QUANTUM ELECTRODYNAMICS

ADVANCED SERIES ON DIRECTIONS IN HIGH ENERGY PHYSICS

ISSN: 1793-1339

This is the best review series in high energy physics today. It comprehensively reviews the most important developments in each sector of high energy physics and is of lasting use to all researchers. All volumes are edited by eminent physicists — researchers who have themselves made substantial contributions to their respective fields of research.

Published

Vol. 29 – *Multiple Parton Interactions at the LHC* (eds. Paolo Bartalini and Jonathan Richard Gaunt)
Vol. 28 – *The State of the Art of Neutrino Physics: A Tutorial for Graduate Students and Young Researchers* (ed. A. Ereditato)
Vol. 27 – *Technology Meets Research 60 Years of CERN Technology: Selected Highlights* (eds. C. Fabjan, T. Taylor, D. Treille and H. Wenninger)
Vol. 26 – *The Standard Theory of Particle Physics: Essays to Celebrate CERN's 60th Anniversary* (eds. L. Maiani and L. Rolandi)
Vol. 25 – *Massive Neutrinos: Flavor Mixing of Leptons and Neutrino Oscillations* (ed. H. Fritzsch)
Vol. 24 – *The High Luminosity Large Hadron Collider: The New Machine for Illuminating the Mysteries of Universe* (eds. O. Brüning and L. Rossi)
Vol. 23 – *60 Years of CERN Experiments and Discoveries* (eds. H. Schopper and L. Di Lella)
Vol. 22 – *Perspectives on String Phenomenology* (eds. B. Acharya, G. L. Kane and P. Kumar)
Vol. 21 – *Perspectives on Supersymmetry II* (ed. G. L. Kane)
Vol. 20 – *Lepton Dipole Moments* (eds. B. Lee Roberts and William J. Marciano)
Vol. 19 – *Linear Collider Physics in the New Millennium* (eds. K. Fujii, D. J. Miller and A. Soni)
Vol. 18 – *Perspectives on Supersymmetry* (ed. G. L. Kane)
Vol. 17 – *Perspectives on Higgs Physics II* (ed. G. L. Kane)
Vol. 16 – *Electroweak Symmetry Breaking and New Physics at the TeV Scale* (eds. T. L. Barklow, S. Dawson, H. E. Haber and J. L. Siegrist)
Vol. 15 – *Heavy Flavours II* (eds. A. J. Buras and M. Lindner)
Vol. 14 – *Precision Tests of the Standard Electroweak Model* (ed. P. Langacker)
Vol. 13 – *Perspectives on Higgs Physics* (ed. G. L. Kane)
Vol. 12 – *Advances of Accelerator Physics and Technologies* (ed. H. Schopper)
Vol. 11 – *Quantum Fields on the Computer* (ed. M. Creutz)
Vol. 10 – *Heavy Flavours* (eds. A. J. Buras and M. Lindner)
Vol. 9 – *Instrumentation in High Energy Physics* (ed. F. Sauli)

The complete list of titles in the series can be found at
http://www.worldscientific.com/series/asdhep

Advanced Series on
Directions in High Energy Physics — Vol. 7

QUANTUM ELECTRODYNAMICS

Editor

T. Kinoshita
Newman Laboratory
Cornell University

World Scientific
Singapore • New Jersey • London • Hong Kong

Published by

World Scientific Publishing Co. Pte. Ltd.
5 Toh Tuck Link, Singapore 596224
USA office: 27 Warren Street, Suite 401-402, Hackensack, NJ 07601
UK office: 57 Shelton Street, Covent Garden, London WC2H 9HE

British Library Cataloguing-in-Publication Data
A catalogue record for this book is available from the British Library.

Advanced Series on Directions in High Energy Physics — Vol. 7
QUANTUM ELECTRODYNAMICS

ISBN-13 978-981-02-0213-2
ISBN-10 981-02-0213-X
ISBN-13 978-981-02-0214-9 (pbk)
ISBN-10 981-02-0214-8 (pbk)

FOREWORD

Quantum Electrodynamics (QED) occupies a unique position in the hierarchy of theoretical physics. As the basic theory of the electromagnetic interaction, it provides the foundation for the quantum mechanics of atoms and molecules as well as condensed matter physics, not to mention their numerous applications in chemistry, biology, etc. At the same time, it is the first theory that has been applied successfully to high energy phenomena. As such it has become the prototype for a comprehensive (but not yet fully established) theory of unified electromagnetic, weak, and strong interactions of elementary particles, the so-called standard model. Examining the validity of QED is important not only because it is the fundamental theory of the electromagnetic interaction but also because it is one of the basic building blocks of the standard model. Any sign of a departure of experimental observation from the predictions of QED could have a profound impact on our understanding of the laws of nature.

This is why QED has been tested as rigorously as possible over the last 40 years. Sometimes this has been achieved by pushing the available state-of-the-art technologies and computational facilities to the limits of feasibility. Occasionally breakthroughs come as a consequence of discoveries of new phenomena and ensuing development of new technologies. It is important to keep pushing the limits; new physics might reveal itself in the next decimal place in theory and measurement.

This volume deals mainly with high precision tests of the "pure" QED aspect of the standard model. The exceptions are the muon anomalous magnetic moment and e^+e^- collisions at large momentum transfers, which push us somewhat beyond pure QED. Such tests are necessarily restricted to systems for which precise theoretical calculations are available and precise measurements are feasible. At present such requirements are satisfied only in a small number of cases such as the anomalous magnetic moments of the electron and muon, and fine and hyperfine structures of positronium, muonium, and the hydrogen and helium atoms. Other important but less critical tests of QED are not included for lack of space.

Through the necessity to access the most precise value of the fine structure constant α, high precision tests of QED are also related to high precision determinations of α based on the ac Josephson and quantized Hall effects discovered in the domain of condensed matter physics. A potential complication concerning the theoretical basis of condensed matter physics is pointed out and its resolution outlined in this volume.

As a beginning physics student I had the good fortune to watch at close range the exciting development of the renormalization theory of QED by Tomonaga and his collaborators. Since then QED has gone through subtle metamorphoses as a consequence of the explosion of new experimental discoveries in high energy physics and the development of the gauge theory of unified interactions. I have always been fascinated by the remarkable resilience of QED against the challenge of rigorous experimental tests. For more than twenty years I myself have participated in such tests, working out theoretical predictions for the anomalous magnetic moments of the electron and muon. I was therefore thrilled to be given an opportunity to assemble a comprehensive review of high precision tests of QED. Thanks are due to all the authors who have participated in this worthwhile enterprise. Without their enthusiastic support and hard work this volume would not have become a reality.

It is my sincere hope that this volume will prove useful both as a comprehensive document of the current status of high precision tests of QED and as a guide for future research.

Toichiro Kinoshita
Ithaca, New York
May 1990

CONTENTS

QUANTUM ELECTRODYNAMICS

High Precision Tests of Quantum Electrodynamics—An Overview

Toichiro Kinoshita and Donald R. Yennie
Newman Laboratory of Nuclear Studies, Cornell University
Ithaca, NY 14853

CONTENTS

1. Introduction

The "modern" era of quantum electrodynamics (QED) dates from the late 1940's, when the experimental discoveries of the anomalous moment of the electron[1] and the Lamb shift[2] stimulated the evolution of quantum electrodynamics to its present precise form. This era has been one of interplay between experiment and theory, in which each has encouraged increasing refinement in the other. On the experimental side, this evolution has led to the development of very sophisticated techniques; while on the theoretical side it has required the development of more and more powerful methods of calculation. Over the years, theory and experiment have been more often in agreement than in disagreement. Although there have been occasions when there seemed to be real conflicts between them, new experiments, better calculations, or better values of the fine structure constant, ultimately led to resolution. At the present time there seem to be no definitive conflicts, although there are situations in which further improvements of theory and experiment are desirable to obtain more stringent tests of the basic theory.

The purpose of this introductory article is to provide some perspective on the more detailed technical discussions which follow. This section gives some general remarks about the nature of QED. Sec. 2 describes some of the tests of QED which have occupied the attention of both theorists and experimentalists for the past four decades. It is also noteworthy that precision QED has become intimately related to condensed-matter physics phenomena through the need for a precise value of the fine structure constant. This is discussed in Sec. 3.

The success of QED as a quantized field theory owes to a large extent to the fact that it is constructed on a solid foundation of classical electrodynamics—Maxwell-Lorentz theory—which is one of the most successful theories of physics. The latter is applicable to most electromagnetic phenomena and describes them quite accurately. Its disagreement with observations manifests itself only at microscopic scales, the spectrum of the black body radiation and the photo-electric effect being the earliest evidences. Also, it is incapable of explaining the structure of atomic spectra or the stability of atoms and molecules. From a more theoretical point of view, it suffers from internal inconsistencies such as the infinite energy of self-interaction of charged particles. As is well-known, efforts to resolve some of these difficulties led to the discovery of quantum mechanics. The success of quantum mechanics is phenomenal. It describes emission and absorption of photons by atoms and molecules correctly. The structure of the spectra of atoms and molecules are understood for the first time. Modern physics and associated technologies including lasers, semiconductors, and superconductors are built upon the foundation of quantum mechanics. High precision realizations of practical electrical units now depend on the ac Josephson effect and the quantized Hall effect, both of which are nothing but macroscopic manifestations of quantum mechanics.

Nevertheless, the conventional (nonrelativistic) quantum mechanics is not completely satisfactory in the sense that it leaves out relativity which is one of the pillars of modern physics. The effort to make quantum mechanics compatible with relativity was initiated almost as soon as quantum mechanics was formulated. However, it took more than twenty years before it was reformulated in a way that satisfied the requirement of relativity in (an at least superficially) consistent manner. Renormalized quantum electrodynamics is the outcome of this effort.

QED thus formulated turns out to be much more than a mere marriage of quantum mechanics and relativity. The wave-particle duality of quantum mechanics is now fully incorporated into the theory; and charged particles (such as electrons) and photons are treated as quantized fields, placing them on the same footing. Quantum electrodynamics is the theory of the interaction of charged leptons with photons, in which all observable effects can be expressed in terms of measured charge and mass. Given its tremendous success in predicting experimental facts, ranging from very refined details of the properties of electrons and muons and atomic spectra to interactions in the multi-GeV range, it must represent very accurately the true nature of these interactions. It is inconceivable that any theory which is conceptually less sophisticated could produce the same results.

On the other hand, recent progress in particle physics shows clearly that QED is no more than a part of a larger, deeper theory involving the weak, as well as the strong and possibly gravitational, interactions. Many aspects of the electroweak theory seem already well understood. While the leptons might be composites of more elementary entities, there is no present evidence for such structure, down to scales of 10^{-16} cm (equivalently, up to energies of over 100 GeV). Thus, although QED should ultimately be regarded as a low energy phenomenology to some more fundamental theory, we can in practice often ignore that fact and treat QED as a self-contained theory, occasionally incorporating small corrections from the more complete theory, the so-called standard model of the electroweak and strong interactions. To the extent that pure QED predictions agree with experiment, the deeper theory has not yet revealed itself in low energy experiments; but it is important to continue to push QED to more and more stringent tests in the expectation that at some point there will appear a genuine conflict between predictions and experiment.

If QED is merely a phenomenology, how can one account for its remarkable quantitative success? The answer is intimately related to one of its most characteristic features, renormalizability. Because of this, short distance, high energy effects in QED can be absorbed into a finite number of *measurable* masses and charges. For the first time in the history of physics, there exists a theory which has no obvious intrinsic limitation and enables us in principle to calculate physical phenomena to any accuracy we need in terms of a few measurable parameters such as the elementary electric charge e and the electron mass m. Thus the detailed high energy structure of the ultimate theory is irrelevant to the analysis of low energy phenomena except insofar as it determines these parameters. As originally conceived, the renormalization program was to deal with the fact that the high energy behavior of pure QED leads to divergent integrals in perturbation theory. This occurs because QED inherits some diseases of classical electrodynamics such as the infinite self energy and also generates additional infinities of its own. Whether or not these infinities reflect the incompleteness of QED, or are just an artifact of perturbation theory, all divergent integrals are made convergent "simply" by expressing observable quantities in terms of physical masses and charges. The results of this intricate renormalization procedure have been outstandingly successful.

It is clearly impossible to cover all achievements of QED in a single volume. Instead, we shall concentrate on experimental high precision tests of the validity of QED and the development of theoretical infrastructures required to carry out such tests. We assume that the reader has a textbook knowledge of QED. For further information we attach at the end of this volume an extensive bibliography on QED assembled by Kubo and Yokoyama, together with their concise review of the development of QED.[a] In order to be up-to-date about the theoretical foundation of QED, however, two articles written specifically for this volume are included: It is still unknown whether or not QED is a mathematically viable theory, not to mention the question of whether it is self-contained as a physical theory. Ito's article[b] is an attempt to throw some light on this problem, albeit for a much simpler case of the ϕ_4^4 theory; it is based on a renormalization group method. Within the confine of the conventional QED, there still remain several technically sticky problems. Some of them are discussed in the article by Nakanishi.

The rest of this volume is devoted to the survey of those topics which come under the framework of conventional quantum electrodynamics. With the exception of Martyn's article on the test of QED in high energy electron-positron collisions, it concentrates on low-energy high-precision tests. Other interesting areas which could not be included are scattering and collisions at all energies, recent developments in heavy atoms, heavy ion collisions, and the emerging field of laser-cooled atomic spectroscopy.

From the theoretical and technical points of view, our topics break down into two quite different categories, each requiring its own special types of expertise:

(a) *Perturbative realm.* To compute lepton anomalous magnetic moments, or scattering amplitudes for leptons or photons, it is necessary to evaluate a series of Feynman graphs. Each graph contributes to a definite order in α, and the results are given as a (truncated) power series in α. To accomplish this, a thorough mastery of the renormalization program, with its intricate subtraction of nested and overlapping divergences, is necessary. The case of the electron anomalous moment represents the greatest refinement of this renormalization analysis. In some cases, for example in the infrared contribution

[a]This is a translation by Kubo of an article published by the Physical Society of Japan in Japanese, and is included here with the permission of the authors and the Physical Society of Japan.

[b]References to articles refer to contributions to this volume.

to scattering amplitudes and cross sections, the perturbation expansions can be partially summed; in the infrared example, they are found to "exponentiate."

(b) *Non-perturbative realm.* In bound state problems, a given graph does not correspond to a definite order in α. Often infinite sets of graphs must be summed to evaluate a contribution to a particular order; this situation is usually signalled by results which contain factors of $\ln \alpha$. These non-perturbative effects are found to be associated with the photons which are responsible for the binding, rather than with those which are involved in the self interaction of the electron. It is conventional to emphasize this distinction by labeling the α-dependence associated with binding photons by $Z\alpha$, even though Z is unity in many cases of interest. So far renormalization needed in the bound state problem has been relatively simple, being only at the one- and two-loop levels. However, this is overshadowed by the complication that there is really no uniquely agreed upon procedure for incorporating higher-order corrections in the binding because contributions can readily be shifted between the structure of the wave function and that of the interaction kernels. Of course, only the details of intermediate steps, not the final result, are affected by this ambiguity.

2. Survey of Tests of QED

High Energy Tests of QED. In recent years QED has been tested extensively using high energy collisions of electrons and positrons in storage rings; these experiments explore the validity and limitation of theory in the region of very short distances or, equivalently, large momentum transfers. Reactions used for such tests include

$$\begin{aligned} e^+e^- &\rightarrow \gamma\gamma, \\ e^+e^- &\rightarrow e^+e^-, \\ e^+e^- &\rightarrow \mu^+\mu^-, \\ e^+e^- &\rightarrow \tau^+\tau^-. \end{aligned} \tag{1}$$

The latest of these tests carried out at the high energy e^+e^- colliding rings PEP, PETRA, and TRISTAN are reviewed by Martyn. The measurement of the differential cross section for the reaction $e^+e^- \rightarrow \gamma\gamma$ is in agreement with the $O(\alpha^3)$ QED cross section including virtual and bremsstrahlung radiative corrections at the one percent level. This means that possible existence of excited heavy electrons of mass less than 80 GeV/c^2 is ruled out. The differential cross section for the Bhabha scattering $e^+e^- \rightarrow e^+e^-$ confirms the $O(\alpha^3)$ QED calculations at the level of a few percent, but indicates the presence of the electroweak effect. Similarly for the other processes. Lower limits on the lepton QED cut-off parameters Λ exceed values of 250 GeV and suggest that the electron, muon, and tauon behave as point-like particles down to distances smaller than 10^{-16} cm. Further refinement of these results is expected shortly from the LEP e^+e^- colliding beam experiments at CERN.

If one recalls that momentum scales of concern at the time QED was first formulated is below a few MeV at most, it is truly remarkable that the theory holds unmodified up to momentum scales of order 100 GeV/c. The fact that weak and strong interaction effects must be taken into account in interpreting the data at larger momentum scales does not mean that QED fails at short distances, but simply that the real world has forces besides electromagnetism and QED must be regarded as part of a more comprehensive theory that accomodates these forces. Fortunately, for high precision tests on low-energy systems, which are our main concern, the effect of non-QED forces can be treated as a small calculable deviation from the description of the pure QED.

Anomalous Moment of the Electron. The anomalous magnetic moment $a(e)$ of an electron in a weak magnetic field, which is one-half the deviation of the g factor from the value 2 predicted by the Dirac theory, is the simplest quantity that can be calculated from QED. This quantity was first measured accurately by Kusch and Foley.[3] Schwinger[4] then showed that it can be calculated from QED, which was just being developed. Since then there has been enormous progress both experimentally and theoretically. The latest experiment of Dehmelt and his collaborators, in which $a(e)$ of an isolated electron suspended in a Penning trap is measured, provides an opportunity to check the theoretical calculation very closely. The most recent experimental values of the magnetic moment anomaly of the electron and positron are[5]

$$\begin{aligned} a(e^-) &= 1\ 159\ 652\ 188.4\ (4.3) \times 10^{-12} \\ a(e^+) &= 1\ 159\ 652\ 187.9\ (4.3) \times 10^{-12}\ , \end{aligned} \tag{2}$$

respectively. The experiments are discussed in the article by Van Dyck. The experimental error comes from several sources: The statistical error is 0.62 $\times 10^{-12}$. The error due to the microwave power shift is 1.3 $\times 10^{-12}$. The radiative interaction of an electron with the metallic walls of the Penning trap is the largest source of uncertainty at present.[6] The uncertainty from this cavity shift effect is estimated to be 4×10^{-12}. An instructive discussion of this effect is presented in the article by Gabrielse, Tan, and Brown.

The theoretical expression for the electron anomaly $a(e)$ can be written in the form:

$$a(e) = C_1 \left(\frac{\alpha}{\pi}\right) + C_2 \left(\frac{\alpha}{\pi}\right)^2 + C_3 \left(\frac{\alpha}{\pi}\right)^3 + C_4 \left(\frac{\alpha}{\pi}\right)^4 + \ldots + \delta a(e) \tag{3a}$$

where

$$\begin{aligned} C_1 &= 0.5\ , \\ C_2 &= \ -\ 0.328\ 478\ 965\ \ldots\ , \\ C_3 &= \ 1.176\ 11\ (42)\ , \\ C_4 &= \ -\ 1.434\ (138)\ , \\ \delta a(e) &= \ 4.46 \times 10^{-12}\ . \end{aligned} \tag{3b}$$

The coefficients C_1 and C_2 are known analytically,[4,7] while the evaluation of C_3 is partly analytic and partly numerical.[8] The present status of the analytic evaluation of C_3 is described in the article by Remiddi, Roskies, and Levine, while the determination of C_3 and C_4 by numerical means is discussed in Kinoshita's article. The term $\delta a(e)$ is a sum of contributions due to the muon and tauon loops, hadronic effects, and the electroweak effect.[9]

If one uses the value of α given later in (21) rather than the one quoted in (13), [c] one

[c] The reason for avoiding use of α given in (13) is that it depends partly on an earlier determination of α which included the electron anomaly and hence is not appropriate for the testing QED by means of $a(e)$.

obtains from (3)

$$a(e) = 1\ 159\ 652\ 140\ (5.3)\ (4.1)\ (27.1) \times 10^{-12}\ , \tag{4}$$

where the first and second errors come from the numerical errors on C_3 and C_4 in (3b) while the third reflects the error on α quoted in (21).

Agreement between theory and measurement of $a(e)$ is at the relative level of 10^{-7}. Note, however, that the error on $a(e)$ is dominated by that of α. This means that a rigorous test of QED using the electron anomaly $a(e)$ must wait until a better value of α is found. Pending an improved measurement of α, however, one can calculate α from theory and the experimental value of $a(e)$ to a precision better than any other means available at present:

$$\alpha^{-1}(g-2) = 137.035\ 992\ 22\ (94)\ \ (0.0069\ \text{ppm}) \tag{5}$$

where the experimental and theoretical errors contribute 0.0037 ppm and 0.0058 ppm, respectively.

Anomalous Moment of the Muon. The best measured values of the muon anomalous magnetic moments thus far are those obtained at CERN:

$$\begin{aligned} a(\mu^-) &= 1\ 165\ 937(12) \times 10^{-9}, \\ a(\mu^+) &= 1\ 165\ 911(11) \times 10^{-9}, \end{aligned} \tag{6}$$

for negative and positive muons, respectively.

The present theoretical prediction for a_μ is

$$a(\mu) = 1\ 165\ 919\ 18(191) \times 10^{-11}, \tag{7}$$

which is in good agreement with (6). The theoretical result consists of the QED contribution evaluated by partly analytic and partly numerical means up to the α^4 order and estimated numerically in the α^5 order, as well as the contributions of the hadronic vacuum-polarization term, the hadronic light-by-light scattering effect, and the weak interaction effect of the standard electroweak theory.

The muon $g-2$ experiment is discussed in detail in the article by Farley and Picasso and the theory in the article by Kinoshita and Marciano. The theory of the muon's anomalous moment is of course very similar to that of the electron. However, the physics of the muon anomalous moment is quite different from that of the electron due to the fact that the internal momenta of the muon's structure scale as the muon mass rather than the electron mass. This makes closed electron loops quite important, and leads to logarithms of m_μ/m_e in the coefficient of $(\alpha/\pi)^2$ and higher powers. Also, the contribution of hadronic vacuum polarization to $a(\mu)$ is rather large. Although the contribution of the weak interaction is about 2×10^{-9} which is five times smaller than the current experimental uncertainties, it certainly becomes observable if the experiment is improved significantly. This will provide one of the most important tests for the validity of the standard model of the electroweak interaction. A new muon $g-2$ experiment is in progress at the Brookhaven National Laboratory; it will have sufficient precision to make such a test feasible. In view of the recent measurements[10,11] of the mass and width of the Z^0 boson, which give convincing confirmation of the standard model with three generations of lepton families with light neutrinos, the calculated electroweak contribution to $a(\mu)$ is very likely to be confirmed in such an experiment. If that happens, comparisons of theory and experiment on $a(\mu)$ may enable us to impose a very strict lower bound on the mass of not-yet-discovered particles in the energy range of the SSC. In this sense, the importance of the Brookhaven $a(\mu)$ experiment is further enhanced by recent results from LEP.

Lamb Shift. The experimental discovery of the Lamb shift of the hydrogen atom[2] showed that the $2S_{1/2}$ level lies about 1060 MHz above the $2P_{1/2}$ level, contrary to the Dirac theory which predicts the degeneracy of these levels. This ushered in the new era of renormalized QED.[12] Since then the theory and experiment of the Lamb shift have gone through numerous improvements. The most recent measurements of the hydrogen atom Lamb shift $\Delta E(2S_{1/2} - 2P_{1/2})$ are

$$\begin{aligned} &1\,057\,862\ (20)\ \text{kHz},^{13]} \\ &1\,057\,845\ (9)\ \text{kHz},^{14]} \\ &1\,057\,851\ (2)\ \text{kHz}.^{15]} \end{aligned} \tag{8}$$

The latest theoretical values are[d]

$$\begin{aligned} &1\,057\,853\ (13)\ \text{kHz} \quad \text{if the rms radius of proton is } 0.805\ (11)\ \text{fm}, \\ &1\,057\,871\ (13)\ \text{kHz} \quad \text{if the rms radius of proton is } 0.862\ (12)\ \text{fm}. \end{aligned} \tag{9}$$

The main difficulty in comparing QED theory and experiment is the lack of an agreed upon value for the radius of the proton; however, it is seen that there is no conflict at the level of about 20 ppm. This also means that the theory and experiment of the hydrogen Lamb shift may be used to determine α to similar precision. (The main contribution to the theory of the Lamb shift is proportional to $\alpha^3 R_\infty$, where the Rydberg R_∞ is known to very high accuracy; therefore, the fractional uncertainty in α determined this way is one-third of that of the Lamb shift itself.) In separate articles, the Lamb shift experiments for hydrogen and other atoms are discussed by Pipkin, and the corresponding theory is discussed by Sapirstein and Yennie.

Hydrogen Hyperfine Structure. The validity of QED can be tested even more precisely from various spectral intervals of simple systems such as hydrogen and helium atoms. In many cases, however, it turns out to be very difficult to go beyond the precision of a few ppm for experimental and/or theoretical reasons. A good example of theoretical difficulty is the hyperfine structure of the ground state of the hydrogen atom which is one of the most precisely measured numbers in physics:[16,17]

$$\Delta\nu(\text{hydrogen}) = 1\,420.405\,751\,766\,7\ (9)\ \text{MHz}\ . \tag{10}$$

The article by Ramsey gives a historical survey of these experiments. The theoretical description of this system, outlined in the article by Sapirstein and Yennie, has several parts: the main (non-recoil) term incorporates those radiative corrections in which the proton mass enters only through the proton's magnetic moment and an overall reduced

[d] A recent revision is given in the article by Sapirstein and Yennie in this volume.

mass dependence; the remaining (recoil) terms are sensitive to the structure of the proton and depend not only on the electromagnetic radius of the proton but also on its polarizability, which (the latter in particular) are rather poorly known at present. For this reason the test of QED based on the hydrogen hfs is limited to a precision of only a few ppm, although the actual magnitude of the proton polarizability correction might be much smaller.[18]

Helium Fine Structure. Since the helium atom has two orbiting electrons, it has not been possible to solve its Schrödinger equation analyatically. Thus it may appear surprising that QED can be tested with high precision for such a system. It turns out, however, that energy levels and wave functions of some lowest-lying states can be calculated very accurately by the variational method. Furthermore, transitions between $2\,^3P_J$ (J = 0, 1, 2) states have been measured very accurately, verifying the validity of theory to a few ppm. These measurements and comparison with theory are reviewed in the article by Hughes and Pichanick.

Because of the hadronic complications, it is customary for a more precise test of QED to avoid systems containing hadrons (such as protons and atomic nuclei) as primary components. The spectra of muonium (a hydrogen-like bound state of an electron and a positive muon) and positronium and the anomalous magnetic moments of the electron and muon are the only systems available that satisfy this criterion. Of these the hyperfine structure of the muonium ground state, together with the anomalous magnetic moment of the electron discussed earlier, is particularly blessed with very precise measurements and extensive theoretical calculations. In muonium the effects of strong and weak interactions are extremely small and under tight control.

Muonium Hyperfine Structure. Muonium is very similar to the hydrogen atom in many respects. For instance, as in hydrogen, the muonium has different energies depending on whether the magnetic moments of e^- and μ^+ are aligned parallel or antiparallel. This hyperfine splitting between the spin-0 and spin-1 levels of the muonium ground state has been measured very accurately:[19]

$$\Delta\nu(\text{exp}) \;=\; 4\,463\,302.88\,(16)\ \text{kHz}. \tag{11}$$

The experiments are described in the article by Hughes and zu Putlitz. As is well known, the bulk of the hyperfine splitting is given by the Fermi formula

$$E_F = \frac{16}{3}(Z\alpha)^2 R_\infty \frac{\mu_\mu}{\mu_B}\left[1+\frac{m_e}{m_\mu}\right]^{-3}, \tag{12}$$

where Z is the charge of the muon in units of the electron charge, R_∞ is the Rydberg constant for infinite nuclear mass, μ_μ is the muon magnetic moment, μ_B is the Bohr magneton, and m_e and m_μ are the electron and muon masses, respectively. Of course Z is equal to one for the muon, but it is useful to keep it in the formula in order to distinguish contributions arising from binding effects ($Z\alpha$) and radiative corrections (α). Many correction terms (of both α and $Z\alpha$ type) have been calculated since the work of Fermi. Conceptually, the theoretical result is separated into three main types of contributions: non-recoil terms which include radiative corrections, pure recoil corrections, and radiative-recoil corrections. The details of these different contributions are contained in the article by Sapirstein and Yennie.

If one uses the values

$$R_\infty = 10\,973\ \ 731.534\ (13)\ \mathrm{m}^{-1},^{20]}$$
$$\frac{m_\mu}{m_e} = 206.768\,262\,0\ (617),^{19,21]} \tag{13}$$
$$\alpha^{-1} = 137.035\,989\,5\ (61),^{20]}$$

one finds that

$$\Delta\nu(\text{theory}) = 4\,463\,303.11\ (1.33)\ (0.40)\ (1.0)\ \text{kHz}\ , \tag{14}$$

where the first and second errors reflect the uncertainties in the measurements of m_μ and α^{-1} listed in (13), and the third is an order of magnitude estimate of an uncalculated radiative correction contribution. The difference between theory and measurement is thus given by

$$\Delta\nu(\text{theory}) - \Delta\nu(\text{exp}) = 0.23\ (0.16)\ (1.33)\ (0.40)\ (1.0)\ \text{kHz} \tag{15}$$

where the first error comes from the $\Delta\nu$ measurement and the rest are carried over from (14).

We wish to emphasize here that the muonium hfs is potentially one of the cleanest ways to determine α, although the best value available at present

$$\alpha^{-1}(\mu\ \text{hfs}) = 137.035\,993\,0\ (224)\ \ (0.164\ \text{ppm})\ . \tag{16}$$

is less accurate than desired. In order to make further progress, one must evaluate the uncalculated term mentioned above and measure m_μ more accurately. When theory and experiment are improved sufficiently it will provide one of the most stringent tests of QED.

Positronium Fine and Hyperfine Structure. Positronium is an effectively pure QED system, with weak and strong interaction contributions relatively small because of the low momenta which are being probed. One complication of the theoretical analysis is that the absence of a large mass ratio requires one to avoid certain approximations which are valid for atoms with heavier nuclei. This shows up most notably in the fact that the hyperfine structure is of the same order of magnitude as the fine structure. Another complication is the presence of "annihilation diagrams", which contribute to the hyperfine splitting in S-states at the same level as the usual spin-spin interactions. Unfortunately, these complications prevent positronium from giving as crucial a test of some of the details of the theory as the other systems described above. Nevertheless, it is a very challenging system to study and it certainly merits further effort. Experiments are described in the article by Mills and Chu, and the theory is reviewed by Sapirstein and Yennie.

The energies of a postronium atom depend on four quantum numbers. First, the electron and positron spins can be combined to produce a total spin of zero or one (singlet or triplet, respectively). This total spin can then be combined with the orbital angular momentum L to produce the total angular momentum J. In addition to these angular momentum quantum numbers, there is of course the principal quantum number n. One measurement is the analogue of the usual fine structure in which transitions are made between states of different orbital angular momentum with the total spin unchanged. The current experimental and theoretical results for three of these transitions (in MHz) are shown as follows:

	$2\,^3S_1 \to 2\,^3P_0$	$2\,^3S_1 \to 2\,^3P_1$	$2\,^3S_1 \to 2\,^3P_2$
experiment	18504.1(10.0)	13001.3(3.9)	8619.6(2.7)
theory	18496.1	13010.9	8625.2

The experimental errors shown are statistical. The agreement is reasonable considering that the present theory incorporates only two orders of α, namely $\alpha^4 m_e$ and $\alpha^5 m_e$. Only the S-states are affected by annihilation to this order. The next order could contribute at the few MHz level.

The present experimental value for the hyperfine structure in the ground state of positronium is

$$\nu = 203\ 389.10(74)\ \text{MHz} \tag{17}$$

The theory is similar to that for the muonium hyperfine structure, with the additional complications mentioned above. At present, calculations of relative order α^2 contributions are nearly completed, and the resulting theoretical value is 203 404.5(9.3) MHz, where the error estimate arises from an estimate of the size of uncalculated contributions. Considering the theoretical uncertainty, the agreement is satisfactory; but obviously the theory, and hopefully the experiment, should be pushed further.

Orthopositronium Decay. This process gives a possible blemish on the generally superb agreement between theory and experiment. For some years the measured orthopositronium decay rate has been somewhat larger than expected from theory. Because the experiments were relatively inaccurate until recently, this was only a two standard deviation effect, and not a source of concern. Now, however, a significant improvement in accuracy has been achieved,[22] as is described in the article by Mills and Chu. The new measurement is compared with theory in Table 1.

Table 1. Comparison of theory and experiment for orthopositronium decay rate Γ.

Term	Coefficient	Rate contribution (μsec^{-1})
$m\alpha^6$	$\frac{2}{9\pi}(\pi^2 - 9) =$ 0.0615119	7.2112
$m\alpha^7$	−0.20130(6)	−0.1722
$m\alpha^8$ ($\log\alpha$ term only)	−0.1008791	−0.0006
Sum		7.0384
Γ (exp)		7.0516(13)
$m\alpha^8$ (inferred constant)	2.12(20)	

Since no discussion of the theoretical implication of this result is given elsewhere in this volume, let us describe it here. Adkins[23] surveyed the $m\alpha^7$ terms in 1983. They have been improved by analytic evaluation of one-loop self-energy and vertex corrections.[24] As is seen from the last row of Table 1, a relatively large (nonlogarithmic) constant in order $m\alpha^8$ is required to remove the discrepancy between theory and experiment. This constant appears to be much larger if written in terms of $(\alpha/\pi)^2 m\alpha^6$, and even larger if written in terms of $(\alpha/\pi)^2\Gamma$, ~21 and ~340, respectively. The question of which expansion is more natural is open. With respect to powers of α/π, it is certainly true that QED perturbation theory is more properly expanded in terms of this quantity rather than only α. However, in the bound state problem, factors of π frequently occur in the numerator: an example of this is the large first order binding correction to the Lamb shift, where such factors can be seen in the value of A_{50} in Eq. (2.5b) in the article by Sapirstein and Yennie. Of course, in the present calculation there is a squared power of π already in lowest order, but it is made small numerically by having 9 subtracted from it. This situation is frequently encountered also in analytic evaluations of electron $g-2$ graphs. With respect to the question of whether it is appropriate to expand in terms of Γ or $m\alpha^6$, we again consider the question open. The small coefficient of $m\alpha^6$ in Γ comes from integrating the square of the matrix element for annihilation into three photons over three-body phase space. Radiative corrections to this process are to be integrated over the same three-body phase space, but with a significantly altered matrix element. Because of this ambiguity, we do not consider orthopositronium decay an example of a breakdown of QED: it is, however, clearly important to complete the calculation of the constant term in order $m\alpha^8$, after which more definite conclusions can be drawn. At the same time, an independent high precision measurement of the orthopositronium decay rate is highly desirable.

3. Condensed-matter Physics Determination of α

The very good agreement between the values of α given by (16) and (5) shows that QED is valid to the level of 0.17 ppm. At the same time (16) must obviously be improved by a factor of 20 to reach the precision achieved by (5) so that one can test the internal consistency of QED at the level of 0.01 ppm or better.

The uncertainties in (16) come mainly from the muon mass and the theory of muonium hfs. In particular the calculation of the uncalculated radiative correction is urgently needed. No obstacle is in sight for improving the theoretical prediction by at least an order of magnitude although it will certainly require an extensive effort. A new experiment for a more accurate measurement of the muon mass, which is the largest source of experimental uncertainty, is being prepared.[25] The same experiment will improve $\Delta\nu$ of (11), too.

While the internal consistency of QED is testable to only 0.17 ppm at present, it is possible to test the validity of QED to a higher precision by completely independent means based on macroscopic quantum effects discovered in the realm of condensed matter physics, namely the quantum Hall effect and the ac Josephson effect. The nature of these two measurements will now be described briefly.

Quantum Hall Effect. When the electrons participating in a Hall effect measurement are confined to a very thin, almost two-dimensional, layer with a strong perpendicular magnetic field and the effect is observed at very low temperatures, there is a tendency for the Hall resistance to be quantized at certain particular values. That is, as some external parameter such as the magnetic field or a gate voltage is varied (depending on the type of device), one observes certain 'plateaus' in the Hall resistance. The Hall resistance (R_H) is defined as the ratio of the voltage (V_H) across the sample to the current (I_H) flowing along the sample. With the use of the exactly defined constants μ_0, the permeability of the vacuum, and c, the speed of light, the integer steps in the quantum Hall resistance are simply related to the fine structure constant by

$$\alpha^{-1} = \frac{2R_H}{\mu_0 c}. \tag{18}$$

Aside from uncertainties inherent in the measurement itself, this requires precision reference voltage and current standards, as well as an assumption, or proof, that the theory is exact at the required level of interest. As a practical matter, one replaces the need for the precision voltage and current standards by a precision resistance standard. In turn, the resistance standard is calibrated by reference to a 'calculable capacitor'.[26] Any error in this calibration is reflected directly in α^{-1}. The exactness of the theoretical expression is asserted by the work of Laughlin.[27] A discussion for non-specialists is given by Yennie.[28]

ac Josephson Effect. A chain of four separate precision quantities are combined to give a value for α by

$$\alpha^{-2} = \frac{c}{4R_\infty \gamma_p'} \frac{\mu_p'}{\mu_B} \frac{2e}{h}. \tag{19}$$

The most critical two are the ac Josephson effect, which gives a precise value for the combination $2e/h$, and the gyromagnetic ratio of the proton in water (called γ_p'). In the ac Josephson effect, a pair of electrons tunnels through a Josephson junction separating superconductors at two different potentials. Their energy change $2eV$ is related to the frequency ν associated with the transition through $h\nu = 2eV$. The frequency can of course be determined very precisely, and the voltage across the sample can be related to a voltage standard. The gyromagnetic ratio is the spin-flip frequency of the proton divided by the magnetic field. The frequency can be measured quite exactly, and the magnetic field is obtained by constructing a precision coil with a current known from a precision current standard. Two other important quantities are the Rydberg, R_∞, and the proton's magnetic moment (in water) in units of the Bohr magneton, μ_p'/μ_B.

Combination Method. It is also possible to determine α from the formula

$$\alpha^{-3} = \frac{R_H}{2\mu_0 R_\infty \gamma_p'} \frac{\mu_p'}{\mu_B} \frac{2e}{h}, \tag{20}$$

which is obtained by combining (18) and (19). The advantage of this form is that the dependence on the resistance standard cancels out.

The value of α quoted in (13) is the 1986 recommended value determined by analyzing numerous high precision data including those determined from the ac Josephson effect and the quantized Hall effect available before the end of 1986. This will have to be modified significantly in view of recent developments. In particular, two new measurements of α based on (18) and (19) have substantially improved uncertainties. They are[29]

$$\alpha^{-1}(\text{QHE}) = 137.035\,997\,9\,(32) \qquad (0.024\ \text{ppm}) \tag{21}$$

and[30]

$$\alpha^{-1}(\text{acJ}\&\gamma_p') = 137.035\,977\,0\,(77) \qquad (0.056\ \text{ppm})\,. \tag{22}$$

Corresponding to (20) one also finds

$$\alpha^{-1}(\text{Eq.}(20)) = 137.035\,984\,0\,(51) \qquad (0.037\ \text{ppm})\,. \tag{23}$$

Measurements of (21) and (22) depend on a calculable capacitor and are thus sensitive to its uncertainty. In (23), however, such a dependency cancels out and its error is mainly due to that of γ'_p.

Accepting the results (21), (22), and (23) as the best values available at present, a few comments are in order: It is generally believed that the α's determined by the ac Josephson effect and quantized Hall effect have no theoretical uncertainty, although this has never been proved. The difference of 21×10^{-6} between (21) and (22), which is about 3 times larger than the error in (22), might be the first indication that this belief is open to question. Before accepting this view, however, one must examine carefully the possible errors in the measurements, in particular those of the calculable capacitor and γ'_p. If the difference between (21) and (22) persists even after the experimental errors are fully understood and improved, it may become necessary to examine the theoretical basis of formulas (18) and (19) closely and evaluate correction terms, if any, to (18) and/or (19). Strictly speaking, until the internal consistency of condensed matter physics is firmly established, it is not possible to push the test of QED to the level of 10^{-8}.

In summary, α's determined by QED and condensed matter physics methods agree with each other at the level of 10^{-7}. However, it is still an open question whether the good agreement can be maintained down to the level of 10^{-8}. One cannot exclude the possibility that the theories of the ac Josephson effect and/or the quantized Hall effect are not exact and require further refinement. It is also possible that we are at last at the threshold of breakdown of the old reliable QED. Work in the next few years might be crucial for clarifying the situation. In this context, Kinoshita and Lepage reexamine quantum mechanics as the starting point of the ac Josephson effect and the quantized Hall effect from the viewpoint of relativistic QED adapted to nonrelativistic systems.

Finally, we comment on the often-expressed view that the fine structure constant has no fundamental significance since it is just a particular value of the running coupling constant describing a more fundamental theory, say a grand unified theory. In that point of view, α is merely the value of one branch of the running coupling constant at $q^2 = 0$. Although this may well be true, it does not dimish the importance of a precise determination of α. This is not only because such a fundamental theory is still far from being discovered at present but also, even when it becomes finally available, because α certainly remains the most accurately determinable value of the universal function describing the running coupling constant. This particular value provides a well-defined anchor point for the running coupling constant, and hence the entire theoretical structure.

Acknowledgements

We wish to thank J. Sapirstein and G. S. Adkins for useful comments. This work is supported in part by the National Science Foundation.

References

1. J. E. Nafe, E. B. Nelson, and I. I. Rabi, Phys. Rev. **71**, 914 (1947); D. E. Nagle, R. S. Julian, and J. R. Zacharias, Phys. Rev. **72**, 971 (1947).
2. W. E. Lamb and R. C. Retherford, Phys. Rev. **72**, 241 (1947).
3. P. Kusch and H. M. Foley, Phys. Rev. **72**, 1256 (1947).

4. J. Schwinger, Phys. Rev. **73**, 416L (1948).
5. R. S. Van Dyck, Jr., P. B. Schwinberg, and H. G. Dehmelt, Phys. Rev. Lett. **59**, 26 (1987).
6. L. S. Brown and G. Gabrielse, Rev. Mod. Phys. **58**, 233 (1986).
7. C. Sommerfield, Phys. Rev. **107**, 328 (1957); A. Petermann, Helv. Phys. Acta **30**, 407 (1957).
8. P. Cvitanovic and T. Kinoshita, Phys. Rev. **D10**, 4007 (1974); M. J. Levine, H. Y. Park, and R. Z. Roskies, Phys. Rev. **D25**, 2205 (1982).
9. See, for instance, T. Kinoshita in "New Frontiers in High Energy Physics", B. Kursunoglu *et al.*, eds. (Plenum, 1978), 127.
10. G. S. Abrams *et al.*, Phys. Rev. Lett. **63**, 2173 (1989).
11. L3 Collab., B. Adeva *et al.*, Phys. Lett. B **231**, 509 (1989); ALEPH Collab., D. Decamp *et al.*, Phys. Lett. B **231**, 519 (1989); OPAL Collab., M. Z. Akrawy *et al.*, Phys. Lett. B **231**, 530 (1989); DELPHI Collab., P. Aarnio *et al.*, Phys. Lett. B **231**, 539 (1989).
12. H. A. Bethe, Phys. Rev. **72**, 339 (1947).
13. D. A. Andrews and G. Newton, Phys.Rev. Lett. **37**, 1254 (1976).
14. S. R. Lundeen and F. M. Pipkin, Phys. Rev. Lett. **46**, 232 (1981).
15. V. G. Pal'chikov, Yu. L. Sokolov, and V. P. Yakovlev, Pis'ma Zh. Eksp. Teor. Fiz. **38**, 347 (1983) [JETP Lett. **38**, 418 (1983)].
16. H. Helwig *et al.*, IEEE Trans. Instrum. Meas. **19**, 200 (1970).
17. L. Essen *et al.*, Nature **229**, 110 (1971).
18. G. T. Bodwin and D. R. Yennie, Phys. Rev. **D37**, 498 (1988).
19. F. G. Mariam *et al.*, Phys. Rev. Lett. **49**, 993 (1982).
20. E. R. Cohen and B. N. Taylor, Rev. Mod. Phys. **59**, 1121 (1987).
21. E. Klempt *et al.*, Phys. Rev. **D25**, 652 (1982).
22. C. I. Westbrook, D. W. Gidley, R. S. Conti, and A. Rich, Phys. Rev. Lett. **58**, 1328 (1987); erratum: Phys. Rev. Lett. **58**, 2153 (1987).
23. G. S. Adkins, Ann. Phys. (N.Y.) **146**, 78 (1983).
24. M. A. Stroscio, Phys. Rev. Lett. **48**, 571 (1982); G. S. Adkins, Phys. Rev. A **27**, 530 (1983), *ibid.*, **31**, 1250 (1983).
25. V. W. Hughes and G. Z. Putlitz, Comm. Nucl. Part. Phys. **12**, 259 (1984).
26. A. M. Thompson and D. G. Lampard, Nature (London) **177**, 888 (1956).
27. R. Laughlin, Phys. Rev. **B23**, 5632 (1981).
28. D. R. Yennie, Rev. Mod. Phys. **59**, 781 (1987).
29. M. E. Cage *et al.*, IEEE Trans. Instrum. Meas. **38**, 284 (1989).
30. E. R. Williams *et al.*, IEEE Trans. Instrum. Meas. **38**, 233 (1989).

CONSTRUCTION OF FOUR DIMENSIONAL QUANTUM FIELD MODELS: Φ_4^4 AND QED_4

Keïichi R.Ito

Department of Mathematics, College of Liberal Arts

Kyoto University, Kyoto 606, Japan

and

Department of Mathematics and Information Science,

Konan College of Women, Takaya-Cho, Konan 483

Contents

1. Present Status of Construction of Field Models

Quantum electrodynamics is certainly a great achievement in field theory [1-2]. Its predictions (by renormalization) are so accurate that many physicists believed that QED was established. However, soon after that, Landau conjectured that QED (in 4 dimensions) may not exist as a complete theory in the high energy limit[3]. (Conjectured no-go theorem for QED_4.) Though there have been many failed attempts to overcome this difficulty, there has not been any serious attempts to check or confirm Landau's no-go theorem.

In my opinion, it is dangerous to believe conventional (perturbative) renormalization group arguments. In fact the recent studies of two-dimensional statistical models with non-abelian symmetry suggest that the conventional renormalization group method is unreliable unless renormalization group flow is followed to fairly high orders[4-9].

The second possibility, which is more or less related to the first idea, is that there is a non-trivial zero point in the β-function of the renormalization group equation. This is related to the phase transition in QED (breaking of the chiral gauge transformation)[10-11]. Though it is very hard to prove the existence of this transition, one may be able to construct QED around this point.

In this note, I originally planned to construct QED_4 as the continuum limit of euclidean lattice QED. One of the recent achievements in this field is that the perturbation series in (euclidean) QED_4 are locally Borel summable. This follows from a non-trivial fact that the n-th term of the perturbation series is bounded by $AR^n(n!)^{1/2}$, and implies that the Borel transform of the series is analytic in a neighborhood of the origin[2]. But, alas, this does not imply that the resultant theory is non-trivial and meaningful !

To argue this problem, one has to define block spin transformations for fermions or has to investigate the Matthews-Salam determinant. I have found, however, that these problems are beyond the current technologies of mathematical physics. Therefore (though I am afraid that this topics may not match the editor's original intention) I decided to argue a construction of four-dimensional ϕ^4 model with a coupling constant of *unusual sign* (i.e. negative) because such a model shares many *common features* with QED_4: these models are not asymptotically free at short distances (at least for ϕ_4^4) and thus believed not to exist without assuming something unphysical (e.g. negative coupling constant.) See also the arguments in the final section.

However, as is discussed below, once negative coupling ϕ_4^4 is introduced, then its continuum limit exists and non-trivial. One can also prove that the perturbation expansions are asymptotic to the Green's functions (up to the third order at least). Moreover there is a possibility to find analytic continuations of Green's functions with respect to the coupling constant and to study whether or not physically sensible quantum electrodynamics exists.

Therefore the present study will cast a light on new and constructive study of 4D quantum electrodynamics which is believed trivial by most of modern physicists. ("Trivial" means that the full theory of QED does not exist and the perturbation expansion has nothing to do with the physical phenomena in the world of electrons and photons.)

Remark 1. It is not yet established rigorously that ϕ_4^4 is trivial. See Refs.12 and 13. But following Refs.15 and 16, one finds that the continuum theory is trivial and gaussian if the bare coupling constant is positive and small.

2. Construction of the ϕ_4^4 Sytem and Main Results

A difficulty in constructing ϕ_4^4 (as well as QED_4) is that the system is (perturbatively) asymptotically free at *long distance scales* if λ_0(=bare coupling constant) is > 0. This means that if we start with a bare action (e.g. action with distance scale = δ << Plank's length)

$$\upsilon_0 = \frac{1}{2} \sum_{\substack{|x-y|=1 \\ x,y\in\Lambda}} (\phi(x)-\phi(y))^2 + \frac{1}{2} m_0^2 \sum_{x\in\Lambda} \phi^2(x) + \lambda_0 \sum_{x\in\Lambda} \phi^4(x) \tag{2.1}$$

and if $m_0^2=m_{cr}^2$ (defined later), then the effective action υ_n at distance scale $L^n\delta$ defined by

$$\exp[-\upsilon_n(\Phi)] = \int \exp[-\upsilon_0(\phi)] \prod_{x\in\Lambda_n} \delta[\Phi(x)-(C^n\phi)(x)] \prod_{z\in\Lambda} d\phi(z) \tag{2.2}$$

looks like that of free (gaussian) systems in the limit of $n \to \infty$. Here L is an even integer and $\Lambda = \Lambda_0$ is a rectangular set of integer points in four dimensions:

$$\Lambda = \{(x_1,..,x_4);\ x_i= -L^K/2,\ -L^K/2 +1,..,\ L^K/2 -1\}, \tag{2.3a}$$

$$\Lambda_n= \{(x_1,..,x_4);\ x_i= -L^{K-n}/2,\ -L^{K-n}/2 + 1,..,\ L^{K-n}/2 -1\}, \tag{2.3b}$$

(K is an arbitralily large integer) and the block spin operator C is defined as an averaging operator which maps functions defined on Λ_n to functions defined on Λ_{n+1}:

$$C: R^{\Lambda_k} \rightarrow R^{\Lambda_{k+1}}$$

$$f(x) \rightarrow (Cf)(x) = L^{-3} \sum_{\substack{z \\ -L/2 \le z_\mu \le L/2-1}} f(Lx+z) \quad . \qquad (2.4)$$

Namely (Cf)(x) is the arithmatic average of f(z) over the points of square of size L×..×L with center at Lx, except for the additional scaling factor L. Therefore (2.2) is the integrations over the fluctuations around the fixed block spins { $\Phi(x)$ } which will be integrated out in later inductive steps.

The procedure (2.2) has a semi-group structure, and can be iterated many times. It is usually a hard work to follow the trajectory of effective actions { υ_n }, and our main efforts are devoted to integrate (2.2) recursively without losing control of the flow of the trajectory.

Let us explain how to obtain the continuum theory within our framework.

Prepare the bare mass $m_0^2=m_0^{2(N)}$ and the coupling constant $\lambda_0=\lambda_0^{(N)}$ at the distance scale $\delta=L^{-N}$ (though we regard δ as 1) and iterate the recursion formulas N times to obtain the theory at the distance scale 1. Insist that all physical quantities should be non-zero and finite at this scale. Then we let $N \rightarrow \infty$ keeping these quantities non-zero and finite, and check whether or not λ_0 and m_0^2 stay bounded.

There have been several attempts to construct a non-trivial 4D ϕ^4 model. Among them, Gawedzki and Kupiainen showed[14)] that one can construct a non-trivial continuum euclidean ϕ_4^4 model if one starts with the bare lattice action of negative coupling constant, within the *hierarchical approximation* invented by Wilson and Ma and deeply studied by Gawedzki and Kupiainen themselves subsequently.

The trick used by them is to rotate $\phi(x)\in R$ by an angle $\alpha \in (\pi/8,\pi/4)$ in the complex plane : $\phi(x)\rightarrow e^{i\alpha}\phi(x)$. Then $\mathrm{Re}\ e^{4i\alpha}\lambda_0>0$ and the integral (2.2) is stabilized. Moreover $\mathrm{Re}\ e^{2i\alpha} > 0$, which means that the gaussian integral $\langle P\rangle \equiv\int P\ d\mu_1(e^{i\alpha}z)$ is well defined whenever P is a polynomial of z, and thus perturbation expansions are left invariant. The reader should note that nothing is changed by this trick as far as the perturbation expansion is concered. See Fig.1.

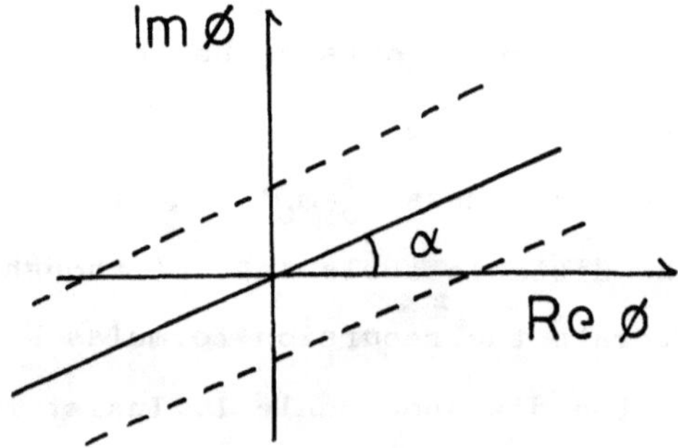

Fig.1. Complex contour $e^{i\alpha}\phi(x)$. v_n is analytic in the strip region $|\mathrm{Im}\ e^{-i\alpha}\phi|< B|\lambda_n|^{-1/4}$. λ_n is the coupling constant at the distance scale $L^n\delta$. See sec.3.

Remark 2. It is widely believed that $(\mathrm{QED})_4$ belongs to the same class of models as ϕ_4^4. But to my best knowledge, there is no rigorous proof for this claim. If one starts with lattice $(QED)_4$ with fermions, it is hard to carry out the block spin transformation because of the non-local nature of fermions and essentially nothing is known.

Our main task in this note is to extend this program to a *real model* and show that we can learn much about typical field models which can be obtained by perturbation theory. In fact we can show that the perturbation expansion is asymptotic with respect to renormalized coupling constant up to the third order. (Presumably they are Borel summable. But so far we do not have a proof.) However, as it will turn out soon, the resultant theory may be unphysical and may not be analytically continued to the Minkowski space.

Before we state our main theorem, we have to explain our notation following Refs.15 and 16:

$$\phi_n(x)=(C\,\phi_{n-1})(x)=L^{-3}\sum_{-L/2\le\xi_\mu\le L/2-1}\phi_{n-1}(Lx+\xi) \tag{2.5}$$

$$G_n(x,y)=(CG_{n-1}C^+)(x,y) \tag{2.6a}$$

where $\phi_n(x)$ is the block spin variable at distance scale L^n. L^{-3} is chosen so that their covariances (two point funcions) $G_n(x,y)$ are essentially equal to the original $G_0(x,y)$:

$$\begin{aligned} G_0(x,y)&=\text{Green's function of the lattice Laplacian}\\ &=(-\Delta)^{-1}(x,y)\sim (x-y)^{-2} \end{aligned} \tag{2.6b}$$

and thus

$$G_n(x,y) = \text{propagator (Green's function) of } \phi_n$$
$$\equiv L^{-6n} \sum_{\zeta,\xi} G_0(L^n x+\zeta, L^n y+\xi)$$
$$\sim (x-y)^{-2}, \qquad (2.7)$$

where ζ and ξ are points in the square of size $L^n\times..\times L^n$ with center at the origin. This means that $\{\phi_n\}$ are very similar to the original $\{\phi=\phi_0\}$ and reflects the idea of *scaling* in critical phenomena. We introduce two operators though they will not be used explicitly in this note.

The first one is Q which maps $f(x)\in R^{\Lambda_n\setminus L\Lambda_{n+1}}$ (functions defined only for $x\in\Lambda_n\setminus L\Lambda_{n+1}$(set of points which are in Λ_n but not in $L\Lambda_{n+1}$)) to $(Qf)(x)$ which is defined for all $x\in\Lambda_n$. Here $(Qf)(x)=f(x)$ if $x\in\Lambda_n\setminus L\Lambda_{n+1}$ and the values of $(Qf)(x)$ for $x\in L\Lambda_{n+1}$ are defined so that $C(Qf)=0$:

$$Q: R^{\Lambda_n\setminus L\Lambda_{n+1}} \to R^{\Lambda_n}$$
$$f(x) \to (Qf)(x) = \begin{cases} f(x) \text{ if } x\in\Lambda_n\setminus L\Lambda_{n+1} \\ -\sum\limits_{-\frac{L}{2}\le\zeta_\mu\le\frac{L}{2}-1} f(x+\zeta) \text{ if } x\in L\Lambda_{n+1} \end{cases} \qquad (2.8)$$

Roughly speaking, Q is an operator which gets fluctuaion fields from the field variables.

The second one is the projection R:

$$R : R^{\Lambda_n} \to R^{\Lambda_n\setminus L\Lambda_{n+1}}$$
$$f(x) \to (Rf)(x) = \begin{cases} f(x) \text{ if } x\in\Lambda_n\setminus L\Lambda_{n+1} \\ 0 \quad \text{otherwise.} \end{cases} \qquad (2.9)$$

Thus for any function $f(x)$ defined on Λ_n, we see that $g(x)\equiv (QRf)(x)$ satisfies $Cg=0$.

Using these operators, we can naturally decompose $\{\phi_n(x);\ x\in\Lambda_n\}$ into block spin variables $\{\phi_{n+1}\}$ of next distance scale and fluctuation fields $\{\xi_n(x);\ x\in\Lambda_n\setminus L\Lambda_{n+1}\}$:

$$\phi_n(x)= (A_n\phi_{n+1})(x)+(Q\xi_n)(x) \tag{2.10}$$

where $A_n:R^{\Lambda n+1} \to R^{\Lambda n}$ is given by

$$A_n(x,y)= (G_n\, C^+ G_{n+1}^{-1})(x,y) \tag{2.11}$$

and $\{\xi_n(x);x\in\Lambda_n\setminus L\Lambda_{n+1}\}$ are gaussian random variables of zero mean and covariance

$$\Gamma_n(x,y)=R(G_n-G_nC^+G_{n+1}^{-1}G_n)R^+. \tag{2.12}$$

[Multiply C to (2.10) from left hand side, and use (2.11) to check that (2.10) is true. Consider the two point functions of the both sides of (2.10) and use (2.11) to confirm that (2.12) is true.]

It is not difficult to see that

$$|A_n(Lx+\tilde{x},y)| \leqq C_1\exp[-\beta|x-y|], \tag{2.13a}$$

$$|\Gamma_n(x,y)| \leqq C_2\exp[-\beta|x-y|], \tag{2.13b}$$

where $|\tilde{x}|< L/2$. For the later purposes, it is convenient to define

$$\mathcal{A}_n(x,y)= L^nA_0A_1\cdots A_{n-1}(L^nx,y) \tag{2.14}$$

$x\in L^{-n}\Lambda$ (lattice width=L^{-n}), $y\in\Lambda$ (lattice width=1)

and introduce two types of gaussian random variables:

$$\psi_n(x)= (\mathcal{A}_n\phi_n)(x), \tag{2.15a}$$

$$z_n(x)= (\mathcal{A}_nQ\xi_n)(x)\equiv (\mathcal{A}_nQ\Gamma_n^{1/2}\, z_n)(x), \tag{2.15b}$$

where $x \in L^{-n}\Lambda$. We also define

$$\int F(\psi_n(x),z_n(x))dx \equiv L^{-4n} \sum_x F(\psi_n(x),z_n(x)) .$$

Thus eq.(2.10) is simplified:

$$\psi_n(x)= \frac{1}{L}\psi_{n+1}(x/L) + z_n(x) \tag{2.16}$$

where $x\in L^{-n}\Lambda$. Their means are zero, and their covariances are respectively given by

$$\mathscr{G}_n(x,y) = (\mathscr{A}_n G_n \mathscr{A}_n^+)(x,y) \sim (x-y)^{-2} \tag{2.17a}$$

and

$$\mathscr{T}_n(x,y) = (\mathscr{A}_n Q\Gamma_n Q^+ \mathscr{A}_n^+)(x,y) \sim e^{-\beta|x-y|} \,, \tag{2.17b}$$

where both x and y $\in L^{-n}\Lambda$. We also define

$$\begin{aligned}\mathscr{Q}_n(x,y) &= \sum_{k=0}^{n-1} L^{2(n-k)} \mathscr{T}_k(L^{n-k}x, L^{n-k}y) \\ &= L^{2n} G_0(L^n x, L^n y) - \mathscr{G}_n(x,y)\end{aligned} \tag{2.17c}$$

and

$$\mathscr{S}_n(x_1,x_2,x_3) = \mathscr{Q}_n(x_1,x_2)\mathscr{Q}_n(x_2,x_3). \tag{2.17d}$$

$\mathscr{T}_n$, $\mathscr{Q}_n$ and $\mathscr{S}_n$ have exponential decay property uniform in n and the difference between G_n and $\mathscr{G}_n$ is marginal.

To calculate the renormalization recursion formulas (2.2), we separate $(\phi,(-\Delta)\phi) = \Sigma\,(\phi(x)-\phi(y))^2$ ($|x-y|=1$) from $v_0 = v_0^{(N)}$ and represent it as

$$\begin{aligned}&\prod_n [\ \exp[-(\xi_n,\Gamma_n^{-1}\xi_n)/2]\ \Pi d\xi_n(x)\ ] \\ &\equiv \prod_n d\mu_{\Gamma_n}(\xi)\end{aligned}$$

or as $\prod_n d\mu_1(z_n)$, after introducing new variables $z_n = (\mathscr{A}_n Q\Gamma_n^{1/2} z_n)$ where $d\mu_1$ is the gaussian measure of zero mean and covariance 1. Therefore our recursion formulas is now

$$\exp[-v_{n+1}(\psi(.))] = \mathscr{N}^{-1} \int \exp[-v_n(\psi(./L)/L + z_n(.))]\ d\mu_1(z_n). \tag{2.18}$$

We are now ready to state our theorem. Note, however, that the following theorem is oversimplified at the cost of accuracy. In fact $\{\psi_n(x); x\in L^{-n}\Lambda\}$ are connected to each other by functions $\mathscr{T}_n$, $\mathscr{S}_n$ and etc : they are not local though they are approximately local. Namely neither small field configuration nor large field configuration is a local notion, and they cannot be sharply separated. So I am slightly cheating the reader by pretending as if all fields were simultaneously large or small. See Ref.17 for the details, but the essential structure of the proof is seen in Refs.15 and 16. We just mention that we need to estimate v_n up to ψ^8 (or $O(|\lambda_n|^{3+\varepsilon})$) to fix the flow.

Theorem 1. Let the bare lattice action on $\Lambda=\Lambda^{(K)}$ be

$$v_0 = v_0^{(N)} = \frac{1}{2} m_0^2 \int \psi_0^2(x)dx - 6\lambda_0 \int \mathcal{G}_0(x,x)\psi_0^2(x)dx + \lambda_0 \int \psi_0^4(x)dx \tag{2.19}$$

where

$$\frac{1}{\lambda_0} = \frac{1}{\lambda_{phys}} - \beta_2 N + c_3 \log(1-\beta_2\lambda_{phys}N) \tag{2.20a}$$

$$\lambda_{phys} < 0 \tag{2.20b}$$

and β_2 (>0) and c_3 are constants specified later. Then for $|\lambda_{phys}| << 1$, there exists $m_0^2 \in [-|\lambda_0|^{3/2}, \ |\lambda_0|^{3/2}]$ such that $\exp[-v_n(\psi)]$ exists for all $n < N$ and the continuum limit $\lim_{N\to\infty} \exp[-v_{N-1}^{(N)}]$ exists. The series $\{ v_n = v_n^{(N)} \}$ satisfy:

(i) *Analyticity in the small field region*

There exist constants m_n^2, λ_n, γ_n and η_n such that

$$\lambda_n \in [-c_-/(N+n_0-n), -c_+/(N+n_0-n)], \tag{2.21a}$$

$(c_\pm > 0, \ n_0 > 0)$

$$m_n^2 \in [-c_1|\lambda_n|^{3/2}, \ c_1|\lambda_n|^{3/2}], \tag{2.21b}$$

$$\gamma_n \in [8\lambda_n^2 - O(\lambda_n^{2+\varepsilon}),\ 8\lambda_n^2 + O(\lambda_n^{2+\varepsilon})], \tag{2.21c}$$

$$\eta_n \in [96\ \lambda_n^3 - O(\lambda_n^{3+\varepsilon}),\ 96\ \lambda_n^3 + O(\lambda_n^{3+\varepsilon})]. \tag{2.21d}$$

Then v_n is analytic in $\mathcal{X}_n = \{\ \psi_n(x) \in C\ ;\ |\psi_n| < B|\lambda_n|^{-1/4}\ \}$ and admits the following expansion there:

$$\begin{aligned} v_n = {} & \frac{1}{2}\, m_n^2 \int dx\ \psi_n^2(x) - 6\lambda_n \int dx\ \mathcal{G}_n(x,x)\psi_n^2(x) \\ & + \lambda_n \int dx\ \psi_n^4(x) + \gamma_n \int dx dy\ \mathcal{D}_n(x,y)\psi_n^3(x)\psi_n^3(y) \\ & + \eta_n \int dx_1 dx_2 dx_3\ \mathcal{G}_n(x_1,x_2,x_3)\ \psi_n^3(x_1)\psi_n^2(x_2)\psi_n^3(x_3) \\ & + (\text{irrelevant terms}) \\ & + \tilde{v}_n(\psi), \end{aligned} \tag{2.22}$$

where

$$\partial^k\ \tilde{v}_n / \partial\psi_n^k\ |_{\psi=0} = 0 \text{ for } k=1,\ldots,8,$$
$$|\tilde{v}_n| \leqq \text{const}\ |\lambda_n|^{3/2}\ .$$

(ii) *Uniform Boundedness of the Gibbs Factor*

Let $\mathcal{D} = \{\ |\mathrm{Im}\ e^{-i\alpha}\psi\ | < B|\lambda_n|^{-1/4}\ \}$, where $\alpha \in (\pi/8, \pi/4)$. Then the Gibbs factor $\exp[-v_n(\psi)]$ is analytic there and satisfies

$$|\exp[-v_n]| < \exp[\ -|\lambda_n|^{1/2}|\psi_n|^2 + |\lambda_n||\mathrm{Im}\ \psi_n|^4 + D] \tag{2.23}$$

with a uniform constant D.

We must add some remarks and notes here:

(1) For simplicity we neglected the wave function renormalization which comes from the quadratic terms in v_n.

(2) Since $\{\psi_n(x)\}$ are extended to $L^{-n}\Lambda$ and are related to each other (non-local effects), one cannot pick out $\psi_n(x)$ and discuss $\psi_n(x)$ only. One would rather defines $\mathcal{X}_n$ and $\mathcal{D}_n$

in a much refined (and then complicated) way so that the cluster expansion can be inductively applied [14-17].

3. Outline of the Proof

We prove the theorem through several steps:

[*Small Field Region*]

If $|\psi_n(x)| < B|\lambda_n|^{-1/4}$, we have the expansion (2.22). So set $\psi_n(x)=\psi_{n+1}(x/L)/L + z_n(x)$, and substitute it into the right hand side of eq.(2.22). Then one finds

$$v_n = v_n^{(0)} + \delta v_n$$

where

$$\begin{aligned} v_n^{(0)} = \ & \tfrac{1}{2}L^2 m_n^2 \int \psi_{n+1}^2(x)dx \\ & +\lambda_n \int \psi_{n+1}^4(x)dx - 6L^2\lambda_n \int \mathscr{G}_n(Lx,Lx)\psi_{n+1}^2(x)dx \\ & +L^2\gamma_n \int \mathscr{D}_n(Lx,Ly)\psi_{n+1}^3(x)\psi_{n+1}^3(y)\, dxdy \\ & +L^4\eta_n \int \mathscr{S}_n(Lx_1,Lx_2,Lx_3) \times \\ & \quad \times \psi_{n+1}^3(x_1)\psi_{n+1}^2(x_2)\psi_{n+1}^3(x_3)\, \Pi dx_i, \end{aligned} \tag{3.1}$$

and

$$\delta v_n = \delta v_n^{(1)} + \delta v_n^{(2)} + \delta v_n^{(3)} + \tilde{v}_n \ . \tag{3.2}$$

Here $\delta v_n^{(i)}$ are polynomials of ψ_{n+1} and z_n, and $\delta v_n^{(1)} = O(\lambda_n, m_n^2)$, $\delta v_n^{(2)} = O(\lambda_n^2, \gamma_n)$, $\delta v_n^{(3)} = O(\lambda_n^3, \eta_n)$. Finally $\tilde{v}_n$ is the remainder which may contain irrelevant terms.

Explicitly writing , we have

$$\begin{aligned} \delta v_n^{(1)} = \ & (m_n^2/L) \int \psi_{n+1}(x/L)z_{n+1}(x)\, dx \\ & +\lambda_n \int \Big[\tfrac{4}{L}\psi_{n+1}^3(x/L)z_{n+1}(x) + \tfrac{6}{L^2}\psi_{n+1}^2(x/L)z_{n+1}^2(x) \\ & \qquad + \tfrac{4}{L^3}\psi_{n+1}(x/L)z_{n+1}^3(x)\Big]\, dx \\ & -12\lambda_n \int \mathscr{G}_n(x,x)\, \tfrac{1}{L}\psi_{n+1}(x/L)z_{n+1}(x)dx \end{aligned} \tag{3.3a}$$

where we omit z_n^4 which have no effect on the renormalization of the coefficients. In the same way

$$\delta v_n^{(2)} = \gamma_n \int \mathcal{Q}_n(x,y) \{ \frac{3}{L^3} [\psi_{n+1}^3(x/L)\psi_{n+1}^2(y/L)z_{n+1}(y) + (x \leftrightarrow y)] + \frac{1}{L^4} [9\psi_{n+1}^2(x/L)\psi_{n+1}^2(y/L)z_{n+1}(x)z_{n+1}(y) + 3\psi_{n+1}^3(x/L)\psi_{n+1}(y/L)z_n^2(y) + (x \leftrightarrow y)] + \ldots \} dxdy, \tag{3.3b}$$

$$\delta v_n^{(3)} = \eta_n \int \mathcal{G}_n(x_1,x_2,x_3) \{ \frac{3}{L} [\psi_{n+1}^2(x_1/L)\psi_{n+1}^2(x_2/L) \times \psi_{n+1}^3(x_3/L)z_{n+1}(x_1)+(x_1 \leftrightarrow x_3)]+\ldots\} \Pi dx_i . \tag{3.3c}$$

We now have

$$\begin{aligned} v_{n+1} &= v_n^{(0)} - \log [\int \exp[-\delta v_n] d\mu_1(z_n)] \\ &= v_n^{(0)} + \{ \langle\delta v_n\rangle - (2!)^{-1} \langle\delta v_n,\delta v_n\rangle \\ &\qquad + (3!)^{-1}\langle\delta v_n,\delta v_n,\delta v_n\rangle + \text{remainder} \}, \end{aligned} \tag{3.4}$$

where

$$\langle . \rangle \equiv \int e^{i\alpha_R} (.) d\mu_1(z), \quad \langle\delta v,\delta v\rangle \equiv \langle\delta v\delta v\rangle - \langle\delta v\rangle^2 \tag{3.5}$$

and so on.

Even if $|\psi_{n+1}|$ are small, $\psi_{n+1}/L + z_n$ may be large and thus the expansion (2.20) may fail there. In this case we must use the bound (ii) of section 2. For simplicity we skip all these difficulties. The results are:

(1) mass terms

$$\begin{aligned} &\tfrac{1}{2} L^2 m_n^2 \int \psi_{n+1}^2(x)\, dx \\ &- \tfrac{1}{2} L^2 \lambda_n^2 \int dxdy\, \psi_{n+1}(x)\psi_{n+1}(y) [144\, \mathcal{G}_n(Lx,Lx)\mathcal{G}_n(Ly,Ly) \\ &\mathcal{T}_n(Lx,Ly) + 96\mathcal{T}_n^3(Lx,Ly) - 288\mathcal{G}_n(Lx,Lx)\mathcal{T}_n(Ly,Ly)\mathcal{T}_n(Lx,Ly)] \\ &+ 24L^6\lambda_n m_n^2 \int dxdy\, \psi_{n+1}(x)\psi_{n+1}(y)\, \mathcal{T}_{n+1}(Lx,Ly) \\ &\times [-\mathcal{G}_n(Lx,Ly) + \mathcal{T}_n(Lx,Ly)] + O(\lambda_n^3, m_n^2\lambda_n, \gamma_n, \eta_n). \end{aligned}$$

In the last terms, we decompose kernel functions depending

on x and y. The diagonal components are just mass terms and off diagonal terms are absorbed as a wave function re-normalization $(\sim(\partial\psi)^2)$ or as irrelevant terms.
They are all of order $O(\lambda_n^2)$:

$$m_{n+1}^2 = L^2 m_n^2 - \alpha_1\lambda_n^2 + O(\lambda_n^{2+\varepsilon}). \tag{3.6}$$

(2) ψ^4 and Wick order terms

After some calculations, we find that they are:

$$\begin{aligned}
&6L^2\lambda_n \int dx\, [\mathcal{G}_n(Lx,Lx)-\mathcal{T}_n(Lx,Lx)]\psi_{n+1}^2(x)+\lambda_n \int \psi_{n+1}^4(x)dx \\
&+ L^4\lambda_n^2 \{ -72 \int dxdy\, \mathcal{D}_n(Lx,Ly)\mathcal{T}_n(Lx,Ly)\psi_{n+1}^2(x)\psi_{n+1}^2(y) \\
&\qquad -48 \int dxdy\, \mathcal{D}_n(Lx,Ly)\mathcal{T}_n(Ly,Ly)\psi_{n+1}^3(x)\psi_{n+1}(y) \\
&\qquad -36 \int dxdy\, \mathcal{T}_n(Lx,Ly)^2\, \psi_{n+1}^2(x)\psi_{n+1}^2(y) \\
&\qquad -48 \int dxdy\, \mathcal{T}_n(Lx,Ly)\mathcal{T}_n(Ly,Ly)\, \psi_{n+1}^3(x)\psi_{n+1}(y) \\
&\qquad +48 \int dxdy\, \mathcal{T}_n(Lx,Ly)\mathcal{G}_n(Ly,Ly)\, \psi_{n+1}^3(x)\psi_{n+1}(y) \} \\
&-8L^4\lambda_n m_n^2 \int dxdy\, \mathcal{T}_n(Lx,Ly)\psi_{n+1}^2(x)\psi_{n+1}^2(y) \\
&+O(\lambda_n\gamma_n)+O(\lambda_n^3)+O(\eta_n).
\end{aligned}$$

We estimate the contributions to the coefficient of $\int\psi_{n+1}^4(x)$, for example, by examining

$$-72L^4\lambda_n^2 \int_{\square_0} dx \int dy\, \mathcal{D}_n(Lx,Ly)\mathcal{T}_n(Lx,Ly)$$

as for the first term in the second order terms, where $\square_0$ is the square of size $L\times..\times L$ with center at the origin. As was proved in [15], the main contribution comes from

$$L^4 \int_{\square_0} dx \int dy\, \mathcal{T}_n^2(Lx,Ly) \sim O(\ln L).$$

Thus (since $L^2 (\mathcal{G}_n(Lx,Lx)-\mathcal{T}_n(Lx,Lx))= \mathcal{G}_{n+1}(x,x)$)

$$\lambda_{n+1}=\lambda_n-\beta_2\lambda_n^2 -\beta_3\lambda_n^3 -\beta_4 m_n^2\lambda_n+ O(\lambda_n^4 , m_n^4 , m_n^2\lambda_n^2), \tag{3.7a}$$

$$\beta_2 > 0. \tag{3.7b}$$

(3) ψ^6 and ψ^8 terms

We first set $\gamma_n=8\lambda_2^2$ and $\eta_n=96\lambda_n^3$. Then these terms are

$$-8L^2\lambda_n^2 \int dxdy\ [\mathcal{Q}_n(Lx,Ly)+\mathcal{T}_n(Lx,Ly)]\psi_{n+1}^3(x)\psi_{n+1}^3(y)$$
$$+96\lambda_n^3 \int \pi\ dx_i\ L^4\ [\ \mathcal{G}_n(Lx_1,Lx_2,Lx_3)$$
$$+(\mathcal{Q}_n(Lx_1,Lx_2)\mathcal{T}_n(Lx_2,Lx_3)+(x_1\leftrightarrow x_3))$$
$$+\ \mathcal{T}_n(Lx_1,Lx_2)\mathcal{T}_n(Lx_2,Lx_3)]\psi_{n+1}^3(x_1)\psi_{n+1}^2(x_2)\psi_{n+1}^3(x_3)$$
$$+O(\lambda_n^3)\times(\text{ Polynomials of order} \leq 6\ )$$

where $L^2[..] = \mathcal{Q}_{n+1}(x,y)$ as for the first square brakette and $L^4[..]=\mathcal{G}_{n+1}(x_1,x_2,x_3)$ as for the second square brakette. See eqs.(2.17b-d).

(4) Irrelevant terms

The remaining terms not absorbed by these main terms are either of higher order or connected by functions which decrease exponentially as their arguments are separated.

(5) Solving recursion relations and tuning of m_n^2

Now we can solve the recursion relations:

R.1) $m_{n+1}^2 = L^2m_n^2 - \alpha_1\lambda_n^2 + O(\lambda_n^3\ ,\ m_n^2\lambda_n)$,

R.2) $\lambda_{n+1} = \lambda_n-\beta_2\lambda_n^2 -\beta_3\lambda_n^3 -\beta_4 m_n^2\lambda_n +O(\lambda_n^4,\ m_n^2\lambda_n^2)$,

R.3) $\gamma_{n+1} =8\lambda_{n+1}^2 +O(\lambda_{n+1}^3,\ m_n^2\lambda_n)$,

R.4) $\eta_{n+1} = 96\lambda_{n+1}^3+O(\lambda_{n+1}^4\ ,m_n^2\lambda_{n+1}^2)$,

R.5) $\tilde{v}_{n+1} =O((|\lambda_n|+|m_n^2|)^4)$.

The parameters γ_n and η_n are completely determined. [λ_{n+1} and m_{n+1}^2 of course have feedback from γ_n and η_n which appear in R.1) and R.2) as $O(\lambda_n^3)$. Thus as will be seen, their effects are well controlled.]

First we consider the flow of $\xi_n={}^t(m_n^2, \lambda_n)$ defined by R.1) and R.2) or roughly by

$$\xi_{n+1} = \begin{pmatrix} L^2 & -\alpha_1\lambda \\ -\beta_4\lambda & 1-\beta_2\lambda-\beta_3\lambda^2 \end{pmatrix} \xi_n ,$$

$$\lambda = \lambda_n \ (\text{ or } = \lambda_{n_0} \text{ for some } n_0 < n),$$

and insist that $\{ |\xi_n| \}$ stay as $O(1)$ for $n < N$. Then we find that $m_n^2=[\alpha_1/(L^2-1)]\lambda_n^2 +O(\lambda_n^3)$. Thus the third and fourth terms in the right hand side of R.2) are replaced by $-\tilde{\beta}_3\lambda_n^3$. Deviding both sides of R.2) by $\lambda_n\lambda_{n+1}$, we get:

$$\begin{aligned} \frac{1}{\lambda_n} - \frac{1}{\lambda_{n+1}} &= -\beta_2 \frac{\lambda_n}{\lambda_{n+1}} -\tilde{\beta}_3 \frac{\lambda_n^2}{\lambda_{n+1}} +O(\lambda_n^2) \\ &= -\beta_2 -(\beta_2^2+\tilde{\beta}_3)\frac{\lambda_n^2}{\lambda_{n+1}} +O(\lambda_n^2) \\ &= -\beta_2-(\beta_2^2+\tilde{\beta}_3)\lambda_n+O(\lambda_n^2). \end{aligned}$$

Now assume that λ_N is the observed physical coupling constant $\lambda_{phys} < 0$ ($|\lambda_{phys}| << 1$). Thus $\lambda_0=\lambda_0^{(N)}$ should satisfy

$$\begin{aligned} \frac{1}{\lambda_0} - \frac{1}{\lambda_{phys}} &= -\beta_2 N -(\beta_2^2+\tilde{\beta}_3) \sum_{k=0}^{N-1} [\frac{1}{\lambda_{phys}} -k\beta_2]^{-1}+O(1) \\ &= -\beta_2 N + \frac{\beta_2^2 + \tilde{\beta}_3}{\beta_2} \log[1-\lambda_{phys}\beta_2 N]+ O(1). \end{aligned}$$

This is nothing but the relation (2.20).

Finally let me explain how to choose m_n^2. This is the famous fine tuning problem considered by Bleher and Sinai[15].

Assume m_n^2 changes in I_n containing the origin. Then as a function of m_n^2 , m_{n+1}^2 is continuous and $I_{n+1} \equiv \mathrm{Range}(m_{n+1}^2)$ contains I_n. Thus we can choose m_0^2 as our requirements hold.

[*Large Field Region*]

This is not so difficult. See 14) for the hierarchical model case and 17) for the present case. (See also Ref.15.)

4. Several Properties of the Resultant Theory

First of all the resultant theory is non-trivial (differs from a free theory). In fact we have ψ^6 and ψ^8 terms . Secondly, the perturbation expansion is asymptotic to the Green's (Schwinger's) functions with respect to the renormalized coupling constant at least up to the third order. (In fact they are written in terms of $\lambda_N = \lambda_{phys}$, not of λ_0.) Moreover, though we did not explain explicitly, the wave function renormalization constant is non-zero and finite. In fact the correction to the kinetic term at the n-th step is $O(\lambda_n^2)$, and $0 < \sum \lambda_n^2 < \infty$.

Therefore the resultant theory satisfies almost all properties which should be satisfied by field theory models (explained in many standard text books of field theory !)

However, since the contour is complexified, the reflection positivity is lost and the resultant theory may not be analytically continued to our Minkowski space. (If so, this is related to the probable lack of the ground state of the Hamiltonian.)

5. What We Can Learn from the Model

So how about the answer to the construction (or destruction) of 4D QED ? As we mentioned already, there are several possibilities:

(i) The theory exists as a complete theory. Landau's no-go theorem is not true. This may take the form of QED constructed around a non-trivial zero point of the beta function.

(ii) The theory does not exist at all though it may exist just as a subtheory of a unified theory.

(iii) Same as (ii). But negative coupling theory may exist since this seems to recover the asymptotic freedom at short distances.

What we can do best now is to check Landau's no-go theorem which is still a conjecture. If the answer is true, then we have to give up the theory. But some recent calculations [11)], though some of them are speculative, seem to suggest that there is a chance for the theory to survive as a physically sensible one. Finally, as for the negative coupling QED, it should be mentioned that the renormalized effective interaction of photons obtained from the Matthews-Salam determinant, seems to behave like the negative coupling ϕ^4-model discussed here[18)]. Thus the discussion in this note may play an important role in the construction of 4D QED, more than what we think.

Acknowledgements: The idea in this note arose from the discussions with Drs.T.Hara and K.Kondo while the author attended the BiBoS (Bielefeld-Bochm Stochastics) in 1986. The author also benefited very much from the discussion with Prof.K.Gawedezki in I.H.E.S.

References

1) "Quantum Electrodynamics", J.Schwinger, ed. (Dover, New York, 1958)

2) J.S.Feldman, T.R.Hurd, L.Rosen, and J.D.Wright, "QED: A Proof of Renormalizability", in "Lecture Notes in Physics", vol.312 (Springer-Verlag, Heidelberg 1989).

3) L.D.Landau, "On the Quantum Theory of Fields", in "Niels Bohr and the Developement of Physics", W.Pauli ed. (Pergmon, London, 1955).

4) J.Kogut and K.Wilson, Phys. Rep. 12(1974),75. S.K.Ma, "Modern Theory of Critical Phenomena" (Benjamin, Reading, MA., 1976).

5) A.M.Polyakov, Phys. Letters 59B (1975), 79.

6) K.R.Ito, Phys. Rev. Letters 58 (1987), 439.

7) E.Seiler, I.O.Stamatesacu, I.O.Linke, and A.Patrasciou, Nucl.Phys. B405 (1988), 623.

8) A.Patrasciou, Phys. Rev. Letters 58 (1987), 2285.

9) K.R.Ito, "Kosterlitz-Thouless Transitions in 2 Dimensional Non-Abelian Field Models", preprint (1989).

10) V.A.Miransky, Nuovo Cimento 90A (1985), 149.

11) "Proceedings of the 1988 International Workshop on New Trends in Strong Coupling Gauge Theories" , K.Yamawaki et al, ed. (World Scientific, Singapore, 1988).

12) M.Aizenmann, Commun. Math. Phys. 86 (1982), 1.

13) J.Frohlich, Nucl.Phys. B200 (1982), 281.

14) K.Gawedzki and A.Kupiainen, Nucl.Phys. B257 [FS14] (1985), 474.

15) K.Gawedzki and A.Kupiainen, Commun. Math. Phys. 99 (1985), 197.

16) H.Hara, Journal of Stat. Phys. 47 (1987), 57.

17) K.R.Ito, "Construction of Non-Trivial Four-Dimensional Φ^4-theory with Negative Coupling Constant", preprint (1989).

18) Z.Haba, private communications.

CRITICAL REVIEW OF THE THEORY OF QUANTUM ELECTRODYNAMICS

Noboru Nakanishi
Research Institute for Mathematical Sciences
Kyoto University, Kyoto 606, Japan

Contents

1. INTRODUCTION

Quantum electrodynamics (QED) is a prototype of quantum field theory. It is most well-established both theoretically and experimentally. There are a lot of excellent textbooks and review articles on QED. It is, therefore, not very profitable to add another orthodox review on QED to the bibliography. The present author wishes to present an eccentric review on subtle problems of QED, which is perhaps biased by his own prejudice. It is assumed here that the readers have standard knowledge about QED.

Seven topics are discussed in the following seven sections. Each section is almost independent of others. Those topics are selected by the author's interest. They by no means cover all important theoretical problems of QED. But topics are limited to those of the proper framework of QED, that is, we discuss no such related theories as massive QED, Higgs model, electroweak theory, Schwinger model, Chern-Simons gauge theory, lattice QED, strong-coupling QED, etc.

We denote the Minkowski metric by $\eta_{\mu\nu}$ with the metric convention (+,-,-,-). Summation over repeated indices should be understood. References are not intended to be complete, but limited to those which are directly related to our topics.

2. DIFFICULTY OF THE AXIAL-TYPE GAUGES

Before considering the covariant quantization, we point out that there is an intrinsic difficulty in quantizing the electromagnetic field in such non-covariant gauges as axial, light-cone, and temporal gauges, which we collectively call axial-type gauges.*

Let A_μ be the electromagnetic field; the field strength $F_{\mu\nu}$ is defined by

$$F_{\mu\nu} \equiv \partial_\mu A_\nu - \partial_\nu A_\mu. \tag{2.1}$$

The Lagrangian density of the electromagnetic field is given by

$$\mathcal{L}_{em} = -\frac{1}{4} F^{\mu\nu} F_{\mu\nu}, \tag{2.2}$$

to which the charged-field Lagrangian density $\mathcal{L}_{ch}$ should be added. The electromagnetic current j_μ, defined by

$$ej^\mu \equiv \partial \mathcal{L}_{ch}/\partial A_\mu, \tag{2.3}$$

is conserved:

* For an extensive review on the axial-type gauge, see [Leibbrandt 1987].[1)]

$$\partial_\mu j^\mu = 0. \tag{2.4}$$

Then the Maxwell equation is

$$\Box A_\mu - \partial_\mu \partial^\nu A_\nu = -ej_\mu. \tag{2.5}$$

As is well known, it is impossible to quantize a gauge field in a gauge-invariant way; one must make gauge fixing. Usually, the gauge-fixing condition is set up by introducing a gauge-fixing term into the Lagrangian density. But the axial-type gauge-fixing condition,

$$n^\mu A_\mu = 0, \tag{2.6}$$

can be taken into account directly at the level of the operator equation. Here n_μ is a c-number constant vector; the gauge fixing is called axial, light-cone, and temporal according as n_μ is spacelike, light-like, and timelike, respectively. The axial-type gauges are of particular interest in the non-abelian gauge theory because there the Coulomb gauge becomes very complicated and because the Faddeev-Popov (FP) ghosts decouple in the axial-type gauges. However, the axial-type gauges are quite subtle ones in spite of their simple-looking appearance.

For definiteness, we consider the axial gauge

$$A_3 = 0. \tag{2.7}$$

The canonical variables are A_1 and A_2, and their canonical conjugates are

$$\pi^i \equiv \partial^i A_0 - \partial_0 A^i \qquad (i=1,2). \tag{2.8}$$

From the zeroth component of (2.5) and from (2.8), we obtain

$$A_0 = \partial_3^{-2}(\partial_i \pi^i + ej_0). \tag{2.9}$$

Here a *non-local* operator ∂_3^{-2} must be introduced. If we define ∂_3^{-1} by

$$\partial_3^{-1} f(x^3) \equiv \frac{1}{2} \int_{-\infty}^{+\infty} dy^3 \, \epsilon(x^3-y^3) f(y^3) \tag{2.10}$$

then $(\partial_3^{-1})^2$ is *divergent* [Schwinger 1963][2]. Hence instead of identifying ∂_3^{-2} with $(\partial_3^{-1})^2$, we set

$$\partial_3^{-2} f(x^3) \equiv \frac{1}{2} \int_{-\infty}^{+\infty} dy^3 |x^3-y^3| f(y^3) \tag{2.11}$$

by dropping a divergent term proportional to

$$\int_{-\infty}^{+\infty} dy^3 f(y^3). \tag{2.12}$$

Accordingly, if the time-independent operator

$$G(x^1,x^2) \equiv \int_{-\infty}^{+\infty} dx^3 (\partial_i \pi^i + e j_0) \tag{2.13}$$

could be identified with zero, the above trouble would be evaded. The operator $G(x^1,x^2)$ is essentially the generator of the so-called residual local gauge transformation $A_i \to A_i' = A_i + \partial_i \Lambda$ with $\Lambda \equiv \Lambda(x^1,x^2)$, that is,

$$i[G(x^1,x^2), A_i(y)] = \partial_i \delta^2(x-y). \tag{2.14}$$

Accordingly, $G(x^1,x^2)$ <u>cannot</u> be identified with zero. It is customary, therefore, to set up a supplementary condition

$$G(x^1,x^2)|f\rangle = 0. \tag{2.15}$$

Then $G(x^1,x^2)$ cannot be hermitian in the usual sense (for further discussion, see Sec. 3).

As mentioned above, ∂_3^{-2} is defined by dropping a divergent term, that is, it is defined as a *finite part*. But, in general, to take a finite part is incompatible with the positivity of the Hilbert-space metric. In the following, we prove the indispensability of indefinite metric in the axial-type gauge quite generally, *provided that the usual framework* (of course, except for manifest covariance) *of quantum field theory is maintained* [Nakanishi 1983].[3)]

From the Lorentz transformation properties of A_μ, we can write [Frenkel et al. 1976, West 1983][4),5)]

$$\langle 0|A_\mu(x)A_\nu(y)|0\rangle = (2\pi)^{-3}\int d^4k e^{-ik(x-y)}\theta(k_0)\rho_{\mu\nu}(k), \tag{2.16}$$

where

$$\rho_{\mu\nu}(k) \equiv \rho^{(1)}(k^2,k_L)[-\eta_{\mu\nu}k_L{}^2+(n_\mu k_\nu+n_\nu k_\mu)k_L-n^2k_\mu k_\nu]k_L{}^{-2} + \rho^{(2)}(k^2,k_L)(\eta_{\mu\nu}n^2-n_\mu n_\nu) \tag{2.17}$$

with $k_L \equiv n^\mu k_\mu$. Since the canonical commutation relation implies a sum rule

$$\int_0^\infty dk^2\rho^{(1)}(k^2,k_L) = 1, \tag{2.18}$$

$\rho^{(1)}(k^2,0)$ cannot identically vanish. Accordingly, we must define $k_L{}^{-2}$ appearing in (2.17) in the sense of a distribution. Since $k_L{}^{-2}$ is real and satisfies $k_L{}^2k_L{}^{-2} = 1$, we can write

$$k_L{}^{-2} \equiv \mathrm{Re}(k_L-i0)^{-2}+c_1\delta(k_L)+c_2\delta'(k_L), \tag{2.19}$$

where c_1 and c_2 are undetermined (real) constants.*

---

* In the case $n_\mu = (0,0,0,1)$, $k_L{}^{-2} = k_3{}^{-2}$ is the Fourier transform of $-\partial_3{}^{-2}$. The definition (2.11) means to set $c_1=c_2=0$; more precisely, (2.10) implies $c_2=0$, and it implies $c_1=0$ to drop the term proportional to (2.12).

It is very crucial to observe that k_L^{-2} *is not a positive-definite quantity*. Mathematically, *it does not define a measure*; it is not well-defined in a closed interval $[0,a]$ $(a > 0)$.

Let m_μ be a constant c-number vector which is neither parallel nor orthogonal to n_ν. Then, since $m^\mu k_\mu$ is linearly independent of k_L, we see from (2.17) that $m^\mu m^\nu \rho_{\mu\nu}(k)$ is *not a measure*.

We prove the indispensability of indefinite metric by reduction to absurdity; so, we assume that the state-vector space has positive-definite metric. Then there is an orthonormal basis $\{|n\rangle\}$ such that

$$\sum_n |n\rangle\langle n| = 1. \tag{2.20}$$

Inserting (2.20) into $\langle 0|m^\mu A_\mu \cdot m^\nu A_\nu|0\rangle$, we find that

$$m^\mu m^\nu \rho_{\mu\nu}(k) \geqq 0 \tag{2.21}$$

owing to the positivity. There is, however, a general theorem in the theory of distributions [Schwartz 1950],[6] which states that *any positive-semidefinite distribution is a measure*. We thus encounter a contradiction. Therefore, metric cannot be positive definite. Evidently, this proof is applicable to any theory involving the momentum-space singularities *non-integrable in the sense of functions*.*

The above pathology already arises in the definition of the transverse projection, which is needed to specify physical states. The transverse unit vectors $e_{i\mu}$ $(i=1,2)$ are defined by

$$n^\mu e_{i\mu} = k^\mu e_{i\mu} = 0, \quad e_{i\mu} e_j{}^\mu = \eta_{ij} \tag{2.22}$$

on mass shell $k^2 = 0$. We then have

$$\sum_{i=1}^{2} e_{i\mu} e_{i\nu} = [-\eta_{\mu\nu} k_L{}^2 + (n_\mu k_\nu + n_\nu k_\mu) k_L - n^2 k_\mu k_\nu] k_L{}^{-2}. \tag{2.23}$$

---

* Dimensional regularization resolves nothing in this respect because it cannot be realized in terms of state vectors.

When n_μ is spacelike or lightlike, (2.23) is self-contradictory [Nakanishi 1982].[7] Indeed, let m_μ be a real vector orthogonal to n_μ; then multiplying (2.23) by $m^\mu m^\nu$, we find that the l.h.s. is positive definite, while the r.h.s. is indefinite owing to the non-measure nature of k_L^{-2}. Although the canonical formulation of the light-cone gauge [Bassetto et al. 1985][8] is nicely constructed, it is not free of this difficulty.

As noted at the beginning, however, the indispensability of indefinite metric is proved under the postulate that the usual axioms (except for manifest covariance) of quantum field theory are satisfied. Hence, for example, if translational invariance is violated, the above theorem is no longer applicable. Quite interestingly, it has been found from the direct calculation of the Wilson loop (to fourth order) that translational invariance should be violated in the temporal gauge [Caracciolo et al. 1982, Dahmen et al. 1982].[9),10] For $n_\mu = (1,0,0,0)$, the longitudinal propagator factorizes and its time factor $D(y^0,x^0)$ satisfies

$$(\partial_0{}^y)^2 D(y^0,x^0) = -\delta(y^0-x^0). \tag{2.24}$$

Assuming that $D(y^0,x^0)$ is symmetric in $x^0 \leftrightarrow y^0$, we obtain

$$D(y^0,x^0) = -\frac{1}{2}|y^0-x^0| + \frac{1}{2}\,a(y^0+x^0) + \text{const}. \tag{2.25}$$

Here the first term corresponds to the momentum-space principal value. In order to reproduce the correct result (to fourth order)* obtained by means of the covariant gauge, it is seen that we should set $a = \pm 1$. Thus *translational invariance is violated*, though only for gauge non-invariant quantities.

The violation of translational invariance can be evaded if k_L^{-2} is replaced by $(k_L{}^2+\delta^2)^{-1}$ with $\delta > 0$ and if *the limit* $\delta \to 0$ *is taken at the end of calculation* [Landshoff 1986].[11] In this case, of course, $(k_L{}^2+\delta^2)^{-1}$ is positive definite, whence no indefinite

---

* The general validity is yet unknown.

metric is encountered. Although gauge invariant quantities are well-defined for $\delta \to 0$, non-invariant ones are *divergent* for $\delta \to 0$. Thus this procedure does not define a theory at $\delta = 0$.

For $\delta > 0$, the above replacement can be realized (apart from some inessential terms) by adopting a certain non-local gauge fixing depending on δ [Steiner 1986].[12] Although the original axial-type gauge is formally recovered by taking $\delta \to 0$, such a procedure cannot be regarded as a resolution of the difficulty, to the same extent as the introduction of a mass term cannot be regarded as a resolution of the infrared problem in a massless theory.

In summary, the theory of the axial-type gauges suffers from either of the following difficulties:

1. Indefinite metric is present, whence unitarity is violated.
2. Spontaneous breakdown of translational invariance is introduced by hand (in the temporal gauge).

Otherwise, we must be contented with changing the gauge-fixing condition into a complicated one.

Finally, we note that even the most time-honored Coulomb gauge is not free of trouble [Cheng et al. 1986].[13] If A_0 is not eliminated beforehand, the 00-component of the photon propagator behaves like $1/\mathbf{k}^2$, whence the integration over k_0 becomes non-convergent owing to the singularity at $k_0 = \infty$.

3. INDEFINITE METRIC AND SUBSIDIARY CONDITION

We now proceed to the quantization of the electromagnetic field in the covariant gauges corresponding to the classical Lorentz condition $\partial_\mu A^\mu = 0$.

Since the scalar field, the Dirac field, and even the massive vector field can be quantized in the positive-metric Hilbert space, one might hope that the electromagnetic field could be quantized covariantly without using indefinite metric. Indeed, in 1960's, there appeared a number of papers claiming the success of realizing this hope. Of course, all of them are wrong.

The reason for the necessity of indefinite metric is very simple: The little group on a lightlike vector is a *non-compact* group E(2) (two-dimensional Euclidean group). Although the compact subgroup SO(2) of the little group E(2) characterizes the electromagnetic field as spin-one, it is impossible to discard the remaining degrees of freedom in the field-theoretical formulation. The proof of the following proposition can be made by analyzing the Lorentz transformation property of the two-point function [Mathews et al. 1974]:[14)]
Any non-trivial massless field transforming as a covariant vector cannot be quantized without using indefinite metric.

The introduction of indefinite metric, however, endangers the probabilistic interpretability of the theory. The only successful way-out of this trouble is to set up a subsidiary condition. The states satisfying the subsidiary condition are called physical states. The totality of physical states are called a physical subspace, and denoted by $\mathcal{V}_{phys}$. It should satisfy the following requirements:

R1. $\mathcal{V}_{phys}$ should be *closed* not only with respect to taking linear combination of states but also with respect to taking a limiting state.

R2. $\mathcal{V}_{phys}$ should be time-independent, that is, it should be an invariant subspace of the total Hamiltonian H. (In the covariant theory, $\mathcal{V}_{phys}$ should be Poincaré invariant.)

R3. $\mathcal{V}_{phys}$ should contain no states of negative norm, though it may contain zero-norm states.

Then the physical S-matrix defined on the quotient space $\mathcal{V}_{phys}/\mathcal{V}_0$ is shown to be genuinely unitary, i.e., probabilistically interpretable, where $\mathcal{V}_0$ stands for the totality of the zero-norm physical states.

The requirement R1 is most naturally satisfied by a subsidiary condition of the form*

---

* More generally, $G_i|f\rangle = 0$ $(i=1,\ldots,n)$.

$$G|f\rangle = 0, \tag{3.1}$$

where G is a particular operator. Such a subsidiary condition as $\langle f|G|f\rangle = 0$ does not satisfy R1, whence it cannot guarantee the unitarity.

The requirement R2 is trivially satisfied if G is a conserved quantity. More generally, it is all right if $i[H,G]$ vanishes *modulo* G. For example, if G is a local operator, $i[H,G]$ equals $\partial_0 G$, which vanishes for $G=0$.

The requirement R3 is the most crucial one; its satisfaction must be checked explicitly.

When R3 is not taken care of, a subsidiary condition is often called a supplementary condition, which usually arises in the quantum treatment of a first-class constraint in the Dirac method of quantization [Dirac 1964].[15] A typical example of a supplementary condition is the Gauss law (2.15) in the temporal gauge. An occasionally taken approach for resolving the inconsistency between (2.15) with (2.14) and the hermiticity of G seems to be that of adopting a continuous set of linearly independent states, in the same way as one adopts momentum eigenstates for representing the canonical commutation relation $[p,q] = -i$ [Hossein Partovi 1984, Rossi et al. 1984, Bialynicki-Birula et al. 1984].[16)-18)]

If the above standpoint were admitted, then one could revive the old subsidiary condition [Fermi 1932][19]

$$\partial_\mu A^\mu |f\rangle = 0. \tag{3.2}$$

It was abandoned because it is inconsistent with any non-trivial free-field four-dimensional commutation relation

$$[A_\mu(x),\ A_\nu(y)] = i(c_1\eta_{\mu\nu} + c_2\partial_\mu\partial_\nu)D(x-y) \qquad (c_1\neq 0) \tag{3.3}$$

together with the hermiticity of $\partial_\mu A^\mu$:

$$0 = \langle f | [\partial_\mu A^\mu(x), A_\nu(y)] | f \rangle = ic_1 \partial_\nu D(x-y) \tag{3.4}$$

for $\langle f|f\rangle = 1$. But if the continuous-spectrum states are used for $|f\rangle$, it is forbidden to consider (3.4), whence no contradiction would be encountered.

The use of a continuous set of linearly independent states, however, contradicts a basic principle (separability of the state-vector space) of quantum field theory [Streater et al. 1964].[20] Although we often consider continuous energy eigenstates, we always have the understanding that those states can be reconsidered in terms of normalizable wave-packet states if necessary. That is, we postulate that the number of linearly independent states is essentially countable. Without this postulate, we cannot safely discuss symmetry generators, because then two eigenstates whose eigenvalues are infinitesimally different are *finitely different*, whence an *infinitesimal symmetry transformation* can induce a *finite change* of a state.

In order to have the usual particle interpretation of quantum field theory, it is of course natural to postulate the existence of a Poincaré-invariant vacuum state $|0\rangle$. Furthermore, it is postulated that $|0\rangle$ is cyclic, that is, any state can be arbitrarily accurately approximated by $\Phi|0\rangle$, where Φ is a polynomial in field operators smeared by many-particle wave-packet functions. We then have the following important theorem [Streater et al. 1964]:[21]

Theorem RS: If $\varphi(x)$ is a local operator,

$$\varphi(x)|0\rangle = 0 \Longleftrightarrow \varphi(x) = 0. \tag{3.5}$$

This theorem holds even in the indefinite-metric case because it is essentially a consequence of cyclicity, locality, and analyticity.

Now, we introduce one more requirement for $\mathcal{V}_{phys}$:

R4. The vacuum $|0\rangle$ is a physical state, i.e.,

$$G|0\rangle = 0. \tag{3.6}$$

From R4 and Theorem RS, we find that $G = 0$ if G is a local operator. Thus G must be a *non-local* operator.* This result essentially excludes the Dirac quantization having first-class constraints. If, therefore, one employs the Dirac quantization, one should explicitly introduce a gauge-fixing condition so as to convert first-class constraints into second-class ones.

As is well known, the correct subsidiary condition for the electromagnetic field in the Feynman gauge was first proposed by Gupta [Gupta 1950, Bleuler 1950].[23),24)] In this gauge, since $\partial_\mu A^\mu \neq 0$ but $\Box\partial_\mu A^\mu = 0$, we can define the positive-frequency part $(\partial_\mu A^\mu)^{(+)}$ in the Lorentz-invariant way and set up a subsidiary condition

$$[\partial_\mu A^\mu(x)]^{(+)}|f> = 0. \tag{3.7}$$

It is important to note that $(\partial_\mu A^\mu)^{(+)}$ is not only non-hermitian but also non-local.

As described in Sec. 4, the Gupta subsidiary condition (3.7) is rewritten into a more convenient expression

$$B^{(+)}(x)|f> = 0 \tag{3.8}$$

in the B-field formalism. It is valid for the linear covariant gauges including the Landau gauge.

The Gupta subsidiary condition can no longer be set up for *nonlinear* gauges. For the general gauge fixing, as in the non-abelian gauge theory, we must introduce an FP ghost term in such a way that the gauge-fixing plus FP-ghost Lagrangian density be invariant under the Becchi-Rouet-Stora (BRS) transformation [Becchi et al. 1976].[25)] We then set up the Kugo-Ojima subsidiary condition [Kugo et al. 1979],[26)]

$$Q_B|f> = 0, \tag{3.9}$$

---

* It is proved in the framework of the axiomatic quantum field theory that something non-local must be introduced [Strocchi 1967].[22)]

where Q_B denotes the BRS charge. Here the B-field formalism should be employed so as to ensure the important property $Q_B{}^2 = 0$. Although Q_B is hermitian, we encounter no trouble because Q_B is spacetime-independent.

It is quite remarkable that the one-dimensional condition (3.9) is sufficient for guaranteeing R3 [Kugo et al. 1979].[26] In the linear covariant gauges, the FP ghosts become free fields, whence we can set up an additional subsidiary condition requiring the absence of the FP ghosts in $\mathcal{V}_{phys}$. If this condition is combined with (3.9), then (3.8) is reproduced. In this sense, (3.9) is a generalization of (3.8). Even in the linear covariant gauges, however, (3.9) is better than (3.8) in the following respects:

1. If (3.8) is employed, the notion of gauge invariance of an observable $\mathcal{O}$ must be classified into gauge independence ($\langle f_1+g_1 | \mathcal{O} | f_2+g_2 \rangle = \langle f_1 | \mathcal{O} | f_2 \rangle$ for $|f_i\rangle \in \mathcal{V}_{phys}$ and $|g_i\rangle \in \mathcal{V}_0$), weak gauge invariance ($\mathcal{O}\mathcal{V}_0$, $\mathcal{O}^\dagger \mathcal{V}_0 \subset \mathcal{V}_0$), and gauge invariance ($\mathcal{O}\mathcal{V}_{phys}$, $\mathcal{O}^\dagger \mathcal{V}_{phys} \subset \mathcal{V}_{phys}$) [Strocchi et al. 1974].[27] If (3.9) is employed, however, no such distinction is necessary because they are all equivalent [Kugo et al. 1979].[26]

2. When the system has a non-trivial boundary condition, an unwanted Casimir energy arising from unphysical photons is encountered if (3.8) is employed, but canceled by FP ghosts if (3.9) is employed [Ambjørn et al. 1983].[28]

3. When gravity couples with A_μ, i.e., when the theory is generally-covariantized, (3.9) is the only correct subsidiary condition [Nakanishi 1980].[29]

4. DIPOLE GHOSTS AND THE B-FIELD FORMALISM

In the linear covariant gauges, the free photon propagator in the momentum space is

$$iD_{F\mu\nu}(k) = \frac{\eta_{\mu\nu}}{k^2+i0} - (1-\alpha)\frac{k_\mu k_\nu}{(k^2+i0)^2} \,. \tag{4.1}$$

For $\alpha \neq 1$, it has a double pole, whence there are dipole-ghost one-photon states. This simple fact was, however, not recognized explicitly until 1966 [Nakanishi 1966].[30] Indeed, it had, or even have, been a custum to write the photon propagator as

$$\frac{1}{k^2+i\varepsilon}\,[\eta_{\mu\nu} - (1-\alpha)\frac{k_\mu k_\nu}{k^2}\,], \qquad (\varepsilon \to +0). \tag{4.2}$$

Evidently, (4.2) is mathematically meaningless.

The definition of a dipole ghost* is as follows. Let H be a Hamiltonian. A state $|d\rangle$ is called a dipole ghost if

$$(H-E)|d\rangle \neq 0 \qquad \text{but} \qquad (H-E)^2|d\rangle = 0. \tag{4.3}$$

It follows from (4.3) that the state $|a\rangle \equiv (H-E)|d\rangle$ is a zero-norm eigenstate of H. Although $|d\rangle$ is not an eigenstate, a complete set needs both $|d\rangle$ and $|a\rangle$.

It is straightforward to define a multipole ghost by extending (4.3). The appearance of multipole ghosts is quite a normal phenomenon in the indefinite-metric theory; indeed, even in a finite-dimensional space, there can exist multipole ghosts, as is understood by remembering the Jordan form of a square matrix [Nakanishi 1972].[32]

When a *continuous spectrum* is considered, a subtlety arises because $\delta(H-E)$ can be changed into its derivative by partial integration with respect to E. For example, a subsidiary condition stating that $\mathcal{V}_{phys}$ is spanned by all *eigenstates* of H does not satisfy R1 (see Sec. 3) because a dipole-ghost state can be constructed from eigenstates by using the above trick [Nakanishi 1971].[33] This fact invalidiates Heisenberg's proposal for securing the unitarity by

---

* This notion was first introduced by Heisenberg in his unified theory and then in the Lee model [Heisenberg 1957].[31]

using dipole ghosts [Heisenberg 1957].[31] It is therefore important to avoid the explicit use of continuous-spectrum states (See Sec. 3). For one-particle states, the above subtlety does not arise if one requires manifest covariance.

Multiple poles, and therefore multipole ghosts, are inevitably encountered in the theory of the Bethe-Salpeter (BS) equation [Nakanishi 1965, Nakanishi 1969].[34),35] In the generic situation, the occurrence of multiple poles in the scattering BS amplitude is as follows. Let P_μ be the total momentum of the system, and $s \equiv P^2$. The location, $s = s(\lambda)$, of a pole is an analytic function of the coupling constant squared λ. Its residue is expressed in terms of the bound-state BS amplitudes. Although the physical bound-state value of s is restricted to the interval $0 \leqq s \leqq s_0$, $\sqrt{s_0}$ being the elastic threshold, the scattering BS amplitude and therefore its poles exist even for $s < 0$. Since the little group on P_μ for $s < 0$ is a non-compact group SO(2,1), some of "bound states" must be of negative norm. Furthermore, the existence of the wrong-sign residue is "analytically" continued to the $s > 0$ region. Now, an interesting thing happens at $s = 0$. Since $s = 0$ is realized by $P_\mu = 0$ and since the little group on $P_\mu = 0$ is a larger group SO(3,1), pole trajectories collide at $s = 0$. But, from the continuity, the *bound-state amplitudes* for $s = 0$ should be characterized by *lightlike* P_μ. What happens here is the "conspiracy"* of the colliding right-sign and wrong-sign pole trajectories; N colliding simple poles generically yield an N-ple pole. From this discussion, it can be inferred that a massless bound state having orbital angular momentum $\ell = N$ is generically accompanied with an N-pole ghost.

To be concrete, we consider the simplest non-trivial example. In the scalar-scalar system, the bound-state BS amplitudes are characterized by four quantum numbers n, ℓ, m, and κ ($n \geqq 1$, $0 \leqq \ell \leqq n-1$, $|m| \leqq \ell$, $\kappa \geqq 0$; states with $\kappa > 0$ have no nonrelativistic counterpart). For $s > 0$, there are a positive-norm P-wave

---

* This word was invented for a similar phenomenon found later in the Regge-pole theory [Freedman et al. 1967].[36]

trajectory ($n = 2$, $\ell = 1$, $m = 0, \pm 1$, $\kappa = 0$) and a negative-norm S-wave one ($n = 2$, $\ell = 0$, $m = 0$, $\kappa = 1$), which collide at $s = 0$. The $m = \pm 1$ amplitudes go without anything, while the two $m = 0$ states ($\kappa = 0$ and $\kappa = 1$) change into a dipole ghost and a zero-norm state through their rearrangement. Then, for $s < 0$, we obtain an SO(2,1) "P"-wave trajectory consisting of two positive-norm and one negative-norm amplitudes and an SO(2,1) "S"-wave trajectory having positive norm.

It is natural to expect that a similar phenomenon takes place also for the case of elementary particles; to formulate a massless spin-1 theory we introduce a set of massless vector field A_μ and *massless scalar field* B. This idea is realized by the B-field formalism [Nakanishi 1966, Lautrup 1967].[30),37)] Indeed, in the massive extension of the B-field formalism [Nakanishi 1972],[39)] the B-field has negative norm and its mass is different from that of A_μ except for $\alpha = 1$. In the Feynman gauge $\alpha = 1$, we encounter no dipole ghost; the corresponding situation exists also in the BS equation: In the unequal-mass Wick-Cutkosky model (*massless*-scalar-exchange scalar-scalar ladder model) [Nakanishi 1988],[40)] there are no multiple poles at $s = 0$ owing to the fact that its Wick-rotated equation is SO(4) symmetric for *all* values of s.

The Lagrangian density of the B-field formalism* consists of the local gauge invariant one and the gauge-fixing one,

$$\mathcal{L}_{GF} = B\partial_\mu A^\mu + \tfrac{1}{2}\alpha B^2. \tag{4.4}$$

For $\alpha \neq 0$, it is possible to eliminate B, but it is not legitimate to do so for the most important Landau gauge $\alpha = 0$, especially in the non-perturbative treatment. The field equations are

$$\partial_\mu F^{\mu\nu} - \partial^\nu B = -ej^\nu, \tag{4.5}$$

$$\partial_\mu A^\mu + \alpha B = 0. \tag{4.6}$$

---

* A thorough review of the B-field formalism is given in a book [Nakanishi et al. —].[38)]

From (4.5) and (2.4), it follows that

$$\Box B = 0. \tag{4.7}$$

The canonical quantization can be carried out consistently if B is not regarded as a canonical variable. The four-dimensional commutation relations,

$$[A_\mu(x), B(y)] = -i\partial_\mu D(x-y), \tag{4.8}$$

$$[B(x), B(y)] = 0, \tag{4.9}$$

hold even for $j_\mu \neq 0$. Clearly, (4.9) shows that B is of zero norm. Furthermore, (4.7) shows that we can consistently define $B^{(+)}$ and set up the subsidiary condition (3.8). Note, especially, that R2 is satisfied.

In order to see the dipole-ghost character, we consider the free (or asymptotic) fields by setting $j_\mu = 0$. We define the dipole-ghost extension of the Pauli-Jordan D-function by

$$\begin{aligned} E(x) &\equiv -(\partial/\partial m^2)\Delta(x;m^2)\big|_{m^2=0} \\ &= -i(2\pi)^{-3}\int d^4k\, \epsilon(k_0)\delta'(k^2)e^{-ikx} \\ &= -(8\pi)^{-1}\,\epsilon(x^0)\theta(x^2), \end{aligned} \tag{4.10}$$

$$\Box E(x) = D(x). \tag{4.11}$$

Then the four-dimensional commutation relation,

$$[A_\mu(x), A_\nu(y)] = -i\eta_{\mu\nu}D(x-y)+i(1-\alpha)\partial_\mu\partial_\nu E(x-y), \tag{4.12}$$

holds. Clearly, (4.12) shows that the longitudinal part of A_μ is a dipole ghost for $\alpha \neq 1$.

When a dipole ghost is present, it is impossible to consider the three-dimensional Fourier transform of $A_\mu(x)$. It is not legitimate to gauge-transform A_μ into the Feynman-gauge one because this procedure necessarily violates manifest covariance. In order to define the positive/negative frequency part $A_\mu^{(\pm)}(x)$, we first decompose $iE(x)$ into $E^{(+)}(x)$ and $E^{(-)}(x)$:

$$E^{(\pm)}(x) \equiv -(\partial/\partial m^2)\Delta^{(\pm)}(x;m^2)\big|_{m^2=0}$$

$$= \pm(2\pi)^{-3}\int d^4k\,\theta(\pm k_0)\delta'(k^2)e^{-ikx}$$

$$= \mp[(4\pi)^{-2}\log(-x^2\pm i0x^0)+\text{const}], \tag{4.13}$$

$$\Box E^{(\pm)}(x) = D^{(\pm)}(x). \tag{4.14}$$

Here, an additional constant in (4.13) is actually *infinite*, but *it is harmless*. Indeed, the expression

$$A_\mu^{(\pm)}(x) \equiv -i\int d^3z\,\{\partial_0^z D^{(\pm)}(x-z)\cdot A_\mu(z)-D^{(\pm)}(x-z)\partial_0 A_\mu(z)$$

$$+(1-\alpha)[\partial_0^z E^{(\pm)}(x-z)\cdot\partial_\mu B(z)-E^{(\pm)}(x-z)\partial_0\partial_\mu B(z)]\} \tag{4.15}$$

is independent of this constant because it is multiplied by*

$$\int d^3z\,\partial_0\partial_\mu B(z) = 0. \tag{4.16}$$

By using $A_\mu^{(-)}$ and $B^{(-)}$, we can construct a Fock space and confirm that $\mathcal{V}_{\text{phys}}$ is positive semi-definite because the dipole ghost is excluded by (3.8) with (4.8).

Finally, we again emphasize the impossibility of *q-number* gauge transformation in the framework of the B-field formalism. In order to realize it, it is necessary to introduce the freedom of

---

* Use (4.7) for $\mu = 0$.

"gaugeon" field. The Lagrangian density of the gaugeon formalism [Yokoyama 1974][41] is obtained by further adding the gaugeon one,

$$\mathcal{L}_{\text{gaugeon}} = -\partial^\mu G\cdot\partial_\mu C - \frac{1}{2}\epsilon(C+\gamma B)^2, \tag{4.17}$$

where G and C are new scalar fields, γ being a real parameter and $\epsilon = \pm 1$. It is easy to see that G is a massless dipole ghost. In this formalism, the effective gauge parameter $\tilde{\alpha}$ (twice the coefficient of B^2) is given by*

$$\tilde{\alpha} = \alpha - \epsilon\gamma^2. \tag{4.18}$$

The total action of the gaugeon formalism having γ and ϵ is converted into the one having γ' and $\epsilon' = \epsilon$ by the following transformation of fields:

$$A_\mu = A'_\mu - (\gamma'-\gamma)\partial_\mu G', \qquad B = B',$$

$$G = G', \quad C = C' + (\gamma'-\gamma)B',$$

$$\psi = [\exp ie(\gamma-\gamma')G']\psi', \tag{4.19}$$

where ψ denotes a generic charged field (having a charge e for simplicity). Since (4.19) is nothing but field redefinition, the physical contents of the theory remain unchanged. Nevertheless, it precisely reproduces the effect of the q-number gauge transformation at the level of the Green's function.

5. GOTO-IMAMURA DIFFICULTY AND THE SCHWINGER TERM

QED involves various kinds of pathologies, that is, we encounter

---

* In the original work, the $\alpha = 0$ case is considered.

the problems of photon self-mass, anomaly, and the Schwinger term in addition to the infrared and ultraviolet divergences.

If we calculate the photon self-energy naively, we encounter a non-vanishing, quadratically divergent self-mass, violating gauge invariance. But if we calculate it in a gauge-invariant way (e.g., by using dimensional regularization), the photon self-mass, of course, vanishes exactly. Thus there remains no pathology in this problem. In contrast to it, the problem of anomaly is more serious because there is no resolution which is consistent with all of chiral invariance, gauge invariance, and Lorentz invariance. But since anomaly is not a problem within the proper framework of QED and since we are interested in anomaly in the non-abelian gauge theory rather than in the abelian one, we do not discuss this problem here.

The subtlest pathology is the problem of the Schwinger term. Historically, the existence of inconsistency was found first by Goto and Imamura [Goto et al. 1955][42] in the discussion of criticizing Källén's work (cf. Sec. 8). Later, it was rediscovered by Schwinger [Schwinger 1959][43] by using a somewhat different reasoning and he proposed to introduce a non-canonical term in the equal-time current-current commutator. Therefore, it is not appropriate to call this term "Goto-Imamura-Schwinger term". Here we call the inconsistency "Goto-Imamura difficulty" and the non-canonical term "Schwinger term".

We consider a Dirac field $\psi(x)$, which may be free or interacting with the electromagnetic field. The conserved current is defined by

$$j^{\mu}(x) \equiv \bar{\psi}(x)\gamma^{\mu}\psi(x) \tag{5.1}$$

in the standard notation. If we *naively* calculate the equal-time current-current commutator by means of the canonical anticommutation relations for ψ and $\psi^{\dagger}$, we find that

$$[j^{0}(x^{0}, \mathbf{x}), j^{k}(x^{0}, \mathbf{y})] = 0. \tag{5.2}$$

On the other hand, from such general principles as Poincaré

invariance, locality, and the current conservation (2.4), we obtain the spectral representation

$$\langle 0|[j^{\mu}(x), j^{\nu}(y)]|0\rangle$$

$$= -i\int_0^{\infty} ds\,\pi(s)(s\eta^{\mu\nu} + \partial^{\mu}\partial^{\nu})\Delta(x-y;\ s). \tag{5.3}$$

Here the spectral function $\pi(s)$ must satisfy the condition

$$\pi(s) \geq 0, \quad \pi(s) \not\equiv 0 \tag{5.4}$$

because of the positivity of $\langle 0|j^0 j^0|0\rangle$. It is important to note that this is true in spite of the use of indefinite metric for the covariant quantization of A_{μ}, because j^{μ} commutes with B, whence $j^{\mu}|0\rangle \in \mathcal{V}_{phys}$. Now, we set $y^0 = x^0$; then for $\mu = 0$, $\nu = k$, (5.3) reduces to

$$\langle 0|[j^0(x^0,\mathbf{x}), j^k(x^0, \mathbf{y})]|0\rangle = i\partial^k\delta(\mathbf{x}-\mathbf{y})\int_0^{\infty} ds\,\pi(s)$$

$$\neq 0 \tag{5.5}$$

in contradiction to (5.2). This inconsistency is the Goto-Imamura difficulty.

The standard way of evading this trouble is to distrust the naive result (5.2). The basis for this belief is that (5.2) is essentially to assert $\infty - \infty = 0$ because the product of field operators at the same spacetime point is singular. Indeed, if one recalculates $[j^0, j^k]$ by using the point-splitting definition or the Wilson operator product expansion of j^{μ}, one finds that the limiting value is non-vanishing. Such a term is called the Schwinger term.

The Schwinger term is, however, quite an unwelcome guy in the following respects:*

---

* For its mathematical detail, see [Orzalesi —].[44)]

1. There are many ways of calculating the Schwinger term (e.g., point-splitting, regularized δ-function, covariant perturbation, etc.), and the results are not necessarily identical.

2. The presence of the Schwinger term violates the Jacobi identity because the two equal-time limiting processes are non-commutative [Johnson et al. 1966].[45)]

3. The Schwinger term yields no physically observable effects.

If the presence of the Schwinger term is seriously considered, the same rule must be applied to any equal-time commutator of composite operators. Although it is calculable by means of the covariant *perturbation* theory, one is then forgetting for what purpose one considers the equal-time commutator: The equal-time commutation relations are indispensable to the *non-perturbative* approach in the Heisenberg picture.

It should be noted also that the Schwinger term disappears if all calculations are made by employing the Pauli-Villars regulator method [Moffat 1960].[46)] Of course, since the regulator mass must finally go to infinity, the negative contribution to $\pi(s)$ arises from $s = \infty$, but it is unacceptable mathematically to consider $\pi(\infty)$ literally.

At this point, we must remember an important fact: Neither the free Dirac theory nor QED cannot be a physically closed theory. Since those theories have an energy-momentum tensor, they necessarily couple with gravity, that is, *we must explicitly take account of quantum gravity*. In the usual calculation of particle physics, we always neglect the effect of quantum gravity; this is quite legitimate *as an approximation* because of the extreme largeness of the Planck mass squared, κ^{-1}. This is no longer true, however, in such an exact statement as (5.4), because κ^{-1} *is certainly smaller than infinity*. In the following, we show that if quantum gravity is taken into account seriously, we cannot conclude (5.4) nor, therefore, (5.5) [Nakanishi 1980].[47)]

The manifestly covariant canonical formalism of quantum gravity

[Nakanishi 1983, Nakanishi et al. —],[48),38)] which was formulated in 1978-1985, is an outstandingly beautiful theory. Although it does not bring any improvement to calculational technique, the theoretical foundation of quantum gravity has become quite transparent. We can now make non-perturbative discussion of quantum gravity in a reliable way. Here it must be noted that this theory is heavily based on the Lagrangian canonical formalism. We emphasize that the axiomatic approach is inappropriate in quantum gravity because of the following reasons:

1. Indefinite metric is indispensable.
2. Microcausality is meaningless at the operator level because $g_{\mu\nu}$ is q-number.
3. The contravariant metric tensor $g^{\lambda\rho}(x)$ is a complicated function of the covariant one $g_{\mu\nu}(x)$ at the same spacetime point.

Especially, the last one implies that it is catastrophic if one cannot naively treat operator products at the same spacetime point. In this sense, we must rescue (5.2) from the Goto-Imamura difficulty.

Since general coordinate invariance is a local symmetry, to quantize gravity we introduce gauge fixing and FP-ghost terms in such a way that their sum be BRS invariant. Let Q_b be the BRS charge of quantum gravity, and set up the Kugo-Ojima subsidiary condition*

$$Q_b |f\rangle = 0. \tag{5.6}$$

Then the unitarity of the physical S-matrix is guaranteed.

The generally covariant Dirac theory can be formulated by means of the vierbein formalism. Let $h^\mu{}_a$ (a=0,1,2,3) be the vierbein field and $h \equiv \det h^\mu{}_a$. Then the generally-covariantized electric

* Of course, if the theory has other BRS charges, the corresponding equations should also be imposed to define $\mathcal{V}_{phys}$.

current is defined by

$$j^{\mu} \equiv hh^{\mu}{}_{a}\bar{\psi}\gamma^{a}\psi, \tag{5.7}$$

where γ^a denotes the flat-space Dirac matrix. Canonical quantization is carried out consistently, and (5.2) remains unmodified. Furthermore, in the manifestly covariant canonical formalism of quantum gravity, the Poincaré invariance of particle physics is realized at the level of the representation of field operators, whence (5.3) also remains valid. But a crucial change arises in (5.4).

In contrast to gauge invariance, general coordinate invariance is a *spacetime* symmetry, and correspondingly the BRS charge Q_b does *not* (anti)commute with *any* local operator. For example, we have

$$\begin{aligned} i[Q_b, j^{\mu}] &= \kappa\partial_{\lambda}c^{\mu}\cdot j^{\lambda} - \kappa\partial_{\lambda}(c^{\lambda}j^{\mu}) \\ &\neq 0, \end{aligned} \tag{5.8}$$

where c^{μ} denotes the gravitational FP ghost. It is important to note that (5.6) and (5.8) imply

$$j^{\mu}(x)|0\rangle \notin \mathcal{V}_{\text{phys}} \tag{5.9}$$

in contrast to the usual QED. Hence $\pi(s)$ can receive the contribution from unphysical intermediate states, which are of indefinite metric. Thus (5.4) is no longer valid if quantum gravity is taken into account; $\pi(s)$ can become negative for $s \gtrsim \kappa^{-1}$. Thus *the Goto-Imamura difficulty has been resolved without introducing the Schwinger term.* We now have a sum rule

$$\int_0^{\infty} ds\,\pi(s) = 0. \tag{5.10}$$

For example, if $\pi(s)$ behaves like

$$\frac{1-(\kappa s)^2}{[1+(\kappa s)^2]^2}\,, \tag{5.11}$$

both (5.4) for $s < \kappa^{-1}$ and (5.10) are satisfied.

Finally, we note that the above consideration is no longer valid in the two-dimensional spacetime because of the absence of Einstein gravity. Indeed, the boson-fermion equivalence $(j_\mu \sim \partial_\mu \phi)$ could not hold without introducing the Schwinger term.

6. INFRARED PROBLEM

QED suffers from both ultraviolet and infrared divergences. The latter is caused by the masslessness of photons. Since QED is a correct theory in low energies, infrared catastrophe must be superficial.

Real or virtual photons having $k_\mu \simeq 0$ are called soft photons, though this notion is dependent on the Lorentz frame. Since real soft photons are not identifiable as particles experimentally, the observed cross section receives contributions also from the processes accompanied with soft-photon emission. Although the real-photon's propagator,

$$2\pi\theta(k_0)\,\delta(k^2), \tag{6.1}$$

is different from the virtual-photon's one,

$$i(k^2+i0)^{-1} \tag{6.2}$$

plus gauge term, the difference,

$$i(k^2-ik_0 0)^{-1}, \tag{6.3}$$

can be made free of infrared singularity by deforming the k_0 contour, *provided that* there is no obstruction for doing so.

We consider a real electron which emits or absorbs a real or virtual photon. Apart from inessential factors, we have two products of the electron's propagators:

$$\delta(p^2-m^2)[(p+k)^2-m^2+i0]^{-1}, \tag{6.4}$$

$$(p^2-m^2+i0)^{-1}\delta((p+k)^2-m^2). \tag{6.5}$$

Evidently, each of (6.4) and (6.5) has infrared singulairty of order 1 if it is multiplied by (6.1).* Since we encounter two such factors in the calculation of the cross section, there arises a logarithmic infrared divergence when integrated over k_μ.

If, however, we consider the *sum* of (6.4) and (6.5), it can be rewritten as

$$(2\pi)^{-1}i\{(p^2-m^2+i0)^{-1}[(p+k)^2-m^2+i0]^{-1}$$
$$-(p^2-m^2-i0)^{-1}[(p+k)^2-m^2-i0]^{-1}\}. \tag{6.6}$$

Evidently, each term of (6.6) is free of infrared singularity when multiplied by (6.1).** This is the essential mechanism of the infrared-divergence cancellation [Kinoshita 1950].[49)]

The above cancellation mechanism can be generalized as follows [Nakanishi 1958].[50)] We introduce a double-cut diagram by connecting a pair of Feynman graphs not only at the final state but also at the initial state. An electron line is called "essentially external" if its momentum can be different from the on-shell value only by soft-photon momentum. Given a double-cut diagram, we cut all its

---

* There is no such singularity if (6.1) is not multiplied.

** It is unnecessary to rewrite (6.6) in terms of a derivative of the δ function. The same applies also to (6.8). The author would like to thank Professor T. Kawai for pointing this out.

essentially external lines, in any admissible way, so as to obtain a set of pairs of Feynman graphs. On each essentially external line, we encounter the following sum if we consider all pairs of Feynman graphs arising from a double-cut diagram:

$$\sum_{n=0}^{N} \prod_{i=0}^{n-1} [(p+k_i)^2-m^2-i0]^{-1} \cdot \delta((p+k_n)^2-m^2) \times \prod_{j=n+1}^{N} [(p+k_j)^2-m^2+i0]^{-1}, \tag{6.7}$$

where the propagator having $-i0$ arises from the complex-conjugate amplitude. Now, it is easy to prove by mathematical induction that for any real quantities $A_1,\ldots,A_N$, the following identity holds:

$$\sum_{n=0}^{N} \prod_{i=0}^{n-1} (A_i-i0)^{-1} \cdot \delta(A_n) \cdot \prod_{j=n+1}^{N} (A_j+i0)^{-1} = (2\pi)^{-1} i [\prod_{n=0}^{N} (A_n+i0)^{-1} - \prod_{n=0}^{N} (A_n-i0)^{-1}]. \tag{6.8}$$

Setting $A_n = (p+k_n)^2-m^2$ in (6.8), we see how the infrared singularities cancel on each essentially external line.*

The above analysis clarifies the cancellation mechanism of infrared divergence quite generally and independently of the speciality of QED. Unfortunately, however, it has a serious drawback: The contour deformation of (6.3) assumed at the beginning is not always possible. If a soft photon is exchanged between two *distinct* electrons, its energy sign cannot be assigned naturally. Then the artificial decomposition of (6.2) into (6.1) and (6.3) induces a pinch between (6.3) and either of the electron's propagators. Thus we must further investigate the terms arising from the retarded propagators like (6.3).

In QED, the infrared divergences of this type are summed up into a multiplicative phase factor, called the Coulomb phase factor,

---

* The identity (6.8) was proposed twenty years earlier than the proposal of "Libby-Sterman identity" [Libby et al. 1979].[51] The latter is the lowest-order approximation of the former [Nakanishi 1983].[52]

whence they are totally harmless. In quantum chromodynamics (QCD), however, because gluons are not colorless, such simple exponentiation no longer holds; two infrared singularities of the type (6.1) and the type (6.3) can interfere so that the infrared cancellation actually breaks down [Doria et al. 1980, Andraši et al. 1981, Yoshida 1981].[53)-55)]

As far as QED is concerned, all infrared contributions are shown to be factored out [Yennie et al. 1961].[56)] The proof can be made more elegantly by explicitly using the gauge-invariant structure of QED [Grammer et al. 1973],[57)] though this analysis is not extendable to QCD.

The infrared exponentiation in QED can be seen more directly by the following operator analysis [Murota 1960].[58)]

Let $H(t)$ be the QED interaction Hamiltonian in the interaction picture:

$$H(t) = -e\int d\mathbf{x}\colon \bar{\psi}(x)\gamma^{\mu}\psi(x)\colon A_{\mu}(x), \quad (t=x^0). \tag{6.9}$$

We decompose it into the soft and hard parts:

$$H(t) = H_s(t) + H_h(t), \tag{6.10}$$

where $H_s(t)$ is the part of $H(t)$ in which the photon's energy $k_0 = |\mathbf{k}|$ is smaller than an infrared cutoff λ (<< m), and $H_h(t)$ is the remainder. The Dyson S-matrix is then written as

$$\begin{aligned} S &\equiv T\exp\left[-i\int_{-\infty}^{\infty} dtH(t)\right] \\ &= \sum_{n=0}^{\infty} \frac{(-i)^n}{n!}\int dt_1\cdots dt_n\ T[U_s(\infty,-\infty)H_h(t_1)\cdots H_h(t_n)], \end{aligned} \tag{6.11}$$

where

$$U_s(t,t_0) \equiv T\exp\left[-i\int_{t_0}^{t} d\tau H_s(\tau)\right]. \tag{6.12}$$

In the soft-photon approximation, we can write

$$H_s(t) = -\frac{e}{(2\pi)^{3/2}}\int d\mathbb{p}\, d\mathbb{k}\, \theta(\lambda - k_0)\frac{p^\mu}{p_0}\rho(\mathbb{p})$$

$$\times \frac{1}{\sqrt{2k_0}} a_\mu(\mathbb{k})\exp(-i\frac{kp}{p_0}t) + \text{hermitian conjugate.} \qquad (6.13)$$

where $\rho(\mathbb{p})$ and $a_\mu(\mathbb{k})$ denote the charge-density operator (electron number minus positron number) and the photon's annihilation operator (in the Feynman gauge), respectively. It is important to note that the commutator,

$$Q_s(t_1,t_2) \equiv [H_s(t_1),\, H_s(t_2)], \qquad (6.14)$$

commutes with $H_s(t)$. Accordingly, operator calculus yields that

$$U_s(t,t_0) = \exp[-i\int_{t_0}^{t} d\tau H_s(\tau) - \frac{1}{2}\int_{t_0}^{t} d\tau \int_{t_0}^{\tau} d\tau' Q_s(\tau,\tau')]. \qquad (6.15)$$

Physically, Q_s consists of the electron's self-energy and the Coulomb energy.

Now, since the t_j dependence of U_s may be neglected compared with that of $H_h(t_j)$ in (6.11), S can be approximated by

$$S = U_s(\infty,0)S_h U_s(0,-\infty) \qquad (6.16)$$

with

$$S_h \equiv T\exp[-i\int_{-\infty}^{\infty} dt\, H_h(t)]. \qquad (6.17)$$

Since S_h contains no soft photons, it should be infrared finite. All infrared contributions are contained in $U_s(0,-\infty)$ and $U_s(\infty,0) = [U_s(0,-\infty)]^\dagger$. A Fock state $|h>$ having no soft photons is an eigenstate of $U_s(0,-\infty)$. Let $U_s(h)$ be its eigenvalue, though it is still an operator containing soft-photon's annihilation and creation operators. Then the transition matrix element is given by

$$<\text{final} + \text{soft photons}\ |S|\ \text{initial}>$$

$$= <\text{soft photons}\ |[U_s(\text{final})]^\dagger U_s(\text{initial})|0>$$

$$<\text{final}\ |S_h|\ \text{initial}>. \tag{6.18}$$

From the first factor of (6.18) together with (6.15), we can explicitly calculate the infrared correction factor in the differential cross section.*

It is clear from the above analysis that all infrared divergences are contained in $U_s(0,-\infty)$ and its hermitian conjugate. Accordingly, if we *redefine* the S matrix

$$S = \lim_{t\to-\infty,\, t'\to\infty} U(t',t) \tag{6.19}$$

by

$$S_{coh} = \lim_{t\to-\infty,\ t'\to\infty} Z^\dagger(t')U(t',t)Z(t), \tag{6.20}$$

where $Z(t)$ is such a unitary operator that

$$\lim_{t\to-\infty} U_s(0,t)Z(t) = \text{infrared finite}, \tag{6.21}$$

then the new S-matrix S_{coh} is infrared finite.** Of course, Z(t) should be chosen in such a way that S_{coh} be Lorentz invariant. A desired one can be constructed by means of the $|t| \to \infty$ asymptotic dynamics [Kulish et al. 1970].[60] The result is given by

---

* We note that the above analysis was made ten years earlier than the Kulish-Faddeev theory described below.

** The infrared finiteness can be shown perturbatively [Stapp 1983].[59]

$$Z(t) = \lim_{\lambda\to\infty} U_s(t,t_0) \tag{6.22}$$

with the understanding

$$\int_{t_0}^{t} e^{i\alpha\tau}\, d\tau = \frac{1}{i\alpha}\, e^{i\alpha t}. \tag{6.23}$$

That is, from (6.15) we have

$$Z(t) = \exp R(t)\cdot \exp i\Phi(t), \tag{6.24}$$

where

$$R(t) = \frac{e}{(2\pi)^{3/2}} \int d\mathbf{p} d\mathbf{k}\, \frac{p_\mu}{pk}\, \rho(\mathbf{p})\, \frac{1}{\sqrt{2k_0}} [a_\mu{}^{\dagger}(\mathbf{k}) \exp(i\, \frac{kp}{p_0}\, t)$$

$$- a_\mu(\mathbf{k}) \exp(-i\, \frac{kp}{p_0}\, t)], \tag{6.25}$$

$$\Phi(t) \sim \frac{e^2}{8\pi} \int d\mathbf{p} d\mathbf{q}\, \frac{pq}{\sqrt{(pq)^2 - m^4}}\ :\rho(\mathbf{p})\rho(\mathbf{q}):\ \epsilon(t) \log |t|. \tag{6.26}$$

Here $\Phi(t)$ is the Coulomb phase factor, whose existence is a consequence of the long-range nature of the Coulomb potential, as is known in the nonrelativistic quantum mechanics.

The Fock space $\mathcal{V}_F$ is not an appropriate operand on which $R(t)$ acts. We should consider a space

$$\mathcal{V}_{coh} = \exp(-R_f) \cdot \mathcal{V}_F \tag{6.27}$$

with

$$R_f = \frac{e}{(2\pi)^{3/2}} \int d\mathbf{p}\, \frac{d\mathbf{k}}{\sqrt{2k_0}}\, \rho(\mathbf{p}) [f^\mu(\mathbf{k},\mathbf{p}) a_\mu{}^{\dagger}(\mathbf{k}) - f^{\mu *}(\mathbf{k},\mathbf{p}) a_\mu(\mathbf{k})], \tag{6.28}$$

where $f_\mu(\mathbf{k},\mathbf{p})$ is a function having a singularity of p_μ/pk for small k_0. The state vectors of $\mathcal{V}_{coh}$ are called coherent states.

Because of the singularity structure of f_μ, $k^\mu f_\mu$ is necessarily non-vanishing in the low-energy limit if Poincaré covariance should be maintained. Accordingly, in $\mathcal{V}_{coh}$ we have

$$\lim_{k_0 \to 0} k^\mu a_\mu(\mathbf{k})/\sqrt{2k_0} = -(2\pi)^{-3/2} Q, \tag{6.29}$$

where Q denotes the charge operator. Since the Fourier transform of the l.h.s. of (6.29) before taking limit is $(\partial^\mu A_\mu)^{(+)}$, (6.29) contradicts the Gupta subsidiary condition (3.7) in a charged sector. Thus, if we adhere to the coherent-state formalism, we must abandon (3.7), or equivalently, (3.8) (we are exclusively using the Feynman gauge) [Zwanziger 1976].[61] In the coherent-state formalism, the subsidiary condition is modified into

$$B^{(+)}|f\rangle = b^{(+)}|f\rangle, \tag{6.30}$$

where $b^{(+)}(x)$ is the positive-frequency part of a *c-number* function $b(x)$, which satisfies $\Box b(x)=0$ and

$$\int d\mathbf{x}\partial_0 b(x) = q, \tag{6.31}$$

q being an eigenvalue of Q.

The modified subsidiary condition (6.30) is, however, necessarily *time-dependent* because of (6.31), that is, requirement R2 (see Sec. 3) is violated. Thus the coherent-state formalism remains unsatisfactory.

7. ELECTRON'S ASYMPTOTIC FIELD AND THE GAUSS LAW

Asymptotic fields are very important in formulating quantum field theory in the Heisenberg picture. They are the t $\pm\infty$ *weak* limits of primary or composite local fields. Although asymptotic fields satisfy free-field equations and free-field (anti)commutation relations, they are *non-local* fields in the Heisenberg picture.

The S-matrix is defined as the transformation operator between in-fields ($t \to -\infty$ asymptotic fields) and out-fields ($t \to +\infty$ ones). It is important to postulate the *asymptotic completeness*: Any (well-defined) operator which (anti)commutes with all asymptotic fields is a c-number only. The unitarity of the S-matrix can be established only under this postulate.

Asymptotic fields are determined in the following way. Let $\Phi_j(x)$ be a (renormalized) local field and

$$\Phi_j(x) \longrightarrow \Phi_j^{as}(x) \qquad (x^0 \to \pm\infty), \tag{7.1}$$

where "as" means "in" for $x^0 \to -\infty$ and "out" for $x^0 \to +\infty$. The (anti)commutator between Φ_i^{as} and Φ_j^{as} is given by

$$[\Phi_i^{as}(x),\ \Phi_j^{as}(y)]_{\mp}$$
$$= \text{discrete spectrum of } \langle 0|[\Phi_i(x),\ \Phi_j(y)]_{\mp}|0\rangle. \tag{7.2}$$

The field equations, $f_k(\Phi^{as}(x)) = 0$, for asymptotic fields can be established by showing

$$[f_k(\Phi^{as}(x)),\ \Phi_j^{as}(y)] = 0 \qquad \text{for any } j, \tag{7.3}$$

together with $\langle 0|f_k|0\rangle = 0$, on the basis of the asymptotic completeness.

Now, we wish to apply the above general consideration to QED. In this case, we encounter the infrared problem as seen in Sec. 6. Since the electron is accompanied by a soft-photon cloud, its two-point function does not have the expected discrete spectrum in contradiction to (7.2). This difficulty was found by means of the renormalization group analysis [Bogoliubov et al. 1959][62] much earlier than the development of the coherent-state formalism. The electron's two-point function can be shown to behave like

$$(\gamma^\mu p_\mu + m)(p^2 - m^2 + i0)^{-1+\beta} \tag{7.4}$$

near the mass shell, where*

$$\beta = e^2(\alpha-3)/8\pi^2, \tag{7.5}$$

α being the gauge parameter. Thus the corresponding commutator function does not have the usual mass-shell δ function *except for the Yennie gauge* $\alpha = 3$.

The behavior similar to (7.4) is encountered also in various exactly solvable *two-dimensional* models such as the Thirring model, the Schwinger model (i.e., massless two-dimensional QED), the pre-Schwinger model (i.e., Schwinger model with massive A_μ), the Schroer model, etc. Since the Schroer model is simplest and has a massive fermion, it is very instructive to discuss it in some detail.

The vector-coupling theory of the spinor-scalar system is exactly solvable in any dimensions, but renormalizable only in two dimensions. The Schroer model is the two-dimensional vector-coupling theory of a massive spinor ψ and a *massless* scalar ϕ [Schroer 1963].[63] The field equations are

$$(i\gamma^\mu \partial_\mu - m)\psi = -e\gamma^\mu\psi\partial_\mu\phi, \tag{7.6}$$

$$\Box\phi = 0. \tag{7.7}$$

To derive (7.7), use has been made of the current conservation. Let $\psi^{(0)}$ be an auxiliary free massive spinor satisfying

$$(i\gamma^\mu\partial_\mu - m)\psi^{(0)} = 0; \tag{7.8}$$

then we have the exact solution

$$\psi(x) = Z^{1/2} :\exp ie\phi(x): \psi^{(0)}(x), \tag{7.9}$$

---

* The expression for β receives no higher-order correction [Yennie et al. 1961][56] if parameters m, e, and α are renormalized.

where Z is a renormalization constant. By using (7.9), it is easy to calculate the two-point function, which behaves like

$$(\gamma^\mu p_\mu + m)(p^2 - m^2 + i0)^{-1+e^2/2\pi} \tag{7.10}$$

near $p^2 = m^2$. More generally, it is proved that there is no massive discrete spectrum in this model if quantized in the positive-metric Hilbert space. On the basis of this fact, Schroer introduced the notion of "infraparticle", which is a particle interacting with a massless field so that it loses its discrete spectrum. According to him, the QED electron is an infraparticle.

In Schroer's analysis, ψ and $\psi^{(0)}$ cannot be represented in a common Hilbert space. The difficulty arises from the impossibility of quantizing the two-dimensional massless scalar field in the positive-metric Hilbert space owing to the infrared divergence of its two-point function. This divergence can, however, be evaded, without violating Poincaré invariance, if quantized in the *indefinite-metric* Hilbert space [Nakanishi 1977].[64] The zero-frequency mode of ϕ becomes of negative norm, but it is excluded from $\mathcal{V}_{\text{phys}}$ by setting up a subsidiary condition

$$\int dx^1\ \partial_0 \phi^{(+)}(x)|f\rangle = 0. \tag{7.11}$$

If one works in this indefinite-metric formalism, then both ψ and $\psi^{(0)}$ are represented in a *common* indefinite-metric Hilbert space [Nakanishi 1977].[65] Although ψ has no discrete spectrum, $\psi^{(0)}$ does have one, that is, $\psi^{(0)}$ is an *asymptotic field of a composite field* $e^{-ie\phi}\psi$. In contrast to Schroer's analysis, (7.9) now shows that ψ is expressible in terms of asymptotic fields. Similar phenomenon arises also in other exactly solvable two-dimensional models listed above.* It is important to note that the absence of discrete spectrum in the two-point function of the primary spinor field is quite consistent with the existence of the asymptotic field

---

* For further details, see [Nakanishi et al. —].[38]

satisfying the asymptotic completeness.

It is quite natural to expect that a similar thing takes place also in QED. Indeed, we can realize this expectation if we employ the gaugeon formalism presented in Sec 4. The gaugeon G plays the role of ϕ of the two-dimensional models. The composite field ψ' defined by (4.19) with γ' satisfying $\alpha-\epsilon'\gamma'^2 = 3$ does have a discrete spectrum.

Finally, we make a comment to the proposition that the Gauss law would imply the absence of charged physical states. If $|f\rangle \in \mathcal{V}_{\rm phys}$, (4.5) together with (3.8) yields

$$\langle f \,|\, \partial_\mu F^{\mu\nu}+ej^\nu \,|\, f\rangle = 0. \tag{7.12}$$

This result is usually regarded as the reproduction of the classical Maxwell equation in QED. The zeroth component of (7.12),

$$\langle f \,|\, \partial_k F^{k0}+ej^0 \,|\, f\rangle = 0, \tag{7.13}$$

is nothing but the Gauss law. If $|f\rangle$ is a charged physical state, then we should have

$$\int d\mathbf{x} \langle f\,|j^0|\,f\rangle \neq 0. \tag{7.14}$$

Hence (7.13) implies

$$\int d\mathbf{x}\ \partial_k \langle f\,|F^{k0}|\,f\rangle \neq 0. \tag{7.15}$$

But (7.15) is impossible to hold. The proof is given in the following form. Let

$$\lim_{R\to\infty} \int d^4x\ R^{-2}\varphi_R(x)\langle f\,|F_{\mu\nu}(x)|\,f\rangle = f_{\mu\nu}(\varphi), \tag{7.16}$$

where $\varphi_R(x)$ is an arbitrary test function having a compact support in the spacelike region of the origin, R being the diameter of the support and $\varphi = \lim_{R\to\infty} \varphi_R$. Then, with some technical assumption concerning the sense of the limit, it can be rigorously proved that $f_{\mu\nu}$ must identically vanish [Buchholz 1986].[66)] Thus the Gauss law would imply the absence of charged physical states.

The above difficulty is caused by the identification of (7.12) with the classical Maxwell equation. The distinction between them can be seen by considering

$$\langle f|[\partial_\mu F^{\mu\nu}(x)+ej^\nu(x)]A_\lambda(y)|f\rangle. \tag{7.17}$$

The genuine classical Maxwell equation indicates also the vanishing of (7.17), but it does *not* vanish in general; for example, it equals $\partial^\nu\partial_\lambda D^{(+)}(x-y)$ for $|f\rangle=|0\rangle$.

From (4.5), we have

$$Q \equiv e\int d\mathbf{x}\, j^0(x) = \int d\mathbf{x}\, \partial^0 B(x) - \int d\mathbf{x}\, \partial_k F^{k0}. \tag{7.18}$$

Since the second term is converted into a surface integral, local commutativity implies that it commutes with any local operator. Accordingly, the quantity

$$Q' \equiv \int d\mathbf{x}\, \partial^0 B(x) \tag{7.19}$$

plays the role of the U(1) generator in the same way as Q does. But it is extremely important to note that Q' *cannot* be a well-defined operator in QED, that is, one can consider neither its eigenvalue nor its expectation value.

In what follows we prove that if Q' is well-defined then the asymptotic completeness must be violated, that is, we show that Q' commutes with all asymptotic fields. Firstly, since $B^{as} = B$, (4.9) implies that Q' commutes with B^{as}. Secondly, (4.8) and (7.2) imply

$$\begin{aligned}[A_\mu{}^{as}(x),\, Q'] &= -i\int d\mathbf{y}(\partial_0)^x(\partial_0)^y D(x-y) \\ &= -i\int d\mathbf{y}(\partial_k\partial^k)^y D(x-y) = 0. \end{aligned} \tag{7.20}$$

Finally, from the canonical commutation relations together with (4.7), we can derive

$$[\psi(x),\ B(y)] = e\psi(x)D(x-y). \tag{7.21}$$

Then, in the Yennie gauge, (7.2) together with $<0\,|\psi|\,0> = 0$ implies

$$[\psi^{as}(x),\ B(y)] = 0, \tag{7.22}$$

whence Q' commutes with ψ^{as}. In the general gauge, we must take account of G and C. But since they commute with B, the above reasoning remains valid.

Thus Q' is not well-defined under the postulate of asymptotic completeness. Since Q should be well-defined, (7.18) implies that $\int d\mathbf{x}\ \partial_k F^{k0}$ is *not well-defined*. This means that *it is forbidden to consider the l.h.s. of (7.15)*. Thus the existence of charged physical states yields no contradiction.

We remark that the mathematically rigorous analysis often yields misleading results in the indefinite-metric theory even if one restricts oneself to considering the physical subspace. One must be extremely careful for the mathematical setting-up of the starting point of the analysis.

8. FINITENESS OF THE RENORMALIZATION CONSTANTS

QED suffers from the ultraviolet divergence difficulty, but in the actual calculation it can be bypassed by means of the renormalization procedure. Here, perhaps the following two points must be emphasized:

1. Renormalization does not intrinsically resolve the divergence difficulty.
2. Renormalization is necessary even if there were no ultraviolet divergences.

In QED, we have four renormalization constants δm, Z_1, Z_2, and Z_3, though $Z_1 = Z_2$ (Ward identity). Let $A^{(0)}{}_\mu$, $B^{(0)}$, $\psi^{(0)}$, $m^{(0)}$, $e^{(0)}$, and $\alpha^{(0)}$ be unrenormalized quantities. Then the corresponding renormalized ones are defined as follows:

$$A_\mu = Z_3{}^{-1/2} A^{(0)}{}_\mu, \tag{8.1}$$

$$B = Z_3{}^{1/2} B^{(0)}, \tag{8.2}$$

$$\psi = Z_2{}^{-1/2} \psi^{(0)}, \quad \bar{\psi} = Z_2{}^{-1/2} \bar{\psi}^{(0)}, \tag{8.3}$$

$$m = m^{(0)} + \delta m, \tag{8.4}$$

$$e = Z_1{}^{-1} Z_2 Z_3{}^{1/2} e^{(0)}, \tag{8.5}$$

$$\alpha = Z_3{}^{-1} \alpha^{(0)}. \tag{8.6}$$

The QED Lagrangian density is given by

$$\begin{aligned} \mathcal{L} = &-\frac{1}{4} F^{(0)\mu\nu} F^{(0)}{}_{\mu\nu} + B^{(0)} \partial_\mu A^{(0)\mu} + \frac{1}{2} \alpha^{(0)} B^{(0)2} \\ &+ e^{(0)} \bar{\psi}^{(0)} \gamma^\mu \psi^{(0)} A^{(0)}{}_\mu + \bar{\psi}^{(0)} (i\gamma^\mu \partial_\mu - m^{(0)}) \psi^{(0)}. \end{aligned} \tag{8.7}$$

We substitute (8.1) - (8.6) into (8.7) and set

$$\mathcal{L}_0 \equiv -\frac{1}{4} F^{\mu\nu} F_{\mu\nu} + B \partial_\mu A^\mu + \frac{1}{2} \alpha B^2 + \bar{\psi} (i\gamma^\mu \partial_\mu - m) \psi. \tag{8.8}$$

Then the remainder

$$\mathcal{L}_I \equiv \mathcal{L} - \mathcal{L}_0 \tag{8.9}$$

is the interaction Lagrangian density of the renormalized QED. As is well known, it is free of ultraviolet divergences in every order of perturbation theory if the renormalization constants are appropriately chosen perturbatively. The field equations (4.5) - (4.7) hold for the renormalized quantities, where

$$j^{\mu} \equiv Z_1 Z_3^{-1}\bar{\psi}\gamma^{\mu}\psi - e^{-1}(Z_3^{-1}-1)\partial^{\mu}B. \qquad (8.10)$$

Note the existence of the second term in (8.10).

There is an opinion that unrenormalized quantities are meaningless, and therefore one should consider renormalized quantities only. But then $\mathcal{L}$ becomes too complicated and it is inappropriate to regard it as the starting point of QED. It is quite natural to expect that (8.1) - (8.6) are meaningful in some sense rather than purely formal. If (8.1) - (8.6) are taken seriously, we should have

$$Z_3 \geqq 0 \qquad (8.11)$$

from the hermiticity of A_μ and

$$Z_2 \geqq 0 \qquad (8.12)$$

from the self-consistency of (8.3). Although the perturbation theory guarantees the reality of all renormalization constants, it can tell us nothing about their signatures because the perturbation series seems to be non-convergent (at best asymptotically convergent). In the Heisenberg picture, however, we can discuss the spectral representations of two-point functions. Then we can show the validity of (8.11) by using the facts that $j_\mu|0\rangle \in \mathcal{V}_{phys}$ and that $\mathcal{V}_{phys}$ is of positive semi-definite. On the other hand, one cannot prove (8.12) because Z_2 is not a gauge-invariant quantity.

In order for (8.1) - (8.6) to be meaningful, the renormalization constants δm, Z_1, Z_2, and Z_3 should be all finite. Although Källén once claimed that at least one of them must be infinite [Källén 1954],[67] his proof was criticized: He did not correctly take account of the infrared factor [Yennie et al. 1961].[56]. At present, it is fair to say that the internal consistency of QED is neither proved nor disproved. Of course, it is purely academic to discuss QED alone, because it is now established that QED must be regarded as a part of the electroweak theory, and furthermore, one must take QCD into account. As suggested in Sec. 5, it is perhaps more crucial to take account of quantum gravity. It is optimistically expected that the Planck mass may play the role of a natural ultraviolet cutoff. Since all ultraviolet divergences encountered in QED are logarithmic, the extreme largeness of the Planck mass does not make the renormalization constants extremely large. Since the electromagnetic mass differences of hadrons are experimentally finite and of order e^2, the effect of the QED self-interaction must be finite and moderate.

Johnson and his collaborators studied the possibility of making all renormalization constants of QED explicitly finite in a self-consistent way.* The gauge-dependent renormalization constants Z_2 and Z_1 $(= Z_2)$ can be made finite by choosing the gauge parameter α appropriately (Landau gauge in lowest order) [Johnson et al. 1964].[69] The divergence of δm can be evaded if $m^{(0)} = 0$, that is, if m is generated in a self-consistent way. Assuming the finiteness of Z_3 so that the photon's two-point function asymptotically behaves like the free one, one solves the Schwinger-Dyson equation for the electron's two-point function asymptotically and finds that the electron's proper self-energy part can behave like $(p^2)^{-\sigma}$ $(\sigma > 0)$ as $p^2 \to \infty$, so that no ultraviolet divergences are encountered [Baker et al. 1971].[70] In this case, although chiral invariance is broken, they claim that the Nambu-Goldstone boson does not arise because of the anomalous behavior (nowadays known as anomaly) of the pseudoscalar density [Maris et al. 1966].[71]**

Thus the central problem is the finiteness of Z_3. In the calculation of the photon's proper self-energy part, one encounters ultraviolet divergences, but to all orders they are of single

* Their work is summarized in [Johnson et al. 1973].[68]

** For the recent discussions on the chiral symmetry breaking from the viewpoint of the strong-coupling QED, see [Miransky 1985].[72]

logarithm, i.e., $\sim\log \Lambda^2/m^2$ but not $\sim(\log \Lambda^2/m^2)^n$ with $n \geqq 2$, where Λ denotes the ultraviolet cutoff. This is because there are no internal divergences owing to the self-consistency assumption of finiteness. Accordingly, Z_3 is finite if and only if the coefficient of $\log \Lambda^2/m^2$ vanishes. Because of the dimensional reason, it is a function of $\hat{\alpha}^{(0)} \equiv (e^{(0)})^2/4\pi$ only and is written as $\hat{\alpha}^{(0)} f(\hat{\alpha}^{(0)})$. Thus QED would be a consistent theory if and only if a transcendental equation*

$$f(x) = 0 \tag{8.13}$$

has a positive root $x = x_0$.

Some remarks should be added [Adler 1972].[73] Contrary to the naive expectation, the root x_0 is not a simple zero of $f(x)$ but a zero of infinite order, i.e., an essential singularity of an analytic function $f(z)$. There is an alternative for having a consistent theory: The root x_0 may be identified with $\hat{\alpha} \equiv e^2/4\pi$ instead of $\hat{\alpha}^{(0)}$. Such ambiguity arises from taking different orders of summations.

The above scenario of having a finite QED is very interesting if true, but does not seem to be very promising. It is practically impossible to calculate $f(x)$ explicitly beyond the first several orders, and it is hopeless to write $f(x)$ in a closed form. More fundamental difficulty of the above analysis is that the order of summations is freely interchanged in spite of the fact that absolute convergence is never guaranteed. Remember that a power series is absolutely convergent if the expansion parameter is *less* than its convergence radius. Although it is unlikely that the perturbation series of QED has a non-zero convergence radius, its validity is implicitly postulated in the renormalization theory. Even under this postulate, in the present case, x_0 cannot be less than the (supposed) convergence radius because it is an essential singularity of $f(z)$. The ambiguity of identification stated above ($x_0 = \hat{\alpha}^{(0)}$ or $\hat{\alpha}$) may be regarded as a manifestation of

---

* It is possible to replace $f(x)$ by a simpler function $f_1(x)$, which is calculable by taking single-loop graphs only.

the lack of absolute convergence.

As is suggested by the discussion of the Schwinger term (see Sec. 5), it is perhaps not the right way to try to find a resolution of the high-energy consistency problem of QED within its own framework, but we should take account of the effect of quantum gravity.

REFERENCES

1) Leibbrandt, G., Rev. Mod. Phys. 59, 1067 (1987).
2) Schwinger, J., Phys. Rev. 130, 402 (1963).
3) Nakanishi, N., Phys. Lett. 131B, 381 (1983).
4) Frenkel, J. and Taylor, J. C., Nucl. Phys. B109, 439 (1976).
5) West, G. B., Phys. Rev. D27, 1878 (1983).
6) Schwartz, L., *Théorie des distributions* (Hermann et Cie, 1950), Chap. 1.
7) Nakanishi, N., Prog. Theor. Phys. 67, 965 (1982).
8) Bassetto, A., Dalbosco, M., Lazzizzera, I. and Soldati, R., Phys. Rev. D31, 2012 (1985).
9) Caracciolo, S., Curci, G. and Menotti, P., Phys. Lett. 113B, 311 (1982).
10) Dahmen, H. D., Scholz, B and Steiner, F., Phys. Lett. 117B, 339 (1982).
11) Landshoff, P. V., Phys. Lett. 169B, 69 (1986).
12) Steiner, F., Phys. Lett. 173B, 321 (1986).
13) Cheng, H. and Tsai, E.-C., Phys. Rev. Lett. 57, 511 (1986).
14) Mathews, P. M., Seetharaman, M. and Simon, M. T., Phys. Rev. D9, 1706 (1974).
15) Dirac, P. A. M., *Lectures on Quantum Mechanics* (Belfer Graduate School of Science, Yeshiva Univ., 1964).
16) Hossein Partovi, M., Phys. Rev. D29, 2993 (1984).
17) Rossi, G. C. and Testa, M., Phys. Rev. D29, 2997 (1984).
18) Bialynicki-Birula, I. and Kurzepa, P., Phys. Rev. D29, 3000 (1984).
19) Fermi, E., Rev. Mod. Phys. 4, 87 (1932).

20) Streater, R. F. and Wightman, A. S., *PCT, Spin & Statistics, and All That* (W. A. Benjamin, 1964), pp.86-87.

21) Streater, R. F. and Wightman, A. S., *ibid.* pp.139 & 163.

22) Strocchi, F., Phys. Rev. 162, 1429 (1967).

23) Gupta, S. N., Proc. Phys. Soc. A63, 681 (1950).

24) Bleuler, M., Helv. Phys. Acta 23, 567 (1950).

25) Becchi, C., Rouet, A. and Stora, R., Ann. of Phys. 98, 287 (1976).

26) Kugo, T. and Ojima, I., Prog. Theor. Phys. Suppl. 66, 1 (1979).

27) Strocchi, F. and Wightman, A. S., J. Math. Phys. 15, 2198 (1974).

28) Ambjørn, J. and Hughes, R. J., Nucl. Phys. B217, 336 (1983).

29) Nakanishi, N., Prog. Theor. Phys. 63, 656 (1980).

30) Nakanishi, N., Prog. Theor. Phys. 35, 1111 (1966).

31) Heisenberg, W., Nucl. Phys. 4, 532 (1957).

32) Nakanishi, N., Prog. Theor. Phys. Suppl. 51, 1 (1972).

33) Nakanishi, N., Phys. Rev. D3, 1343 (1971).

34) Nakanishi, N., Phys. Rev. B140, 947 (1965).

35) Nakanishi, N., Prog. Theor. Phys. 41, 233 (1969).

36) Freedman, D. Z. and Wang, J.-M., Phys. Rev. 153, 1596 (1967).

37) Lautrup, B., Mat. Fys. Medd. Dan. Vid. Selsk. 35, No.11 (1967).

38) Nakanishi, N. and Ojima, I., *Covariant Operator Formalism of Gauge Theories and Quantum Gravity* (World Scientific), to be published.

39) Nakanishi, N., Phys. Rev. D5, 1324 (1972).

40) Nakanishi, N., Prog. Theor. Phys. Suppl. 95, 1 (1988).

41) Yokoyama, K., Prog. Theor. Phys. 52, 1669 (1974).

42) Goto, T. and Imamura T., Prog. Theor. Phys. 14, 396 (1955).

43) Schwinger, Phys. Rev. Lett. 3, 296 (1959).

44) Orzalesi, C. A., *Lectures on Field Theoretic Aspects of Current Algebra* (1968, unpublished).

45) Johnson, K. and Low, F. E., Prog. Theor. Phys. Suppl. 37 & 38, 74 (1966).

46) Moffat, J., Nucl. Phys. 16, 304 (1960).

47) Nakanishi, N., Prog. Theor. Phys. 63, 1823 (1980).

48) Nakanishi, N., Publications RIMS (Kyoto) 19, 1095 (1983).

49) Kinoshita, T., Prog. Theor. Phys. 5, 1045 (1950).

50) Nakanishi, N., Prog. Theor. Phys. 19, 159 (1958).

51) Libby, S. B. and Sterman, G., Phys. Rev. D19, 2468 (1979).

52) Nakanishi, N., Phys. Rev. D28, 2112 (1983).

53) Doria, R., Frenkel, J. and Taylor, J. C., Nucl. Phys. B168, 93 (1980).

54) Andraši, A., Day, M., Doria, R., Frenkel, J. and Taylor, J. C., Nucl. Phys. B182, 104 (1981).

55) Yoshida, N., Prog. Theor. Phys. 66, 269 (1981).

56) Yennie, D. R., Frauschi, S. C. and Suura, H., Ann. of Phys. 13, 379 (1961).

57) Grammer Jr., G. and Yennie, D. R., Phys. Rev. D8, 4332 (1973).

58) Murota, T., Prog. Theor. Phys. 24, 1109 (1960).

59) Stapp, H. P., Phys. Rev. D28, 1386 (1983).

60) Kulish, P. P. and Faddeev, L. D., Teor. Mat. Fiz. 4, 153 (1970) [transl. Theor. Math. Phys. 4, 745 (1971)].

61) Zwanziger, D., Phys. Rev. D14, 2570 (1976).

62) Bogoliubov, N. N. and Shirkov, D. V., *Introduction to the Theory of Quantized Fields* (Interscience, 1959), §§41-44.

63) Schroer, B., Fortschr. Phys. 17, 1 (1963).

64) Nakanishi, N., Prog. Theor. Phys. 57, 269 (1977).

65) Nakanishi, N., Prog. Theor. Phys. 57, 1079 (1977).

66) Buchholz, D., Phys. Lett. B174, 331 (1986).

67) Källén, G., Kgl. Dan. Vid. Selsk. Mat.-fys. Medd. 27, No.12 (1954).

68) Johnson, K. and Baker, M., Phys. Rev. D8, 177 (1973).

69) Johnson, K., Baker, M. and Willey, R., Phys. Rev. B136, 1111 (1964).

70) Baker, M. and Johnson, K., Phys. Rev. D3, 2516 (1971).

71) Maris, Th. A. J. and Jacob, G., Phys. Rev. Lett. 17, 1300 (1966).

72) Miransky, V. A., Nuovo Cim. 90A, 149 (1985).

73) Adler, S. L., Phys. Rev. D5, 3021 (1972).

Quantum Electrodynamics for Nonrelativistic Systems and High Precision Determinations of α

Toichiro Kinoshita and G. Peter Lepage
Newman Laboratory, Cornell University
Ithaca, New York 14853

CONTENTS

1. Introduction

Nonrelativistic quantum mechanics is a very successful theory. It is the starting point for theories of atoms, molecules, and condensed-matter physics. Two of its most precise predictions are for the ac Josephson effect and for the quantized Hall effect, both of which have been confirmed by very accurate measurements. Indeed, the measurements of these effects are so precise that they are among the best ways of determining the fine structure constant α. These determinations are based upon two formulas derived from nonrelativistic quantum mechanics. The first,

$$\alpha^{-2} = \frac{c}{4R_\infty \gamma_p'} \frac{\mu_p'}{\mu_B} \frac{2e}{h}, \tag{1}$$

relates α to measurements of the ac Josephson frequency-to-voltage quotient $2e/h$ and of the proton's gyromagnetic ratio γ_p'. The second,

$$\alpha^{-1} = \frac{2R_H}{\mu_0 c}, \tag{2}$$

relates α to the quantized Hall resistance R_H.

The most recent measurements give[1,2]

$$\alpha^{-1}(\text{acJ}) = 137.035\,977\,0(77) \quad (0.056 \text{ ppm}), \tag{3}$$
$$\alpha^{-1}(\text{QHE}) = 137.035\,997\,9(32) \quad (0.024 \text{ ppm}). \tag{4}$$

These results may be contrasted with the best alternate determination of α, obtained by comparing the measured anomalous magnetic moment a_e of the electron[3] with the prediction from quantum electrodynamics (QED)[4]:

$$\alpha^{-1}(a_e) = 137.035\,992\,22(94) \quad (0.0069 \text{ ppm}). \tag{5}$$

These three values of α agree with each other to a part in 10^7, providing compelling confirmation both of QED and of the condensed-matter analyses leading to predictions for α(acJ) and α(QHE). Beyond a part in 10^7, however, there are problems. First, our ability to test QED is limited by the fact that neither α(acJ) nor α(QHE) is as accurately determined as $\alpha(a_e)$. Furthermore, the determinations based upon condensed-matter physics differ from each other by almost three standard deviations, and lie on the opposite sides of the QED determination.

Some possible causes of this disagreement are:

- statistical fluctuations;
- errors in the measurement of the proton's gyromagnetic ratio γ_p';
- errors in the calibration of the laboratory ohm standard;
- errors in the measurement of the electron anomaly;
- dependence of the quantized Hall effect on such things as the composition, quantum state, or temperature of the sample;
- corrections to the standard theories of the ac Josephson effect and quantized Hall effect;
- errors in the QED calculation of the electron anomaly;
- a breakdown of QED at the level of 10^{-8}.

The experimental issues in this list are best discussed by experimentalists. We note only that the $\alpha(\text{acJ})^{-2}$ is proportional to $\Omega_{\text{SI}}/\Omega_{\text{lab}}$, where Ω_{SI} is the SI ohm and Ω_{lab} is a laboratory realization of Ω_{SI}. This means that part of the uncertainty in α(acJ) is due to calibration errors in Ω_{lab}. On the other hand, $\alpha(\text{QHE})^{-1}$ is inversely proportional to Ω_{SI}/Ω_{lab}. Thus the combination $\alpha(\text{acJ})^{-2}\alpha(\text{QHE})^{-1}$ is independent of the resistance standard provided the same Ω_{lab} is used for both measurements. This leads to a fourth determination of the fine structure constant:[2]

$$\alpha^{-1}(\text{comb}) = 137.035\,984\,0(51) \quad (0.037 \text{ ppm}). \tag{6}$$

As to theoretical problems, it is generally believed that the theoretical expressions for α based upon the ac Josephson and quantized Hall effects are exact. As is well known, the first effect results from pair condensation of conduction electrons, while the second is due to the formation of a two-dimensional electric current perpendicular to an applied magnetic field. Although the underlying mechanisms differ in detail, the exactness of the α-determination in both cases derives primarily from the single-valuedness of the wave function and from the gauge invariance of the theory (that is, nonrelativistic quantum mechanics for a system of electrons and nuclei interacting through the Coulomb force[5,6]). It does not depend upon detailed properties of the system such as its geometric shape or size, its material composition, or the distribution of impurities.

This orthodox view will need reexamination should the disagreement between the condensed-matter experiments persist as these experiments are refined. Correction terms as small as a part in 10^8 will be important in such a study. This poses a problem for traditional nonrelativistic quantum mechanics. This theory ignores effects of order $(v/c)^2$, as well as radiative corrections due to the emission and reabsorption of virtual photons. Each of these effects leads to corrections that are significant given the high level of precision required; and each, when treated as a perturbation, leads to infinities in the theory. So the nonrelativistic theory, as usually posed, is either incomplete or undefined, or both. The inadequacy of the underlying theory calls into question the exactness of the condensed-matter determinations of α.

These considerations point to the need for a more reliable theory to serve as a basis for the analysis of the condensed-matter systems. A natural candidate is QED, which is relativistic and renormalizable, and which has been tested to very high precision. Unfortunately the theory of electrons is greatly complicated by the inclusion of relativity, and thus QED proves far more cumbersome for many-electron systems than the traditional nonrelativistic theory. Luckily there is an alternative formulation of QED that corresponds closely to the usual nonrelativistic theory, but where relativistic and radiative corrections can be systematically included. This new formulation is called nonrelativistic QED or NRQED.[7]

The infinities in the nonrelativistic theory due to relativistic and radiative effects result from the mistreatment of relativistic virtual states that couple to the nonrelativistic states of interest in second-order perturbation theory (and in higher orders). One way to deal with this problem is to remove all relativistic states from the quantum theory by in effect cutting off all momentum integrations in perturbation theory at $p \sim m$, where m is the electron's mass. Of course the relativistic states do affect nonrelativistic physics, but such effects can be taken into account by adding simple perturbative correction terms to the nonrelativistic Hamiltonian. NRQED is basically just the ordinary nonrelativistic theory but with a cutoff that makes the theory finite, and with correction terms in the Hamiltonian that compensate for the effects of the cutoff.

In this note we will outline the derivation of NRQED. The theory was first developed for high-precision analyses of QED-atoms like positronium or muonium. We will discuss its application to such problems, as well as to general problems in condensed-matter physics.

2. The Cutoff in QED

A key ingredient in NRQED is the cutoff that removes relativistic states from the theory. Thus we begin our formulation of the theory by examining the effects of the ultraviolet regulator in ordinary QED. According to conventional renormalization theory, QED is defined by a Lagrangian,

$$\mathcal{L}^{(\Lambda)} = \overline{\Psi}\left(i\partial \cdot \gamma - e(\Lambda) A \cdot \gamma - m(\Lambda)\right)\Psi - \tfrac{1}{4}F_{\mu\nu}F^{\mu\nu}, \tag{7}$$

together with an ultraviolet regulator that cuts off momentum integrations in perturbation theory around cutoff Λ. Given the correct numerical values for $e(\Lambda)$ and $m(\Lambda)$, the "bare" charge and mass of the electron, this cut-off theory is correct up to errors of order p/Λ, where p is a momentum typical of the process under study. It is traditional to remove the cutoff at the end of a calculation by taking the limit $\Lambda \to \infty$, thereby removing the p/Λ errors. However there is an alternative to taking Λ to infinity. In formulating NRQED, we prefer to use the cutoff to exclude relativistic-electron states from the theory, by setting

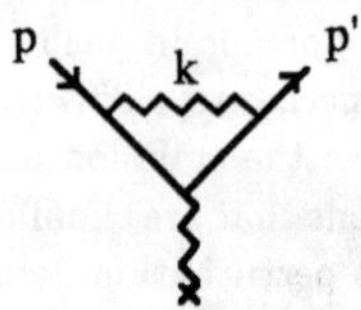

Figure 1: A one-loop radiative correction to the amplitude for an electron to scatter off an external field.

$\Lambda \sim m$. The errors that result from the finite cutoff are small, of order $p/\Lambda \sim v/c$, for nonrelativistic systems, but not sufficiently small for high precision analyses. So the theory must be modified to remove the errors of order p/Λ and higher.

To see how this is done, consider the one-loop radiative corrections to the renormalized amplitude for scattering a low-energy electron off an external field (see Fig. 1). The difference between the correct amplitude T and the amplitude $T^{(\Lambda)}$ in the cut-off theory involves internal photons with momentum k only of order Λ or larger. Given that the external electrons have small momenta, p and p', relative to Λ, we can expand the difference $T - T^{(\Lambda)}$ in a Taylor series in p/Λ and p'/Λ to obtain[a]

$$\begin{aligned} T - T^{(\Lambda)} &= \frac{e(\Lambda)m(\Lambda)c_1(\Lambda)}{\Lambda^2}\,\overline{u}A^{\mu}_{\text{ext}}\sigma_{\mu\nu}(p-p')^{\nu}u \\ &+ \frac{ie(\Lambda)c_2(\Lambda)}{\Lambda^2}\,(p-p')^2\,\overline{u}A_{\text{ext}}\cdot\gamma u \\ &+ \mathcal{O}(1/\Lambda^3), \end{aligned} \tag{8}$$

where coefficients c_1 and c_2 are dimensionless, and where the structure of the amplitude is constrained by the need for current conservation, and for chiral invariance (in the limit $m(\Lambda) = 0$). The first term dominates the errors in $T^{(\Lambda)}$ when $p, p' \ll \Lambda \sim m$, since it is then first order in p/Λ. We can incorporate this term into the cut-off theory by adding a new local interaction to the Lagrangian,

$$\delta\mathcal{L}^{(\Lambda)} = \frac{e(\Lambda)m(\Lambda)c_1(\Lambda)}{2\Lambda^2}\,\overline{\Psi}\sigma_{\mu\nu}F^{\mu\nu}\Psi. \tag{9}$$

This interaction introduces a "bare" anomalous magnetic moment for the electron into the theory. It is nonrenormalizable, but that causes no problem since the (finite) cutoff prevents infinities.

By replacing Lagrangian $\mathcal{L}^{(\Lambda)}$ with $\mathcal{L}^{(\Lambda)} + \delta\mathcal{L}^{(\Lambda)}$, we remove the order p/Λ errors from the theory, at least for the process $e\,A_{\text{ext}} \to e$ to one-loop order. Simple dimensional analysis shows that only three-point amplitudes can generate terms of order p/Λ, and so adding in $\delta\mathcal{L}^{(\Lambda)}$ removes *all* errors of order p/Λ from the theory. Furthermore this result can be extended to all orders in perturbation theory: $c_1(\Lambda)$ has a perturbative expannsion in α that is designed so that $\mathcal{O}(p/\Lambda)$ errors are removed order by order in α.

[a] There are no $\mathcal{O}(1/\Lambda^0)$ terms in $T - T^{(\Lambda)}$ since each of T and $T^{(\Lambda)}$ is a renormalized amplitude.

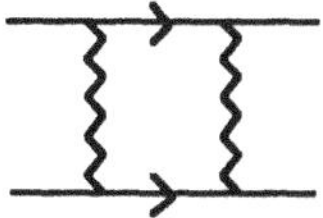

Figure 2: One-loop amplitude contributing to $ee \to ee$.

Similarly, errors of order $(p/\Lambda)^2$ can be removed by adding terms like

$$\frac{e(\Lambda)c_2(\Lambda)}{2\Lambda^2}\overline{\Psi}i\partial_\mu F^{\mu\nu}\gamma_\nu\Psi \tag{10}$$

to correct the amplitude for $e\,A_{\text{ext}} \to e$, and other terms like

$$\frac{d_2(\Lambda)}{\Lambda^2}\left(\overline{\Psi}\gamma_\mu\Psi\right)^2 \tag{11}$$

to correct for the omission of the $k > \Lambda$ contributions to amplitudes like that in Fig. 2 for $ee \to ee$. By dimensional arguments, one need only consider three- and four-point amplitudes to second order in p/Λ, and so again we find that all $(p/\Lambda)^2$ errors can be removed by adding a finite number of nonrenormalizable interactions to $\mathcal{L}^{(\Lambda)}$.

This process can be generalized to higher orders in p/Λ, and the theory made as accurate as desired without increasing the cutoff. Since contributions omitted as a consequence of the cutoff can be Taylor-expanded in powers of p/Λ, the correction terms in $\delta\mathcal{L}^{(\Lambda)}$ are all local—that is, they are polynomials in the fields and in derivatives of the fields. This is intuitively reasonable since these terms correct for contributions from intermediate states that are highly virtual and thus quite local in extent. The relative importance of the different interactions is determined by the number of powers of $1/\Lambda$ in their coefficients, and that is determined by the dimensions of the operators involved: each term in $\delta\mathcal{L}^{(\Lambda)}$ must have total (energy) dimension four, and so if the operator in a particular term has dimension $n+4$ then the coefficient must contain a factor $(1/\Lambda)^n$. One need only include interactions with operators of dimension $n+4$ or less to achieve accuracy through order $(p/\Lambda)^n$. On the other hand, all such interactions that respect the symmetries of the theory (and regulator) must be included. Thus, for example, the only dimension five operator that is local, gauge invariant, Lorentz invariant, and conserves parity is $\overline{\Psi}\sigma_{\mu\nu}F^{\mu\nu}\Psi$, and it was the only interaction needed to remove $\mathcal{O}(p/\Lambda)$ errors.

We have explained how to build a theory with a finite cutoff that approximates QED as closely as one wishes. By choosing a cutoff of $\mathcal{O}(m)$, we can make the final transformation to a truly nonrelativistic theory that greatly simplifies the analysis of nonrelativistic systems.

3. Nonrelativistic Quantum Electrodynamics (NRQED)

The cut-off version of QED developed in the previous section is not yet particularly useful in the study of nonrelativistic systems. Although restricted by the cutoff to nonrelativistic electrons, the theory is still basically a Dirac theory and therefore substantially more complicated than the nonrelativistic theory. The utility of the theory is greatly enhanced if one transforms the Dirac field so as to decouple its upper components, representing the electron, from its lower components, representing the positron. This can be achieved using the Foldy-Wouthuysen-Tani transformation[8]. It transforms the Dirac Lagrangian into a nonrelativistic Lagrangian. In QED it leads to

$$\begin{aligned}\overline{\Psi}(iD\cdot\gamma - m)\Psi \quad \rightarrow \quad & \psi^\dagger iD_0\psi + \psi^\dagger \frac{\mathbf{D}^2}{2m}\psi + \psi^\dagger \frac{\mathbf{D}^4}{8m^3}\psi \\ & + \frac{e}{2m}\,\psi^\dagger \boldsymbol{\sigma}\cdot\mathbf{B}\psi + \frac{e}{8m^2}\,\psi^\dagger \mathrm{div}\mathbf{E}\psi + \cdots \end{aligned} \tag{12}$$

where $D_\mu = \partial_\mu + ieA_\mu$ is the gauge-covariant derivative, $\mathbf{E}$ and $\mathbf{B}$ are the electric and magnetic fields, and ψ is a two-component Pauli spinor representing the electron part of the original Dirac field. The lower components of the Dirac field lead to analogous terms that specify the electromagnetic interactions of positrons. The Foldy-Wouthuysen-Tani transformation generates an infinite expansion of the action in powers of $1/m$. In an ordinary $\Lambda \to \infty$ field theory this expansion is a disaster: the renormalizability of the theory is completely disguised since it relies upon a delicate conspiracy involving terms of all orders in $1/m$. However, setting $\Lambda \sim m$ implies that the Foldy-Wouthuysen-Tani expansion is an expansion in $1/\Lambda$, and from our general analysis of cut-off theories we know that only a finite number of terms need be retained in the expansion if we want to work to some finite order in $p/\Lambda \sim p/m \sim v/c$. Thus, to work through order $(v/c)^2$, we can replace relativistic QED by a nonrelativistic theory (NRQED) with the Lagrangian

$$\begin{aligned}\mathcal{L}_{\mathrm{NRQED}} \quad = \quad & -\tfrac{1}{4}(F^{\mu\nu})^2 + \psi^\dagger\Big\{ i\partial_t - eA_0 + \frac{\mathbf{D}^2}{2m} + \frac{\mathbf{D}^4}{8m^3} \\ & + c_1\frac{e}{2m}\,\boldsymbol{\sigma}\cdot\mathbf{B} + c_2\frac{e}{8m^2}\,\mathrm{div}\mathbf{E} \\ & + c_3\frac{ie}{8m^2}\,\boldsymbol{\sigma}\cdot(\mathbf{D}\times\mathbf{E} - \mathbf{E}\times\mathbf{D})\Big\}\psi \\ & + \frac{d_1}{m^2}\,(\psi^\dagger\psi)^2 + \frac{d_2}{m^2}\,(\psi^\dagger\boldsymbol{\sigma}\psi)^2 \\ & + \text{positron and positron-electron terms.}\end{aligned} \tag{13}$$

The coupling constants e, m, $c_1 \ldots$ are all specified for a particular cutoff with $\Lambda \sim m$. Renormalization theory tells us that there exists a choice for the coupling constants in this theory such that NRQED reproduces all of the results of QED up to corrections of order $(p/m)^3$.

The values of the coupling constants depend upon the choice of cutoff or regulator, since the interactions must compensate for the finiteness of the cutoff. Generally in field theory, one has tremendous freedom in designing regulators, and that freedom should be used to tailor the cutoff to the particular problem of interest. For perturbative calculations in NRQED it is usually most convenient if the regulator respects such symmetries of the theory as gauge invariance, rotation invariance, and parity conservation; breaking one of these symmetries with the cutoff results in a more complicated Lagrangian. For example, if the regulator breaks the gauge symmetry of NRQED then extra gauge noninvariant interactions must be added to $\mathcal{L}_{\mathrm{NRQED}}$ to cancel gauge noninvariant effects induced by the cutoff. While it is possible to work with such a regulator, it is usually simpler to maintain gauge invariance, thereby avoiding additional interaction vertices. Insofar as

$\mathcal{L}_{\rm NRQED}$ is not Lorentz invariant, it is not critical that the regulator be Lorentz invariant, although a Lorentz invariant cutoff does simplify the calculation of some of the coupling constants: for example, the $\psi^\dagger(\mathbf{D}^4/8m^3)\psi$ term in $\mathcal{L}_{\rm NRQED}$ has no further coefficient if the cutoff is Lorentz invariant.

Obvious choices for the regulator include dimensional regularization and the Pauli-Villars regulator. One regulator that has been used successfully with NRQED is based upon a subtraction scheme: the divergences in any diagram (or subdiagram) are removed by subtracting counterterms that are identical in structure to the diagram, or derivatives of the diagram, but evaluated on-shell and at threshold.[7]. This regulator is particularly convenient for numerical work because the subtractions can be made in the integrands of Feynman integrals, before integrating.

Once a regulator has been specified for NRQED, the coupling constants are readily calculated. One simple procedure is to compute scattering amplitudes both in QED and in NRQED, and to adjust the NRQED coupling constants so that the two calculations agree (to a given order in α and v/c). For example, the coefficient c_1 of the $\psi^\dagger\boldsymbol{\sigma}\cdot\mathbf{B}\psi$ interaction can be determined by comparing calculations in the two theories of the spin-flip amplitude for an electron to scatter off an external magnetic field. Once c_1 is tuned for this process, it is correct for all other processes. One finds from such calculations that each of the coupling constants has a perturbative expansion in α; at tree level, $c_1 = c_2 = c_3 = 1$ and $d_1 = d_2 = 0$. The electric charge e is not renormalized in NRQED since vacuum polarization corrections to the low-energy photon propagator involve highly virtual electron-positron states that are omitted from the nonrelativistic theory; there is no vacuum polarization in NRQED. The effects of vacuum polarization on the photon's dynamics are of order $\alpha(v/c)^2$ and smaller, and can be incorporated into NRQED by adding new local interactions to the photon's Lagrangian.[b]

4. NRQED for Positronium and Other Simple Atoms

Simple atoms like positronium, hydrogen, and muonium (a muon-electron atom) have long played an important role in high-precision tests of QED. Although most issues of principle concerning nonrelativistic bound states have long been resolved, there remain significant technical problems connected with the study of these systems. Central among these is the problem of too many energy scales. Typically a nonrelativistic system has at least three important energy scales: the masses m of the constituents, their three-momenta $p \sim mv$, and their kinetic energies $K \sim mv^2$. These scales are widely different in a nonrelativistic system since $v \ll 1$ (where the speed of light $c = 1$), and this complicates any analysis of such a system.

To illustrate the problems that arise, consider a traditional Bethe-Salpeter calculation of the energy levels of positronium. The potential in the Bethe-Salpeter equation is given by a sum of two-particle irreducible Feynman amplitudes. One generally solves the problem for some approximate potential and then incorporates corrections using time-independent perturbation theory. Unfortunately, perturbation theory for a bound state is far more complicated than perturbation theory for, say, the electron's g-factor. In the

[b]The leading correction involves an operator $(\partial_\mu F^{\mu\nu})^2$. The equations of motion imply that such a term can be replaced by four-fermion interactions such as those already present in $\mathcal{L}_{\rm NRQED}$. In higher orders there are four-photon vertices of the sort computed (long ago) by Euler and Heisenberg.[9]

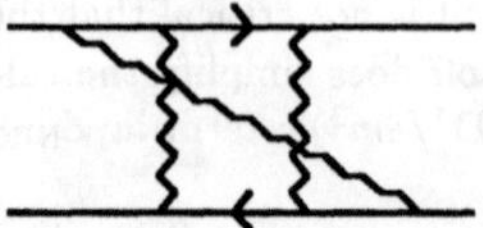

Figure 3: A two-loop kernel contributing to the Bethe-Salpeter potential for positronium.

latter case a diagram with three photons contributes only in order α^3. In positronium a kernel involving the exchange of three photons (e.g., Fig. 3) can also contribute to order α^3, but the same kernel will contribute to all higher orders as well:

$$\langle V_3 \rangle = \alpha^3 m \left\{ a_0 + a_1\alpha + a_2\alpha^2 + \ldots \right\}. \tag{14}$$

So in the bound state calculation there is no simple correlation between the importance of an amplitude and the number of photons in it. Such behavior is at the root of the complexities in high-precision analyses of positronium or other QED bound states, and it is a direct consequence of the multiple scales in the problem. Any expectation value like that in Eq. (14) will be some complicated function of ratios of the three scales in the atom:

$$\langle V_3 \rangle = \alpha^3 m \; F(\langle p \rangle / m, \langle K \rangle / m). \tag{15}$$

Since $\langle p \rangle / m \sim \alpha$ and $\langle K \rangle / m \sim \alpha^2$, a Taylor expansion of F in powers of these ratios generates an infinite series of contributions just as in Eq. (14). A similar series does not occur in the g-factor calculation because there is but one scale in that problem, the mass of the electron.[c]

Traditional methods for analyzing these bound states fail to take advantage of the nonrelativistic character of these systems; and atoms like positronium are very nonrelativistic: the probability for finding relative momenta of $\mathcal{O}(m)$ or larger is roughly $\alpha^5 \approx 10^{-11}$! Clearly NRQED is very well suited for studying these atoms. In NRQED, states with relativistic energies are explicitly excluded, thereby removing the most troublesome of the energy scales in the bound state problem, the constituent's mass. The essentially nonperturbative character of the binding is primarily a nonrelativistic effect. Thus the relativistic physics can be included perturbatively, to any desired precision, simply by adding local interactions to the cut-off Lagrangian of NRQED.

NRQED is far simpler to use than the original relativistic theory when studying nonrelativistic atoms like positronium. The analysis falls into two parts. First one must

[c]The bound state calculation is actually even worse than indicated because QED is a gauge theory. The contribution from a particular kernel is highly dependent upon gauge. For example, the coefficients a_0 and a_1 in Eq. (14) vanish in Coulomb gauge but not in Feynman gauge. The Feynman gauge result is roughly 10^4 times larger, and it is largely spurious: the bulk of the contribution comes from unphysical retardation effects in the Coulomb interaction that cancel when an infinite number of other diagrams are included. The Coulomb interaction is instantaneous in Coulomb gauge and so this gauge generally does a better job describing the fields created by slowly moving charged particles. On the other hand contributions coming from relativistic momenta are more naturally handled in a covariant gauge like Feynman gauge; in particular renormalization is far simpler in Feynman gauge than it is in Coulomb gauge. Unfortunately most Bethe-Salpeter kernels have contributions coming from both nonrelativistic and relativistic momenta, and so there is no optimal choice of gauge. This is again a problem due to the multiple scales in the system.

determine the coupling constants in the NRQED Lagrangian. This is easily done by comparing simple scattering amplitudes computed in both QED and NRQED, as discussed in the previous section. Since the couplings contain the relativistic physics, this part of the calculation involves only scales of order m; it is similar in character to a calculation of the g-factor. Furthermore there is no need to deal with complicated bound states at this stage. Having solved the high-energy part of QED by computing the coupling constants in $\mathcal{L}_{\rm NRQED}$, one goes on to solve NRQED for any nonrelativistic process or system. To study positronium one uses the Bethe-Salpeter equation for this theory, which is just the Schrödinger equation, and ordinary time-independent perturbation theory. One of the main virtues of this approach is that it builds directly on the simple results of nonrelativistic quantum mechanics, leaving our intuition intact. Even more important for high-precision calculations is that only two dynamical scales remain in the problem, the momentum and the kinetic energy, and these are easily separated on a diagram-by-diagram basis. As a result infinite series in α can be avoided in calculating the contributions due to individual diagrams, and thus it is trivial to separate, say, $\mathcal{O}(\alpha^6)$ contributions from $\mathcal{O}(\alpha^5)$ contributions.[d]

Positronium is composed entirely of leptons; ordinary atoms like hydrogen can also be analyzed using NRQED. The major difference is that the nucleus in such atoms has structure. NRQED allows us to incorporate the effects of this structure in a systematic and rigorous fashion. The proton in the hydrogen atom, for example, is treated as a point-like particle described by a Lagrangian that is identical in form to the electron's Lagrangian (Eq. (13)). The proton's structure, being very short-ranged, enters only through the coupling constants $c_1, c_2 \ldots$. Lacking a complete theory of proton structure, we are unable to compute these coupling constants. However, we can measure them: for example, c_1, c_2, and c_3 are completely determined by the proton's magnetic moment and charge radius, and by relativistic invariance. Thus proton structure is for our purposes completely specified by a small number of coupling constants, the actual number depending upon the accuracy needed. Given these couplings, one can forget that the proton has structure and treat it as we treat electrons.

5. NRQED for Condensed-Matter Systems

In the previous section we described the utility of NRQED in analyzing simple atoms. NRQED is immediately applicable to many-electron systems as well, because it is formulated as a second-quantized field theory. Going to Coulomb gauge, one obtains the standard nonrelativistic quantum field theory for electrons, but with new interactions that correct for relativity and photon-electron interactions. This theory can be made as precise as is desired, and so it provides a rigorous framework for high-precision analyses in condensed-matter physics. It would be very interesting to explore the consequences of the "new" interactions in the theory for condensed-matter systems. As regards the condensed-matter determinations of α, the most relevant feature of NRQED is its exact gauge invariance. The gauge symmetry is unaffected by the new interactions, and con-

[d]Furthermore the choice of gauge is now obvious. In computing the coupling constants for NRQED, one uses a covariant gauge like Feynman gauge since this calculation involves only relativistic momenta. These couplings are gauge invariant provided the cutoff used in defining NRQED is gauge invariant. Thus when we use NRQED to compute the properties of positronium, for example, we are free to choose Coulomb gauge, the optimal choice for nonrelativistic systems.

sequently the conventional analyses[5,6] of the ac Josephson effect and the quantized Hall effect remain intact.

It follows therefore that any discrepancy between α(acJ) and α(QHE) cannot be attributed to a breakdown of gauge invariance; the cause must lie elsewhere. Various theoretical corrections to the ac Josephson effect have been examined and evaluated in the literature[10]. In general these corrections seem to have a negligible effect upon α(acJ), at least to a part in 10^8. One complication in the quantized Hall effect is the possibility of crosstalk between edge states on opposite sides of the sample.[11]. At present there are vigorous experimental efforts under way to understand such effects.[e] Hopefully these issues will be clarified to the extent that uncertainties in the quantized Hall effect become smaller than a part in 10^8 for well-prepared samples.

Acknowledgements

One of us (T.K.) thanks B.I. Halperin and A.D. Stone for illuminating discussions concerning the current state of our understanding of the quantized Hall effect. This work was supported in part by a grant from the National Science Foundation.

References

1. E. R. Williams, G. R. Jones, Jr., S. Ye, R. Liu, H. Sasaki, P. T. Olsen, W. D. Phillips, and H. P. Layer, IEEE Trans. Instrum. Meas., vol. IM-**38**, 233 (1989).

2. M. E. Cage, R. F. Dziuba, R. E. Elmquist, B. F. Field, G. R. Jones, Jr., P. T. Olsen, W. D. Phillips, J. Q. Shields, R. L. Steiner, B. N. Taylor, and E. R. Williams, IEEE Trans. Instrum. Meas., vol. IM-**38**, 284 (1989).

3. See the article by Van Dyck in this volume.

4. See the article by Kinoshita in this volume.

5. F. Bloch, Phys. Rev. Lett. **21**, 1241 (1968); Phys. Rev. B **2**, 109 (1970); T. A. Fulton, Phys. Rev. B **7**, 981 (1973).

6. R. B. Laughlin, Phys. Rev. B **23**, 5632 (1981); B. I. Halperin, Phys. Rev. B **25**, 2185 (1982).

7. W. E. Caswell and G. P. Lepage, Phys. Lett. **167B**, 437 (1986).

8. M. H. L. Pryce, Proc. Roy. Soc. **195**, 62 (1948); S. Tani, Soryushiron Kenkyu, **1**, 15 (1949) (in Japanese); Prog. Theor. Phys **6**, 267 (1951); L. L. Foldy and S. A. Wouthuysen, Phys. Rev. **78**, 29 (1950).

9. W. Heisenberg and H. Euler, Z. Physik **98**, 714 (1936).

10. R. H. Koch, D. J. Van Harlingen, and J. Clarke, Phys. Rev. Lett. **45**, 2132 (1980); A. O. Caldeira and A. J. Leggett, Phys. Rev. Lett. **46**, 211 (1981); A. Widom and T. D. Clark, Phys. Rev. Lett. **48**, 1572 (1982); V. Ambegaokar, U. Eckern, and G. Schon, Phys. Rev. Lett. **48**, 1745 (1982); M. R. Arai, Cornell University Ph. D. thesis, 1983.

---

[e]See Ref. 12, for example.

11. B.I. Halperim, Phys. Rev. B **25**, 2185 (1982); P. Streda, J. Kucera, and A.H. MacDonald, Phys. Rev. Lett. **59**, 1973 (1987); J.K. Jain and S.A. Kivelson, Phys. Rev. Lett. **60**, 1542 (1988); M. Büttiker, Phys. Rev. B **38**, 9375 (1988).

12. P.L. McEuen, A. Szater, C.A. Richter, B.W. Alphenaar, J.K. Jain, A.D. Stone, R.G. Wheeler, and R.N. Sachs, Phys. Rev. Lett. **64**, 2062 (1990).

Test of QED by High Energy Electron–Positron Collisions

Hans – Ulrich Martyn
I. Physikalisches Institut, RWTH Aachen
Sommerfeldstraße, D 5100 Aachen, Germany

August 1989

Contents

1 Introduction

High energy electron–positron collision experiments have provided a large amount of information and contributed considerably to our understanding of particle physics. Despite its high potential to search for new phenomena, one of the main objectives to construct e^+e^- colliders was and still is to test the validity of Quantum Electrodynamics (QED) at large momentum transfers. The particularly clean initial state offers an ideal laboratory to investigate the following basic lowest order processes

- photon pair production
$$e^+e^- \rightarrow \gamma\gamma\,, \tag{1}$$
- lepton pair production ($l = e,\ \mu,\ \tau$, or new kind)
$$e^+e^- \rightarrow l^+l^-\,, \tag{2}$$
- hadron formation through quark pair production ($q = u,\ d,\ s,\ c,\ b$, or new kind)
$$e^+e^- \rightarrow q\bar{q}\,, \tag{3}$$
- fermion pair production through photon–photon scattering $\gamma\gamma \rightarrow f\bar{f}$ ($f = l,\ q$) via
$$e^+e^- \rightarrow e^+e^- f\bar{f}\,. \tag{4}$$

These processes have been extensively studied at all e^+e^- colliders, covering the large range of centre of mass energy from 1 to 60 GeV, which will very soon be extended to 100 GeV. With the operatión of high energy machines, like PETRA, PEP and TRISTAN, it became apparent that it is not sufficient to describe these processes by QED alone. The weak interaction mediated through the intermediate vector boson Z^0 and the strong interaction among the final state hadrons become important with increasing centre of mass energy. Gauge theories propose a unified description of electromagnetic and weak interactions, Electroweak Interaction or Quantum Flavourdynamics (QFD), and strong interactions between quarks and gluons, Quantum Chromodynamics (QCD). Various attempts to arrive at a unified theory have been made, the most successful one up to now being the so called Standard Theory based on the $SU(3) \times SU(2) \times U(1)$ symmetry.

QED remains a consistent theory within the embracing standard theory and can be tested by itself provided the following conditions are fullfilled [1]:

1. The characteristic invariants of a process, the centre of mass energy squared s and the space-like momentum transfer squared Q^2, have to be small compared to m_Z^2, the scale given by the Z^0 mass.

2. The experimental precision should be lower than $\mathcal{O}(s/m_Z^2)$ or $\mathcal{O}(Q^2/m_Z^2)$.

Both criteria are not always satisfied in high energy e^+e^- experiments, the momentum transfers reach $s/m_Z^2 \sim 0.35$ while the typical experimental precision is a few percent. Large momentum transfers imply that only special reactions or certain distributions can be used for pure QED tests, where QFD and QCD effects are either neglibible or can be treated as small corrections. With regard to the experimental precision of e^+e^- experiments it is often sufficient to perform QED calculations up to the next to leading order in the fine structure constant α, which is $\mathcal{O}(\alpha^3)$ for the two body reactions. These higher orders can be checked explicitly by detecting additional photons in the basic reactions (1) – (4). This has to be compared to low energy high precision measurements like the anomalous magnetic moments of leptons, the Lamb shift and the hyperfine structure splittings (discussed in separate articles of this book), where QED corrections up to the forth and fifth order in α have been calculated. Therefore high energy e^+e^- experiments are complementary by testing QED at very large momentum transfers or equivalently at very small distances.

A major task of QED is concerned with the detailed and precise calculation of higher order radiative effects. Large efforts have been made to implement them into Monte Carlo programs with realistic detector and analysis simulation in order to verify them with experimental data. These tests are of prime importance for the validity of QED and can be considered as a prerequisite to disentangle QFD and QCD effects. They can also be used to search for or set limits on a possible substructrure of the elementary fermions.

The photon pair production $e^+e^- \to \gamma\gamma$, reaction (1), provides a very clean QED test. It proceeds in lowest order α^2 via single electron exchange. Also higher order contributions can be entirely described by QED alone, the first order weak corrections being completely negligible.

The e^+e^- annihilation into a fermion pair $e^+e^- \to f\bar{f}$, reactions (2) and (3), belongs to the most simple processes. The production cross section in lowest order receives contributions from photon and Z^0 exchange resulting in a sum of pure QED, pure weak and interference terms with a characteristic energy dependence

$$\frac{d\sigma}{d\Omega} = \frac{d\sigma}{d\Omega}^{QED} (\sim \frac{\alpha^2}{s}) + \frac{d\sigma}{d\Omega}^{interference} (\sim \alpha \cdot G) + \frac{d\sigma}{d\Omega}^{weak} (\sim G^2 \cdot s)$$

(G = Fermi coupling constant). For quark pair production additional QCD corrections have to be applied. The QED cross section decreases as $1/s$ while the pure weak cross section rises with s. At the PETRA, PEP and TRISTAN e^+e^- colliders, even at the highest energies, QED is still the dominating force, the purely weak contributions being only $\sim 2\%$ for lepton pair production but reaching $\sim 50\%$ for quark pair production. However, the electroweak interference between the photon and the Z^0 leads to appreciable measurable asymmetries in the fermion angular distributions, which are partly counteracted by practically energy independent higher order QED contributions. At SLC and LEP the fermion pair production will be completely dominated by Z^0 formation.

In recent years photon–photon scattering of reaction (4) which proceeds via forth order in α has received special attention. QED tests concentrate on two–photon lepton pair production while hadron final states offer the possibility to study QCD effects and resonance production. The momentum transfers are moderate compared to the annihilation diagrams, thus providing less stringent QED bounds. However, as an $\mathcal{O}(\alpha^4)$ process it has an interest on its own right.

In this article the phenomenology and experimental results of QED tests in high energy e^+e^- collisions will be presented. Since QED can only be seen in the context of QFD a brief review on electroweak phenomena will be given as well. The material comprises results from the experiments at PETRA, PEP and TRISTAN as published until July 1989.

2 Interaction Lagrangians and couplings

The standard theory [2,3,4] combines the electromagnetic, weak and strong interactions between elementary particles into an unified renormalizable theory. It is based on local gauge invariance, which is then spontaneously broken in order to generate the masses of the particles. Although known to be unsatisfactory (e.g. 17 free parameters in the minimal electroweak model) and leaving many questions open, it is still the most successful unified theory and has survived all experimental tests so far. For recent reviews on the status of the electroweak standard model see e.g. [5,1].

The basic theoretical foundations can be found in modern text books. Here only the most important ingredients of the electroweak standard theory relevant for the understanding of e^+e^- reactions will be briefly outlined [6].

2.1 The electromagnetic interaction

Quantum Electrodynamics describes the electromagnetic interaction of charged particles with photons. The dimensionless coupling strength is given by the electric charge ($\alpha = e^2/4\pi = 1/137.0359895(61)$ in the low energy Thomson limit). Of special interest is the $ff\gamma$ coupling of a charged fermion (Dirac field $\psi(x)$, charge $e_f = Q_f\, e$) to the photon field $A_\mu(x)$ with the interaction Lagrangian

$$\mathcal{L}_I^{em} = -e_f\, \bar{\psi}(x)\, \gamma_\mu\, \psi(x)\, A^\mu(x)\,. \tag{5}$$

The Lagrangian is invariant under parity and charge conjugation. It is also invariant under a local U(1) gauge transformation which implies a conserved vector current

$$J_\mu^{em}(x) = \bar{\psi}(x)\, \gamma_\mu\, \psi(x) \tag{6}$$

and completely fixes the interaction. The electromagnetic interaction Lagrangian can now being written as

$$\mathcal{L}_I^{em} = -e_f\, J_\mu^{em}(x)\, A^\mu(x)\,. \tag{7}$$

2.2 The electroweak interaction

Quantum Flavourdynamics describes the electromagnetic and weak interactions of fermions and bosons with the gauge fields and among the gauge bosons themselves.

It is a renormalizable gauge theory which was first proposed by Glashow [2], Salam [3] and Weinberg [4] for the leptonic sector and has later been extended to the quark sector [7]. The minimal model is based on the group $SU(2) \times U(1)$ with a triplet of gauge fields for the weak isospin group of $SU(2)$ and a singlet gauge field for the weak hypercharge group of $U(1)$. Mass generation of all particles is provided through symmetry breaking by the Higgs mechanism, which introduces an additional scalar Higgs doublet and results in the four physical gauge fields, the massless photon γ and the massive intermediate vector bosons Z^0, W^+ and W^-. The symmetry is spontaneously broken from $SU(2) \times U(1)$ to $U(1)_{em}$ of electromagnetism. The coupling constants are related by the unification condition

$$e = g \sin\theta_W \,, \tag{8}$$

where θ_W is called the weak mixing angle. The relation to the charged weak interaction (Fermi coupling constant $G = 1.16637(2)\,10^{-5}\,GeV^{-2}$ from μ decay) is given in the tree approximation by

$$\frac{G}{\sqrt{2}} = \frac{g^2}{8\,m_W^2} = \frac{e^2}{8\,\sin^2\theta_W\,m_W^2} \,. \tag{9}$$

Together with

$$m_W = m_Z \cos\theta_W \tag{10}$$

the masses of the intermediate bosons m_W and m_Z are determined by a measurement of α, G and $\sin^2\theta_W$.

In the standard theory all fermions are represented as left-handed doublets and right-handed singlets of weak isospin I^f. However, there exist no right-handed neutrinos, i.e. neutrinos are massless. There are three generations of leptons, (ν_e, e^-), (ν_μ, μ^-) and (ν_τ, τ^-), and three generations of quarks, (u, d'), (c, s') and (t, b'), where the top quark t still awaits its experimental discovery. The quarks q' represent the weak eigenstates. Transitions between lepton families are not allowed, while the quark generations are mixed through the weak charged current.

The interaction ffZ between a fermion and the Z^0 gauge field $Z_\mu(x)$ can be written analogous to eq. (7) as the weak neutral current interaction Lagrangian

$$\mathcal{L}_I^Z = -\frac{g}{2\,\cos\theta_W}\,J_\mu^Z(x)\,Z^\mu(x) \tag{11}$$

with the weak neutral current

$$J_\mu^Z(x) = \bar{\psi}(x)\,\gamma_\mu\,(g_V^f - g_A^f\,\gamma^5)\,\psi(x) \,. \tag{12}$$

This Lagrangian violates parity and charge conjugation. The vector and axial vector couplings are defined as

$$g_V^f = I_3^f - 2\,Q_f\,\sin^2\theta_W \,, \tag{13}$$
$$g_A^f = I_3^f \,, \tag{14}$$

Table 1: Quantum numbers and coupling constants of the elementary fermions to the electroweak neutral current. Numerical values are given for $\sin^2\theta_W = 0.23$

Fermion	Q_f	I_3^f	g_V^f		g_A^f
ν_e, ν_μ, ν_τ	0	$+\frac{1}{2}$	$+\frac{1}{2}$	+0.50	+0.50
e^-, μ^-, τ^-	−1	$-\frac{1}{2}$	$-\frac{1}{2} + 2\sin^2\theta_W$	−0.04	−0.50
u, c, t	$+\frac{2}{3}$	$+\frac{1}{2}$	$+\frac{1}{2} - \frac{4}{3}\sin^2\theta_W$	+0.19	+0.50
d, s, b	$-\frac{1}{3}$	$-\frac{1}{2}$	$-\frac{1}{2} + \frac{2}{3}\sin^2\theta_W$	−0.35	−0.50

where I_3^f is the third component of the weak isospin. The experimentally determined value of the weak mixing angle using all available information (assuming a top quark mass of $m_t = 45\,GeV$ and a Higgs mass of $m_H = 100\,GeV$) is [8]

$$\sin^2\theta_W = 0.230 \pm 0.0048 \ .$$

The quantum numbers and coupling constants of the fermions are summarized in Table 1.

The complete electroweak neutral current interaction Lagrangian then becomes

$$\mathcal{L}_I^{NC} = \mathcal{L}_I^{em} + \mathcal{L}_I^Z \ . \tag{15}$$

3 Cross sections in lowest order

At the energies discussed in this article electrons and positrons couple to the neutral gauge bosons ($e^+e^-\gamma$, e^+e^-Z). Couplings to the Higgs boson are suppressed by m_e/m_W factors and can be safely neglected. In this section cross section formulas for the standard reactions will be given in lowest order in the general frame of QFD, and also for the embedded special case of QED. Only unpolarized electron and positron beams will be considered, since so far experiments with transverse polarized beams did not surpass an exploratory study. This situation may change at the next generation of e^+e^- colliders, SLC and LEP, where experiments hope to profit from longitudinal polarization. Higher order radiative corrections will be treated in detail in section 4.

3.1 $e^+e^- \to \gamma\gamma$

Two photon production $e^+e^- \to \gamma\gamma$ is a pure QED process mediated by the two diagrams of t and u channel electron exchange of Fig. 1. Defining θ as the polar angle between the electron and photon the differential cross section for $e^+e^- \to \gamma\gamma$ in second order QED is

$$\frac{d\sigma}{d\Omega} = \frac{\alpha^2}{s}\frac{1+\cos^2\theta}{\sin^2\theta} \ . \tag{16}$$

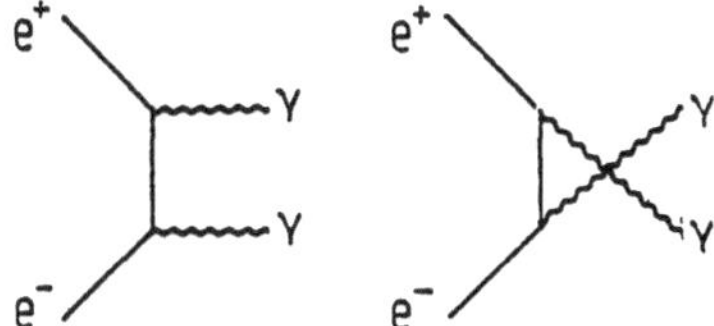

Figure 1: Lowest order diagrams for $e^+e^- \to \gamma\gamma$

Typically for t channel exchange the cross section is steeply rising in the forward and backward region and appears to be singular as θ approaches 0 or π. This is due to the fact that the electron mass has been neglected. The denominator $\sin^2\theta$ should be replaced by $1-\beta_e^2\cos^2\theta \simeq \sin^2\theta + 4\,m_e^2/s$, where $\beta_e = p_e/E_e = \sqrt{1-4\,m_e^2/s}$ is the velocity of the electron. The correction term $4\,m_e^2/s$ reaches the level 10^{-3} only in the direction very close to the beams if θ or $\pi - \theta \leq 30/\sqrt{s}$ $[GeV]$ $mrad$. For detectors with limited acceptance $\theta > \theta_{min}$ the integrated total cross section becomes

$$\begin{aligned}\sigma(e^+e^- \to \gamma\gamma) &= \int_{\theta_{min}}^{\pi/2} \frac{d\sigma}{d\Omega} d\Omega \\ &= \frac{2\,\pi\,\alpha^2}{s}\left\{\ln\frac{1+\cos\theta_{min}}{1-\cos\theta_{min}} - \cos\theta_{min}\right\}\,. \end{aligned} \tag{17}$$

The integration is performed over one hemisphere to take into account that both photons are indistinguishable. Apart from QED radiative corrections this reaction should also at very high energies be completely described by electron exchange.

3.2 $e^+e^- \to f\bar{f}$

The e^+e^- annihilation into a fermion pair $e^+e^- \to f\bar{f}$ with f being a μ, τ or quark proceeds in lowest order via photon and Z^0 exchange as sketched in Fig. 2 (for $f = e$ additional t channel diagrams have to be added, see below). Defining θ as the polar angle between the incoming electron and the outgoing fermion the electroweak differential cross section for $e^+e^- \to f\bar{f}$ including mass effects is given in lowest order by

$$\begin{aligned}\frac{d\sigma}{d\Omega} &= \frac{\alpha^2}{4\,s} N_f^c \beta_f \left\{C_1(1+\cos^2\theta) + \beta_f\,C_2\cos\theta + (1-\beta_f^2)\,C_3\,\sin^2\theta\right\}\,, \qquad (18)\\ C_1 &= Q_f^2 - 2\,g_V^e\,g_V^f\,Q_f\,\Re(\chi) + ({g_V^e}^2 + {g_A^e}^2)({g_V^f}^2 + {g_A^f}^2)\,|\chi|^2\,,\\ C_2 &= \quad -4\,g_A^e\,g_A^f\,Q_f\,\Re(\chi) + 8\,g_V^e\,g_V^f\,g_A^e\,g_A^f\,|\chi|^2\,,\\ C_3 &= Q_f^2 - 2\,g_V^e\,g_V^f\,Q_f\,\Re(\chi) + ({g_V^e}^2 + {g_A^e}^2)\,{g_V^f}^2\,|\chi|^2\,.\end{aligned}$$

Here N_f^c is a colour factor (1 for leptons and 3 for quarks), $\beta_f = p_f/E_f = \sqrt{1-4\,m_f^2/s}$ is the velocity of the fermion and g_V^e, g_V^f, g_A^e, g_A^f are the vector and axial vector couplings of the electron and fermion given in Table 1.

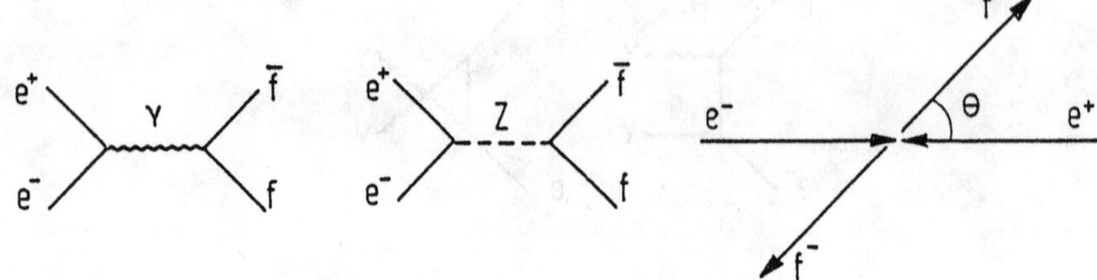

Figure 2: Lowest order diagrams for $e^+e^- \to f\bar{f}$ and definition of the polar angle θ

The function χ is given by the coupling strengths and the Z^0 propagator and depends on the renormalization scheme. Choosing α, m_Z and $\sin^2\theta_W$ (on-shell scheme) one has

$$\chi_I = \frac{1}{4\,\sin^2\theta_W\,\cos^2\theta_W}\,\frac{s}{s-m_Z^2+im_Z\Gamma_Z}\;. \tag{19}$$

In this scheme the measured cross sections can be directly compared to the lowest order Born term calculations, once 'reduced QED' radiative corrections are applied to the data (see section 4). The disadvantage of this parametrization is that it needs the weak mixing angle and the Z^0 mass as input, where the latter has still large experimental errors. Alternatively one can use the parameters α, G and m_Z (Z-mass scheme) and obtains

$$\chi_{II} = \frac{G\,(1-\Delta r)\,m_Z^2}{2\,\sqrt{2}\,\pi\,\alpha}\;\frac{s}{s-m_Z^2+im_Z\Gamma_Z}\;. \tag{20}$$

The factor $(1-\Delta r)$ accounts for the radiative corrections of the Fermi coupling constant derived from μ decay. It is predicted to be $\Delta r = 0.0713 \pm 0.0013$ [5] (assuming a top quark mass $m_t = 45\,GeV$ and a Higgs mass $m_H = 100\,GeV$). In this scheme only well known quantities enter and the dependence on the Z^0 mass is weak in the interference region. However, the data have to be corrected for the complete electroweak radiative corrections (see section 4).

In many cases it is sufficient to consider the relativistic limit $\beta_f \to 1$, thus

$$\frac{d\sigma}{d\Omega} = \frac{\alpha^2}{4\,s}\,N_f^c\left\{C_1(1+\cos^2\theta)+C_2\cos\theta\right\}\;. \tag{21}$$

In QED the lowest order differential cross section for $e^+e^- \to f\bar{f}$ via single photon exchange simplifies to

$$\frac{d\sigma}{d\Omega} = \frac{\alpha^2}{4\,s}\,N_f^c\,Q_f^2\,\left\{1+\cos^2\theta\right\}\;, \tag{22}$$

and the total cross section becomes

$$\sigma(e^+e^- \to f\bar{f}) = \frac{4\,\pi\,\alpha^2}{3\,s}\,N_f^c\,Q_f^2 = \sigma^0\,N_f^c\,Q_f^2\;. \tag{23}$$

The quantity

$$\sigma^0 = \frac{4\pi\,\alpha^2}{3\,s} = \frac{86.8\,nb}{s\,[GeV^2]}\,. \tag{24}$$

is called the pointlike QED cross section for asymptotic lepton pair production and is often used to present normalized total cross section data as

$$R_{f\bar{f}} = \frac{\sigma(e^+e^- \to f\bar{f})}{\sigma^0} = N_f^c \cdot C_1\,. \tag{25}$$

The QFD differential cross section eq. (21) has a term symmetric in $\cos\theta$ and another term linear in $\cos\theta$. The symmetric part is a sum of purely electromagnetic, interference and purely weak contributions, while the asymmetric part contains only interference and purely weak terms. Note that the asymmetry is not a parity violating effect but solely due to the interference of the vector and axial vector currents. Parity violation can only be observed by using longitudinally polarized $e^\pm$ beams or by analyzing the final state fermion polarization.

In the total cross section the term linear in $\cos\theta$ drops out and the rate is proportional to the coefficient C_1 of eq. (18), which is sensitive to the vector couplings in the interference region. Since the weak mixing angle $\sin^2\theta_W$ is close to 1/4 and therefore the electron vector coupling is close to zero, the weak contributions are still small up to centre of mass energies around $\sqrt{s} \leq m_Z/2$. This means that total cross section measurements could serve as a test of QED for not too high energies. The behaviour of $R_{f\bar{f}}$, the total cross section of fermion pair production normalized to the pointlike QED cross section, as a function of the centre of mass energy is displayed in Fig. 3 for μ (or τ) pair production and hadron production $e^+e^- \to \sum q\bar{q} \to hadrons$. $R_{\mu\mu}$ is very close to one up to energies around $\sqrt{s} = 55\,GeV$ and then begins to rise towards a value of ~ 200 at the Z^0 resonance. For hadron production one notices the charm and bottom thresholds (araound 4 and 10 GeV) and a subsequently fairly constant level. Beyond $\sqrt{s} = 35\,GeV$ R_{had} starts to increase, by about 50% at 60 GeV and reaching finally values of ~ 4000 at the Z^0 pole.

On the other hand the differential cross section exhibits a sizeable asymmetry at PETRA, PEP and TRISTAN energies. A convenient quantity is the forward backward charge asymmetry defined as

$$\begin{aligned} A_{f\bar{f}} &= \frac{N_F - N_B}{N_F + N_B}\,, \\ &= \frac{3}{8}\frac{C_2}{C_1}\,, \\ &\simeq -\frac{3}{2}\,\frac{g_A^e\,g_A^f}{Q_f}\,\Re(\chi)\,. \end{aligned} \tag{26}$$

Here N_F is the number of fermions produced in the forward direction ($\cos\theta \geq 0$) and N_B is the number of fermions produced in the backward direction ($\cos\theta \leq 0$). The approximation holds for the interference region where $s \ll m_Z^2$ and indicates that the asymmetries are larger by a factor $1/Q_f$ for quarks than for leptons. The expected fermion asymmetries $A_{f\bar{f}}$ as a function of the centre of mass energy are shown in Fig.

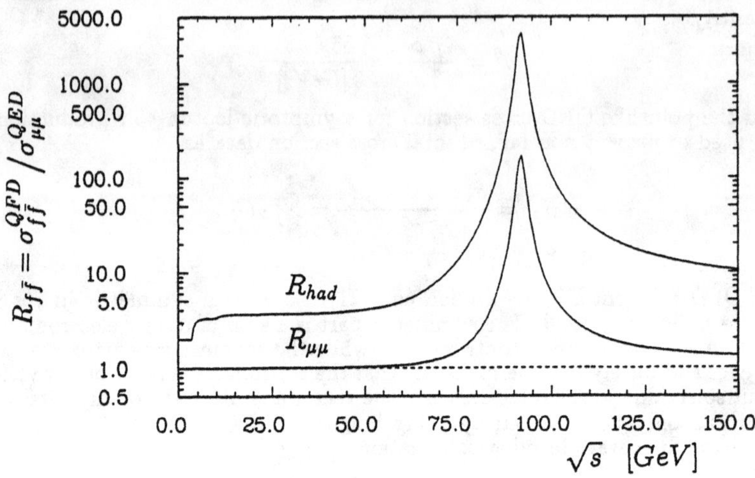

Figure 3: Normalized total cross sections $R_{f\bar{f}}$ as a function of the centre of mass energy $\sqrt{s}$. The curves show $R_{\mu\mu}(e^+e^- \to \mu^+\mu^-)$ and $R_{had}(e^+e^- \to \sum q\bar{q} \to hadrons)$ for 5 quarks

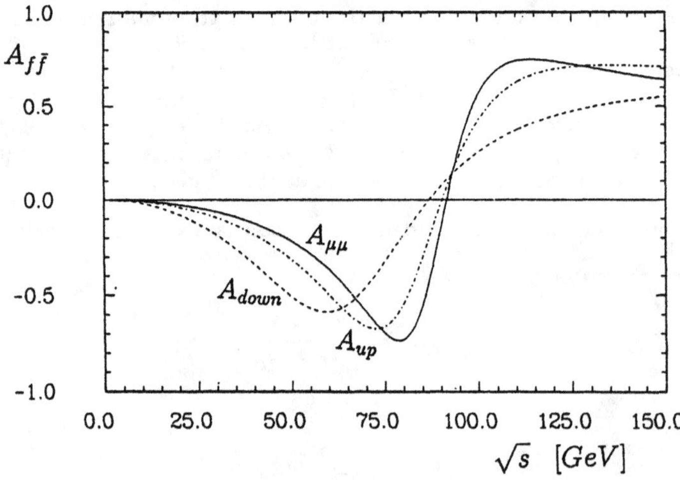

Figure 4: Forward backward charge asymmetries $A_{f\bar{f}}$ as a function of the centre of mass energy $\sqrt{s}$ for $e^+e^- \to f\bar{f}$. The curves show $A_{\mu\mu}$ for μ or τ pair production, A_{up} for *up* type quarks and A_{down} for *down* type quarks

4. They are negative below the Z^0 resonance and after passing through a maximimum they change their sign at energies close to the Z^0 mass. The shape of the differential cross section essentially determines the axial vector couplings and cannot be used for QED tests.

3.3 $e^+e^- \rightarrow e^+e^-$

In Bhabha scattering $e^+e^- \rightarrow e^+e^-$ there are in addition to the annihilation graphs two space-like diagrams for t channel exchange (Fig. 5). Defining θ as the polar angle between the incoming and outgoing electron the differential cross section for $e^+e^- \rightarrow e^+e^-$ in second electroweak order was calculated in ref. [9]

$$\begin{aligned}
\frac{d\sigma}{d\Omega} &= \frac{\alpha^2}{8\,s}\left\{4\,B_1 + B_2\,(1-\cos\theta)^2 + B_3\,(1+\cos\theta)^2\right\}\,, \qquad (27)\\
B_1 &= \left(\frac{s}{t}\right)^2 \left|1 + ({g_V}^2 - {g_A}^2)\,\chi(t)\right|^2\,,\\
B_2 &= \left|1 + ({g_V}^2 - {g_A}^2)\,\chi(s)\right|^2\,,\\
B_3 &= \frac{1}{2}\left|1 + \frac{s}{t} + (g_V + g_A)^2\,(\frac{s}{t}\,\chi(t) + \chi(s))\right|^2\\
&\quad + \frac{1}{2}\left|1 + \frac{s}{t} + (g_V - g_A)^2\,(\frac{s}{t}\,\chi(t) + \chi(s))\right|^2\,.
\end{aligned}$$

Here $t = -s/2\,(1-\cos\theta)$ is the space-like momentum transfer squared, the function $\chi(s)$ is one of equations (19) or (20) and $\chi(t)$ is the same function with s replaced by t, g_V and g_A denote the vector and axial vector couplings of the electron as given in Table 1.

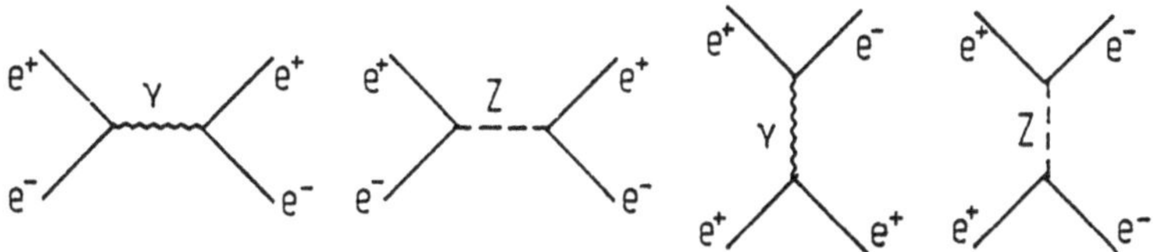

Figure 5: Lowest order diagrams for $e^+e^- \rightarrow e^+e^-$

The differential cross section for Bhabha scattering in second order QED via photon exchange is

$$\frac{d\sigma}{d\Omega} = \frac{\alpha^2}{4\,s}\,\frac{(3+\cos^2\theta)^2}{(1-\cos\theta)^2}\,. \qquad (28)$$

The differential cross section is dominated by large t channel contributions and shows therefore a strong forward peaking (the total cross section is infinite because of the long range force due to photon exchange). Bhabha scattering is thus extremely useful to determine with high rates the luminosity in e^+e^- experiments at very small angles ($\theta \simeq 30\,mrad$), where electroweak effects can be completely neglected as the propagator term $\chi(t) \to 0$ for $t \to 0$.

In the central and backward region the differential cross section is sensitive to electroweak interference. However, due to cancelations of s and t diagrams and due to the fact that the electron vector coupling is close to zero, the expected effects are rather small. As an example Fig. 6 shows the ratio of the electroweak to the QED cross section at a typical PETRA centre of mass energy $\sqrt{s} = 35\,GeV$. The maximum deviation is about -2.5% in the backward hemisphere, close to the achieved experimental resolution. This means that after applying the small corrections due to the standard model Bhabha scattering could be used for QED tests.

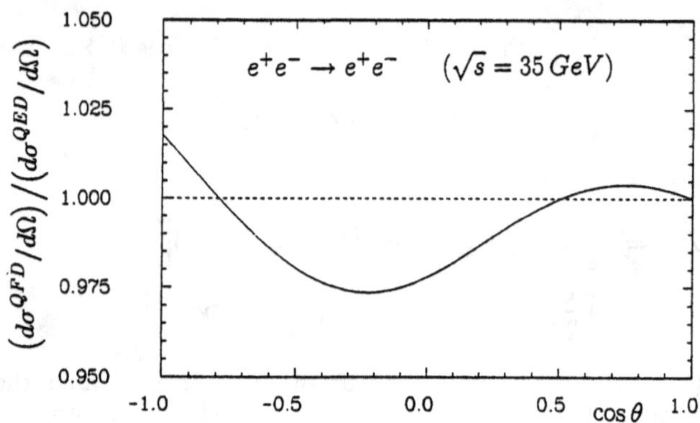

Figure 6: The electroweak differential Bhabha cross section divided by the QED cross section at $\sqrt{s} = 35\,GeV$

4 Radiative corrections

The second order e^+e^- two body reactions described in section 3 represent an ideal situation. Any real experiment has to deal with higher order processes involving either more than two particles in the final state or virtual effects (loop diagrams). Examples of the first type are photon bremsstrahlung off fermions with a probability increasing with decreasing mass, leading e.g. to the final states

$$
\begin{aligned}
e^+e^- &\to \gamma\gamma\gamma,\ \gamma\gamma\gamma\gamma,\ \ldots \\
e^+e^- &\to f\bar{f}\gamma,\ f\bar{f}\gamma\gamma,\ \ldots
\end{aligned}
$$

Examples of the second type are loop corrections appearing in boson or fermion propagators, vertices and box diagrams. The size of these so called radiative corrections depends on the actual experimental set up and event selection criteria.

Both QED and weak contributions modify the measured cross section which can formally be written as

$$\frac{d\sigma}{d\Omega}^{meas} = \frac{d\sigma}{d\Omega}^{0} \left(1 + \delta^{QED} + \delta^{weak}\right) , \tag{29}$$

where $d\sigma^0/d\Omega$ is the Born cross section. The radiative corrections are calculable order by order. Complete calculations for the e^+e^- processes (1), (2) and (3) have been performed up to $\mathcal{O}(\alpha^3)$, which means first order corrections to the lowest order cross sections. In some special cases $\mathcal{O}(\alpha^4)$ cross sections have been computed.

Let us consider the first order corrections. The radiative corrections can be divided in two parts: virtual corrections to the non radiative lowest order diagrams and photon bremsstrahlung.

The virtual corrections are obtained by adding loops to the internal lines, known as vacuum polarization, and by adding an internal photon line in all possible ways to the external lines, known as vertex corrections and box diagrams. These diagrams are of $\mathcal{O}(\alpha^4)$, their interference with the lowest order diagrams leads to $\mathcal{O}(\alpha^3)$ cross sections.

The bremsstrahlung diagrams are obtained by adding an external photon line to the lowest order diagrams, thus giving $\mathcal{O}(\alpha^3)$ contributions.

Corrections involving photons and massive gauge bosons have to be treated differently. Virtual QED corrections δ_v^{QED} exhibit infrared divergences caused by the photon propagator $1/k$ in the soft photon limit $k \to 0$. They can be regularized by introducing a fictitious photon mass λ, which is of course unphysical but drops out after summing up all divergences. Similarly, divergences arise for the bremsstrahlung, where an infinite number of photons are emitted as $k \to 0$, and in collinear configurations where a photon is emitted along the direction of one of the initial or final state particles. In most cases the experiments do not detect radiated photons, one reason being the limited solid angular acceptance in particular along the beam directions. But even if they had photon detection capabilities the finite experimental resolution ΔE does not allow to distinguish between single and multiple photon emission if $k < \Delta E$. The bremsstrahlung calculations are therefore split in an infrared divergent soft photon part δ_s^{QED} and a finite hard photon part δ_h^{QED}. The renormalizability of QED assures that each order of the perturbation expansion is calculable and finite. The divergences from the virtual corrections exactly cancel the divergences from the soft photon bremsstrahlung, such that the final QED correction

$$\delta^{QED} = \delta_v^{QED} + \delta_s^{QED} + \delta_h^{QED}$$

remains finite.

The evaluation of the weak radiative corrections is much more elaborate. Although numerous calculations exist they are sometimes difficult to compare since no common choice of basic parameters and/or renormalization scheme has been adopted by the various authors. However, there seems to be agreement that the weak corrections are small compared to the QED corrections (at least at energies below the Z^0 pole) when using the physical Z^0 mass and the coupling constants renormalized at the mass of the Z^0.

Therefore it seems appropriate to separate the radiative correction calculations at the one-loop level into three categories with the following definitions of the notion:

1. Electromagnetic corrections to the γ or electron exchange diagrams consisting of virtual photon and bremsstrahlung contributions and the fermionic vacuum polarization of the photon. They are called 'reduced QED corrections' and are model independent but depend on the experimental conditions.
2. Electromagnetic corrections to the Z^0 exchange diagrams consisting of virtual and bremsstrahlung contributions in all possible combinations. They are dependent on the model parameters as well as on the experimental cuts. Together with the above corrections they are called 'full QED corrections'.
3. Weak (non-photonic) corrections to both γ and Z^0 exchange diagrams. They contain dynamical aspects of the electroweak model and depend sensitively on the renormalization scheme.

A convenient way to compare measurements with the lowest order theory is to unfold the data by using the calculated radiative corrections, thus arriving at 'lowest order data'. This method to present results was applied by almost all the experiments at PETRA, PEP and TRISTAN. At these energies the most important contributions come from QED effects and the electroweak radiative corrections play a minor rôle. This situation may be different at SLC and LEP, where the 'lowest order data' depend on the electroweak theory which one wants to test in detail. Here it may be more sensible to compare the acceptance corrected data with the specific electroweak model. However, in both cases it is very important to have event generators including higher order radiative corrections in the form of Monte Carlo programs which allow to simulate the experimental conditions as precisely as possible.

In the following the QED and weak radiative corrections will be discussed separately for the basic two body reactions. A detailed discussion of radiative corrections in e^+e^- lepton pair production including cross section formulas can be found in [10].

4.1 $e^+e^- \to \gamma\gamma,\ \gamma\gamma\gamma$

The complete set of diagrams leading to $\mathcal{O}(\alpha^3)$ cross sections for $e^+e^- \to \gamma\gamma,\ \gamma\gamma\gamma$ is shown in Fig. 7. Besides the lowest order graphs (a) they consist of virtual corrections to the vertices (b), to the electron propagator (c), box diagrams (d) and photon bremsstrahlung (e).

The differential cross section formulas are rather involved and can be found, e.g., in ref. [10]. The total $\mathcal{O}(\alpha^3)$ cross section is

$$\begin{aligned} \sigma(e^+e^- \to \gamma\gamma, \gamma\gamma\gamma) &= \sigma^{2\gamma} + \sigma^{3\gamma} \qquad (30) \\ &= \sigma^0 \left\{ 1 + \frac{\alpha}{\pi} \frac{1}{2v-1} \left(\frac{4}{3} v^3 - v^2 + \left(\frac{2\pi^2}{3} - 2 \right) v + 2 - \frac{\pi^2}{12} \right) \right\}. \end{aligned}$$

$\sigma^0 = 2\pi\alpha^2 (2v-1)/s$ is the lowest order cross section given in section 3.1 and $v = 1/2 \ln(s/m_e^2)$. The correction to σ^0 increases with energy and amounts to appreciable values of 18.8% and 20.6% at $\sqrt{s} = 30$ and 50 GeV.

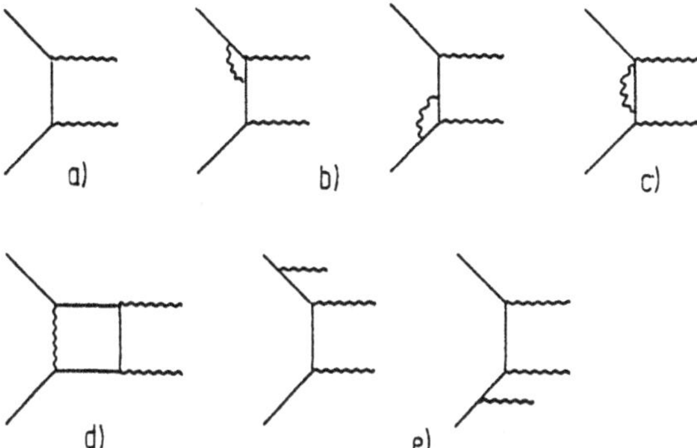

Figure 7: Diagrams leading to $\mathcal{O}(\alpha^3)$ cross section for the reaction $e^+e^- \to \gamma\gamma$. a) Lowest order, b) vertex corrections, c) electron propagator corrections, d) box diagram, e) photon bremsstrahlung. The diagrams with the two external photons interchanged have to be added

The first order weak corrections turn out to be very small, below 1%. Therefore the reaction $e^+e^- \to \gamma\gamma$ can be considered for all practical purposes as a pure QED process even at Z^0 energies.

4.2 $e^+e^- \to f\bar{f},\ f\bar{f}\gamma$

The lowest order radiative corrections to the reaction $e^+e^- \to f\bar{f}$ are conveniently split into QED corrections and weak corrections. The first order QED correction diagrams to γ and Z^0 exchange are shown in Figs. 8 and 9. Some first order electroweak correction diagrams are shown in Fig. 10.

The first order QED correction diagrams to γ exchange (Fig. 8) consist of the vacuum polarization of the photon (b), vertex corrections (c), box diagrams (d) and photon bremsstrahlung (e). The largest effect of the virtual QED corrections comes from the vacuum polarization of the photon. The electric charge polarizes the surrounding vacuum into fermion antifermion pairs and is thus screened to a lesser degree the closer the distance or the higher the momentum transfer of the scattering particles. This leads to a modification of the photon propagator which can be absorbed in the definition of the now running electromagnetic coupling constant α at momentum transfer Q^2

$$\alpha^{-1}(Q^2) = \alpha^{-1}\left\{1 - \frac{\alpha}{3\pi} N_f^c \sum_f Q_f^2 \ln\frac{Q^2}{m_f^2}\right\} . \tag{31}$$

Here α is the fine structure constant obtained from low energy measurements ($Q^2 \sim m_e^2$) and the sum extends over all fermions (leptons and quarks) with $m_f^2 < Q^2$. Inserting the known lepton masses and effective quark masses, e.g. $0.1\,GeV$ for the

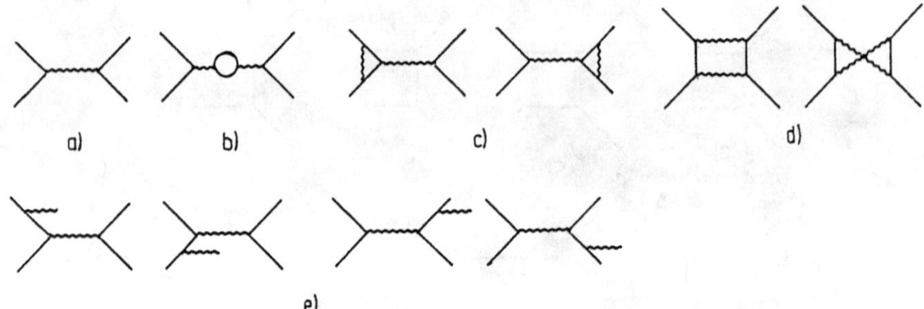

Figure 8: First order QED diagrams to γ exchange leading to $\mathcal{O}(\alpha^3)$ cross section for the reaction $e^+e^- \to f\bar{f}$. **a)** Lowest order, **b)** photon vacuum polarization, **c)** vertex corrections, **d)** box diagrams, **e)** bremsstrahlung diagrams

light quarks u, d, s, one obtains for a typical PETRA energy of $35\,GeV$ an increased effective coupling constant $\alpha \sim 1/129$ compared to the low energy value of $\sim 1/137$. The above formula is valid for the continuum region away from the vector meson resonances ρ, ϕ, J/ψ, Υ, etc. A more careful analysis using dispersion relations together with measured cross section data from $e^+e^- \to$ *hadrons* leads to consistent results. The vertex corrections, box diagrams and bremsstrahlung diagrams involve an additional photon with the quantum numbers $J^{PC} = 1^{--}$. The parity of these diagrams is therefore opposite to the parity of the Born graph and their interference induces a forward backward charge asymmetry. As mentioned above the vertex corrections, box diagrams and soft photon bremsstrahlung terms are each divergent by themselves, but the total corrections are finite after summing up all diagrams. Expressions for the individual contributions are rather involved and can be found in [10] and references therein.

For the hard photon bremsstrahlung the largest effect is the radiation off the initial state electrons and positrons. Emission of a photon with energy k leads to a reduced centre of mass energy $s' = 4\,E\,(E-k)$ and thus the rate increases due to the $1/s$ dependence of the cross section. The bremsstrahlung spectrum can be given separately for the initial electron and the final fermion

$$\frac{d\sigma^e}{dk} = \sigma^0(s')\,N_f^c\,\frac{\alpha}{\pi}\left(\ln\frac{s}{m_e^2}-1\right)\left(1+\left(\frac{s'}{s}\right)^2\right)\frac{1}{k}\,, \tag{32}$$

$$\frac{d\sigma^f}{dk} = \sigma^0(s)\,N_f^c\,Q_f^2\,\frac{\alpha}{\pi}\left(\ln\frac{s}{m_f^2}-1+\ln\frac{s'}{s}\right)\left(1+\left(\frac{s'}{s}\right)^2\right)\frac{1}{k}\,. \tag{33}$$

$\sigma^0(s) = 4\,\pi\,\alpha^2/3\,s$ is the lowest order QED total cross section given in section 3.2, and $\sigma^0(s')$ is the total cross section at a reduced centre of mass energy s'. Integrating over the photon spectrum for $k > k_0$ gives

$$\sigma^e = \sigma^0\,N_f^c\,\frac{2\,\alpha}{\pi}\left(\ln\frac{s}{m_e^2}-1\right)\left(\ln\frac{E}{k_0}+\frac{1}{2}\ln\frac{s}{m_f^2}-\frac{4}{3}\right)\,, \tag{34}$$

$$\sigma^f = \sigma^0 N_f^c Q_f^2 \frac{2\alpha}{\pi} \left\{ \left(\ln \frac{s}{m_f^2} - 1 \right) \left(\ln \frac{E}{k_0} - \frac{3}{4} \right) + \frac{5}{8} - \frac{\pi^2}{6} \right\} . \tag{35}$$

These expressions diverge for $k_0 \to 0$, but the whole cross section remains finite after adding the soft photon part and the virtual corrections.

The total cross section for the QED corrections to the photon exchange diagrams alone takes the form

$$\begin{aligned}
\sigma(e^+e^- \to f\bar{f}, f\bar{f}\gamma) &= \sigma^{f\bar{f}} + \sigma^{f\bar{f}\gamma} , \\
&= \sigma^0 (1 + \delta_e + \delta_f + \delta_{vp}) , \qquad (36) \\
\delta_e &= \frac{2\alpha}{\pi} \left\{ \frac{1}{2} \ln \frac{s}{m_e^2} \ln \frac{s}{m_f^2} - \frac{7}{12} \ln \frac{s}{m_e^2} - \frac{1}{2} \ln \frac{s}{m_f^2} + \frac{\pi^2}{6} + \frac{1}{3} \right\} , \\
\delta_f &= \frac{3\alpha}{4\pi} , \\
\delta_{vp} &= \sum_i \frac{2\alpha}{\pi} \left\{ -\frac{5}{9} + \frac{1}{3} \ln \frac{s}{m_i^2} \right\} .
\end{aligned}$$

δ_e contains the virtual, soft and hard bremsstrahlung corrections, while δ_f is a similar term for the final fermion, and δ_{vp} is the vacuum polarization of the photon, where the sum runs over the three leptons and six quark flavours (suitable quark masses are $m_u = m_d = 0.032\,GeV$, $m_s = 0.15\,GeV$, $m_c = 1.5\,GeV$, $m_b = 4.5\,GeV$, $m_t = 30\,GeV$).

The corrections discussed so far are called the 'reduced QED corrections' and constitute the major part of the radiative corrections below the Z^0 resonance. It is obvious that the electron contribution is by far the most important one. Due to the double log term $\ln(s/m_e^2)\,\ln(s/m_f^2)$ the radiative corrections become large if hard photon emission along the $e^\pm$ beams cannot be rejected experimentally. To illustrate the effect, take as example the reaction $e^+e^- \to \mu^+\mu^-$ at $\sqrt{s} = 35\,GeV$. The δ_e term contributes a 53% correction, while the vacuum polarization is about 12% and the muonic contribution is entirely negligible. The complete corrections to the total cross sections appear to be extremely large. In practice, however, experimental conditions restrict the acceptable phase space and the corrections become less dramatic.

In order to control the hard photon radiation and the available phase space, experimental cuts on the energy E_f of the final state fermion and the acollinearity angle ξ between the fermion momenta $\vec{p_f}$ are applied, where $\xi = \arccos(-\vec{p_{\bar{f}}} \cdot \vec{p_f})$. To illustrate the effect of radiative corrections for the reaction $e^+e^- \to \mu^+\mu^-$ at $\sqrt{s} = 35\,GeV$, standard experimental conditions of $\xi < 10^0$ and $E_\mu > \sqrt{s}/4$ lead to a total correction of about 25%. The interference of the Born amplitude with the box diagrams and soft photon radiation results in a positive forward backward charge asymmetry, while the hard photons produce a negative asymmetry. The net effect on the observable asymmetry within the polar angle acceptance of $|\cos\theta_\mu| < 0.80$ is a positive correction term of about $+(1.5 - 2.0)\%$ from QED $\mathcal{O}(\alpha^3)$ radiative corrections.

The diagrams of first order QED corrections to the Z^0 exchange graphs are shown in Fig. 9. They contain vertex corrections (b), box diagrams involving γ and Z^0 lines

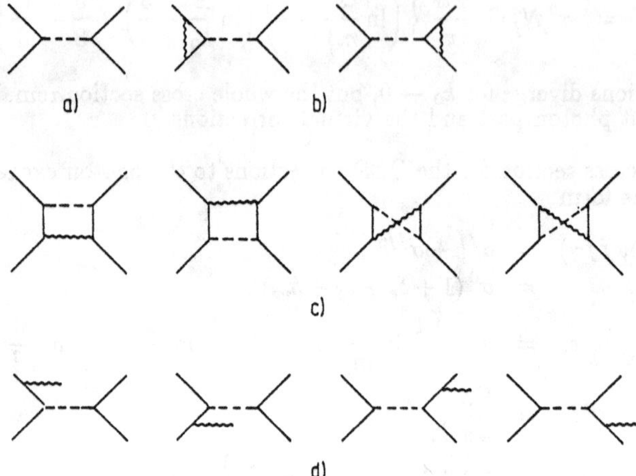

Figure 9: First order QED diagrams to Z^0 exchange leading to $\mathcal{O}(\alpha^3)$ cross section for the reaction $e^+e^- \to f\bar{f}$. **a**) Lowest order, **b**) vertex corrections, **c**) box diagrams, **d**) bremsstrahlung diagrams

(c) and photon bremsstrahlung to the Z^0 Born graph (d). Again there appear infrared divergences in the vertex corrections and box diagrams which cancel against those of the soft photon bremsstrahlung. Hard photon initial state radiation further reduces the observable electroweak charge asymmetry in absolute magnitude as it shifts the available centre of mass energy towards lower values away from the Z^0 resonance. The size of the effect depends on the experimental selection criteria. Taking all QED corrections to the γ and Z^0 exchange diagrams as depicted in Figs. 8 and 9 one arrives at the so called 'full QED corrections'.

In order to complete the radiative corrections the purely weak terms have to be included, some of the virtual weak loop diagrams are shown in Fig. 10. Among those the most important one is the self energy diagram to the Z^0 (a). Since the standard theory is renormalizable the higher orders are calculable. Due to the finite mass of the weak bosons they are infrared finite. Although a large number of calculations have been performed, the results are difficult to compare since the authors prefer different choices of the input parameters and of the renormalization schemes. This does not apply to QED where the on-shell scheme with α and the fermion masses as parameters has been commonly accepted. For the weak corrections two popular schemes are the on-shell scheme using renormalized parameters α, m_Z and $\sin^2\theta_W$ (or equivalently m_W via the relation $\sin^2\theta_W = 1 - m_W^2/m_Z^2$) and the Z-mass scheme using α, G and m_Z. In addition some assumptions on the masses of the yet undiscovered top quark and Higgs (in most cases $m_t \simeq 40\,GeV$ and $m_H = 100\,GeV$) enter in the calculations. The results of the two schemes differ at the Born level (as a consequence of the two different Z^0 couplings of eqs. (19) and (20) in section 3.2) as well as in the QED and weak radiative corrections, but arrive at the same results once all corrections are taken into account. This is illustrated in Table 2 for the observable charge asymmetry in the reaction $e^+e^- \to \mu^+\mu^-$ at two typical PETRA energies of

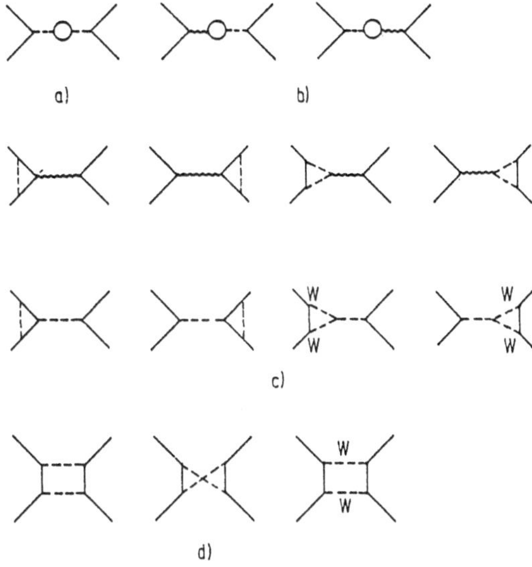

Figure 10: Some first order weak correction diagrams for the reaction $e^+e^- \rightarrow f\bar{f}$. a) Z^0 self energy, b) $\gamma - Z^0$ self energy, c) vertex corrections, d) box diagrams

Table 2: Effects of radiative corrections on the observable charge asymmetry $A_{\mu\mu}[\%]$ in the reaction $e^+e^- \rightarrow \mu^+\mu^-$ at $\sqrt{s} = 34.5\,GeV$ and $44.0\,GeV$. Event selection criteria correspond to a polar angle acceptance $|\cos\theta_\mu| < 0.80$, acollinearity angle $\xi < 10^0$, radiated photon energy $k < \sqrt{s}/4$. Values are given for the on-shell scheme and the Z-mass scheme using $m_Z = 92\,GeV$ and $\sin^2\theta_W = 0.23$

	$\sqrt{s} = 34.5\,GeV$ renormalization scheme		$\sqrt{s} = 44.0\,GeV$ renormalization scheme	
	on-shell	Z-mass	on-shell	Z-mass
Born term	−7.62	−8.16	−13.63	−14.66
+ reduced QED	−5.80	−6.31	−11.82	−12.86
+ full QED	−5.28	−5.82	−10.85	−11.98
+ weak corr.	−5.89	−5.86	−11.93	−11.89

$\sqrt{s} = 34.5\,GeV$ and $\sqrt{s} = 44\,GeV$ for 'reasonable' experimental cuts [11]. It turns out that at PETRA energies there is the accidental situation that the QED corrections to the photon Born term using the on-shell scheme (i.e. the Z^0 coupling of eq. (19)) coincide with the full corrections applied to the Z^0 diagrams. The small differences are below the experimental precision. Therefore it became customary that PETRA, PEP and TRISTAN experiments present 'lowest order data' with only 'reduced QED corrections' applied. This procedure need not to be applicable at higher energies, where different strategies have to be employed.

4.3 $e^+e^- \to e^+e^-,\ e^+e^-\gamma$

The first order radiative corrections to Bhabha scattering $e^+e^- \to e^+e^-$ are completely analogous to fermion pair production. There are twice as many diagrams, besides the γ and Z^0 annihilation graphs for the s channel a similar set of diagrams occur in the t channel (see Figs. 8, 9). The calculations of the QED radiative corrections are again split into virtual and soft bremsstrahlung corrections, which by themselves are divergent but, when summed up, give finite results, and hard photon bremsstrahlung. 'Reduced QED corrections' to the γ diagrams are the dominant part. QED corrections to the Z^0 graphs have been calculated and turn out to be less important. They are therefore not applied to the data. Also the first order weak corrections have been computed. In the on-shell scheme they are completely negligible at present energies and achieved experimental resolutions. Therefore all data are published as 'lowest order data' unfolded by detector acceptance and 'reduced QED corrections'.

4.4 Event generators

Experimental selection criteria cannot easily be translated into analytic cross section formulas. Therefore a tremendous amount of work has been done to provide Monte Carlo event generators for the previously discussed elementary reactions including first order radiative corrections. Event generators exist for the two photon production $e^+e^- \to \gamma\gamma$ [12], for lepton pair production $e^+e^- \to \mu^+\mu^-$ [13] and $e^+e^- \to \tau^+\tau^-$ [14], which can be easily modifyed to quark pair production, and Bhabha scattering $e^+e^- \to e^+e^-$ [15]. In general they contain the 'reduced QED corrections', the 'full QED corrections' and the most important weak corrections, adequate for energies below the Z^0 resonance. Recently also programs including weak corrections for high precision electroweak tests around the Z^0 pole are being developed.

Contrary to analytic formulas the Monte Carlo event generators cannot be exact in all regions of the phase space. For example the cross sections exhibit extremely peaking structures in the soft photon limit $k \to 0$ and in collinear configurations where a soft photon is emitted in the direction of the charged fermions. Hard bremsstrahlung peaks also occur in fermion production and Bhabha scattering from $1/s'$, $1/t$ and $1/t'$ terms. None of the peaks get singular. However, numerical methods have to be devised in order to perform the integrations with the required precision.

An example where finer details of an event sample cannot be reproduced is the photon energy spectrum. Within an event generator all events with a soft photon below a certain energy threshold $k < k_0$ (typically below 1% of the beam energy) are simulated as two body reactions, while for photons with $k > k_0$ three body final states are produced. Thus there appears a discontinuity in the photon spectrum which also reflects in the acollinearity angle distribution ξ of the final state particles for very small values of ξ. However, for all practical purposes it is possible to avoid the peaking regions and limitations of the event generators. Thus the Monte Carlo event generator programs are of prime importance and allow to simulate the experiments as precisely as possible and necessary and extract from the data cross sections which can be confronted with QED and the electroweak standard theory.

In addition to the first order radiative correction generators of the basic two body reactions there exist satisfactory approximations to higher order four particle final states such as $e^+e^- \to f\bar{f}\gamma\gamma$ and four lepton final states $e^+e^- \to e^+e^-l^+l^-(\gamma)$ [16], which are interesting by themselves and have to be accounted for as background sources to the two body reactions.

5 Photon–photon collision cross sections

The four lepton production $e^+e^- \to e^+e^-l^+l^-$ can be envisaged as a two stage process. In the first step the initial electron and positron radiate virtual photons, which in the second step produce the final state X, in this case the lepton pairs $\gamma\gamma \to X \to l^+l^-$. In this picture it is obvious that e^+e^- colliders are an ideal place to study photon–photon collisions. The Feynman diagram and relevant kinematic quantities for the reaction $e^+e^- \to e^+e^-l^+l^-$ are shown in Fig. 11. It is a pure QED process of order α^4 and the cross section can be calculated in a straighforward but tedious way, giving lengthy and not very handy formulas.

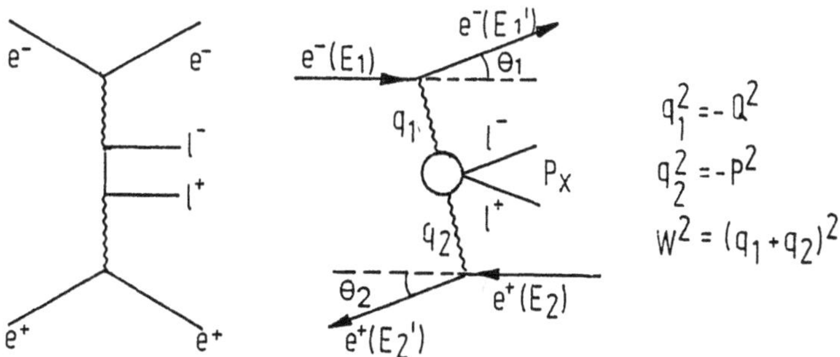

Figure 11: The two–photon process $e^+e^- \to e^+e^-X \to e^+e^-l^+l^-$. Feynman diagram and kinematics

Fortunately simple and reliable approximations exist which will be used in the following discussion. A very successful approach is the Weizsäcker-Williams approximation. The spectrum of the bremsstrahlung photons is given by

$$\frac{dN}{d\omega} = \frac{\alpha}{2\pi} \frac{1+(1-\omega)^2}{\omega} \eta \, , \tag{37}$$

where $\omega = E_\gamma / E$ is the fractional energy of the photon in terms of the beam energy E. The function η accounts for the direction of the electron/positron and depends on the momentum transfer q^2 or virtual mass of the photon. The momentum transfer is experimentally controlable. A vanishing momentum transfer, $q^2 \to 0$, means that the electron/positron is not deflected, $\theta \simeq 0^0$, this is called the no-tag case. If the momentum transfer becomes large, $q^2 \neq 0$, the electron/positron will be scattered at a polar angle θ. If $\theta > \theta_{min}$ falls inside the detector acceptance the so called tag case occurs. For small angles $-q^2 = Q^2 \simeq E\,E'\,\theta^2$ and the function η takes for the two situations the form

$$\begin{aligned} \eta &= \ln \frac{E^2}{m_e^2} && \text{no} - \text{tag} \, , \\ &= \ln \frac{Q^2}{Q^2_{min}} = \ln \frac{\theta^2}{\theta^2_{min}} && \text{tag} \, . \end{aligned}$$

The convoluted photon spectra at a given invariant $\gamma\gamma$ mass W or fractional mass $\tilde{s} = W/\sqrt{s}$ can be translated into a $\gamma\gamma$ luminosity function $L_{\gamma\gamma}$

$$\begin{aligned} \frac{d\,L_{\gamma\gamma}(\tilde{s})}{d\,\tilde{s}} &= \left(\frac{\alpha}{\pi}\right)^2 \frac{f(\tilde{s})}{\tilde{s}} \eta^2 \, , \\ f(\tilde{s}) &= (2+\tilde{s}^2)^2 \ln \frac{1}{\tilde{s}} - (1-\tilde{s}^2)(3+\tilde{s}^2) \, . \end{aligned} \tag{38}$$

The observable two–photon cross section can then be written as a convolution of the cross section $\sigma_{\gamma\gamma}$ for the sub-process $\gamma\gamma \to l^+l^-$ with the $\gamma\gamma$ luminosity $L_{\gamma\gamma}$

$$\sigma(e^+e^- \to e^+e^-l^+l^-) = \int L_{\gamma\gamma}(\tilde{s})\, \sigma_{\gamma\gamma}(\tilde{s})\, d\tilde{s} \, . \tag{39}$$

First the no-tag case will be considered, where both photons are nearly on mass shell, i.e. $q_1^2 \to 0$ and $q_2^2 \to 0$. Using for the specific sub process $\gamma\gamma \to \mu^+\mu^-$ the very rough approximation $\sigma_{\gamma\gamma} = 4\pi\alpha^2/W^2$ the total cross section for $e^+e^- \to e^+e^-\mu^+\mu^-$ becomes

$$\sigma(e^+e^- \to e^+e^-\mu^+\mu^-) = \frac{8\,\alpha^4}{\pi} \ln^2\left(\frac{E}{m_e}\right) \ln\left(\frac{E}{m_\mu}\right) \frac{1}{m_\mu^2} \, . \tag{40}$$

The surprising result is that the $\mathcal{O}(\alpha^4)$ photon–photon collision cross section rises logarithmically with the beam energy and becomes even larger then the $\mathcal{O}(\alpha^2)$ annihilation cross section at not too high energy. This is shown in Fig. 12, where the total cross sections for one-photon and two-photon μ pair production are compared.

It should be noted, however, that most of the cross section is unobservable in the experiments. First, the initial state photons are strongly aligned along the beam

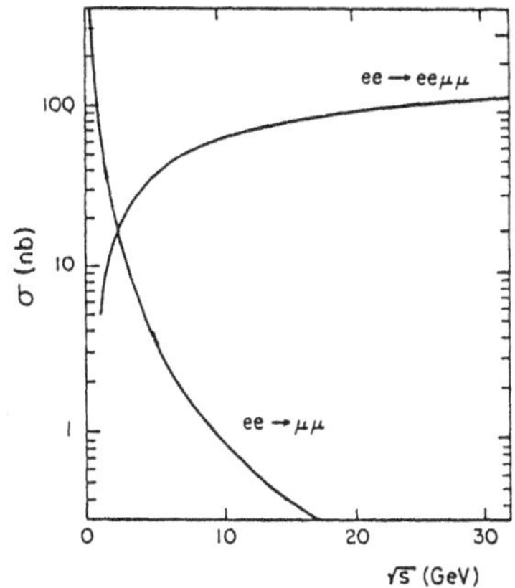

Figure 12: The two–photon cross section for $e^+e^- \rightarrow e^+e^-\mu^+\mu^-$ compared to the annihilation cross section for $e^+e^- \rightarrow \mu^+\mu^-$ as a function of the centre of mass energy $\sqrt{s}$

direction and the $\gamma\gamma$ centre of mass system is moving along that direction. Second, the lepton pair production cross section $\gamma\gamma \rightarrow l^+l^-$, which is obviously the reverse of $e^+e^- \rightarrow \gamma\gamma$ production discussed in section 3.1 modified according to the involved leptons, is in the $\gamma\gamma$ centre of mass system strongly peaked along the photon direction with a polar angle θ^* distribution of

$$\frac{d\sigma(\gamma\gamma \rightarrow l^+l^-)}{d\cos\theta^*} \propto \frac{1+\cos^2\theta^*}{\sin^2\theta^*} . \tag{41}$$

The observable cross section is therefore only a few percent of the total cross section falling rapidly with the $\gamma\gamma$ centre of mass energy W. The Weizsäcker-Williams approximation overestimates the exact luminosity calculations by about 20% for $\tilde{s} < 0.8$, but reproduces the shape of the distribution well.

Now the tag case will be discussed. As explained above one can tune the masses of the photons by a measurement of the electron/positron scattering angle θ . The limit of q_1^2 large and $q_2^2 \rightarrow 0$ can be interpreted as deep inelastic electron scattering off an (almost) real photon target $e\gamma \rightarrow e\mu^+\mu^-$. The factorization into a $\gamma\gamma$ cross section and the $\gamma\gamma$ luminosity function is no longer valid. Instead one has to sum over the transverse and longitudinal photon polarizations weighted with the appropriate spin densities. The formalism is completely analoguous to deep inelastic lepton nucleon scattering. The differential cross section for $e\gamma$ scattering is parametrized by two structure functions F_1 and F_2 as [17]

$$\frac{d^2\sigma}{dx\,dQ^2} = \frac{4\pi\alpha^2}{Q^4 x}\left\{(1-y)F_2 + x\,y^2 F_1\right\} , \tag{42}$$

with the kinematic variables

$$\begin{aligned} Q^2 &= 4\,E\,E'\sin^2\frac{\theta}{2} , \\ x &= \frac{Q^2}{Q^2+W^2} , \\ y &= 1-\frac{E'}{E}\cos^2\frac{\theta}{2} . \end{aligned}$$

E' is the energy of the scattered electron/positron and θ its scattering angle. The structure functions depend on the scaling variables x and Q^2. F_1 is under standard experimental conditions not accessible since the coefficient $x\,y^2$ is too small. The photon structure function $F_2(x, Q^2)$ describes the pointlike coupling of the photon to the muon pair. It is calculable in QED [17] and is for $W^2 \gg 4\,m_\mu^2$ given by

$$F_2(x, Q^2) = \frac{\alpha}{\pi}\,x\left\{\left(x^2 + (1-x)^2\right)\ln\frac{Q^2\,(1/x-1)}{m_\mu^2} - 1 + 8\,x\,(1-x)\right\}. \qquad (43)$$

The above formulas can be easily applied to other leptons with obvious modifications regarding the lepton mass. For quarks an additional factor $N_q^c \sum_q Q_q^4$ has to be added containing a colour factor $N_q^c = 3$ and the quark charges Q_q. The corresponding hadronic structure function is theoretically less well known and is an interesting object for QCD studies. The leptonic reactions $e^+e^- \to e^+e^-e^+e^-$ and $e^+e^- \to e^+e^-\mu^+\mu^-$ serve as prototypes for QED photon–photon collisions. Only if they are understood one can hope to learn something from two–photon quark pair or hadron production.

Photon–photon collisions are extremely difficult to analyze without detailed acceptance simulation of the detector, because of the moving centre of mass system and the rapidly falling $\gamma\gamma$ luminosity. Monte Carlo event generators for four lepton production $e^+e^- \to e^+e^-l^+l^-(\gamma)$ have been provided [16] without and with hard photon radiation. The lepton event generators have also been extended to two–photon production of quarks.

6 Experiments

One of the most critical design parameter of a collider is the luminosity $\mathcal{L}$ $[cm^{-2}\,s^{-1}]$ determining the rate at which an experiment will take data. The number of recorded events for a given reaction is

$$N = \sigma \cdot \epsilon \cdot \int \mathcal{L}\,dt\,.$$

Here σ is the observable cross section, ϵ an efficiency factor and $\int\mathcal{L}\,dt$ the time integrated luminosity of the e^+e^- collider.

In the past decade the most significant e^+e^- experiments to test QED and the electroweak theory have been performed at the storage rings PETRA (at DESY in Hamburg) and PEP (at SLAC in Stanford).

PETRA has been operated from 1978 till 1986 and covered a large centre of mass energy region from 12 to 46.8 GeV, thereby extending the energy scale of previous QED tests from SPEAR (SLAC) and DORIS (DESY) by almost an order of magnitude. Much of the data were taken in a scanning mode, where the energy was varied in small step sizes of 20 or 30 MeV. This mode of operation was motivated by the search for new open quark or bound quark states, notably the top quark. Five experiments have been operated during the years: CELLO which succeeded PLUTO, JADE, MARK J and TASSO. A typical history of an experiment in terms of luminosity and energy is shown in Fig. 13. As can be seen the energy was continuously upgraded, only the last year run was at a fixed energy of $35.0\,GeV$. Each experiment has collected a total integrated luminosity of about $250\,pb^{-1}$. For the analysis energy intervals were grouped together. Statistically significant data samples are available for average energies around $\sqrt{s} \sim 14\,GeV$, $22\,GeV$, $35\,GeV$ ($\sim 180\,pb^{-1}$), $38\,GeV$ ($\sim 15\,pb^{-1}$), and $44\,GeV$ ($\sim 40\,pb^{-1}$).

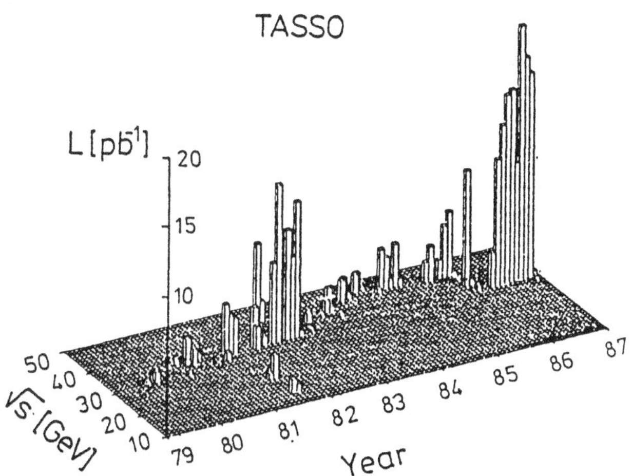

Figure 13: History of the TASSO experiment in terms of accumulated integrated luminosity as function of the centre of mass energy and date of running

The storage ring PEP operated from 1980 on at a fixed energy of $\sqrt{s} = 29\,GeV$. The experiments HRS, MAC, MARK II and TPC (and previously DELCO and ASP) took data until the end of 1986 and have each collected about $220 - 300\,pb^{-1}$ of integrated luminosity, comparable with what has been accumulated by the PETRA experiments. The lower energy yields larger event samples due to the $1/s$ dependence of the cross sections. On the other hand the sensitivity to QED and to the electroweak interference as well as to phenomena beyond the standard theory is rising with s, in fact sometimes faster than linear in s. Larger effects also mean that systematic experimental biases become less important. In this respect the PETRA experiments were in a more favourable position than the PEP experiments.

Immediately after the PETRA shut down at the end of 1986 the TRISTAN storage ring at KEK, Japan, started its physics program at energies between 50 and 60 GeV. By the end of 1988 a maximum energy of $60\,GeV$ was reached and the experiments AMY, TOPAZ and VENUS have reported results based on an integrated luminosity of about $20\,pb^{-1}$ for each experiment. The upgrade program foresees to extend the energy range up to $66\,GeV$.

In 1989 the two new colliders SLC (at SLAC) and LEP (at CERN) came into operation working in the energy region around $\sqrt{s} \simeq m_Z$. They aim at studying the properties of the Z^0 resonance with high precision ('Z^0 factories'). The MARK II experiment at SLC has gathered the first Z^0 events in the spring of 1989, while the LEP experiments just recorded the first events in summer 1989.

Most e^+e^- experiments have been built as general multi-purpose detectors in order to exploit the large variety of physics questions. They should be able to detect the different topologies of low multiplicity (leptons, photons) and high multiplicity (hadrons) events from conventional γ and Z^0 exchange, from two-photon processes which are boosted along the beam directions and unexpected exotic reactions. This requires hermetically closed '$4\,\pi$ detectors' with momentum analysis of charged particles through bending in a magnetic field and particle identification of electrons, muons and photons. Charged track measurements are usually provided by drift chambers over a solid angle of more than 80 % of $4\,\pi$, whereas electromagnetic calorimetry by lead sandwhich or lead glass counters extends up to 95 % of the solid angle. Muon identification is typically done over 80 % of $4\,\pi$ with wire chambers situated inside or behind the iron magnetic flux return which absorbs the hadrons. The luminosity is monitored by shower counters using the high rate Bhabha scattering at very low angles ($\theta \sim 30\,mrad$).

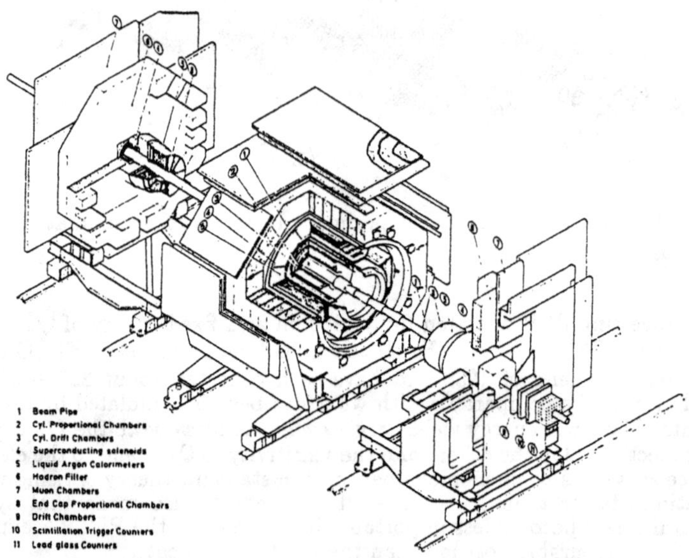

Figure 14: The CELLO detector at the PETRA e^+e^- storage ring

A typical representative is the CELLO detector at PETRA shown in Fig. 14. The central detector consists of drift and proportional chambers placed inside a thin superconducting coil (0.47 radiation length thick, 1.5 m diameter) providing a magnetic field of 1.3 Tesla. The coverage extends over 91% of $4\,\pi$ and the achieved momentum resolution is $\sigma/p = 0.013\,p\,[GeV]$ at $\theta = 90^0$. Outside the coil is a 20 radiation length thick liquid argon electromagnetic calorimeter (octagonal barrel and end caps) covering 86% of the solid angle. A fine grained strip read out with 19 samplings in depth allows electron/photon identification with an energy resolution of $\sigma/E = 0.13/\sqrt{E\,[GeV]}$ and an angular resolution of $4\,mrad$. Muon detection is done with one set of planar wire chambers behind the flux return (80 cm iron or 5.5 nuclear interaction lengths at $\theta = 90^0$) over 92% of $4\,\pi$.

The limited space here does not permit to go into more details and to present other detectors as well. More information can be found in the original literature or, e.g., in the review article on e^+e^- detectors by Lynch in ref. [18]. In the following the general features of measurement and analysis of the basic e^+e^- reactions will be briefly described.

The experimental signature for the reaction $\underline{e^+e^- \rightarrow \gamma\gamma}$ is two energetic clusters in the electromagnetic calorimeter with no additional charged track in the detector. The two clusters should point to the common interaction region and are required to be back-to-back within a typical acollinearity angle of $\xi < 20^0$. The energy of each cluster should be close to the beam energy, which poses no experimental problems since the resolution of shower counters improves with energy as $\sigma/E \sim 1/\sqrt{E}$. The most serious background comes from Bhabha scattering $e^+e^- \rightarrow e^+e^-$, where both particles leave no signature in the tracking chambers. These inefficiencies have to be corrected for in the analysis. To further reduce background from e.g. $e^+e^- \rightarrow e^+e^-\pi^0$, $e^+e^-\eta$ with $\pi^0 \rightarrow \gamma\gamma$ and $\eta \rightarrow \gamma\gamma$ and undetected $e^\pm$ a typical cut on $E_\gamma > 0.25\,E_{beam}$ is applied. Multiple photon emission at large polar angles is in most cases easily detected through the fine spatial granularity of the electromagnetic calorimeter.

The experimental signature for the reaction $\underline{e^+e^- \rightarrow \mu^+\mu^-}$ is two oppositely charged tracks emerging from a common interaction vertex. The tracks have to be acollinear with typical cuts of $\xi < 20^0$. The particles are identified as muons by their penetration through iron of about 6 - 8 interaction lengths, which efficiently absorbs all other particles. The momentum is either measured by drift chambers within a solenoidal magnetic field with a typical resolution of $\sigma/p \sim 1-2\,\%\cdot p\,[GeV]$ or through bending in magnetized iron (MAC, MARK J) with a resolution of $\sigma/p \sim 30\%$. The achieved resolutions are in general sufficient to determine the charge of the muon, with a probability of at most a few percent that both tracks have a wrong charge assignment simultaneously. The experiments have shown in detailed investigations that the charge measurement is not affected by any instrumental polar angle dependent effects in a systematic way. The systematic uncertainties of the measured charge asymmetry are of the order of 0.5% or less. A typical momentum cut $p_\mu > 0.25\,p_{beam}$ reduces background from the two-photon reaction $e^+e^- \rightarrow e^+e^-\mu^+\mu^-$, cosmic rays (plus additional time of flight cuts) and τ pair production where both particles decay into $\tau \rightarrow \mu\nu_\mu\nu_\tau$ (branching ratio $17.8 \pm 0.4\,\%$). The residual background is less than a percent. The radiative process $e^+e^- \rightarrow \mu^+\mu^-\gamma$ with a photon at large polar angle is again detected with the help of the electromagnetic shower counters.

The reaction $\underline{e^+e^- \rightarrow \tau^+\tau^-}$ at high energies has a very distinct signature. The decay multiplicities are small. The τ decays after a mean life time of $(3.04 \pm 0.09) \cdot 10^{-13}\,s$ either in a lepton plus two neutrinos or in one to three hadrons plus neutrino; decays into 5 or more charged particles are negligible. The topological branching fractions of the τ decaying into 1 charged particle plus neutrals (neutrinos, photons) is 85.7 ± 0.4 % and of the τ decaying into 3 charged particles plus neutrals is 14.3 ± 0.4 %. At high energies the decay particles are boosted in a narrow cone around the original flight direction of the τ. Requiring, as most experiments do, a one-prong decay recoiling against a narrow jet like three-prong decay with certain momentum and geometrical cuts leaves a very clean τ pair event sample without any particle identification. An example of such a topology is shown in Fig. 15. However, the price is that only a quarter of all decays are used. The background from low multiplicity hadron events, Bhabha events where an accompanying photon converts in the beam pipe into an e^+e^- pair faking a three-prong jet, and the two-photon reaction $e^+e^- \rightarrow e^+e^-\tau^+\tau^-$ can be kept low, of the order of a few percent. In principle all τ decays can be used for analysis except those where both τ decay into electrons or muons, thereby raising the acceptable decay modes to about 90%. In this case, however, refined particle identification is mandatory. This has been achieved only by few experiments. Small corrections to the direction measurement have to be applied, since the momentum vector of the one- or three-prong decay does not coincide with the direction of the produced τ.

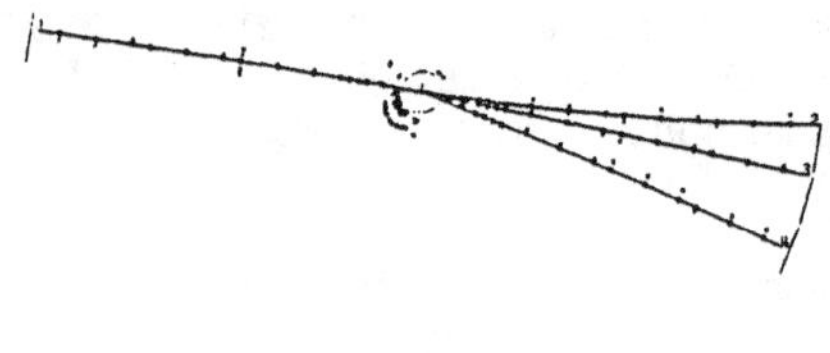

Figure 15: An $e^+e^- \rightarrow \tau^+\tau^-$ event observed in the TASSO detector at $\sqrt{s} = 35\,GeV$ (view perpendicular to the beams). The one-prong decay is identified as $\tau^- \rightarrow \mu^-\bar{\nu}_\mu\nu_\tau$, the three-prong decay as $\tau^+ \rightarrow (3\,hadrons)^+\,\nu_\tau\,(n\,\gamma)$

The reaction $\underline{e^+e^- \rightarrow e^+e^-}$ has a signature very similar to the two photon pair production above, except that a charged track has to be associated with each calorimetric energy cluster. Geometrical matching requires that the charged track points to the centre of the energy cluster. If magnetic momentum analysis is available, a track momentum energy balance within the experimental resolutions is performed. Note that at high energies the energy measurement is superior to the momentum measurement. As mentioned previously the angular distribution in Bhabha scattering is extremely peaked in the forward direction, the differential cross section at $\cos\theta = +0.8$ is about 80 times larger than at $\cos\theta = -0.8$. This implies that extremely good momentum resolution is needed in order not to affect the angular distribution measurements in the backward region by charge confusion. Some experiments therefore prefer to show angular distributions folded around $\cos\theta = 0$. The selection criteria are typically an acollinearity cut of $\xi < 20^0$ and an energy cut for each particle of $E_e > 0.25\,E_{beam}$. The background from hadron production, τ pair production and two-photon reactions can be easily separated since almost all centre of mass energy is released in the electromagnetic calorimeter and well measured. A remaining small background comes from $e^+e^- \rightarrow \gamma\gamma$, where both photons convert in

122

the beam pipe into highly asymmetric e^+e^- pairs. Thes into account in the acceptance calculations. Since Bhabha largest cross section of all reactions leading to two charged pa the TASSO collaboration did an analysis without any partick somewhat more stringent selection criteria on high momentum events, their Bhabha event sample contains about 5.5% overall and τ pairs. This background is statistically subtracted from the ang the correction reaching substantial values of about 25% in the backw

The experimental signature of the photon-photon reactions e^+e^- (l = e or μ) is two oppositely charged tracks which appear to be ba (coplanar) in the plane perpendicular to the $e^\pm$ beams but are acollinea the moving $\gamma\gamma$ centre of mass system. The event selection is just opposite charged two-prong annihilation events, which are discriminated by requiring a mum acollinearity angle in space and a cut on the maximum visible energy. Since $\gamma\gamma$ luminosity drops rapidly with the $\gamma\gamma$ centre of mass energy the bulk of the even appears at low energies. Therefore the track momentum resolution poses no problems. Momentum balance in the plane perpendicular to the $e^\pm$ beam direction between the two charged tracks (and the electron/positron if it is tagged) assures that it is an exclusive process with no further undetected particles. A more severe problem is the electron and muon identification at low momenta around $1\,GeV$ against hadrons, which are copiously produced either directly or via resonance decays. A possible background comes from beam gas events where the electron/positron hits a nucleus inside the vacuum chamber thereby producing acoplanar low energy event topologies. But their interaction vertex is uniformly distributed along the beam direction and they can be statistically subtracted from the two-photon event sample.

7 Experimental tests of QED

In this section the experimental tests of QED will be discussed. It is divided in two particle final states, hard photon radiation processes and photon-photon collisions. Due to the overwhelming amount of material not all the beautiful experimental results can be presented. Instead a few examples are chosen which demonstrate the achievements of the experimental QED tests.

7.1 Two particle final states

7.1.1 $e^+e^- \to \gamma\gamma$

The reaction $e^+e^- \to \gamma\gamma$ provides the cleanest QED test at the one loop level, since hadronic vacuum polarization and electroweak effects only enter in the second order corrections. This process has been studied by many experiments.

Fig. 16 shows the differential cross section distributions at four energies and the total cross section integrated over $|\cos\theta| < 0.76$ between 12 and $44\,GeV$ from the JADE experiment. The agreement with QED is very good.

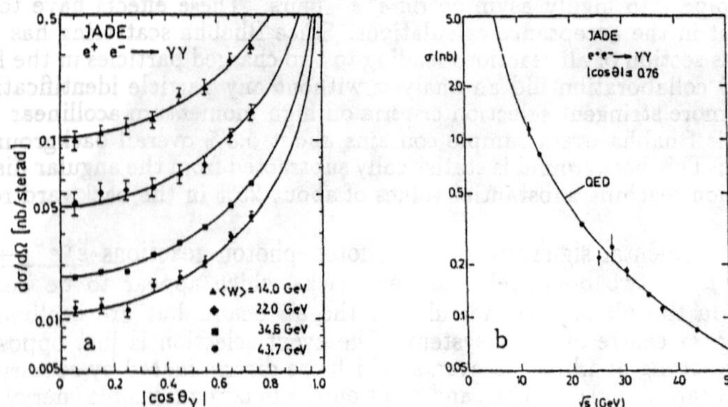

Figure 16: Cross section distributions for the reaction $e^+e^- \to \gamma\gamma$ from JADE. a) Differential cross sections at $\sqrt{s} = 14$, 22, 34.5 and 43.1 GeV, b) total cross section as a function of the energy. The curves show the QED prediction

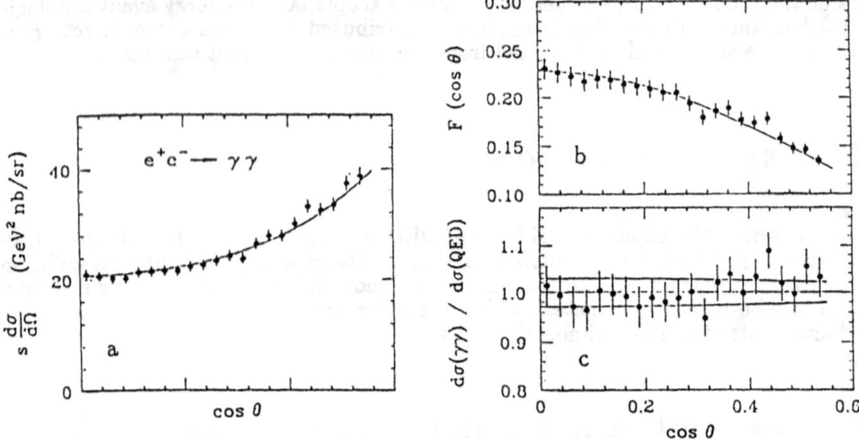

Figure 17: Cross section distributions for the reaction $e^+e^- \to \gamma\gamma$ from HRS at $\sqrt{s} = 29\,GeV$. a) The differential cross section, b) the differential cross section divided by the Bhabha cross section, c) the differential cross section divided by the QED cross section. The curves are the QED prediction, the upper and lower curves in c) represent experimental limits for $\Lambda > 59\,GeV$

The most precise measurement has been performed by the HRS collaboration, which collected 14880 γ pairs at large angles $|\cos\theta| < 0.55$ at an energy of $\sqrt{s} = 29\,GeV$. The differential cross section is shown in Fig. 17a, and the same data normalized to the QED prediction are plotted in Fig. 17c. Each data point has an error of about 4%, while the overall systematic uncertainty is 0.6%. The average value is very close to one with an overall error of less than 2%, significantly lower than the first order radiative corrections, which amount on average to about 8%. The measurement clearly demonstrates the necessity and correctness of the $\mathcal{O}(\alpha^3)$ QED calculations. In order to overcome the problems related to an absolute luminosity measurement the $\gamma\gamma$ data were compared bin by bin to a high statistics Bhabha event sample (84423 events) selected under similar conditions. This is shown in Fig. 17b, where the function $F(\cos\theta)$ represents the $\gamma\gamma$ angular distribution divided by the Bhabha cross section folded around $\cos\theta$ and corrected for small electroweak effects. In this quantity common systematic uncertainties cancel. As can be seen the agreement with QED is excellent. Defining Σ as the integral of $F(\cos\theta)$ over $|\cos\theta| < 0.55$ a fit to the QED prediction yields

$$\frac{\Sigma^{meas}}{\Sigma^{QED}} = 1.007 \pm 0.009 \pm 0.008 \ ,$$

where the first error is statistical and the second systematic. This most precise measurement shows that $\mathcal{O}(\alpha^3)$ QED calculations have been experimentally tested at the level of one percent.

A compilation of results on $R_{\gamma\gamma} = \sigma^{meas}/\sigma^{QED}$ as a function of the centre of mass energy squared s is shown in Fig. 18. All measurements confirm QED within a few percent.

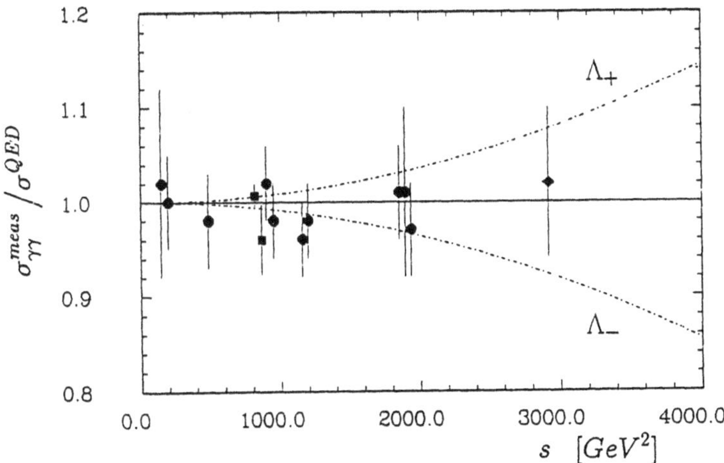

Figure 18: Cross sections normalized to QED for the reaction $e^+e^- \to \gamma\gamma$ as a function of the centre of mass energy squared s. Data are from PETRA (circles), PEP (squares) and TRISTAN (diamond) experiments. The full line shows the QED expectation, the dashed curves show the expected deviation from QED for cut-off parameters $\Lambda = 80$ GeV

A quantitative non-gauge invariant formulation of a possible break down of QED has been proposed by [19] long time before the invention of the electroweak theory. In this scheme the photon propagator is modified according to

$$\frac{1}{q^2} \to \frac{1}{q^2} - \frac{1}{q^2 - \Lambda^2} = \frac{1}{q^2}\left(1 - \frac{q^2}{q^2 - \Lambda^2}\right),$$

where the cut-off parameter Λ can be interpreted as the mass of an additional heavy photon with coupling strength α. For high values $\Lambda \to \infty$ the original theory is restored. Such a heavy photon would also modify the vertex and the fermion propagator.

All these contributions can be absorbed in a so-called form factor, which alters the observable differential cross section [20] of the reaction $e^+e^- \to \gamma\gamma$ as

$$\frac{d\sigma}{d\Omega} = \frac{\alpha^2}{s}\frac{1+\cos^2\theta}{\sin^2\theta}\left(1 \pm \frac{s^2}{2\,\Lambda_\pm^4}\frac{\sin^4\theta}{1+\cos^2\theta}\right). \tag{44}$$

The term with Λ_- has no simple physical interpretation, but is added for convenience to also account for a lower cross section.

Another possible deviation from QED could arise through the exchange of a heavy or excited electron E^* with mass m^* and coupling strength $\lambda\, e$. Electromagnetic current conservation allows only a coupling via the magnetic moments with the interaction Lagrangian

$$\mathcal{L}_{int} = \frac{\lambda\, e}{2\, m^*}\,\bar{\psi}^{(*)}\,\sigma_{\mu\nu}\,F^{\mu\nu}\,\psi\,,$$

where $\psi^{(*)}$ is the Dirac field of the spin 1/2 heavy electron, $\sigma_{\mu\nu} = i/2\,(\gamma_\mu\gamma_\nu - \gamma_\nu\gamma_\mu)$ and $F^{\mu\nu}$ is the elctromagnetic field tensor. Replacing the mass by Λ and assuming $\lambda = 1$ the modified cross section becomes

$$\frac{d\sigma}{d\Omega} = \frac{\alpha^2}{s}\frac{1+\cos^2\theta}{\sin^2\theta}\left(1 \pm \frac{s^2}{2\,\Lambda_\pm^4}\frac{\sin^4\theta}{1-\cos^2\theta}\right). \tag{45}$$

Again the term proportional to Λ_- is introduced for convenience to allow for a lower cross section.

Both form factor extensions to the cross section for $e^+e^- \to \gamma\gamma$ vary as s^2/Λ^4 and are identical at $\theta = 90^0$, where the sensitivity reaches its maximum. In the static limit the modified photon propagator can be interpreted as a modification of the r dependence of the Coulomb potential $1/r \to 1/r\,(1 - e^{-\Lambda r})$, i.e. the cut-off parameter Λ is a measure of the pointlike nature of the $e\gamma$ interaction. In the second case the derived Λ values set limits on the mass of an additional heavy electron. Most experiments analyze their data using eq. (45) and quote lower limits on $\Lambda_\pm$. However, the difference with eq. (44) is small, of the order of a few GeV.

As an example Fig. 17b shows the effect on the differential cross section for the lower limits (95% confidence level) of $\Lambda_+ > 59\,GeV$ and $\Lambda_- > 59\,GeV$ derived by the HRS experiment. The results from other experiments on the cut-off parameters Λ are summarized in Table 3. The sensitivity is experimentally limited by the systematic error of the luminosity measurement (typically 3%) but rises with the centre of mass energy. This is evident from the higher values obtained by the TRISTAN experiments at $\sqrt{s} \simeq 55\,GeV$, although they are less precise and have only a tenth of the luminosity compared to the high precision HRS data at $\sqrt{s} = 29\,GeV$. The effect of a hypothetical cut-off parameter of $\Lambda = 80\ GeV$ can also be seen in the total cross section data of Fig. 18.

Table 3: Lower limits (95 % confidence level) on QED cut-off parameters for the reaction $e^+e^- \to \gamma\gamma$

Experiment	Λ_+ $[GeV]$	Λ_- $[GeV]$
HRS [21,22]	59	59
MAC [23]	66	67
CELLO [24]	84	44
JADE [25]	66	75
MARK J [26]	70	70
PLUTO [27]	46	36
TASSO [28]	61	56
AMY [29]	65	—
TOPAZ [30]	94	59
VENUS [31]	81	82

In summary QED is well tested up to $\mathcal{O}(\alpha^3)$ in the reaction $e^+e^- \to \gamma\gamma$ at the level of one percent. Lower limits on the cut-off parameters of $\Lambda_+ > 94\,GeV$ and $\Lambda_- > 82\,GeV$ have been obtained excluding a heavy electron of this mass.

7.1.2 $e^+e^- \to e^+e^-$

Bhabha scattering $e^+e^- \to e^+e^-$ serves two purposes. At small scattering angles or low momentum transfers, where QED is well established, it serves as a luminosity monitor with high rates, while at large angles or high momentum transfers tests of QED and the electroweak theory can be performed.

The precision of the luminosity determination is limited by systematic effects, which are caused by the steep rise of the angular distribution and the non perfect knowledge of the geometrical acceptance of the luminosity counters. The systematic uncertainties typically amount to 2 - 4 %.

Wide angle measurements of $e^+e^- \to e^+e^-$ have been performed by several experimental groups. An example of the differential cross section at $\sqrt{s} = 34.6\,GeV$ from the MARK J collaboration is shown in Fig 19. The detector covers polar angles of $12^0 < \theta < 168^0$, but, due to the lack of charged particle momentum analysis, the data are folded around $\theta = 90^0$. Another example is shown in Fig 20a, where the differential cross sections measured over $-0.8 < \cos\theta < 0.8$ at energies between $\sqrt{s} = 14$ and $43.6\,GeV$ by the TASSO collaboration are displayed.

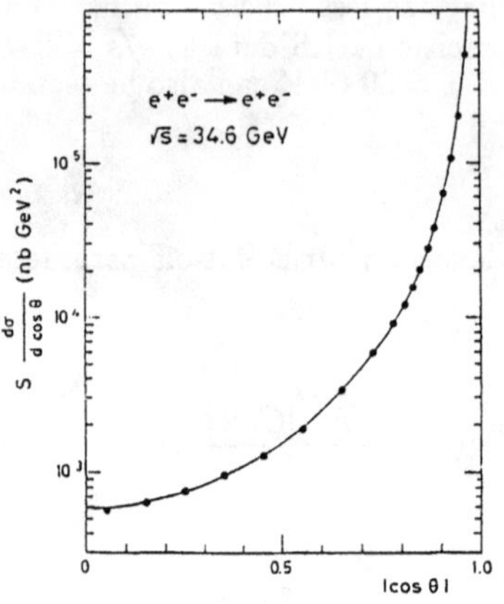

Figure 19: Differential cross section for the reaction $e^+e^- \to e^+e^-$ from MARK J at $\sqrt{s} = 34.6\,GeV$. The curve shows the QED prediction

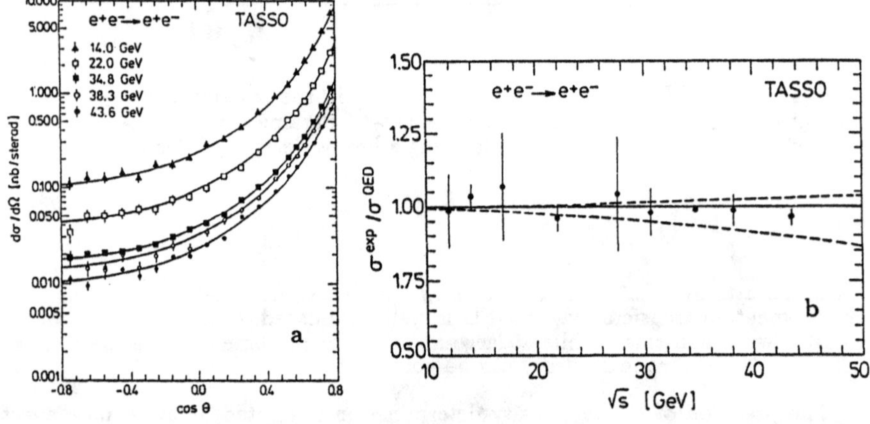

Figure 20: Cross section measurements for the reaction $e^+e^- \to e^+e^-$ from TASSO. a) Differential cross sections at $\sqrt{s} = 14$, 22, 34.8 and $43.6\,GeV$, b) total cross section integrated over $|\cos\theta| < 0.8$ divided by QED as a function of the centre of mass energy. The full curves show the QED prediction, the dashed curves show the deviation from QED for cut-off parameters $\Lambda_+ = 370\,GeV$ and $\Lambda_- = 190\,GeV$

At first sight Bhabha scattering seems to be well described by QED. A closer look at the data is provided by plotting the ratio of the measured over the QED predicted differential cross sections on a linear scale, as shown for some experiments in Fig. 21. Although in general good agreement with QED, the data are slightly better described if electroweak effects are included. To prove the presence of the small electroweak contributions is just marginally feasible for most experiments. Nevertheless, the measurements put stringent bounds on the parameters of the standard theory.

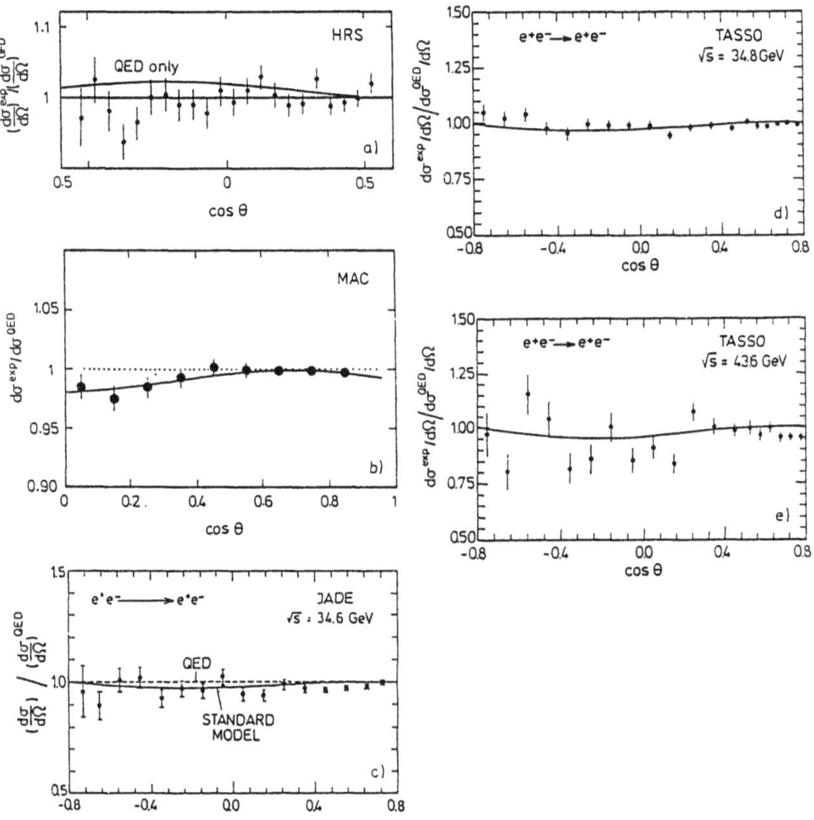

Figure 21: Differential cross section measurements for the reaction $e^+e^- \to e^+e^-$ divided by QED. a) HRS at $\sqrt{s} = 29\,GeV$ normalized to QFD, b) MAC at $\sqrt{s} = 29\,GeV$, c) JADE at $\sqrt{s} = 34.5\,GeV$, d) TASSO at $\sqrt{s} = 34.8\,GeV$, e) TASSO at $\sqrt{s} = 43.6\,GeV$. The curves show the QFD prediction

In order to perform quantitative tests of QED one can either use the total cross section, which is dominated by the forward region with negligible electroweak contributions, or apply small electroweak corrections to the differential cross sections. Fig.

20b shows the total Bhabha cross section integrated over $|\cos\theta| < 0.8$ as a function of energy from the TASSO experiment. The agreement with $\mathcal{O}(\alpha^3)$ QED calculations is at the level of 3% up to energies of $\sqrt{s} = 44\,GeV$.

In pure QED a breakdown can be formulated by inserting form factors

$$F(q^2) = 1 \mp \frac{q^2}{q^2 - \Lambda_\pm^2} \tag{46}$$

at the time-like, $q^2 = s$, and space-like, $q^2 = t = -s/2\,(1 - \cos\theta)$, vertices, which leads to the modified differential cross section

$$\frac{d\sigma}{d\Omega} = \frac{\alpha^2}{2\,s}\left\{\left(\frac{s}{t}\right)^2 F(t)^2 + \left(\frac{t}{s}\right)^2 F(s)^2 + \left(1+\frac{s}{t}\right)^2\left(1+\frac{t}{s}\right)^2\left(\frac{s}{t}F(t) + F(s)\right)^2\right\}. \tag{47}$$

The space-like, time-like and interference terms are apparent. Note that in Bhabha scattering the sensitivity is proportional to q^2/Λ^2. The data are analyzed after applying the small electroweak standard theory corrections (assuming $\sin^2\theta_W = 0.23$ and $m_Z = 92\,GeV$) and any residual deviation is attributed to the QED cut-off parameters Λ. The results of such fits are summarized in Table 4. Again the lower bounds are limited by systematic uncertainties of the luminosity and rise with the centre of mass energy. Lower limits of $\Lambda_+ > 435\,GeV$ and $\Lambda_- > 590\,GeV$ have been obtained. The effect of QED cut-off parameters on the total cross section can also be seen in Fig. 20b for the TASSO data.

Table 4: Lower limits (95 % confidence level) on QED cut-off parameters for the reaction $e^+e^- \to e^+e^-$

Experiment	Λ_+ $[GeV]$	Λ_- $[GeV]$
HRS [32,22]	154	220
MAC [33]	256	179
CELLO [34]	83	155
JADE [35]	267	200
MARK J [36]	173	177
PLUTO [37]	162	184
TASSO [38]	435	590
AMY [29]	154	342
TOPAZ [30]	115	236
VENUS [39]	180	255

To summarize Bhabha scattering experiments are well described by $\mathcal{O}(\alpha^3)$ QED calculations at the level of a few percent with first hints of electroweak contributions. Lower limits on QED cut-off parameters Λ exceed 400 GeV, which can be interpreted that the $ee\gamma$ interaction behaves as pointlike down to distances of less than $10^{-16}\,cm$.

7.1.3 $e^+e^- \rightarrow \mu^+\mu^-$ and $e^+e^- \rightarrow \tau^+\tau^-$

The lepton pair production reactions $e^+e^- \rightarrow \mu^+\mu^-$ and $e^+e^- \rightarrow \tau^+\tau^-$ are mediated at high energies through γ and Z^0 annihilation diagrams. As discussed in section 3.2 their interference leads to considerable forward backward asymmetries, thus the differential cross section measurements are not suited for QED tests. However, at the energies considered here the total cross section can be well described by QED alone, because the electroweak interference effects cancel after integration over the polar angle. The weak contributions are expected to be very small due to the fact that the electron, muon and tau vector couplings are close to zero in the standard theory (see section 3.2).

The $\mathcal{O}(\alpha^3)$ radiative corrections are experimentally verified, as shown e.g. in Fig. 22 for the acollinearity angle distribution of μ pair production measured by MARK J. Data and QED simulations agree up to very large acollinerity angles. For the total cross section measurements the first order virtual and bremsstrahlung radiative corrections are of the order of 10%, depending on the selection criteria.

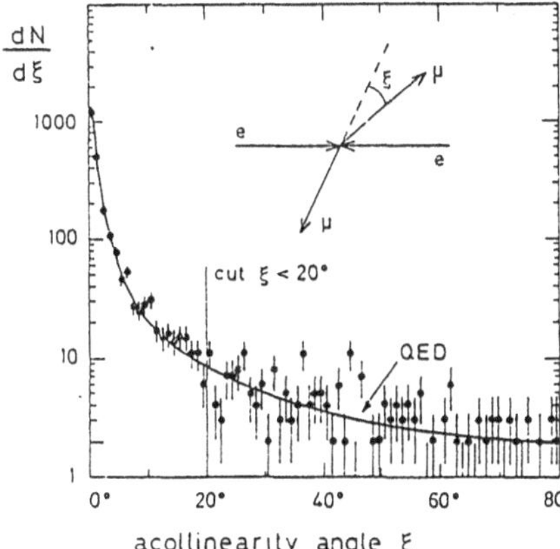

Figure 22: Acollinearity angle distribution for the reaction $e^+e^- \rightarrow \mu^+\mu^-$ by MARK J at $\sqrt{s} = 34.6\,GeV$. The curve shows the QED prediction

The $e^+e^- \rightarrow \mu^+\mu^-$ and $e^+e^- \rightarrow \tau^+\tau^-$ total cross sections normalized to the asymptotic QED prediction, $R_{\mu\mu,\tau\tau} = \sigma^{\mu\mu,\tau\tau}/\sigma^0$ with $\sigma^0 = 4\,\pi\,\alpha^2/3\,s$, are displayed in Fig. 23 as a function of the centre of mass energy squared. For both reactions the measured R values are close to one with partially substantial errors. In the larger data samples the errors are dominated by systematic uncertainties on the luminosity of the order of 3 - 5 %. In addition τ pairs suffer from lower statistics and uncertainties in the decay branching ratios. The measurements are certainly consistent with QED, which is within the presently achieved accuracy practically indistinguishable from the QFD prediction. Only at TRISTAN energies around $s =\simeq 3000\,GeV^2$ deviations of a few percent are expected. The rise is due to the purely weak terms in the cross section formula of eq. (18). However, the statistical significance of the data is still unsufficient to observe weak contributions to the total leptonic cross sections.

Within QED any deviation is parametrized by a time-like polar angle independent

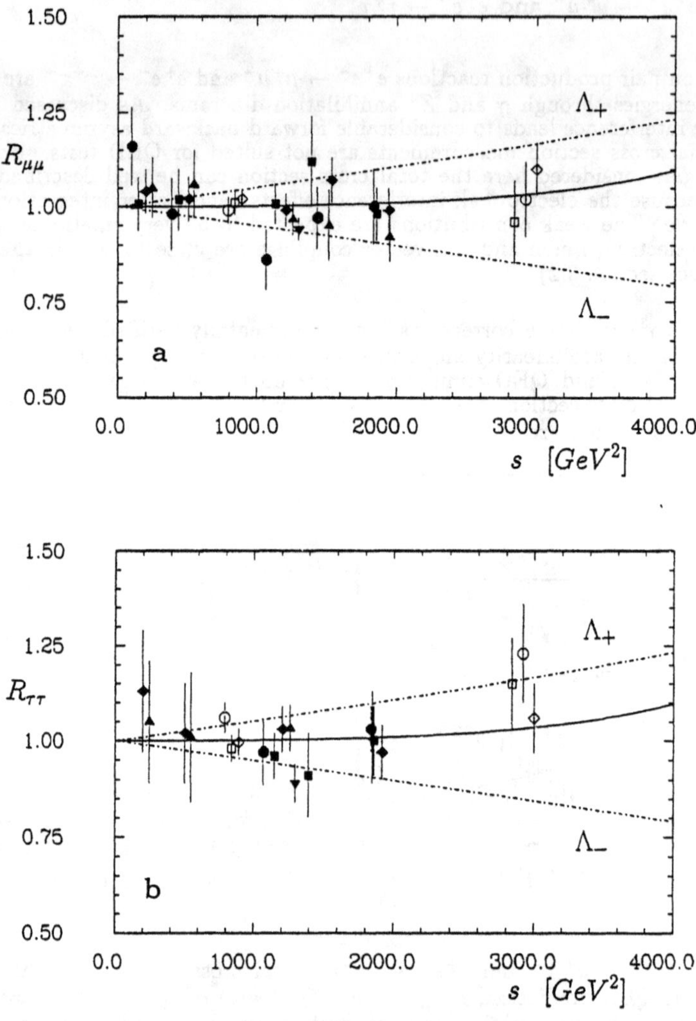

Figure 23: Total cross sections normalized to QED as a function of the centre of mass energy squared s. a) $R_{\mu\mu}$ for $e^+e^- \to \mu^+\mu^-$, b) $R_{\tau\tau}$ for $e^+e^- \to \tau^+\tau^-$. Data are from PETRA (full symbols), PEP (open symbols) and TRISTAN (open symbols) experiments and contain systematic uncertainties. The full curves show the QFD expectation. The dashed curves show the deviation from QED for cut-off parameters $\Lambda = 200\,GeV$

form factor, which modifies the total cross section according to

$$\sigma(e^+e^- \to l^+l^-) = \frac{4\pi\alpha^2}{3s}\left(1 \mp \frac{s}{s-\Lambda_\pm^2}\right)^2 . \tag{48}$$

After correcting for the small electroweak effects (assuming the validity of the standard theory with $\sin^2\theta_W = 0.23$ and $m_Z = 92\,GeV$), fits to the data yield lower limits on the QED cut-off parameters for the reactions $e^+e^- \to \mu^+\mu^-$ and $e^+e^- \to \tau^+\tau^-$ summarized in Table 5. They reach values above 250 GeV. The effect of a cut-off parameter of $\Lambda = 200\,GeV$ on the total cross section measurements is illustrated in Figs. 23. At very high energies such an interpretation of the data certainly becomes doubtful and should instead be compared to the QFD predictions.

Table 5: Lower limits (95 % confidence level) on QED cut-off parameters for the reactions $e^+e^- \to \mu^+\mu^-$ and $e^+e^- \to \tau^+\tau^-$

	$e^+e^- \to \mu^+\mu^-$		$e^+e^- \to \tau^+\tau^-$	
Experiment	Λ_+ [GeV]	Λ_- [GeV]	Λ_+ [GeV]	Λ_- [GeV]
HRS [40,41]	170	146	92	246
MAC [42]	172	172	—	—
CELLO [43,44]	230	171	142	121
JADE [35]	230	245	285	210
MARK J [36,45]	355	209	235	205
PLUTO [37,46]	206	141	79	63
TASSO [47,48]	325	150	184	221
AMY [29]	265	220	—	—
TOPAZ [39]	330	128	—	—
VENUS [39]	188	265	—	—

Summary on two particle final states. The cleanest QED reaction $e^+e^- \to \gamma\gamma$ has been measured very precisely. The $\mathcal{O}(\alpha^3)$ differential cross section including virtual and bremsstrahlung radiative corrections is found to be in agreement with QED at the one percent level. QED cut-off parameters of $\Lambda_+ > 98\,GeV$ and $\Lambda_- > 80\,GeV$ exclude excited heavy electrons of this mass. The differential Bhabha cross section $e^+e^- \to e^+e^-$ confirms $\mathcal{O}(\alpha^3)$ QED calculations at the level of a few percent, but indicates the presence of electroweak effects. The total cross section measurements for the reactions $e^+e^- \to \mu^+\mu^-$ and $e^+e^- \to \tau^+\tau^-$ are well described by $\mathcal{O}(\alpha^3)$ QED within a few percent up to energies of 55 GeV. Lower limits on the lepton QED cut-off parameters Λ exceed values of 250 GeV and suggest that the electron, the muon and the tau behave as pointlike particles down to distances smaller than $10^{-16}\,cm$. This is very remarkable, since the lepton masses differ by more than 3 orders of magnitude and the τ is about twice as heavy as the proton.

7.2 Hard photon radiation processes

In section 4 on radiative corrections it was pointed out that first order corrections to the basic two body processes can become quite large. It is therefore very important to test these calculations experimentally. Among the $\mathcal{O}(\alpha^3)$ QED corrections the initial and final state photon bremsstrahlung give the largest contributions. An example of collinear photon emission is the acollinearity angle distribution of μ pair production shown in Fig. 22. In this section experimental results on processes involving hard photon radiation will be presented.

Experimental studies have been performed for the $\mathcal{O}(\alpha^3)$ and $\mathcal{O}(\alpha^4)$ final states listed in Table 6. The signature is that one or two photons are scattered under a large polar angle inside the detector acceptance and are well separated from the other particles. The event samples consist of the order of hundred to thousand three particle final states and of the order of ten to hundred four particle final states for each experiment. Since the event selection criteria are very specific to each experiment, a comparison of data with QED can only be done with the help of detailed Monte Carlo simulations. For e^+e^- annihilation into lepton pairs obviously the lowest order Z^0 exchange diagrams have to be included. Qualitative and quantitative comparisons with QED are provided through a number of characteristic distributions and by comparing absolute rates.

Fig. 24 shows the energy and polar angle distributions of the photons in the reaction $e^+e^- \to \gamma\gamma\gamma$ measured by JADE. While most of the photons have an energy around $E_\gamma/E_{beam} \simeq 1$ the spectrum rises towards low energies exhibiting the typical bremsstrahlung behaviour. One also observes that the angular distribution becomes flatter compared to the $e^+e^- \to \gamma\gamma$ reaction (Fig. 16a), because the additional photon is produced almost isotropically. The agreement of the three and four photon final state data with $\mathcal{O}(\alpha^3)$ and $\mathcal{O}(\alpha^4)$ QED simulations is generally good concerning the shapes of the distributions as well as the rates. Similar conclusions have been reached by MAC (see Table 6).

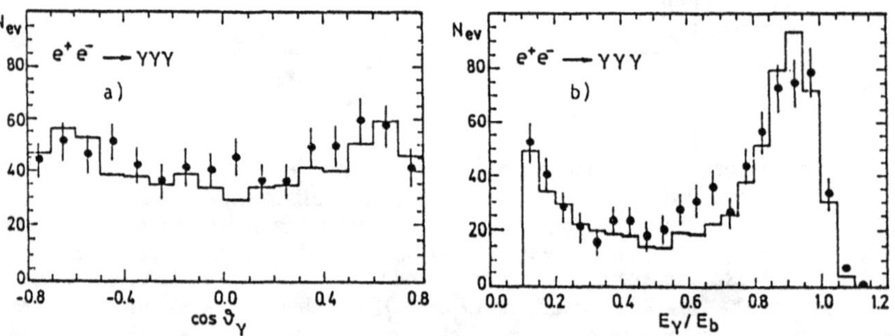

Figure 24: Distributions for the reaction $e^+e^- \to \gamma\gamma\gamma$ measured by JADE at $\sqrt{s} = 34.4\,GeV$ (3 entries/event). a) Polar angle θ_γ, b) photon energy E_γ/E_{beam}. The histograms show the $\mathcal{O}(\alpha^3)$ QED predictions

Table 6: Data samples on photonic and leptonic $\mathcal{O}(\alpha^3)$ and $\mathcal{O}(\alpha^4)$ processes involving one or two additional photons in the final state and comparison with QED predictions

Experiment	Final state	$\sqrt{<s>}$ $[GeV]$	No. events observed	No. events predicted
JADE [49]	$\gamma\gamma\gamma$	34.4	246	235 ± 4
	$\gamma\gamma\gamma\gamma$		10	9.0 ± 0.3
MAC [23]	$\gamma\gamma\gamma$	29	284	310.6
CELLO [50]	$e^+e^-\gamma$	$14 - 46.8$	654	661.5
JADE [49]	$e^+e^-\gamma$	34.4	3227	3320 ± 9
	$e^+e^-\gamma\gamma$		176	170 ± 5
CELLO [50]	$\mu^+\mu^-\gamma$	$14 - 46.8$	62	59.8
JADE [49]	$\mu^+\mu^-\gamma$	$27 - 36.6$	270	290
	$\mu^+\mu^-\gamma\gamma$		16	11.7
MAC [51]	$\mu^+\mu^-\gamma$	29	273	280
	$\mu^+\mu^-\gamma\gamma$		17	14
MARK J [52]	$\mu^+\mu^-\gamma$	$28.7 - 46.0$	795	811
JADE [49]	$\tau^+\tau^-\gamma$	$30 - 46.8$	112	100 ± 3
	$\tau^+\tau^-\gamma\gamma$		9.4 ± 3.3	9.7

Fig. 25 shows photon and electron distributions for the reactions $e^+e^- \to e^+e^-\gamma$ and $e^+e^- \to e^+e^-\gamma\gamma$ measured by JADE. The photon energy spectra show the typical bremsstrahlung nature. They peak at low values but have long tails extending up to the maximum energy of $E_\gamma/E_{beam} \simeq 1$. The photon angular distribution in the reaction $e^+e^- \to e^+e^-\gamma\gamma$ becomes relatively flat except for angles close to the $e^\pm$ beams. The photons are emitted almost isotropically in the central detector region. The interference of the initial and final state bremsstrahlung diagrams leads to a negative forward backward asymmetry in the lepton polar angle distribution (see section 4). This effect is very large in the reaction $e^+e^- \to e^+e^-\gamma$ (Fig. 25b). The forward region is drastically depopulated compared to the pure Bhabha scattering (Figs. 19 and 20a). Similar investigations have been done by CELLO. The agreement between the data and $\mathcal{O}(\alpha^3)$ and $\mathcal{O}(\alpha^4)$ QED simulations is in all cases satisfactory at the level of a few percent (see Table 6).

The reactions $e^+e^- \to \mu^+\mu^-\gamma$ and $e^+e^- \to \mu^+\mu^-\gamma\gamma$ have been studied by several experiments. Fig. 26 shows some distributions for the reactions $e^+e^- \to \mu^+\mu^-\gamma$ and $e^+e^- \to \mu^+\mu^-\gamma\gamma$ from the MAC experiment. The $\mu\gamma$ invariant mass distributions are well described by phase space and give no hints for an excited muon decaying into a muon and a photon. The CELLO collaboration reported a two standard deviation excess of events in the reaction $e^+e^- \to \mu^+\mu^-\gamma$ at high invariant mass $m^2_{\mu\gamma}/s > 0.8$ for

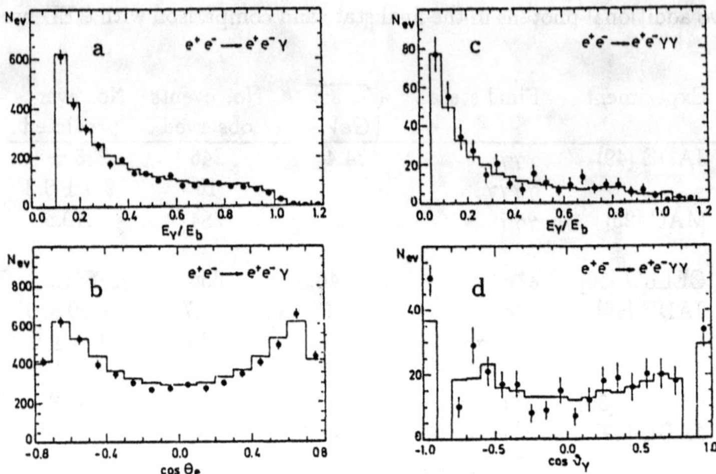

Figure 25: Distributions for the reactions $e^+e^- \to e^+e^-\gamma$ and $e^+e^- \to e^+e^-\gamma\gamma$ measured by JADE at $\sqrt{s} = 34.4\,GeV$. $e^+e^- \to e^+e^-\gamma$: a) photon energy E_γ/E_{beam}, b) electron polar angle θ_e (2 entries/event); $e^+e^- \to e^+e^-\gamma\gamma$ (2 entries/event): c) photon energy E_γ/E_{beam}, d) photon polar angle θ_γ. The histograms show the $\mathcal{O}(\alpha^3)$ and $\mathcal{O}(\alpha^4)$ QED predictions

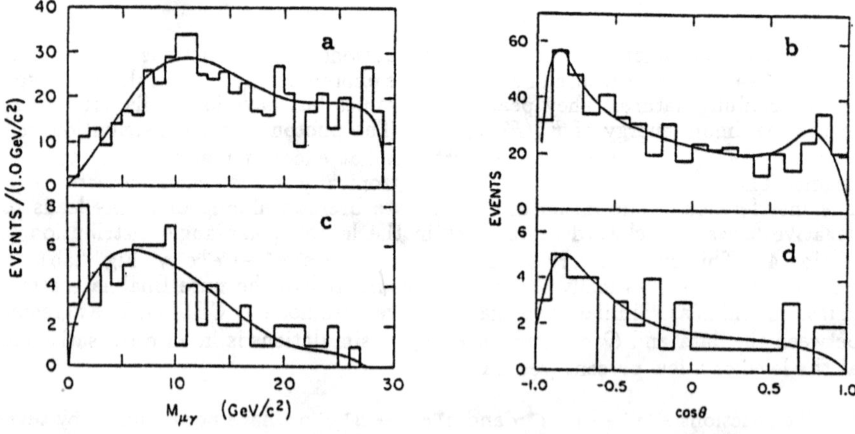

Figure 26: Distributions for the reactions $e^+e^- \to \mu^+\mu^-\gamma$ and $e^+e^- \to \mu^+\mu^-\gamma\gamma$ measured by MAC at $\sqrt{s} = 29\,GeV$. $e^+e^- \to \mu^+\mu^-\gamma$: a) $\mu\gamma$ invariant mass (2 entries/event), b) μ polar angle θ_μ; $e^+e^- \to \mu^+\mu^-\gamma\gamma$: c) $\mu\gamma$ invariant mass (4 entries/event), b) μ polar angle θ_μ. The curves show the $\mathcal{O}(\alpha^3)$ and $\mathcal{O}(\alpha^4)$ QED predictions

centre of mass energies above 34 GeV. This observation was, however, not confirmed by other experiments [49,53]. The muon polar angle distributions (Figs. 26b,d) exhibit large negative forward backward asymmetries caused by the interference of the initial and final state bremsstrahlung diagrams. The results on the asymmetry in the reaction $e^+e^- \to \mu^+\mu^-\gamma$ obtained by MAC, MARK J and JADE are summarized in Table 7. They are in good agreement with QED as well as with QFD. Note that the major part of the asymmetry is due to QED and that its value is very large and negative. This has to be compared with the positive QED induced asymmetry of about +2% in the analysis of $e^+e^- \to \mu^+\mu^-$ requiring an acollinearity angle of $\xi < 10^0$ (see section 4). The bremsstrahlung contributions can be varied by choosing different photon energy cuts. JADE and MARK J found that the absolute magnitude of the asymmetry decreases with rising photon energy cut in accordance with QED. The distributions and rates for both reactions $e^+e^- \to \mu^+\mu^-\gamma$ and $e^+e^- \to \mu^+\mu^-\gamma\gamma$ are in good agreement with $\mathcal{O}(\alpha^3)$ and $\mathcal{O}(\alpha^4)$ QED calculations.

Table 7: Charge asymmetries in the reactions $e^+e^- \to \mu^+\mu^-\gamma$ and $e^+e^- \to \tau^+\tau^-\gamma$ compared with QED and QFD predictions

Experiment	Final state	$\sqrt{s}$	A^{meas} [%]	A^{QED} [%]	A^{QFD} [%]
JADE [49]	$\mu^+\mu^-\gamma$	$27 - 36.6$	-39 ± 8	-34 ± 1	-40 ± 1
MAC [51]	$\mu^+\mu^-\gamma$	29	-21.6 ± 4.1	-21.1 ± 1.3	
MARK J [52]	$\mu^+\mu^-\gamma$	$28.7 - 46.0$	-14.7 ± 3.6	-11.6 ± 1.3	-18.4 ± 1.3
JADE [49]	$\tau^+\tau^-\gamma$	$30 - 46.8$	-27.6 ± 8.7	-31.4 ± 2.2	-36.2 ± 4

The reactions $e^+e^- \to \tau^+\tau^-\gamma$ and $e^+e^- \to \tau^+\tau^-\gamma\gamma$ are experimentally more difficult to access, since photons from τ decays have to be efficiently removed. Photon energy and τ angular distributions of the reaction $e^+e^- \to \tau^+\tau^-\gamma$ measured by JADE are presented in Fig. 27 . Again the photon bremsstrahlung spectrum is well described by QED. This also holds for the τ angular distribution, which shows a large negative forward backward asymmetry. The result on the asymmetry is given in Table 7, it is consistent with QED as well as with the standard theory. The data of both reactions $e^+e^- \to \tau^+\tau^-\gamma$ and $e^+e^- \to \tau^+\tau^-\gamma\gamma$ are in good agreement with the QED predictions (see Table 6).

Summary on hard photon radiation processes. Radiative processes with one or two additional hard photons have been studied in great detail. Characteristic distributions as well as absolute rates are in good agreement with $\mathcal{O}(\alpha^3)$ and $\mathcal{O}(\alpha^4)$ QED calculations at the level of 5 – 10 %. The available Monte Carlo event generators are able to reproduce the experimental data in almost all kinematical regions with sufficient accuracy. Finally the quality of the QED tests allows to set lower limits on the production of excited heavy leptons decaying into a lepton and a photon.

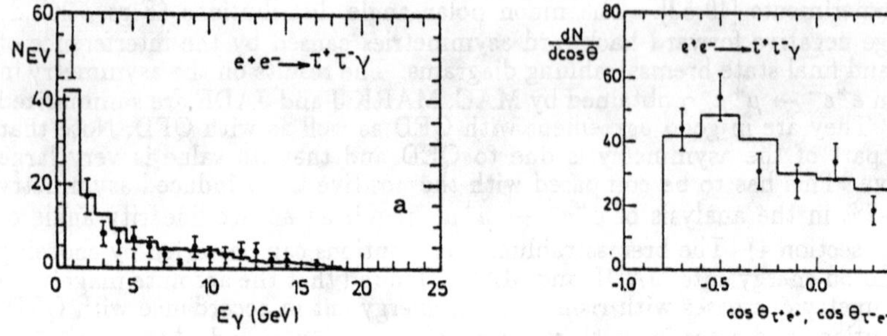

Figure 27: Distributions for the reaction $e^+e^- \rightarrow \tau^+\tau^-\gamma$ measured by JADE at $\sqrt{s} = 34.4\,GeV$. a) Photon energy E_γ, b) τ polar angle θ_τ. The histograms show the $\mathcal{O}(\alpha^3)$ QED predictions

7.3 Photon–photon collisions

Experimental QED tests in photon–photon collisions have been performed by studying the lepton pair production $e^+e^- \rightarrow e^+e^-e^+e^-$ and $e^+e^- \rightarrow e^+e^-\mu^+\mu^-$. Three types of $\mathcal{O}(\alpha^4)$ QED diagrams contribute to four lepton final states as sketched in Fig. 28. The dominant process is the photon–photon interaction or multiperipheral graph involving two space-like photons with $q_{\gamma 1}^2 < 0$ and $q_{\gamma 2}^2 < 0$. In addition there occur single conversions with one space-like and one time-like photon, $q_{\gamma 1}^2 < 0$ and $q_{\gamma 2}^2 > 0$, and double conversions with two time-like photons, $q_{\gamma 1}^2 > 0$ and $q_{\gamma 2}^2 > 0$. The latter two contributions are small, of the order of a few percent compared to the multiperipheral graph, and can, in principle, be disentangled by searching for low mass di-lepton pairs.

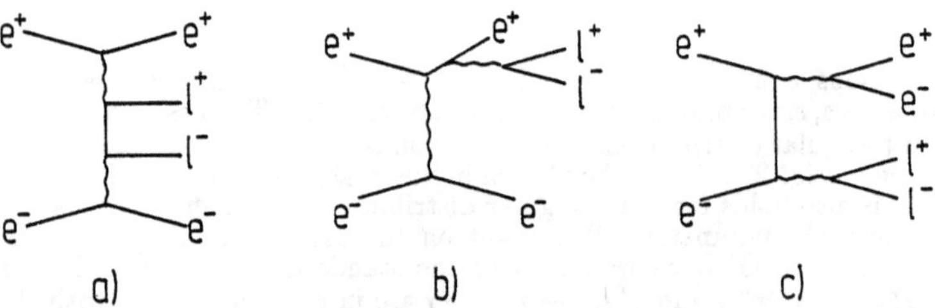

Figure 28: Feynman graphs contributing to the reaction $e^+e^- \rightarrow e^+e^-l^+l^-$. a) Photon–photon collision or multiperipheral process, b) single photon conversion, c) double photon conversion

As discussed in section 5 photon–photon collisions can be subdivided into the following classes: (i) no-tag events where both photons are quasi real and the electron/positron remains almost undeflected in the beam pipe, (ii) single-tag events where one photon is highly virtual and the electron/positron is scattered into the detector, (iii) double-tag events where both photons are highly virtual and both the electron and the positron are detected in the apparatus. The electron/positron tagging is usually done with shower counters in the very forward or backward regions or in the central detector, where the rate decreases considerably with increasing scattering angle. The available data samples on photon–photon experiments are compiled in Table 8 together with the experimental conditions and a comparison with the QED predictions. The kinematical quantitites are defined in section 5. The accessible momentum transfers are quite low compared to the annihilation processes. Therefore it is not very meaningful to analyze the data in terms of QED cut-off parameters, which cannot compete with those derived in e^+e^- annihilation events.

Table 8: Data samples of photon–photon collision experiments for the reactions $e^+e^- \rightarrow e^+e^-l^+l^-$ and comparison with QED predictions

Experiment	$\sqrt{s}$ [GeV]	Q^2 [GeV^2]	Observed particles	Tag	Events observed	Events predicted
MAC [54]	29	$\simeq 0$	$\mu\mu$	no	4849 ± 122	5090 ± 260
PEP-9 [55]	29	< 1.7	$e\mu\mu$	single	3400	
CELLO [56]	35 - 46.8	0.5 - 800	eee	single	3358	3330 ± 113
			$e\mu\mu$	single	1415	1387 ± 56
			$eeee$	double	45	48
			$ee\mu\mu$	double	28	30
JADE [57]	29 - 46.8	9 - 1200	$eeee$	double	13	12.0 ± 0.6
			$ee\mu\mu$	double	8	8.6 ± 0.5
MARK J [52]	30 - 46.8	0.7 - 166	$\mu\mu$	no	3671 ± 61	3834 ± 31
			$e\mu\mu$	single	283 ± 17	256 ± 8
			$ee\mu, ee\mu\mu$	double	43 ± 7	39 ± 3
PLUTO [58]	35	0.1 - 100	$e\mu\mu$	single	643 ± 40	674 ± 32

Studies of the process $e^+e^- \rightarrow e^+e^-\mu^+\mu^-$ with untagged electrons/positrons have been presented by MAC and MARK J. Data on the reaction $e^+e^- \rightarrow e^+e^-e^+e^-$ in the no-tag mode do not exist, because they are very difficult to discriminate against the high background from radiative Bhabha scattering events $e^+e^- \rightarrow e^+e^-\gamma$. Muon pairs from two-photon events can easily be separated from one-photon annihilation due to their low momenta. Furthermore transverse momentum balance requires that the two tracks are coplanar, i.e. appear to be back-to-back in the plane perpendicular to the $e^\pm$ beam direction, while the moving centre of mass system causes an almost flat acollinearity angle distribution. The invariant di-muon mass distribution is shown in Fig. 29a for the MARK J data taken at $\sqrt{s} = 44\,GeV$. One notices the rapidly falling spectrum towards high masses. The low mass part is affected by detector acceptance. There is no indication of a state with charge conjugation $C = +1$ decaying into a $\mu^+\mu^-$ pair. The data are well described by the QED Monte Carlo simulation.

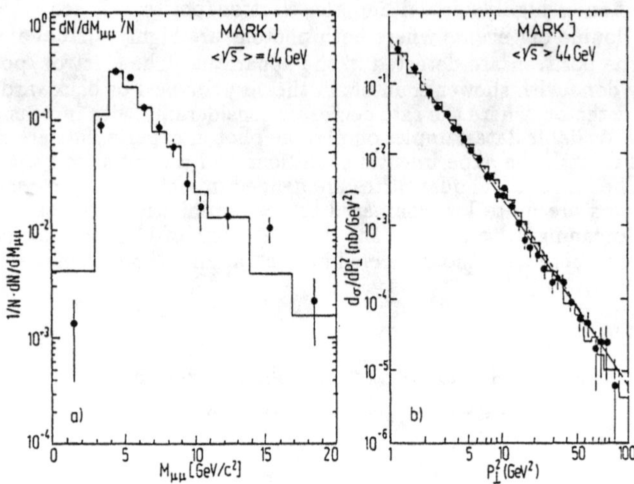

Figure 29: Distributions for the reaction $e^+e^- \to e^+e^-\mu^+\mu^-$ from MARK J at $\sqrt{s} = 44\,GeV$. **a)** The observed invariant di-muon mass $M_{\mu\mu}$, **b)** the differential cross section as a function of the muon transverse momentum squared $p_\perp^2$. The histograms show the $\mathcal{O}(\alpha^4)$ QED predictions. The broken line shows the Weizsäcker-Williams approximation and the full line a fit to the $p_\perp^2$ distribution

Fig. 29b shows the differential cross section as a function of the transverse momentum squared $d\sigma/dp_\perp^2$, where $p_\perp$ is measured with respect to the $e^\pm$ beam direction. Also shown is the $\mathcal{O}(\alpha^4)$ QED prediction. Both the exact formula and the Weizsäcker–Williams approximation describe the data equally well. The differential cross section can be well fitted by a power law behaviour

$$d\sigma/dp_\perp^2 = A \cdot p_\perp^n$$

with $n = 4.47 \pm 0.15$. For peripheral processes with massless particles and infinite centre of mass energy one expects [17] the exponent to be $n = 4$, close to the observed value. The MAC collaboration has performed a similar analysis and obtained with a slightly different parametrization $n = (4.00 \pm 0.12) + (0.019 \pm 0.004)\,p_\perp^2$ for the $p_\perp^2$ range of $3 - 30\,GeV^2/c^2$. It is interesting to note that n approaches the theoretical value of 4 as $p_\perp^2 \to 0$. To substantiate this hypothesis will require much more data.

The total cross section for the reaction $e^+e^- \to e^+e^-\mu^+\mu^-$ as a function of the e^+e^- centre of mass energy is displayed in Fig. 30 as measured by MARK J. The cross section of the untagged data rises logarithmically with the energy and becomes even larger than the annihilation cross section as expected. The measurements exhibit a remarkable agreement with the $\mathcal{O}(\alpha^4)$ QED predictions.

Tagged photon–photon collisions have been studied in both reactions $e^+e^- \to e^+e^-e^+e^-$ and $e^+e^- \to e^+e^-\mu^+\mu^-$. Through a measurement of the scattered electron/positron energy and direction the squared momentum transfer Q^2 can be deter-

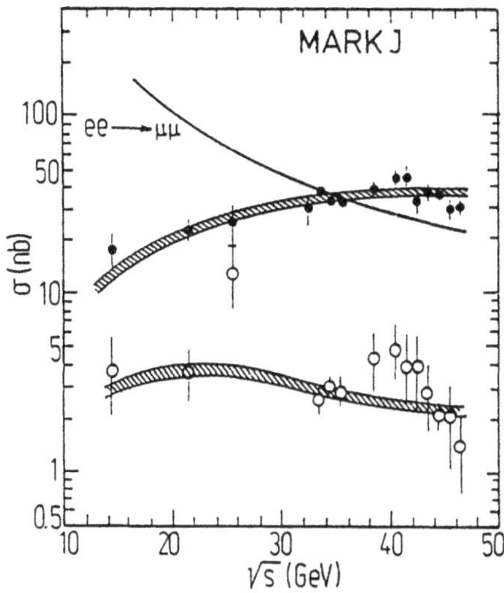

Figure 30: The total cross section for the reaction $e^+e^- \rightarrow e^+e^-\mu^+\mu^-$ from MARK J as a function of the e^+e^- centre of mass energy. The dots represent the untagged data, the circles the tagged data. The curves show the $\mathcal{O}(\alpha^4)$ QED predictions and the $e^+e^- \rightarrow \mu^+\mu^-$ cross section for comparison

mined. The experiments prefer to show distributions of various kinematic quanitites like Q^2, W, x, momenta, invariant masses, etc., for limited Q^2 regions as given by the geometrical postion and acceptance of the tagging devices. In general the distributions agree well with $\mathcal{O}(\alpha^4)$ QED simulations concerning the shape as well as the absolute rate. Fig. 30 shows the total cross section for tagged two-photon muon pair production as a function of the e^+e^- centre of mass energy measured by MARK J. The observed cross sections are about an order of magnitude lower than those of the tagged data and are well described by the QED predictions.

Of particular interest is the deep inelastic scattering of electrons off an almost real photon target. This process is described by two structure functions, eq.(42), of which under usual experimental conditions only one, namely$F_2(x, Q^2)$ is accessible.

The PEP-9/$\gamma\gamma$ collaboration has made a careful study of the reaction $e\gamma \rightarrow e\mu^+\mu^-$ at $\sqrt{s} = 29\,GeV$. The electron/positron is tagged at angles between 30 and 90 mrad, thus covering a momentum transfer range of $0 < Q^2 < 1.7\,GeV^2$. Fig. 31 shows differential cross sections as function of the muon scattering angle θ^*_μ (measured in the $\mu^+\mu^-$ centre of mass system) and of the scaling variable x. One clearly observes a peaking of θ^*_μ along the virtual photon directions which coincide with the $e^\pm$ beams. Both distributions agree well with the QED Monte Carlo simulation including all diagrams of Fig. 28. After integration over $\cos\theta^*_\mu$ the formula (42) can be applied. A further cut on the relative energy transfer $y < 0.25$ suppresses contributions from F_1 and allows to use the approximation of eq.(43) for $F_2(x, Q^2)$. The structure function $F_2(x)$, i.e. $F_2(x, Q^2)$ averaged over Q^2, is extracted by comparing the data to a Monte

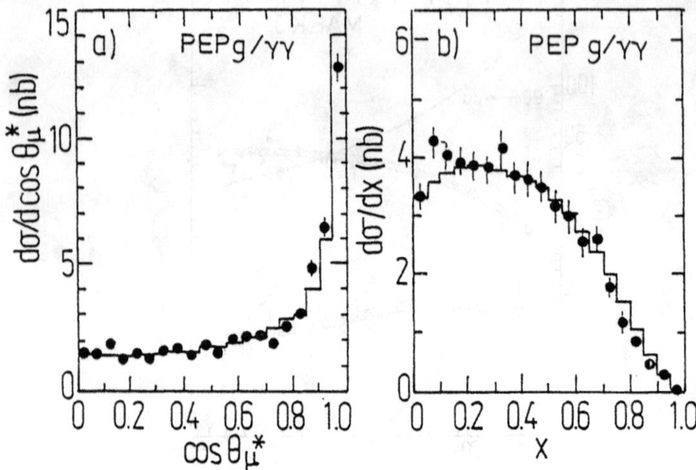

Figure 31: Differential cross section distributions for single tag $e^+e^- \rightarrow e^+e^-\mu^+\mu^-$ events from PEP-9/$\gamma\gamma$ at $\sqrt{s} = 29\,GeV$. a) $d\sigma/d\cos\theta^*_\mu$, b) $d\sigma/dx$. The histograms show the $\mathcal{O}(\alpha^4)$ QED simulations

Carlo event sample generated with $F_2 = 1$ and $F_1 = 0$. The result is shown in Fig. 32. One observes that $F_2(x)$ rises and remains large with increasing x values, characteristic for a poinlike interaction. The acceptance corrected data can be directly compared to the QED predictions taken at average values of $Q^2 \simeq 0.5\,GeV^2$. It is interesting to note that the structure function F_2 is very sensitive to the muon mass and would allow a mass determination within about 10%. It is essentially this mass dependence which makes the photon structure function F_2 attractive for a determination of the QCD mass scale parameter Λ_{QCD} in two–photn hadron production (see [59] and Kolanoski and Zerwas [18]). These results are a very beautiful confirmation of the $\mathcal{O}(\alpha^4)$ QED calculations in photon–photon collisions.

The CELLO collaboration has carried out an analysis of four lepton final states in the double tagged reactions $e^+e^- \rightarrow e^+e^-e^+e^-$ and $e^+e^- \rightarrow e^+e^-\mu^+\mu^-$ over a large Q^2 range. The data can be well described by QED once all graphs of Fig. 28 are included in the Monte Carlo simulation. The complete matrix element can be written as $\mathcal{M}(eell) = \alpha_0\mathcal{M}_0 + \alpha_1\mathcal{M}_1 + \alpha_2\mathcal{M}_2$, where the indices 0, 1 and 2 denote the multiperipheral, single conversion and double conversion contributions, respectively. The conversion diagrams essentially contribute to high the x values covered by this experiment. From a simultaneous fit to all data they find for the coefficients $\alpha_0 = +0.94 \pm 0.08$, $\alpha_1 = +1.02 \pm 0.08$ and $\alpha_2 = +1.05 \pm 0.30$ in agreement with the expectation. The relative signs occur through interferences of the different graphs. This result provides clear evidence for the (although small) non–multiperipheral contributions and again support the $\mathcal{O}(\alpha^4)$ QED calculations.

To summarize lepton pair production in untagged and tagged photon–photon collisions is found to be in good agreement with QED and confirm the $\mathcal{O}(\alpha^4)$ cal-

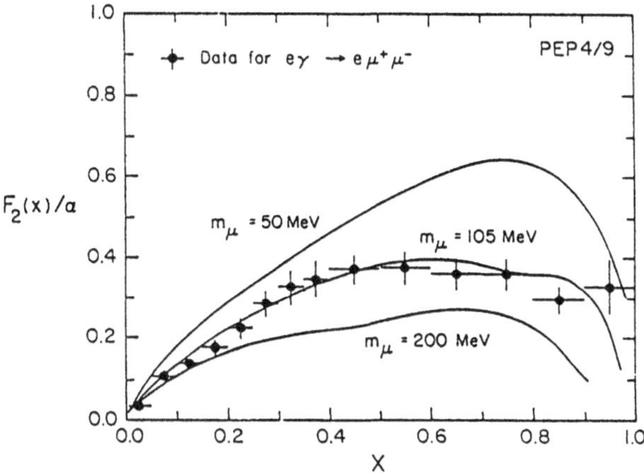

Figure 32: The structure function $F_2(x)$ for the reaction $e^+e^- \to e^+e^-\mu^+\mu^-$ from PEP-9/$\gamma\gamma$ at $\sqrt{s} = 29\,GeV$. The curves show the sensitivity to different values of the muon mass

culations at the level of ten percent. The 'muonic' photon structure function $F_2(x)$ demonstrates the pointlike interaction of the photon with the muon pair and is very sensitive to the muon mass.

8 Experimental tests of QFD

A review on experimental tests of the validity of QED in high energy electron–positron collisons would be incomplete without mentioning its 'natural' break down due to the weak interaction. Meanwhile the gauge invariant electroweak standard theory or QFD, unifying QED and the weak interaction, has been theoretically and experimentally well established. At the PETRA, PEP and TRISTAN e^+e^- colliders the electroweak interaction manifests itself in $\gamma - Z^0$ interference phenomena leading to appreciable asymmetries in fermion pair production as well as in the rise of the normalized total hadronic cross section at centre of mass energies above $\sqrt{s} \simeq 35\,GeV$ due to the tail of the Z^0 resonance.

In the following section lepton and quark angular asymmetries and the total hadronic cross section will be discussed briefly with emphasis on the determination of the electroweak parameters of the standard theory. More detailed information can be found in [18], a collection of articles on e^+e^- physics published in the same series as this volume, or in the recent compilation of electroweak interactions by [1].

8.1 Lepton asymmetries

The differential cross section distributions of Bhabha scattering $e^+e^- \to e^+e^-$ are shown in Fig. 21. The electroweak interference effects are very small and often comparable to the achieved experimental precision. Therefore the significance of individual experiments is generally poor. In Fig. 33 the results of a two parameter fit to the vector and axial vector couplings obtained by HRS and MAC are shown in the ${g_V^e}^2 - {g_A^e}^2$ plane. The two parameters are highly correlated (correlation coefficient of 0.5 to 1) and depend sensitively on the absolute normalization. Both experiments claim to exclude pure QED at the level of 2 standard deviations. Results on ${g_V^e}^2$ and ${g_A^e}^2$ are summarized in Table 9. The individual experiments are certainly consistent

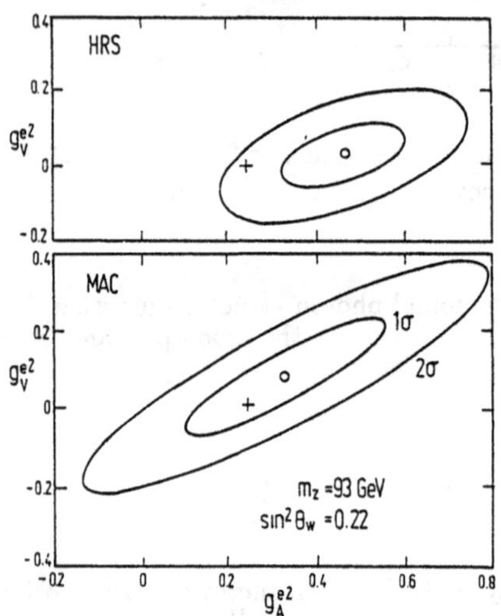

Figure 33: Correlation of vector coupling versus axial vector coupling constants for the reaction $e^+e^- \to e^+e^-$. a) HRS at $\sqrt{s} = 29\,GeV$, b) MAC at $\sqrt{s} = 29\,GeV$. The QFD prediction is marked by the crosses

Table 9: Results on ${g_V^e}^2$ and ${g_A^e}^2$ from the reaction $e^+e^- \to e^+e^-$

Experiment	${g_V^e}^2$	${g_A^e}^2$
HRS [32]	0.03 ± 0.09	0.46 ± 0.14
MAC [33]	0.09 ± 0.14	0.33 ± 0.24
JADE [60]	0.02 ± 0.10	0.22 ± 0.17
PLUTO [61]	0.09 ± 0.12	0.39 ± 0.20
TASSO [38]	-0.08 ± 0.04	0.14 ± 0.12
Average	-0.035 ± 0.032	0.285 ± 0.071
Standard theory	0.002	0.25

with QFD. The average values do not take the correlations into account, which were not always published. They are in good agreement with the standard theory.

The weak mixing angle $\sin^2\theta_W$ cannot unambiguously be extracted from Bhabha scattering. The small value of $g_V^{e\,2} = (-1/2 + 2\sin^2\theta_W)^2$ implies a value of $\sin^2\theta_W$ close to 1/4 with a twofold ambiguity, which could only be resolved with a considerably improved precision. Consequently low cross section measurements resulting in negative values of $g_V^{e\,2}$ lead to unrealistically small errors on $\sin^2\theta_W$.

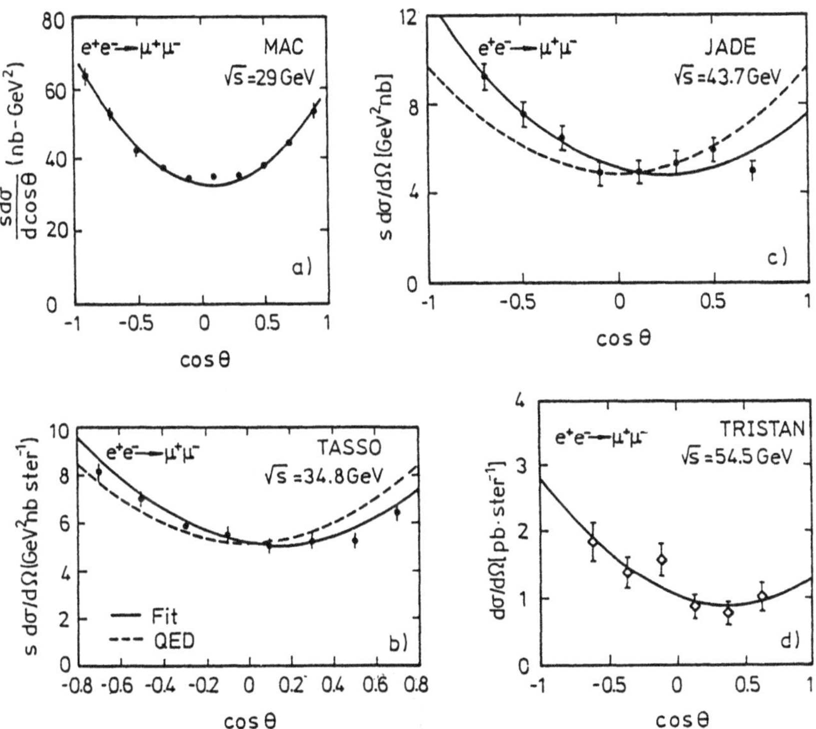

Figure 34: Differential cross sections for the reaction $e^+e^- \to \mu^+\mu^-$. a) MAC at $\sqrt{s} = 29\,GeV$, b) TASSO at $\sqrt{s} = 34.8\,GeV$, c) JADE at $\sqrt{s} = 43.7\,GeV$, d) combined TRISTAN experiments at $\sqrt{s} = 54.5\,GeV$

The reaction $e^+e^- \to \mu^+\mu^-$ was the first process where electroweak $\gamma - Z^0$ interference was observed at e^+e^- colliders. QFD predicts a sizeable negative charge asymmetry in the differential cross section distribution at high energy (see Fig. 4). The first significant evidence on a forward backward charge asymmetry in the angular distributions was announced by PETRA experiments in 1981 [62]. Since then extensive measurements have been performed at PETRA, PEP and TRISTAN. Present experimental data are not sufficient to detect electroweak effects in the total lepton pair production cross section (see section 7).

Fig. 34 shows the differential cross section distributions for the reaction $e^+e^- \rightarrow \mu^+\mu^-$ measured by MAC, TASSO, JADE and TRISTAN experiments. All experiments measure a clear deviation from QED with a statistical significance of several standard deviations. The experimental results of the observed charge asymmetries $A_{\mu\mu}$ are plotted in Fig. 35 as a function of the centre of mass energy squared s. One notices a clear rise in absolute magnitude roughly proportional to s in good agreement with the standard theory.

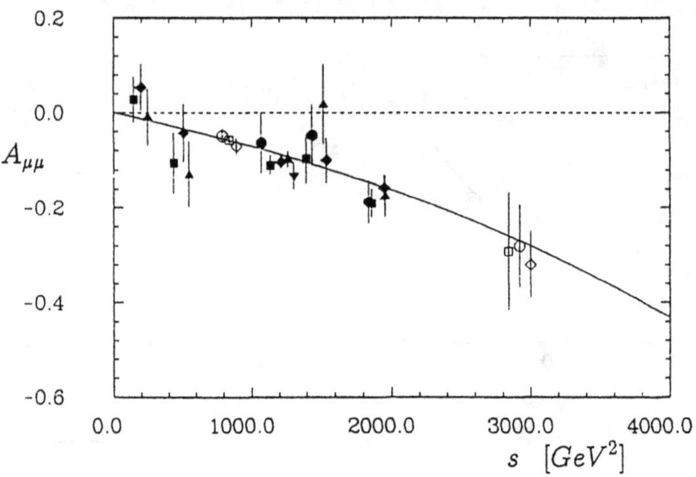

Figure 35: Measurements of the forward backward asymmetry $A_{\mu\mu}$ in the reaction $e^+e^- \rightarrow \mu^+\mu^-$ as a function of the centre of mass energy squared s. The curve shows the standard theory expectation for $\sin^2\theta_W = 0.23$ and $m_Z = 91.9\,GeV$

Since the asymmetry is a firm prediction of QFD, another way to compare the different experiments independent of the centre of mass energy is to extract the electroweak coupling constants. The charge asymmetry depends essentially on the axial vector couplings, while the total cross section depends essentially on the vector coupling constants. Both measurements can be used to determine the products of $g_A^e\, g_A^\mu$ and $g_V^e\, g_V^\mu$ with almost no correlation. The results are summarized in Table 10 together with average values. The data do not provide a precise measurement of the vector couplings, they rather put strong constraints on them. The axial vector couplings, however, deviate significantly from zero. The measured values are in very good agreement with the electroweak theory within one standard deviation or to better than 6%.

In the reaction $e^+e^- \rightarrow \tau^+\tau^-$ the same electroweak $\gamma - Z^0$ interference effects are expected as in μ pair production, since in QFD all charged leptons have identical vector and axial vector couplings and the mass of the τ is negligible. The data samples are in general smaller and have larger systematic uncertainties because only few experiments analyze all accessible τ decay modes.

Differential cross section measurements for the reaction $e^+e^- \rightarrow \tau^+\tau^-$ are shown

Table 10: Results on $g_V^e g_V^\mu$ and $g_A^e g_A^\mu$ from the reaction $e^+e^- \to \mu^+\mu^-$

Experiment	$g_V^e g_V^\mu$	$g_A^e g_A^\mu$
HRS [40]	0.03 ± 0.10	0.208 ± 0.067
MAC [63]	-0.02 ± 0.09	0.25 ± 0.03
MARK II [64]	—	0.32 ± 0.07
CELLO [65]	—	0.28 ± 0.08
JADE [35]	0.01 ± 0.08	0.325 ± 0.043
MARK J [52]	0.04 ± 0.04	0.280 ± 0.035
PLUTO [66]	—	0.35 ± 0.08
TASSO [47]	0.09 ± 0.06	0.264 ± 0.037
AMY [67]	0.01 ± 0.08	0.26 ± 0.08
TOPAZ [67]	0.07 ± 0.10	0.28 ± 0.12
VENUS [67]	-0.05 ± 0.06	0.29 ± 0.06
Average	0.025 ± 0.022	0.276 ± 0.015
Standard theory	0.002	0.25

Table 11: Results on $g_V^e g_V^\tau$ and $g_A^e g_A^\tau$ from the reaction $e^+e^- \to \tau^+\tau^-$

Experiment	$g_V^e g_V^\tau$	$g_A^e g_A^\tau$
HRS [41]	—	0.28 ± 0.11
MAC [68]	0.06 ± 0.10	0.24 ± 0.05
MARK II [64]	—	0.19 ± 0.09
CELLO [44]	—	0.28 ± 0.14
JADE [35]	0.08 ± 0.07	0.185 ± 0.055
MARK J [52]	0.07 ± 0.07	0.223 ± 0.064
TASSO [48]	-0.05 ± 0.10	0.18 ± 0.09
AMY [67]	-0.15 ± 0.12	0.32 ± 0.09
TOPAZ [67]	-0.11 ± 0.10	0.19 ± 0.13
VENUS [67]	-0.03 ± 0.07	0.18 ± 0.10
Average	-0.003 ± 0.029	0.231 ± 0.022
Standard theory	0.002	0.25

in Fig. 36 at four different energies around 29 GeV, 35 GeV, 44 GeV and 55 GeV. They exhibit a significant deviation from the QED prediction and show a negative forward backward charge asymmetry growing in modulus with energy.

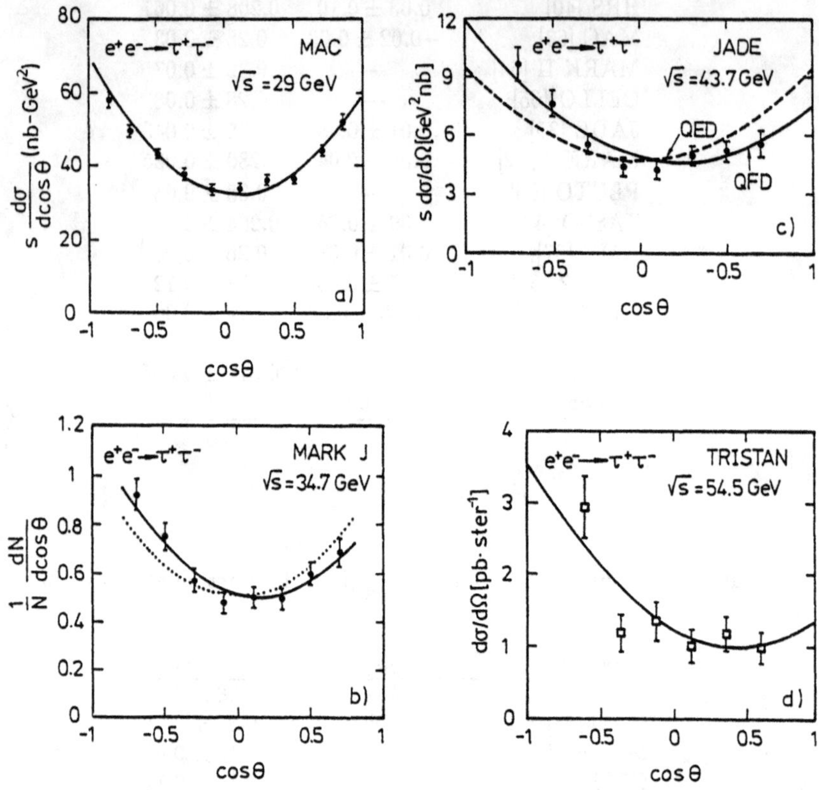

Figure 36: Differential cross sections for the reaction $e^+e^- \to \tau^+\tau^-$. a) MAC at $\sqrt{s} = 29\,GeV$, b) MARK J at $\sqrt{s} = 34.6\,GeV$, c) JADE at $\sqrt{s} = 43.7\,GeV$, d) combined TRISTAN experiments at $\sqrt{s} = 54.5\,GeV$

The measured asymmetries $A_{\tau\tau}$ from all experiments are displayed in Fig. 37 as a function of the centre of mass energy squared s. The values rise in absolute magnitude with s and closely follow the QFD curve. The charge asymmetry in the reaction $e^+e^- \to \tau^+\tau^-$ is experimentally well established.

The results on the product of the vector couplings $g_V^e\, g_V^\tau$ and the axial vector couplings $g_A^e\, g_A^\tau$ are compiled in Table 11. Again they are in good agreement with QFD. The axial vector couplings of the τ and the μ differ by about 2 standard deviations, ${g_A^\tau}^2$ being lower and ${g_A^\mu}^2$ being higher than the standard theory prediction. This 'discrepancy' has been discussed for quite some time, regarding, e.g., whether

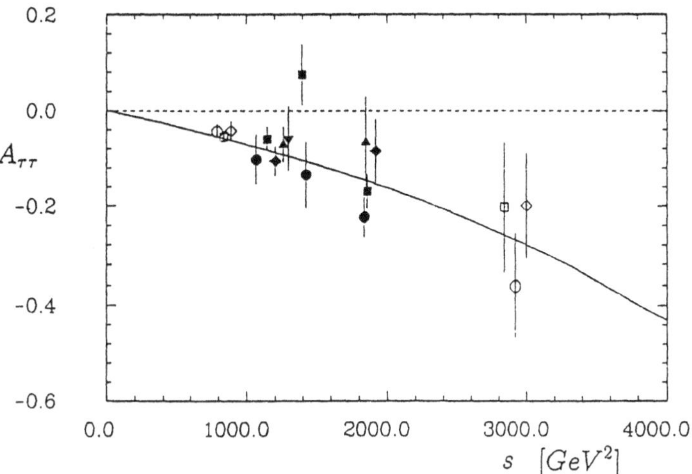

Figure 37: Measurements of the forward backward asymmetry $A_{\tau\tau}$ in the reaction $e^+e^- \rightarrow \tau^+\tau^-$ as a function of the centre of mass energy squared s. The curve shows the standard theory expectation for $\sin^2\theta_W = 0.23$ and $m_Z = 91.9\,GeV$

systematic effects in the more problematic τ analysis have not been correctly taken into account. No explanations other than statistical fluctuations have been found. Both reactions are certainly compatible with each other and confirm the standard theory prediction.

The electroweak couplings of the three leptonic reactions are in agreement with each other and support the concept of $e - \mu - \tau$ universality. It thus makes sense to assume lepton universality. A combined fit yields for the lepton coupling constants

$$
\begin{aligned}
{g_V^l}^2 &= 0.010 \pm 0.015 \,, \\
{g_A^l}^2 &= 0.258 \pm 0.012 \,.
\end{aligned}
$$

In a single parameter fit one can determine the weak mixing angle $\sin^2\theta_W$. Using all data one finds for the leptonic sector in e^+e^- experiments

$$\sin^2\theta_W = 0.216 \pm 0.014 \,.$$

This value is in agreement with the world average from all experiments and should in particular be compared to that obtained in low momentum transfer purely leptonic $\nu_\mu(\bar{\nu}_\mu)\, e \rightarrow \nu_\mu(\bar{\nu}_\mu)\, e$ scattering experiments [5], $\sin^2\theta_W = 0.223 \pm 0.018$.

8.2 Quark asymmetries

The electroweak standard theory predicts large asymmetries for quark pair production $e^+e^- \rightarrow q\bar{q} \rightarrow hadrons$ proportional to the inverse of the quark charge. Hadronic annihilation events are experimentally readily separable from lepton final states and photon–photon collisions. The difficulty is to disentangle the individual quark flavours being produced proportional to their charge squared with comparable rates in the debris of hadrons. The heavy quarks c and b are essentially produced as primary quarks, while the light quarks are as well copiously produced in the fragmentation process and via hadron decays. Two methods have been developed to tag the heavy quark flavours: (i) the study of semileptonic decays with subsequent assignment of the original quark from the lepton charge; (ii) the reconstruction of the parent hadrons, in particular D and D^* mesons.

Prompt leptons from the heavy quarks c and b emerge in the following sequence (neglecting suppressed decay modes)

$$\begin{aligned} e^+e^- &\rightarrow c\bar{c} \\ &\quad \hookrightarrow s\, l^+ \nu_l \,, \\ e^+e^- &\rightarrow b\bar{b} \\ &\quad \hookrightarrow c\, l^- \bar{\nu}_l \,. \end{aligned}$$

The quark flavour is uniquely assigned to the lepton charge. (Here and in the following only states containing the heavy quark will be considered, charge conjugate states can be treated accordingly.) The analysis relies on two ingredients. Firstly, the heavy quarks are essentially produced as primary quarks with high fractional energy, thus they are aligned along the direction of the jet formed by the decay particles. Secondly, the quark masses are heavy, thus the decay leptons have high transverse momenta with respect to the quark or jet direction, up to half the quark mass. The light u, d, s quarks decay semileptonically into low transverse momentum leptons with respect to the jet axis and can be easily separated. The charmed event sample has to be corrected for wrong sign leptons from b quark decays. The observable bottom quark asymmetry will be reduced by a factor of about 1.25 ± 0.1 due to $B^0\bar{B}^0$ mixing [69].

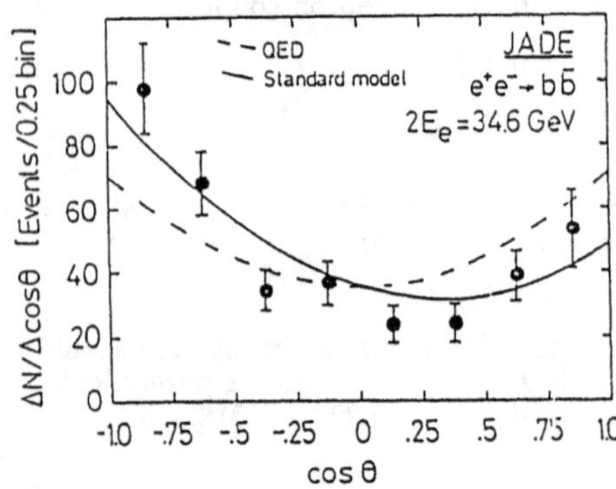

Figure 38: Angular distribution for the reaction $e^+e^- \rightarrow b\bar{b}$ using inclusive muons from JADE at $\sqrt{s} = 34.6\,GeV$. The full curve shows the QFD prediction, while the dashed curve shows the QED expectation

Several experiments on inclusive electron and muon spectra from heavy quark decays have been performed. Fig. 38 shows the b quark angular distribution derived from inclusive muon spectra measured by JADE at $\sqrt{s} = 34.6\,GeV$, the most precise data. A distinct asymmetry of $A_{bb} = -22.8 \pm 6.5\,\%$ is seen in good agreement with the QFD prediction of $A_{b\bar{b}}^{QFD} = -25.6\,\%$ neglecting $B^0\bar{B}^0$ mixing. The asymmetry is about 3 times larger as for μ pairs at the same centre of mass energy, reflecting the charge $Q_b = -1/3$ of the b quark.

A more direct method to tag heavy quarks is to reconstruct the parent mesons. At high energies this has so far only been accomplished for charmed mesons produced via the following chain

$$\begin{aligned} e^+e^- \;\rightarrow\; & c\bar{c} \\ & \hookrightarrow D^{*+}X \\ & \qquad \hookrightarrow D^0\pi^+ \\ & \qquad\qquad \hookrightarrow K^-\pi^+,\; K^-\pi^+\pi^0,\; K^-\pi^+\pi^+\pi^-,\; \ldots \end{aligned}$$

The experimental resolution is considerably improved by making use of the small mass difference $\Delta m = m_{D^{*+}} - m_{D^0}$, which can be measured much more accurately than the masses of the D^{*+} or D^0 mesons themselves. The event samples are almost free of background, the main source being b decays, but are statistically limited by the small branching ratios of low multiplicity hadronic D meson decays. Fig. 39 shows the angular distribution of D^* mesons measured by HRS at $\sqrt{s} = 29\,GeV$, which is the only experiment capable of identifying the decays $D^0 \rightarrow K^-\pi^+$ and $D^+ \rightarrow K^-\pi^+\pi^+$ directly without the above mentioned kinematical trick. The D^* asymmetry is $-6.1 \pm 3.9\,\%$ to be compared with the QFD prediction of $-8.6\,\%$.

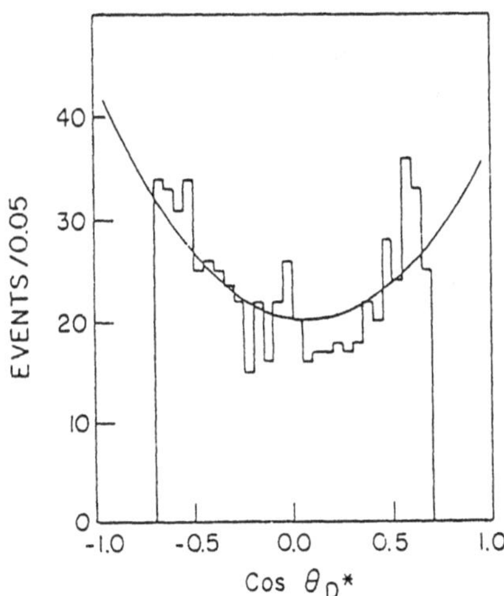

Figure 39: Angular distribution of D^* mesons in the reaction $e^+e^- \rightarrow c\bar{c}$ from HRS at $\sqrt{s} = 29\,GeV$. The curve shows the measured asymmetry

Results on the axial vector couplings of the charm and bottom quarks deduced from the asymmetries of the most accurate experiments are compiled in Tables 12 and 13 (assuming $g_A^e = -1/2$ for the electron axial vector charge). The analysis methods are listed as well. With a few exceptions the signals of individual measurements are generally poor. The averaged values of the heavy quark axial vector couplings, however, have a significance of about 5 standard deviations and definitely support the standard theory predictions of $g_A^c = 1/2$ and $g_A^b = -1/2$. Note that the measured

Table 12: Results on the charm quark axial vector coupling g_A^c from the reaction $e^+e^- \to c\bar{c}$

Experiment	Method	g_A^c
HRS [70]	D^*, D	0.58 ± 0.16
TPC [71]	e, μ, D^*	0.95 ± 0.50
JADE [72]	D^*	0.55 ± 0.25
MARK J [36]	μ	0.60 ± 0.30
TASSO [73]	D^*	0.55 ± 0.15
Average		0.58 ± 0.09

Table 13: Results on the bottom quark axial vector coupling g_A^b from the reaction $e^+e^- \to b\bar{b}$. The MARK II and JADE data are not corrected for $B^0\bar{B}^0$ mixing

Experiment	Method	g_A^b
MARK II [74]	e, μ	-0.75 ± 0.38
CELLO [75]	e, μ	-0.60 ± 0.22
JADE [76]	μ	-0.45 ± 0.13
TASSO [77]	e, μ, jet charge	-0.62 ± 0.25
Average		-0.53 ± 0.10

axial vector coupling of the bottom quark requires that it is a member of a weak isospin doublet. This is the most significant argument that the still undiscovered top quark has to exist.

If no flavour tagging is possible, as e.g. for the light quarks, one can try to measure the elctroweak induced charge asymmetry of all quarks simultaneously. The asymmetries of the positively charged quarks and antiquarks tend to cancel each other and the residual asymmetry summed over 5 quarks is expected to have a very small value of

$$< A_q > \;=\; -\frac{3}{11} A_{\mu\mu} \, .$$

The charge of the primary leading quark should be reflected in the jet charge, which is the sum of all particle charges in a jet weighted with some function according to their momenta, favouring high momentum tracks. The axial vector couplings can then be deduced by a comparison with model calculations, but requires high statistics and excellent understanding of the detector as well as of the underlying hadronization model.

The first analysis of this type was carried out by MAC [78] at a centre of mass energy of $\sqrt{s} = 29\,GeV$. The jet angular distribution is shown in Fig. 40. The observed asymmetry is small but significant, which is best seen after subtracting the symmetric QED term and plotting the term linear in $\cos\theta$. The measured asymmetry is $A_{jet} = +2.8 \pm 0.5\,\%$ while the expected asymmetry including detector effects is $+2.0 \pm 0.5\,\%$. Similar analyses have been done subsequently by TASSO [79] and JADE [80] at 35 and 44 GeV.

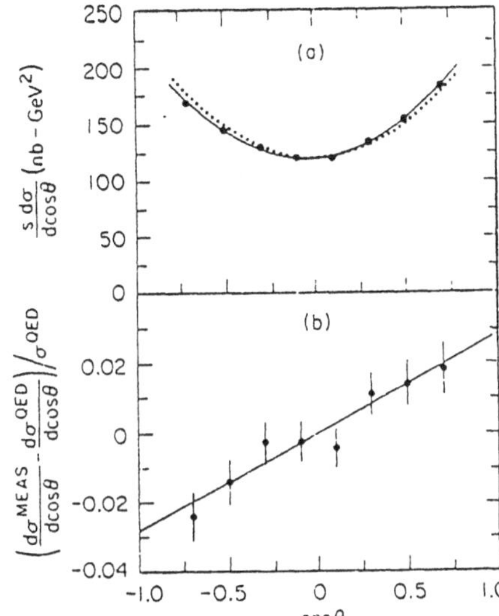

Figure 40: Jet angular distribution for the reaction $e^+e^- \to q\bar{q}$ from MAC at $\sqrt{s} = 29\,GeV$. The polar angle is measured between the positive jet and the e^- direction. **a**) Differential cross section, **b**) difference between measurement and QED divided by QED. The full curves are the QFD prediction, the dashed curve is the QED prediction

The combined data of the three experiments yield for the average quark axial vector coupling

$$< g_A^q > \;=\; 0.62 \pm 0.10$$

in good agrement with QFD, where $g_A^q = g_A^{up} = -g_A^{down}$ has been assumed for the up and down type quarks. This is clear evidence that also the light u, d, s quarks carry the axial vector charges assigned to them in the standard theory.

8.3 Total hadronic cross section

The total hadronic cross section normalized to the asymptotic QED cross section for lepton pair production neglecting quark masses can be written as

$$R_{had} \;=\; \sum_q R_{q\bar{q}} \left\{ 1 + \frac{\alpha_s}{\pi} + 1.41 \left(\frac{\alpha_s}{\pi}\right)^2 + 64.8 \left(\frac{\alpha_s}{\pi}\right)^3 + \ldots \right\} . \qquad (49)$$

The sum runs over all quark flavours. The notation is taken from section 3.2 with the electroweak part

$$R_{q\bar{q}} = 3\left\{Q_q^2 - 2g_V^e g_V^q Q_q \Re(\chi) + ({g_V^e}^2 + {g_A^e}^2)({g_V^q}^2 + {g_A^q}^2)|\chi|^2\right\}.$$

The factor 3 accounts for the colour and the function χ contains the Z^0 propagator (eqs. (19) or (20)). The terms in curly brackets of eq. (49) represent the QCD corrections due to gluon radiation to third order in the strong coupling constant α_s in the $\overline{MS}$ renormalization scheme assuming 5 quarks.

The bulk of the cross section is given by QED quark pair production, in the naive quark parton model (QPM) $R^{QPM} = 3\sum Q_q^2 = 11/3$ for energies well above the bottom threshold. The QCD corrections decrease with rising energy as a consequence of the running strong coupling constant. The total hadronic cross section offers a clean possibility to measure α_s since the data do not depend on any fragmentation model and the QCD dependence is theoretically calculable. However, recent calculations have shown that the third order corrections are comparable to the second order corrections. The electroweak contributions increase with energy. The QCD and QFD terms have opposite energy behaviours and can thus be separated with very little correlation.

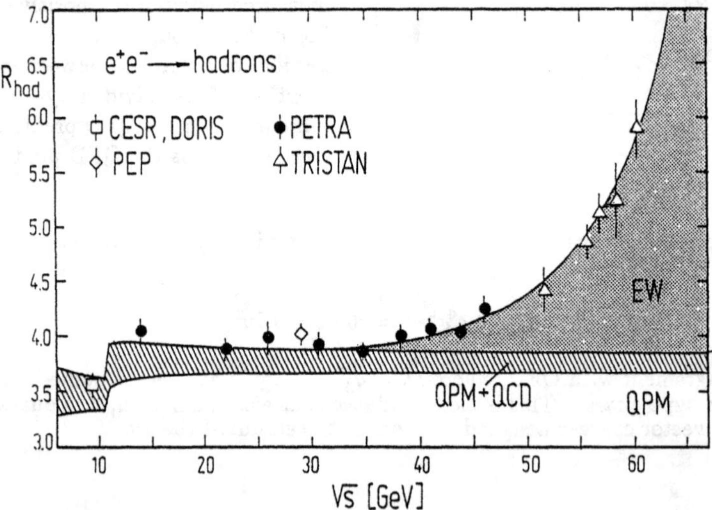

Figure 41: The total hadronic cross section R_{had} as a function of the centre of mass energy. The errors contain statistical and systematic uncertainties and their correlations. The curves represent the fits to the data

The combined total hadronic cross section measurements are shown in Fig. 41 as a function of the centre of mass energy. The rise in R_{had} observed at PETRA between 35 and 46.8 GeV was the first evidence for a direct purely weak Z^0 pole contribution, i.e. not a $\gamma - Z^0$ interference effect, in e^+e^- experiments. This observation is verified by the TRISTAN measurements [84], showing a 50% increase in R_{had} at $\sqrt{s} = 60\,GeV$. The CELLO collaboration [81] and Marshall [82] have performed an analysis using all available cross section data and taking the individual systematic errors and known correlations properly into account. Much care has been given to apply the same higher order radiative corrections to all data. The analysis was recently repeated by [83] including the latest TRISTAN results. A 3 paramter fit to the total hadronic cross section data was performed to determine α_s, $\sin^2\theta_W$ and m_Z (assuming a top mass $m_t = 60\,GeV$). The QPM, QCD and electroweak (EW) contributions to the fit are shown in Fig. 41. The strong coupling constant was found to be $\alpha_s(\sqrt{s} = 34\,GeV) = 0.141 \pm 0.019$. The electroweak parameters are

$$\begin{aligned} \sin^2\theta_W &= 0.221\,^{+0.029}_{-0.022}\,, \\ m_Z &= 89.4 \pm 1.9\,GeV\,. \end{aligned}$$

The result on the weak mixing angle $\sin^2\theta_W$, determined at momentum transfers up to $\sim 3500\,GeV^2$, is in very good agreement with the values found in deep inelastic neutrino nucleon scattering and intermediate vector boson production. It is also in remarkable agreement with neutrino electron scattering, polarized electron deuteron scattering and atomic parity violation experiments [5], demonstrating the universality of the weak neutral current coupling over many orders of magnitude in the momentum transfer squared. The mass of the Z^0 turns out to be somewhat lower than the values obtained by the $p\bar{p}$ collider experiments at CERN and FNAL. A definite precise answer will be provided soon by the SLC and LEP experiments.

Finally the total cross section data can be used to look for quark substructures, similarly as done for lepton pair production (section 7). Assuming the QFD and QCD terms are given by the previous parameters, any further deviation can be described by a QED like form factor modifying the normalized total hadronic cross section as

$$R'_{had} = R_{had}\left\{1 \mp \frac{s}{s - \Lambda_\pm^2}\right\}^2 . \tag{50}$$

From a fit to the data lower limits (95% confidence level) on the QED cut-off parameters of $\Lambda_+ > 435\,GeV$ and $\Lambda_- > 335\,GeV$ have been obtained [85].

8.4 Beyond the standard theory

The electroweak standard theory has been very successful in describing the e^+e^- experiments to a high degree of precision. No new particles or interactions have been found. Limits on the production of new sequential leptons and quarks, supersymmetric particles, neutral and charged Higgs particles have been set with lower limits on the masses close to or beyond the $e^\pm$ beam energy (see refernces in [18]). The known leptons and quarks behave as pointlike particles with no substructure beyond $\sim 10^{-16}\,cm$.

However, it is still conceivable that the electroweak theory is the low energy limit of a more fundamental interaction observable only at very high energy. If 'new physics' would exist at a mass scale Λ^C much larger than the presently accessible momentum transfers Q, the interference with the γ and Z^0 fields would lead to residual contact interactions which may result in deviations from the standard model. A general parametrization of such contact terms, originally intended for substructure or compositeness, has been proposed by [86]. Assuming helicity conservation an effective Lagrangian of the form

$$\mathcal{L}_{eff} = \pm \frac{g^2}{2\,\Lambda^{C\,2}_{\pm}} \{\eta_{LL}\, j_L\, j_L + \eta_{RR}\, j_R\, j_R + 2\,\eta_{LR}\, j_L\, j_R\} \tag{51}$$

has to be added to the neutral current QFD interaction Lagrangian $\mathcal{L}_I^{NC}$ of section 2. The coupling constant is arbitrarily fixed to $g^2/4\pi = 1$, leaving the mass scale Λ^C as the only free parameter. j_L and j_R are left-handed and right-handed currents. The coefficients η are chosen as 0 or ± 1, allowing to construct left-handed, right-handed, vector and axial vector currents. For the LL coupling $\eta_{LL} = 1$, $\eta_{RR} = \eta_{LR} = 0$, for the RR coupling $\eta_{RR} = 1$, $\eta_{LL} = \eta_{LR} = 0$, for the VV coupling $\eta_{LL} = \eta_{RR} = \eta_{LR} = 1$, and for the AA coupling $\eta_{LL} = \eta_{RR} = -\eta_{LR} = 1$.

A particularly simple example for compositeness is Bhabha scattering. Since initial and final particles are the same no assumptions on the constitutents have to be made. The differential cross section for the reaction $e^+e^- \to e^+e^-$ becomes

$$\begin{aligned}
\frac{d\sigma}{d\Omega} &= \frac{\alpha^2}{8\,s}\left\{4\,B_1 + B_2\,(1-\cos\theta)^2 + B_3\,(1+\cos\theta)^2\right\}\,, \qquad (52)\\
B_1 &= \left(\frac{s}{t}\right)^2 \left|1 + (g_V{}^2 - g_A{}^2)\,\chi(t) \pm \frac{\eta_{LR}\, t}{\alpha\, \Lambda^{C\,2}_{\pm}}\right|^2 ,\\
B_2 &= \left|1 + (g_V{}^2 - g_A{}^2)\,\chi(s) \pm \frac{\eta_{LR}\, s}{\alpha\, \Lambda^{C\,2}_{\pm}}\right|^2 ,\\
B_3 &= \frac{1}{2}\left|1 + \frac{s}{t} + (g_V + g_A)^2\,(\frac{s}{t}\,\chi(t) + \chi(s)) \pm \frac{2\,\eta_{RR}\, s}{\alpha\, \Lambda^{C\,2}_{\pm}}\right|^2\\
&\quad + \frac{1}{2}\left|1 + \frac{s}{t} + (g_V - g_A)^2\,(\frac{s}{t}\,\chi(t) + \chi(s)) \pm \frac{2\,\eta_{LL}\, t}{\alpha\, \Lambda^{C\,2}_{\pm}}\right|^2 .
\end{aligned}$$

The cross section formula for s channel lepton pair production can be simply obtained by setting all terms proprtional to s/t to zero and dropping the factor 2 in front of η_{RR} and η_{LL}. For quarks the fractional charges have to be considered.

The sensitivity of the Bhabha and μ pair differential cross sections on various values of the cut-off parameters Λ^C are shown in Fig. 42 for TASSO data at $\sqrt{s} = 34.8$ GeV. The cross sections have been normalized to the QFD prediction. The angular dependences on the chirality structures are markedly different for the two processes.

Typical values of lower limits on the scale parameters Λ^C for leptons and quarks are summarized in Table 14. They are of the order of a few TeV depending on the chirality structure. The traditional QED form factor is a special case of the vector type contact interaction, related by $\Lambda^{QED} \simeq \sqrt{\alpha}\,\Lambda^{VV}$. Note that LL and RR interactions cannot be distinguished in leptonic reactions. This is different for quarks, where up and down type quarks couple differently to left-handed and right-handed currents. Summing over five quarks therefore leads to different LL and RR coupling limits. All quark flavours are assumed to contribute with the same Λ^C parameter. However, substructures need not to show up at the same scale for all quark families. If only the u and d quarks are allowed to have a structure while the heavier quarks behave still pointlike the limits are slightly less stringent.

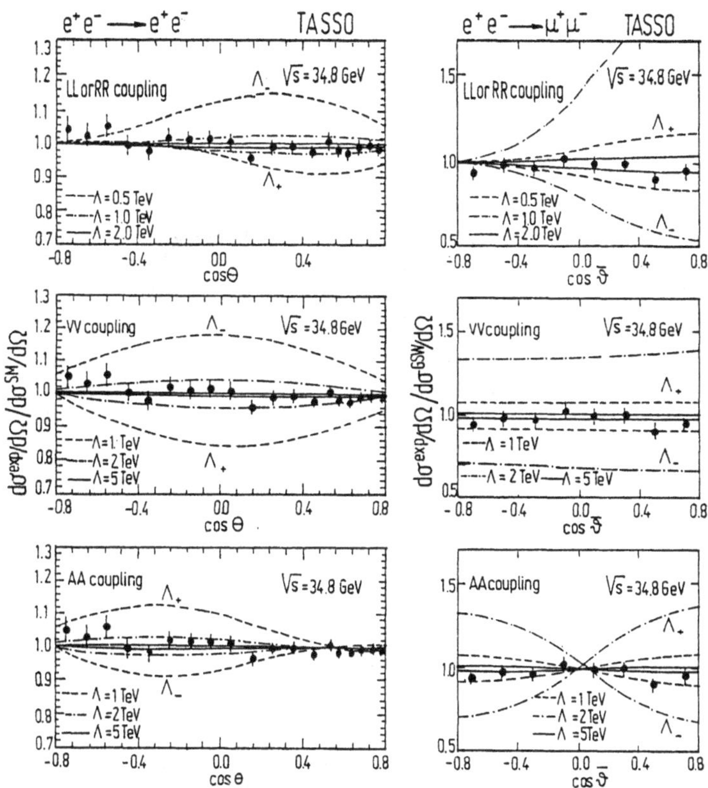

Figure 42: The differential cross sections normalized to QFD for the reations $e^+e^- \to e^+e^-$ and $e^+e^- \to \mu^+\mu^-$ from TASSO at $\sqrt{s} = 34.8\,GeV$. The curves show possible contributions of contact terms for left-handed (LL) or right-handed (RR) coupling, vector (VV) coupling, and axial vector (AA) coupling

Table 14: Lower limits (95% confidence level) of mass scale parameters Λ^C $[TeV]$ in contact interactions for left-handed (LL), right-handed (RR), vector (VV), and axial vector (AA) couplings

Reaction	Λ_+^{LL}	Λ_-^{LL}	Λ_+^{RR}	Λ_-^{RR}	Λ_+^{VV}	Λ_-^{VV}	Λ_+^{AA}	Λ_-^{AA}
$e^+e^- \to e^+e^-$ [38]	1.4	3.3	1.4	3.3	3.6	7.1	2.8	2.4
$e^+e^- \to \mu^+\mu^-$ [47]	2.3	1.3	2.3	1.3	3.5	1.8	3.2	2.9
$e^+e^- \to \tau^+\tau^-$ [49]	2.2	3.2	2.2	3.2	2.7	5.7	4.1	5.7
$e^+e^- \to q\bar{q}$ [85]	1.2	1.8	3.4	2.7	5.1	3.9	1.7	2.4

9 Conclusion

The impressive amount of data accumulated by high energy electron–positron collision experiments at PETRA, PEP and TRISTAN during the past decade allows to test the fundamental interactions very precisely. Tests of Quantum Electrodynamics, the paradigm of all field theories, at unprecedented high time-like momentum transfers of up to $s \sim 3000\,GeV^2$ confirm its validity to $\mathcal{O}(\alpha^3)$ and $\mathcal{O}(\alpha^4)$ within a few percent, but also clearly unvealed the presence of electroweak effects in lepton and hadron final states. The quality of the QED tests gives confidence in extracting the electroweak parameters of the standard theory. The embracing theory Quantum Flavourdynamics provides a coherent description of all observations at the ten percent level. The e^+e^- experiments contribute in particular information on the heavy lepton and quark families which cannot be obtained in other experiments.

The main achievements of electron–positron collisions with respect to QED and QFD tests are:

- The reaction $e^+e^- \to \gamma\gamma$, the only process where QED can be tested as a self consistent theory, has been measured very precisely and confirms QED including radiative corrections to $\mathcal{O}(\alpha^3)$ at the one percent level. QED cut-off parameters exceeding $\Lambda > 80\,GeV$ have been derived.

- The total cross sections for lepton pair production $e^+e^- \to e^+e^-$, $e^+e^- \to \mu^+\mu^-$ and $e^+e^- \to \tau^+\tau^-$ have been measured at the level of a few percent and agree with $\mathcal{O}(\alpha^3)$ QED calculations. QED cut-off parameters larger than $\Lambda > 250\,GeV$ have been obtained, indicating that all three leptons behave as pointlike particles down to a distance of about $10^{-16}\,cm$.

- Hard photon radiation with one and two additional photons accompanying the basic two particle final states has been studied and verifies the $\mathcal{O}(\alpha^3)$ and $\mathcal{O}(\alpha^4)$ QED calculations to better than 10%. It has to be stressed that large efforts have been devoted to the development of Monte Carlo event generators including the higher order virtual and bremsstrahlung radiative corrections, which are able to reproduce the experimental results reliably.

- Photon–photon collisons $e^+e^- \rightarrow e^+e^-e^+e^-$ and $e^+e^- \rightarrow e^+e^-\mu^+\mu^-$ have been investigated and show that the total cross section for untagged $e^+e^- \rightarrow e^+e^-\mu^+\mu^-$ events becomes larger than the annihilation cross section for $e^+e^- \rightarrow \mu^+\mu^-$ at high energy. Although the momentum transfers are moderate compared to the annihilation processes photon–photon collisions demonstrate the nice agreement with $\mathcal{O}(\alpha^4)$ QED calculations.

- Electroweak $\gamma - Z^0$ interference effects have been observed in lepton and quark production. Sizeable forward backward charge asymmetries have been established for the reactions $e^+e^- \rightarrow \mu^+\mu^-$ and $e^+e^- \rightarrow \tau^+\tau^-$. The electroweak parameters of the three charged leptons agree with each other and with the standard theory and support the concept of $e - \mu - \tau$ lepton universality. The axial vector charges of the heavy quarks as well as of the light quarks have been investigated and found to be in good agreement with the QFD assignments.

- The definite rise of the normalized total hadronic cross section at centre of mass energies above 35 GeV is the first evidence for direct contributions from the Z^0 resonance in e^+e^- collisions. The electroweak mixing angle derived from these measurements is comparable and competitive to the values found in other experiments.

The next generation of e^+e^- experiments at SLC and LEP just start their physics program at energies around the Z^0 resonance. They aim at precision tests of the electroweak theory, which rely sensitively on the understanding of higher order electroweak radiative corrections. QED can therefore only be seen in the context of QFD and QCD. The only area which will remain a domain of QED for quite some time will be photon pair production $e^+e^- \rightarrow \gamma\gamma$ and lepton pair production by photon–photon collisions $e^+e^- \rightarrow e^+e^-l^+l^-$.

Acknowledgements. I would like to thank my colleagues from the TASSO collaboration and from the other experiments at PETRA and PEP for the numerous dicussions which helped in understanding the physics of electron–positron collisions. My special thanks go to D. Haidt for critical reading of the manuscript.

References

[1] D. Haidt and H. Pietschmann. *Electroweak Interactions.* Volume I/10 of *Landolt - Börnstein, New Series*, Springer Verlag, Berlin Heidelberg, 1988.

[2] S.L. Glashow. *Nucl. Phys.* **22** (1961) 579 and *Rev. Mod. Phys.* **52** (1980) 539.

[3] A. Salam. *Phys. Rev.* **127** (1962) 331 and *Rev. Mod. Phys.* **52** (1980) 525.

[4] S. Weinberg. *Phys. Rev. Lett.* **19** (1967) 1264 and *Rev. Mod. Phys.* **52** (1980) 515.

[5] P. Langacker. Proc. xxiv Int. Conf. on High Energy Physics, Munich (1988), 190.

[6] F. M. Renard. *Basics of Electron Positron Collisions.* Edition Frontières, Gif sur Yvette, France, 1981.

[7] S.L. Glashow, J. Iliopoulos, and L. Maiani. *Phys. Rev.* **D2** (1970) 1285.

[8] Particle Data Group. *Phys. Lett.* **204B** (1988) 1.

[9] R. Budny. *Phys. Lett.* **55B** (1975) 227.

[10] F.A. Berends and A. Böhm. Lepton pair production, radiative corrections and electroweak parameters. In A. Ali and P. Söding, editors, *High Energy Electron–Positron Physics*, World Scientific Publishing, Signapore, 1988.

[11] M. Böhm and W. Hollik. *Phys. Lett.* **139B** (1984) 213.

[12] F.A. Berends and R. Kleiss. *Nucl. Phys.* **B186** (1981) 22.

[13] F.A. Berends, R. Kleiss, and S. Jadach. *Nucl. Phys.* **B202** (1982) 63.

[14] S. Jadach and Z. Was. *Acta Phys. Pol.* **B15** (1984) 1151.

[15] F.A. Berends and R. Kleiss. *Nucl. Phys.* **B228** (1983) 537.

[16] F.A. Berends, P.H. Daverveldt, and R. Kleiss. *Comp. Phys. Comm.* **40** (1986) 271, 285, 309.

[17] V.M. Budnev, I.F. Ginzburg, G.V. Meledin, and V.G. Serbo. *Phys. Rep.* **15** (1975) 181.

[18] A. Ali and P. Söding, editors. *High Energy Electron–Positron Physics.* Volume 1 of *Advanced Series on Directions in High Energy Physics*, World Scientific Publishing, Signapore, 1988.

[19] S.D. Drell. *Ann. Phys.* **41** (1958) 75.

[20] A. Litke. Ph.D. Thesis, Harvard University, unpublished (1970).

[21] HRS Coll., M. Derrick *et al.* *Phys. Lett.* **166B** (1986) 468.

[22] HRS Coll., M. Derrick *et al.* *Phys. Rev.* **D34** (1986) 3286.

[23] MAC Coll., E. Fernandez *et al.* *Phys. Rev.* **D35** (1987) 1.

[24] CELLO Coll., H.J. Behrend *et al.* *Phys. Lett.* **168B** (1986) 420.

[25] JADE Coll., W. Bartel *et al.* *Z. Phys.* **C19** (1983) 197.

[26] MARK J Coll., B. Adeva *et al.* *Phys. Rev. Lett.* **53** (1984) 134.

[27] PLUTO Coll., C. Berger *et al.* *Phys. Lett.* **94B** (1980) 87.

[28] TASSO Coll., M. Althoff *et al.* *Z. Phys.* **C26** (1984) 337.

[29] AMY Coll., S.K. Kim *et al.* *Phys. Lett.* **223B** (1989) 476.

[30] TOPAZ Coll., I. Adachi *et al.* *Phys. Lett.* **200B** (1988) 391.

[31] VENUS Coll., K. Abe *et al.* subm. to *Z. Phys.* **C** (1989).

[32] HRS Coll., M. Derrick *et al.* *Phys. Lett.* **166B** (1986) 463.

[33] MAC Coll., E. Fernandez *et al.* *Phys. Rev.* **D35** (1987) 10.

[34] CELLO Coll., H.J. Behrend *et al.* *Phys. Lett.* **103B** (1981) 148.

[35] JADE Coll., W. Bartel *et al.* *Z. Phys.* **C30** (1986) 371.

[36] MARK J Coll., B. Adeva *et al.* *Phys. Rep.* **109** (1984) 133.

[37] PLUTO Coll., C. Berger *et al.* *Z. Phys.* **C27** (1985) 341.

[38] TASSO Coll., W. Braunschweig *et al.* *Z. Phys.* **C37** (1988) 171.

[39] Y. Unno. Proc. xxiv Int. Conf. on High Energy Physics, Munich (1988), 860.

[40] HRS Coll., M. Derrick *et al.* *Phys. Rev.* **D31** (1985) 2352.

[41] HRS Coll., K.K. Gan *et al.* *Phys. Lett.* **153B** (1985) 116.

[42] MAC Coll., D. Bender *et al.* *Phys. Rev.* **D30** (1985) 515.

[43] CELLO Coll., H.J. Behrend *et al.* *Phys. Lett.* **183B** (1987) 400.

[44] CELLO Coll., H.J. Behrend *et al.* *Phys. Lett.* **114B** (1982) 282.

[45] MARK J Coll., B. Adeva *et al.* *Phys. Lett.* **179B** (1986) 177.

[46] PLUTO Coll., L. Criegee and G. Knies. *Phys. Rep.* **83** (1982) 153.

[47] TASSO Coll., W. Braunschweig *et al.* *Z. Phys.* **C40** (1988) 163.

[48] TASSO Coll., W. Braunschweig *et al.* DESY 89-035, subm. to *Z. Phys.* C.

[49] B. Naroska. *Phys. Rep.* **148** (1986) 67.

[50] CELLO Coll., H.J. Behrend *et al.* *Phys. Lett.* **158B** (1985) 536.

[51] MAC Coll., W.T. Ford *et al.* *Phys. Rev. Lett.* **51** (1983) 257.

[52] MARK J Coll., B. Adeva *et al.* *Phys. Rev.* **D38** (1988) 2665.

[53] MARK J Coll., B. Adeva *et al.* *Phys. Rev. Lett.* **55** (1985) 665.

[54] MAC Coll., E. Fernandez *et al.* *Phys. Rev.* **D28** (1983) 2721.

[55] PEP-9 $\gamma\gamma$ Coll., M.P. Cain *et al.* *Phys. Lett.* **147B** (1984) 232.

[56] CELLO Coll., H.J. Behrend *et al.* *Z. Phys.* **C43** (1989) 1.

[57] JADE Coll., W. Bartel *et al.* *Z. Phys.* **C30** (1986) 545.

[58] PLUTO Coll., C. Berger *et al.* *Z. Phys.* **C27** (1985) 249.

[59] Ch. Berger and W. Wagner. *Phys. Rep.* **146** (1987) 1.

[60] JADE Coll., W. Bartel *et al.* *Phys. Lett.* **161B** (1985) 188.

[61] PLUTO Coll., C. Berger *et al.* *Z. Phys.* **C28** (1985) 1.

[62] J.G. Branson. Proc. Int. Symp. on Lepton and Photon Interactions, Bonn (1981), 279.

[63] MAC Coll., W.W. Ash *et al.* *Phys. Rev. Lett.* **55** (1985) 1831.

[64] MARK II Coll., M.E. Levi *et al.* *Phys. Rev. Lett.* **51** (1983) 1941.

[65] CELLO Coll., H.J. Behrend *et al.* *Phys. Lett.* **191B** (1987) 209.

[66] PLUTO Coll., C. Berger *et al.* *Z. Phys.* **C21** (1983) 53.

[67] S.L. Olsen. Proc. xxiv Int. Conf. on High Energy Physics, Munich (1988), 868.

[68] MAC Coll., E. Fernandez *et al.* *Phys. Rev. Lett.* **54** (1985) 1620.

[69] ARGUS Coll., H. Albrecht *et al.* *Phys. Lett.* **192B** (1987) 245.

[70] HRS Coll., P. Baringer *et al.* *Phys. Lett.* **206B** (1988) 551.

[71] TPC Coll., H. Aihara *et al.* *Phys. Rev.* **D34** (1986) 1945.

[72] JADE Coll., F. Ould-Saada *et al.* DESY 89-063, subm. to *Z. Phys.* C.

[73] TASSO Coll., W. Braunschweig *et al.* DESY 89-053, subm. to *Z. Phys.* C.

[74] N.S. Lockyer. SLAC-PUB 3245 (1983).

[75] C. Kiesling. Proc. xxiv ème Rencontre de Moriond (1989).

[76] JADE Coll., W. Bartel *et al.* *Phys. Lett.* **146B** (1984) 437.

[77] P. Rehders. Ph.D. Thesis, Hamburg University, unpublished (1989).

[78] MAC Coll., W.W. Ash *et al.* *Phys. Rev. Lett.* **58** (1987) 1080.

[79] TASSO Coll., W. Braunschweig *et al.* *Z. Phys.* **C41** (1988) 385.

[80] JADE Coll., T. Greenshaw *et al.* *Z. Phys.* **C42** (1989) 1.

[81] CELLO Coll., H.J. Behrend *et al.* *Phys. Lett.* **183B** (1987) 400.

[82] R. Marshall. *Z. Phys.* **C43** (1989) 607.

[83] G. D'Agostini, W. de Boer, and G. Grindhammer. DESY 89-057.

[84] T. Kamae. Proc. xxiv Int. Conf. on High Energy Physics, Munich (1988), 156.

[85] H.-U. Martyn. Proc. of the HERA workshop, Hamburg, (1987) Vol. 2, 811.

[86] E.J. Eichten, K.D. Lane, and M.E. Peskin. *Phys. Rev. Lett.* **50** (1983) 811.

ANALYTIC EVALUATION OF SIXTH-ORDER CONTRIBUTIONS TO THE ELECTRON'S g FACTOR

Ralph Z. Roskies
Department of Physics and Astronomy
University of Pittsburgh, Pittsburgh, PA 15260

Ettore Remiddi
Department of Physics
Università di Bologna, and INFN, Sezione di Bologna,
Bologna Italy

Michael J. Levine
Department of Physics
Carnegie Mellon University, Pittsburgh, PA 15213

Contents

1 INTRODUCTION

The electron's g factor is the best testing ground for perturbative quantum field theory. The phenomenal agreement between theory and experiment are testament to the validity of the renormalization prescription of quantum electrodynamics (QED). The electron's g factor is the physical parameter which one can both calculate theoretically and measure to the greatest precision. The experiments are accurate to a few parts in 10^{12} [1]. In fact the 1989 Nobel prize in physics was awarded to Professor Hans Dehmelt for these experiments. Although there is no closed form for the theoretical value of the electron's g factor, it can be computed to higher and higher order in perturbation theory. The perturbative parameter is the fine structure constant α(roughly 1/137), and the perturbation series appears to converge quite rapidly. One can think of the comparison between theory and experiment as a test of the whole formalism of quantum electrodynamics. Or else, one can accept the formalism, and use these calculations to obtain very accurate estimates of the fine structure constant. There are good grounds for believing that the perturbation series is only an asymptotic one, rather than a convergent power series [2]. But for comparison with experiment this complication can so far be ignored.

The first term in the expansion, called the second order moment, is the famous $\alpha/(2\pi)$ computed by Schwinger [3] in 1947, which helped establish the formalism of quantum electrodynamics. In modern parlance, this involves just one Feynman diagram (See Fig. 1), and can now be assigned as a homework problem in an introductory quantum field theory course.

The second term of the expansion, called the fourth order moment, involves 5 different Feynman diagrams (Fig. 2).

It was not computed correctly until 1957 [4]. These were intricate calculations, carried out analytically by hand.

The third term, the sixth order moment, is still not known completely analytically.

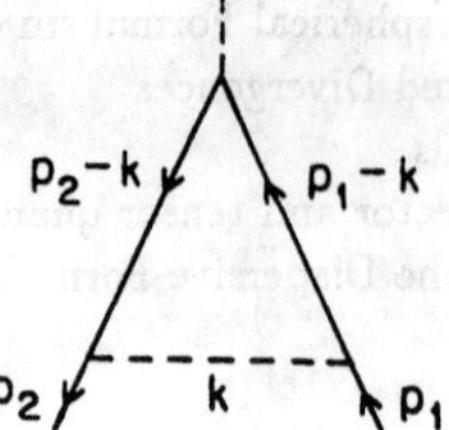

Figure 1: Second order Feynman graph contributing to electron g factor.

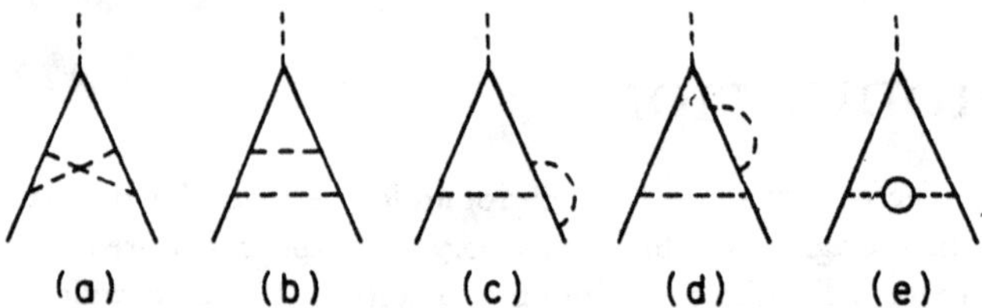

Figure 2: Distinct fourth order Feynman graphs contributing to electron g factor. The contributions of graphs (c) and (d) must be multiplied by 2 to take account of their mirror images.

It involves the sum of 40 different 3 loop Feynman diagrams (Fig. 3). The value of the sixth order moment was first computed numerically by three different groups [5,6,7] in 1973-4. But the estimated error of these numerical integrations exceeded the estimated experimental error of the best measurements. To permit a better test of quantum electrodynamics, it was necessary to improve the numerical accuracy of the calculations. This was achieved primarily over a 10 year period 1972-82, by computing many of the graphs analytically. Some analytic work is still continuing. It is this set of analytic calculations on which we concentrate in this review article. For each graph we indicate in Table 1 which technique has been used in its analytic evaluation and we collect the analytic results for these graphs in Appendix 6.6.

The numerical approaches to the three-loop Feynman diagrams were all carried out using Feynman parameters. These are 7 dimensional integrals, carried out using Monte Carlo or Gaussian integration methods. They were heroic numerical calculations at the time. In Feynman parameter space, the integrand is a rational function, so that the first of the 7 dimensional integrals could easily be done analytically. But the complexity grew rapidly, and no one has seen their way through a complete 7 dimensional analytic calculation using Feynman parameters.

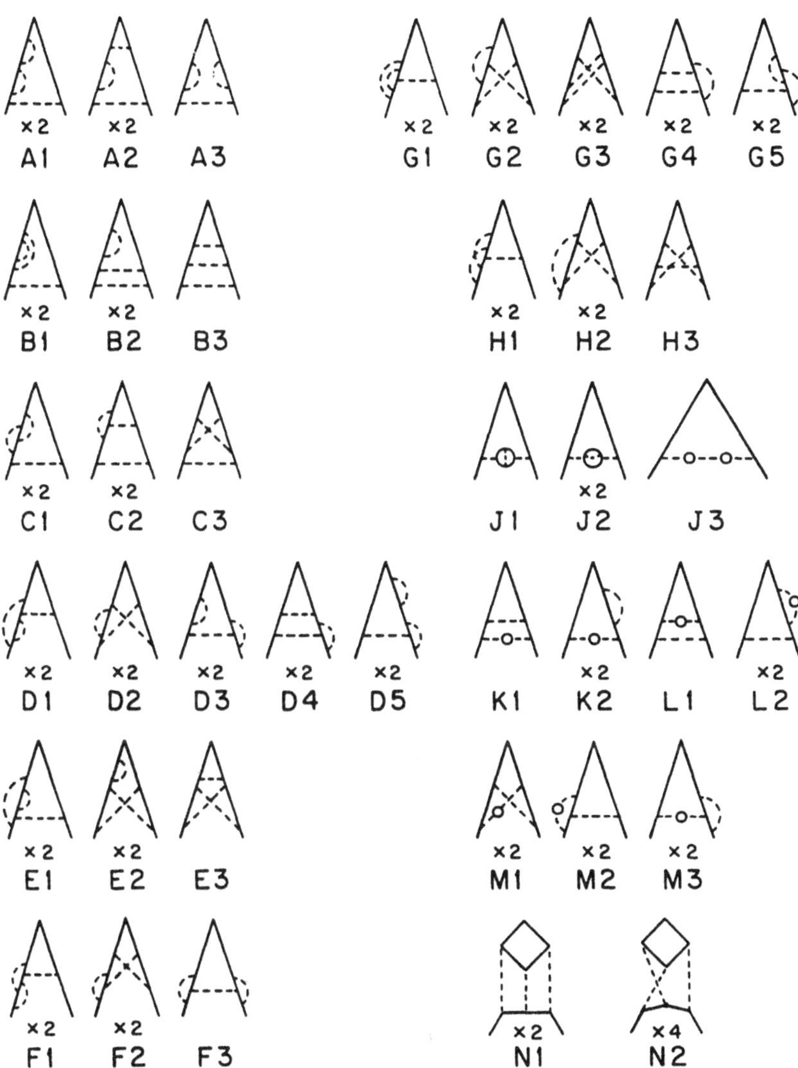

Figure 3: 40 distinct Feynman graphs contributing to the electron g factor in sixth order, organized by self energy sets. The first 28 graphs are labelled as in [6].

TABLE 1: Techniques and references for the analytic evaluation of the Feynman graphs of Figure 3. "Numerical" lists the best numerical evaluation of the graphs. In these cases, no analytic result is known.

Graph Number	Technique	Reference
A1 - A3	Hyperspherical	[12]
	Schwinger Mass Operator	[16]
B1 - B3	Hyperspherical	[12]
C1 - C3	Dispersive	[17,27]
	Hyperspherical	[13]
D1 - D5	Hyperspherical	[13]
E1 - E3	Hyperspherical	[32]
	Dispersive	[27]
F1 - F3	Hybrid	[29]
G1 - G3	Numerical	[33]
H1 - H3	Numerical	[34]
J1 - J3	Dispersive	[22]
K1 - K2	Dispersive	[23,25]
	Schwinger Mass Operator	[35]
L1 - L2	Dispersive	[24,25]
	Schwinger Mass Operator	[35]
M1 - M3	Dispersive	[26]
N1 - N2	Numerical	[36]

The numerical evaluations will always yield a result with the brute force application of more computing time. Analytic evaluations require some insight or trick for their success. The analytic evaluation of the subset of graphs involving vacuum polarizations was done using a dispersive representation for vacuum polarization. Of the 40 graphs contributing to the sixth order moment, 10 (sets J,K,L,M of Fig. 3) involve vacuum polarization. Two of the remaining graphs (set N) involve photon-photon scattering diagrams, also known as light-by-light graphs. To date these two graphs are not known analytically, although work is proceeding. The remaining 28 graphs have no fermion loops. The first subsets of these graphs were evaluated analytically using the hyperspherical formalism, which establishes an ordering of these graphs by difficulty. The first few of them (the easiest)could be done straighforwardly using this formalism. In the course of evaluating the more complicated graphs, it was discovered that the dispersive approach or a hybrid of the dispersive and hyperspherical methods was the more fruitful. The methods we discuss apply in principle to all graphs except the "triple cross" graphs (set H). No method suitable for the analytic calculation of these three graphs has yet been found. The methods we describe apply also to eighth or higher order graphs with the same simple topology.

Apart from the analytic integration problems, the sheer algebraic complications of all these graphs is daunting. Almost all evaluations use algebraic computer techniques just to keep track of the hundreds and even thousands of terms. Even the numerical evaluations used these computer algebra techniques to write down the integrand in Feynman parameter language.

The hyperspherical or dispersive methods reduce the dimension of the integrations to be done to at most three. We will show how this comes about, and how the remaining three dimensional integrals can sometimes be carried out completely analytically. In other cases, the remaining three dimensional integral has only been carried out numerically. Because the dimensionality has been reduced from 7 to 3, accurate numerical evaluation is sometimes easier.

We organize this paper as follows. First we outline the general approach to the magnetic moment, and discuss ultraviolet and infrared divergences. We then present the techniques behind the analytic results. We introduce the hyperspherical formalism, show how it organizes the graphs without fermion loops by topology, and indicate how the analytic evaluation of the simplest of these proceeds. As the graphs become more difficult, we show how the dispersive techiques simplify the calculation. Finally, we indicate how far the mixed hyperspherical-dispersive methods have taken us. In each case, we illustrate the relevant techniques by a complete computation of the second order moment contribution.

More technical questions are relegated to appendices. These include the explicit evaluation of infrared divergences and difficult angular integrals in the hyperspherical formalism, extensions of the dispersive formalism to vector and tensor quantities, the subtleties of infrared divergences in this formalism, and a discussion of polylogarithms, the special functions which occur in the analytic evaluation of these higher order Feynman graphs.

Graph by graph analytic results are collected in the final section of the Appendix.

2 GENERAL FORMALISM

2.1 Magnetic Moment Projection Operators

Let $M_\mu(p_1, p_2)$ be the QED vertex amplitude corresponding to an electron of initial and final four-momenta p_1, p_2. Define further (See Fig. 4)

$$p \equiv \frac{1}{2}(p_1 + p_2), \qquad \Delta \equiv p_1 - p_2, \qquad t \equiv -\Delta^2,$$

$$p_1 = p + \frac{1}{2}\Delta, \qquad p_2 = p - \frac{1}{2}\Delta \ . \tag{2.1.1}$$

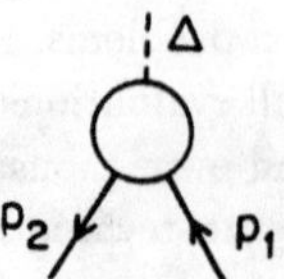

Figure 4: Generic vertex graph.

(We use the metric $\Delta^2 = \vec{\Delta}^2 - \Delta_0^2$). The mass shell conditions for the electron are

$$p_1^2 = p_2^2 = -m^2 \ , \tag{2.1.2}$$

so that

$$(p \cdot \Delta) = 0 \ , \qquad p^2 = \frac{1}{4}(t - 4m^2) \ . \tag{2.1.3}$$

If $u_1, \bar{u}_2$, are the spinors describing the on shell initial and final electron states, on invariance grounds one finds that M_μ can be expressed in full generality in terms of three form factors

$$\bar{u}_2 M_\mu u_1 = u_2 \left[F_1(t)\gamma_\mu - \frac{i}{4m} F_2(t)(\gamma_\mu \not{\Delta} - \not{\Delta}\gamma_\mu) - \frac{i}{m} F_3(t)\Delta_\mu \right] u_1 \ . \tag{2.1.4}$$

Such an expression is somewhat redundant, because $F_3(t)$ in fact vanishes, as required by the conservation of the electromagnetic current (which does not apply, however, to each graph separately)

$$\Delta_\mu \bar{u}_2 M_\mu u_1 = 0 \ . \tag{2.1.5}$$

$F_1(t)$ is known as the charge form factor, $F_2(t)$ as the magnetic form factor; the anomalous magnetic moment of the electron a is given by

$$a \equiv \frac{1}{2}(g-2) = F_2(0). \tag{2.1.6}$$

In order to extract the form factors from Eq.(2.1.4), one can tentatively introduce a projector operator

$$N_\mu = (-i\not{p}_1 + m) \left[g_1 \gamma_\mu + \frac{i}{m} g_2 p_\mu + \frac{i}{m} g_3 \Delta_\mu \right] (-i\not{p}_2 + m) \ , \tag{2.1.7}$$

where the parameters g_i are unspecified, and evaluate the trace of $N_\mu M_\mu$; note that the factors $(-i\not{p}_i + m)$, present in Eq.(2.1.7), guarantee together with Eq.(2.1.2) that the mass shell conditions are enforced. (The convention for the gamma matrices are $\gamma_\mu\gamma_\nu + \gamma_\nu\gamma_\mu = 2\delta_{\mu\nu}, \gamma_\mu^\dagger = \gamma_\mu$).

One finds

$$\mathrm{Tr}\,(N_\mu M_\mu) = \left(4(t+2m^2)g_1 + 2(t-4m^2)g_2\right) F_1(t) + t\left(6g_1 + \frac{t-4m^2}{2m^2}g_2\right) F_2(t)$$

$$+2t(t-4m^2)g_3F_3(t). \tag{2.1.8}$$

Note that at this very first step already the unwanted form factor $F_3(t)$ is decoupled from the other two. With a suitable choice of the coefficients g_i one can extract [8] the two electromagnetic form factors $F_1(t), F_2(t)$ from the general amplitude M_μ,

$$F_1(t) = \frac{1}{4(t-4m^2)}\mathrm{Tr}\left[(-i\not{p}_1+m)\left(\gamma_\mu - i\frac{12m}{t-4m^2}p_\mu\right)(-i\not{p}_2+m)M_\mu\right]$$

$$F_2(t) = \frac{m^2}{t(t-4m^2)}\mathrm{Tr}\left[(-i\not{p}_1+m)\left(-\gamma_\mu + 2i\frac{t+2m^2}{m(t-4m^2)}p_\mu\right)(-i\not{p}_2+m)M_\mu\right]. \tag{2.1.9}$$

When one is interested in the anomaly a only, one can expand M_μ in Eq.(2.1.9) to first order in Δ_μ

$$M_\mu(p,\Delta) \simeq M_\mu(p,0) + \Delta_\nu\frac{\partial}{\Delta_\nu}M_\mu(p,\Delta)|_{\Delta=0} \equiv V_\mu(p) + \Delta_\nu T_{\nu\mu}(p)\ , \tag{2.1.10}$$

average over the spatial directions of Δ, with the formulas

$$\int\frac{d\Omega(\Delta)}{4\pi}\Delta_\mu\Delta_\nu = \frac{1}{3}\Delta^2\left(\delta_{\mu\nu} - \frac{p_\mu p_\nu}{p^2}\right),$$

$$\int\frac{d\Omega(\Delta)}{4\pi}\Delta_\mu = 0\ , \tag{2.1.11}$$

and then take the $t \to 0$ limit; the result is [9]

$$a = \frac{1}{12m^2}\mathrm{Tr}\left[(m^2\gamma_\mu + 3imp_\mu + 4\not{p}p_\mu)V_\mu(p) - \frac{i}{4}(-i\not{p}+m)(\gamma_\mu\gamma_\nu - \gamma_\nu\gamma_\mu)(-i\not{p}+m)T_{\nu\mu}(p)\right], \tag{2.1.12}$$

to be evaluated at $p^2 = -m^2$.

2.2 Second Order Example

Let us denote by $\left(\frac{\alpha}{\pi}\right)M_\mu^{(2)}$ the vertex graph amplitude of second order in the electron charge in the perturbative expansion of QED, and the corresponding second order form factors by

$$\left(\frac{\alpha}{\pi}\right)\bar{u}_2M_\mu^{(2)}u_1 = \left(\frac{\alpha}{\pi}\right)u_2\left[F_1^{(2)}(t)\gamma_\mu - \frac{i}{4m}F_2^{(2)}(t)(\gamma_\mu\not{\Delta} - \not{\Delta}\gamma_\mu)\right]u_1\ . \tag{2.2.1}$$

One has from perturbation theory (See Fig. 1)

$$\left(\frac{\alpha}{\pi}\right) M_\mu^{(2)}(p,\Delta) = e^2\, i \int \frac{d^4k}{(2\pi)^4}\, \gamma_\nu \frac{-i\not{p}_2 + i\not{k} + m}{(p_2-k)^2+m^2-i\epsilon} \gamma_\mu \frac{-i\not{p}_1 + i\not{k} + m}{(p_1-k)^2+m^2-i\epsilon} \gamma_\nu$$

$$\times \left(\frac{1}{k^2+\lambda^2-i\epsilon} - \frac{1}{k^2+\Lambda^2-i\epsilon}\right) . \qquad (2.2.2)$$

Note the in Eq.(2.2.2) we have used, instead of the simple photon propagator

$$\frac{1}{k^2-i\epsilon} , \qquad (2.2.3)$$

its regularized version

$$\frac{1}{k^2+\lambda^2-i\epsilon} - \frac{1}{k^2+\Lambda^2-i\epsilon} . \qquad (2.2.4)$$

The use of Eq.(2.2.3) would give in fact a meaningless k integral, diverging both for small k (infrared divergence) and for large k (ultraviolet divergence), which becomes however mathematically well defined with the regularized propagator Eq.(2.2.4); λ, to be considered small, is a fictitious "photon mass" which regularizes the infrared divergences, while Λ, to be considered larger than any other mass or energy, is the Pauli-Villars mass which regularizes the ultraviolet divergences. Neither λ nor Λ can appear in the final result for any physical quantity, such as the anomaly a (but it is to be recalled that the charge form factor $F_1(t)$ is infrared divergent): the λ dependence disappears when all the contributions to a physical quantity are suitably combined, while the Λ dependence is removed by renormalization. When trying to evaluate analytically some quantities, it can happen that different contributions must be integrated by different techniques, so that the infrared or ultraviolet regularizing parameters must sometimes be kept in the intermediate steps of the calculations.

By using Eq.(2.1.9) with $M_\mu^{(2)}$ given by Eq.(2.2.2), one finds

$$\left(\frac{\alpha}{\pi}\right) F_i^{(2)}(t) = e^2\, i \int \frac{d^4k}{(2\pi)^4} \frac{1}{(p_1-k)^2+m^2-i\epsilon} \frac{1}{(p_2-k)^2+m^2-i\epsilon}$$

$$\times \left(\frac{1}{k^2+\lambda^2-i\epsilon} - \frac{1}{k^2+\Lambda^2-i\epsilon}\right) f_i^{(2)}(k) , \qquad (2.2.5)$$

with

$$f_1^{(2)}(k) = 4m^2 - t + \frac{2}{t-4m^2}\left(4m^2k^2 + (\Delta.k)^2 + 4(t-2m^2)(p.k)\right) - 8\frac{t+8m^2}{(t-4m^2)^2}(p.k)^2 ,$$

$$f_2^{(2)}(k) = -\frac{8m^2}{t-4m^2}\left(2(p.k) + k^2 + \frac{1}{t}(\Delta.k)^2\right) + 96\frac{m^2(p.k)^2}{(t-4m^2)^2} . \qquad (2.2.6)$$

$F_2^{(2)}(t)$ turns out to be infrared convergent, thanks to the presence of a power of k in all the terms of Eq.(2.2.6), so that $\lambda = 0$ is allowed from the very beginning and, although the terms quadratic in k in the numerator are separately ultraviolet divergent, as a matter of fact the result is finite and the Pauli-Villars regularization can also be ignored.

In the case of the vertex amplitude given by Eq.(2.2.2), the Δ-expansion of Eq.(2.1.10) would require in principle also the terms

$$\frac{1}{(p_1-k)^2+m^2-i\epsilon} \simeq \frac{1}{(p-k)^2+m^2-i\epsilon}\left(1+\frac{(\Delta.k)}{(p-k)^2+m^2-i\epsilon}\right) ,$$

$$\frac{1}{(p_2-k)^2+m^2-i\epsilon} \simeq \frac{1}{(p-k)^2+m^2-i\epsilon}\left(1-\frac{(\Delta.k)}{(p-k)^2+m^2-i\epsilon}\right) ; \qquad (2.2.7)$$

as a matter of fact only the symmetric combination of the two factors occurs, so that up to first order in Δ one has simply

$$\frac{1}{(p_1-k)^2+m^2-i\epsilon}\frac{1}{(p_2-k)^2+m^2-i\epsilon} \simeq \frac{1}{((p-k)^2+m^2-i\epsilon)^2} . \qquad (2.2.8)$$

The lowest order anomaly can then be written as

$$\left(\frac{\alpha}{\pi}\right) a_2 = e^2\, i \int \frac{d^4k}{(2\pi)^4}\frac{1}{((p-k)^2+m^2-i\epsilon)^2}$$
$$\times\left(\frac{1}{k^2+\lambda^2-i\epsilon}-\frac{1}{k^2+\Lambda^2-i\epsilon}\right)\left[\frac{16(p.k)^2}{3m^2}+\frac{4}{3}k^2+4(p.k)\right] , \qquad (2.2.9)$$

to be evaluated at $p^2=-m^2$; one finds *a posteriori*, as in the case of $F_2^{(2)}(t)$, that the regularized photon propagator Eq.(2.2.4) can be replaced in Eq.(2.2.9) by its simpler form Eq.(2.2.3).

2.3 Extraction of Divergences

As illustrated by the second order example, the diagrammatic expansion and many of the procedures based upon it are plagued by the existence of nonintegrable singularities in the integrands. Because of these singularities, many of the operations which one would perform in the simplification of these expressions are at best only formal. The singularities are of two types: the ultraviolet (UV) divergences associated with the short-distance behaviour of the theory and the infrared (IR) divergences associated with the long-range behaviour of the photon propagator. These singularities are well understood [10]. Analytically, they may be handled by regularization of integrals using UV and IR cutoffs (use (2.2.4) as the photon propagator rather than (2.2.3)).

Although this prescription can in fact be followed for textbook examples, in both numerical and analytic integration of high order graphs, it is preferable to minimize the requirement of the extraction of limits in explicit cutoff parameters. One is usually able to introduce counter terms and to group expressions so that those grouped

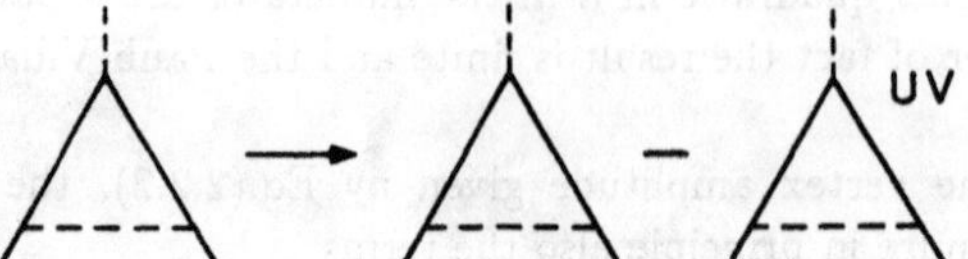

Figure 5: Second-order vertex insertion and the renormalizing UV counterterm.

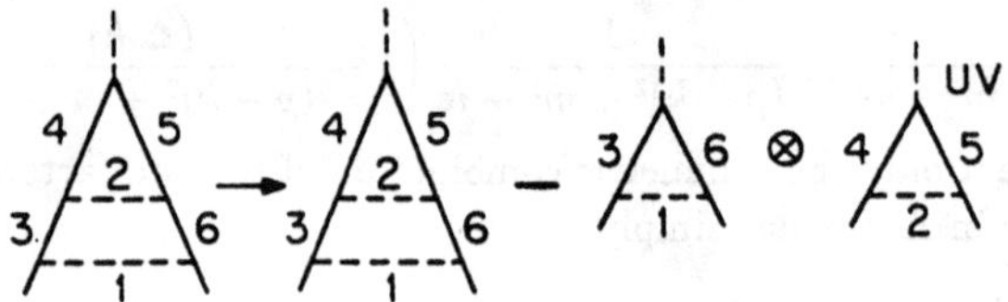

Figure 6: A fourth-order vertex function before and after the introduction of renormalizing vertex UV insertions. The symbol ⊗ means multiplication.

expressions are well behaved in the appropriate limit. This technique is quite delicate and requires great care in avoiding illegal operations on individual subexpressions which are, themselves, divergent. Using this technique, one can eliminate all ultraviolet divergences and isolate those infrared divergences which exist in a diagrammatic expansion in a particular gauge. If it is desired to produce expressions, graph by graph, the infrared divergences which have been isolated can then be evaluated by other techniques.

2.3.1 Extraction of Ultraviolet Divergences

For the ultraviolet divergences we have introduced the usual renormalization counterterms. The vertex function requires one subtraction

$$\Lambda_\mu(p_2, p_1) \to \Lambda^r_\mu(p_2, p_1) = \Lambda_\mu(p_2, p_1) - \Lambda_\mu(p_0, p_0) \tag{2.3.1}$$

where p_0 is a mass-shell momentum. This may be represented graphically for the second-order vertex as shown in Fig. 5.

In Fig. 6 we draw the main and UV counterterm for the fourth-order ladder graph. Here we ignore the overall subtraction which does not contribute to the magnetic moment. Although we may consider the counterterm numerator to be factored into two second-order pieces, we label lines and route momenta in the counterterm as we did for the main graph. In this way we produce a counterterm which when subtracted

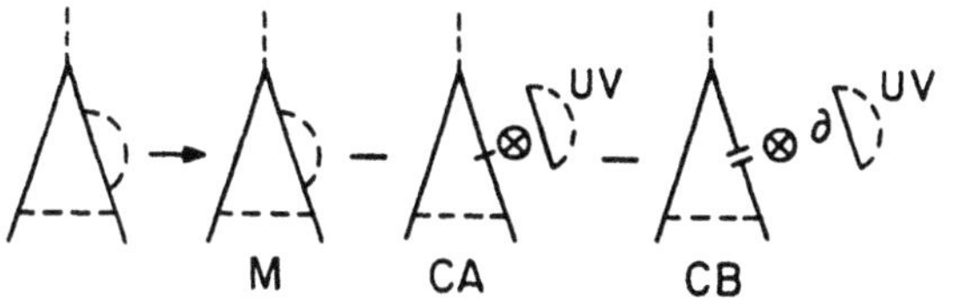

Figure 7: A vertex graph before and after renormalizing electron propagator UV insertions.

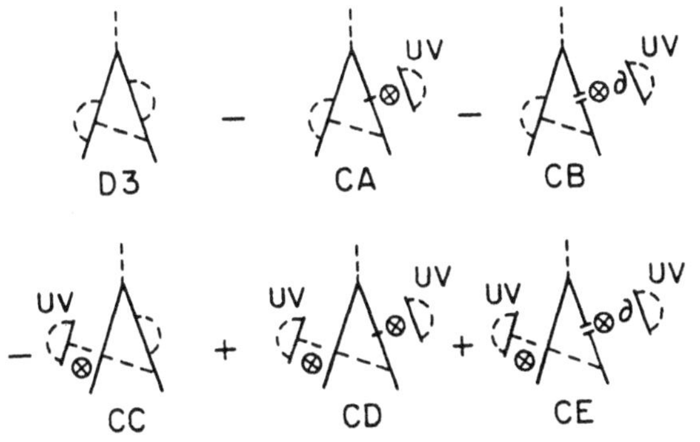

Figure 8: Sixth-order graph D3 with UV counterterms. Both vertex and electron propagator renormalizations are included.

from the main term leaves a function free from UV divergences point by point in momentum space. We treat the electron self-energy insertions in a similar manner:

$$\Sigma(p) \to \Sigma^r(p) = \Sigma(p) - \Sigma(p_0) - (\not{p} - \not{p}_0)(\partial\Sigma(p)/\partial p)\mid_{p_0} \qquad (2.3.2)$$

In Fig. 7 we give an example for a fourth-order graph. The bar on the electron line represents the operator "1" (i.e., there is a propagator on either side of the bar). The electron line with the double bar denotes the factors $(i\not{p}+m)^{-1}(i\not{p}+m)(i\not{p}+m)^{-1}$.

The sixth-order graph of Fig. 8 (graph D3 of Fig. 3) contains both vertex and propagator insertions. Using this procedure, all ultraviolet divergences are removed and the necessary renormalization is performed.

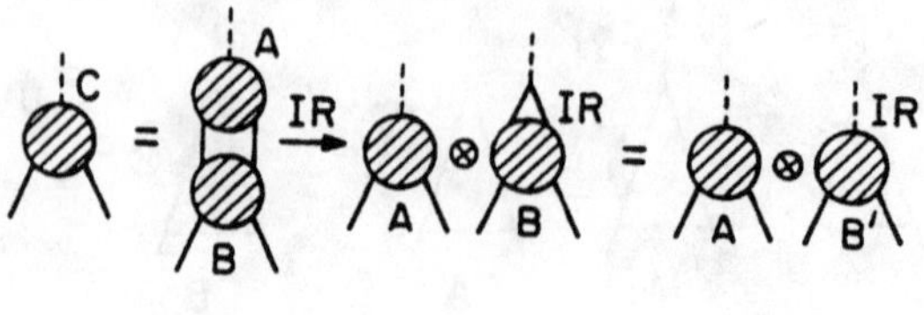

Figure 9: General fermion vertex graph (C) separated into a vertex (A) and scattering (B) graph. In the IR limit, $k_B \longrightarrow 0$, this factors into a vertex graph (A) multiplied by the IR limit of the vertex graph (B′) corresponding to the scattering graph.

2.3.2 Extraction of Infrared Divergences

We have used an analogous procedure to handle the IR divergences [5]. It is based upon the factorization property used to show general, analytical cancellation of IR divergences [11]. In momentum space, the infrared divergences arise when some photon momenta in a graph go to zero. If, as in Fig. 9, a vertex graph (C) can be written as a combination of a vertex graph (A) and a proper two-particle scattering graph (B), then when all photon momenta in B (k_B) vanish, we have an infrared divergence. If we are interested in $F_2(0)$, then A must be of higher than first order. For $F_1(0)$, the case where A is the bare vertex must also be considered. Such situations arise in the original graphs and in the UV counter graphs, including the second subtraction terms for electron-propagator renormalizations. While the latter are equivalent to vertex graphs because of Ward's identity, these quantities are formally divergent and great care must be taken in handling expressions containing them.

In the infrared limit, $k_B \to 0$, the fermion lines joining A and B go to the mass shell and the graph factors. The resultant expression for the B part of the graph is the infrared limit of the $F_1(0)$ part of the corresponding vertex graph (B′). We represent it graphically as shown in Fig. 9. The infrared limit involves deleting all references to the k_B in A and replacing the fermion propagators in B, $\dfrac{1}{i(\not{p} + \not{k}_B) + m}$ with the expression $\dfrac{-i\not{p} + m}{(p + k_B)^2 + m^2}$. In these expressions all numerator terms involving the k_B are missing.

As an example, consider the fourth-order ladder graph. In Fig. 10 we have drawn the main graph M and UV counterterm CA, and subtracted (CB) and added (CC) the IR limits of graphs M and CA, respectively. If we label and route momenta in the subtracted counterterm (CB) in parallel with the main graph, as we did for the UV counterterms, the difference (M-CB) will be IR-divergence-free in momentum space.

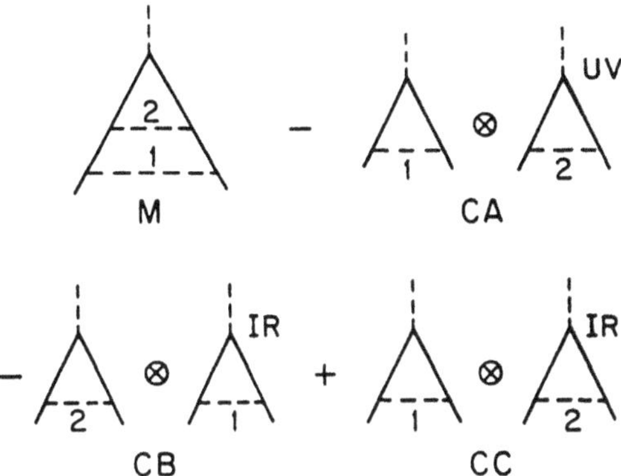

Figure 10: Fourth-order ladder vertex graph with UV counterterm (CA) and IR counterterms (CB) and (CC). (M - CA) is UV-convergent. (M - CB) and (CA - CC) are IR-convergent. CB = CC so that the sum of the IR counterterms is zero.

However, unlike the UV case, this counterterm is a computational artifice rather than a requirement of the theory and must finally be reincluded. If desired, this may be done analytically or numerically to obtain a term like $(a + b\ln\frac{\lambda^2}{m^2})$ in the limit $\lambda \to 0$. This is easier since we have now factored the graph into lower order pieces. The UV counterterm (CA) for this graph is also infrared divergent. The necessary IR counterterm (CC) is, in this case, the same as for the main graph but of opposite sign. The only difference between the two IR counterterms is in the labeling of some variables of integration. For the graphs in Fig. 10 the UV divergences cancel between M and CA, while the IR divergences cancel between M and CB and between CA and CC. CB and CC cancel identically after integration. There is no need to calculate even lower-order IR-divergent terms for this graph.

In general such infrared divergence cancellations occur between graphs rather than within a single graph. Because of this cancellation property, no explicit IR-divergent terms need be calculated unless one wants a cutoff-dependent result for a single graph. For $F_2(0)$ there are subsets of graphs that are gauge-invariant corresponding to sets that have the same number of virtual photon lines crossing the external field vertex. The IR counterterms cancel within each subset. As a further example consider the main graph and counterterms for diagrams C3 and E3 as shown in Fig. 11. In this case all IR counterterms cancel between the two graphs. After introducing the counterterms, the resultant functions are everywhere integrable and no divergent expressions need to be evaluated. Explicit cutoffs are avoided at all stages.

Ignoring graphs with vacuum polarization, there is the well-known second-order counterterm and four irreducible fourth-order infrared counterterms needed. They are

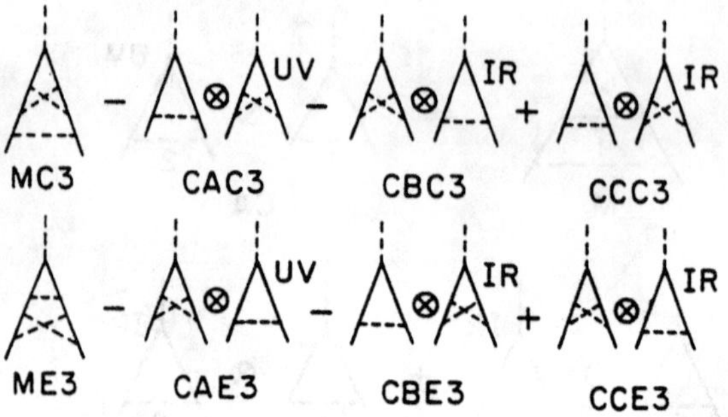

Figure 11: Main, UV, and IR counterterms for sixth-order graphs C3 and E3. (MC3 - CAC3) and (ME3 - CAE3) are UV-convergent. (MC3 - CBC3), (CAC3 - CCC3), (ME3 - CBE3), and (CAE3 - CCE3) are IR-convergent. CBC3 = CCE3 and CCC3 = CBE3 so that the total added IR counterterm is zero.

the first four graphs of Fig. 2 with special rules for their evaluation. The graphs are evaluated with all photon momenta deleted from numerators of propagators with the exception of graph d, where we found it convenient to keep the photon momentum for the self-energy insertion, in which case the renormalization terms are also included. In Table 2 we list the value of the infrared counterterms associated with these four graphs[1] and in terms of these, in Table 3 we list the counterterms for each sixth order vertex graph without vacuum polarization. To these we must add pieces involving the second order infrared counterterm which is simply $\frac{1}{2}\ln\lambda^2$ without any additive constant.

TABLE 2: Values for fourth-order infrared counterterms for the first four graphs shown in Fig. 2. A corresponds to graph a, B to graph b, C to graph c, D to graph d. The electron mass has been set to 1.

A	=	$3/2 - \pi^2/4 - \frac{1}{2}\ln\lambda^2$
B	=	$3\pi^2/8 - 1 + \frac{1}{2}\ln\lambda^2 + \frac{1}{8}(\ln\lambda^2)^2$
C	=	$\pi^2/8 - 1/2 + \frac{1}{8}(\ln\lambda^2)^2$
D	=	$3/2 - \pi^2/3 + \frac{1}{4}\ln\lambda^2 + \frac{1}{8}(\ln\lambda^2)^2$

[1]The analytic values in Table 2 have, to our knowledge, not been published before.

TABLE 3: IR counterterms required by graphs without electron loops

Graph	Counterterm
A2	D
B1	2D + B
B2	-D
C1	2C + A
C2	-C
C3	$-\frac{1}{2}A$
D1	-D
D4	C
E1	-D
E3	$\frac{1}{2}A$
F2	-C
G1	-B
G5	-C
H1	-A

3 HYPERSPHERICAL FORMALISM

The Feynman graph for the lowest order graph contributing to the second order electron magnetic moment anomaly is a one loop graph. In fourth and sixth orders, these become 2 and 3 loop graphs respectively (see Figs. 2, 3). The problem is to carry out the evaluation of the associated Feynman integrals. In this section, we discuss the hyperspherical approach to this problem [12,13]. In the next section, we present the dispersive approach and then discuss their combination into a hybrid approach which has proven useful.

Given a Feynman graph, we begin by analytically continuing the Feynman integral in momentum space into the Euclidean region. This is easily accomplished by letting the energy associated with each internal and external line acquire the same phase ϕ, which is then varied from 0 to $\frac{1}{2}\pi$. The $i\epsilon$ prescription of each propagator assures that no poles are crossed. However, now the external momenta are all spacelike, so that we will eventually have to analytically continue the external momentum back to the timelike region. This will be done in Sec. 3.2.

3.1 Performing the Angular Integrals

Having a Euclidean integrand, we introduce four-dimensional spherical coordinates for each loop momentum. The first problem is to perform the angular integrations.

Suppose that it is possible to choose the independent loop momenta so that the momentum of each line is a linear combination of at most two distinct momenta (including both loop and external momenta). Then, in the Euclidean region, each propagator is of the form

$$\frac{1}{K^2+m^2} \quad \text{or} \quad \frac{1}{(K-L)^2+m^2} \,, \tag{3.1.1}$$

where K, L are independent Euclidean four-momenta and m is the mass of the line. (When there is no ambiguity, we denote the magnitude of K by K.) In the first case, the propagator has no angular dependence, while in the second the angular dependence is made explicit by the formula

$$\frac{1}{(K-L)^2+m^2} = \frac{Z_{KL}}{KL} \sum_{n=0}^{\infty} Z_{KL}^n C_n(\hat{K}.\hat{L}) \,, \tag{3.1.2}$$

where $\hat{K}, \hat{L}$ are unit vectors along K, L, C_n is the Gegenbauer polynomial, [14] and

$$Z_{KL} = \frac{K^2+L^2+m^2-[(K^2+L^2+m^2)^2-4K^2L^2]^{\frac{1}{2}}}{2KL} \,. \tag{3.1.3}$$

After performing the appropriate traces, the numerator of the Feynman integrand, in theories with spin, is a polynomial in the scalar products of the momenta, which can be written in terms of the Gegenbauer polynomials and the lengths of the (Euclidean) momenta. Thus the entire angular dependence of the integrand is in the Gegenbauer polynomials, which are the orthogonal functions on the four-dimensional sphere.

To see when it is possible to label each line with no more than two momenta, we now restrict our attention to self-energy graphs with external momentum P, where P is spacelike. We imagine joining the two external lines and consider the topological properties of the resulting graph. For reasons outlined in the next paragraph, we insist that the line corresponding to the external fermion be on the boundary of the graph.

If the graph is planar, then considering it as a map, we associate a circulating loop momentum with each country. The momentum of each line is given by the algebraic sum of the momenta associated with the countries (at most two) which it borders (see Fig. 12). We have insisted that the external line be one of the outer boundaries of the map. It then borders only one country, whose loop momentum is assigned to be P, so that the magnitude of the momentum of the line is unchanged when doing the angular integrations. Thus, planarity is a sufficient condition for being able to label each line with at most two momenta.

In up to sixth order self-energy graphs, there are at most three loop momenta and one external momentum, i.e. at most four momenta, which means at most four countries. If there is at least one country which does not border all three remaining

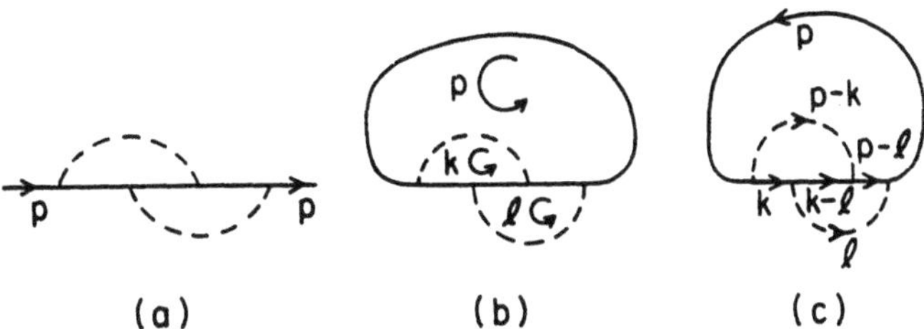

Figure 12: (a) A typical self-energy graph. In (b), it is converted to a map, with loop momenta shown. The subsequent assignment of momenta to the lines is shown in (c).

countries, the angular integrals can always be performed, as we now show. The following relations suffice to perform all the angular integrals

$$\int C_n(\hat{a}.\hat{b})C_m(\hat{b}.\hat{c})\frac{d\Omega_b}{2\pi^2} = \frac{\delta_{mn}}{n+1}C_n(\hat{a}.\hat{c}) \ , \tag{3.1.4}$$

$$C_n(x)C_m(x) = \sum_{j=0}^{\mathrm{Min}(m,n)} C_{m+n-2j}(x). \tag{3.1.5}$$

First, ignore the numerator trace of the Feynman diagram. Then because at least one loop momentum (call it M) does not border all three remaining loop momenta, the angular dependence of the integrand on M appears in the dot product of M with at most two other vectors, and so the M angular integral can always be done using (3.1.4). The result of this integration may involve the dot products of the remaining three loop momenta. But by using (3.1.5) we can write the integrand as a sum of products of terms, none of which contains two Gegenbauer polynomials with the same argument. The remaining angular integrals can then be done by using (3.1.4).

Returning to theories with spin, the numerator trace is a polynomial of finite degree in the dot products of the loop and external momenta, with coefficients which are products of the lengths of the loop and external momenta. The angular dependence can be rewritten in terms of Gegenbauer polynomials. When the numerator trace is multiplied by the infinite sums which come from the propagators, by repeatedly using (3.1.5) the complexity of the angular integrals is not increased by terms involving the same dot products already present in the propagators. The only increased complexity comes when the trace term contains a dot product which does not appear in the propagators. Because such terms are low degree polynomials, we can deal with them as we explain in Appendix 6.2.

The argument we have just given for self-energy graphs is easily extended to vertex graphs or their derivatives at $\Delta^2 = 0$. where Δ is the external photon momentum. For example, the magnetic-moment calculation corresponds to keeping only linear terms in Δ. Thus propagators of the form $1/[(k+\Delta)^2+m^2]$ can be expanded as

$$\frac{1}{(k^2+m^2)^2}[(k-\Delta)^2+m^2] = \frac{1}{(k^2+m^2)^2}[k^2+m^2-2k.\Delta] \quad \text{to order } \Delta. \tag{3.1.6}$$

Thus, the denominators which appear are just as in the self-energy case, except that they may appear squared. But the generalization of (3.1.2),

$$\left(\frac{1}{(K-L)^2+m^2}\right)^2 = \frac{Z_{KL}^2}{K^2L^2}\frac{1}{(1-Z_{KL}^2)}\sum_{n=0}^{\infty}(n+1)Z_{KL}^n C_n(\hat{K}.\hat{L}), \tag{3.1.7}$$

is easily obtained, and the rest of the analysis is unchanged.

It turns out that of the 28 distinct graphs without fermion loops in the sixth order magnetic moment calculation, all but three of them (set H of Fig. 3), are planar. In the planar graphs, the momenta can always be chosen so that there is at least one loop momentum which does not appear paired with all other three in the propagators. So in all but the three non-planar graphs, the angular integrals can be carried out analytically. The remaining integrand is at most three dimensional. The infinite sums associated with the propagators (See 3.1.2) turn out either to be finite sums because of the Kronecker deltas in the angular integrations, or to be simple power series in product of Z's which can be summed in closed form to give logarithms or dilogarithms.

3.2 Performing the "Radial" Integrations, General Approach

Having performed the angular integrations, the integral is now of the form

$$\int dK_1^2 dK_2^2 dK_3^2 f(K_1^2, K_2^2, K_3^2, P^2), \tag{3.2.1}$$

where K_i^2 are the lengths of the (Euclidean) loop momenta, and P^2 is the external momentum, which is spacelike. Since we are ultimately interested in the answer for timelike P^2, we have two options. The first is to evaluate (3.2.1) for arbitrary spacelike P^2, and analytically continue the answer. The second is to perform the continuation before integration. The latter is the simpler alternative; it is mandatory if part of the integral is to be evaluated numerically. We now discuss its implementation.

Except for singularities coming from the divergences in the infinite sums in (3.1.2), which does not affect this analysis, the integrand is meromorphic in the variables K_i^2, P^2 except for the square root in the Z variables defined in (3.1.3). We now study the behavior of Z_{PK} as P^2 is varied from a positive quantity to the negative

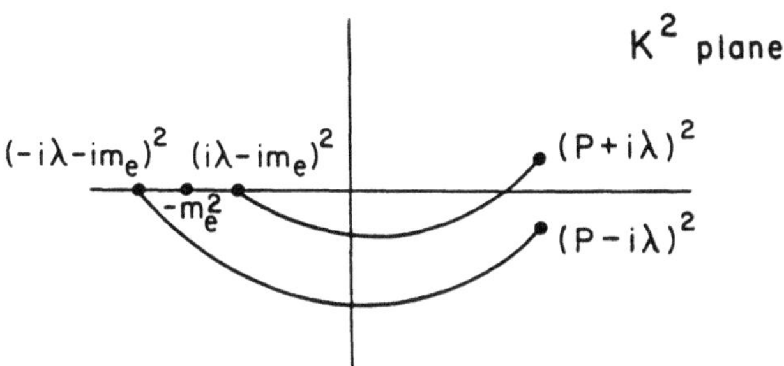

Figure 13: Location of the branch points in Z_{PK}, in the K^2 plane. The path shows how these branch points move as P^2 is varied from a positive value to $-m_e^2$. m_e is the electron mass.

on-mass-shell value $P^2 = -m_e^2$. In the K^2 plane, the square root has its branch points at

$$K^2 = (P \pm im)^2 \tag{3.2.2}$$

and to begin with, P is real and positive, and K^2 is to be integrated from 0 to ∞. If m is the electron mass m_e, then we can continue P^2 to $-m_e^2$ without distorting the K^2 contour. However, if m is the infinitesimal photon mass, which we denote by λ, in continuing P^2 to $-m_e^2$ the branch point crosses the positive K^2 axis, as shown in Fig. 13. To get the correct continuation of the integral, we must deform the contour to avoid the branch point. This can be done if we write the contour as shown in Fig. 14, i.e., we write

$$\int_0^\infty dK^2 = \int_{\substack{0 \\ \text{above cut}}}^{-m_e^2} dK^2 + \int_{\substack{-m_e^2 \\ \text{below cut}}}^{\infty} dK^2 , \tag{3.2.3}$$

where, below the cut, we have

$$\frac{Z_{KP}}{KP} = \frac{K^2 - m_e^2 + \lambda^2 - \mid (K^2 - m_e^2 + \lambda^2)^2 + 4K^2 m_e^2 \mid^{\frac{1}{2}}}{-2K^2 m_e^2} , \tag{3.2.4}$$

whereas above the cut

$$\frac{Z_{KP}}{KP} = \frac{K^2 - m_e^2 + \lambda^2 + \mid (K^2 - m_e^2 + \lambda^2)^2 + 4K^2 m_e^2 \mid^{\frac{1}{2}}}{-2K^2 m_e^2} . \tag{3.2.5}$$

Having established the branch of the square root, we can let $\lambda \to 0$, and obtain

$$\left(\frac{Z_{KP}}{KP}\right)_{\text{above}} = \frac{K^2 - m_e^2 + \mid K^2 + m_e^2 \mid}{-2K^2 m_e^2} = -\frac{1}{m_e^2} , \tag{3.2.6}$$

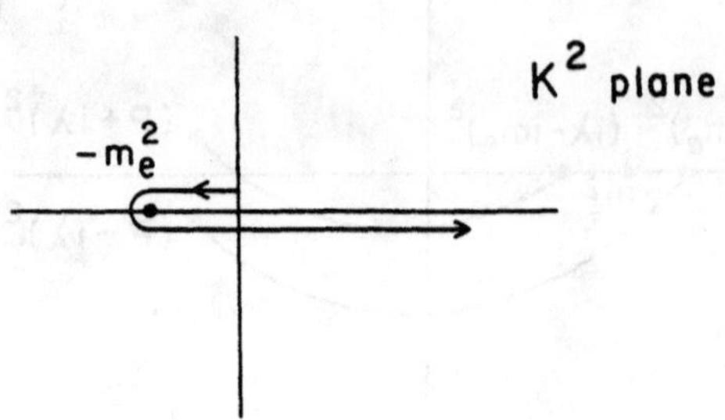

Figure 14: Integration contour in the K^2 plane necessary to avoid the branch points of Z_{PK} shown in Fig. (13).

whereas

$$\left(\frac{Z_{KP}}{KP}\right)_{\text{below}} = \frac{K^2 - m_e^2 - |\ K^2 + m_e^2\ |}{-2K^2 m_e^2} = \frac{1}{K^2}\ , \tag{3.2.7}$$

Thus, if the external momentum appears in a photon line, then analytically continuing back to the external mass shell implies that the radial integral for the other momentum in that photon line is dragged to $-m_e^2$. This effect can propagate as illustrated in the graph in Fig. 15. As $P^2 \to -m_e^2$, the K^2 variable in the outer photon is dragged to $-m_e^2$. But as K^2 becomes negative, it drags the L^2 variable in the inner photon to K^2. Thus, for $K^2 < 0$, the L^2 integration contour is shown in Fig. 16, where above the cut

$$\frac{Z_{KL}}{KL} = \frac{1}{K^2}\ , \tag{3.2.8}$$

while below

$$\frac{Z_{KL}}{KL} = \frac{1}{L^2}\ . \tag{3.2.9}$$

The contour distortion is necessitated by the existence of lines (photons) with mass less than m_e. However, the vanishing of the photon mass leads to a great simplification, for in this case the associated Z becomes a rational function of the lengths of the momenta.

Complications in the radial integrations differ markedly depending on the topology of the graphs. We will deal explicitly with the details of different subsets of graphs in subsequent sections.

Figure 15: A fourth-order self-energy graph, with momentum assignments.

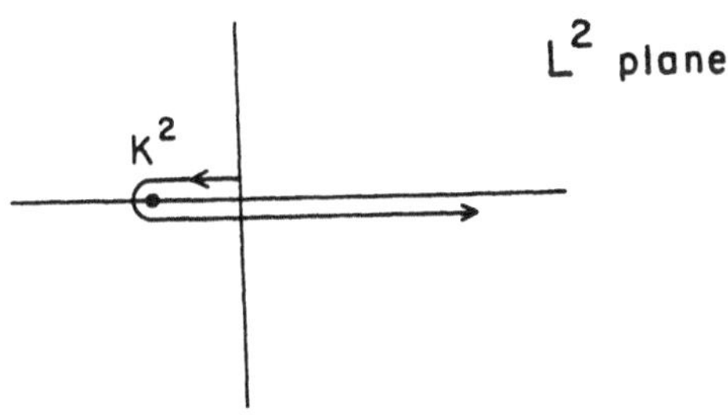

Figure 16: Integration contour in the L^2 plane required to avoid the branch points in Z_{KL}, when K^2 is negative.

3.3 Second Order Example

Before proceeding further, and as an illustration of our technique, we shall work out the second-order contribution to the g factor, first calculated by Schwinger. With lines labeled as in Fig. 1, using equation (2.2.9) but replacing k by $p-k$, and setting the electron mass to 1, the anomaly is given by

$$\frac{\alpha}{\pi}a_2 = 4e^2 i\int\frac{d^4k}{(2\pi)^4}\frac{1}{(k^2+1-i\epsilon)^2}\frac{1}{(p-k)^2-i\epsilon}\left[p.k+\frac{1}{3}k^2+\frac{4}{3}(p.k)^2\right] \qquad (3.3.1)$$

We now continue this integral into the Euclidean region. As explained in the previous section, we temporarily discard the constraint $p^2=-1$, and perform the Wick rotation with p spacelike,

$$p^2 = P^2, \qquad (3.3.2)$$

and define

$$I(P^2) = -\frac{4e^2}{(2\pi)^4}\int d^4K\frac{P.K+\frac{1}{3}K^2+\frac{4}{3}(P.K)^2}{(P-K)^2(K^2+1)^2}. \qquad (3.3.3)$$

The anomaly is determined from I by

$$\frac{\alpha}{\pi}a_2 = \lim_{P^2\to -1} I(P^2). \qquad (3.3.4)$$

Here P, K are four-dimensional Euclidean vectors, and initially P^2 is taken positive.

Rewriting the numerator of the integrand as

$$\frac{1}{2}PKC_1(\hat{P}.\hat{K})+\frac{1}{3}K^2+\frac{1}{3}P^2K^2C_2(\hat{P}.\hat{K})+\frac{1}{3}P^2K^2 \qquad (3.3.5)$$

and expanding

$$\frac{1}{(P-K)^2} = \frac{Z_{PK}}{PK}\sum_{n=0}^{\infty}Z_{PK}^n C_n(\hat{P}.\hat{K}), \qquad (3.3.6)$$

where

$$Z_{PK} = \mathrm{Min(P/K, K/P)}, \qquad (3.3.7)$$

we can perform the angular integrations using (3.1.4) and (3.1.5) to obtain

$$I(P^2) = \frac{2\alpha}{\pi}\int_0^{\infty}\frac{K^3dK}{(K^2+1)^2}\frac{Z_{PK}}{PK}\left(\frac{1}{3}K^2-\frac{1}{3}P^2K^2-\frac{1}{3}P^2K^2Z_{PK}^2-\frac{1}{2}PKZ_{PK}\right). \quad (3.3.8)$$

Here

$$\alpha = e^2/4\pi. \qquad (3.3.9)$$

We now continue P^2 to -1, deforming the contour as in Fig. 14 to obtain

$$\frac{\alpha}{\pi}a_2 = \frac{\alpha}{\pi}\int_0^{-1}\frac{K^2dK^2}{(K^2+1)^2}\left[\frac{(K^2+1)^2(2K^2-1)}{6K^2}\right]+\frac{\alpha}{\pi}\int_0^{\infty}\frac{K^2dK^2}{(1+K^2)^2}\frac{1}{6K^2} = \frac{\alpha}{2\pi}. \qquad (3.3.10)$$

3.4 Details of the Radial Integrals

To understand the complexity of the radial integrals (of the four-dimensional Euclidean vectors), and how it depends on the topology of the graph, we introduce the notion of the truncated dual of a graph. The usual dual of a planar graph is obtained by associating a vertex with every face of the original graph, and joining two vertices if the corresponding faces have an edge in common [15]. For this purpose mathematicians include the infinite (outer) face as one of the faces of the graph. We define the truncated dual to be the graph obtained by the same construction, but omitting the infinite face.

In the 28 graphs without fermion loops occurring in the sixth order g-2 calculation, all but 3 come from planar self-energy diagrams. 6 arise from self-energy graphs whose truncated dual is a tree, 8 from truncated duals with one loop, and 11 from truncated duals with two loops. As the complexity of the truncated dual increases, so does the complexity of analytically evaluating the radial integrals.

We begin with the six graphs whose associated self-energy graphs (obtained by omitting the external photon) have tree graphs as truncated duals. They are sets A and B of Fig. 3. All the angular integrals can be done trivally, and the infinite sums arising from the expansion of the propagators (3.1.2) collapse to finite sums. The point is that each vertex of the truncated dual corresponds to a loop momentum and every tree graph has at least one vertex i of degree 1, i.e., which is connected to only one other vertex, call it j. If the loop momenta associated with vertices i and j are K_i and K_j, respectively, then ignoring the numerator trace, the complete angular dependence on K_i is given by

$$\sum_{n_i} Z_i^{n_i+1} C_{n_i}(\hat{K}_i.\hat{K}_j). \tag{3.4.1}$$

This is trivally integrated over $d\Omega_{K_i}/2\pi^2$ to give

$$\sum_{n_i} (Z_i)^{n_i+1} \delta_{n_i 0} = Z_i. \tag{3.4.2}$$

Thus, for considerations of the angular integral it is as though the edge from i to j were deleted. Thus, ignoring the numerator trace, the result follows by induction. But the numerator trace is a polynomial of finite degree in the dot products of the loop momenta. By orthogonality, taking this numerator into account gives at most a finite sum. It is easily seen that in these graphs, it is always possible to have the fermion lines carry one momentum. As a result, the radial integrand is a rational function of the square of the lengths of the independent loop momenta. This enables us to do the radial integrals in a straightforward way.

Consider the integration over M^2, the innermost loop in Fig. 17. Because the integrand is rational, the result is itself rational plus at most a logarithm. The end point of the M^2 integral can either be $0, \infty$ or L^2, because of the contour dragging

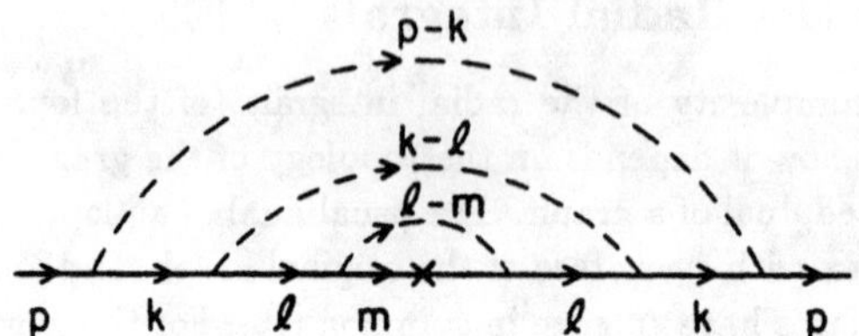

Figure 17: Graph B3, with momentum assignments after the external photon momentum Δ has been set to 0.

explained in the section 3.2. So after the M^2 integral, the L^2 dependence of the integrand is made up of rational function of L^2, as well as possibly $\log(L^2)$ and $\log(1+L^2)$. (We have set the electron mass to 1). The L^2 integral can then be done, yielding rational functions, logarithms and at worst dilogarithms of the remaining variable K^2. This final integral can then be done using the table of integrals for such functions (see Appendix 6.5).

Graphs of set A have also been computed with less algebra using Schwinger's mass-operator formalism [16], but this technique has not been generalized to other sixth order graphs without vacuum polarization.

The next most complicated set of graphs have truncated duals which have one loop. These are sets C and D of Fig. 3. In these graphs, it is not possible to have all fermion lines carry a single momentum. This means that there will be square roots in the integrands associated with the Z factors as explained in the sections 3.1 and 3.2.

Of set D, consider graph D4 with the lines labeled as in Fig. 18. After the angular integrations are done, the remaining integral is

$$\int K^2 dK^2 \int L^2 dL^2 \int M^2 dM^2 \frac{1}{(1+K^2)^2} \frac{1}{(1+M^2)^2} \frac{Z_{KM}}{KM} \frac{Z_{KP}}{KP} \frac{Z_{PL}}{PL} \frac{Z_{KL}}{KL} \frac{1}{L^4}$$

$$\times F(K^2, L^2, M^2, Z_{KM}, Z_{KP}, Z_{KL}, Z_{PL}), \tag{3.4.3}$$

where

$$P^2 = -1,$$

$$Z_{KL} = \{K^2 + L^2 + 1 - [(K^2 + L^2 + 1)^2 - 4K^2L^2]^{1/2}\}/(2KL), \tag{3.4.4}$$

$$Z_{PL} = [L^2 - (L^4 + 4L^2)^{1/2}]/(2PL).$$

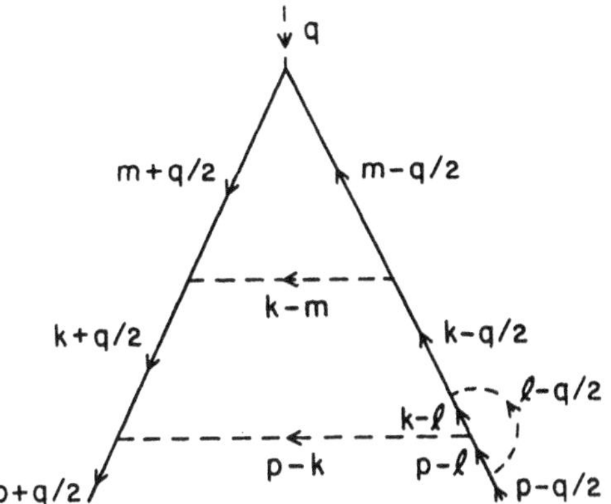

Figure 18: Graph D4, a graph whose truncated dual has one loop.

The L^2 integral is from 0 to ∞, while the K^2, M^2 integrations and the values of Z_{KP}, Z_{KM} can be divided into the following regions:

$$\text{region I}: \quad Z_{PK} = P/K, \quad Z_{KM} = K/M, \quad 0 < K^2 < M^2 < \infty\ ;$$

$$\text{region II}: \quad Z_{PK} = P/K, \quad Z_{KM} = M/K, \quad 0 < M^2 < K^2 < \infty\ ;$$

$$\text{region III}: \ Z_{KM} = K/M, \quad -1 < K^2 < 0, \quad K^2 < M^2 < \infty\ ;$$

$$\text{region IV}: \ Z_{KM} = M/K, \quad -1 < K^2 < 0, \quad K^2 < M^2 < 0.$$

In each of the regions III and IV, the integrands are evaluated with $Z_{PK} = P/K$, then with $Z_{PK} = K/P$, and one takes the difference between the two values.

F is a polynomial in its variables, except that some terms also have a factor $\ln(1 - Z_{KL}Z_{PK}Z_{PL})$ arising from the angular integrals. The M^2 integral is essentially trivial, since the integrand is a rational function of M^2. Performing this integration leaves a two-dimension integral, whose integrand is a polynomial in $K^2, L^2, Z_{KP}, Z_{KL}, Z_{PL}$ divided by powers of $(1 + K^2), K^2, L^2$, some terms of which also have a factor $\ln(1 - Z_{KL}Z_{PK}Z_{PL})$ and some have a factor $\ln(1 + K^2)$ from performing the M^2 integration. The regions of integration are now

$$\text{region A}: \quad Z_{PK} = K/P, \quad -1 < K^2 < 0, \quad 0 < L^2 < \infty\ ;$$

$$\text{region B}: \quad Z_{PK} = P/K, \quad -1 < K^2 < \infty, \quad 0 < L^2 < \infty.$$

At this point, one looks for a variable change to eliminate the square roots in the Z's, leaving a rational function as integrand, apart from the two logarithmic factors. The variables chosen were

$$w = -Z_{PK}Z_{KL}Z_{PL}, \tag{3.4.5}$$

$$u = \frac{1}{2}[-L^2 + (L^4 + 4L^2)^{1/2}].$$

Since Z_{PK} is a different function of K and P in regions A and B, the expressions for K^2 and L^2 in terms of u, w will differ in these regions.

With this choice of variables it is easily verified that regions A and B are mapped into the same u, w region $0 < w < 1 - u$, $0 < u < 1$, that all denominators can be factored into linear functions of w, and that the arguments of the logarithms are also linear functions of w. Consequently, the w integrals can be evaluated with the observation that

$$\int dw(c + dw)^{-1} \ln(a + bw)$$

involves, at worst, Spence functions (dilogarithms) while

$$\int dw(e + fw)^{-1} \ln(a + bw) \ln(c + dw)$$

involves, at worst, trilogarithms.

After evaluating the w integrals, we are left with a one-dimensional u integral, whose integrand contains rational functions, logarithms, Spence functions, and trilogarithms. The only denominator factors which appear are $u, (1-u), (2-u), (1-u+u^2)$. The final integration can then be performed.

The discussion so far has avoided the problem of ultraviolet or infrared divergences. The ultraviolet divergences are canceled by the appropriate renormalization counterterms. To assure pointwise cancellation of these divergences the change of variable from K, L space to u, w space must also be done on the counterterm. It is only upon such a pointwise convergent function that one can legitimately perform that required change of variables. As usual, the infrared divergences were trickier. We discuss the extraction of infrared divergences in Appendix 6.1.

Turning now to a typical graph of set C, label the momenta of graph C3 as shown in Fig. 19.

We can easily perform the traces and the angular integrations. The remaining integral is then

$$\int K^2 dK^2 \int L^2 dL^2 \int M^2 dM^2 \frac{1}{(K^2+1)^2} \frac{1}{(L^2+1)^2} \frac{Z_{LM}^2}{L^2 M^2 (1 - Z_{LM}{}^2)} \tag{3.4.6}$$

$$\times \frac{Z_{KM}^2}{K^2 M^2 (1 - Z_{KM}^2)} \frac{Z_{KL}}{KL} \frac{Z_{KP}}{PK} F(K^2, L^2, M^2, Z_{KM}, Z_{LM}, Z_{KL}, Z_{KP}),$$

where

$$Z_{LM} = \{L^2 + M^2 + 1 - [(L^2 + M^2 + 1)^2 - 4L^2 M^2]^{1/2}\}/(2LM),$$

$$Z_{KM} = \{K^2 + M^2 + 1 - [(K^2 + M^2 + 1)^2 - 4K^2 M^2]^{1/2}\}/(2KM). \tag{3.4.7}$$

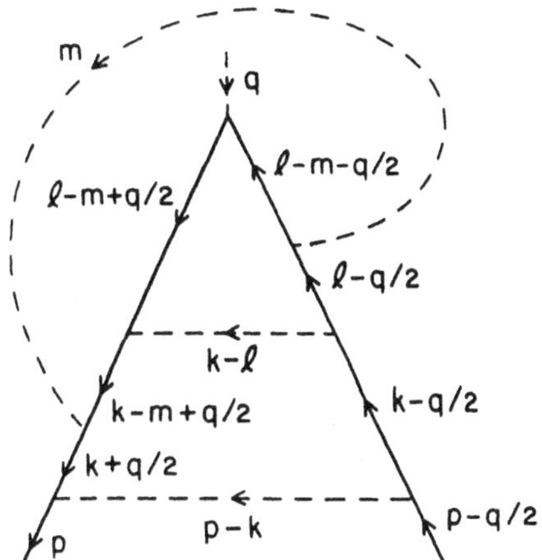

Figure 19: Graph C3, another graph whose truncated dual has one loop.

The M^2 integration is from 0 to ∞, while the K^2, L^2 integrations and the values of Z_{KL}, Z_{KP} can be divided into four regions:

$$\text{region 1}: \quad Z_{KL} = K/L, \quad Z_{KP} = P/K, \quad 0 < K^2 < L^2 < \infty \;;$$

$$\text{region 2}: \quad Z_{KL} = L/K, \quad Z_{KP} = P/K, \quad 0 < L^2 < K^2 < \infty \;;$$

$$\text{region 3}: \quad Z_{KL} = K/L, \quad -1 < K^2 < 0, K^2 < L^2 < \infty \;;$$

$$\text{region 4}: \quad Z_{KL} = L/K, \quad -1 < K^2 < 0, K^2 < L^2 < 0.$$

In each of regions 3 and 4, the integrands are evaluated at $Z_{KP} = P/K$, then at $Z_{KP} = K/P$, and the two values are subtracted. F is a polynomial in its variables, except that some terms have a factor $\ln(1 - Z_{KL}Z_{LM}Z_{KM})$ which arises from doing the angular integrations.

The main difference between the integrand of (3.4.6) (set C) and the integrand of (3.4.3) (set D) is:

1. In (3.4.6), there is no trivial integral in the sense that the M^2 integral was trivial in (3.4.3).

2. Both square roots (3.4.7) involve two integration variables. In (3.4.4), one of the square roots depended only on L^2.

It is still possible to eliminate all square roots from the integrand by defining the variables

$$u = LZ_{LM}/M,$$
$$v = KZ_{KM}/M,$$
$$w = -Z_{KL}Z_{KM}Z_{LM}.$$

The integrand now consisted of rational functions of u, v, w with some terms also multiplied by $\ln(1+w)$.

For definiteness, we concentrate on region 1, although the results for other regions are similar. In this region, it is easily verified that all denominators in the integrand can be written as a product of linear functions of u, without introducing any square roots. Thus, after performing partial fractions in u, the u integration is trivial.

The problem is that the denominators in the remaining v, w integrals involve quadratic functions of v and w, which can not be factored without introducing square roots. In general, for a given term, one can either evaluate the v integral or the w integral straightforwardly or one can find another variable change involving both v and w which allows the evaluation of one integral easily. Thus one can reduce the integrand to one-dimensional form, with limits 0 and 1. The set of one-dimensional integrals is larger than those encountered in evaluating (3.4.3). It also turns out that the extraction of the infrared divergences is much harder.

The complications encountered in this set suggest that this approach is not the simplest possible. Historically, while Levine, Perisho and Roskies were calculating these graphs with the hyperspherical method being presented here [13], Barbieri, Caffo and Remiddi [17] were calculating them using the dispersive methods they had developed. It turns out that the dispersive method is easier for these graphs. We therefore suspend our discussion of the hyperspherical approach and develop some of the general formalism for the dispersive approach.

4 DISPERSION RELATIONS FORMALISM

Let us recall that if $f(z)$ is an analytic function satisfying the condition $f(z^*) = f^*(z)$, with a cut along the real positive axis starting at $z = z_0$, so that for $x > z_0$ one has $f(x+i\epsilon) - f(x-i\epsilon) = 2i\mathrm{Im}f(x)$, and if $f(z)$ vanishes for $z \to \infty$, a straightforward application of the theorem of the residues gives, for any value of z not on the cut, the dispersion relation

$$f(z) = \frac{1}{\pi}\int_{z_0}^{\infty} \frac{dx}{x-z}\mathrm{Im}f(x)\,. \tag{4.1}$$

By subtracting the same equation written for $z = z_1$, one obtains

$$f(z) - f(z_1) = \frac{1}{\pi}\int_{z_0}^{\infty} dx \left(\frac{1}{x-z} - \frac{1}{x-z_1}\right)\mathrm{Im}f(x)\,, \tag{4.2}$$

which can be written in the form of a dispersion relation subtracted at z_1,

$$f(z) = f(z_1) + \frac{1}{\pi}\int_{z_0}^{\infty} dx \frac{(z-z_1)}{(x-z_1)(x-z)}\mathrm{Im}f(x)\,. \tag{4.3}$$

Due to the extra factor $(z-z_1)/(x-z_1)$ appearing in Eq.(4.3), the convergence of the x−integration is faster than in Eq.(4.1); in particular the integral might be convergent in Eq.(4.2), but divergent in Eq.(4.1).

Dispersion relations hold for Feynman graph scalar amplitudes or scalar quantities or form factors which can be extracted from a Feynman graph. The imaginary parts are obtained by means of the Cutkosky cutting rule [18], which amounts to replacing, within a graph, suitable combinations of propagators by the corresponding mass-shell δ−functions. The integration of the δ−functions over the loop variables is almost trivial, giving typical physical phase space factors (usually square roots); as a rule, the number of non trivial integrations, including the last dispersion relation integration, is strongly reduced.

Dispersion relations for Feynman graphs can be derived by analyticity properties (which can be assumed, or somehow inferred from first principles); it is to be noted, however, that the Cutkosky rule can be immediately derived by using Veltman's greatest time equation, whose proof is elementary and relies only on the properties of Feynman propagators in the coordinate space [19].

The general formulation of the cutting rules is quite complicated, but their practical application remarkably simple; to illustrate the dispersion relation approach, consider a vertex amplitude. It depends on the three (not independent) external four vectors p_1, p_2 and $\Delta = p_1 - p_2$. (See Fig. 4) Disregarding for the moment the algebraic complications due to loop momenta in the numerator, any scalar amplitude A associated with it is a function of the three independent invariant variables

$$t \equiv -\Delta^2 = \Delta_0^2 - \vec{\Delta}^2 , \tag{4.4}$$

and p_1^2, p_2^2; for simplicity let us consider the mass shell case

$$p_1^2 = p_2^2 = -m^2 . \tag{4.5}$$

so that $A = A(t)$. One gets the imaginary part for the t-dispersion relation by cutting the graph (or rather the internal propagator lines) in all possible ways such that the algebraic sum of the momenta flowing on the cut propagators is Δ; if the orientation of Δ is, say, upwards, one must change the sign, in the algebraic sum, of the momenta flowing downwards. The cut propagators are then replaced by the mass shell δ−function according to

$$\frac{1}{q^2+m^2-i\epsilon} \rightarrow 2\pi i\theta(+q_0)\delta(q^2+m^2) , \tag{4.6}$$

if q flows upwards, or

$$\frac{1}{q^2+m^2-i\epsilon} \rightarrow 2\pi i\theta(-q_0)\delta(q^2+m^2) , \tag{4.7}$$

if q flows downwards; in so doing one obtains $2i$ times the required imaginary part of $A(t)$.

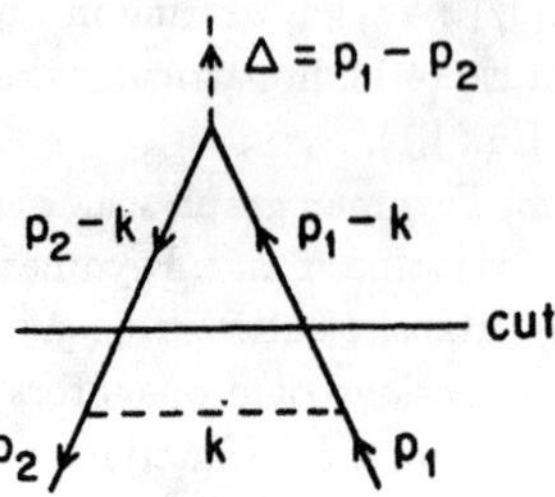

Figure 20: The only cut in the lowest order vertex diagram.

4.1 Second Order Example

The purely scalar part of the lowest order vertex graph depicted in Fig. 20 is

$$A(-\Delta^2) \equiv e^2 i \int \frac{d^4k}{(2\pi)^4} \frac{1}{(p_1-k)^2+m^2-i\epsilon} \frac{1}{(p_2-k)^2+m^2-i\epsilon} \frac{1}{k^2+\lambda^2-i\epsilon} \tag{4.1.1}$$

where the "small photon mass" λ has been kept, as the scalar amplitude would be otherwise infrared divergent. The cutting rule gives for the imaginary part

$$2i\mathrm{Im}A(t) = e^2 i \int \frac{d^4k}{(2\pi)^4}(2\pi i)\theta(p_{10}-k_0)\delta((p_1-k)^2+m^2)$$

$$\times(2\pi i)\theta(k_0-p_{20})\delta((p_2-k)^2+m^2)\frac{1}{k^2+\lambda^2-i\epsilon}\ ; \tag{4.1.2}$$

note the choice of the signs inside the θ functions: (p_1-k) flows upwards, giving $\theta(p_{10}-k_0)$, (p_2-k) downwards, giving $\theta(-(p_{20}-k_0)) = \theta(k_0-p_{20})$. It is convenient to introduce the new loop variable $q = p_1 - k$, so that Eq.(4.1.2) becomes

$$\mathrm{Im}A(t) = -e^2\frac{1}{2}\int d\Phi(t,m^2,m^2;q)\frac{1}{(p_1-q)^2+\lambda^2}\ , \tag{4.1.3}$$

where we have introduced the invariant two body phase space

$$d\Phi(t,m_1^2,m_2^2;q) \equiv \int \frac{d^4q}{(2\pi)^4}(2\pi)\theta(+q_0)\delta(q^2+m_1^2)(2\pi)\theta(\Delta_0-q_0)\delta((\Delta-q)^2+m_2^2). \tag{4.1.4}$$

The two θ functions imply $q_0 \geq \sqrt{m_1^2 + \vec{q}^2}$, $\Delta_0 - q_0 \geq \sqrt{m_2^2 + (\vec{\Delta} - \vec{q})^2}$, i.e. Δ must be timelike, $\Delta_0 \geq m_1 + m_2$ and

$$t \geq (m_1 + m_2)^2 \ , \tag{4.1.5}$$

otherwise the two body phase space would vanish. Once Eq.(4.1.5) is established, the θ functions in Eq.(4.1.4) can be forgotten. To evaluate the two body phase space, exploit the Lorentz invariance of the phase space integral, go to the rest frame of Δ, where $\vec{\Delta} = 0, \Delta_0 = \sqrt{t}$, use spherical coordinates

$$d^4q = dq_0 Q^2 dQ d\Omega(\hat{q}), \tag{4.1.6}$$

$\hat{q}$ being the direction of $\vec{q}$, and integrate the δ functions as follows:

$$Q^2 dQ \delta(q^2 + m_1^2) = Q^2 dQ \delta(-q_0^2 + Q^2 + m_1^2) = \frac{1}{2}Q, \tag{4.1.7}$$

with the mass shell condition $Q = \sqrt{q_0^2 - m_1^2}$;

$$dq_0 \delta((\Delta - q)^2 + m_2^2) = dq_0 \delta(-t + 2\sqrt{t}q_0 - m_1^2 + m_2^2) = \frac{1}{2\sqrt{t}}, \tag{4.1.8}$$

with the condition

$$q_0 = \frac{t + m_1^2 - m_2^2}{2\sqrt{t}}, \tag{4.1.9}$$

and therefore

$$Q = \frac{1}{2\sqrt{t}} R(t, m_1^2, m_2^2), \tag{4.1.10}$$

where the familiar phace space square root has been introduced

$$R(a,b,c) \equiv \sqrt{(a-b-c)^2 - 4bc} = \sqrt{a^2 + b^2 + c^2 - 2ab - 2bc - 2ca} \ . \tag{4.1.11}$$

Some particular cases are

$$R(t, m^2, m^2) = \sqrt{t(t - 4m^2)}, \qquad R(u, m^2, 0) = u - m^2, \qquad R(u, 0, 0) = u \ . \tag{4.1.12}$$

As a result

$$d\Phi(t, m_1^2, m_2^2) = \frac{1}{(2\pi)^2}\theta(t - (m_1 + m_2)^2)\frac{1}{8t}R(t, m_1^2, m_2^2) \int d\Omega(\hat{q}) \ . \tag{4.1.13}$$

By using that result and $\alpha = e^2/4\pi$, Eq.(4.1.3) becomes

$$\mathrm{Im}A(t) = -\left(\frac{\alpha}{\pi}\right)\theta(t - 4m^2)\frac{\sqrt{t(t-4m^2)}}{16t}\int \frac{d\Omega(\hat{q})}{(p_1 - q)^2 + \lambda^2} \ . \tag{4.1.14}$$

In the Δ-center of mass system

$$(p_1 \cdot q) = (-t + (t - 4m^2)(\hat{p} \cdot \hat{q}))/4, \tag{4.1.15}$$

so that finally, for $\lambda^2 << t - 4m^2$,

$$\frac{1}{\pi}\mathrm{Im}A(t) = -\left(\frac{\alpha}{\pi}\right)\theta(t - 4m^2)\frac{1}{4\sqrt{t(t-4m^2)}}\log\left(\frac{t - 4m^2}{\lambda^2}\right). \tag{4.1.16}$$

From the definition Eq.(4.1.1) it is seen that $A(t)$ vanishes for $t \to \infty$; assuming for definiteness Δ to be spacelike, one can then obtain $A(-\Delta^2)$, which is a real quantity, by means of the unsubtracted dispersion relation

$$A(-\Delta^2) = \frac{1}{\pi}\int_{(m_1+m_2)^2}^{\infty}\frac{dt}{t + \Delta^2}\mathrm{Im}A(t)\ . \tag{4.1.17}$$

In general, the two body phase space square root $R(t, m_1^2, m_2^2)$ can be removed by the corresponding rationalizing variable

$$x = x(t) = \frac{t - m_1^2 - m_2^2 - R(t, m_1^2, m_2^2)}{4m_1m_2}, \tag{4.1.18}$$

in terms of which

$$t = t(x) = m_1^2 + m_2^2 + m_1m_2\frac{1 + x^2}{x}, \qquad R(t, m_1^2, m_2^2) = m_1m_2\frac{1 - x^2}{x}, \tag{4.1.19}$$

and

$$\int_{(m_1+m_2)^2}^{\infty}\frac{dt}{R(t, m_1^2, m_2^2)}f(t) = \int_0^1\frac{dx}{x}\ f(t(x)). \tag{4.1.20}$$

4.2 Form Factors

Inspecting equations (2.2.5) and (2.2.6) for the form factors, one needs dispersion relations not only for the scalar amplitude (4.1.1) but also for terms which contain k_μ or $k_\mu k_\nu$ in the numerator . The dispersion relations for such terms are worked out in Appendix 6.3. The result is

$$\left(\frac{\alpha}{\pi}\right)\mathrm{Im}F_2^{(2)}(t) = -e^2\frac{1}{2}\int d\Phi(t, m^2, m^2; q)\frac{f_2^{(2)}(p_1 - q)}{(p_1 - q)^2}, \tag{4.2.1}$$

and $F_2^{(2)}$ satisfies the unsubtracted dispersion relation

$$F_2^{(2)}(-\Delta^2) = \frac{1}{\pi}\int\frac{dt}{t + \Delta^2}\mathrm{Im}F_2^{(2)}(t). \tag{4.2.2}$$

By using the explicit form (2.2.6) for $f_2^{(2)}$,one finds

$$\mathrm{Im}F_2^{(2)}(t) = \theta(t-4m^2)\frac{m^2}{\sqrt{t(t-4m^2)}}. \tag{4.2.3}$$

$F_2^{(2)}(0)$ can be obtained directly from the equation

$$a_2 = \frac{1}{\pi}\int \frac{dt}{t}\mathrm{Im}F_2^{(2)}(t) = \int_{4m^2}^{\infty} \frac{dt}{t}\frac{m^2}{\sqrt{t(t-4m^2)}} = \int_0^1 \frac{dx}{(1+x)^2} = \frac{1}{2}, \tag{4.2.4}$$

where we have used (4.1.18) for $m_1^2 = m_2^2 = m^2$.

The form of the t-dispersion relations remains the same in higher orders, so that one can write, in general

$$a = \frac{1}{\pi}\int \frac{dt}{t}\mathrm{Im}F_2(t)\ , \qquad r = \frac{1}{\pi}\int \frac{dt}{t^2}\mathrm{Im}F_1(t)\ , \tag{4.2.5}$$

where r, also called the electromagnetic radius of the electron, is the slope at $t=0$ of the charge form factor $F_1(t)$; r, which is infrared divergent at order $\left(\frac{\alpha}{\pi}\right)$, infrared finite at higher orders, is used in the evaluation of radiative corrections to atomic levels.

Dispersion relations are easily established also for the self-mass like expression for the anomaly of Eq. (2.2.2), (2.2.9). As a starting point, consider the integral

$$\left(\frac{\alpha}{\pi}\right) S^{(2)}(-p^2,a^2) \equiv e^2 i \int \frac{d^4k}{(2\pi)^4}\frac{1}{(p-k)^2+a^2-i\epsilon} \times \left(\frac{1}{k^2+\lambda^2-i\epsilon} - \frac{1}{k^2+\Lambda^2-i\epsilon}\right) . \tag{4.2.6}$$

The Cutkosky-Veltman rule gives, for large but finite Λ,

$$\frac{1}{\pi}\mathrm{Im}S^{(2)}(u,a^2) = -\left(\theta(u-(a+\lambda)^2)\frac{R(u,a^2,\lambda^2)}{4u} - \theta(u-\Lambda^2)\frac{u-\Lambda^2}{4u}\right), \tag{4.2.7}$$

and the unsubtracted dispersion relation

$$S^{(2)}(-p^2,a^2) = \frac{1}{\pi}\int \frac{du}{u+p^2-i\epsilon}\mathrm{Im}S^{(2)}(u,a^2). \tag{4.2.8}$$

In the anomaly one is often interested in quantities with squared electron denominators, such as for instance

$$\left(\frac{\alpha}{\pi}\right) T^{(2)}(-p^2) \equiv e^2 i \int \frac{d^4k}{(2\pi)^4} \frac{1}{((p-k)^2+m^2-i\epsilon)^2}$$
$$\times \left(\frac{1}{k^2-\lambda^2-i\epsilon} - \frac{1}{k^2-\Lambda^2-i\epsilon}\right) . \tag{4.2.9}$$

One begins with

$$\left(\frac{\alpha}{\pi}\right) S^{(2)}(-p^2, a^2) \equiv e^2 i \int \frac{d^4k}{(2\pi)^4} \frac{1}{(p-k)^2-a^2-i\epsilon}$$
$$\times \left(\frac{1}{k^2-\lambda^2-i\epsilon} - \frac{1}{k^2-\Lambda^2-i\epsilon}\right). \tag{4.2.10}$$

and writes

$$T^{(2)}(-p^2) = \left[-\frac{\partial}{\partial a^2} S^{(2)}(-p^2, a^2)\right]_{a^2=m^2} \tag{4.2.11}$$

In order to obtain a dispersion relation for $T^{(2)}(-p^2)$, one can differentiate the dispersion relation for $S^{(2)}$, with respect to a^2. At $p^2 = -m^2$, in particular, we find

$$T^{(2)}(m^2) = -\frac{1}{4}\int_{(m+\lambda)^2}^{\infty} \frac{du}{R(u, m^2, \lambda^2)} \frac{1}{u} = -\frac{1}{4}\int_0^1 \frac{dx}{x} \frac{x}{(m+\lambda x)(mx-\lambda)} = \frac{1}{4}\log\frac{\lambda}{m} . \tag{4.2.12}$$

where use has been made of the change of variable Eq.(4.1.18). The result is of course the value of the scalar vertex amplitude $A(-\Delta^2)$, (4.1.1), at $\Delta = 0$. The integrals involving $k_\mu, k_\mu k_\nu$ are dealt with as in the vertex case discussed in Appendix 6.3 (the projections are actually simpler). On invariance grounds, one can write

$$\left(\frac{\alpha}{\pi}\right) H_\mu^{(2)}(p, a^2) \equiv e^2 i \int \frac{d^4k}{(2\pi)^4} \frac{k_\mu}{((p-k)^2+a^2-i\epsilon)^2} \left(\frac{1}{k^2+\lambda^2-i\epsilon} - \frac{1}{k^2+\Lambda^2-i\epsilon}\right)$$
$$= \left(\frac{\alpha}{\pi}\right) H^{(2)}(-p^2, a^2) p_\mu ,$$

$$\left(\frac{\alpha}{\pi}\right) L_{\mu\nu}^{(2)}(p, a^2) \equiv e^2 i \int \frac{d^4k}{(2\pi)^4} \frac{k_\mu k_\nu}{((p-k)^2+a^2-i\epsilon)^2} \left(\frac{1}{k^2+\lambda^2-i\epsilon} - \frac{1}{k^2+\Lambda^2-i\epsilon}\right)$$
$$= \left(\frac{\alpha}{\pi}\right) L_0^{(2)}(-p^2, a^2)\delta_{\mu\nu} + \left(\frac{\alpha}{\pi}\right) L_1^{(2)}(-p^2, a^2) p_\mu p_\nu ;$$

$$\left(\frac{\alpha}{\pi}\right) (H^{(2)}(-p^2, a^2), L_i^{(2)}(-p^2, a^2)) = e^2 i \int \frac{d^4k}{(2\pi)^4} \frac{1}{((p-k)^2+a^2-i\epsilon)^2}$$
$$\times \left(\frac{1}{k^2+\lambda^2-i\epsilon} - \frac{1}{k^2+\Lambda^2-i\epsilon}\right) (h(k), l_i(k)) \tag{4.2.13}$$

with

$$h(k) = \frac{(p.k)}{p^2},\ l_0(k) = \frac{1}{3}\left(k^2 - \frac{(p.k)^2}{p^2}\right),\ l_1(k) = \frac{1}{3p^2}\left(4\frac{(p.k)^2}{p^2} - k^2\right) . \tag{4.2.14}$$

The scalar functions $H^{(2)}(-p^2, a^2), L_i^{(2)}(-p^2, a^2)$, *i.e.* the coefficients of $p_\mu, \delta_{\mu\nu}$ and $p_\mu p_\nu$, are free from kinematical singularities and, for finite Λ, satisfy unsubtracted dispersion relations of the form of Eq. (4.2.8).

Eq.(2.2.9) for the second-order moment can now be rewritten as

$$a_2 = 4m^2 \left[-\frac{\partial}{\partial a^2} \left(L_1^{(2)}(m^2, a^2) - H^{(2)}(m^2, a^2) \right) \right]_{a^2=m^2} . \tag{4.2.15}$$

With the obvious simplifications due to the fact that the above expression is both infrared and ultraviolet finite, one has

$$\begin{aligned} a_2 &= 4m^2 \left[-\frac{\partial}{\partial a^2} \int_{a^2}^{\infty} \frac{du}{u - m^2} \frac{1}{\pi} \left(\mathrm{Im} L_1^{(2)}(u, a^2) - \mathrm{Im} H^{(2)}(u, a^2) \right) \right]_{a^2=m^2} \\ &= m^2 \left[\frac{\partial}{\partial a^2} \int_{a^2}^{\infty} \left(\frac{(u-a^2)^3}{3u^3} - \frac{(u-a^2)^2}{2u^2} \right) \right]_{a^2=m^2} \\ &= m^4 \int_{m^2}^{\infty} \frac{du}{u^3} = \frac{1}{2} , \end{aligned} \tag{4.2.16}$$

recovering once more the electron anomaly to first order in $\left(\frac{\alpha}{\pi}\right)$.

Self-mass-like dispersion relations analogous to Eq.(4.2.15) can be worked out to all orders by suitably projecting out terms with powers of the loop variables in the numerators into scalar amplitudes free from kinematical singularities, as in Eq.(4.2.13).

The t-dispersion relation Eq.(4.2.5) has been used in [20] to reproduce the known value of the fourth order (two loops) electron anomaly. The same approach has been followed in [8,28], which gave the explicit analytic expression of Im $F_1(t)$, Im $F_2(t)$, up to 4th order and, by using (4.2.5), reproduced once more the 4th order anomaly and gave the first correct analytic expression of the electron electromagnetic radius in 4th order.

Infrared divergences are quite tricky in the dispersive formalism, and are briefly discussed in Appendix 6.4.

4.3 Applications of dispersion relations to graphs with vacuum polarization

Dispersion relations (also referred to, in this context, as spectral representations) are the natural tool for dealing with vacuum polarization insertions in lowest order graphs. The vacuum polarization tensor $\Pi_{\mu\nu}(k)$ can be expressed in terms of a single form factor $\Pi(-k^2)$,

$$\Pi_{\mu\nu}(k) = (k_\mu k_\nu - k^2 \delta_{\mu\nu}) \Pi(-k^2) , \tag{4.3.1}$$

which satisfies, after renormalization, the subtracted dispersion relation

$$\Pi(-k^2) = -k^2\frac{1}{\pi}\int\frac{du}{u(u+k^2-i\epsilon)}\mathrm{Im}\Pi(u)\ . \tag{4.3.2}$$

The explicit analytic value of $\mathrm{Im}\Pi(u)$ up to 4th order has been known [21] for a long time. When (4.3.1) and (4.3.2) are used, a vacuum polarization insertion in some lower order graph can be considered, apart from trivial algebra, as that lower order graph with a "massive" photon of mass $\sqrt{u}$, to be averaged in u with weight $\frac{1}{\pi}\mathrm{Im}\Pi(u)$.

The above representation for the vacuum polarization has been systematically used in the analytic evaluation of the contributions to the electron anomaly in 6th order due to vacuum polarization insertions. Ref. [22], using the t-dispersion relation Eq.(4.2.5) for the electron anomaly, gives the analytic value of the 6th order contributions due to the insertion of the 4th order vacuum polarization into the lowest order vertex graph (set J of Fig. 3); by the same approach, some 6th order contributions due to 2nd order vacuum polarization insertion into a 4th order graph are evaluated in [23] (graph K2), [24] (graph L2). The analytic calculation of all the remaining 6th order contributions due to 2nd order vacuum polarization insertions into 4th order vertex graphs was completed in [25] (graphs K1,L1) and in [26] (set M) by using the "self-mass" approach Eq.(2.1.12) and the suitable generalization to two loops of the self-mass like dispersion relation of Eq. (4.2.15). The same self-mass approach is followed in [27], which gives the analytic values of set C and E of Fig. 3.

5 HYBRID DISPERSIVE AND HYPERSPHERICAL APPROACH

The number of the different ways of cutting a graph for obtaining all the contributions to its imaginary part increases quickly with the order of the graph and with its topological complexity; to keep the number of the cuts down, dispersive and hyperspherical approaches have been combined in [29] to provide high precision semi-analytic values of the "double corner" graphs , set F. The combined approach has also been used in [30] in a first attempt to show the feasibility of the analytic calculation of the "light-by-light" graphs, set N, and "corner ladder" graphs, set G. (The scalar terms have been evaluated, but the calculation is still in progress). The idea in either case is the following: once the $\Delta \to 0$ limit is taken, any contribution X to the anomaly take the self-mass form depicted in Fig. 21 and can be written as

$$X = i\int\frac{d^4k}{(2\pi)^4}\frac{1}{k^2-i\epsilon}\frac{1}{(p-k)^2+m^2-i\epsilon}V(-(p-k)^2,-k^2,-p^2)\ , \tag{5.1}$$

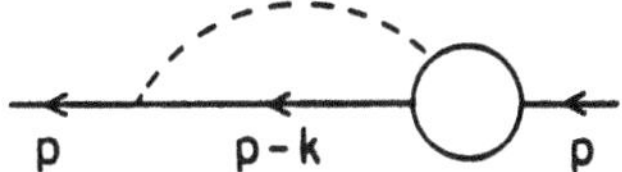

Figure 21: Self-mass like representation of a generic anomaly contribution.

where $V(-q^2, -k^2, -p^2)$ represents the inserted vertex-like amplitude (as a matter of fact $p^2 = -m^2$). One can write for it a dispersion relation in q^2 at fixed k^2, p^2,

$$V(-q^2, -k^2, -p^2) = \frac{1}{\pi} \int \frac{da}{a + q^2 - i\epsilon} \mathrm{Im}V(a, -k^2, -p^2) \ . \tag{5.2}$$

Insert the above dispersion relation in Eq.(5.1) and introduce hyperspherical coordinates for k; the resulting angular integrations are then easily carried out to give

$$X = \frac{1}{32\pi^3} \int \frac{da}{a - m^2} \int_0^\infty \frac{dK^2}{K^2} \left(\sqrt{K^2(K^2 + 4m^2)} + a - m^2 - R(a, -K^2, m^2) \right)$$
$$\times \mathrm{Im}V(a, -K^2, m^2) \ . \tag{5.3}$$

For graph F3, it is straightforward to write a dispersion relation for the other second order vertex insertion.

These methods apply in principle to all graphs except the "triple-cross" graphs, set H. No method suitable for the analytic calculation of these graphs has yet been found. The methods developed for the topologically simple cases apply also to 8th or higher order graphs with the same simple topology. For the "light-by-light" graphs considered in [30], $\mathrm{Im}V(a, -K^2, m^2)$ has the form of a two dimensional integral, the integrand being essentially a logarithm (of a somewhat complicated argument), so that one has as the starting point for the analytic integration a four dimensional integral of a known logarithm.

6 APPENDIX

6.1 Divergences in the Hyperspherical Formalism

The rules we have developed in the chapter 3 in the hyperspherical formalism would suffice if all the integrals existed, that is, if there were no divergences. In this section, we discuss how the divergences affect the rules.

The ultraviolet divergences are easily handled. From renormalization theory, we have a prescription for constructing counterterms, which when combined with the original integrand, give rise to the desired finite quantity. As explained in section 2.2, the counterterms cancel the original divergences in each graph pointwise in momentum space.

The counterterm for self-energy insertions, which are of course linearly divergent, requires special mention. It is usually stated that if one writes the self-energy and vertex pieces as

$$\sum(p) - \delta m = B(-i\not{p} + m) + \text{finite}, \tag{6.1.1}$$

$$\Gamma_\mu(p,p) = (1 + L)\gamma_\mu + \text{finite}, \tag{6.1.2}$$

then the Ward identity implies that the divergent constants B and L are related by

$$B = -L. \tag{6.1.3}$$

This is true only if the external momentum p follows the fermion line throughout. Because the self-energy graph is linearly divergent, this equation does not remain true if the external momentum is routed through the photon line. For example, $L^{(2)}$ and $B^{(2)}$ defined from the graphs in Fig. 22 are given by

$$L^{(2)} = \frac{\alpha}{\pi}(\frac{9}{8} + \frac{1}{2}\ \ln\ \lambda^2 + \frac{1}{4}\ \ln\ \Lambda^2), \tag{6.1.4}$$

$$B^{(2)} = -\frac{\alpha}{\pi}(\frac{7}{8} + \frac{1}{2}\ \ln\ \lambda^2 + \frac{1}{4}\ \ln\ \Lambda^2). \tag{6.1.5}$$

This nonconventional routing of the external momentum p is often more convenient because we may wish to label the fermion lines by a single momentum. The correct prescription for such counterterms is to evaluate B by the equation

$$-iB\gamma_\mu = \frac{\partial}{\partial p^\mu}\Big[\sum(p)\Big]\Big|_{i\not{p}=m}. \tag{6.1.6}$$

The removal of infrared divergences is much trickier. The main problem is that our method for performing the radial integrations was dependent on setting the photon mass to zero. But to treat the infrared divergence properly requires setting the photon mass to λ and keeping those pieces which contribute as $\lambda \to 0$. This introduces

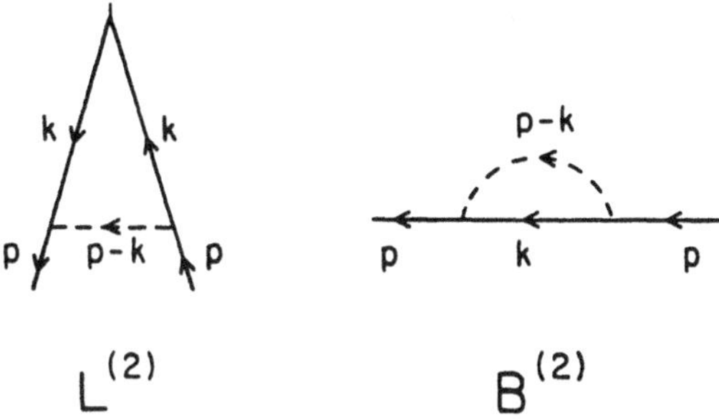

Figure 22: Graphs defining the divergent constants $L^{(2)}$ and $B^{(2)}$. The assignment of internal momenta is unorthodox since p does not follow the fermion line.

complicated square roots into the integrands. Using the technique of section 2.2, it is possible to isolate those pieces of the graph which contribute to the infrared singularity, and these are treated explicitly in the following section.

Although we are primarily interested in the magnetic-moment calculation, which has no overall infrared divergence, we have to compute the infrared pieces because we want to compare our results with the numerical values graph by graph. Secondly, the set of graphs we are able to evaluate analytically do not form a gauge invariant set, and so their sum may possess an infrared divergent piece.

6.1.1 Removal of Infrared Divergences

To illustrate the technique used to isolate the infrared divergences, let us suppose we had to perform the following integration:

$$I = -\frac{i}{\pi^2}\int \frac{d^4k}{(k-p)^2+\lambda^2-i\epsilon}\frac{1}{(k^2+m^2-i\epsilon)^2} \tag{6.1.1.1}$$

where $p^2 = -m^2$, and λ^2 is the photon mass. Setting λ^2 to zero at the outset gives an infrared divergence. This integration is easily done using Feynman techniques, but suppose we use the hyperspherical technique outlined in this paper. Taking p^2 spacelike, transforming to Euclidean variables, and doing the angular integration gives rise to the following integral (setting $m = 1$):

$$I = \int_0^\infty \frac{K}{P}\frac{Z_{KP}}{(1+K^2)^2}dK^2, \tag{6.1.1.2}$$

where

$$Z_{KP} = \frac{K^2 + P^2 + \lambda^2 - [(K^2 + P^2 + \lambda^2)^2 - 4P^2K^2]^{1/2}}{2PK}. \qquad (6.1.1.3)$$

Here we must continue P^2 to the mass-shell value (-1). This can be done by first transforming the K^2 contour to the one shown in Fig. 14 and setting $P^2 = -1$ in the integrand. This gives the result

$$I = I_1 + I_2, \qquad (6.1.1.4)$$

$$I_1 = -\frac{1}{2}\int_0^\infty \frac{dK^2}{(1+K^2)^2}\{(K^2 - 1 + \lambda^2) - [(K^2 + \lambda^2 - 1)^2 + 4K^2]^{1/2}\}, \qquad (6.1.1.5)$$

$$I_2 = \int_{-(1-\lambda)^2}^0 \frac{dK^2}{(1+K^2)^2}[(K^2 + \lambda^2 - 1)^2 + 4K^2]^{1/2}. \qquad (6.1.1.6)$$

Clearly, we can set λ to zero in I_1. This simply reflects the fact that the infrared divergence occurs near the mass shell, i.e., near $K^2 = -1$.

By making the transformation

$$K^2 = -1 + 2\eta\lambda, \qquad (6.1.1.7)$$

I_2 becomes

$$\int_{1-\frac{\lambda}{2}}^{\frac{1}{2\lambda}} \frac{d\eta}{2\eta^2}\left[4\eta^2 - 4 + 4\eta\lambda + \lambda^2\right]^{1/2} \qquad (6.1.1.8)$$

The limit $\lambda \to 0$ can be taken under the integral sign, and in the lower limit. The result is

$$\int_1^{\frac{1}{2\lambda}} \frac{d\eta}{2\eta^2}\left[4\eta^2 - 4\right]^{1/2} \qquad (6.1.1.9)$$

which is easily evaluated to give

$$I_2 = -1 - \ln\lambda + O(\lambda). \qquad (6.1.1.10)$$

If we had set λ to zero at the outset, we would have had to calculate

$$I_2 = \int_{-1}^0 \frac{dK^2}{1+K^2}. \qquad (6.1.1.11)$$

If one is sure that the infrared singularity arises from exactly the type of integral in (6.1.1.1), one can then formally set

$$\int_{-1}^0 \frac{dK^2}{1+K^2} = -1 - \ln\lambda \qquad (6.1.1.12)$$

in the calculation.

But one must be very careful. Consider an infrared singularity of the form

$$I_3 = \frac{i}{\pi^2} \int d^4k \frac{1}{[(k-p)^2 + \lambda^2]^2} , \tag{6.1.1.13}$$

which becomes

$$I_3 = - \int \frac{K}{P} Z_{KP} \frac{1}{[(K^2 + P^2 + \lambda^2)^2 - 4P^2K^2]^{1/2}} dK^2 , \tag{6.1.1.14}$$

where Z_{KP} is given by (6.1.1.3). Again deforming the contour and setting $P^2 = -1$, the singular term becomes

$$I_4 = \int_{-(1-\lambda)^2}^{0} \frac{dK^2}{[(K^2 + \lambda^2 - 1)^2 + 4K^2]^{1/2}} . \tag{6.1.1.15}$$

In the limit $\lambda \to 0$, we would have set

$$I_4 \sim \int_{-1}^{0} \frac{dK^2}{1 + K^2} , \tag{6.1.1.16}$$

but a careful evaluation gives

$$I_4 = - \ln \lambda \tag{6.1.1.17}$$

instead of (6.1.1.10). Thus there is no general formula for

$$\int_{-1}^{0} \frac{dK^2}{1 + K^2} , \tag{6.1.1.18}$$

unless the exact origin of the singularity is understood. It turns out that there are only two forms of single loop logarithmically divergent infrared pieces, and which of the two to use depends only on whether the photon line in the loop has been differentiated [use (6.1.1.17)] or not [use (6.1.1.10)]. (Photon lines can be differentiated either because they carry part of the external-photon momentum q or if in the subtraction constant of a fermion self-energy piece they carry the external-fermion momentum.)

Similarly, there turn out to be several types of divergences of the form $\ln^2 \lambda$. For example

$$\int_{-(1-\lambda)^2}^{0} dK^2 \frac{[(K^2 - 1 + \lambda^2)^2 + 4K^2]^{1/2}}{(1 + K^2)^3} \left[\int_{-(\sqrt{-K^2} - \lambda)^2}^{0} dL^2 \frac{[(K^2 + L^2 + \lambda^2)^2 - 4K^2L^2]^{1/2}}{1 + L^2} \right.$$

$$\left. - \int_{-(1-\lambda)^2}^{0} dL^2 \frac{[(L^2 + \lambda^2 - 1)^2 + 4L^2]^{1/2}}{1 + L^2} \right] = -\frac{1}{8} \ln^2 \lambda^2 + \frac{1}{2} + \frac{1}{12}\pi^2 , \tag{6.1.1.19}$$

$$\int_{-(1-\lambda)^2}^{0} dK^2 \frac{1}{(1+K^2)[(K^2+\lambda^2-1)^2+4K^2]^{1/2}} \Bigg[\int_{-(\sqrt{-K^2}-\lambda)^2}^{0} dL^2 \frac{[(K^2+L^2+\lambda^2)^2-4K^2L^2]^{1/2}}{1+L^2}$$

$$- \int_{-(1-\lambda)^2}^{0} dL^2 \frac{[(L^2+\lambda^2-1)^2+4L^2]^{1/2}}{1+L^2} \Bigg] = -\frac{1}{8}\ln^2\lambda^2 + \frac{1}{2}\ln\lambda^2 + \frac{5}{24}\pi^2 . \tag{6.1.1.20}$$

Having set λ to zero initially, the left-hand side of these expressions would have been

$$\int_{-1}^{0} \frac{dK^2}{1+K^2}[-1+\ln(1+K^2)] .$$

(6.1.1.19) and (6.1.1.20) arise respectively in the infrared divergent counter terms to graph B2 (corresponding to graph d of Fig. 2) and graph B1 (in view of the unorthodox routing of the external momentum along the photon lines). Similarly

$$\int_{-(1-\lambda)^2}^{0} \frac{dK^2}{K^2(1+K^2)^2}\Big[(K^2+\lambda^2-1)^2+4K^2\Big]^{1/2} \int_{-(\sqrt{-K^2}-\lambda)^2}^{0} \frac{dL^2}{(1+L^2)^2}\Big[(L^2+K^2+\lambda^2)^2-4K^2L^2\Big]^{1/2}$$

$$= -\frac{1}{8}\ln^2\lambda^2 - \ln\lambda^2 - 1 - \frac{5\pi^2}{24} \tag{6.1.1.21}$$

arises in the infrared counter term corresponding to graph b of Fig. 2.

The infrared counterterms corresponding to graphs a and c of Fig. 2 are more complicated, because the natural hyperspherical variables (See Equation (3.4.5)) are quite different functions of the momenta above and below the cut, and considerable care must be taken in evaluating the difference across the cut.

6.2 Difficult Angular Integrals

Suppose that the loop momenta in Euclidean space are P, K, L, M and that the denominator of the Feynman integrand does not involve $M.L$. So the denominator depends on the direction of M only through the combinations $(M.P)$ and $(M.K)$. If the numerator is also independent of $M.L$, the angular integral over M can be done using the techniques of section 3.1. But suppose the numerator has a term $M.L$. We then have to perform

$$\int \frac{d\Omega_M}{2\pi^2} \frac{M.L}{\mathit{Polynomial\ in\ }(M.K, M.P)} \tag{6.2.1}$$

The techniques in section 3.1 do not suffice. But they are easily adapted. Write (6.2.1) as $L_\alpha I_\alpha$ with

$$I_\alpha = \int \frac{d\Omega_M}{2\pi^2} \frac{M_\alpha}{\mathit{Polynomial\ in\ }(M.K, M.P)} \tag{6.2.2}$$

By four-dimensional rotational invariance, I_α must be a linear combination of K_α and P_α, with scalar coefficients.

$$I_\alpha = a\, K_\alpha + b\, P_\alpha \tag{6.2.3}$$

Introduce two orthonormal vectors R_α, S_α in the 2-dimensional space spanned by K_α, P_α e.g.

$$R_\alpha = \frac{P_\alpha}{\sqrt{P.P}} \qquad S_\alpha = \frac{K_\alpha - (K.P)\, P_\alpha/P.P}{\sqrt{K^2 - (K.P)^2/P.P}} \tag{6.2.4}$$

Using the completeness of R, S in the space spanned by K, P we can rewrite M_α in (6.2.2) as

$$M_\alpha = (R_\alpha R_\beta + S_\alpha S_\beta) M_\beta \tag{6.2.5}$$

$$= R_\alpha (R.M) + S_\alpha (S.M) \tag{6.2.6}$$

so that the free index no longer appears on the variable of integration M. Now the techniques in the text allow one to carry out the angular M integral.

If the numerator is quadratic in M, so that we need to evaluate

$$I_{\alpha\beta} = \int \frac{d\Omega_M}{2\pi^2} \quad \frac{M_\alpha M_\beta}{Polynomial\ in\ (M.K, M.P)} \tag{6.2.7}$$

then one needs a complete set of orthonormal symmetric tensors in the R, S space. These are, for example,

$$T_{1\alpha\beta} = R_\alpha R_\beta, T_{2\alpha\beta} = S_\alpha S_\beta, T_{3\alpha\beta} = (R_\alpha S_\beta + S_\alpha R_\beta)/\sqrt{2}, T_{4\alpha\beta} = (\delta_{\alpha\beta} - R_\alpha R_\beta - S_\alpha S_\beta)/\sqrt{2} \tag{6.2.8}$$

Then writing

$$M_\alpha M_\beta = \sum_{i=1}^{4} T_{i\alpha\beta}\; T_{i\mu\nu} M_\mu M_\nu \tag{6.2.9}$$

again transforms the free index dependence away from the variable of integration, and one can carry out the integral.

6.3 Dispersion relation for vector and tensor quantities

Consider

$$B(-\Delta^2) \equiv e^2 i \int \frac{d^4k}{(2\pi)^4} \frac{1}{(p_1 - k)^2 + m^2 - i\epsilon} \frac{1}{(p_2 - k)^2 + m^2 - i\epsilon} \frac{(p_1 \cdot k)}{k^2 - i\epsilon} \,. \tag{6.3.1}$$

Due to the presence of a power of k in the numerator, infrared regularization is not needed. By following the same steps as for $A(-\Delta^2)$, Eq.(4.1.1), one has

$$\mathrm{Im} B(t) = e^2 \frac{1}{2} \int d\Phi(t, m^2, m^2; q) \frac{m^2 + (p_1 \cdot q)}{(p_1 - q)^2} \,, \tag{6.3.2}$$

i.e.,

$$\frac{1}{\pi}\text{Im}B(t) = -\left(\frac{\alpha}{\pi}\right)\theta(t-4m^2)\frac{\sqrt{t(t-4m^2)}}{8t} . \qquad (6.3.3)$$

As opposed to $A(t)$, $\text{Im}B(t)$ does not vanish for $t \to \infty$, so that a dispersion relation as simple as Eq.(4.1.17) cannot be written. In order to evaluate $B(t)$ by means of a dispersion relation, one has to consider the integral

$$C_\mu(p_1,p_2) \equiv e^2 i \int \frac{d^4k}{(2\pi)^4}\frac{1}{(p_1-k)^2+m^2-i\epsilon}\frac{1}{(p_2-k)^2+m^2-i\epsilon}\frac{k_\mu}{k^2-i\epsilon} ; \qquad (6.3.4)$$

On invariance grounds, one can write

$$C_\mu(p_1,p_2) = C_1(-\Delta^2)p_{1\mu} + C_2(-\Delta^2)p_{2\mu} . \qquad (6.3.5)$$

By contracting Eq.(6.3.5) with $p_{1\mu}, p_{2\mu}$ one obtains two linear equations involving the two invariant amplitudes $C_i(-\Delta^2)$; solving them for $C_i(-\Delta^2)$ one finds

$$C_i(-\Delta^2) = e^2 i \int \frac{d^4k}{(2\pi)^4}\frac{1}{(p_1-k)^2+m^2-i\epsilon}\frac{1}{(p_2-k)^2+m^2-i\epsilon}\frac{c_i(k)}{k^2-i\epsilon} , \qquad (6.3.6)$$

where, in full generality (i.e. without imposing mass shell constraints on p_1, p_2),

$$c_1(k) = \frac{p_2^2(p_1\cdot k)-(p_1\cdot p_2)(p_2\cdot k)}{p_1^2p_2^2-(p_1\cdot p_2)^2} ,$$

$$c_2(k) = \frac{p_1^2(p_2\cdot k)-(p_1\cdot p_2)(p_1\cdot k)}{p_1^2p_2^2-(p_1\cdot p_2)^2} . \qquad (6.3.7)$$

A direct study of Eq.(6.3.5) shows that the $C_i(-\Delta^2)$ have the same analytic properties and asymptotic behaviour as $A(-\Delta^2)$. To see this, one can for instance introduce standard Feynman parameters and perform the loop integral; when that is done, the scalar functions $A(-\Delta^2), C_i(-\Delta^2)$ take the form of integrals over the Feynman parameters, and their integrands have the same denominator (which contains Δ^2 and determines the analytic and asymptotic properties in that variable) and different numerators, consisting however of simple polynomials in the Feynman parameters, which do not change those properties. The expressions in (6.3.7) involve the denominator

$$\frac{1}{p_1^2p_2^2-(p_1\cdot p_2)^2} \qquad (6.3.8)$$

which, when imposing the mass shell conditions for p_1^2, p_2^2,becomes

$$\frac{1}{p_1^2p_2^2-(p_1\cdot p_2)^2} = -\frac{4}{t(t-4m^2)} . \qquad (6.3.9)$$

The presence of such a factor seems to cause, at first sight, the appearance of poles at $t = 0, t = 4m^2$ (known, when actually present, as kinematical singularities); a more careful analysis shows however that the poles are not there, as the integrals vanish when $t = 0$ or $t = 4m^2$, in full agreement, of course, with the results of the Feynman parameter analysis.

As a result of the discussion, the $C_i(-\Delta^2)$, (6.3.6) do satisfy unsubtracted dispersion relations of the form of Eq.(4.1.17)

$$C_i(-\Delta^2) = \frac{1}{\pi}\int_{4m^2}^{\infty} \frac{dt}{t+\Delta^2}\mathrm{Im}C_i(t) \ , \tag{6.3.10}$$

and their imaginary parts are given by the obvious extensions of (4.1.3)

$$\mathrm{Im}C_i(t) = -e^2\frac{1}{2}\int d\Phi(t, m^2, m^2; q)\frac{c_i(p_1 - q)}{(p_1 - q)^2} \ . \tag{6.3.11}$$

To obtain $B(-\Delta^2)$, Eq.(6.3.1), one has

$$B(-\Delta^2) = p_{1\mu}\cdot C_\mu(p_1, p_2) = p_1^2 C_1(-\Delta^2) + (p_1\cdot p_2)C_2(-\Delta^2) \ , \tag{6.3.12}$$

and, on account of the mass shell conditions,

$$B(t) = -m^2 C_1(t) + \frac{1}{2}(t - 4m^2)C_2(t) \ . \tag{6.3.13}$$

The last term to consider is

$$D_{\mu\nu}(p_1, p_2) \equiv e^2 i\int \frac{d^4k}{(2\pi)^4}\frac{1}{(p_1-k)^2+m^2-i\epsilon}\,\frac{1}{(p_2-k)^2+m^2-i\epsilon} \left(\frac{1}{k^2-i\epsilon} - \frac{1}{k^2+\Lambda^2-i\epsilon}\right)k_\mu k_\nu \ ; \tag{6.3.14}$$

the Pauli-Villars regularization must be kept to parametrize the ultraviolet divergence due to the presence of two powers of k in the numerator. On invariance grounds and showing explicitly the Λ dependence one can write

$$D_{\mu\nu}(p_1, p_2) = D_0(-\Delta^2, \Lambda^2)\delta_{\mu\nu} + D_1(-\Delta^2, \Lambda^2)p_{1\mu}p_{1\nu} + D_2(-\Delta^2, \Lambda^2)(p_{1\mu}p_{2\nu} + p_{2\mu}p_{1\nu}) + D_3(-\Delta^2, \Lambda^2)p_{2\mu}p_{2\nu}. \tag{6.3.15}$$

Obtaining the invariant amplitudes $D_i(-\Delta^2, \Lambda^2)$ is straightforward, although somewhat cumbersome; the calculation can be speeded up by using an orthonormal basis built in terms of the orthogonal vectors $p_{1\mu}$ and $\left(p_{2\mu} - \frac{(p_1\cdot p_2)}{p_1^2}p_{1\mu}\right)$; the result can be written as

$$D_i(-\Delta^2,\Lambda^2) \equiv e^2 i \int \frac{d^4k}{(2\pi)^4} \frac{1}{(p_1-k)^2+m^2-i\epsilon} \frac{1}{(p_2-k)^2+m^2-i\epsilon}$$

$$\left(\frac{1}{k^2-i\epsilon} - \frac{1}{k^2+\Lambda^2-i\epsilon}\right) d_i(k) , \tag{6.3.16}$$

where, for arbitrary p_1, p_2,

$$d_0(k) = \frac{1}{2}\left(k^2 - \frac{p_1^2(p_2\cdot k)^2 - 2(p_1\cdot p_2)(p_1\cdot k)(p_2\cdot k) + p_2^2(p_1\cdot k)^2}{p_1^2p_2^2 - (p_1\cdot p_2)^2}\right) ,$$

$$d_1(k) = \frac{1}{2}\left(3p_2^2\frac{p_1^2(p_2\cdot k)^2 - 2(p_1\cdot p_2)(p_1\cdot k)(p_2\cdot k) + p_2^2(p_1\cdot k)^2}{(p_1^2p_2^2 - (p_1\cdot p_2)^2)^2} - \frac{p_2^2k^2 + 2(p_2\cdot k)^2}{p_1^2p_2^2 - (p_1\cdot p_2)^2}\right) ,$$

$$d_2(k) = 3\frac{2p_1^2p_2^2(p_1\cdot k)(p_2\cdot k) - (p_1\cdot p_2)(p_1^2(p_2\cdot k)^2 + p_2^2(p_1\cdot k)^2)}{2(p_1^2p_2^2 - (p_1\cdot p_2)^2)^2}$$

$$+\frac{(p_1\cdot p_2)k^2 - 4(p_1\cdot k)(p_2\cdot k)}{2(p_1^2p_2^2 - (p_1\cdot p_2)^2)} ,$$

$$d_3(k) = \frac{1}{2}\left(3p_1^2\frac{p_2^2(p_1\cdot k)^2 - 2(p_1\cdot p_2)(p_1\cdot k)(p_2\cdot k) + p_1^2(p_2\cdot k)^2}{(p_1^2p_2^2 - (p_1\cdot p_2)^2)^2} - \frac{p_1^2k^2 + 2(p_1\cdot k)^2}{p_1^2p_2^2 - (p_1\cdot p_2)^2}\right) . \tag{6.3.17}$$

The unsubtracted dispersion relations Eq.(4.1.17) and (6.3.10) which are valid for $A(-\Delta^2)$ and the $C_i(-\Delta^2)$ hold also for the $D_i(-\Delta^2,\Lambda^2)$, as long as Λ is kept finite. The explicit expressions Eq.(6.3.17) shows that for $i = 1,2,3$ the $\Lambda \to \infty$ limit is trivial, i.e. one can define

$$\lim_{\Lambda\to\infty} D_i(-\Delta^2,\Lambda^2) = D_i(-\Delta^2) , \tag{6.3.18}$$

where the $D_i(-\Delta^2)$ are obtained by simply dropping the Λ-term from Eq.(6.3.16) and satisfy the unsubtracted dispersion relations

$$D_i(-\Delta^2) = \frac{1}{\pi}\int \frac{dt}{t+\Delta^2}\mathrm{Im}D_i(t) . \tag{6.3.19}$$

For $D_0(-\Delta^2,\Lambda^2)$ one has

$$\lim_{\Lambda\to\infty} \mathrm{Im}D_0(t,\Lambda^2) = \mathrm{Im}D_0(t) , \tag{6.3.20}$$

where $\mathrm{Im}D_0(t)$, again, is obtained by simply dropping the Λ-term from Eq.(6.3.16), but $\mathrm{Re}D_0(-\Delta^2,\Lambda^2)$ diverges for $\Lambda \to \infty$; for large Λ one can define $D_{0f}(-\Delta^2)$ through

$$D_0(\Delta^2,\Lambda^2) = D_0(0,\Lambda^2) + D_{0f}(-\Delta^2) \tag{6.3.21}$$

and $D_{0f}(-\Delta^2)$ satisfies the subtracted dispersion relation

$$D_{0f}(-\Delta^2) = -\Delta^2\frac{1}{\pi}\int \frac{dt}{t(t+\Delta^2)}\mathrm{Im}D_0(t) . \tag{6.3.22}$$

As a result, all the Λ-dependence of $D_{\mu\nu}(p_1,p_2)$ is buried in the constant $D_0(0,\Lambda^2)$.

6.4 Infrared Divergences in the Dispersive Formalism

Among the problems encountered in the analytic integration when using dispersion relations, infrared divergences deserve a special comment. Although the anomaly is a physical quantity, free from infrared divergences, separate graphs are not in general infrared finite, as their infrared divergences cancel only in the total sum (and also within suitably chosen subsets of graphs; see Section 2.2). When using dispersion relations, one has further to consider separately the contributions from the different cuts of a same graph (or group of graphs). That implies the use of different integration variables or of a different order of integration for a same set of integration variables, so that the compensation of infrared divergences between the various contributions is postponed to a later stage and a larger variety of infrared divergences can appear in the intermediate results. It is therefore necessary to regularize all the potentially infrared divergent terms and to carry out their complete calculation with the infrared regularization.

When one uses as regularization a small photon mass λ, a photon with energy ω has momentum $k = \sqrt{\omega^2 - \lambda^2}$, as opposed to $k = \omega$ in the "true" massless case; that introduces one more square root in the phase space, changing completely the pattern of the analytic integration. The typical integral which appears is of the type $\int_\lambda \frac{d\omega}{\sqrt{\omega^2-\lambda^2}} F(\omega,\lambda)$; the basic technique for evaluating it consists in finding the simplest (and hopefully easy to integrate) function $\bar{F}(\omega,\lambda)$ which approximates $F(\omega,\lambda)$ uniformly for $\omega \to \lambda$, so that

$$\int_\lambda \frac{d\omega}{\sqrt{\omega^2-\lambda^2}} F(\omega,\lambda) = \int_\lambda \frac{d\omega}{\sqrt{\omega^2-\lambda^2}} \left[\bar{F}(\omega,\lambda) + F(\omega,\lambda) - \bar{F}(\omega,\lambda)\right]$$

$$\simeq \int_\lambda \frac{d\omega}{\sqrt{\omega^2-\lambda^2}} \bar{F}(\omega,\lambda) + \int_0 \frac{d\omega}{\omega}\left[F(\omega,0) - \bar{F}(\omega,0)\right] , \qquad (6.4.1)$$

and only the supposedly simpler first term depends still on λ.

Usually the naive $\lambda \to 0$ limit gives as a warning an explicitly divergent integral; but cases are known in the literature in which the naive $\lambda \to 0$ limit leads to a finite but wrong result: see [8], footnote 2 and Sec. 1.3 of [28].

6.5 Polylogarithms

The result of the analytic evaluation of some contribution to the electron anomaly is often written as the combination of a few mathematical constants times rational coefficients; those elegant and relatively simple formulae are in most cases the values taken at some suitable points by particular special functions belonging to the class of the Nielsen polylogarithms, [31] (see also [8]); the same functions appear quite naturally also in the intermediate steps leading to the analytic results.

The Nielsen polylogarithms, a generalization of the Euler dilogarithm, are defined by

$$S_{n,p}(y) \equiv \frac{(-1)^{n+p-1}}{(n-1)!p!} \int_0^1 \frac{dt}{t} \ln^{n-1} t \, \ln^p(1-yt) \; ; \tag{6.5.1}$$

they are real for real $y \leq 1$ and develop an imaginary part for real $y > 1$. The dilogarithm $\mathrm{Li}_2(y)$ (sometimes called also Spence function) corresponds to the $n = p = 1$ case,

$$\mathrm{Li}_2(y) \equiv S_{1,1}(y) = -\int_0^1 \frac{dt}{t} \ln(1-yt) = -\int_0^y \frac{dt}{t} \ln(1-t) \; ; \tag{6.5.2}$$

it appears naturally when integrating a logarithm multiplied by a rational expression, in the same way as logarithms appear when integrating rational expressions. From Eq.(6.5.2) one has the expansion

$$\mathrm{Li}_2(y) = \sum_{m=1}^{\infty} \frac{y^m}{m^2} \, , \tag{6.5.3}$$

which converges for $y \leq 1$, and the value at $y = 1$

$$\mathrm{Li}_2(1) = -\int_0^1 \frac{dt}{t} \ln(1-t) = \zeta(2) \, , \tag{6.5.4}$$

where

$$\zeta(p) \equiv \sum_{m=1}^{\infty} \frac{1}{m^p} \tag{6.5.5}$$

is the Riemann ζ-function of argument p.

From the definition Eq.(6.5.1) one can easily establish the important relations

$$S_{n+1,p}(y) = \int_0^y \frac{dt}{t} S_{n,p}(t), \qquad \frac{d}{dy} S_{n,p}(y) = \frac{1}{y} S_{n-1,p}(y) \, . \tag{6.5.6}$$

It is convenient for classification purposes to define the PolyLogarithmic Degree (PLD for short) of a product of logarithms and polylogarithms with the rules: the logarithm has PLD 1; the polylogarithm of indices n, p has PLD $(n+p)$; the PLD of a product is the sum of the PLD of its factors. The dilogarithm then has PLD 2, the product $\mathrm{Li}_2(y) \ln y$ has PLD 3, etc. As an obvious consequence of the definition and of Eq.(6.5.6), the derivative with respect to some variable of a term with PLD m is a combination of terms all having PLD $(m-1)$, times the derivatives of the arguments; if the arguments are algebraic or rational functions of that variable, the derivatives of the arguments are still algebraic or rational functions.

A great number of identities involving polylogarithms of related arguments, such as for instance $y, 1/y, (1-y), y/(1-y)$, etc, can be found in the literature. All the identities can be easily established by checking their validity at some convenient particular value of y and then differentiating with respect to y, thus obtaining an identity between functions with lower PLD. The procedure can be repeated until one obtains an identity between logarithms, whose validity is trivial to ascertain. As an example, consider the relation

$$\mathrm{Li}_2(1-y) = -\mathrm{Li}_2(y) - \ln(1-y)\ln y + \zeta(2) \ ; \tag{6.5.7}$$

it is satisfied at $y = 0$; by differentiating with respect to y it becomes

$$\frac{1}{1-y}\ln y = \frac{1}{y}\ln(1-y) + \frac{1}{1-y}\ln y - \frac{1}{y}\ln(1-y) \ ;$$

which is obvious, so that Eq.(6.5.7) is established. At $y = 1/2$ it gives in particular

$$\mathrm{Li}_2\left(\frac{1}{2}\right) \equiv a_2 = \frac{1}{2}\zeta(2) - \frac{1}{2}\ln^2 2 \ . \tag{6.5.8}$$

Similarly, one establishes

$$\frac{1}{2}\mathrm{Li}_2(y^2) = \mathrm{Li}_2(y) + \mathrm{Li}_2(-y) \ , \tag{6.5.9}$$

from which, at $y = -1$,

$$\int_0^1 \frac{dt}{t}\ln(1+t) = -\mathrm{Li}_2(-1) = \frac{1}{2}\zeta(2); \tag{6.5.10}$$

for real $y \geq 0$ one then obtains

$$\mathrm{Li}_2(-y) = -\mathrm{Li}_2\left(-\frac{1}{y}\right) - \frac{1}{2}\ln^2 y - \zeta(2) \ , \tag{6.5.11}$$

where the constant is fixed by the value at $y = 1$; by analytic continuation to $y = -1$, as $\ln^2(-1) = -\pi^2$, one obtains the known relation

$$\zeta(2) = \frac{\pi^2}{6} \ . \tag{6.5.12}$$

One can easily extend the various identities obtained for the dilogarithm, (Eqs. (6.5.7), (6.5.9), (6.5.11)), and its values for arguments $\pm 1, 1/2$, (Eqs. (6.5.4), (6.5.8), (6.5.10)), to the polylogarithms of higher order. For those values of the arguments, the polylogarithms are expressed in terms of the Riemann ζ-function $\zeta(p)$, ($\zeta(p)$ for even p is known to be equal to π^p times a rational coefficient), and of the constants a_p defined by

$$a_p \equiv \sum_{m=1}^{\infty} \frac{1}{m^p 2^m} \ . \tag{6.5.13}$$

($a_1 = \ln 2$, a_2 is given by (6.5.8) and similarly one finds also $a_3 = \frac{7}{8}\zeta(3) - \frac{1}{2}\zeta(2)\ln 2 + \frac{1}{6}\ln^3 2$, but no similar expressions seem to exist for $p > 3$). It is natural to extend the definition of PLD, introduced for the polylogarithms of arbitrary arguments, to the above mathematical constants too, assigning to $\zeta(p)$ and a_p a PLD value of p. The values of the polylogarithms of argument $\pm 1, 1/2$, in analogy with (6.5.4),(6.5.8),(6.5.10), can then be seen as a table of definite integrals, whose left hand side is the integral between 0 and 1 in some variable, say t, of an integrand containing typically one of the rational factors $1/t, 1/(1 \pm t)$ times a product of logarithms and polylogarithms of simple arguments with PLD n, while the r.h.s is a combination of products of rational fractions times mathematical constants with PLD $(n+1)$. Further entries in the table can be generated by integrations by parts, changes of the integration variables, use of identities between polylogarithms etc. Ref. [8] contains a fairly complete table of integrals of the kind described above, whose left hand side involves practically all the integrands with PLD of 2 and 3, and the right hand side consists of mathematical constants with PLD of 3 and 4. A number of definite integrals with left hand side having PLD 4 and right hand side having PLD 5 can be found in [29] (the first coefficient in the third entry of Table IV is to be corrected into 99/16); a complete table for PLD=5 is however still missing. Definite integrals containing a function with PLD n but different powers of the rational factors $t, (1 \pm t)$ can be simplified by integrating the rational factors by parts, so obtaining a simpler definite integral involving a product of polylogarithms with lower PLD.

The concept of PLD is also useful in working out algorithms for the analytic integration of integrands depending on many variables. Consider for instance a definite integral like

$$\int_{x_1}^{x_2} \frac{dx}{R(x,y,z)} \frac{1}{x-a} G_p(x,y,...) ,$$

where $R(x,y,z)$ is the usual square root as in Eq.(4.1.11) and the subscript p in $G_p(x,y,...)$ indicates that G_p has PLD equal to p. One finds that

$$F_{p+1}(y,...) \equiv \int_{x_1}^{x_2} \frac{dx}{R(x,y,z)} \frac{R(a,y,z)}{x-a} G_p(x,y,...) , \tag{6.5.14}$$

thanks to the introduction of the factor $R(a,y,z)$ in the numerator, has PLD exactly equal to $(p+1)$; that is, its derivative with respect to y has PLD equal to p. The result can be verified by introducing for instance the rationalizing variable Eq.(4.1.18) and then carrying out the required algebra. Then (6.5.14) becomes

$$F_{p+1}(y,...) = \int_{u(x_1)}^{u(x_2)} du\left(\frac{1}{u - 1/u(a)} - \frac{1}{u - u(a)}\right) G_p(x,y,...) \tag{6.5.15}$$

where

$$u(x) = \frac{x - y - z - R(x,y,z)}{2\sqrt{yz}} \tag{6.5.16}$$

Differentiating the right side of (6.5.15) with respect to y gives end point contributions, an explicit term involving $\frac{\partial G_p}{\partial y}$, and terms arising from, for example,

$$\frac{\partial}{\partial y}\left(\frac{1}{u-u(a)}\right)=\frac{1}{(u-u(a))^2}\frac{\partial u(a)}{\partial y} .$$

After integrating by parts with respect to u, the u derivative can act on G_p, reducing its PLD to $(p-1)$. In the same way one finds that the integrals of the kind

$$\int_{x_1}^{x_2}\frac{dx}{R(x,y,z)}G_p(x,y,...) , \qquad \int_{x_1}^{x_2}\frac{dx}{x-a}G_p(x,y,...) , \tag{6.5.17}$$

also have PLD exactly equal to $(p+1)$, without the introduction of any extra factor. Higher powers of the denominator $(x-a)$, on the other hand, can be worked out by using (repeatedly, when needed) the formula

$$\int_{x_1}^{x_2}\frac{dx}{R(x,y,z)}\frac{G_p(x,...)}{(x-a)^n}=-\frac{1}{n-1}\frac{1}{R^2(a,y,z)}\int_{x_1}^{x_2}\frac{dx}{(x-a)^{n-1}}\Bigg[R(x,y,z)$$
$$\left(\delta(x-x_2)-\delta(x-x_1)-\frac{\partial}{\partial x}\right)+\frac{(n-2)(x-a)+(3-2n)((y+z)-a)}{R(x,y,z)}\Bigg]G_p(x,...) ; \tag{6.5.18}$$

when that is done, one obtains a combination of integrals of the form of (6.5.14),(6.5.17) plus end-point values or simpler integrals, which involve derivatives of $G_p(x,...)$ and therefore have PLD equal to $(p-1)$. Similar formulae for higher powers of $R(x,y,z)$ and no powers of $(x-a)$ are easily established, while products of higher powers of R and $(x-a)$ can be decomposed by partial fractions into combinations of the previous terms; it turns out, therefore, that the nontrivial integrals are all of the form of (6.5.14),(6.5.17).

As $F_{p+1}(y,...)$ has PLD equal to $(p+1)$, its derivatives with respect to any of the arguments are equal to a combination of rational functions times polylogarithms with PLD equal to p. When the derivative is explicitly carried out, the result is a combination of end-point values and integrals whose integrand are derivatives of $G_p(x,y,...)$ and therefore have PLD equal to $(p-1)$. By elementary means one can for instance obtain, Ref. [30],

$$\frac{\partial}{\partial y}F_{p+1}(y,...)=\int_{x_1}^{x_2}\frac{dx}{R(x,y,z)}\Bigg[\frac{R(a,y,z)}{x-a}\frac{\partial}{\partial y}$$
$$+\frac{xy-3y^2+2yz-zx+z^2+a(x+y-z)}{2yR(a,y,z)}\left(\delta(x-x_2)-\delta(x-x_1)-\frac{\partial}{\partial x}\right)\Bigg]G_p(x,y,...) , \tag{6.5.19}$$

where it has been assumed, for simplicity, that the end-points x_1, x_2 and a are independent of y. Note that $G_p(x, y, ...)$ itself might also be given as the definite integral of a simpler function with PLD equal to $(p-1)$, so that the derivatives of $G_p(x, y, ...)$ can be in turn evaluated by the same algorithm. The integral occurring in (6.5.19) is surely simpler to carry out than the original integral in (6.5.14); if $G_p(x, y, ...)$ is a logarithm, its derivative appearing in the right hand side is a rational function; if it is a dilogarithm the derivative is a logarithm; in any case the resulting expressions are easier to handle. Once the y-derivative of $F_{p+1}(y, ...)$ is obtained in a convenient closed analytic form, $F_{p+1}(y, ...)$ itself can be evaluated by means of a quadrature, to be performed with the aid of (6.5.6) and their generalizations.

6.6 Analytic Results

In this section, we gather all the analytic results which have been obtained for the graphs of Fig. 3. For each graph, we give the coefficient of $(\frac{\alpha}{\pi})^3$ evaluated in the Feynman gauge. The contribution of the mirror image graph is included where appropriate. Terms proportional to $\ln \lambda^2$ and $(\ln \lambda^2)^2$ have been dropped. These results are taken from the references shown in Table 1.

$$\mathbf{A1} = \tfrac{51}{32} - \tfrac{19}{48}\pi^2 - \tfrac{1}{8}\zeta(3)$$

$$\mathbf{A2} = \tfrac{95}{144} + \tfrac{139}{216}\pi^2 - \tfrac{17}{12}\zeta(3)$$

$$\mathbf{A3} = \tfrac{169}{576} - \tfrac{143}{432}\pi^2 - \tfrac{1}{3}\zeta(3)$$

$$\mathbf{B1} = \tfrac{33}{32} + \tfrac{1}{8}\pi^2$$

$$\mathbf{B2} = -\tfrac{595}{432} + \tfrac{11}{324}\pi^2 - \tfrac{7}{18}\zeta(3)$$

$$\mathbf{B3} = \tfrac{733}{1728} + \tfrac{59}{648}\pi^2 + \tfrac{7}{18}\zeta(3)$$

$$\mathbf{C1} = -\tfrac{5}{2} + \tfrac{25}{12}\zeta(2) + \tfrac{71}{24}\zeta(3) - 3\zeta(2)\ln 2 - \tfrac{5}{4}\zeta^2(2) - \tfrac{5}{3}\zeta(2)\ln^2 2 + \tfrac{20}{3}\left(a_4 + \tfrac{1}{24}\ln^4 2\right)$$

$$\mathbf{C2} = \tfrac{235}{288} - \tfrac{1429}{144}\zeta(2) + \tfrac{11}{2}\zeta(3) + 8\zeta(2)\ln 2 - \tfrac{5}{6}\zeta^2(2) - \tfrac{19}{3}\zeta(2)\ln^2 2 + \tfrac{40}{3}\left(a_4 + \tfrac{1}{24}\ln^4 2\right)$$

$$\mathbf{C3} = -\tfrac{173}{288} + \tfrac{287}{16}\zeta(2) - \tfrac{17}{3}\zeta(3) - 18\zeta(2)\ln 2 - \tfrac{37}{24}\zeta^2(2) + 11\zeta(2)\ln^2 2 - 14\left(a_4 + \tfrac{1}{24}\ln^4 2\right)$$

$$\mathbf{D1} = -\tfrac{47}{48} + \tfrac{3}{8}\pi^2 - \tfrac{23}{12}\zeta(3) + \tfrac{5}{6}\pi^2\ln 2 - \tfrac{11}{432}\pi^4 - \tfrac{4}{9}\pi^2\ln^2 2 - \tfrac{4}{3}\left(a_4 + \tfrac{1}{24}\ln^4 2\right)$$

$$\mathbf{D2} = -\tfrac{415}{144} + \tfrac{74}{27}\pi^2 + \tfrac{89}{12}\zeta(3) - 6\pi^2\ln 2 - \tfrac{223}{2160}\pi^4 + \tfrac{20}{9}\pi^2\ln^2 2 + \tfrac{20}{3}\left(a_4 + \tfrac{1}{24}\ln^4 2\right)$$

$$\mathbf{D3} = -\tfrac{349}{288} - \tfrac{239}{144}\pi^2 - \tfrac{33}{8}\zeta(3) + \tfrac{38}{9}\pi^2\ln 2 + \tfrac{127}{1080}\pi^4 - \tfrac{16}{9}\pi^2\ln^2 2 - \tfrac{16}{3}\left(a_4 + \tfrac{1}{24}\ln^4 2\right)$$

$$\mathbf{D4} = -4 + \tfrac{463}{216}\pi^2 + \tfrac{25}{8}\zeta(3) - \tfrac{53}{18}\pi^2\ln 2 - \tfrac{11}{80}\pi^4 + \tfrac{4}{3}\pi^2\ln^2 2 + 4\left(a_4 + \tfrac{1}{24}\ln^4 2\right)$$

$$\mathbf{D5} = -\tfrac{763}{288} + \tfrac{19}{48}\pi^2 + \tfrac{5}{12}\zeta(3) - \tfrac{1}{9}\pi^2\ln 2 - \tfrac{1}{90}\pi^4$$

$$\mathbf{E1} = \tfrac{155}{144} - \tfrac{493}{432}\pi^2 - \tfrac{79}{12}\zeta(3) + \tfrac{5}{6}\pi^2\ln 2 + \tfrac{71}{432}\pi^4 - \tfrac{4}{9}\pi^2\ln^2 2 - \tfrac{4}{3}\left(a_4 + \tfrac{1}{24}\ln^4 2\right)$$

$$\mathbf{E2} = -\tfrac{7}{144} + \tfrac{221}{144}\pi^2 + \tfrac{26}{3}\zeta(3) - \tfrac{37}{18}\pi^2\ln 2 - \tfrac{191}{1080}\pi^4 + \tfrac{8}{9}\pi^2\ln^2 2 + \tfrac{8}{3}\left(a_4 + \tfrac{1}{24}\ln^4 2\right)$$

$$\mathbf{E3} = \tfrac{53}{36} + \tfrac{313}{432}\pi^2 - \tfrac{13}{3}\zeta(3) - \tfrac{17}{18}\pi^2\ln 2 - \tfrac{11}{4320}\pi^4 + \tfrac{2}{9}\pi^2\ln^2 2 + \tfrac{2}{3}\left(a_4 + \tfrac{1}{24}\ln^4 2\right)$$

$$\mathbf{F1} = \tfrac{439}{144} + \tfrac{311}{144}\pi^2 - \tfrac{1213}{24}\zeta(3) - \tfrac{11}{18}\pi^2\ln 2 + \tfrac{467}{1080}\pi^4 - \tfrac{32}{9}\pi^2\ln^2 2 - \tfrac{32}{3}\left(a_4 + \tfrac{1}{24}\ln^4 2\right)$$
$$-\tfrac{347}{96}\zeta(5) + \tfrac{539}{48}\pi^2\zeta(3) + \textstyle\int_0^1 db\,(-23I_1 - 65I_2 - 157I_3 + 487I_4)$$

$$\mathbf{F2} = -\tfrac{7}{48} - \tfrac{295}{72}\pi^2 + \tfrac{629}{12}\zeta(3) + \tfrac{1}{3}\pi^2\ln 2 - \tfrac{37}{432}\pi^4 + \tfrac{4}{9}\pi^2\ln^2 2 + \tfrac{4}{3}\left(a_4 + \tfrac{1}{24}\ln^4 2\right)$$
$$-\tfrac{1153}{96}\zeta(5) - \tfrac{307}{48}\pi^2\zeta(3) + \textstyle\int_0^1 db\,(23I_1 + 29I_2 + 121I_3 - 523I_4)$$

$$\mathbf{F3} = \tfrac{2929}{576} + \tfrac{431}{432}\pi^2 + 6\zeta(3) - \tfrac{11}{9}\pi^2\ln 2 - \tfrac{53}{216}\pi^4 + \tfrac{16}{9}\pi^2\ln^2 2 + \tfrac{16}{3}\left(a_4 + \tfrac{1}{24}\ln^4 2\right)$$
$$+\tfrac{689}{96}\zeta(5) - \tfrac{383}{144}\pi^2\zeta(3) + \textstyle\int_0^1 db\,(I_1 + 19I_2 + 23I_3 - 5I_4)$$

where

$$I_1 = Li_2(-b)\ln^2(1-b)/b + \ln(1-b+b^2)\times$$
$$\Big([2\ln(b)\ln^2(1-b) + 3\ln(b)\ln(1-b)\ln(1+b)$$
$$-2Li_2(b)\ln(1-b) - 3Li_2(-b)\ln(1-b) - 3\ln^2(1-b)\ln(1+b)]/3b$$
$$+[2Li_2(b)\ln(b) + 3Li_2(-b)\ln(b) - 4Li_2(b)\ln(1-b) - 6Li_2(-b)\ln(1-b)]/3(1-b)\Big)$$

$$I_2 = \ln(1-b+b^2)[4Li_2(b)\ln(b) + 6Li_2(-b)\ln(b) - \ln^3(1-b) - 3\ln^2(b)\ln(1+b)]/12b$$

$$I_3 = \ln(1-b+b^2)\ln^3(b)/6b$$

$$I_4 = \ln(1-b+b^2)\ln^2(b)\ln(1-b)/12b$$

$$\mathbf{J1} = \tfrac{1145}{432} + \tfrac{161}{27}\zeta(2) + \tfrac{49}{18}\zeta(3) - \tfrac{44}{3}\zeta(2)\ln 2 - \tfrac{14}{15}\zeta(2)^2 - \tfrac{8}{3}\zeta(2)\ln^2(2) + \tfrac{32}{3}\left(a_4 + \tfrac{1}{24}\ln^4 2\right)$$

$$\mathbf{J2} = \tfrac{1547}{432} - 9\zeta(2) - 2\zeta(3) + 12\zeta(2)\ln 2$$

$$\mathbf{J3} = -\tfrac{943}{324} - \tfrac{8}{45}\zeta(2) + \tfrac{8}{3}\zeta(3)$$

$$\mathbf{K1} = \tfrac{73}{216} - \tfrac{25}{6}\zeta(2) - \tfrac{31}{9}\zeta(3) + \tfrac{28}{3}\zeta(2)\ln 2$$

$$\mathbf{K2} = \tfrac{11}{24} + \tfrac{5}{3}\zeta(2) + \tfrac{4}{9}\zeta(3) - \tfrac{10}{3}\zeta(2)\ln 2$$

$$\mathbf{L1} = \tfrac{133}{216} - \tfrac{353}{54}\zeta(2) + \tfrac{403}{144}\zeta(3) + 6\zeta(2)\ln 2$$

$$\mathbf{L2} = -\tfrac{623}{144} + \tfrac{32}{45}\zeta(2) + \tfrac{39}{16}\zeta(3)$$

$$M1 = \frac{2005}{324} + \frac{854}{81}\zeta(2) - \frac{197}{54}\zeta(3) - \frac{154}{9}\zeta(2)\ln 2 + \frac{2}{15}\zeta^2(2)$$

$$M2 = \frac{599}{162} - \frac{205}{81}\zeta(2) + \frac{97}{54}\zeta(3) - \frac{16}{9}\zeta(2)\ln 2 + \frac{2}{15}\zeta^2(2)$$

$$M3 = -\frac{2641}{648} + \frac{305}{81}\zeta(2) - \frac{251}{54}\zeta(3) - \frac{10}{9}\zeta(2)\ln 2 + \frac{26}{15}\zeta^2(2)$$

where

$$\zeta(2) = \sum \frac{1}{n^2} = \frac{1}{6}\pi^2$$

$$\zeta(3) = \sum \frac{1}{n^3} = 1.202\,0569\ldots$$

$$\zeta(5) = \sum \frac{1}{n^5} = 1.036\,927\ldots$$

$$a_4 = \sum \frac{1}{2^n n^4} = 0.517\,479\ldots$$

References

[1] R. S. Van Dyck, Jr., P. B. Schwinberg, and H. G. Dehmelt, Phys. Rev. Lett. 59, 26 (1987)

[2] F. J. Dyson, Phys. Rev. 85, 631 (1952)

[3] J. Schwinger, Phys. Rev. 73, 416 (1948); 76, 790 (1949).

[4] A. Petermann, Helv. Phys. Acta 30, 407 (1957); C. M. Sommerfield, Ann. Phys. (N.Y.) 5, 26 (1958).

[5] M. J. Levine and J. Wright, Phys. Rev. D 8, 3171 (1973).

[6] P. Cvitanovic and T. Kinoshita, Phys. Rev. D10, 4007 (1974).

[7] R. Carroll and Y. P. Yao, Phys. Lett. 48*B*, 125 (1974). R. Carroll, Phys. Rev. D12, 2344 (1975).

[8] R. Barbieri, J. A. Mignaco, and E. Remiddi, Nuovo Cimento 11*A*, 824 (1972).

[9] R. Barbieri and E. Remiddi, Nucl. Phys. *B*90, 233 (1975).

[10] See for example J. D. Bjorken and S. D. Drell, *Relativistic Quantum Fields*, McGraw Hill, 1965 or N. N. Bogoliubov and D. V. Shirkov, *Introduction to the Theory of Quantized Fields*, Wiley-Interscience, 1959.

[11] D. R. Yennie, S. C. Frautschi, and H. Suura, Ann. Phys. (N.Y.) 13, 379 (1961).

[12] M. J. Levine and R. Roskies, Phys. Rev. D. 9, 421 (1974).

[13] M. J. Levine, R. C. Perisho and R. Roskies, Phys. Rev. D 13, 997 (1976).

[14] Properties of the Gegenbauer polynomials (usually denoted by C_n^1) can be found in the text by W. Magnus, F. Oberhettinger, and R. P. Soni, *Formulas and Theorems for the Special Functions of Mathematical Physics* (Springer, New York, 1966), p. 218.

[15] See, e.g., J. W. Essam and Michael E. Fisher, Rev. Mod. Phys. 42 271 (1970).

[16] K. A. Milton, W. Tsai and L. L. DeRaad, Jr., Phys. Rev. D9 1809 (1974).

[17] R. Barbieri, M. Caffo, and E. Remiddi, Phys. Lett. 57*B*, 460 (1975).

[18] See [19] for an elementary treatment.

[19] E. Remiddi, Helv. Phys. Acta 54, 364 (1981) and references cited therein.

[20] M. V. Terent'ev, Sov. Phys. JETP 16, 444 (1963).

[21] G. Källén and A. Sabry, Dan. Mat. Fys. Medd. 29, n. 17 (1955). See also [22].

[22] J. A. Mignaco and E. Remiddi, Nuovo Cimento 60*A*, 519 (1969).

[23] D. Billi, M. Caffo and E. Remiddi, Nuovo Cimento Lett. 4, 657 (1972).

[24] R. Barbieri, M. Caffo and E. Remiddi, Nuovo Cimento Lett. 5, 769 (1972).

[25] R. Barbieri, M. Caffo and E. Remiddi, Nuovo Cimento Lett. 9, 690 (1974).

[26] R. Barbieri and E. Remiddi, Phys. Lett. 49*B*, 468 (1974).

[27] R. Barbieri, M. Caffo, E. Remiddi, S. Turrini and D. Oury, Nucl. Phys. *B*144, 329 (1978).

[28] R. Barbieri, J. A. Mignaco, and E. Remiddi, Nuovo Cimento 11*A*, 865 (1972).

[29] M. J. Levine, E. Remiddi, and R. Roskies, Phys. Rev. D 20, 2068 (1979).

[30] M. Caffo, E. Remiddi, and S. Turrini, Nuovo Cimento 79*A*, 220 (1984).

[31] N. Nielsen, Nova Acta 90, 125 (Halle 1909).

[32] M. J. Levine and R. Roskies, Phys. Rev. D 14, 2191 (1976).

[33] M. J. Levine, H. Y. Park and R. Roskies, Phys. Rev. D 25, 2205 (1982).

[34] See T. Kinoshita's article in this volume.

[35] K. A. Milton, W. Tsai and L. L. DeRaad Jr., Phys. Rev. D 9, 1814 (1974).

[36] T. Engelman and M. J. Levine, unpublished.

Theory of the Anomalous Magnetic Moment of the Electron – Numerical Approach

Toichiro Kinoshita
Newman Laboratory, Cornell University
Ithaca, New York 14853

CONTENTS

1. Introduction and Summary

According to Dirac's theory of electrons, the electron has an intrinsic magnetic moment accompanying its spin whose value, when expressed in the form $\frac{geh}{4\pi mc}$, is given by $g = 2$. In 1947, however, a tiny deviation from this prediction was discovered.[1] The excess $a_e = \frac{1}{2}(g-2)$ is called the anomalous magnetic moment. Together with the discovery of the Lamb shift of the hydrogen atom in the same year, this presented a timely stimulus to the renormalization theory of quantum electrodynamics (QED) just being developed. Schwinger's calculation[2] of a_e to order α provided one of the most important milestones in the development of QED. Since then, both experiment and theory of a_e have been improved by several orders of magnitude, and keep providing the most precise and rigorous test for the validity of QED.

Instead of reviewing the history of the electron anomalous magnetic moment, which can be found elsewhere[3], let me just quote the latest Penning trap measurement of the electron and positron anomalies obtained by the University of Washington group:[4],[a]

$$\begin{aligned} a_{e^-} &= 1\ 159\ 652\ 188.4(4.3) \times 10^{-12}, \\ a_{e^+} &= 1\ 159\ 652\ 187.9(4.3) \times 10^{-12}. \end{aligned} \tag{1.1}$$

As is obvious from

$$\left(\frac{\alpha}{\pi}\right)^4 \simeq 29 \times 10^{-12},$$

the very high precision of these measurements demands knowledge of the eighth-order contribution to a_e. It is to meet this challenge that the theoretical calculation of the anomaly has been pushed to the order α^4.

[a]See Van Dyck's article in this volume.

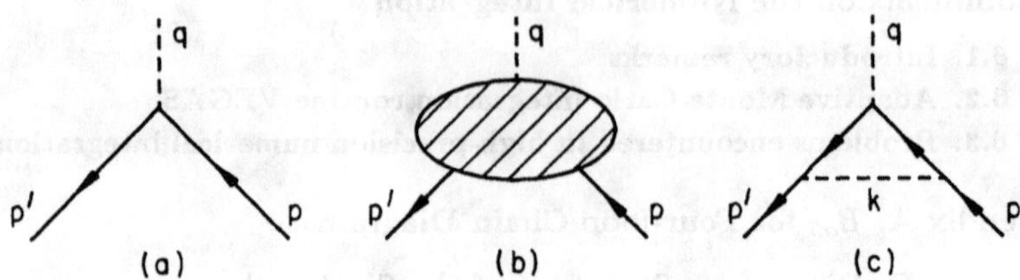

Figure 1: (a) Lowest-order Feynman diagram describing scattering of an electron by an external magnetic field. (b) Schematic diagram representing an infinite set of Feynman diagrams contributing to a_e. (c) Second-order vertex diagram.

It is important to note that, at the present stage of development of QED (or its generalizations including the standard model of electroweak and strong interactions), neither the mass nor the charge of the electron is calculable from the theory itself. They must be treated as input parameters. Because of this the simplest quantity that can be actually calculated from first principles is the anomalous magnetic moment a_e of the electron. This is why the study of a_e occupies a particularly important niche in the high-precision test of QED.

The magnetic property of an electron can be studied most conveniently by examining its scattering by a static magnetic field. If the interaction with virtual photons is ignored, this can be expressed, in the weak-field limit (which is the case under normal experimental conditions) by the Feynman diagram shown in Fig. 1(a). Application of Feynman-Dyson rules to this diagram leads to the amplitude (apart from a factor $-2\pi i\delta(p'_0 - p_0)$ for energy conservation)

$$e\bar{u}(p')\gamma^{\mu}u(p)A_{\mu}{}^{e}(\vec{q}), \tag{1.2}$$

with

$$A_{\mu}{}^{e}(\vec{q}) = \frac{1}{(2\pi)^3}\int d^3x e^{-i\vec{q}\cdot\vec{x}}A_{\mu}{}^{e}(\vec{x}), \tag{1.3}$$

where $A_{\mu}{}^{e}(\vec{x})$ is the vector potential representing the external static magnetic field. The electric current $\bar{u}(p')\gamma^{\mu}u(p)$ can be decomposed into convection and spin currents:

$$\bar{u}(p')\gamma^{\mu}u(p) = \frac{1}{2m}\bar{u}(p')(p'+p)^{\mu}u(p) + \frac{i}{2m}\bar{u}(p')q_{\nu}\sigma^{\mu\nu}u(p), \tag{1.4}$$

where $q = p' - p$ and $\sigma^{\mu\nu} = \frac{i}{2}(\gamma^{\mu}\gamma^{\nu} - \gamma^{\nu}\gamma^{\mu})$. The latter exhibits the well-known fact that the Landé g factor g of a free electron is equal to 2 in Dirac's theory of the electron.

Because of the interaction with the virtual photon field surrounding the charge, however, the diagram in Fig. 1(a) must be replaced by an infinite set of Feynman diagrams, all having the structure schematically represented by Fig. 1(b).

Taking account of Lorentz, C, P, and T invariances, the corresponding amplitude can be written as a sum of two terms

$$e\bar{u}(p')[\gamma^{\mu}F_1(q^2) + \frac{i}{2m}\sigma^{\mu\nu}q_{\nu}F_2(q^2)]u(p)A_{\mu}{}^{e}(\vec{q}). \tag{1.5}$$

F_1 and F_2 are called the charge and magnetic form factors, respectively. The charge form factor is normalized so that $F_1(0) = 1$. Thus the first term reduces to the amplitude (1.2) in the static limit and contributes a factor 2 to the g factor. The magnetic moment anomaly a_e is the static limit of $F_2(q^2)$, and, rewriting p' and p as $p + \frac{1}{2}q$ and $p - \frac{1}{2}q$, can be expressed as

$$a_e = F_2(0) = Z_2 M, \tag{1.6}$$

with

$$M = \lim_{q=0} \frac{m}{4p^4 q^2} Tr[(m\gamma^\nu p^2 - (m^2 + \frac{1}{2}q^2)p^\nu)(\not{p} + \frac{1}{2}\not{q} + m)\Gamma^\nu(\not{p} - \frac{1}{2}\not{q} + m)], \tag{1.7}$$

where $p^2 = m^2 - \frac{1}{4}q^2, p \cdot q = 0$, Z_2 is the wave-function renormalization constant, and Γ^ν is the proper vertex part represented by Fig. 1(b).

As a consequence of the renormalizability of QED, the magnetic moment anomaly a_e can be written as a power series in $\frac{\alpha}{\pi}$:

$$a_e = C_1 \left(\frac{\alpha}{\pi}\right) + C_2 \left(\frac{\alpha}{\pi}\right)^2 + C_3 \left(\frac{\alpha}{\pi}\right)^3 + \dots , \tag{1.8}$$

where the coefficients C_1 , C_2 , ... are finite calculable quantities.

The lowest-order contribution C_1 can be calculated from the Feynman diagram of Fig. 1(c). The scattering amplitude corresponding to this diagram is readily given by the standard Feynman-Dyson rules:

$$\frac{-ie^3}{(2\pi)^4} \int d^4k \frac{1}{k^2} \bar{u}(p')\gamma^\lambda \frac{1}{\not{p}' + \not{k} - m} \gamma^\mu \frac{1}{\not{p} + \not{k} - m} \gamma_\lambda u(p) A_\mu{}^e(\vec{q}), \tag{1.9}$$

where $\not{p} = \gamma^\mu p_\mu$. Carrying out the integration over the 4-momentum k, one finds

$$C_1 = \frac{1}{2} \quad \text{or} \quad a_e^{(2)} = \frac{\alpha}{2\pi}, \tag{1.10}$$

which was first obtained by Schwinger[2].

Before discussing higher-order corrections to a_e, let us consider the second-order radiative correction in an arbitrary constant magnetic field B, which can be calculated using an exact integral representation for the electron propagator in a constant magnetic field.[5] For $eB/m^2 \ll 1$ the shift in the ground-state energy due to B is found to be[6,7]

$$\begin{aligned} \Delta E = \frac{\alpha}{2\pi} m \Bigg[-\frac{eB}{2m^2} + \left(\frac{eB}{m^2}\right)^2 \left(\frac{4}{3}\ln\frac{m^2}{2eB} - \frac{13}{18}\right) \\ + \left(\frac{eB}{m^2}\right)^3 \left(\frac{14}{3}\ln\frac{m^2}{2eB} - \frac{32}{5}\ln 2 + \frac{83}{90}\right) + \dots \Bigg], \end{aligned} \tag{1.11}$$

where the magnetic field B points in the + z direction. From this one can estimate the importance of nonlinear effects in B. For the Seattle Penning trap experiment[4], in which $B \simeq 5$ tesla, we find $eB/m^2 \approx 1.1 \times 10^{-9}$. In view of the great precision of this experiment,

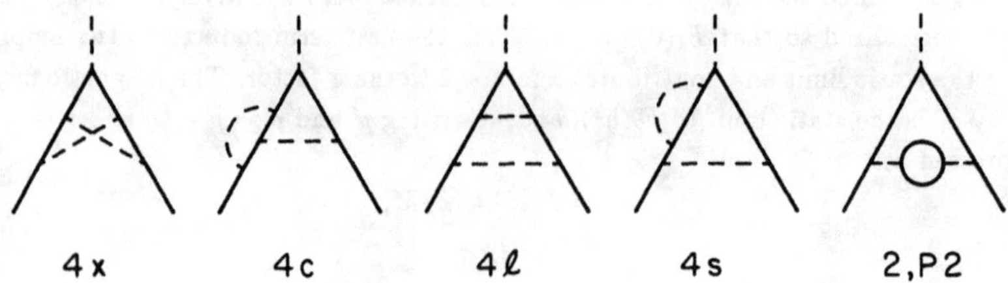

Figure 2: Examples of fourth-order vertex diagrams contributing to C_2.

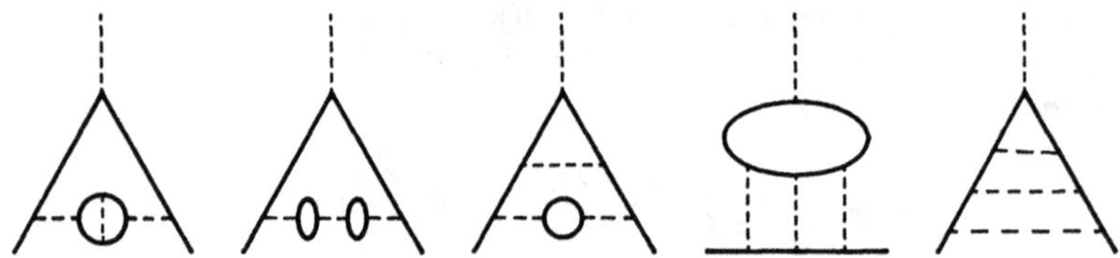

Figure 3: Examples of sixth-order vertex diagrams contributing to C_3.

the B^2 term could be important: however, being spin independent, it does not affect the $g-2$ experiment, which measures energy differences between different spin states. The spin-dependent effect comes from the B^3 term. This is of the relative order of 2×10^{-16}, and is clearly negligible. It is thus legitimate to keep only the first term in Eq. (1.11), which is equivalent to working with the diagram in Fig. 1(c).

The fourth order coefficient C_2 is represented by seven Feynman diagrams, some of which are shown in Fig. 2 . This contribution was analytically evaluated in the Feynman gauge by Petermann[8] and Sommerfield[9]:

$$\begin{aligned} C_2 &= \frac{197}{144} + \left(\frac{1}{2} - 3\ln 2\right)\zeta(2) + \frac{3}{4}\zeta(3) \\ &= -\,0.328\,478\,965\ldots\,, \end{aligned} \tag{1.12}$$

where ζ is the Riemann ζ function. The same result was obtained recently in the Fried-Yennie gauge.[10]

The computation of the sixth order coefficient C_3 , represented by 72 Feynman diagrams, some of which are shown in Fig. 3, is considerably more complicated. Although direct analytic evaluation has been carried out for some diagrams[11], there is a clear need for developing systematic methods for handling these complicated integrals. Also, because of the size of the integrands, it is not practical to handle them manually. Thus computers are used extensively for both analytic and numerical work. The following are four major approaches that have been applied to this problem:

A. Dispersion-theoretical approach.[12]

B. Partial wave expansion in 4-dimensional Euclidean space.[13,14]

C. Straightforward numerical integration of ordinary Feynman integrals.[15,16,17]

D. Numerical integration of Feynman integrals summed with the help of the Ward-Takahashi identity.[17]

Method A is based on the fact that the integral of an analytic function can be evaluated by dispersion theory once the absorptive part, which is generally easier to evaluate, is known.

In Method B one utilizes the fact that $F_2(0)$, which was originally an integral over Minkowski 4-momenta, is invariant under 4-dimensional Euclidean rotation once they are Wick-rotated. One can then expand all propagators into "spherical" harmonics in the 4-dimensional space and carry out the angular integration using their orthogonality relations. In this manner the original 12-dimensional momentum space integration (in the sixth-order case) is reduced to an integral over three radial variables, which is then evaluated either analytically (where possible) or numerically.

Method C relies on numerical integration applied to integrals obtained by straightforward application of Feynman-Dyson rules.

In Method D the emphasis is on minimizing the amount of algebraic work needed to set up the integrals. It is not particularly suitable for analytic integration (up to 7 variables for C_3 in contrast to 3 variables in the Method B) and one must rely on numerical integration instead. This aim is achieved by the following steps:

i) Development of Feynman-Dyson rules expressed in Feynman parameters.[18]

ii) Use of the Ward-Takahashi identity to combine topologically related Feynman diagrams into one integral.[17] This allows one to cut down dramatically the number of integrals to be evaluated.

iii) Development of a renormalization procedure that applies directly to Feynman-parametric integrals. Infrared divergences, although they have a different origin, are handled in a similar manner.[19]

For diagrams for which analytical results are available, numerical results are useful only to the extent that they provide a check of analytical results. In other cases, the numerical approach is the only one available. All 72 diagrams of sixth order have been evaluated by Methods C and D. (See Refs. 17, 20, 21.) At present 51 of 72 diagrams have also been evaluated analytically by Methods A and/or B. Of the remaining 21 diagrams, 16 have been reduced to 3-dimensional integrals by Method B and then evaluated numerically. At present the values of the remaining five diagrams are known only by Methods C and D.

The best value of C_3 now available is[22]

$$C_3 = 1.176\,11\,(42), \tag{1.13}$$

where the error comes from the numerical evaluation of the 21 diagrams mentioned above.

We also need the eighth-order coefficient C_4 in order to take full advantage of the experimental precision in (1.1) in testing the validity of QED. The number of contributing

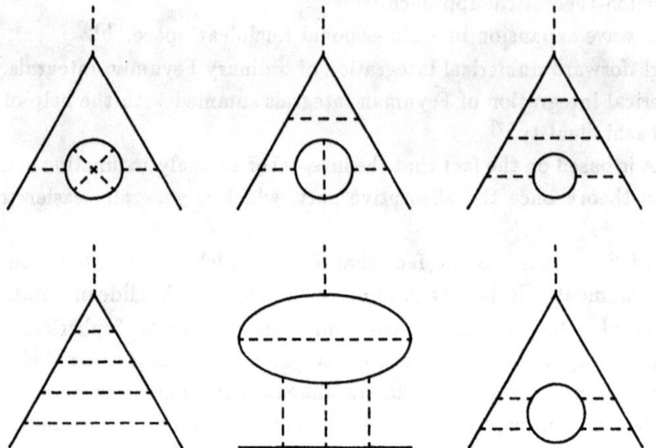

Figure 4: Examples of eighth-order vertex diagrams contributing to C_4.

Feynman diagrams is 891. Typical eighth-order diagrams are shown in Fig. 4. Thus far they have been evaluated only by Method D, except for a few of the simplest cases.[23] The results are reported in Refs. 22, 24, 25, 26, 27. The latest value for C_4 is[22]

$$C_4 \;=\; -\;1.434\,(138). \tag{1.14}$$

No terms of order 10 or higher have been evaluated thus far except for some cases involving multiple insertion of simple vacuum-polarization-loops.[28] An estimate of terms of very large order has been attempted based on the steepest descent method.[29] However, further study is needed to see whether this gives a good asymptotic estimate or not.

To compare with the experimental data, it is also necessary to include the vacuum-polarization contributions of muon, tauon, and hadron loops as well as the contribution of the electroweak effect. They are:[30]

$$\begin{aligned} \Delta a_e(muon) &= 2.804 \times 10^{-12}, \\ \Delta a_e(tauon) &= 0.010 \times 10^{-12}, \\ \Delta a_e(hadron) &= 1.6(2) \times 10^{-12}, \\ \Delta a_e(weak) &= 0.05 \times 10^{-12}. \end{aligned} \tag{1.15}$$

Finally, we need an accurate value of the fine structure constant α. If one uses the most recent measurement of α based on the quantized Hall effect:[31]

$$\alpha^{-1} \;=\; 137.035\,997\,9(32), \tag{1.16}$$

one obtains from (1.10),(1.12), (1.13), (1.14), and (1.15) the theoretical value[22]

$$a_e(theory) \;=\; 1\,159\,652\,140(5.3)(4.1)(27.1) \times 10^{-12}, \tag{1.17}$$

where the first and second uncertainties come from the numerical uncertainties in C_3 and C_4 while the third reflects the uncertainty in α quoted in (1.16).

The result (1.17) agrees within 1.7 standard deviations with the latest Penning trap measurements.[4] Note, however, that the uncertainty in $a_e(theory)$ is dominated by that of α. This means that a test of QED using the electron anomaly a_e must be postponed until a better value of α is found. Pending improved measurement of α, however, one can calculate α from theory and the experimental value of a_e with a better precision than by any other means presently available:

$$\alpha^{-1}(a_e) = 137.035\ 992\ 22(94) \qquad (0.0069\ ppm) \tag{1.18}$$

where the experimental and theoretical uncertainties contribute 0.0037 ppm and 0.0058 ppm, respectively.

At present it is not known where the remaining discrepancy between (1.16) and (1.18) comes from. Further improvement of the quantized Hall α as well as better measurements of α^{-1} by means of the ac Josephson effect[32] and other methods[33] is urgently needed.

Other possible sources of discrepancy include some unknown interaction involving particles more massive than the weak bosons W and Z, and possible substructure of the electron.[34]. However, they are unlikely to contribute significantly at this level since their effects will scale as $(m/M)^2$ and are already very small for M of the order of the W or Z mass as is seen from (1.15). Clearly the electron anomaly is rather insensitive to heavy mass particle contribution. In fact this is why it is such a good tool for testing the validity of "pure" QED.

On the other hand, the electron anomaly responds very sensitively if a very light particle interacts with the electron. The near agreement of (1.16) and (1.18) enables us to impose a tight upper bound on the possible interaction strength of such a particle with the electron. For example, the possibility that the object reported[35] to be peaking around 1.7 MeV is an elementary particle can be ruled out in view of our result and the estimated cross section for producing such a particle.[36]

Finally we note that comparison of different determinations of α, such as (1.16) and (1.18) provides an opportunity for re-examining the theoretical basis of the quantized Hall and ac Josephson effects. An aspect of this problem is discussed in the article by Kinoshita and Lepage in this volume.

As was noted above, Method D is the only method that has been applied to the calculation of the entire C_4. Since the analytic approach is described in detail elsewhere[b], let me concentrate in this article on the theory and numerical analysis of the electron anomalous magnetic moment based on Method D.

In order to develop Method D, it is convenient to formulate the perturbative QED in terms of Feynman parameters from the very beginning. A parametric formulation of Feynman-Dyson rules and their application to self-energy and vertex diagrams are described in Section 2. Section 3 is devoted to an analysis of the divergence structure

---

[b]See the article in this volume by Levine, Remiddi, and Roskies

of the renormalization constants. Section 4 and Section 5 describe the calculation of a_e without and with closed electron loops, respectively. Problems encountered in the numerical integration are discussed in Section 6.

In view of the enormous amount of information generated in the calculation of a_e, it is not possible to describe the details of the calculation in full within the limited space available. Instead I shall concentrate on presenting the key ideas and steps that are required to construct a workable algorithm for the numerical evaluation of a_e. Most of the rules and equations are presented without proof. Original articles should be consulted for further details. In order to help elucidate our approach, however, all formulas up to the fourth order are shown explicitly. Sixth- and eighth-order terms are too lengthy to be included here. FORTRAN programs of these terms are available on request.

Finally I note that this is the first time that all the articles I have written on the electron anomalous moment have been put together in a self-contained article in which a uniform convention is employed throughout. In so doing, I had to rectify some confusing inconsistencies in the notation of original articles, written over a period of 25 years in a somewhat evolving manner. Wherever there is an apparent discrepancy between the original article and this one, it is the latter that should be regarded as preferred.

2. A Parametric Formulation of Perturbation Theory for QED

2.1. Introductory remarks

Over the years a number of parametric formulas for Feynman integrals have been introduced. Some of these formulas can be expressed concisely in terms of auxiliary functions rather than Feynman parameters themselves. For instance, a significant simplification has been achieved in Ref. 21 by expressing the integrand in terms of the so-called Kirchhoff currents. These auxiliary functions themselves can be expressed in various forms (e.g., as determinants, sums over loops[37], sums over cut-sets[38,39], etc.), some of which look simple but become very lengthy when worked out explicitly. Such differences are irrelevant for the second-order calculation, but for higher-order calculations it is crucial to write the integrals in as simple and explicit a form as possible.

For practical calculations it is therefore useful to assemble simple expressions for these functions in a systematic and coherent manner. For this purpose it is most convenient to present them as Feynman-Dyson rules in a Feynman-parametric space, which enables us to write down S-matrix elements directly in terms of parametric functions. As is shown below, our parametric representation[18] is systematic and economical, and minimizes the labor of setting up a computer program for numerical integration. By stressing common features of related diagrams through parametric functions defined directly by the topology of diagrams, it also reduces redundancy of effort in writing programs for different diagrams and provides a means of cross-checking related diagrams.

Such a formulation would be deficient, however, if it does not incorporate a procedure

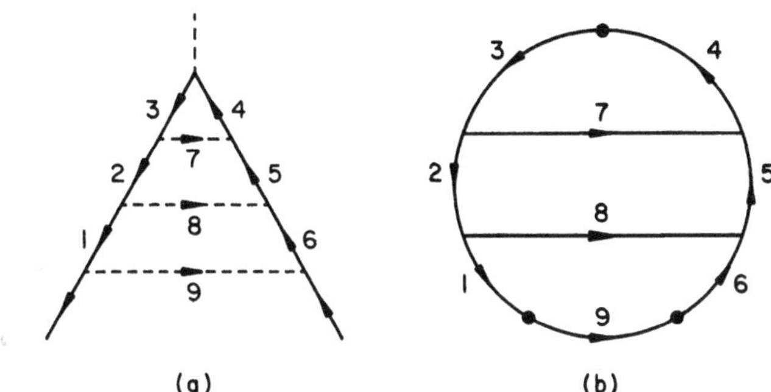

Figure 5: (a) A sixth-order vertex diagram. (b) The chain diagram corresponding to the vertex diagram (a) which consists of 5 chains (1,6,9), (2,5), (3,4), (7), (8).

for locating and subtracting the ultraviolet (UV) and infrared (IR) divergences beforehand. Our solution to this requirement is discussed in Sec. 3.

2.2. Feynman-Dyson rules for parametric integrals

In this subsection we give a set of rules for writing down an invariant amplitude $\mathcal{M}$ for a Feynman diagram G directly in terms of parametric functions without going through the Feynman-Dyson rules in momentum or coordinate variables. We follow the notation and convention of Bjorken and Drell.[40] It is sufficient to consider only proper diagrams G, namely those that cannot be separated into two disconnected parts by cutting an internal line.

Let G be a proper diagram that consists of N internal lines labeled 1, 2, ..., N, some of which are electron lines and others of which are photon lines. Momentum flow directions are indicated by the arrows assigned to both electron and photon lines. An example is shown in Fig. 5(a). Feynman parameters $z_1, z_2, \ldots, z_N$ satisfying $\sum_{i=1}^{N} z_i = 1, z_i \geq 0$, are assigned to the internal lines. In this formulation, loop momenta are completely integrated out and do not appear at all. Each line j carries a fixed momentum q_j, which satisfies the momentum conservation law at each vertex and hence depends linearly on the external momenta.

The basic building blocks of our formulation are parametric functions $B_{\alpha\beta}$, where α, β refer to chains α, β of the chain diagram corresponding to G. Here chain α is a set of all internal lines $i, j, \ldots, k$ carrying the same loop momentum, and the chain diagram is obtained from G by amputating all external lines and disregarding the distinction between electron and photon lines. The Greek letter α will also be used to denote the Feynman parameter for the chain α: $\alpha \equiv z_i + z_j + \ldots + z_k$. Fig. 5(b) shows the chain diagram corresponding to the vertex diagram of Fig. 5(a). A crucial feature of $B_{\alpha\beta}$ is that it is a homogeneous form of degree $n-1$, n being the number of independent loops of G, whose structure is completely determined by the topology of the chain diagram.

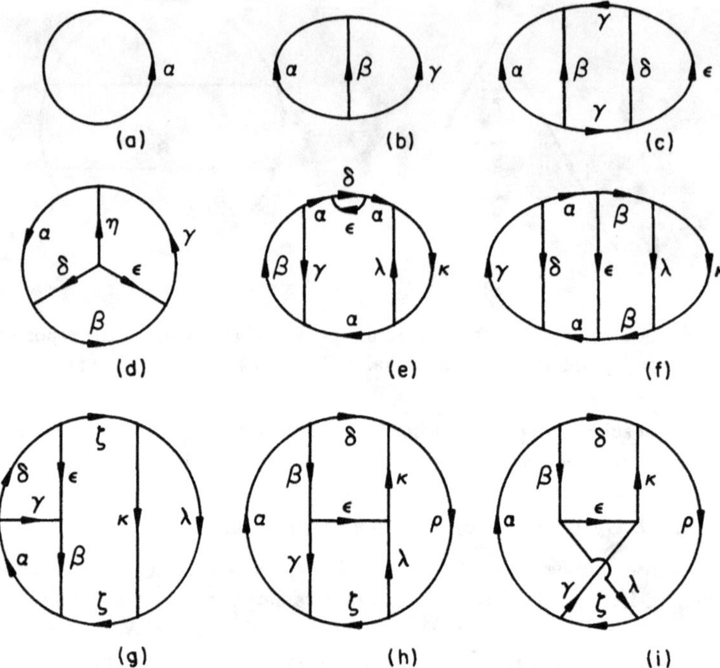

Figure 6: Chain diagrams for one- to four-loop Feynman diagrams.

In any given order the number of topologically distinct chain diagrams is quite small. All chain diagrams needed to calculate processes of up to four loops are shown in Fig. 6. For each chain diagram the number of topologically inequivalent $B_{\alpha\beta}$ is also very small. It is therefore easy to prepare beforehand a complete list of $B_{\alpha\beta}$ up to a given order. All $B_{\alpha\beta}$ needed for the calculation of *any* Feynman diagram with up to three loops are shown in Table I. $B_{\alpha\beta}$ for diagrams with four internal loops are listed in Appendix A.

Once $B_{\alpha\beta}$ is known, it is easy to find B_{ij}, where i and j refer to the internal lines i and j of G. For instance, if G is a single-loop diagram consisting of lines $i, j, \ldots, k$, and if all lines point in the same direction, one finds

$$B_{ij} = B_{\alpha\alpha} = 1, \quad \text{for all } i,\ j\ \in\ \alpha.$$

In general one can construct B_{ij} successively using the formulas[18]

$$B_{ij} = \sum_c \eta_{ic}\eta_{jc}U_c, \tag{2.1}$$

and

$$\eta_{ks} U = \sum_{j=1}^{N} \eta_{js} z_j B_{jk}, \qquad \text{any } k \in s, \tag{2.2}$$

where the summation in (2.1) is over all (not necessarily independent) self-nonintersecting loops c that contain both lines i and j, and η_{ic} is the projection $(\pm 1, 0)$ of line i along the loop c. U_c is the U function for the reduced diagram obtained from G by shrinking the loop c to a point. In (2.2) s is an arbitrary closed loop in G. The number of terms actually contributing to U in (2.2) can be minimized by choosing the shortest loop in G. Note that the overall sign of B_{ij} changes if the direction of either line i or j changes.

Although it is not obvious from (2.2), U is also completely determined by the chain structure. For instance, if G is a single-loop diagram consisting of lines $i, j, \ldots, k$, one finds

$$U = \alpha, \tag{2.3}$$

where $\alpha = z_i + z_j + \ldots + z_k$.

Table I. Parametric functions $B_{\alpha\beta}$. Functions not listed can be obtained by relabeling the suffixes. Consult Eq.(2.1) to determine the sign.

Number of loops	Chain diagram	$B_{\alpha\beta}$
1	Fig. 6(a)	$B_{\alpha\alpha} = 1$
2	Fig. 6(b)	$B_{\alpha\alpha} = \beta + \gamma$
		$B_{\alpha\beta} = -\gamma$
3	Fig. 6(c)	$B_{\alpha\alpha} = (\beta+\gamma)(\delta+\epsilon)+\delta\epsilon$
		$B_{\gamma\gamma} = (\alpha+\beta)(\delta+\epsilon)$
		$B_{\alpha\beta} = -\gamma(\delta+\epsilon)-\delta\epsilon$
		$B_{\alpha\gamma} = -\beta(\delta+\epsilon)$
3	Fig. 6(d)	$B_{\alpha\alpha} = \epsilon(\beta+\gamma+\delta+\eta)+(\beta+\delta)(\gamma+\eta)$
		$B_{\alpha\beta} = \delta(\gamma+\epsilon+\eta)+\epsilon\eta$
		$B_{\alpha\epsilon} = \beta\eta-\gamma\delta$

We are now ready to state the parametric Feynman-Dyson rules:

1. Find parametric functions B_{ij} from Table I, Appendix A, or (2.1).
2. Construct the parametric function U using (2.2).
3. Construct the parametric functions $Q_i'^{\mu}$ and V:

$$Q_i'^{\mu} = -\frac{1}{U} \sum_{j=1}^{N} q_j^{\mu} z_j B_{ij}', \tag{2.4}$$

$$V = \sum_{j=1}^{N} z_j (m_j^2 - q_j \cdot Q_j'), \tag{2.5}$$

where m_i is the mass of the ith line and

$$B'_{ij} = B_{ij} - \delta_{ij}\frac{U}{z_i}. \tag{2.6}$$

4. Construct the parametric integral

$$\left(\frac{i}{16\pi^2}\right)^n (-1)^N (N-2n-1)! \int (dz)_G \frac{1}{U^2(V-i\epsilon)^{N-2n}}, \tag{2.7}$$

where

$$(dz)_G = \delta(1-\sum_{i=1}^{N} z_i)\prod_{i=1}^{N} dz_i, \quad z_i \geq 0 \text{ for all } i, \tag{2.8}$$

and n is the total number of independent loops.

5. Multiply Eq. (2.7) by the following factors associated with the remaining elements of the diagram G:[c]

(a) for each external electron line entering the diagram a factor $\sqrt{Z_2}u(p,s)$ or a factor $\sqrt{Z_2}v(p,s)$ depending on whether the line is in the initial or final state; likewise, for each electron line leaving the diagram a factor $\sqrt{Z_2}\bar{v}(p,s)$ or $\sqrt{Z_2}\bar{u}(p,s)$;

(b) for each external photon line a factor $\sqrt{Z_3}\epsilon_\mu$, ϵ_μ being the photon polarization vector;

(c) for each internal electron line j a factor $i(\not{D}_j + m_j)$, where $D_j{}^\mu$ is defined by

$$D_j^\mu \equiv \frac{1}{2}\int_{m_j^2}^{\infty} dm_j^2 \frac{\partial}{\partial q_{j\mu}}; \tag{2.9}$$

(d) for each internal photon line j a factor $-ig_{\mu\nu}$ (in Feynman gauge only);

(e) for each vertex a factor $-ie_0\gamma_\mu$;

(f) for each mass counterterm a factor $i\delta m$;

(g) for each closed electron loop a factor -1.

6. Denote by $\mathbf{F}$ the product of γ^μ from vertices, $\not{D}_j + m_j$ from electron lines, and appropriate spinor factors generated by rule 5. Then, the action of $\mathbf{F}$ on the integral in Eq. (2.7) produces terms of the form

$$\mathbf{F}\frac{1}{V^m} = \frac{F_0}{V^m} + \frac{F_1}{UV^{m-1}} + \frac{F_2}{U^2V^{m-2}} + \cdots, \tag{2.10}$$

where the subscript k of F_k stands for the number of contractions. By contraction we mean picking a pair of $\not{D}_i + m_i$, $\not{D}_j + m_j$ from $\mathbf{F}$, making the substitution

$$\not{D}_i + m_i, \quad \not{D}_j + m_j \implies \gamma^\mu, \; \gamma_\mu, \tag{2.11}$$

multiplying the result by a factor $-\frac{1}{2}B_{ij}$, and summing the results of this operation over all distinct pairs. Remaining uncontracted D_i are then replaced by Q'_i. For $k \geq 1$, F_k includes an overall factor $(m-1)^{-1}\cdots(m-k)^{-1}$.

For details of the derivation of these rules see Ref. 18. Once an integral is constructed, it must be regularized to make the integral well-defined. For this purpose, one may use Feynman cut-off, Pauli-Villars regularization, dimensional regularization, *etc.*

[c] These rules are quoted from Appendix B of Ref. 40, with minor modifications.

Kirchhoff's laws

By construction the momenta Q'^{μ}_j satisfy the conservation law (analogue of Kirchhoff's junction law for electrical circuits)

$$\sum_{j=1}^{N} \epsilon_{vj} Q'^{\mu}_j + p^{\mu}_v = 0 \tag{2.12}$$

at each vertex v, where ϵ_{vj} is an incidence matrix defined by

$$\epsilon_{vj} = \begin{cases} 1 & \text{if the line } j \text{ enters the vertex } v \text{ ,} \\ -1 & \text{if the line } j \text{ leaves the vertex } v \text{ ,} \\ 0 & \text{otherwise,} \end{cases}$$

and p^{μ}_v is the external momentum entering the vertex v. Furthermore the Q'^{μ}_j satisfy an analogue of Kirchhoff's loop law

$$\sum_{i=1}^{N} \eta_{is} z_i Q'^{\mu}_i = 0, \tag{2.13}$$

where η_{is} is defined in (2.1). Substituting Q'^{μ}_i of (2.4) in this equation and equating the coefficients of q_j to 0 we find another useful relation:

$$\sum_{i=1}^{N} \eta_{is} z_i B'_{ij} = 0. \tag{2.14}$$

Note that (2.2) follows from (2.6) and (2.14).

Scalar currents

In many applications it is convenient to replace the vectors Q'^{μ}_i by scalar functions. Suppose an external momentum p^{μ} enters a graph G at point A and leaves it at point B. Then we may write

$$Q'^{\mu}_i = A^{(AB)}_i p^{\mu} + \text{terms linear in other external momenta.} \tag{2.15}$$

Substituting this in (2.12) and (2.13), one finds that $A^{(AB)}_i$ is completely determined by Kirchhoff's laws and represents a fractional "current" associated with the external current p^{μ} flowing through the line i. It will be called the scalar current. In order to find an explicit formula for $A^{(AB)}_i$, let us note that one choice of constant momenta q_j consistent with momentum conservation is

$$q^{\mu}_j = \eta_{jP} p^{\mu} + \text{other terms,} \tag{2.16}$$

where $P = P(AB)$ is any self-nonintersecting path starting at A and ending at B, and $\eta_{jP} = (1, -1, 0)$ according to whether the line j lies (along, against, outside of) the path P. Substituting (2.15) and (2.16) in (2.4), one obtains

$$A_i^{(AB)} = -\frac{1}{U} \sum_{j=1}^{N} \eta_{jP} z_j B'_{ji}, \qquad P = P(AB). \tag{2.17}$$

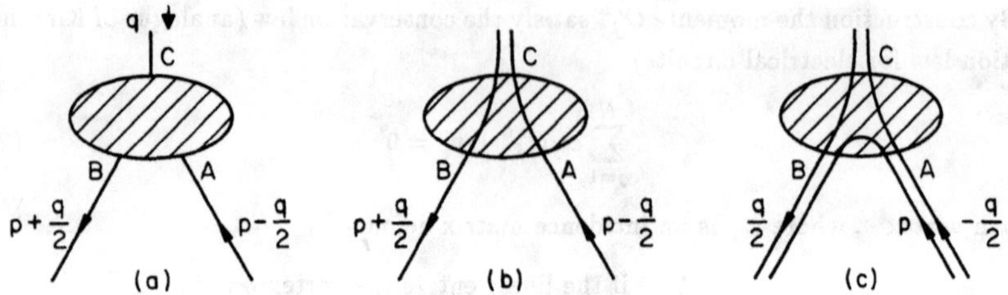

Figure 7: A vertex diagram with given external momenta is shown in (a). Some possible paths of external momenta are indicated in (b) and (c).

By choosing $P(AB)$ to be the shortest possible path, one obtains a very compact expression for $A_i^{(AB)}$.

To illustrate the usefulness of scalar currents, we shall compute $Q_i'^{\mu}$ of (2.4) for a vertex diagram of Fig. 7(a). Alternate definitions of external momenta and their possible paths through the diagram are indicated in Fig. 7(b) and 7(c). The most efficient choice depends on the particular diagram. If one follows the paths in Fig. 7(b), the constant momenta are given by

$$\begin{aligned} q_i &= p - \frac{1}{2}q, \qquad i \in P(AC), \\ q_i &= p + \frac{1}{2}q, \qquad i \in P(CB), \end{aligned}$$

and zero otherwise. Kirchhoff's current $Q_i'^{\mu}$ then becomes

$$Q_i'^{\mu} = A_i^{(AC)}\left(p - \frac{1}{2}q\right)^{\mu} + A_i^{(CB)}\left(p + \frac{1}{2}q\right)^{\mu}. \tag{2.18}$$

If one chooses the paths of Fig. 7(c), one obtains instead

$$Q_i'^{\mu} = A_i^{(AB)}p^{\mu} - A_i^{(AC)}\frac{1}{2}q^{\mu} + A_i^{(CB)}\frac{1}{2}q^{\mu}. \tag{2.19}$$

Comparing these equations one finds

$$A_i^{(AB)} = A_i^{(AC)} + A_i^{(CB)},$$

which is obvious from the definition (2.17).

2.3. Amplitudes for proper self-energy and vertex diagrams

In this article we are primarily interested in the diagrams of self-energy and vertex type. According to the Feynman-Dyson rules of Sec. 2.2, the amplitude for a $2n$-th order contribution to the electron proper self-energy G, aside from factors such as Z_2, $\bar{u}$, and u, can be expressed in the form

$$\left(\frac{\alpha}{\pi}\right)^n \Sigma^{(2n)} = -\left(\frac{-\alpha}{4\pi}\right)^n (n-2)!\, \mathbf{F} \int \frac{(dz)_G}{U^2 V^{n-1}}, \tag{2.20}$$

where $(dz)_G$ is given by (2.8) with $N = 3n-1$. The operator $\mathbf{F}$ is given by

$$\mathbf{F} = \gamma^{\alpha_1}(\not{D}_1 + m_1)\gamma^{\alpha_2}(\not{D}_2 + m_2)\cdot\cdot\cdot\ \gamma_{\alpha_i}(\not{D}_{2n-1} + m_{2n-1})\gamma_{\alpha_j}, \tag{2.21}$$

where i and j refer to the internal photon lines arriving at the $(2n-1)$-st and $2n$-th vertices along the electron line. Explicit forms of U, V, etc. are given later.

From now on we set the physical electron mass m equal to 1 for simplicity. The mass renormalization constant $\delta m^{(2n)}$ is obtained by evaluating (2.20) for $\not{p} = 1$. However, the mass m_i of the i-th electron line is kept as a parameter until all manipulations with m_i are completed.

In order to obtain the wave function renormalization constant $B^{(2n)}$, where B is related to Z_2 by $Z_2 = 1 + B$, we must evaluate $\partial\Sigma^{(2n)}/\partial p^\mu$. This leads to

$$\left(\frac{\alpha}{\pi}\right)^n B^{(2n)} = -\left(\frac{-\alpha}{4\pi}\right)^n (n-2)! \int \frac{(dz)_G}{U^2}\left(\mathbf{E}\frac{1}{V^{n-1}} + (n-1)\mathbf{N}\frac{1}{V^n}\right), \tag{2.22}$$

where $\mathbf{N} = 2G\mathbf{F}$, $2G \equiv -p^\nu(\partial V/\partial p^\nu)|_{q=0,p^2=1}$, and

$$\mathbf{E} = \sum_i A_i \mathbf{F}_i, \quad \text{(summation over electron lines only)}. \tag{2.23}$$

$\mathbf{F}_i$ is derived from $\mathbf{F}$ by the replacement in the electron line i:

$$\not{D}_i + m_i \Longrightarrow \not{p},$$

and A_i is the scalar current defined by (2.17).

Similarly, the amplitude for a $2n$-th order contribution to the proper electron vertex part, with the external photon vertex inserted in the k-th electron line, can be expressed in the form

$$\left(\frac{\alpha}{\pi}\right)^n \Gamma_\nu{}^{(2n)} = \left(\frac{-\alpha}{4\pi}\right)^n (n-1)!\mathbf{F}_\nu \int \frac{(dz)_G}{U^2 V^n}, \tag{2.24}$$

where $(dz)_G$ is of the form (2.8) with $N = 3n$ and $\mathbf{F}_\nu$ is obtained from (2.21) by the substitution in the electron line k:

$$(\not{D}_k + m_k) \Longrightarrow (\not{D}_k + m_k)\gamma_\nu(\not{D}_{k'} + m_{k'}),$$

k' being the new electron line created by insertion of a vertex. Aside from minor complications arising from the renormalization constant Z_2, this amplitude can be decomposed into the charge form factor F_1 and the magnetic form factor F_2, as was discussed in (1.5), where

$$F_1^{(2n)}(q) = \frac{1}{4}Tr[(\not{p}+1)p^\nu \Gamma_\nu^{(2n)}], \tag{2.25}$$

and

$$F_2^{(2n)}(q) = \frac{1}{4p^4q^2} Tr[(\gamma^\nu p^2 - (1+\frac{1}{2}q^2)p^\nu)(\not p + \frac{1}{2}\not q + 1)\Gamma_\nu^{(2n)}(\not p - \frac{1}{2}\not q + 1)]. \quad (2.26)$$

The $2n$-th order vertex renormalization constant $L^{(2n)}$, related to Z_1 by $Z_1 = 1 - L$, is obtained by evaluating (2.25) in the limit $q = 0$, $\not p = 1$. The anomalous magnetic moment a_e is the static limit of the magnetic form factor $F_2(q)$ in (2.26) and is given by (1.7). Parametric integrals for a_e are more complicated than those for the charge form factor because of the more elaborate trace projection and the appearance of two scalar currents associated with the momenta p and q.

We also need integrals obtained by inserting a δm vertex in an electron line of a proper vertex diagram of order $2n$:

$$-\delta m \left(\frac{\alpha}{\pi}\right)^n \Gamma_\nu{}^{(2n)*} = \delta m \left(\frac{-\alpha}{4\pi}\right)^n n!\, \mathbf{F}_\nu^* \int \frac{(dz)_G}{U^2 V^{n+1}}. \quad (2.27)$$

Similar expressions for $\Sigma^{(2n)*}$, etc., can be written down using the rules of Sec. 2.2.

Of course these formulas are meaningless without regularization. As is shown later, regularized Feynman integrals can be decomposed into UV-divergent (i.e., regulator-dependent) parts and UV-finite parts. The latter can further be decomposed into IR-divergent and IR-finite parts.

For simplicity we shall omit the factor $(\alpha/\pi)^n$ in the following.

2.4. Magnetic moment from a set of proper vertices summed by the Ward-Takahashi identity

Although the magnetic moment anomaly can be computed from (2.26) for any diagram, it is useful, in particular for higher order terms, to note that a set of vertex diagrams, derived from a self-energy diagram by inserting an external vertex representing the static magnetic field in all possible ways, share many properties. In the limit $q = 0$, they have functions U, V, B_{ij}, and A_i in common so that it is natural to treat them collectively. (Only the scalar currents A_i associated with q are not common.) We can go even further and amalgamate these integrals into a single one using the Ward-Takahashi identity. This approach enables us to reduce the number of independent integrals substantially, leading to considerable savings in time and effort of computation.

As is well-known, the proper vertex and self-energy parts are related by the Ward-Takahashi identity

$$q_\mu \Lambda^\mu(p, q) = -\Sigma(p + \frac{1}{2}q) + \Sigma(p - \frac{1}{2}q), \quad (2.28)$$

where we have set $\Gamma^\mu = \gamma^\mu + \Lambda^\mu$. This identity holds not only for the exact Σ and Λ but also for perturbation-theoretical Σ_G and Λ_G, where Σ_G represents the electron self-energy diagram G and Λ_G is the sum of vertex diagrams obtained by inserting an external vertex in G in all possible ways. Differentiating both sides of (2.28) with respect to q^μ

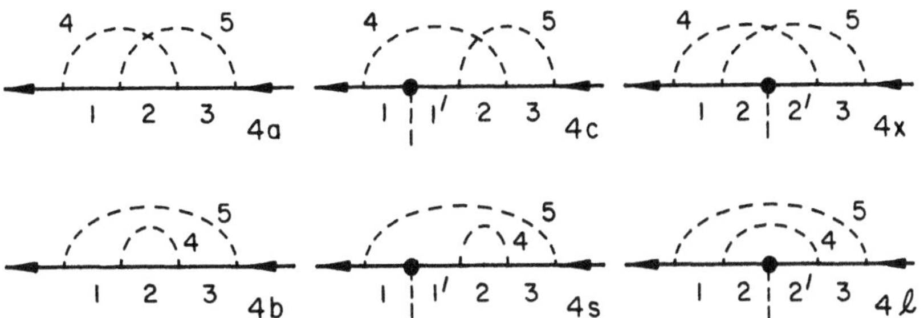

Figure 8: Line numbering of fourth-order diagrams. Upper half shows the electron self-energy diagram $4a$ and associated vertex diagrams $4x$ (crossed-ladder) and $4c$ (corner). Lower half shows the electron self-energy diagram $4b$ and associated vertex diagrams $4l$ (ladder) and $4s$ (self-energy).

and dropping terms quadratic or higher in q^μ, we obtain

$$\Lambda^\nu(p,q) \simeq \; - q^\mu \left[\frac{\partial \Lambda_\mu(p,q)}{\partial q_\nu}\right]_{q=0} - \frac{\partial \Sigma(p)}{\partial p_\nu}. \tag{2.29}$$

This is the starting point of our consideration.

In order to take advantage of (2.29), it is convenient to carry out the differentiation with respect to q_ν first and then parametrize the resulting integral. This can be achieved with the help of the identities:[17]

$$\left[\frac{\partial}{\partial q_\nu} \frac{1}{\not{p} \pm \frac{1}{2}\not{q} - m}\right]_{q=0} = \mp \frac{(\not{p}+m)\gamma^\nu(\not{p}+m)}{2(p^2-m^2)^2}, \tag{2.30}$$

$$\left[\frac{\partial}{\partial q_\nu}\left(\frac{1}{\not{p}+\frac{1}{2}\not{q}-m}\gamma^\mu \frac{1}{\not{p}-\frac{1}{2}\not{q}-m}\right)\right]_{q=0} = -\frac{\gamma^\mu\gamma^\nu(\not{p}+m) - (\not{p}+m)\gamma^\nu\gamma^\mu}{2(p^2-m^2)^2}, \tag{2.31}$$

$$\frac{(\not{p}_i + m_i)\gamma^\nu(\not{p}_i+m_i)}{(p_i^2-m_i^2)^2} = 2D_i^\nu(\not{D}_i + m_i)\frac{1}{(p_i^2-m_i^2)^2}. \tag{2.32}$$

As a concrete example, let us consider the fourth-order vertex diagram Λ_1 in Fig. 8(4c) in which an external photon line is inserted in the electron line 1. For this diagram one finds

$$\left[\frac{\partial}{\partial q_\nu}\left(\gamma^\alpha \frac{1}{\not{p}_1+\frac{1}{2}\not{q}-m_1}\gamma^\mu \frac{1}{\not{p}_1-\frac{1}{2}\not{q}-m_1}\gamma^\beta \frac{1}{\not{p}_2-\frac{1}{2}\not{q}-m_2}\gamma_\alpha \frac{1}{\not{p}_3-\frac{1}{2}\not{q}-m_3}\gamma_\beta\right)\right]_{q=0}$$

$$= -\left(\mathbf{Z}_1^{\mu\nu} + 2\sum_{j=1}^{3}\epsilon_{1j}D_1^\mu D_j^\nu \mathbf{F}\frac{1}{p_j^2-m_j^2}\right)\frac{1}{(p_1^2-m_1^2)\prod_{i=1}^{3}(p_i^2-m_i^2)},$$

where

$$\mathbf{F} = \gamma^\alpha(\not{D}_1+m_1)\gamma^\beta(\not{D}_2+m_2)\gamma_\alpha(\not{D}_3+m_3)\gamma_\beta$$

is identical with $\mathbf{F}$ for the corresponding electron self-energy diagrams of Fig.8(4a),

$$\epsilon_{ij} = -\epsilon_{ji} = 1 \quad \text{if } i < j,$$
$$\epsilon_{ii} = 0,$$

and $\mathbf{Z}_1^{\mu\nu}$ is obtained from $\mathbf{F}$ by the substitution:

$$\not{D}_1 + m_1 \Longrightarrow \frac{1}{2}[\gamma^\mu\gamma^\nu(\not{D}_1 + m_1) - (\not{D}_1 + m_1)\gamma^\nu\gamma^\mu]. \tag{2.33}$$

We are now ready to parametrize $q_\mu(\partial\Lambda^\mu/\partial q_\nu)$ by a slight adaptation of the method described in Sec. 2.2. The result is

$$-q_\mu\left[\frac{\partial\Lambda_1^\mu}{\partial q_\nu}\right]_{q=0} = \frac{1}{16}\,q_\mu\int(dz)\left(z_1\mathbf{Z}_1^{\mu\nu}\frac{1}{U^2V^2} + 4\sum_{j=1}^{3}\epsilon_{1j}D_1^\mu D_j^\nu\mathbf{F}\frac{z_1z_j}{U^2V^3}\right).$$

Similar results are obtained for the diagram Λ_2 of Fig. 8(4x) and the other corner diagram Λ_3. Summing up these contributions we obtain

$$-q_\mu\left[\frac{\partial\Lambda_a^\mu}{\partial q_\nu}\right]_{q=0} = \frac{1}{16}\,q_\mu\int(dz)\left(\mathbf{Z}^{\mu\nu}\frac{1}{U^2V^2} + 4\sum_{i,j=1}^{3}\epsilon_{ij}D_i^\mu D_j^\nu\mathbf{F}\frac{z_iz_j}{U^2V^3}\right), \tag{2.34}$$

where $\Lambda_a^\mu = \Lambda_1^\mu + \Lambda_2^\mu + \Lambda_3^\mu$ and

$$\mathbf{Z}^{\mu\nu} = \sum_{j=1}^{3} z_j\mathbf{Z}_j^{\mu\nu}. \tag{2.35}$$

In projecting out the magnetic moment component of (2.34), it is seen that the only contribution to a_e arises from the case where D_i^μ and D_j^ν are both contracted with the $\not{D}_k$ operators within $\mathbf{F}$. Thus the second term within the parentheses of (2.34) can be written as

$$\frac{1}{2U^2}\sum_{i,j,k,l}^{1,2,3}\epsilon_{ij}z_iz_jB'_{ik}B'_{jl}\mathbf{F}_{kl}^{\mu\nu}\frac{1}{U^2V},$$

where $\mathbf{F}_{kl}^{\mu\nu}$ is obtained from $\mathbf{F}$ by the substitution:

$$\not{D}_k + m_k\ ,\ \not{D}_l + m_l \implies \gamma^\mu\ ,\ \gamma^\nu \tag{2.36}$$

in the k-th and l-th electron lines.

This result can be readily generalized to any set of vertex diagrams generated by insertion of an external photon vertex in a self-energy diagram G in all possible ways. For simplicity let us define

$$\mathbf{C}^{\mu\nu} = \sum_{i<j} C_{ij}\mathbf{F}_{ij}^{\mu\nu}, \tag{2.37}$$

where

$$C_{ij} = \frac{1}{U^2}\sum_{k=1}^{2n-2}\sum_{l=k+1}^{2n-1} z_kz_l(B'_{ik}B'_{jl} - B'_{il}B'_{jk}).$$

Then we find

$$-q_\mu \left[\frac{\partial \Lambda_G^\mu}{\partial q_\nu}\right]_{mag.} = \left(\frac{-1}{4}\right)^n (n-1)!\, q_\mu \int (dz) \left(\mathbf{Z}^{\mu\nu}\frac{1}{U^2V^n} + \mathbf{C}^{\mu\nu}\frac{1}{(n-1)U^2V^{n-1}}\right),$$

putting $\mathbf{F}_{kl}^{\mu\nu} = -\mathbf{F}_{lk}^{\mu\nu}$ (which holds only for magnetic moment terms). If we now project out the magnetic moment term from this equation and the second term of (2.29) with the help of (2.26), we obtain

$$M_G^{(2n)} = \left(\frac{-1}{4}\right)^n (n-1)! \int (dz) \left(\frac{\mathbf{E}+\mathbf{C}}{n-1}\frac{1}{U^2V^{n-1}} + (\mathbf{N}+\mathbf{Z})\frac{1}{U^2V^n}\right), \tag{2.38}$$

as the contribution to the electron anomaly from all vertex diagrams associated with the self-energy diagram G of order $2n$. Here

$$\begin{aligned} \mathbf{N} &= \frac{1}{4}Tr[P_1^\nu p_\nu (2G\mathbf{F})], \\ \mathbf{E} &= \frac{1}{4}Tr[P_1^\nu \mathbf{E}_\nu], \\ \mathbf{C} &= \frac{1}{4}Tr[P_2^{\mu\nu}\mathbf{C}_{\mu\nu}], \\ \mathbf{Z} &= \frac{1}{4}Tr[P_2^{\mu\nu}\mathbf{Z}_{\mu\nu}]. \end{aligned} \tag{2.39}$$

The factors P_1^ν and P_2^ν are obtained from the magnetic moment projection in (2.26) by averaging over the directions of q_μ subject to the constraint $q^\mu p_\mu = 0$, and are given by

$$\begin{aligned} P_1^\nu &= \frac{1}{3}\gamma^\nu - \left(1 + \frac{4}{3}\not{p}\right) p^\nu, \\ P_2^{\mu\nu} &= \frac{1}{3}(\not{p}+1)(g^{\mu\nu} - \gamma^\mu\gamma^\nu + p^\mu\gamma^\nu - p^\nu\gamma^\mu). \end{aligned} \tag{2.40}$$

The operator $\mathbf{F}$ is the numerator factor for the self-energy diagram G:

$$\mathbf{F} = \gamma^{\alpha_1}(\not{D}_1 + m_1)\gamma^{\alpha_2}(\not{D}_2 + m_2)\cdots\gamma^{\alpha_i}(\not{D}_{2n-1} + m_{2n-1})\gamma_{\alpha_j},$$

where i and j refer to the internal photon lines arriving at the $(2n-1)$-th and $2n$-th vertices along the electron line (which depend on the diagram G), $\mathbf{E}^\nu$ is defined by

$$\mathbf{E}^\nu \equiv \frac{\partial \mathbf{F}}{\partial p_\nu} = \sum_{electron\ only} A_i \mathbf{F}_i^\nu, \tag{2.41}$$

where $\mathbf{F}_i^\nu$ is obtained from $\mathbf{F}$ by the substitution in the i-th electron line:

$$\not{D}_i + m_i \implies \gamma^\nu, \tag{2.42}$$

and G in (2.39) is defined in the sentence following (2.22) and can be written as

$$G = \sum_{electron\ only} z_i A_i. \tag{2.43}$$

Note that the troublesome q^2 in the denominator of (2.26) is absent in the final formula (2.38) and the limit $q = 0$ can be taken without running into difficulty. As a consequence, (2.38) depends only on one external momentum p, and the only scalar currents needed are A_i of the electron lines associated with p. If these A_i are expressed in terms of B_{ij}'s, they have the same form for all diagrams irrespective of how virtual photons attach to the electron line. Finally V in (2.38) can be written as

$$\begin{aligned} V &= \sum_{electron\ only} z_i(1 - A_i), \\ &= \sum_{electron\ only} z_i - G, \end{aligned} \tag{2.44}$$

which follows from the general formula (2.5) if one lets the external momentum p flow through the electron lines only. This form is again independent of how virtual photons are attached to the electron line.

3. Divergence Structure of Renormalization Constants

From the computational point of view the standard renormalization procedure is not necessarily the most desirable one, since renormalization terms themselves are generally infrared singular and introduce spurious IR divergences, making numerical evaluation more difficult. It is to circumvent this problem as much as possible that an alternative scheme, a kind of intermediate renormalization, was developed. The first step is to understand how UV and IR divergences appear in the renormalization constants. This section is devoted to the analysis of this problem.

3.1. $\mathbf{K}_S$ and $\mathbf{I}_R$ operations

Suppose $\mathcal{M}_G$ is a proper electron self-energy part or a proper vertex part of $2n$-th order defined by (2.20) or (2.24). Carrying out the D-operations in the $\mathbf{F}$ or $\mathbf{F}_\nu$ factor in these formulas, one finds an expression of the form

$$\begin{aligned} \mathcal{M}_G = (-1)^s \left(\frac{-1}{4}\right)^n (N - 2n - 1)! \int (dz)_G \Bigg(&\frac{F_0}{U^2 V^{N-2n}} + \frac{F_1}{U^3 V^{N-2n-1}} \\ &+ \frac{F_2}{U^4 V^{N-2n-2}} + \cdots + \frac{F_{m_G}}{U^{m_G+2} V^{N-2n-m_G}} \Bigg), \end{aligned} \tag{3.1}$$

where $s = 1$, $N = 3n - 1$ if G is a self-energy part and $s = 0$, $N = 3n$ if G is a vertex part. m_G is the maximum number of contractions of D operators, which is equal to $n - 1$ or n according to whether G is a self-energy part or a vertex part.

Suppose we want to find out whether $\mathcal{M}_G$ diverges when all loop momenta of a subdiagram S consisting of N_S lines and n_S closed loops go to infinity. In the parametric formulation, this corresponds to the vanishing of the denominator U when all $z_i \in S$ vanish simultaneously. To find a criterion for a UV divergence from S, consider the part of the integration domain where the $z_i \in S$ satisfy $\sum_{i\in S} z_i \leq \epsilon$. In the limit $\epsilon \to 0$, one finds

$$\begin{aligned} V &= O(1), \qquad U = O(\epsilon^{n_S}), \\ B_{ij} &= O(\epsilon^{n_S-1}) \ \text{ if } \ i,j \in S, \\ B_{ij} &= O(\epsilon^{n_S}) \qquad \text{otherwise.} \end{aligned} \tag{3.2}$$

Let m_S be the maximum number of contractions of D operators within S. Simple power counting shows that the $(m+1)$-th term of M_G in (3.1), whose numerator has at most m_S factors of B_{ij}, with $i,j \in S$, is divergent in the limit $\epsilon \to 0$ if and only if

$$N_S - 2n_S \leq \min[m, m_S] \leq m_S, \tag{3.3}$$

where $\min[m, m_S]$ means the smaller of m and m_S.

If S is a vertex part, we have $N_S = 3n_S$ and $m_S = n_S$. If S is an electron self-energy part, we have $N_S = 3n_S - 1$ and $m_S = n_S - 1$. Thus, in both cases, (3.3) is satisfied only for $\min[m, m_S] = m_S$, namely $m \geq m_S$. This means that the UV divergence from S is restricted to the terms with m_S contractions within S in the last $m_G - m_S + 1$ terms of (3.1).

For $S = G$ the above argument must be modified since $\sum_{i\in G} z_i \leq \epsilon$ is in conflict with $\sum_{i\in G} z_i = 1$. However, it is easy to remove this constraint[18] and show that the *UV* divergence is confined to the last term of (3.1). Since this divergence corresponds to the case where all loop momenta go to infinity, let us call it an *overall UV* divergence of $\mathcal{M}_G$, and denote it as $\hat{\mathcal{M}}_G$. The difference $\mathcal{M}_G - \hat{\mathcal{M}}_G$ will often be denoted as $\tilde{\mathcal{M}}_G$.

In order to deal with subdiagram divergences systematically, it is useful to introduce a reduced diagram R. If a subdiagram S is a vertex part, we write the reduced diagram R as G/S which is obtained by shrinking S to a point. If S is a self-energy part connected to the rest of G by electron lines i and j, shrinking S to a point produces a reduced diagram R, which we denote as $G/S(i^*)$, where i^* indicates that it has a two-point vertex between lines i and j. The diagram obtained by shrinking both S and line j to points will be denoted as $G/[S,j]$.

Let us now introduce a $\mathbf{K}_S$ operation, which extracts the overall UV-divergent part (due to the subdiagram S) of the Feynman integral

$$\mathcal{M}_G \equiv \int (dz)_G J_G \tag{3.4}$$

in an analytically factorizable way. This can be achieved by the following steps:

(a) In the limit (3.2) keep only terms with the lowest power of ϵ in the parametric functions U, B_{ij}, A_i, V. [This means that U factorizes into a product of U_S and U_R, where U_S is U defined on S and U_R is defined on R. Similarly for B_{ij}. V is reduced to V_R. Factorization of A_i is more subtle since it has a factor of U in its denominator. However, only those A_i's that factorize unambiguously are needed in the end.]

(b) Replace V_R obtained in (a) by $V_S + V_R$, where V_S is defined on S. [Since $V_S = O(\epsilon)$ while $V_R = O(1)$, this does not affect the leading singularity of the integrand in the $\epsilon \to 0$ limit.]

(c) Rewrite J_G in terms of the redefined parametric functions, drop all terms except those with m_S contractions within S, and call the result $\mathbf{K}_S J_G$. The corresponding integral is referred to as $\mathbf{K}_S \mathcal{M}_G$.

Since we deal in practice with logarithmic divergences only, the steps (a), (b) and (c) are sufficient to ensure that $(1 - \mathbf{K}_S)\mathcal{M}_G$ is convergent for $\epsilon \to 0$. The inclusion of V_S in (b) enables us to decompose $\mathbf{K}_S \mathcal{M}_G$ analytically into lower-order factors as

$$\mathbf{K}_S \mathcal{M}_G = \hat{L}_S \mathcal{M}_{G/S} \tag{3.5}$$

if S is a vertex subdiagram, and

$$\mathbf{K}_S \mathcal{M}_G = \delta\hat{m}_S \mathcal{M}_{G/S(i^\bullet)} + \hat{B}_S \mathcal{M}_{G/S(i')} \tag{3.6}$$

if S is an electron self-energy inserted between consecutive lines i and j of G. Here $\hat{L}_S$, $\delta\hat{m}_S$, and $\hat{B}_S$ are the overall UV-divergent parts (terms with maximal contraction) of the renormalization constants L_S, δm_S, and B_S. The second term $\mathcal{M}_{G/S(i')}$ in (3.6) is related to $\mathcal{M}_{G/[S,j]}$ by Nakanishi's identity.[37] Suppose $\mathcal{M}_{G/[S,j]}$ is written as

$$\mathcal{M}_{G/[S,j]} = \left(\frac{-1}{4}\right)^n (n-1)! \int (dz)_{G/[S,j]} \left(\frac{F_0}{U^2 V^n} + \frac{F_1}{U^3 V^{n-1}} + \cdots\right).$$

Then, it turns out that $\mathcal{M}_{G/S(i')}$, which has a complicated expression starting with a V^{-n-1} term, can always be transformed into the form

$$\mathcal{M}_{G/S(i')} = \left(\frac{-1}{4}\right)^n (n-1)! \int (dz)_{G/[S,j]} \left(-z_i \frac{\partial}{\partial z_i}\right) \left(\frac{F_0}{U^2 V^n} + \frac{F_1}{U^3 V^{n-1}} + \cdots\right). \tag{3.7}$$

Thus integration by part leads to

$$\mathcal{M}_{G/S(i')} = \mathcal{M}_{G/[S,j]}, \tag{3.8}$$

and we can rewrite (3.6) as

$$\mathbf{K}_S \mathcal{M}_G = \delta\hat{m}_S \mathcal{M}_{G/S(i^\bullet)} + \hat{B}_S \mathcal{M}_{G/[S,j]}. \tag{3.9}$$

We shall call $\mathcal{M}_{G/S(i')}$ a "derivative amplitude" since it is derived from $M_{G/[S,j]}$ by applying the operation $-z_i(\partial/\partial z_i)$ on its integrand. This operation gives a new parametric integrand, whose divergence structure can be analyzed by a method similar to the one described above. Note that its divergence structure is different from that of $\mathcal{M}_{G/[S,j]}$. In particular one finds

$$\hat{\mathcal{M}}_{G/[S,j]} \neq \hat{\mathcal{M}}_{G/S(i')},$$

although (3.8) must be satisfied as a whole.

If i runs through the set of all vertex and self-energy subdiagrams S_i of G, the integral

$$\prod_i (1 - \mathbf{K}_{S_i}) \mathcal{M}_G \tag{3.10}$$

is UV-divergence-free by construction. This is a kind of intermediate renormalization. Note that, although the $\mathbf{K}_S$ operation deals with the UV divergence arising from the subdiagram S, $\mathbf{K}_S J_G$ is defined in the entire domain of integration, and the factorizations in (3.5) and (3.6) are exact.

An infrared divergence, which is caused by vanishing virtual photon momenta, arises from the part of integration domain where the z_i's assigned to these photons take the largest possible values under the constraint $\sum z_i = 1$. This means that all the other z_i's are pushed to zero in the IR limit. Furthermore, the IR singularity, in order that it actually becomes divergent, must be enhanced by vanishing denominators of electron propagators adjacent to the infrared photon. In parametric language this corresponds to the vanishing of V in the integration domain characterized by

$$z_i = \begin{cases} O(\delta) & \text{if } i \text{ is an electron line in } R, \\ O(1) & \text{if } i \text{ is a photon line in } R, \\ O(\epsilon),\ \epsilon \simeq \delta^2, & \text{if } i \in S\ . \end{cases} \tag{3.11}$$

Let us first consider the case $R = G$. Then, from (3.11) we have

$$U = O(1), \quad V = O(\delta^2), \quad F_i = O(1), \quad i = 1,\ 2,\ \ldots,\ m_G.$$

Thus, a naive power counting of the $(m+1)$-st term of $\mathcal{M}_G$ in (3.1) leads to

$$\frac{(dz)_G F_m}{U^{2+m} V^{N-2n-m}} = O\left(\frac{\delta^{2n}}{\delta^{2(N-2n-m)}}\right) = O(\delta^{2m}),$$

if G is a vertex part, and

$$\frac{(dz)_G F_m}{U^{2+m} V^{N-2n-m}} = O\left(\frac{\delta^{2n-1}}{\delta^{2(N-2n-m)}}\right) = O(\delta^{1+2m}),$$

if G is a self-energy part. Thus only the $m = 0$ term of the vertex part has logarithmic IR divergence. This will be called the overall IR-divergent term of $\mathcal{M}_G$.

This argument can be extended to any self-energy or vertex subdiagram S. Fortunately, for the discussion of the magnetic moment, we need not worry about a subdiagram S with four or more external lines that may develop complicated IR-divergence structure dependent on the momentum configuration.

Taking these remarks into account, let us introduce the $\mathbf{I}_R$ operation on a vertex diagram G by the following procedure:

(a′) Keep only the lowest powers of ϵ, δ in parametric functions U, B_{ij} and A_i.

(b′) Make the following replacements:

$$U \to U_S U_R, \quad V \to V_S + V_R, \quad \mathbf{F} \to F_0[L_R]\mathbf{F}_S,$$

where $F_0[L_R]$ is the no-contraction term of the vertex renormalization constant defined on R, and $\mathbf{F}_S$ is the product of γ matrices and D_i^μ operators for the diagram S.

(c′) Rewrite the integrand J_G in terms of redefined parametric functions, keep only the IR-divergent terms, and denote the result as $\mathbf{I}_R J_G$. The corresponding integral will be denoted as $\mathbf{I}_R \mathcal{M}_G$.

Step (a′) is a twofold limit; $\epsilon \to 0$ by itself is just the UV limit for the subdiagram S, and $\delta \to 0$ corresponds to all photons of the reduced diagram R going to the IR limit. This means that most (but not all) parametric functions defined to take care of the UV divergence from the subdiagram S can be used for IR divergence, too. When an IR-divergent term $\mathbf{I}_R \mathcal{M}_G$ contains a UV-divergent subdiagram S, it is useful to treat $\mathbf{I}_R \mathcal{M}_G$ and $\mathbf{I}_R \mathbf{K}_S \mathcal{M}_G$ together by introducing a new notation:

$$\tilde{\mathbf{I}}_R \equiv \mathbf{I}_R(1 - \mathbf{K}_S). \tag{3.12}$$

$\tilde{\mathbf{I}}_R \mathcal{M}_G$ has a much simpler IR structure than either $\mathbf{I}_R \mathcal{M}_G$ or $\mathbf{I}_R \mathbf{K}_S \mathcal{M}_G$ since most of the IR-singular terms cancel out when combined. Step (b′) is chosen to guarantee analytic factorization of the subtraction term.

As a specific example, consider a vertex diagram $G(k)$ which is obtained from a self-energy diagram G by inserting an external photon (magnetic field) line in the electron line k. Suppose a vertex subdiagram $S(k)$, which contains the electron line k, is connected to the rest of $G(k)$ through the electron lines i and j. Then one finds

$$\tilde{\mathbf{I}}_R \mathcal{M}_{G(k)} = L_{R(i)}[F_0] \mathcal{M}_{S(k)}, \tag{3.13}$$

where $L_{R(i)}[F_0]$ is the no-contraction term of $\mathcal{M}_{R(i)}$. This factorization is valid in the entire domain of integration, not just near the infrared divergent region.

Amplitudes containing mass-vertex insertions have IR divergences stronger (linear or worse) than those described above. For instance, a vertex with a mass-vertex insertion, such as $\Gamma_\nu^{(2n)*}$ of (2.27), generates an integral of the form

$$\mathcal{M}_G^* = -\left(\frac{-1}{4}\right)^n n! \int (dz)_G \left(\frac{F_0^*}{U^2 V^{n+1}} + \frac{F_1^*}{U^3 V^n} + \ldots \right). \tag{3.14}$$

Because of the increased power of V in the denominator, the degree of IR divergence is increased and the preceding analysis must be revised accordingly. Although these IR divergences cancel out in the final result, we have to learn how to deal with them since all of them appear somewhere in our calculation. They will be discussed case by case as the need arises.

To carry out the $\mathbf{I}_R$ operation on amplitudes such as (2.38) that have been summed via the Ward-Takahashi identity, it is necessary to exercise extra care. This is due to the fact that the sum may include terms in which S is a self-energy part and that Eq. (2.29) blurs the distinction between denominators and numerators, thereby making infrared power counting more subtle. Let $\mathcal{M}_G$ be the Ward-Takahashi summed magnetic moment projection of all vertex diagrams $G(k)$ associated with the self-energy diagram G. We then find

$$\tilde{\mathbf{I}}_R M_G = L_{R(i)}[F_0] M_S + M_{R,i^*}[I] \Delta\delta\tilde{m}_S, \tag{3.15}$$

where M_S is the Ward-Takahashi summed magnetic moment projection of all vertex diagrams determined by S, and $L_{R(i)}[F_0]$ and $\Delta\delta\tilde{m}_S$ are parts of renormalization constants. For the meaning of $M_{R,i^*}[I]$, see later sections.

Formula (3.13) applies to the magnetic moment projection ($M_{G(k)}$) of an individual (namely, un-summed) vertex graph, the associated vertex renormalization constant ($L_{G(k)}$), and the mass counterterm (δm_G).

Formula (3.15) applies to the Ward-Takahashi summed magnetic moment projection (M_G) as well as the wave-function renormalization constant (B_G), which may be regarded as the Ward-Takahashi sum of vertex renormalization constants. For the latter (3.15) takes the form

$$\tilde{\mathbf{I}}_R B_G = L_{R(i)}[F_0]B_S + B_{R,i^*}[I]\Delta\delta\tilde{m}_S. \tag{3.16}$$

If S is a second-order self-energy subdiagram, we have

$$\Delta\delta\tilde{m}_S \equiv \Delta\delta m_2 = 0,$$

as is seen in the next subsection. Thus it is not until we deal with a sixth-order diagram G that the additional terms of (3.15) and (3.16) become visible.

Let i run through the set of all UV-divergent subdiagrams S_i and j through all reduced diagrams R_i which are IR-divergent. Then the integral

$$\Delta\mathcal{M}_G = \prod_{i,j}(1 - \mathbf{I}_{R_i})(1 - \mathbf{K}_{S_j})\mathcal{M}_G \tag{3.17}$$

is free from both UV and IR divergences, if S_j are all vertex diagrams. If some S_j are self-energy diagrams, more careful treatment of the IR divergence is needed. This will be discussed when we run into such cases.

In the following we often identify $\mathbf{K}_S$ or $\mathbf{I}_R$ not by S or R but by the names of electron lines contained in S or R. [For instance, if S contains two electron lines 1 and 2, we write $\mathbf{K}_{12}$ for $\mathbf{K}_S$.] Also, the overall UV-divergent term of $\mathcal{M}_G$ will be denoted as $\mathbf{K}_{all}\mathcal{M}_G$ ($= \hat{\mathcal{M}}_G$), while the overall IR-divergent term of $\mathcal{M}_G$ will be denoted as $\mathbf{I}_{all}\mathcal{M}_G$.

3.2. Second-order renormalization constants

Diagrams needed for constructing second-order renormalization constants are shown in Fig. 9. As was mentioned previously, we omit powers of α/π, since it is not difficult to retrieve them from the context.

Mass renormalization constant δm_2

In this order, only one diagram (Σ_2 of Fig. 9) contributes to the electron mass operator

$$\begin{aligned} \Sigma_2(p) &= \frac{1}{4}\mathbf{F}\int(dz)\int_{\lambda^2}^{\Lambda^2} z_2 dm_2^2 \frac{1}{U^2V}, \\ (dz) &= \delta(1 - z_1 - z_2)dz_1 dz_2, \\ \mathbf{F} &= \gamma^\mu(\not{D}_1 + m_1)\gamma_\mu. \end{aligned} \tag{3.18}$$

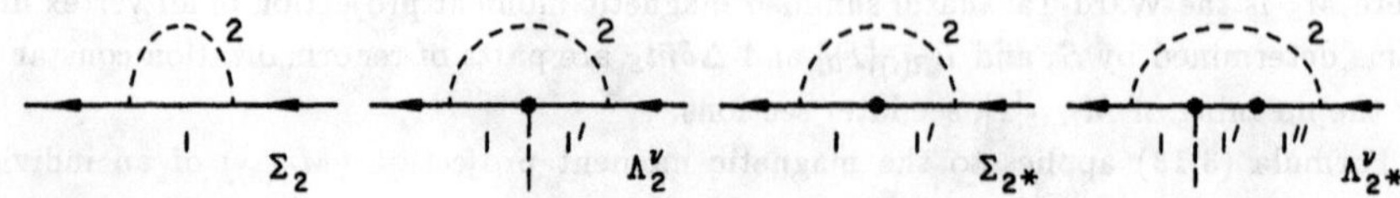

Figure 9: Second-order electron self-energy and vertex diagrams and corresponding diagrams with mass insertion.

The operator D_i^μ is defined by (2.9). Λ is the Feynman cut-off for the photon (line 2 in Fig. 9), and λ is an infinitesimal photon mass introduced to avoid possible IR divergence. Carrying out the $\mathbf{F}$ operation and setting $\not{p} = 1$, one obtains

$$\begin{aligned} \delta m_2 &= \frac{1}{4}\int (dz) \int_{\lambda^2}^{\Lambda^2} z_2 dm_2^2 \frac{F_0}{U^2 V}, \\ F_0 &= 2(2 - A_1). \end{aligned} \tag{3.19}$$

To construct the integrand explicitly, one first looks at Table I and finds

$$B_{11} = 1. \tag{3.20}$$

Then, from (2.2), (2.4), (2.5), and (2.43), one obtains [d]

$$\begin{aligned} U &= z_1 + z_2 \equiv z_{12}, \\ A_1 &= 1 - z_1 B_{11}/U = z_2/z_{12}, \\ V &= z_1 + m_2^2 z_2 - z_1 A_1 \qquad (p^2 = 1). \end{aligned} \tag{3.21}$$

The F_0 term contributes to the overall UV divergence, but has no IR divergence. Thus, the intermediate renormalization constant $\delta\hat{m}_2 \equiv \mathbf{K}_{all}\delta m_2$ can be identified with δm_2:

$$\delta\hat{m}_2 = \delta m_2. \tag{3.22}$$

Carrying out the integration in (3.19) one finds

$$\delta m_2 = \frac{3}{2}(\ln\Lambda + \frac{1}{4}). \tag{3.23}$$

Vertex renormalization constant L_2

According to (2.24) the second-order vertex part (Λ_2^ν of Fig. 9) is given by

$$\Lambda_2^\nu = -\frac{1}{4}\mathbf{F}^\nu \int (dz) \int_{\lambda^2}^{\Lambda^2} z_2 dm_2^2 \frac{1}{U^2 V^2}, \tag{3.24}$$

with

$$\mathbf{F}^\nu = \gamma^\mu(\not{D}_1 + m_1)\gamma^\nu(\not{D}_{1'} + m_{1'})\gamma_\mu.$$

[d]For brevity we use the notation $z_{12\cdots n}$ for $z_1 + z_2 + \cdots + z_n$ from here on.

Carrying out the D operations, we can reduce (3.24) to

$$\Gamma_2^\nu = -\frac{1}{4}\int (dz) \int_{\lambda^2}^{\Lambda^2} z_2 dm_2^2 \left(\frac{F_0^\nu}{U^2V^2} + \frac{F_1^\nu}{U^3V}\right), \tag{3.25}$$

where

$$\begin{aligned} (dz) &= \delta(1 - z_{11'2}) dz_1 dz_{1'} dz_2 \,, \\ F_0^\nu &= \gamma^\mu(\not{Q}'_1 + m_1)\gamma^\nu(\not{Q}'_{1'} + m_{1'})\gamma_\mu \,, \\ F_1^\nu &= -\frac{1}{2} B_{11'}\gamma^\mu\gamma^\lambda\gamma^\nu\gamma_\lambda\gamma_\mu \,. \end{aligned}$$

The vertex renormalization constant L_2 is obtained by evaluating (3.25) for $q = 0$, $\not{p} = 1$. Note that $\not{Q}'_i = A_i\not{p}$ by (2.15), A_i being the scalar current associated with p. Since $A_{i'} = A_i, B_{11'} = B_{11}$ for $q = 0$, one obtains

$$L_2 = -\frac{1}{4}\int (dz) \int_{\lambda^2}^{\Lambda^2} z_2 dm_2^2 \left(\frac{F_0}{U^2V^2} + \frac{F_1}{U^3V}\right), \tag{3.26}$$

where

$$F_0 = -\,2(1 - 4A_1 + A_1^2), \quad F_1 = -\,2B_{11}.$$

Of course this F_0 is different from F_0 of (3.19). Functions A_1, B_{11}, U, and V are obtained from (3.20) and (3.21) by the substitution $z_1 \to z_{11'}$.

Since z_1 and $z_{1'}$ appear in the combined form $z_{11'}$ in the integrand of (3.26), one can perform one of the z integrations and reduce it to

$$L_2 = -\frac{1}{4}\int z_{11'}(dz) \int_{\lambda^2}^{\Lambda^2} z_2 dm_2^2 \left(\frac{F_0}{U^2V^2} + \frac{F_1}{U^3V}\right), \tag{3.27}$$

where (dz) is redefined as

$$(dz) = \delta(1 - z_{11'} - z_2) dz_{11'} dz_2.$$

For economy of notation, let us write the integrals over F_0 and F_1 in (3.27) as

$$L_2 = L_2[F_0] + L_2[F_1],$$

or, even more symbolically, as

$$L_2 = \mathcal{F}_0 + \mathcal{F}_1, \tag{3.28}$$

whenever it is clear from the context. In the intermediate renormalization scheme of Sec. 3.1, we treat only the UV-divergent part of L_2 as the renormalization constant. Since only $\mathcal{F}_1$ satisfies the overall UV divergence criterion (3.3), we choose it as the intermediate vertex renormalization constant $\hat{L}_2$:

$$\hat{L}_2 = \mathbf{K}_{all} L_2 = \mathcal{F}_1.$$

Let $\tilde{L}_2$ be the difference of L_2 and $\hat{L}_2$. Then we have $\tilde{L}_2 = \mathcal{F}_0$. Since $\mathcal{F}_0$ is UV-finite, we can remove the Feynman cut-off and write it as

$$\tilde{L}_2 = \mathcal{F}_0 = -\frac{1}{4}\int z_{11'}(dz)\frac{F_0}{U^2V}\,. \tag{3.29}$$

Carrying out the m_2 and z integrations, we find

$$\hat{L}_2 = \frac{1}{2}(\ln \Lambda - \frac{1}{4}), \tag{3.30}$$

$$\tilde{L}_2 = \ln \lambda + \frac{5}{4} \equiv I_2 . \tag{3.31}$$

$\tilde{L}_2$ is IR-divergent and can be identified with $\mathbf{I}_{all}L_2$. From these we obtain the standard result

$$L_2 = \hat{L}_2 + \tilde{L}_2 = \frac{1}{2}\ln \Lambda + \ln \lambda + \frac{9}{8} . \tag{3.32}$$

Note that the photon mass dependence of V is presented somewhat sloppily: The "photon" mass in V of (3.27) is m_2 while that of (3.29) is λ. This is done to avoid excessive notational complication.

Wave-function renormalization constant B_2

According to (2.22) the wave-function renormalization constant B_2 is given by

$$B_2 = \frac{1}{4}\int (dz) \int z_2 dm_2^2 \left(\mathbf{E}\frac{1}{U^2V} + \mathbf{N}\frac{1}{U^2V^2}\right), \tag{3.33}$$

where (dz) and $\mathbf{F}$ are defined by (3.18), $G = z_1 A_1$, $\mathbf{N} = 2G\mathbf{F}$, and

$$\mathbf{E} = [\gamma^\mu (A_1 \not{p})\gamma_\mu]_{\not{p}=1} .$$

This reduces to

$$\begin{aligned} B_2 &= \frac{1}{4}\int (dz) \int z_2 dm_2^2 \left(\frac{E_0}{U^2V} + \frac{N_0}{U^2V^2}\right) \\ &\equiv \mathcal{E}_0 + \mathcal{N}_0 \end{aligned} \tag{3.34}$$

where $N_0 = 2GF_0$, F_0 being given by (3.19), and $E_0 = -2A_1$.

Note that, if we identify z_1 in (3.34) and $z_{11'}$ in (3.26), all parametric functions defining B_2 and L_2 become identical. In general, parametric functions unrelated to the external photon momentum q are common for a self-energy diagram and corresponding set of vertex diagrams.

Since $\mathcal{E}_0$ is the only UV-divergent term, the intermediate renormalization constant $\hat{B}_2 \equiv \mathbf{K}_{all}B_2$ is given by

$$\hat{B}_2 = \mathcal{E}_0 = -\frac{1}{2}(\ln \Lambda + \frac{5}{4}). \tag{3.35}$$

The difference between B_2 and $\hat{B}_2$:

$$\tilde{B}_2 = \mathcal{N}_0 = \frac{1}{4}\int (dz)\, \frac{N_0}{U^2V} \tag{3.36}$$

is UV-finite but IR-divergent. The intermediate renormalization scheme becomes somewhat simpler if we express this IR divergence in terms of I_2 of (3.31). With this in mind let us rewrite the integrand of $\tilde{B}_2$ as

$$N_0 = -2z_1(1 - 4A_1 + A_1^2) + 2z_1(1 - A_1^2).$$

Note that the first term is z_1 times F_0 of (3.26). The second term vanishes in the IR limit $A_1 \to 1$ and hence gives an IR-finite integral. Thus we can write

$$\tilde{B}_2 = -I_2 + \Delta B_2,$$

where

$$\Delta B_2 \equiv \frac{1}{4}\int (dz)\, \frac{2z_1(1-A_1^2)}{U^2 V} = \frac{3}{4}. \tag{3.37}$$

From these results and (3.32) follows the second-order Ward identity:

$$B_2 + L_2 = 0. \tag{3.38}$$

3.3. Second-order graphs with mass insertions

In discussing the renormalization of the higher-order terms of the electron anomaly, it becomes necessary to deal explicitly with renormalization constants arising from diagrams with mass insertion, such as $\Sigma_{2^\bullet}$ and $\Lambda^\nu_{2^\bullet}$ of Fig. 9.

Mass renormalization constant $\delta m_{2^\bullet}$

The parametric form of $\delta m_{2^\bullet}$, the self-mass part of $\Sigma_{2^\bullet}$, can be written as

$$\begin{aligned} \delta m_{2^\bullet} &= -\frac{1}{4}\int (dz) \int_{\lambda^2}^{\Lambda^2} z_2 dm_2^2 \left(\frac{F_0^*}{U^2V^2} + \frac{F_1^*}{U^3 V}\right) \\ &\equiv \mathcal{F}_0^* + \mathcal{F}_1^*, \end{aligned} \tag{3.39}$$

where

$$F_0^* = 4(1 - A_1 + A_1^2), \quad F_1^* = -8B_{11},$$

U, V, B_{11}, and A_1 being obtained from those of (3.19) by the substitution $z_1 \to z_{11'}$.

It is easily seen that $\mathcal{F}_0^*$ is IR-divergent while $\mathcal{F}_1^*$ is UV-divergent. Thus we write

$$\begin{aligned} \delta \hat{m}_{2^\bullet} &= \mathcal{F}_1^* \\ \delta \tilde{m}_{2^\bullet} &= \mathcal{F}_0^* \equiv I_2 + \Delta \delta m_{2^\bullet}, \end{aligned} \tag{3.40}$$

where I_2 is given by (3.31) and $\Delta\delta m_{2^\bullet} = -\frac{3}{4}$.

Vertex renormalization constant $L_{2^\bullet}$

The parametric form for $L_{2^\bullet}$, the vertex renormalization constant derived from $\Lambda^\nu_{2^\bullet}$, is

$$\begin{aligned} L_{2^\bullet} &= -\frac{1}{4}\int (dz) \left(\frac{F_0^*}{U^2V^2} + \frac{F_1^*}{U^3 V}\right) \\ &\equiv \mathcal{F}_0^* + \mathcal{F}_1^*, \end{aligned} \tag{3.41}$$

where U and V are obtained from those of (3.39) by the substitution $z_{11'} \to z_{11'1''}$, and

$$\begin{aligned} (dz) &= \delta(1 - z_{11'1''} - z_2) dz_1 dz_{1'} dz_{1''} dz_2 \\ &= \delta(1 - z_{11'1''} - z_2) \frac{1}{2} z_{11'1''}^2 dz_{11'1''} dz_2, \end{aligned}$$

$$F_0^* = 2(-1 + 6A_1 - 3A_1^2 + 2A_1^3), \quad F_1^* = -12 A_1 B_{11}.$$

L_{2*} is UV-finite. Thus $\hat{L}_{2*} = 0$. This is a general feature of vertex and (by the Ward identity) wave-function renormalization constants for diagrams having self-mass insertions. On the other hand, IR divergences become worse: $\mathcal{F}_0^* \equiv \mathbf{I}_{all} L_{2*}$ has a linear overall IR singularity

$$I_{2*} \equiv \mathcal{F}_0^* = \lim_{\lambda \to 0} \left(\frac{\pi}{4\lambda} + \frac{3}{2} \ln \lambda + \frac{3}{4} \right). \tag{3.42}$$

The second term $\mathcal{F}_1^*$ is finite. On integration one finds

$$\Delta L_{2*} = -\frac{3}{4}. \tag{3.43}$$

Wave-function renormalization constant B_{2*}

The wave-function renormalization constant B_{2*} (from Σ_{2*} of Fig. 9) is given by

$$\begin{aligned} B_{2*} &= -\frac{1}{4} \int (dz) \left(\frac{N_0^*}{U^2 V^2} + \frac{N_1^*}{U^3 V} + \frac{E_0^*}{U^2 V} \right) \\ &\equiv \mathcal{N}_0^* + \mathcal{N}_1^* + \mathcal{E}_0^*, \end{aligned} \tag{3.44}$$

where

$$N_0^* = 2GF_0^*, \quad N_1^* = 2GF_1^*, \quad G = z_{11'} A_1, \quad E_0^* = 4A_1(2A_1 - 1),$$

and U, V, F_0^*, F_1^*, and (dz) are as in (3.39).

All terms of B_{2*} are UV-finite. However, $\mathcal{N}_0^*$ and $\mathcal{E}_0^*$ are overall IR divergent. Writing $\mathcal{N}_0^* + \mathcal{E}_0^* (\equiv \mathbf{I}_{all} B_{2*})$ as

$$\mathcal{N}_0^* + \mathcal{E}_0^* = -2I_{2*} + H_{2*},$$

where I_{2*} is given by (3.42), one finds that

$$H_{2*} = -\frac{1}{4} \int (dz) \left(\frac{z_{11'}(2 - 4A_1 - 2A_1^2 + 4A_1^3)}{U^2 V^2} + \frac{4A_1(2A_1 - 1)}{U^2 V} \right)$$

is finite. If one defines ΔB_{2*} by

$$\begin{aligned} \Delta B_{2*} &= B_{2*} + 2I_{2*}, \\ &= \mathcal{N}_1^* + H_{2*}, \end{aligned}$$

one obtains $\Delta B_{2*} = \frac{3}{2}$, which is consistent with the extended Ward identity

$$2L_{2*} + B_{2*} = 0.$$

3.4. Fourth-order renormalization constants

Diagrams needed for constructing fourth-order renormalization constants are shown in Fig. 8. The numbering of electron lines and photon lines shown there will be adhered to throughout Sec. 3.4.

Mass renormalization constant δm_{4a}

The self-mass term corresponding to the diagram 4a of Fig. 8 is given by

$$\begin{aligned}\delta m_{4a} &= -\frac{1}{16}\int(dz)\int z_4 dm_4^2 \int z_5 dm_5^2 \left(\frac{2F_0}{U^2V^3} + \frac{F_1}{U^3V^2}\right) \\ &\equiv \mathcal{F}_0 + \mathcal{F}_1,\end{aligned} \tag{3.45}$$

where

$$\begin{aligned}F_0 &= 4(-1 + 2A_2 + A_1A_2 + A_1A_3 + A_2A_3 - 2A_1A_2A_3), \\ F_1 &= 4[B_{12}(A_3 - 2) + B_{13}(4A_2 - 2) + B_{23}(A_1 - 2)].\end{aligned} \tag{3.46}$$

Applying the rules of Sec. 2, one finds

$$\begin{aligned}&B_{12} = z_{35}, \quad B_{13} = -z_2, \quad B_{23} = z_{14}, \\ &B_{11} = z_{235}, \quad B_{22} = z_{1345}, \quad B_{33} = z_{124}, \quad U = z_{14}B_{12} + z_2B_{22}, \\ &A_i = 1 - (z_1B_{1i} + z_2B_{2i} + z_3B_{3i})/U, \quad i = 1,\ 2,\ 3, \\ &G = z_1A_1 + z_2A_2 + z_3A_3, \quad V = z_{123} + m_4^2z_4 + m_5^2z_5 - G, \\ &(dz) = \delta(1 - z_{12345})dz_1dz_2dz_3dz_4dz_5.\end{aligned}$$

The integral $\mathcal{F}_1$ has not only an overall UV-divergence but also UV-divergences from subdiagrams S = [1,2,4] and S' = [2,3,5]. Since the overall UV-divergent term will be subtracted in its entirety in our intermediate renormalization procedure, no further analysis of subdiagram divergences is necessary for $\mathcal{F}_1$. Both $\mathcal{F}_0$ and $\mathcal{F}_1$ are IR-finite. The divergence structure is therefore

$$\begin{aligned}\delta m_{4a} &= \mathcal{F}_0 + \mathcal{F}_1 \\ &= \Delta\delta m_{4a} + \delta\hat{m}_{4a}.\end{aligned}$$

Here, and in all analogous formulas that follow, there is a one-to-one correspondence between terms of the first and second lines. Note also that, following the convention introduced earlier, we identify UV- and IR-finite quantities by the prefix Δ.

Renormalization constants B_{4a}, L_{4x}, and L_{4c}

The wave-function renormalization constant associated with diagram 4a of Fig. 8 can be written as

$$\begin{aligned}B_{4a} &= -\frac{1}{16}\int(dz)\int z_4 dm_4^2 \int z_5 dm_5^2 \left(\frac{2E_0}{U^2V^3} + \frac{E_1}{U^3V^2} + \frac{6N_0}{U^2V^4} + \frac{2N_1}{U^3V^3}\right) \\ &\equiv \mathcal{E}_0 + \mathcal{E}_1 + \mathcal{N}_0 + \mathcal{N}_1,\end{aligned} \tag{3.47}$$

where $N_{0,1} = 2GF_{0,1}$ with $F_{0,1}$ given in (3.46), and

$$\begin{aligned}E_0 &= 4[A_1 + A_2 + A_3 + 2(A_1A_2 + A_1A_3 + A_2A_3) - 6A_1A_2A_3], \\ E_1 &= 4(B_{12}A_3 + 4B_{13}A_2 + B_{23}A_1).\end{aligned}$$

The overall UV divergence is confined to $\mathcal{E}_1$. Thus we define the intermediate wave-function renormalization constant as

$$\hat{B}_{4a} \equiv \mathbf{K}_{all}B_{4a} = \mathcal{E}_1.$$

For the remaining terms of (3.47) one has to analyze subdiagram UV divergences. As is easily seen, they are confined to $\mathcal{N}_1$. Applying the $\mathbf{K}_{12}$ and $\mathbf{K}_{23}$ operations, where suffixes 12 and 23 refer to the electron lines in S and S', one finds

$$\mathbf{K}_{12}\mathcal{N}_1 = \mathbf{K}_{23}\mathcal{N}_1 = \hat{L}_2\tilde{B}_2.$$

If we define the UV-finite part of B_{4a}, which no longer needs regularization, by

$$\Delta' B_{4a} = -\frac{1}{16}\int (dz)\left(\frac{E_0}{U^2V} + \frac{N_0}{U^2V^2} + (1 - \mathbf{K}_{12} - \mathbf{K}_{23})\frac{N_1}{U^3V}\right), \tag{3.48}$$

the UV divergence structure of B_{4a} may be summarized as follows:

$$\begin{aligned} B_{4a} &= [\mathcal{E}_0 + \mathcal{N}_0 + (1 - \mathbf{K}_{12} - \mathbf{K}_{23})\mathcal{N}_1] + (\mathbf{K}_{12} + \mathbf{K}_{23})\mathcal{N}_1 + \mathcal{E}_1 \\ &= \quad \Delta' B_{4a} \quad + \quad 2\hat{L}_2\tilde{B}_2 \quad + \quad \hat{B}_{4a}. \end{aligned}$$

Although $\Delta' B_{4a}$ is UV-finite, it is divergent in the (overall) IR limit where the momenta of *both* internal photons vanish. In the parameter space this limit corresponds to

$$z_4 + z_5 = 1 - \delta, \quad \delta \to 0.$$

Examining the integrand in this limit, one finds that only $\mathcal{N}_0$ is IR-divergent:

$$I_{4a} \equiv \mathbf{I}_{all} B_{4a} = \mathcal{N}_0 = -\frac{1}{16}\int (dz)\frac{N_0}{U^2V^2}. \tag{3.49}$$

Before analyzing this integral further, let us examine the overall IR divergence of the associated vertex diagrams, which are generated by inserting an external vertex in the electron lines of diagram $4a$. To begin with let us consider the vertex renormalization constant L_{4x} for the diagram of Fig. 8(4x):

$$\begin{aligned} L_{4x} &= \frac{1}{16}\int (dz)\int z_4 dm_4^2 \int z_5 dm_5^2 \left(\frac{6F_0}{U^2V^4} + \frac{2F_1}{U^3V^3} + \frac{F_2}{U^4V^2}\right) \\ &\equiv \mathcal{F}_0 \quad + \quad \mathcal{F}_1 \quad + \quad \mathcal{F}_2, \end{aligned} \tag{3.50}$$

where

$$\begin{aligned} F_0 &= 4[(1 + A_2^2)(1 + A_1 + A_3 - 2A_1A_3) + 2A_2(-2 + A_1 + A_3 + A_1A_3)], \\ F_1 &= -4[B_{12}(1 + 4A_2 + A_3 - 2A_2A_3) + B_{22}(-1 - A_1 - A_3 + 2A_1A_3) \\ &\quad + B_{23}(1 + A_1 + 4A_2 - 2A_1A_2) + 4B_{13}(-1 + A_2 - A_2^2)], \\ F_2 &= 8(B_{12}B_{23} + 2B_{12}B_{13} + 2B_{13}B_{23}). \end{aligned} \tag{3.51}$$

The overall UV divergence is confined to $\mathcal{F}_2$ while the overall IR divergence is confined to $\mathcal{F}_0$. Thus we denote

$$\hat{L}_{4x} = \mathcal{F}_2, \tag{3.52}$$

$$I_{4x} = \mathcal{F}_0. \tag{3.53}$$

$\mathcal{F}_1$ is UV- and IR-finite. Writing

$$\Delta L_{4x} = \mathcal{F}_1,$$

the divergence structure of L_{4x} can therefore be expressed as

$$\begin{aligned} L_{4x} &= \mathcal{F}_0 + \mathcal{F}_1 + \mathcal{F}_2 \\ &= I_{4x} + \Delta L_{4x} + \hat{L}_{4x}. \end{aligned}$$

For the "corner" diagram in Fig. 8, the integral L_{4c} has the same form as (3.50) except that F_0, F_1, F_2 are now

$$\begin{aligned} F_0 &= 4[1 - 4A_1 + A_1^2 + (A_2 + A_3)(1 + A_1)^2 - 2A_2A_3(1 - A_1 + A_1^2)], \\ F_1 &= 4[B_{11}(1 + A_2 + A_3 - 2A_2A_3) + B_{12}(-1 - 4A_1 - A_3 + 2A_1A_3) \\ &\quad + B_{13}(-1 - 4A_1 - A_2 + 8A_1A_2) + B_{23}(1 - 4A_1 + A_1^2)], \\ F_2 &= 4(B_{11}B_{23} - 4B_{12}B_{13}). \end{aligned} \tag{3.54}$$

The term $\mathcal{F}_2$ again defines an overall UV-divergent part. Now, however, $\mathcal{F}_1$ has a UV-divergence arising from the subdiagram S = [2,3,5], which is isolated by the $\mathbf{K}_{23}$ operator:

$$\mathbf{K}_{23}\mathcal{F}_1 = \hat{L}_2\tilde{L}_2.$$

The UV-finite piece

$$\Delta' L_{4c} = \mathcal{F}_0 + (1 - \mathbf{K}_{23})\mathcal{F}_1$$

can be split further into IR-divergent and IR-finite parts:

$$I_{4c} = \mathcal{F}_0, \qquad \Delta L_{4c} = (1 - \mathbf{K}_{23})\mathcal{F}_1.$$

Thus the divergence structure of L_{4c} can be expressed in the form

$$\begin{aligned} L_{4c} &= \mathcal{F}_0 + (1 - \mathbf{K}_{23})\mathcal{F}_1 + \mathbf{K}_{23}\mathcal{F}_1 + \mathcal{F}_2 \\ &= I_{4c} \; + \; \Delta L_{4c} \; + \; \hat{L}_2\tilde{L}_2 + \hat{L}_{4c}. \end{aligned}$$

The same result holds for the other corner diagram obtained by reversing the direction of arrows of the electron lines.

The renormalization constants B_{4a}, L_{4x}, and L_{4c} satisfy the Ward identity

$$B_{4a} + L_{4x} + 2L_{4c} = 0. \tag{3.55}$$

This means in particular that the sum $I_{4a} + I_{4x} + 2I_{4c}$ is free from IR divergences. In order to evaluate this sum, note that, if we replace $z_{22'}$ by z_2 in all parametric functions for the crossed diagram $4x$, they reduce to the corresponding functions for the self-energy diagram $4a$. This is why we have not distinguished the parametric functions in (3.47) and (3.50). Similar comment applies to the corner diagram $4c$. The only difference among these diagrams is in their phase spaces:

$$(dz)_{4c} = z_1(dz)_{4a} \;\text{ or }\; z_3(dz)_{4a}, \quad (dz)_{4x} = z_2(dz)_{4a}.$$

Thus we can write

$$I_{4x} + 2I_{4c} = \frac{1}{16}\int (dz)_{4a} \frac{z_1F_0^{(1)} + z_2F_0^{(2)} + z_3F_0^{(3)}}{U^2V^2}, \tag{3.56}$$

where $F_0^{(1)}$, $F_0^{(2)}$, $F_0^{(3)}$ are the F_0 functions for 4c, 4x and the second 4c, respectively.

This integral is very similar in form to I_{4a} of (3.49). In fact, using the identity (for $p^2 = 1$)

$$2A_i(A_i\not{p}+1) = (A_i\not{p}+1)\not{p}(A_i\not{p}+1) - (1-A_i^2)\not{p}, \quad i = 1,2,3,$$

one can rewrite the numerator N_0 of (3.49) as

$$N_0 = \sum_{i=1}^{3} z_i F_0^{(i)} + H,$$

where

$$H = -\sum_{i=1}^{3} z_i(1-A_i^2)H^{(i)},$$

with

$$\begin{aligned} H^{(1)} &= 4(1+A_2+A_3-2A_2A_3), \\ H^{(2)} &= 4(1+A_1+A_3-2A_1A_3), \\ H^{(3)} &= 4(1+A_1+A_2-2A_1A_2). \end{aligned}$$

From (3.49) and (3.56) one obtains

$$H_{4a} \equiv I_{4a} + I_{4x} + 2I_{4c} = -\frac{1}{16}\int (dz)_{4a}\frac{H}{U^2V^2}.$$

Since $A_i \to 1$ and $H \to 0$ in the IR limit, this is IR-finite as expected.

Defining the UV- and IR-finite part of B_{4a} by

$$\Delta B_{4a} = \mathcal{E}_0 + H_{4a} + (1 - \mathbf{K}_{12} - \mathbf{K}_{23})\mathcal{N}_1,$$

one can finally rewrite B_{4a} as

$$B_{4a} = \Delta B_{4a} - I_{4x} - 2I_{4c} + 2\hat{L}_2\tilde{B}_2 + \hat{B}_{4a}. \tag{3.57}$$

Mass renormalization constant δm_{4b}

The self-mass term corresponding to the self-energy diagram 4b of Fig. 8 can be written in the same form as in (3.45), where, however,

$$\begin{aligned} F_0 &= 4[4(1-A_1+A_1^2) + A_2(1-4A_1+A_1^2)], \\ F_1 &= -4[3B_{12}A_1 + 8(B_{11}-B_{12})]. \end{aligned} \tag{3.58}$$

The parametric functions are

$$\begin{aligned} &B_{11} = z_{24}, \quad B_{12} = z_4, \quad U = z_{135}B_{11} + z_2B_{12}, \\ &(dz) = \delta(1 - z_{12345})z_{13}dz_{13}dz_2dz_4dz_5, \end{aligned}$$

A_i, G, and V having the same form as in (3.45).

The integral $\mathcal{F}_1$ can be identified with the overall UV-divergent part $\delta\hat{m}_{4b}$. $\mathcal{F}_0$ contains a UV divergence arising from the $z_{24} \to 0$ region ([2,4] being the self-energy subdiagram).

We must therefore apply the $\mathbf{K}_2$ operation to extract the divergent part. This leads to the UV- and IR-finite part

$$\Delta\delta m_{4b} = (1 - \mathbf{K}_2)\mathcal{F}_0.$$

Instead of computing $\mathbf{K}_2\mathcal{F}_0$ directly, let us first consider

$$\mathbf{K}_2\delta m_{4b} = \delta m_2 \delta m_{2^*} + \hat{B}_2 \delta m_{2(1')}, \tag{3.59}$$

where δm_{2^*} is the self-mass contribution of the diagram Σ_{2^*} of Fig. 9, and $\delta m_{2(1')}$ is the derivative amplitude corresponding to δm_2. [See (3.7).] To avoid heavy notation, let us introduce, for the second-order amplitudes only, the following simplifications:

$$\delta m_{2(1')} \equiv \delta m'_2, \quad B_{2(1')} \equiv B'_2, \quad L_{2(1')} \equiv L'_2, \quad \text{etc.} \tag{3.60}$$

Similar calculation gives

$$\mathbf{K}_2\delta \hat{m}_{4b} = \delta m_2 \delta \hat{m}_{2^*} + \hat{B}_2 \delta \hat{m}'_2, \tag{3.61}$$

where $\delta\hat{m}_{2^*}$ and $\delta\hat{m}'_2$ are the UV-divergent parts of δm_{2^*} and $\delta m'_2$. $\mathbf{K}_2\mathcal{F}_0$ is the difference of (3.59) and (3.61). Thus the divergence structure can be written as

$$\begin{aligned}\delta m_{4b} &= (1 - \mathbf{K}_2)\mathcal{F}_0 + \mathbf{K}_2\mathcal{F}_0 + \mathcal{F}_1 \\ &= \Delta\delta m_{4b} + [\delta m_2 \delta\tilde{m}_{2^*} + \hat{B}_2 \delta\tilde{m}'_2] + \delta\hat{m}_{4b},\end{aligned}$$

where $\delta\tilde{m} = \delta m - \delta\hat{m}$. $\delta\tilde{m}_{2^*}$ is IR-divergent while $\delta\tilde{m}'_2$ is completely finite.

Renormalization constants B_{4b}, L_{4l}, and L_{4s}

The wave-function renormalization constant associated with the diagram $4b$ of Fig. 8 can be written in the same form as (3.47) except that F_0 and F_1 are given by (3.58) while

$$\begin{aligned}E_0 &= 4[4(-A_1 + 2A_1^2) + A_2(1 - 8A_1 + 3A_1^2)], \\ E_1 &= -12B_{12}A_1.\end{aligned}$$

The UV divergence structure of B_{4b} is given by

$$\begin{aligned}B_{4b} &= (1 - \mathbf{K}_2)\mathcal{N}_0 + (1 - \mathbf{K}_2)(\mathcal{E}_0 + \mathcal{N}_1) + \mathbf{K}_2(\mathcal{N}_0 + \mathcal{N}_1 + \mathcal{E}_0) + \mathcal{E}_1 \\ &= \quad \check{I}_{4b} \quad + \quad \overline{\Delta B_{4b}} \quad + \quad (\delta m_2 B_{2^*} + \hat{B}_2 \tilde{B}'_2) \quad + \quad \hat{B}_{4b}.\end{aligned} \tag{3.62}$$

$\check{I}_{4b}$ satisfies the divergence criteria for both overall IR divergence, isolated by the $\mathbf{I}_{all}$ operation, and IR divergence associated with the subdiagram $S = [2,4]$, isolated by the $\tilde{\mathbf{I}}_{13}$ operation. $\overline{\Delta B_{4b}}$ is finite. B_{2^*} is UV-finite but IR-divergent. $\tilde{B}'_2 = B'_2 - \hat{B}'_2$. See (3.60) for the meaning of B'_2.

Before going into more detail on B_{4b}, let us examine the vertex diagrams generated by inserting an external vertex in the electron lines of the diagram $4b$. Let us first consider the vertex renormalization constant L_{4l} for the diagram of Fig. 8($4l$). This can be written in the form (3.50) where

$$\begin{aligned}F_0 &= 4[8A_2(1 - A_1 + A_1^2) + (1 + A_2^2)(1 - 4A_1 + A_1^2)], \\ F_1 &= 4[B_{11}(1 - 16A_2 + A_2^2) + B_{22}(1 - 4A_1 + A_1^2) + 4B_{12}(1 - 2A_1 + 4A_2 - 2A_1A_2)], \\ F_2 &= 4(B_{11}B_{22} + 5B_{12}^2).\end{aligned} \tag{3.63}$$

The divergence structure of L_{4l} is

$$\begin{aligned} L_{4l} &= \mathcal{F}_0 + (1-\mathbf{K}_{22'})\mathcal{F}_1 + \mathbf{K}_{22'}\mathcal{F}_1 + \mathcal{F}_2 \\ &= \check{I}_{4l} + \Delta L_{4l} + \hat{L}_2\tilde{L}_2 + \hat{L}_{4l}. \end{aligned}$$

$\check{I}_{4l}$ satisfies the divergence criteria for both $\mathbf{I}_{all}$ and $\mathbf{I}_{13}$.

The vertex renormalization constant L_{4s} for the diagram of Fig. 8(4s) has the form (3.50), but now

$$\begin{aligned} F_0 &= 4[(4A_1 - 2A_2)(1 - A_1 + A_1^2) + (-2 + A_1A_2)(1 - 4A_1 + A_1^2)], \\ F_1 &= 12[B_{11}A_1(-4 + A_2) + B_{12}(-1 + 6A_1 - 2A_1^2)], \\ F_2 &= -12B_{11}B_{12}. \end{aligned} \tag{3.64}$$

The divergence structure of L_{4s} is given by

$$\begin{aligned} L_{4s} &= (1-\mathbf{K}_2)\mathcal{F}_0 + (1-\mathbf{K}_2)\mathcal{F}_1 + \mathbf{K}_2(\mathcal{F}_0 + \mathcal{F}_1) + \mathcal{F}_2 \\ &= \check{I}_{4s} + \Delta L_{4s} + (\delta m_2 L_{2*} + \hat{B}_2\tilde{L}_2) + \hat{L}_{4s}. \end{aligned}$$

$\check{I}_{4s}$ is the only IR divergent term. Note that L_{2*} is UV-finite. $\tilde{L}_2' = L_2' - \hat{L}_2'$. See (3.60) for the meaning of L_2'.

Returning to the analysis of the IR structure of B_{4b}, let us introduce the sum of F_0 terms of L_{4s}, L_{4l}, and B_{4b}:

$$H_{4b} \equiv B_{4b}[N_0] + 2L_{4s}[F_0] + L_{4l}[F_0], \tag{3.65}$$

which has the structure

$$H_{4b} = -\frac{1}{16}\int (dz)_{4b} \sum_{i=1}^{3} [-z_i(1 - A_i^2)H^{(i)}](UV)^{-2},$$

where the $H^{(i)}$ are given by

$$\begin{aligned} H^{(1)} &= H^{(3)} = 4(-2 + 4A_1 - 2A_2 + A_1A_2), \\ H^{(2)} &= 4(1 - 4A_1 + A_1^2). \end{aligned}$$

Using (3.65), we can rewrite $\check{I}_{4b}$ defined in (3.62) as

$$\begin{aligned} \check{I}_{4b} &= (1-\mathbf{K}_2)B_{4b}[N_0] \\ &= (1-\mathbf{K}_2)(H_{4b} - 2L_{4s}[F_0]) - L_{4l}[F_0] \\ &= (1-\mathbf{K}_2)H_{4b} - 2\check{I}_{4s} - \check{I}_{4l}. \end{aligned} \tag{3.66}$$

The introduction of H_{4b} enables us to avoid dealing directly with a fairly malevolent IR structure ($\check{I}_{4b}$ and $\check{I}_{4s}$ have linear divergences associated with the self-energy subdiagram S= [2,4]). In the above method only the IR structure of H_{4b} has to be examined. The integral $(1-\mathbf{K}_2)H_{4b}$ has a logarithmic singularity whose appearance can be traced to the fact that $\mathbf{K}_2$ acts on H_{4b} and L_{4s} but not on L_{4l}, as is seen from (3.66). This IR singularity can be isolated by $\tilde{\mathbf{I}}_{13}$ [see (3.12)] as

$$(1-\mathbf{K}_2)H_{4b} = (1-\mathbf{K}_2-\tilde{\mathbf{I}}_{13})H_{4b} + \tilde{\mathbf{I}}_{13}H_{4b},$$

where $\tilde{\mathbf{I}}_{13} \equiv \mathbf{I}_{13}(1-\mathbf{K}_2)$. The first term is finite. The second term simplifies as

$$\begin{aligned}\tilde{\mathbf{I}}_{13}H_{4b} &= \tilde{\mathbf{I}}_{13}H_{4b}[-z_2(1-A_2^2)H^{(2)}] \\ &= I_2\Delta B_2.\end{aligned}$$

Collecting these results one finds

$$B_{4b} = -2\check{I}_{4s} - \check{I}_{4l} + I_2\Delta B_2 + \Delta B_{4b} + \delta m_2 B_{2*} + \hat{B}_2\tilde{B}_2' + \hat{B}_{4b}, \tag{3.67}$$

where

$$\Delta B_{4b} \equiv \overline{\Delta B_{4b}} + (1-\mathbf{K}_2-\tilde{\mathbf{I}}_{13})H_{4b}.$$

For higher-order graphs it is still useful to introduce a combination H similar to (3.65) in which linear IR singularities cancel each other. Remaining IR singularities in H are logarithmic, all of the subdiagram type. Unfortunately, it becomes more difficult to decipher the IR structure of H directly in terms of lower-order amplitudes. In fact it is easier to go back to the constituent amplitudes $L[F_0]$ and $B[N_0]$ and study their structure piece by piece. The advantage of introducing H is not totally lost, however, since we need examine only those subdiagram IR divergences of the constituent amplitudes that contribute to the divergence of H.

Although such an additional complication is not actually necessary in fourth order, we shall work out the alternate method on the fourth-order coefficients just for illustration. Let us first note that the subdiagram IR divergence of L_{4s} cancels out with that of B_{4b} in the sum H_{4b} of (3.65). Thus we do not have to examine the subdiagram IR structure of $\check{I}_{4s}$ further. We write it simply as I_{4s}. On the other hand, $\check{I}_{4l}$ has a subdiagram IR divergence isolated by $\mathbf{I}_{13}$ that contributes to H:

$$\begin{aligned}\check{I}_{4l} &= (1-\mathbf{I}_{13})L_{4l}[F_0] + \mathbf{I}_{13}L_{4l}[F_0] \\ &\equiv I_{4l} + \mathbf{I}_{13}L_{4l}[F_0].\end{aligned}$$

I_{4l} is still overall IR-divergent. We find

$$\mathbf{I}_{13}L_{4l}[F_0] = I_2\tilde{L}_2 = (I_2)^2.$$

Analogously, $\check{I}_{4b}$ can be written as

$$\begin{aligned}\check{I}_{4b} &= (1-\mathbf{K}_2-\tilde{\mathbf{I}}_{13})B_{4b}[N_0] + \tilde{\mathbf{I}}_{13}B_{4b}[N_0] \\ &\equiv I_{4b} + I_2\tilde{B}_2.\end{aligned}$$

Substituting these results in (3.67) we obtain the final form for B_{4b}:

$$B_{4b} = -2I_{4s} - I_{4l} + I_2\tilde{B}_2 + \Delta B_{4b} + \delta m_2 B_{2*} + \hat{B}_2\tilde{B}_2' + \hat{B}_{4b}. \tag{3.68}$$

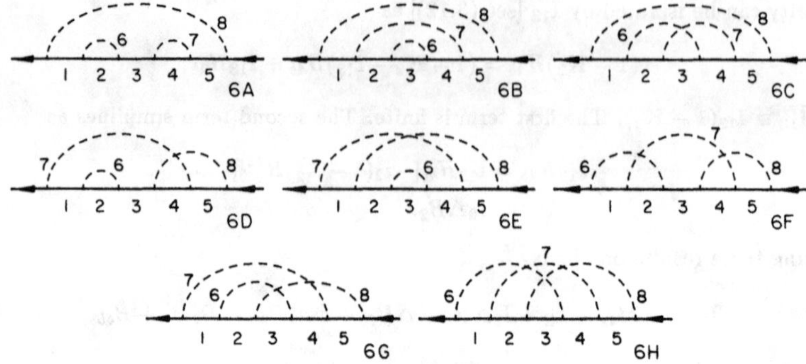

Figure 10: Sixth-order self-energy diagrams. There are ten such diagrams, of which eight are shown. Two diagrams are related to $6D$ and $6G$ by time-reversal. These diagrams also represent 50 sixth-order vertex diagrams of three-photon-exchange type obtained by insertion of an external magnetic field in their electron lines.

3.5. Sixth-order renormalization constants

Fig. 10 displays the labels and line numbering for the three-photon-exchange diagrams contributing to the sixth-order self-energy amplitude. The same line numbering is used for the vertex diagrams that are obtained from those of Fig. 10 by inserting an external magnetic field vertex in their electron lines. We shall discuss in detail the divergence structure for the diagrams $6B$ and $6D$ only. Together they exhibit all the new types of divergence structure not found in lower-order renormalization constants. The divergence structure of all sixth-order renormalization constants are listed in Appendix B. We begin with the $6D$-type diagrams.

Mass renormalization constant δm_{6D}

The self-mass term corresponding to the diagram $6D$ of Fig. 10 can be written in the form

$$\begin{aligned}\delta m_{6D} &= \frac{1}{64}\int (dz) \int z_6 dm_6^2 \int z_7 dm_7^2 \int z_8 dm_8^2 \left(\frac{4!F_0}{U^2V^5} + \frac{3!F_1}{U^3V^4} + \frac{2!F_2}{U^4V^3}\right) \\ &\equiv \mathcal{F}_0 + \mathcal{F}_1 + \mathcal{F}_2,\end{aligned} \tag{3.69}$$

where F_0, F_1, F_2 are expressed in terms of appropriate parametric functions generated, for instance, by the algebraic manipulation program SCHOONSCHIP.[41] They are too lengthy to be listed here.

Table II lists, underneath individual terms of δm_{6D}, the various UV divergence criteria they satisfy. The $\mathcal{F}_2$ term satisfies the overall UV divergence criterion and thus defines the intermediate renormalization constant

$$\delta \hat{m}_{6D} = \mathbf{K}_{all}\delta m_{6D} = \mathcal{F}_2.$$

Since the parametric functions for $\mathbf{K}_{all}$ are the same as for the complete diagram, this term is isolated in its entirety by $\mathbf{K}_{all}$ and we need not be concerned with its $\mathbf{K}_2$ and $\mathbf{K}_{45}$ structures.

Table II. Divergence structure of δm_{6D}.

$\mathcal{F}_0$	$\mathcal{F}_1$	$\mathcal{F}_2$
		$\mathbf{K}_{all}$
	$\mathbf{K}_{45}$	$\mathbf{K}_{45}$
$\mathbf{K}_2$	$\mathbf{K}_2$	$\mathbf{K}_2$

The terms $\mathcal{F}_0$ and $\mathcal{F}_1$ have UV divergences isolated by the $\mathbf{K}_2$ and $\mathbf{K}_{45}$ operations. After subtracting these divergences, we obtain

$$\Delta\delta m_{6D} = (1-\mathbf{K}_2)\mathcal{F}_0 + (1-\mathbf{K}_2)(1-\mathbf{K}_{45})\mathcal{F}_1,$$

which is UV- and IR-finite. The divergent pieces isolated by $\mathbf{K}_2$ and $\mathbf{K}_{45}$ are related to lower-order renormalization constants as

$$\begin{aligned}
\mathbf{K}_2(\mathcal{F}_0+\mathcal{F}_1) &= \delta m_2 \delta\tilde{m}_{4a(1^*)} + \hat{B}_2\delta\tilde{m}_{4a(1')},\\
\mathbf{K}_{45}\mathcal{F}_1 &= \hat{L}_2\delta\tilde{m}_{4b},\\
\mathbf{K}_2\mathbf{K}_{45}\mathcal{F}_1 &= \hat{L}_2(\delta m_2\delta\tilde{m}_{2^*} + \hat{B}_2\delta\tilde{m}_2').
\end{aligned}$$

The divergence structure of δm_{6D} is thus

$$\begin{aligned}
\delta m_{6D} &= \Delta\delta m_{6D} + \delta m_2\delta\tilde{m}_{4a(1^*)} + \hat{B}_2\delta\tilde{m}_{4a(1')}\\
&\quad + \hat{L}_2\delta\tilde{m}_{4b} - \hat{L}_2(\delta m_2\delta\tilde{m}_{2^*} + \hat{B}_2\delta\tilde{m}_2') + \delta\hat{m}_{6D}.
\end{aligned}$$

Note that $\delta\tilde{m}_{4a(1^*)}$ and $\delta\tilde{m}_{2^*}$, which are defined by reduced diagrams, are IR-divergent even though such divergences do not show up explicitly in Table II.

Renormalization constants L_{6Dj} and B_{6D}

Each sixth-order vertex renormalization constant L_{6Dj} has the form

$$\begin{aligned}
L_{6Dj} &= -\frac{2}{64}\int(dz)\int z_6 dm_6^2\int z_7 dm_7^2\int z_8 dm_8^2\left(\frac{5!F_0}{2U^2V^6} + \frac{4!F_1}{U^3V^5} + \frac{3!F_2}{U^4V^4} + \frac{2!F_3}{U^5V^3}\right)\\
&\equiv \mathcal{F}_0 + \mathcal{F}_1 + \mathcal{F}_2 + \mathcal{F}_3,
\end{aligned} \tag{3.70}$$

where the third suffix j is the label of the electron line to which the external magnetic field is attached.[e] The F_i's are, again, obtained using SCHOONSCHIP. The divergence structure of the L_{6Dj} ($j = 1, 2, .., 5$) are listed in Table III.

The term $\mathcal{F}_3$ for each j satisfies the overall UV divergence criterion. Thus we define the intermediate renormalization constant by

$$\hat{L}_{6Dj} = \mathbf{K}_{all}L_{6Dj} = \mathcal{F}_3 \quad \text{for} \quad j = 1,\ 2,\ \ldots,\ 5.$$

[e]The electron line sharing the external vertex with the line j is denoted as j'.

Table III. Divergence structure of L_{6Dj}.

Diagram	$\mathcal{F}_0$	$\mathcal{F}_1$	$\mathcal{F}_2$	$\mathcal{F}_3$
L_{6D1},L_{6D3}	$\mathbf{I}_{all}$			$\mathbf{K}_{all}$
	$\tilde{\mathbf{I}}_{123}$	$\mathbf{K}_{45}$	$\mathbf{K}_{45}$	$\mathbf{K}_{45}$
	$\mathbf{K}_2$	$\mathbf{K}_2$	$\mathbf{K}_2$	$\mathbf{K}_2$
L_{6D2}	$\mathbf{I}_{all}$			$\mathbf{K}_{all}$
	$\tilde{\mathbf{I}}_{123}$	$\mathbf{K}_{45}$	$\mathbf{K}_{45}$	$\mathbf{K}_{45}$
	$\tilde{\mathbf{I}}_{1345}$	$\mathbf{K}_{22'}$	$\mathbf{K}_{22'}$	$\mathbf{K}_{22'}$
L_{6D4}	$\mathbf{I}_{all}$			$\mathbf{K}_{all}$
	$\mathbf{K}_2$	$\mathbf{K}_2$	$\mathbf{K}_2$	$\mathbf{K}_2$
L_{6D5}	$\mathbf{I}_{all}$			$\mathbf{K}_{all}$
	$\tilde{\mathbf{I}}_5$	$\tilde{\mathbf{I}}_5$	$\mathbf{K}_{1234}$	$\mathbf{K}_{1234}$
	$\mathbf{K}_2$	$\mathbf{K}_2$	$\mathbf{K}_2$	$\mathbf{K}_2$

The IR structure of the L_{6Dj} is quite complicated. In addition to the overall IR divergences, the terms $L_{6Dj}[F_0]$ have subdiagram IR divergences of various kinds, including some linear ones. Fortunately it is not necessary to examine them in great detail. As far as the $\mathcal{F}_0$ terms are concerned, it is sufficient to examine the IR property of

$$H_{6D} \equiv \sum_{j=1}^{5} L_{6Dj}[F_0] + B_{6D}[N_0], \tag{3.71}$$

where B_{6D} is defined later. This sum has a simple divergence structure since overall IR divergences as well as subdiagram IR divergences isolated by the $\tilde{\mathbf{I}}_{123}$ and $\tilde{\mathbf{I}}_5$ operations cancel one another. The only IR divergences left are those associated with the $\tilde{\mathbf{I}}_{1345}$ operation in $L_{6D2}[F_0]$ and $B_{6D}[N_0]$, in addition to $\tilde{\mathbf{I}}_5[F_1]$ inL_{6D5}.

Keeping these remarks in mind, we obtain for L_{6D1} the result

$$\begin{aligned} L_{6D1} &= (1-\mathbf{K}_2)\mathcal{F}_0 + (1-\mathbf{K}_2)(1-\mathbf{K}_{45})(\mathcal{F}_1+\mathcal{F}_2) + \mathbf{K}_2(\mathcal{F}_0+\mathcal{F}_1+\mathcal{F}_2) \\ &\quad + \mathbf{K}_{45}(\mathcal{F}_1+\mathcal{F}_2) - \mathbf{K}_2\mathbf{K}_{45}(\mathcal{F}_1+\mathcal{F}_2) + \mathcal{F}_3 \\ &= I_{6D1} + \Delta L_{6D1} + (\delta m_2 L_{4c(1^*)} + \hat{B}_2\tilde{L}_{4c(1')}) \\ &\quad + \hat{L}_2\tilde{L}_{4s} - \hat{L}_2(\delta m_2 L_{2^*} + \hat{B}_2\tilde{L}_2') + \hat{L}_{6D1}, \end{aligned}$$

where ΔL_{6D1} is finite and $L_{4c(1^*)}$ and $\tilde{L}_{4c(1')}$ are defined as in (3.59). For L_{6D2} we find

$$\begin{aligned} L_{6D2} &= (1-\tilde{\mathbf{I}}_{1345})\mathcal{F}_0 + \tilde{\mathbf{I}}_{1345}\mathcal{F}_0 + (1-\mathbf{K}_{22'})(1-\mathbf{K}_{45})(\mathcal{F}_1+\mathcal{F}_2) \\ &\quad + \mathbf{K}_{22'}(\mathcal{F}_1+\mathcal{F}_2) + \mathbf{K}_{45}(\mathcal{F}_1+\mathcal{F}_2) - \mathbf{K}_{22'}\mathbf{K}_{45}(\mathcal{F}_1+\mathcal{F}_2) + \mathcal{F}_3 \\ &= I_{6D2} + I_{4c}\tilde{L}_2 + \Delta L_{6D2} + \hat{L}_2\tilde{L}_{4c} + \hat{L}_2\tilde{L}_{4l} - (\hat{L}_2)^2\tilde{L}_2 + \hat{L}_{6D2}. \end{aligned}$$

L_{6D3} has the same structure as L_{6D1}:

$$L_{6D3} = I_{6D3} + \Delta L_{6D3} + (\delta m_2 L_{4c(1^\bullet)} + \hat{B}_2 \tilde{L}_{4c(1')}) \\ + \hat{L}_2 \tilde{L}_{4s} - \hat{L}_2(\delta m_2 L_{2^\bullet} + \hat{B}_2 \tilde{L}_2') + \hat{L}_{6D3},$$

L_{6D4} has a simple structure:

$$L_{6D4} = I_{6D4} + \Delta L_{6D4} + \delta m_2 L_{4x(1^\bullet)} + \hat{B}_2 \tilde{L}_{4x(1')} + \hat{L}_{6D4}.$$

Applying the $\tilde{\mathbf{I}}_5$ operation to isolate the IR divergence in $L_{6D5}[F_1]$, we find

$$\tilde{\mathbf{I}}_5 L_{6D5}[F_1] = I_2 L_{4s}[F_1].$$

Because of our procedure of defining I_{6D5} and H_{6D}, we need not consider the effect of $\tilde{\mathbf{I}}_5$ on the F_0 term of L_{6D5} (as H_{6D} is finite in this limit). Thus we find

$$\begin{aligned} L_{6D5} &= (1 - \mathbf{K}_2)\mathcal{F}_0 + ((1 - \tilde{\mathbf{I}}_5)(1 - \mathbf{K}_2)\mathcal{F}_1 + (1 - \mathbf{K}_2)(1 - \mathbf{K}_{1234})\mathcal{F}_2) \\ &\quad + \mathbf{K}_2(\mathcal{F}_0 + \mathcal{F}_1 + \mathcal{F}_2) + \mathbf{K}_{1234}\mathcal{F}_2 - \mathbf{K}_2\mathbf{K}_{1234}\mathcal{F}_2 + \tilde{\mathbf{I}}_5(1 - \mathbf{K}_2)\mathcal{F}_1 + \mathcal{F}_3 \\ &= I_{6D5} + \Delta L_{6D5} + (\delta m_2 L_{4c(3^\bullet)} + \hat{B}_2 \tilde{L}_{4c(3')}) \\ &\quad + \hat{L}_{4s}\tilde{L}_2 - \hat{B}_2 \hat{L}_2' \tilde{L}_2 + I_2 \Delta L_{4s} + \hat{L}_{6D5}, \end{aligned}$$

where

$$\Delta L_{4s} = (1 - \mathbf{K}_2) L_{4s}[F_1]$$

is finite.

Table IV. Divergence structure of B_{6D}.

$\mathcal{N}_0$	$\mathcal{N}_1$	$\mathcal{N}_2$	$\mathcal{E}_0$	$\mathcal{E}_1$	$\mathcal{E}_2$
$\mathbf{I}_{all}$					$\mathbf{K}_{all}$
$\tilde{\mathbf{I}}_5$	$\tilde{\mathbf{I}}_5$	$\mathbf{K}_{1234}$			$\mathbf{K}_{1234}$
$\tilde{\mathbf{I}}_{123}$	$\mathbf{K}_{45}$	$\mathbf{K}_{45}$		$\mathbf{K}_{45}$	$\mathbf{K}_{45}$
$\tilde{\mathbf{I}}_{1345}$,$\mathbf{K}_2$	$\mathbf{K}_2$	$\mathbf{K}_2$	$\mathbf{K}_2$	$\mathbf{K}_2$	$\mathbf{K}_2$

The sixth-order wave-function renormalization constant B_{6D} has the form

$$\begin{aligned} B_{6D} &= \frac{1}{64}\int (dz) \int z_6 dm_6^2 \int z_7 dm_7^2 \int z_8 dm_8^2 \\ &\quad \times \left(2\left[\frac{5!N_0}{2!U^2V^6} + \frac{4!N_1}{U^3V^5} + \frac{3!N_2}{U^4V^4}\right] + \frac{4!E_0}{U^2V^5} + \frac{3!E_1}{U^3V^4} + \frac{2!E_2}{U^4V^3}\right) \\ &\equiv \mathcal{N}_0 + \mathcal{N}_1 + \mathcal{N}_2 + \mathcal{E}_0 + \mathcal{E}_1 + \mathcal{E}_2. \end{aligned} \tag{3.72}$$

Table IV indicates the nature of the singularities of B_{6D}. The complete $\mathcal{E}_2$ term is isolated by $\mathbf{K}_{all}$, the operation defining the overall UV-divergent piece:

$$\hat{B}_{6D} = \mathbf{K}_{all} B_{6D} = \mathcal{E}_2.$$

Because of (3.71) we need to isolate only the $\tilde{\mathbf{I}}_{1345}$-related divergence of $\mathcal{N}_0$ and the $\tilde{\mathbf{I}}_5$-related divergence of $\mathcal{N}_1$. The latter has the form

$$\tilde{\mathbf{I}}_5\mathcal{N}_1 = \tilde{B}_2 L_{4s}[F_1].$$

Thus we obtain

$$\begin{aligned}
B_{6D} &= (1-\mathbf{K}_2-\tilde{\mathbf{I}}_{1345})\mathcal{N}_0 \\
&+ (1-\tilde{\mathbf{I}}_5-\mathbf{K}_{1234})(1-\mathbf{K}_{45})(1-\mathbf{K}_2)(\mathcal{N}_1+\mathcal{N}_2+\mathcal{E}_0+\mathcal{E}_1) \\
&+ \mathbf{K}_2(\mathcal{N}_0+\mathcal{N}_1+\mathcal{N}_2+\mathcal{E}_0+\mathcal{E}_1)+\mathbf{K}_{45}(\mathcal{N}_1+\mathcal{N}_2+\mathcal{E}_1) \\
&+ \mathbf{K}_{1234}\mathcal{N}_2-\mathbf{K}_2\mathbf{K}_{45}(\mathcal{N}_1+\mathcal{N}_2+\mathcal{E}_1)-\mathbf{K}_2\mathbf{K}_{1234}\mathcal{N}_2 \\
&+ \tilde{\mathbf{I}}_{1345}\mathcal{N}_0+\tilde{\mathbf{I}}_5(1-\mathbf{K}_2)\mathcal{N}_1+\mathcal{E}_2 \\
&= I_{6D}+\overline{\Delta B_{6D}}+(\delta m_2 B_{4a(1^*)}+\hat{B}_2\tilde{B}_{4a(1')}) \\
&+ \hat{L}_2\tilde{B}_{4b}+\hat{L}_{4s}\tilde{B}_2-\hat{L}_2(\delta m_2 B_{2^*}+\hat{B}_2\tilde{B}_2') \\
&- \hat{B}_2\hat{L}_2'\tilde{B}_2+I_{4c}\tilde{B}_2+\tilde{B}_2\Delta L_{4s}+\hat{B}_{6D},
\end{aligned}$$

where, in $\overline{\Delta B_{6D}}$, the $\mathbf{K}$ and $\mathbf{I}$ operations produce null results acting on some of $\mathcal{N}_1,\ldots,\mathcal{E}_1$.

Following (3.71) we rewrite I_{6D} as

$$I_{6D} = (1-\mathbf{K}_2-\tilde{\mathbf{I}}_{1345})H_{6D}-\sum_{j=1}^{5} I_{6Dj}.$$

The first term is finite. Adding it to $\overline{\Delta B_{6D}}$ we define

$$\Delta B_{6D} \equiv (1-\mathbf{K}_2-\tilde{\mathbf{I}}_{1345})H_{6D}+\overline{\Delta B_{6D}}.$$

This leads to the structure of B_{6D} given in Appendix B. Note that

$$\tilde{\mathbf{I}}_{1345}H_{6D} = I_{4c}\Delta B_2.$$

The divergence structures of the renormalization constants for diagrams 6A, 6E, 6F, 6G, and 6H are similar to or simpler than that for 6D and are listed in Appendix B. The structures for 6B and 6C are similar to each other. We illustrate them with 6B.

Mass renormalization constant δm_{6B}

The self-mass term corresponding to the diagram $6B$ of Fig. 10 can be written in the same form as (3.69). Table V indicates the singularity structure of δm_{6B}. The analysis of this term is straightforward:

$$\begin{aligned}
\delta m_{6B} &= ((1-\tilde{\mathbf{I}}_{15})(1-\mathbf{K}_3)\mathcal{F}_0+(1-\mathbf{K}_3)(1-\mathbf{K}_{234})\mathcal{F}_1)+\tilde{\mathbf{I}}_{15}(1-\mathbf{K}_3)\mathcal{F}_0 \\
&+ \mathbf{K}_3(\mathcal{F}_0+\mathcal{F}_1)+\mathbf{K}_{234}\mathcal{F}_1-\mathbf{K}_3\mathbf{K}_{234}\mathcal{F}_1+\mathcal{F}_2 \\
&= \Delta\delta m_{6B}+I_2(\delta\tilde{m}_{4b}-\delta m_2\delta\tilde{m}_{2^*}-\hat{B}_2\delta\tilde{m}_2') \\
&+ (\delta m_2\delta\tilde{m}_{4(2^*)}+\hat{B}_2\delta\tilde{m}_{4b(2')})+(\delta\hat{m}_{4b}\delta\tilde{m}_{2^*}+\hat{B}_{4b}\delta\tilde{m}_2') \\
&- (\delta m_2\delta\hat{m}_{2^*}\delta\tilde{m}_{2^*}+\hat{B}_2(\delta\hat{m}_2'\delta\tilde{m}_{2^*}+\hat{B}_2'\delta\tilde{m}_2'))+\delta\hat{m}_{6B}.
\end{aligned}$$

Table V. Divergence structure of δm_{6B}.

$\mathcal{F}_0$	$\mathcal{F}_1$	$\mathcal{F}_2$
		$\mathbf{K}_{all}$
$\tilde{\mathbf{I}}_{15}$	$\mathbf{K}_{234}$	$\mathbf{K}_{234}$
$\mathbf{K}_3$	$\mathbf{K}_3$	$\mathbf{K}_3$

Renormalization constants L_{6Bj} and B_{6B}

The sixth-order vertex renormalization constant L_{6Bj} has the same form as (3.70). The divergence structure of L_{6Bj} $(j = 1,2,3)$ is outlined in Table VI. The UV divergences can be isolated in the same way as before.

Table VI. Divergence structure of L_{6Bj}.

Diagram	$\mathcal{F}_0$	$\mathcal{F}_1$	$\mathcal{F}_2$	$\mathcal{F}_3$
L_{6B1}	$\mathbf{I}_{all}$			$\mathbf{K}_{all}$
	$\tilde{\mathbf{I}}_{15}$	$\tilde{\mathbf{I}}_{15}$, $\mathbf{K}_{234}$	$\mathbf{K}_{234}$	$\mathbf{K}_{234}$
	$\mathbf{K}_3$	$\mathbf{K}_3$	$\mathbf{K}_3$	$\mathbf{K}_3$
L_{6B2}	$\mathbf{I}_{all}$			$\mathbf{K}_{all}$
	$\tilde{\mathbf{I}}_{15}$	$\tilde{\mathbf{I}}_{15}$	$\mathbf{K}_{234}$	$\mathbf{K}_{234}$
	$\mathbf{K}_3$	$\mathbf{K}_3$	$\mathbf{K}_3$	$\mathbf{K}_3$
L_{6B3}	$\mathbf{I}_{all}$			$\mathbf{K}_{all}$
	$\tilde{\mathbf{I}}_{15}$	$\tilde{\mathbf{I}}_{15}$	$\mathbf{K}_{234}$	$\mathbf{K}_{234}$
	$\tilde{\mathbf{I}}_{1245}$	$\mathbf{K}_{33'}$	$\mathbf{K}_{33'}$	$\mathbf{K}_{33'}$

The limiting process associated with the $\tilde{\mathbf{I}}_{15}$ operation indicates the appearance of a linear IR divergence in L_{6B1} in the $\mathcal{F}_0$ and $\mathcal{F}_1$ terms. Our definition of I_{6B1} and H_{6B} ensures that we do not have to worry about the linear divergent piece in $\mathcal{F}_0$. We handle the linear divergence in $L_{6B1}[F_1]$ by simply removing the offending piece. This piece can be conveniently written as

$$L_{6B1}[f_1] \equiv -\frac{1}{32}\int (dz)\frac{f_1}{U^3V^2},$$

where

$$\begin{aligned} f_1 &= -16(3A_3B_{22} + 8(B_{22} - B_{23})), \\ (dz) &= dz_1dz_{1'}dz_5dz_2dz_3dz_4dz_8dz_9dz_0\delta(1 - z_{11'} - z_5 - \cdots - z_9 - z_0). \end{aligned} \tag{3.73}$$

Then the difference

$$L_{6B1}[F_1] - L_{6B1}[f_1] = -\frac{1}{32}\int (dz)\frac{F_1 - f_1}{U^3V^2}$$

is at most logarithmically divergent. As a matter of fact, we find that

$$\tilde{\mathbf{I}}_{15}\left(-\frac{1}{32}\int (dz)\frac{F_1 - f_1}{U^3V^2}\right) = 0.$$

[The logarithmic divergence appears in both the $\mathbf{I}_{15}$ and $\mathbf{K}_{234}\mathbf{I}_{15}$ operations but cancels in the combination $\tilde{\mathbf{I}}_{15} = \mathbf{I}_{15}(1 - \mathbf{K}_{234})$.] With these preliminaries and anticipating the IR structure of H_{6B} we investigate only the IR singularities isolated by $\tilde{\mathbf{I}}_{15}$ in L_{6B2} and L_{6B3} and by $\tilde{\mathbf{I}}_{1245}$ in L_{6B3}.

We find the following divergence structure: For L_{6B1} we have

$$\begin{aligned} L_{6B1} &= (1 - \mathbf{K}_3)\mathcal{F}_0 + (1 - \mathbf{K}_{234})(1 - \mathbf{K}_3)(\mathcal{F}_1 - f_1 + \mathcal{F}_2) \\ &\quad + (1 - \mathbf{K}_{234})(1 - \mathbf{K}_3)f_1 + \mathbf{K}_3(\mathcal{F}_0 + \mathcal{F}_1 + \mathcal{F}_2) \\ &\quad + \mathbf{K}_{234}(\mathcal{F}_1 + \mathcal{F}_2) - \mathbf{K}_3\mathbf{K}_{234}(\mathcal{F}_1 + \mathcal{F}_2) + \mathcal{F}_3 \\ &= I_{6B1} + \Delta L_{6B1} + \frac{1}{2}J_{6B} + (\delta m_2 L_{4s(2^*)} + \hat{B}_2\tilde{L}_{4s(2')}) + (\delta\tilde{m}_{4b}L_{2^*} + \hat{B}_{4b}\tilde{L}_2') \\ &\quad - (\delta m_2\delta\tilde{m}_{2^*}L_{2^*} + \hat{B}_2(\delta\tilde{m}_2' L_{2^*} + \hat{B}_2'\tilde{L}_2')) + \hat{L}_{6B1}, \end{aligned}$$

where we have defined

$$\frac{1}{2}J_{6B} = (1 - \mathbf{K}_{234})(1 - \mathbf{K}_3)\left(-\frac{1}{32}\int (dz)\frac{f_1}{U^3V^2}\right). \tag{3.74}$$

For L_{6B2} and L_{6B3} we find

$$\begin{aligned} L_{6B2} &= (1 - \tilde{\mathbf{I}}_{15})(1 - \mathbf{K}_3)\mathcal{F}_0 + (1 - \mathbf{I}_{15})(1 - \mathbf{K}_{234})(1 - \mathbf{K}_3)(\mathcal{F}_1 + \mathcal{F}_2) \\ &\quad + \tilde{\mathbf{I}}_{15}(\mathcal{F}_0 + \mathcal{F}_1) - \tilde{\mathbf{I}}_{15}\mathbf{K}_3(\mathcal{F}_0 + \mathcal{F}_1) + \mathbf{K}_3(\mathcal{F}_0 + \mathcal{F}_1 + \mathcal{F}_2) \\ &\quad + \mathbf{K}_{234}\mathcal{F}_2 - \mathbf{K}_3\mathbf{K}_{234}\mathcal{F}_2 + \mathcal{F}_3 \\ &= I_{6B2} + \Delta L_{6B2} + I_2\tilde{L}_{4s} - I_2(\delta m_2 L_{2^*} + \hat{B}_2\tilde{L}_2') \\ &\quad + (\delta m_2 L_{4l(2^*)} + \hat{B}_2\tilde{L}_{4l(2')}) + \hat{L}_{4s}\tilde{L}_2 - \hat{B}_2\hat{L}_2'\tilde{L}_2 + \hat{L}_{6B2}, \end{aligned}$$

and

$$\begin{aligned} L_{6B3} &= (1 - \tilde{\mathbf{I}}_{15})(1 - \tilde{\mathbf{I}}_{1245})\mathcal{F}_0 + (1 - \mathbf{I}_{15})(1 - \mathbf{K}_{234})(1 - \mathbf{K}_{33'})(\mathcal{F}_1 + \mathcal{F}_2) \\ &\quad + \tilde{\mathbf{I}}_{15}(\mathcal{F}_0 + \mathcal{F}_1) - \tilde{\mathbf{I}}_{15}\mathbf{K}_{33'}\mathcal{F}_1 + \tilde{\mathbf{I}}_{1245}(1 - \tilde{\mathbf{I}}_{15})\mathcal{F}_0 \\ &\quad + \mathbf{K}_{33'}(\mathcal{F}_1 + \mathcal{F}_2) + \mathbf{K}_{234}\mathcal{F}_2 - \mathbf{K}_{33'}\mathbf{K}_{234}\mathcal{F}_2 + \mathcal{F}_3 \\ &= I_{6B3} + \Delta L_{6B3} + I_2\tilde{L}_{4l} - I_2\hat{L}_2\tilde{L}_2 + I_{4l}\tilde{L}_2 \\ &\quad + \hat{L}_2\tilde{L}_{4l} + \hat{L}_{4l}\tilde{L}_2 - (\hat{L}_2)^2\tilde{L}_2 + \hat{L}_{6B3}. \end{aligned}$$

The sixth-order wave-function renormalization constant B_{6B} has the same form as (3.72). Table VII indicates the nature of the singularities in B_{6B}. The $\mathcal{E}_2$ term is isolated by $\mathbf{K}_{all}$, defining the overall UV-divergent piece:

$$\hat{B}_{6B} = \mathcal{E}_2.$$

Table VII. Divergence structure of B_{6B}.

$\mathcal{N}_0$	$\mathcal{N}_1$	$\mathcal{N}_2$	$\mathcal{E}_0$	$\mathcal{E}_1$	$\mathcal{E}_2$
$\mathbf{I}_{all}$					$\mathbf{K}_{all}$
$\tilde{\mathbf{I}}_{15}$	$\tilde{\mathbf{I}}_{15}, \mathbf{K}_{234}$	$\mathbf{K}_{234}$	$\tilde{\mathbf{I}}_{15}$	$\mathbf{K}_{234}$	$\mathbf{K}_{234}$
$\tilde{\mathbf{I}}_{1245}, \mathbf{K}_3$	$\mathbf{K}_3$	$\mathbf{K}_3$	$\mathbf{K}_3$	$\mathbf{K}_3$	$\mathbf{K}_3$

The divergence structure of B_{6B} is as follows:

$$\begin{aligned}
B_{6B} &= (1-\tilde{\mathbf{I}}_{15})(1-\tilde{\mathbf{I}}_{1245}-\mathbf{K}_3)\mathcal{N}_0 \\
&+ (1-\tilde{\mathbf{I}}_{15}-\mathbf{K}_{234})(1-\mathbf{K}_3)(\overline{\mathcal{N}_1}+\mathcal{N}_2+\mathcal{E}_0+\mathcal{E}_1) \\
&+ (1-\mathbf{K}_{234})(1-\mathbf{K}_3)(-z_{15}f_1)+\mathbf{K}_3(\mathcal{N}_0+\mathcal{N}_1+\mathcal{N}_2+\mathcal{E}_0+\mathcal{E}_1) \\
&+ \mathbf{K}_{234}(\mathcal{N}_1+\mathcal{N}_2+\mathcal{E}_1)-\mathbf{K}_3\mathbf{K}_{234}(\mathcal{N}_1+\mathcal{N}_2+\mathcal{E}_1) \\
&+ \tilde{\mathbf{I}}_{15}(1-\mathbf{K}_3)(\mathcal{N}_0+\overline{\mathcal{N}_1}+\mathcal{E}_0)+\tilde{\mathbf{I}}_{1245}(1-\tilde{\mathbf{I}}_{15})\mathcal{N}_0+\mathcal{E}_2 \\
&= I_{6B}+\overline{\Delta B_{6B}}-J_{6B}+(\delta m_2 B_{4b(2^*)}+\hat{B}_2\tilde{B}_{4b(2')}) \\
&+ (\delta\hat{m}_{4b}B_{2^*}+\hat{B}_{4b}\tilde{B}_2')-(\delta m_2\delta\hat{m}_{2^*}B_{2^*}+\hat{B}_2(\delta\hat{m}_2'B_{2^*}+\hat{B}_2'\tilde{B}_2')) \\
&+ (I_2(\tilde{B}_{4b}-\delta m_2B_{2^*}-\hat{B}_2\tilde{B}_2')+B_{2^*}[I]\Delta\delta m_{4b})+I_{4l}\tilde{B}_2+\hat{B}_{6B}.
\end{aligned} \tag{3.75}$$

Here

$$\overline{\mathcal{N}_1} \equiv \frac{1}{32}\int (dz)\frac{\overline{N_1}}{U^3V^2} \equiv \frac{1}{32}\int (dz)\frac{N_1-z_{15}f_1}{U^3V^2},$$

with

$$(dz) = z_{15}dz_{15}dz_2dz_3dz_4dz_8dz_9dz_0\delta(1-z_{15}-z_2-\cdots-z_0), \tag{3.76}$$

and f_1 is given by (3.73). This f_1 is chosen (other than for its suggestive structure – compare f_1 and F_1 of B_{4b}) because it performs the same function for the $z_{15}A_1F_1$ terms of N_1 in B_{6B} that it does in L_{6B1}. In fact the combination $\overline{\mathcal{N}_1}$ has no linear IR divergence. Notice, however, that, due to different definitions of (dz) for (3.76) and (3.73) and a difference in a minus sign in the definitions of B_{6B} and L_{6B1}, the linear IR divergence isolated from B_{6B} is $-J_{6B}$ rather than $\frac{1}{2}J_{6B}$. Introducing

$$H_{6B} \equiv B_{6B}[N_0]+\sum_{j=1}^{5}L_{6Bj}[F_0],$$

one can rewrite $I_{6B}+\overline{\Delta B_{6B}}$ in (3.75) as

$$\begin{aligned}
I_{6B}+\overline{\Delta B_{6B}} &= -\sum_{j=1}^{5}I_{6Bj}+(1-\tilde{\mathbf{I}}_{15})(1-\tilde{\mathbf{I}}_{1245}-\mathbf{K}_3)H_{6B}+\overline{\Delta B_{6B}} \\
&\equiv -2I_{6B1}-2I_{6B2}-I_{6B3}+\Delta B_{6B},
\end{aligned}$$

which defines ΔB_{6B}.

In (3.75) we note the appearance of the term $B_{2\bullet}[I]\Delta\delta m_{4b}$:

$$\tilde{I}_{15}B_{6B} = I_2 B_{4b} + B_{2\bullet}[I]\Delta\delta m_{4b},$$

as is expected from (3.16). The factor $B_{2\bullet}[I]$ has the explicit form

$$B_{2\bullet}[I] = \lim_{\lambda\to 0}\int (dz)\frac{(1-4A_1+A_1^2)}{2U^2}\left(\frac{1}{V} - \frac{z_{15}(1-A_1)}{V^2}\right), \tag{3.77}$$

where

$$(dz) = z_{15}dz_{15}dz_8\delta(1-z_{15}-z_8),$$
$$A_1 = z_8/z_{158}, \quad U = z_{158}, \quad V = z_{15}(1-A_1)+\lambda^2 z_8.$$

Carrying out the integration and taking the limit $\lambda \to 0$ one finds

$$B_{2\bullet}[I] = -\frac{1}{2}.$$

Note that individual terms of $B_{2\bullet}[I]$ in (3.77) are IR divergent. The final form of B_{6B} is listed in Appendix B together with all other sixth-order renormalization constants.

4. Electron Anomalous Moment without Closed Electron Loops

In this section we consider those diagrams that contain no closed electron loop. One diagram of second order, six diagrams of fourth order, 50 diagrams of sixth order, and 518 diagrams of eighth order belong to this category. Diagrams with closed electron loops are considered in Sec. 5.

4.1. Second-order term

In second order only one vertex diagram contributes to a_e. Our construction of the magnetic moment starts from the operator

$$\mathbf{F} = \gamma^\alpha(\not{D}_1 + m_1)\gamma_\alpha$$

for the electron self-energy diagram of Fig. 9. From (2.41), (2.37), and (2.35) we obtain

$$\mathbf{E}^\nu = \gamma^\alpha\gamma^\nu\gamma_\alpha, \quad \mathbf{C}^{\mu\nu} = 0,$$
$$\mathbf{Z}^{\mu\nu} = \frac{1}{2}\gamma^\alpha[\gamma^\mu\gamma^\nu(\not{D}_1 + m_1) - (\not{D}_1 + m_1)\gamma^\nu\gamma^\mu]\gamma_\alpha.$$

Since $\mathbf{F}$ has only one D operator, no contraction of D's is possible and we can replace it by $A_1\not{p}$. Thus, from (2.39) we obtain the no-contraction terms:

$$N_0 = 2G(2A_1-4), \quad E_0 = 0, \quad C_0 = 0, \quad Z_0 = 4z_1A_1, \quad G = z_1A_1,$$

and the magnetic moment (2.38) reduces to

$$M_2 = -\frac{1}{4}\int (dz)\frac{4z_1A_1(A_1-1)}{U^2V}, \tag{4.1}$$

where

$$(dz) = dz_1 dz_2 \delta(1 - z_1 - z_2),$$

$$B_{11} = 1, \quad U = z_{12} B_{11}, \quad A_1 = z_2/U, \quad V = z_1 - z_1 A_1,$$

according to (2.8), Table I, (2.3), (2.17), and (2.5).

Since M_2 is UV- and IR-finite and since $Z_2 = 1$ to this order, it is equal to C_1 of (1.10). Carrying out the integration, one finds the familiar result

$$C_1 = M_2 = \frac{1}{2}. \tag{4.2}$$

Of course this result can be obtained with much less effort by a straightforward evaluation of the ordinary Feynman integral. Note, however, that the above presentation contains all the steps, except for the UV and IR divergence subtractions, that calculation of higher order terms must follow.

4.2. Fourth-order terms

There are six fourth-order vertex diagrams contributing to a_e that contain no electron loops. They give rise to two magnetic projections M_{4a} and M_{4b} corresponding to the self-energy diagrams of Figs. 8(4a) and 8(4b). M_{4a} has UV divergences arising from the vertices $S' = [1,2,4]$ and $S'' = [2,3,5]$. M_{4b} has a UV divergence from $S = [2,4]$. The standard subtractive renormalization gives

$$a_{4a} = M_{4a} - 2L_2 M_2, \tag{4.3}$$

$$a_{4b} = M_{4b} - \delta m_2 M_{2*} - B_2 M_2, \tag{4.4}$$

as renormalized quantities, where renormalization constants δm_2, L_2, B_2 are defined by (3.23), (3.32), and (3.34), respectively. M_{2*} is the magnetic moment contribution of the mass-insertion diagram Σ_{2*} of Fig. 9. Carrying out the integration one finds that

$$M_{2*} = 1.$$

The fourth-order term of the multiplicatively-renormalized magnetic moment $Z_2 M$ in (1.6), excluding a vacuum-polarization contribution to be discussed later, can be written in the form

$$a_e(\text{Fig. } 8) = M_{4a} + M_{4b} - \delta m_2 M_{2*} + B_2 M_2.$$

As is expected, this is equal to the sum of (4.3) and (4.4) because of the Ward identity (3.38). In the following we adhere to the subtractive renormalization.

The starting point of our intermediate renormalization scheme is

$$M_{4a} = \frac{1}{16} \int (dz) \left(\frac{E_0 + C_0}{U^2 V} + \frac{N_0 + Z_0}{U^2 V^2} + \frac{N_1 + Z_1}{U^3 V} \right), \tag{4.5}$$

obtained from (2.38), where the suffixes 0 and 1 refer to the no-contraction and one-contraction terms. In (4.5) we have

$$\begin{aligned}
(dz) &= dz_1 dz_2 dz_3 dz_4 dz_5 \delta(1 - z_{12345}), \\
B_{11} &= z_{235}, \quad B_{12} = z_{35}, \quad B_{13} = -z_2, \quad B_{23} = z_{14}, \\
B_{22} &= z_{1345}, \quad B_{33} = z_{124}, \quad U = z_2 B_{12} + z_{14} B_{11}, \\
A_i &= 1 - (z_1 B_{1i} + z_2 B_{2i} + z_3 B_{3i})/U, \quad i = 1,\ 2,\ 3, \\
G &= z_1 A_1 + z_2 A_2 + z_3 A_3, \quad V = z_{123} - G,
\end{aligned} \tag{4.6}$$

and

$$\begin{aligned}
E_0 &= 8(2A_1A_2A_3 - A_1A_2 - A_1A_3 - A_2A_3), \quad C_0 = -24 z_4 z_5/U, \\
N_0 &= G(E_0 - 8(2A_2 - 1)), \\
Z_0 &= 8z_1(-A_1 + A_2 + A_3 + A_1A_2 + A_1A_3 - A_2A_3) \\
&\quad + 8z_2(1 - A_1A_2 + A_1A_3 - A_2A_3 + 2A_1A_2A_3) \\
&\quad + 8z_3(A_1 + A_2 - A_3 - A_1A_2 + A_1A_3 + A_2A_3), \\
N_1 &= 8G[B_{12}(2 - A_3) + 2B_{13}(1 - 2A_2) + B_{23}(2 - A_1)], \\
Z_1 &= -8z_1[B_{12}(1 - A_3) + B_{13} + B_{23}A_1] \\
&\quad + 8z_2[B_{12}(1 - A_3) - 4B_{13}A_2 + B_{23}(1 - A_1)] \\
&\quad - 8z_3[B_{12}A_3 + B_{13} + B_{23}(1 - A_1)].
\end{aligned}$$

In our approach, the central role is played by the finite integral

$$\Delta M_{4a} = (1 - \mathbf{K}_{12} - \mathbf{K}_{23}) M_{4a}, \tag{4.7}$$

where the operation of $\mathbf{K}_{12}$ on M_{4a} produces the integral

$$\mathbf{K}_{12} M_{4a} = \frac{1}{16} \int (dz) \frac{N_1 + Z_1}{U^3 V}, \tag{4.8}$$

with

$$\begin{gathered}
U = z_{124} B_{12}, \quad A_3 = z_5/z_{35}, \quad G = z_3 A_3, \quad V = z_{123} - G, \\
N_1 + Z_1 = 8 G B_{12}(1 - A_3),
\end{gathered}$$

B_{12} being the same as in (4.6). $\mathbf{K}_{23} M_{4a}$ can be constructed similarly. $\mathbf{K}_{12} M_{4a}$ and $\mathbf{K}_{23} M_{4a}$ factorize exactly as

$$\mathbf{K}_{12} M_{4a} = \mathbf{K}_{23} M_{4a} = \hat{L}_2 M_2,$$

where $\hat{L}_2$ is the UV-divergent part of the vertex renormalization constant L_2 given by (3.30). Thus one can write

$$M_{4a} = \Delta M_{4a} + 2\hat{L}_2 M_2. \tag{4.9}$$

Substituting (4.9) in (4.3) one obtains

$$a_{4a} = \Delta M_{4a} - 2\tilde{L}_2 M_2, \tag{4.10}$$

where $\tilde{L}_2$ is defined by (3.31). While the first term is finite, the second term is IR-divergent, reflecting the IR divergence introduced by L_2 of the standard renormalization.

The contribution M_{4b} of the diagram of Fig. 8(4b) to the electron moment can be expressed in exactly the same form as (4.5) except that now

$$B_{11} = z_{24}, \quad B_{12} = z_4, \quad B_{22} = z_{1345}, \quad U = z_{135}B_{11} + z_2B_{12},$$
$$G = z_{13}A_1 + z_2A_2, \quad A_1 = z_5B_{11}/U, \quad A_2 = z_5B_{12}/U, \quad V = z_{123} - G,$$

and

$$\begin{aligned} E_0 &= 8A_1[4(A_2 - A_1) - A_1A_2], \quad C_0 = -8A_2, \\ N_0 &= -8G[4(1 - A_1 + A_1^2) + A_2(1 - 4A_1 + A_1^2)], \\ Z_0 &= 8z_{13}[4A_1 - A_2(1 + A_1^2)] + 8z_2A_2(1 + A_1^2), \\ N_1 &= 8G[8(B_{11} - B_{12}) + 3A_1B_{12}], \\ Z_1 &= 24(z_{13} - z_2)A_1B_{12}. \end{aligned}$$

The intermediate renormalization of M_{4b} starts from

$$\Delta' M_{4b} \equiv (1 - \mathbf{K}_2)M_{4b}, \tag{4.11}$$

where $\mathbf{K}_2$ isolates the UV divergence arising from the subdiagram $S = [2,4]$. The subtraction term $\mathbf{K}_2M_{4b}$ is constructed by dropping all terms in the numerators of M_{4b} which are explicitly proportional to z_2. (z_2 and z_4 hidden within A_i's and B_{ij}'s must be kept to the leading order in the $\mathbf{K}_2$ limit.) This construction leads to an exact relation

$$\mathbf{K}_2M_{4b} = \delta m_2 M_{2*} + \hat{B}_2M_2, \tag{4.12}$$

where δm_2 is given in (3.23) and $\hat{B}_2$ is the UV-divergent part of the wave-function renormalization constant B_2 given by (3.35).

From (4.11) and (4.12) one obtains

$$M_{4b} = \Delta' M_{4b} + \delta m_2 M_{2*} + \hat{B}_2M_2. \tag{4.13}$$

The integral $\Delta' M_{4b}$ is UV-finite but contains an IR divergence which can be extracted by an $\mathbf{I}_R$ operation as described in Sec. 2.5:

$$\Delta M_{4b} = (1 - \mathbf{I}_2)\Delta' M_{4b}.$$

The term $\mathbf{I}_2\Delta' M_{4b}$ factorizes analytically as

$$\mathbf{I}_2\Delta' M_{4b} = I_2M_2,$$

where I_2 is defined by (3.31). Putting these results in (4.4) one finds

$$a_{4b} = \Delta M_{4b} + (2I_2 - \Delta B_2)M_2. \tag{4.14}$$

The fourth-order contribution to a_e from the diagrams of Figs. 8(4a) and 8(4b) is the sum of (4.10) and (4.14):

$$a_e(\text{Fig. } 8) = \Delta M_{4a} + \Delta M_{4b} - \Delta B_2M_2, \tag{4.15}$$

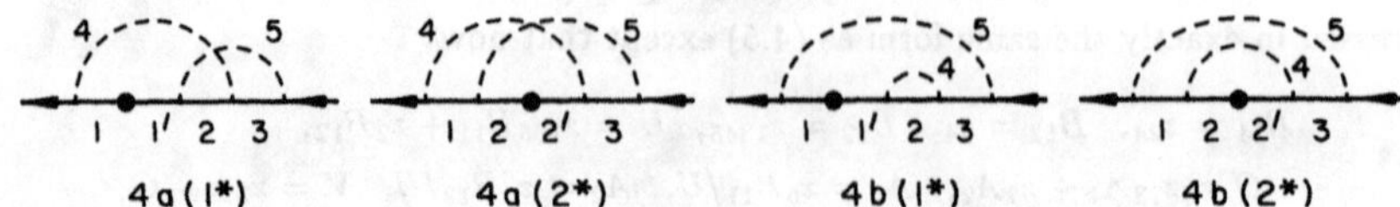

Figure 11: Fourth-order self-energy diagrams with a mass-vertex insertion. They represent the Ward-Takahashi-summed vertex diagrams discussed in Sec. 4.3.

where ΔB_2 is given by (3.37). Numerical integration by VEGAS[42] gives

$$\Delta M_{4a} = 0.219\,36(24), \quad \Delta M_{4b} = -0.188\,22(30), \quad a_e(\text{Fig. 8}) = -0.343\,86(39),$$

where the value of ΔM_{4a} is obtained after 30 iterations with 4×10^7 function calls per iteration and that of ΔM_{4b} is a consequence of 33 iterations with 10^7 function calls per iteration. The result a_e(Fig. 8) is in good agreement with the analytical value $-0.344\,166\ \ldots$.

4.3. Fourth-order terms with mass insertion

In dealing with the sixth- and higher-order terms, it is necessary to know the divergence structure of the Ward-Takahashi-summed vertex graphs with a mass-vertex insertion, represented by the self-energy diagrams of Fig. 11. As is seen from (3.14), these diagrams have overall IR divergences that are not suppressed by the magnetic moment projection. These overall singularities, isolated in the parametric form, are what are needed to cancel *subdiagram* IR divergences appearing in various higher-order diagrams that contain self-energy subdiagrams.

The parametric integral for the Ward-Takahashi-summed magnetic moment projections of vertex graphs related to the diagrams of Fig. 11 is of the form

$$\begin{aligned} M_{4\alpha(i^*)} = -\frac{1}{8}\int (dz)\bigg(&\frac{E_0+C_0}{2U^2V^2}+\frac{E_1+C_1}{2U^3V} \\ &+\frac{N_0+Z_0}{U^2V^3}+\frac{N_1+Z_1}{U^3V^2}+\frac{N_2+Z_2}{U^4V}\bigg), \end{aligned} \tag{4.16}$$

where $\alpha = a,\ b\ ;\ i = 1,\ 2$. Divergence structures for these amplitudes are shown in Table VIII. Isolation of the UV divergences by the $\mathbf{K}_2$ and $\mathbf{K}_{23}$ operations is straightforward, and needs no further discussion. However, it is necesssary to isolate all IR divergences explicitly as shown below.

The effect of $\mathbf{I}_{all}$ operation on $M_{4a(1^*)}$ can be written as

$$\mathbf{I}_{all}M_{4a(1^*)} \equiv \lim_{\lambda\to 0}\int (dz)\frac{F_0[L_{4c}]}{4U^2}\left(\frac{1}{V^2}-\frac{V_t}{V^3}\right), \tag{4.17}$$

where $F_0[L_{4c}]$ is given by (3.54) and

$$V_t = z_{11'}(1-A_1)+z_2(1-A_2)+z_3(1-A_3), \quad V = V_t + \lambda^2 z_{45}.$$

Note that $\mathbf{I}_{all}$ operation does not uniquely give the factor $F_0[L_{4c}]$ in (4.17). Any function having the same IR limit could be used. Our choice has been made based on a (superficial) similarity of the structures of $M_{4a(1^*)}$ and L_{4c}. All parametric functions of $\mathbf{I}_{all}M_{4a(1^*)}$ are identical with those of $M_{4a(1^*)}$.

Table VIII. Divergence structure of $M_{4a(i^*)}$ and $M_{4b(i^*)}$.

Diagram	$\mathcal{N}_0, \mathcal{Z}_0$	$\mathcal{N}_1, \mathcal{Z}_1$	$\mathcal{N}_2, \mathcal{Z}_2$	$\mathcal{E}_0, \mathcal{C}_0$	$\mathcal{E}_1, \mathcal{C}_1$
$M_{4a(1^*)}$	$\mathbf{I}_{all}$			$\mathbf{I}_{all}$	
	$\tilde{\mathbf{I}}_1$	$\tilde{\mathbf{I}}_1$, $\mathbf{K}_{23}$	$\mathbf{K}_{23}$	$\tilde{\mathbf{I}}_1$	$\mathbf{K}_{23}$
$M_{4a(2^*)}$	$\mathbf{I}_{all}$			$\mathbf{I}_{all}$	
$M_{4b(1^*)}$	$\mathbf{I}_{all}$			$\mathbf{I}_{all}$	
	$\tilde{\mathbf{I}}_{13}$,$\mathbf{K}_2$	$\mathbf{K}_2$	$\mathbf{K}_2$	$\mathbf{K}_2$	$\mathbf{K}_2$
$M_{4b(2^*)}$	$\mathbf{I}_{all}$			$\mathbf{I}_{all}$	
	$\tilde{\mathbf{I}}_{13}$	$\tilde{\mathbf{I}}_{13}$, $\mathbf{K}_{22'}$	$\mathbf{K}_{22'}$	$\tilde{\mathbf{I}}_{13}$	$\mathbf{K}_{22'}$

Application of the subdiagram $\tilde{\mathbf{I}}_1$ operation on $M_{4a(1^*)}$ isolates the term

$$\tilde{\mathbf{I}}_1 M_{4a(1^*)} = \lim_{\lambda\to 0} -\frac{1}{8}\int (dz)\left(\frac{E_0+C_0}{2U^2V^2} + \frac{N_0+Z_0}{U^2V^3} + \frac{N_1+Z_1}{U^3V^2}\right), \tag{4.18}$$

where

$$N_0 + Z_0 = -16z_{11'}(1-A_1)(1-4A_2+A_2^2) + 16z_{23}(1-A_2)(1+2A_2-A_2^2),$$
$$N_1 + Z_1 = 8z_{23}A_2B_{23}, \quad E_0 + C_0 = 16(1-5A_2);$$

all parametric functions being defined under the $\tilde{\mathbf{I}}_1$ limit.

Application of $\mathbf{I}_{all}\tilde{\mathbf{I}}_1$, which isolates the N_0+Z_0 and E_0+C_0 terms of (4.18), leads to

$$\mathbf{I}_{all}\tilde{\mathbf{I}}_1 M_{4a(1^*)} = \lim_{\lambda\to 0}\int (dz)\frac{(-2)(1-4A_2+A_2^2)}{U^2}\left(\frac{1}{V^2} - \frac{V_t}{V^3}\right), \tag{4.19}$$

where all functions are again defined under the $\tilde{\mathbf{I}}_1$ limit.

We do not bother to express (4.18) and (4.19) as recognizable products of lower-order graphs. It is enough to define

$$I_{4a(1^*)} \equiv (\mathbf{I}_{all} + \tilde{\mathbf{I}}_1(1-\mathbf{I}_{all}))M_{4a(1^*)}.$$

Then we can write the divergence structure of $M_{4a(1^*)}$ as

$$\begin{aligned}M_{4a(1^*)} &= (1-\mathbf{I}_{all})(1-\tilde{\mathbf{I}}_1-\mathbf{K}_{23})M_{4a(1^*)}+\mathbf{K}_{23}M_{4a(1^*)}+(\mathbf{I}_{all}+\tilde{\mathbf{I}}_1(1-\mathbf{I}_{all}))M_{4a(1^*)}\\ &= \Delta M_{4a(1^*)}+\hat{L}_2M_{2^*}+I_{4a(1^*)}.\end{aligned}$$

The integral $M_{4a(2^*)}$ has only one IR divergence, isolated by $\mathbf{I}_{all}$. In analogy with (4.17) we write it as

$$\begin{aligned}I_{4a(2^*)} &\equiv \mathbf{I}_{all}M_{4a(2^*)}\\ &= \lim_{\lambda\to 0}\int(dz)\frac{F_0[L_{4x}]}{4U^2}\left(\frac{1}{V^2}-\frac{V_t}{V^3}\right),\end{aligned} \tag{4.20}$$

where $F_0[L_{4x}]$ is given by (3.51) and

$$V_t = z_1(1-A_1)+z_{22'}(1-A_2)+z_3(1-A_3), \quad V = V_t+\lambda^2 z_{45}.$$

Parametric functions for $I_{4a(2^*)}$ are identical with those for $M_{4a(2^*)}$. The divergence structure of $M_{4a(2^*)}$ can thus be expressed as

$$\begin{aligned}M_{4a(2^*)} &= (1-\mathbf{I}_{all})M_{4a(2^*)}+\mathbf{I}_{all}M_{4a(2^*)}\\ &= \Delta M_{4a(2^*)}+I_{4a(2^*)}.\end{aligned}$$

The integral $M_{4b(1^*)}$ has an IR divergence associated with the $\mathbf{I}_{13}$ operation which "power counting" shows to be linear. This is difficult to isolate by the effective operation $\tilde{\mathbf{I}}_{13}\equiv\mathbf{I}_{13}(1-\mathbf{K}_2)$. Thus we follow an alternate route. Namely we subtract the integral

$$M_{4b(1^*)}[f]\equiv -\frac{1}{8}\int(dz)\frac{f}{U^2V^3}$$

from $M_{4b(1^*)}$, where

$$f = -8z_2A_2(1-A_2)(-1+6A_1-3A_1^2+2A_1^3).$$

Then $(1-\mathbf{I}_{all})(M_{4b(1^*)}-M_{4b(1^*)}[f])$ is free from the divergence related to $\tilde{\mathbf{I}}_{13}$. The term isolated by the $\mathbf{I}_{all}$ operation can be written as

$$\begin{aligned}&\mathbf{I}_{all}M_{4b(1^*)}[N_0+Z_0-f, E_0+C_0]\\ &= \lim_{\lambda\to 0}\int(dz)\frac{F_0[L_{4s}]}{4U^2}\left(\frac{1}{2V^2}-\frac{z_{13}(1-A_1)}{V^3}\right),\end{aligned} \tag{4.21}$$

where $F_0[L_{4s}]$ is given by (3.64) and all parametric functions are defined as for the complete amplitude $M_{4b(1^*)}$. The term $\mathbf{K}_2\mathbf{I}_{all}M_{4b(1^*)}[N_0+Z_0-f,...]$ is also given by the r.h.s. of (4.21) but functions are now defined in the $\mathbf{K}_2$ limit.

We can therefore write the divergence structure of $M_{4b(1^*)}$ as

$$\begin{aligned}M_{4b(1^*)} &= (1-\mathbf{K}_2)(1-\mathbf{I}_{all})(M_{4b(1^*)}-M_{4b(1^*)}[f])+\mathbf{K}_2M_{4b(1^*)}\\ &\quad+(\mathbf{I}_{all}(1-\mathbf{K}_2)M_{4b(1^*)}[N_0+Z_0-f,E_0+C_0]+M_{4b(1^*)}[f])\\ &\equiv \Delta M_{4b(1^*)}+(\delta m_2M_{2^{**}}+\hat{B}_2M_{2^*})+I_{4b(1^*)},\end{aligned}$$

which also defines $I_{4b(1^\bullet)}$.

Since IR divergences in $M_{4b(2^\bullet)}$ are logarithmic, the terms isolated by the $\tilde{\mathbf{I}}_R$ operations can be easily expressed as products of lower-order amplitudes. The term isolated by the $\mathbf{I}_{all}$ operation is

$$\begin{aligned} I_{4b(2^\bullet)} &\equiv \mathbf{I}_{all} M_{4b(2^\bullet)} \\ &= \lim_{\lambda \to 0} \int (dz) \frac{F_0[L_{4l}]}{4U^2} \left(\frac{1}{V^2} - \frac{V_t}{V^3} \right), \end{aligned} \tag{4.22}$$

where $F_0[L_{4l}]$ is given by (3.63), and

$$V_t = z_{13}(1 - A_1) + z_{22'}(1 - A_2), \quad V = V_t + \lambda^2 z_{45}.$$

The term isolated by the $\tilde{\mathbf{I}}_{13}$ operation can be written as

$$\tilde{\mathbf{I}}_{13} M_{4b(2^\bullet)} = \lim_{\lambda \to 0} \int (dz) \frac{(1 - 4A_1 + A_1^2)}{16} \left(\frac{E_0 + C_0}{2U^2V^2} + \frac{N_0 + Z_0}{U^2V^3} + \frac{N_1 + Z_1}{U^3V^2} \right), \tag{4.23}$$

with

$$\begin{aligned} N_0 + Z_0 &= 32[z_{13}(1 - A_1)(1 - A_2 + A_2^2) + z_{22'} A_2^2 (1 - A_2)], \\ N_1 + Z_1 &= 32 z_{22'} A_2 B_{22}, \quad E_0 + C_0 = -32(1 - A_2 + 2A_2^2). \end{aligned}$$

Eq. (4.23) is recognizable as the sum

$$\tilde{\mathbf{I}}_{13} M_{4b(2^\bullet)} = I_2 M_{2^\bullet} + M_{2^\bullet}[I] \delta \tilde{m}_{2^\bullet},$$

where the second term is expected from (3.15). $M_{2^\bullet}[I]$ is given by the integral

$$M_{2^\bullet}[I] = \lim_{\lambda \to 0} \int (dz)(1 - 4A_1 + A_1^2) \left(\frac{1}{U_1^2 V_1} - \frac{z_{15}(1 - A_1)}{U_1^2 V_1^2} \right), \tag{4.24}$$

with

$$(dz) = dz_1 dz_3 dz_5 \delta(1 - z_{135}), \quad U_1 = z_{135}, \quad A_1 = z_5 / U_1, \quad V_1 = z_{13}(1 - A_1) + \lambda^2 z_5.$$

On integration one finds $M_{2^\bullet}[I] = -1$. The combined operation $\mathbf{I}_{all}\tilde{\mathbf{I}}_{13}$ gives

$$\mathbf{I}_{all} \tilde{\mathbf{I}}_{13} M_{4b(2^\bullet)} = \lim_{\lambda \to 0} \int (dz) \frac{(1 - 4A_1 + A_1^2)(1 - 4A_2 + A_2^2)}{U^2} \left(\frac{1}{V^2} - \frac{V_t}{V^3} \right), \tag{4.25}$$

where V_t is as in (4.22) but all functions are now defined in the $\mathbf{I}_{13}$ limit. Eq. (4.25) is recognizable as

$$\mathbf{I}_{all} \tilde{\mathbf{I}}_{13} M_{4b(2^\bullet)} = 2 M_{2^\bullet}[I] I_2.$$

Thus the divergence structure of $M_{4b(2^\bullet)}$ can be written as

$$\begin{aligned} M_{4b(2^\bullet)} &= (1 - \mathbf{I}_{all})(1 - \tilde{\mathbf{I}}_{13} - \mathbf{K}_{22'}) M_{4b(2^\bullet)} + \mathbf{K}_{22'} M_{4b(2^\bullet)} \\ &\quad + \mathbf{I}_{all} M_{4b(2^\bullet)} + \tilde{\mathbf{I}}_{13} M_{4b(2^\bullet)} - \mathbf{I}_{all} \tilde{\mathbf{I}}_{13} M_{4b(2^\bullet)} \\ &= \Delta M_{4b(2^\bullet)} + \delta \tilde{m}_{2^\bullet} M_{2^\bullet} + I_{4b(2^\bullet)} \\ &\quad + (I_2 M_{2^\bullet} + M_{2^\bullet}[I] \delta \tilde{m}_{2^\bullet}) - 2 M_{2^\bullet}[I] I_2. \end{aligned}$$

4.4. Sixth-order terms

Out of 72 vertex diagrams contributing to a_e in sixth order, 50 are without closed electron loops. They are represented by the eight self-energy diagrams of Fig. 10 whose magnetic moment projections (2.38) have the form

$$M_{6\alpha} = -\frac{2}{64}\int (dz)\left(\frac{E_0+C_0}{2U^2V^2} + \frac{E_1+C_1}{2U^3V} + \frac{N_0+Z_0}{U^2V^3} + \frac{N_1+Z_1}{U^3V^2} + \frac{N_2+Z_2}{U^4V}\right), \tag{4.26}$$

where $\alpha = A, B, \ldots, H$, and

$$A_i = 1 - (z_1B_{1i} + z_2B_{2i} + z_3B_{3i} + z_4B_{4i} + z_5B_{5i})/U,$$
$$G = z_1A_1 + z_2A_2 + z_3A_3 + z_4A_4 + z_5A_5, \quad V = z_{12345} - G.$$

The B_{ij} and U are diagram-dependent. For instance, for the diagram $6B$ we have

$$B_{11} = z_{247}z_{36} + z_3z_6, \quad B_{12} = z_{36}z_7, \quad B_{13} = z_6z_7,$$
$$U = z_{158}B_{11} + z_{24}B_{12} + z_3B_{13}.$$

The trace calculation and integrations over loop momenta needed to express E_n, C_n, N_n, and Z_n in terms of these parametric functions are carried out by SCHOONSCHIP.[41] The resulting expressions are too lengthy to be listed here.

In the standard subtractive renormalization scheme, the renormalized contribution of the diagrams of Fig. 10 to the anomalous moment can be written in the form

$$a_e^{(6)}(\text{Fig.10}) = \sum_{\alpha=A}^{H} \eta_\alpha a_{6\alpha}, \tag{4.27}$$

where

$$a_{6\alpha} = M_{6\alpha} - \text{renormalization terms}$$
$$\eta_\alpha = 2, \quad \alpha = D,\ G,$$
$$= 1, \quad \text{otherwise},$$

which takes into account the time-reversed contributions of Figs. 10(6D), (6G).

Each renormalized $a_{6\alpha}$ is IR-divergent in general and hence not numerically calculable. In order to relate $a_{6\alpha}$ to a finite, calculable quantity $\Delta M_{6\alpha}$, we apply the method of intermediate renormalization, which employes $\mathbf{K}_S$ and $\mathbf{I}_R$ operations to isolate and subtract the UV and IR divergences from $M_{6\alpha}$, and express them as products of intermediate renormalization constants and magnetic moment projections of lower-order graphs.

Let us examine this procedure more closely, choosing Fig. 10(6B) as an example. Table IX lists the divergence structure of M_{6B}. The standard renormalized contribution of this diagram can be written as

$$a_{6B} = M_{6B} - (\delta m_2 M_{4b(2^*)} + B_2 M_{4b}) - (\delta m_{4b} M_{2^*} + B_{4b} M_2) + \delta m_2(\delta m_{2^*} M_{2^*} + B_{2^*} M_2) + B_2(\delta m_2 M_{2^*} + B_2 M_2). \tag{4.28}$$

Table IX. Divergence structure of M_{6B}.

$\mathcal{N}_0, \mathcal{Z}_0$	$\mathcal{N}_1, \mathcal{Z}_1$	$\mathcal{N}_2, \mathcal{Z}_2$	$\mathcal{E}_0, \mathcal{C}_0$	$\mathcal{E}_1, \mathcal{C}_1$
$\tilde{\mathbf{I}}_{15}$	$\tilde{\mathbf{I}}_{15}, \mathbf{K}_{234}$	$\mathbf{K}_{234}$	$\tilde{\mathbf{I}}_{15}$	$\mathbf{K}_{234}$
$\tilde{\mathbf{I}}_{1245}, \mathbf{K}_3$	$\mathbf{K}_3$	$\mathbf{K}_3$	$\mathbf{K}_3$	$\mathbf{K}_3$

In the intermediate renormalization scheme, we introduce

$$\begin{aligned} \Delta M_{6B} &= (1 - \mathbf{I}_{15})(1 - \mathbf{K}_{234})(1 - \mathbf{I}_{1245})(1 - \mathbf{K}_3) M_{6B} \\ &\equiv (1 - \tilde{\mathbf{I}}_{15} - \mathbf{K}_{234})(1 - \tilde{\mathbf{I}}_{1245} - \mathbf{K}_3) M_{6B}, \end{aligned} \tag{4.29}$$

which is free from all divergences. As is apparent from (4.29), ΔM_{6B} is related to M_{6B} in the following manner:

$$\begin{aligned} M_{6B} = \Delta M_{6B} &+ \mathbf{K}_3 M_{6B} + \mathbf{K}_{234} M_{6B} - \mathbf{K}_3 \mathbf{K}_{234} M_{6B} \\ &+ \tilde{\mathbf{I}}_{15}(1 - \tilde{\mathbf{I}}_{1245} - \mathbf{K}_3) M_{6B} + \tilde{\mathbf{I}}_{1245} M_{6B} - \tilde{\mathbf{I}}_{1245} \mathbf{K}_{234} M_{6B}. \end{aligned} \tag{4.30}$$

Applying (3.6) to M_{6B} and noting that $\hat{B}_{2^*} = 0$, one obtains

$$\begin{aligned} \mathbf{K}_3 M_{6B} &= \delta m_2 M_{4b(2^*)} + \hat{B}_2 M_{4b}, \\ \mathbf{K}_{234} M_{6B} &= \delta \hat{m}_{4b} M_{2^*} + \hat{B}_{4b} M_2, \\ \mathbf{K}_3 \mathbf{K}_{234} M_{6B} &= \delta m_2 \delta \hat{m}_{2^*} M_{2^*} + \hat{B}_2 (\delta \hat{m}'_2 M_{2^*} + \hat{B}'_2 M_2). \end{aligned}$$

Considering the terms in (4.30) involving IR divergences, one finds

$$\begin{aligned} \tilde{\mathbf{I}}_{15} M_{6B} &= I_2 M_{4b} + M_{2^*}[I] \delta \tilde{m}_{4b}, \\ \tilde{\mathbf{I}}_{15} \tilde{\mathbf{I}}_{1245} M_{6B} &= (I_2)^2 M_2, \\ \tilde{\mathbf{I}}_{15} \mathbf{K}_3 M_{6B} &= I_2(\delta m_2 M_{2^*} + \hat{B}_2 M_2) + M_{2^*}[I](\delta m_2 \delta \tilde{m}_{2^*} + \hat{B}_2 \delta \tilde{m}'_2). \end{aligned}$$

From these results one obtains

$$\tilde{\mathbf{I}}_{15}(1 - \tilde{\mathbf{I}}_{1245} - \mathbf{K}_3) M_{6B} = I_2 \Delta M_{4b} + M_{2^*}[I] \Delta \delta m_{4b}.$$

The integral $\tilde{\mathbf{I}}_{1245} M_{6B}$ factorizes exactly as

$$\tilde{\mathbf{I}}_{1245} M_{6B} = L_{4l}[F_0] M_2 = [I_{4l} + (I_2)^2] M_2.$$

Since the divergences separately isolated by $\mathbf{K}_{234}$ and $\tilde{\mathbf{I}}_{1245}$ have no overlap, the oversubtraction correction vanishes:

$$\mathbf{K}_{234} \tilde{\mathbf{I}}_{1245} M_{6B} = 0.$$

Substitution of these results in (4.30) leads to the intermediate renormalization relation listed in Appendix C. From this and the standard renormalization (4.28) one obtains the relation between ΔM_{6B} and a_{6B}:

$$\begin{aligned} a_{6B} = \Delta M_{6B} &+ (I_2 - \tilde{B}_2) \Delta M_{4b} + \Delta \delta m_{4b} (M_{2^*}[I] - M_{2^*}) \\ &+ M_2 (2 I_{4s} + 2 I_{4l} - \Delta B_{4b} + (\tilde{B}_2)^2 + (I_2)^2 - 2 \tilde{B}_2 I_2). \end{aligned} \tag{4.31}$$

Similar analysis of divergence structure can be carried out for all diagrams of Fig. 10. In Appendix C we list the formulas that relate the standard-renormalized amplitudes $a_{6\alpha}$ to the unrenormalized integrals $M_{6\alpha}$ ($\alpha = A$ to H), which in turn are expressed in terms of intermediate renormalization quantities. Summing over all diagrams of this group, we obtain

$$\begin{aligned} a_e^{(6)}(\text{Fig.10}) = \sum_{\alpha=A}^{H} \eta_\alpha \Delta M_{6\alpha} - 3\Delta B_2 \Delta M^{(4)} \\ + (M_{2^*}[I] - M_{2^*})\Delta\delta m^{(4)} - M_2[\Delta B^{(4)} + 2\Delta L^{(4)} - 2(\Delta B_2)^2], \end{aligned} \tag{4.32}$$

where

$$\begin{aligned} \Delta M^{(4)} &= \Delta M_{4a} + \Delta M_{4b}, \quad \Delta\delta m^{(4)} = \Delta\delta m_{4a} + \Delta\delta m_{4b}, \\ \Delta B^{(4)} &= \Delta B_{4a} + \Delta B_{4b}, \quad \Delta L^{(4)} = \Delta L_{4x} + 2\Delta L_{4c} + \Delta L_{4l} + 2\Delta L_{4s}. \end{aligned} \tag{4.33}$$

Table X lists the latest values of ΔM_α evaluated by VEGAS specifically for this article. Note that ΔM_{6B} and ΔM_{6C} have values different from those of Ref. 17 because they are defined differently. Other terms in (4.32) are listed in Table XI. From (4.32), Table X, and Table XI we obtain

$$a_e^{(6)}(\text{Fig.10}) = 0.905\ 1(86), \tag{4.34}$$

which is in good agreement with the best result available at present obtained by combining the analytic and semianalytic results[14] for the diagrams $6A$ through $6G$ with $6H$ from Table X for which no analytical result exists:

$$a_e^{(6)}(\text{Fig.10}) = 0.899\ 87(42). \tag{4.35}$$

Table X. Magnetic moment contribution of the sixth-order diagrams of Fig. 10. $\eta_\alpha = 1$ (2) for time-reversal symmetric (asymmetric) diagrams. All integrals are evaluated in double precision except in a small domain of $6G$ where the integral is evaluated in quadruple precision (using 10^6 function calls, 37 iterations).

Diagram	$\eta_\alpha \Delta M_\alpha$	No. of function calls (in units of 10^6)	No. of iterations
6A	−1.353 5(27)	10	37
6B	3.015 3(33)	10	37
6C	−0.337 6(34)	10	36
6D	0.930 2(52)	10	38
6E	1.202 3(19)	10	37
6F	0.753 5(27)	10	37
6G	2.466 3(18)	10,1	37,37
6H	−2.206 32(19)	900	30

Table XI. Auxiliary integrals.

Integral	Value	Integral	Value
M_2	0.5	ΔB_2	0.75
M_{2^*}	1.	$M_{2^*}[I]$	−1.
$\Delta M^{(4)}$	0.030 834	$\Delta\delta m^{(4)}$	1.905 42(65)
$\Delta B^{(4)}$	−0.435 79(47)	$\Delta L^{(4)}$	0.465 32(35)

4.5. Eighth-order terms

There are 518 vertex diagrams without closed electron loops that contribute to a_e in eighth order. They are represented by the 47 self-energy diagrams shown in Fig. 12 whose magnetic moment projections have the form

$$M_{8\alpha} = \frac{3!}{256}\int(dz)\left(\frac{E_0+C_0}{3U^2V^3}+\frac{E_1+C_1}{3U^3V^2}+\frac{E_2+C_2}{3U^4V}\right.$$
$$\left.+\frac{N_0+Z_0}{U^2V^4}+\frac{N_1+Z_1}{U^3V^3}+\frac{N_2+Z_2}{U^4V^2}+\frac{N_3+Z_3}{U^5V}\right), \tag{4.36}$$

where $\alpha = 1, \ldots, 47$. Parametric functions are appropriately defined for each amplitude, with N_0 through C_2 being generated by SCHOONSCHIP.[41]

Standard subtractive renormalization relates the $M_{8\alpha}$ to the the renormalized contribution by

$$a_e^{(8)}(\text{Fig.12}) = \sum_{\alpha=1}^{47}\eta_\alpha a_{8\alpha}, \tag{4.37}$$

where

$$a_{8\alpha} = M_{8\alpha} - \text{renormalization terms},$$

and $\eta_\alpha = 1$ (2) if diagram α is time-reversal symmetric (asymmetric). For simplicity we shall denote these amplitudes as M_α instead of $M_{8\alpha}$ in the following. To avoid possible confusion with M_2 defined elsewhere, however, let us denote $M_1, M_2, \ldots, M_9$ as M_{01}, $M_{02}, \ldots, M_{09}$.

The intermediate renormalization of eighth-order terms is generally not much more difficult than the discussions given in the preceding sections. We therefore restrict our consideration to the two most troublesome of the 47 amplitudes, whose line numberings are shown in Fig. 13. In Appendix D we give a list of the intermediate renormalization terms of M_{01} through M_{47}.

The amplitude M_{47} provides an instructive lesson on nested IR divergences each of which, in general, consists of two distinct pieces. (See (3.15).) Table XII indicates the singularity structure of M_{47}.

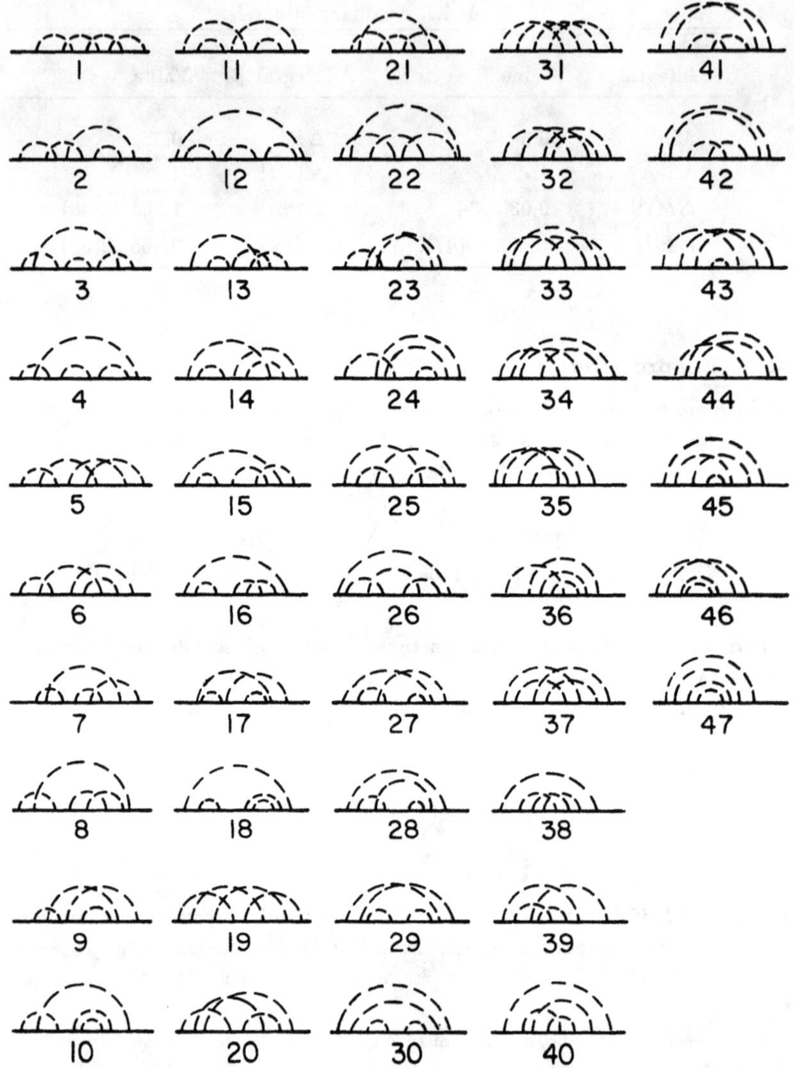

Figure 12: Self-energy diagrams representing 518 eighth-order diagrams of four-photon exchange type that contribute to a_e. There are 74 such diagrams, of which 47 are shown. Other diagrams are related to some of them by time reversal.

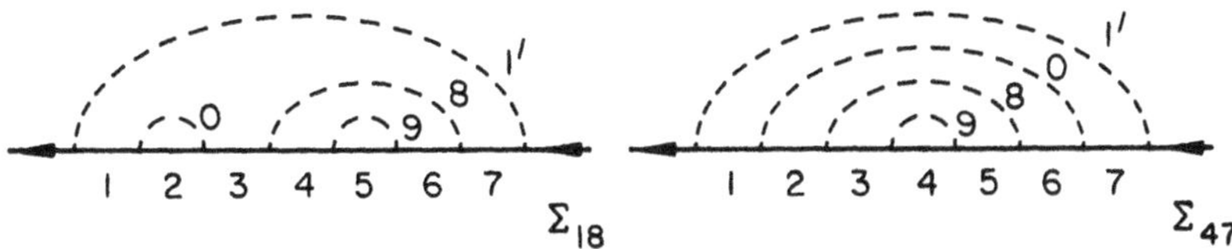

Figure 13: Eighth-order diagrams Σ_{18} and Σ_{47}, showing the line numbering.

Table XII. Divergence structure of M_{47}.

$\mathcal{N}_0, \mathcal{Z}_0$	$\mathcal{N}_1, \mathcal{Z}_1$	$\mathcal{N}_2, \mathcal{Z}_2$	$\mathcal{N}_3, \mathcal{Z}_3$	$\mathcal{E}_0, \mathcal{C}_0$	$\mathcal{E}_1, \mathcal{C}_1$	$\mathcal{E}_2, \mathcal{C}_2$
$\tilde{\mathbf{I}}_{17}$	$\tilde{\mathbf{I}}_{17}$	$\tilde{\mathbf{I}}_{17}, \mathbf{K}_{23456}$	$\mathbf{K}_{23456}$	$\tilde{\mathbf{I}}_{17}$	$\tilde{\mathbf{I}}_{17}$	$\mathbf{K}_{23456}$
$\tilde{\mathbf{I}}_{1267}$	$\tilde{\mathbf{I}}_{1267}, \mathbf{K}_{345}$	$\mathbf{K}_{345}$	$\mathbf{K}_{345}$	$\tilde{\mathbf{I}}_{1267}$	$\mathbf{K}_{345}$	$\mathbf{K}_{345}$
$\tilde{\mathbf{I}}_{123567}, \mathbf{K}_4$	$\mathbf{K}_4$	$\mathbf{K}_4$	$\mathbf{K}_4$	$\mathbf{K}_4$	$\mathbf{K}_4$	$\mathbf{K}_4$

We find that

$$\Delta M_{47} = (1 - \tilde{\mathbf{I}}_{17} - \mathbf{K}_{23456})(1 - \tilde{\mathbf{I}}_{1267} - \mathbf{K}_{345})(1 - \tilde{\mathbf{I}}_{123567} - \mathbf{K}_4)M_{47} \tag{4.38}$$

is finite and numerically integrable. Let us write the parts of (4.38) dealing with the IR-divergence-isolating operations as

$$\begin{aligned} &- \tilde{\mathbf{I}}_{17}(1 - \tilde{\mathbf{I}}_{1267} - \mathbf{K}_{345})(1 - \tilde{\mathbf{I}}_{123567} - \mathbf{K}_4)M_{47} \\ &- \tilde{\mathbf{I}}_{1267}(1 - \tilde{\mathbf{I}}_{123567} - \mathbf{K}_4)M_{47} \\ &- \tilde{\mathbf{I}}_{123567}M_{47}, \end{aligned}$$

and analyze the structure of these three terms one by one.

As $\delta\tilde{m}_2 = 0$, $\tilde{\mathbf{I}}_{123567}$ acting on M_{47} generates only terms proportional to M_2:

$$\begin{aligned} \tilde{\mathbf{I}}_{123567}M_{47} &= L_{6B3}[F_0]M_2 \\ &= (I_{6B3} + 2I_{4l}I_2 + (I_2)^3)M_2. \end{aligned}$$

The term $\tilde{\mathbf{I}}_{1267}(1 - \tilde{\mathbf{I}}_{123567} - \mathbf{K}_4)M_{47}$ consists of two pieces, the second proportional to $\Delta\delta m_{4b}$, its coefficient being recognizable as $I_{4b(2^*)} \equiv \mathbf{I}_{all}M_{4b(2^*)}$:

$$\tilde{\mathbf{I}}_{1267}(1 - \tilde{\mathbf{I}}_{123567} - \mathbf{K}_4)M_{47} = (I_{4l} + (I_2)^2)\Delta M_{4b} + I_{4b(2^*)}\Delta\delta m_{4b}.$$

For $\tilde{\mathbf{I}}_{17}$ we find the structure

$$\tilde{\mathbf{I}}_{17}M_{47} = I_2 M_{6B} + M_{2^*}[I]\Delta\delta\tilde{m}_{6B}.$$

The potentially tricky operation $\tilde{\mathbf{I}}_{17}\tilde{\mathbf{I}}_{1267}$ actually turns out to be quite tame:

$$\tilde{\mathbf{I}}_{1267}\tilde{\mathbf{I}}_{17}M_{47} = I_2(I_2 M_{4b} + 2M_{2^*}[I]\Delta\delta\tilde{m}_{4b}).$$

Combining these results we find

$$\tilde{\mathbf{I}}_{17}(1-\tilde{\mathbf{I}}_{1267}-\mathbf{K}_{345})(1-\tilde{\mathbf{I}}_{123567}-\mathbf{K}_4)M_{47} = I_2\Delta M_{6B} + M_{2^*}[I]\Delta\delta m_{6B}.$$

Including the terms isolated by the $\mathbf{K}_S$ operations only, which we have not shown explicitly, we obtain the intermediate renormalization equation for M_{47}:

$$\begin{aligned}
M_{47} &= \Delta M_{47} + \delta m_2 M_{6B(3^*)} + \hat{B}_2 M_{6B} + \delta\hat{m}_{4b} M_{4b(2^*)} + \hat{B}_{4b} M_{4b} \\
&+ \delta\hat{m}_{6B} M_{2^*} + \hat{B}_{6B} M_2 - \delta m_2 \delta\hat{m}_{2^*} M_{4b(2^*)} - \hat{B}_2(\delta\hat{m}_2' M_{4b(2^*)} + \hat{B}_2' M_{4b}) \\
&- \delta m_2 \delta\hat{m}_{4b(2^*)} M_{2^*} - \hat{B}_2(\delta\hat{m}_{4b(2')} M_{2^*} + \hat{B}_{4b(2')} M_2) \\
&- \delta\hat{m}_{4b}\delta\hat{m}_{2^*} M_{2^*} - \hat{B}_{4b}(\delta\hat{m}_2' M_{2^*} + \hat{B}_2' M_2) \\
&+ \delta m_2(\delta\hat{m}_{2^*})^2 M_{2^*} + \hat{B}_2 \delta\hat{m}_2' \delta\hat{m}_{2^*} M_{2^*} + \hat{B}_2 \hat{B}_2'(\delta\hat{m}_2' M_{2^*} + \hat{B}_2' M_2) \\
&+ (I_{6B3} + 2I_{4l}I_2 + (I_2)^3) M_2 + (I_{4l} + (I_2)^2)\Delta M_{4b} \\
&+ I_{4b(2^*)}\Delta\delta m_{4b} + I_2 \Delta M_{6B} + M_{2^*}[I]\Delta\delta m_{6B},
\end{aligned} \tag{4.39}$$

which is to be compared with the standard renormalization relating M_{47} to a_{47}:

$$\begin{aligned}
a_{47} &= M_{47} - \delta m_2 M_{6B(3^*)} - B_2 M_{6B} - \delta m_{4b} M_{4b(2^*)} - B_{4b} M_{4b} \\
&- \delta m_{6B} M_{2^*} - B_{6B} M_2 + \delta m_2(\delta m_{2^*} M_{4b(2^*)} + B_{2^*} M_{4b}) \\
&+ B_2(\delta m_2 M_{4b(2^*)} + B_2 M_{4b}) + \delta m_2(\delta m_{4b(2^*)} M_{2^*} + B_{4b(2^*)} M_2) \\
&+ B_2(\delta m_{4b} M_{2^*} + B_{4b} M_2) + \delta m_{4b}(\delta m_{2^*} M_{2^*} + B_{2^*} M_2) \\
&+ B_{4b}(\delta m_2 M_{2^*} + B_2 M_2) - \delta m_2 \delta m_{2^*}(\delta m_{2^*} M_{2^*} + B_{2^*} M_2) \\
&- \delta m_2 B_{2^*}(\delta m_2 M_{2^*} + B_2 M_2) - B_2 \delta m_2(\delta m_{2^*} M_{2^*} + B_{2^*} M_2) \\
&- (B_2)^2(\delta m_2 M_{2^*} + B_2 M_2).
\end{aligned} \tag{4.40}$$

The insertion of (4.39) into (4.40) provides the working relation between a_{47} and the numerically finite ΔM_{47}. The amplitudes M_{41}, M_{42}, and M_{46} can be analyzed in a manner analogous to M_{47}.

The second amplitude we discuss, M_{18}, provides another instructive lesson, which we can illustrate best by first considering the sixth-order diagram M_{6A}.

The prescription for identifying IR singularities is to let all possible combinations of photons "go soft" and "power count" under the rules of Sec. 3 for each combination. The IR-divergence-isolating operators corresponding to various soft photon limits for M_{6A} thus identified are as follows:

$$\begin{aligned}
\mathbf{I}_{1345} &: \quad z_1,\ z_3,\ z_4,\ z_5 = O(\delta), \quad z_2,\ z_6 = O(\delta^2), \quad z_7,\ z_8 = O(1), \\
\mathbf{I}_{1235} &: \quad z_1,\ z_2,\ z_3,\ z_5 = O(\delta), \quad z_4,\ z_7 = O(\delta^2), \quad z_6,\ z_8 = O(1), \\
\mathbf{I}_{135} &: \quad z_1,\ z_3,\ z_5 = O(\delta), \quad z_2,\ z_4,\ z_6,\ z_7 = O(\delta^2), \quad z_8 = O(1),
\end{aligned}$$

Thus

$$\Delta M_{6A} = (1-\mathbf{I}_{135})(1-\mathbf{I}_{1345})(1-\mathbf{K}_2)(1-\mathbf{I}_{1235})(1-\mathbf{K}_4)M_{6A} \tag{4.41}$$

is finite. The terms in (4.41) in which $\mathbf{I}_{135}$ appear cancel. This is easily seen since the limiting procedures of the $\mathbf{I}_{135}$ operation define a subset of the overlap of the $\mathbf{K}_2$ and $\mathbf{K}_4$ operations and hence

$$\mathbf{I}_{135}(1-\mathbf{K}_2)(1-\mathbf{K}_4)M_{6A}=0.$$

Thus (4.41) simplifies to

$$\Delta M_{6A}=(1-\tilde{\mathbf{I}}_{1345}-\mathbf{K}_2)(1-\tilde{\mathbf{I}}_{1235}-\mathbf{K}_4)M_{6A}.$$

The terms M_{16} and M_{18} are the first amplitudes where an IR subtraction analogous to $\mathbf{I}_{135}(1-\mathbf{K}_2)(1-\mathbf{K}_4)$ above produces a non-zero result.

The IR singularities of M_{18} are isolated by $\mathbf{I}_\alpha$ where $\alpha=137$, 1237, 13467, 134567, or 123467. $\mathbf{I}_{13467}$ cancels just as $\mathbf{I}_{135}$ for M_{6A}:

$$\mathbf{I}_{13467}(1-\mathbf{K}_2)(1-\mathbf{K}_5)M_{18}=0,$$

and we can write

$$\overline{\Delta M_{18}}\equiv(1-\mathbf{I}_{137})\overline{\Delta' M_{18}},$$

where

$$\overline{\Delta' M_{18}}\equiv(1-\tilde{\mathbf{I}}_{1237}-\mathbf{K}_{456})(1-\tilde{\mathbf{I}}_{134567}-\mathbf{K}_2)(1-\tilde{\mathbf{I}}_{123467}-\mathbf{K}_5)M_{18}.$$

"Power counting" in the limit defined by $\mathbf{I}_{137}$ indicates a linear IR divergence. We deal with this unpleasant divergence by avoiding it. Rather than working with $\overline{\Delta M_{18}}$ directly it is found to be easier to deal with the quantity

$$\Delta M_{18}\equiv\overline{\Delta' M_{18}}-\frac{1}{2}J_{6B}M_2-(1-\mathbf{I}_{all}(1-\mathbf{K}_2))M_{4b(1^*)}[f]\Delta\delta m_{4b}+I_{2^*}\Delta\delta m_{4b}M_2, \quad (4.42)$$

where J_{6B} is defined in (3.74). ΔM_{18} is IR-finite. The last three terms of (4.42) can be included in the computer code for $\overline{\Delta' M_{18}}$ to give ΔM_{18}, which is then the quantity numerically calculated. I_{2^*} is defined by (3.42).

Carrying out the divergence isolating operations in $\overline{\Delta' M_{18}}$ we obtain the intermediate renormalization relation between M_{18} and ΔM_{18} :

$$\begin{aligned}
M_{18}=\;&\Delta M_{18}+\delta m_2 M_{6B(1^*)}+\hat{B}_2 M_{6B}+\delta m_2 M_{6A(4^*)}+\hat{B}_2 M_{6A}\\
&+\delta\tilde{m}_{4b}M_{4b(3^*)}+\hat{B}_{4b}M_{4b}-\delta m_2(\delta m_2 M_{4b(1^*,2^*)}+\hat{B}_2 M_{4b(1^*)})\\
&-\hat{B}_2(\delta m_2 M_{4b(2^*)}+\hat{B}_2 M_{4b})-\delta m_2(\delta\tilde{m}_{4b}M_{2^{**}}+\hat{B}_{4b}M_{2^*})\\
&-\hat{B}_2(\delta\tilde{m}_{4b}M_{2^*}+\hat{B}_{4b}M_2)-\delta m_2\delta\tilde{m}_{2^*}M_{4b(3^*)}\\
&-\hat{B}_2(\delta\tilde{m}_2' M_{4b(3^*)}+\hat{B}_2' M_{4b})\\
&+(\delta m_2)^2\delta\tilde{m}_{2^*}M_{2^{**}}+\delta m_2\hat{B}_2(\delta\tilde{m}_2' M_{2^{**}}+\hat{B}_2' M_{2^*})\\
&+\hat{B}_2\delta m_2\delta\tilde{m}_{2^*}M_{2^*}+(\hat{B}_2)^2(\delta\tilde{m}_2' M_{2^*}+\hat{B}_2' M_2)\\
&+I_{6B1}M_2+(I_{6A2}+I_{4s}I_2)M_2+I_{4s}\Delta M_{4b}\\
&+\frac{1}{2}J_{6B}M_2+I_{4b(1^*)}\Delta\delta m_{4b}-I_{2^*}M_2\Delta\delta m_{4b},
\end{aligned}$$

while standard renormalization relates M_{18} to a_{18} by

$$\begin{aligned}
a_{18} = M_{18} &- \delta m_2 M_{6B(1^*)} - B_2 M_{6B} - \delta m_2 M_{6A(4^*)} - B_2 M_{6A} \\
&- \delta m_{4b} M_{4b(3^*)} - B_{4b} M_{4b} + \delta m_2(\delta m_2 M_{4b(1^*,2^*)} + B_2 M_{4b(1^*)}) \\
&+ B_2(\delta m_2 M_{4b(2^*)} + B_2 M_{4b}) + \delta m_2(\delta m_{4b} M_{2^{**}} + B_{4b} M_{2^*}) \\
&+ B_2(\delta m_{4b} M_{2^*} + B_{4b} M_2) + \delta m_2(\delta m_{2^*} M_{4b(3^*)} + B_{2^*} M_{4b}) \\
&+ B_2(\delta m_2 M_{4b(3^*)} + B_2 M_{4b}) \\
&- (\delta m_2)^2(\delta m_{2^*} M_{2^{**}} + B_{2^*} M_{2^*}) - \delta m_2 B_2(\delta m_2 M_{2^{**}} + B_2 M_{2^*}) \\
&- B_2 \delta m_2(\delta m_{2^*} M_{2^*} + B_{2^*} M_2) - (B_2)^2(\delta m_2 M_{2^*} + B_2 M_2).
\end{aligned}$$

The procedure for M_{16} is analogous to that for M_{18}. The problematic IR divergence in M_{16} is again the one that is isolated by $\mathbf{I}_{137}$. Here, we find that

$$\Delta M_{16} \equiv \overline{\Delta' M_{16}} - \frac{1}{2} J_{6C} M_2 - (1 - \mathbf{I}_{all}(1 - \mathbf{K}_2)) M_{4b(1^*)}[f] \Delta\delta m_{4a} + I_{2^*} \Delta\delta m_{4a} M_2 \quad (4.43)$$

is free from IR divergences (and finite), where

$$\overline{\Delta' M_{16}} \equiv (1 - \tilde{\mathbf{I}}_{1237} - \mathbf{K}_{456})(1 - \tilde{\mathbf{I}}_{134567} - \mathbf{K}_2)(1 - \mathbf{K}_{45})(1 - \mathbf{K}_{56}) M_{16},$$

and

$$\frac{1}{2} J_{6C} = (1 - \mathbf{K}_{234})(1 - \mathbf{K}_{23})(1 - \mathbf{K}_{34}) \left(-\frac{1}{32} \int (dz) \frac{f_1}{U^3 V^2} \right), \quad (4.44)$$

in which (dz) is given by (3.73) and

$$f_1 = -16[B_{23}(2 - A_4) + 2B_{24}(1 - 2A_3) + B_{34}(2 - A_2)].$$

For M_{08} and M_{10}, the IR divergence associated with $\mathbf{I}_{37}$ is linear. We avoid dealing with it in a manner analogous to M_{16} and M_{18} by considering

$$\begin{aligned}
\Delta M_{08} &\equiv \overline{\Delta' M_{08}} - \tilde{\mathbf{I}}_1 (1 - \mathbf{I}_{all}) M_{4a(1^*)} \Delta\delta m_{4a} \\
\Delta M_{10} &\equiv \overline{\Delta' M_{10}} - \tilde{\mathbf{I}}_1 (1 - \mathbf{I}_{all}) M_{4a(1^*)} \Delta\delta m_{4b},
\end{aligned}$$

where

$$\begin{aligned}
\overline{\Delta' M_{08}} &= (1 - \tilde{\mathbf{I}}_{1237} - \mathbf{K}_{456})(1 - \mathbf{K}_{12})(1 - \mathbf{K}_{45})(1 - \mathbf{K}_{56})(1 - \mathbf{K}_{234567}) M_{08}, \\
\overline{\Delta' M_{10}} &= (1 - \tilde{\mathbf{I}}_{1237} - \mathbf{K}_{456})(1 - \tilde{\mathbf{I}}_{123467} - \mathbf{K}_5)(1 - \mathbf{K}_{12})(1 - \mathbf{K}_{234567}) M_{10}.
\end{aligned}$$

Collecting a_1 through a_{47} we find

$$\begin{aligned}
a_e^{(8)}(\text{Fig.12}) = \Delta M^{(8)} &- 5\Delta M^{(6)} \Delta B_2 \\
&- \Delta M^{(4)}[4\Delta L^{(4)} + 3\Delta B^{(4)} - 9(\Delta B_2)^2] - \Delta M^{(4^*)} \Delta\delta m^{(4)} \\
&- M_2[2\Delta L^{(6)} + \Delta B^{(6)} - \Delta\delta m^{(4)}(4\Delta L_{2^*} + \Delta B_{2^*} - B_{2^*}[I])] \quad (4.45) \\
&- (M_{2^*} - M_{2^*}[I])[\Delta\delta m^{(6)} - \Delta\delta m^{(4)}(5\Delta B_2 + \Delta\delta m_{2^*})] \\
&+ M_2 \Delta B_2[10\Delta L^{(4)} + 6\Delta B^{(4)} - 5(\Delta B_2)^2],
\end{aligned}$$

where

$$\Delta M^{(8)} = \sum_{i=1}^{47} \eta_i \Delta M_i, \quad \Delta M^{(6)} = \sum_{\alpha=A}^{H} \eta_\alpha \Delta M_{6\alpha},$$
$$\Delta M^{(4^*)} = 2\Delta M_{4a(1^*)} + \Delta M_{4a(2^*)} + 2\Delta M_{4b(1^*)} + \Delta M_{4b(2^*)},$$
$$\Delta L^{(6)} = \sum_{\alpha=A}^{H} \eta_\alpha \sum_{i=1}^{5} \Delta L_{6\alpha i}, \quad \Delta B^{(6)} = \sum_{\alpha=A}^{H} \eta_\alpha \Delta B_{6\alpha}, \tag{4.46}$$
$$\Delta \delta m^{(6)} = \sum_{\alpha=A}^{H} \eta_\alpha \Delta \delta m_{6\alpha}.$$

In addition, $\Delta M^{(4)}$, $\Delta\delta m^{(4)}$, $\Delta B^{(4)}$, $\Delta L^{(4)}$ are defined in (4.33), and $\Delta\delta m_{2^*}$, $B_{2^*}[I]$, $M_{2^*}[I]$ are defined by (3.40), (3.77), and (4.24), respectively.

The programming and numerical evaluation of the 47 integrals of group V are by far the most difficult and time-consuming part of the whole calculation. The integrands, including UV and IR subtraction terms, are generated by SCHOONSCHIP. All of them are constructed in more than two different forms and checked against each other by numerical spot samplings. Furthermore, subtraction terms generated by $\mathbf{K}_S$ and $\mathbf{I}_R$ operations from different integrands must agree with each other, providing a valuable cross check.

The largest integrands consist of more than 20,000 complicated rational functions constructed from ten integration variables, their FORTRAN codes occupying up to 350 kilobytes of memory. This bulkiness prevents us from evaluating the integrals with high precision except on the fastest computers, and is responsible for the enormous amount of computing time required, spread out over a period of more than seven years. Numerically integrated values of all quantities in (4.45) and (4.46) are listed in Tables XIII and XIV and XI.

Most integrals have been evaluated in double precision. However, in some cases (M12, M16, M18), it was necessary to evaluate the integral in quadruple precision in a small domain around some singularity, and in double precision in the rest of the domain. For instance, for M_{12} in Table XIII, this is indicated in the column 7 by "10,1.5" which means that 10^7 double precision and 1.5×10^6 quadruple precision integrand samplings were made. Such a caution was required in order to deal with severe round-off errors. This problem is intrinsic to our particular approach to renormalization in which individual terms may diverge on some boundaries of the integration domain and the integral as a whole is made finite only as a consequence of point-by-point cancellation of divergences by carefully tailored counterterms. This would cause no problem if each step of the computation were carried out with infinite precision. In reality we have to perform calculations in finite precision. The intended cancellation may fail occasionally because cancelling terms have no more than 12 or 13 significant digits (in double precision) and their difference tends to be dominated near a singularity by round-off errors causing undesirable fluctuations. As a matter of fact, it is fortunate that this problem has not been a severe one in most cases, requiring adoption of quadruple precision arithmetic only in a small number of cases. However, some other integrals in Table XIII may have to be scrutinized for possible signs of round-off errors if one should want to reduce their errors further.

Substituting the results of Tables XIII and XIV in (4.45) one finds

$$a_e^{(8)}(\text{Fig.12}) = -1.934\,4(1370). \tag{4.47}$$

At present there is no analytic result (except for M_{12}) to compare with this value.

Table XIII. Magnetic moment contribution of eighth-order diagrams of Fig. 12. $\eta_\alpha = 1$ (2) for time-reversal symmetric (asymmetric) diagram. N is the number of sampling points per iteration in units of 10^6. I is the number of iteration. All calculations are in double precision except for the cases where a pair of numbers are assigned to N and I. In both columns the second numbers refer to quadruple precision. See text for details.

Diagram	$\eta_\alpha \Delta M_\alpha$	N	I	Diagram	$\eta_\alpha \Delta M_\alpha$	N	I
M01	−0.085 1(235)	8	27	M02	−0.351 8(264)	22	30
M03	−0.764 3(254)	10	27	M04	4.643 3(349)	13	36
M05	2.333 1(153)	8	22	M06	−1.396 1(128)	8	29
M07	0.119 5(108)	8	33	M08	−6.726 3(295)	15	33
M09	0.841 8(299)	10	35	M10	18.062 6(339)	40	30
M11	2.471 8(182)	8	30	M12	−4.146 1(179)	10,1.5	35,30
M13	−6.620 1(177)	10	20	M14	2.592 4(183)	8	33
M15	1.048 7(176)	10	33	M16	3.266 2(341)	18,1.2	26,32
M17	3.137 5(195)	16	43	M18	12.370 8(188)	18,2	36,26
M19	−0.832 1(63)	4	27	M20	0.691 2(77)	4	29
M21	0.218 9(48)	4	27	M22	−0.236 7(135)	4	27
M23	−4.379 7(191)	8	32	M24	2.141 9(187)	8	31
M25	0.078 9(52)	4	27	M26	1.785 8(168)	8	31
M27	0.950 6(233)	10	31	M28	−4.769 3(261)	20	34
M29	3.048 2(120)	8	33	M30	−3.774 8(273)	10	27
M31	3.011 6(93)	4	27	M32	−2.359 9(76)	4	25
M33	−1.188 1(57)	4	27	M34	1.311 9(95)	4	33
M35	−0.524 7(96)	4	25	M36	−0.258 4(160)	10	27
M37	0.487 0(51)	4	27	M38	−3.041 6(132)	8	34
M39	−0.622 3(98)	4	34	M40	5.095 7(213)	10	24
M41	−1.453 1(225)	8	27	M42	−4.757 4(258)	16	41
M43	−2.370 2(108)	4	31	M44	3.463 4(169)	12	28
M45	0.740 3(128)	10	45	M46	−6.698 2(118)	10	29
M47	8.243 5(223)	20	25				

Table XIV. Auxiliary integrals.
In this table we denote $\Delta M_{4a^*} \equiv 2\Delta M_{4a(1^*)} + \Delta M_{4a(2^*)}$. Same for ΔM_{4b^*}.
$A_{LBD} \equiv \sum_{i=1}^{5} \Delta L_{6Ai} + \frac{1}{2}\Delta B_{6A} + 2\Delta\delta m_{6A}$. Similarly for B_{LBD}, etc.

Integral	Value	Integral	Value
ΔB_{2^*}	1.5	ΔL_{2^*}	−0.75
$B_{2^*}[I]$	−0.5	$\Delta\delta m_{2^*}$	−0.75
ΔM_{4a^*}	3.646 5(88)	ΔM_{4b^*}	10.145 4(52)
A_{LBD}	1.366 1(82)	B_{LBD}	3.560 9(86)
C_{LBD}	−6.768 8(113)	$2D_{LBD}$	−8.762 2(190)
E_{LBD}	−1.592 8(94)	F_{LBD}	2.618 1(139)
$2G_{LBD}$	3.869 6(124)	H_{LBD}	−0.232 2(63)
$\Delta M^{(6)}$	4.465 01(144)		

5. Electron Anomalous Moment with Closed Electron Loops

In this section we consider diagrams that have closed electron loops. One diagram of fourth order, 22 diagrams of sixth order, and 373 diagrams of eighth order belong to this category. In dealing with terms of up to eighth order, only two types of closed electron loops have to be considered; one is of the vacuum-polarization (abbreviated as v-p) type, being attached to two photon lines, and the other of the light-by-light scattering (abbreviated as l-b-l) type, being attached to four photon lines.

5.1. Feynman-parametric form of the photon spectral function

The effect of v-p loops on the electron anomaly can be calculated most easily using the spectral representation of the photon propagator. This is because insertion of a v-p loop in a photon line amounts to giving a (variable) mass to the photon. To take advantage of this fact, let us begin by writing the contribution to a_e from a second-order vertex diagram containing a virtual photon of mass μ in the form

$$M^{(2)}(\mu) = \int_0^1 dy(1-y)\frac{1}{W},$$

where

$$W = 1 + \mu^2 \frac{1-y}{y^2}.$$

The effect of insertion of a gauge-invariant set of closed electron loops in the internal photon line is expressed by the renormalized vacuum polarization tensor

$$\Pi^{\mu\nu}(q) = (q^\mu q^\nu - g^{\mu\nu} q^2)\Pi(q^2), \tag{5.1}$$

with

$$\Pi(q^2) = -q^2 \int_0^1 dt \frac{\rho(t)}{q^2 - 4(1-t^2)^{-1}}, \quad \Pi(0) = 0. \tag{5.2}$$

Since this can be regarded as a superposition of propagators of mass $4(1-t^2)^{-1}$, its contribution to a_e can be written in the form

$$M_{2,P} = \int_0^1 dy(1-y) \int_0^1 dt \frac{\rho(t)}{W_t}, \tag{5.3}$$

where

$$W_t = 1 + \frac{4}{1-t^2} \frac{1-y}{y^2}.$$

If n such loops are inserted sequentially in the same photon line one obtains

$$M_{2,P:n} = \int_0^1 dy(1-y) \left(\int_0^1 dt \frac{\rho(t)}{W_t} \right)^n. \tag{5.4}$$

These formulas enable us to calculate the contribution of v-p loops to a_e once the spectral function $\rho(t)$ is known. For second order we have

$$\rho_2(t) = \frac{t^2(1-\frac{1}{3}t^2)}{1-t^2}. \tag{5.5}$$

The fourth-order spectral function is also known analytically.[43] Unfortunately no explicit formula is known for higher orders. The only practical way to deal with the sixth-order spectral function at present is to make use of the parametric integral representation.

In order to see how to construct such parametric forms, let us first examine the parametric forms of second- and fourth-order v-p loops and their contribution to a_e when inserted in a second-order vertex. Although analytic results are known in these cases, this is a useful exercise for understanding the general procedure. Furthermore, these parametric formulas are needed later in the renormalization of integrals of higher orders. For the second-order electron loop of Fig. 14(a) we can write the renormalized vacuum polarization tensor in the form (5.1) where

$$\Pi(q^2) = \Pi^{(2)}(q^2) = \int (dz) \frac{D_0}{U^2} \ln\left(\frac{V_0}{V}\right), \tag{5.6}$$

with

$$(dz) = dz_1 dz_2 \delta(1 - z_{12}), \quad z_1,\ z_2 \geq 0,$$
$$U = z_{12}, \quad A_1 = z_2/U, \quad V_0 = z_{12},$$
$$G = z_1 A_1, \quad V = V_0 - q^2 G, \quad D_0 = 2A_1(1-A_1).$$

Eq. (5.6) can be recast in the form

$$\Pi^{(2)}(q^2) = q^2 \int (dz) \int_0^1 dt \frac{D_0 G}{U^2 V_t}, \tag{5.7}$$

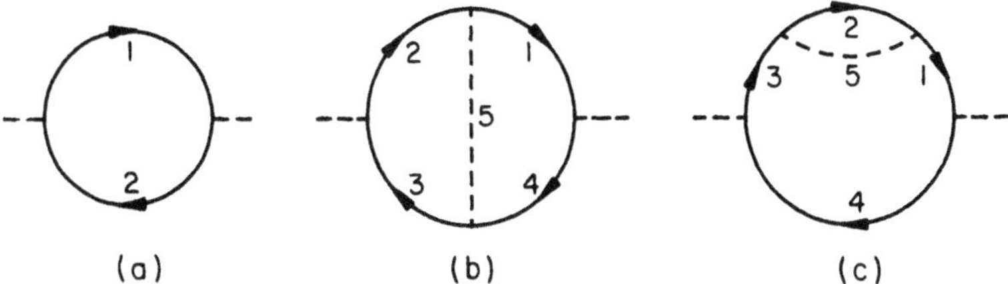

Figure 14: Line numbering for the vacuum polarization diagrams of second and fourth orders.

with

$$V_t = V_0 - tq^2 G.$$

Noting the similarity of (5.7) and (5.2), one may readily conclude that the contribution of (5.7) to the electron anomaly is

$$M_{2,P2} = \int_0^1 dy(1-y) \int (dz) \frac{D_0}{U^2} \ln\left(\frac{W}{W-1}\right), \tag{5.8}$$

where

$$W = 1 + \frac{V_0}{G}\frac{1-y}{y^2}.$$

With the change of variable $z_1 = \frac{1}{2}(1+t)$ and integration by parts, (5.8) becomes (5.3). In the form (5.8) the relation between $M_{2,P2}$ and $\Pi^{(2)}$ is transparent. Furthermore it can be readily generalized to higher-order cases.

The fourth-order vacuum polarization tensor has contibutions from the diagram of Fig. 14(b) and from two diagrams of the type shown in Fig. 14(c). The former contribution, normalized at $q = 0$ but with subvertex divergences not yet removed, can be written as

$$\Pi^{(4a)}(q^2) = \int (dz) \left[\frac{D_0}{U^2}\left(\frac{1}{V} - \frac{1}{V_0}\right) + \frac{D_1}{U^3} \ln\left(\frac{V_0}{V}\right)\right], \tag{5.9}$$

where

$$\begin{aligned}
&B_{11} = z_{235}, \quad B_{12} = z_5, \quad B_{22} = z_{145}, \quad U = z_{14}B_{11} + z_{23}B_{12}, \\
&A_1 = (z_3 B_{12} + z_4 B_{11})/U, \quad A_2 = (z_3 B_{22} + z_4 B_{12})/U, \\
&A_3 = A_2 - 1, \quad A_4 = A_1 - 1, \\
&V_0 = z_{1234}, \quad G = z_1 A_1 + z_2 A_2, \quad V = V_0 - q^2 G, \\
&D_0 = (A_1 + A_4)(A_2 + A_3) - A_1 A_4 - A_2 A_3, \\
&D_1 = (A_1 A_2 + A_3 A_4) B_{12} - A_1 A_4 B_{22} - A_2 A_3 B_{11}.
\end{aligned} \tag{5.10}$$

This integral contains UV divergences arising from the vertex subdiagrams [2,3,5] and [1,4,5]. The divergent terms can be isolated by applying $\mathbf{K}_{23}$ and $\mathbf{K}_{14}$ operations to the integrand of (5.9). This leads to

$$\mathbf{K}_{23}\Pi^{(4a)} = \int (dz) \frac{D_1'}{U'^3} \ln \frac{V_0'}{V'}, \tag{5.11}$$

where

$$\begin{aligned} &B'_{22} = z_{14}, \quad U' = z_{235}B'_{22}, \quad V'_0 = z_{14} + z_{23}^2/z_{235}, \\ &A'_1 = z_4/z_{14}, \quad A'_4 = A'_1 - 1, \\ &D'_1 = -A'_1A'_4B'_{22}, \quad G' = z_1A'_1, \quad V' = V'_0 - q^2G'. \end{aligned} \tag{5.12}$$

A similar formula holds for $\mathbf{K}_{14}\Pi^{(4a)}$. It is easily seen that (5.11) factorizes as

$$\mathbf{K}_{23}\Pi^{(4a)} = \hat{L}_2\Pi^{(2)}, \tag{5.13}$$

where $\hat{L}_2$, given by (3.30), is the UV-divergent part of the renormalization constant L_2. Similarly for $\mathbf{K}_{14}\Pi^{(4a)}$.

Making use of (5.13) we can rewrite the standard renormalization equation

$$\Pi^{(4a)}_{ren} = \Pi^{(4a)} - 2L_2\Pi^{(2)}$$

as

$$\Pi^{(4a)}_{ren} = \Delta\Pi^{(4a)} - 2\tilde{L}_2\Pi^{(2)}, \tag{5.14}$$

where $\tilde{L}_2$ is given by (3.31) and

$$\Delta\Pi^{(4a)} = (1 - \mathbf{K}_{23} - \mathbf{K}_{14})\Pi^{(4a)}$$

is free from UV divergences.

The renormalized vacuum polarization term due to Fig. 14(c) is given by

$$\Pi^{(4b)}_{ren} = \Delta\Pi^{(4b)} - \tilde{B}_2\Pi^{(2)}, \tag{5.15}$$

where $\tilde{B}_2$ is defined in (3.36) and

$$\Delta\Pi^{(4b)}(q^2) = \int (dz)(1 - \mathbf{K}_2)\left[\frac{D_0}{U^2}\left(\frac{1}{V} - \frac{1}{V_0}\right) + \frac{q^2C_0}{U^2V} + \frac{D_1}{U^3}\ln\left(\frac{V_0}{V}\right)\right], \tag{5.16}$$

with

$$\begin{aligned} &B_{11} = z_{25}, \quad B_{12} = z_5, \quad B_{22} = z_{1345}, \quad U = z_{134}B_{11} + z_2B_{12}, \\ &A_1 = z_4B_{11}/U, \quad A_2 = z_4B_{12}/U, \quad A_3 = A_1, \quad A_4 = A_1 - 1, \\ &G = z_{13}A_1 + z_2A_2, \quad V = V_0 - q^2G, \\ &D_0 = (4A_1 - A_2)A_4, \quad C_0 = -A_1^2A_2A_4, \quad D_1 = B_{12}A_1(A_1 + 3A_4). \end{aligned}$$

V_0 and (dz) are the same as in (5.9).

It is now straightforward to write down the parametric integrals for the sixth-order v-p loop diagrams of Fig. 15. We have

$$\Pi^{(6)}_{ren} = \Delta\Pi^{(6)}(q^2) + \text{ residual renormalization terms,}$$

where

$$\begin{aligned} \Delta\Pi^{(6)}(q^2) = \frac{1}{2}\int (dz)\prod_S(1 - \mathbf{K}_S)&\left[\frac{D_0}{U^2}\left(\frac{1}{V^2} - \frac{1}{V_0^2}\right) + \frac{q^2}{U^2V^2}(B_0 + q^2C_0)\right. \\ &\left. + \frac{D_1}{U^3}\left(\frac{1}{V} - \frac{1}{V_0}\right) + \frac{q^2B_1}{U^3V} + \frac{D_2}{U^4}\ln\left(\frac{V_0}{V}\right)\right] \end{aligned} \tag{5.17}$$

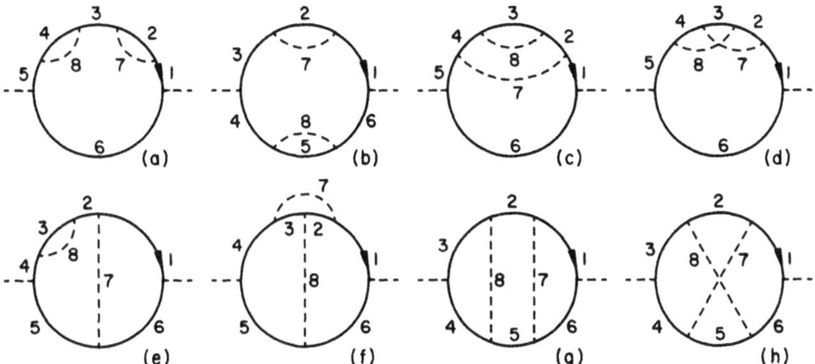

Figure 15: Line numbering for the vacuum polarization diagrams of sixth order.

with

$$V_0 = z_{123456}, \quad V = V_0 - q^2 G,$$

G being determined by the individual diagram. The functions D_0, D_1, D_2, B_0, B_1, and C_0 are obtained using the algebraic manipulation program SCHOONSCHIP[41] and are listed in the Appendix of Ref. 45. The $\mathbf{K}_S$ operations for all UV-divergent subdiagrams S must be carried out in (5.17). The incorporation of these results in the electron anomaly will be discussed in Sec. 5.4.

5.2. Fourth-order term

In fourth order the only contribution to a_e arises from the diagram shown at the right end of Fig. 2, obtained by inserting a second-order electron vacuum polarization loop in the second-order electron vertex. This contribution, $M_{2,P2}$, can be calculated from (5.3) with the spectral function ρ given by (5.5). The numerical evaluation, listed in Table XV, is in agreement with the analytic result:

$$\begin{aligned} a_{2,P2} &= \left(\frac{119}{36} - 2\zeta(2)\right)\left(\frac{\alpha}{\pi}\right)^2 \\ &= 0.015\,687\,\cdots\left(\frac{\alpha}{\pi}\right)^2, \end{aligned} \tag{5.18}$$

where ζ is the Riemann ζ function.

5.3. Sixth-order terms

The 22 sixth-order diagrams contributing to a_e fall naturally into four types of gauge-invariant sets according to the way closed electron loops appear in them:

Type a. Three diagrams containing a fourth-order v-p subgraph.

Type b. **A diagram containing two second-order v-p loops.**
Type c. **Twelve diagrams containing a second-order v-p loop.**
Type d. **Six diagrams containing a light-by-light scattering subgraph.**

Table XV. Magnetic moment contributions from sixth-order vertex diagrams with various closed electron loop insertions. $\eta_\alpha = 1$ (2) for time-reversal symmetric (asymmetric) diagrams.

Diagram	$\eta_\alpha \Delta M_\alpha$	No. of function calls (in units of 10^6)	No. of iterations
$M_{2,P2}$	0.015 688(3)	1	8
$M_{2,P4a}$	0.034 550(7)	2	20
$M_{2,P4b}$	0.041 861(15)	2	20
$M_{2,P2:2}$	0.002 560(1)	1	20
$M_{4a,P2}$	0.039 53(19)	2	30
$M_{4b,P2}$	−0.146 45(7)	2	30
M_{6LL}	0.371 12(28)	100	30

The type *a* contribution can be readily calculated from (5.3) using the Källén-Lehmann photon spectral function of fourth order. Instead, we shall pursue the parametric approach here. Going through the steps in parallel with (5.14) we find the contribution of type *a* coming from the fourth-order vacuum polarization loop of Fig. 14(b) to be

$$a_{2,P4a} = \Delta M_{2,P4a} - 2\tilde{L}_2 M_{2,P2}, \tag{5.19}$$

with

$$\begin{aligned}\Delta M_{2,P4a} = \int_0^1 dy(1-y)\int(dz)\Bigg[&\frac{D_0}{U^2 V_0 W} + \frac{D_1}{U^3}\ln\left(\frac{W}{W-1}\right) \\ &-\frac{D_1'}{U'^3}\ln\left(\frac{W'}{W'-1}\right) - \frac{D_1''}{U''^3}\ln\left(\frac{W''}{W''-1}\right)\Bigg],\end{aligned} \tag{5.20}$$

where most functions are defined in (5.10) and (5.12), and

$$W = 1 + \frac{V_0}{G}\frac{1-y}{y^2}, \quad W' = 1 + \frac{V_0'}{G'}\frac{1-y}{y^2}, \quad W'' = 1 + \frac{V_0''}{G''}\frac{1-y}{y^2}. \tag{5.21}$$

The double-primed functions are defined in analogy with (5.12) for the $\mathbf{K}_{14}$ operation.

The contribution to a_e of the two diagrams of the type Fig. 14(c) is given by

$$a_{2,P4b} = \Delta M_{2,P4b} - 2\tilde{B}_2 M_{2,P2}, \tag{5.22}$$

where

$$\Delta M_{2,P4b} = 2\int_0^1 dy(1-y)\int(dz)(1-\mathbf{K}_2)\left[\frac{D_0}{U^2V_0W}+\frac{C_0}{U^2GW}+\frac{D_1}{U^3}\ln\left(\frac{W}{W-1}\right)\right],$$

the functions in the integrand being identical with those of (5.16).

The total contribution of the diagrams of type a to the electron anomaly is thus

$$a_{2,P4} = a_{2,P4a} + a_{2,P4b} = \Delta M_{2,P4a} + \Delta M_{2,P4b} - 2\Delta B_2 M_{2,P2}, \tag{5.23}$$

where $\Delta B_2(=\frac{3}{4})$ is defined by (3.37). We have evaluated $\Delta M_{2,P4a}$ and $\Delta M_{2,P4b}$ numerically. The results are listed in Table XV. The value of $a_{2,P4}$ calculated from the formula (5.23) [0.052 88(3)] is in good agreement with the exact result 0.052 87 ⋯.[12]

The magnetic moment contribution from the diagram of type b (two second-order v-p loops inserted in a second-order vertex) can be readily computed using formula (5.4) for $n = 2$. The result, $M_{2,P2:2}$, is listed in Table XV.

The type c contribution of 12 vertex diagrams can be calculated once the fourth-order contributions M_{4a} and M_{4b} corresponding to the diagrams of Fig. 8 are known. For instance, the contribution of three diagrams obtained by inserting a v-p loop in "photon line 4" of M_{4a} can be written in the form

$$M_{4a,P2(4)} = \frac{1}{16}\int_0^1 dt\rho_2(t)\int(dz)\left(\frac{E_0+C_0}{U^2V}+\frac{N_0+Z_0}{U^2V^2}+\frac{N_1+Z_1}{U^3V}\right), \tag{5.24}$$

where

$$V = z_{123} - G + z_4\frac{4}{1-t^2},$$

all other quantities being the same as those for (4.5). UV divergences arising from the subdiagrams [1,2,4] and [2,3,5] can be subtracted as before, leading to a finite quantity

$$\Delta M_{4a,P2(4)} = (1-\mathbf{K}_{12}-\mathbf{K}_{23})M_{4a,P2(4)},$$

with

$$\mathbf{K}_{12}M_{4a,P2(4)} = \hat{L}_{2,P2}M_2, \quad \mathbf{K}_{23}M_{4a,P2(4)} = \hat{L}_2 M_{2,P2},$$

where $\hat{L}_{2,P2}$ is the UV-divergent part of the vertex renormalization constant $L_{2,P2}$. A similar result is obtained for insertion of a v-p loop in "photon line 5". [Replace (4) by (5) in the above formulas.] In terms of these results the standard-renormalized contribution to the electron anomaly can be written as

$$a_{4a,P2} = \Delta M_{4a,P2} - 2\tilde{L}_2 M_{2,P2} - 2\tilde{L}_{2,P2}M_2, \tag{5.25}$$

where $\Delta M_{4a,P2}$ is the sum of "photon 4" and "photon 5" insertions.

The contribution of the remaining six diagrams of type c obtained by inserting a v-p loop in "photon lines 4 and 5" of M_{4b} can be written in the same form as (5.24). After a similar divergence analysis one finds

$$a_{4b,P2} = \Delta M_{4b,P2} + (2\tilde{L}_2 - \Delta B_2)M_{2,P2} + (2\tilde{L}_{2,P2} - \Delta B_{2,P2})M_2, \tag{5.26}$$

where $\Delta M_{4b,P2}$ is the sum of "photon 4" and "photon 5" insertions. We therefore obtain

$$\begin{aligned} a_{4,P2} &= a_{4a,P2} + a_{4b,P2} \\ &= \Delta M_{4a,P2} + \Delta M_{4b,P2} - \Delta B_2 M_{2,P2} - \Delta B_{2,P2} M_2 . \end{aligned} \tag{5.27}$$

Numerically evaluated values of these integrals are listed in Table XV. They are in good agreement with the analytical results. Note that $\Delta B_{2,P2}$ here corresponds to $\Delta L_{2,P2} + \Delta B_{2,P2}$ in (4.7) of Ref. 24 and has the value $\frac{41}{24} - \frac{\pi^2}{6} = 0.063\,399\cdots$.

The contribution of the six diagrams containing a light-by-light scattering subdiagram (see Fig. 3) can be expressed in the form

$$M_{6LL} = 2 \times \frac{1}{32} \int (dz) \left(\frac{C_0}{2U^2V^2} + \frac{C_1}{2U^3V} + \frac{Z_0}{U^2V^3} + \frac{Z_1}{U^3V^2} + \frac{Z_2}{U^4V} \right) .. \tag{5.28}$$

Note that the factor 2 in (5.28) accounts for the contribution of charge-conjugated electron loops. Note also that N- and E-terms in the general formula (2.38) are absent in (5.28) since the second term in (2.29), which is derived from the Ward-Takahashi identity, vanishes because of Furry's theorem. Furthermore, differentiation with respect to q_ν on the right-hand-side of (2.29) means that the degree of UV divergence arising from the l-b-l scattering loop is effectively reduced by one. This is why the Pauli-Villars regularization is removed in (5.28). The renormalization is now trivial (the subtraction term vanishes by gauge invariance). The convergent integral (5.28) is evaluated by VEGAS and its value is listed in Table XV. It is in good agreement with the semi-analytic result 0.370 986 (20) obtained by Levine and Engelmann.[44]

The sum of the sixth-order contributions of the types $a,\ b,\ c,\ d$ is given by

$$a_e^{(6)}(a,b,c,d) = 0.276\ 18(35), \tag{5.29}$$

which is consistent with 0.276 243(20) obtained by analytic[12] and semi-analytic[44] means.

5.4. Eighth-order terms

There are 373 eighth-order diagrams that contribute to a_e. They fall naturally into four (gauge-invariant) groups according to the way closed electron loops appear in them:

Group I. Second-order vertex diagrams containing v-p loops of second, fourth, and sixth orders.

Group II. Fourth-order vertex diagrams containing v-p loops of second and fourth orders.

Group III. Sixth-order vertex diagrams containing a v-p loop of second order.

Group IV. Vertex diagrams containing a l-b-l scattering subdiagram with further radiative corrections of various kinds.

Group I

These diagrams can be classified further into the following gauge-invariant subgroups:

Subgroup I(a). This subgroup consists of one diagram obtained by inserting three second-order v-p loops in a second-order vertex. The contribution of this diagram to the

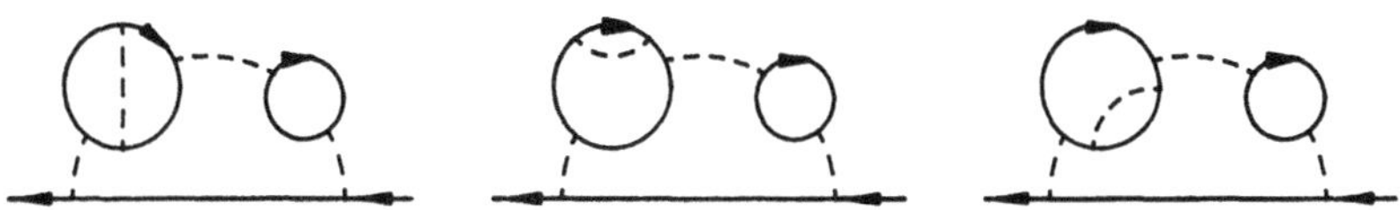

Figure 16: Three of the diagrams contributing to subgroup I(b). The other three are obtained from these by time reversal.

electron anomaly, denoted as $a^{(8)}_{I(a)} \equiv M_{2,P2:3}$, is given by the formula (5.4) with $n = 3$. The result of numerical evaluation is listed in Table XVI.

Subgroup I(b). Diagrams obtained by inserting one second-order and one fourth-order v-p loop in a second-order vertex. Six diagrams belong to this group. Some of them are shown in Fig. 16. The contribution of this subgroup to the electron anomaly, denoted as $a^{(8)}_{I(b)} \equiv M_{2,P2,P4}$, can be obtained from (5.4) choosing $n = 2$ and $\rho = \rho_2 + \rho_4$, and picking up the $\rho_2\rho_4$ term. The result of numerical integration is listed in Table XVI.

Subgroup I(c). Vertex diagrams containing two closed electron loops – one within the other. Three diagrams belong to this group. Some of them are shown in Fig. 17. The contributions from these diagrams can be readily calculated once the sixth-order contributions $a_{2,P4a}$ and $a_{2,P4b}$ of (5.19) and (5.22) are known. Consulting the formulas (5.3) and (5.20) we find that the diagram of Fig. 17(a) gives the renormalized contribution

$$a_{2,P(4a,P2)} = \Delta M_{2,P(4a,P2)} - 2\tilde{L}_{2,P2}M_{2,P2},$$

where

$$\begin{aligned}\Delta M_{2,P(4a,P2)} = \int_0^1 dy(1-y)\int_0^1 dt\rho_2(t)\int(dz)\Bigg[\frac{D_0}{U^2V_0W} \\ +\frac{D_1}{U^3}\ln\left(\frac{W}{W-1}\right) - \frac{D_1'}{U'^3}\ln\left(\frac{W'}{W'-1}\right) - \frac{D_1''}{U''^3}\ln\left(\frac{W''}{W''-1}\right)\Bigg],\end{aligned} \tag{5.30}$$

with

$$V_0 = z_{1234} + \frac{4}{1-t^2}z_5, \quad W = 1 + \frac{V_0}{G}\frac{1-y}{y^2}.$$

Similarly two diagrams of the type Fig. 17(b) contribute

$$a_{2,P(4b,P2)} = \Delta M_{2,P(4b,P2)} - 2\tilde{B}_{2,P2}M_{2,P2},$$

where

$$\begin{aligned}\Delta M_{2,P(4b,P2)} = 2\int_0^1 dy(1-y)\int_0^1 dt\rho_2(t)\int(dz)(1-\mathbf{K}_2) \\ \times\left[\frac{D_0}{U^2V_0W} + \frac{C_0}{U^2GW} + \frac{D_1}{U^3}\ln\left(\frac{W}{W-1}\right)\right].\end{aligned} \tag{5.31}$$

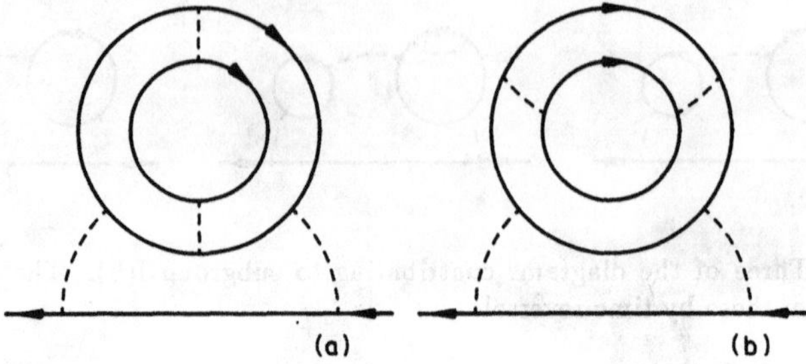

Figure 17: Two of the vertex diagrams containing two closed electron loops, one within the other. The third is obtained from the second by charge conjugation.

The total contribution of subgroup I(c) is

$$a^{(8)}_{I(c)} = \Delta M_{2,P(4a,P2)} + \Delta M_{2,P(4b,P2)} - 2\Delta B_{2,P2} M_{2,P2}.$$

The results of numerical integration are listed in Table XVI.

Table XVI. Magnetic moment contributions from the eighth-order diagrams of Group I. $\eta_\alpha = 1(2 \text{ or } 4)$ for time-reversal symmetric (asymmetric) diagrams.

Diagram	$\eta_\alpha \Delta M_\alpha$	No. of function calls (in units of 10^6)	No. of iterations
$M_{2,P2:3}$	0.000 876 8(2)	0.4	10
$M_{2,P2,P4}$	0.015 325(2)	0.4	10
$M_{2,P(4a,P2)}$	0.011 41(1)	1	20
$M_{2,P(4b,P2)}$	0.001 72(1)	1	20
$M_{2,P6A}$	0.044 44(3)	2	40
$M_{2,P6B}$	0.028 57(5)	1	20
$M_{2,P6C}$	−0.038 35(4)	1	20
$M_{2,P6D}$	−0.027 50(2)	2	20
$M_{2,P6E}$	0.179 29(8)	2	30
$M_{2,P6F}$	−0.061 96(4)	1	20
$M_{2,P6G}$	0.038 88(4)	1	20
$M_{2,P6H}$	0.023 66(2)	2	20

Subgroup I(d). This subgroup consists of diagrams obtained by inserting sixth-order (single electron loop) vacuum polarization loops in a second-order vertex. Fifteen diagrams

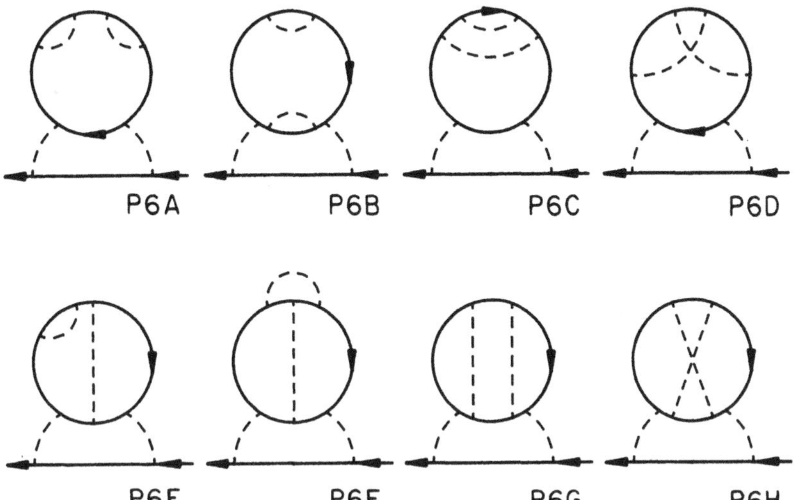

Figure 18: Eighth-order vertices obtained by insertion of sixth-order (single electron loop) vacuum polarization subdiagrams in a second-order vertex.

belong to this subgroup. Eight are shown in Fig. 18. Each of A, C, D, E, and F and the time-reversed diagram for E has a charge conjugated counterpart.

Given the structure of the sixth-order vacuum polarization (5.17), the contributions of these diagrams to the electron anomaly can be written in the form

$$
\begin{aligned}
a_{2,P6i} &= \Delta M_{2,P6i} \ + \text{ residual renormalization terms,} \\
\Delta M_{2,P6i} &= \frac{1}{2}\int_0^1 dy(1-y)\int (dz)\prod_S (1-\mathbf{K}_S) \\
&\quad \left[\frac{-1}{U^2W^2}\left(\frac{D_0}{V_0^2}(1-2W)+\frac{B_0}{GV_0}(1-W)+\frac{C_0}{G^2}\right)\right. \\
&\quad \left. + \ \frac{1}{U^3W}\left(\frac{D_1}{V_0}+\frac{B_1}{G}\right)+\frac{D_2}{U^4}\ln\left(\frac{W}{W-1}\right)\right]
\end{aligned}
\tag{5.32}
$$

where W is given by (5.21) in terms of the appropriate V_0 and G.

The integrals $\Delta M_{2,P6i}$ are free from divergences and can be evaluated numerically. Their values (multiplied by appropriate multiplicity factors η_α) are listed in Table XVI.

Summing up the contributions of diagrams A to H of Fig. 18 yields

$$
\begin{aligned}
a^{(8)}_{I(d)} = &\sum_{i=A}^{H}\eta_i\Delta M_{2,P6,i} - 4\Delta B_2\Delta M_{2,P4} + 5(\Delta B_2)^2 M_{2,P2} \\
&- 2(\Delta L^{(4)} + \Delta B^{(4)})M_{2,P2} - 2\Delta\delta m^{(4)} M_{2,P2^*},
\end{aligned}
\tag{5.33}
$$

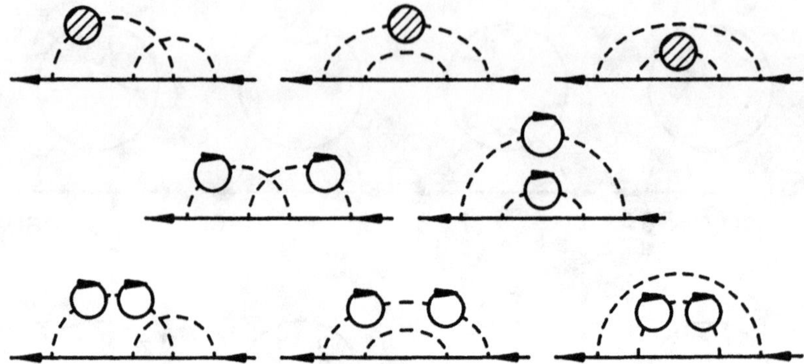

Figure 19: Eighth-order electron self-energy diagrams generated from the fourth-order vertex diagrams by insertion of various vacuum-polarization loops. Shaded circles represent the sum of all fourth-order vacuum polarization loops.

where

$$\eta_B = \eta_G = \eta_H = 1, \quad \eta_A = \eta_C = \eta_D = \eta_F = 2, \quad \eta_E = 4,$$
$$\Delta M_{2,P4} = \Delta M_{2,P4a} + \Delta M_{2,P4b},$$

and other quantities are defined in (4.46) and their values are listed in Tables XI and XV.

The total contribution of Group I diagrams is given by

$$a_I^{(8)} = 0.076\ 73(15). \tag{5.34}$$

Group II

These diagrams can be classified further into the following gauge-invariant subgroups:

Subgroup II(a). Diagrams obtained by inserting fourth-order vacuum polarization loops in fourth-order vertices. Twelve diagrams belong to this group. They are shown schematically in the top row of Fig. 19. The contribution of this subgroup to the electron anomaly, denoted as $a_{4,P4}$, can be obtained from (5.24) by replacing ρ_2 with ρ_4. The renormalized contribution to the electron anomaly from these diagrams can be written as

$$a_{II(a)}^{(8)} = a_{4a,P4} + a_{4b,P4},$$

where

$$a_{4a,P4} = \Delta M_{4a,P4} - 2\tilde{L}_2 M_{2,P4} - 2\tilde{L}_{2,P4} M_2, \tag{5.35}$$

$$a_{4b,P4} = \Delta M_{4b,P4} + (2\tilde{L}_2 - \Delta B_2) M_{2,P4} + (2\tilde{L}_{2,P4} - \Delta B_{2,P4}) M_2, \tag{5.36}$$

in parallel with (5.25) and (5.26). The results of numerical integration are listed in Table XVII.

Subgroup II(b). Diagrams obtained by inserting two second-order vacuum polarization loops in different photon lines of fourth-order vertices. Two diagrams belong to this group. They are shown in the middle row of Fig. 19. The contribution of the diagram on the left-end of this row can be expressen in the form

$$M_{4a,P2,P2} = \frac{1}{16}\int_0^1 ds\rho_2(s)\int_0^1 dt\rho_2(t)\int(dz)\left(\frac{E_0+C_0}{U^2V}+\frac{N_0+Z_0}{U^2V^2}+\frac{N_1+Z_1}{U^3V}\right),$$

where

$$V = z_{123} + \frac{4}{1-s^2}z_4 + \frac{4}{1-t^2}z_5 - z_1A_1 - z_2A_2 - z_3A_3,$$

the functions U, E_0, . . ., Z_1 being identical with those for M_{4a} of (4.5). A similar formula holds for $M_{4b,P2,P2}$, which corresponds to the diagram at the right-end of the row.

The contribution to the electron anomaly from this subgroup is thus given by

$$a^{(8)}_{II(b)} = \Delta M_{4a,P2,P2} + \Delta M_{4b,P2,P2} - \Delta B_{2,P2}M_{2,P2},$$

where

$$\Delta M_{4a,P2,P2} = (1-\mathbf{K}_{12}-\mathbf{K}_{23})M_{4a,P2,P2}, \quad \Delta M_{4b,P2,P2} = (1-\mathbf{K}_2-\tilde{\mathbf{I}}_{13})M_{4b,P2,P2}.$$

The results of numerical integration are listed in Table XVII.

Subgroup II(c). Diagrams obtained by inserting two second-order vacuum polarization loops in the same photon line of fourth-order vertices. Four diagrams belong to this group. Three of them are shown in the bottom row of Fig. 19. The contribution of these diagrams takes a slightly different form. For the diagram at the left-end we find

$$M_{4a,P2:2} = \frac{1}{16}\int_0^1 ds\rho_2(s)\int_0^1 dt\rho_2(t)\int(dz) \times\left[\left(\frac{E_0+C_0}{U^2}+\frac{N_1+Z_1}{U^3}\right)\frac{V}{V_sV_t}+\frac{N_0+Z_0}{U^2V_s^2V_t^2}\left(V^2-\frac{4}{1-s^2}\frac{4}{1-t^2}z_4^2\right)\right],$$

where all quantities are as for M_{4a} of (4.5) except that

$$V_s = V + \frac{4}{1-s^2}z_4, \quad V_t = V + \frac{4}{1-t^2}z_4.$$

Similar formulas hold for other diagrams of the row.

The eighth-order electron anomaly arising from the diagrams of this subgroup is then

$$a^{(8)}_{II(c)} = 2\Delta M_{4a,P2:2} + \Delta M_{4b,P2:2(4)} + \Delta M_{4b,P2:2(5)} - \Delta B_2M_{2,P2:2} - \Delta B_{2,P2:2}M_2.$$

where (4) and (5) refer to the photon lines in which the insertion have been made. The results of numerical integration are listed in Table XVII. Auxiliary integrals are listed in Table XVIII. From these one obtains the total contribution of Group II:

$$a^{(8)}_{II} = -0.521\ 68(24). \tag{5.37}$$

Note that $\Delta M_{4b,P4}$, $\Delta M_{4b,P2,P2}$, $\Delta M_{4b,P2:2}$ are defined differently from those of Ref. 25.

Table XVII. Magnetic moment contributions from the eighth-order diagrams of Group II. $\eta_\alpha = 1(2)$ for time-reversal symmetric (asymmetric) diagrams.

Diagram	$\eta_\alpha \Delta M_\alpha$	No. of function calls (in units of 10^6)	No. of iterations
$M_{4a,P4}$	0.131 61(19)	10	36
$M_{4b,P4}$	−0.420 62(10)	10	32
$M_{4a,P2,P2}$	−0.004 37(2)	2	32
$M_{4b,P2,P2}$	−0.022 34(1)	2	30
$M_{4a,P2:2}$	0.007 91(9)	4	32
$M_{4b,P2:2}$	−0.065 47(4)	2	30

Table XVIII. Auxiliary integrals – Group II.

Integral	Value	Integral	Value
$M_{2,P2}$	0.015 687 ⋯	$M_{2,P2^*}$	−0.012 702 ⋯
$\Delta B_{2,P2}$	0.063 399 ⋯	$\Delta B_{2,P4}$	0.183 76(3)
$M_{2,P2:2}$	0.002 558 ⋯	$\Delta B_{2,P2:2}$	0.027 92(1)

Group III

Diagrams belonging to this group can be generated from the 50 sixth-order diagrams (represented by the eight self-energy diagrams of Fig. 10) by insertion of a second-order vacuum-polarization loop in all possible ways. These 150 diagrams can therefore be represented by the eight independent integrals of the form

$$M_{6\alpha,P2} = -\frac{2}{64}\int_0^1 dt\, \rho_2(t) \int (dz) \sum_{i=6,7,8} \left(\frac{E_0 + C_0}{2U^2 V_i(t)^2} + \frac{E_1 + C_1}{2U^3 V_i(t)} + \frac{N_0 + Z_0}{U^2 V_i(t)^3} + \frac{N_1 + Z_1}{U^3 V_i(t)^2} + \frac{N_2 + Z_2}{U^4 V_i(t)} \right), \tag{5.38}$$

where $\alpha = A, B, , , , H$, and

$$V_i(t) = V + \frac{4}{1-t^2} z_i, \quad i = 6,\ 7,\ 8.$$

All other parametric functions are as given in (4.26).

Renormalization of $M_{6\alpha,P2}$ proceeds in the same way as that of $M_{6\alpha}$, and hence will not be repreated here. Summing over all diagrams in group III we can write the total contribution to the eighth-order anomalous moment of the electron as a sum of finite pieces:

$$
\begin{aligned}
a_{III}^{(8)} = & \sum_{\alpha=A}^{H} \eta_\alpha \Delta M_{6\alpha,P2} - 3\Delta B_{2,P2}\Delta M^{(4)} - 3\Delta B_2 \Delta M_{P2}^{(4)} \\
& + (M_{2^*,P2}[I] - M_{2^*,P2})\Delta\delta m^{(4)} + (M_{2^*}[I] - M_{2^*})\Delta\delta m_{P2}^{(4)} \\
& - M_{2,P2}(\Delta B^{(4)} + 2\Delta L^{(4)} - 2(\Delta B_2)^2) - M_2(\Delta B_{P2}^{(4)} + 2\Delta L_{P2}^{(4)} - 4\Delta B_2 \Delta B_{2,P2}),
\end{aligned}
\tag{5.39}
$$

where $\Delta M^{(4)}$, $\Delta\delta m^{(4)}$, $\Delta B^{(4)}$, and $\Delta L^{(4)}$ are given in Table XI and

$$
\begin{aligned}
\Delta M_{P2}^{(4)} &= \Delta M_{4a,P2(4)} + \Delta M_{4a,P2(5)} + \Delta M_{4b,P2(4)} + \Delta M_{4b,P2(5)}, \\
\Delta\delta m_{P2}^{(4)} &= \Delta\delta m_{4a,P2(4)} + \Delta\delta m_{4a,P2(5)} + \Delta\delta m_{4b,P2(4)} + \Delta\delta m_{4b,P2(5)}, \\
\Delta B_{P2}^{(4)} &= \Delta B_{4a,P2(4)} + \Delta B_{4a,P2(5)} + \Delta B_{4b,P2(4)} + \Delta B_{4b,P2(5)}, \\
\Delta L_{P2}^{(4)} &= \Delta L_{4x,P2(4)} + 2\Delta L_{4c,P2(4)} + \Delta L_{4x,P2(5)} + 2\Delta L_{4c,P2(5)} \\
&\quad + \Delta L_{4l,P2(4)} + 2\Delta L_{4s,P2(4)} + \Delta L_{4l,P2(5)} + 2\Delta L_{4s,P2(5)}.
\end{aligned}
$$

The numerals 4 and 5 within the parentheses refer to the photon lines in which vacuum-polarization loops are inserted. The results of numerical integration are listed in Table XIX. In the row for $M_{6G,P2}$, the entries 10,1 and 42,20 mean that one part of the integration domain is evaluated in double precision with 10^7 function calls for each of 42 iterations and the rest in quadruple precision with 10^6 function calls for each of 20 iterations.

From these results one obtains the total contribution of Group III:

$$
a_{III}^{(8)} = 1.418\,6(34). \tag{5.40}
$$

Table XIX. Magnetic moment contributions from the eighth-order diagrams of group III. $\eta_\alpha = 1$ (2) for time-reversal symmetric (asymmetric) diagrams.

Diagram	$\eta_\alpha \Delta M_\alpha$	No. of function calls (in units of 10^6)	No. of iterations
$M_{6A,P2}$	−0.440 3(3)	10	30
$M_{6B,P2}$	0.690 7(7)	10	40
$M_{6C,P2}$	0.638 2(7)	10	40
$M_{6D,P2}$	0.442 0(7)	10	35
$M_{6E,P2}$	0.432 1(4)	10	36
$M_{6F,P2}$	0.441 9(8)	10	36
$M_{6G,P2}$	1.140 4(10)	10,1	42,20
$M_{6H,P2}$	−0.797 5(12)	10	40

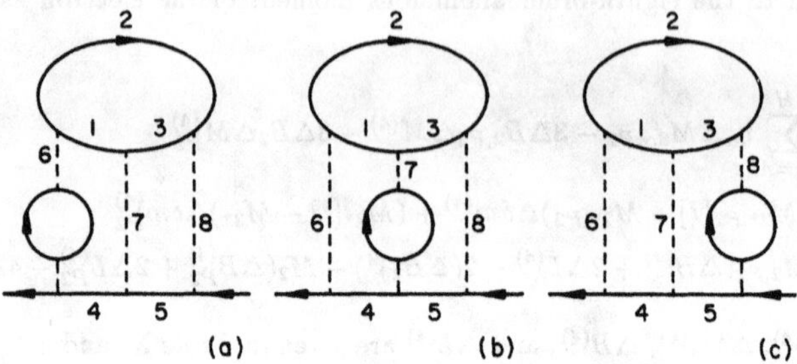

Figure 20: Line numbering for the self-energy diagrams, which generically represent the diagrams of subgroup IVa.

Table XX. Auxiliary integrals – Group III.

Integral	Value	Integral	Value
$\Delta B_{2,P2}$	0.063 399	$\Delta B^{(4)}_{P2} + 2\Delta L^{(4)}_{P2}$	0.091 3(22)
$M_{2*,P2}$	0.044 079(5)	$M_{2*,P2}[I]$	0.010 291(42)
$\Delta M^{(4)}_{P2}$	−0.106 9(8)	$\Delta\delta m^{(4)}_{P2}$	0.679 3(7)

Group IV

Group IV consists of all diagrams that contain a light-by-light scattering subdiagram. These diagrams can be conveniently divided into four subgroups. Each subgroup is gauge invariant and is composed of two equivalent sets of graphs, the first set being related to the second by charge conjugation (reversal of the direction of momentum flow) in the electron loop of the light-by-light scattering subdiagram.

Subgroup IVa consists of 18 diagrams that are obtained from the sixth-order diagrams containing a light-by-light scattering subdiagram by inserting a second-order vacuum-polarization loop in one of the three internal photon lines. Typical diagrams are shown in Fig. 20.

The contribution of the sum of all 18 diagrams of subgroup IVa can therefore be expressed in the form

$$
\begin{aligned}
M_{6LL,P2} &= 2 \times \frac{1}{32} \sum_{i=6,7,8} \int_0^1 dt \rho_2(t) \int (dz) \\
&\times \left(\frac{C_0}{2U^2 V_i(t)^2} + \frac{C_1}{2U^3 V_i(t)} + \frac{Z_0}{U^2 V_i(t)^3} + \frac{Z_1}{U^3 V_i(t)^2} + \frac{Z_2}{U^4 V_i(t)} \right),
\end{aligned}
\tag{5.41}
$$

where $V_i(t)$ is defined in the same way as in (5.38). The result of numerical integration of this integral is given in Table XXI.

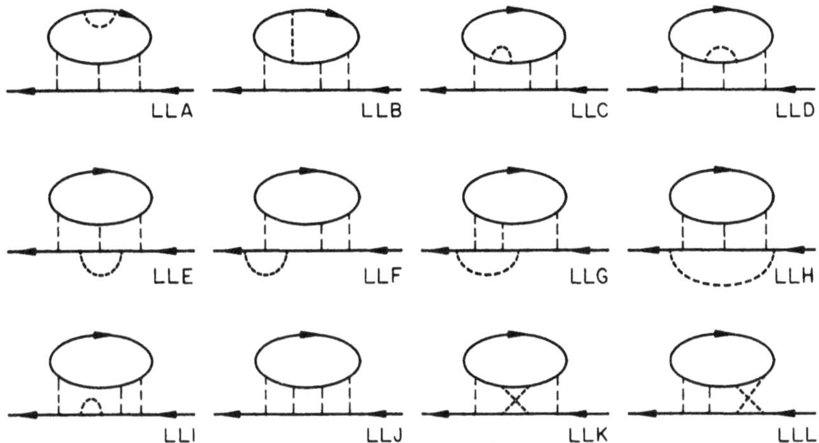

Figure 21: Self-energy diagrams that generically represent the diagrams of subgroups IVb, IVc, and IVd.

Subgroup IVb consists of 60 diagrams, each containing a sixth-order light-by-light subdiagram. They are generically represented by the self-energy diagrams LLA, LLB, LLC, LLD of Fig. 21. Subgroup IVc consists of 48 diagrams which are represented by the diagrams LLE, LLF, LLG, LLH, LLI of Fig. 21. In both subgroups contributions of N terms and E terms are absent because of Furry's theorem. Thus the parametric integrals of the unrenormalized magnetic moment projections for the subgroups IVb and IVc can be expressed in the form

$$\begin{aligned} M_{8LL\alpha} = -2 \times \frac{3!}{256} \int (dz) \Big(& \frac{C_0}{3U^2V^3} + \frac{C_1}{3U^3V^2} + \frac{C_2}{3U^4V} \\ & + \frac{Z_0}{U^2V^4} + \frac{Z_1}{U^3V^3} + \frac{Z_2}{U^4V^2} + \frac{Z_3}{U^5V} \Big), \end{aligned} \tag{5.42}$$

where $\alpha = A,\ B,\ ,\ ,\ ,\ , I$.

The divergence structure of diagrams of group IVb and IVc is straightforward. For instance, for the diagram LLA, the renormalized amplitude a_{8LLA} can be written as

$$a_{8LLA} = \Delta M_{8LLA} - \tilde{B}_2 M_{6LL},$$

where

$$\Delta M_{8LLA} = (1 - \mathbf{K}_3) M_{8LLA}.$$

Similarly, we find

$$a_{8LLB} = \Delta M_{8LLB} - \tilde{L}_2 M_{6LL},$$
$$a_{8LLC} = \Delta M_{8LLC} - \tilde{B}_2 M_{6LL},$$

$$a_{8LLD} = \Delta M_{8LLD} - \tilde{L}_2 M_{6LL},$$
$$a_{8LLE} = \Delta M_{8LLE} - \tilde{L}_2 M_{6LL},$$
$$a_{8LLF} = \Delta M_{8LLF} - \tilde{L}_2 M_{6LL},$$
$$a_{8LLI} = \Delta M_{8LLI} - \tilde{B}_2 M_{6LL}.$$

The integral M_{8LLG} is finite, hence

$$a_{8LLG} = M_{8LLG} = \Delta M_{8LLG}.$$

The integral M_{8LLH} has an IR divergence, which is isolated by the $\mathbf{I}_{47}$ operation. Thus we find

$$a_{8LLH} = \Delta M_{8LLH} + \tilde{L}_2 M_{6LL},$$

where

$$\Delta M_{8LLH} = (1 - \mathbf{I}_{47}) M_{8LLH}$$

is finite and

$$\mathbf{I}_{47} M_{8LLH} = \tilde{L}_2 M_{6LL}$$

holds exactly. The results of numerical integration of group IVb and IVc contributions are listed in Table XXI. The sum of all contributions of group IVb is given by

$$a^{(8)}_{IVb} = \sum_{\alpha=A}^{D} \eta_\alpha \Delta M_{8LL\alpha} - 3\Delta B_2 M_{6LL}.$$

Similarly for group IVc:

$$a^{(8)}_{IVc} = \sum_{\alpha=E}^{I} \eta_\alpha \Delta M_{8LL\alpha} - 2\Delta B_2 M_{6LL}.$$

Parametric integrals for the diagrams in subgroup IVd have the form

$$\begin{aligned} M_{8LL\alpha} = -2 \times \frac{3!}{256} \int (dz) \bigg(& \frac{E_0 + C_0}{3U^2V^3} + \frac{E_1 + C_1}{3U^3V^2} + \frac{E_2 + C_2}{3U^4V} \\ & + \frac{N_0 + Z_0}{U^2V^4} + \frac{N_1 + Z_1}{U^3V^3} + \frac{N_2 + Z_2}{U^4V^2} + \frac{N_3 + Z_3}{U^5V} \bigg), \end{aligned} \tag{5.43}$$

where $\alpha = J,\ K,\ L$. As is readily seen by power counting, the integrals M_{8LLJ}, M_{8LLK}, M_{8LLL} of subgroup IVd have UV divergences associated not only with the electron loop T =[4,5,6,7] but also with the sixth-order vertex subdiagrams R=[2,3,4,5,6,7,8,9,3′] and S=[1,2,4,5,6,7,8,9,1′]. It is necessary to renormalize all these divergences. It turns out that there is no advantage in resorting to the intermediate renormalization for the diagrams of subgroup IVd. Thus we have carried out the renormalization of divergences associated with the subdiagrams T, R, and S according to the standard prescription. Explicit construction of these renormalization terms was one of the most tedious and difficult problems we had to tackle throughout the entire calculation of the eighth-order term. When the contributions of renormalized amplitudes for LLJ, LLK,and LLL are summed up, the effects of renormalization terms cancel out and we obtain

$$a_{IVd}^{(8)} = \sum_{\alpha=J}^{L} \eta_\alpha a_{8LL\alpha}.$$

Summing all these results one finds the contribution of group IV diagrams to be

$$a_{IV}^{(8)} = \Delta M_{6LL,P2} + \sum_{\alpha=A}^{I} \eta_\alpha \Delta M_{8LL\alpha} - 5\Delta B_2 M_{6LL} + \sum_{\alpha=J}^{L} \eta_\alpha a_{8LL\alpha}. \quad (5.44)$$

Numerical values of these integrals have been evaluated in double precision using VEGAS. The results are listed in Table XXI. Substituting them in (5.44) we find

$$a_{IV}^{(8)} = -0.473(15). \quad (5.45)$$

Table XXI. Magnetic moment contributions from eighth-order vertex diagrams of group IV. $\eta_\alpha = 1$ (2) for time-reversal symmetric (asymmetric) diagrams.

Diagram	$\eta_\alpha \Delta M_\alpha$	No. of function calls (in units of 10^6)	No. of iterations
$6LL, P2$	0.599 9(18)	2	28
LLA	0.167 4(35)	4	34
LLB	−1.290 7(49)	8	30
LLC	2.832 1(55)	8	35
LLD	−0.055 9(30)	4	32
LLE	−0.265 3(21)	4	32
LLF	−0.761 1(43)	8	35
LLG	−0.549 8(62)	8	36
LLH	0.189 4(60)	12	36
LLI	0.802 1(41)	8	38
LLJ	2.125 1(30)	4	29
LLK	−2.098 0(39)	4	30
LLL	−0.777 4(34)	4	29

6. Comments on the Numerical Integration

6.1. Introductory remarks

Although analytic evaluation has been successful for the fourth-order terms and some of the sixth-order terms, the prospect for applying it to the eighth-order terms seems to be very dim in view of the enormous complexity and size of the eighth-order integrands. Thus, it is inevitable to rely on numerical integration techniques in order to extract useful information from the QED theory of the electron anomalous moment. The parametric formulation described in this article was developed to be particularly adaptable to numerical integration. In this sense the numerical integration algorithm is an integral component of our theory.

Of course there are many integration routines available. Some of them converge very rapidly but are applicable only to a very limited class of functions. Others are not effective in handling multi-dimensional integrals. What we need is an algorithm capable of handling very large integrands (some of them occupy several hundred kilobytes of source code) and large numbers of independent variables (up to 10 for the eighth-order terms). An algorithm satisfying such requirements is an iterative-adaptive Monte-Carlo integration routine that was originally devised by Sheppy and developed subsequently by Lautrup and Lepage into versions such as RIWIAD[46] and VEGAS[42]. Although I used RIWIAD extensively in the early stages of my work, I have relied on VEGAS more recently. This is mainly because a version of VEGAS that works on a vector processor is available.[f] In other respects RIWIAD and VEGAS work equally well, even though they are based on somewhat different adaptation criteria.

In view of the essential role VEGAS has played in providing a highly accurate evaluation of gigantic integrals, let me begin by describing how VEGAS works, borrowing heavily from Ref. 42.

6.2. Adaptive Monte-Carlo integration routine VEGAS

To see how VEGAS works let us first consider a one-dimensional integral:

$$I = \int_0^1 dx\ f(x). \tag{6.1}$$

If N points x are randomly chosen from a set of points distributed in the interval [0, 1] with density $p(x)$, the integral I is approximated by

$$S^{(1)} = \frac{1}{N} \sum_x \frac{f(x)}{p(x)} \tag{6.2}$$

which converges to I as $N \to \infty$, where the probability density function is normalized to

[f] I thank G. P. Lepage for providing me with a vectorized version of VEGAS.

unity:

$$\int_0^1 dx\; p(x) = 1.$$

The quantity $S^{(1)}$ will fluctuate about the true value of the integral as different sets of N random points are chosen. The variance of this fluctuation is given by

$$\sigma^2 = \frac{1}{N}\left(\int_0^1 dx\, \frac{f^2(x)}{p(x)} - \left[\int_0^1 dx\; f(x)\right]^2\right). \tag{6.3}$$

For large N, this quantity is approximated by

$$\sigma^2 \simeq \frac{S^{(2)} - (S^{(1)})^2}{N-1}, \tag{6.4}$$

where

$$S^{(2)} = \frac{1}{N}\sum_x \left(\frac{f(x)}{p(x)}\right)^2. \tag{6.5}$$

The standard deviation σ indicates the accuracy of $S^{(1)}$ as an estimate of I. Note that reliable estimates of the variance are possible only if the integral

$$\int_0^1 dx\, \frac{f^2(x)}{p(x)} \tag{6.6}$$

is finite.

There are a number of techniques used to reduce the variance σ for fixed N. One strategy is to subdivide the integration volume into Q smaller volumes of varying sizes. Then Monte Carlo integration is performed in each subvolume using N/Q random points. The variance is varied by changing the relative sizes and locations of the subvolumes and is minimized when the contributions to σ^2 from each subvolume are identical ($=\sigma^2/Q$). Thus, function evaluations are concentrated where the potential error is largest – i.e., where the integrand is both large and rapidly changing.

This and other methods of variance reduction appear to be inappropriate for general-purpose algorithms as they require detailed knowledge of the integrand's behavior prior to implementation.

However, Sheppy has devised an iterative algorithm that uses information generated about the integrand during a Monte Carlo integration to reduce the variance in subsequent integrations. In the absence of any prior information on the integrand, this algorithm proceeds as follows: Initially an N-point Monte Carlo integration is performed with a uniform probability density ($p(x) = 1$). Besides providing estimates of the integral and the possible error (Eqs. (6.2) and (6.3)), the N integrand evaluations can also be used to define an improved probability density for use in the next (N-point) Monte Carlo integration. In this approach an empirical variance reduction can be gradually introduced over several iterations.

In VEGAS, rather than dealing with a density function $p(x)$ of arbitrary shape, one deals with a step function with Q steps such that the probability of a random number being chosen from any given step is a constant, equal to $1/Q$ for all steps ($0 = x_0 \leq \cdots \leq x_Q = 1$, $\Delta x_i = x_i - x_{i-1}$):

$$p(x) = \frac{1}{Q\Delta x_i}, \quad x_i - \Delta x_i \leq x < x_i, \; i = 1, \ldots, Q,$$

where

$$\sum_{i=1}^{Q} \Delta x_i = 1.$$

The probability distribution is tailored to a particular integrand simply by adjusting the incremental sizes Δx_i.

Given N integrand evaluations, the probability distribution or, equivalently, the increment density is refined by subdividing each increment Δx_i into $m_i + 1$ subincrements, where

$$m_i = K \frac{\bar{f}_i \Delta x_i}{\sum_j \bar{f}_j \Delta x_j}$$

and

$$\bar{f}_i \equiv \sum_{x \in x_i - \Delta x_i}^{x_i} |f(x)| \propto \frac{1}{\Delta x_i} \int_{x_i - \Delta x_i}^{x_i} dx |f(x)|. \tag{6.7}$$

Thus each increment is subdivided into as many as $(K+1)$ subincrements, and its contribution to the weight function increased in proportion to its contribution to the integral of $|f(x)|$, which is the expected behavior of the optimal variance:

$$p(x) = \frac{|f(x)|}{\int dx |f(x)|}.$$

Since it is desirable to restore the number of increments to its original value ($=Q$), groups of the new increments are amalgamated into larger increments, the number of subincrements in each group being constant (to preserve the relative increment density). The net effect is to alter the increment sizes, while keeping the total number constant, so that the smallest increments occur where $|f(x)|$ is largest. The new grid is used and further refined in subsequent iterations until the optimal grid has been obtained.

The values $\bar{f}_i$ must be discarded after each iteration because of storage limitations. However, a cumulative estimate of the integral and its error can be made that uses every evaluation of the integrand:

$$\begin{aligned} \bar{I} &= \sigma_I^2 \sum_i \frac{I_i}{\sigma_i^2} \\ \sigma_I &= \left[\sum_i \frac{1}{\sigma_i^2} \right]^{-1/2}. \end{aligned} \tag{6.8}$$

Here I_i and σ_i^2 are the integral and standard deviation estimated in the i-th iteration using Eqs. (6.2) and (6.3). The quality of the result of integration can be estimated by the value of χ^2:

$$\chi^2 \simeq \sum_i \frac{(I_i - \bar{I})^2}{\sigma_i^2},$$

which should not greatly exceed the number of iterations (minus one).

The number of iterations and the number of integrand samplings per iteration needed clearly depend upon the complexity of the integrand and the accuracy being sought. Usually, initial few iterations give large errors, but they are essential for establishing a relatively stable grid structure. Once the (approximately) optimal grid has been found, the uncertainty (σ_I) in the subsequent iterations becomes roughly proportional to $N^{-1/2}$, N being the number of integrand evaluations.

The algorithm described above is easily generalized to integrals of arbitrary dimensionality. To illustrate the necessary modifications consider

$$I = \int_0^1 dx \int_0^1 dy\ f(x,y).$$

VEGAS adopts a separable probability density function in order to limit data storage requirements:

$$p(x,y) = p_x(x) p_y(y).$$

In this case the optimal densities are easily shown to be

$$p_x(x) = \frac{\left[\int_0^1 dy\ \frac{f^2(x,y)}{p_y(y)}\right]^{1/2}}{\int_0^1 dx \left[\int_0^1 dy\ \frac{f^2(x,y)}{p_y(y)}\right]^{1/2}}$$

with a similar result for $p_y(y)$. Thus the one-dimensional algorithm can be applied along each axis but with $\bar{f}_i$ in (6.7) replaced by

$$(\bar{f}_i)^2 \equiv \sum_{x \in x_i - \Delta x_i}^{x_i} \sum_y \frac{f^2(x,y)}{p_y^2(y)} \propto \frac{1}{\Delta x_i} \int_{x_i - \Delta x_i}^{x_i} dx \int_0^1 dy\ \frac{f^2(x,y)}{p_y(y)} \tag{6.9}$$

for the x axis and with an analogous definition for the y axis. The generalization to an integral over a unit hypercube of n dimension is obvious. In order to apply this to parametric integrals defined on a hyperplane of the form

$$z_1 + z_2 + \cdots + z_{n+1} = 1,$$

it is of course necessary to map it onto a unit n-dimensional hypercube:

$$0 \le x_i \le 1, \quad i = 1, 2, \ldots, n, \tag{6.10}$$

which can be realized in many ways.

6.3. Problems encountered in high-precision numerical integration

In this section we shall consider in some detail salient features of the numerical work specific to our evaluation of the electron anomalous moment. In particular, it is of utmost importance to understand thoroughly the nature of error estimates generated by VEGAS in order to assure the credibility of numerical results.

Errors encountered in this work arise primarily from the following three features of our integrands:

(a) In general our integrands are singular on some boundaries of the integration domain, whether renormalized or not, because of the vanishing of the denominator factors U and/or V.

(b) In our scheme, renormalization is performed numerically, which means mutual cancellation of ∞'s at every singular point in the integration domain.

(c) The sheer size of the integrands, particularly in the eighth order, makes it difficult to accumulate adequate number of integrand samplings for high-precision calculation of the integrals.

At first sight, the feature (a) seems to indicate that it is hopeless to obtain a reliable result for such a singular integrand. Note, however, that we are dealing with a multi-dimensional integral and the size of the integration measure at singular points is so small that the integral itself is well defined and convergent once it is renormalized. To see this clearly, let M_G be the unrenormalized integral corresponding to a Feynman diagram G with UV-divergent subdiagrams S, S', S'', ... Then the integral

$$(1 - \mathbf{K}_{S'})(1 - \mathbf{K}_{S''}) \cdots M_G \tag{6.11}$$

where the $\mathbf{K}$ operations are defined in Sec. 3, is free from all divergences except a logarithmic one associated with the subdiagram S. Suppose S consists of the lines 1, 2, . . . , m, and z_1, z_2, . . . , z_m are the corresponding Feynman parameters. Thus the divergence of the integral (6.11) arises from the subdomain

$$0 \leq z_1 + z_2 + \cdots + z_m \leq z_S, \tag{6.12}$$

where $z_S \to 0$ at the singularity. In this subdomain the "phase space" is of order z_S^m and the integrand of (6.11) must behave as z_S^{-m} in order to generate a logarithmic divergence in (6.11). If we now apply the $(1 - \mathbf{K}_S)$ operation to (6.11), the resulting integrand of the convergent integral

$$(1 - \mathbf{K}_S)(1 - \mathbf{K}_{S'})(1 - \mathbf{K}_{S''}) \cdots M_G \tag{6.13}$$

behaves as z_S^{-m+1} in the domain (6.12). This means that the contribution to the integral (6.13) from the domain (6.12) is of order z_S and hence tends linearly to 0 as $z_S \to 0$.

Nevertheless, the singular nature of the integrand may be a cause for concern in the sense that the interval adaptation by VEGAS might not reach the optimal stage as rapidly as one would wish. This problem can be alleviated, however, by "stretching" the integration variables. Suppose VEGAS finds after several iterations that the integrand samplings are highly concentrated at one end of the integration domain, say $x_1 = 0$, of the axis x_1 of the hypercube (6.10). In such a case, if one maps x_1 into x_1' as

$$x_1' = x_1^{a_1}, \tag{6.14}$$

where a_1 is some real number greater than 1, the domain near the point $x_1 = 0$ is stretched out and random samplings in the x_1' variable give more attention to the region near $x_1 = 0$

from the beginning of iteration. Also, the Jacobian $a_1 x^{a_1-1}$ of the transformation (6.14) has the effect of reducing the peaking of the integrand. Similarly, the singularity at $x_1 = 1$ can be weakened by the stretching

$$x_1' = 1 - (1 - x_1)^{b_1}, \tag{6.15}$$

where b_1 is a real number greater than 1. Such stretching may be applied to all integration variables $x_1, \cdots, x_n$ independently. By an appropriate choice of the values of $a_1, \cdots, a_n$ and/or $b_1, \cdots, b_n$, one can accelerate the speed of convergence of the iteration considerably. Note that all these stretchings are one-to-one mappings of a unit hypercube onto another.

In practice, based on the result of a few iterations by VEGAS, one decides to stretch some axes by choosing appropriate trial values of stretch parameters a_i and b_i, and then run several iterations to see what happens to the structure of the integrand along each axis. One then readjusts the stretch parameters and tries again. In this manner one can find, after a few trials, a set of stretch parameters that gives a reasonable speed of convergence, although this method will never be perfect in view of the complex structure of integrands.

It will be obvious from this argument that there is a great deal of freedom in the choice of integration variables even if one restricts oneself to the unit hypercube. In particular, the criterion of square integrability (6.6), which seems to be necessary for reliable estimates of variance, is actually dependent on the choice of integration variables. For the type of integrals we are interested in, we can always make an appropriate stretch mapping so that the resulting integral is square integrable.

Let us now discuss feature (b). In our approach, the integrand, generated by an algebraic manipulation program such as SCHOONSCHIP, whether renormalized or not, actually diverges on some boundaries of the integration domain. In cases that require no renormalization, the integral converges in spite of this singularity because it is compensated by a small enough integration measure. When renormalization is required, the integral is made convergent as a consequence of point-by-point cancellation of divergences by carefully tailored counter-terms created by the intermediate renormalization procedure.

All this would pose no problem if each step of computation were carried out with infinite precision. In reality we have to perform calculations in finite precision, typically in double precision arithmetics. The intended cancellation of divergences may fail occasionally because cancelling terms have no more than 12 or 13 significant digits (in double precision) and their difference tends to be dominated near a singularity by roundoff errors causing undesirable fluctuations. The difficulty is compounded further by register overflow. To the extent that this takes place in very small regions like (6.12), it does not cause appreciable error in most integrals since the error is proportional to the fractional volume of the integration domain involved. However, in a few cases where this is a serious problem, it is found that the error can be reduced to a manageable level by adopting quadruple precision arithmetic, although it slows down the computation substantially. In practice quadruple precision is needed only in the neighborhood of some singularities. Thus one may evaluate the bulk of the integral in double precision, resorting to quadruple precision only where it is absolutely needed. In this manner we have been able to reduce the impact of computation slowdown considerably.

Let us next discuss feature (c). The huge size of our integrands (particularly in the eighth order) means that it requires a large amount of computing time in order to accumulate adequate sampling statistics. Indeed, this combined with the unavailability (until the last few years) of adequate computing power has been the main reason why we have not been able to finish our calculation sooner. Some of our eighth-order integrands consist of more than 20,000 complicated rational functions constructed from the parametric functions B_{ij}, A_i, C_{ij}, U, and V, which in turn are constructed from 10 integration variables, resulting in FORTRAN source codes of 300 to 500 kilobytes in size. This bulkiness prevents us from performing a sufficient number of integrand samplings per iteration except on the fastest computers. This also makes it very time-consuming to experiment with alternate choices of variable stretching which are often required to test the reliability of numerical results.

In order to appreciate the nature of the difficulties caused by the size of the integrand, suppose we sample it at N randomly chosen points in each iteration. If N is larger than a certain critical value, which unfortunately is not known *a priori* and varies from integrand to integrand, the error will decrease steadily as $N^{-1/2}$ as is expected for statistically significant samplings. If N is not large enough, however, VEGAS may not be able to explore the integrand thoroughly and may give deceptively optimistic error estimates. In such a case the estimated error may even increase as N increases, an effect we have had occasion to observe. The only practical way to know that the error estimate obtained is reliable is to repeat iteration sequences several times, each time increasing N, until the $N^{-1/2}$ behavior in the estimated error is observed.

Acknowledgments

This work is supported in part by the National Science Foundation. I should like to thank P. Cvitanovic and W. B. Lindquist whose contributions were essential for the computation of the C_3 and C_4 terms, respectively. I am grateful to W. B. Lindquist who read this article critically and made numerous useful suggestions. I am indebted for assistance in computing to numerous colleagues, G. P. Lepage in particular. Thanks are due to R. F. Peierls, H. Sugawara, and H. Miyazawa for their encouragement and support at various phases of this work, which extended over more than 20 years. Support by the computer centers of the Brookhaven National Laboratory, the National Laboratory for High Energy Physics, Japan (KEK), Tokyo University, and CERN, where parts of the numerical work were carried out, is gratefully acknowledged. The last phase of this research was conducted using the Cornell National Supercomputer Facility, which receives funding from the National Science Foundation and the IBM Corporation, with additional support from New York State and members of the Corporate Research Institute.

Appendix A. $B_{\alpha\beta}$ for Four-loop Chain Diagrams

We list parametric functions $B_{\alpha\beta}$ for the four-loop chain diagrams of Fig. 6. A chain is a set of all internal lines carrying the same loop momentum. Lines may be consecutive or separated by other parts of the chain diagram. The Greek letter for each chain also represents the sum of Feynman parameters of lines belonging to the chain. Functions not listed can be obtained by relabeling the names of chains taking the symmetry property of the chain diagram into account. Consult Eq. (2.1) to determine their overall signs.

Fig. 6(e):

$$
\begin{aligned}
B_{\alpha\alpha} &= (\beta+\gamma)(\delta+\epsilon)(\kappa+\lambda), \\
B_{\beta\beta} &= (\alpha+\gamma)(\delta+\epsilon)(\kappa+\lambda)+\delta\epsilon(\kappa+\lambda)+(\delta+\epsilon)\kappa\lambda, \\
B_{\alpha\beta} &= \gamma(\delta+\epsilon)(\kappa+\lambda), \\
B_{\beta\gamma} &= \alpha(\delta+\epsilon)(\kappa+\lambda)+\delta\epsilon(\kappa+\lambda)+(\delta+\epsilon)\kappa\lambda, \\
B_{\beta\delta} &= \gamma\epsilon(\kappa+\lambda).
\end{aligned}
$$

Fig. 6(f):

$$
\begin{aligned}
B_{\alpha\alpha} &= (\gamma+\delta)((\beta+\epsilon)(\kappa+\lambda)+\kappa\lambda), \\
B_{\gamma\gamma} &= (\alpha+\delta+\epsilon)(\beta(\kappa+\lambda)+\kappa\lambda)+(\alpha+\delta)\epsilon(\kappa+\lambda), \\
B_{\epsilon\epsilon} &= (\alpha+\beta)(\gamma+\delta)(\kappa+\lambda)+\gamma\delta(\kappa+\lambda)+(\gamma+\delta)\kappa\lambda, \\
B_{\alpha\beta} &= (\gamma+\delta)\epsilon(\kappa+\lambda), \\
B_{\alpha\gamma} &= \delta((\beta+\epsilon)(\kappa+\lambda)+\kappa\lambda), \\
B_{\alpha\epsilon} &= (\gamma+\delta)(\beta(\kappa+\lambda)+\kappa\lambda), \\
B_{\alpha\kappa} &= (\gamma+\delta)\epsilon\lambda, \\
B_{\gamma\delta} &= (\alpha+\epsilon)(\beta(\kappa+\lambda)+\kappa\lambda)+\alpha\epsilon(\kappa+\lambda), \\
B_{\gamma\epsilon} &= \delta(\beta(\kappa+\lambda)+\kappa\lambda), \\
B_{\gamma\kappa} &= \delta\epsilon\lambda.
\end{aligned}
$$

Fig. 6(g):

$$
\begin{aligned}
B_{\alpha\alpha} &= (\gamma+\delta+\epsilon)((\beta+\zeta)(\kappa+\lambda)+\kappa\lambda)+(\gamma+\delta)\epsilon(\kappa+\lambda), \\
B_{\gamma\gamma} &= (\alpha+\beta+\delta+\epsilon)(\zeta(\kappa+\lambda)+\kappa\lambda)+(\alpha+\delta)(\beta+\epsilon)(\kappa+\lambda), \\
B_{\zeta\zeta} &= ((\alpha+\beta)(\gamma+\delta+\epsilon)+\gamma(\delta+\epsilon))(\kappa+\lambda), \\
B_{\kappa\kappa} &= ((\alpha+\beta)(\gamma+\delta+\epsilon)+\gamma(\delta+\epsilon))(\zeta+\lambda) \\
&\quad +(\beta+\epsilon)(\alpha(\gamma+\delta)+\gamma\delta)+\beta\epsilon(\alpha+\delta), \\
B_{\alpha\beta} &= (\gamma+\delta+\epsilon)(\zeta(\kappa+\lambda)+\kappa\lambda)+\delta\epsilon(\kappa+\lambda), \\
B_{\alpha\gamma} &= (\delta+\epsilon)(\zeta(\kappa+\lambda)+\kappa\lambda)+\delta(\beta+\epsilon)(\kappa+\lambda), \\
B_{\alpha\delta} &= \gamma(\zeta(\kappa+\lambda)+\kappa\lambda)+(\gamma(\beta+\epsilon)+\beta\epsilon)(\kappa+\lambda), \\
B_{\alpha\epsilon} &= \gamma(\zeta(\kappa+\lambda)+\kappa\lambda)-\beta\delta(\kappa+\lambda), \\
B_{\alpha\zeta} &= (\beta(\gamma+\delta+\epsilon)+\gamma\epsilon)(\kappa+\lambda), \\
B_{\alpha\kappa} &= (\beta(\gamma+\delta+\epsilon)+\gamma\epsilon)\lambda,
\end{aligned}
$$

$$B_{\gamma\zeta} = (\beta\delta - \alpha\epsilon)(\kappa + \lambda),$$
$$B_{\gamma\kappa} = (\beta\delta - \alpha\epsilon)\lambda,$$
$$B_{\zeta\kappa} = ((\alpha + \beta)(\gamma + \delta + \epsilon) + \gamma(\delta + \epsilon))\lambda,$$
$$B_{\kappa\lambda} = ((\alpha + \beta)(\gamma + \delta + \epsilon) + \gamma(\delta + \epsilon))\zeta + (\beta + \epsilon)(\alpha(\gamma + \delta) + \gamma\delta) + \beta\epsilon(\alpha + \delta).$$

Fig. 6(h):

$$B_{\alpha\alpha} = (\kappa + \lambda + \rho)(\epsilon(\beta + \delta + \gamma + \zeta) + (\beta + \delta)(\gamma + \zeta)) + \kappa\lambda(\beta + \delta + \gamma + \zeta) + \rho(\epsilon(\kappa + \lambda) + \kappa\lambda + \lambda(\beta + \delta) + \kappa(\gamma + \zeta)),$$
$$B_{\delta\delta} = (\kappa + \lambda + \rho)((\alpha + \beta + \gamma)(\epsilon + \zeta) + (\alpha + \beta)\gamma) + \lambda(\alpha + \beta + \gamma)(\kappa + \rho),$$
$$B_{\alpha\beta} = (\kappa + \lambda + \rho)(\delta(\gamma + \epsilon + \zeta) + \epsilon\zeta) + \kappa\lambda(\delta + \zeta) + \rho(\kappa(\gamma + \epsilon + \zeta) + \lambda(\delta + \epsilon + \kappa)),$$
$$B_{\alpha\delta} = (\kappa + \lambda + \rho)(\beta(\gamma + \epsilon + \zeta) + \gamma\epsilon) + \lambda(\kappa(\beta + \gamma) + \beta\rho),$$
$$B_{\alpha\epsilon} = (\kappa + \lambda + \rho)(\gamma\delta - \beta\zeta) + (\gamma\kappa - \beta\lambda)\rho,$$
$$B_{\alpha\kappa} = \lambda(\gamma\delta - \beta\zeta) - \rho(\beta(\gamma + \epsilon + \zeta + \lambda) + \gamma\epsilon),$$
$$B_{\alpha\rho} = (\kappa + \lambda)(\epsilon(\beta + \gamma) + \beta\gamma) + \kappa\lambda(\beta + \gamma) + \beta\zeta\kappa + \gamma\delta\lambda,$$
$$B_{\delta\epsilon} = (\alpha + \beta + \gamma)(\zeta(\kappa + \lambda + \rho) + \lambda\rho) + \alpha\gamma(\kappa + \lambda + \rho).$$

Fig. 6(i):

$$B_{\alpha\alpha} = (\gamma + \zeta + \kappa + \rho)((\beta + \delta)(\epsilon + \lambda) + \epsilon\lambda) + (\gamma + \zeta)(\kappa(\beta + \delta + \lambda + \rho) + \rho(\beta + \delta + \epsilon)) + \kappa\rho(\epsilon + \lambda),$$
$$B_{\alpha\beta} = (\gamma + \epsilon + \zeta + \lambda)(\delta(\kappa + \rho) + \kappa\rho) + \delta(\gamma + \zeta)(\epsilon + \lambda) + \epsilon\gamma\rho + \zeta\kappa\lambda,$$
$$B_{\alpha\epsilon} = \delta\lambda(\gamma + \zeta + \kappa + \rho) + \kappa\zeta(\beta + \delta + \lambda + \rho) + \rho(\delta\zeta + \kappa\lambda - \beta\gamma).$$

Appendix B. Divergence Structure of the Sixth-order Renormalization Constants

We list the divergence structure (relationship between the standard and intermediate renormalization schemes) of the sixth-order renormalization constants defined by the graphs of Fig. 10. For any integral A, $\hat{A}$ represents the overall UV-divergent part of A and $\tilde{A} \equiv A - \hat{A}$. Note, however, that corresponding parts of the self-mass integral δm are denoted as $\delta\hat{m}$ and $\delta\tilde{m}$, respectively. δm_{2**} in δm_{6A} is obtained from δm_2 by inserting two self-mass vertices in the electron line. Similarly for B_{2**} in B_{6A}. $L_{2**\dagger}$ and $L_{2*\dagger*}$ in L_{6A1} and L_{6A3} are obtained from L_2 by inserting two self-mass vertices in one electron line and two electron lines, respectively. To be precise, $\delta\tilde{m}'_{2*}$ and $\delta\tilde{m}''_2$ in δm_{6A} and L'_{2*} and $\tilde{L}''_2$ in L_{6A1} and L_{6A3} should also be distinguished similarly. Fortunately, this does not matter since these terms cancel out in the end.

Mass renormalization constants:

$$\begin{aligned}
\delta m_{6A} &= \Delta\delta m_{6A} + 2(\delta m_2 \delta\tilde{m}_{4b(1^*)} + \hat{B}_2 \delta\tilde{m}_{4b(1')}) \\
&\quad - \delta m_2(\delta m_2 \delta m_{2^{**}} + \hat{B}_2 \delta\tilde{m}'_{2^*}) - \hat{B}_2(\delta m_2 \delta\tilde{m}'_{2^*} + \hat{B}_2 \delta\tilde{m}''_2) + \delta\hat{m}_{6A}, \\
\delta m_{6B} &= \Delta\delta m_{6B} + \delta m_2 \delta\tilde{m}_{4b(2^*)} + \hat{B}_2 \delta\tilde{m}_{4b(2')} + \delta\hat{m}_{4b} \delta\tilde{m}_{2^*} + \hat{B}_{4b} \delta\tilde{m}'_2 \\
&\quad - \delta m_2 \delta\hat{m}_{2^*} \delta\tilde{m}_{2^*} - \hat{B}_2(\delta\hat{m}'_2 \delta\tilde{m}_{2^*} + \hat{B}'_2 \delta\tilde{m}'_2) \\
&\quad + I_2 \delta\tilde{m}_{4b} - I_2(\delta m_2 \delta\tilde{m}_{2^*} + \hat{B}_2 \delta\tilde{m}'_2) + \delta\hat{m}_{6B}, \\
\delta m_{6C} &= \Delta\delta m_{6C} + 2\hat{L}_2 \delta\tilde{m}_{4b} + \delta\hat{m}_{4a} \delta\tilde{m}_{2^*} + \hat{B}_{4a} \delta\tilde{m}'_2 \\
&\quad - 2\hat{L}_2(\delta m_2 \delta\tilde{m}_{2^*} + \hat{B}_2 \delta\tilde{m}'_2) + I_2 \delta\tilde{m}_{4a} + \delta\hat{m}_{6C}, \\
\delta m_{6D} &= \Delta\delta m_{6D} + \delta m_2 \delta\tilde{m}_{4a(1^*)} + \hat{B}_2 \delta\tilde{m}_{4a(1')} \\
&\quad + \hat{L}_2 \delta\tilde{m}_{4b} - \hat{L}_2(\delta m_2 \delta\tilde{m}_{2^*} + \hat{B}_2 \delta\tilde{m}'_2) + \delta\hat{m}_{6D}, \\
\delta m_{6E} &= \Delta\delta m_{6E} + \delta m_2 \delta\tilde{m}_{4a(2^*)} + \hat{B}_2 \delta\tilde{m}_{4a(2')} + \delta\hat{m}_{6E}, \\
\delta m_{6F} &= \Delta\delta m_{6F} + 2\hat{L}_2 \delta\tilde{m}_{4a} + \delta\hat{m}_{6F}, \\
\delta m_{6G} &= \Delta\delta m_{6G} + \hat{L}_2 \delta\tilde{m}_{4a} + \delta\hat{m}_{6G}, \\
\delta m_{6H} &= \Delta\delta m_{6H} + \delta\hat{m}_{6H}.
\end{aligned}$$

Vertex renormalization constants:

$$\begin{aligned}
L_{6A1} &= I_{6A1} + \Delta L_{6A1} + 2\delta m_2 L_{4s(1^*)} + 2\hat{B}_2 \tilde{L}_{4s(1')} \\
&\quad - \delta m_2(\delta m_2 L_{2^{**}\dagger} + \hat{B}_2 L'_{2^*}) - \hat{B}_2(\delta m_2 L'_{2^*} + \hat{B}_2 \tilde{L}''_2) + \hat{L}_{6A1}, \\
L_{6A2} &= I_{6A2} + \Delta L_{6A2} + \hat{L}_2 \tilde{L}_{4s} + \delta m_2 L_{4l(1^*)} + \hat{B}_2 \tilde{L}_{4l(1')} \\
&\quad - \hat{L}_2(\delta m_2 L_{2^*} + \hat{B}_2 \tilde{L}'_2) + I_{4s} \tilde{L}_2 + \hat{L}_{6A2}, \\
L_{6A3} &= I_{6A3} + \Delta L_{6A3} + 2(\delta m_2 L_{4s(1^*)} + \hat{B}_2 \tilde{L}_{4s(1')}) \\
&\quad - \delta m_2(\delta m_2 L_{2^*\dagger^*} + \hat{B}_2 L'_{2^*}) - \hat{B}_2(\delta m_2 L'_{2^*} + \hat{B}_2 \tilde{L}''_2) + \hat{L}_{6A3}, \\
L_{6B1} &= I_{6B1} + \Delta L_{6B1} + \delta m_2 L_{4s(2^*)} + \hat{B}_2 \tilde{L}_{4s(2')} + \delta\hat{m}_{4b} L_{2^*} + \hat{B}_{4b} \tilde{L}'_2 \\
&\quad - \delta m_2 \delta\hat{m}_{2^*} L_{2^*} - \hat{B}_2(\delta\hat{m}'_2 L_{2^*} + \hat{B}'_2 \tilde{L}'_2) + \frac{1}{2} J_{6B} + \hat{L}_{6B1}, \\
L_{6B2} &= I_{6B2} + \Delta L_{6B2} + \delta m_2 L_{4l(2^*)} + \hat{B}_2 \tilde{L}_{4l(2')} + \hat{L}_{4s} \tilde{L}_2 \\
&\quad - \hat{B}_2 \hat{L}'_2 \tilde{L}_2 + I_2 \tilde{L}_{4s} - I_2(\delta m_2 L_{2^*} + \hat{B}_2 \tilde{L}'_2) + \hat{L}_{6B2}, \\
L_{6B3} &= I_{6B3} + \Delta L_{6B3} + \hat{L}_2 \tilde{L}_{4l} + \hat{L}_{4l} \tilde{L}_2 - (\hat{L}_2)^2 \tilde{L}_2 \\
&\quad + I_2 \tilde{L}_{4l} - I_2 \hat{L}_2 \tilde{L}_2 + I_{4l} \tilde{L}_2 + \hat{L}_{6B3}, \\
L_{6C1} &= I_{6C1} + \Delta L_{6C1} + 2\hat{L}_2 \tilde{L}_{4s} + \delta\hat{m}_{4a} L_{2^*} + \hat{B}_{4a} \tilde{L}'_2 \\
&\quad - 2\hat{L}_2(\delta m_2 L_{2^*} + \hat{B}_2 \tilde{L}'_2) + \frac{1}{2} J_{6C} + \hat{L}_{6C1}, \\
L_{6C2} &= I_{6C2} + \Delta L_{6C2} + \hat{L}_2 \tilde{L}_{4l} + \hat{L}_{4c} \tilde{L}_2 - (\hat{L}_2)^2 \tilde{L}_2 \\
&\quad + I_2 \tilde{L}_{4c} - I_2 \hat{L}_2 \tilde{L}_2 + \hat{L}_{6C2}, \\
L_{6C3} &= I_{6C3} + \Delta L_{6C3} + \hat{L}_{4x} \tilde{L}_2 + I_2 \tilde{L}_{4x} + \hat{L}_{6C3},
\end{aligned}$$

$$
\begin{aligned}
L_{6D1} &= I_{6D1} + \Delta L_{6D1} + \delta m_2 L_{4c(1^*)} + \hat{B}_2 \tilde{L}_{4c(1')} + \hat{L}_2 \tilde{L}_{4s} \\
&\quad - \hat{L}_2(\delta m_2 L_{2^*} + \hat{B}_2 \tilde{L}_2') + \hat{L}_{6D1}, \\
L_{6D2} &= I_{6D2} + \Delta L_{6D2} + \hat{L}_2 \tilde{L}_{4c} + \hat{L}_2 \tilde{L}_{4l} - (\hat{L}_2)^2 \tilde{L}_2 + I_{4c} \tilde{L}_2 + \hat{L}_{6D2}, \\
L_{6D3} &= I_{6D3} + \Delta L_{6D3} + \delta m_2 L_{4c(1^*)} + \hat{B}_2 \tilde{L}_{4c(1')} + \hat{L}_2 \tilde{L}_{4s} \\
&\quad - \hat{L}_2(\delta m_2 L_{2^*} + \hat{B}_2 \tilde{L}_2') + \hat{L}_{6D3}, \\
L_{6D4} &= I_{6D4} + \Delta L_{6D4} + \delta m_2 L_{4x(1^*)} + \hat{B}_2 \tilde{L}_{4x(1')} + \hat{L}_{6D4}, \\
L_{6D5} &= I_{6D5} + \Delta L_{6D5} + \delta m_2 L_{4c(3^*)} + \hat{B}_2 \tilde{L}_{4c(3')} + \hat{L}_{4s} \tilde{L}_2 \\
&\quad - \hat{B}_2 \hat{L}_2' \tilde{L}_2 + I_2 \Delta L_{4s} + \hat{L}_{6D5}, \\
L_{6E1} &= I_{6E1} + \Delta L_{6E1} + \delta m_2 L_{4c(2^*)} + \hat{B}_2 \tilde{L}_{4c(2')} + \hat{L}_{4s} \tilde{L}_2 \\
&\quad - \hat{B}_2 \hat{L}_2' \tilde{L}_2 + I_2 \Delta L_{4s} + \hat{L}_{6E1}, \\
L_{6E2} &= I_{6E2} + \Delta L_{6E2} + \delta m_2 L_{4x(2^*)} + \hat{B}_2 \tilde{L}_{4x(2')} + \hat{L}_{6E2}, \\
L_{6E3} &= I_{6E3} + \Delta L_{6E3} + I_{4x} \tilde{L}_2 + \hat{L}_2 \tilde{L}_{4x} + \hat{L}_{6E3}, \\
L_{6F1} &= I_{6F1} + \Delta L_{6F1} + \hat{L}_2 \tilde{L}_{4c} + \hat{L}_{4c} \tilde{L}_2 - (\hat{L}_2)^2 \tilde{L}_2 + I_2 \Delta L_{4c} + \hat{L}_{6F1}, \\
L_{6F2} &= I_{6F2} + \Delta L_{6F2} + \hat{L}_2 \tilde{L}_{4x} + \hat{L}_{6F2}, \\
L_{6F3} &= I_{6F3} + \Delta L_{6F3} + 2\hat{L}_2 \tilde{L}_{4c} - (\hat{L}_2)^2 \tilde{L}_2 + \hat{L}_{6F3}, \\
L_{6G1} &= I_{6G1} + \Delta L_{6G1} + \hat{L}_2 \tilde{L}_{4c} + \hat{L}_{4c} \tilde{L}_2 - (\hat{L}_2)^2 \tilde{L}_2 + I_2 \Delta L_{4c} + \hat{L}_{6G1}, \\
L_{6G2} &= I_{6G2} + \Delta L_{6G2} + \hat{L}_{6G2}, \\
L_{6G3} &= I_{6G3} + \Delta L_{6G3} + \hat{L}_{6G3}, \\
L_{6G4} &= I_{6G4} + \Delta L_{6G4} + \hat{L}_2 \tilde{L}_{4x} + \hat{L}_{6G4}, \\
L_{6G5} &= I_{6G5} + \Delta L_{6G5} + \hat{L}_2 \tilde{L}_{4c} + \hat{L}_{4l} \tilde{L}_2 - (\hat{L}_2)^2 \tilde{L}_2 + I_2 \Delta L_{4l} + \hat{L}_{6G5}, \\
L_{6H1} &= I_{6H1} + \Delta L_{6H1} + \hat{L}_{4x} \tilde{L}_2 + I_2 \Delta L_{4x} + \hat{L}_{6H1}, \\
L_{6H2} &= I_{6H2} + \Delta L_{6H2} + \hat{L}_{6H2}, \\
L_{6H3} &= I_{6H3} + \Delta L_{6H3} + \hat{L}_{6H3}.
\end{aligned}
$$

Wave-function renormalization constants:

$$
\begin{aligned}
B_{6A} &= -\sum_{i=1}^{5} I_{6Ai} + \Delta B_{6A} + 2(\delta m_2 B_{4b(1^*)} + \hat{B}_2 \tilde{B}_{4b(1')}) \\
&\quad - \delta m_2(\delta m_2 B_{2^{**}} + \hat{B}_2 B_{2^*}') - \hat{B}_2(\delta m_2 B_{2^*}' + \hat{B}_2 \tilde{B}_2'') + 2 I_{4s} \tilde{B}_2 + \hat{B}_{6A}, \\
B_{6B} &= -\sum_{i=1}^{5} I_{6Bi} + \Delta B_{6B} + \delta m_2 B_{4b(2^*)} + \hat{B}_2 \tilde{B}_{4b(2')} \\
&\quad + \delta \hat{m}_{4b} B_{2^*} + \hat{B}_{4b} \tilde{B}_2' - \delta m_2 \delta \hat{m}_{2^*} B_{2^*} - \hat{B}_2(\delta \hat{m}_2' B_{2^*} + \hat{B}_2' \tilde{B}_2') \\
&\quad + I_2 \tilde{B}_{4b} - I_2(\delta m_2 B_{2^*} + \hat{B}_2 \tilde{B}_2') + I_{4l} \tilde{B}_2 + B_{2^*}[I] \Delta \delta m_{4b} - J_{6B} + \hat{B}_{6B},
\end{aligned}
$$

$$
\begin{aligned}
B_{6C} = &-\sum_{i=1}^{5} I_{6Ci} + \Delta B_{6C} + 2\hat{L}_2\tilde{B}_{4b} + \delta m_{4a}B_{2*} + \hat{B}_{4a}\tilde{B}_2' \\
&- 2\hat{L}_2(\delta m_2 B_{2*} + \hat{B}_2\tilde{B}_2') \\
&+ I_2\tilde{B}_{4a} - 2I_2\hat{L}_2\tilde{B}_2 + B_{2*}[I]\Delta\delta m_{4a} - J_{6C} + \hat{B}_{6C}, \\
B_{6D} = &-\sum_{i=1}^{5} I_{6Di} + \Delta B_{6D} + \delta m_2 B_{4a(1^*)} + \hat{B}_2\tilde{B}_{4a(1')} \\
&+ \hat{L}_2\tilde{B}_{4b} + \hat{L}_{4s}\tilde{B}_2 - \hat{L}_2(\delta m_2 B_{2*} + \hat{B}_2\tilde{B}_2') \\
&- \hat{B}_2\hat{L}_2'\tilde{B}_2 + I_{4c}\tilde{B}_2 + \tilde{B}_2\Delta L_{4s} + \hat{B}_{6D}, \\
B_{6E} = &-\sum_{i=1}^{5} I_{6Ei} + \Delta B_{6E} + \delta m_2 B_{4a(2^*)} + \hat{B}_2\tilde{B}_{4a(2')} \\
&+ 2\hat{L}_{4s}\tilde{B}_2 - 2\hat{B}_2\hat{L}_2'\tilde{B}_2 + I_{4x}\tilde{B}_2 + 2\tilde{B}_2\Delta L_{4s} + \hat{B}_{6E}, \\
B_{6F} = &-\sum_{i=1}^{5} I_{6Fi} + \Delta B_{6F} + 2\hat{L}_2\tilde{B}_{4a} + 2\hat{L}_{4c}\tilde{B}_2 \\
&- 3(\hat{L}_2)^2\tilde{B}_2 + 2\tilde{B}_2\Delta L_{4c} + \hat{B}_{6F}, \\
B_{6G} = &-\sum_{i=1}^{5} I_{6Gi} + \Delta B_{6G} + \hat{L}_2\tilde{B}_{4a} + \hat{L}_{4c}\tilde{B}_2 \\
&+ \hat{L}_{4l}\tilde{B}_2 - 2(\hat{L}_2)^2\tilde{B}_2 + \tilde{B}_2(\Delta L_{4l} + \Delta L_{4c}) + \hat{B}_{6G}, \\
B_{6H} = &-\sum_{i=1}^{5} I_{6Hi} + \Delta B_{6H} + 2\hat{L}_{4x}\tilde{B}_2 + 2\tilde{B}_2\Delta L_{4x} + \hat{B}_{6H}.
\end{aligned}
$$

Appendix C. Renormalization Scheme for Sixth-order Moment of the Three-photon Exchange Type

We list the formulas that relate the standard-renormalized amplitudes $a_{6\alpha}$ to the unrenormalized integrals $M_{6\alpha}$ (α = A to H) and those that express $M_{6\alpha}$ in terms of intermediate renormalization quantities. Substitution of the latter in the former provides a working relation for our renormalization procedure.

$$
\begin{aligned}
a_{6A} =& M_{6A} - 2\delta m_2 M_{4b(1^*)} - 2B_2 M_{4b} + \delta m_2(\delta m_2 M_{2**} + B_2 M_{2*}) \\
&+ B_2(\delta m_2 M_{2*} + B_2 M_2), \\
M_{6A} =& \Delta M_{6A} + 2\delta m_2 M_{4b(1^*)} + 2\hat{B}_2 M_{4b} - \delta m_2(\delta m_2 M_{2**} + \hat{B}_2 M_{2*}) \\
&- \hat{B}_2(\delta m_2 M_{2*} + \hat{B}_2 M_2) + 2I_{4s}M_2\ , \\
a_{6B} =& M_{6B} - \delta m_2 M_{4b(2^*)} - B_2 M_{4b} - \delta m_{4b} M_{2*} - B_{4b} M_2 \\
&+ \delta m_2(\delta m_{2*} M_{2*} + B_{2*} M_2) + B_2(\delta m_2 M_{2*} + B_2 M_2), \\
M_{6B} =& \Delta M_{6B} + \delta m_2 M_{4b(2^*)} + \hat{B}_2 M_{4b} + \delta\hat{m}_{4b} M_{2*} + \hat{B}_{4b} M_2 - \delta m_2 \delta\hat{m}_{2*} M_{2*} \\
&- \hat{B}_2(\delta\hat{m}_2' M_{2*} + \hat{B}_2' M_2) + I_2\Delta M_{4b} + M_{2*}[I]\Delta\delta m_{4b} + (I_{4l} + (I_2)^2)M_2\ ,
\end{aligned}
$$

$$
\begin{aligned}
a_{6C} =& M_{6C} - 2L_2M_{4b} - \delta m_{4a}M_{2^*} - B_{4a}M_2 + 2L_2(\delta m_2M_{2^*} + B_2M_2), \\
&M_{6C} = \Delta M_{6C} + 2\hat{L}_2M_{4b} + \delta\hat{m}_{4a}M_{2^*} + \hat{B}_{4a}M_2 - 2\hat{L}_2(\delta m_2M_{2^*} + \hat{B}_2M_2) \\
&\qquad + I_2M_{4a} - 2I_2\hat{L}_2M_2 + M_{2^*}[I]\Delta\delta m_{4a} , \\
a_{6D} =& M_{6D} - L_2M_{4b} - L_{4s}M_2 - \delta m_2M_{4a(1^*)} - B_2M_{4a} \\
&\quad + L_2(\delta m_2M_{2^*} + B_2M_2) + B_2L_2M_2 + \delta m_2L_{2^*}M_2, \\
&M_{6D} = \Delta M_{6D} + \hat{L}_2M_{4b} + \hat{L}_{4s}M_2 + \delta m_2M_{4a(1^*)} + \hat{B}_2M_{4a} \\
&\qquad - \hat{L}_2(\delta m_2M_{2^*} + \hat{B}_2M_2) - \hat{B}_2\hat{L}_2'M_2 + I_{4c}M_2 , \\
a_{6E} =& M_{6E} - 2L_{4s}M_2 - \delta m_2M_{4a(2^*)} - B_2M_{4a} + 2B_2L_2M_2 + 2\delta m_2L_{2^*}M_2, \\
&M_{6E} = \Delta M_{6E} + 2\hat{L}_{4s}M_2 + \delta m_2M_{4a(2^*)} + \hat{B}_2M_{4a} - 2\hat{B}_2\hat{L}_2'M_2 \\
&\qquad + I_{4x}M_2 , \\
a_{6F} =& M_{6F} - 2L_2M_{4a} - 2L_{4c}M_2 + 3(L_2)^2M_2 , \\
&M_{6F} = \Delta M_{6F} + 2\hat{L}_2M_{4a} + 2\hat{L}_{4c}M_2 - 3(\hat{L}_2)^2M_2 , \\
a_{6G} =& M_{6G} - L_{4l}M_2 - L_{4c}M_2 - L_2M_{4a} + 2(L_2)^2M_2 , \\
&M_{6G} = \Delta M_{6G} + \hat{L}_{4l}M_2 + \hat{L}_{4c}M_2 + \hat{L}_2M_{4a} - 2(\hat{L}_2)^2M_2 , \\
a_{6H} =& M_{6H} - 2L_{4x}M_2 , \\
&M_{6H} = \Delta M_{6H} + 2\hat{L}_{4x}M_2 .
\end{aligned}
$$

Appendix D. Intermediate Renormalization of Eighth-order Moment of the Four-photon Exchange Type

We list the intermediate renormalization procedure for amplitudes M_{01} through M_{47}. Standard renormalization terms are not listed to save space. They can be obtained from the formulas below by removing IR subtraction terms and replacing $\hat{L}_2$, etc. by L_2, etc., following examples in Appendix C. M_{2^*}, $M_{2^{**}}$, $M_{2^{***}}$ are obtained by inserting one, two, three mass-vertices (two-point vertices) in the electron line of M_2. $M_{4a(1^*2^*)}$ is obtained by inserting mass-vertices in the electron lines 1 and 2 of M_{4a}. $L_{4c((1')')}$ in M_{03} is a derivative amplitude obtained by applying $-z_{1'}\frac{\partial}{\partial z_{1'}}$ on the integrand of L_{4c} of Fig. 8.

$$
\begin{aligned}
M_{01} =& \Delta M_{01} + 2\hat{L}_2M_{6F} + 2\hat{L}_{4c}M_{4a} + 2\hat{L}_{6F1}M_2 - 3(\hat{L}_2)^2M_{4a} - 6\hat{L}_2\hat{L}_{4c}M_2 + 4(\hat{L}_2)^3M_2 , \\
M_{02} =& \Delta M_{02} + \delta m_2M_{6F(1^*)} + \hat{B}_2M_{6F} + \hat{L}_2M_{6D} + \hat{L}_{4s}M_{4a} + \hat{L}_{4c}M_{4b} + \hat{L}_{6D5}M_2 \\
&- \hat{L}_2(\delta m_2M_{4a(1^*)} + \hat{B}_2M_{4a}) - \hat{B}_2\hat{L}_2'M_{4a} - \hat{L}_{4c}(\delta m_2M_{2^*} + \hat{B}_2M_2) - \hat{B}_2\hat{L}_{4c(3')}M_2 \\
&- 2\hat{L}_2\hat{L}_{4s}M_2 - (\hat{L})^2M_{4b} + 2\hat{L}_2\hat{B}_2\hat{L}_2'M_2 + (\hat{L}_2)^2(\delta m_2M_{2^*} + \hat{B}_2M_2) \\
&+ I_{6F1}M_2 + I_2\Delta L_{4c}M_2 , \\
M_{03} =& \Delta M_{03} + 2\hat{L}_2M_{6D} + \delta m_2M_{6F(3^*)} + \hat{B}_2M_{6F} + 2\hat{L}_{6D1}M_2 \\
&- 2\hat{L}_2(\delta m_2M_{4a(1^*)} + \hat{B}_2M_{4a}) - (\hat{L}_2)^2M_{4b} \\
&- 2\hat{L}_2\hat{L}_{4s}M_2 - 2\hat{B}_2\hat{L}_{4c((1')')}M_2 + (\hat{L}_2)^2(\delta m_2M_{2^*} + \hat{B}_2M_2) + 2\hat{L}_2\hat{B}_2\hat{L}_2'M_2 \\
&+ I_{6F3}M_2 ,
\end{aligned}
$$

$$
\begin{aligned}
M_{04} &= \Delta M_{04} + \delta m_2 M_{6D(3^*)} + \hat{B}_2 M_{6D} + \delta m_2 M_{6D(1^*)} + \hat{B}_2 M_{6D} + \hat{L}_2 M_{6A} + \hat{L}_{6A1} M_2 \\
&\quad - \delta m_2(\delta m_2 M_{4a(1^{**})} + \hat{B}_2 M_{4a(1^*)}) - \hat{B}_2(\delta m_2 M_{4a(1^*)} + \hat{B}_2 M_{4a}) \\
&\quad - 2\hat{L}_2(\delta m_2 M_{4b(1^*)} + \hat{B}_2 M_{4b}) - \hat{B}_2(\hat{L}_{4s(3')} + \hat{L}_{4s((1')')}) M_2 \\
&\quad + \hat{L}_2 \delta m_2(\delta m_2 M_{2^{**}} + \hat{B}_2 M_{2^*}) + \hat{L}_2 \hat{B}_2(\delta m_2 M_{2^*} + \hat{B}_2 M_2) + (\hat{B}_2)^2 \hat{L}_2'' M_2 \\
&\quad + I_{6D3} M_2 + I_{6D1} M_2 \,, \\
M_{05} &= \Delta M_{05} + \hat{L}_2 M_{6H} + \hat{L}_{4x} M_{4a} + \hat{L}_{6F2} M_2 + \hat{L}_{6H1} M_2 - 3\hat{L}_2 \hat{L}_{4x} M_2 \,, \\
M_{06} &= \Delta M_{06} + \hat{L}_2 M_{6G} + \hat{L}_2 M_{6F} + \hat{L}_{4l} M_{4a} + \hat{L}_{6F3} M_2 + \hat{L}_{6G5} M_2 \\
&\quad - 2(\hat{L}_2)^2 M_{4a} - 2\hat{L}_2 \hat{L}_{4l} M_2 - 3\hat{L}_2 \hat{L}_{4c} M_2 + 3(\hat{L}_2)^3 M_2 \,, \\
M_{07} &= \Delta M_{07} + \hat{L}_2 M_{6G} + \hat{L}_2 M_{6F} + \hat{L}_{4c} M_{4a} + \hat{L}_{6D2} M_2 + \hat{L}_{6G1} M_2 \\
&\quad - 2(\hat{L}_2)^2 M_{4a} - 4\hat{L}_2 \hat{L}_{4c} M_2 - \hat{L}_2 \hat{L}_{4l} M_2 + 3(\hat{L}_2)^3 M_2 \,, \\
M_{08} &= \Delta M_{08} + \hat{L}_2 M_{6C} + 2\hat{L}_2 M_{6D} + \delta\hat{m}_{4a} M_{4a(1^*)} + \hat{B}_{4a} M_{4a} + \hat{L}_{6C1} M_2 \\
&\quad - 2(\hat{L}_2)^2 M_{4b} - \hat{L}_2(\delta\hat{m}_{4a} M_{2^*} + \hat{B}_{4a} M_2) - 2\hat{L}_2(\delta m_2 M_{4a(1^*)} + \hat{B}_2 M_{4a}) \\
&\quad - 2\hat{L}_2 \hat{L}_{4s} M_2 - \hat{B}_{4a} \hat{L}_2' M_2 + 2(\hat{L}_2)^2(\delta m_2 M_{2^*} + \hat{B}_2 M_2) + 2\hat{L}_2 \hat{B}_2 \hat{L}_2' M_2 \\
&\quad + I_{4c} \Delta M_{4a} + I_{4a(1^*)} \Delta\delta m_{4a}, \\
M_{09} &= \Delta M_{09} + \delta m_2 M_{6F(2^*)} + \hat{B}_2 M_{6F} + \hat{L}_2 M_{6E} + \hat{L}_{4s} M_{4a} + \hat{L}_{6E1} M_2 + \hat{L}_{6D3} M_2 \\
&\quad - \hat{L}_2(\delta m_2 M_{4a(2^*)} + \hat{B}_2 M_{4a}) - \hat{B}_2 \hat{L}_2' M_{4a} - \hat{B}_2 \hat{L}_{4c(2')} M_2 \\
&\quad - \hat{B}_2 \hat{L}_{4c(1')} M_2 - 3\hat{L}_2 \hat{L}_{4s} M_2 + 3\hat{L}_2 \hat{B}_2 \hat{L}_2' M_2 \\
&\quad + I_{6F2} M_2, \\
M_{10} &= \Delta M_{10} + \delta m_2 M_{6D(2^*)} + \hat{B}_2 M_{6D} + \delta\hat{m}_{4b} M_{4a(1^*)} + \hat{B}_{4b} M_{4a} \\
&\quad + \hat{L}_2 M_{6B} + \hat{L}_{6B1} M_2 - \delta m_2 \delta\hat{m}_{2^*} M_{4a(1^*)} - \hat{B}_2(\delta\hat{m}_2' M_{4a(1^*)} + \hat{B}_2' M_{4a}) \\
&\quad - \hat{L}_2(\delta m_2 M_{4b(2^*)} + \hat{B}_2 M_{4b}) \\
&\quad - \hat{B}_2 \hat{L}_{4s(2')} M_2 - \hat{L}_2(\delta\hat{m}_{4b} M_{2^*} + \hat{B}_{4b} M_2) - \hat{B}_{4b} \hat{L}_2' M_2 \\
&\quad + \hat{L}_2 \delta m_2 \delta\hat{m}_{2^*} M_{2^*} + \hat{L}_2 \hat{B}_2(\delta\hat{m}_2' M_{2^*} + \hat{B}_2' M_2) + \hat{B}_2 \hat{B}_2' \hat{L}_2' M_2 \\
&\quad + I_{4c} \Delta M_{4b} + I_{4c} I_2 M_2 + I_{6D2} M_2 + I_{4a(1^*)} \Delta\delta m_{4b}, \\
M_{11} &= \Delta M_{11} + 2\delta m_2 M_{6D(5^*)} + 2\hat{B}_2 M_{6D} + 2\hat{L}_{4s} M_{4b} \\
&\quad - \delta m_2(\delta m_2 M_{4a(1^*3^*)} + \hat{B}_2 M_{4a(1^*)}) - \hat{B}_2(\delta m_2 M_{4a(1^*)} + \hat{B}_2 M_{4a}) \\
&\quad - 2\hat{L}_{4s}(\delta m_2 M_{2^*} + \hat{B}_2 M_2) - 2\hat{B}_2 \hat{L}_2' M_{4b} + 2\hat{B}_2 \hat{L}_2'(\delta m_2 M_{2^*} + \hat{B}_2 M_2) \\
&\quad + 2I_{6D5} M_2 + 2I_2 \Delta L_{4s} M_2, \\
M_{12} &= \Delta M_{12} + 2\delta m_2 M_{6A(1^*)} + 2\hat{B}_2 M_{6A} + \delta m_2 M_{6A(3^*)} + \hat{B}_2 M_{6A} \\
&\quad - 2\delta m_2(\delta m_2 M_{4b(1^{**})} + \hat{B}_2 M_{4b(1^*)}) - 2\hat{B}_2(\delta m_2 M_{4b(1^*)} + \hat{B}_2 M_{4b}) \\
&\quad - \delta m_2(\delta m_2 M_{4b(1^*3^*)} + \hat{B}_2 M_{4b(1^*)}) - \hat{B}_2(\delta m_2 M_{4b(1^*)} + \hat{B}_2 M_{4b}) \\
&\quad + (\delta m_2)^2(\delta m_2 M_{2^{***}} + \hat{B}_2 M_{2^{**}}) + 2\delta m_2 \hat{B}_2(\delta m_2 M_{2^{**}} + \hat{B}_2 M_{2^*}) \\
&\quad + (\hat{B}_2)^2(\delta m_2 M_{2^*} + \hat{B}_2 M_2) \\
&\quad + 2I_{6A1} M_2 + I_{6A3} M_2,
\end{aligned}
$$

$$\begin{aligned}
M_{13} = {} & \Delta M_{13} + \delta m_2 M_{6H(1^*)} + \hat{B}_2 M_{6H} + \hat{L}_{4x} M_{4b} + \hat{L}_{6D4} M_2 \\
& - \hat{L}_{4x}(\delta m_2 M_{2^*} + \hat{B}_2 M_2) - \hat{B}_2 \hat{L}_{4x(1')} M_2 \\
& + I_{6H1} M_2 + I_2 \Delta L_{4x} M_2, \\
M_{14} = {} & \Delta M_{14} + \delta m_2 M_{6G(5^*)} + \hat{B}_2 M_{6G} + \hat{L}_2 M_{6D} + \hat{L}_{4l} M_{4b} + \hat{L}_{6D3} M_2 \\
& - \hat{L}_2(\delta m_2 M_{4a(1^*)} + \hat{B}_2 M_{4a}) - \hat{L}_{4l}(\delta m_2 M_{2^*} + \hat{B}_2 M_2) - \hat{B}_2 \hat{L}_{4c(1')} M_2 \\
& - (\hat{L}_2)^2 M_{4b} - \hat{L}_2 \hat{L}_{4s} M_2 + (\hat{L}_2)^2(\delta m_2 M_{2^*} + \hat{B}_2 M_2) + \hat{L}_2 \hat{B}_2 \hat{L}'_2 M_2 \\
& + I_{6G5} M_2 + I_2 \Delta L_{4l} M_2, \\
M_{15} = {} & \Delta M_{15} + \delta m_2 M_{6G(1^*)} + \hat{B}_2 M_{6G} + \hat{L}_2 M_{6D} + \hat{L}_{4c} M_{4b} + \hat{L}_{6A2} M_2 \\
& - \hat{L}_2(\delta m_2 M_{4a(1^*)} + \hat{B}_2 M_{4a}) - \hat{L}_{4c}(\delta m_2 M_{2^*} + \hat{B}_2 M_2) - \hat{B}_2 \hat{L}_{4l(1')} M_2 \\
& - (\hat{L}_2)^2 M_{4b} - \hat{L}_2 \hat{L}_{4s} M_2 + (\hat{L}_2)^2(\delta m_2 M_{2^*} + \hat{B}_2 M_2) + \hat{L}_2 \hat{B}_2 \hat{L}'_2 M_2 \\
& + I_{6G1} M_2 + I_2 \Delta L_{4c} M_2, \\
M_{16} = {} & \Delta M_{16} + \delta m_2 M_{6C(1^*)} + \hat{B}_2 M_{6C} + \delta\hat{m}_{4a} M_{4b(1^*)} + \hat{B}_{4a} M_{4b} + 2\hat{L}_2 M_{6A} \\
& - \delta m_2(\delta\hat{m}_{4a} M_{2^{**}} + \hat{B}_{4a} M_{2^*}) - \hat{B}_2(\delta\hat{m}_{4a} M_{2^*} + \hat{B}_{4a} M_2) \\
& - 4\hat{L}_2(\delta m_2 M_{4b(1^*)} + \hat{B}_2 M_{4b}) + 2\hat{L}_2 \delta m_2(\delta m_2 M_{2^{**}} + \hat{B}_2 M_{2^*}) \\
& + 2\hat{B}_2 \hat{L}_2(\delta m_2 M_{2^*} + \hat{B}_2 M_2) \\
& + I_{6C1} M_2 + I_{4s} \Delta M_{4a} + \frac{1}{2} J_{6C} M_2 + I_{4b(1^*)} \Delta\delta m_{4a} - I_{2^*} \Delta\delta m_{4a} M_2, \\
M_{17} = {} & \Delta M_{17} + \delta m_2 M_{6E(1^*)} + \hat{B}_2 M_{6E} + \delta m_2 M_{6D(4^*)} + \hat{B}_2 M_{6D} + \hat{L}_{4s} M_{4b} + \hat{L}_{6A3} M_2 \\
& - \delta m_2(\delta m_2 M_{4a(1^*2^*)} + \hat{B}_2 M_{4a(1^*)}) - \hat{B}_2(\delta m_2 M_{4a(2^*)} + \hat{B}_2 M_{4a}) \\
& - \hat{L}_{4s}(\delta m_2 M_{2^*} + \hat{B}_2 M_2) - 2\hat{B}_2 \hat{L}_{4s(1')} M_2 - \hat{B}_2 \hat{L}'_2 M_{4b} \\
& + \delta m_2 \hat{B}_2 \hat{L}'_2 M_{2^*} + (\hat{B}_2)^2(\hat{L}'_2 + \hat{L}''_2) M_2 \\
& + I_{6E1} M_2 + I_{6D4} M_2 + I_2 \Delta L_{4s} M_2, \\
M_{18} = {} & \Delta M_{18} + \delta m_2 M_{6B(1^*)} + \hat{B}_2 M_{6B} + \delta m_2 M_{6A(2^*)} + \hat{B}_2 M_{6A} \\
& + \delta\hat{m}_{4b} M_{4b(1^*)} + \hat{B}_{4b} M_{4b} \\
& - \delta m_2(\delta m_2 M_{4b(1^*2^*)} + \hat{B}_2 M_{4b(1^*)}) - \hat{B}_2(\delta m_2 M_{4b(2^*)} + \hat{B}_2 M_{4b}) \\
& - \delta m_2(\delta\hat{m}_{4b} M_{2^{**}} + \hat{B}_{4b} M_{2^*}) - \hat{B}_2(\delta\hat{m}_{4b} M_{2^*} + \hat{B}_{4b} M_2) \\
& - \delta m_2 \delta\hat{m}_{2^*} M_{4b(1^*)} - \hat{B}_2(\delta\hat{m}'_2 M_{4b(1^*)} + \hat{B}'_2 M_{4b}) \\
& + \delta m_2 \delta\hat{m}_{2^*}(\delta m_2 M_{2^{**}} + \hat{B}_2 M_{2^*}) + \hat{B}_2 \delta\hat{m}'_2(\delta m_2 M_{2^{**}} + \hat{B}_2 M_{2^*}) \\
& + \hat{B}_2 \hat{B}'_2(\delta m_2 M_{2^*} + \hat{B}_2 M_2) \\
& + I_{6B1} M_2 + I_{4s} \Delta M_{4b} + I_{6A2} M_2 + I_{4s} I_2 M_2 \\
& + \frac{1}{2} J_{6B} M_2 + I_{4b(1^*)} \Delta\delta m_{4b} - I_{2^*} \Delta\delta m_{4b} M_2, \\
M_{19} = {} & \Delta M_{19} + 2\hat{L}_{6H2} M_2 \ , \\
M_{20} = {} & \Delta M_{20} + \hat{L}_2 M_{6H} + \hat{L}_{6F2} M_2 + \hat{L}_{6G4} M_2 - 2\hat{L}_2 \hat{L}_{4x} M_2 \ , \\
M_{21} = {} & \Delta M_{21} + 2\hat{L}_{6G2} M_2 \ ,
\end{aligned}$$

$$
\begin{aligned}
M_{22} &= \Delta M_{22} + \hat{L}_2 M_{6G} + \hat{L}_{4c} M_{4a} + \hat{L}_{6F1} M_2 + \hat{L}_{6C2} M_2 \\
&\quad - (\hat{L}_2)^2 M_{4a} - 3\hat{L}_2 \hat{L}_{4c} M_2 - \hat{L}_2 \hat{L}_{4l} M_2 + 2(\hat{L}_2)^3 M_2 \,, \\
M_{23} &= \Delta M_{23} + \delta m_2 M_{6H(2^*)} + \hat{B}_2 M_{6H} + \hat{L}_{6E2} M_2 + \hat{L}_{6D4} M_2 - \hat{B}_2(\hat{L}_{4x(1')} + \hat{L}_{4x(2')}) M_2 \\
&\quad + I_{6H2} M_2 \,, \\
M_{24} &= \Delta M_{24} + \delta m_2 M_{6G(2^*)} + \hat{B}_2 M_{6G} + \hat{L}_{4s} M_{4a} + \hat{L}_{6B2} M_2 + \hat{L}_{6D5} M_2 \\
&\quad - \hat{B}_2 \hat{L}'_2 M_{4a} - \hat{B}_2 \hat{L}_{4l(2')} M_2 - 2\hat{L}_{4s} \hat{L}_2 M_2 - \hat{B}_2 \hat{L}_{4c(3')} M_2 + 2\hat{L}_2 \hat{B}_2 \hat{L}'_2 M_2 \\
&\quad + I_{6G2} M_2, \\
M_{25} &= \Delta M_{25} + 2\hat{L}_2 M_{6G} + 2\hat{L}_{6D2} M_2 \\
&\quad - (\hat{L}_2)^2 M_{4a} - 2\hat{L}_2 \hat{L}_{4l} M_2 - 2\hat{L}_2 \hat{L}_{4c} M_2 + 2(\hat{L}_2)^3 M_2 \,, \\
M_{26} &= \Delta M_{26} + 2\hat{L}_2 M_{6C} + 2\hat{L}_{4c} M_{4b} + \delta\hat{m}_{6F} M_{2^*} + \hat{B}_{6F} M_2 - 3(\hat{L}_2)^2 M_{4b} \\
&\quad - 2\hat{L}_2(\delta\hat{m}_{4a} M_{2^*} + \hat{B}_{4a} M_2) - 2\hat{L}_{4c}(\delta m_2 M_{2^*} + \hat{B}_2 M_2) \\
&\quad + 3(\hat{L}_2)^2(\delta m_2 M_{2^*} + \hat{B}_2 M_2) \\
&\quad + I_2 \Delta M_{6F} + M_{2^*}[I] \Delta\delta m_{6F}, \\
M_{27} &= \Delta M_{27} + \delta m_2 M_{6G(4^*)} + \hat{B}_2 M_{6G} + \hat{L}_2 M_{6E} + \hat{L}_{6D1} M_2 + \hat{L}_{6A2} M_2 - \hat{B}_2 \hat{L}_{4l(1')} M_2 \\
&\quad - \hat{B}_2 \hat{L}_{4c((1')')} M_2 - 2\hat{L}_2 \hat{L}_{4s} M_2 - \hat{L}_2(\delta m_2 M_{4a(2^*)} + \hat{B}_2 M_{4a}) + 2\hat{L}_2 \hat{B}_2 \hat{L}'_2 M_2 \\
&\quad + I_{6G4} M_2, \\
M_{28} &= \Delta M_{28} + \delta m_2 M_{6C(2^*)} + \hat{B}_2 M_{6C} + \hat{L}_2 M_{6B} + \hat{L}_{4s} M_{4b} + \delta\hat{m}_{6D} M_{2^*} + \hat{B}_{6D} M_2 \\
&\quad - \hat{L}_2(\delta m_2 M_{4b(2^*)} + \hat{B}_2 M_{4b}) - \hat{B}_2 \hat{L}'_2 M_{4b} - \delta m_2 \delta\hat{m}_{4a(1^*)} M_{2^*} \\
&\quad - \hat{B}_2(\delta\hat{m}_{4a(1')} M_{2^*} + \hat{B}_{4a(1')} M_2) - \hat{L}_2(\delta\hat{m}_{4b} M_{2^*} + \hat{B}_{4b} M_2) \\
&\quad - \hat{L}_{4s}(\delta m_2 M_{2^*} + \hat{B}_2 M_2) + \hat{L}_2 \delta m_2 \delta\hat{m}_{2^*} M_{2^*} \\
&\quad + \hat{L}_2 \hat{B}_2(\delta\hat{m}'_2 M_{2^*} + \hat{B}'_2 M_2) + \hat{B}_2 \hat{L}'_2(\delta m_2 M_{2^*} + \hat{B}_2 M_2) \\
&\quad + I_{6C2} M_2 + I_2 I_{4c} M_2 + I_2 \Delta M_{6D} + M_{2^*}[I] \Delta\delta m_{6D}, \\
M_{29} &= \Delta M_{29} + 2\delta m_2 M_{6E(2^*)} + 2\hat{B}_2 M_{6E} + 2\hat{L}_{6A1} M_2 \\
&\quad - \delta m_2(\delta m_2 M_{4a(2^*2^*)} + \hat{B}_2 M_{4a(2^*)}) - \hat{B}_2(\delta m_2 M_{4a(2^*)} + \hat{B}_2 M_{4a}) \\
&\quad - 2\hat{B}_2(\hat{L}_{4s(3')} + \hat{L}_{4s((1')')}) M_2 + 2(\hat{B}_2)^2 \hat{L}''_2 M_2 \\
&\quad + 2I_{6E2} M_2, \\
M_{30} &= \Delta M_{30} + 2\delta m_2 M_{6B(2^*)} + 2\hat{B}_2 M_{6B} + \delta\hat{m}_{6A} M_{2^*} + \hat{B}_{6A} M_2 \\
&\quad - \delta m_2(\delta m_2 M_{4b(2^*2^*)} + \hat{B}_2 M_{4b(2^*)}) - \hat{B}_2(\delta m_2 M_{4b(2^*)} + \hat{B}_2 M_{4b}) \\
&\quad - 2\delta m_2 \delta\hat{m}_{4b(1^*)} M_{2^*} - 2\hat{B}_2(\delta\hat{m}_{4b(1')} M_{2^*} + \hat{B}_{4b(1')} M_2) \\
&\quad + 2\delta m_2 \hat{B}_2 \delta\hat{m}'_{2^*} M_{2^*} + (\hat{B}_2)^2(\delta\hat{m}''_2 M_{2^*} + \hat{B}''_2 M_2) \\
&\quad + 2I_{6B2} M_2 + 2I_2 I_{4s} M_2 + I_2 \Delta M_{6A} + M_{2^*}[I] \Delta\delta m_{6A}, \\
M_{31} &= \Delta M_{31} + 2\hat{L}_{6H3} M_2 \\
M_{32} &= \Delta M_{32} + \hat{L}_{6G3} M_2 + \hat{L}_{6H2} M_2 \\
M_{33} &= \Delta M_{33} + 2\hat{L}_{6G3} M_2
\end{aligned}
$$

$$M_{34} = \Delta M_{34} + \hat{L}_{4x} M_{4a} + \hat{L}_{6C3} M_2 + \hat{L}_{6H1} M_2 - 2\hat{L}_{4x}\hat{L}_2 M_2 \ ,$$

$$M_{35} = \Delta M_{35} + \hat{L}_2 M_{6H} + \hat{L}_{6E3} M_2 + \hat{L}_{6G4} M_2 - 2\hat{L}_{4x}\hat{L}_2 M_2 \ ,$$

$$\begin{aligned} M_{36} = {} & \Delta M_{36} + \hat{L}_2 M_{6G} + \hat{L}_{4l} M_{4a} + \hat{L}_{6B3} M_2 + \hat{L}_{6G5} M_2 \\ & - (\hat{L}_2)^2 M_{4a} - 3\hat{L}_2\hat{L}_{4l} M_2 - \hat{L}_2\hat{L}_{4c} M_2 + 2(\hat{L}_2)^3 M_2 \ , \end{aligned}$$

$$M_{37} = \Delta M_{37} + 2\hat{L}_{6G2} M_2$$

$$\begin{aligned} M_{38} = {} & \Delta M_{38} + 2\hat{L}_{4x} M_{4b} + \delta\hat{m}_{6H} M_{2^*} + \hat{B}_{6H} M_2 - 2\hat{L}_{4x}(\delta m_2 M_{2^*} + \hat{B}_2 M_2) \\ & + I_2 \Delta M_{6H} + M_{2^*}[I]\Delta\delta m_{6H}, \end{aligned}$$

$$\begin{aligned} M_{39} = {} & \Delta M_{39} + \hat{L}_2 M_{6G} + \hat{L}_{4c} M_{4a} + \hat{L}_{6G1} M_2 + \hat{L}_{6C2} M_2 \\ & - (\hat{L}_2)^2 M_{4a} - 3\hat{L}_2\hat{L}_{4c} M_2 - \hat{L}_2\hat{L}_{4l} M_2 + 2(\hat{L}_2)^3 M_2 \ , \end{aligned}$$

$$\begin{aligned} M_{40} = {} & \Delta M_{40} + \hat{L}_2 M_{6C} + \hat{L}_{4l} M_{4b} + \hat{L}_{4c} M_{4b} + \delta\hat{m}_{6G} M_{2^*} + \hat{B}_{6G} M_2 \\ & - 2(\hat{L}_2)^2 M_{4b} - \hat{L}_2(\delta\hat{m}_{4a} M_{2^*} + \hat{B}_{4a} M_2) - \hat{L}_{4l}(\delta m_2 M_{2^*} + \hat{B}_2 M_2) \\ & - \hat{L}_{4c}(\delta m_2 M_{2^*} + \hat{B}_2 M_2) + 2(\hat{L}_2)^2(\delta m_2 M_{2^*} + \hat{B}_2 M_2) \\ & + I_2 \Delta M_{6G} + M_{2^*}[I]\Delta\delta m_{6G}, \end{aligned}$$

$$\begin{aligned} M_{41} = {} & \Delta M_{41} + 2\hat{L}_2 M_{6E} + \delta\hat{m}_{4a} M_{4a(2^*)} + \hat{B}_{4a} M_{4a} + 2\hat{L}_{6C1} M_2 \\ & - 2\hat{L}_2(\delta m_2 M_{4a(2^*)} + \hat{B}_2 M_{4a}) - 4\hat{L}_2\hat{L}_{4s} M_2 - 2\hat{B}_{4a}\hat{L}'_2 M_2 + 4\hat{L}_2\hat{B}_2\hat{L}'_2 M_2 \\ & + I_{4x}\Delta M_{4a} + I_{4a(2^*)}\Delta\delta m_{4a}, \end{aligned}$$

$$\begin{aligned} M_{42} = {} & \Delta M_{42} + 2\hat{L}_2 M_{6B} + \delta\hat{m}_{4a} M_{4b(2^*)} + \hat{B}_{4a} M_{4b} + \delta\hat{m}_{6C} M_{2^*} + \hat{B}_{6C} M_2 \\ & - 2\hat{L}_2(\delta m_2 M_{4b(2^*)} + \hat{B}_2 M_{4b}) - 2\hat{L}_2(\delta\hat{m}_{4b} M_{2^*} + \hat{B}_{4b} M_2) \\ & - \delta\hat{m}_{4a}\delta\hat{m}_{2^*} M_{2^*} - \hat{B}_{4a}(\delta\hat{m}'_2 M_{2^*} + \hat{B}'_2 M_2) + 2\hat{L}_2\delta m_2\delta\hat{m}_{2^*} M_{2^*} \\ & + 2\hat{L}_2\hat{B}_2(\delta\hat{m}'_2 M_{2^*} + \hat{B}'_2 M_2) \\ & + I_{4l}\Delta M_{4a} + (I_2)^2\Delta M_{4a} + I_2\Delta M_{6C} + M_{2^*}[I]\Delta\delta m_{6C} \\ & + I_{4b(2^*)}\Delta\delta m_{4a}, \end{aligned}$$

$$\begin{aligned} M_{43} = {} & \Delta M_{43} + \delta m_2 M_{6H(3^*)} + \hat{B}_2 M_{6H} + 2\hat{L}_{6E2} M_2 - 2\hat{B}_2\hat{L}_{4x(2')} M_2 \\ & + I_{6H3} M_2, \end{aligned}$$

$$\begin{aligned} M_{44} = {} & \Delta M_{44} + \delta m_2 M_{6G(3^*)} + \hat{B}_2 M_{6G} + \hat{L}_{4s} M_{4a} + \hat{L}_{6B2} M_2 + \hat{L}_{6E1} M_2 \\ & - \hat{B}_2\hat{L}'_2 M_{4a} - \hat{B}_2\hat{L}_{4l(2')} M_2 - \hat{B}_2\hat{L}_{4c(2')} M_2 - 2\hat{L}_{4s}\hat{L}_2 M_2 + 2\hat{L}_2\hat{B}_2\hat{L}'_2 M_2 \\ & + I_{6G3} M_2, \end{aligned}$$

$$\begin{aligned} M_{45} = {} & \Delta M_{45} + \delta m_2 M_{6C(3^*)} + \hat{B}_2 M_{6C} + 2\hat{L}_{4s} M_{4b} + \delta\hat{m}_{6E} M_{2^*} + \hat{B}_{6E} M_2 \\ & - 2\hat{B}_2\hat{L}'_2 M_{4b} - \delta m_2\delta\hat{m}_{4a(2^*)} M_{2^*} - \hat{B}_2(\delta\hat{m}_{4a(2')} M_{2^*} + \hat{B}_{4a(2')} M_2) \\ & - 2\hat{L}_{4s}(\delta m_2 M_{2^*} + \hat{B}_2 M_2) + 2\hat{B}_2\hat{L}'_2(\delta m_2 M_{2^*} + \hat{B}_2 M_2) \\ & + I_{6C3} M_2 + I_2 I_{4x} M_2 + I_2\Delta M_{6E} + M_{2^*}[I]\Delta\delta m_{6E}, \end{aligned}$$

$$\begin{aligned} M_{46} = {} & \Delta M_{46} + \delta m_2 M_{6E(3^*)} + \hat{B}_2 M_{6E} + \delta\hat{m}_{4b} M_{4a(2^*)} + \hat{B}_{4b} M_{4a} + 2\hat{L}_{6B1} M_2 \\ & - \delta m_2\delta\hat{m}_{2^*} M_{4a(2^*)} - \hat{B}_2(\delta\hat{m}'_2 M_{4a(2^*)} + \hat{B}'_2 M_{4a}) - 2\hat{B}_2\hat{L}_{4s(2')} M_2 \\ & - 2\hat{B}_{4b}\hat{L}'_2 M_2 + 2\hat{B}_2\hat{B}'_2\hat{L}'_2 M_2 \\ & + I_{6E3} M_2 + I_{4x} I_2 M_2 + I_{4x}\Delta M_{4b} + I_{4a(2^*)}\Delta\delta m_{4b}, \end{aligned}$$

$$
\begin{aligned}
M_{47} = {} & \Delta M_{47} + \delta m_2 M_{6B(3^*)} + \hat{B}_2 M_{6B} + \delta\hat{m}_{4b} M_{4b(2^*)} + \hat{B}_{4b} M_{4b} + \delta\hat{m}_{6B} M_{2^*} + \hat{B}_{6B} M_2 \\
& - \delta m_2 \delta\hat{m}_{2^*} M_{4b(2^*)} - \hat{B}_2(\delta\hat{m}'_2 M_{4b(2^*)} + \hat{B}'_2 M_{4b}) - \delta m_2 \delta\hat{m}_{4b(2^*)} M_{2^*} \\
& - \hat{B}_2(\delta\hat{m}_{4b(2')} M_{2^*} + \hat{B}_{4b(2')} M_2) - \delta\hat{m}_{4b}\delta\hat{m}_{2^*} M_{2^*} \\
& - \hat{B}_{4b}(\delta\hat{m}'_2 M_{2^*} + \hat{B}'_2 M_2) + \delta m_2 (\delta\hat{m}_{2^*})^2 M_{2^*} + \hat{B}_2 \delta\hat{m}'_2 \delta\hat{m}_{2^*} M_{2^*} \\
& + \hat{B}_2 \hat{B}'_2 (\delta\hat{m}'_2 M_{2^*} + \hat{B}'_2 M_2) \\
& + I_{6B3} M_2 + 2 I_{4l} I_2 M_2 + (I_2)^3 M_2 + I_{4l} \Delta M_{4b} + (I_2)^2 \Delta M_{4b} \\
& + I_2 \Delta M_{6B} + M_{2^*}[I] \Delta\delta m_{6B} + I_{4b(2^*)} \Delta\delta m_{4b}.
\end{aligned}
$$

References

1. P. Kusch and H. M. Foley, Phys. Rev. **72**, 1256 (1947).
2. J. Schwinger, Phys. Rev. **73**, 416L (1948).
3. A. Rich and J. C. Wesley, Rev. Mod. Phys. **44**, 250 (1972); T. Kinoshita, *Shelter Island II*, Eds., R. Jackiw, N. N. Khuri, S. Weinberg, and E. Witten (MIT Press, Cambridge, London, 1985), pp.278-297.
4. R. S. Van Dyck, Jr., P. B. Schwinberg, and H. G. Dehmelt, Phys. Rev. Lett. **59**, 26 (1987).
5. J. Schwinger, Phys. Rev. **82**, 664 (1951).
6. R. G. Newton, Phys. Rev. **96**, 523 (1954).
7. W.-Y. Tsai and A. Yildiz, Phys. Rev. D **8**, 3446 (1973).
8. A. Petermann, Helv. Phys. Acta **30**, 407 (1957).
9. C. M. Sommerfield, Phys. Rev. **107**, 328 (1957).
10. G. S. Adkins, Phys. Rev. D **39**, 3798 (1989).
11. K. A. Milton, W. Y. Tsai, and L. L. De Raad, Jr., Phys. Rev. D **9**, 1809 (1974); L. L. De Raad, Jr., K. A. Milton, and W. Y. Tsai, Phys. Rev. D **9**, 1814 (1974).
12. J. A. Mignaco and E. Remiddi, Nuovo Cimento **60A**, 519 (1969); R. Barbieri and E. Remiddi, Nucl. Phys. **B90**, 233 (1975); R. Barbieri, M. Caffo, and E. Remiddi, Phys. Lett. **57B**, 460 (1975).
13. M. J. Levine and R. Z. Roskies, Phys. Rev. D **9**, 421 (1974).
14. M. J. Levine, H. Y. Park, and R. Z. Roskies, Phys. Rev. D **25**, 2205 (1982).
15. M. J. Levine and J. Wright, Phys. Rev. D **8**, 3171 (1973).

16. R. Carroll, Phys. Rev. D **12**, 2344 (1975).

17. P. Cvitanovic and T. Kinoshita, Phys. Rev. D **10**, 4007 (1974).

18. P. Cvitanovic and T. Kinoshita, Phys. Rev. D **10**, 3978 (1974).

19. P. Cvitanovic and T. Kinoshita, Phys. Rev. D **10**, 3991 (1974).

20. S. J. Brodsky and T. Kinoshita, Phys. Rev. D **3**, 356 (1971).

21. J. Aldins, S. J. Brodsky, A. Dufner, and T. Kinoshita, Phys. Rev. Lett. **23**, 441, (1969); Phys. Rev. D **1**, 2378 (1970).

22. T. Kinoshita and W. B. Lindquist, to be published in Phys. Rev. D. A preliminary value was reported in T. Kinoshita, IEEE Trans. Instrum. Meas., **38**, 172 (1989).

23. M. Caffo, S. Turrini, and E. Remiddi, Phys. Rev. D **30**, 483 (1984); E. Remiddi and S. P. Sorella, Lett. Nuovo Cimento **44**, 231 (1985).

24. T. Kinoshita and W. B. Lindquist, Phys. Rev. D **27**, 867 (1983).

25. T. Kinoshita and W. B. Lindquist, Phys. Rev. D **27**, 877 (1983).

26. T. Kinoshita and W. B. Lindquist, Phys. Rev. D **27**, 886 (1983).

27. T. Kinoshita and W. B. Lindquist, Phys. Rev. D **39**, 2407 (1989).

28. B. Lautrup, Phys. Lett. **69B**, 109 (1977); M. Samuel, Nuovo Ciment Lett., **21**, 227 (1978); M. Caffo, S. Turrini, and E. Remiddi, Nuclear Phys. **B141**, 302 (1978); M. L. Laursen and M. A. Samuel, Oklahoma State University preprint, (1979).

29. C. Itzykson, G. Parisi, and J.-B. Zuber, Phys. Rev. D **16**, 996 (1977).

30. T. Kinoshita, *New Frontiers in High Energy Physics*, Eds., B. Kursunoglu, A. Perlmutter, and L. F. Scott (Plenum, New York, 1978), pp.127-143.

31. M. E. Cage *et al.*, IEEE Trans. Instrum. Meas., **38**, 284 (1989).

32. E. R. Williams *et al.*, IEEE Trans. Instrum. Meas., **38**, 284 (1989).

33. F. G. Mariam *et al.*, Phys. Rev. Lett. **49**, 993 (1982).

34. S. Brodsky and S. Drell, Phys. Rev. D **22**, 2236 (1980); M. E. Peskin, in *Proc. of Intl. Symp. on Lepton Photon Interactions at High Energies*, Bonn, 1981, ed. by W. Pfeil (Bonn 1981), p.880; L. Lyons, Prog. Part. Nucl. Phys. **10**, 227 (1983).

35. T. Cowan *et al.*, Phys. Rev. Lett. **56**, 444 (1986).

36. A. Schaefer *et al.*, J. Phys. G **11**, L69 (1985); A. Schaefer *et al.*, Phys. Lett. **149B**, 455 (1984); A. B. Balantekin *et al.*, Phys. Rev. Lett. **55**, 461 (1985); A. Chodos and L. C. R. Wijewardhana, Phys. Rev. Lett. **56**, 302 (1986).

37. N. Nakanishi, Prog. Theor. Phys. **17**, 401 (1957).

38. T. Kinoshita, J. Math. Phys. **3**, 650 (1962).

39. Y. Shimamoto, Nuovo Cimento **25**, 1292 (1962).

40. J. D. Bjorken and S. D. Drell, *Relativistic Quantum Fields* (McGraw-Hill, New York, 1965).

41. H. Strubbe, Compt. Phys. Commun. **8**, 1 (1974); **18**, 1 (1979).

42. G. P. Lepage, J. Comput. Phys. **27**, 192 (1978).

43. G. Källén and A. Sabry, K. Dan. Vidensk. Selsk. Mat.-Fys. Medd. **29**, No. 17 (1955).

44. T. Engelmann and M. J. Levine (unpublished), quoted in M. J. Levine, H. Y. Park, and R. Z. Roskies, Phys. Rev. D **25**, 2205 (1982).

45. T. Kinoshita and W. B. Lindquist, Phys. Rev. D **27**, 853 (1983).

46. B. E. Lautrup, in *Proceedings of the Second Colloquium in Advanced Computer Methods in Theoretical Physics*, ed. by A. Visconti (Univ. of Marseille, Marseille, 1971).

Anomalous Magnetic Moment of Single Electrons and Positrons: Experiment

Robert S. Van Dyck, Jr.*
Department of Physics, FM-15
University of Washington
Seattle, Washington 98195

Contents

---

* Supported by a grant from the National Science Foundation.

1. Introduction

1.1 Historical Overview

The lofty goal to which some metrologists aspire is the possibility of making measurements of fundamental constants which are so precise that they can stand the test of time until another more brilliant technique and/or technology emerges which will push precision well beyond its present limitations. The g-2 experiment at the University of Washington has fulfilled this dream far beyond one's reasonable expectations. A number of previous reviews of this fascinating experiment can be found in the literature with various depths of treatment.[1,2,3,4,5,6,7] The history of this work, however, really begins in 1947 with the first precision measurement[8] of the hfs in hydrogen and deuterium, which generated a puzzling 0.2% discrepancy between the measured and predicted values. The theoretical prediction was based on the accepted Dirac g-factor of 2. We now realize the limitations of this theory. The first experiment specifically designed to measure this g-factor was conducted[9] by Louisell, Pidd, and Crane in 1953. Their reported accuracy of 1% for the magnetic moment of 420 keV electrons did not however provide an experimental value for the anomalous part "a_e" of the free electron's g-factor given by

$$g = 2(1 + a_e)\,. \tag{1}$$

For free electrons, the first *direct* measurement[10] of the anomaly (with a 3% accuracy) was obtained by Dehmelt at the University of Washington using an experiment in which electrons are stored in the field of a positive-ion cloud diffusing slowly in a dense inert gas. The substitution of the quadrupole trap, similar to that shown in Fig. 1, for the optical pumping cell began the illustrious history of the "Penning trap."

The first high precision measurements on the g-factor, using this Penning trap and the ion-storage-collision technique, which had been previously developed[11] by Dehmelt and Major, was conducted by Gräff and associates in 1968. The Penning trap was now located in the uniform magnetic field region of a standard atomic beam machine[12]. The state-selected sodium beam then interacts via charge exchange collisions with the trapped electrons in order to polarize the cloud. Subsequent spin-dependent energy-transfer collisions with the polarized atomic beam then leads to an observable *change* in the number of electrons still trapped after a fixed interaction time. By applying a microwave field at the spin precession frequency ω_s and subsequently an rf field at the spin-cyclotron difference frequency ω_a, the degree of polarization of the electron cloud could be varied. As a result,

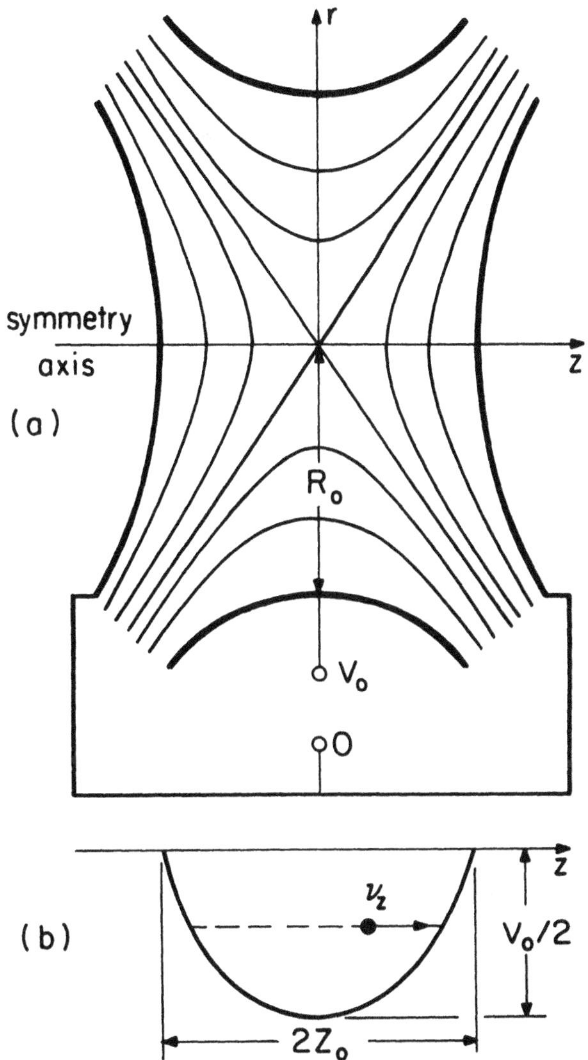

Figure 1: *Electric trapping potentials defined by the Penning trap. (a) Electrode surfaces (two endcaps and one ring) are placed upon a given set of equipotentials, amongst the family that satisfies Laplace's equation, by choosing a particular R_0 and Z_0. The z-axis is the rotational axis of symmetry. (b) An axial frequency is obtained by choosing the appropriate bias potential V_0 for the ring relative to the endcaps (from Ref. 3).*

the free electron g-factor was measured[13] to an accuracy of 0.3 ppm:

$$\frac{1}{2}g(e^-) = \frac{\omega_s}{\omega_s + \omega_a} = 1.001\,159\,660(300)\,. \tag{2}$$

This same experiment was then modified to use the non-destructive bolometric detection technique developed[14] at the University of Washington by Dehmelt and Walls whereby one observes the mean square value of the (thermal) currents induced into the trapping electrodes. This modified experiment yielded the g-factor with approximately the same precision.[15] The major limitations in this method were the relatively large linewidths attributable to relativistic kinematics, second-order Doppler effects, and magnetic field inhomogeneity over the large electron cloud. The ultimate limitation to this particular method is the electrostatic cloud shifts[16] (associated with space charge) of the anomaly resonance at ω_a.

In contrast, the method introduced by Crane at the University of Michigan (see Wilkinson and Crane[17] and Rich[18]) had enjoyed the most success prior to the present sequence of "geonium"[19] experiments at the University of Washington. Typically, bunches containing $< 10^4$ energetic (≈ 100 keV) electrons were localized (for several milliseconds) in an evacuated ($\sim 2 \times 10^{-8}$ Torr) magnetic mirror trap, formed by a slightly inhomogeneous 100–1000 Gauss magnetic field. Referred to as the free precession method, the scheme directly observed the relative orientation between the precessing spin and the electron's orbital momentum after the fixed containment time. The spin dependence inherent in Mott scattering was utilized for both producing the initial polarization and for analyzing the final relative orientation, which then yielded the direct observation of the spin-cyclotron-beat frequency $\omega_a = \omega_s - \omega_c$. This technique was dramatically refined[20] by Wesley and Rich at Michigan, achieving a final precision of 3 ppb in the electron's g-factor:

$$\frac{1}{2}g(e^-) = 1.001\,159\,657\,7(3\,5)\,. \tag{3}$$

The free-precession method succumbed at this point to some potentially serious limitations (see for instance, Ford and Granger[21]) such as finite observation time, relativistic mass corrections, space-charge shifts and large containment volumes with the subsequent need for precise determination of the average magnetic field over this trapping volume. The most serious limitation is the need to map the magnetic field in a separate experiment, and to relate the electron cyclotron frequency by means of NMR measurements using known water sample configurations. Proposals[22] for a 10 to 100 fold improvement of this experiment did not materialize.

The work initiated by Dehmelt in 1959 led to the first non-destructive detection of electron clouds in the Penning trap at very low pressures ($< 10^{-11}$

Torr), cooled by axial coupling to a resonant tuned circuit. Using the bolometric technique,[14] either the excitation of the cyclotron resonance or an alternating excitation of spin and anomaly frequencies could be applied to produce an increase in the monitored axial temperature. This scheme produced the reported g-factor measurement:[23]

$$\frac{1}{2}g(e^-) = 1.001\,159\,658(80)\,. \tag{4}$$

whose precision appeared to be limited by electrostatic cloud perturbations. Then, noting this limitation on space charge, Dehmelt and colleagues leaped to the logical conclusion that the cloud should be reduced to its irreducible limit, i.e. a single electron, capable of being continuously observed for several days (and now months). The successfully observed[24] "monoelectron oscillator" subsequently made it attractive to complete the development of the second critical aspect of the geonium experiment, that is the axial frequency shift detector,[25] also referred to as the continuous Stern-Gerlach effect.[26] Thus, in striking contrast to other methods, these geonium experiments are conducted on the *same* individual cold (< 1 meV) electron, confined to a volume $< 10^{-7}$ cm^3 in a Penning trap with extremely well-defined electric fields and a homogeneous 50 kG magnetic field with superimposed weak (< 20 G deep) magnetic bottle. In addition, both cyclotron and anomaly frequencies are measured on this *same* electron by high precision rf resonance techniques to yield the recent result:[27]

$$\frac{1}{2}g(e^-) = 1 + \frac{\omega_a}{\omega_c} = 1.001\,159\,652\,188\,4(4\,3)\,. \tag{5}$$

It is expected that this technique will yield another order of magnitude in precision which will have a two-fold purpose. First, since experiment and theory can not be directly compared unless a more accurate independent measurement of the fine structure constant, α, is made, the QED theory and a_e experiment can be used to define a more accurate value of α which is presently

$$\alpha^{-1}(\text{QED}) = 137.035\,991\,4(1\,1)\,. \tag{6}$$

This precision is ~ 3 times more precise than existing direct measurements of α (see Sec. 7.2). Secondly, the experiment is intrinsically identical for positrons and electrons, except for the change in sign of the ring voltage and the initial loading/centering in a storage trap and subsequent transferring of positrons into the experiment trap (see Sec. 2.3.) The most recent measurement[27] of the positron's g-factor is

$$\frac{1}{2}g(e^+) = 1.001\,159\,652\,187\,9(4\,3)\,. \tag{7}$$

Because common systematics cancel out, the ratio of these measured g-factors yield the more accurate matter/antimatter comparison:

$$g(e^-)/g(e^+) = 1 + (0.5 \pm 2.1) \times 10^{-12} \tag{8}$$

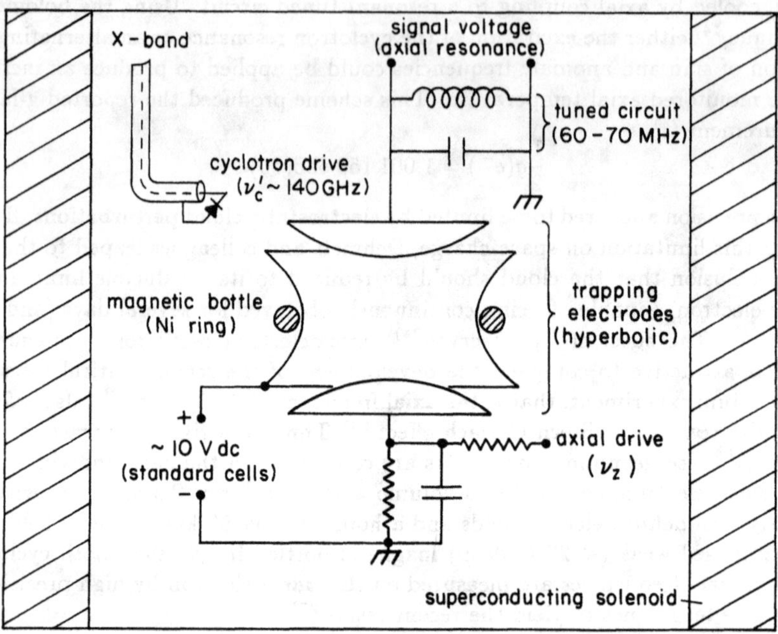

Figure 2: *Schematic of the geonium apparatus. The appropriately biased hyperbolic endcaps and ring electrodes trap the charge axially while coupling the driven harmonic motion to an external LC circuit tuned to the driven axial frequency. Radial trapping of the charge is produced by the strong magnetic field obtained from a superconducting solenoid (from Ref. 2).*

1.2 Basic Description

The Penning trap which isolates the charged particles consists of a combination of static electric and magnetic fields (see Fig. 1 and 2). The magnetic field B_0 which provides the confinement in the radial direction is applied along the z-axis of symmetry in Fig. 1 and is usually obtained from a very stable and uniform superconducting magnet. The electric field which produces the axial restoring force is obtained from the stable voltage V_0, provided by the standard cells shown in Fig. 2. This voltage is applied between the ring electrode and two common endcap electrodes as defined by the following hyperboloids of revolution (in cylindrical coordinates):

$$\begin{aligned} z^2 &= Z_0^2 + r^2/2 \qquad (endcaps) \\ r^2 &= R_0^2 + 2z^2 \qquad (ring) \end{aligned} \tag{9}$$

where $2Z_0$ and $2R_0$ represent the minimum endcap separation and the minimum ring diameter respectively. The resulting harmonic potential (shown in Fig. 1b), which arises from a selection of R_0 and Z_0 from the family of equipotentials shown in Fig. 1a, is given (also in cylindrical coordinates) by:

$$V(r,z) = V_0 \frac{r^2 - 2z^2}{4d^2} \tag{10}$$

where d is the characteristic trap dimension defined by $4d^2 = 2Z_0^2 + R_0^2$. A symmetric trap is described by $2Z_0^2 = R_0^2$ which yields $d = Z_0$.

Actual devices (examples of which will be discussed in Sec. 2) achieve their ultra high vacuum because of the liquid helium environment within which they are placed. This low temperature also improves the signal/noise achieved by the preamplifier which is located (within the cryogenic bath) as close to the signal endcap as possible. (See Sec. 3.2 for a discussion of this system.) The opposite endcap is used to drive the axial motion. and the resulting axial motion induces an image current into the external tuned circuit (resonant at the axial frequency) which then developes a voltage which can be amplified by the preamp. The magnetic bottle, shown schematically as a nickel wire in Fig. 2, is used to couple the total magnetic moment to the axial resonance. This will be described in Sec. 3.4. The microwave power required to excite the cyclotron resonance is obtained from an X-band source which is multiplied by a Schottky diode in order to obtain a harmonic at the appropriate frequency.

The motion of the trapped electron in the ideal Penning trap is described by the following set of equations:

$$\omega_z^2 = \frac{eV_0}{m_e d^2} \tag{11a}$$

$$\omega_c = \frac{eB_0}{m_e c} \tag{11b}$$

$$2\omega_\pm = \omega_c \pm (\omega_c^2 - 2\omega_z^2)^{1/2} \,. \tag{11c}$$

The first frequency in Eq. 11 (for ω_z) is associated with the axial harmonic oscillation of the trapped particle and is the subject of Sec. 3. The second is its unperturbed cyclotron frequency (in the absence of the Penning trap). The last equation represents the radial motion; in the ideal case, $\omega_- \equiv \delta_e$ is the slow magnetron frequency (the subject of Sec. 4) and $\omega_+ \equiv \omega_c'$ is the perturbed cyclotron frequency (the subject of Sec. 5). Thus, it follows that

$$\omega_c' = \omega_c - \delta_e \quad \text{and} \quad \delta_e = \omega_z^2/2\omega_c' \,. \tag{12}$$

However, in a less-than-ideal Penning trap, one finds that the observed magnetron frequency, ω_m, is slightly shifted from its value δ_e given by Eq. 12. But, because of its relatively small magnitude for electrons (by construction), one may ignore the difference between δ_e and ω_m; they may be used interchangeably in the remainder of this paper. (However, see Sec. 4.3.)

The last of the important frequencies which needs to be introduced is associated with the spin magnetic moment of the electron. It is purely quantum mechanical in nature and is given by

$$\hbar\omega_s = g\mu_B B_0 \tag{13}$$

where μ_B is the Bohr magneton ($\mu_B \equiv e\hbar/2m_e c$). Producing an adequate strength oscillating magnetic field directly at ω_s turns out to be quite difficult and inefficient at 140 GHz. Since ω_s and ω_c' are nearly equal, differing only by 0.1%, a far more convenient frequency to use is $\omega_a' = \omega_s - \omega_c'$ which is referred to as the anomaly frequency and is the subject of Sec. 6. Thus, in a Penning trap, the anomaly is obtained from

$$a_e = \frac{\omega_a' - \delta_e}{\omega_c' + \delta_e} \tag{14}$$

which turns out to be accurate against all perturbations that have been considered so far, at least at the 1 part in 10^{13} level.

2. Apparatus

2.1 Various Traps

The first compensated trap used in the geonium g-2 experiments[1,28] is shown in Fig. 3. It was constructed on a 20-lead tungsten-to-glass flatpress pin base which was sealed into one end of a 4.0 cm o.d. glass tube. All trap electrodes were made of molybdenum since it is commonly believed that molybdenum oxide would be uniform and thus not be subject to large electrical patch effects. The endcap closest to the pin base is designated as the signal cap, thus leaving the far endcap for rf drives. In this early version of the compensated trap, roughly flat guards were placed between the hyperbolic endcaps and ring electrodes in order to tune out the fourth order term in the potential. Evidence for the trap's ability to compensate the potential well was demonstrated[29] in 1975 with nearly a hundred fold reduction in the driven axial linewidth. A field emission point (FEP) is shown in the drive cap and is negatively biased relative to the endcap to produce an ionizing electron beam along the trap's symmetry axis. Finally, a small nickel ring symmetrically placed in the hyperbolic ring electrode is used to observe the magnetic state via the precise harmonic oscillation frequency, where the coupling is determined solely by geometry, the trap's potential, and the charge-to-mass ratio of the trapped charge.

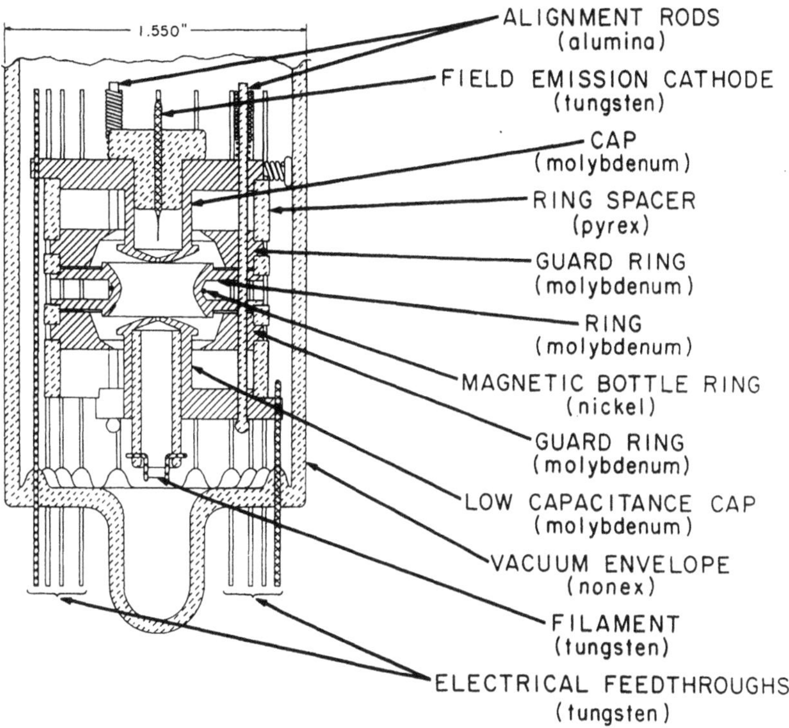

Figure 3: *First high-precision compensated Penning trap made of molybdenum. The usual endcap-ring-endcap combination is enhanced by the addition of two guard rings placed symmetrically between each endcap and the ring electrode. By carefully adjusting the bias on these guards rings, the trapping potential can be made more than 100 times more harmonic. Additional features include a field emission cathode, back-up tungsten filament, and a nickel wire loop which generates the magnetic bottle (from Ref. 1).*

A standard Penning getter-ion discharge pump was constructed, mounted onto a second 20-lead pin base, and then sealed into the other end of the glass cylinder to produce a closed vacuum tube. The entire apparatus was then baked under vacuum at 450°C for several days before being sealed off. However, the assembly process for these glass systems was very time consuming and certainly did not encourage changes in the basic apparatus. In addition, they are fragile and sometimes leaked. Thus, the reusable metal envelope with indium seals was developed as shown in Fig. 4. The outer envelope is made of beryllium copper with threaded rings (of the same material) that are used to apply pressure to the indium which is melted into each end of the vacuum tube. The pin bases are made

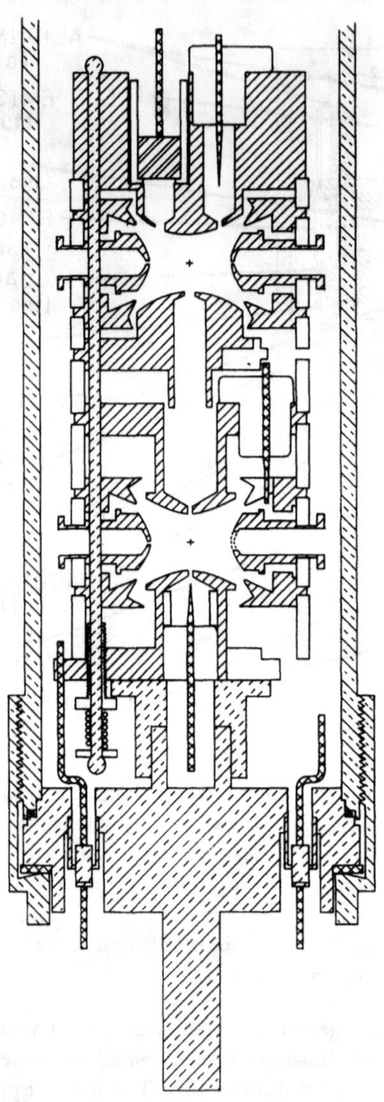

Figure 4: *Cross-section of the double trap apparatus used in the electron-positron comparison experiment. An OFHC copper pin base with non-magnetic feedthrus is sealed to an all-metal beryllium copper envelope via a compressed indium O-ring. The lower trap is a well compensated Penning trap used for precision measurements; the upper trap is used only for the storage of positrons which can be transferred into the lower "experiment" trap when needed (from Ref. 2).*

of OFHC copper and each has several ceramic-to-constantan feedthrus silver-soldered into the base. The upper pin base (not shown) is also used for the sputter-ion pump whereas the lower base contains the trap located at the solenoid's field center. This particular apparatus is the double trap for the positron/electron comparison.[2,27,30,31] This all-metal envelope requires substantially less time for assembly than the previous glass tubes, is more reliable and robust, and has been reused several times.

2.2 Compensation Feature

One of the main features of the new style Penning traps is the use of compensation electrodes and Fig. 4 illustrates the type which is incorporated into all of our present traps. It has guard rings that protrude into the truncation region between endcap and ring electrodes (compared to the flat style shown in Fig. 3). The early reasoning for constructing this style involved getting more control over the anharmonic terms in the potential. Even though in principal there should be very little dependence on shape[32] as far as tunability is concerned, the sharp style used in present traps can tune out substantially larger perturbations simply because they penetrate more deeply into the trap.

Several methods are available to tune up the trap. The most common method uses the amplitude asymmetry versus frequency as shown in Fig. 5. The procedure involves symmetrizing the axial resonance using increasingly stronger drives. Until the 6*th*-order perturbation enters, symmetry should occur at the same guard setting. A second method involves minimizing the noise in the correction signal of the axial-frequency-shift detector described in the Sec. 3.4. Upon varying the compensation potential, a dc shift in the well depth is also observed by means of this axial-frequency-shift detector. The size of this effect was observed to depend exponentially on how close the compensation electrodes are placed in the truncation region relative to trap center and the size of the gap at truncation. In addition, the ability to tune the trap was observed to scale with this shift in well depth for symmetric Penning traps. Recently, these observations were verified[32] theoretically by Gabrielse using a numerical relaxation program which computes the exact fields in the real Penning trap. For Fig. 5, the normalized 4*th*-order coefficient, C_4, of the anharmonic potential is negative when pulled to the low frequency side, indicating that the guard potential was set too low.

2.3 Adaptations for Positrons

Since the basic geonium apparatus can accept either charge sign simply by applying the appropriately signed potential, the full sensitivity of the experiment is available for positrons as well as for electrons with all the systematics that are discussed in the following sections being the same for both. Thus, the principal challenge

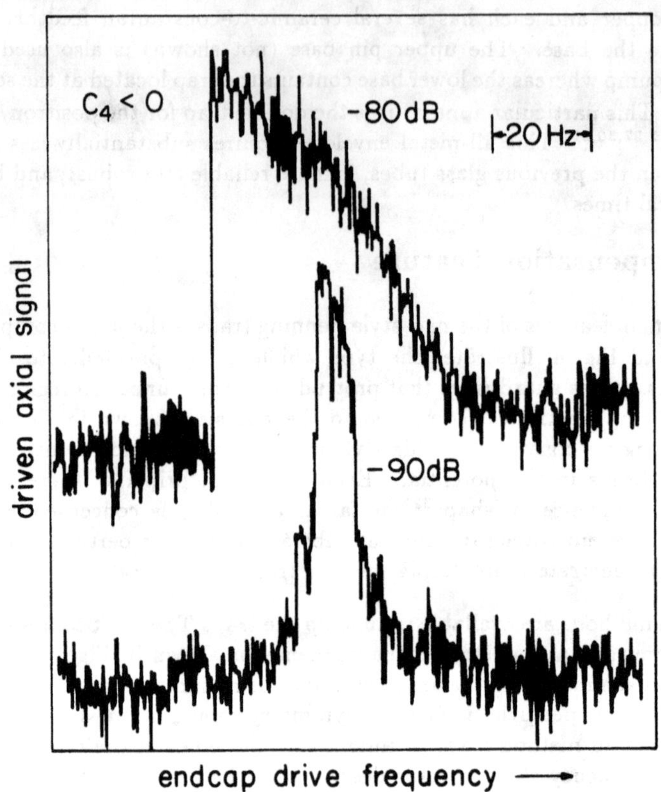

Figure 5: *Typical anharmonically pulled axial resonance. When the normalized coefficient, C_4, of the 4th-order term in the potential is negative (i.e. the guard potential is set too low), the resonance is pulled to low frequency side. Note that 10 dB less drive nearly restores the completely symmetrical lineshape.*

for this variation of the geonium experiment is the trapping of a single positron. The loading of large clouds (~ 100) of positrons was demonstrated[33)] by Schwinberg in 1979 and only a short description of the trapping process is available in the literature.[30,31,34)] [However, the full description is available from University Microfilms.[33)]] Radiation damping is the basic mechanism for losing some of the positron's energy, such that it might fall into a potential well. For the axial motion, this comes in the form of the resistance obtained from the *LC* circuit tuned to ω_z. For the radial motion, synchrotron radiation quickly damps the 50–100 keV initial energy in the cyclotron motion within a few seconds down to the thermal range.

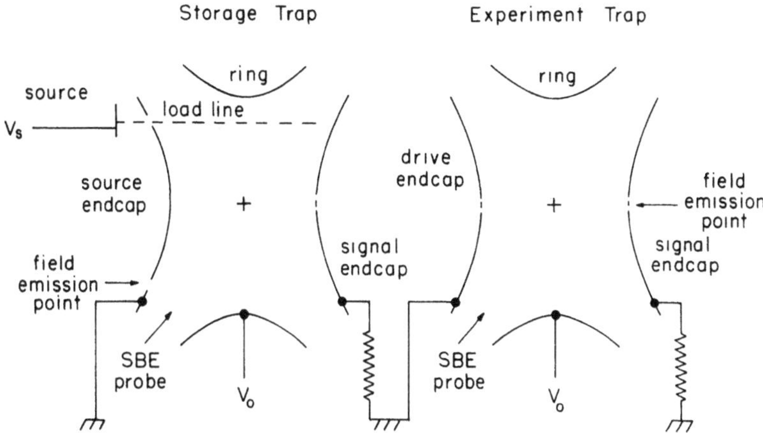

Figure 6: *Schematic of the double-trap configuration. The storage trap contains a sealed ^{22}Na positron emitter located at $\frac{3}{4}R_0$. The sideband excitation (SBE) probes are used in the radial centering process for off-axis-loaded positrons and position stabilizing in the experiment trap. Signal endcaps are tuned to the axial frequency via an external inductor in order to observe the axial motion driven by an rf signal applied to the opposite endcap (from Ref. 30).*

Since the requirements imposed on the trapping electrodes for high precision measurements are inconsistent with those imposed in order to achieve an adequate capture rate, it was decided to have two traps, one optimized to capture the positrons and one designed to carry out the high precision g-2 measurements. This double trap configuration is shown schematically in Fig. 6. The source for the positrons consists of a 2.6 yr half-life sodium-22 salt in a sealed container and is visible in Fig. 4. located inside the largest endcap (farthest from the pin-base). In order to fit both a field emission point (FEP) and the source within the same endcap at the same radius, as well as to improve the trapping rate. each is mounted at $\frac{3}{4}R_0 \sim 3.55$ mm from the central axis. The FEP is required in order to optimize the centering process using an adequate number of easily obtainable charges. The source hole through this endcap is 1.2 mm in diameter and the line, parallel to the trap axis, which passes through this source hole is referred to as the "load line".

The positrons of interest (with 50–100 keV in the radial motion) will travel down this load line guided by the axial magnetic field, reflect at the opposite endcap, reflect again at the source cap because the magnetron motion has rotated it away from the load line, and finally lose enough energy after one magnetron period to be permanently trapped. Only a very small fraction of those emitted satisfy all these criteria, typically $\sim 10^{-9}$. In fact, it was estimated[33)] that about 90 positrons/h should have been trappable in 1979, but experimentally, only about

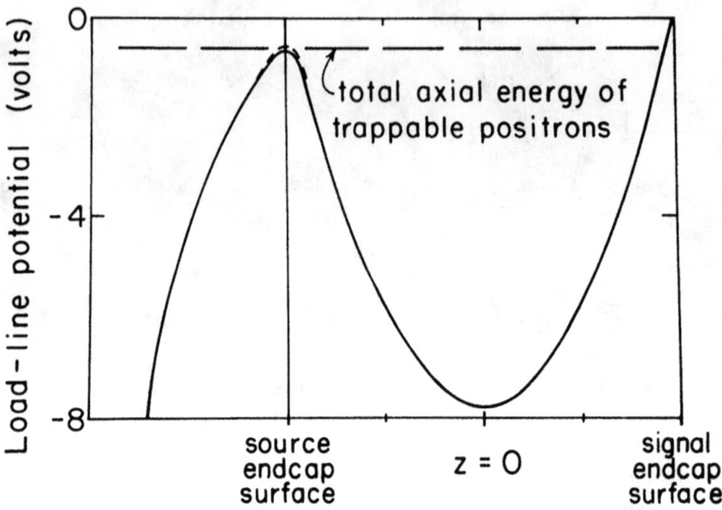

Figure 7: *Potential along the load-line. The source is biased at −300 V relative to the endcap in order to produce this potential variation along the line parallel to the z-axis that passes through the hole in the source endcap. The broken line greatly exaggerates the slightly higher potential (∼ 1 mV) seen by a positron on the first return to the entry hole (shifted off the load-line by the magnetron motion) (from Ref. 34).*

20 positrons/h were observed for the 0.5 mCi source.

Figure 7 illustrates the applied potential along the load line. The shape of the repulsive potential hill (top endcap surface in Fig. 4) is established by the −300 V bias applied to the source relative to the grounded cylindrical walls leading to the source hole in the endcap. The potential well between the endcaps is established as usual by the Penning trap ring electrode, biased in this case to −9.1 volts in order to provide the well into which the positrons fall. These potentials then establish the antitrapping hill at the center of the source hole, with a depth ∼ −0.7 volts. Those positrons with axial energies less than 0.7 eV will reflect at the signal cap and most of those will exit right back to the source. Some, however, will have a total axial energy at this entry point only marginally above the potential hill (say 0 to 1 meV) and these are the tiny fraction which have any chance of being trapped.

Once loaded, we seldom have enough charge that the positrons can be observed at the $\frac{3}{4}R_0$ position directly. This general lack of sensitivity requires typically 50 positrons in number before we can say they are definitely trapped there.

This loss of S/N is due to the extreme anharmonic nature of the trap in which large holes occur in the source and signal endcaps, but without symmetric counterparts placed in opposite endcaps. One set of guard electrodes in general can not compensate all perturbations at different locations simultaneously. Thus, the requirement for centering is often done without the aid of knowing exactly where the positrons are in frequency space. The problem of shifting frequencies while centering is minimized, however, by tuning the guards for a relatively flat spectrum between $\frac{3}{4}R_0$ and the center, as determined using electrons loaded at $\frac{3}{4}R_0$.

On occasion, the loading process would be found to be very inefficient. One cause for this is the presence of background gas. In the early stages of the experiment with a glass envelope, we had a leaky feedthru and the gas pressure (probably above 10^{-10} Torr) was scattering some of the initial high-energy of the radial motion into the axial motion via random small angle scattering. Later, we observed that any strong drive seen by the positrons during loading can also inhibit the process. Clearly, very little excitation is needed to exceed the 100 to 1000 μeV of axial kinetic energy allowed during the critical, single axial cycle of the first turn around. As a result of this mechanism, there are times, in the present noisy environment, in which positrons load only about one a day, though they are expected to load ten times faster from simple scaling of the source strength.

Finally, once a positron cloud is observed to be centered, as indicated by a clean synchronously detected axial resonance located at the correct frequency and ring voltage, we should be ready for the transfer into the experiment trap. However, we have observed yet another dirty effect which must be taken care of first. During the positron loading process, positive ions have been self-loading into the trap from either the γ-ray or positron flux. We know that the γ-rays are capable of doing this since electrons load automatically into the experiment trap with a rate that has varied over the last five years from once every few hours to about once a day. The ions which are centered with the positron cloud appear to cause large enough space-charge perturbations to make the centered positron cloud invisible except for moderately strong axial drives. The cure for this problem consists of applying very strong axial drives at the ions' axial frequency while the positrons are strongly cooled by the tuned circuit and the centering drive. When the ions are removed, the cloud becomes visible with weak axial drives. The penalty for not cleaning these out first is a very low transfer efficiency. In fact on many occasions, entire positron clouds have been lost without transferring a single positron to the experiment trap. The actual transfer is accomplished by pulsing the two adjacent endcaps (storage trap signal cap and experiment trap drive cap) to the common ring potential for a few microseconds. Typical transfer efficiencies between 25 and 50% have been observed; however with pulses much longer than 10 μs, the transfer tends to be very inefficient, possibly due to radial drifting during the passage between traps.

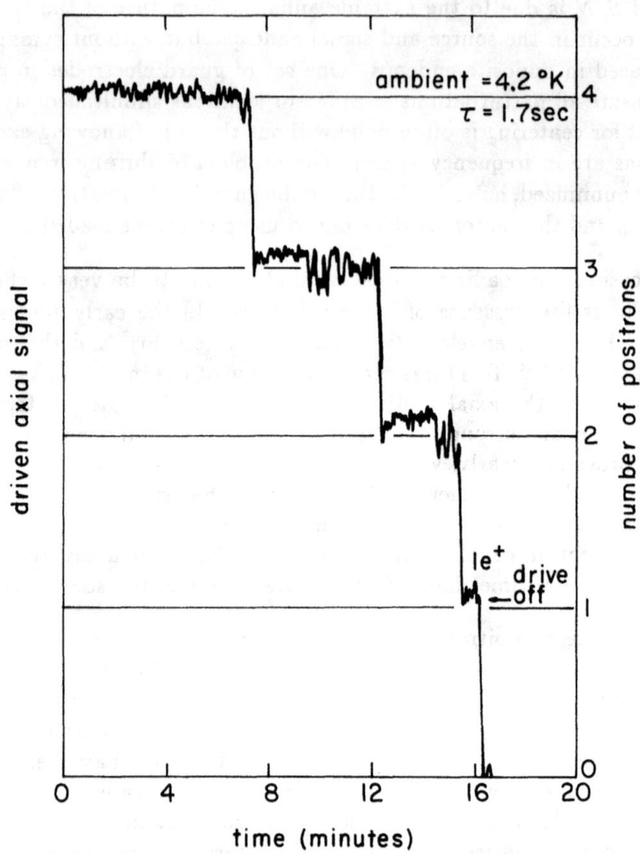

Figure 8: *Positron ejection record. Positrons transferred from storage to the experiment trap are detected via strong off-resonance drives (with fixed amplitude and frequency). Then, intense rf pulses at* $\omega_z + \omega_m$ *are applied to the SBE probe in order to systematically eject one charge at a time until only one remains (from Ref. 30).*

Once identified in the experiment trap, the exact number can be determined rapidly by two methods: the weakly driven axial linewidth and the hard off-resonance driven amplitude both scale as the number (see Sec. 3.3). Having previously calibrated the off-resonance drive, the cloud in Fig. 8 was immediately identified to contain four positrons. Subsequently, the excess beyond one were systematically ejected by using intense rf pulses, also at the sideband cooling frequency. The rf amplitude of the ejection/cooling pulse is carefully adjusted such that at least 10 consecutive pulses are required in order to eject one positron from the cloud. Note however that the probability for being thrown out increases as the number drops since the individual charges have larger amplitudes as the number reduces (i.e. center-of-mass amplitude scales inversely with the number trapped for fixed drives). Once a single positron is isolated, the drive signal is reduced by orders of magnitude in order to resolve the narrow (4–6 Hz) axial resonance, typical of a well-compensated trap.

3. Axial Resonance

3.1 Stable Ring-Endcap Potential

As indicated in Eq. 11a, the axial frequency depends on the applied ring potential relative to the grounded endcaps. The heart of this source of potential is a set of ten oven-stabilized standard cells which achieve an overall stability of 10 parts per billion. Also, the internal impedance of the batteries is low enough that Johnson noise is not important. To achieve the required resolution, there were several effects which had to be considered when applying V_0 such as thermal electric, contact, and patch potentials. In order to control the residual thermal electric potentials, and thus further stabilize the applied ring voltage, the liquid helium level is controlled as well. The contact and common average patch potentials, δV_+, have been investigated[1] on at least one occasion by varying V_0 and measuring ν_z. By fitting the data to the linear relation of ν_z^2 vs V_0, it has been determined that δV_+ is less than 60 mV out of a total of 10 volts applied to the ring electrode. This particular offset is not a serious problem because it can be compensated by an appropriate adjustment of the ring potential. An antisymmetric offset, $\pm\delta V_-$, has the effect of moving the center of the electron's motion away from trap center. By applying an external antisymmetric potential to the endcaps, δV_- is likewise determined[2] to be less than 60 mV and will not move the electron by much more than 0.001 cm from trap center. A shift this size is not significant since the tolerance on fabricating the electrodes is no better than this amount. Nevertheless, in early Penning traps, molybdenum was always used to minimize patch effects on the electrodes.

3.2 Detection of Trapped Charge

As indicated in Fig. 2, the trapped particle is driven on one endcap and detected on the other cap. A fairly straight forward calculation shows that driving the trapped electron is equivalent to exciting a series ℓc tuned circuit.[35] Assume an electron with charge e and mass m_e is bound between the plates of a parallel-plate capacitor (with separation, D) by a electrically neutral, massless spring with elastic

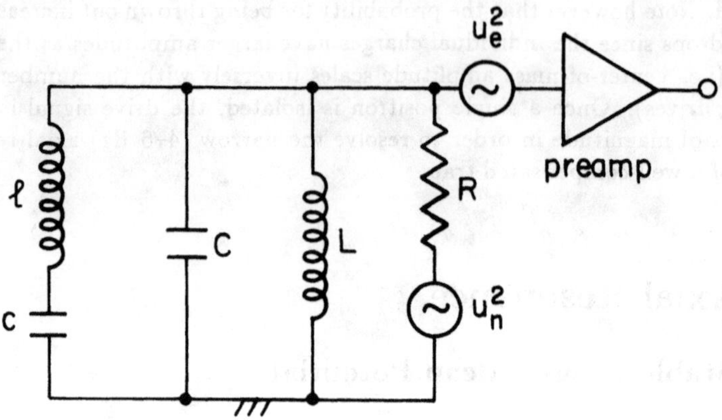

Figure 9: *Input detection circuit with equivalent ℓc representing the trapped electron. The trap's net capacity, C, is tuned out with an external inductor, L, yielding a large parallel resistor, R, with equivalent noise generator,* u_n^2*, presumably at 4 K. The preamp is represented as ideal except for the equivalent series noise generator depicted by* u_e^2*, which is presumably not at 4 K (adapted from Ref. 35).*

constant $k = m_e\omega_z^2$. For this calculation, a drive voltage V is applied across the parallel plates. The equation of motion for this single electron is given by

$$\ddot{z} + \gamma_{z,1}\dot{z} + \omega_z^2 = \frac{e}{m_e}\left(\frac{C_1 V}{D}\right). \tag{15}$$

where $\gamma_{z,1}$ represents the damping coefficient for the harmonic motion at ω_z and C_1 is the dimensionless constant of order unity which represents the finite geometry of the electrodes.[36)] For an infinite parallel plate arrangement, $C_1 = 1$, but for a Penning trap, $C_1 < 1$ because the net field at trap center is reduced by the attraction of some lines of force onto the ring electrode. (It is theoretically predicted[36)] to be 0.8 for the Penning trap where D is set equal to $2Z_0$.) Using the appropriate change of variable, $i = C_1 e\dot{z}/D$, one achieves

$$\ell_1\frac{di}{dt} + Ri + \frac{1}{c_1}\int idt = V \tag{16}$$

where ℓ_1 is associated with the combination $m_e D^2/(C_1 e)^2$, c_1 with $(C_1 e)^2/kD^2$ and R with $\gamma_{z,1}\ell_1$. Note that the reduction in field strength by trap geometry has been transferred to an effective charge $C_1 e$ at trap center. The dc trapping potentials on the Penning trap electrodes are therefore replaced by the fictitious capacitor c_1 in series with fictitious inductance ℓ_1 such that $\omega_z^2 = (\ell_1 c_1)^{-1}$. As shown in Fig. 9, the ℓc series circuit is effectively connected in parallel with the net trap capacitance, C, which is then put in parallel with an external inductor L such that the LC combination is also resonant at ω_z. This external parallel tuned circuit then provides the series R that damps the axial motion.

Now, if V is associated with an rf electric field, then the driven motion of the trapped charge will induce image currents in the signal endcap or equivalent current $i = C_1 e\dot{z}/D$ will flow through the ℓc series combination. To be observed, one would like to drop this current across a *large* resistor R. This is achieved by making the external tuned circuit have as high a Q as possible. As summarized in Table 1 for the recent molybdenum trap, some typical parameters are $C \sim 15$pf, $\nu_z \sim 64$ MHz, $Q \sim 1000$, and the equivalent parallel resistance is $R = Q/\omega_z C \sim 170\text{k}\Omega$. Now, the steady-state axial motion, driven by an electric field at frequency ω with arbitrary phase ϕ and amplitude E_{rf}, is given by.

$$z = \frac{(eE_{\text{rf}}/m_e)\sin(\omega t + \phi)}{\{(\omega_z^2 - \omega^2)^2 + \gamma_{z,1}^2\omega^2\}^{1/2}} \tag{17}$$

where $E_{\text{rf}} = C_1 V_{\text{rf}}/2Z_0$ for a given rf voltage applied to the drive endcap only. From Eq. 17, it follows that the relative driven on-resonance amplitude Z_d is

$$\frac{Z_d}{Z_0} = \left(\frac{C_1 V_{\text{rf}}}{2V_0}\right)\left(\frac{\omega_z}{\gamma_{z,1}}\right) \approx 4 \times 10^5 V_{\text{rf}} \tag{18}$$

where typically $V_{\text{rf}} \sim 10$ nV. From the expression for total energy of a harmonic oscillator $(m_e\omega_z^2 Z_d^2)$, the relative energy for a single electron in a Penning trap is:

$$\frac{W_k}{eV_0} = \left(\frac{Z_d}{Z_0}\right)^2 = \left(\frac{2V_{\text{rf}}}{C_1 eR\omega_z}\right)^2 \tag{19}$$

or its absolute axial energy is given by

$$W_k = \frac{V_{\text{rf}}^2}{R\gamma_{z,1}} \sim 10^{-4}\ \text{eV} \tag{20}$$

It should be noted that the formal definition of Q_z of the axial resonance $(Q_z \equiv \omega_z/\gamma_z)$ is the energy stored (ℓi^2) divided by the energy lost/cycle (power dissipated $\times$ period $= i^2 R/\omega_z$), from which it also follows that the axial linewidth (FWHM) is $\gamma_z = R/\ell$. For a $2\pi(6\text{ Hz})$ wide line in the experiment trap shown in Fig. 4, this relation predicts $\ell_1 \sim 4,500$ h for a single electron.

Table 1: *Summary of parameters characteristic of experiment trap in double trap configuration shown in Fig. 4.*

symbol	definition	typical value
ω_z	axial frequency	$2\pi(64.0\ \mathrm{MHz})$
ω_m	magnetron frequency	$2\pi(14.5\ \mathrm{kHz})$
ω_c'	cyclotron frequency	$2\pi(141\ \mathrm{GHz})$
ω_a'	anomaly frequency	$2\pi(164\ \mathrm{MHz})$
γ_z	axial linewidth	$2\pi(6\ \mathrm{Hz})$
ω_{mod}	ring modulation frequency	$2\pi(1.00\ \mathrm{MHz})$
δ	magnetic bottle step-size	$2\pi(1.3\ \mathrm{Hz})$
B_0	magnetic field strength	50.5 kG
B_2	magnetic bottle coefficient	$155(4)\ \mathrm{G/cm}^2$
V_0	applied ring-endcap potential	10.2 volts
$\delta V_{\mathrm{cap}}(\mathrm{offset})$	common endcap offset potential	21 mV
$2Z_0$	minimum endcap-endcap separation	6.70 mm
$2R_0$	minimum ring diameter	9.47 mm
R_m	typical minimum magnetron radius	1.4×10^{-3} cm
R_c	cyclotron radius for $n = 0$	114 Å
T.C.	trap constant	$20.0\ \mathrm{MHz}/\sqrt{\mathrm{Hz}}$
Q	tuned circuit quality factor	~ 1000
C_1	electric field coupling constant	0.78(4)
ℓ_1	single lepton inductance	4500 H
C	total tuned circuit capacitance	15 pf
R	equivalent tuned circuit resistance	170 kΩ
C_4	normalized anharmonic coefficient	$\sim 5 \times 10^{-5}$
$B_2 Z_0^2/B_0$	relative maximum magnetic bottle	3.44×10^{-4}
$eV_0/2m_ec^2$	ratio of well depth to rest energy	1.0×10^{-5}
$\hbar\omega_c^2/m_ec^2$	relativistic shift per Landau level	$2\pi(140\ \mathrm{Hz})$
kT_z	thermal energy, assuming $T_z = 4.2$ K	3.6×10^{-4} eV
$\hbar\omega_c'$	quantum of cyclotron energy	5.8×10^{-4} eV
$\hbar\omega_z$	quantum of axial energy	2.6×10^{-7} eV
$\hbar\omega_m$	quantum of magnetron energy	6×10^{-11} eV
$(eV_0/2)(R_m/R_0)^2$	magnetron energy at minimum orbit	4.5×10^{-5} eV

The equivalent current on resonance can be found by taking the time derivative of Eq. 17 (using $D = 2Z_0$):

$$i = \frac{C_1 e^2 E_{\mathrm{rf}}}{2\gamma_z m_e Z_0} = \frac{V_{\mathrm{rf}}}{R} \tag{21}$$

where the expression for ℓ_1 has been used. For a moderate drive applied to the endcaps which is effectively -160 dB down from a 13 dBm (or 1 volt rms) drive source, one finds a typical current of $\sim 10^{-13}$ amps for a single electron with ~ 6 Hz axial linewidth. This current also corresponds to a drive energy that is about equal to kT_z when $T_z = 4$ K. As a result, the on-resonance signal amplitude is about 17 nV in this case. If $\Delta\nu_{\mathrm{det}}$ is the detection bandwidth, then the signal voltage can be compared to the rms noise voltage associated with a resistance R given by

$$V_N/(\Delta\nu_{\mathrm{det}})^{1/2} = \left(4kT_z R\right)^{1/2} \tag{22}$$

which, for a 170 kΩ resistor in liquid helium, yields 6 nV/$\sqrt{Hz}$. As one can see in Fig. 9, if the trap is ideal and the ℓc circuit is tuned to make it resonate at the parallel tuned circuit frequency, the series circuit would totally short out the resistor R and its equivalent noise generator u_n^2. The remaining noise in the experiment would then come from the series equivalent input noise generator u_e^2 associated with the imperfect preamplifier. However, the anharmonic part of the real trapping potential keeps the finite temperature electron from totally shorting the trap. Experimentally, we see the noise reduce by about 50% when on-resonance relative to the off-resonance noise.

From here, the signal is post amplified and mixed down to 1 MHz before being observed in a phase sensitive detector (with a 1 MHz reference). One would normally have difficulty observing such small signals in the presence of a strong local oscillator (for which it is difficult to obtain adequate shielding of the tuned circuit) if not for the usual trick of modulating the well depth[24)] at $\omega_{\mathrm{mod}} = 2\pi(1.0\,\mathrm{MHz})$. Because of the application of an rf field of amplitude V_{mod} to the ring electrode, the instantaneous ring-to-endcap potential is given by

$$V = V_0 + V_{\mathrm{mod}} \cos \omega_{\mathrm{mod}} t \tag{23}$$

and this leads to frequency modulated sidebands of the driven motion with a modulation index, $(\omega_z/\omega_{\mathrm{mod}})(V_{\mathrm{mod}}/2V_0)$, which is chosen to be between 0.01 and 0.02. The axial motion is thus driven on either the upper or lower sideband and detected at the fundamental frequency, ω_z.

3.3 Isolating the Single Electron

It can be shown that Eq. 15 applies to the center of mass of N charges if the damping constant $\gamma_{z,1}$ is replaced by $\gamma_{z,N}$. This linewidth will scale as $N\gamma_{z,1}$ since it is proportional to $(\text{charge})^2/\text{mass}$, unlike all other coefficients which enter as charge/mass. The total current of the center-of-mass motion is then given by

$$i_{\rm cm} = \frac{NC_1 e\dot{z}_{\rm cm}}{2Z_0} \tag{24}$$

with D again replaced by $2Z_0$. The ability to isolate a single quanta of charge arises from the nature of our off-resonance drive technique. One notes from Eq. 21 that on resonance and under ideal conditions, the observable voltage signal (iR) across the tuned circuit is simply $V_{\rm rf}$, the value of the rf potential applied only to the drive endcap. There is no number dependence in this case. However, again taking the derivative of Eq. 17, and specifying a detuning $\delta\omega$ such that $\omega_z \gg \delta\omega \gg \gamma_{z,1}$, one finds that

$$\dot{z}_{\rm cm}(\text{off}) = \left(\frac{e}{m_e}\right)\frac{E_{\rm rf}}{2\delta\omega} \tag{25}$$

which, in this case, is the same for N electrons or a single. Thus, using Eq. 24, the observed off-resonance signal voltage becomes

$$V_{\rm sig}(\text{off}) = N\left[\frac{V_{\rm rf}R}{2\ell_1\delta\omega}\right]. \tag{26}$$

If the rf amplitude and the detuning are held constant, the signal voltage is proportional to the number of charge quanta in the trap.

To load these quanta of charge, a current of ~ 0.1 nA is emitted from the field emission electrode for 10–20 seconds. After each load, the detection system is switched on and the signal observed. Then, the trap is dumped by reversing the sign of the trapping potential. The experiment is repeated and the results are shown in Fig. 10. Note that there is a minimum voltage step, of which all others are integer multiples. The drive is then reduced to the nominal on-resonance power and the frequency is swept through the resonance, obtaining the result shown in Fig. 11, when the minimum voltage step is achieved. At multiples of this step, γ_z is the same multiple of $\gamma_{z,1}$. The amplitude signal (which is the Lorentzian lineshape) is used to monitor the lock signal (see Sec. 3.4) which is itself derived from the quadrature component (or dispersion shape also shown in Fig. 11). The resolution of this resonance is $\sim 1\times10^{-8}$ or about 0.5 Hz out of 60 MHz. Such a well-resolved axial resonance typically shows some noise shorting as described earlier.

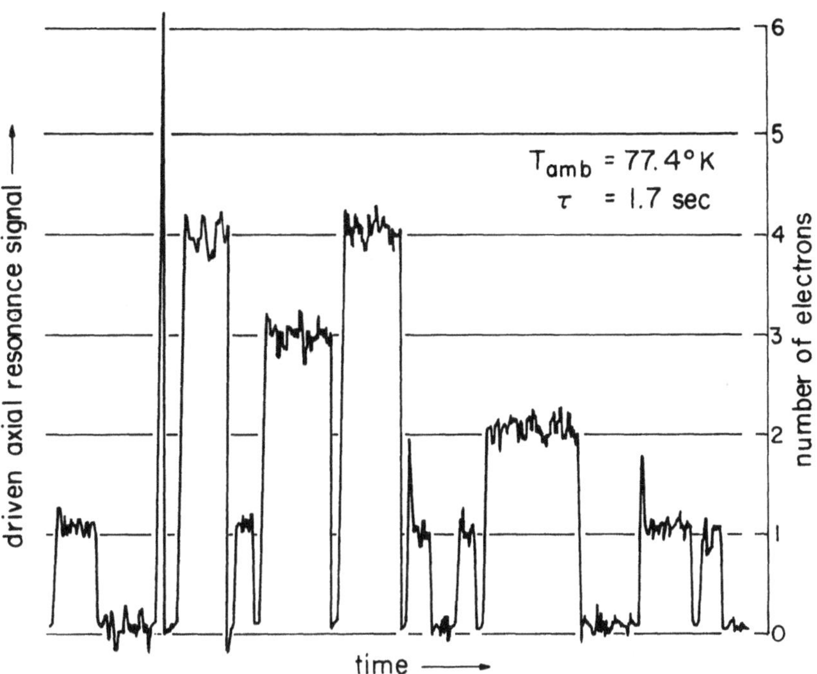

Figure 10: *Identification of a single electron. A strong off-resonance signal (with fixed frequency and amplitude) is applied continuously to the drive endcap of the trap. Then, the FEP is turned on for* ~ 10 *sec with* < 1 *nA of current until a non-zero signal is observed. The trap is occasionally dumped and reloaded until the smallest quantized signal is observed (from Ref. 1).*

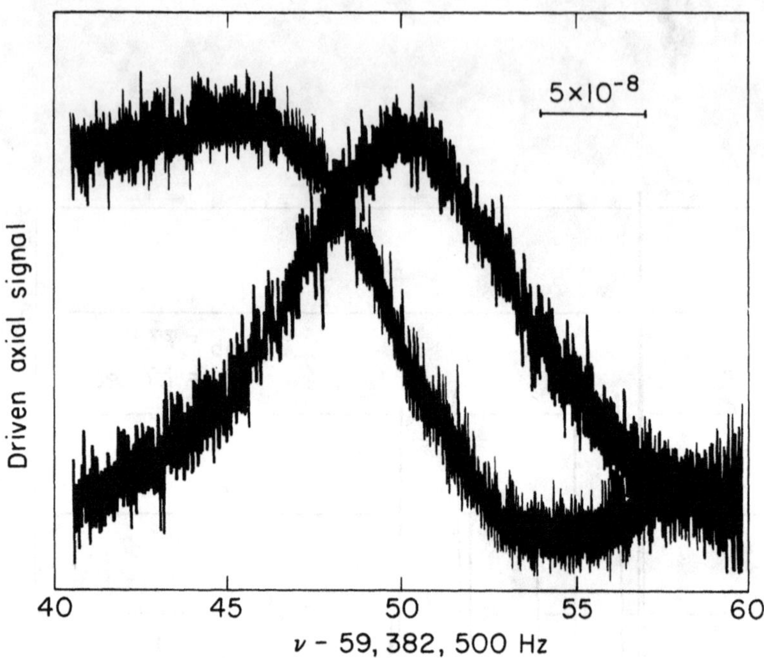

Figure 11: *Axial resonance signals at ≈ 60 MHz. The signal-to-noise ratio of this ≈ 8 Hz wide line corresponds to a frequency resolution of 10 ppb. Both absorption and dispersion modes are shown with the latter mode appropriate for the frequency shift detection scheme employed in these geonium experiments (from Ref. 1).*

3.4 Magnetic Coupling

The ultimate goal of this experiment is to observe the magnetic resonances, ω_c' and ω_s (or ω_a') but observation must come indirectly through the purely electrostatic axial resonance. In order to provide a weak coupling to this frequency, a small magnetic bottle is produced from either a nickel wire wound around the ring electrode and placed in the symmetry plane (as in our first compensated Penning trap shown in Fig. 3), or from nickel screws, placed equally spaced within the ring electrode, and again threaded so as to provide 4-fold symmetry in the ring symmetry plane and inversion symmetry along the z-axis (as in the case of the experiment trap in Fig. 4). The magnetic material used in the rings of these traps has an effective radius a and generates a quadratic gradient B_2, which for the latter trap is $\sim$ 155(15) G/cm^2. The net bottle field (in cylindrical coordinates r, z) is thus described by

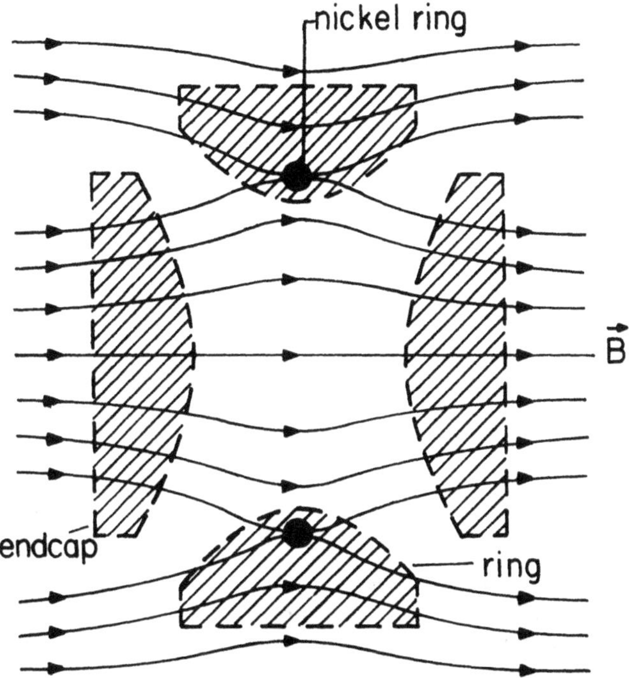

Figure 12: *Magnetic bottle from saturated nickel wire loop. When placed in a strong magnetic field, the nickel's magnetic moment aligns in such a way as to weaken the applied field in a cylindrically symmetric fashion (along the z-axis).*

$$\begin{aligned} B_z &= (B_s - \frac{2}{9}B_2a^2) + B_2\big(z^2 - \frac{r^2}{2}\big) \\ B_r &= -B_2rz \end{aligned} \tag{27}$$

where B_s is the applied solenoid field. Figure 12 illustrates how this bottle shape would appear due to a symmetric ring of magnetic dipoles.

The coupling to the axial frequency follows from the orientational potential associated with the total magnetic moment $\vec{\mu}$ of the electron relative to the total field $\vec{B}$ (ignoring radial terms):

$$W_m = -\vec{\mu}\cdot\vec{B} = -\mu_z(B_0 + B_2z^2)\,. \tag{28}$$

This can now be added to the electrostatic potential energy $W_e = \frac{1}{2}m\omega_z^2z^2$ to yield the resultant axial frequency:

$$\omega_z = \omega_{z,0} - \frac{B_2\mu_z}{m_e\omega_{z,0}}\,. \tag{29}$$

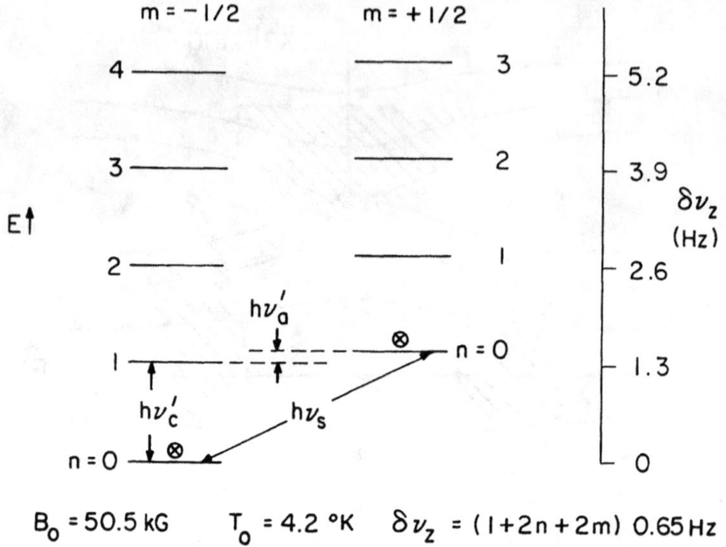

Figure 13: *Lowest Rabi-Landau levels for a geonium atom. The axial frequency (shown in the right-hand scale) corresponds to the coupling via the axial magnetic bottle field. The lowest state ($n = 0$) which is occupied by the electron or positron 80–90% of the time differs by 1.3 Hz depending on the exact spin state. This is the signature used to indicate that a spin has flipped (from Ref. 2).*

The observable magnetic moments are the cyclotron moment, $-2\mu_B n$, the spin moment, $-g\mu_B m$, and the magnetron moment, $-2(\omega_m/\omega_c')\mu_B q$, where n, m and q are cyclotron, spin and magnetron quantum numbers respectively. One can now see the intrinsic importance of the axial shift detection technique, which can be summarized in the quantum mechanically correct expression (assuming $g \approx 2$):

$$\omega_z = \omega_{z,0} + (n + m + \frac{1}{2} + \frac{\omega_m}{\omega_c'}q)\delta \tag{30}$$

where $\delta = 2\mu_B B_2/(m_e\omega_z) \sim 2\pi(1.3\,\text{Hz})$ for electrons in the nickel-screw bottle for the experiment trap in Fig. 4. Figure 13 shows the Landau level scheme and the associated shifts in the axial frequency for this particular trap.

In order to observe these shifts, the axial motion is incorporated into a feedback loop as illustrated in Fig. 14. In this apparatus, the dispersion-shaped axial resonance (shown in Fig. 11) is used as an error signal applied to an integrator, whose output fine-tunes the total ring voltage. If something causes ω_z to shift relative to the synthesizer output, then a nonzero error is integrated until ω_z again coincides with the synthesizer's reference frequency and the error returns to zero.

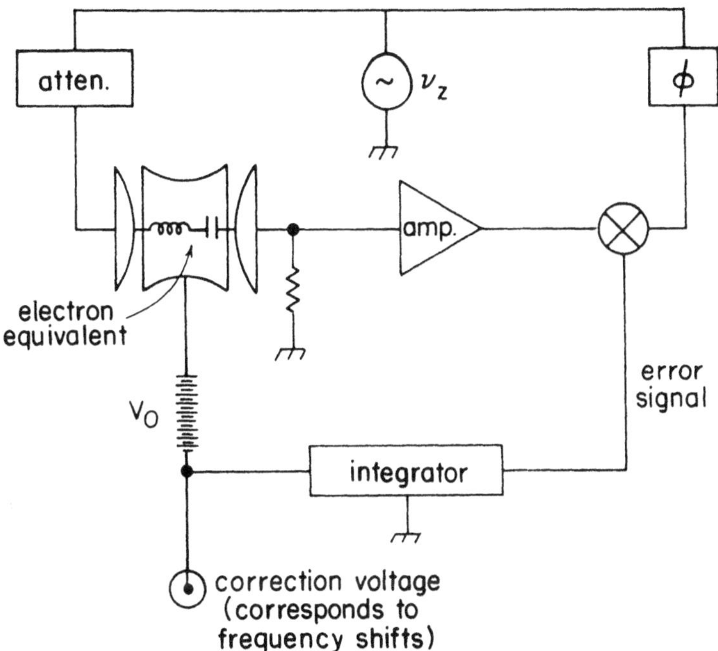

Figure 14: *A schematic representation of the superheterodyne system used to detect the electron's motion. The upper half is the standard axial drive and synchronous detection whereas the lower half generates the frequency shift information. The feedback signal is generated by mixing the phase-shifted drive synthesizer with the amplified electron signal to produce an error voltage. This error is integrated and fed back as a correction voltage to the ring bias circuitry, thereby producing a frequency lock (from Ref. 1).*

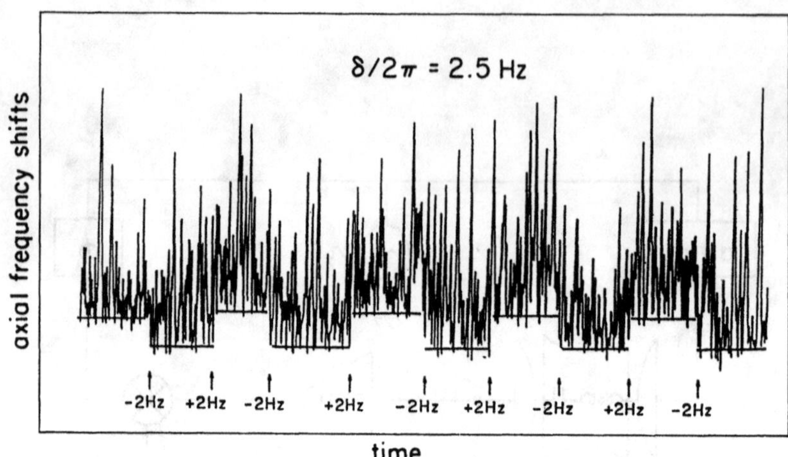

Figure 15: *Control and calibration obtained with an earlier 2.5 Hz magnetic bottle. The asymmetric noise spikes are due to thermal excitation of the cyclotron motion (at ~ 4 K). The 2 Hz changes in the floor were obtained by changing the generator frequency by 2 Hz (from Ref. 1).*

The output of the integrator is thus used to note the frequency shifts which have occurred. An example of the correction signal is shown in Fig. 15 in which the drive synthesizer is shifted deliberately by 2 Hz every minute. This calibration pattern indicates that we have 10^{-8} resolution for an earlier 2.5 Hz bottle. The correction signal will now reflect, via the magnetic bottle, any shift in the axial frequency due to all sources (i.e. noisy voltage, anharmonic shifts with change in stored energy, cyclotron and/or spin quantum jumps, etc.). This description thus embodies the essence of the axial-frequency-shift detection technique.

3.5 Other Axial Shifts

By far the largest perturbation is associated with the inability to perfectly align the magnetic field with the z-axis of symmetry for the trap. This results in a coupling amongst the normal modes via the angle of tilt, θ, between $\vec{B}$ and $\hat{z}$. In 1982, Brown and Gabrielse investigated this problem[37] and worked out the perturbations to the quadratic trapping potential associated with both the misalignment angle, θ, and an asymmetry-of-electrodes parameter, ϵ, which is defined according to the ellipticity generated in the normally circular projections of equipotentials onto the xy-plane. Thus, when $\theta \ll 1$ and $\epsilon \ll 1$, the result predicted for the observed axial frequency $\overline{\omega}_z$ is

$$\overline{\omega}_z^2 \approx \omega_{z,0}^2\left(1 - \frac{3}{2}\theta^2\right) \tag{31}$$

which through second order does not depend on ϵ. (Thus. maximizing ω_z will remove the angle θ.) Typically, the relative shift in the axial frequency is $-3\theta^2/4 \sim 10^{-5}$ for $\theta \sim 0.2°$. The constraint that the axial frequency is kept locked causes a shift in the dc voltage according to $(\delta\omega/\omega)_\theta = -(\delta V/2V_0)$; therefore, both ω_m and ω_c' are affected. The effect of such a lock-induced shift for ω_c' in particular is

$$\frac{\delta\omega_c'}{\omega_c'} = \left(\frac{\omega_z}{\omega_c'}\right)^2 \left(\frac{-\delta V}{2V_0}\right). \tag{32}$$

However, since the angle of tilt is expected to remain fixed (at least at the 0.1% level because axial stability is 10 ppb), the impact on ω_c' is less than 2×10^{-15}.

There are also a few minor perturbations associated with the non-uniformity of the magnetic field, the non-quadratic terms in the electric potential, and the relativistic corrections. Their effects on the normal modes have been studied by Brown and Gabrielse. From their analysis,[4)] one can show that for our traps and energies less than 0.01 eV in the respective motions, similar feedback-forced relative shifts in ω_c' are less than 10^{-13}.

Finally however, there is one large shift in axial frequency which is used for calibration purposes. When very strong anomaly drives are applied to one endcap or to the split guard rings, the axial frequency is found to shift proportional to the drive power. For anomaly power applied to the drive endcap only, as in very early Geonium experiments, the shift was a direct measure of the amplitude of the electric field used as an off-resonance axial drive. When applied to the guard rings on the other hand, the circulating current in a LC circuit, tuned to ω_a', develops a differential voltage across the loop inductance of the guards which is then capacitively coupled over to the endcaps. Since the drive cap typically has a very low impedance, unlike the signal cap, the rf drive is again asymmetrically applied. The shift in axial frequency itself is due to the production of a pseudo-potential associated with the rf trapping field as routinely observed in Paul traps.[38)]

4. Magnetron Motion

4.1 The "Resonance"

The next normal mode which is relatively easy to observe is the slow guiding center frequency in the radial plane, referred to as the magnetron frequency. This particular "resonance" is useful because it can reflect asymmetries between the electric and magnetic fields as well as perturbations relative to the ideal quadratic potential (see Sec. 5.1). Actually, this motion is not a true resonance since it does not have a binding potential well to restore equilibrium. The motion, however, is

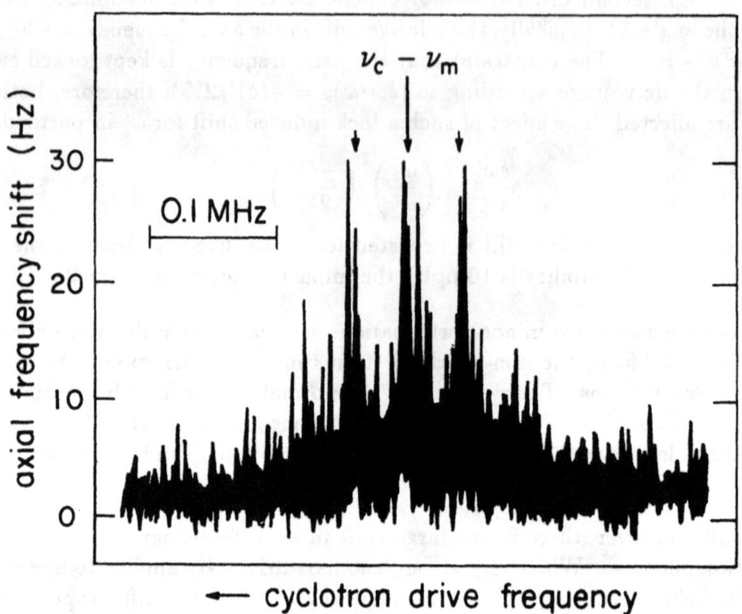

Figure 16: *Externally excited cyclotron resonance using the 1.0 Hz bottle in the first compensated Penning trap and the axial-frequency-shift detection scheme. Clearly noticeable are the upper and lower magnetron sidebands which are produced by the magnetron motion in a radially non-uniform magnetic field. For this resonance,* $\nu_c - \nu_m = 51,073,965$ *kHz (from Ref. 28).*

metastable in that the decay out of the trap by all dissipative forces is very slow, as experimentally verified to be $\gg 10^5$ sec. The real problem with this motion is that electrons are typically loaded with initial orbits which can vary significantly from electron to electron. Because of imperfections in both trapping potential and uniformity of magnetic field, it was imperative that a reproducible set of conditions be developed with which to measure the magnetic moments. By necessity, this means shrinking the magnetic orbit to as close to zero as possible and this is discussed in Sec. 4.2.

Historically, one of the earliest geonium examples of the magnetron resonance was as a sideband frequency on the cyclotron resonance as illustrated in Fig. 16. Since the electron's motion has a component of its guiding center along the x-direction, (i.e. $x = R_m \sin\omega_m t$), if $B = B_0 + B_1 x$, then from Eq. 12,

$$\begin{aligned} \omega_c' &= \omega_{c,0}(1 + \alpha \sin\omega_m t) - \omega_{m,0}(1 - \alpha \sin\omega_m t) \\ &= \omega_{c,0}' + (\omega_{c,0} + \omega_{m,0})\alpha \sin\omega_m t \end{aligned} \tag{33}$$

where $\alpha = B_1 R_m/B_0$. This is an example of frequency modulation with sideband strength determined by first order Bessel functions with modulation index equal to frequency deviation/modulation frequency $\approx \alpha\omega'_{c,0}/\omega_m$ (which is therefore proportional to the radius of the magnetron motion, R_m, as well as the linear gradient, B_1). In Fig. 16, one can even discern a hint of second sidebands in the spectrum, a case which is characteristic of modulation indexes between 0.5 and 1.0. Due to an unforeseen asymmetry in the nickel wire loop forming the trap's 1.0 Hz magnetic bottle, the relative field inhomogeneity was about 10^{-4}/cm. Thus, for $\omega'_c/2\pi = 51$ GHz and $\omega_m/2\pi = 35$ kHz, one can estimate that the magnetron orbit in this trap was $\lesssim 1\%$ of R_0.

In 1975, magnetron sidebands were observed on the axial frequency. Using the frequency shift detector (described in Sec. 3.4) for observation, an additional drive field at ω' is applied to a guard probe that protrudes slightly into the gap between endcap and main electrode to produce the sideband excitation (SBE). This drive field is not uniform at the center of the trap, and will contain an axial component of the form $\vec{E}' = Ay\hat{z}$ where coordinate axes are preferentially chosen such that the probe is on the y-axis. Thus if $A = A_0 \sin(\omega_z + \omega_m)t$ and $y = R_m \sin\omega_m t$ for the instantaneous position of the electron, then, the sideband drive becomes:

$$\begin{aligned}\vec{E}' &= [A_0 \sin(\omega_z + \omega_m)t][R_m \sin\omega_m t]\hat{z} \\ &= \frac{A_0 R_m}{2}\cos[\omega_z t]\,\hat{z} - \frac{A_0 R_m}{2}\cos[(\omega_z + 2\omega_m)t]\,\hat{z}\,.\end{aligned} \tag{34}$$

Thus, relative to the magnetron motion, the first term appears to be a constant axial drive, similar to the one used to lock up the axial resonance. Note that again in this case, the strength of the drive field is proportional to R_m. Now suppose this auxiliary drive is not exactly resonant, i.e. $\omega' = \omega_z + \omega_m + \delta$; then, an image of the axial resonance will appear δ above the fundamental when sweeping the normal endcap drive with frequency. This image will have essentially the same phase condition which characterizes the fundamental, relative to the 2π(1 MHz) standard modulation frequency. Thus, if phase of the normal axial drive is adjusted correctly for the axial lock, then the phase of the image is also correct for locking up the axial motion. If the SBE drive is swept such that $\delta \to 0$, then, assuming the signal-to-noise is adequate (i.e. strong enough SBE drive amplitude), the lock loop will then follow the sharp image until S/N decreases to the point that lock is lost by the image and is regained by the main resonance. In this way, resonances of the type shown in Fig. 17 are obtained. Once lock is lost, a ringing is observed that is associated with a beat note from the excited motion and the applied magnetron drive. By sweeping both directions, the onset of image-lock allows a determination of ν_m to better than 0.01 Hz or 1 ppm.

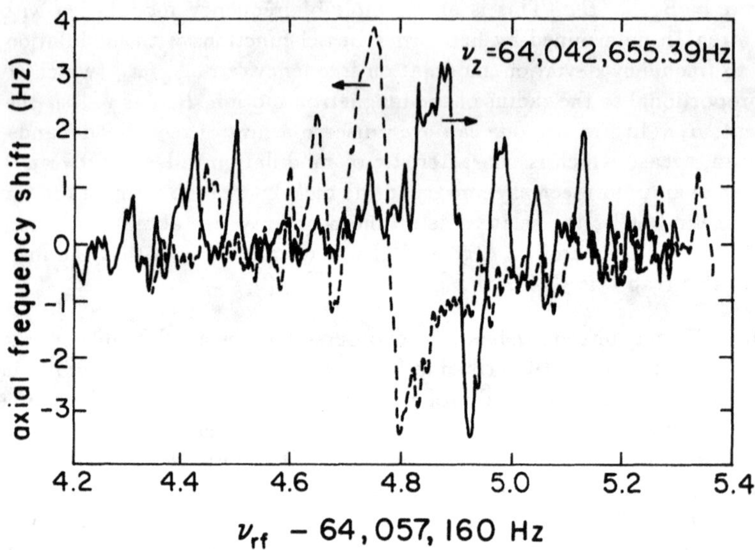

Figure 17: *The $\omega_z - \omega_m$ cooling resonance using SBE probes. The magnetron sideband of the axial resonance is observed as a pulling of the locked axial resonance and a subsequent beating between the excited magnetron motion and the applied magnetron drive. Such resonances allow ω_m to be determined to 1 ppm and the magnetron radius to be reduced because of the absorbed energy $\hbar\omega_m$ (from Ref. 2).*

4.2 Changing the Magnetron Orbit

The nature of the centering process is best visualized by considering conservation of energy. When a photon of energy, $\hbar(\omega_z + \omega_m)$, is absorbed from the SBE probe drive, $\hbar\omega_m$ is absorbed by the magnetron motion while the remaining part, $\hbar\omega_z$, is harmlessly added to the axial motion which quickly damps away because of the strongly coupled tuned circuit, held at some fixed temperature T_z. Now, the magnetron energy is strongly dominated by the negative radial potential hill given by $-m_e\omega_z^2 R_m^2/4$, compared to the very small kinetic energy $-m_e\omega_m^2 R_m^2/2$. Thus, adding a positive amount of energy $\hbar\omega_m$ to this negative hill will reduce the magnetron radius, R_m. This we designate as a "cooling" or "centering" drive, in contrast to the application of a sideband at $\omega_z - \omega_m$ which drives the electron radially out of the trap. This technique is referred to as "motional sideband cooling".

In principal, cooling will continue to occur by application of this upper sideband drive until the occupation quantum numbers (k for axial, q for magnetron) in the 2 separate motions become equal,[4)] i.e. $q = k$. This represents the statement that the absorption rate equals the stimulated emission rate for photons of $\hbar(\omega_z+\omega_m)$. In terms of the thermodynamic temperature T_z of the axial frequency, the average axial energy can be written $\hbar\omega_z(k+\frac{1}{2}) = k_B T_z$. If the driven energy, used to lock the axial frequency, is much less than $k_B T_z$, then T_z is associated with the thermal reservoir produced by the strongly coupled tuned circuit in a liquid helium bath with attached real (somewhat hotter) preamplifier. Otherwise, T_z is larger and is associated with the drive source. In either case, the minimum magnetron energy is determined by the condition:

$$-m_e\omega_z^2 R_m^2/4 = -\hbar\omega_m(q+\frac{1}{2})\,|_{q=k} \tag{35}$$

and the resulting theoretical minimum orbit, R_{min}, can then be written

$$R_{\text{min}} = \left(\frac{2ck_B T_z}{e\omega_z B_0}\right)^{1/2} \tag{36}$$

which explicitly shows the dependence on magnetic field. It is also insensitive to well depth varying as the inverse 4*th* root.

As indicated in Eq. 34, the ability to change the magnetron orbit by applying a fixed amplitude axial sideband drive will undoubtedly be proportional to the size of the orbit, R_m. This is equivalent to stating that whether one adds or subtracts Δq to the initial quantum number q_0, the magnitude of the rate will be proportional to $q(t)$ and the rate of approach to the minimum will be exponential as well as the rate of blowing the electron out from the center. The latter possibility has been used in the past[1)] to estimate q_0. Recalling that the magnetic bottle couples ω_z to the magnetic moment for the magnetron motion, the locked axial frequency will have the following time dependence when subjected to an axial sideband at $\omega_z - \omega_m$ (see Eq. 35 where $q \neq k$):

$$\omega_z = \omega_{z,0} + \frac{\omega_m}{\omega_c'} q_0 \delta e^{t/\tau} \tag{37}$$

where τ can be experimentally measured as illustrated in Fig. 18. For this example, the magnetic field was at 18 kG and the minimum orbit was found to be 0.0014 cm, though many runs suggested that R_{min} was probably half this value. Nevertheless, Eq. 36 would predict $R_{\text{min}} \approx 0.0001$ cm which suggests that either one can not realistically achieve the minimum orbit due to some uncontrolled heating mechanism or else the effective axial drive temperature is much greater than 4°K, or a combination of both.

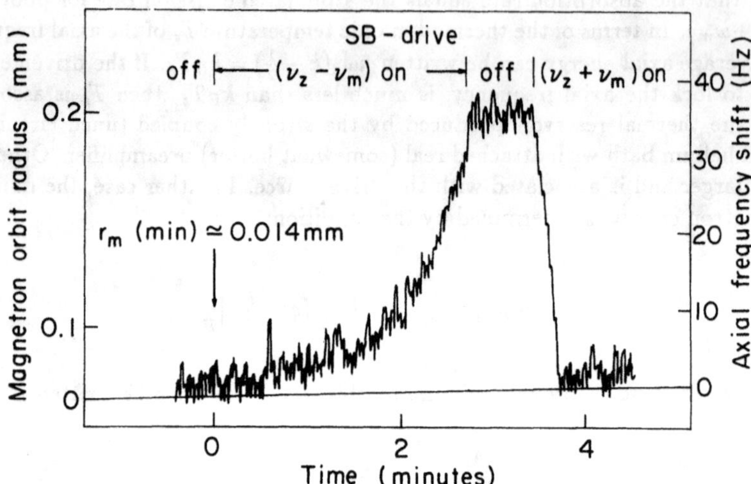

Figure 18: *Expanding and shrinking of magnetron orbit radius. By driving the axial motion on the sidebands $\omega_z \pm \omega_m$, it is possible to force the magnetron motion at $\omega_m = 2\pi\nu_m$ to absorb/emit the energy balance $\hbar\omega_m$ and thereby shrink/expand the magnetron orbit (from Ref. 1).*

4.3 Frequency Shifts

As suggested in Sec. 3.5, the magnetron frequency is very sensitive to the angle of tilt, θ, that occurs between $\vec{B}$ and $\hat{z}$ and the asymmetry-of-electrodes parameter, ϵ. The obvious question that arises concerns how these perturbations may affect one's ability to determine the spin anomaly. To see how serious the problem could be, in 1978 the following frequencies were reported:[1] the measured magnetron frequency, $\nu_m = 34,471.9 \pm 0.1$ Hz compared to the calculated frequency $\delta_e = 34,468.18 \pm 0.1$ Hz. The resulting discrepancy was 3.7 Hz, and relative to the anomaly frequency represented a 62 ppb shift. Though for this early 200 ppb measurement, the discrepancy was not serious, it might have become important as precision increased.

In their 1982 analysis,[37] Brown and Gabrielse predicted that the observed magnetron frequency $\bar{\omega}_m$ would be given approximately by the following formula (when $\epsilon \ll 1$ and $\theta \ll 1$):

$$\bar{\omega}_m \approx \bar{\delta}_e\left[1 - \frac{\epsilon^2}{2} + \frac{9}{4}\theta^2\right] \qquad (38)$$

where $\bar{\delta}_e$ is defined in terms of the observed axial and cyclotron frequencies: $\bar{\delta}_e = \bar{\omega}_z^2/2\bar{\omega}_c$. It is believed that ϵ is $\ll 0.01$ in these high precision Penning traps (by construction) and thus may be ignored in most cases in comparison to the effect of the tilt angle θ. All large compensated Penning traps used in electron g-2 work have always had $\bar{\omega}_m > \bar{\delta}_e$ which is consistent with the dominance of θ. In the 1978 result, $(\bar{\omega}_m - \bar{\delta}_e)/\bar{\delta}_e = 0.0001$ and corresponds to an angle $\theta \approx 0.4°$. Such typical angles did pose a problem on one occasion when the double trap scheme (shown in Fig. 6) was being used to transfer positrons to the high precision experiment trap. Transfer was made possible only by minimizing $\bar{\omega}_m - \bar{\delta}_e$.

Again, there are a few minor perturbations associated with the non-uniformity of the magnetic field, the non-quadratic terms in the electric potential, and the relativistic corrections. From the 1986 analysis[4)] by Brown and Gabrielse, the most significant corrections to the magnetron frequency would result in a relative magnetron shift less than 10^{-6} in existing g-2 traps, assuming that all energies were less than 0.01 eV. Thus, when compared with the cyclotron frequency, the probable effect on a g-2 measurement is below the 10^{-16} level of accuracy.

5. Cyclotron Resonance

5.1 Microwave System

The excitation of the cyclotron resonance is accomplished by means of a clean X-band source, transported by 1 mm diameter, semi-rigid coax to the dc-biased Schottky diode multiplier at 4 K (see Fig. 2). The amount of power delivered to the diode is related to the increase in dc current, obtained by rectification of the X-band drive. Initially, the diode was placed outside the glass envelope, but with the new metal envelopes, it is sealed inside the vacuum envelope and now radiates from inside of a small circular tube which acts as a high pass filter to attenuate the strong X-band field. The X-band source consisted principally of a Klystron, stabilized to a frequency standard at 10 MHz, and amplified by a travelling-wave amplifier. This simple system was adequate for multiplication to 140 GHz at the ppb level of precision primarily because the classical decay time was $\sim$ 1 sec or 10 times the corresponding free space value. This long decay time made it possible to very weakly tickle the cyclotron resonance, exciting no more than 1-3 quantum levels (see Fig. 13), thus keeping the noise pedestal of the multiplied X-band field from significantly exciting the cyclotron resonance.

5.2 Lineshape

The shape of the cyclotron resonance is almost exclusively determined by the magnetic bottle and the thermal Boltzmann distribution of axial states. In addition, the relativistic pulling effect has some influence on the lineshape, causing the

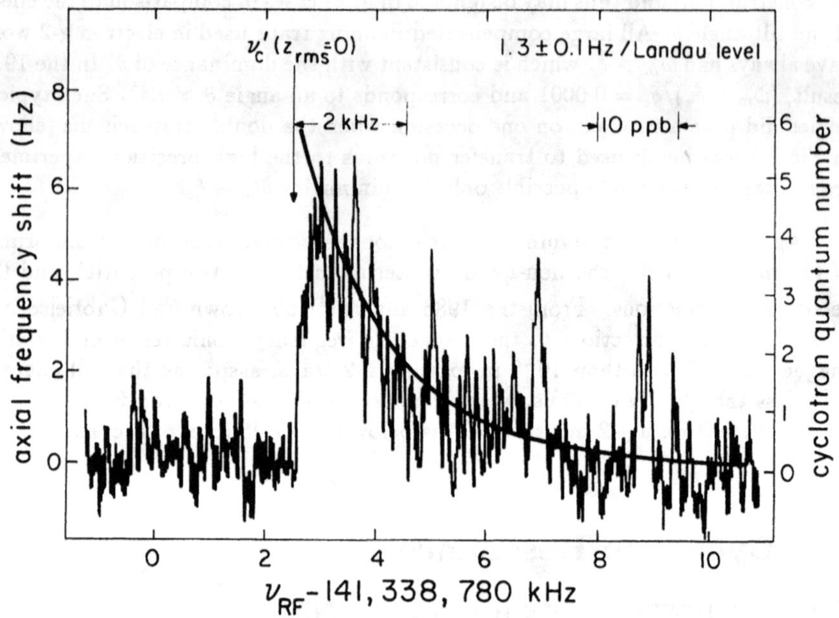

Figure 19: *Single electron cyclotron resonance for the 1.3 Hz bottle in the most recent compensated Penning trap. The characteristic magnetic lineshape has an exponential tail that reflects the Boltzmann distribution of axial states coupled via the z^2 term in the magnetic field. The sharp edge feature corresponds to the field at the trap center and the solid line represents a fit to the data with a width corresponding to a 5 K axial temperature (from Ref. 2).*

resonance to appear somewhat broader than expected.[39] It is also imperative that the axial locking drive not be applied when determining the true cyclotron lineshape. This follows from the z^2 term in the magnetic bottle which will show a cross term of the form $Z_{\rm th} Z_d$ that will also broaden the resonance ($Z_{\rm th}$ is the thermal amplitude of the axial motion). Necessarily then, the detection and excitation for this resonance are alternated and the observed ω_z shift signals the onset of the cyclotron resonance, assuming that the frequency is swept up in magnitude for a positive magnetic bottle. Figure 19 shows a typical cyclotron resonance taken with this scheme in the trap shown in Fig. 4. The main requirement of the alternating scheme is that the alternating rate not be too slow compared to $1/\tau_c \sim$ 1 Hz, but must allow time for axial energy to decay away. This latter condition is satisfied with a delay $\gtrsim$100 ms before onset of excitation and the alternating rate is optimized at 0.5 Hz (1 sec on, 1 sec off). In addition, the axial damping time ($\tau_z \sim$ 50 ms) needs to be short compared to $\tau_c \sim$ 1 sec so that some signal is

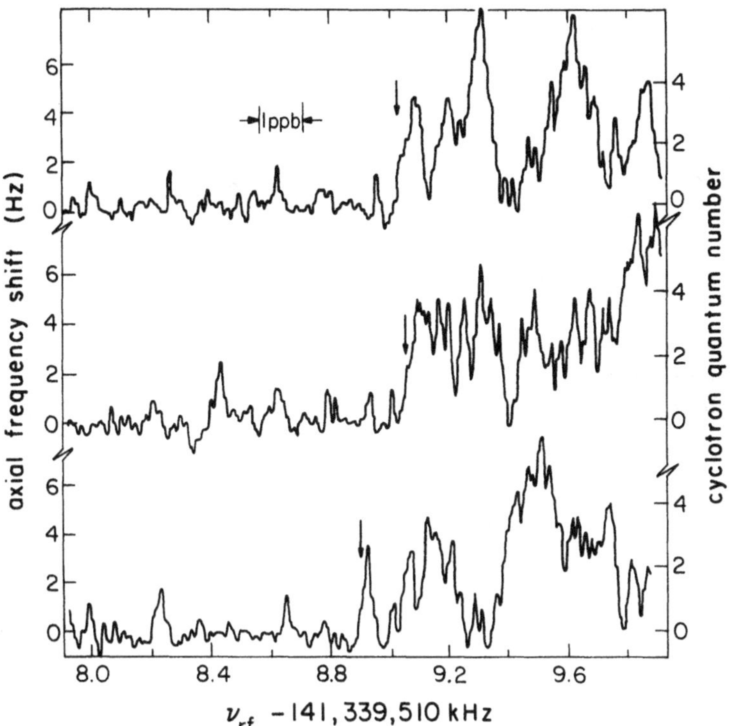

Figure 20: *High resolution of the low frequency edge. The magnification of three successive traces indicates that the magnetic field at the trap center can be measured to have a short-term precision ≈ 1 ppb (from Ref. 2).*

registered in the detection part of the cycle. It is worth noting however, that if the axial locking drive is scaled down significantly such that lock is just barely possible, a continuous cyclotron resonance can be swept out whose low-frequency edge agrees very well with the alternating low-frequency edge, but without quite as good resolution. Under alternating conditions, it is possible to achieve 1 ppb short-term resolution of this low frequency edge (see Fig. 20) which represents the $Z_{\rm rms} = 0$ position in the magnetic bottle, and thus the center of the quadratic field (assuming no linear gradients).

The high frequency tail of the cyclotron resonance represents the Boltzmann distribution of axial states, where the 1/e linewidth should correspond to

$$\Delta\omega_c' = \frac{eB_2 k_B T_z}{m_e^2 c\omega_z^2} = \frac{k_B T_z}{\hbar\omega_z}\delta \tag{39}$$

where δ is defined by Eq. 30. As one can see, it is expected that this linewidth will be proportional to T_z, yet often this relation would indicate that $T_z > 4$ K. This could be due to the equilibration to an amplifier whose temperature is greater than ambient, but some of the discrepancy is believed to be due to the relativistic pulling effect.[39] Here, the onset of excitation of the edge immediately pulls the resonance edge to lower frequency, thus artificially reducing the peak response. Since this reduction will be more significant (in terms of absolute signal) near the peak than on the tail of the resonance, the shift has the effect of broadening the 1/e resonance linewidth.

5.3 Perturbations to the Cyclotron Frequency

One possible source of shifts in ω_c' that was considered in earlier sections was associated with the angle of tilt θ between $\vec{B}$ and the z-axis of symmetry, as well as possible ellipticity ϵ in the trapping electrodes. In terms of $\bar{\delta}_e$ defined in Sec. 4.3, the observed cyclotron frequency $\bar{\omega}_c'$ can be written in the form

$$\bar{\omega}_c' \approx \omega_c - \bar{\delta}_e - \frac{\bar{\delta}_e}{\bar{\omega}_c'}(\bar{\omega}_m - \bar{\delta}_e) \tag{40}$$

which differs from Eq. 12 only by the discrepancy between measured and calculated magnetron frequencies, scaled by the ratio of magnetron and cyclotron frequencies. The relative shift is given by

$$\frac{\Delta\omega_c'}{\bar{\omega}_c'} = \frac{1}{4}\left(\frac{\omega_z}{\omega_c'}\right)^4 \left(\frac{\bar{\omega}_m - \bar{\delta}_e}{\bar{\delta}_e}\right) \tag{41}$$

where the quantity $(\omega_z/\omega_c')^4 \approx 10^{-13}$ for parameters chosen for our experiments (see Table 1) and the last factor is usually no larger than 10^{-4}. Thus, this shift is totally negligible for precision in the foreseeable future. It should be noted that the above correction is *not* negligible for ions since it scales as the square of the mass. However, under this type of perturbation, the true free space cyclotron frequency can always be retrieved[37] from the three measured frequencies $\overline{\omega}_c$, $\overline{\omega}_z$, and $\overline{\omega}_m$ according to

$$\omega_c^2 = \overline{\omega}_c{}^2 + \overline{\omega}_z^2 + \overline{\omega}_m^2 \tag{42}$$

which is referred to as an "invariance theorem". For high precision, stored-ion mass spectroscopy,[40] this theorem will be useful.

Another possible perturbation is the electrostatic shift due to the intrinsic constraint of a real charged particle surrounded entirely by conducting surfaces. The simple model of a point charge located inside a grounded sphere of radius a has been used to predict[41] the electrostatic shifts due to the image charge located outside the sphere. Using the experimental condition that $\overline{\omega}_z$ is held constant by the feedback loop (described in Sec. 3.4), the observed frequencies become

$$\begin{aligned}\overline{\omega}_{m,N} &= \omega_{m,0} + \frac{3\Delta_N}{2\overline{\omega}_{c,N}} \\ \overline{\omega}_{c,N} &= \omega'_{c,0} - \frac{3\Delta_N}{2\overline{\omega}_{c,N}}\end{aligned} \tag{43}$$

where $\omega_{m,0}$ and $\omega'_{c,0}$ represent the magnetron and cyclotron frequencies respectively in the limit that the number of trapped charges $N \to 0$ for $\Delta_N = Ne^2/m_e a^3$. By fitting the observed shifts caused by changing the number of ions trapped in the 3 times smaller quadring Penning trap[42] with the prediction described by Eq. 43, it is found that $a \approx R_0$. Thus, for a single electron in a g-2 trap,

$$\begin{aligned}\frac{\delta\omega'_c}{\overline{\omega}'_c} &= \frac{3m_e c^2}{a^3 B_0^2} \\ &\approx 5 \times 10^{-15}\,.\end{aligned} \tag{44}$$

This is quite negligible and may be ignored for all the g-2 experiments.

Next, the shifts in the cyclotron frequency due to the non-uniformity of the magnetic field, the non-quadratic (electrostatic) terms in the potential, and the relativistic corrections must be considered. Assuming W_n, W_k, and W_q represent actual energies in cyclotron, axial, and magnetron motions respectively, it can be shown[4] that

$$\frac{\Delta\omega'_c}{\omega'_c} = \left\{\frac{-eV_0}{m_e c^2}\right\}\frac{W_n}{eV_0} + \left\{\frac{B_2 Z_0^2}{B_0} - \frac{eV_0}{2m_e c^2}\right\}\frac{W_k}{eV_0} + \left\{\frac{2B_2 Z_0^2}{B_0}\right\}\frac{W_q}{eV_0} \tag{45}$$

where the conspicuous absence of the C_4 term occurs because its contribution is reduced by the factor $(\omega_z/\omega'_c)^2 \sim 2 \times 10^{-7}$. For the energies, one can obtain W_k from Eq. 20, W_q from $-(eV_0/2)(R_m/R_0)^2$ and W_n from $(n+\frac{1}{2})\hbar\omega'_c$. Using some of the parameters listed in Table 1, this equation becomes

$$\frac{\Delta\omega'_c}{\omega'_c} = \{-0.2W_n + 3.4W_k + 6.8W_q\} \times 10^{-5} \tag{46}$$

where all energies are measured in eV. The first term represents the relativistic correction and corresponds to $\sim$ 1.2 ppb per integer change in the cyclotron quantum number. For a positive bottle, swept up in frequency, it should not affect the edge resolution but only the height of the excitation, unless the noise pedestal heats up the cyclotron temperature and allows the resonance to be pulled into excitation. Normally, the average occupation level in a 50 kG magnetic field is 0.2 at 4 K. The second term is responsible for the shape of the cyclotron resonance and was the subject of the last section. Note that there appears to be a special positive bottle size B_2' which would cancel the relativistic shift, $B_2' = (B_0/Z_0^2)(eV_0/2m_ec^2) \approx 5\,\text{Gauss/cm}^2$, but the detectable signal will include the same B_2' (in Eq. 30) and thus will go to zero also. The last term generates the most concern for us in our g-2 experiment since a change in the metastable magnetron orbit yields a corresponding change in ω_c'. It illustrates the importance of cooling the magnetron motion. Assume that $R_m \sim 0.0014$ cm (as shown in Fig. 18); then $W_q = -4.5 \times 10^{-5}$ eV. Thus, a -3 ppb relative shift occurs in the cyclotron frequency which must be *stable* over time if a 1 ppb accuracy is to be maintained. Ideally, if one could reach the cooling limit, the relative shift could be kept below 0.02 ppb. This term generates the motivation to reduce or eliminate the magnetic bottle in these measurements.

5.4 Cavity Effects

The effect of the Penning trap microwave cavity was first noted in these geonium experiments with the determination that the optimum alternation rate for observing cyclotron resonances in g-2 trap shown in Fig. 4 (at 50.5 kG) was about 0.5 Hz (i.e. 1 second on and 1 second off). It was therefore apparent that the excitation to higher cyclotron quantum states was not decaying as quickly as expected from the classical free-space damping time of 0.1 sec at this field.

To observe the inhibition of spontaneous emission, the initial amplitude of the cyclotron edge is measured for various fixed time delays after turning off microwave power and turning on the axial detection drive. The response time for the axial motion was less than 40 ms and thus did not affect these measurements. The relative initial amplitude is then plotted on a semi-log graph versus the fixed delay time as shown in Fig. 21 in order to determine[2)] that the classical decay time was 1.0 ± 0.1 s, or about 10 times longer than τ_c predicted by free space radiation damping. It was quite fortuitous that these early experiments exhibited such long decay times. As suggested earlier, it meant that much less microwave power was required to see a sharp ν_c'-edge frequency, thus putting less demand on how clean the microwave source needed to be if the noise pedestal of the 16*th* harmonic of the Klystron was not to dominate the carrier frequency. The inhibition was first reported[43)] by Gabrielse and Dehmelt for a Penning trap which did not have a magnetic bottle, but did have a much more open structure. This separate experiment found a decay time τ_c which ranged from 86 to 347 ms for B_0 near 60 kG.

Specifically, the present interest in the inhibition of spontaneous emission arose because of the associated effect of cavity-pulling[5)] on a resonant frequency by the oscillators which represent the natural modes of the cavity. The presence of shifts in the cyclotron frequency could only be detected by changing the magnetic field but even with this, the true unshifted frequency would be difficult to deduce

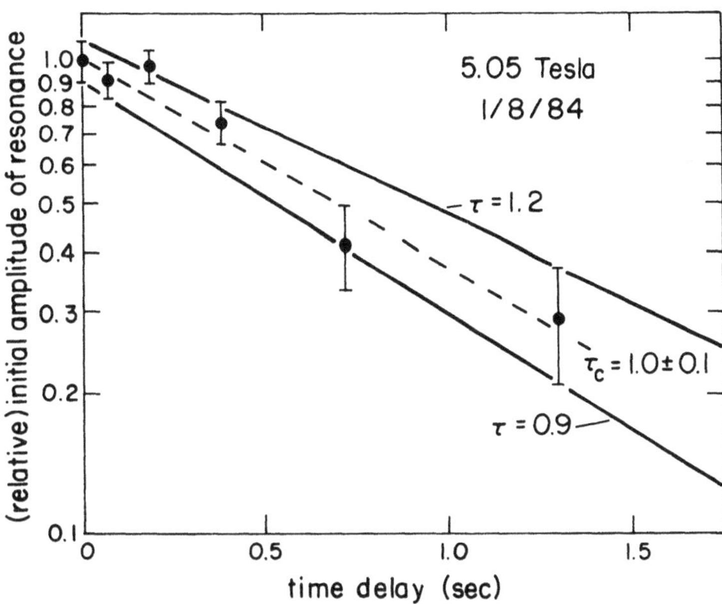

Figure 21: *Cyclotron damping time for a single electron in a Penning trap cavity. By delaying the onset of axial detection drive in the normal alternation scheme, the initial resonance amplitude is found to decay exponentially with a time constant which is ten times longer than that predicted for free space radiation (from Ref. 2).*

without some theoretical model as a guide. Since the g-factor is related to the ratio of spin precession frequency divided by the cyclotron frequency, a shift of several parts in 10^{12} in ω_c without a corresponding shift in ω_s would produce a comparable systematic error in the measured g-factor (and 10^3 times larger shift in a_e). Since present experimental and theoretical precision for the electron g-factor is on the order of several parts in 10^{12}, this systematic effect becomes very important.

The first complete theoretical treatment[44)] of the effect of a microwave cavity on the cyclotron motion of the trapped electron was based on a cylindrical model which was chosen at the time because of the ease of theoretical treatment that the geometry afforded. It predicted that for a cavity with a $Q \approx 1000$, shifts in the cyclotron frequency could exceed 70 parts in 10^{12}. However, a closer look[45)] at the model suggested that for this same $Q \approx 1000$ with $\tau_c(\text{cavity}) \approx 10\tau_c(\text{vacuum})$, the maximum and probable shifts would be on the order of 8 and 4 parts in 10^{12} respectively. The latter uncertainty thus represents the experimental precision of

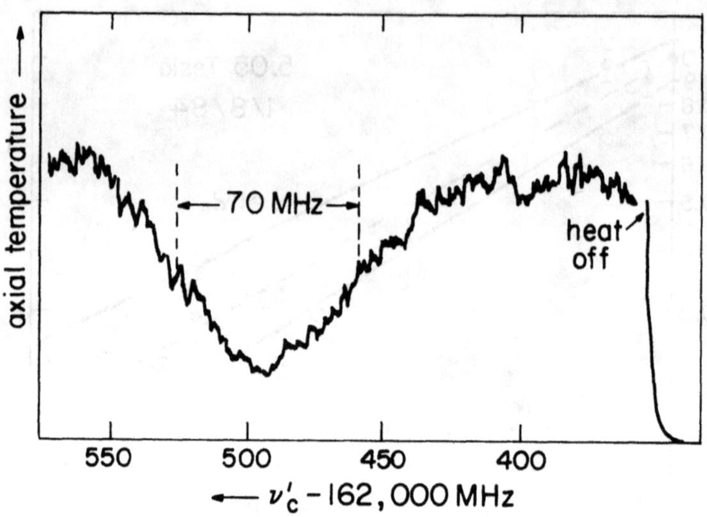

Figure 22: *Axial temperature of small electron cloud vs magnetic field (converted to equivalent cyclotron frequency) for a single cavity mode near 162 GHz. The point where heat is turned off indicates the detection response time and corresponding base line. FWHM (full width at half maximum) linewidth yields relative line Q of* ≈ 2300 *(from Ref. 43).*

the present electron g-2 experiment. To see if a $Q \sim 1000$ was reasonable, an experimental technique was devised to observe the cavity modes[46] in the trap shown in Fig. 3. The method was based on the well-established bolometric technique[14,35] for clouds of several hundred electrons. The axial noise temperature is monitored (without any axial drive at ω_z) by means of the noise currents induced in the parallel resonant LC circuit. To raise the cloud's temperature above the 4 K ambient, a $2\omega_z$ drive is applied to one endcap in order to excite the dominant, non-center-of-mass axial motion. Due to fast internal collision coupling, the entire cloud is heated, causing a corresponding rise in temperature of the center-of-mass as well, which is monitored via a square-law detector whose signal is proportional to temperature. By sweeping the magnetic field, cavity modes which can be coupled to the cloud at trap center are made visible because of the strong coupling of the cyclotron motion (via the internal collisions) to the cold microwave cavity, which subsequently reduces the equilibrium temperature of the cloud, as observed through the center-of-mass motion. Figure 22 is an example of axial temperature versus magnetic field for a particular mode near 162 MHz (one of the narrowest observed) whose line Q was estimated at 2300. Using the same technique for the trap in Fig. 4, a lower than expected line Q (less than 500) was observed which makes the estimate of the probable cavity shift (based on a $Q \sim 1000$) look quite reasonable.

It should be noted at this point that this work has stimulated further theoretical interest in pursuing a numerical calculation of the mode structure in an ideal Penning trap[47]. Unfortunately, subtle variations in slits and holes in the structure as well as deviations from the perfect family of hyperboloids did not allow for any reasonable agreement between the observed and the predicted mode structure for this particular g-2 Penning trap. Future efforts to account for the mode structure will either concentrate on using very low Q cavities that can not exhibit significant frequency pulling or else on measuring the few nearest modes on each side of our operating frequency in hopes of predicting[45,47,48] the true cyclotrón frequency. The latter approach requires much more experimental effort, but does have the advantage of putting less dependence on an absolutely clean microwave source and may be necessary for the future relativistic mass shift approach of detecting spin flips.

5.5 Observed Shift vs Cyclotron Drive

In the early years of the g-2 experiment, power shifts in ω_c' were not of major concern, but as precision increased, there were several observed (i.e. ω_c' versus anomaly, axial and cyclotron drive powers). The first two of these were eliminated by alternating the excitation and detection processes in which either the drives were turned off or else kept on with same power level, but shifted to a non-resonant frequency. This alternation process was expected to eliminate the third shift as well, yet it has persisted. Figure 23 shows this residual shift for the ν_c'-resonance edge versus applied X-band microwave power in the g-2 trap shown in Fig. 4. For the diode used in this trap, the rectified current through the multiplier diode varies approximately linearly with applied X-band power according to 0.0175 db/μA. The limit of 310 μA appears to represent the "zero" for the sixteenth harmonic, which is related in some non-linear way to the required small rf signal which generates this harmonic. This conclusion arises from the observation that the cyclotron resonances appear to abruptly disappear near this limit (i.e. the $n = 1$ level is not excited). The exact nature of this shift is still unknown, but the prevailing evidence suggests that this shift is a resonant phenomenon associated with observing the cyclotron edge frequency and depends directly on the strength of the magnetic bottle. A possible mechanism for the shift would involve a forced increase in the metastable magnetron orbit, possible due to radial coupling of the normal mode frequencies by the slightly anharmonic potential well. Having observed this shift, an adjustment can be made in the cyclotron frequency and thus the g-2 results. This has been done for the 1987 data as well as some previously published data such as the 1984 data which was initially reported without this correction. The results are described in the sext section.

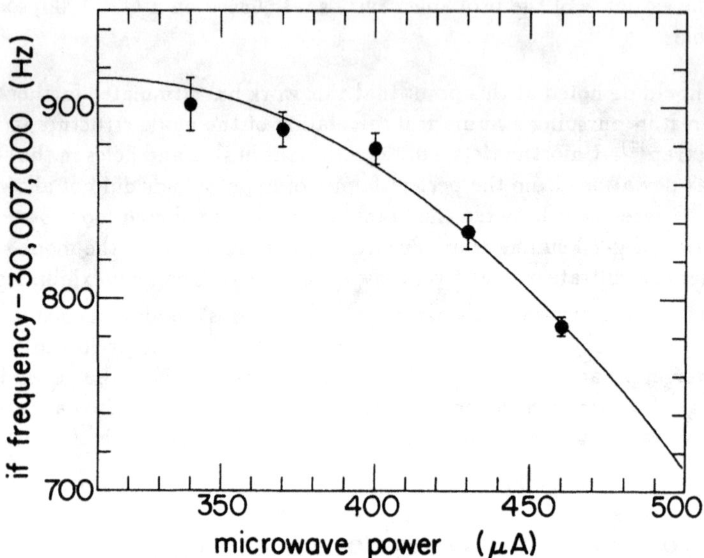

Figure 23: *Residual systematic shift of $\nu_c'(e^-)$ versus applied X-band microwave power. The vertical scale denotes the intermediate frequency where* $\nu_c' = 16(\nu_{\rm if} +$ 8 803 699 296 *Hz) and the horizontal scale is in units of rectified X-band current in the multiplier diode. The solid curve represents a quadratic fit to these data with the constraint that the "zero" for the sixteenth harmonic be fixed at 310 μA (from Ref. 27).*

6. Anomaly Resonance

6.1 The Clean "Beat-Note"

The anomaly resonance is actually a beat-note between the precession rate ω_s of the magnetic moment about $\vec{B}_0$ and the cyclotron rotation at $\omega_c \approx \omega_s$ about the same axis. This beating was dramatically illustrated by the work[20,49)] on electrons and positrons by Rich and Crane completed in 1971, and the work[50)] on positive and negative muons by the CERN collaboration completed in 1977. Excitation of the resonance is produced by a simultaneous two photon process, in which the spin is flipped and the cyclotron state changes by one unit (see Fig. 13). This yields an advantage of 3 orders of magnitude in precision for g-factor measurements over any method which independently measures ω_s and ω_c since major perturbations are reduced accordingly by the simultaneous transitions.

To give a particular example, the relative shifts in ω_s due to the bottle field are (but not including terms of order $\omega_z^2/\omega_c'^2$):

$$\frac{\Delta\omega_s}{\omega_s} = +\left\{\frac{B_2 Z_0^2}{B_0}\right\}\frac{W_k}{eV_0} + \left\{\frac{2B_2 Z_0^2}{B_0}\right\}\frac{W_q}{eV_0} \tag{47}$$

where again W_k and W_q are axial and magnetron energies respectively. As with the cyclotron frequency, terms associated with the anharmonic potential do not appear at this order of precision. Upon comparing this equation with corresponding terms for the relative cyclotron perturbation, given in Eq. 45, the difference frequency, $\omega_s - \omega_c'$, then has the same functional form shown in Eq. 47 but now for $\Delta\omega_a'/\omega_a'$.

The shifts associated with the relativistic mass increase are a little trickier to find because one must now specify to which state the cyclotron frequency is referenced. To be specific, the cyclotron resonance is referenced[3)] to $\omega_c'(n, m) = \omega_c'(0, -\frac{1}{2})$ (i.e. excited from the $n = 0$ ground state with "spin down") such that

$$\omega_c'(0, -\tfrac{1}{2}) = \omega_{c,0}' - \frac{\hbar\omega_c^2}{2m_e c^2} - \frac{\omega_c}{2} \cdot \frac{W_k}{m_e c^2} \tag{48}$$

where $\omega_{c,0}'$ incorporates all other shifts common to both spin and cyclotron frequencies. If the spin happens to be "up" when the cyclotron frequency is measured, then measurements must be corrected according to

$$\omega_c'(0, -\tfrac{1}{2}) = \omega_c'(0, +\tfrac{1}{2}) + \frac{\hbar\omega_c^2}{m_e c^2}, \tag{49}$$

where the correction, $\hbar\omega_c^2/m_e c^2$, corresponds to $\approx 2\pi(140\ \text{Hz})$ in the present 50 kG field or equivalently, 1 ppb. This correction is fairly reliable since excitation occurs primarily from the $n = 0$ state (due to the high field at 4 K); the basic determination of spin state follows from the direction of the axial frequency shift for the first observed spin flip.

The relativistically correct spin precession frequency is given[3)] by

$$\omega_s = \omega_{s,0} - \omega_{s,0}\frac{W_k}{2m_e c^2} - \frac{\hbar\omega_c^2}{2m_e c^2}. \tag{50}$$

By taking the difference between ω_s and ω_c', the anomaly beat frequency becomes

$$\omega_a' = \omega_{a,0}\left(1 - \frac{W_k}{2m_e c^2}\right) \tag{51}$$

where the large 1-ppb dependence on cyclotron energy is now absent. Thus, if W_k is a coherent axial energy comparable to kT_z, then, $\Delta\omega_a'/\omega_a' \sim -0.35$ ppb.

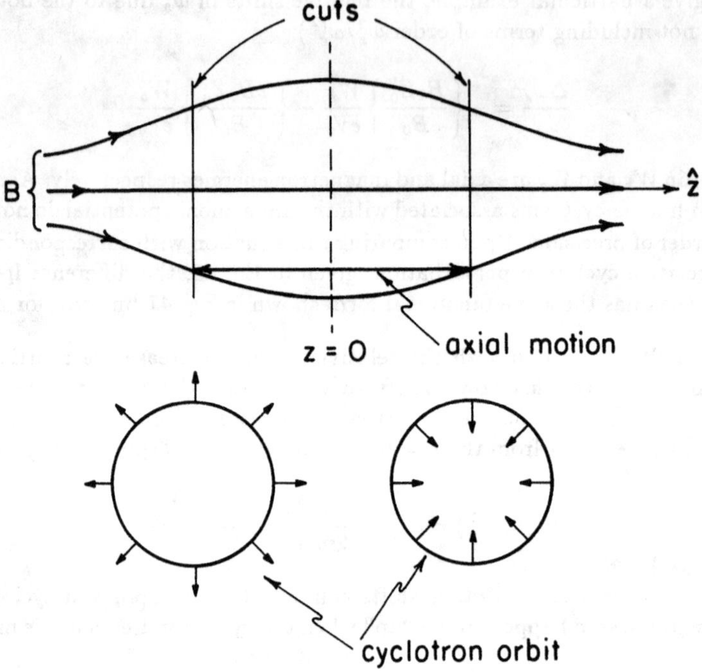

Figure 24: *Mechanism for inducing spin flips by the electron's axial motion in the magnetic bottle. From the electron's frame of reference, a magnetic field is seen to be rotating at* ω'_c. *This field is then modulated by the axial motion at* ω'_a, *thus yielding sidebands at* $\omega'_c \pm \omega'_a$ *where* $\omega_s = \omega'_c + \omega'_a$ *(from Refs. 1 and 28).*

6.2 Axial Excitation Technique

An added benefit of using a magnetic bottle for detection is that it can be used fortuitously to generate the appropriate perpendicular spin flipping field, readily associated with standard NMR experiments. The technique is based on axial amplitude modulation at the anomaly frequency through the non-homogeneous magnetic field, whose component in the radial plane has the form: $b_r = -B_2 rz$ where B_2 is the usual quadratic coefficient as described earlier. (Existence of this component is required by $\nabla \cdot \vec{B} = 0$.) Figure 24 illustrates the motion through this bottle at the two extremes of the axial motion, where for this discussion the magnetron motion can be safely ignored. From the frame of reference of the electron, going around its cyclotron orbit, it sees a magnetic field rotating about its axis, which points in opposite directions at the two extremes of the motion as shown. Thus, from this rotating frame, the radial magnetic field is given by

$$b_r = -B_2 R_c Z_a \sin(\omega'_c t) \sin(\omega'_a t) = b_r(1) + b_r(2) \qquad (52)$$

where

$$b_r(1) = \frac{B_2 R_c Z_a}{2} \cos(\omega_c' + \omega_a')t$$
$$b_r(2) = -\frac{B_2 R_c Z_a}{2} \cos(\omega_c' - \omega_a')t\,. \tag{53}$$

The cyclotron radius, R_c, in a 50.5 kG field and for the $n = 0$ ground state corresponds to ≈ 114 Å. The axial amplitude Z_a, produced by the electric field $E_{\rm rf}$ applied to one endcap, is given by

$$\frac{Z_a}{Z_0} = \left(\frac{Z_0 E_{\rm rf}}{V_0}\right)\left[\left(\frac{\omega_a'}{\omega_z}\right)^2 - 1\right]^{-1}$$
$$\approx \left(\frac{Z_0 E_{\rm rf}}{5V_0}\right) \tag{54}$$

according to Eq. 17 when far from the axial resonance. For the conditions described in Table 1, Eq. 54 yields $Z_a \sim 0.003$ cm for a 1 volt rf drive. The radial component $b_r(1)$ corresponds to the appropriate perpendicular spin flipping field at frequency $\omega_c' + \omega_a' = \omega_c + \omega_a = \omega_s$, whereas the second component will correspond to an insignificant non-resonant perturbation. From standard NMR theory, the $b_r(1)$ component will then produce spin flips with a Rabi frequency

$$\omega_{\rm Ra} = \frac{2\mu_B}{\hbar} b_r(1) \tag{55}$$

which is $\sim 2\pi(0.7$ Hz) for a 1 volt rf endcap drive in the experiment trap shown in Fig. 4. The average time between flips is then given[3)] by

$$\langle T \rangle = 2\pi \frac{\Delta\omega_a'}{\omega_{\rm Ra}^2} \tag{56}$$

which, on resonance, is about 6 sec for a 3 Hz wide anomaly line.

Unfortunately, applying a volt of anomaly power would also produce a significant amount of liquid helium evaporation from a 50Ω terminated drive line. Such time varying heat loads then upsets the equilibrium of the axial frequency lock loop and reduces the effective S/N. However, a reduction in anomaly power would then require much longer data-taking periods. This problem was alleviated somewhat by applying a simultaneous microwave drive in the tail of the cyclotron resonance in order to heat the cyclotron motion into higher excited states on the average with correspondingly larger values of R_c and thus larger $b_r(1)$.

6.3 Guard-Ring Excitation Technique

Every major systematic shift encountered so far appears to be related to the strength of the detection magnetic bottle. To get around these shifts, the bottle must be reduced which means some other mechanism is needed to generate the

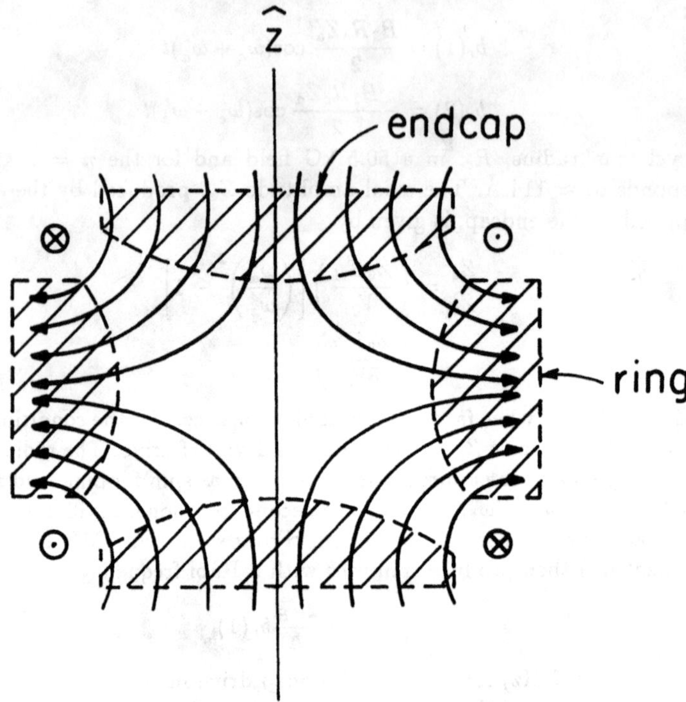

Figure 25: *Guard-ring anomaly excitation. RF current loops above and below the central ring plane, formed by splitting the compensation guard-rings on one side only, produces a constant radial rf magnetic field gradient* B_1. *However, the ideal shape of this field as shown applies only in the absence of the metal electrodes. Coupled with the cyclotron motion, this rf field at* ω_a' *produces motional sidebands, one of which will be at the spin precession frequency* ω_s *(from Ref. 2).*

required spin-flip field. As a first step, a direct current loop is made possible by splitting the guard rings on one side only. The injection of current to these guards is arranged to yield currents flowing in opposite directions for each guard. With this bucking coil drive, the field lines will be as shown in Fig. 25. The amplitude of the rf magnetic field near the center of the trap can be obtained from the expansion of the magnetic field near the axis of the current loop. Of course, the electrodes are expected to somewhat alter the shape of this field and to strongly attenuate it as well, but still effectively producing a constant radial magnetic gradient (near trap center) of the form

$$B_1 \approx 1.3 I_{\rm rf} f_s \text{ Gauss/cm} \tag{57}$$

where $I_{\rm rf}$ is in amps and f_s is the rf electrode shielding factor. The measured spin flipping rate indicates that $f_s \sim 0.1$. A tuned circuit attached to the guard rings

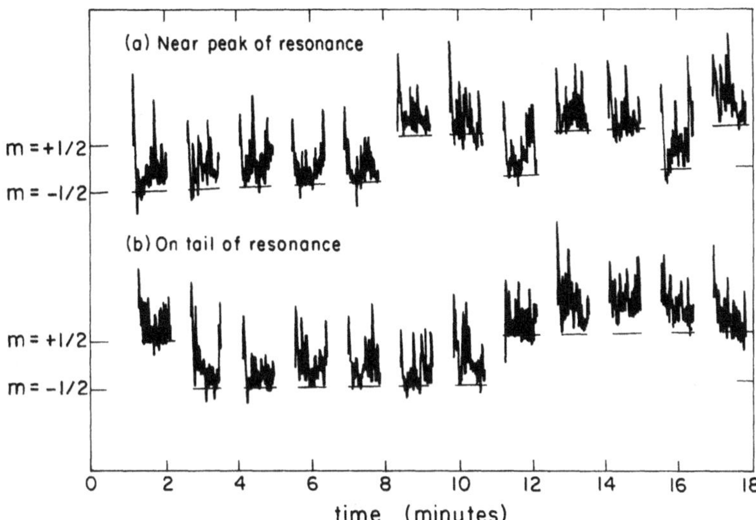

Figure 26: *Example of spin-flip data. The alternation between detection drive and anomaly excitation is much slower than for cyclotron excitation in order to minimize the power in the anomaly field. The excitation frequency is kept constant for each 12–25 alternation cycles while spin-flips are being detected in the correction signal. Note the increased rate of flipping near the peak of the resonance shown in (a) compared to the rate on the exponential tail shown in (b) (from Ref. 2).*

also enhances the production of rf current. The radial field then becomes:

$$b_r = R_c B_1 \sin(\omega_a' t)\sin(\omega_c' t) = b_r(1) + b_r(2)\,. \tag{58}$$

One immediately notes the similarity to the two components described in Eq. 52, in which $B_2 R_c Z_a$ will be replaced by $B_1 R_c$. If R_c is again associated with $n = 0$, this amplitude becomes approximately $0.74 I_{rf} f_s$(in μG), assuming I_{rf} is in amps.

6.4 The Resonance Lineshape

The excitation cycle for anomaly resonances involves turning off the axial drive at ω_z, and turning on the anomaly drive at ω_a' with the microwave drive also applied, but tuned into the tail of the exponential lineshape of the cyclotron resonance. The subsequent detection cycle then involves turning off both anomaly and cyclotron drives while the ω_z drive registers the floor of the lock-loop correction signal Examples of the type of anomaly data obtainable with this method are shown in Fig. 26. The anomaly drive was kept fixed in frequency for some specified number of excitation/detection cycles and then the number of successful flips is plotted versus the applied drive frequency. An example of such early resonances is shown in Fig. 27a for run 109. The dotted line is added only to aid in recognizing the resonance shape. Before a theoretical lineshape was available with which to fit this data, an estimate had to be made as to where the leading edge would be for $Z_{rms} = 0$, typically taken at one-third of peak height. The lack of a sharp edge is due to the noise modulation (or statistical fluctuation) of the anomaly resonance by the axial motion within the inhomogeneous magnetic field. The estimation process produced a serious "apparent" systematic shift of the anomaly with anomaly power. This shift is related to the tendency for this type of resonance to saturate (or to be power broadened). A maximum spin flipping rate is 0.5 (with equal probability of being left in spin-up or spin-down state). As more anomaly power is applied, the resonance edge is broadened and the "apparent" anomaly frequency (and therefore a_e) is shifted to a slightly lower frequency. However, the extrapolation to zero power should be consistent with our presently accepted value of a_e.

The basis for this unusual lineshape is found in Eq. 27 for which the magnetic field depends weakly on the instantaneous value of z^2. This dependence is shown explicitly for ω_c' in the second term of Eq. 45 (not including the relativistic correction). As described in Sec. 5.2, the magnetic width, which represents the $1/e$ linewidth, will be proportional to the average axial energy, and will reflect the Boltzmann distribution of axial states at some temperature T_z:

$$\Delta\omega_a' = \omega_a' \frac{B_2 Z_0^2}{B_0} \cdot \frac{k_B T_z}{eV_0} . \tag{59}$$

In other words, the weak magnetic coupling to the axial frequency also causes the random thermal axial fluctuations to appear in all magnetic resonances. This effect was accurately described[51)] by Brown in 1985 by writing the stochastic average of the axial motion as a functional integral which transcribes it into a soluable one-dimensional, quantum-mechanical barrier penetration problem. The resulting lineshape profile $\chi(\omega)$ as a function of frequency is given for very weak rf drives by

$$\chi(\omega) = \frac{4}{\pi}\mathrm{Re}\left\{ \frac{\gamma'\gamma_z}{(\gamma'+\gamma_z)^2} \sum_{n=0}^{\infty} \frac{(\gamma'-\gamma_z)^{2n}(\gamma'+\gamma_z)^{-2n}}{(n+\frac{1}{2})\gamma' - \gamma_z/2 + \gamma_c/2 - i(\omega-\omega_0)} \right\} \tag{60}$$

where Re denotes the real part, $\gamma' \equiv (\gamma_z^2 + 4i\gamma_z\Delta\omega_a')^{1/2}$, γ_z is the corresponding axial linewidth, and γ_c is the natural cyclotron damping linewidth in the absence of the magnetic bottle, i.e. associated with the decay of cyclotron excitation via synchrotron radiation. However, moderately hard anomaly drives are typically used such that near the peak, the probability for a spin flip approaches $\frac{1}{2}$ as indicated earlier. If t_e represents the period of time that the excitation is on, then the net probability for observing a spin flip is given by

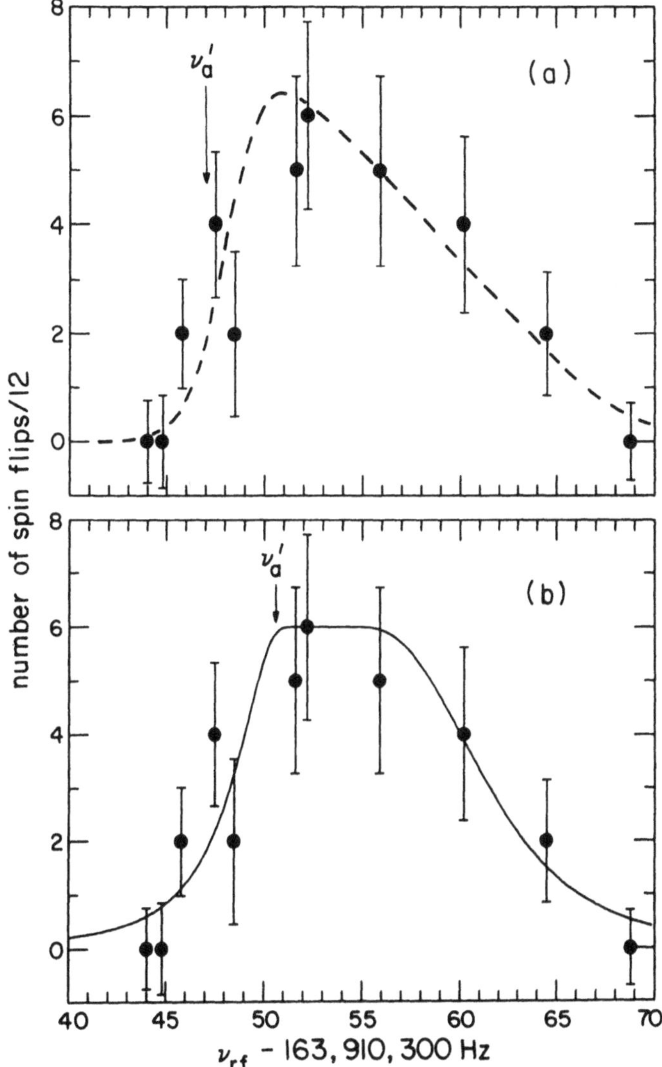

Figure 27: *Early anomaly resonance (run 109) showing strong saturation on the peak. These early runs, obtained by plotting the number of flips observed out of a fixed number of tries vs anomaly frequency, were found to show a systematic shift towards lower* $\nu_a'(Z_{\mathrm{rms}} = 0)$ *with increasing applied power. As shown in (a), this arose from the artificial choice of determining the resonance edge at* $\approx 1/3$*rd of peak height. As the new lineshape fitting shows in (b), the correct* $Z_{\mathrm{rms}} = 0$ *is near the top of the peak (adapted from Ref. 2).*

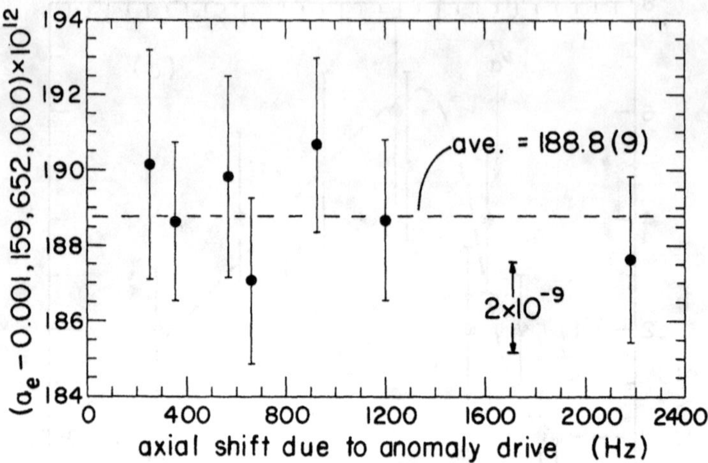

Figure 28: *Summary of 1983-84 anomaly runs using guard-ring excitation and lineshape fitting. Power of the anomaly drive is measured relatively as an axial frequency shift due to rf deepening of the effective potential well. The residual anomaly-power dependence is < 1 ppb over the range of drives used, but there was an undetected (at that time) −4 ppb error due to the residual systematic shift of the cyclotron frequency (from Ref. 2).*

$$\wp(t_e) = \tfrac{1}{2}\{1 - \exp[-\pi\omega_{\mathrm{Ra}}^2 t_e \chi(\omega)]\} \tag{61}$$

where the strength of the drive is contained in the Rabi frequency ω_{Ra}. Note that when ω_{Ra} is sufficiently large, $\wp \to \frac{1}{2}$ (i.e. saturated). If run 109, shown in Fig. 27a, is fitted to Eq. 61, one sees in Fig. 27b an excellent example of this saturation effect where the $Z_{rms} = 0$ position is near the peak instead. In fact, if all the runs taken at that time under similar conditions (and having a sufficient amount of data) are fitted to Eq. 61, one sees that there is no longer a serious systematic shift of a_e with anomaly power (see Fig. 28). However, it was later learned that these runs still contained a small systematic shift with cyclotron power as described in Sec. 5.5

The ability to resolve the smeared low-frequency edge obviously depends directly on the size of the magnetic linewidth, $\Delta\omega_a'$, relative to the width of the axial resonance. It is clearly desirable to reduce this smearing and increase accuracy by reducing the width of the axial resonance during the anomaly excitation. By detuning the voltage on the endcaps by a small amount (30–50 mV), it is possible to shift the axial resonance far enough from the peak of the detection tuned circuit that the axial damping is substantially reduced. An example of data, for which the axial linewidth is reduced to $\gamma_z/2\pi \sim 1$ Hz, is given in Fig. 29 for run 146.

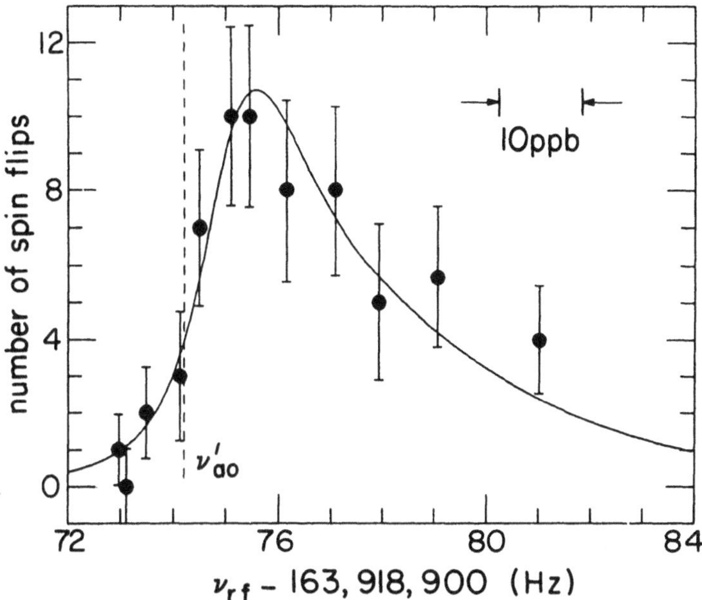

Figure 29: *Single-electron anomaly resonance showing additional narrowing by detuning the axial resonance. Each point represents 25 measurements of the spin state. The line shape now clearly shows the expected exponential tail, but with the low-frequency edge still smeared by the axial noise modulation, however at a reduced amount due to a factor of six narrower axial linewidth. Error bars are derived from the binomial distribution, and are computed with the least-squares-fitted curve as the true parent distribution. A typical fitted ($Z_{\rm rms} = 0$) edge is uncertain by 0.2 Hz (from Ref. 27).*

The magnetic width is $\sim$ 3 Hz for this run, corresponding to an effective axial temperature of $\sim$ 6 K, which follows from Eq. 59. Typical precision of such fits is about 1 ppb.

7. Results, Conclusions, and Goals

7.1 The g-Factor Results

New electron g-factor results were reported at four different times in the past dozen years. The first of these was the introduction in 1977 of the basic technique which utilized continuous observation of the cyclotron resonances and some alternating excitation/detection for the anomaly resonances. The accuracy of this initial observation was almost 20 times better than any previous measurement of the electron's g-factor. The work reported in 1979 gives the g-factor at three different magnetic fields. This was done at the time as a very simple expedience for finding an accurate g-factor since systematics were expected to depend on magnetic field. The 1984 results noted the use of a new trap and several improvements in technique, most notably the use of guard-ring drive and the use of line shape fitting to eliminate the major systematic error. The work published in 1987 uncovered the presence of systematic dependencies on power, estimated the microwave cavity shift and established the present precision comparison of the electron with the positron.

7.1.1 The 1977 work: At this time, the first ever g-2 measurement using the geonium system was reported. This work was characterized initially by the use of continuously observed cyclotron and anomaly resonances, though the beneficial aspects of the alternating excitation/detection method for observing narrower anomaly resonances was quickly noted. The magnetic field was chosen to be 18 kG, primarily because of the near degeneracy of the anomaly and axial frequencies. This choice was necessitated by the need for very strong b_r fields according to Eq. 53 which was accomplished by having $\omega'_a \sim \omega_z$ in order to obtain large axial amplitudes at ω'_a. This allowed the initial observation of the first spin flipping by saturating the anomaly resonance. Typical widths under weak drives with these conditions were about 0.25 ppm, and at that time no bottle-dependent lineshape could be detected for either cyclotron or anomaly resonances.

The one important feature, which was first introduced with this work, was the use of SBE magnetron cooling which maintained a reproducible position in the trap from electron to electron and with time. Also, the measurement of ω_m and its comparison and close agreement with $\delta_e = \omega_z^2/2\omega'_c$ was used as experimental proof that the dc trapping potential was effectively axially symmetric in these Penning traps as assumed in the derivation of Eq. 11. Thus, using the alternating anomaly resonances, for the first time, the g-factor anomaly was reported[1] as

$$a_e(1977) = 1\,159\,652\,410(200) \times 10^{-12} \tag{62}$$

where the uncertainty was based on estimates of all possible systematics and the half linewidths of the observed resonances.

7.1.2 The 1979 work: The work reported here involved making $\sim$ 16 measurements at three different magnetic fields, 18.6, 32.0 and 51.5 kG. This data now exclusively used alternating detection and excitation periods for both cyclotron and anomaly resonances and, since the bottle-dependent lineshape could now be observed, the estimate for ν'_a was taken to be about one-third the way up the low-frequency edge from the base line. However, the scatter for these runs was quite large, typically 1-3 standard deviations, and precision was not adequate at this time to even observe the apparent anomaly shift. Because of the somewhat non-statistically nature of these runs, the anomaly was reported[52] as

$$a_e(1979) = 1\,159\,652\,200(40) \times 10^{-12} \tag{63}$$

where the uncertainty represents the maximum displacement from the average. The reason for the large scatter was believed to be due to magnetic field wander in some non-linear fashion. The experiment could not be improved until a more stable magnet was installed.

7.1.3 <u>The 1984 work:</u> The data reported here involved a number of important changes in addition to the magnet. First of all, a previous vacuum leak into the new all-metal vacuum envelope was traced to the lack of cold-welding to the beryllium copper envelope. This was remedied by pre-wetting the indium to the metal parts prior to sealing the tube. The effect of a slightly higher background of helium gas probably accounted for the jumping of the cyclotron frequency edge as the particle's position in the inhomogeneous magnetic field was occasionally altered. Another improvement was the lowering of the amount of anomaly power required to flip the spin by using a weak microwave drive-assist to heat up the electron's cyclotron orbit and thus increase the spin flipping rate according to Eq. 53 or 58. This "assist" reduced the amount of anomaly power dissipated into the liquid helium environment, thus allowing the liquid level to remain stable and the axial S/N to remain near optimum for both cyclotron and anomaly data-taking periods. For symmetry purposes, both anomaly power and cyclotron power remain on during each such period, but one drive was always detuned from resonance while the other was being varied through resonance.

Along the same line of reasoning for reducing the amount of anomaly power, the new guard-ring excitation method was introduced which utilized the direct magnetic field excitation at the anomaly frequency. The assumption was that this mechanism would not depend on proximity to ω_z for excitation amplitude and thus would be more efficient. This assumption has not yet been conclusively verified by experimentation, though it is certainly expected to be valid.

By far the most significant improvement in technique was the utilization of the lineshape fitting program to predict the "true" location of the $Z_{\rm rms} = 0$ anomaly edge. By taking power broadening into account, the fitting process successfully eliminated the systematic shift in (older) past data that had an adequate amount of the lineshape with which to fit to the theoretical shape. This effect was discussed in Sec. 6.5. The apparent systematic is clearly not present in the runs shown in Fig. 28 which comprises the 1984 data. Because of these improvements in technique, the 1984 runs show an order of magnitude reduction in uncertainty, and were reported[2] to yield:

$$a_e(1984) = 1\,159\,652\,193(4) \times 10^{-12}\,. \tag{64}$$

The uncertainty at this point was still limited by apparent magnetic field fluctuations (but at least an order of magnitude less than for the 1979 runs). The statistical error was about half of the reported uncertainty. Because the same microwave diode was used for this data as that reported in the next section, we are now justified in correcting each run for the shift in cyclotron frequency associated with the strength of the microwave field as described in Fig. 23. The data shown in Fig. 28 contain this correction. As a result,

$$a_e(1984)_{\mathrm{corr.}} = 1\,159\,652\,189(4) \times 10^{-12}\,. \tag{65}$$

7.1.4 **The 1987 work:** At this point, the improved technique of microwave "assist" was refined to include a small shift in trapping potential that is applied commonly to both endcaps during the ω_c', ω_a' excitation phases. (This practice had just been introduced at the end of the 1984 run session.) This shift is done to reduce the axial damping linewidth, γ_z. The Brownian-motion lineshape theory[51] has confirmed that the anomaly resonance will have a sharper[5] $Z_{\mathrm{rms}} = 0$ edge if this axial linewidth is reduced well below the magnetic resonance width. The ability to alter the axial damping follows from the ability to change the effective impedance of the detection tuned circuit presented to the electron. Thus, if the equivalent ℓc series circuit of the electron is detuned by $\delta\omega$ with respect to the center frequency, the linewidth, which is proportional to the effective damping resistor, can be shown to equal γ_z' such that

$$\gamma_z' = \frac{\gamma_z}{1 + \left(\frac{2\delta\omega}{\Delta\Omega}\right)^2} \tag{66}$$

where $\Delta\Omega$ represents the FWHM for the detection circuit. For the parameters shown in Table 1, this equation predicts a reduction factor of 0.19 and an axial linewidth of 1.1 Hz. This is compared to the magnetic linewidth, fitted on the average to be 2.8 ± 0.3 Hz. The effect of this narrowing when $\gamma_c \sim 2\pi(0.16\ \mathrm{Hz})$ is a reduction of about one half for the low frequency edge width and thus an improved resolution. Typical fitted uncertainty due to the lineshape fitting program is typically 0.2 Hz for this set of data. However, as indicated in Sec. 5.5, our precision had improved to the point that we could now observe a residual shift of the cyclotron frequency with cyclotron drive power. The effect is shown in Fig. 23 for this data and the evidence suggests that the shift is due to a forced change in the magnetron orbit by the applied cyclotron drive. After correcting the cyclotron resonances for this effect, the data is again displayed in a plot of δa_e versus microwave power (in Fig. 30b) which is directly related to the rectified current through the multiplier diode and similarly versus rf anomaly power (in Fig. 30a)

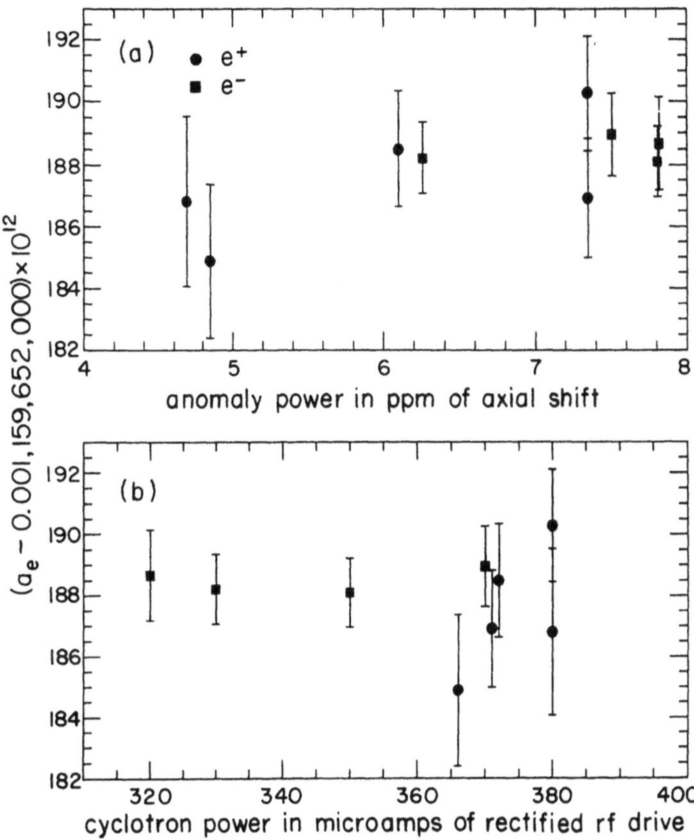

Figure 30: *Summary of recent high precision data used in the electron/positron comparison. Single electron and single positron runs are distinguished by squares and circles respectively. In (a), the data are plotted vs anomaly power measured in ppm of the axial shift when anomaly power is applied. In (b), the same data are plotted vs cyclotron power measured in μA of rectified rf drive. Any residual systematic shift is at or below the 1 ppb level of uncertainty (data obtained from Ref. 27).*

which is again related directly to the axial frequency shift produced by the anomaly drive as discussed in Sec. 3.5. At this level of the statistical fluctuations, no further dependence is apparent for these runs.

Finally, our improved precision has brought considerable theoretical interest in possible cavity shifts due to the mode-pulling effects. Using the model of a cylindrical cavity as utilized[44)] by Brown et al. (1985), and the constraint that the classical decay time is inhibited by a factor of 10 (shown in Fig. 21) in the neighborhood of mode-structure (believed to exist for a comparable sized cylindrical cavity at 50 kG), it was estimated that a probable shift of no more than 4 for δa_e (in units of 10^{-12}) would exist. This estimate was also made using the assumption that the cavity Q was about 1000. Subsequently, having observed cavity modes in the same trap to have a cavity $Q < 500$, the following g-factor anomaly could be quoted[27)] with confidence:

$$a_e(1987) = 1\,159\,652\,188.4(4.3) \times 10^{-12} \tag{67}$$

where a statistical error of 0.62×10^{-12} in the weighted average of all electron runs was combined in quadrature with our estimate of 1.3×10^{-12} for the uncertainty in the residual microwave power shift and the 4×10^{-12} potential cavity-mode shift discussed above. Also mentioned at the beginning of this section, similar runs were taken at this same time using positrons in place of electrons. These runs are also shown in Fig. 30 and give essentially the same result within their uncertainties (see Eq. 7).

7.2 "a_e" and QED

The intrinsic interest in this very precise measurement of the electron's g-factor resides in the ability of theorists to calculate the same quantity to roughly the same accuracy via the powerful theoretical technique of Quantum Electrodynamics (QED). Since early experiments suggested that the electron's g-factor could not be exactly 2, a strong motivation existed at that time to develop the revolutionary new QED theory. The discrepancy which had existed was found to have its basis in the processes of virtual emission and absorption of photons and the polarization of the vacuum by electron/positron pairs. These radiative corrections for this simple lepton system can be arranged into an infinite power series of the quantity α/π where α is the fine-structure constant, and C_i are the corresponding coefficients of the power series. This particular lepton system is also fortunately blessed with understandable and calculable effects due to the strong and weak interactions such that they do not limit the present accuracy of the g-factor prediction.

The leading term has one Feynman diagram due to a single virtual photon exchange and its coefficient C_1 was first shown[53)] by J.S. Schwinger to be exactly 0.5; it alone was adequate at the 1 part in 10^3 level of accuracy at accounting for the observed discrepancy of that early hfs work.[8] The second term allows for two virtual photon exchanges or a single vacuum polarization bubble containing an e^+/e^- pair (7 total Fynman diagrams). The C_2 coefficient in front of this term is known analytically to be $-0.328,478,965,\ldots$ (most recently predicted[54]) by C. Sommerfeld and A. Petermann). The next order of complexity contains 72 Fynman diagrams and the corresponding C_3 coefficient has been evaluated in part analytically[55]) and in part numerically:[56]) $C_3 = 1.175,62(56)$. As if this was not difficult enough, T. Kinoshita undertook the mammoth job of evaluating[57]) the 891 Feynman diagrams associated with the $8th$ order term, yielding $C_4 = -1.472(152)$. The remaining contributions due to the other-lepton loops, hadronic terms, and the electroweak effect have all been collected[58]) into the term: $\delta a_e' = 4.46 \times 10^{-12}$.

In order to make connection between theory and experiment, one needs to have an accurate measurement of the fine-structure constant, α. Unfortunately, the tremendous improvement in precision of the g-2 experiment and theory have both exceeded the present improvements in the accuracy of α. At present, the most accurate independent measure of α is by the quantum Hall effect:[59])

$$\alpha^{-1}(\mathrm{QHE}) = 137.035\,997\,9(3\,3) \tag{68}$$

which can be compared to the fine-structure constant computed from the ac Josephson value of $2e/h$ combined with the most recent low field determination[60]) of the gyromagnetic ratio γ_p' (for protons in a water sample):

$$\alpha^{-1}(\mathrm{acJ}\&\gamma_p') = 137.035\,984\,0(5\,1)\,. \tag{69}$$

These two values do not even agree within their combined uncertainties. Using the first of these (because it is more accurate), the theoretical value for a_e is

$$a(\mathrm{theory}) = 1\,159\,652\,133(29) \times 10^{-12} \tag{70}$$

where the error is almost exclusively due to the measurement error of α. Because of this limitation, it is preferable to determine α from the g-2 experiment and QED theory which is repeated from Eq. 6,

$$\alpha^{-1}(\mathrm{QED}) = 137.035\,991\,4(1\,1) \tag{71}$$

where the uncertainty is still primarily determined by the theoretical error. However, it is anticipated that one day there will be comparisons of all three of these major ingredients [α, a_e(expt), and a_e(theory)] at the one part per billion level of precision. This will provide a far more stringent test of QED than is presently possible.

7.3 Potential Future Efforts

The present efforts involve the use of a new phosphor bronze trap with $R_0 = 3.8$ mm and $Z_0 = 3.2$ mm. This trap is designed to have a lower cavity Q at 50 kG and its primary function is to verify that previous work was not plagued by large cavity shifts. This work will not be reviewed here because it is still in the development stage. However, the preliminary results are consistent with all past g-2 results within the same 4 ppb uncertainty in a_e.

The best method for handling the cavity problem for the future has not yet been determined. It has been suggested by Brown, et al.[47] and Dehmelt[45,48] that the characteristics of several neighboring modes can be investigated and the theoretical model can then be used to predict the region in frequency space which is free of shifts. Such regions do exist, but considerable effort may be required to find them. However, it is always preferable, if a choice exists, to design the experiment such that all regions of space are relatively shift free. One approach to achieving this might involve using smaller traps or lower magnetic fields such that the frequency of interest is well below the lowest mode where shifts can not exist. More than likely, both low cavity Q traps will be used and the nearest modes will be studied by actually measuring g-2 at the maximum shift positions, just to be assured of the accuracy between modes.

The next task will be the improvement of axial sensitivity in order to allow still smaller axial frequency shifts to be observed, with the intent of reducing the size of the magnetic bottle. The clear evidence is that all major systematic effects, other than cavity shifts, that have been studied so far, depend directly on the strength of this bottle in some way. A more stable voltage source (possibly from a Josephson junction array) would translate immediately into improved signal-to-noise, since voltage fluctuations in the standard cells are believed to produce the the present limit of 10 ppb resolution. A higher Q tuned circuit would also generate a larger signal for the same oscillation amplitude in the trap.

However, since the real object is to achieve an almost perfectly uniform magnetic field (with some means of detecting $\vec{\mu}$), two possible methods have been suggested for which this condition is approached at least by a factor of 10 improvement in field homogeneity. One method involves utilizing the existing relativistic bottle in some way. The primary impediment to this approach is axial sensitivity which presently requires 10–20 times larger bottles just to see a spin flip. Several elegant methods have been proposed[39,61,62,63] which might enhance the effect in order to yield the required S/N, but no technique has yet demonstrated that it will work.

The method favored by the author involves the use of the variable magnetic bottle.[64] Here, the B_2-term can be modulated[65] at some slow rate and a PSD can be used on the axial-frequency-shift detector to pick out the corresponding modulated magnetic moment (see Eq. 30). This technique has been attempted, but has so far achieved only partial success. This effort is presently languishing because of much needed technical improvements in equipment and trapping apparatus.

In all cases, the primary objective is improvement in precision, not necessarily accuracy, since the test to be performed is an ultra high precision comparison of electron and positron g-factors. As stated earlier, common systematics such as cavity shifts or B_2-related shifts will be common to both. Thus, a statistical improvement is all that is required for the next improved comparison of matter and antimatter.

References

1. Van Dyck, Jr., R.S., Schwinberg, P.B., and Dehmelt, H.G., "Electron Magnetic Moment from Geonium Spectra," in *New Frontiers in High-Energy Physics*, edited by B. Kursunoglu, A. Permutter, and L.F. Scott (Plenum Publishing, N.Y., 1978) p. 159.

2. Van Dyck, Jr, R.S., Schwinberg, P.B., and Dehmelt, H.G., "The Electron and Positron Geonium Experiments," in *Atomic Physics 9*, edited by R.S. Van Dyck, Jr. and E.N. Fortson (World Scientific, Singapore, 1984), p. 53.

3. Van Dyck, Jr., R.S., Schwinberg, P.B., and Dehmelt, H.G., "Electron Magnetic Moment from Geonium Spectra: Early Experiments and Background Concepts," Phys. Rev. D34, 722 (1986).

4. Brown, L.S. and Gabrielse, G., "Geonium Theory: Physics of an Electron or Ion in a Penning Trap," Rev. Mod. Phys. 58, 233 (1986).

5. Dehmelt, H.G., "Invariant Frequency Ratios in Electron and Positron Geonium Spectra Yield Refined Data on Electron Structure," in *Atomic Physics 7*, edited by D. Kleppner and F.M. Pipkin (Plenum Publishing, N.Y. 1981) p. 337.

6. Wineland, D.J., Itano, W.M., and Van Dyck, Jr., R.S., "High-Resolution Spectroscopy of Stored Ions," in *Advances in Atomic and Molecular Physics*, Vol 19, edited by B. Bederson NS D. Bates (Academic Press, N.Y. 1983) p. 135.

7. Dehmelt, H.G., "A Single Atomic Particle Forever Floating at Rest in Free Space: New Value for Electron Radius," Physica Scripta T22, 102 (1988).

8. Nafe, J.E., Nelson, E.B., and Rabi, I.I., "The Hyperfine Structure of Atomic Hydrogen and Deuterium," Phys. Rev. 71, 914 (1947).

9. Louisell, W.H., Pidd, R.W., and Crane, H.R., "An Experimental Measurement of the Gyromagnetic Ratio of the Free Electron," Phys. Rev. 91, 475 (1953); 94, 7 (1954).

10. Dehmelt, H.G., "Spin Resonance of Free Electrons Polarized by Exchange Collisions," Phys. Rev. 109, 381 (1958); also Science 124, 1039 (1956).

11. Major, F.G., and Dehmelt, H.G., "Exchange-Collision Technique for the rf Spectroscopy of Stored Ions," Phys. Rev. 170, 91 (1968).

12. Gräff, G., Major, F.G., Roeder, R.W., and Werth, G., "Method For Measuring the Cyclotron and Spin Resonance of Free Electrons," Phys. Rev. Lett. 21, 340 (1968).

13. Gräff, G., Klempt, E., and Werth, G., "Method For Measuring the Anomalous Magnetic Moment of Free Electrons," Z. Phys. 222, 201 (1968).

14. Dehmelt H.G. and Walls, F.L., "Bolometric Technique for the rf Spectroscopy of Stored Ions," Phys. Rev. Lett. 21, 127 (1968).

15. Church, D.A. and Mokri, B., "Polarization Detection of Trapped Electrons via Interaction with Polarized Atoms," Z. Phys. 244, 6 (1971).

16. Wineland, D.J. and Dehmelt, H.G., "Line Shifts and Widths of Axial, Cyclotron, and g-2 Resonances in Tailored, Stored Electron (Ion) Cloud," Int. J. Mass Spectrom. Ion Phys. 16, 338 (1975) and erratum: 19, 251 (1976).

17. Wilkinson, D.T. and Crane, H.R., "Precision Measurement of the g Factor of the Free Electron," Phys. Rev. 130, 852 (1963).

18. Rich, A., "Corrections to the Experimental Value for the Electron g-Factor Anomaly," Phys. Rev. Lett. 20, 967 (1968) and Phys. Rev. Lett. 20, 1221 (1968)

19. We look upon the single electron bound via magnet and trap structure to the earth as a metastable pseudo-atom, called "geonium".

20. Wesley, J.C. and Rich, A., "High-Field Electron g-2 Measurement," Phys. Rev. A4, 1341 (1971).

21. Granger, S. and Ford, G.W., "Lepton Spin Motion in a Weak Electromagnetic Trap," Phys. Rev. D13, 1897 (1976).

22. Newman, D., Sweetman, E., Conti, R. and Rich, A., "Magnetic Moments of Electrons and Positrons," in *Atomic Masses and Fundamental Constants 6*, edited by J.A. Nolen, Jr. and W. Benenson (Plenum, New York, 1980), p.183.

23. Walls, F.L., Ph.D. thesis, "Determination of the Anomalous Magnetic Moment of the Free Electron from Measurements Made on an Electron Gas at 80°K Using a Bolometric Technique," University of Washington, 1970; also Walls, F.L., and Stein, T.S., Phys. Rev. Lett. 31, 975 (1973).

24. Wineland, D.J., Ekstrom, P., and Dehmelt, H., "Monoelectron Oscillator," Phys. Rev. Lett. 31 1279 (1973).

25. Dehmelt, H., Ekstrom, P., Wineland, D., Van Dyck, R., "Landau Level Dependent ν_z-Shifts in the Monoelectron Oscillator," Bull. Am. Phys. Soc. 19, 572 (1974).

26. Dehmelt, H., "Continuous Stern-Gerlach Effect: Principle and Idealized Apparatus," Proc. Natl. Acad. Sci. USA 83, 2291 (1986).

27. Van Dyck, Jr, R., Schwinberg, P., and Dehmelt, H., "New High-Precision Comparison of Electron and Positron g Factors," Phys. Rev. Lett. 59, 26 (1987).

28. Van Dyck, Jr, R.S., Schwinberg, P.B., and Dehmelt, H.G., "Precise Measurements of Axial, Magnetron, Cyclotron, and Spin-Cyclotron-Beat Frequencies on an Isolated 1-meV Electron," Phys. Rev. Lett 38, 310 (1977).

29. Van Dyck, R.S., Jr., Wineland, D.J., Ekstrom, P.A., and Dehmelt, H.G., "High Mass Resolution with a New Variable Anharmonicity Penning Trap," Appl. Phys. Lett. 28, 446 (1976).

30. Schwinberg, P.B., Van Dyck, Jr, R.S., and Dehmelt, H.G., "New Comparison of the Positron and Electron g Factors," Phys. Rev. Lett. 47, 1679 (1981).

31. Schwinberg, P.B., Van Dyck, Jr, R.S., and Dehmelt, H.G., "Preliminary Comparison of the Positron and Electron Spin Anomalies," in *Precision Measurement and Fundamental Constants II*, edited by B.N. Taylor and W.D. Phillips, Natl. Bur. Stand. (U.S.) Spec. Publ. 617 (1984) p. 21.

32. Gabrielse, G., "Relaxation Calculation of the Electrostatic Properties of Compensated Penning Traps with Hyperbolic Electrodes," Phys. Rev. A27, 2277 (1983).

33. Schwinberg, P.B., Ph.D. thesis, "A Technique for Catching Positrons in a Penning Trap via Radiation Damping," University of Washington, 1979 [available from University Microfilms International, Ann Arbor, MI.].

34. Schwinberg, P.B., Van Dyck, Jr, R.S., and Dehmelt, H.G., "Trapping and Thermalization of Positrons for Geonium Spectroscopy," Phys. Lett. 81A, 119 (1981).

35. Wineland, D.J. and Dehmelt, H.G., "Principles of the Stored Ion Calorimeter," J. Appl. Phys. 46, 919 (1975).

36. Gabrielse, G., "Detection, Damping, and Translating the Center of the Axial Oscillation of a Charged Particle in a Penning Trap with Hyperbolic Electrodes," Phys. Rev. A29, 462 (1984).

37. Brown, L.S. and Gabrielse, G., "Precision Spectroscopy of a Charged Particle in an Imperfect Penning Trap," Phys. Rev. A25, 2423 (1982).

38. McLachlan, N.W., "Theory and Applications of Mathieu Functions," (Oxford University Press, N.Y., 1947), p. 20.

39. Dehmelt, H., Mittleman, R., and Liu, Y., "Relativistic Cyclotron Resonance Shape in Magnetic Bottle Geonium," Proc. Natl. Acad. Sci. USA 85, 7041 (1988).

40. Van Dyck, Jr., R.S., "Summary of the Physics in Traps Panel," Physica Scripta T22, 228 (1988).

41. Van Dyck, Jr., R.S., Moore, F.L., Farnham, D.L., and Schwinberg, P.B., "Number Dependency in the Compensated Penning Trap," to be published in Phys. Rev. A.

42. Van Dyck, Jr., R.S., Schwinberg, P.B., and Bailey, S.H., "High Resolution Penning Trap as a Precision Mass-Ratio Spectrometer," in *Atomic Masses and Fundamental Constants 6*, edited by J.A. Nolen, Jr. and W. Benenson (Plenum, New York, 1980), p. 173.

43. Gabrielse, G. and Dehmelt, H., "Observation of Inhibited Spontaneous Emission," Phys. Rev. Lett. 55, 67 (1985).

44. Brown, L.S., Gabrielse, G., Helmerson, K., and Tan, J., "Cyclotron Motion in a Microwave Cavity: Lifetime and Frequency Shifts," Phys. Rev A32, 3204 (1985); also Phys. Rev. Lett. 55, 44 (1985).

45. Dehmelt, H., et al. "Practical Zero-Shift Tuning in Geonium," (unpublished).

46. Van Dyck, Jr., R., Moore, F., Farnham, D., Schwinberg, P., and Dehmelt, H. "Microwave-Cavity Modes directly Observed in a Penning Trap," Phys. Rev. A (Brief Reports) 36, 3455 (1987).

47. Brown, L.S., Gabrielse, G., Tan, J., and Chan, K.C.D., "Cyclotron Motion in a Penning Trap Microwave Cavity," Phys. Rev. A37, 4163 (1988).

48. Dehmelt, H., "Single Atomic Particle at Rest in Free Space: Shift-Free Suppression of the Natural Line Width?" in *Laser Spectroscopy VIII*, edited by W. Persson and S. Svanberg (Springer-Verlag, 1987), p. 39.

49. Gilleland, J.R. and Rich, A., "Precision Measurement of the g Factor of the Free Positron," Phys. Rev. A5, 38 (1972).

50. Bailey, J., et al., "Final Report on the CERN Muon Storage Ring Including the Anomalous Magnetic Moment and Electric Dipole Moment of the Muon, and a Direct Test of Relativistic Time Dilation," Nuc. Phys. B150, 1 (1979).

51. Brown, L.S., "Geonium Lineshape," Ann. Phys. (NY) 159, 62 (1985); also Phys. Rev. Lett. 52, 2013 (1984).

52. Van Dyck, Jr., R.S., Schwinberg, P.B., and Dehmelt, H.G., "Progress of the Electron Spin Anomaly Experiment," Bull. Am. Phys. Soc. 24, 758 (1979).

53. Schwinger, J., "On Quantum-Electrodynamics and the Magnetic Moment of the Electron," Phys. Rev. 73, 416 (1948).

54. Sommerfield, C., "Magnetic Dipole Moment of the Electron," Phys. Rev. 107, 328 (1957); A. Petermann, "Fourth Order Magnetic Moment of the Electron," Helv. Phys. Acta 30, 407 (1957).

55. Levine, M.J., Park, H.Y., and Roskies, R.Z., "High-Precision Evaluation of Contributions to g-2 of the Electron in Sixth Order," Phys. Rev. 25, 2205 (1982).

56. Cvitanovic, P. and Kinoshita, T., "Sixth-Order Magnetic Moment of the Electron," Phys. Rev. D10, 4007 (1974); Kinoshita, T. and Lindquist, W.B., "Improving the Theoretical Prediction of the Electron Anomalous Magnetic Moment," Cornell preprint CLNS-374, 1977.

57. Kinoshita, T., "Fine-Structure Constant Derived from Quantum Electrodynamics," Metrologia 25, 233 (1988); see also "Accuracy of the Fine-Structure Constant," IEEE Trans. Instrum. Meas. 38, 172 (1989).

58. See, for instance, Kinoshita, T., in *New Frontiers in High Energy Physics,*, edited by B. Kursunoglu, A. Permutter, and L.F. Scott (Plenum Publishing, N.Y., 1978) p. 127; also Kinoshita, T. and Lindquist, W.B., "Eighth-Order Anomalous Magnetic Moment of the Electron," Phys. Rev. Lett. 47, 1573, (1981).

59. Cage, M.E., et al., "NBS Determination of the Fine-Structure Constant, and of the Quantized Hall Resistance and Josephson Frequency to Voltage Quotient in SI Units," IEEE Trans. Instrum. Meas. 38, 284 (1989).

60. Williams, E.R., et al., "A Low Field Determination of the Proton Gyromagnetic Ratio in Water," IEEE Trans. Instrum. Meas. 38, 233 (1989).

61. Dehmelt, H., Van Dyck, R., and Schwinberg, P., "Proposal for Detection of Geonium Spectra via Radial Displacement," Bull. Am. Phys. Soc. 24, 491 (1979).

62. Dehmelt, H., Van Dyck, R., Schwinberg, P., and Gabrielse, G., "Proposal to Detect Spin Flips in Geonium via Linked Axial Excitation," Bull. Am. Phys. Soc. 24, 675 (1979).

63. Dehmelt, H. and Gabrielse, G., "Faster, Simpler Schemes to Distinguish $n = 0, 1$ in Geonium," Bull. Am. Phys. Soc. 26, 797 (1981); "Comb Excitation Scheme for Resolving the Cyclotron Spectrum of Geonium," Bull. Am. Phys. Soc. 29, 44 (1984); "Quasi-Thermal, Multi-Step Excitation Scheme for Geonium Cyclotron Spectroscopy," Bull. Am. Phys. Soc. 29, 926 1984).

64. Van Dyck, Jr., R.S., Moore, F.L., Farnham, D.L., and Schwinberg, P.B., "Variable Magnetic Bottle for Precision Geonium Experiments," Rev. Sci. Instrum. 57, 593 (1986).

65. Schwinberg, P.B. and Van Dyck, Jr., R.S., "Geonium Spectroscopy Using a Modulated Magnetic Bottle," Bull. Am. Phys. Soc. 26, 598 (1981).

CAVITY SHIFTS OF MEASURED ELECTRON MAGNETIC MOMENTS

Gerald Gabrielse* and Joseph Tan*
Department of Physics
Harvard University
Cambridge, MA 02138

Lowell S. Brown**
Department of Physics
University of Washington
Seattle, WA 98195

Contents

* Supported by the U.S. National Science Foundation.
** Supported by the U.S. Department of Energy.

1. Introduction and History

The measured magnetic moment of the electron provides the most accurate test of quantum electrodynamics.[1,2] Until a few years ago, the measurements were conceptually simple, with no large systematic corrections required. However, the accuracy of the experiments has improved to the extent that this is no longer true. Presently, the measurements utilize line splitting based upon an intricate theory of the observed line shape[3] . The experimental accuracy is limited by a lack of control over the interaction of a single electron oscillator with the electromagnetic modes of a surrounding microwave cavity. (The unavoidable cavity in experiments so far is formed by the metal electrodes of the Penning trap used to confine the electron.) Efforts are underway to remove the reliance upon the theory of the observed resonance line. One possibility is to switch on and off the inhomogeneous magnetic field which both makes possible and limits the current experiments.[2] Another possibility is to utilize relativistic couplings which have been observed[4] to avoid the noise broadened lineshape which is an unpleasant consequences of the magnetic field inhomogeneity. Whatever approach succeeds, possible cavity shifts of the measured magnetic moment (the subject of this chapter) must be controlled or eliminated if further experimental progress is to be made.

The possibility of cavity shifts of the measured cyclotron frequency used to determine an electron's magnetic moment was first demonstrated when the coupling of the electron oscillator and a trap cavity was directly observed by Gabrielse and Dehmelt.[5] The first of the reported observations was made almost a year and one half before they were published. The reason for the delay illustrates a difficulty in performing these measurements. A proper study required shifting the electron cyclotron frequency by changing the current in a superconducting solenoid and this was delayed to avoid the long time (up to a month) required for the field to restabilize sufficiently to allow more measurements. During this year and a half, estimates with electrical analogs[6] and an explicit model calculation[7] were undertaken to estimate the size of possible cavity shifts. Eventually, the observation of a coupling between an electron and a trap cavity was repeated at one electron frequency in another trap.[2] In hindsight, it is now clear that the cavity coupling is responsible for a slower than expected decay of the electron's cyclotron motion which had earlier been attributed to electronic causes.

Although the importance of cavity shifts for measurements of the electron magnetic moment was realized only recently, the basic notion that the couplings of two oscillators can shift both the damping rate and oscillation frequency of the oscillations is certainly very familiar. (The electron in its cyclotron motion and the electromagnetic cavity modes are the coupled oscillators here.) Long ago, for example, it was mentioned that the spontaneous emission of an atom placed in a cavity could be inhibited.[8] Further discussions of cavity-induced modifications to atom damping rates came later,[9] with clear realization of the problems that the frequency shifts would present for precise measurements of resonance frequencies.[10] In fact, the observation of inhibited spontaneous emission in a trap was the first observation of inhibited spontaneous emission in a cavity. Soon after, similar effects were observed with Rydberg atoms and now a variety of such studies have been carried out.[11]

The electron cyclotron oscillator and the observed coupling to a trap cavity are introduced in Sec. 2. Only a pure cyclotron motion is considered because other aspects of the motion of a single particle in a Penning trap are well understood[12] and do not complicate our examination of damping and frequency shifts. A simple model, wherein an electron oscillator is coupled to an LCR circuit which represents a single cavity mode, is developed in Secs. 3 and 4. The simple model is useful for understanding and for estimating changes in the damping rate and the resonance frequency of the electron oscillator. It is qualitatively useful as well, when the cyclotron oscillator is near to resonance with a cavity mode. In Sec. 4, eigenmodes of cylindrical, hyperbolic and spherical cavities are examined. Calculated coupling coefficients are presented which indicate how well each mode couples to an electron cyclotron motion around the symmetry axis of the cavity at its center. Sufficient information is provided so that the coupling of any of the cavity modes to the electron oscillator can be treated using the simple model. The way that cavity shifts limit the current measurements of the electron magnetic moment is discussed in Sec. 5. Only the simple model is used, given the limited experimental information which makes a more detailed theoretical analysis to be unwarranted.

Unfortunately, the simple model is not a complete description, especially when the electron cyclotron oscillator is not near to resonance with a high Q cavity mode. In fact, the sum of frequency shifts computed with the simple model diverges if contributions from too many cavity modes are included. Rigorous calculations are thus required to establish the validity and limitations of the simple model.[13] Detailed, renormalized calculations for a cylindrical cavity[14] and a spherical cavity[15] are reviewed in Sec. 6. These calculations are classical since it has been shown that within a high level of accuracy, the exact apparatus of quantum electrodynamics yields the classical results.[16] A comparison of the exact cylindrical calculation with the simple model has been used to justify an approximate treatment of the hyperbolic cavity, for which a more exact calculation cannot be done.[13] Finally, future possibilities for dealing with cavity shifts are discussed in Sec. 7. Particular attention is paid to a cylindrical microwave cavity[17] since it has recently been demonstrated that a single electron can be accurately studied with trap electrodes which approximate this configuration,[18] just as well as within the customary hyperbolic electrodes.

The present state of experimental information about the cavity modes and how much they affect the measurements of the electron magnetic moment is far from satisfactory. Ideally, the magnetic moment would be measured over a wide range of electron cyclotron frequencies, a range over which the location, Q and symmetry of cavity modes is well known (instead of at only a single frequency). Although there is now some experimental indication of the location and Q of cavity modes,[2,19] the estimates in this article are badly in need of more experimental information. Hopefully, the estimates will be replaced by measurements.

2. One-electron Cyclotron Oscillator

For our purposes it suffices to deal with a cyclotron motion of an electron (of charge e and mass m) in a uniform magnetic field B. In free space, the electron orbits a magnetic field line at angular frequency (CGS units)

$$\omega_c = \frac{eB}{mc} \tag{1}$$

This cyclotron motion is damped via synchrotron radiation with a damping width

$$\gamma_c = \frac{4e^2\omega_c^2}{3mc^2} \tag{2}$$

which is just the ratio of the power radiated from the circular motion (given by the familiar Larmor formula) divided by the instantaneous energy stored in this motion. Typically, the cyclotron motion is damped only very weakly so that the radiation width is very much smaller than the cyclotron frequency. For $B = 5.9$ Tesla, for example, $\omega_c/2\pi = 164$ GHz and $\gamma_c/2\pi = 1.3$ Hz so that

$$\frac{\gamma_c}{\omega_c} = 8 \times 10^{-12}. \tag{3}$$

Without line splitting, therefore, the cyclotron frequency could (in principle) be measured to 8×10^{-12}. The corresponding decay time is $\tau_c = \gamma_c^{-1} = 100$ ms.

To relate the uncertainty $\Delta\omega_c$ in measuring the cyclotron frequency to the resulting uncertainty Δa in the measured anomalous magnetic moment a which is so important for tests of quantum electrodynamics, we note that

$$a = \frac{\omega_s - \omega_c}{\omega_c} \tag{4}$$

can be regarded as a definition of a, where ω_s is the electron's spin precession frequency. Since $a \approx 10^{-3}$ is small,

$$\frac{\Delta a}{a} \approx \frac{1}{a}\frac{\Delta\omega_c}{\omega_c}. \tag{5}$$

Setting $\Delta\omega_c$ equal to the free space line width yields $\Delta a/a \approx 8 \times 10^{-9}$. The experimental error currently quoted[2] is half this amount.

Both the cyclotron frequency and the damping rate can be shifted when the one electron cyclotron oscillator is located within a microwave cavity rather than in free space. Fig. 1 shows the cavity within which the the inhibition of spontaneous emission was first observed. Fig. 2 shows measured cyclotron damping times[5] of 86 ± 2 and 347 ± 64 ms for cyclotron frequencies which differ by only 0.5%. These are the first observations of damping times which differ from the $\tau_c = 100$ ms expected for radiation into free space from Eq. (2). The damping clearly varies as the electron cyclotron frequency (and hence the detuning from the nearest eigenmode of the trap cavity) is changed. The longest damping time observed in this particular trap was about three times longer than the damping time for free space. This first observation was subsequently repeated in another trap[2], but with a damping time $\tau = \gamma^{-1} = 1.0 \pm 0.1s$, about 10 times larger than in free space.

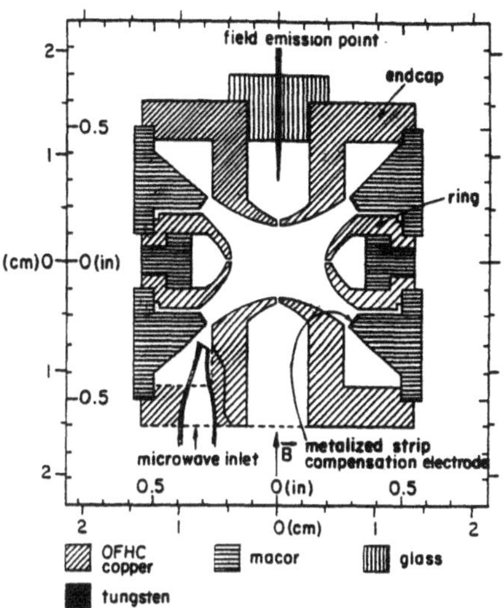

Fig. 1. Hyperbolic Penning Trap in which the inhibited spontaneous emission of a one-electron cyclotron oscillator was first observed. (From Ref. 5.)

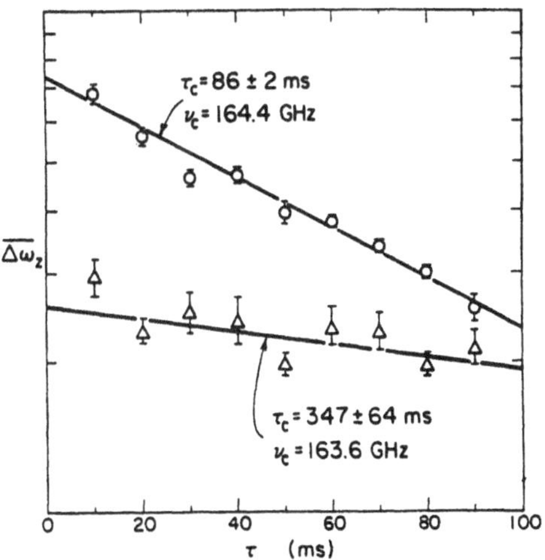

Fig. 2. Measured decay of the energy of a one-electron cyclotron oscillator in the trap of Fig. 1 as a function of time for two electron cyclotron frequencies which differ by only 0.5 %. Both damping times differ from the free space value $(\gamma_c)^{-1} = 0.1$ *s because of the microwave cavity formed by the trap electrodes. (From Ref. 5.)*

3. Simple Model

To illustrate the characteristic way that the damping rate and resonant frequency of an electron cyclotron oscillator is modified by its interaction with a microwave cavity, we use the simple model in Fig. 3a. In this model, the Nth cavity mode is represented to the right by an LCR tuned circuit.[20] Such a circuit is resonant at angular frequency $\omega_N = (LC)^{-1/2}$ with damping width $\Gamma_N = (RC)^{-1}$. The electron oscillator is represented as a charge e and mass m on a spring with spring constant $m\omega_c^2$. The interaction between the LCR and the electron is via the force $-eV/d$ on the charge which occurs when a potential V is present across two extended plates (separated by an effective distance d) between which the particle is located. (The effective capacitance of the two plates is included in C.) The unperturbed resonance frequency of the two oscillators are related by a detuning δ defined by

$$\omega_c = \omega_N + \frac{1}{2}\Gamma_N \delta. \tag{6}$$

Presently we shall provide explicit expressions for the combination of the parameters L,C,R and d which will make the simple model quantitatively useful as well.

To emphasize that the cavity mode is in fact an oscillator, at any given frequency we can substitute an equivalent series $L_s C_s R_s$ circuit for the parallel LCR circuit as shown in Fig. 3b. We choose the same resonant frequency $\omega_N = (L_s C_s)^{-1/2}$, and the parallel and series circuits are indistinguishable at any given frequency provided that

$$L_s = -\frac{R_s}{\Gamma_N} \tag{7}$$

and

$$R_s = \frac{R}{1+\delta^2}. \tag{8}$$

The negative series inductance (and capacitance) signify an effective phase change and to simplify the latter expression we have assumed that $\delta << (2\omega_N/\Gamma_N)$. Both oscillators now explicitly satisfy harmonic oscillator equations

$$\ddot{z} + \omega_c^2 z = -\frac{eV}{md} \tag{9}$$

$$\ddot{q} + \frac{R_s}{L_s}\dot{q} + \frac{1}{L_s C_s} q = \frac{V}{L_s}, \tag{10}$$

where q is the charge stored in the capacitor and $\dot{q}$ denotes that q is differentiated with respect to time. The two equations are easily combined to eliminate V which couples them.

The remaining equation can be solved by inspection if the instantaneous power transfered from circuit to electron $-\dot{q}V$ is equated to the power dissipated by the electron $-eV\dot{z}/d$, yielding the Schotky formula

$$\dot{q} = \left(\frac{e}{d}\right)\dot{z}. \tag{11}$$

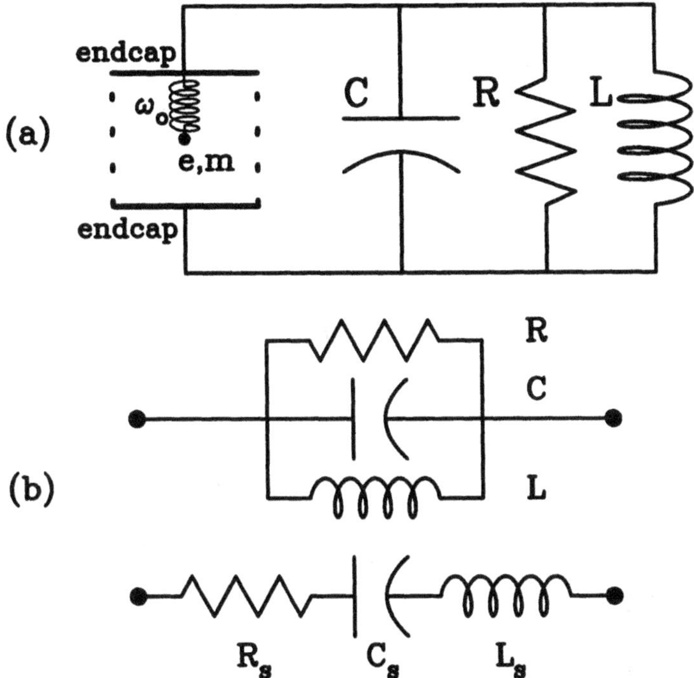

Fig. 3. (a) Simple model of the interaction of a one-electron cyclotron oscillator (represented as a charge e and mass m on a spring with spring constant $m\omega_c^2$) and a electromagnetic mode of a microwave cavity (represented as a parallel LCR circuit). (b) At any one frequency, the parallel LCR circuit may be represented as a series $L_sC_sR_s$ circuit.

We choose the origin of the coordinate system to coincide with the center of the trap so that $q = (e/d)z$ as well. The q dependence can then be eliminated to obtain a single damped harmonic oscillator equation

$$\ddot{z} + \gamma\dot{z} + (\omega_c + \Delta\omega)^2 z = 0 \tag{12}$$

which describes the motion of the coupled electron. The electron is damped via its coupling to the cavity mode at a rate

$$\gamma = \frac{\gamma_N}{1+\delta^2}. \tag{13}$$

The maximum damping rate γ_N (discussed in the next paragraph) pertains on resonance when ω_c and ω_N coincide (i.e. $\delta = 0$). The electron frequency is shifted from ω_c to $\omega_c + \Delta\omega$

with

$$\Delta\omega = \frac{\gamma_N}{2}\frac{\delta}{1+\delta^2} = \frac{1}{2}\gamma\delta. \tag{14}$$

If the electron oscillator and the LCR circuit are tuned to the same unperturbed resonance frequency (i.e. $\delta = 0$) there is no frequency shift. The maximum frequency shifts $\pm\gamma_N/4$ occur near resonance, at detunings $\delta = \pm 1$. The characteristic shapes for γ and $\Delta\omega$ are shown in Fig. 4 and these will be clearly evident in more detailed calculations which follow as well.

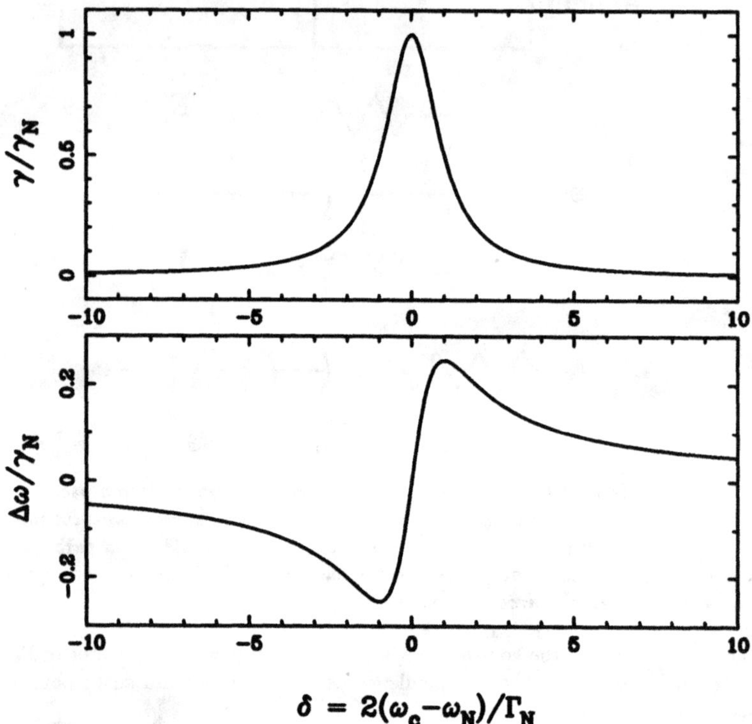

Fig. 4. Characteristic dependence of an oscillator's damping rate γ in (a) and frequency shift $\Delta\omega$ in (b) as a function of its detuning δ from the resonant frequency of an LCR circuit (or a cavity mode).

The maximum of the damping rate and the maximum of the frequency shift are both determined by

$$\gamma_N = \frac{e^2}{d^2}\frac{R}{m}. \tag{15}$$

Since $\gamma_N \sim R \sim Q_N$, the maximum damping and maximum frequency shift are larger when the quality factor $Q_N = \omega_N/\Gamma_N$ is larger. To display the Q dependence explicitly we write

$$\frac{\gamma_N}{\omega_N} = 2Q_N \left(\frac{\lambda_N}{\omega_N}\right)^2 \tag{16}$$

thereby defining the coupling strength[21] λ_N. This definition also allows the use of a simple form for the electron's frequency shift and damping rate

$$\Delta\omega - i\frac{\gamma}{2} = \frac{\omega(\lambda_N)^2}{\omega^2 + i\omega\Gamma_N - \omega_N^2}. \tag{17}$$

Using the parameters of the LCR model,

$$(\lambda_N)^2 = \frac{1}{2}\frac{r_e c^2}{Cd^2}, \tag{18}$$

where we have utilized the classical electron radius $r_e = e^2/mc^2$. Coupling strengths for cavities of interest will be specified in the next section.

To illustrate the results of the simple model, the solid lines in Fig. 5 represent the damping γ and frequency shift $\Delta\omega$ as a function of detuning δ (upper abcissa), for an electron cyclotron oscillator coupled to a mode of a cylindrical microwave cavity. (The mode eigenfrequency is within the frequency range where experiments are being done and the coupling strength used is described in the next section.) For comparison, the results of a more complete calculation (described in Sec. 6) are represented by the dotted line. For this relatively high Q value, $Q = 10^4$, the simple model describes both the damping and frequency shifts near the mode so well as to make the two curves almost indistinguishable.

Fig. 6 shows a similar comparison for a lower $Q = 1000$. Again the simple, one-mode model works very well near the mode. The larger frequency spread being viewed, however, reveals significant deviations between the one-mode model (solid lines) and the exact calculation (dotted lines). The major difficulty is the coupling of the electron to additional cavity modes which are nearby, as is illustrated at the left of the figure because of a nearby mode which is strongly coupled. These contributions can be largely accounted for by summing the contributions to γ and $\Delta\omega$ over additional, nearby modes. We thus replace Eq. (17) by the mode sum

$$\Delta\omega - i\frac{\gamma}{2} = \omega - \omega_c - i\frac{\gamma_c}{2} = \omega \sum_N \frac{\lambda_N^2}{\omega^2 + i\omega\Gamma_N - \omega_N^2}, \tag{19}$$

where the frequency of the cyclotron oscillator, ω, is now taken to be complex. The dashed line in Fig. 6 shows what happens when contributions are included from one additional cavity mode on either side. The agreement with exact values is much better, particularly at the left of the figure.

More than 3 modes are clearly required to describe the region between cavity modes with accuracy comparable to the experimental precision. The contributions to $\Delta\omega$ from off-resonant modes are clearly important, going as $\frac{1}{2}(\lambda_N)^2/(\omega_c - \omega_N)$, independent of Q_N.

Including the contribution of more modes to the mode sum does improve the agreement with the exact values. A careful study for $Q = 10^3$ shows that the agreement gets better as more modes are included up to approximately 7 modes to either side of the frequency being considered.[13] If more modes are included, however, the error in the frequency shift (i.e. in the real part of the mode sum) begins to increase. In fact, the frequency shift from the mode sum diverges if the mode sum includes all the cavity modes which couple to the electron cyclotron motion. This divergence, and how it can be correctly removed by renormalization in certain cases, is discussed in Sec. 6. A truncated mode sum was used to calculate the damping and frequency shifts of an electron cyclotron oscillator within a hyperbolic cavity,[13] since an exact calculation has not been feasible in this geometry.

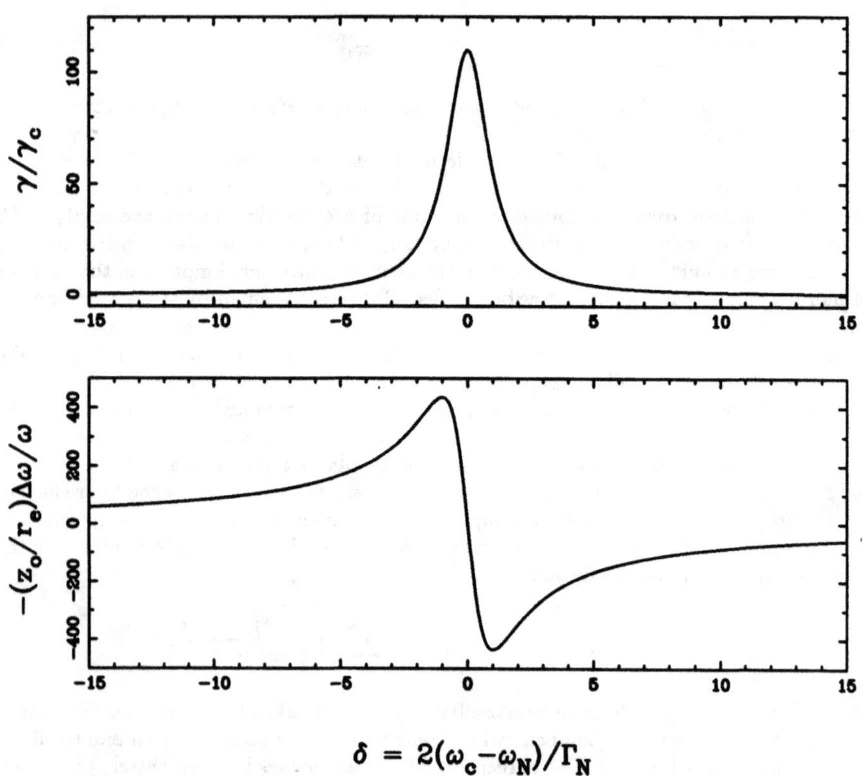

Fig. 5 The simple model and the exact calculation of Sec. 6 yield indistinguishable damping rate (above) and frequency shift (below) near a cylindrical cavity mode at $\xi_N \approx 3.781$ *with* $Q = 10^4$.

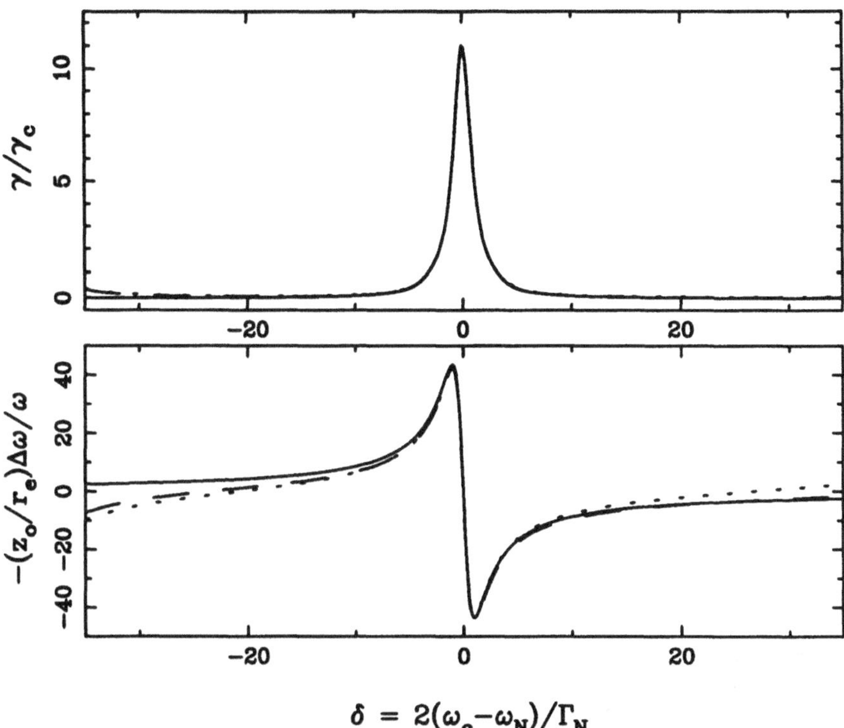

Fig. 6. Comparison of simple model (solid line) and the exact calculation of Sec. 6 (dotted line) for a cylindrical cavity mode at $\xi_N \approx 3.781$ with $Q = 10^3$. Adding the contributions of an additional cavity mode to either side (dashed line) improves the agreement.

4. Cylindrical, Hyperbolic and Spherical Cavities

To use the simple model to quantitatively describe the interaction of the electron cyclotron oscillator with a cavity mode, we must of course specify the resonant frequency ω_N and the quality factor $Q_N = \omega_N/\Gamma_N$ for the mode. Also, the coupling constant $(\lambda_N)^2$, which indicates the strength of the interaction between an electron and a cavity mode, must be specified. This is equivalent to specifying an effective value of Cd^2. We focus upon three cavity geometries of interest, considering only modes which couple to cyclotron motion about the symmetry axis of a single electron located at the center of the cavity. For each cavity geometry, the size of the cavity is indicated by an appropriately defined constant z_o. The use of a normalized frequency

$$\xi = \frac{2z_o}{\lambda} = \frac{z_o\omega}{\pi c} \tag{20}$$

which is independent of the overall size, makes it possible to separate out the dependence of the cavity properties on the cavity size.

The first cavity considered is a cylindrical cavity (Fig. 7) which is of particular interest because a one-electron cyclotron oscillator is being studied within a Penning trap whose electrodes approximate the idealized cavity shown in the figure. Also, a detailed calculation for this geometry (Sec. 6), makes it possible to establish the limitations of simple models. The second cavity discussed is a simple spherical cavity (Fig. 8). This cavity is included because its high degree of symmetry allows a particularly simple rigorous calculation of its effects upon a one-electron cyclotron oscillator located at its center. The third is a hyperbolic cavity (Fig. 9). The traps used in the past to study one-electron cyclotron oscillators (including the trap used for the most precise electron magnetic moment measurement[2]) have electrodes which approximate equipotentials of an electric quadrupole potential which they are used to produce.

At the outset it is possible to estimate possible Q values. Up to a geometrical factor (which depends upon the specific field configuration of a particular cavity mode) the Q value is simply the ratio

$$Q \sim \frac{V}{A\delta_s} \sim \frac{z_0}{\delta_s}, \tag{21}$$

where V is the volume of the cavity and $A\delta_s$ is the penetration volume of the fields into the cavity walls. Here A is the interior surface area for the cavity and δ_s is the skin depth. The skin depth is related to the conductivity of the cavity walls and the angular frequency ω by

$$\delta_s = .c/\sqrt{2\pi\omega\sigma}. \tag{22}$$

For a pure material like copper at low temperature ($T < 30\ K$), the electrical conductivity is very high ($\sigma > 10^{20}\ s^{-1}$) and the skin depth is very small ($\delta_s < 10^{-6}$ cm). This yields an estimate of $Q > 10^5$.

In practice, the Q values of the hyperbolic traps used so far seem to be lower. Holes and slits in the electrodes (to admit particles and to allow potentials to be applied to the trap electrodes) can allow radiation to leak out of the cavity, lowering the Q. In the only hyperbolic trap in which Q values were measured, Q values on the order of 1000 were indicated.[2] The Q value in the cylindrical trap now being used may well be higher, since choke flanges were incorporated to minimize the losses of radiation through the slits between electrodes. Unfortunately, the Q values are not yet well known and hence cannot be treated very accurately in any calculation.

a. Cylindrical Cavity

The cylindrical cavity represented in Fig. 7 is invariant under rotations about z-axis, with radial dimension ρ_o and axial dimension z_o. For numerical examples, we choose a relative geometry of $\rho_o/z_o = 1.186$ as indicated in Fig. 7, the choice which is used experimentally to provide optimum electrostatic properties for the Penning trap.[14] The well known mode eigenfrequencies are obtained by solving the boundary-value problem for perfectly conducting surfaces. To find the analytic expressions for C and d, however, we

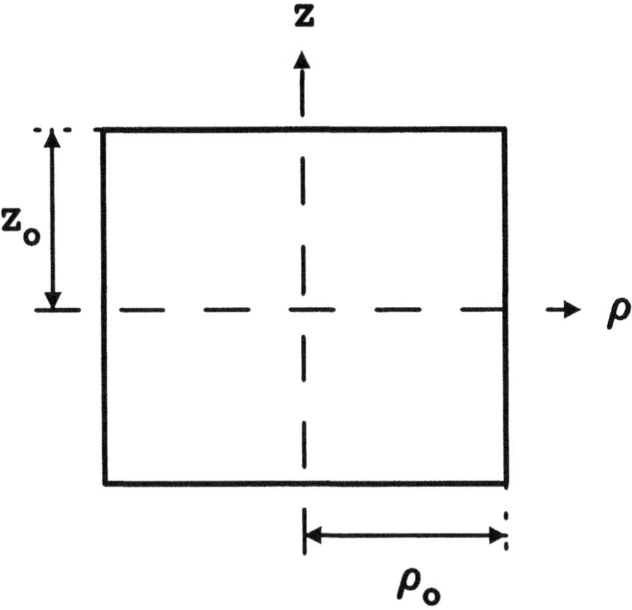

Fig. 7. Geometry of a cylindrical microwave cavity used to calculate the damping rate and frequency shift of a one-electron cyclotron oscillator which is located at the center of the cavity and oscillates about the symmetry axis of the cavity. A trap with relative geometry given by $\rho_o/z_o = 1.186$ *and* $z_o = 0.385$ *cm is presently being used for experiments.*

have to solve the system of coupled differential equations describing the interaction of the cyclotron motion with the standing waves excited in the cavity by the accelerating charged particle at the cavity center. Comparing the solution with the simple model, we choose

$$C = \frac{\rho_o^2}{8z_o} \tag{23}$$

and

$$d = \frac{2z_o}{\kappa_{n,l}}, \tag{24}$$

to maintain a close analogy when the oscillator in the model is driven. The dimensionless constant $\kappa_{n,l}$ (discussed in the next paragraph) depends upon the particular mode N, which we identify here with the two quantum numbers $n = 0, 1, 2, \ldots$ and $l = 1, 2, 3, \ldots$. When the magnetic field is along the symmetry axis of the cavity, these two indices identify the subset of cavity modes which couple to a cyclotron oscillator located at the center of the cavity. The coupling constant is thus given by

$$\lambda_{n,l}^2 = \frac{r_e c^2}{z_o \rho_o^2} (\kappa_{n,l})^2 \tag{25}$$

from which $\gamma_N = \gamma_{n,l}$ can be calculated.

Two types of modes couple to the cyclotron motion and for both it is convenient to use $k_n = (n + 1/2)\pi / z_o$. For TE (transverse electric) modes

$$\kappa_{n,l}^2 = \frac{2}{\alpha_l^2 - 1} \frac{\alpha_l^2}{J_1(\alpha_l)^2} \tag{26}$$

$$\omega_{n,l}^2 = \left(k_n^2 + \frac{\alpha_l^2}{\rho_o^2} \right) c^2 \tag{27}$$

where α_l, defined by

$$J_1'(\alpha_l) = 0, \tag{28}$$

is the lth zero of the derivative of the first order Bessel function. For TM (transverse magnetic) modes which couple to electron cyclotron motion

$$\kappa_{n,l}^2 = \frac{2k_n^2 c^2}{\omega_{n,l}^2} \frac{1}{J_0(\beta_l)} \tag{29}$$

$$\omega_{n,l}^2 = \left(k_n^2 + \frac{\beta_l^2}{\rho_o^2} \right) c^2 \tag{30}$$

where β_l, given by

$$J_1(\beta_l) = 0, \tag{31}$$

is the lth zero of the first order Bessel function. The quantum numbers n and l which we use to label the cavity modes which couple to the electron are simply related to common conventions for labeling all the modes of a cylindrical cavity. For example, in the textbook by Jackson the origin of the coordinate system is translated to the center of the bottom endcap,[22] and the TE and TM modes identified above are labeled as $TE_{1,l,2n+1}$ and the $TM_{1,l,2n+1}$, respectively. Table I gives specific values calculated for the cylindrical trap in Fig. 3.

b. Spherical Cavity

Fig. 8 represents a spherical cavity of radius ρ_o. The mode structure is particularly simple owing to the high degree of symmetry for a sphere. Only one quantum number suffices to label the modes, since only the partial waves which rotate as vectors couple to a cyclotron oscillator at the center of the cavity. Eqs.(19)-(24) still apply for the spherical cavity with $z_o = \rho_o$. The coupling parameter is given by

$$(\kappa_N)^2 = \frac{(1-\chi_N^2)^2 + \chi_N^2}{\chi_N^2 - 2}, \tag{32}$$

and the mode eigenfrequencies by

$$\omega_N = \chi_N \frac{c}{\rho_o}, \tag{33}$$

where χ_N is the Nth root of the equation

$$0 = \left(1 - \chi_N^2\right) \tan(\chi_N) - \chi_N. \tag{34}$$

The eigenfrequencies and couplings of the first 25 modes are given in Table II.

c. Hyperbolic Cavity

The g-2 experiments so far have been performed in Penning traps formed from hyperbolic surfaces of revolution. Unfortunately, the microwave properties of a hyperbolic cavity are not known like those of the cylindrical or spherical cavities. The eigenfrequencies ω_N and the coupling parameter λ_N can only be obtained numerically.[13] A computer code called URMELT-T developed for accelerator studies was used.[23] The numerically calculated values for the mode frequencies and couplings in Table III are for a hyperbolic cavity closed as shown in Fig. 9. So-called endcap electrodes and ring electrodes lie approximately along hyperbola of revolution generated, respectively, by

$$z^2 = z_o^2 + \rho^2/2 \tag{35}$$

and

$$\rho^2 = \rho_o^2 + 2z^2. \tag{36}$$

An aspect ratio $\rho_o/z_o = \sqrt{2}$ was chosen, corresponding to the trap used for the measurement of the electron magnetic moment. Flat compensation ring electrodes terminate the truncation region, perpendicular to the asymptote with the perpendicular distance from the origin given by $h/z_o = 2.80$.

The numerical method was tested in computations of the eigenfrequencies and coupling parameters for the cylindrical cavity with aspect ratio $\rho_0/z_0 = 1.186$. Comparisons with the exact, analytic expressions showed that the errors in the eigenfrequencies are approximately 0.2 % on the average and do not exceed 0.4 %. The errors in the numerical

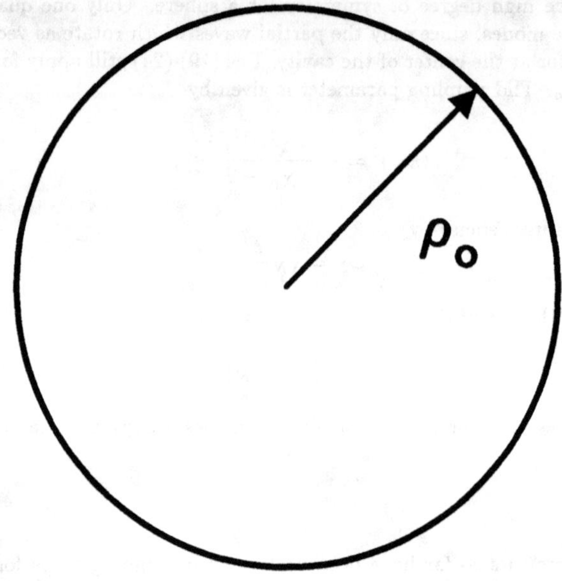

Fig. 8. Spherical cavity.

evaluation of the coupling constants are about two orders of magnitude larger. The coupling parameter for the highest frequency mode calculated (see Table III) has an 87% error because the parameters of the numerical method were not optimized. A more accurate calculation is not warranted since the actual dimensions must be known to better than 1% for the model to be directly useful. Hence, the measured values of these parameters are needed.

5. Implication for current experiments

Unfortunately, the frequencies and quality factors of the electromagnetic modes of the Penning trap used for the most accurate, electron g-2 measurements have not been measured.[2] Also, the highest precision experiment was only done for a single oscillation frequency of the electron oscillator (i.e. at only a single value of the magnetic field).

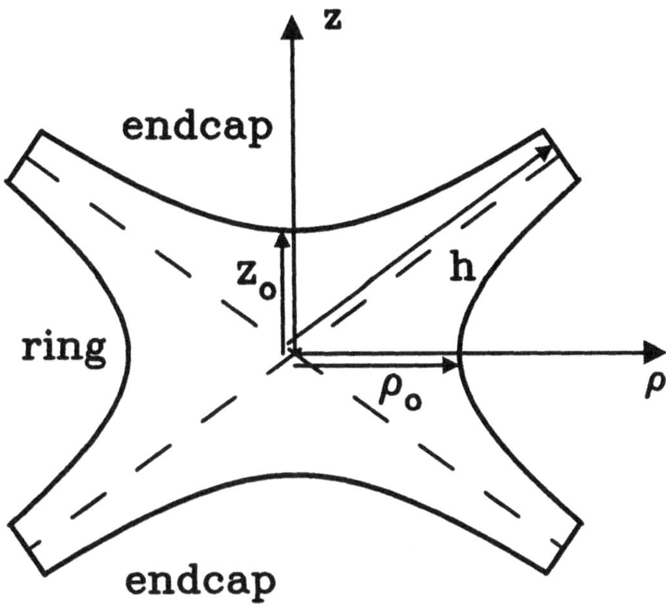

Fig. 9 Hyperbolic cavity. (From Ref. 13.)

This means that the detuning between the oscillator and the nearest cavity mode is not known at all. For an unfortunate choice of detuning, near $\delta = \pm 1$, the frequency shift could be as large as $\gamma_N/4$. From Table III we see that $\gamma_N/\gamma_c Q_N$ can be as large as 0.027 in the frequency range of interest, corresponding to a fractional frequency shift $\Delta\omega/\omega_N \approx (Q_N/20) \times 10^{-12}$. For the "reasonable" value of $Q_N = 1000$ suggested by measurements[2], this gives a possible frequency shift of 50×10^{-9}, or ten times the claimed uncertainty. For larger Q_N the maximum shift is proportionately higher.

As it happens, the experimenters were not so unlucky. A small damping rate $\gamma \approx \gamma_c/10$ was measured. Inverting Eq. (13) shows that such a shift could be caused by a detuning

$$\delta = \sqrt{\frac{\gamma_N}{\gamma} - 1} = \sqrt{\frac{\gamma_N}{\gamma_c Q_N} \frac{\gamma_c}{\gamma} Q_N - 1}. \tag{37}$$

For the illustrative mode used above, a detuning $\delta \approx \sqrt{Q_N}/2$ is required to attain the measured damping. The corresponding fractional frequency shift is $\Delta\omega/\omega \approx 0.2\sqrt{Q_N} \times$

10^{-12}. If we again use the suggested $Q_N = 1000$, this gives $\delta \approx 16$ and $\Delta\omega/\omega \approx 6 \times 10^{-12}$. From the calculated mode density for a hyperbolic cavity (see Table III) this detuning is reasonable. This fractional frequency shift is only slightly larger than the 4×10^{-12} quoted,[2] but would be higher for Q_N greater than 1000.

As illustrated in Fig. 6, including the effect of neighboring cavity modes tends to make the damping rate slightly higher than when only one mode is considered. This means that the frequency shift could tend to be slightly smaller than estimated if the cavity modes happen to be uniformly distributed in frequency. However, given the lack of knowledge of the frequencies and quality factors for the cavity modes of the real trap, we stress that these estimates are rather unreliable. We have also pointed out earlier that the coupling constants were obtained numerically with rather low accuracy. Hopefully, subsequent measurements will be done as a function of the electron's cyclotron frequency (by changing the magnetic field), within a trap cavity whose microwave properties are well measured, so that such estimates of probable measurement error are no longer required.

6. Renormalized calculation of cavity shifts

a. Renormalization

In Sec. 4, the coupling constants reported were obtained by solving the two-dimensional equation of motion for the velocity $\mathbf{v}$ of a one-electron cyclotron oscillator within a microwave cavity,

$$m\dot{\mathbf{v}} - (e/c)\mathbf{B} \times \mathbf{v} = e\mathbf{E}^{\mathrm{T}}. \tag{38}$$

The transverse radiation field within the cavity $\mathbf{E}^{\mathrm{T}}$ was taken to be the standing wave field of a particular cavity mode, generated as the accelerating particle radiates into the cavity. In this equation of motion the standing wave field acts back upon the cyclotron oscillator.

As illustrated earlier, a simple mode sum is actually not valid except for oscillator frequencies in close proximity to the resonant frequency of a high Q cavity mode. To understand this, we note that the standing wave field is actually composed of two contributions

$$\mathbf{E}^{\mathrm{T}} = \mathbf{E}_{self} + \mathbf{E}', \tag{39}$$

the self-field $\mathbf{E}_{self}$ radiated directly by the oscillator (as if into free space) and the reflected field $\mathbf{E}'$ which is reflected from the cavity walls. The back reaction of a self-field upon the accelerating charge which is radiating, is well known to lead to difficulties and divergences in classical electricity and magnetism.[24] In our particular situation, the real part of the mode sum [Eq. (19)] diverges when the sum includes all the cavity modes.

Given the usefulness of the simple model and mode sum in certain circumstances, it is instructive to see how the divergences arise and can be partially circumvented in this model. Consider first the case that the cyclotron oscillator's frequency is near enough to the eigenfrequencies of several cavity modes to be dominated by these modes. In the standing waves of these modes, the self-field is much smaller than the built up reflected wave in the cavity and can be neglected. The simple model thus describes quite well the contributions of these dominant modes. Cavity modes far from resonance with the cyclotron oscillator contribute relatively little to the damping or to the frequency shift of the oscillator and thus can be neglected. Neglecting the off-resonant contributions is essential here, since an unrenormalized calculation of these contributions overstates them and leads to a divergent result when the mode sum includes all cavity modes.

The situation is quite different when the cyclotron oscillator is not close in frequency to the resonant frequency of any cavity mode. A number of off-resonant cavity modes make small contributions to the damping of the oscillator as well as to shifting its frequency. Typically, including the contributions of more nearest neighbor modes in the mode sum initially improves the accuracy of the calculated frequency shift. The difficulty is that for off resonant modes, the Fourier component of the standing wave fields at the electron frequency contains a non-negligible amount of self-field. As mentioned above, these self-fields are the problem. For any particular off-resonant mode, the contribution to the frequency shift of the electron oscillator is slightly overstated due to self-field in the standing wave. The overstated contributions add up as the contributions from many modes are included. Optimal use of the simple, mode sum model thus requires a careful choice of the number of cavity modes included in the sum. Beyond a certain number of terms, the real part of the mode sum will start to diverge. Eventually, the mode sum over an infinite number of such small contributions diverges. It is difficult to establish the optimum number of terms or the accuracy of the truncated mode sum except by comparison to a calculation which avoids the divergences entirely and which is the subject of the remainder of this section.

For a correctly renormalized calculation, the self-field term is replaced by a radiation damping term for radiation into free space (with damping time γ_c discussed earlier) and only the transverse reflected field $\mathbf{E}'$ acts back upon the cyclotron oscillator,

$$\dot{\mathbf{v}} - \omega_c \times \mathbf{v} + \frac{1}{2}\gamma_c \mathbf{v} = (e/m)\mathbf{E}'. \tag{40}$$

Only in special cases is it possible to separate the reflected field and the self-field which together make up the standing wave. The high degree of symmetry for a spherical cavity (Sec. 6.1) makes the removal of the self-field relatively simple because the free-space radiation from the oscillator at the center contains only outgoing spherical waves, easily distinguished from the reflected waves. A cylindrical cavity (Sec. 6.2) has less symmetry, but the separation can still be accomplished by using image charges to satisfy the cavity boundary conditions. The reflected field is thus clearly distinguished as the field of the images. Unfortunately, for a hyperbolic cavity (which corresponds to the trap within which the electron's magnetic moment was measured) a separation of self and reflected fields is completely intractable. Finite mode sums as done in the simple model are the only possibility. Comparisons of a mode sum and a complete calculation for the cylindrical cavity are used to estimate the optimal number of terms to be included in the finite mode sum. To set up a framework for specific calculations, we use the radiation gauge. The transverse electric fields $\mathbf{E}^{\mathrm{T}}$, $\mathbf{E}_{\mathbf{self}}$ and $\mathbf{E}'$ are thus time derivatives of vector potentials. These vector potentials, in turn, can be written in terms of Green's functions. For example, the standing wave field $\mathbf{E}^{\mathrm{T}}$ can be written as

$$\mathbf{E}^{\mathrm{T}}_{\mathbf{k}}(t, \mathbf{r}) = -\frac{\partial}{\partial t} \int dt' \sum_{l=1}^{3} D_{kl}(t - t'; \mathbf{r}, \mathbf{r}(t')) e v_l(t')/c^2. \tag{41}$$

It is convenient to Fourier transform $D_{kl}(t-t';\mathbf{r},\mathbf{r}')$, according to

$$D_{kl}(t-t';\mathbf{r},\mathbf{r}') = \int \frac{d\omega}{2\pi} e^{-i\omega(t-t')} \tilde{D}_{kl}(\omega;\mathbf{r},\mathbf{r}'). \tag{42}$$

From the wave equation for the vector potential in the radiation gauge, with a point current source, we obtain

$$\left(-\nabla^2 - k^2\right) \tilde{D}_{kl}(\omega;\mathbf{r},\mathbf{r}') = 4\pi \left[1 - \nabla \frac{1}{\nabla^2} \nabla\right]_{kl} \delta(\mathbf{r}-\mathbf{r}'). \tag{43}$$

The gradient operator ∇ and the inverse $1/\nabla^2$ in the square brackets insure that only transverse radiation fields are being considered. Of course, the Green's function must satisfy the cavity boundary conditions as well.

The self-field $\mathbf{E_{self}}(t,\mathbf{r})$ can be similarly written in terms of the free space Green's function $D_{kl}^{self}(t-t';\mathbf{r},\mathbf{r}')$ or its Fourier transform $\tilde{D}_{kl}^{self}(\omega;\mathbf{r},\mathbf{r}')$. The latter is a solution to the same wave equation (43), but with boundary conditions appropriate to free space. What we are most interested in, however, is the reflected field $E'(t,\mathbf{r})$ which is needed to solve the renormalized equation of motion (40) for the cyclotron oscillator in a cavity. As above, we can write the reflected wave in terms of $D'_{kl}(t-t';\mathbf{r},\mathbf{r}')$ and its Fourier transform $\tilde{D}'_{kl}(\omega;\mathbf{r},\mathbf{r}')$. This Fourier transform is a solution of the homogeneous equation

$$\left(-\nabla^2 - k^2\right) \tilde{D}'_{kl}(\omega;\mathbf{r},\mathbf{r}') = 0, \tag{44}$$

since the sources of the reflected waves are not located within the cavity. $\tilde{D}'_{kl}(\omega;\mathbf{r},\mathbf{r}')$ is thus the unique modification which must be added to the free space Green's function to obtain the Green's function for the standing waves which satisfies the cavity boundary conditions.

The desired damping rates and frequency shifts for the cyclotron oscillator can be simply expressed in terms of the modification to the Green's function. We use the Fourier expansion of $D'_{kl}(t-t';\mathbf{r},\mathbf{r}')$ in the renormalized equation of motion and use the complex notation for the velocities $v(t) = v_x(t) - iv_y(t) \sim e^{-i\omega t}$, which includes the assumption that the oscillator has a well defined frequency. If we restrict our attention to a cyclotron oscillator located at the center of the cavity ($\mathbf{r} = 0 = \mathbf{r}'$) we obtain

$$\omega - \omega_c + i\gamma_c/2 = -\omega r_0 \tilde{D}'_{xx}(\omega;0,0). \tag{45}$$

The simplicity of the right-hand side of Eq.(45) results from the axial symmetry which implies that $\tilde{D}'(\omega;0,0)$ is proportional to the unit dyadic in the xy plane, with the proportionality constant $\tilde{D}'_{xx}(\omega;0,0) = \tilde{D}'_{yy}(\omega;0,0)$. In general, the Green's function modification $\tilde{D}'_{xx}(\omega;0,0)$ is a complex number, and thus the presence of the cavity modifies the cyclotron decay constant away from its free-space value γ_c. In the limit of a perfect cavity with perfectly conducting walls, the imaginary part of $\tilde{D}'_{xx}(\omega;0,0)$ cancels γ_c exactly. In this limit there is no decay of the cyclotron motion because there is no dissipative process to absorb the energy.

b. Spherical cavity

A notable feature of the spherical cavity is that the effects of dissipation can be treated exactly. Because of the spherical geometry, application of the boundary conditions appropriate for cavity walls with finite conductivity,[15] does not mix the modes and leads to

$$\omega - \omega_c' + \frac{1}{2}i\gamma_c = \frac{1}{2}i\gamma_c\left[\frac{(1-i\eta k\rho_o)h_1(k\rho_o)+(k\rho_o)h_1'(k\rho_o)}{(1-i\eta k\rho_o)j_1(k\rho_o)+(k\rho_o)j_1'(k\rho_o)} - \frac{3}{2}i\frac{1}{(k\rho_o)^3}\right], \tag{46}$$

where $k = \omega/c$ is the wave number at frequency ω. Here, η relates the electric and magnetic field components just outside the surface of the lossy conductor

$$\mathbf{E}_{\|} = \eta\mathbf{n}\times\mathbf{B}_{\|}, \tag{47}$$

where $\mathbf{n}$ is the outward unit normal to the surface. In terms of the magnetic permeability of the conductor μ and the skin depth δ_s,

$$\eta = \mu(1-i)k\delta_s/2. \tag{48}$$

In Eq. (46) we use the usual notation $j_1(x)$ for the spherical Bessel function and $h_1(x)$ for the outgoing wave spherical Hankel function, both of order one. The prime denotes a derivative with respect to the argument.

Since the dissipative effects given in Eq.(46) only approximately model an actual trap, and since dissipative effects are generally very small except near a cavity resonance, it suffices to adopt the simple procedure of replacing ω by the complex extension $\omega \to \omega(1+i/2Q)$, where Q is an average quality factor for the frequency range of interest. In this way, we obtain from Eq.(46)

$$\omega - \omega_c' = \Delta\omega - \frac{i}{2}\gamma = -\frac{1}{2}i\gamma_c + \frac{1}{2}i\gamma_c\left[\frac{(1-z^2)\cos z + z\sin z}{(1-z^2)\sin z - z\cos z} + \frac{3}{2z^3}\right] \tag{49}$$

where

$$z = \frac{\omega\rho_o}{c}\left[1+\frac{i}{2Q}\right] \tag{50}$$

and explicit forms for the Bessel functions have been used.

In Fig. 10, we highlight the important features of this result and compare directly with those of the cylindrical cavity (to be discussed). The cavity shifts are plotted against the dimensionless variable $\xi = 2\rho_o/\lambda$ for the spherical cavity, while for the cylindrical cavity we use $\xi = 2z_o/\lambda$ where $2z_o$ is the length of the cylinder. We take ρ_o to be also the radius of the cylinder, and set $z_o = 2\rho_o/3$ so that both cavities have the same volume. We use a rather low value, $Q = 100$, so as to more clearly exhibit the structure. The regularity of the spherical modes (thick line) is set against the complicated mode structure of the cylindrical cavity (thin line). The cylindrical-cavity mode spacing steadily decreases as the frequency increases, with considerable merging of the modes giving rise to a complicated fine structure. In contrast, the modes of the spherical cavity are quite evenly spaced with a larger spacing, and although there is a broadening of the widths with increasing frequency, with an accompanying decrease in the height of the peaks, the dispersive and absorptive structure maintain a distinct corelation with one another and to the normal resonant frequencies. But because the spherical-cavity mode spacing is larger than that of the cylinder, while the average total "oscillator strength" of the two cavities must be the same, the effects are much larger for the sphere. Roughly speaking, the cavity reflection field from a spherical surface is optimally focused to the center of the trap and thus produces a stronger effect.

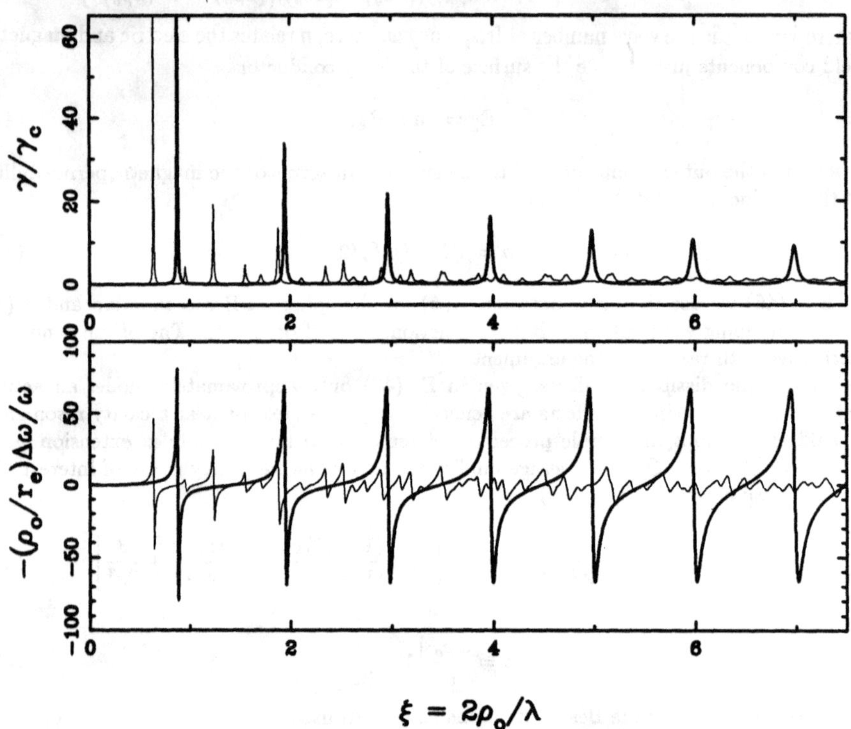

Fig. 10. Damping time and frequency shift as a function of normalized frequency ξ, for a spherical cavity (thick line) and for a cylindrical cavity of the same volume (thin line), both with $Q = 100$. (From Ref.15.)

6.3 Cylindrical cavity

To get the renormalized alteration for the cylindrical cavity, we note that the limit in which the cavity radius goes to infinity yields a geometry with two parallel, infinite

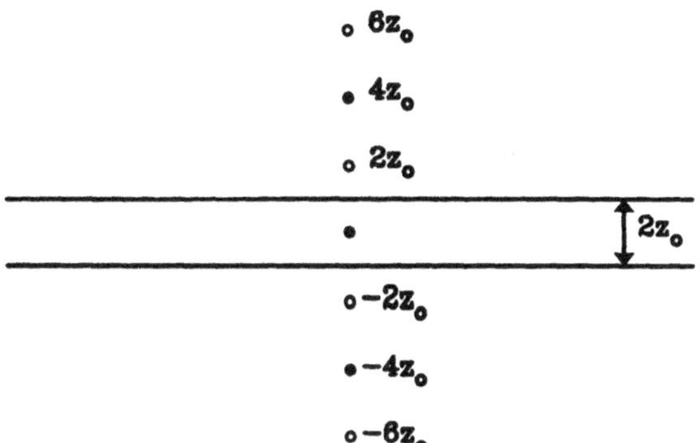

Fig. 11. Image charges for an electron midway between two parallel plates.

conducting planes a distance $2z_o$ apart. Thus we express the Green's function as the sum of the Green's function for the parallel plate problem plus the solution to the homogeneous wave equation which corrects for the presence of the cylindrical side wall. This gives

$$\omega - \omega_c' = \Delta\omega - \frac{i}{2}\gamma = -\frac{i}{2}\gamma_c + \omega\left[\Sigma_P(\omega + \frac{i}{2}\Gamma) + \Sigma_S(\omega + \frac{i}{2}\Gamma)\right] \tag{51}$$

where Σ_P is the parallel-plate contribution to the renormalized green's function alteration, and Σ_S is the correction due to the cylindrical side of the cavity. Since the Green's function for the two-parallel plate geometry can be expressed as an infinite sum of image contributions as shown in Fig. 11, the removal of the self-field term is now trivial: the direct contribution is omitted from the sum. Using the method of images we obtain

$$\Sigma_P(\omega) = \frac{r_e}{z_o}\ln\left[1 + e^{2i\omega z_o/c}\right] - \frac{r_e}{z_o}\sum_{n=1}^{\infty}(-1)^n\left[e^{2in\omega z_o/c}\left(\frac{ic}{2n^2 z_o\omega} - \frac{c^2}{4n^3 z_o^2\omega^2}\right) + \frac{c^2}{4n^3 z_o^2\omega^2}\right]. \tag{52}$$

The alteration of the Green's function brought about by the presence of the circular side can be expressed in terms of an infinite sum over the axial standing waves that fit between the two endcap planes. The wave numbers of the waves that do not vanish at the midplane location of the charged particle are given by

$$k_n = (n + \frac{1}{2})\pi/z_o, \tag{53}$$

where $n = 0, 1, 2, \ldots$. With ω below the first axial threshold, $\xi < \frac{1}{2}$, the radial waves are exponentially damped with the damping constant

$$\mu_n = (k_n^2 - \omega^2/c^2)^{1/2}. \tag{54}$$

In terms of this decomposition, the cavity side addition to the complex frequency shift (47) is given by

$$\Sigma_S(\omega) = -\frac{r_e}{z_o}\sum_{n=0}^{\infty}\left\{\frac{K_1'(\mu_n\rho_o)}{I_1'(\mu_n\rho_o)} + \frac{k_n^2c^2}{\omega^2}\left[\frac{K_1(\mu_n\rho_o)}{I_1(\mu_n\rho_o)} - \frac{K_1(k_n\rho_o)}{I_1(k_n\rho_o)}\right]\right\}, \tag{55}$$

where the prime denotes a derivative. The first ratio of Bessel functions is the TE contribution, the terms in the large parentheses are the TM contribution. When ω is near the nth threshold, μ_n becomes small, and the nth term in the sum (55) has a large logarithmic contribution that cancels the large logarithm in the parallel plate term. As ω passes the threshold, μ_n becomes a negative imaginary number. In the limit of vanishing dissipation ($\Gamma = 0$), the imaginary part of the Bessel function ratios cancels the imaginary part of the parallel plate term. Past a threshold, the Bessel functions in the denominator can vanish, producing poles corresponding to the normal modes of the cavity. The replacement $\omega \to \omega + i\Gamma/2$ changes these poles into Lorentzian forms of width Γ. The sum in (55) converges very rapidly; for large n, $\mu_n \sim k_n \sim n\pi/z_o$, and it is exponentially damped. Thus the sum is easily calculated on a digital computer. Adding the result to the previous parallel plate contribution gives the complete shift of Eq.(51).

We use the aspect ratio $\rho_o/z_o = 1.186$ for the cylindrical cavity, which has been employed to demonstrate the possibility of the cylindrical geometry for a Penning trap. Experiments are generally performed in the region $3.5 < \xi < 4.5$, and Fig. 12 illustrates the result in this region. But before passing to this, we note that the various cavity modes have different quality factors. In the region that we are considering, the quality factors for perfect cylindrical geometry for the TE modes Q_E are , to within about 10 percent, twice the quality factors Q_M of the TM modes. We approximately account for this difference in cavity widths by using the complex frequency $\omega(1+i/2Q_E)$ in the TE denominator function $I_1'(\mu_n\rho_o)$ in Eq.(55), while all the other complex frequencies are given by $\omega(1+i/2Q_M)$, with $Q_M = Q_E/2$. (By keeping all the other complex frequencies the same, we preserve the threshold cancellation discussed above, as we must.) Actual cylindrical penning traps contain holes and slits, and the quality factors are difficult to calculate accurately. Thus, although one could alter the individual modes in the interval $3.5 < \xi < 4.5$ by putting in the exact widths, this is not warranted by the uncertainty in our knowledge of the widths of the experimental traps. Thus we use the simple substitution described above to compute the decay constants and frequency shifts plotted in Fig. 12 for a typical quality factor $Q_E = 1000$. A $Q \gtrsim 1000$ is required to make possible the decrease in the damping constant by the factor of 10 which has been observed.[2]

7. Summary and future prospects

Cavity shifts of the cyclotron frequency of a one-electron cyclotron oscillator presently limit the accuracy of measurements of the anomalous magnetic moment of the electron. A substantial change in the electron's damping rate has been observed,[5] but a frequency shift has not been observed directly. In fact, little is known experimentally about the microwave

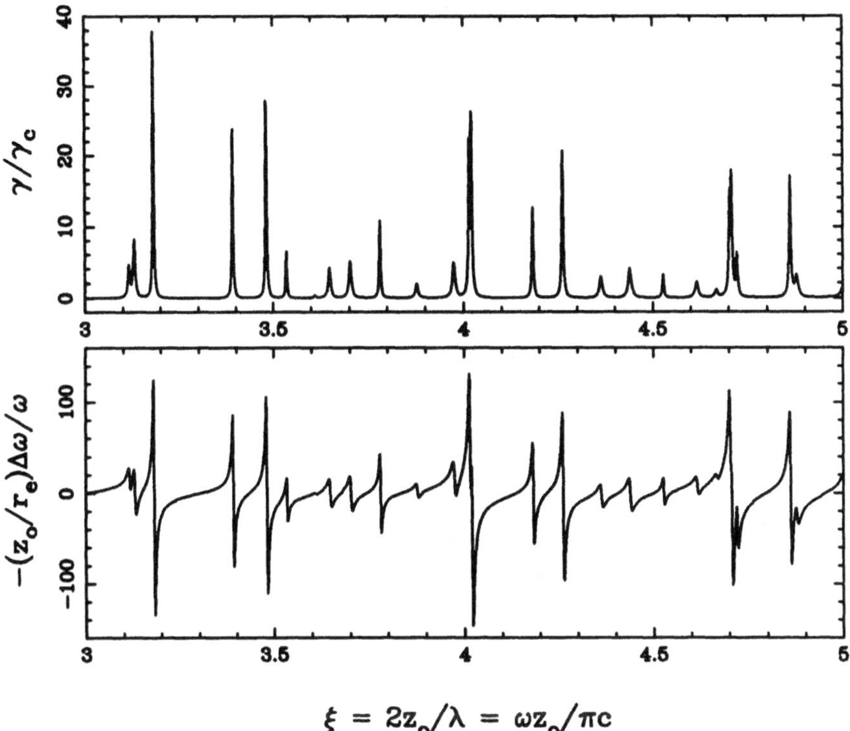

Fig. 12. Damping time and frequency shift as a function of normalized frequency ξ of the cyclotron oscillator, for the cylindrical cavity in Fig. 7 with $Q = 1000$.

properties of the cavities in which magnetic moment measurements have been done. On the theoretical side, a simple model whereby the electron is coupled to a single cavity mode illustrates the characteristic damping and frequency shift of the electron oscillator. Moreover, the simple model can be quantitatively useful when the electron oscillator and a cavity mode are nearly resonant. When the electron oscillator is not resonant with a cavity mode, however, a properly renormalized calculation is required for a precise description. Such a calculation can be done for cylindrical and spherical cavities, but not for hyperbolic cavities (which correspond most closely to traps used for precise measurements so far).

Two experimental directions are being pursued. One is to deliberately make the trap of lossy materials[2] to minimize the radiation which reflects from the cavity walls and acts back upon the oscillator. A possible alternative might be to make trap electrodes which lie along a cylinder, but with completely open ends[25] through which radiation can escape. The second direction is to use the cavity to control the radiation, by tuning the electron oscillator's frequency as close as desired to the eigenfrequency of a high Q cavity mode. A cylindrical trap cavity[14] is promising here (compared to the hyperbolic traps) because a proper renormalized calculation can be done for this geometry and because it is easier to build a high Q cavity of this geometry. Moreover, it has recently been demonstrated that a single electron can be confined and interrogated within such a cavity just as well as in the traditional hyperbolic geometry. Hopefully the perturbations due to slits in the cavity (needed to make a particle trap) are very small, with radiation losses kept small by choke flanges, but this remains to be seen. Prospects seem good for a systematic experimental study of the damping and frequency shifts of a one-electron cyclotron oscillator within a well-characterized microwave cavity.

8. References

1. The theory is described in the article by T. Kinoshita in this volume.
2. The experiment is described in the article by R.S. Van Dyck, Jr. in this volume.
3. L.S. Brown, "Line Shape for a Precise Measurement of the Electron's Magnetic Moment", Phys. Rev. Lett. **52**, 2013 (1984). "Geonium Lineshape", Ann. Phys. **159**, 62 (1985).
4. G. Gabrielse, H. Dehmelt and W. Kells, "Observation of a Relativistic, Bistable Hysteresis in the Cyclotron Motion of a Single Electron", Phys. Rev. Lett. **54**. 537 (1985).
5. G. Gabrielse and H.G. Dehmelt, "Observation of Inhibited Spontaneous Emission", Phys. Rev. Lett. **55**, 67 (1985).
6. H.G. Dehmelt, "g-Factor of Electron Centered in a Symmetric Cavity", Proc. Nat. Acad Sci. U.S.A. **81**, 8037 (1984). See also the erratum in Proc. Nat. Acad. Sci. U.S.A. **82**, 6366 (1985).
7. L.S. Brown, G. Gabrielse, K. Helmerson and J. Tan, "Cyclotron Motion in a Microwave Cavity: Possible Shifts of the Measured Electron g Factor", Phys. Rev. Lett. **55**, 44 (1985).
8. E.M. Purcell, "Spontaneous Emission Probabilities at Radio Frequencies", Phys. Rev. **69**, 681 (1946).
9. D. Kleppner, "Inhibited Spontaneous Emission", Phys. Rev. Lett. **47**, 233 (1981).
10. D. Kleppner, in *Atomic Physics and Astrophysics*, edited by M. Chretien and E. Lipworth (Gordon & Breach, NY, 1971).
11. Reviewed by S. Haroche and D. Kleppner, "Cavity Quantum Electrodynamics", Physics Today, Jan. 1989, p. 24. See also E.A. Hinds, "Cavity Quantum Electrodynamics", in Adv. At. Mol. Opt. Phys. **87** (1990).
12. L.S. Brown and G. Gabrielse, "Geonium Theory: Physics of an Electron or Ion in a Penning Trap", Rev. Mod. Phys. **58**, 233 (1986).
13. L.S. Brown, G. Gabrielse, J. Tan and K.C.D. Chan, "Cyclotron Motion in a Penning-trap Microwave Cavity", Phys. Rev. A **37**, 4163 (1988).
14. L.S. Brown, G. Gabrielse, K. Helmerson, J. Tan, "Cyclotron Motion in a Microwave Cavity: Lifetime and Frequency Shifts", Phys. Rev. A **32**, 3204 (1985).
15. L.S. Brown, K. Helmerson and J. Tan, "Cyclotron Motion in a Spherical Microwave Cavity", Phys. Rev. A. **34**, 2638 (1986).

16. D.G. Boulware, L.S. Brown and T. Lee, "Apparatus-dependent contributions to $g - 2$?", Phys. Rev. D **32**, 729 (1985).
17. G. Gabrielse and F.C. MacKintosh, "Cylindrical Penning Traps with Orthogonalized Anharmonicity Compensation", Int. J. of Mass Spec. and Ion Proc. **57**, 1 (1984).
18. J. Tan and G. Gabrielse,"One Electron in an Orthogonalized Cylindrical Penning Trap," Appl. Phys. Lett. **55**, 2144 (1989).
19. R.S. Van Dyck, Jr., F.L. Moore, D.L. Farnham, P.B. Schwinberg and H.G. Dehmelt, "Microwave-cavity Modes Directly Observed in a Penning Trap", Phys. Rev. A **36**, 3455 (1987).
20. It is customary in microwave electronics to present information in the form of equivalent circuits. See for example Robert Beringer, "Resonant Cavities as Microwave Circuit Elements", in *Principles of Microwave Circuits* pp. 207 - 239, edited by C.G. Montgomery, R.H. Dicke, E.M. Purcell (MIT Rad. Lab. Series) (McGraw, NY, 1948).
21. This is the definition of $(\lambda_N)^2$ used in Ref. 14 [Phys. Rev. A **32**, 3204 (1985)]. It is smaller than the dimensionless $(\lambda_N)^2$ used in Ref. 13 [Phys. Rev. A **37**, 4163 (1988)] by a factor $\omega_N^2 r_e / z_o$.
22. J.D. Jackson, *Classical Electrodynamics*, 2nd Edition (Wiley, New York).
23. The computer code is the Los Alamos National Laboratory-Deutsches Elecktronen-Synchotron (LANL-DESY) program URMELT-T. The LANL-DESY program called URMEL is described by T. Weiland, DESY report No. 82-24, 1982 (unpublished).
24. See e.g., Panovsky and M. Phillips, *Classical Electricity and Magnetism*, 2nd edition, (Addison-Wesley, Reading, MA., 1962) chapters 21 and 22 , or Ref. 19, Chapter 17.
25. G. Gabrielse, L. Haarsma and S.L. Rolston, "Open-endcap Penning Traps for Precision Experiments", Intl. J. of Mass Spec. and Ion Proc. **88**, 319 (1989).

Table I. Mode structure for a cylindrical cavity with $\rho_0/z_0 = 1.186$

$N = (n,l)$	$\xi_N = z_o\omega_N/\pi c$	$(z_o/r_e)(\lambda_N/\omega_N)^2$	$\gamma_N/(Q_N\gamma_c)$	TYPE
0, 1	0.702985	1.2213	0.8295	TE
0, 1	1.143497	0.1299	0.0542	TM
0, 2	1.515745	0.5425	0.1709	TE
1, 1	1.579300	0.2420	0.0732	TE
1, 1	1.818677	0.1827	0.0480	TM
0, 2	1.948165	0.0278	0.0068	TM
1, 2	2.073037	0.2900	0.0668	TE
0, 3	2.344983	0.3556	0.0724	TE
1, 2	2.407353	0.1072	0.0213	TM
2, 1	2.548370	0.0929	0.0174	TE
2, 1	2.703254	0.1039	0.0184	TM
1, 3	2.738420	0.2608	0.0455	TE
0, 3	2.775854	0.0097	0.0017	TM
2, 2	2.880535	0.1502	0.0249	TE
1, 3	3.115344	0.0552	0.0085	TM
2, 2	3.129752	0.1042	0.0159	TM
0, 4	3.181306	0.2634	0.0395	TE
2, 3	3.391009	0.1701	0.0239	TE
1, 4	3.481481	0.2200	0.0302	TE
3, 1	3.534712	0.0483	0.0065	TE
0, 4	3.610725	0.0044	0.0006	TM
3, 1	3.647956	0.0614	0.0080	TM
2, 3	3.702076	0.0769	0.0099	TM
3, 2	3.781201	0.0872	0.0110	TE
1, 4	3.877800	0.0301	0.0037	TM
3, 2	3.974336	0.0785	0.0094	TM
2, 4	4.015060	0.1654	0.0197	TE
0, 5	4.020443	0.2089	0.0248	TE
3, 3	4.183174	0.1118	0.0128	TE
1, 5	4.261919	0.1859	0.0208	TE
2, 4	4.363179	0.0521	0.0057	TM
3, 3	4.439073	0.0729	0.0078	TM
0, 5	4.448730	0.0024	0.0003	TM
4, 1	4.527051	0.0295	0.0031	TE
4, 1	4.616014	0.0396	0.0041	TM
1, 5	4.668105	0.0177	0.0018	TM
3, 4	4.703266	0.1205	0.0122	TE
2, 5	4.707861	0.1524	0.0155	TE
4, 2	4.722021	0.0559	0.0057	TE
0, 6	4.860961	0.1730	0.0170	TE

Table II. Mode structure for a spherical cavity

N	$\xi_N = \rho_o\omega_N/\pi c$	$(\rho_o/r_e)(\lambda_N/\omega_N)^2$	$\gamma_N/(Q_N\gamma_c)$
1	0.873349	1.606573	0.878323
2	1.947027	1.371989	0.336450
3	2.965571	1.349238	0.217231
4	3.974397	1.342053	0.161228
5	4.979597	1.338849	0.128374
6	5.983033	1.337140	0.106708
7	6.985475	1.336119	0.091325
8	7.987301	1.335461	0.079831
9	8.988719	1.335012	0.070913
10	9.989851	1.334691	0.063792
11	10.990776	1.334455	0.057972
12	11.991547	1.334275	0.053127
13	12.992198	1.334135	0.049030
14	13.992757	1.334024	0.045520
15	14.993240	1.333935	0.042480
16	15.993663	1.333862	0.039820
17	16.994036	1.333802	0.037475
18	17.994368	1.333751	0.035390
19	18.994665	1.333708	0.033525
20	19.994932	1.333671	0.031847
21	20.995173	1.333640	0.030329
22	21.995393	1.333613	0.028949
23	22.995593	1.333589	0.027690
24	23.995777	1.333568	0.026535
25	24.995946	1.333550	0.025473

Table III. Mode structure for a hyperbolic cavity

N	$\xi_N = z_o\omega_N/\pi c$	$(z_o/r_e)(\lambda_N/\omega_N)^2$	$\gamma_N/(Q_N\gamma_c)$
1	0.443	0.30	0.32
2	0.663	0.71	0.51
3	0.926	0.12	0.061
4	1.184	0.47	0.19
5	1.291	0.049	0.018
6	1.337	0.043	0.015
7	1.469	0.065	0.021
8	1.565	0.27	0.083
9	1.693	0.042	0.012
10	1.729	0.040	0.011
11	1.740	0.32	0.087
12	1.898	0.071	0.018
13	1.971	0.0029	0.00070
14	2.021	0.034	0.0081
15	2.079	0.34	0.079
16	2.147	0.011	0.0025
17	2.237	0.018	0.0039
18	2.299	0.12	0.025
19	2.352	0.079	0.016
20	2.405	0.082	0.016
21	2.471	0.048	0.0093
22	2.534	0.0032	0.00060
23	2.562	0.066	0.012
24	2.669	0.0052	0.00093
25	2.693	0.28	0.050
26	2.730	0.14	0.024
27	2.739	0.0058	0.0010
28	2.836	0.11	0.019
29	2.894	0.0034	0.00056
30	2.954	0.080	0.013
31	2.988	0.11	0.017
32	3.075	0.043	0.0066
33	3.105	0.015	0.0022
34	3.137	0.057	0.0087
35	3.161	0.0015	0.00022
36	3.177	0.053	0.0079
37	3.285	0.00015	0.000021
38	3.308	0.18	0.026
39	3.324	0.13	0.018
40	3.374	0.018	0.0025
41	3.446	0.072	0.010
42	3.458	0.00050	0.000069
43	3.514	0.11	0.014
44	3.571	0.20	0.027
45	3.711	0.025	0.0032

Theory of the Muon Anomalous Magnetic Moment

Toichiro Kinoshita
Newman Laboratory, Cornell University
Ithaca, New York 14853
and
William J. Marciano
Physics Department, Brookhaven National Laboratory
Upton, New York 11973

CONTENTS

1. Introduction and Summary

The anomalous magnetic moment, a_μ, of the muon is one of the fundamental quantities in elementary particles physics. Its importance derives from the fact that it is one of the few quantities that can be measured very precisely and, at the same time, is calculable from first principles. The relative sensitivity of the muon anomaly to short distance scales makes it a useful tool for studying physics involving high-mass particles. This is in strong contrast to the electron anomalous magnetic moment, a_e, which is rather insensitive to strong and weak interactions, and hence offers a better testing ground for "pure" quantum electrodynamics (QED).

The analytic structure of the QED contribution is identical for both electrons and muons. However, they have completely different values due to the fact that the effect of the electron on the muon anomaly, a_μ, and that of the muon on the electron anomaly, a_e, via vacuum polarization, are quite asymmetric. The electron, being much less massive than the muon, cannot readily create a virtual muon-antimuon pair. Thus muons (and all heavier particles) have little effect on a_e. Muons, on the other hand, can create a virtual electron-positron pair with relative ease. Indeed, in the fourth and higher orders, diagrams containing electron loops dominate. For the same reason, effects due to strong and weak interactions are much more important in a_μ than in a_e.

In a sequence of increasingly more precise measurements at CERN, the value of a_μ was determined for both positive and negative muons. The best values obtained thus far are[1]

$$\begin{aligned} a_{\mu^-}^{exp} &= 1\ 165\ 937(12) \times 10^{-9} \ , \\ a_{\mu^+}^{exp} &= 1\ 165\ 911(11) \times 10^{-9} \ . \end{aligned} \tag{1.1}$$

These results are in good agreement with theoretical calculations that include the QED contributions up to the eighth-order[2], the effects of hadronic vacuum polarization[3], and weak interactions[4].

Actually, the uncertainties in (1.1) are 5 times larger than the weak loop effect. This means that (1.1) is not accurate enough to comfirm the existence of the weak loop effect. Further improvement in both the precision of measurement and theory is of great interest because it provides one of the most stringent tests of the renormalization program of the standard model, the unified electroweak sector in particular. A new muon $g-2$ experiment is in progress at the Brookhaven National Laboratory (BNL) which will have sufficient precision to make such a test feasible. At the same time, experiments to make better measurements of the hadron production cross sections in electron-positron collisions are in various stages of progress. This is necessary to reduce the uncertainties in the hadronic vacuum polarization contribution to a_μ to a level at which the weak loop contribution can be identified from the new BNL experiment.

When these experiments are completed, the prediction of the standard model can be tested at the one-loop level. Although masses of the weak bosons Z^0 and $W^\pm$ have sizable one-loop corrections[5], these contributions arise from intermediate states quite different than those contributing to a_μ. Thus we should study Z^0 and $W^\pm$ masses, a_μ, and any other quantities that can be precisely measured in order to fully test the standard model of the electroweak interactions. It should also be noted that a_μ may have contributions from particles too massive to be studied directly at high-energy accelerators. Consequently, precision measurements of a_μ will enable us to impose strong constraints on appendages to the standard model and theoretical model building.

Let us now present an up-to-date summary of the theoretical calculation of the muon anomaly. This includes the calculation of up to the tenth order in QED[6], the contribution of the hadronic vacuum polarization effect[3,7,8], and the effect of weak interactions.[4] Details of these calculations are presented in the subsequent sections.

Let us first discuss the status of the QED calculation. It is convenient to write the QED part of the muon anomaly a_μ as $a_e + (a_\mu - a_e)$. To the tenth order the QED contribution to the second term, which represents the contribution of all diagrams containing closed electron and tauon loops, can be written as[6]

$$\begin{aligned} a_\mu^{QED} - a_e^{QED} &= 1.094\,337\,0\,(\alpha/\pi)^2 \\ &+ 22.867\,6(33)\,(\alpha/\pi)^3 \\ &+ 127.00(41)\,(\alpha/\pi)^4 \\ &+ 570(140)\,(\alpha/\pi)^5. \end{aligned} \tag{1.2}$$

Here the fourth-order (i.e. α^2) term is known analytically, and consists of terms obtained by inserting an electron vacuum polarization loop in the second-order muon vertex, plus a similar contribution in which the electron is replaced by the tauon, minus a contribution from a muon vacuum polarization loop in a second-order electron vertex. The contribution of the tauon loop to the second-order electron vertex is negligible. Some parts of the sixth-order term are known analytically, but the numerically dominant contribution arising from a sixth-order diagram containing a light-by-light scattering subdiagram is known only by numerical integration. The eighth-order term is known mostly by numerical integration. So is the tenth-order term, which is a crude estimate at best.

If one uses the latest value of the fine structure constant determined in terms of the quantized Hall effect[9]

$$\alpha^{-1} = 137.035\,997\,9(32), \tag{1.3}$$

one finds from (1.2) that

$$a_\mu^{QED} - a_e^{QED} = 6\,194\,815(45) \times 10^{-12}\,. \tag{1.4}$$

Adding the value of a_e which has been calculated to the eighth order,[a] one finds that the pure QED contribution to the muon anomaly is given by

$$a_\mu^{QED} = 1\,165\,846\,955(46)(28) \times 10^{-12}\,. \tag{1.5}$$

The first error is an estimate of the theoretical uncertainty and the second reflects the measurement uncertainty of α in (1.3). We note that the theoretical uncertainty in (1.2)

[a] See Ref. 10 and Kinoshita's article on the electron anomaly in this volume.

is mostly due to that of the α^3 term. There is no intrinsic difficulty in reducing this uncertainty further. This will be attempted in the near future.

The sensitivity of the muon anomaly to the hadronic structure of the photon was first pointed out by Bouchiat and Michel[11] and Durand[12], who showed the importance of the hadronic vacuum-polarization, in particular the dramatic enhancement effect of low-lying resonances such as the ρ meson resonance on a_μ. A recent estimate, including the hadronic vacuum-polarization contribution to the second- and fourth-order QED diagrams and the hadronic light-by-light scattering contribution, gives[3]

$$a_\mu^{had} = 7027(175) \times 10^{-11} , \tag{1.6}$$

while the lowest order weak interaction contribution calculated within the framework of the Weinberg-Salam model is[4]

$$a_\mu^{weak} = 195(10) \times 10^{-11} . \tag{1.7}$$

Here the uncertainty comes mostly from a rough estimate of the potential two-loop electroweak correction. (See Sec. 5.) The current uncertainty in the Weinberg angle has a negligible effect, contributing only 0.05×10^{-11} to the error in (1.7).

Summing up the contributions (1.5) through (1.7), we obtain the overall theoretical prediction for the muon anomaly

$$a_\mu^{theory} = 116\,591\,918(176) \times 10^{-11} . \tag{1.8}$$

This is in good agreement with the current experimental values (1.1).

Note that the theoretical QED-type error of a_μ is less than 3 per cent of the error in the hadronic term (1.6). The latter happens to be of the same order of magnitude as the electroweak effect (1.7). Thus, for experimental detection of the electroweak effect, it is necessary to improve the accuracy of the hadronic contribution (1.6) by a factor of 5 or more. This requires more accurate measurements of R, the ratio of the hadron production cross section and muon pair production cross section in e^+e^- collisions. See Ref. 13 for a discussion of the feasibility of such an improvement. The new experiment at the Brookhaven National Laboratory aims to measure the a_μ about 20 times more accurately than those of (1.1). Together with improved measurements of R, this will enable us to detect the electroweak effect (1.7) within an uncertainty of 30 per cent or better.

The rest of this article is organized as follows: In Sec. 2 we give an outline of the calculation of the QED contribution to the muon anomaly. After a brief preliminary remark in Sec. 2.1, we discuss the fourth-, sixth-, eighth-, and tenth-order calculations in Sec. 2.2 through Sec. 2.5. Sec. 3 is devoted to the discussion of the $\ln(m_\mu/m_e)$ dependence of a_μ based on the renormalization group technique. Sec. 4 is concerned with the hadronic effect on the muon anomaly. Sec. 4.2 discusses the effect of hadronic vacuum polarization on a_μ. Sec. 4.3 describes the effect of hadronic light-by-light scattering subdiagrams on the muon anomaly. Sec. 5 is devoted to the discussion of the weak loop effects on a_μ. Finally, we discuss possible contributions from various postulated "new" particles or interactions in Sec. 6.

2. QED Contribution to the Muon Anomaly

2.1. Preliminary remarks

The contribution of QED diagrams to the anomalous magnetic moment of the muon, being a dimensionless quantity, can be written in the general form

$$a_\mu = A_1 + A_2(m_\mu/m_e) + A_2(m_\mu/m_\tau) + A_3(m_\mu/m_e, m_\mu/m_\tau), \tag{2.1}$$

where m_e, m_μ, and m_τ are the masses of the electron, muon, and tauon, respectively. Throughout this article we shall use the values $m_e = 0.510\,999\,06(15)$ MeV/c^2, $m_\mu = 105.658\,39(36)$ MeV/c^2, and $m_\tau = 1\,784.1(+2.7/-3.6)$ MeV/c^2, respectively.[14]

Renormalizability of QED guarantees that the functions A_1, A_2, and A_3 can be expanded in power series in α/π with finite calculable coefficients:

$$A_i = A_i^{(2)}\left(\frac{\alpha}{\pi}\right) + A_i^{(4)}\left(\frac{\alpha}{\pi}\right)^2 + A_i^{(6)}\left(\frac{\alpha}{\pi}\right)^3 + \ldots, \quad i = 1, 2, 3. \tag{2.2}$$

The value of A_1 has been evaluated to the eighth (i. e., α^4) order[b] in the calculation of the electron anomaly a_e. As for A_2 and A_3, it is easy to see that $A_2^{(2)} = A_3^{(2)} = A_3^{(4)} = 0$ since they have no corresponding Feynman diagram. Of the remaining terms, some low-order contributions are known analytically up to terms of order m_e/m_μ while others are known only by numerical integration. We shall discuss all terms that have been evaluated thus far, starting with the fourth order.

2.2. Fourth-order terms

In the fourth order there are contributions to a_μ from two Feynman diagrams. One is from a diagram obtained by inserting an electron vacuum-polarization loop in the second-order muon vertex. (See Fig. 1(a).) The other is from the same diagram except that the electron loop is replaced by a tauon loop. Both are known analytically. Up to terms of the order $(m_e/m_\mu)^2$, the first contribution is given by [15]

$$\begin{aligned} A_2^{(4)}(m_\mu/m_e) &= \frac{1}{3}\ln\frac{m_\mu}{m_e} - \frac{25}{36} + \frac{3}{2}\frac{m_e}{m_\mu}\zeta(2) \\ &\quad - 4\left(\frac{m_e}{m_\mu}\right)^2 \ln\frac{m_\mu}{m_e} + 3\left(\frac{m_e}{m_\mu}\right)^2 + O\left[\left(\frac{m_e}{m_\mu}\right)^3\right] \\ &= 1.094\,259\,6\ldots \end{aligned} \tag{2.3}$$

where $\zeta(n)$ is the Riemann zeta function of argument n. The second contribution is [16]

$$\begin{aligned} A_2^{(4)}(m_\mu/m_\tau) &= \frac{1}{45}\left(\frac{m_\mu}{m_\tau}\right)^2 + O\left[\left(\frac{m_\mu}{m_\tau}\right)^4 \ln\frac{m_\mu}{m_\tau}\right] \\ &= 7.794(32)\times 10^{-5}. \end{aligned} \tag{2.4}$$

Clearly, the contribution of (2.3) is the dominant one.

---

[b]See Kinoshita's article in this volume.

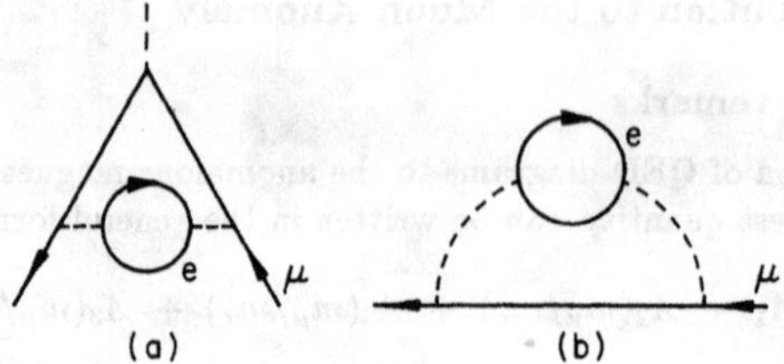

Figure 1: (a) Fourth-order muon vertex obtained by insertion of an electron vacuum-polarization loop in a second-order muon vertex. (b) Self-energy diagram (in which electrons and muons propagate in a constant magnetic field) which is equivalent to the diagram in (a) in the limit of the weak magnetic field.

The α^2 term in (1.2) also contains a term $-\ A_2^{(4)}(m_e/m_\mu)$ which contributes to the electron anomaly a_e. This contribution has the same analytic form as (2.4). Numerically one finds

$$A_2^{(4)}(m_e/m_\mu) \;=\; 5.198 \times 10^{-7}. \tag{2.5}$$

In this article we frequently use "self-energy" diagrams such as Fig. 1(b) as graphic representations of the magnetic moment anomaly. One may regard electron and muon lines in such diagrams as propagating in a constant magnetic field. Then, the diagram in Fig. 1(a) is identical with the contribution of Fig. 1(b) in the limit of a weak magnetic field. The representation of the magnetic moment in terms of "self-energy" diagrams is particularly useful in higher orders since we can simplify the corresponding Feynman-parametric integrals considerably with the help of the Ward-Takahashi identity.

2.3. Sixth-order terms

In the sixth order there are 24 Feynman diagrams contributing to $A_2^{(6)}(m_\mu/m_e)$. These include 18 diagrams containing electron vacuum-polarization loops and 6 diagrams containing light-by-light scattering subdiagrams. Typical diagrams of both kinds are shown in Fig. 2 and Fig. 3, respectively.

We list analytic results for the first group for each subgroup represented by the diagrams of Fig. 2 separately:[17,18,19,20]

$$\begin{aligned}
A_2^{(6)}[Fig.2(a)] &= \frac{1}{4}\ln\frac{m_\mu}{m_e} + \frac{1}{2}\zeta(3) - \frac{5}{12} + \ldots, \\
A_2^{(6)}[Fig.2(b)] &= \frac{2}{9}\left(\ln\frac{m_\mu}{m_e}\right)^2 - \frac{25}{27}\ln\frac{m_\mu}{m_e} + \frac{317}{324} + \frac{2}{9}\zeta(2) + \ldots, \\
A_2^{(6)}[Fig.2(c)] &= \left(\frac{119}{27} - \frac{8}{3}\zeta(2)\right)\ln\frac{m_\mu}{m_e} - \frac{61}{162} + \frac{2}{9}\zeta(2) + \ldots, \\
A_2^{(6)}[Fig.2(d,e)] &= \left(-\frac{31}{12} + \frac{10}{3}\zeta(2) - 4\zeta(2)\ln 2 + \zeta(3)\right)\ln\frac{m_\mu}{m_e} \\
&\quad + \frac{115}{24} - \frac{79}{9}\zeta(2) + 10\zeta(2)\ln 2 - \frac{7}{2}\zeta(3) + 3C_4 + \ldots,
\end{aligned} \tag{2.6}$$

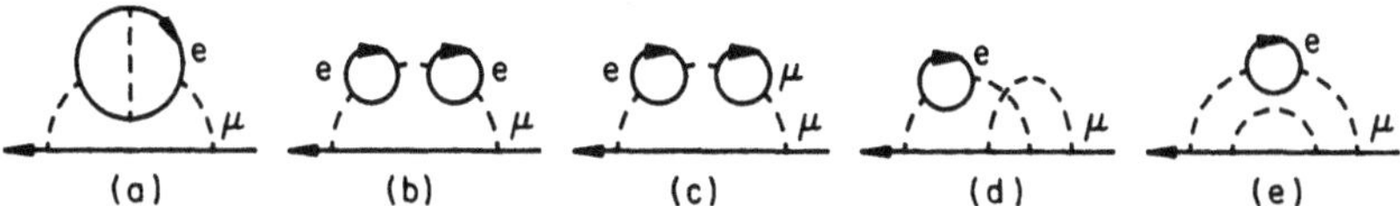

Figure 2: Typical sixth-order muon vertices obtained by insertion of electron vacuum polarization loops in second- and fourth-order muon vertices. The number of vertex diagrams represented by the self-energy diagrams (a), (b), (c), (d), and (e) are 3, 1, 2, 6, and 6, respectively.

Table I. Integrals contributing to the sixth-order muon magnetic moment.

Integral	Value	Reference
M_2	0.5	$A_1^{(2)}$ in (2.2)
$M_{2,P2}^{(\mu,e)}$	1.094 259 6	Fig.1(b)
$M_{2,P4}^{(\mu,e)}$	1.494 3(6)	Fig.2(a)
$M_{2,P2:2}^{(e,e)}$	2.720 1(3)	Fig.2(b)
$M_{2,P2:2}^{(\mu,e)}$	0.050 28(1)	Fig.2(c)
$\Delta M_{4a,P2}^{(\mu,e)}$	1.728 7(10)	Fig.2(d)
$\Delta M_{4b,P2}^{(\mu,e)}$	−2.359 8(6)	Fig.2(e)
ΔB_2	0.75	residual renormalization
$\Delta B_{2,P2}^{(\mu,e)}$	1.886 33(8)	Fig.1(b)

where C_4 is a known constant with the value 0.833 768 09 ..., and the uncalculated terms represented by ... are of the order of $(m_e/m_\mu)\ln(m_\mu/m_e)$.

Unfortunately, because of the uncalculated terms, these formulas do not have enough precision to satisfy the need of forthcoming measurements. More precise values can be obtained by numerical integration. The results obtained using the iterative and adaptive Monte Carlo integration routine VEGAS[21] are listed in Table I, where the anomalies are related to the table entries as follows:

$$\begin{aligned}
A_2^{(6)}[Fig.2(a)] &= M_{2,P4}^{(\mu,e)}, \\
A_2^{(6)}[Fig.2(b)] &= M_{2,P2:2}^{(e,e)}, \\
A_2^{(6)}[Fig.2(c)] &= M_{2,P2:2}^{(\mu,e)} + M_{2,P2:2}^{(e,\mu)} = 2M_{2,P2:2}^{(\mu,e)}, \\
A_2^{(6)}[Fig.2(d,e)] &= \Delta M_{4a,P2}^{(\mu,e)} + \Delta M_{4b,P2}^{(\mu,e)} - \Delta B_2 M_{2,P2}^{(\mu,e)} - \Delta B_{2,P2}^{(\mu,e)} M_2.
\end{aligned} \tag{2.7}$$

Quantities on the right-hand side of these equations are defined in Ref. 6. In particular,

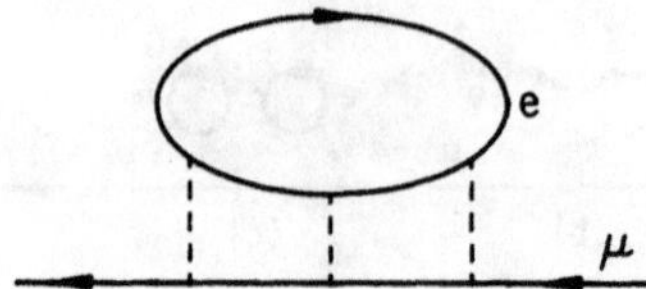

Figure 3: Sixth-order muon vertex obtained by insertion of an electron light-by-light scattering subdiagram. This diagram represents six vertex diagrams.

ΔM_α is the finite part of M_α obtained by applying the $\mathbf{K}_S$ renormalization procedure.[c] Each integral in (2.7) is evaluated using 40 million randomly chosen sampling points and by iterating the procedure from 30 to 40 times. The sum of contributions from all diagrams of Fig. 2 thus evaluated is given by

$$A_2^{(6)}[Fig.2] \;=\; 1.920\,0(14). \tag{2.8}$$

The contribution of Fig. 3 is surprisingly large.[22] Although it is not yet known analytically and the origin of its large size is somewhat of a mystery, it is clearly related to the logarithmic mass singularity at $m_e = 0$. The coefficient of $\ln(m_\mu/m_e)$ is known[23] to be $2\pi^2/3$. In view of the fact that this is the most important sixth-order term and its numerical precision is crucial in providing a good theoretical prediction for a_μ, it has been re-evaluated recently using VEGAS. The latest result is[24]

$$A_2^{(6)}[Fig.3] \;=\; 20.947\,1(29), \tag{2.9}$$

which is obtained by evaluating the integrand at 140 million sampling points each for the first ten iterations and 280 million sampling points each for the next twenty iterations.

There is another sixth-order term $A_3^{(6)}(m_\mu/m_e, m_\mu/m_\tau)$ which is obtained by replacing the muon vacuum-polarization loop of Fig. 2(c) by a tauon loop. This contribution has been evaluated by VEGAS with the result[6]

$$A_3^{(6)}(m_\mu/m_e, m_\mu/m_\tau) \;=\; 5.24(1) \times 10^{-4}. \tag{2.10}$$

Contributions of other sixth-order terms to $A_3^{(6)}$ are completely negligible at present. From (2.8), (2.9), and (2.10), we find the coefficient of the α^3 term given in (1.2).

2.4. Eighth-order terms

In the eighth order there are altogether 469 Feynman diagrams contributing to the term $A_2(m_\mu/m_e)$. They all have subdiagrams of the vacuum-polarization type and/or light-by-light scattering type and can be classified into four (gauge-invariant) groups:

Group I. Second-order muon vertex diagrams containing electron vacuum-polarization loops of second, fourth, and sixth orders. This group consists of 49 diagrams.

Group II. Fourth-order vertex diagrams containing electron vacuum-polarization loops of second and fourth orders. This group consists of 90 diagrams.

[c]See Sec. 3 of Kinoshita's article on the electron anomaly in this volume.

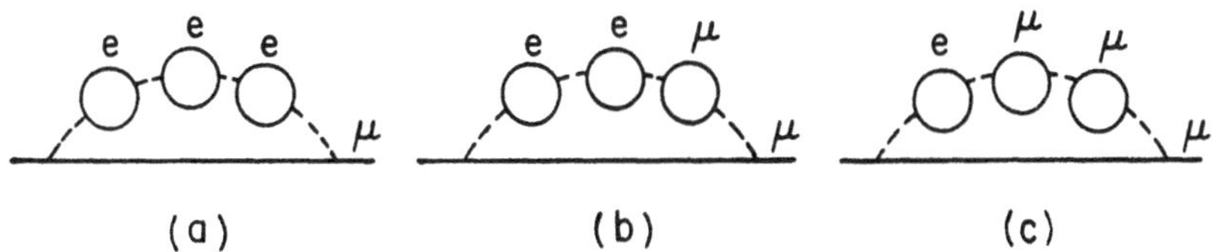

Figure 4: Three of the diagrams contributing to subgroup I(a).

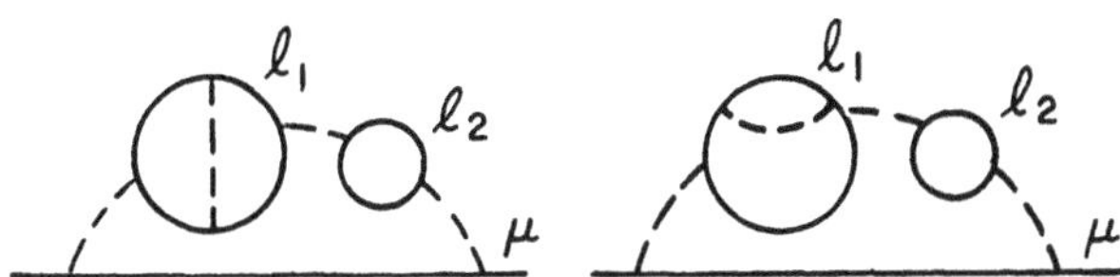

Figure 5: Diagrams contributing to subgroup I(b). $(l_1, l_2) = (e, e), (e, \mu)$, or (μ, e).

Group III. Sixth-order muon vertex diagrams containing a second-order electron vacuum-polarization loop. This group consists of 150 diagrams.

Group IV. Muon vertex diagrams containing a light-by-light scattering subdiagram due to an electron loop with further radiative corrections of various kinds. This group consists of 180 diagrams.

In addition there are also eighth-order contributions to $A_3(m_\mu/m_e, m_\mu/m_\tau)$. The only significant contributions to this term arise from a muon vertex that contains an electron light-by-light scattering subdiagram and a tauon vacuum-polarization loop and another in which the roles of electron and tauon are interchanged.

2.4.1. Group I diagrams

The diagrams of Group I can be classified further into the following gauge-invariant subgroups:

Subgroup I(a). Diagrams obtained by inserting three second-order vacuum-polarization loops in a second-order muon vertex. Seven diagrams belong to this subgroup. Three are shown in Fig. 4. The other four are obtained from diagrams (b) and (c) of this figure by permuting electron and muon loops along the photon line.

Subgroup I(b). Diagrams obtained by inserting one second-order and one fourth-order vacuum-polarization loop in a second-order muon vertex. See Fig. 5. Eighteen diagrams belong to this subgroup.

Subgroup I(c). Diagrams containing two closed lepton loops one within the other. See Fig. 6. There are nine diagrams that belong to this subgroup.

Subgroup I(d). Diagrams obtained by insertion of sixth-order (single electron loop) vacuum-polarization subdiagrams in a second-order muon vertex. Fifteen diagrams belong to this subgroup. Eight are shown in Fig. 7. Each of A, C, D, E and F and the time-reversed diagram for E has a charge-conjugated counterpart.

The evaluation of contributions of subgroups I(a) and I(b) is greatly facilitated by the analytic formulas available for the second- and fourth-order Källén-Lehmann spectral representations of the renormalized photon propagator.[25]

The contribution to a_μ from the diagram obtained by sequential insertion of m k-th order electron and n l-th order muon vacuum-polarization loops into a second-order muon vertex is given by

$$a = \int_0^1 dy(1-y)\left[\int_0^1 ds \frac{\rho_k(s)}{1+\frac{4}{1-s^2}\frac{1-y}{y^2}\left(\frac{m_e}{m_\mu}\right)^2}\right]^m \left[\int_0^1 dt \frac{\rho_l(t)}{1+\frac{4}{1-t^2}\frac{1-y}{y^2}}\right]^n , \tag{2.11}$$

where ρ_k is the k-th order spectral function. Explicit forms of ρ_2 and ρ_4 can be found in Ref. 26.

As a special case of (2.11) the contribution of the diagram in Fig. 4(a) can be written as

$$a[Fig.4(a)] = \int_0^1 dy(1-y)\left[\int_0^1 ds \frac{\rho_2(s)}{1+\frac{4}{1-s^2}\frac{1-y}{y^2}\left(\frac{m_e}{m_\mu}\right)^2}\right]^3 . \tag{2.12}$$

The contributions of the diagrams in Figs. 4(b) and (c) are given by similar expressions. Evaluating these integrals numerically using the integration routine RIWIAD[27] with 1.6×10^5 subcubes and 12 iterations, it is found that

$$a[Fig.4(a)] = 7.223\ 7(13), \tag{2.13}$$

$$a[Fig.4(b)] = 0.494\ 2(2), \tag{2.14}$$

$$a[Fig.4(c)] = 0.028\ 0(1). \tag{2.15}$$

Thus the total contribution of the diagrams of subgroup I(a) is

$$a^{(8)}_{I(a)} = 7.745\ 9(13). \tag{2.16}$$

The contributions of Fig. 5 for $(l_1, l_2) = (e,e), (e,\mu)$, and (μ, e) can be written down in a similar fashion. Numerical integration by RIWIAD using 1.6×10^5 subcubes and 10 iterations gives

$$a[Fig.5(e,e)] = 7.128\ 9(23), \tag{2.17}$$

$$a[Fig.5(e,\mu)] = 0.119\ 5(1), \tag{2.18}$$

$$a[Fig.5(\mu,e)] = 0.333\ 7(1). \tag{2.19}$$

Summing up the results (2.17) through (2.19), one finds the contribution of the subgroup I(b) to be

$$a^{(8)}_{I(b)} = 7.582\ 1(23). \tag{2.20}$$

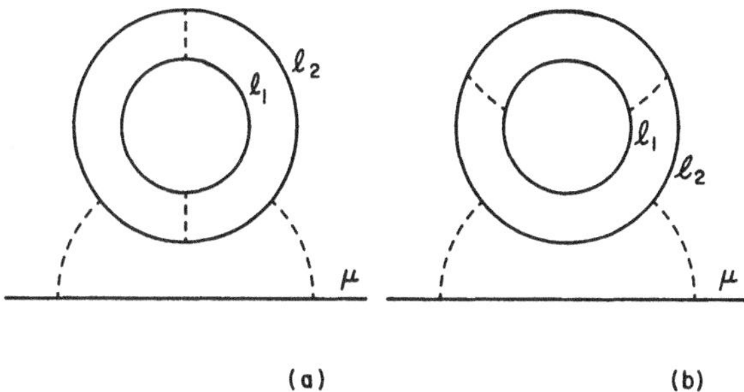

Figure 6: Diagrams contributing to subgroup I(c). $(l_1, l_2) = (e,e), (e,\mu)$, or (μ, e).

In order to evaluate the contribution to a_μ coming from the 9 Feynman diagrams of subgroup I(c), one makes use of the parametric integral representation of the sixth-order QED vacuum-polarization terms. These contributions can be written in the form

$$a_{I(c)}^{(8)} = \sum_{(l_1,l_2)} \left(\Delta M_{2,P(P4,P2)}^{(l_1,l_2)} - 2\Delta B_{2,P2}^{(l_1,l_2)} M_{2,P2}^{(\mu,l_2)}\right), \tag{2.21}$$

where

$$(l_1, l_2) = (e,\ e),\ (e,\ \mu)\ or\ (\mu,\ e),$$

and

$$\Delta M_{2,P(P4,P2)}^{(l_1,l_2)} = \Delta M_{2,P(P4a,P2)}^{(l_1,l_2)} + \Delta M_{2,P(P4b,P2)}^{(l_1,l_2)} \tag{2.22}$$

are finite integrals obtained by the $\mathbf{K}_S$ renormalization procedure described in Ref. 26. The suffix $P2$ stands for the second-order vacuum-polarization diagram while $P4$ represents the complete fourth-order vacuum polarization, which receives contributions from two distinct diagrams P_{4a} and P_{4b}, $P4 = P_{4a} + P_{4b}$. For the meaning of other notations see Ref. 26. Terms appearing on the right-hand side of (2.21) are finite, numerically calculable integrals.

The results of numerical evaluation of (2.22), obtained by RIWIAD using 10^5 subcubes and 10 iterations, as well as the corresponding residual renormalization terms expressing the difference between the standard and $\mathbf{K}_S$ renormalizations, are listed in Table II. Numerical values of lower-order Feynman integrals, in terms of which the residual renormalization terms are expressed, are given in Table III. From these tables it is found that

$$a[Fig.6(e,e)] = 1.441\,6(18), \tag{2.23}$$

$$a[Fig.6(e,\mu)] = 0.172\,7(2), \tag{2.24}$$

$$a[Fig.6(\mu,e)] = 0.021\,6(1). \tag{2.25}$$

Summing up (2.23), (2.24) and (2.25), one finds the contribution of the subgroup I(c) to the muon anomaly to be

$$a_{I(c)}^{(8)} = 1.635\,9(19). \tag{2.26}$$

The contribution to a_μ from the 15 diagrams of subgroup I(d) (see Fig. 7) can be written as

$$a_{2,P6i} = \Delta M_{2,P6i} + \text{residual renormalization terms}\ , \quad (i = A, \ldots, H). \tag{2.27}$$

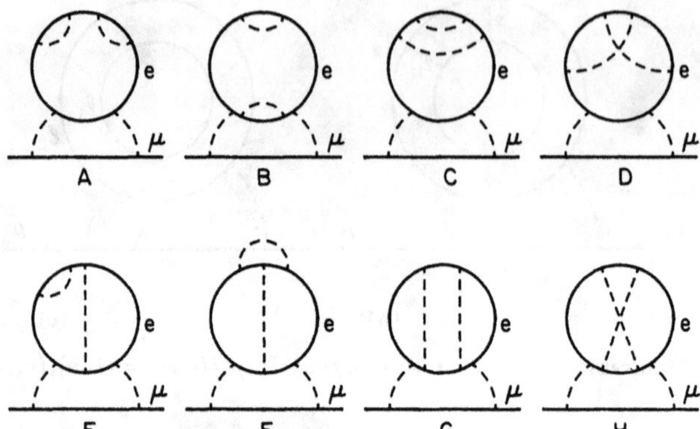

Figure 7: Eighth-order vertices of subgroup I(d) obtained by insertion of sixth-order (single electron loop) vacuum-polarization diagrams in a second-order muon vertex.

Table II. Contributions of diagrams of Figs. 6 and 7.
(η_i is a multiplicity factor.)

Figure	η_i	$\eta_i \Delta M_i$	Residual renormalization terms
6(e, e)	1	1.580 3(18)	$-2\Delta B_{2,P2}^{(e,e)} M_{2,P2}^{(\mu,e)}$
6(e, μ)	1	0.231 9(2)	$-2\Delta B_{2,P2}^{(\mu,e)} M_{2,P2}^{(\mu,\mu)}$
6(μ, e)	1	0.021 6(1)	$-2\Delta B_{2,P2}^{(e,\mu)} M_{2,P2}^{(\mu,e)}$
7A	2	5.411 4(74)	$-4\Delta' B_2 M_{2,P4b}^{(\mu,e)} + 2(\Delta' B_2)^2 M_{2,P2}^{(\mu,e)}$
7B	1	2.900 2(63)	$-2\Delta' B_2 M_{2,P4b}^{(\mu,e)} + (\Delta' B_2)^2 M_{2,P2}^{(\mu,e)}$
7C	2	1.365 5(33)	$-2\Delta' B_2 M_{2,P4b}^{(\mu,e)} + 2(\Delta' B_2)^2 M_{2,P2}^{(\mu,e)}$ $-2\Delta' B_{4b} M_{2,P2}^{(\mu,e)} - 2\Delta\delta m_{4b} M_{2,P2^*}^{(\mu,e)}$
7D	2	−3.105 2(59)	$-4\Delta' L_2 \Delta M_{2,P4b}^{(\mu,e)} + 4\Delta' L_2 \Delta' B_2 M_{2,P2}^{(\mu,e)}$ $-2\Delta' B_{4a} M_{2,P2}^{(\mu,e)} - 2\Delta\delta m_{4a} M_{2,P2^*}^{(\mu,e)}$
7E	4	−0.176 5(99)	$-4\Delta' L_2 \Delta M_{2,P4b}^{(\mu,e)} + 8\Delta' L_2 \Delta' B_2 M_{2,P2}^{(\mu,e)}$ $-4\Delta' L_{4s} M_{2,P2}^{(\mu,e)} - 4\Delta' B_2 \Delta M_{2,P4a}^{(\mu,e)}$
7F	2	−4.037 7(71)	$+4(\Delta' L_2)^2 M_{2,P2}^{(\mu,e)} - 4\Delta' L_{4c} M_{2,P2}^{(\mu,e)}$ $-2\Delta' L_2 M_{2,P4a}^{(\mu,e)}$
7G	1	−0.208 4(86)	$+3(\Delta' L_2)^2 M_{2,P2}^{(\mu,e)} - 2\Delta' L_{4l} M_{2,P2}^{(\mu,e)}$ $-2\Delta' L_2 \Delta M_{2,P4a}^{(\mu,e)}$
7H	1	2.834 0(43)	$-2\Delta' L_{4x} M_{2,P2}^{(\mu,e)}$

Table III. Auxiliary integrals. (Suffix $P2$ is sometimes written as P for simplicity.)

Integral	Value	Integral	Value
M_{2*}	1.0	$M_{2*,P}^{(\mu,e)}$	2.350 8(8)
$M_{2,P2}^{(\mu,\mu)}$	0.015 687	$M_{2,P2*}^{(\mu,e)}$	−0.161 09(3)
$\Delta M_{2,P4}^{(\mu,e)}$	3.135 7(6)		
$\Delta B_{2,P2}^{(e,e)}$	0.063 399	$\Delta B_{2,P2}^{(e,\mu)}$	$9.405\,5 \times 10^{-6}$
$\Delta B_{2,P2:2}^{(e,e)}$	5.331 9(15)	$\Delta B_{2,P2:2}^{(e,\mu)}$	0.236 13(6)
$\Delta\delta m_{4a}$	−0.301 5(10)	$\Delta\delta m_{4b}$	2.208 1(4)
B_{axc}	−0.987 3(26)	B_{bls}	1.482 3(9)
$B_{axc,P}^{(\mu,e)}$	−6.195 8(17)	$B_{bls,P}^{(\mu,e)}$	9.009 3(15)
$\Delta B_{4a} + 2\Delta L_{4c} + \Delta L_{4x}$	−0.513 8(17)	$\Delta B_{4b} + 2\Delta L_{4s} + \Delta L_{4l}$	0.542 4(6)
$M_{2*}[I]\Delta\delta m_{4a,P}^{(\mu,e)}$ $+M_{2*,P}^{(\mu,e)}[I]\Delta\delta m_{4a}$	0.185 7(25)	$M_{2*}[I]\Delta\delta m_{4b,P}^{(\mu,e)}$ $+M_{2*,P}^{(\mu,e)}[I]\Delta\delta m_{4b}$	−15.501 8(16)

Divergence-free integrals $\Delta M_{2,P6i}$ are defined by (4.13) of Ref. 26. They are evaluated numerically by RIWIAD with typically 3×10^5 subcubes and numbers of iterations ranging from 10 to 15, depending on the convergence rate. Their numerical values (multiplied by appropriate multiplicity factors η_i accounting for the diagrams related by time-reversal and charge-conjugation symmetries) are listed in the third column of Table II. The residual renormalization terms are listed in the fourth column of the same table.

Summing up the contributions of diagrams A to H of Fig. 7, we obtain the following expression:

$$\begin{aligned} a_{I(d)}^{(8)} &= \sum_{i=A}^{H} \eta_i \Delta M_{2,P6_i} - 4\Delta B_2 \Delta M_{2,P4}^{(\mu,e)} \\ &+ 5(\Delta B_2)^2 M_{2,P2}^{(\mu,e)} - 2(\Delta L^{(4)} + \Delta B^{(4)}) M_{2,P2}^{(\mu,e)} \\ &- 2\Delta\delta m^{(4)} M_{2,P2*}^{(\mu,e)}, \end{aligned} \tag{2.28}$$

where

$$\eta_i = \begin{cases} 1 & \text{for } i = B,\ G,\ H, \\ 2 & \text{for } i = A, C, D, F, \\ 4 & \text{for } i = E, \end{cases}$$

and

$$\begin{aligned} \Delta B_2 &= \Delta' B_2 + \Delta' L_2 = \frac{3}{4}, \\ \Delta M_{2,P4}^{(\mu,e)} &= \Delta M_{2,P4a}^{(\mu,e)} + 2\Delta M_{2,P4b}^{(\mu,e)}, \\ \Delta L^{(4)} &= \Delta L_{4x} + 2\Delta L_{4c} + \Delta L_{4l} + 2\Delta L_{4s}, \\ \Delta B^{(4)} &= \Delta B_{4a} + \Delta B_{4b}, \\ \Delta\delta m^{(4)} &= \Delta\delta m_{4a} + \Delta\delta m_{4b}. \end{aligned} \tag{2.29}$$

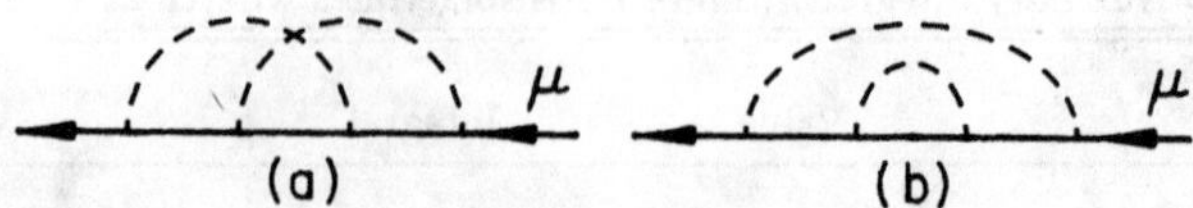

Figure 8: Fourth-order muon self-energy diagrams containing no vacuum-polarization loops.

The quantities in (2.29) are defined in Ref. 26. Their values are given in Table III. From the numerical values listed in Tables II and III we obtain

$$a^{(8)}_{I(d)} = -0.794\,5(202). \tag{2.30}$$

Finally, collecting the results (2.16), (2.20), (2.26) and (2.30), we find the contribution to the muon anomaly from the 49 diagrams of group I to be

$$a^{(8)}_{I} = 16.169(21). \tag{2.31}$$

2.4.2. Group II diagrams

Diagrams of this group are generated by inserting second- and fourth-order vacuum-polarization loops in the photon lines of the fourth-order muon vertex diagrams of Figs. 8(a) and (b).

Use of the analytic expressions for the second- and fourth-order spectral functions for the photon propagator, the Ward-Takahashi identity, and time-reversal symmetry cuts down the number of independent integrals to be evaluated from 90 to 11.

The contribution to a_μ arising from the set of vertex diagrams represented by the diagrams of Fig. 9 can be written in the form

$$a_{4,P_\alpha} = \Delta M_{4,P_\alpha} + \text{residual renormalization terms,}$$

where $\Delta M_{4,P_\alpha}$ are finite integrals.[28] Their numerical values, obtained by RIWIAD using typically 10^5 subcubes and 10 to 15 iterations, are listed in Table IV. The values of auxiliary integrals needed to obtain the total contribution of group II diagrams are given in Table III.

Summing the contributions of diagrams of the first, second, and third rows of Fig. 9, one obtains

$$\begin{aligned} a_{4,P4} &= 2\Delta M^{(\mu,e)}_{4a,P4} + \Delta M^{(\mu,e)}_{4b,P1':4} + \Delta M^{(\mu,e)}_{4b,P0:4} \\ &\quad - \Delta B_2 M^{(\mu,e)}_{2,P4} - \Delta B^{(\mu,e)}_{2,P4} M_2, \end{aligned} \tag{2.32}$$

$$\begin{aligned} a_{4,P2,P2} &= \Delta M^{(e,e)}_{4a,P2,P2} + \Delta M^{(e,e)}_{4b,P1':2,P0:2} \\ &\quad - \Delta B^{(\mu,e)}_{2,P2}\, M^{(\mu,e)}_{2,P2} \\ &\quad + 2\Delta M^{(e,\mu)}_{4a,P2,P2} + \Delta M^{(e,\mu)}_{4b,P1':2,P0:2} + \Delta M^{(\mu,e)}_{4b,P1':2,P0:2} \\ &\quad - \Delta B^{(\mu,\mu)}_{2,P2}\, M^{(\mu,e)}_{2,P2} - \Delta B^{(\mu,e)}_{2,P2}\, M^{(\mu,\mu)}_{2,P2}, \end{aligned} \tag{2.33}$$

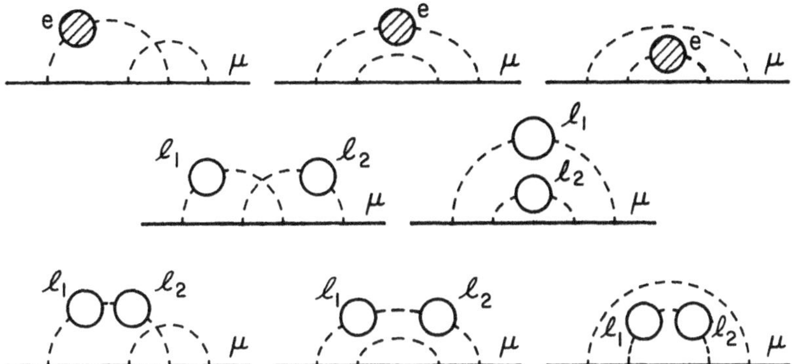

Figure 9: Eighth-order muon self-energy diagrams obtained from the fourth-order diagrams of Fig. 8 by inserting various vacuum-polarization loops. Shaded blobs are proper fourth-order loops. $(l_1, l_2) = (e, e), (\mu, e)$, or (e, μ).

Table IV. Contributions of diagrams of Fig. 9.
(η_i is a multiplicity factor.)

Term	η_i	$\eta_i \Delta M_{4,Pi}$	Residual renormalization terms
$\Delta M^{(\mu,e)}_{4a,P4}$	2	2.050 3(37)	$-2\Delta' L_2 M^{(\mu,e)}_{2,P4} - 2\Delta' L^{(\mu,e)}_{2,P4} M_2$
$\Delta M^{(\mu,e)}_{4b,P1':4}$	1	−2.495 6(24)	$-\Delta' B_2 M^{(\mu,e)}_{2,P4} - \Delta' B^{(\mu,e)}_{2,P4} M_2$
$+\Delta M^{(\mu,e)}_{4b,P0:4}$			$+I_2 M^{(\mu,e)}_{2,P4} + I^{(\mu,e)}_{2,P4} M_2$
$\Delta M^{(e,e)}_{4a,P2,P2}$	1	2.290 7(25)	$-2\Delta' L^{(\mu,e)}_{2,P2} M^{(\mu,e)}_{2,P2}$
$\Delta M^{(e,e)}_{4b,P1':2,P0:2}$	1	−4.254 1(14)	$-\Delta' B^{(\mu,e)}_{2,P2} M^{(\mu,e)}_{2,P2} + I^{(\mu,e)}_{2,P2} M^{(\mu,e)}_{2,P2}$
$\Delta M^{(e,\mu)}_{4a,P2,P2}$	2	0.053 4(9)	$-2\Delta' L^{(\mu,\mu)}_{2,P2} M^{(\mu,e)}_{2,P2} - 2\Delta' L^{(\mu,e)}_{2,P2} M^{(\mu,\mu)}_{2,P2}$
$\Delta M^{(e,\mu)}_{4b,P1':2,P0:2}$	1	−0.485 5(4)	$-\Delta' B^{(\mu,\mu)}_{2,P2} M^{(\mu,e)}_{2,P2} - \Delta' B^{(\mu,e)}_{2,P2} M^{(\mu,\mu)}_{2,P2}$
$+\Delta M^{(\mu,e)}_{4b,P1':2,P0:2}$			$+I^{(\mu,e)}_{2,P2} M^{(\mu,\mu)}_{2,P2} + I^{(\mu,\mu)}_{2,P2} M^{(\mu,e)}_{2,P2}$
$\Delta M^{(e,e)}_{4a,P2:2}$	2	5.151 1(33)	$-2\Delta' L_2 M^{(e,e)}_{2,P2:2} - 2\Delta' L^{(e,e)}_{2,P2:2} M_2$
$\Delta M^{(e,e)}_{4b,P1':2:2}$	1	−8.648 3(15)	$-\Delta' B_2 M^{(e,e)}_{2,P2:2} - \Delta' B^{(e,e)}_{2,P2:2} M_2$
$+\Delta M^{(e,e)}_{4b,P0:2:2}$			$+I_2 M^{(e,e)}_{2,P2:2} + I^{(e,e)}_{2,P2:2} M_2$
$\Delta M^{(e,\mu)}_{4a,P2:2}$	4	0.261 3(6)	$-4\Delta' L_2 M^{(e,\mu)}_{2,P2:2} - 4\Delta' L^{(e,\mu)}_{2,P2:2} M_2$
$\Delta M^{(e,\mu)}_{4b,P1':2:2}$	2	−1.103 6(12)	$-2\Delta' B_2 M^{(e,\mu)}_{2,P2:2} - 2\Delta' B^{(e,\mu)}_{2,P2:2} M_2$
$+\Delta M^{(e,\mu)}_{4b,P0:2:2}$			$+2I_2 M^{(e,\mu)}_{2,P2:2} + 2I^{(e,\mu)}_{2,P2:2} M_2$

$$
\begin{aligned}
a_{4,P2:2} = & \ 2\Delta M^{(e,e)}_{4a,P2:2} + \Delta M^{(e,e)}_{4b,P1':2:2} + \Delta M^{(e,e)}_{4b,P0:2:2} \\
& - \Delta B_2 M^{(e,e)}_{2,P2:2} - \Delta B^{(e,e)}_{2,P2:2} M_2 \\
& + 4\Delta M^{(e,\mu)}_{4a,P2:2} + 2\Delta M^{(e,\mu)}_{4b,P1':2:2} + 2\Delta M^{(e,\mu)}_{4b,P0:2:2} \\
& - 2\Delta B_2 M^{(e,\mu)}_{2,P2:2} - 2\Delta B^{(e,\mu)}_{2,P2:2} M_2,
\end{aligned} \tag{2.34}
$$

respectively, where $M^{(\mu,e)}_{2,P4}$ is equal to $\Delta M^{(\mu,e)}_{2,P4} - 2\Delta B_2 M^{(\mu,e)}_{2,P2}$. Note that the multiplicity factor for each term, which accounts for equivalent diagrams obtained by time-reversal and/or interchange of electron and muon vacuum-polarization loops, is shown explicitly in the above formulas. Thus, entries in Table III do not include the multiplicity factors.

Substituting the data from Tables III and IV into (2.32), (2.33), and (2.34) we obtain

$$
\begin{aligned}
a_{4,P4} &= -2.786\,4(45), \\
a_{4,P2,P2} &= -4.558\,6(31), \\
a_{4,P2:2} &= -9.357\,1(40).
\end{aligned}
$$

Combining these results we find the contribution of the 90 diagrams of group II to be

$$
a^{(8)}_{II} = -16.702(7). \tag{2.35}
$$

2.4.3. Group III diagrams

Diagrams belonging to this group are generated by inserting a second-order vacuum-polarization loop into the photon lines of sixth-order muon vertex diagrams of the three-photon-exchange type. Time-reversal invariance, use of the function ρ_2 (see (2.11)) for the second-order photon spectral function, summation over a set of proper vertex amplitudes that differ only in where the external magnetic field vertex is inserted, and transformation of these sums with the help of the Ward-Takahashi identity reduce the number of independent integrals to be evaluated from 150 to 8. These integrals have a one-to-one correspondence with the self-energy diagrams of Fig. 10 and can be written explicitly in terms of the parametric functions defined for the latter.

Let $M_{6\alpha,P}$ be the Ward-Takahashi-summed magnetic moment projection of the set of 15 diagrams generated from a self-energy diagram α (=A through H) of Fig. 10 by insertion of a second-order electron vacuum-polarization loop and an external vertex.[29] The renormalized contribution due to the group III diagrams can then be written as[d]

$$
a^{(8)}_{III} = \sum_{\alpha=A}^{H} \eta_\alpha a_{6\alpha,P},
$$

where

$$
a_{6\alpha,P} = \Delta M_{6\alpha,P} + \text{residual renormalization terms},
$$

and

$$
\eta_\alpha = \begin{cases} 2 & \text{for } \alpha = D, G, \\ 1 & \text{for } \alpha = A, B, C, E, F, H, \end{cases}
$$

[d]For simplicity we often write P instead of $P2$ if the meaning is obvious from the context.

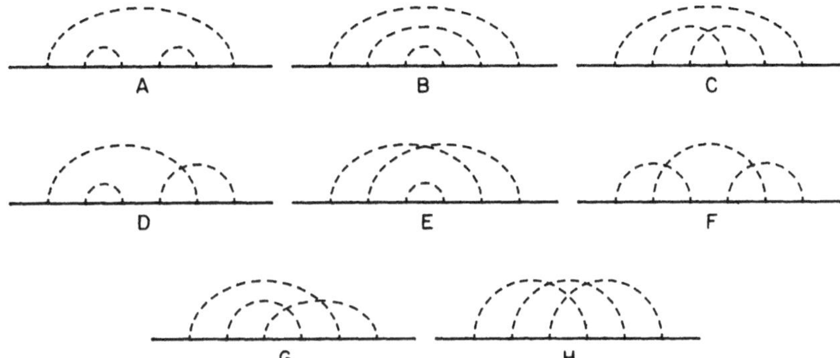

Figure 10: Muon self-energy diagrams of the three-photon-exchange type. Two more diagrams related to D and G by time reversal are not shown.

which takes account of the time-reversed counterparts of the self-energy diagrams D and G of Fig. 10. $\Delta M_{6\alpha,P}$ is the UV- and IR-finite portion of $M_{6\alpha,P}$ where all divergences have been projected out by $\mathbf{K}_S$ and $\mathbf{I}_R$ operations. (See Kinoshita's article on a_e in this volume.) Integrals $\Delta M_{6\alpha,P}$ (α=A through F and H) are evaluated by the integration routine VEGAS[21] with 10^7 subcubes, the number of iterations ranging between 30 and 40 depending on the convergence rate of the particular integral.

The integral $\Delta M_{6G,P}$ required special treatment because double precision arithmetic was not accurate enough to deal with the cancellation of UV-divergences arising from a second-order vertex. This problem was resolved using quadruple precision arithmetic in a small region surrounding the singularity.[30] This region (1 percent of the whole domain) was sampled with 10^6 points per iteration while the rest was sampled with 10^7 points per iteration. The numbers of iterations were 34 and 37, respectively.

The latest results of numerical evaluation of Group III integrals are summarized in Table V. The residual renormalization terms are also shown. Numerical values of auxiliary integrals needed in the renormalization scheme are listed in Table III.

When summed over all the diagrams of group III, the UV- and IR-divergent pieces cancel out and the total contribution to a_μ can be written as a sum of finite pieces:

$$
\begin{aligned}
a_{III}^{(8)} &= \sum_{\alpha=A}^{H} \eta_\alpha \Delta M_{6\alpha,P} \\
&- 3\Delta B_{2,P}^{(\mu,e)}(\Delta M_{4a} + \Delta M_{4b}) - 3\Delta B_2(\Delta M_{4a,P}^{(\mu,e)} + \Delta M_{4b,P}^{(\mu,e)}) \\
&+ M_{2*,P}^{(\mu,e)}[I](\Delta\delta m_{4a} + \Delta\delta m_{4b}) + M_{2*}[I](\Delta\delta m_{4a,P}^{(\mu,e)} + \Delta\delta m_{4b,P}^{(\mu,e)}) \\
&- M_{2*,P}^{(\mu,e)}(\Delta\delta m_{4a} + \Delta\delta m_{4b}) - M_{2*}(\Delta\delta m_{4a,P}^{(\mu,e)} + \Delta\delta m_{4b,P}^{(\mu,e)}) \\
&- M_{2,P}^{(\mu,e)}[B_{axc} + B_{bls} - 2(\Delta B_2)^2] \\
&- M_2(B_{axc,P}^{(\mu,e)} + B_{bls,P}^{(\mu,e)} - 4\Delta B_2 \Delta B_{2,P}^{(\mu,e)}).
\end{aligned}
$$

Table V. Contributions of diagrams of Fig. 10.
($\eta_\alpha = 1(2)$ for symmetric(asymmetric) diagrams.)

Term	$\eta_\alpha \Delta M_{6\alpha,P}$	Residual renormalization terms
A	−12.940 1(130)	$-2(\Delta' B_2 \Delta M_{4b,P}^{(\mu,e)} + 2\Delta' B_{2,P}^{(\mu,e)} \Delta M_{4b})$
		$+M_{2,P}^{(\mu,e)}(2I_{4s} + (\Delta' B_2)^2 - 2I_2\Delta' B_2)$
		$+M_2(2I_{4s,P}^{(\mu,e)} + 2\Delta'B_2\Delta'B_{2,P}^{(\mu,e)} - 2I_2\Delta'B_{2,P}^{(\mu,e)} - 2I_{2,P}^{(\mu,e)}\Delta'B_2)$
B	18.797 9(171)	$+\Delta M_{4b}(I_{2,P}^{(\mu,e)} - \Delta'B_{2,P}^{(\mu,e)}) + \Delta M_{4b,P}^{(\mu,e)}(I_2 - \Delta'B_2)$
		$+\Delta\delta m_{4b}(M_{2*,P}^{(\mu,e)}[I] - M_{2*,P}^{(\mu,e)}) + \Delta\delta m_{4b,P}^{(\mu,e)}(M_{2*}[I] - M_{2*})$
		$+M_{2,P}^{(\mu,e)}(2I_{4s} + 2I_{4l} - \Delta B_{4b} + (\Delta'B_2)^2 + (I_2)^2 - 2\Delta'B_2I_2)$
		$+M_2(2I_{4s,P}^{(\mu,e)} + 2I_{4l,P}^{(\mu,e)} - \Delta B_{4b,P}^{(\mu,e)} + 2\Delta'B_2\Delta'B_{2,P}^{(\mu,e)}$
		$+2I_2I_{2,P}^{(\mu,e)} - 2\Delta'B_{2,P}^{(\mu,e)}I_2 - 2\Delta'B_2I_{2,P}^{(\mu,e)})$
C	4.000 7(178)	$+I_{2,P}^{(\mu,e)}\Delta M_{4a} + I_2\Delta M_{4a,P}^{(\mu,e)} - 2I_{2,P}^{(\mu,e)}\Delta M_{4b} - 2I_2\Delta M_{4b,P}^{(\mu,e)}$
		$+\Delta\delta m_{4a,P}^{(\mu,e)}(M_{2*}[I] - M_{2*}) + \Delta\delta m_{4a}(M_{2*,P}^{(\mu,e)}[I] - M_{2*,P}^{(\mu,e)})$
		$+M_{2,P}^{(\mu,e)}(2I_{4c} + I_{4x} - \Delta B_{4a} - 2(I_2)^2 + 2\Delta'B_2I_2)$
		$+M_2(2I_{4c,P}^{(\mu,e)} + I_{4x,P}^{(\mu,e)} - \Delta B_{4a,P}^{(\mu,e)})$
		$+M_2(-4I_2I_{2,P}^{(\mu,e)} + 2\Delta'B_{2,P}^{(\mu,e)}I_2 + 2\Delta'B_2I_{2,P}^{(\mu,e)})$
D	10.494 0(225)	$-\Delta'B_{2,P}^{(\mu,e)}\Delta M_{4a} - \Delta'B_2\Delta M_{4a,P}^{(\mu,e)}$
		$-I_{2,P}^{(\mu,e)}\Delta M_{4b} - I_2\Delta M_{4b,P}^{(\mu,e)}$
		$+M_{2,P}^{(\mu,e)}(I_{4c} - I_{4s} - \Delta L_{4s} - (I_2)^2 + 2\Delta'B_2I_2)$
		$+M_2(I_{4c,P}^{(\mu,e)} - I_{4s,P}^{(\mu,e)} - \Delta L_{4s,P}^{(\mu,e)})$
		$+M_2(-2I_2I_{2,P}^{(\mu,e)} + 2\Delta'B_{2,P}^{(\mu,e)}I_2 + 2\Delta'B_2I_{2,P}^{(\mu,e)})$
E	11.000 1(121)	$-\Delta'B_{2,P}^{(\mu,e)}\Delta M_{4a} - \Delta'B_2\Delta M_{4a,P}^{(\mu,e)}$
		$+M_{2,P}^{(\mu,e)}(I_{4x} - 2I_{4s} - 2\Delta L_{4s} + 2\Delta'B_2I_2)$
		$+M_2(I_{4x,P}^{(\mu,e)} - 2I_{4s,P}^{(\mu,e)} - 2\Delta L_{4s,P}^{(\mu,e)})$
		$+M_2(2\Delta'B_{2,P}^{(\mu,e)}I_2 + 2\Delta'B_2I_{2,P}^{(\mu,e)})$
F	5.651 8(166)	$-2I_{2,P}^{(\mu,e)}\Delta M_{4a} - 2I_2\Delta M_{4a,P}^{(\mu,e)}$
		$+M_{2,P}^{(\mu,e)}(-2I_{4c} - 2\Delta L_{4c} + 3(I_2)^2)$
		$+M_2(-2I_{4c,P}^{(\mu,e)} - 2\Delta L_{4c,P}^{(\mu,e)} + 6I_2I_{2,P}^{(\mu,e)})$
G	19.742 4(172)	$-I_{2,P}^{(\mu,e)}\Delta M_{4a} - I_2\Delta M_{4a,P}^{(\mu,e)}$
		$+M_{2,P}^{(\mu,e)}(-I_{4l} - I_{4c} - \Delta L_{4l} - \Delta L_{4c} + (I_2)^2)$
		$+M_2(-I_{4l,P}^{(\mu,e)} - I_{4c,P}^{(\mu,e)} - \Delta L_{4l,P}^{(\mu,e)} - \Delta L_{4c,P}^{(\mu,e)} + I_2I_{2,P}^{(\mu,e)})$
H	−18.361 5(141)	$+M_{2,P}^{(\mu,e)}(-2I_{4x} - 2\Delta L_{4x}) + M_2(-2I_{4x,P}^{(\mu,e)} - 2\Delta L_{4x,P}^{(\mu,e)})$

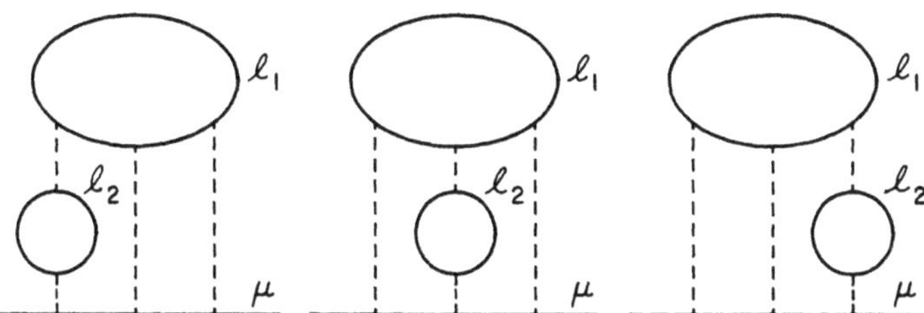

Figure 11: Muon self-energy diagrams representing the external-vertex-summed integrals of subgroup IV(a). $(l_1, l_2) = (e, e), (e, \mu)$, or (μ, e).

Plugging in the values listed in Tables III and V, we obtain

$$a_{III}^{(8)} = 10.793(48). \tag{2.36}$$

2.4.4. Group IV diagrams

Diagrams of this group can be divided into four subgroups. Each subgroup consists of two equivalent sets of diagrams related by charge conjugation (reversal of the direction of momentum flow in the loop of the light-by-light scattering subdiagram).

Subgroup IV(a). Diagrams obtained by inserting a second-order vacuum-polarization loop in the sixth-order light-by-light scattering diagrams. This subgroup is comprised of 54 diagrams. They are all appropriate modifications of the integral $M_{6LL,P}$ defined by (2.4) of Ref. 30. Denote these integrals by $M_{6LL,P}^{(l_1,l_2)}$ where $(l_1, l_2) = (e, e)$, (e, μ) or (μ, e). They are generically represented by the self-energy diagrams shown in Fig. 11.

Subgroup IV(b). Diagrams containing sixth-order light-by-light scattering subdiagrams. Altogether, there are 60 diagrams of this type. Charge-conjugation and time-reversal symmetries and summation over external vertex insertions reduce to 4 the number of integrals to be evaluated. These integrals are generically represented by the self-energy diagrams LLA, LLB, LLC and LLD of Fig. 12.

Subgroup IV(c). Diagrams obtained by attaching a virtual photon line to the muon line of the sixth-order vertex containing a fourth-order electron-loop light-by-light scattering diagram. There are 48 diagrams that belong to this subgroup. Summation over external vertex insertions and use of the interrelations available due to charge-conjugation and time-reversal symmetries leave five independent integrals to be evaluated. They are generically represented by the self-energy diagrams LLE, LLF, LLG, LLH and LLI of Fig. 12.

Subgroup IV(d). Diagrams generated by inserting a fourth-order light-by-light scattering subdiagram internally in a fourth-order vertex diagram. Diagrams of this kind appear for the first time in the eighth order. Charge-conjugation invariance and summation over the external vertex insertion with the help of the Ward-Takahashi identity lead us to three independent integrals. They are represented by the diagrams LLJ, LLK and LLL of Fig. 12.

The renormalized contribution to the muon anomaly arising from the group IV dia-

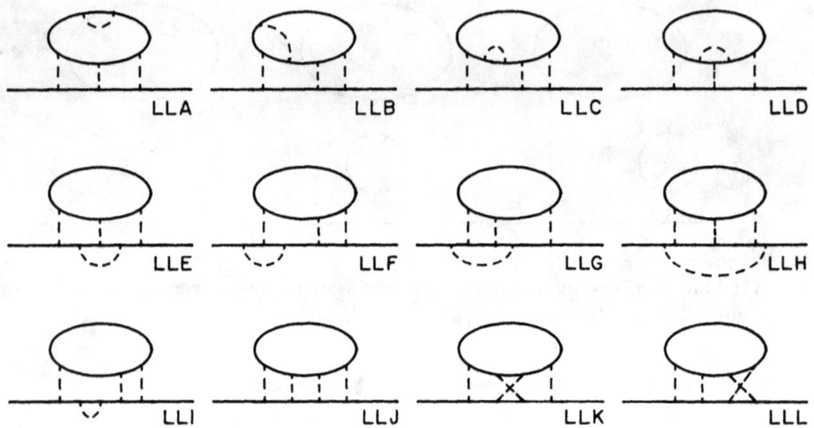

Figure 12: Muon self-energy diagrams representing the external-vertex-summed integrals of subgroup IV(b), IV(c), and IV(d).

grams can be written in the standard renormalization scheme as

$$\begin{aligned} a_{IV}^{(8)} &= a_{IV(a)}^{(8)} + a_{IV(b)}^{(8)} + a_{IV(c)}^{(8)} + a_{IV(d)}^{(8)} \\ &= \sum_{(l_1,l_2)} a_{6LL,P}^{(l_1,l_2)} + \sum_{\alpha=A}^{L} \eta_\alpha a_{8LL\alpha} , \end{aligned}$$

where

$$a_{6LL,P}^{(l_1,l_2)} = M_{6LL,P} - \text{renormalization terms} , \tag{2.37}$$

$$a_{8LL\alpha} = M_{8LL\alpha} - \text{renormalization terms} , \tag{2.38}$$

and

$$\eta_\alpha = \begin{cases} 2 & \text{for } \alpha = B, C, F, G, I, \\ 1 & \text{for } \alpha = A, D, E, H, J, K, L \end{cases}$$

accounts for diagrams related by time reversal. The factor 2 coming from equivalent diagrams obtained by reversing the momentum flow in the electron loop is included in the definitions (2.37) and (2.38).

In the case of subgroups a, b, and c, the UV-divergence arising from the light-by-light scattering subdiagram $\Pi^{\nu\alpha\beta\gamma}(q, k_i, k_j, k_l)$ is taken care of by making use of the identity:

$$\Pi^{\nu\alpha\beta\gamma}(q, k_i, k_j, k_l) = - q_\mu \left[\frac{\partial}{\partial q_\nu} \Pi^{\mu\alpha\beta\gamma}(q, k_i, k_j, k_l) \right] , \tag{2.39}$$

which follows from the Ward-Takahashi identity, and the fact that self-energy diagrams to which vertex diagrams of these subgroups are related vanish by Furry's theorem. On the other hand, the self-energy diagrams from which diagrams of subgroup d are derived are nonzero and the UV-divergence associated with the light-by-light scattering subdiagram must be regularized, e. g., à la Pauli-Villars. For these diagrams it is necessary to carry out explicit renormalizations of the light-by-light subdiagram as well as of the two sixth-order vertex subdiagrams that contain it. For details see Ref. 30.

Making use of (2.39) and the second-order photon spectral function, one finds that integrals $M_{6LL,P}^{(l_1,l_2)}$ are all finite, implying

$$a_{6LL,P}^{(l_1,l_2)} = M_{6LL,P}^{(l_1,l_2)} = \Delta M_{6LL,P}^{(l_1,l_2)},$$

so that the contribution of subgroup IV(a) is given by

$$a_{IV(a)}^{(8)} = \sum_{(l_1,l_2)} \Delta M_{6LL,P}^{(l_1,l_2)}. \tag{2.40}$$

Relating the IR- and UV-divergent $M_{8LL\alpha}$ to the finite, numerically calculable piece $\Delta M_{8LL\alpha}$ defined by the procedure of intermediate renormalization of Ref. 30, one can write (2.38) in the form

$$a_{8LL\alpha} = \Delta M_{8LL\alpha} + \text{ residual renormalization terms }.$$

Specifically, the contributions of the diagrams of subgroups b and c are given by

$$a_{IV(b)}^{(8)} = \sum_{\alpha=A}^{D} \eta_\alpha \Delta M_{8LL\alpha} - 3\Delta B_2 M_{6LL}, \tag{2.41}$$

and

$$a_{IV(c)}^{(8)} = \sum_{\alpha=E}^{I} \eta_\alpha \Delta M_{8LL\alpha} - 2\Delta B_2 M_{6LL}. \tag{2.42}$$

M_{6LL} appearing in (2.41) and (2.42), which is identical with the quantity in (2.9), is the contribution to the muon anomaly from the sixth-order vertex diagrams containing an electron loop light-by-light scattering subdiagram. The contribution of the subgroup d is written as

$$a_{IV(d)}^{(8)} = \sum_{\alpha=J}^{L} \eta_\alpha a_{8LL\alpha}, \tag{2.43}$$

where $a_{8LL\alpha}$ is obtained by the *standard* renormalization instead of the intermediate renormalization (see Ref. 30).

Summing (2.40), (2.41), (2.42) and (2.43), we arrive at

$$a_{IV}^{(8)} = \sum_{(l_1,l_2)} \Delta M_{6LL,P}^{(l_1,l_2)} + \sum_{\alpha=A}^{I} \eta_\alpha \Delta M_{8LL\alpha} - 5\Delta B_2 M_{6LL} + \sum_{\alpha=J}^{L} \eta_\alpha a_{8LL\alpha}. \tag{2.44}$$

Numerical integration of all terms in (2.44) has been carried out using VEGAS.[21] The latest results are listed in Table VI. The number of iterations employed in achieving the stated results was between 22 and 38, while the number of function calls per iteration ranged from 4×10^6 to 16×10^6. In general, the major difficulty in dealing with the diagrams of this group arises from the enormous size of the integrands (up to 5000 terms and 240 kilobytes of FORTRAN source code per integral) and the large number of integration variables (up to 10).[e]

Table VI. Contributions of diagrams of Figs. 11 and 12. $\eta_\alpha = 1$ (2) for symmetric (asymmetric) diagrams. N is the number of sampling points per iteration in units of 10^6. I is the number of iteration.

Diagram	$\eta_\alpha \Delta M_{8LL\alpha}$	N	I
6LL,P(e,e)	116.805 1(609)	10	35
6LL,P(e,μ)	2.701 5(44)	10	22
6LL,P(μ,e)	4.325 7(130)	10	36
LLA	49.882 0(872)	4	38
LLB	−74.485 8(1048)	4	36
LLC	102.370 1(1248)	8	23
LLD	−37.810 6(1024)	4	31
LLE	−21.547 5(688)	4	37
LLF	−75.402 5(2031)	13	29
LLG	−34.942 1(1374)	8	31
LLH	54.091 3(1564)	12	34
LLI	112.662 9(1519)	16	38
LLJ	5.457 5(206)	4	36
LLK	−7.828 9(340)	4	35
LLL	−1.067 3(355)	4	38

From Table VI one finds the contribution from all 180 diagrams of group IV to be

$$a_{IV}^{(8)} = 116.660(405). \tag{2.45}$$

Finally, combining (2.31) with (2.35), (2.36) and (2.45), one obtains the result for the eighth-order QED contribution

$$A_2^{(8)}(m_\mu/m_e) = 126.92(41). \tag{2.46}$$

2.4.5. Other eighth-order terms

The only other nontrivial contribution to the eighth-order term arises from a muon vertex that contains an electron light-by-light scattering subdiagram and a tauon vacuum-polarization loop and another in which the roles of electron and tauon are interchanged.

[e]See Kinoshita's article on the electron anomaly in this volume and Ref. 30 for a discussion of various computational problems encountered in the numerical integration.

(See Fig. 11 with $(l_1, l_2) = (e, \tau), (\tau, e)$.) Their values have been evaluated numerically. The sum is given by

$$A_3^{(8)}(m_\mu/m_e, m_\mu/m_\tau) = 0.079(3). \tag{2.47}$$

Other contributions are insignificant at present. The α^4 term in (1.2) is the sum of (2.46) and (2.47).

2.5. Tenth-order terms – a preliminary estimate

In view of the very large value for the eighth-order coefficient of a_μ, one may naturally wonder how large the tenth-order coefficient might be. To answer this question unambiguously, one has to evaluate all tenth-order terms, a formidable task indeed. To achieve our goal of determining a_μ to a precision of few parts in 10^{-11}, however, it is sufficient if a rough but fairly reliable estimate is available.

Fortunately, it is not difficult to obtain such an estimate. It is based on the following empirical facts accumulated while working on the sixth- and eighth-order terms:

(a) The contribution of a minimal (gauge-invariant) set of diagrams is of order 1 as far as they have no closed electron loop. We mean by "minimal" a gauge-invariant set whose proper subsets are not gauge-invariant.

(b) The contribution of a minimal set of diagrams containing n electron vacuum-polarization loops is of order $(\ln(m_\mu/m_e))^n$ times the term obtained by omitting these loops. In addition, it has a multiplicative factor that depends on the number of ways electron loop insertions can be made.

(c) The contribution of a minimal set of diagrams containing an electron loop light-by-light scattering subdiagram has a $\ln(m_\mu/m_e)$ factor with a large numerical coefficient. If it also contains electron vacuum-polarization loops, it is multiplied by further $\ln(m_\mu/m_e)$ factors.

Summing over a gauge invariant set is a necessity since individual diagrams are UV- and/or IR-divergent in general. As examples of minimal sets, let us list the contributions of eighth-order subgroups IV(b) and IV(c):

$$\begin{aligned} a_{IV(b)}^{(8)} &= -\,7.175(212), \\ a_{IV(c)}^{(8)} &= 3.442(336). \end{aligned} \tag{2.48}$$

As is seen from Table VI, even the convergent parts of individual terms of these subgroups, which are not gauge invariant, may be very large, but their gauge-invariant sums are fractions of the sixth-order light-by-light scattering contribution M_{6LL} discussed in Sec. 2.4 (aside from an extra factor of α/π).

One may conclude from these observations that the most important tenth-order term comes from 36 Feynman diagrams of the type shown in Figs. 13(a) and 13(b), which contain one light-by-light electron loop and two second-order electron vacuum-polarization loops. It is not difficult to write down a FORTRAN program for the sum of all these diagrams, adapting Eqs. (3.13) and (3.19) of Ref. 28 to this case, and evaluate it numerically. The result, based on 28 iterations with 10^7 function calls per iteration, is

$$A_2^{(10)}(Figs.13(a),\ 13(b)) = 569.33(61). \tag{2.49}$$

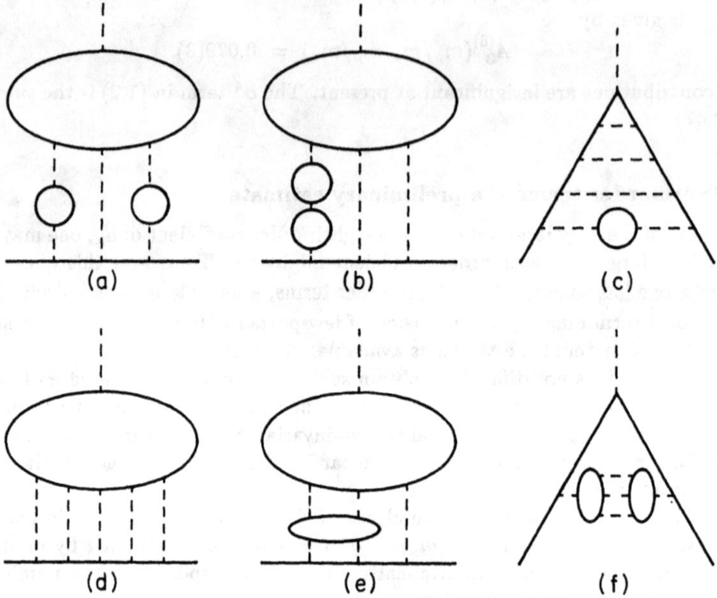

Figure 13: Some tenth-order diagrams. Diagrams (a) and (b) are generated by inserting two electron vacuum-polarization loops in a sixth-order diagram containing a light-by-light scattering subdiagram. There are 36 diagrams of these types. See the text for other diagrams.

Of course, direct evaluation of other terms is much more tedious. Fortunately, it is possible to obtain a rough estimate for a large class of diagrams containing second-order electron loops :[17] First note that such an insertion results in a modification of the photon propagator as is discussed in Sec. 3. Asymptotically this modification can be written as

$$d_R^\infty\left(\frac{q^2}{m_e^2},\, \alpha\right) = 1 + \frac{\alpha}{\pi}\left[\frac{1}{3}\ln\left(\frac{q^2}{m_e^2}\right) - \frac{5}{9}\right] + \ldots, \quad q^2 >> m_e^2. \qquad (2.50)$$

Since the logarithm is a slowly varying function of q^2, one may replace q^2 by an average value $r^2 m_\mu^2$, where r is a constant of order unity. This means that the insertion of a vacuum-polarization loop can be effectively reduced to a multiplication by a factor

$$\frac{\alpha}{\pi}K = \frac{\alpha}{\pi}\left[\frac{2}{3}\ln\left(\frac{rm_\mu}{m_e}\right) - \frac{5}{9}\right]. \qquad (2.51)$$

In order that the approximation (2.51) makes sense, r should be less than $\simeq 1$ which means that K should be less than $\simeq 3$.

Let us now estimate the magnitude of K from the previously calculated results. For example, for the eighth-order diagrams $M_{6LL,P}^{(e,e)}$ of Fig. 11 we will have

$$M_{6LL,P}^{(e,e)} \simeq 3KM_{6LL}. \tag{2.52}$$

The factor 3 accounts for the number of photon lines in which an electron vacuum-polarization loop can be inserted. Similarly we may fix the parameter K from the relation

$$A_2^{(10)}(Figs. 13(a),\ 13(b)) \simeq 6K^2 M_{6LL}. \tag{2.53}$$

The factor 6 arises because two electron vacuum-polarization loops can be inserted in three photon lines in 6 different ways. Using the data from Table VI and (2.49), one finds $K = 1.86$ and $K = 2.13$ from (2.52) and (2.53), respectively. Examination of other diagrams yields K mostly in the range from 2 to 2.5 with the exception of $a_{III}^{(8)}$ of Sec. 2.4.3, which gives $K \simeq 4$. For our purpose it is sufficient to choose

$$K = 2 \sim 4.$$

This shows how poor the approximation (2.51) might actually be. What is important, however, is that these K's are all positive. This means that one can confidently predict the signs of terms obtained by insertion of vacuum-polarization loops.

It is not difficult to turn this heuristic argument into a more rigorous one using the renormalization group technique discussed in Sec. 3. However, it will not be necessary for our present purpose.

As an application of the admittedly very crude method described above, let us estimate the magnitude of the term representing the sum of 2072 Feynman diagrams of the type shown in Fig. 13(c), which are obtained from 518 electron-loop-free eighth-order diagrams by insertion of an electron vacuum-polarization loop in all possible manners. The estimate for this term is

$$4 \times K \times (-1.93) = -(16 \sim 32),$$

where 4 is the number of virtual photons and the factor -1.93 is from Ref. 10.

A similar estimate can be made for each minimal gauge-invariant subgroup discussed in Sec. 2.4. In view of the fact that the results of Sec. 2.4, as well as of (2.48) and other gauge-invariant results calculable from Table VI, are no larger than 17 in magnitude and tend to cancel one another, one finds that the contribution of tenth-order diagrams obtained by insertion of a second-order vacuum-polarization loop in all eighth-order diagrams, excluding the result (2.49), is likely to be substantially less than 100.

Tenth-order diagrams that cannot be estimated by the method discussed above are of the types shown in Figs. 13(d), 13(e), and 13(f). Since they have low powers of $\ln(m_\mu/m_e)$, however, it seems unlikely that they would give large contributions. In view of the fact that we do not know why the sixth-order diagrams containing a light-by-light scattering subdiagram have such a large value, however, it might not be prudent to rule out a surprise in the tenth order. Direct evaluation of contributions of Figs. 13(d) – 13(f) might be interesting.

Our conservative estimate, then, is that the tenth-order term will be found well within the range given by

$$A_2^{(10)} \simeq 570(140).$$

This amounts to $39(10) \times 10^{-12}$ in the final value of a_μ^{QED} given in Sec. 1.

3. Application of the Renormalization Group Method to a_μ

Due to the large mass ratio, $m_\mu/m_e \approx 207$, the leading contribution to the muon anomaly arising from the electron vacuum-polarization insertion is governed by the short-distance behavior of the photon propagator. Exploiting the relationship of this fact and the renormalization procedure it is possible to predict the $\ln(m_\mu/m_e)$ structure for a large number of diagrams contributing to the muon anomaly.[17] Refining this technique further, Lautrup and de Rafael[31] were able to predict the complete $\ln(m_\mu/m_e)$ structure of a large class of eighth-order diagrams contributing to a_μ. The purpose of this section is to compare the numerical results obtained in the earlier sections with the corresponding formulas derived from the renormalization group method. We start with a brief review of the renormalization group method.

Let us denote by $a(m_\mu/m_e, \alpha)$ the contribution to the muon anomaly from the class of vertex diagrams obtained by replacing all the internal photon lines in a renormalized muon vertex by a dressed renormalized photon propagator whose fermion loops are of the electron type only. (See Fig. 14.) The starting set of muon vertices may contain muon loops to which any number of photon lines are attached. It may even include electron loops to which four or more photon lines are attached. For simplicity, however, let us first consider the case where the starting set is the largest set containing all types of muon loops but no electron loop at all. Actually, for some smaller starting sets, it is possible to extract more information. Some of them will be considered later.

Following the procedure of Ref. 31, let us define the asymptotic part $a_\infty(m_\mu/m_e, \alpha)$ of $a(m_\mu/m_e, \alpha)$ as follows: In each order of perturbation theory drop terms that vanish as $m_\mu/m_e \to \infty$, while retaining terms that are constant or increase logarithmically. It can be shown that a_∞ obeys the homogenous Callan-Symanzik equation:

$$\left(m_e\frac{\partial}{\partial m_e} + \beta(\alpha)\alpha\frac{\partial}{\partial\alpha}\right) a_\infty\left(\frac{m_\mu}{m_e}, \alpha\right) = 0. \tag{3.1}$$

The solution to this equation is

$$\begin{aligned} a_\infty\left(\frac{m_\mu}{m_e}, \alpha\right) &= \left(\frac{m_\mu}{m_e}\right)^{\beta(\alpha)\alpha\frac{\partial}{\partial\alpha}} B(\alpha) \\ &= \sum_{n=0}^{\infty}\frac{1}{n!}\left(\ln\frac{m_\mu}{m_e}\right)^n\left(\beta(\alpha)\alpha\frac{\partial}{\partial\alpha}\right)^n B(\alpha), \end{aligned} \tag{3.2}$$

where $\beta(\alpha)$ is the Callan-Symanzik function and

$$B(\alpha) = a_\infty\left(\frac{m_\mu}{m_e}, \alpha\right)_{m_\mu=m_e} \tag{3.3}$$

is the nonlogarithmic part of a_∞. As is seen from Eq. (3.2), knowledge of $\beta(\alpha)$ and $B(\alpha)$ determines a_∞ completely. Expanding $\beta(\alpha)$ and $B(\alpha)$ in powers of α/π as

$$\beta(\alpha) = \sum_{k=1}^{\infty}\beta_k\left(\frac{\alpha}{\pi}\right)^k, \tag{3.4}$$

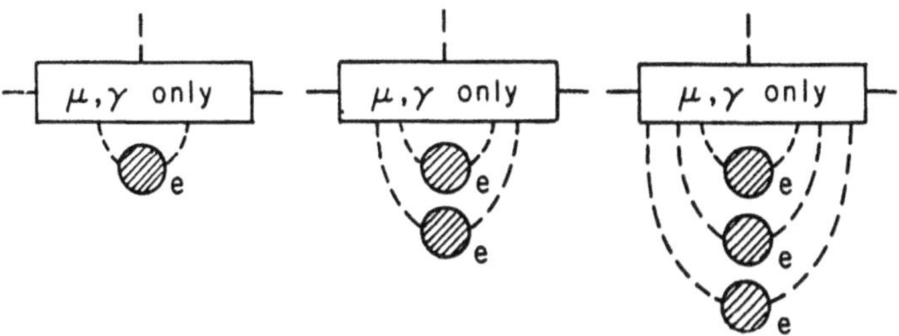

Figure 14: A class of diagrams generated by inserting electron vacuum-polarization loops into a muon vertex.

$$B(\alpha) = \sum_{k=1}^{\infty} B_k \left(\frac{\alpha}{\pi}\right)^k , \tag{3.5}$$

and substituting them into Eq. (3.2), we find that

$$\begin{aligned} a_\infty\left(\frac{m_\mu}{m_e}, \alpha\right) &= \left(\frac{\alpha}{\pi}\right) B_1 \\ &+ \left(\frac{\alpha}{\pi}\right)^2 \left(B_2 + C_2 \ln\frac{m_\mu}{m_e}\right) \\ &+ \left(\frac{\alpha}{\pi}\right)^3 \left(B_3 + C_3 \ln\frac{m_\mu}{m_e} + D_3 \ln^2\frac{m_\mu}{m_e}\right) \\ &+ \left(\frac{\alpha}{\pi}\right)^4 \left(B_4 + C_4 \ln\frac{m_\mu}{m_e} + D_4 \ln^2\frac{m_\mu}{m_e} + E_4 \ln^3\frac{m_\mu}{m_e}\right) \\ &+ \ldots , \end{aligned} \tag{3.6}$$

where

$$\begin{aligned} C_2 &= \beta_1 B_1, \\ C_3 &= \beta_2 B_1 + 2\beta_1 B_2, \\ C_4 &= \beta_3 B_1 + 2\beta_2 B_2 + 3\beta_1 B_3, \\ D_3 &= \beta_1^2 B_1, \\ D_4 &= \frac{5}{2}\beta_1\beta_2 B_1 + 3\beta_1^2 B_2, \\ E_4 &= \beta_1^3 B_1. \end{aligned} \tag{3.7}$$

The functions $\beta(\alpha)$ and $B(\alpha)$ are both known up to the sixth order in perturbation theory. The coefficients β_k (k = 1, 2, 3) are given by[32]

$$\beta_1 = \frac{2}{3}, \quad \beta_2 = \frac{1}{2}, \quad \beta_3 = -\frac{121}{144}, \tag{3.8}$$

while the values of B_k (k = 1, 2, 3) are[10,16,17,33]

$$B_1 = \frac{1}{2}, \quad B_2 = -1.022\,923\ldots, \quad B_3 = 2.741\,64(56). \tag{3.9}$$

Substituting (3.7), (3.8), and (3.9) in (3.6) we find

$$\begin{aligned}
a_\infty^{(2)}\left(\frac{m_\mu}{m_e},\alpha\right) &= 0.5\left(\frac{\alpha}{\pi}\right), \\
a_\infty^{(4)}\left(\frac{m_\mu}{m_e},\alpha\right) &= 0.765\,78\left(\frac{\alpha}{\pi}\right)^2, \\
a_\infty^{(6)}\left(\frac{m_\mu}{m_e},\alpha\right) &= 3.119\,67(56)\left(\frac{\alpha}{\pi}\right)^3, \\
a_\infty^{(8)}\left(\frac{m_\mu}{m_e},\alpha\right) &= (17.067\,6(60) + B_4)\left(\frac{\alpha}{\pi}\right)^4, \\
&\ldots,
\end{aligned} \tag{3.10}$$

where $a_\infty^{(n)}(m_\mu/m_e,\alpha)$ represents the n-th order contribution to a_∞.

Let us now concentrate on $a_\infty^{(8)}$ which contains an unknown term B_4 and thus cannot be fully determined from known results. Of the 469 eighth-order Feynman diagrams that contribute to $A_2^{(8)}(m_\mu/m_e)$, 304 belong to the class represented schematically by the diagrams shown in Fig. 14. It is the contribution of these diagrams to a_∞ that is given by (3.10). These 304 diagrams consist of i) all diagrams of group I with the exception of those shown in Fig. 6($(l_1,l_2) = (\mu,e)$), ii) all diagrams of group II, iii) all diagrams of group III, iv) diagrams of group IV that are represented by the self-energy diagrams of Fig. 11 with $(l_1,l_2) = (\mu,e)$.

Having evaluated the contribution to the muon anomaly from these 304 diagrams "exactly", we are now in a position to extract the value of the previously undetermined coefficient B_4. Summing the results (2.16), (2.20), (2.23), (2.24), (2.30), (2.35), (2.36), and $M_{6LL,P}^{(\mu,e)}$ (third entry in Table VI), we find

$$a^{(8)}\left(\frac{m_\mu}{m_e},\alpha\right) = 14.565(55)\left(\frac{\alpha}{\pi}\right)^4, \tag{3.11}$$

which, when compared with the fourth line of (3.10), means that

$$B_4 = -2.503(55) + O\left(\frac{m_e}{m_\mu}\ln\frac{m_\mu}{m_e}\right). \tag{3.12}$$

The method used above was such that, because of lumping together of all the diagrams at a given order, information about individual diagrams was lost. Also, for the eighth-order calculation, it was possible to determine only the logarithmic mass dependence of $a_\infty^{(8)}$. The mass-independent term B_4 in (3.6) could not be determined until direct evaluation (3.11) was made. There is, however, a class of eighth-order diagrams whose contributions to $a_\infty^{(8)}$, including mass-independent terms in some cases, can be obtained by making use of the asymptotic form of the vacuum-polarization terms and lower-order anomalies. Diagrams that belong to this class are those that are generated by inserting an electron vacuum-polarization loop into a single photon line of an arbitrary muon vertex. (See Fig. 15.) In order to consider the contribution of this class of diagrams we start with a brief discussion of the asymptotic photon propagator.

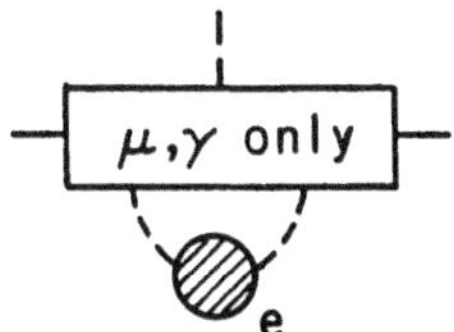

Figure 15: A class of diagrams generated by inserting electron vacuum-polarization loops into a single photon line of an arbitrary muon vertex.

The general expression for the renormalized photon propagator is of the form

$$D_R^{\mu\nu}(q) = -i\frac{g^{\mu\nu}}{q^2}d_R\left(\frac{q^2}{m_e^2},\ \alpha\right) + \text{the } q_\mu q_\nu \text{ term.} \tag{3.13}$$

The asymptotic part of the renormalized photon propagator is defined as follows: In each order of perturbation theory drop terms that vanish in the limit $-q^2/{m_e}^2 \rightarrow \infty$ while keeping divergent and constant terms. The asymptotic propagator satisfies the following Callan-Symanzik equation:[34]

$$\left(m_e\frac{\partial}{\partial m_e} + \beta(\alpha)\alpha\frac{\partial}{\partial\alpha}\right)\alpha d_R^\infty\left(-\frac{q^2}{{m_e}^2},\ \alpha\right) = 0. \tag{3.14}$$

The solution of this equation may be expanded in the form

$$\begin{aligned} d_R^\infty\left(-\frac{q^2}{{m_e}^2},\ \alpha\right) &= 1-(a_1+b_1L)\left(\frac{\alpha}{\pi}\right) \\ &+ \left[-(a_2+b_2L)+(a_1+b_1L)^2\right]\left(\frac{\alpha}{\pi}\right)^2 \\ &+ \left[-(a_3+b_3L+c_3L^2)+2(a_1+b_1L)(a_2+b_2L)\right. \\ &- \left.(a_1+b_1L)^3\right]\left(\frac{\alpha}{\pi}\right)^3 + \ldots, \end{aligned} \tag{3.15}$$

where

$$L = \ln\left(-\frac{q^2}{{m_e}^2}\right).$$

The first few coefficients are known from perturbation theory:[31]

$$\begin{aligned} b_1 &= -\frac{1}{3},\quad a_1 = \frac{5}{9}, \\ b_2 &= -\frac{1}{4},\quad a_2 = \frac{5}{24} - \zeta(3), \\ c_3 &= -\frac{1}{24},\quad b_3 = \frac{47}{96} - \frac{1}{3}\zeta(3). \end{aligned} \tag{3.16}$$

The coefficient a_3 has not been determined thus far. We are now ready to evaluate it.

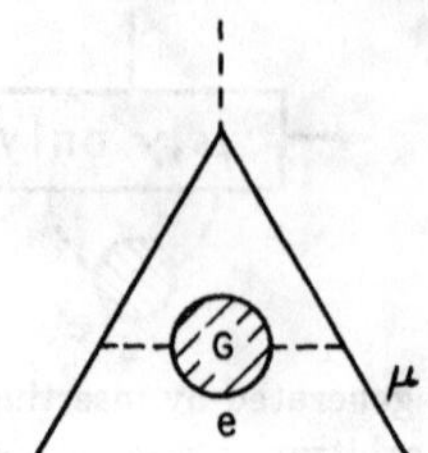

Figure 16: Diagrams generated by inserting electron vacuum-polarization loops into the second-order muon vertex.

From (3.15) one can see which contributions come from proper vacuum-polarization diagrams, which come from improper diagrams consisting of two proper parts, and which come from improper diagrams consisting of three proper diagrams, etc. Having found the asymptotic form of the photon propagator, we now turn to the discussion of the contribution to a_∞ from the diagrams that belong to the class represented by the diagram shown in Fig. 15.

We start by considering the diagrams shown in Fig. 16, generated by inserting the electron vacuum-polarization diagram G into the lowest-order muon vertex. An exact expression for the contribution to the muon anomaly from these diagrams is given by[31]

$$a_{(2;G)} = \frac{\alpha}{\pi}\int_0^1 dx(1-x)\left[d_R\left(\frac{-x^2}{1-x}\frac{m_\mu{}^2}{m_e{}^2}\right) - 1\right]_{(G)} . \tag{3.17}$$

Since we are interested in the asymptotic contribution to the anomaly, we have

$$a^\infty_{(2;G)} = \frac{\alpha}{\pi}\int_0^1 dx(1-x)\left[d_R^\infty\left(\frac{-x^2}{1-x}\frac{m_\mu{}^2}{m_e{}^2}\right) - 1\right]_{(G)} . \tag{3.18}$$

Eighth-order diagrams that belong to the class represented in Fig. 16 are those shown in Figs. 4, 5, 6, and 7 having only closed electron loops. From (3.15), (3.16), and (3.18) one finds[31]

$$\begin{aligned} a^\infty_{(2;2:2:2)} &= a^\infty[Fig.4(three\ electron\ loops)] \\ &= \left[-\frac{8609}{5832} - \frac{25}{27}\zeta(2) - \frac{2}{9}\zeta(3)\right. \\ &\quad + \left(\frac{317}{162} + \frac{4}{9}\zeta(2)\right)\ln\frac{m_\mu}{m_e} \\ &\quad \left.-\frac{25}{27}\ln^2\frac{m_\mu}{m_e} + \frac{4}{27}\ln^3\frac{m_\mu}{m_e}\right]\left(\frac{\alpha}{\pi}\right)^4 \\ &= 7.196\,7\left(\frac{\alpha}{\pi}\right)^4 , \end{aligned} \tag{3.19}$$

$$\begin{aligned} a^{\infty}_{(2;2:4)} &= a^{\infty}[Fig.5((l_1,l_2)=(e,e))] \\ &= \left[\frac{509}{432} + \frac{1}{3}\zeta(2) - \frac{25}{18}\zeta(3)\right. \\ &+ \left.\left(-\frac{5}{4} + \frac{2}{3}\zeta(3)\right)\ln\frac{m_\mu}{m_e} + \frac{1}{3}\ln^2\frac{m_\mu}{m_e}\right]\left(\frac{\alpha}{\pi}\right)^4 \\ &= 7.140\,4\left(\frac{\alpha}{\pi}\right)^4. \end{aligned} \tag{3.20}$$

Note that (3.19) and (3.20) have no undetermined constants. On the other hand, numerically integrating the exact expressions obtained in Sec. 2, one finds:

$$a_{(2;2:2:2)} = 7.223\,7(13)\left(\frac{\alpha}{\pi}\right)^4, \tag{3.21}$$

$$a_{(2;2:4)} = 7.128\,9(23)\left(\frac{\alpha}{\pi}\right)^4, \tag{3.22}$$

which are consistent with (3.19) and (3.20), respectively, within the uncertainty of the order of $(m_e/m_\mu)\ \ln\,(m_\mu/m_e)$.

The total contribution to a_∞ from the proper sixth-order vacuum-polarization insertions (eighteen diagrams of Fig. 6($(l_1,l_2)=(e,e)$) and Fig. 7) is given in terms of one unknown parameter a_3, the constant term in the asymptotic vacuum polarization in the sixth-order:[31]

$$\begin{aligned} a^{\infty}_{(2;6)} &= \left[-\frac{1}{2}a_3 + \frac{287}{384} + \frac{1}{12}\zeta(2) - \frac{5}{12}\zeta(3)\right. \\ &+ \left.\left(-\frac{67}{96} + \frac{1}{3}\zeta(3)\right)\ln\frac{m_\mu}{m_e} + \frac{1}{12}\ln^2\frac{m_\mu}{m_e}\right]\left(\frac{\alpha}{\pi}\right)^4 \\ &= \left(-\frac{1}{2}a_3 + 1.167\,729\right)\left(\frac{\alpha}{\pi}\right)^4. \end{aligned} \tag{3.23}$$

Comparing this with the exact result obtained in Sec. 2 :

$$a_{(2;6)} = 0.647\,1(203)\left(\frac{\alpha}{\pi}\right)^4, \tag{3.24}$$

one finds the unknown coefficient a_3 to be

$$a_3 = 1.041(41) + O\left(\frac{m_e}{m_\mu}\ln\frac{m_\mu}{m_e}\right). \tag{3.25}$$

We now consider the contribution to a_∞ from the set of 42 eighth-order diagrams generated by inserting proper fourth-order electron vacuum-polarization diagrams into fourth-order muon vertices. The diagrams that belong to this set are those shown in Fig. 5 (for $(l_1,l_2)=(e,\mu)$) and those represented by the eighth-order muon self-energy diagrams in the top row of Figs. 9.

Let us denote by $K^{(4)}(t)$ the anomaly due to fourth-order muon vertex diagrams with one heavy photon of mass squared t. Analytic properties of $K^{(4)}(t)$ allow one to write[31]

$$K^{(4)}(t) = \int_{-\infty}^{0} dt' \frac{1}{t' - t}\frac{1}{\pi}\,\mathrm{Im}K^{(4)}(t'). \tag{3.26}$$

Introducing the notation

$$\frac{1}{\pi}\,\mathrm{Im}K^{(4)}(t) = -k^{(4)}\left(-\frac{t}{m_\mu{}^2}\right)\left(\frac{\alpha}{\pi}\right)^2, \quad (t<0), \tag{3.27}$$

the contribution to a_∞ from the class of diagrams under consideration can be written as

$$a^{\infty}_{(4;4)} = \left(\frac{\alpha}{\pi}\right)^2 \int_0^\infty \frac{dy}{y} k^{(4)}(y)\left[d_R^\infty\left(-y\frac{m_\mu^2}{m_e{}^2}\right) - 1\right]_{(4)}. \tag{3.28}$$

Inserting the fourth-order contribution to the asymptotic propagator and defining the integrals

$$I_N = \int_0^\infty \frac{dy}{y} k^{(4)}(y)\ln^N y, \quad (N=0,\ 1) \tag{3.29}$$

we may write Eq. (3.28) as

$$a^{\infty}_{(4;4)} = \left(-a_2 I_0 - b_2 I_1 - 2b_2 I_0 \ln\frac{m_\mu}{m_e}\right)\left(\frac{\alpha}{\pi}\right)^4. \tag{3.30}$$

The integral I_0 is 2 times the fourth-order electron anomaly $a_e^{(4)}$:

$$I_0 = 2\left(\frac{197}{144} + \frac{1}{2}\zeta(2) - 3\zeta(2)\ln 2 + \frac{3}{4}\zeta(3)\right). \tag{3.31}$$

The value of I_1 is not known analytically. However, it can be determined from

$$a^{\infty}_{(4;2)} = \left(-a_1 I_0 - b_1 I_1 - 2b_1 I_0 \ln\frac{m_\mu}{m_e}\right)\left(\frac{\alpha}{\pi}\right)^3, \tag{3.32}$$

where $a_{(4;2)}$ is the sixth-order contribution to the muon anomaly obtained by inserting a second-order electron vacuum-polarization loop into the fourth-order muon vertices. Using the values listed in Tables I and III and a result reported in Ref. 18, we obtain

$$a_{(4;2)} = -\,2.294\,4(12)\left(\frac{\alpha}{\pi}\right)^3, \tag{3.33}$$

which leads to

$$I_1 = -\,0.972\,9(36). \tag{3.34}$$

Substituting (3.31) and (3.34) into (3.30), and taking (3.16) into account, we find

$$a^{\infty}_{(4;4)} = -2.647\,4(9)\left(\frac{\alpha}{\pi}\right)^4, \tag{3.35}$$

which is consistent with the numerically obtained "exact" result:

$$a_{(4;4)} = -2.666\,9(45)\left(\frac{\alpha}{\pi}\right)^4 \tag{3.36}$$

within the uncertainty of order $(m_e/m_\mu)\ln(m_\mu/m_e)$.

4. Hadronic Contributions to the Muon Anomaly

4.1. Introductory remarks

The importance of the hadronic vacuum-polarization insertions in the second-order muon vertex was first pointed out by Bouchiat and Michel[11] and by Durand.[12] This is based on the observation that typical momenta of virtual photons emitted by a muon is not much smaller than the masses of important hadron resonances. In fact the muon anomaly a_μ has a substantial contribution ($\simeq 7 \times 10^{-8}$) from the hadronic effect. (On the other hand, this effect is not important for the electron anomaly.) It arises from two types of diagrams: Hadronic vacuum-polarization diagrams and hadronic light-by-light scattering diagrams shown in Figs. 17 and 18, respectively. The dominant contribution, which also has the largest uncertainty, comes from the diagram in Fig. 17(a), and it is this uncertainty that is the most serious obstacle for improvement of theoretical prediction. In principle this is calculable by quantum chromodynamics (QCD), which is widely accepted as the correct theory of the strong interaction. At the present stage of development, however, QCD is powerless to handle this problem since we are dealing here with the processes dominated by momenta of order m_μ where perturbative QCD is not expected to be reliable. Nonperturbative approaches based on the lattice QCD method have not been developed far enough either.

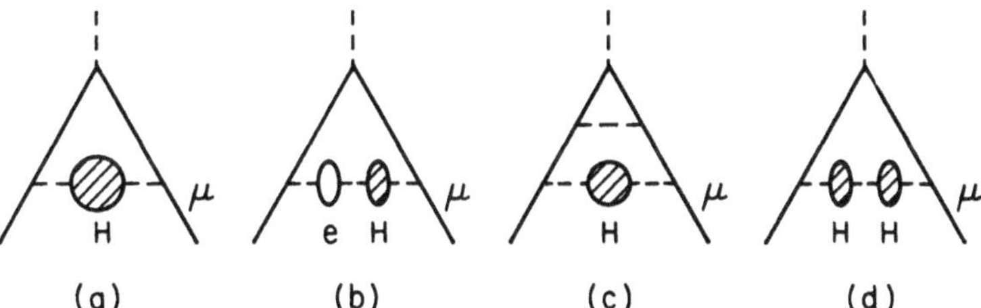

Figure 17: Various hadronic vacuum-polarization contributions to a_μ.

It is thus very fortunate that the hadronic vacuum-polarization contribution to the muon anomaly can be evaluated without detailed knowledge of the underlying theory of strong interactions: it can be directly related to the observed e^+e^- annihilation cross section measured in symmetric and/or asymmetric colliding beam experiments. Using this information the contribution of the hadronic vacuum polarization has been evaluated by several groups. The latest evaluations are[3,7,8]

$$a_\mu(Fig.17(a)) = 7068(59)(164) \times 10^{-11}, \tag{4.1}$$

$$= 684(11) \times 10^{-10}, \tag{4.2}$$

$$= 7100(105)(49) \times 10^{-11}, \tag{4.3}$$

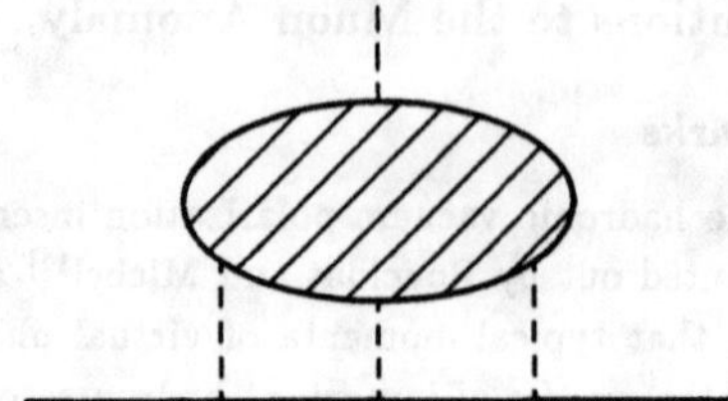

Figure 18: Hadronic light-by-light scattering contribution to a_μ.

where the first error is statistical and the second is systematic. (Only the total error is given in (4.2).) These results agree with each other within their errors. The apparent difference in error estimates seems to be due to somewhat different theoretical approaches and different handlings of available data, experimental systematic errors in particular. More on this in Sec. 4.2.

There are also small contributions coming from the higher-order hadronic terms arising from the diagrams of Figs. 17(b), (c), and (d):[3,35]

$$a_\mu(Figs.17(b)-(d)) \;=\; -90(5)\times 10^{-11}. \tag{4.4}$$

As for the hadronic light-by-light contribution, it is unfortunately not possible to use experimental data directly. Instead we have to evaluate this contribution using the theory of strong interactions. Therefore, at present, the hadronic light-by-light contribution to the muon anomaly can be evaluated only in a model-dependent way. One approach is to assume that the blob in Fig. 18 can be approximated by quark loops of various flavors and colors.[35] Another approach is based on the assumption that the blob in Fig. 18 can be approximated by charged pion loops and various low-energy resonances. The values of the hadronic contribution obtained by these methods are[3]

$$\begin{aligned} a_\mu(Fig.18) &= 60(4)\times 10^{-11} \quad \text{(quark loop)} \\ &= 49(5)\times 10^{-11} \quad \text{(pion loop and resonances)}, \end{aligned} \tag{4.5}$$

which are not inconsistent with each other. However, the ambiguity in the choice of quark mass values makes the result obained by the first method rather unreliable. Only the order of magnitude may be correct. What is important is the fact that (4.5) is relatively small in magnitude and thus does not contribute significantly to the theoretical uncertainty, at least as far as the next experiment is concerned.

4.2. Hadronic vacuum-polarization contribution to a_μ

If one assumes that the dispersion integral for the hadronic vacuum polarization requires only one subtraction (charge renormalization), one can express the hadronic contribution to a_μ as an integral over the annihilation cross section measured in the e^+e^-

colliding beam experiments:

$$a_\mu(Fig.17(a)) = \frac{1}{4\pi^3}\int_{4m_\pi^2}^{\infty} ds K(s)\sigma_H(s), \tag{4.6}$$

where

$$\begin{aligned} K(s) &= \int_0^1 \frac{y^2(1-y)dy}{y^2 + (1-y)(s/m_\mu^2)} \\ &= x^2\left(1 - \frac{x^2}{2}\right) + \frac{(1+x)}{(1-x)}x^2 \ln x \\ &+ (1+x)^2(1+x^{-2})\left(\ln(1+x) - x + \frac{x^2}{2}\right), \end{aligned}$$

$$x = \frac{1 - (1 - 4m_\mu^2/s)^{\frac{1}{2}}}{1 + (1 - 4m_\mu^2/s)^{\frac{1}{2}}},$$

and $\sigma_H(s)$ is the total cross section for e^+e^- annihilation into hadrons. Since $K(s)$ is positive definite in the integration domain $4m_\pi^2 \leq s < \infty$, a_μ(Fig.17(a)) must be positive. Furthermore, $K(s)$ decreases as $m_\mu^2/(3s)$ for large s. Thus the most important contribution to a_μ(Fig. 17(a)) comes from the region of low-energy π-π resonances, the ρ resonance in particular.

The $e^+e^- \to \pi^+\pi^-$ cross section in this energy range can be expressed in terms of the pion form factor $F_\pi(s)$ as

$$\sigma(e^+e^-) = \frac{8\pi}{3}\frac{\alpha^2 q^3}{s^{5/2}}|F_\pi(s)|^2, \tag{4.7}$$

where

$$q = \left(\frac{s}{4} - m_\pi^2\right)^{\frac{1}{2}}.$$

There are several ways to determine F_π from the experimental data. Let us first discuss the simplest approach in which one parametrizes $F_\pi(s)$ using the modified Gounaris-Sakurai formula[36]

$$F_\pi = \left(\frac{A_1 - m_\pi^2 A_2}{A_1 + A_2 q^2 + f(s)} + A_3 e^{iA_4}\frac{m_\omega^2}{s - m_\omega^2 + im_\omega\Gamma_\omega}\right) G(s), \tag{4.8}$$

which deals with the energy region including the $\rho - \omega$ resonances. Here

$$f(s) = \frac{1}{\pi}\left(m_\pi^2 - \frac{s}{3}\right) + \frac{2}{\pi}\frac{q^3}{\sqrt{s}}\ln\left(\frac{\sqrt{s} + 2q}{2m_\pi}\right) - i\frac{q^3}{\sqrt{s}}, \tag{4.9}$$

$$G(s) = \left(\frac{M^2}{s - M^2 + iM\Gamma}\right)^n. \tag{4.10}$$

In (4.8) only the real part is kept for $\sqrt{s} < m_\pi + m_\omega$. The first term in the brackets in (4.8) is the standard Gounaris-Sakurai formula.[37] The second term accounts for the

$\rho - \omega$ interference.[38] The factor $G(s)$ was introduced in Ref. 36 to crudely incorporate the effects of the $\rho - \omega$ inelastic channel. For the numerical work the parameters are chosen as $M = 1.2$ GeV, $\Gamma = 0.15$ GeV, and $n = 0.22$. Using the data for $|F_\pi|^2$ from Refs. 36,39,40,41, and 42, it is found that the parameters of (4.8) are given by

$$A_1 = 0.290(2)(\text{GeV})^2, \quad A_2 = -2.30(1),$$

$$A_3 = -0.012(1), \quad A_4 = 1.84(9).$$

The χ^2 is 175.7 with 95 degrees of freedom. The integration region was taken to be $2m_\pi \leq \sqrt{s} \leq 1.197\,6$ GeV.

From this result one can calculate the contribution of σ_H to the muon anomaly. The result is

$$a_\mu(\rho,\omega) = 5\,063.9(21.5)(150) \times 10^{-11}, \tag{4.11}$$

where the first error is statistical and the second is an estimate of the systematical error. The systematic error in the measurement of $|F_\pi(s)|^2$ in this region is about 2 percent. It is, however, not clear, especially after fitting the parameters, how the systematic error of a_μ depends on those of the experimental data for $|F_\pi(s)|^2$. In fact the latter is the dominant source of uncertainty in evaluating the hadronic contribution to a_μ. For a better determination of this contribution it is crucial that the next generation of measurements of σ_H pay particular attention to the reduction of systematic errors.

The ω and ϕ resonances are treated by the Breit-Wigner formula

$$\sigma_{BW} = \frac{3\pi}{s} \frac{\Gamma_{total}\Gamma_{e^+e^-}}{(\sqrt{s} - M_R)^2 + \frac{1}{4}\Gamma^2_{total}}, \tag{4.12}$$

where M_R is the mass of ω or ϕ, Γ_{total} is the total width of ω or ϕ, and $\Gamma_{e^+e^-}$ is the partial width of ω or ϕ decaying into an e^+e^- pair. The statistical error was estimated using the statistical errors in the measurements of Γ_{total} and $\Gamma_{e^+e^-}$. The systematic error is again somewhat unclear. We use 3.2 percent, which is the systematic error in the measurements of Γ_{total} and $\Gamma_{e^+e^-}$.[43]

The other resonances have been treated using a narrow width approximation. The error estimates are made in the same way as for the ω and ρ resonances.

The background contribution to (4.6) from the region $1.197\,6 \text{ GeV} \leq \sqrt{s} \leq 30.8$ GeV has been evaluated by the trapezoidal rule, using the experimental data for R, where

$$R(s) = \frac{\sigma_{total}(e^+e^- \to hadrons)}{\sigma_{total}(e^+e^- \to \mu^+\mu^-)}.$$

This treatment is permissible because the background R is more or less constant (i.e., $\approx 3 \sum e_q^2$) for most of s. Since a_μ depends on R linearly, we can estimate the systematic error of a_μ by that of the measurement of R. Finally, the contribution from the region $\sqrt{s} \geq 30.8$ GeV is estimated by the lowest-order QCD with six quarks.

More details of the above analysis can be found in Ref. 3. The total contribution from the diagram of Fig. 17(a) is given by (4.1). Note that the uncertainty in (4.1) is dominated by that of the systematic error of 164×10^{-11}, which in turn is mostly due to the uncertainty in the ρ and ω contribution (4.11). This is why it is crucial to carry out a new measurement of R in the region below 1 GeV with particular attention to the reduction of systematic errors.

The result (4.2) is based on more or less the same data set as that of (4.1) and is treated in a similar way. The small difference in these results may reflect the difficulty in assessing errors in diverse sets of data.

The result (4.3) is obtained by a somewhat different method[7] in which both time-like and space-like data for the pion form factor are utilized. In this calculation F_π is represented by a 15-parameter function written as a product of the Omnes function and the inelastic part which is parametrized in terms of higher vector meson contributions, a three-parameter background function, and a function providing the asymptotic behavior of $F_\pi(s)$. The error assessment appears to be more subtle in this approach than others. In particular, the relatively small systematic error in (4.3) may require a rather optimistic asumption about the randomness of the experimental systematic error distribution, which may be difficult to justify. Any significant improvement in the precision of $a_\mu(Fig.17(a))$ can come only from measurements of $R(s)$ with better statistical and systematic errors.

4.3. Hadronic light-by-light scattering contribution to a_μ

The contribution of the hadronic light-by-light scattering to the muon anomaly was first discussed by Calmet *et al.*[35] Their calculation is based on the assumption that it is effectively given by the sum of quark loop contributions of various colors and flavors (see Fig. 19). Furthermore, they assumed that quark masses m_q are larger than the muon mass m_μ so that an expansion in m_μ/m_q is justified. Assuming the quark mass values of $m_u = m_d = 0.3$, $m_s = 0.5$, and $m_c = 1.5$ GeV/c^2, they obtained the value

$$a_\mu(had2) \; = \; -\,26(10) \times 10^{-10}, \tag{4.13}$$

where the error is due to the numerical integration procedure only. The negative sign in (4.13) was rather unexpected since it is known[22] to be positive for $m_q = m_\mu$.

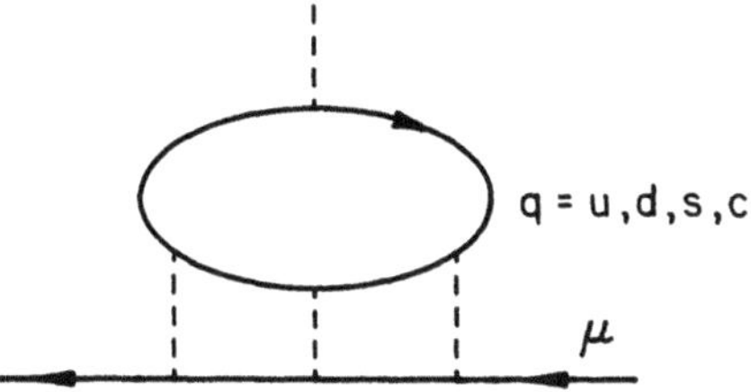

Figure 19: Quark-loop light-by-light scattering contribution to a_μ.

Let $a(m_\mu/m_q)$ be the naive quark loop contribution of Fig. 19 due to a quark of mass m_q and electric charge e_q. Then, adapting from the pure QED calculation of Sec. 2.3, we can write

$$a(m_\mu/m_q = 1) \;=\; 0.370\,986(20) \times (3e_q^4) \times \left(\frac{\alpha}{\pi}\right)^3 . \qquad (4.14)$$

For $e_q = \frac{2}{3}$ this is equal to 27.6×10^{-10}. Comparison of (4.13) and (4.14) raises two questions:

(i) Is it reasonable that $a(m_\mu/m_q)$ changes sign as $(m_\mu/m_q)^2$ decreases from 1 to 0.1 ?

(ii) Why is (4.13) of the same order of magnitude as (4.14) instead of being an order of magnitude smaller, which would be the case if $a(m_\mu/m_q)$ behaves as $(m_\mu/m_q)^2$?

To answer these questions the contribution of Fig. 19 was reexamined in Ref. 3. The starting point is to integrate the exact expression, rather than the leading term of the expansion in m_μ/m_q , for the light-by-light contribution for various values of quark mass ranging from 0.3 to 1000 GeV/c^2. It is found that the results are positive and are roughly proportional to $(m_\mu/m_q)^2$ for all values of m_q. It is also seen that the relative numerical accuracy deteriorates steadily as m_q increases, indicating that the value of this integral reflects a delicate cancellation between positive and negative contributions and the cancellation becomes more and more difficult as m_q increases. To verify this observation the behavior of the integrand was examined analytically in the limit of large m_q. It is found that the integral contains various terms that behave as $(m_\mu/m_q)^2 \ln(m_q/m_\mu)$ and regains the expected $(m_\mu/m_q)^2$ behavior only as a consequence of delicate cancellation of logarithmic terms from different parts of the integration domain. This means that the coefficient of $(m_\mu/m_q)^2$ in the expansion of $a(m_\mu/m_q)$, which is an integral over Feynman parameters, is not pointwise integrable, contrary to the implicit assumption of Ref. 35, and hence cannot be evaluated reliably by a Monte Carlo procedure.

As a consequence of the work in Ref. 3, the quark loop light-by-light contribution to the muon anomaly is found to be

$$a_\mu(had2) \;=\; 60(4) \times 10^{-11} \qquad (4.15)$$

for $m_u = m_d = 0.3$ GeV/c^2, $m_s \;=\; 0.5$ GeV/c^2, and $m_c \;=\; 1.5$ GeV/c^2, which is positive and substantially smaller than (4.13).

The calculation described above, however, is not satisfactory in several respects. First of all, the result depends strongly on m_q. Because of quark confinement, however, there is an ambiguity in the definition of the quark mass m_q. Second, the contribution (4.15) is governed by the low-energy behavior of the virtual quark loop whose typical momenta are of order $m_\mu c$. It is not clear to what extent this approximation represents the correct physical picture in the low-energy region. In order to test the validity of the approximation discussed above it was evaluated in another picture in which the hadronic part of the diagram in Fig. 19 is approximated by a loop of pion, treated as an elementary particle, and various low-energy resonances. Since typical momenta of virtual photons attached to the hadron loop are of order m_μ, these photons are not hard enough to resolve the internal structure of the hadron. Therefore, to a reasonably good approximation,

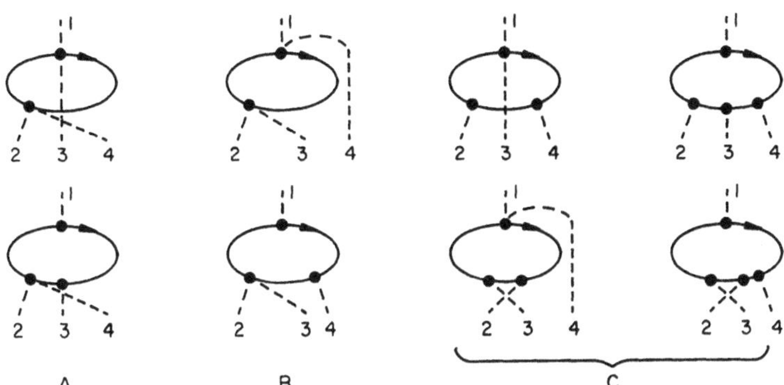

Figure 20: Fourth-order diagrams contributing to the light-by-light scattering amplitude in scalar QED. The subsets A, B, and C consist of 3, 6, and 12 diagrams, respectively, and are gauge-invariant by themselves with respect to the photon 1.

the pion may be treated as an elementary field and scalar QED may be used to describe the photon-pion interaction. Two versions of this treatment are discussed in the following subsections. To make this picture more realistic, it is also necessary to incorporate the vector-meson-dominance (VMD) approximation. Finally, the contribution of other resonances is considered in Sec. 4.3.3.

4.3.1. Charged-pion-loop contribution

As was mentioned above, we treat the pion as an elementary field and use scalar QED to describe the pion-photon interaction.

There are altogether 21 Feynman diagrams that contribute to the lowest-order light-by-light scattering amplitude in scalar QED. They are shown in Fig. 20. The total amplitude is given in terms of the fourth-rank vacuum-polarization tensor $\Pi_{\nu\rho\lambda\sigma}(k_1, k_2, k_3, k_4)$, where k_i, $i = 1, 2, 3, 4$, are the momenta of photons attached to the pion loop, and the Pauli-Villars regularization is understood. Because of gauge invariance we have

$$\begin{aligned} k_1^\nu \Pi_{\nu\rho\lambda\sigma}(k_1, k_2, k_3, k_4) &= 0, \\ k_2^\rho \Pi_{\nu\rho\lambda\sigma}(k_1, k_2, k_3, k_4) &= 0, \text{ etc.} \end{aligned} \tag{4.16}$$

As is indicated in Fig. 20, the set of 21 Feynman diagrams can be classified into subsets A, B, and C, comprised of 3, 6, and 12 diagrams, respectively, each being gauge-invariant with respect to the photon 1. Relations (4.16) hold for each of the three subsets separately. As a consequence the total amplitude, as well as the partial amplitudes corresponding to the subsets A, B, and C, are finite, even though each contributing diagram is logarithmically divergent according to power counting.

Attaching the photon lines 2, 3, and 4 of these diagrams to the muon line, we obtain 21 sixth-order vertex diagrams. Charge-conjugation invariance (invariance under reversal

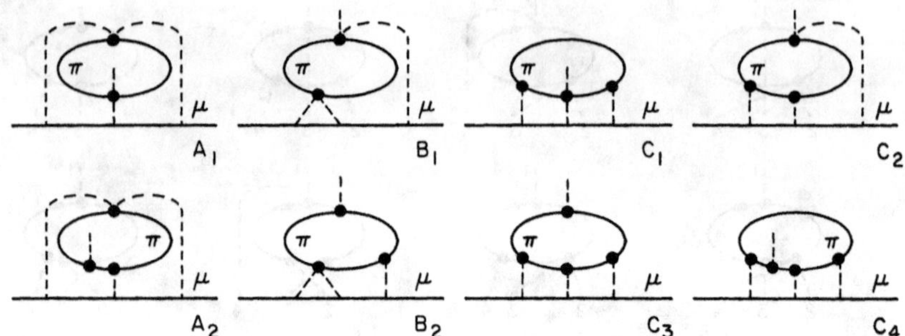

Figure 21: Vertex diagrams containing the scalar QED light-by-light scattering subdiagrams. The subsets A, B, and C consist of 3, 6, and 12 diagrams, respectively, and are gauge-invariant by themselves.

of the direction of momentum flow in the pion loop) and time-reversal invariance reduce the number of independent diagrams to be evaluated to eight. They are shown in Fig. 21 together with the corresponding multiplicity factors that account for the diagrams related by the symmetries mentioned above.

To set up the Feynman integrals it is convenient to use the Feynman-Dyson rules in parameter space described in Kinoshita's article on the electron anomaly in this volume, modified slightly for theories with derivative coupling such as scalar QED.

Omitting the factor $(\alpha/\pi)^3$, the magnetic moment projection of the amplitudes corresponding to the diagram α $(=A_1, A_2, \cdots, C_4)$ of Fig. 21 can be written in the form

$$\begin{aligned} M_\alpha &= -\frac{1}{16}\int(dz)\frac{F_0}{U^2V}, \quad \text{for } \alpha = A_1,\ B_1\ , \\ M_\alpha &= -\frac{1}{32}\int(dz)\left(\frac{F_0}{U^2V^2}+\frac{F_1}{U^3V}\right), \quad \text{for } \alpha = A_2,\ B_2,\ C_1,\ C_2\ , \\ M_\alpha &= -\frac{1}{32}\int(dz)\left(\frac{F_0}{U^2V^3}+\frac{F_1}{U^3V^2}+\frac{F_2}{U^4V}\right), \quad \text{for } \alpha = C_3,\ C_4\ . \end{aligned} \tag{4.17}$$

The details of the parametric functions U, V, F_0, F_1, and F_2 for each diagram in Fig. 21 are given in Ref. 3. The overall divergence of the vertex function is absent in the magnetic moment projection. The integral M_α depends logarithmically on the Pauli-Villars regularization mass of the light-by-light scattering subdiagram S_α. However, in the case of scalar QED, the regularization terms can be replaced with those obtained by applying the $\mathbf{K}_S$ operation[f] to the integrands of (4.17). The quantity

$$\Delta M_\alpha = (1 - \mathbf{K}_{S_\alpha})M_\alpha \tag{4.18}$$

[f] See Sec. 3 of Kinoshita's article on the electron anomaly in this volume.

is finite and can be evaluated numerically.

The results of numerical evaluation of individual integrals by RIWIAD are summarized in Table VII. The quoted uncertainties represent the 90 percent confidence limit estimated by the integration routine. Summing up the contribution listed in Table VII we find the total contribution due to the Feynman diagrams of Fig. 21 to be

$$a_{HLL} = -0.043\,7(36)\left(\frac{\alpha}{\pi}\right)^3 . \tag{4.19}$$

Table VII. Numerical results for ΔM_α of (4.18).

Diagram	$\eta_\alpha \Delta M_\alpha$	Subcubes ($\times 10^5$)	Iterations
A_1	$-0.392\,7(5)$	5	31
A_2	$0.276\,7(6)$	5	35
B_1	$-0.461\,7(4)$	5	26
B_2	$0.160\,3(4)$	5	25
C_1	$0.402\,4(10)$	5	26
C_2	$0.830\,9(12)$	5	35
C_3	$-0.301\,7(10)$	5	13
C_4	$-0.557\,9(30)$	5	10

As a check of (4.19), a_{HLL} is also evaluated by an alternative method based on the Ward-Takahashi identity. We start from the observation that the three gauge-invariant sets of vertex diagrams in Fig. 21 can be obtained from the self-energy diagrams shown in Fig. 22 by inserting an external magnetic vertex in the pion lines in all possible ways. As is well known, proper vertex and self-energy parts are related by the Ward-Takahashi identity

$$q_\mu \Lambda^\mu_G = \Sigma_G(p+q/2) - \Sigma_G(p-q/2),$$

where Σ_G is calculated from the self-energy diagram G and Λ_G is the sum of vertex diagrams obtained by inserting an external vertex in G in all possible ways. In our case all self-energy diagrams shown in Fig 22 vanish by (generalized) Furry's theorem so that

$$q_\mu \Lambda^\mu_G(p,\, q) = 0, \quad G = A,\, B,\, C\, . \tag{4.20}$$

Differentiating (4.20) with respect to q_ν and dropping terms quadratic and higher order in q, we obtain

$$\Lambda^\nu_G(p,q)|_{q\to 0} = -\, q_\mu \left(\frac{\partial \Lambda^\mu_G(p,q)}{\partial q_\nu}\right)_{q=0} + O(q^2). \tag{4.21}$$

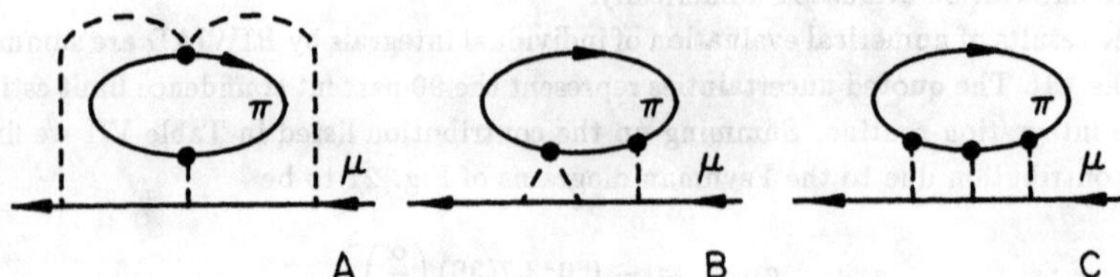

Figure 22: Self-energy diagrams that correspond to the vertex diagrams of the gauge-invariant subgroups A, B, and C.

Parametrizing (4.21) according to the method described in Kinoshita's article on the electron anomaly in this volume and projecting out the magnetic moment term, we obtain integrals that can be evaluated numerically. In this way it is found that the total contribution of diagrams of Fig. 21 is

$$a_{HLL} = -0.038\,3(20)\left(\frac{\alpha}{\pi}\right)^3 . \tag{4.22}$$

This is in reasonable agreement with the result (4.19) obtained by the first method.

4.3.2. Incorporating the vector-meson-dominance picture

The previous discussion is based on the usual scalar QED: We have been treating the pion as elementary. This treatment is not quite satisfactory in the sense that the pion, in fact, has structure. In principle the hadronic structure can be described by QCD. However, the momentum scale of interest is small so that the perturbative QCD does not work here. The lattice QCD is not yet developed enough to apply to our case either. Thus the best available approximation to the actual hadronic picture at this energy scale will be the vector-meson-dominance (VMD) model. In VMD one assumes that hadrons are elementary but the photon has hadronic structure: a photon transforms into a vector meson (such as ρ, ω, or ϕ) and the vector meson in turn couples to structureless hadrons. One benefit of VMD is that it provides a cutoff for momentum integration and makes the resonance contribution to the muon anomaly, which are otherwise logarithmically divergent, finite as discussed later.

As the lowest-order approximation to the VMD picture, we take into account only the lowest-mass constituents of the photon (ρ^0 and ω). One of the diagrams contributing to a_μ in this picture is shown in Fig. 23. The coupling constants of the ρ meson to the pion and photon are $-\ if_\rho$ and igm_ρ^2, respectively.[44] The ρ-photon line which connects the pion loop and the muon line in Fig. 23 can be written as

$$\frac{ef_\rho g m_\rho^2}{(p^2\ -\ m_\rho^2)p^2} = \frac{e^2}{p^2} - \frac{e^2}{p^2\ -\ m_\rho^2}$$

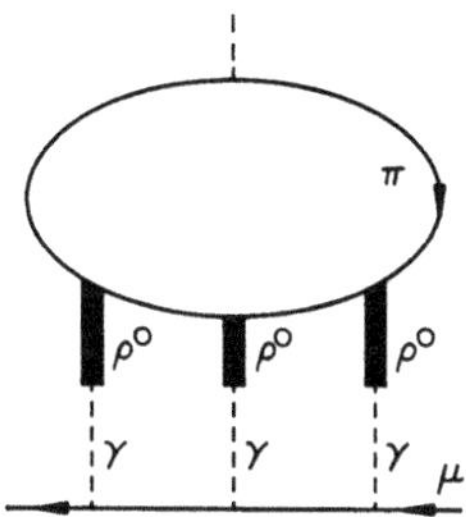

Figure 23: A diagram with vector-meson insertions.

using the identity $g = e/f_\rho$. Thus the ρ-photon line splits into two terms: The first term is exactly the same as that in scalar QED without the VMD, i.e., "bare photon" line, and the second term is the ρ line which provides a momentum cutoff to the scalar QED integral at the ρ mass. Therefore, we have to make only a slight modification to the original scalar QED integrals in order to obtain the integrals of the VMD model. Numerical evaluation by the integration routine VEGAS shows that the contribution to the muon anomaly coming from the 21 Feynam diagrams of Fig. 21 in the VMD picture is given by

$$a_{HLL} = -0.012\,5(19)\left(\frac{\alpha}{\pi}\right)^3 . \tag{4.23}$$

We believe that this value is closer to reality than the values (4.19) and (4.22) obtained without proper attention to the extended structure of hadrons.

4.3.3. Resonance contributions

In this subsection we shall consider the contribution to the muon anomaly arising from various low-energy resonances. The contributing diagrams are depicted in Fig. 24. Let us start with the lowest mass resonance, π^0. The effective Lagrangian for the process $\pi^0 \rightarrow 2\gamma$ is given by

$$L_{\pi^0\rightarrow 2\gamma} = \frac{1}{\pi f_\pi}\alpha F_{\mu\nu}\tilde{F}^{\mu\nu}, \tag{4.24}$$

where $\tilde{F}^{\mu\nu}$ is the dual of $F^{\mu\nu}$, α is the fine-structure constant, and f_π (= 93 MeV) is the pion decay constant. The overall divergence of the integrals corresponding to the diagrams of Fig. 24 is removed by the magnetic-moment projection, but the integrals still contain logarithmic subdivergence, as is seen by power counting. The vector-meson insertion (in the sense of VMD), however, provides the momentum cutoff and renders the integrals finite. Evaluating the resulting integrals numerically by VEGAS, we find

$$a_\mu(\pi^0) = 0.052(5)\left(\frac{\alpha}{\pi}\right)^3, \tag{4.25}$$

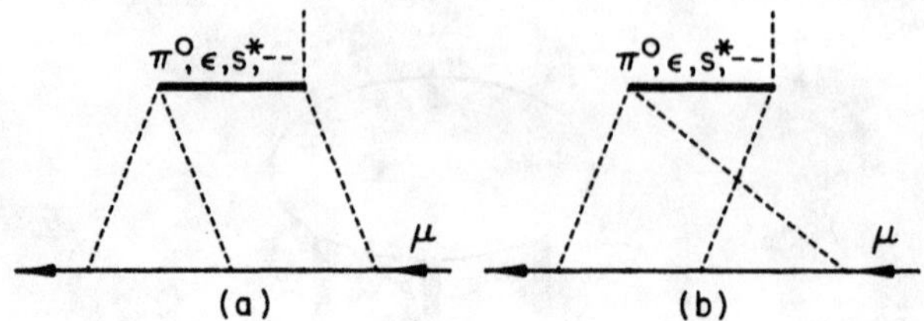

Figure 24: Resonance diagrams contributing to the muon anomaly.

where the number of subcubes used for the integrals is 16×10^3 and the number of iterations is 10.

Since the contribution to a_μ will decrease as the mass of the resonance particle increases, we expect that higher mass resonance contributions to a_μ may be ignored compared with the contribution from the π^0 resonance (4.25). The next-lowest-mass resonances to be taken into account are the scalar resonance ϵ and S^* (at about 1 GeV). Using the effective Lagrangian density[45]

$$L_{\epsilon,S^*\to 2\gamma} = -\sqrt{(8\pi)}\alpha(f\epsilon + f'S^*)F_{\mu\nu}F^{\mu\nu}, \tag{4.26}$$

where $f(f')$ are coupling constant of ϵ (S^*) to the electromagnetic field, we find numerically

$$\begin{aligned} a_\mu(\epsilon) &= 0.13 f^2 a_\mu(\pi^0), \\ a_\mu(S^*) &= 0.13 f'^2 a_\mu(\pi^0). \end{aligned} \tag{4.27}$$

Since $f, f' \approx 0.1$[see ref. 45], $a_\mu(\epsilon)$ and $a_\mu(S^*)$ are in fact negligible compared to $a_\mu(\pi^0)$. Contributions of the higher-mass resonances will be even smaller and may be ignored completely. Combining (4.23) and (4.25), we obtain the following contribution to the muon anomaly due to the pion loop and resonances:

$$a_\mu(had2) = 49(5) \times 10^{-11}, \tag{4.28}$$

which is not inconsistent with (4.15), the result obtained using the quark-loop approximation. The error in (4.28) is our estimate of the model dependence (i. e., dependence on the cutoff mass m_ρ). Note that the π^0 resonance contribution (4.25) dominates (4.28).

5. Weak Contributions to a_μ

The standard $SU(3)_C \times SU(2)_L \times U(1)$ model of strong and electroweak interactions is a fully renormalizable quantum field theory. As such, short-distance ultraviolet divergences can be systematically absorbed into the bare parameters of the theory order by order in perturbation theory. Unambiguous calculations may, therefore, be carried out and confronted by high precision experimental measurements. A clear disagreement between theory and experiment would signal the presence of "new physics" beyond the standard model.

Leptonic anomalous magnetic moments, $a_\ell \equiv \frac{1}{2}(g_\ell - 2)$, $\ell = e, \mu, \tau$, are nice examples of calculable quantities. Renormalizability requires $a_\ell = 0$ in lowest order; so, one loop corrections to a_ℓ must be finite (i.e., there are no counterterms at that order). Higher orders will exhibit ultraviolet infinities, but they will be cancelled by renormalization of the bare parameters (masses and couplings) of the theory. That feature has been well tested in the QED sector, where four-loop (eighth-order) contributions have been calculated (see previous sections) and agreement with experiment is impressive. In the case of weak loop corrections to a_ℓ, they are suppressed by m_ℓ^2/m_W^2, ($m_W \simeq 80$ GeV), relative to QED effects. Hence, they have not been important at the existing level of experimental precision. However, that situation should change for a_μ. A BNL experiment aims to reach an a_μ precision of $\pm\ 40 \times 10^{-11}$ and thus probe weak loop contributions which, as we shall see, would at that level constitute about a 5σ effect. It would, therefore, test the standard model at the weak interaction quantum loop level.

The standard model's one loop weak corrections to a_μ are graphically illustrated in Fig. 25 (for the unitary gauge). They are divided into three contributions

$$a_\mu^{\text{weak}} = a_\mu^W + a_\mu^Z + a_\mu^H \tag{5.1}$$

coming from W boson, Z boson, and Higgs scalar loops respectively. Each contribution is separately finite and given by[4]

$$a_\mu^W = \left(\frac{10}{3}\right)\frac{G_F m_\mu^2}{8\sqrt{2}\pi^2} + \mathcal{O}\left(\frac{m_\mu^4}{m_W^4}\right), \tag{5.2}$$

$$a_\mu^Z = -\left(\frac{5}{3} - \frac{(3-4\cos^2\theta_W)^2}{3}\right)\frac{G_F m_\mu^2}{8\sqrt{2}\pi^2} + \mathcal{O}\left(\frac{m_\mu^4}{m_W^4}\right), \tag{5.3}$$

$$a_\mu^H = 3F\left(\frac{m_H^2}{m_\mu^2}\right)\frac{G_F m_\mu^2}{8\sqrt{2}\pi^2}, \tag{5.4}$$

where $G_F = 1.16637 \times 10^{-5}\ \text{GeV}^{-2}$, $\sin^2\theta_W \simeq 0.233$ and

$$\begin{aligned} F(R) &= \frac{2}{3}\int_0^1 dx \frac{x^2(2-x)}{x^2 + R(1-x)} \\ &= 1 - \frac{2}{3}R + \left(\frac{1}{3}R^2 - R\right)\ln R \\ &\quad + \frac{2}{3}(R-1)\sqrt{4R - R^2}\,\text{Arctg}\left(\frac{4-R}{R}\right)^{\frac{1}{2}}, \quad R < 4 \\ &= 1 - \frac{2}{3}R + \left(\frac{1}{3}R^2 - R\right)\ln R \\ &\quad - \frac{1}{3}(R-1)\sqrt{R^2 - 4R}\,\ln\left(\frac{R + \sqrt{R^2-4R}}{R - \sqrt{R^2-4R}}\right), \quad R \geq 4 \end{aligned} \tag{5.5}$$

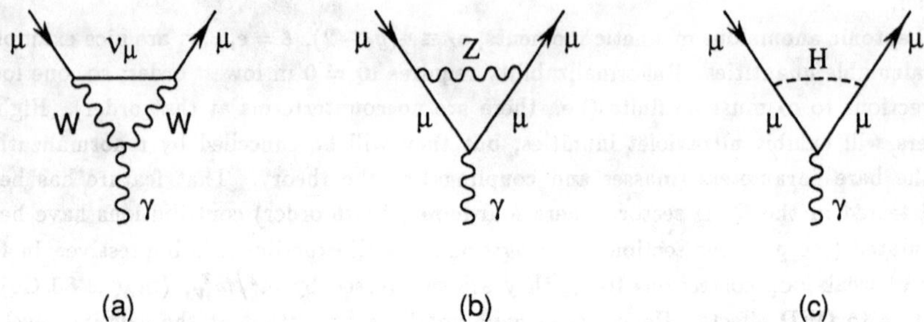

Figure 25: One loop weak corrections to a_μ in the standard model.

with the special values

$$\begin{aligned} F(0) &= 1 \\ F(4) &\simeq 0.182 \\ F(R) &\to \frac{2}{3R}\ln R, \quad \text{large } R. \end{aligned} \tag{5.6}$$

Several interesting features are exhibited by the results in (5.2), (5.3), and (5.4). The W loop contribution, a_μ^W, is finite only because the lowest order gyromagnetic ratio of the W boson, g_W, equals 2 in Yang-Mills theories such as the standard model.[46] We later discuss the effect of $g_W \neq 2$ as a test of possible W boson substructure. For now, it suffices to say that a precision measurement of a_μ directly tests the $g_W = 2$ prediction of the standard model. The Z loop contribution, a_μ^Z, has opposite sign and partially cancels a_μ^W. That cancellation is intimately tied to the renormalizability and well behaved high energy properties of the standard model. Taking[47] $\sin^2\theta_W = 0.233 \pm 0.002$ and estimating an uncertainty of $\pm$ 10×10^{-11} from higher-order (two-loop) uncalculated electroweak corrections, one finds

$$a_\mu^W + a_\mu^Z = 195(10) \times 10^{-11}. \tag{5.7}$$

The rough uncertainty estimate was made as follows: Although a detailed study of the two-loop electroweak corrections to a_μ has not been undertaken, it has been recently pointed out[48] that $\gamma\gamma Z$ electron triangle anomaly diagrams induce a relatively large two-loop contribution

$$-\frac{3\alpha^2}{8\pi^2 \sin^2\theta_W}\left(\frac{m_\mu^2}{m_W^2}\right)\ln\left(\frac{m_Z}{m_\mu}\right) \simeq -10 \times 10^{-11}$$

to a_μ. That effect will be cancelled in part by analogous u and d quark triangle diagrams which cancel the short-distance anomaly. Summing over the three generations of leptons and quarks can bring the total contribution of the two-loop fermion triangle diagrams to $\sim -1 \times 10^{-10}$. There are in addition two-loop weak contributions of relative order $\alpha/\pi\sin^2\theta_W \simeq 0.01$ with respect to the one-loop effects in (5.2) and (5.3). They could potentially add up to $\mathcal{O}(1\times10^{-10})$. A more complete two-loop analysis is clearly warranted.

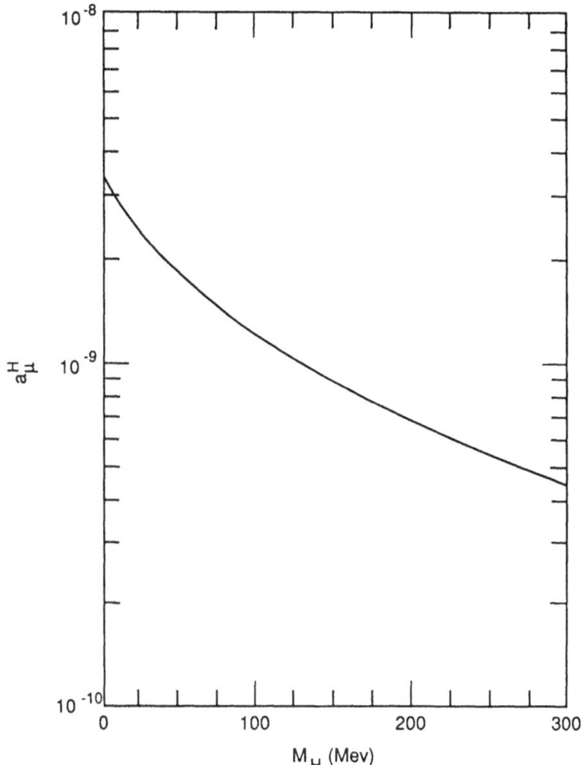

Figure 26: Higgs scalar contribution to a_μ as a function of the Higgs mass.

Until such an analysis is carried out, we employ the central value in (5.7) and assume a rough uncertainty of $\pm\ 10 \times 10^{-11}$ from higher orders. It is, of course, possible that the sum of all two-loop electroweak diagrams could exceed that uncertainty. However, it seems very unlikely that they would exceed the anticipated $\pm\ 40 \times 10^{-11}$ precision of the BNL experiment.

The physical Higgs contribution, a_μ^H, is dependent on the as yet unknown Higgs mass, m_H. For $m_H = 0$, $a_\mu^H = 350 \times 10^{-11}$, a very significant effect which exceeds the weak gauge boson contribution in (5.7). However, for $m_H \geq 3$ GeV, one finds $a_\mu^H < 2 \times 10^{-11}$, which is even less than the higher order uncertainty estimate in (5.7). As illustrated in Fig. 26, the effect of a_μ^H is particularly important for $m_H \leq$ few hundred MeV. Low energy experiments have ruled out most light Higgs mass possibilities in the standard model,[49] but there may be some very small gaps where a light Higgs ($\leq 2m_\mu$) could have escaped detection. In addition, recent searches for the decay $Z \to H\nu\bar{\nu}$ or $H\ell^+\ell^-$, $\ell = e$, μ, have ruled out m_H from about 35 MeV up to about 25 GeV.[50] We assume that a_μ^H is, therefore, likely to be completely negligible and take (5.7) to represent the best estimate for the weak correction

$$a_\mu^{\text{weak}} = 195(10) \times 10^{-11}. \tag{5.8}$$

If, however, a positive deviation in a_μ is experimentally observed, it could signal the effect of a light Higgs and all very low energy constraints on m_H would have to be rescrutinized. Of course, any deviation could also signal the presence of "new physics." In the next section, we survey various "new physics" possibilities and discuss their potential impact on a_μ.

6. "New Physics" Contributions

In this section we examine several "new physics" scenarios and discuss their effect on a_μ. As we shall see, precise a_μ measurements provide stringent constraints on some types of new interactions and particles such as those associated with compositeness, but are in other cases not competitive with existing experimental bounds. In that regard, a_μ complements the broad program of experiments that are testing the standard model.

As a way of gauging the sensitivity of a_μ to different types of "new physics", we will when necessary assume that the proposed BNL experiment reaches its goal of $\pm\, 40 \times 10^{-11}$ and the central value of a_μ coincides with the standard model's prediction (see (1.8))

$$a_\mu^{S.M.} = 116\ 591\ 918 \times 10^{-11}. \tag{6.1}$$

We also assume that uncertainties in the hadronic contributions will be significantly reduced. With these caveats, we will describe bounds on "new physics" mass scales and couplings at the 1σ level. In addition, we assume for simplicity that large cancellations between different types of "new physics" contributions are not occurring. That obviously need not be the case. In fact, for some cutoff dependent loop contributions we expect compensating effects in a more realistic full theory. Also, since our discussion is meant to be a survey, it is necessarily incomplete. For a very detailed recent account that emphasizes constraints on compositeness, we refer the reader to a study by Méry *et al.*[51]

6.1 Extra Gauge Bosons

The $SU(3)_C \times SU(2)_L \times U(1)$ symmetry of the standard model can be easily expanded to a larger gauge group with additional charged and neutral gauge bosons. Many specific examples have been considered in the literature. Here, we focus on two realistic possibilities. The first includes a charged gauge boson, W_R, which couples to right-handed charged currents in generic left-right symmetric models. The second allows for two extra diagonally coupled neutral gauge bosons which arise in superstring inspired E_6 grand unified theories (GUTS).[52] More exotic (colored, fractionally charged, off-diagonal, etc.) gauge bosons also generally occur in GUTS; however, other low energy phenomenology requires their masses to be large and hence their contribution to $g-2$ of the muon to be negligibly small.

A general analysis of one loop contributions to a_μ from extra gauge bosons has been carried out by Leveille.[53] Here we apply his results to the specific cases of interest.

W_R Boson: For the case of a W_R coupled to μ_R and ν_{μ_R} with gauge coupling g_R, one finds in analogy with (5.2) for $m_{\nu_R} << m_{W_R}$

$$a_\mu(W_R) = \frac{10}{3}\frac{G_F m_\mu^2}{8\sqrt{2}\pi^2}\frac{g_R^2}{g^2}\frac{m_W^2}{m_{W_R}^2}. \tag{6.2}$$

If on the other hand $m_{\nu_R} \simeq m_{W_R}$, (6.2) is reduced by a factor of $\frac{17}{20}$. Assuming $g_R \simeq g$, where g is the $SU(2)_L$ gauge coupling constant, (6.2) leads to the potential 1σ bound (i.e., $|\Delta a_\mu| < 40 \times 10^{-11}$)

$$m_W^2/m_{W_R}^2 < 0.10 \quad \text{or} \quad m_{W_R} > 250 \text{ GeV}. \tag{6.3}$$

That bound is not as stringent as existing bounds on m_{W_R} from direct searches for W_R at colliders or indirect low energy constraints which tend to be in the 300 ~ 450 GeV range.

Z' Bosons: Extra neutral gauge bosons (generically called Z' bosons) can be easily appended to the standard model via additional $U'(1)$ gauge symmetries. They arise naturally in grand unified theories and some superstring models. For example, the $SO(10)$ model has one such additional boson which is often denoted by Z_χ while E_6 contains Z_χ as well as a second flavor diagonal neutral boson Z_ψ. Their couplings to fermions are completely specified up to renormalization effects which are in principle calculable.[54]

The interaction Lagrangian for Z_χ and Z_ψ with muons is given by[55]

$$\mathcal{L}_{int} = -\sqrt{\frac{3}{8}}\, g \tan\theta_W \sum_{i=\chi,\psi} \sqrt{\lambda_i} Z_i^\alpha (Q_R^i \bar{\mu}_R \gamma_\alpha \mu_R + Q_L^i \bar{\mu}_L \gamma_\alpha \mu_L), \tag{6.4}$$

where $\sqrt{\lambda_i} \simeq 1$ is a renormalization parameter and

$$Q_L^\chi = 3Q_R^\chi = 1, \quad Q_L^\psi = -Q_R^\psi = \sqrt{\frac{5}{27}}. \tag{6.5}$$

Z_χ and Z_ψ will generally mix with one another such that the mass eigenstates are

$$\begin{aligned} Z(\beta) &= Z_\psi \sin\beta + Z_\chi \cos\beta, \\ Z'(\beta) &= Z_\psi \cos\beta - Z_\chi \sin\beta. \end{aligned} \tag{6.6}$$

(Mixing with the ordinary Z boson is required to be small by Z mass measurements.) The contribution to a_μ from either Z_i is given by

$$\Delta a_\mu = \frac{2G_F m_\mu^2}{8\sqrt{2}\pi^2}\lambda_i \sin^2\theta_W \frac{m_Z^2}{m_{Z_i}^2}(3Q_L^i Q_R^i - Q_L^i Q_L^i - Q_R^i Q_R^i). \tag{6.7}$$

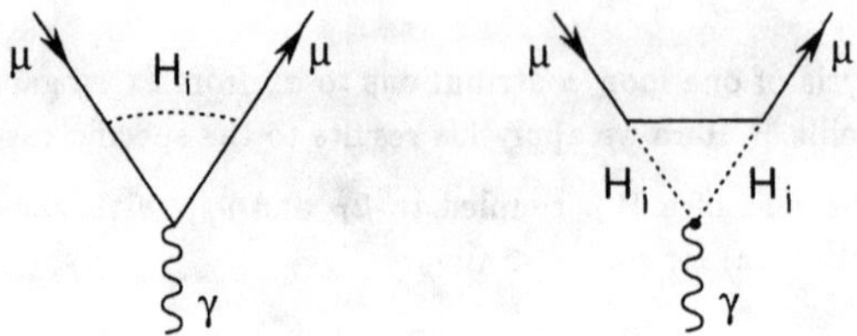

Figure 27: Generic Higgs scalar contribution to a_μ.

Allowing for mixing as in (6.6) then gives (setting $\lambda_i \simeq 1$)

$$\Delta a_\mu = \frac{2G_F m_\mu^2}{8\sqrt{2}\pi^2}\sin^2\theta_W\left(\frac{m_Z^2}{m_{Z(\beta)}^2}\left(-\frac{25}{27}\sin^2\beta - \frac{1}{9}\cos^2\beta - \frac{10}{3}\sqrt{\frac{5}{27}}\sin\beta\cos\beta\right)\right.$$
$$\left. + \frac{m_Z^2}{m_{Z'(\beta)}^2}\left(-\frac{25}{27}\cos^2\beta - \frac{1}{9}\sin^2\beta + \frac{10}{3}\sqrt{\frac{5}{27}}\sin\beta\cos\beta\right)\right). \tag{6.8}$$

In general, the potential bounds derived from the formula (6.8) for $m_{Z(\beta)}$ and $m_{Z'(\beta)}$ are $\mathcal{O}(100 \text{ GeV})$ or smaller and thus not competitive with existing low energy constraints which range from about 120 → 300 GeV.[54,56,57] Of course, for non-GUT Z' bosons the coupling may be much larger than those given above. In that case, larger mass constraints can be obtained.

6.2 Additional Higgs Scalars

Enlargement of the Higgs scalar sector of the standard model can have a potentially important effect on a_μ. That is possible because, unlike the situation for extra gauge bosons, Higgs scalar masses and couplings are not very constrained by phenomenology or symmetry. So, a_μ provides an important probe of expanded Higgs sectors, at least those parts which couple to the muon. Any such probe is extremely useful, since the Higgs mechanism and mass generation in general are the least understood aspects of the standard model and areas where "new physics" surprises could easily arise.

To discuss the effect of a generic Higgs, H_i, with charge Q_{H_i} (in units of $e > 0$), we again follow Leveille[53] and employ the interaction

$$\mathcal{L}_{int} = \overline{\mu}(C_s + C_p\gamma_5)FH_i + h.c. \ , \tag{6.9}$$

where F is an arbitrary fermion with mass m_F and charge $-1 - Q_{H_i}$. F may be a muon, neutrino, quark or even more exotic fermion. Any such interaction gives the following one loop contribution to a_μ (see Fig. 27)

$$
\begin{aligned}
a_\mu(H_i) = & -\frac{Q_{H_i} m_\mu^2}{8\pi^2}\int_0^1 dx \frac{[C_s^2\{x^3 - x^2 + \frac{m_F}{m_\mu}(x^2 - x)\} + C_p^2\{m_F \to -m_F\}]}{m_\mu^2(x^2 - x) + m_H^2 x + m_F^2(1 - x)} \\
& + \frac{(Q_{H_i} + 1)m_\mu^2}{8\pi^2}\int_0^1 dx \frac{[C_s^2\{x^2 - x^3 + \frac{m_F}{m_\mu}x^2\} + C_p^2\{m_F \to -m_F\}]}{m_\mu^2(x^2 - x) + m_H^2(1 - x) + m_F^2 x}.
\end{aligned} \tag{6.10}
$$

If there are several scalars (or color degrees of freedom), one must sum over all individual contributions. Note that, for $Q_{H_i} = 0$, $C_s = gm_\mu/2m_W$, $C_p = 0$ and $m_F = m_\mu$, (6.10) reduces to the standard model's Higgs contribution in (5.4).

There are many examples of specific models with additional Higgs scalars that one can consider. Moore, Whisnant and Young[58] have discussed in some detail the implications of a few popular scenarios. Here, we prefer to keep our discussion general and focus on what conditions are necessary to make $a_\mu(H_i)$ observable, i.e., $\mathcal{O}(40 \times 10^{-11})$, in the forthcoming BNL experiment.

From the form of (6.10), it is clear that $a_\mu(H_i)$ will be sizeable if m_{H_i} is relatively light or if m_F and/or C_s, C_p are large. None of these possibilities occurs for the standard model Higgs where $m_F = m_\mu$, $C_s = gm_\mu/2m_W \simeq 10^{-3}g$, $C_p = 0$ and $m_H \geq 25$ GeV. Enlarging the Higgs sector can, however, lead to larger couplings and, at least for neutral Higgs scalars, reopen the possibility of a very light mass. We consider various scenarios.

If there is a neutral very low mass scalar or pseudoscalar, which couples to the muon and is not ruled out by other phenomenology, we can carry over the discussion in Sec. 5. There it was shown (see Fig. 26) that a $m_H < 3m_\mu$ gives a significant shift in a_μ of $\mathcal{O}(40 \times 10^{-11})$ for C_s and C_p of $\mathcal{O}(10^{-3}g)$. So, the new a_μ measurement is a nice probe of very weakly coupled light neutral scalars or pseudoscalars $\lesssim \mathcal{O}(300$ MeV). Such light particles are still phenomenologically viable in multi-Higgs doublet extensions of the standard model.

Another possibility is that m_H is not so small but C_s or C_p is much larger than $\mathcal{O}(10^{-3}g)$, perhaps even $\mathcal{O}(g)$. Such a large chiral changing coupling seems somewhat unnatural for a fermion as light as the muon, but there is no fundamental argument that rules it out. In that case, for $m_F \lesssim m_\mu$, one finds that a $\pm 40 \times 10^{-11}$ measurement of a_μ probes m_H roughly up to 500 GeV. If $m_F \simeq m_H$, the mass sensitivity goes way up to $\mathcal{O}(10^5$ GeV); but it is difficult to imagine such a large coupling between fermions when their masses are so disparate. Instead, one can envision a coupling of $\mathcal{O}(gm_\mu/m_F)$ as being more plausible. In that case one is again sensitive to $m_F \simeq m_H \lesssim 500$ GeV.

Although the above arguments were rough, the conclusion is fairly definite. A new measurement of a_μ with sensitivity $\pm 40 \times 10^{-11}$ is an extremely good probe of light ($\lesssim \mathcal{O}(300$ MeV)) neutral scalars or pseudoscalars even if their coupling is very weak, $\mathcal{O}(10^{-3}g)$. For stronger coupling $\mathcal{O}(g)$ or very weak coupling to massive fermions $m_F \simeq m_H$, a_μ measurements probe Higgs mass scales up to $\mathcal{O}(500$ GeV).

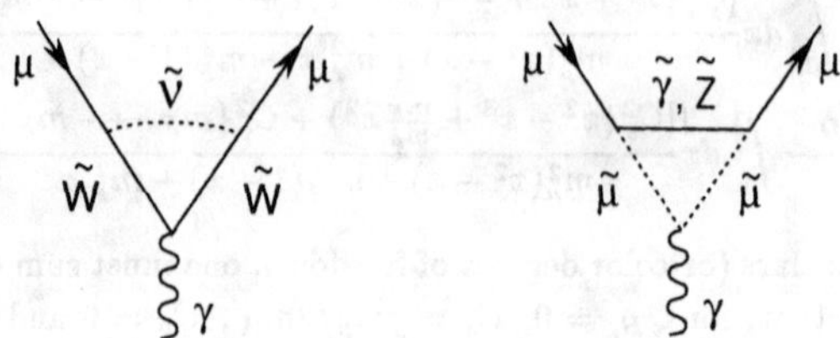

Figure 28: Supersymmetric loop contributions to a_μ.

6.3 Supersymmetry

The basic idea of an underlying supersymmetry which connects fermions and bosons is very attractive. However, so far, that hypothesis is totally lacking in experimental support. It is, therefore, very important to search for indications of supersymmetry wherever possible. In that regard, an improved measurement of the muon's anomalous magnetic moment presents an opportunity to indirectly uncover evidence for supersymmetry or further constrain the parameters of specific supersymmetric models.

The supersymmetric partners of known particles (sparticles) contribute to a_μ via the loops in Fig. 28.[59] The magnitude of their effect depends on the masses and couplings of the internal sparticles (winos, zinos, photinos, smuons and sneutrinos). Those parameters are very model dependent; but they have a few generic features. Generally, the sign of the supersymmetric appendage, $a_\mu(SUSY)$ is negative. Sparticle loops usually tend to cancel ordinary particle loops. In fact, some simple SUSY models can have $a_\mu = 0$ in the limit of exact supersymmetry.[60] The expected magnitude of the correction is of the form

$$|a_\mu(SUSY)| \sim \mathcal{O}\left(\frac{G_F m_\mu^2}{8\pi^2\sqrt{2}}\left(\frac{m_W^2}{m_{SUSY}^2}\right)\right), \tag{6.11}$$

where m_{SUSY} represents some sparticle mass scale. Present constraints from a_μ are around $m_{SUSY} \simeq 30 \sim 40$ GeV. The new BNL experiment is capable of pushing those constraints to about $120 \sim 130$ GeV, an interesting region that competes with the supersymmetry discovery potential of the Fermilab TEVATRON. Of course, the above discussion is meant to roughly estimate the SUSY reach of a new a_μ measurement. More generically, it appears that a signature for SUSY particles in the 100-200 GeV region would be a determination of a_μ below the standard model prediction. That hint of SUSY would then be sorted out by detailed studies at high energy colliders capable of producing sparticles.

6.4 Compositeness

The known fermions and gauge bosons of the standard model are generally assumed to be point-like elementary particles with no underlying substructure. That prevailing view may eventually turn out to be wrong. Indeed, there are numerous precedents in which once perceived "fundamental" objects turned out to be bound states or composites of more elementary constituents. The displacement of protons and neutrons by quarks as the building blocks of matter is a nice example of such a development.

Precision measurements of magnetic moments provide particularly powerful probes of compositeness. For example, the proton's anomalous magnetic moment $a_p \simeq 1.8$ deviates strongly from the QED value of $\alpha/2\pi$, a clear indication of substructure. Although the quark model was not discovered in that way, it was realized early on that the large nucleon magnetic moments could be understood on the basis of the constituent quark model. In an analogous fashion, the muon anomalous magnetic moment allows a direct probe of muon substructure. Furthermore, it is also sensitive to neutrino or gauge boson compositeness via loop effects. We discuss bounds on those possibilities in this subsection.

Muon Substructure: If the muon is a composite object, the underlying physics must be quite different than the bound state dynamics of hadrons. Its compton wavelength = $m_\mu^{-1} \sim 10^{-13}$ cm is much larger than bounds on its possible structure which as we shall see are less than 10^{-17} cm. Equivalently, the composite mass scale, Λ, must be much greater than m_μ. It is, therefore, a challenge for advocates of compositeness to explain why the muon is so light in comparision with Λ. The general approach is to invoke a chiral (left-right) symmetry for the underlying bound state dynamics which keeps the muon light.[61] That chiral symmetry will prevent the induced magnetic moment from being of the form $1/\Lambda$ that would normally be expected and instead lead to

$$a_\mu(\Lambda) \sim \frac{m_\mu^2}{\Lambda^2}. \tag{6.12}$$

In (6.12) we have followed the usual convention of leaving out couplings or other suppression factors; so, Λ is only meant to be a very rough estimate of the composite scale.

Existing a_μ measurements require $|a_\mu(\Lambda)| \leq 10^{-8}$ which gives via (6.12) the bound

$$\Lambda \geq 1 \text{ TeV}. \tag{6.13}$$

That bound is comparable to or in most cases better than bounds on substructure coming from other electroweak phenomena. In those other cases, the typical scale of the interaction is $m_W \simeq 80$ GeV. To see interference between ordinary weak interaction and new composite-induced four Fermi amplitudes of $\mathcal{O}(1/\Lambda^2)$ requires a sensitivity of $\mathcal{O}(m_W^2/\Lambda^2)$. So, to probe $\Lambda \sim 1$ TeV in those experiments requires 1% precision. That is somewhat better than the present sensitivity of, for example, $\nu_\mu N$ scattering experiments by about a factor of 2 or 3.

The new BNL experiment would improve the measurement of a_μ by a factor of 20 and thus probe (via (6.12))

$$\Lambda \geq \mathcal{O}(4 \sim 5 \text{ TeV}) \qquad \text{(Future Sensitivity)}. \tag{6.14}$$

That would nicely complement the SSC effort which has direct discovery potential in the multi-TeV region for many types of "new" physics including compositeness.

<u>Excited Leptons</u>: If the muon is composite, one expects excited muons, μ^*, and very likely excited neutrinos, ν_μ^*, to exist. Strong constraints on $\mu \to e\gamma$, $\tau \to e\gamma$ and $\tau \to \mu\gamma$ as well as $\nu' \to \nu\gamma$ tend to eliminate any of the known leptons as reasonable candidates for excited states. Allowing for new excited muons with mass m_{μ^*}, one can parameterize the $\mu^*\mu\gamma$ amplitude by[51]

$$\frac{e}{2m_{\mu^*}}\bar{\mu}^*\sigma^{\alpha\beta}(c - d\gamma_5)\mu F_{\alpha\beta} + h.c. \,, \tag{6.15}$$

where $F_{\alpha\beta}$ is the electromagnetic field. (Effects of similar couplings for $W^\pm$ and Z bosons and their effect on a_μ have also been studied.[51])

Using that interaction in (6.15) gives at one loop level

$$a_\mu(m_{\mu^*}) \simeq 3\frac{\alpha}{\pi}(c^2 - d^2)\frac{m_\mu}{m_{\mu^*}} \,, \tag{6.16}$$

where the loop integration has been cutoff at m_{μ^*}. For chiral couplings we have $c^2 = d^2$ and the leading term given in (6.16) vanishes. A natural possibility in that case is small chirality violation (or non-leading effects) giving a suppression factor $c^2 - d^2 \simeq \mathcal{O}(m_\mu/m_{\mu^*})$. In that case, one finds (roughly) $m_{\mu*} \geq 100$ GeV from present constraints on a_μ and a sensitivity to $m_{\mu*} \simeq \mathcal{O}(400 \text{ GeV})$ in the new BNL experiment. A much more severe constraint arises if either c or d is $\mathcal{O}(1)$ with little cancellation. In that case, one finds $m_{\mu*} \geq 10^5$ GeV from existing a_μ measurements alone. Of course, this analysis is meant only to be suggestive of the range of excited lepton mass scales probed. An interesting comparison of the sensitivity to $m_{\mu*}$ obtainable from the new BNL measurement of a_μ and future experiments[62] at LEPII (e^+e^- annihilation up to 200 GeV) has been given by Méry *et al.*[51] They conclude that for a given set of input parameters, both types of experiments have comparable sensitivity to $m_{\mu*}$. So, for excited muon searches, a_μ is competitive with very high precision measurements at the next generation of e^+e^- colliders.

<u>Gauge Boson Compositeness</u>: One of the most important tests of the standard model involves measuring the three-gauge-boson coupling. LEPII will allow the direct study of those couplings via $e^+e^- \to W^+W^-$ through γ and Z mediated amplitudes.[62] The $WW\gamma$ couplings can also be tested indirectly through the a_μ contribution in Fig. 25. Generalizing the standard $WW\gamma$ coupling to allow an arbitrary W boson magnetic dipole moment

$$\mu_W = \frac{e}{2m_W}(1 + \kappa + \lambda) \tag{6.17}$$

and electric quadrupole moment

$$Q = -\frac{e}{m_W^2}(\kappa - \lambda), \tag{6.18}$$

where $\kappa = 1$ (or equivalently $g_W = 2$) and $\lambda = 0$ in the standard model or any gauge theory (renormalizability requires that condition), one finds[51,63]

$$a_\mu(\kappa, \lambda) \simeq \frac{G_F m_\mu^2}{4\sqrt{2}\pi^2}\left[(\kappa - 1)\ \ln\frac{\Lambda^2}{m_W^2} - \frac{1}{3}\lambda\right] , \tag{6.19}$$

where a cutoff Λ has been applied to the loop integrations and non-leading terms have been discarded. Assuming $\Lambda \sim \mathcal{O}(1 \text{ TeV})$, the present a_μ value implies

$$\begin{aligned} |g_W - 2| = |\kappa - 1| \leq 0.8, \\ |\lambda| \leq 12. \end{aligned} \tag{6.20}$$

The improvement by a factor of 20 in the measurement of a_μ would improve the bound on $|\kappa - 1|$ to the 0.04 level which is very impressive. Estimates[62] for LEPII experiments indicate it is sensitive to deviations in κ from 1 at the level of 0.4, a factor of 10 worse. (SSC should also constrain $\Delta\kappa$ to roughly $\pm$ 0.4.) So, it seems that determining the anomalous magnetic moment of the W boson is one of the most important benefits of a new a_μ measurement.

To appreciate the implications of measuring $|\kappa - 1|$ at the level of 0.04, we consider what might occur in a composite model for the W boson. In that case, one expects a deviation of $|\kappa - 1|$ from zero by m_W/Λ or m_W^2/Λ^2, with Λ the scale of W boson compositeness. A sensitivity of ± 0.04 then probes $\Lambda \simeq 2$ TeV and 400 GeV for the two cases. W boson compositeness is, therefore, studied at nearly the same level as muon compositeness in a_μ experiments.

6.5 Electric dipole moments

The best bound on the muon's electric dipole moment (e.d.m.) presently comes from a_μ measurements.[1,64] If the muon had an e.d.m., it would have affected spin precession in the CERN experiments and given rise to a tilt in the precession plane. That would have changed the precession frequency and "effectively" increased the measured anomalous magnetic moment by

$$\Delta a_\mu(e.d.m.) = \frac{m_\mu^2}{\alpha^2}|d_\mu|^2, \tag{6.21}$$

where d_μ is the muon e.d.m. Comparison of (6.21) with experiment gives the bound (at 95 per cent CL)[64]

$$|d_\mu| < 7.3 \times 10^{-19} \text{ e-cm}. \tag{6.22}$$

(A similar bound can be directly obtained from measurements of the precession plane.) The new BNL experiment is expected to improve the bound in (6.22) by about an order of magnitude.

The bound on d_μ is in most models not as constraining as the neutron e.d.m. bound $|d_n| \leq 10^{-25}$ e-cm or the electron e.d.m. bound $|d_e| < 1.3 \times 10^{-25}$ e-cm.[g] There can, however, be models where d_n is small and d_ℓ scales as m_ℓ^3 such that d_μ would be about $10^7 d_e$. In that case, an improvement of about an order of magnitude in (6.22) would become a useful constraint.

If the W boson has an e.d.m., $d_W = e\lambda_W/2m_W$, it will induce a muon e.d.m. (via Fig. 25a)[66]

$$d_\mu = -\frac{eG_F m_\mu \lambda_W}{8\sqrt{2}\pi^2} \ln\left(\frac{\Lambda^2}{m_W^2}\right),$$

[g] Ninety five per cent confidence limit calculated from the result of Ref. 65.

where Λ is the loop momentum cutoff. Bounds on $\lambda_W \ln(\Lambda^2/m_W^2)$ and d_W from the neutron e.d.m. then imply $d_\mu < 10^{-25}$ e-cm, which is unobservably small.

6.6 Concluding remarks

We have seen that the standard model makes a rather precise prediction for the muon's anomalous magnetic moment. The largest uncertainty presently comes from hadronic vacuum polarization effects. Ongoing and future measurements of $e^+e^- \to$ hadrons along with renewed theoretical scrutiny should significantly reduce that uncertainty and further refine the prediction. Similar loop effect uncertainties start to affect the interpretation of $W^\pm$ and Z masses as precision measurements approach $\pm$ 20 MeV. Those uncertainties can also be reduced by better measurements of $e^+e^- \to$ hadrons at low energies, and thus provide additional motivation for a new round of accurate experiments.

Theory and experiment currently agree on a_μ. That in itself represents a tremendous success for the QED portion of the radiative corrections in a_μ. A factor of 20 improvement in the measured a_μ anticipated in a new BNL experiment will test the one loop weak corrections of the standard model at about the 5σ level. Confirmation at the level of its electroweak radiative corrections would mark another milestone in the string of tests it has successfully passed.

Table VIII. Estimated upper mass scale sensitivity of some examples of "new physics" made possible by a new measurement of a_μ to the precision of $\pm$ 40×10^{-11}.

"New physics"	Approx. future a_μ sensitivity
W_R	$\sim$ 250 GeV
Z' (E_6 couplings)	$\sim$ 100 GeV
Additional Higgs – standard model couplings	$\sim$ 300 MeV
– $\mathcal{O}$(g) couplings	$\sim$ 500 GeV
Supersymmetry	$\sim$ 130 GeV
Muon substructure	$\sim$ 4–5 TeV
Excited muons – chiral couplings	$\sim$ 400 GeV
– non-chiral couplings	$\sim 1 \times 10^5$ GeV
$W^\pm$ substructure	$\sim$ 2 TeV
W anomalous magnetic moment	$\Delta\left(\frac{g_W-2}{2}\right) \sim 0.02$

In addition, such precise measurements will be sensitive to all sorts of "new physics." Those include (see Table VIII) new gauge bosons, scalars, supersymmetry, etc. In several cases, the sensitivity of a_μ to those effects is competitive or better than other planned experiments. Overall, however, its most impressive capability seems to be as a probe of compositeness. The new measurement will constrain (or find a hint of) compositeness up to scales in the multi-TeV region. At that level it is competitive with the SSC as well as high precision measurements at e^+e^- colliders. The anomalous magnetic moment of the W boson will be studied via a_μ at the level of $\pm\ 0.04e/2m_W$, an impressive unique sensitivity. That tests the standard model's electromagnetic coupling of the W boson and significantly constrains its possible composite structure.

The existence of high precision anomalous magnetic moment measurements for the muon and electron have traditionally provided guidance and constraints for model builders and hurdles any new "potential" preliminary discoveries must overcome. Further improvements in those measurements will allow them to continue to play an active forefront role in elementary particle physics. Together with other high precision and high energy measurements they will contribute to progress in unraveling and deciphering the fundamental laws of nature.

Acknowledgments

This work is supported in part by the U. S. National Science Foundation under contract number PHY-8715272 and the U. S. Department of Energy under contract number DE-AC02-76CH00016. One of us (T. K.) thanks Y. Okamoto and B. Nizic for their contribution to the contents of Sections 2, 3, and 4.

References

1. J. Bailey *et al.*, Phys. Lett. **68B**, 191 (1977); F. J. M. Farley and E. Picasso, this volume.

2. T. Kinoshita, B. Nizic, and Y. Okamoto, Phys. Rev. Lett. **52**, 717 (1984).

3. T. Kinoshita, B. Nizic, and Y. Okamoto, Phys. Rev. D **31**, 2108 (1985).

4. R. Jackiw and S. Weinberg, Phys. Rev. D **5**, 2396 (1972); I. Bars and M. Yoshimura, Phys. Rev. D **6**, 374 (1972); K. Fujikawa, B. W. Lee, and A. I. Sanda, Phys. Rev. D **6**, 2923 (1972); W. A. Bardeen, R. Gastmans, and B. E. Lautrup, Nucl. Phys. B**46**, 319 (1972).

5. A. Sirlin, Phys. Rev. D **22**, 971 (1980); W. J. Marciano and A. Sirlin, Phys. Rev. D **22**, 2695 (1980).

6. T. Kinoshita, B. Nizic, and Y. Okamoto, Phys. Rev. D **41**, 593 (1990).

7. J. A. Casas, C. Lopez, and F. J. Yndurain, Phys. Rev. D **32**, 736 (1985).

8. L. M. Kurdadze *et al.*, Yad. Fiz. **40**, 451 (1984) [Sov. J. Nucl. Phys. **40**, 286 (1984)].

9. M. E. Cage *et al.*, IEEE Trans. Instrum. Meas., IM-**38**, 284 (1989).

10. T. Kinoshita and W. B. Lindquist, Phys. Rev. D. For a preliminary report, see T. Kinoshita, IEEE Trans. Instrum. Meas., IM-**38**, 172 (1989).

11. C. Bouchiat and L. Michel, J. Phys. Radium **22**, 121 (1961).

12. L. Durand, III., Phys. Rev. **128**, 441 (1962).

13. V. W. Hughes and T. Kinoshita, Comments Nucl. Part. Phys. **14**, 341 (1985); M. Greco, Frascati Report No. LNF-88/24(P), 1988 (unpublished).

14. Particle Data Group, Phys. Lett. **B204** (April 1988).

15. H. Suura and E. Wichmann, Phys. Rev. **105**, 1930 (1957); A. Peterman, Phys. Rev. **105**, 1931 (1957); H. H. Elend, Phys. Lett. **20**, 682 (1966); **21**, 720 (1966); G. W. Erickson and H. H. T. Liu, UCD-CNL-81 report (1968).

16. B. E. Lautrup and E. de Rafael, Phys. Rev. **174**, 1835 (1968).

17. T. Kinoshita, Nuovo Cimento **51B**, 140 (1967).

18. B. E. Lautrup and E. de Rafael, Nuovo Cimento **64A**, 322 (1969); B. E. Lautrup, Phys. Lett. **32B**, 627 (1970).

19. S. J. Brodsky and T. Kinoshita, Phys. Rev. D **3**, 356 (1971).

20. R. Barbieri and E. Remiddi, Phys. Lett. **49B**, 468 (1974).

21. G. P. Lepage, J. Comput. Phys. **27**, 192 (1978).

22. J. Aldins, S. Brodsky, A. Dufner, and T. Kinoshita, Phys. Rev. D **1**, 2378 (1970).

23. B. E. Lautrup and M. A. Samuel, Phys. Lett. **72B**, 114 (1977).

24. T. Kinoshita, Phys. Rev. Lett. **61**, 2898 (1988).

25. G. Källén and A. Sabry, K. Dan. Vidensk. Selsk. Mat.-Fys. Medd. **29**, No. 17 (1955).

26. T. Kinoshita and W. B. Lindquist, Phys. Rev. D **27**, 867 (1983).

27. B. E. Lautrup, in "Proceedings of the Second Colloquium in Advanced Computing Methods in Theoretical Physics, Marseille, 1971", ed. A. Visconti (University of Marseille, Marseille, 1971).

28. T. Kinoshita and W. B. Lindquist, Phys. Rev. D **27**, 877 (1983).

29. T. Kinoshita and W. B. Lindquist, Phys. Rev. D **27**, 886 (1983).

30. T. Kinoshita and W. B. Lindquist, Phys. Rev. D **39**, 2407 (1989).

31. B. E. Lautrup and E. de Rafael, Nucl. Phys. **B70**, 317 (1974).

32. E. de Rafael and J. L. Rosner, Ann. Phys. (N. Y.) **82**, 369 (1974).

33. R. Barbieri and E. Remiddi, Nucl. Phys. **B90**, 997 (1975).

34. S. L. Adler, Phys. Rev. D **5**, 3021 (1972).

35. J. Calmet, S. Narison, M. Perrottet, and E. de Rafael, Phys. Lett. **16B**, 283 (12976); Rev. Mod. Phys. **49**, 21 (1977).

36. A. Quenzer *et al.*, Phys. Lett. **76B**, 512 (9178).

37. G. Gounaris and J. Sakurai, Phys. Rev. Lett. **21**, 244 (1968).

38. M. Gourdin, L. Stodolsky, and F. M. Renard, Phys. Lett. **30B**, 347 (1969).

39. S. R. Amendolia *et al.*, Phys. Lett. **138B**, 454 (1984).

40. L. M. Kurdadze *et al.*, Novosibirsk Report No. IYF-82-97, 1982(unpublished).

41. G. V. Anikin *et al.*, in "Proceedings of the 1983 International Symposium on Lepton and Photon Interactions at High Energies, Ithaca, New York", ed. by D. G. Cassel and D. L. Kreinick (Newman Laboratory of Nuclear Studies, Cornell University, Ithaca, 1984).

42. I. B. Vasserman *et al.*, Yad. Fiz. **30**, 999 (1979) [Sov. J. Nucl. Phys. **30**, 519 (179)]; **33**, 709 (1981) [**33**, 368 (1981)].

43. Particle Data Group, Phys. Lett. **111B**,1 (1982).

44. T. H. Bauer, R. D. Spital, D. R. Yennie, and F. M. Pipkin, Rev. Mod. Phys. **50**, 261 (1978).

45. G. Mennessier, Z. Phys. C **16**, 241 (1983).

46. T. Burnett and M. J. Levine, Phys. Lett. **24B**, 467 (1967); S. Brodsky and J. D. Sullivan, Phys. Rev. **156**, 1644 (1967).

47. S. Fanchiotti and A. Sirlin, Phys. Rev. D **41**, 319 (1990).

48. E. A. Kuraev, T. V. Kukhto, and A. Schiller, Novosibirsk preprint 89-153 (1989).

49. J. Gunion, H. Haber, G. Kane, and S. Dawson, "Higgs Hunter's Guide", (Addison-Wesley, New York, 1990).

50. ALEPH Collaboration, D. Decamp *et al.*, CERN preprint EP/89-157 (1989).

51. P.Méry, S.Moubarik, M.Perrottet, and F.Renard, Marseille preprint CPT-89/P.2226.

52. J. Rosner, Comm. Nucl. Part. Phys. **14**, 229 (1985).

53. J. Leveille, Nucl. Phys. B **137**, 63 (1978).

54. L. S. Durkin and P. Langacker, Phys. Lett. **166B**, 436 (1986).

55. W. Marciano, Nucl. Phys. **B11**, 5 (1989)(Proc. Suppl.).

56. W. Marciano and A. Sirlin, Phys. Rev. D **35**, 1672 (1987).

57. U. Amaldi *et al.*, Phys. Rev. D **36**, 1385 (1987).

58. S. Moore, K. Whisnant, and B.-L. Young, Phys. Rev. D **31**, 105 (1985).

59. P. Fayet, in "Unification of Fundamental Particle Interactions", eds. S. Ferrara, J. Ellis, and P. van Nieuwenhuizen (Plenum, N.Y., 1980) p. 587; J. Grifols and A. Mendez, Phys. Rev. D **26**, 1809 (1982); D. Kosower, L. Krauss, and N. Sakai, Phys. Lett. **133B**, 305 (1983); T.-C. Yuan, R. Arnowitt, A. H. Chamseddine, and P. Nath, Zeit. f. Phys. C **26**, 407 (1984).

60. S. Ferrara and E. Remiddi, Phys. Lett. **53B**, 347 (1974).

61. S. Brodsky and S. Drell, Phys. Rev. D **22**, 2236 (1980); G. Shaw, D. Silverman, and R. Slansky, Phys. Lett. **94B**, 57 (1980).

62. P. Méry, M. Perrottet, and F. Renard, Z. Phys. C **36**, 249 (1987); **38**, 579 (1988).

63. F. Herzog, Phys. Lett. **148B**, 355 (1984); M. Suzuki, Phys. Lett. **153B**, 289 (1983); A. Grau and J. Grifols, Phys. Lett. **154B**, 283 (1985).

64. J. Bailey *et al.*, Nucl. Phys. **B150**, 1 (1979).

65. S. A. Murphy *et al.*, Phys. Rev. Lett. **63**, 965 (1989). See also S. M. Barr and W. J. Marciano, in "CP Violation", ed. by C. Jarlskog (World Scientific, Singapore, 1989), p. 455.

66. W. Marciano and A. Queijeiro, Phys. Rev. D **33**, 3449 (1986).

THE MUON g − 2 EXPERIMENTS

F.J.M. Farley
Yale University, New Haven, Conn., USA

and

E. Picasso
CERN, Geneva, Switzerland

CONTENTS

1. INTRODUCTION

It is now thirty years since a group of experimental physicists at CERN, under Leon Lederman, started to study the problem of the muon g-factor. The g-factor relates the magnetic dipole moment to the intrinsic angular momentum of a charged system. Classically, the dipole moments can arise from either charges or currents. For example, the circulating current, which is due to an orbiting particle with an electric charge e and mass m, has associated with it a magnetic dipole moment $\vec{\mu}_L$ given by

$$\vec{\mu}_L = \frac{e}{2mc} \vec{L} , \tag{1.1}$$

where $\vec{L}$ is the orbital angular momentum. Alternatively, the electric dipole moment possessed by certain polar molecules is due to the relative displacement of the centres of the positive and negative charge distributions. Thus we have examples of a magnetic dipole moment and an electric dipole moment both having their origins in electric charge, and it is interesting to note that all electromagnetic phenomena are explained in terms of electric charges and their currents; there is no place, as yet, for magnetic charges. In particular, the intrinsic magnetic dipole moments of all particles can be considered, in the classical picture, to be made up of circulating electric currents and not of distributed magnetic charges [1]. This is just one aspect of the basic asymmetry between electricity and magnetism, which is apparent in Maxwell's equations. The argument, first proposed by Dirac [2], that the existence of magnetic charge would lead naturally to the quantization of both magnetic and electric charge, still stands as a challenge to physicists, both theoretical and experimental, to find a proper place for the magnetic monopole in the electromagnetic theory and to establish its physical reality.

The Dirac equation permits any value of g for the electron or muon, through the possible presence of the Pauli term, but the simplest version without Pauli term implies $g = 2$. To this must be added the corrections due to quantum electrodynamics which it is the purpose of the experiment to measure.

For a particle with both magnetic and electric dipole moments, the electromagnetic interaction Hamiltonian contains a part

$$H = -\vec{\mu}_m \cdot \vec{B} - \vec{\mu}_e \cdot \vec{E} , \tag{1.2}$$

where $\vec{B}$ and $\vec{E}$ are the magnetic and electric field strengths, and $\vec{\mu}_m$ and $\vec{\mu}_e$ are the magnetic and electric dipole moment operators. Following the general form of Eq. (1.1), we write

$$\vec{\mu}_m = g \frac{e}{2mc} \frac{1}{2} \left(\frac{h}{2\pi}\right) \vec{\sigma} , \tag{1.3}$$

$$\vec{\mu}_e = \eta \frac{e}{2mc} \frac{1}{2} \left(\frac{h}{2\pi}\right) \vec{\sigma} , \tag{1.4}$$

where the components of $\vec{\sigma}$ are the three Pauli spin matrices, and for the negative lepton we have to insert the charge $e = -|e|$. Making use of the Bohr magneton $\mu_0 = eh/4\pi mc$, these equations can be simplified to

$$\vec{\mu}_m = g\mu_0 \frac{\vec{\sigma}}{2} \tag{1.5}$$

and

$$\vec{d} \equiv \vec{\mu}_e = \eta\mu_0 \frac{\vec{\sigma}}{2} , \tag{1.6}$$

where we have taken the opportunity to introduce the conventional symbol $\vec{d}$ for the electric dipole moment (EDM).

The expectation value of the electric dipole moment $\vec{d}$ must be zero for a particle described by a state of well-defined parity. The polar molecules that we have referred to above are in a mixture of degenerate states with opposite parities, and so are not covered by this symmetry condition.

Arguments involving the time-reversal operation also require that the electric dipole moment should vanish as a consequence of the different symmetry properties of the magnetic and electric fields. Whilst $\vec{B}$ is an axial vector, $\vec{E}$ is a polar vector. Thus if the Hamiltonian equation (1.2) is to remain invariant with respect to parity inversion P and time reversal T, then $\vec{\mu}_m$ must transform like an axial vector and $\vec{\mu}_e$ must transform like a polar vector; an axial vector changes sign under T but not under P, whilst for a polar vector the opposite is true. From Eqs. (1.5) and (1.6) we see that the dipole moment operators should transform like the spin operator $\vec{\sigma}$. Since this latter behaves like an axial vector, all is consistent for $\vec{\mu}_m$; but in the case of $\vec{\mu}_e$, either of the operations P or T changes the relative sign of the two sides of the equation.

In order to have a satisfactory situation, η must be zero. These arguments can be generalized to show that for a system of definite parity, the odd electric (dipole, sextupole, etc.) and even magnetic (quadrupole, octupole, etc.) moments must be zero. Since the discovery that nature does not respect parity invariance, however, the invariance of interactions with respect to symmetry operations must always be underpinned by experiments.

2. SURVEY OF THE THEORY

If the muon obeys the simple Dirac equation for a particle of its mass (206 times heavier than an electron), then $g = 2$ exactly; but this is modified by the quantum fluctuations in the electromagnetic field around the muon, as specified by the rules of quantum electrodynamics (QED), making g larger by about 1 part in 800. The quantum fluctuations require further correction for the very rare fluctuations, which include virtual pion states and strongly interacting vector mesons.

At present, theory and experiment agree at the level of 1 part in 10^8, and the muon g-factor, together with that of the electron and positron, R_∞, c, and the frequency of the hydrogen maser, are the most accurately known constants in nature.

The muon $(g-2)$ value has played a central role in establishing that the muon obeys QED and behaves like a heavy electron. The experimental value of $(g-2)$ has been determined by three progressively more precise measurements at CERN, the latest one achieving a precision of 7.3 ppm (parts per million) in the anomaly $[a_\mu \equiv (g-2)/2]$. The theoretical value for $(g-2)$ has improved steadily as higher-order QED radiative contributions have been evaluated, and as knowledge of the virtual hadronic contribution to $(g-2)$ has been improved both by further measurements of the experimental ratio $R(s) = \sigma(e^+e^- \rightarrow \text{hadrons})/\sigma(e^+e^- \rightarrow \mu^+\mu^-)$ and by calculations.

The gyromagnetic ratio is increased from its primitive value of 2, arising from the Dirac equation, to $g = 2(1 + a_\mu)$, where a_μ is the anomalous magnetic moment or anomaly. In QED theory, the lepton g-factor may be expressed as a power series in α/π:

$$a_\mu^{th} = A(\alpha/\pi) + B(\alpha/\pi)^2 + C(\alpha/\pi)^3 + \dots . \tag{2.1}$$

Typical Feynman diagrams, which contribute to the calculation of the theoretical value of a_μ for the electron and the muon, are shown in Fig. 1, whilst a complete set of diagrams up to order $(\alpha/\pi)^4$ is given by Kinoshita et al. [3].

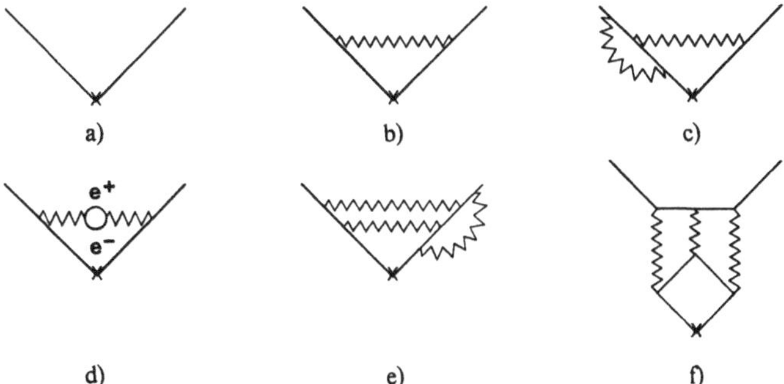

Fig. 1 Feynman diagrams used in calculating a_μ. The solid line represents the muon, which interacts with the laboratory magnetic field at X. The zigzag line represents a virtual photon, which is emitted and later reabsorbed. In (d) and (f) an e^+e^- pair, created and then annihilated, gives rise to the closed loop (solid line).

The calculations [3] of the coefficients in Eq. (2.1) have given the following results

$$
\begin{aligned}
A &= 0.5 \\
B &= 0.765\,858\,6 \\
C &= 24.067\,(3) \\
D &= 124.82\,(65) \\
E &= 570\,(140)\,.
\end{aligned}
\tag{2.2}
$$

Note that the value of B includes a small contribution from τ-loops. Using [4]

$$\alpha^{-1} = 137.035\,989\,5\,(61) \qquad (0.045 \text{ ppm})$$

the result is

$$a_\mu^{QED} = 116\,584\,730\,(5)\,(5) \times 10^{-11}\,, \tag{2.3}$$

where the first error comes from the uncertainty in the coefficients C, D, E and the second reflects the error in α. Combining in quadrature the overall uncertainty in the QED value is 0.06 ppm.

So far, only the change in the gyromagnetic ratio due to the interaction of a particle with its own electromagnetic field has been mentioned. Any other field coupled to the particle should produce a similar effect, and calculations have been made for scalar, pseudoscalar, vector, and axial-vector fields, using a coupling constant $\varkappa$ (assumed to be small) to a boson of mass M [5]. For a vector field, for example,

$$\Delta a_\mu^V = \left(\frac{1}{3\pi}\right)\left(\frac{\varkappa^2}{M^2}\right) m_\mu^2\,. \tag{2.4}$$

A precise measurement of a_μ could therefore reveal the presence of a new field; but first, all known fields, including the weak and the strong interactions, must be taken into account. Strongly interacting particles do not couple directly to the muon, but if they are charged, they couple to the photon. Thus they can appear in the inner loops, as Fig. 1d, for example, with a pion pair replacing the e^+e^- pair. Because of the high mass of the pion, one would initially expect such amplitudes to be small, but there are strong resonances in the $\pi^+\pi^-$ system that enhance the effect. Only a vector resonance can contribute, because it alone can transform directly into the virtual photon which must have $J^{PC} = 1^{--}$ (one unit of angular momentum, negative parity, and negative charge conjugation).

To calculate this contribution it is necessary to specify the overall probability amplitude for a photon of a given q^2 to connect the two muon vertices shown in Fig. 1b, with the effect of virtual hadron loops fully included; that is, one requires the propagator function of Fig. 2a.

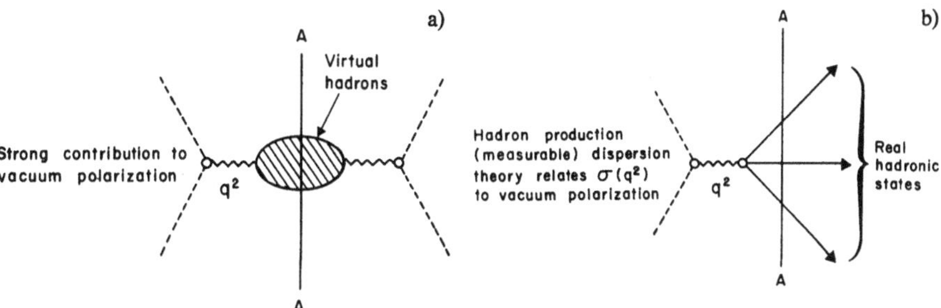

Fig. 2 The photon propagator is modified by the creation of virtual hadrons (a). This is related by the dispersion theory to real hadron production in e^+e^- collisions (b).

This cannot be calculated from theory, because not enough is known about hadrons. But fortunately the propagator in Fig. 2a can, in principle, be cut in half to obtain that of Fig. 2b, which shows an e^+e^- pair annihilating to give real hadronic states. By using dispersion theory, the cross-section for Fig. 2b as a function of s = (total energy)^2 can be related to the propagator shown in Fig. 2a [6], and so to the anomalous moment arising from Fig. 1d with hadron loops:

$$\Delta a_\mu^{\rm had} = \left(\frac{1}{4\pi^3}\right) \int_{4m_\pi^2}^{\infty} \sigma_{\rm H}(s) K(s)\, ds \,, \tag{2.5}$$

where $\sigma_{\rm H}(s)$ is the cross-section for hadron production via one-photon exchange in e^+e^- collisions, s is the square of the centre of mass energy, and

$$K(s) = \left\{ x^2\left(1 - \frac{x^2}{2}\right) + (1+x)^2\left(1 + \frac{1}{x^2}\right)\left[\ln(1+x) - x + \frac{x^2}{2}\right] + \frac{1+x}{1-x} x^2 \ln x \right\} , \tag{2.6}$$

where

$$x = \frac{1-\beta}{1+\beta} ; \qquad \beta = (1 - 4m_\mu^2/s)^{1/2} ;$$

$\sigma_{\rm H}(s)$ is normally obtained by measuring the ratio

$$R(s) = \frac{\sigma(e^+e^- \to \text{hadrons})}{\sigma(e^+e^- \to \mu^+\mu^-)} ,$$

the denominator being known from QED.

The process $e^+e^- \to$ hadrons has been extensively studied in electron–positron colliding beams, and the cross-section is rather well defined from near threshold up to about 3 GeV centre-of-mass energy. Therefore the integration in Eq. (2.5) can be carried out with fair confidence. Three new evaluations were published in 1985 [7] with the following results:

	Δa_μ^{had}
Kinoshita et al.	$703\,(19) \times 10^{-10}$
Casas et al.	$705\,(12) \times 10^{-10}$
Barkov et al.	$679\,(12) \times 10^{-10}$

The first two of these papers are by theorists and use the same experimental data. They analyse separately the contributions of a wide variety of hadronic final states such as 2π, 3π, ..., 6π, K^+K^-, $K_S^0K_L^0$, $K^+K^-\pi^+\pi^-$, ϱ, ω, etc., and include higher-order diagrams with two hadronic loops, and the scattering of light by light via a hadron loop. It is encouraging that these two independent evaluations, using different approaches, agree so well.

The main contribution of Barkov et al. is to present new very precise data on $\sigma_H(s)$ for the $\pi^+\pi^-$ final state, which were unfortunately not available to Kinoshita et al. nor to Casas et al. Barkov et al. use the new data to evaluate Δa_μ^{had}, but they do not say whether all the multiparticle final states mentioned above are included. We cannot tell whether their result is lower because of the new experimental data, or because some states are omitted.

Further measurements are now planned by Barkov et al. and it is felt that a detailed comparison of the calculations should be deferred until their results are available.

Pending further clarification, it seems prudent to split the difference and increase the error to include the whole range of values above. In this case we can state with some confidence

$$\Delta a_\mu^{had} = 69\,(2) \times 10^{-9}\,. \tag{2.7}$$

The contribution of the four-fermion weak interaction is illustrated in Fig. 3a. This is second order in the weak interaction and turns out to be negligibly small ($\sim 10^{-12}$). However, a larger weak contribution arises from single-loop Feynman diagrams involving W, Z, and ϕ, as shown in Figs. 3b–d. Although the W and Z particles have been discovered [8], the Higgs scalar particle ϕ predicted by the Standard Model has not been detected. Calculations based on the Glashow–Salam–Weinberg theory [9] predict this contribution to be

$$\Delta a_\mu^{weak} = 195\,(1) \times 10^{-11}\,, \qquad \text{or} \qquad (1.7 \pm 0.01)\ \text{ppm in } a_\mu^{th}\,. \tag{2.8}$$

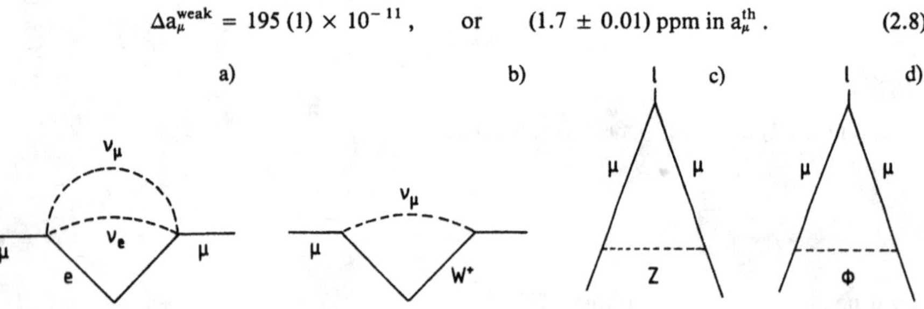

Fig. 3 Contribution to a_μ by the weak interaction: (a) with four-fermion interaction; (b) by the virtual production of an intermediate boson $W^\pm$, (c), (d) single-loop diagrams contributing to the muon g-factor.

The two diagrams with a virtual W and a virtual Z particle contribute at about the same level, but the one involving the Higgs scalar ϕ is relatively negligible if $M_\phi > 7\ \mathrm{GeV}/c^2$, as is now believed [10]. The uncertainty arises from the quoted error in $\sin^2\theta_w$, but it does not include any estimate of higher-order weak interactions. The contribution of the weak interaction is a very small effect, at present masked by the uncertainty in the strong interaction contribution. So in order to utilize fully a higher-precision experimental value of a_μ to determine the electroweak contribution, it is necessary to improve our knowledge of a_μ^{had}, implying better measurements of R(s), particularly in the low-energy region with invariant mass $< 1\ \mathrm{GeV}/c^2$. Combining all the above values the overall theoretical prediction is

$$a_\mu^{th} = 116\,591\,8\,(2) \times 10^{-9}\,. \tag{2.9}$$

Finally, we must consider the effect on a_μ^{th} of a modification of QED. If the muon is not completely point-like in its behavour, but has a form factor $F(q^2) = \Lambda_\mu^2/(q^2 + \Lambda_\mu^2)$, it can be shown that

$$\frac{\Delta a_\mu}{a_\mu} = \frac{-4m_\mu^2}{3\Lambda_\mu^2}, \tag{2.10}$$

implying, for example, a reduction in a_μ of 1 ppm if $\Lambda_\mu = 125\ \mathrm{GeV}/c^2$. Similarly, a modification of the photon propagator by the factor $\Lambda_\gamma^2/(q^2 + \Lambda_\gamma^2)$ implies that

$$\frac{\Delta a_\mu}{a_\mu} = \frac{-2m_\mu^2}{3\Lambda_\gamma^2}. \tag{2.11}$$

This result was first obtained by Berestetskii et al. [5], who emphasized the value of experiments on the muon; the high mass m_μ implies a significant correction to a_μ even when Λ_γ is large. The violation of unitarity suggested by this modification of the photon propagator can be circumvented by introducing a negative metric [11].

It is worth while to emphasize in this brief introduction to the theory that over a period of many years the muon $(g-2)$ has provided a very good constraint on theoretical constructs. Theories that attempt to generalize the Standard Model in one way or another all have their effect on the muon $(g-2)$ value. Such theories include, for example, left-right symmetric theories, Kaluza-Klein theories. Supersymmetric theories introduce many new particles, with masses of the order of m_W, which would contribute to $(g-2)$. Composite models treat the so-called elementary particles as composites of several fundamental entities [12]; the present precision of the muon $(g-2)$ imposes a lower limit, of the order of 1 TeV, on the mass of such constituent particles.

3. MUON PRECESSION AT REST

If a positive muon is brought to rest in a metal, graphite, or other absorber it takes up an interstitial position and keeps its polarization. As the slowing down is accomplished by atomic electric fields which, on the average, are longitudinal, the Lorentz transformation to the rest frame gives no average magnetic field in the rest frame, and so the spin direction does not change. The stationary μ^+ is then influenced by the local magnetic field; the spin rotates at frequency

$$f_s = (g/2)(eB/2\pi mc) \tag{3.1}$$

and the angular distribution of decay electrons must rotate at the same frequency. If decays are counted in a particular direction, the counting rate will be

$$N(t) \propto \exp(-t/\tau)\,[1 + A\cos(2\pi f_s t + \phi)]\,, \tag{3.2}$$

that is, an exponential decay, modulated by the spin frequency f_s. This may be recorded using a time analyser, which sorts the events according to the time delay t between the arrival of the μ^+ and the observation of the e^+.

This precession frequency f_s has been measured many times in fields calibrated in terms of the proton spin precession frequency f_p. The ratio $\lambda = f_s/f_p = \mu_\mu/\mu_p$, where μ_μ and μ_p are the magnetic moments of the muon and proton, has been determined in this way, and also by measurements of the hyperfine splitting in muonium. The current best results are [13], [14]

$$\lambda = 3.183\,344\,(17) \qquad \text{from muon precession}\,,$$

$$\lambda = 3.183\,346\,(11) \qquad \text{from muonium}\,.$$

The mean adjusted value adopted by Cohen and Taylor [4] is

$$\lambda = 3.183\,345\,47\,(47)\,.$$

From Eq. (3.1) the g-factor of the muon could be deduced to a similar accuracy if the mass m_μ of the muon was known. However the only independent measurement of m_μ from X-ray transitions in μ-phosphorus is accurate only to one part in 10^4 [15]. In 1958 this gave

$$g = 2\left(1.0010\begin{matrix}+0.0001\\-0.0002\end{matrix}\right)$$

confirming the theoretical value of a_μ to about 15%.

By studying the spin motion of muons in flight it proves possible to measure a frequency that is proportional not to g, but to (g − 2). This result can then be combined with the magnetic moment measurement, and a precise value of g as well as of the muon mass can then be obtained.

4. MUON PRECESSION IN FLIGHT

4.1 Magnetic Field Only

To calculate the spin motion for a particle rotating at relativistic speeds, it is necessary to consider the influence of the transverse acceleration, which gives rise to an effect known as the Thomas precession [16]. To explain this effect, we follow an argument developed by one of us in the Cargèse Lectures in Physics, 1968 [17].

Consider a gyroscope moving parallel to the x-axis OA, as shown in Fig. 4, and pointing initially along OA. It is assumed that the gyroscope is non-magnetic and is not subject to a

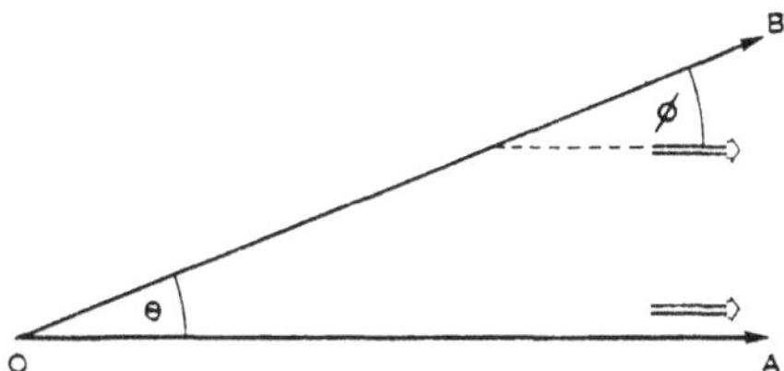

Fig. 4 Thomas precession. Initial motion along OA becomes motion along OB because of transverse acceleration. The arrow represents a gyroscope pointing in a fixed direction. At low speeds we expect $\phi = \theta$, but this is modified at velocities near c.

couple: it just has the property of pointing always in the same direction. Suppose, further, that the gyroscope is accelerated tranversely so that it now moves along OB at an angle θ to the original direction. The angle between the gyroscope axis and its new direction of motion is called ϕ. As the gyroscope keeps its direction, the result expected is $\phi = \theta$. But unfortunately this is not true at relativistic speeds; the direction is modified by the Thomas precession. The result can be derived in a simple way by considering rotations in Minkowski space. This involves some spherical trigonometry, so we first consider a simple geometrical problem in which no velocities are involved.

Figure 5 shows a section of the globe with two lines of longitude separated by the small angle θ. There is an arrow near the pole, pointing due south along the meridian. Let the arrow be displaced parallel to itself until it lies on the line of longitude θ. The angle between the arrow and the new line of longitude is called ϕ. Clearly, near the pole, $\theta = \phi$. However, if the experiment is repeated at the equator the arrow will always point south: $\phi = 0$. At an intermediate latitude, designated by the angle α shown in the figure, the result will be between these two extremes. As only sine and cosine functions are involved in spherical trigonometry, it is not surprising to find that the general law is

$$\phi = \theta \cos \alpha \,. \tag{4.1}$$

Returning now to Minkowski space (Fig. 6) for a gyroscope moving in the x-direction, the time-axis t is rotated to t′ through the angle given by $\cos \alpha = \gamma = (1 - \beta^2)^{-1/2}$, with $\sin \alpha =$

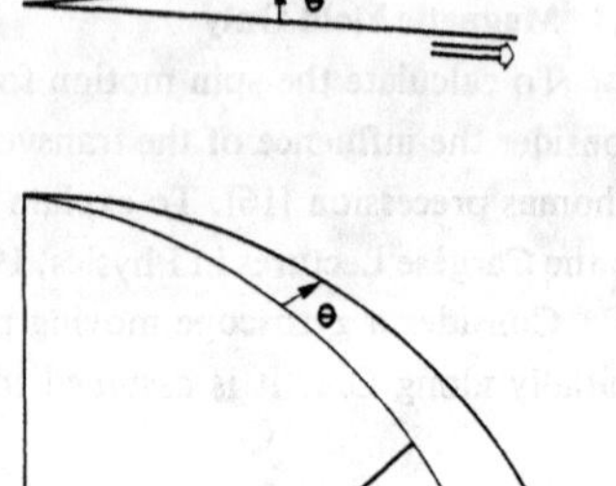

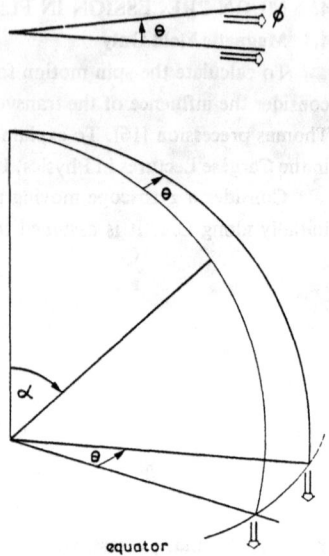

Fig. 5 Spherical trigonometry. When an arrow near the pole is displaced parallel to itself its angle relative to the lines of longitude changes. But at the equator this angle remains fixed.

$i\beta\gamma$, where $\beta = v/c$. An acceleration in the y-direction then implies a rotation of coordinates about the axis Ox ′, bringing the time-axis to t″. Suppose that the gyroscope is now moving at an angle θ to the x-axis; the angle between the gyroscope and the new direction of motion is, according to Eq. (4.1),

$$\phi = \theta \cos \alpha = \gamma\theta \, . \tag{4.2}$$

The angle ϕ is greater than the classical value, because Minkowski space is not really spherical at all, but involves imaginary angles.

In effect, when the direction of the motion is changed by θ, the gyroscope apparently rotates through an angle $(\gamma - 1)\theta$ in the opposite direction. This is called the Thomas precession and is usually thought of as a rotation of the rest-frame axes associated with the transverse

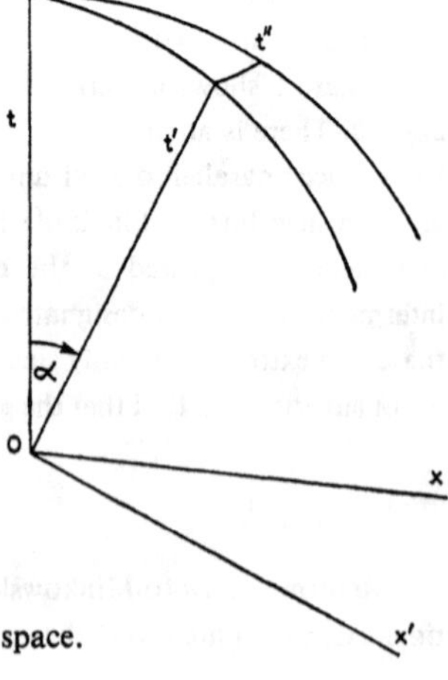

Fig. 6 Derivation of Thomas precession using Minkowski space.

acceleration [16]. A derivation of the Thomas precession using spherical trigonometry in 4-space, broadly similar to the above, was published by Sommerfeld [16]*).

What happens if the gyroscope is a muon and the deflection is due to a magnetic field B? Suppose that in time t the momentum vector is deflected through an angle θ_p. Allowing for the relativistic mass $\gamma \cdot m$,

$$\theta_p = \gamma^{-1}(e/mc)\,Bt\,. \tag{4.3}$$

The rest-frame axes then turn in the opposite direction through the Thomas angle:

$$\theta_t^* = (\gamma - 1)\,\theta_p\,. \tag{4.4}$$

(Starred quantities correspond to the rest-frame, unstarred quantities to the laboratory.) In the rest frame the magnetic field $B^* = \gamma B$ acts on the spin for the time $t^* = t/\gamma$, so the spin turns relative to the axes through

$$\theta_s^* = g\left(\frac{e}{2mc}\right)B^*t^* = g\left(\frac{e}{2mc}\right)Bt = (1 + a_\mu)\left(\frac{e}{mc}\right)Bt\,. \tag{4.5}$$

The overall angle between the spin and the momentum vector is therefore

$$\theta = \theta_s^* - \theta_t^* - \theta_p = a_\mu\left(\frac{e}{mc}\right)Bt\,. \tag{4.6}$$

The frequency at which the spin turns relative to the momentum vector is therefore

$$f_a = a_\mu(eB/2\pi mc)\,. \tag{4.7}$$

Combining with Eq. (3.1) one derives

$$a_\mu = f_a/(f_s - f_a) \tag{4.8}$$

$$= R/(\lambda - R)\,. \tag{4.9}$$

When the magnetic field in which the muons are stored is calibrated in terms of the proton resonance frequency f_p, the experiment becomes a measurement of the ratio $R \equiv f_a/f_p$. The muon spin frequency f_s in the same field if then obtained from the known ratio $\lambda \equiv f_s/f_p$ (see above), and this leads to Eq. (4.9).

Thus the measurement of f_a, the spin frequency of muons turning in a magnetic field allows $a_\mu \equiv (g-2)/2$ to be measured directly, rather than g itself; and the only auxiliary constant required is λ.

---

*) We thank V.L. Telegdi for drawing our attention to this paper.

The value of λ must be corrected for the diamagnetic shielding of the proton in water (~ 26 ppm), to give the field in vacuum seen by the muons.

It must again be stressed that the result (4.7) for the anomalous precession frequency f_a does not contain the relativistic factor γ, essentially because in going to the rest frame, time is shortened by $1/\gamma$, but the magnetic field is increased by γ, so the product Bt remains the same. This is the reason why if relativistic muons are used, the decay lifetime of the particle is lengthened, but the precession frequency remains the same as before: more precession cycles can be measured, leading to increased precision.

More formal derivations of Eq. (4.7) for relativistic particles were given by Mendlowitz and Case and by Carrassi using the Dirac equation, and by Bargmann, Michel and Telegdi using a covariant classical formulation of the spin motion [18].

4.2 Magnetic and Electric Field

If an electric field is used to stabilize the orbit, we must consider whether it upsets the spin motion. In general it does, and in fact stray radial electric fields have caused considerable difficulty in the electron (g – 2) experiments performed at the University of Michigan. It turns out that these difficulties can be avoided by using muons of a carefully chosen 'magic' energy, as has been pointed out in the third (g – 2) experiment performed at CERN (see Section 8). The radial component E_r of the electric field bends the particle orbit; it also induces a magnetic field $B_z^* = \beta\gamma E_r$ in the rest frame of the particle, and this changes the spin precession frequency. To study this in detail, we must repeat the analysis of the relativistic spin precession in the presence of a radial electric field.

The deflection of the momentum vector in time t becomes

$$\theta_p = \gamma^{-1}\left(\frac{e}{mc}\right)(B + E_r/\beta)\,t\,. \tag{4.10}$$

The Thomas angle is again $\theta_t^* = (\gamma - 1)\theta_p$ and $B^* = \gamma B + \beta\gamma E_r$, so that

$$\theta_s^* = (1 + a_\mu)(B + \beta E_r)(e/mc)\,t\,, \tag{4.11}$$

giving

$$\theta = \theta_s^* - \theta_t^* - \theta_p = a_\mu\left(\frac{e}{mc}\right) Bt\,[1 + (\beta - 1/a_\mu\beta\gamma^2)(E_r/B)]\,. \tag{4.12}$$

So the new (g – 2) precession frequency is

$$f_a' = f_a\,[1 + (\beta - 1/a_\mu\beta\gamma^2)(E_r/B)]\,. \tag{4.13}$$

If $\beta^2\gamma^2 = 1/a_\mu$ the second term in Eq. (4.13) is zero and electric fields do not affect the (g – 2) precession frequency. Advantage was taken of this discovery in the third (g – 2) experiment performed at CERN, which used muons of a carefully chosen 'magic energy', 3.096 GeV (see Section 8).

4.3 Pitch Correction

The formulae for the (g − 2) precession frequency f_a derived above assume that the particle velocity is exactly perpendicular to the magnetic field $\vec{B}$, so that the orbit lies in a plane perpendicular to $\vec{B}$. If the velocity has a small angle relative to this plane, the particle will follow a spiral path with pitch angle ψ, and the (g − 2) frequency is altered. In a real storage system, the pitch angle is corrected by the 'vertical' focusing forces, which prevent the particles being lost. The pitch angle changes periodically between positive and negative values, and the correction to the (g − 2) frequency becomes more complex.

All the (g − 2) experiments for electrons and muons are in principle subject to a pitch correction. Early evaluations of the pitch correction by Wilkinson and Crane [19], Henry and Silver [20], and Fierz and Telegdi [18] were superseded by the analysis of Granger and Ford [21] which, for the first time, took proper account of the changes in pitch angle. Farley [22] obtained the same results with a different approach, and extended the analysis to include both electric and magnetic focusing. (See also Field and Fiorentini [23]).

Consider a particle spiralling in the main magnetic field B_z at a small angle ψ (pitch angle) with respect to the x–y plane perpendicular to the z-axis, referred to as the 'axial' direction. Suppose further that the pitch angle varies harmonically according to

$$\psi = \psi_0 \sin \omega_p t \tag{4.14}$$

owing to axial focusing forces (radial component of the magnetic field or the axial electric field). Choose a right-handed Cartesian coordinate frame rotating about the z-axis at the angular frequency

$$\omega = \gamma^{-1}\left(\frac{e}{mc}\right) B_z \tag{4.15}$$

such that the momentum vector lies always in the x–z plane, making an angle ψ to the x-axis. The spin motion will be calculated relative to this frame, and from this the frequency of spin precession relative to the momentum vector, the so-called (g − 2) frequency, follows immediately. We must note that B_z is not necessarily constant, and if it varies with particle position our analysis is still valid but the time average of B_z must be inserted in the final equations. Note that B_x is zero on the average over any closed path, as no current flows through the orbit.

We distinguish two components of the spin motion: i) due to the main field B_z, and ii) due to the axial focusing forces. For (i), the angular velocity of the spin relative to the rotating frame is

$$\omega_a = \gamma^{-1} a_\mu (e/mc) B^* , \tag{4.16}$$

where B^* is the magnetic field in the rest frame of the particle. Resolving B_z parallel and perpendicular to the momentum, transforming to the rest frame, and recombining, one finds the x- and z-components of ω_a:

$$\omega_x = -\omega_0 \frac{\gamma - 1}{\gamma} \psi , \tag{4.17}$$

$$\omega_z = \omega_0 \left(1 - \frac{\gamma - 1}{\gamma} \psi^2\right) \tag{4.18}$$

where $\omega_0 = a_\mu(e/mc)B_z$ is the $(g-2)$ angular frequency for $\psi = 0$, and we use the approximation $\sin \psi = \psi$ for small ψ.

For the axial focusing forces, we remark that when the momentum vector is deflected through the angle ψ in the x–z plane, the spin will rotate about the y-axis through the angle $h\psi$, where

$$h = (1 + \gamma a_\mu) \qquad \text{for magnetic focusing ,} \tag{4.19}$$

and

$$h = (1 + \beta^2 \gamma a_\mu - \gamma^{-1}) \qquad \text{for electric focusing .} \tag{4.20}$$

Thus the harmonic pitch oscillation (4.16) assumed above will induce an instantaneous angular velocity of the spin about the y-axis:

$$\omega_y = -h\omega_p \psi_0 \cos \omega_p t . \tag{4.21}$$

We wish to follow the spin motion under the combined action of the three angular velocities (4.17), (4.18), and (4.21).

If the spin direction is defined by the polar coordinates θ, ϕ such that the projection on the Cartesian axes are

$$x = \cos\theta \cos\phi , \tag{4.22a}$$

$$y = \cos\theta \sin\phi , \tag{4.22b}$$

$$z = \sin\theta , \tag{4.22c}$$

we can calculate $\dot\theta$ and $\dot\phi$ in terms of the Cartesian components of the angular velocity ω_x, ω_y, and ω_z.

Recall that for any vector $\vec{r}$, $\dot{\vec{r}} = \vec{\omega} \times \vec{r}$. Therefore, taking the x-, y-, and z-components,

$$\dot z = \omega_x y - \omega_y x , \text{ etc .} \tag{4.23}$$

Differentiating Eq. (4.22c) then gives

$$\dot\theta = (\omega_x y - \omega_y x)/\cos\theta = \omega_x \sin\phi - \omega_y \cos\phi . \tag{4.24}$$

Furthermore, from Eqs. (4.22a) and (4.22b)

$$\tan\phi = y/x \tag{4.25}$$

and

$$\sec^2\phi = \frac{x^2+y^2}{x^2}\,. \tag{4.26}$$

So differentiating Eq. (4.25) and using Eqs. (4.22a–c) and (4.23), we find, after a little manipulation,

$$\dot{\phi} = \frac{x\dot{y}-\dot{x}y}{x^2+y^2} = \omega_z - (\omega_x\cos\phi + \omega_y\sin\phi)\tan\theta\,. \tag{4.27}$$

Substituting now from Eqs. (4.20) to (4.24), we find

$$\dot{\theta} = -\omega_0\psi_0\gamma^{-1}(\gamma-1)\sin\phi\sin\omega_p t + \omega_p\psi_0 h\cos\phi\cos\omega_p t\,, \tag{4.28}$$

$$\dot{\phi} = \omega_0\{1-\gamma^{-1})(\gamma-1)\psi_0^2\sin^2\omega_p t\}$$

$$+\,\omega_0\psi_0\gamma^{-1}(\gamma-1)\tan\theta\cos\phi\sin\omega_p t + \omega_p\psi_0\,h\tan\theta\sin\phi\cos\omega_p t\,. \tag{4.29}$$

These equations are readily solved if we replace ϕ in the right-hand side with the close approximation $\phi = (\omega_0 t + \xi)$, where ξ is an arbitrary phase. We then find from Eq. (4.28)

$$\theta = \frac{A_1}{\omega_0+\omega_p}\sin\left[(\omega_0+\omega_p)\,t + \xi\right]$$

$$-\,\frac{A_2}{\omega_0-\omega_p}\sin\left[(\omega_0-\omega_p)\,t + \xi\right]\,, \tag{4.30}$$

where

$$A_1 = \frac{1}{2}\psi_0\left[\omega_0\gamma^{-1}(\gamma-1) + h\omega_p\right] \tag{4.31}$$

and A_2 is the same but with a negative sign before the last term.

Note that in general θ oscillates about zero with amplitude $\sim\psi_0$, the spin remaining on the average in the x–y plane. However, the amplitude becomes large if $\omega_p \approx \omega_0$, and if $\omega_p = \omega_0$, $\theta = -A_2 t\cos\xi$, so the spin turns continuously out of the plane. This is an example of the familiar depolarization resonance. It follows from Eq. (4.30) that large values of θ occur only if $(\omega_0-\omega_p)/\omega_0 \lesssim \psi_0/\gamma \lesssim 10^{-2}$ in a typical experiment.

If we are not too close to the resonance, θ remains small, and we may substitute θ as given by Eq. (4.30) for $\tan\theta$ in Eq. (4.29). This gives:

$$\dot{\phi} = \omega_0\left(1 - \frac{\gamma - 1}{2\gamma}\psi_0^2\right) + \frac{A_1^2}{2(\omega_0 + \omega_p)} + \frac{A_2^2}{2(\omega_0 - \omega_p)}$$

$$+ \text{ terms oscillating at frequencies } \omega_p, \omega_0, (\omega_0 - \omega_p), \text{ etc.} \tag{4.32}$$

So the observed (g – 2) frequency is $\omega_a = \bar{\dot{\phi}} = \omega_0(1 - C)$ where, using Eqs. (4.31) and (4.32), the correction factor C is found to be

$$C = \frac{1}{4}\psi_0^2\left[1 - \frac{\omega_0^2}{\gamma^2(\omega_0^2 - \omega_p^2)} - \frac{\omega_p^2(h-1)(h-1+2/\gamma)}{\omega_0^2 - \omega_p^2}\right]. \tag{4.33}$$

Applying this result to three cases that are of experimental interest, we find the following:

a) When the axial focusing forces are magnetic, $h - 1 = \gamma a_\mu$ and

$$C = \frac{1}{4}\psi_0^2\left[1 - \frac{(\omega_0^2 + 2a_\mu\gamma^2\omega_p^2) + a_\mu^2\gamma^4\omega_p^2}{\gamma^2(\omega_0^2 - \omega_p^2)}\right], \tag{4.34}$$

in agreement with Granger and Ford [21]. Except at very high energies, the terms in a_μ will be negligible in view of the uncertainties in determining ψ_0. As ω_p is varied, C changes according to a dispersion curve. When $\omega_p \ll \omega_0$ (the limit of slow pitch changes), $C = 1/4\ \beta^2\phi^2$. As ω_p increase, C falls, eventually becoming negative, but after the resonance at $\omega_p = \omega_0$, C becomes large and positive, finally approaching $1/4\ \psi^2\ (1 + 2a_\mu)$ when $\omega_p \gg \omega_0$ (the limit of rapid pitch changes). Near resonance our approximation $\theta \approx \tan\theta$ is not valid: a more detailed analysis shows that infinitely large corrections suggested by Eq. (4.33) do not, in fact, occur.

b) With axial focusing by electric fields, $h - 1 = \beta^2\gamma a_\mu - \gamma^{-1}$, so

$$C = \frac{1}{4}\psi_0^2\left(\beta^2 - \frac{a_\mu^2\beta^4\gamma^2\omega_p^2}{\omega_0^2 - \omega_p^2}\right). \tag{4.35}$$

At low energy the second term is negligible and the correction is virtually independent of the pitch frequency.

c) With an electric field, using the 'magic' particle energy given by $\beta^2\gamma^2 = a_\mu^{-1}$ so that the electric fields do not affect the (g – 2) precession, then $h = 1$ and the final term in Eq. (4.33) is zero:

$$C = \frac{1}{4}\psi_0^2\left[1 - \frac{\omega_0^2}{\gamma^2(\omega_0^2 - \omega_p^2)}\right]. \tag{4.36}$$

The pitch correction now shows the resonant behaviour discussed above, but it contains no terms in a_μ. Far from resonance, $C = 0.25\psi^2$ at both limits.

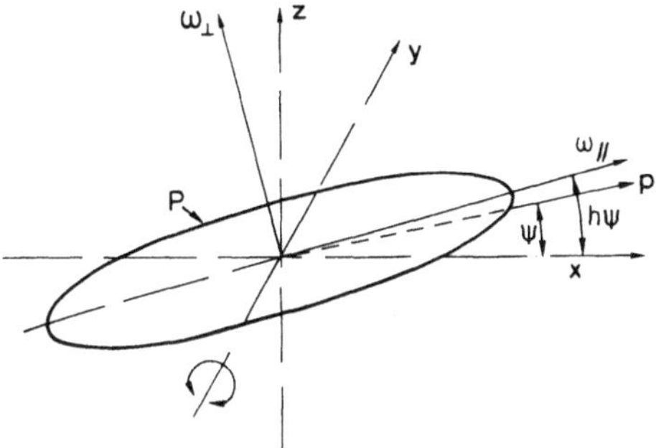

Fig. 7 Axes xyz rotating about z so that the momentum vector p lies always in the x-z plane. Owing to pitch oscillations the pitch angle ψ between p and the x-axis is varying. The spin angle is determined in the plane P (inclined at angle $h\psi$ to the x-axis) because a spin in this plane moves into the xy-plane when ψ passes through zero.

These results may be derived directly by means of a simple physical argument based on Eqs. (4.17) and (4.18). We are interested in the average rotation of the spin about the z-axis, i.e. the projection of the spin on the x–y plane. Let us consider the spin rotation in this plane at instants when the pitch angle is zero, thus neglecting for the average motion the small non-cumulative mutations that occur during the pitch oscillations. The essential step is to recognize that when the pitch angle is ψ, we must determine the progress of the spin in a plane making an angle $h\psi$ with the x-axis, indicated in Fig. 7 as the plane P. This is because any spin direction lying in this plane will be turned into the x–y plane when the pitch angle again becomes zero.

We therefore resolve the angular velocities (or magnetic fields) parallel ($\omega_{\parallel}$) or perpendicular ($\omega_{\perp}$) to the plane P. From Eqs. (4.17) and (4.18),

$$\omega_{\perp} = \omega_z \cos(h\psi) - \omega_x \sin(h\psi) = \omega_0 \left\{1 - \frac{1}{2}\psi^2[1 + (h-1)(h-1+2/\gamma)]\right\} . \quad (4.37)$$

When the pitch oscillation is rapid ($\omega_p \gg \omega_0$), then $\omega_{\parallel}$ changes rapidly in sign and contributes nothing to the net spin precession. So, in this case, the observed spin motion will be determined by $\omega_{\perp}$ alone. Recalling that $\psi^2 = 1/2\ \psi_0^2$, we see that Eqs. (4.33) and (4.37) agree exactly in the limit $\omega_p \gg \omega_0$.

In the case of slow pitching ($\omega_p \ll \omega_0$) however, $\omega_{\parallel}$ remains almost constant over many cycles of (g − 2) precession, and then $\omega_{\perp}$ and $\omega_{\parallel}$ must be added vectorially to give the resultant angular velocity. This is of course equal to the resultant of ω_x and ω_z, so from Eqs. (4.17) and (4.18),

$$\omega_a = \omega_0 \left(1 - \frac{1}{2}\beta^2\psi^2\right), \quad (4.38)$$

agreeing with Eq. (4.33) in the limit $\omega_p \to 0$.

Thus the fast-pitching and slow-pitching limits of Eq. (4.33) are confirmed by an independent argument.

4.4 Experimental Principles

To summarize the principles of the (g − 2) experiment, the essential requirements are a polarized source, an almost uniform magnetic field, and a polarimeter to measure the spin direction as a function of time.

An effective polarimeter would be one that measured $\vec{s} \cdot \vec{\beta}$ as a function of the storage time t spent in the magnetic field. This quantity oscillates with the frequency f_a, and in Fig. 8 we illustrate its measurement, showing the main components of the (g − 2) experiment in block diagram form.

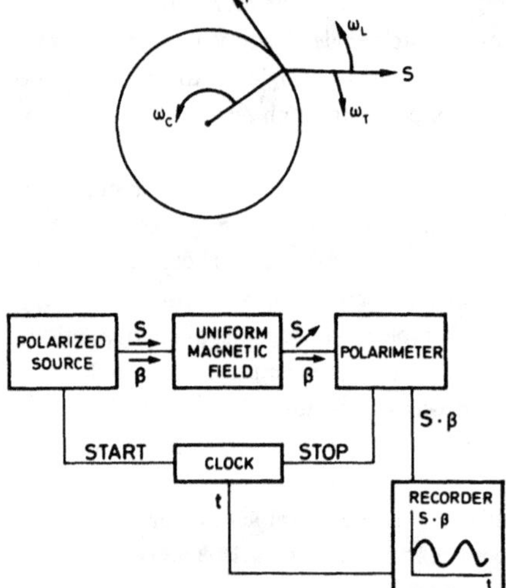

Fig. 8 Relative motion of the spin and momentum in the laboratory frame and the block diagram of an ideal (g − 2) experiment.

The muons are produced by the decay of pions in flight, and are born longitudinally polarized in the pion rest frame. Furthermore, the muon waves a flag to show which way its spin is pointing at the moment of decay; in fact, in the process $\mu^+ \to e^+ + \nu_e + \bar{\nu}_\mu$, the electron angular distribution has its maximum in the direction of the muon spin. By observing a large number of decays from an ensemble of muons, one can measure the average spin direction to any desired accuracy. Thus, the pion-muon decay and the muon-electron decay provide a very simple polarized source and an efficient polarimeter. The ideal (g − 2) experiment consists, then, of trapping longitudinally polarized muons in a uniform magnetic field and measuring the precession frequency f_a of the spin relative to the velocity vector through the decay electron asymmetry.

The observed phenomenon is oscillatory in nature, and so the more oscillation periods there are, the more precise will be the determination of the frequency f_a. This can easily be seen by supposing, for example, that we count n cycles of a wave train, extending over a measured time t, with an uncertainty of one cycle. Then

$$f_a = \frac{n}{t} \pm \frac{1}{t}, \tag{4.39}$$

the error in the frequency decreases as the duration of t increases. It is therefore advantageous to store the charged particle in the uniform-field region for as long as possible. For the electron, this time is limited only by the quality of the trap, and may extend to months. But for the muon, its natural lifetime (2.2 μs) limits the useful storage time, and it is of advantage to dilate the lifetime by going to high energies. In doing this a real gain is made since, as we have seen, the frequency f_a does not depend on the particle energy: through the relativistic increase in the laboratory lifetime of the muon, more (g – 2) modulation cycles can be measured.

In the foregoing we have outlined the general features of the precession experiment under idealized conditions. In real experiments it is not possible to inject particles of a given momentum into a fixed orbit in a static magnetic field. The open curve of the injection trajectory cannot lead to the closed curve of the fixed orbit without some localized change in the orbit parameters, such as the particle velocity or direction. The particles can, of course, be contained within a reasonable volume by allowing quasi-circular orbits to drift slowly through the storage region, but it should be remembered that the less confined the orbits are, the more difficult is the task of obtaining a precise value of the mean magnetic field $\bar{B}$ experienced by the muon sample. Different methods of injection will be discussed when briefly describing the different (g – 2) experiments performed at CERN. A more detailed discussion of some experimental problems encountered in the muon (g – 2) experiments is given in Ref. [24].

5. INITIAL MUON POLARIZATION AND MEASUREMENT OF THE FINAL POLARIZATION

5.1 Generalities

Muons are most readily obtained from the pion decay $\pi^+ \rightarrow \mu^+ + \nu_\mu$, and parity non-conservation in this process leads to their polarization. The relevant decay configuration for the positive pions is shown in Fig. 9. The muons that decay close to the forward direction of the pion momentum have helicity approximately equal to –1, and also laboratory momentum

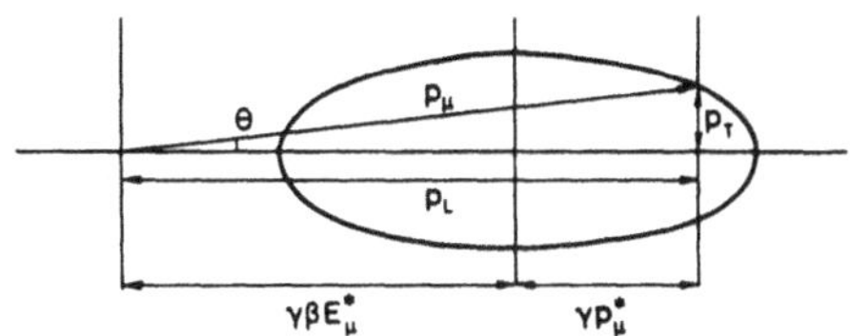

Fig. 9 Pion-muon decay in the laboratory frame.

near the top of the available range. Selection of these maximum-momentum muons will thus yield a sample with high initial longitudinal polarization. By taking a 1% momentum bite around the maximum value of the pion momentum, the longitudinal polarization is about 97%. Thus the muons that have been produced by decays within a momentum-selected pion beam, and have momenta within the acceptance of the beam, will be strongly polarized longitudinally. The degree of polarization will depend on the momentum acceptance of the pion beam and on the decay channel.

The analysis of the longitudinal polarization of this sample as a function of time is effected by making use of the asymmetry in the decay electron angular distribution that arises from parity non-conservation. For the process $\mu^- \to e^- + \bar{\nu}_e + \nu_\mu$, for example, the decay electron angular distribution in the muon rest frame can be expressed as the differential probability

$$dP(y,\theta) = n(y)[1 + A(y)\cos\theta]\, dy\, d\Omega \,, \tag{5.1}$$

where $y = p_e/(p_e)_{max}$, $d\Omega$ is a solid angle, θ is the angle between the electron momentum p_e and the muon spin, and the distribution function and the asymmetry are given by

$$n(y) = 2y^2(3-2y) \,, \qquad A(y) = \frac{2y-1}{3-2y} \,. \tag{5.2}$$

The asymmetry has a maximum value of $+1$ at $y = 1$, changes sign at $y = 1/2$, and reaches a value of $-1/3$ at $y = 0$ (see Fig. 10).

Thus if muons that have emerged from the storage region are stopped in a non-depolarizing target, and the subsequent decay electrons are detected in a counter that is placed, for example, 'forward' with respect to the muon momentum prior to stopping, then the count rate as a function of storage time will be modulated with the $(g-2)$ frequency f_a. The amplitude of this modulation will depend on the polarization of the muon sample, and on the

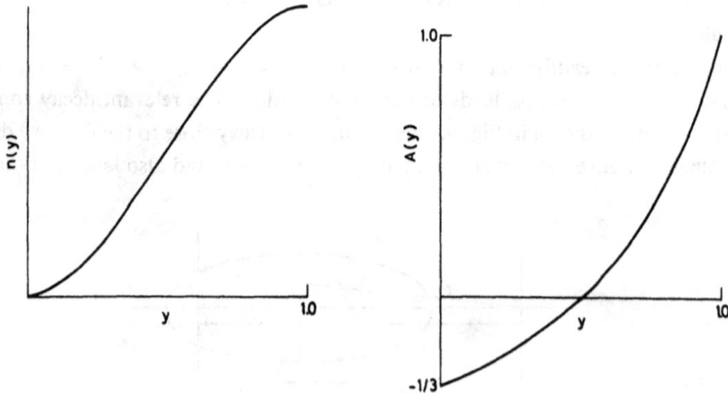

Fig. 10 The decay electron spectrum and asymmetry in the muon rest frame. The quantity y is the electron momentum expressed as a fraction of its maximum value.

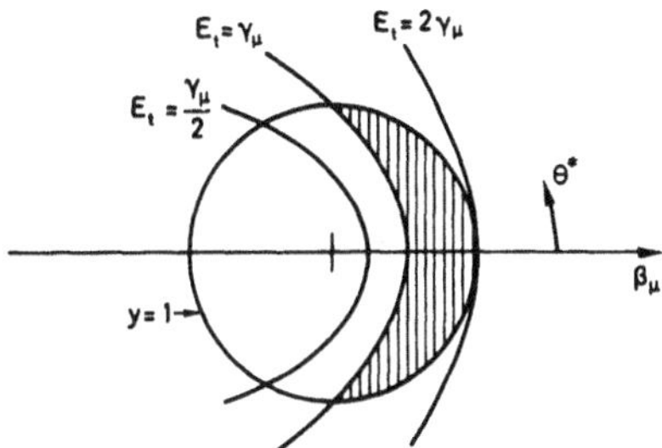

Fig. 11 The decay electron momentum in the muon rest frame. The parabolas illustrate the effect in this frame of an energy cut in the laboratory.

detection efficiency of the counter system as a function of electron angle and momentum. This is essentially the technique employed in the muon experiment carried out at the CERN Synchro-cyclotron (SC), as described in more detail below.

If the muons are allowed to decay within the storage region, then we no longer have direct access to the muon rest frame, and the manner of measuring the final polarization clearly has to be modified. This measurement can be effected by selecting a subset of decay electrons above a given energy threshold in the laboratory. Such a cut can be transformed into the muon rest frame, as illustrated in Fig. 11. This figure shows the decay electron momentum space in the muon rest frame and represents a section, through the axis of symmetry, that coincides with the direction of the muon velocity β_μ. The parabolas illustrate the effect of a laboratory energy threshold E_t in this space, expressed in units of $0.5mc^2$ ($\sim$ 53 MeV). Thus the shaded region represents the decay electron momenta and directions that would be accepted for an energy cut equal to just one half the muon energy in the laboratory ($E_\mu = \gamma_\mu m_\mu c^2$).

This is a general result, which can be derived as follows. Let a moving system have velocity $\beta_c \cdot c$ in the laboratory, and in the moving system suppose particles are emitted with constant momentum p^*, energy E^*, and angular distribution $dN/d\Omega^* = f(\theta^*)$, θ^* being the angle in the moving system relative to the direction of motion. The energy of the particle in the laboratory will be given by

$$E_{lab} = \gamma_e \{E^* + \beta_e p^* \cos\theta^*\}$$

where $\gamma_e = \{1 - \beta_e^2\}^{-1/2}$.

As

$$\Omega^* = 2\pi(1 - \cos\theta^*)$$

$$d\Omega^* = -2\pi\, d(\cos\theta^*)$$

$$dE_{lab} = (\beta_e \gamma_e p^*/2\pi)\, d\Omega^* ,$$

so

$$dN/dE_{lab} = (2\pi/\beta_e\gamma_e p^*)(dN/d\Omega^*)$$

$$= (2\pi/\beta_e\gamma_e p^*)\, f(\theta^*)\,.$$

As $\beta_e\gamma_e$ and p^* are all independent of θ it is clear that the energy distribution in the laboratory has the same functional form as the angular distribution in the moving system. Hence selecting energy in the laboratory is equivalent to selecting angle in the moving system.

In the case of muon decay in flight to electrons this conclusion has to be modified because there is a spectrum of momenta p_e, and this will smear the one-to-one correlation of E_{lab} with θ_e. Now to have high energy in the laboratory the decay electron must a) come from near the top of the spectrum in the moving system (this means also a high asymmetry parameter A) [Eqs. (5.2)] and b) be emitted forwards.

From this simple argument we can see how the energy cut in the laboratory distinguishes between decays that are forward as opposed to backwards with respect to the muon momentum, and how decays at quite large angles in the muon rest frame may be included in the process. If we now envisage that the asymmetric probability distribution given by Eq. (5.1) is rotating in the muon rest frame with angular frequency f_a, we can see that the number of decay electrons within this selected subset will be modulated with the $(g-2)$ frequency and have the form

$$N(t) = N_0 \exp[-t/\tau][1 - A\cos(2\pi f_a t + \phi)]\,, \tag{5.3}$$

where τ is the dilated muon lifetime ($\tau = \gamma\tau_0$), A is the overall asymmetry, and ϕ is the initial phase of the polarized muon sample, that is the initial phase of the spin with respect to the muon velocity vector. Elementary error analysis of the $(g-2)$ signal contained in an observed time spectrum of this form shows that the error in f_a is proportional to $(\tau A\sqrt{N})^{-1}$, where N is the total number of decay electrons detected. This expression indicates that the energy threshold should be set so as to maximize NA^2, and it also demonstrates the importance of the dilated muon lifetime $\tau = \gamma\tau_0$. By integrating over the shaded region shown in Fig. 11 for different values of E_t, we obtain the curves of N_e, A, and N_eA^2 shown in Fig. 12. As can be seen, the

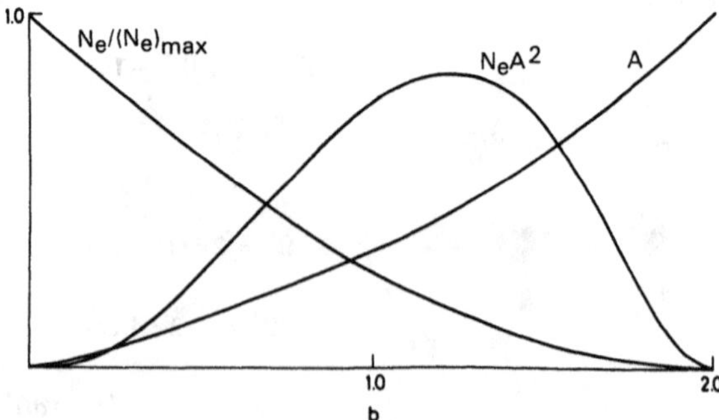

Fig. 12 Dependence of the number of decay electrons N_e, the asymmetry coefficient A and N_eA^2 versus $b = E_t/\gamma_\mu$.

required maximum in NA^2 occurs at a value of E_t that is equivalent to about 0.65 times the highest electron energy. Having established the manner of the final polarization measurement, we will return to the optimization of the (g − 2) signal in the discussion of each actual experimental configuration.

5.2 From an Ideal (g − 2) Experiment to a Real One

In a real (g − 2) experiment, the quality of the storage region (trap) plays an important role because the uncertainty in a_μ is in part inversely proportional to the storage time t, and it is a great advantage to make this time as long as possible.

The first requirement for the quality of the trap is therefore that the particle be trapped or stored for a long period of time. This means the elimination of the characteristics that increase the amplitude of the particle orbits to such an extent that they collide with the material limits of the region and are consequently lost. The most obvious source of such beam blow-up is collisions with gas atoms, so the storage volume must be evacuated. However, since the storage times are relatively short, the vacuum requirements are not as stringent as those of storage rings, in which particles are contained for several hours. Another source of particle loss can be the magnetic field itself. If this field contains local inhomogeneities through which the particles pass at intervals that are simply related to the period of oscillation of the particle about its equilibrium orbit, then the amplitude of these oscillations may grow in time. This distortion of the particle orbit leads to a shift in frequency of the oscillations, so that there is a phase slippage, and the perturbation then decreases the amplitude again. The resulting modulation in amplitude may well take the particle outside the physical limits of the storage volume. This phenomenon is similar to the case of a simple harmonic oscillator driven by a periodic force, and is well known to the designers of particle accelerators and storage rings. So the successful realization of a precession experiment requires a careful study of the stability of the orbits if premature loss of particles is to be avoided.

The second requirement for the quality of the trap concerns the need to measure, with great precision, the average magnetic field seen by the particles. This quantity, $\bar{B}$, must be known to a greater precision than that desired for a_μ itself, since it is just one of the contributions to the overall uncertainty. Any real beam of particles occupies a finite volume of phase space, and consequently the particles do not all follow the same orbits. The precise knowledge of $\bar{B}$ involves, therefore, a precise mapping of the magnetic field and a measurement of the distribution of particle orbits, or at least the calculation of this distribution in a reliable way. The extent to which the uncertainty in this latter leads to an error of $\bar{B}$ is determined by the uniformity of the magnetic field; for example, if the field were perfectly uniform, we would only need to know that the particles were contained within the storage region, but not precisely where.

Another consequence of the finite extension of the phase space of the beam is that it will have some angular divergence, and will thus contain particles whose component of the velocity along the magnetic field lines is not zero. This motion is not restricted by a uniform field. In a precession experiment, the total path length travelled by particles may reach several kilometres, and the confinement of these orbits within a reasonable volume implies some non-uniformity of the magnetic field itself, in the form of a gradient, or perhaps the application of an electric field. A magnetic gradient automatically makes it more difficult to determine $\bar{B}$. Thus there is a conflict between efficient trapping and knowing the average magnetic field, which it is the art of the experimenter to resolve.

6. 1958–1962: THE (g – 2) EXPERIMENT AT THE CERN SYNCHRO-CYCLOTRON

By 1958, QED was an established theory of some 10 years standing, corroborated by accurate measurements of the Lamb shift. The g-factor of the electron was known through electron spin resonance [25] to one part per million (ppm); in 1950, Karplus and Kroll [26] had shown how to calculate the higher-order corrections to g, and a numerical error in their results had been corrected by Petermann [27], Sommerfield [28], and Suura and Wichmann [29], bringing theory into line with experiment at the level of $(\alpha/\pi)^2$.

For the free electron, a direct determination of the anomalous magnetic moment $a_e \equiv (g-2)/2$ was in progress at the University of Michigan [30] using the recently discovered principle of (g – 2) spin motion explained above. Equation (4.7) had been proved to hold for relativistic velocities.

Turning to the muon, the bremsstrahlung cross-section at high energies had been measured with cosmic rays and was shown to agree with a spin assignment of 1/2 rather than 3/2 [31]. A similar conclusion followed from data on neutron production by cosmic-ray muons [32]. Experiments with cosmic-ray and accelerator-generated muons were in progress to compare the electromagnetic scattering of muons and electrons by nuclei.

Thus evidence was accumulating that the muon behaves as a heavy electron of spin 1/2. Berestetskii et al. [5] had emphasized that QED theory implied an anomalous magnetic moment a_μ for the muon, of the same order as for the electron, but as the typical invariant momentum transfer involved was $q^2 \approx m^2$, an experiment for the muon would test the theory at much shorter distances. Feynman [33] felt that the divergences in QED could be limited by a real energy-momentum cut-off Λ, and it seemed reasonable to expect Λ to be of the order of the nucleon mass. This would imply a 0.5% effect in a_μ. On the other hand, it was thought [34] that the muon should have an extra interaction that would distinguish it from the electron and give it its higher mass. This could be a coupling to a new massive field, or some specially mediated coupling to the nucleon. Whatever the source, the new field should have its own quantum fluctuations, and therefore give rise to an extra contribution to the anomalous moment a_μ. The (g – 2) experiment was recognized as a very sensitive test of the existence of such fields, and potentially a crucial signpost to the μ–e problem.

At this stage there was no prospect of such an experiment, but in 1957 parity violation was discovered [35], muon beams were found to be highly polarized, and, better still, it was found that the angular distribution of the decay electrons could indicate the spin direction of the muon as a function of time [36]. The angular distribution of electrons from the decay of polarized muons agreed with spin 1/2 [37] and was inconsistent with spin 3/2 [38]. A wide variety of muon precession and spin-resonance experiments were carried out in the next few years [39]. The (g – 2) principle was invoked in the first paper on muon precession by Garwin et al. [36], who pointed out that g must be within 10% of 2.00, because although the muon trajectory had been deflected through 100° by the cyclotron magnetic field, the muon polarization was still longitudinal.

The possibility of a (g – 2) experiment for muons was envisaged, and groups at Berkeley, Chicago, Columbia, and Dubna started to study the problem [40]. If the muon had a structure that gave a form factor less than one for photon interactions, the value of a_μ should be less than predicted. Compared with the measurement for the electron, the muon (g – 2) experiment was much more difficult because of the low intensity, diffuse nature, and high momentum of available muon sources. This implied large volumes of magnetic field; the lower value of (e/mc) made all precession frequencies 200 times smaller, but the time available for an experiment was limited by the decay lifetime, 2.2 μs. Hence large magnetic fields would be needed to give a reasonable number of precession cycles.

One solution was to scale up the method used at Ann Arbor for the electrons, using a large solenoid and injecting the muons spirally at one end [41]. This was pursued at Berkeley and finally led to a 10% precision measurement [42].

At CERN, the work centred on the belief that it should be possible to store muons in a conventional bending magnet with a more or less uniform vertical field between roughly rectangular pole pieces. In a typical field of 1.5 T, the muon orbit would make 440 turns during the lifetime of 2.2 μs. As $a_\mu \approx \alpha/2\pi \approx 1/800$, the angle between the spin and the momentum vector would develop 800 times more slowly, giving a change in beam polarization of about 180° to be studied.

The polarized muon beam from the CERN Synchro-cyclotron could fairly easily be trapped inside a magnet. The particles were aimed at an absorber in the field; they lost energy and therefore turned more sharply and remained inside the magnet. To prevent them re-entering the absorber after one turn, a small transverse (y-direction) gradient of the magnetic field was introduced, causing the orbits to drift sideways perpendicular to the gradient (x-direction). Vertical focusing was added by means of a parabolic term in the field.

If the field is of the form

$$B_z = B_0 (1 + ay + by^2) , \tag{6.1}$$

where a and b are small, an orbit of radius ϱ moves in the x-direction over a distance $s = a\pi\varrho^2$ per turn (called the step size). On the average, the wavelength of the vertical oscillations is $2\pi/b^{1/2}$. Figure 13 is of historical interest. It shows the first evidence of particles turning several times inside a small experimental magnet. These results gave the laboratory sufficient confidence to order a very long magnet for the experiment.

An overall view of the final storage system [43] is shown in Fig. 14. The magnet pole was 6 m long and 52 cm wide, and the gap was 14 cm. Muons entered on the left through a magnetically shielded iron channel and hit a beryllium absorber in the injection part of the field. Here the step size s was 1.2 cm. Then there was a transition to the long 'storage region', where $s = 0.4$ cm with the field gradient $a = (1/B)(dB/dy) = 3.9 \times 10^{-4}\ \mathrm{cm}^{-1}$. Finally, a smooth transition was made to the ejection gradient, where $s = 11$ cm per turn. After ejection, the muons fell onto the 'polarization analyser' (Fig. 15), where they were stopped and decayed to

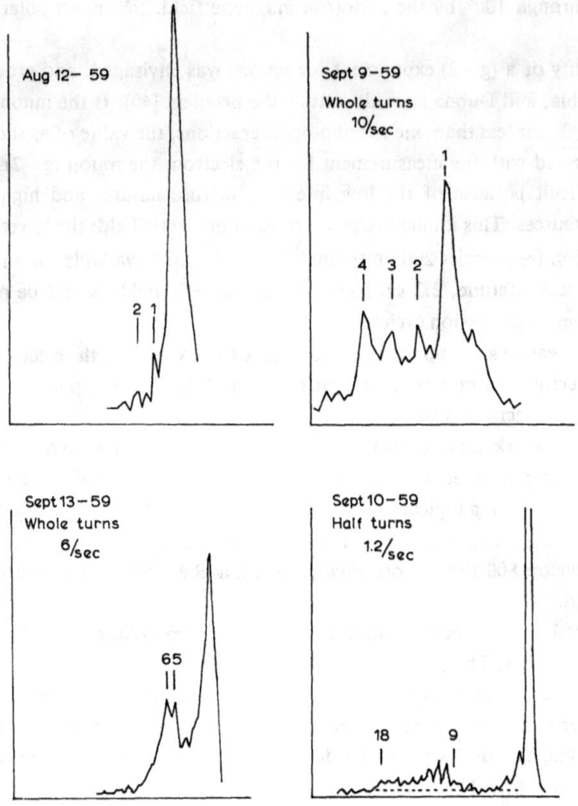

Fig. 13 First evidence of muons making several turns in an experimental magnet. The time of arrival of the particles at a scintillator fixed inside the magnet is plotted horizontally (time increases to the left). The first (right-hand) peak coincides with the moment of injection. The equally spaced later peaks correspond to successive turns. Owing to the spread in orbit diameters and injection angles, some muons hit the counter after nine turns (lower right), while others take 18 turns to reach the same point (Charpak et al., unpublished).

e^+. The time t spent by a muon in the field was determined by recording the coincidences in counters 123 at the input, and counters $4566'\bar{7}$ at the output. The interval was measured with respect to a 10 MHz crystal.

The shimming of this large magnet to produce the correct gradients was a tour de force. This was assisted by the theorem that in weak gradients the flux through a wandering orbit is an invariant of the motion. Therefore, if the field along the centre line of the magnet was constant, unwanted sideways excursions would be avoided, and this could be checked more exactly by moving a flux coil, of the same diameter as the orbit, all along the magnet.

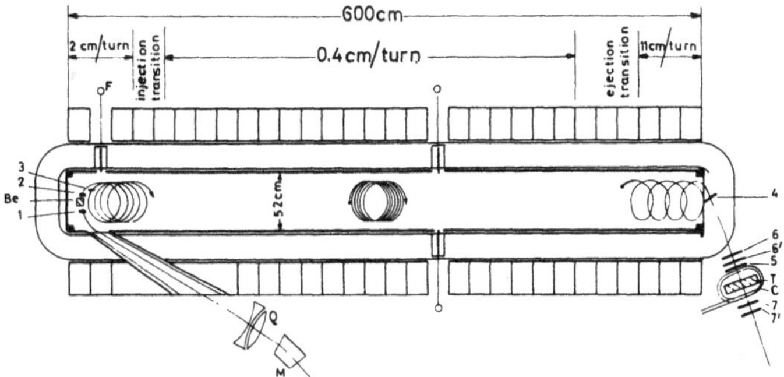

Fig. 14 Storage of muons for up to 2000 turns in a 6 m bending magnet. The field gradient makes the orbit walk to the right. At the end a very large gradient is used to eject the muons so that they are stopped in the polarization analyser. Coincidences 123 and $466'5\bar{7}$, signal an injected and ejected muon, respectively. The coordinates used in the text are x (the long axis of the magnet), y (the transverse axis in the plane of the paper), and z (the axis perpendicular to the paper).

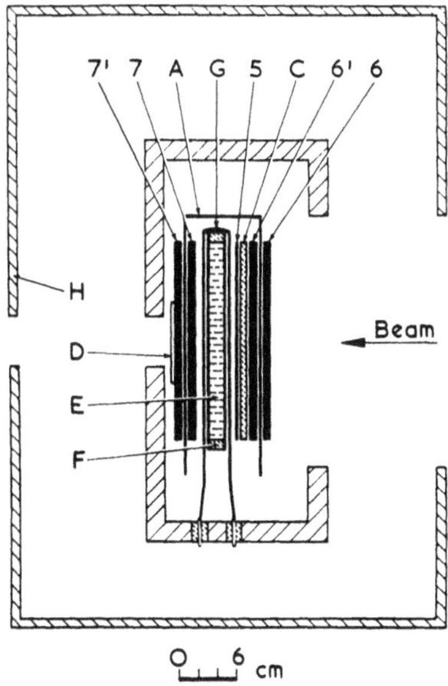

Fig. 15 Polarization analyser. When a muon stops in the liquid methylene iodide (E) a pulse of current in coil G is used to flip the spin through $\pm 90°$. Backward or forward decay electrons are detected in counter telescopes 66′ and 77′. The static magnetic field is kept small by the double iron shield and the mumetal shield A.

However, the constant flux theorem implied that once the particle was trapped inside the magnet it would never emerge. This was seen as a major difficulty, because the final spin direction could only be measured in a weak or zero magnetic field: otherwise, one would lose track of the spin direction while waiting for the muon to decay. For weak gradients and slowly walking orbits, calculations of the orbit confirmed these doubts, and some participants lost faith in the project. Fortunately, it was found that in large gradients, of order ±12% over the orbit diameter, the particles were ejected successfully.

The muons were trapped in the magnet for a time ranging from 2 to 8 μs depending on the location of the orbit centre on the varying gradient given by Eq. (6.1). About one muon per second was stopped finally in the polarization analyser, and the decay electron counting rate was 0.25 per second.

The spin direction can, in principle, be obtained from the ratio of two counting rates measured in different directions. But if two counter telescopes are used (say one forward and one backward relative to the direction of the arriving muons), it is not easy to ensure that they have equal efficiencies and solid angles. Therefore it is more reliable to use only one set of counters, but to move the muon spin direction after it has stopped. This can be done with a small constant magnetic field, but it is more efficient to turn the spin rapidly to a new position by applying a short, sharp magnetic pulse, created by applying a pulse of current to a solenoid wound round the absorber in which the muon is stopped. This flipping was accomplished within 1 μs, before the gate that selected the decay electrons was opened.

In the apparatus shown in Fig. 15, the electron counts c_+ and c_- in the forward telescope 77′ were recorded in separate runs with the spin flipped through $+90°$ and $-90°$, respectively. The asymmetry A of these counts, defined as $(c_+ - c_-)/(c_+ + c_-)$, was then related to the initial direction θ_s of the muon spin (before flipping) relative to the mean electron direction subtended by telescope 77′:

$$A \equiv \frac{(c_+ - c_-)}{(c_+ + c_-)} = A_0 \sin\theta_s \, . \tag{6.2}$$

By flipping instead through 180° and 0°, another ratio proportional to $A_0 \cos\theta_s$ was measured; so θ_s could be determined completely. Similar, but independent, calculations were made for the telescope 66′, which detected decay electrons emitted backwards.

This polarization analyser was first used to study the muon beam that was available for injection. For muons that had been through the magnet, the analyser recorded the asymmetry A as a function of the time t the particle had spent in the field. This showed a sinusoidal variation due to the $(g-2)$ precession in the magnet. Using Eqs. (4.7) and (6.2), it follows that

$$A = A_0 \sin\theta_s = A_0 \sin\left[a_\mu (e/mc) Bt + \phi\right] , \tag{6.3}$$

where ϕ is an initial phase determined by measuring the initial polarization direction and the orientation of the analyser relative to the muon beam.

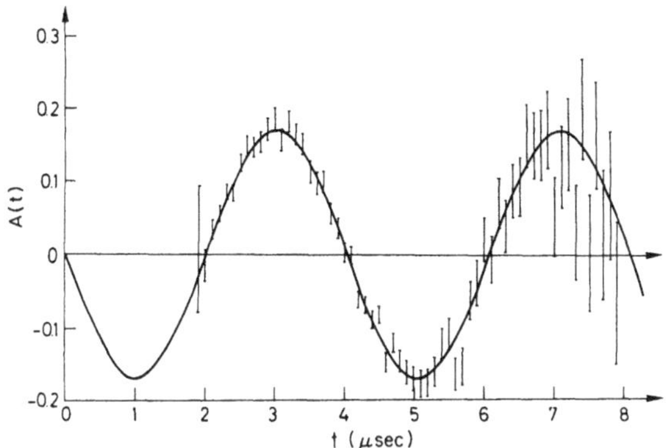

Fig. 16 Asymmetry A of observed decay electron counts as a function of the storage time t. The sinusoidal variation results from the (g − 2) precession; the frequency is measured to ±0.4%.

The experimental data are given in Fig. 16, together with the fitted line obtained by varying A_0 and a_μ in Eq. (6.3). Full discussion of the precautions that are necessary to determine the mean field B seen by the muons, and to avoid systematic errors in the initial phase ϕ, are given in [43].

The results of this experiment are given in Table 1 (see subsection 8.8). The first experiment gave ±2% accuracy in a_μ, and this was later improved to ±0.4%. The figures agreed with theory within experimental errors. The corresponding 95% confidence limit for the photon propagator cut-off [see Eq. (2.11)] was $\Lambda_\gamma > 1.0$ GeV, and for the muon vertex function [Eq. (2.10)] $\Lambda_\mu > 1.3$ GeV.

This was the first real evidence that the muon behaved so precisely like a heavy electron. The result was a surprise to many, because it was confidently expected that g would be perturbed by an extra interaction associated with the muon to account for its larger mass [34, 44]. When nothing was observed at the 0.4% level, the muon became accepted as a structureless point-like QED particle, and the possibility of finding a clue to the μ–e mass difference seemed suddenly much more remote.

7. 1962–1968: MUON STORAGE RING I

7.1 A General View

By now the CERN Proton Synchrotron (PS) and Brookhaven Alternating Gradient Synchrotron (AGS) were operating, and the distinct properties of the two neutrinos ν_e and ν_μ had been established, further emphasizing the parallel but dual behaviour of the muon and the electron [45].

Muon-pair production by 1 GeV gamma rays on carbon was measured by Alberigi-Quaranta et al. [46] in agreement with theory. With this and the (g − 2) data, the evidence for point-like behaviour was now much better for the muon than for the electron. The scattering of muons by lead and carbon [47] agreed with the form factors deduced for electron scattering. Logically this was the best evidence for the point-like behaviour of the electron, but was generally seen as another contribution to our knowledge of the muon. Knock-on electrons from 8 GeV muons confirmed the picture [48]. Muonium formation in high-pressure argon had been oberved by Hughes et al. [49], and the hyperfine splitting of the ground state confirmed the theoretical picture to one part in 2000 [50]. For this and subsequent muonium experiments the $(g-2)$ result was an essential input, not only for the g-factor, but also for deducing the muon mass from the precession frequency at rest, now determined to 16 ppm by Hutchinson et al. [51], see Eqs. (3.1) and (4.7).

The muon $(g-2)$ experiment was now the best test of QED at short distances. For this reason, and to search again for a new interaction, it was desirable to press the accuracy of the experiment to new levels. It would be essential to increase the number of $(g-2)$ cycles observed, either by increasing the field B or by lengthening the storage time. With the CERN PS available, it was attractive to see what could be done by using high-energy muons with relativistically dilated lifetimes. As there is no factor γ in Eq. (4.7), the $(g-2)$ precession frequency would not be reduced, and more cycles would be available before the muons decayed. But to store muons of GeV energy in a magnetic field and measure their polarization required totally new techniques. Farley [52] proposed to measure the anomalous moment using a muon storage ring. As in the cyclotron, if the primary target was placed inside the magnet, muons would be produced by π-μ decay in flight and some of them would remain trapped in the field. With a pulsed accelerator, such as the PS, there should be no continuous background following the injection. Transfer of a pulse of protons from the PS to the muon ring could be achieved with the fast ejected beam already developed for the neutrino experiment. Estimates of the stored muon intensity and polarization looked favourable.

To determine the muon spin direction, decays in flight would need to be observed. The decay electrons would emerge on the inside of the ring and the detectors would respond only to the high-energy particles emitted more or less forward in the muon rest frame. Thus as the spin rotated, the electron counting rate would be modulated at the $(g-2)$ frequency.

It was later realized that at injection the muons would be localized in azimuth (injection time 10 ns, rotation time about 50 ns), so the counting rate would also be modulated at the mean rotation frequency. This would enable the mean radius of the stored muons to be calculated, leading to a precise knowledge of the corresponding magnetic field.

On the evening of 21 October 1963, a random coincidence of time and place influenced the development of the project. The present authors, having first met that morning, found themselves filling a vacant evening in the same bar of the Hawthorne Hotel, Bristol. They drifted into discussing physics, and thus initiated a 26-year collaboration.

The first Muon Storage Ring (Fig. 17) [53] was a weak-focusing ring with $n = 0.13$, orbit diameter 5 m, a useful aperture of 4 cm $\times$ 8 cm (height $\times$ width), a beam momentum of 1.28 GeV/c corresponding to $\gamma = 12$, and a dilated muon lifetime of 27 μs. The mean field at the central orbit was $\bar{B} = 72.852\,7\,(53)$ proton MHz (1.711 T).

The injection of polarized muons was accomplished by the forward decay of pions produced when a target inside the magnetic field was struck by 10.5 GeV/c protons from the

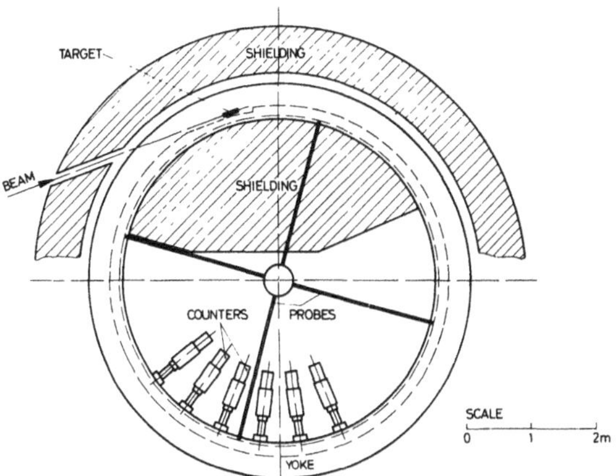

Fig. 17 Muon Storage Ring I: diameter 5 m, muon momentum 1.3 GeV/c, time dilation factor 12. The injected pulse of 10 GeV protons produces pions at the target, which decay in flight to give muons.

CERN PS. The proton beam consisted of either two or three radio-frequency bunches (fast ejection), each ~ 10 ns wide and spaced at ~ 105 ns. As the rotation time in the ring was 52.5 ns, these bunches overlapped exactly inside the ring. Approximately 70% of the protons interacted, creating, among other things, pions of 1.3 GeV/c that started to turn around the ring. The pions made, on an average, four turns before hitting the target again, and in one turn about 20% of the pions decayed. The muons created in the exactly forward decay, together with undecayed pions and stable particles from the target, eventually hit the target and were lost.

However, the decay of pions at small forward angles gave rise to muons of slightly lower momentum, and some of these fell into orbits that missed the target and remained permanently stored in the ring. Thus the perturbation, essential for inflection into any circular machine, was here achieved by the shrinking of the orbit, arising from the change of momentum in π-μ decay, and to some extent by the change in angle at the decay point, which could leave the muon with a smaller oscillation amplitude than that of its parent pion. The muons injected in this way were forward polarized, because they came from the forward decay of pions in flight. About 200 muons were stored per PS cycle. The muon injection was accomplished in a time much shorter than both the dilated muon lifetime (27 μs) and the precession period of the anomalous moment (3.7 μs).

7.2 The Injection System For Muon Storage Rings

The muon capture yield per available proton of given energy is a function of the method of injection and of the ring parameters, that is i) the aperture of the storage ring (b in the horizontal plane, a in the vertical one); ii) the radius r of the ring; iii) the magnetic field; iv) the field index n. It is difficult to express all these parameters in a general formula. However, for a given injection system and inside a limited range of the parameters, the following expression can be accepted:

$$N_\mu = F\frac{ab^3}{r^4}n^{1/2}(1-n)^{5/2}\cdot p_\pi Y$$

$$= F_1\frac{ab^3}{r^3}n^{1/2}(1-n)^{5/2}\,B\cdot Y\,, \tag{7.1}$$

where N_μ is the number of muons stored, Y is the yield of pion production (per GeV/c and steradian), which is a function of the proton and the pion momenta, and F and F_1 are factors depending on the chosen injection system. The formula can be explained simply as the product of the angular acceptance in the vertical plane $(a/r)n^{1/2}$, the angular acceptance in the horizontal plane $(b/r)(1-n)^{1/2}$, the momentum bite of the pions $(b/r)p_\pi(1-n)$, and finally the acceptable fraction of the muon momentum spectrum $\sim(b/r)(1-n)$.

In F, F_1 the decay probability of the pions must be included. This probability is very high in the case of the injection of protons in the ring onto an internal target; it is lower in the case of pion injection through an inflector; and it is very low in the case of backward decay trapping. In F, F_1 the losses due to the decay angles must also be included. There are three methods available for trapping muons in the storage region of the ring:

i) by injecting protons into the ring and having them interact with a target located at the limit of the storage region;

ii) by injecting pions of chosen momentum so that their chance to produce trapped muons is high;

iii) by directly injecting muons of the desired momentum and putting them in stable orbits by means of a very fast inflector.

The first method has been used in this (g – 2) experiment. The pions of the right momentum, produced in the internal target, travel around the ring for about four revolutions before hitting the target again; most of them will decay before, so that the muon capture yield from the circulating pions is rather high. Pions with wrong momenta travel a short distance round the ring, but their much greater number means that a considerable number of extra muons are also trapped. These muons are emitted at large angles in the pion rest frame, so the average longitudinal polarization is only 26% compared with the 95% expected. Furthermore, by injecting protons, the general background is very high around the target, and the electron counters can be located only at the opposite side of the ring (see Fig. 17).

The method of injection used in this first muon storage ring had the advantage of being very simple, but it had the following disadvantages:

i) low muon polarization due to muons from a wide range of pion momenta;

ii) high general background;

iii) contamination with electrons at early times;

iv) low average trapping efficiency.

For some time a magnetic horn was used around the target to concentrate pions of the correct energy in the forward direction. This gave a good muon polarization, but because of increased background it was not finally adopted.

The other methods of injecting muons are discussed below.

7.3 Some General Considerations for Calculating the Number of Muons Stored in a Storage Ring

In order to calculate the number of captured muons in a muon storage ring, the regions of existence of the pions and the muons of different momenta must be traced and compared in phase space, both in the horizontal and in the vertical planes [24].

In a weak-focusing field, the trajectory of any particle can be represented in phase space (with an appropriate choice of scale factors) as a circle (Fig. 18). In the horizontal plane, the centre of the circle is the radial position of the equilibrium orbit, and in the vertical plane it is the vertical position of the median plane. One complete revolution in one of these paths corresponds to a complete betatron oscillation. Taking into account the dependence of the equilibrium orbits on the momentum and the geometrical limits of the chambers, the region of existence of the captured muons is simply limited by circles that are tangential to the given limits. The probability for a pion to produce a captured muon of a given momentum is then evaluated by following its trajectory, which is again a circle centred around the pion equilibrium orbit, inside or outside the chamber, and measuring which fraction of it crosses the region of existence of the chosen muons. When this angle is not small in comparison with the angular acceptance of the ring, a correction must be made to take into account the decay angle between muon and pion.

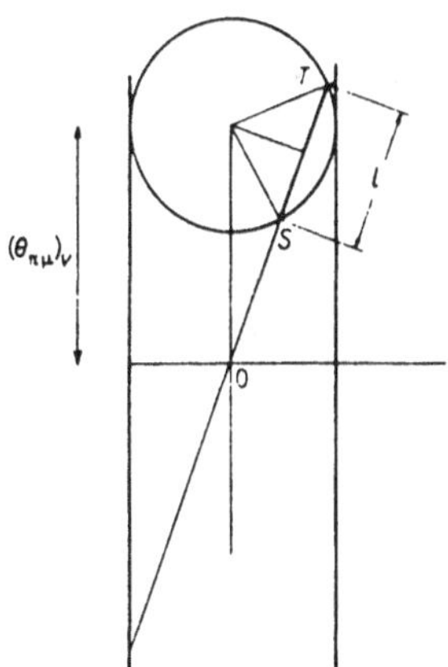

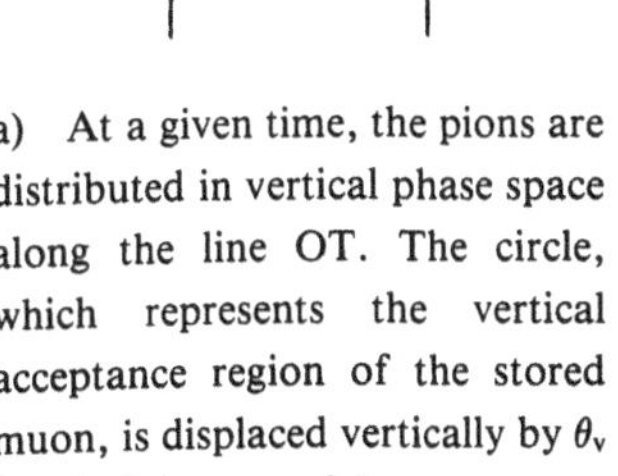

a) At a given time, the pions are distributed in vertical phase space along the line OT. The circle, which represents the vertical acceptance region of the stored muon, is displaced vertically by θ_v (vertical decay angle).

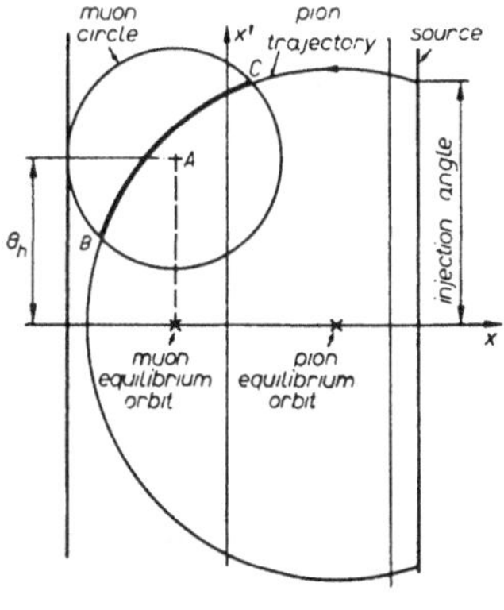

b) In horizontal phase space a pion trajectory originates from a line-source distribution, it is defined by the pion momentum and by the angle of injection. A pion can give a stored muon of momentum p_μ only when it is within the displaced circle, which represents the horizontal acceptance region of the stored muons displaced vertically by θ_h (horizontal decay angle).

Fig. 18

This method gives reliable results when a pion beam with a well-defined phase space, that is, momentum spectrum, and angular and spatial distribution, is injected into the ring through an inflector, in such a way that all pions will accomplish nearly one revolution around the ring before being lost by hitting the injection channel. The ratio between the length of the ring and the decay length of pions then gives a first probability factor; then in the phase space, the fraction of the area which is covered, during their revolution, by pions of given momentum and angular spread, and which is common to the existence area of muons of any chosen momentum, is easily estimated. This operation is repeated for several muon momenta, and then repeated again for several pion momenta and different pion injection angles.

A program based essentially on this method has been developed to compute the number of stored muons versus momentum, as well as the polarization, both longitudinally and radially, as a function of the momenta and of the angle of the pion injection. The programme also gives the number of muons produced per pion (trapping efficiency) as a function of the pion momentum and injection angle.

The program is somewhat less reliable when the protons hit an internal target, since the momentum range and the angular spread of the pions producing captured muons are very large, and the length of the pion trajectory inside the ring is dependent on the initial conditions. Nevertheless, the results gave a stored muon intensity and a muon distribution versus equilibrium radii, in good agreement with the measured intensity and with the reconstructed muon distribution.

It is worth while to point out here that this program was developed primarily to calculate the muon intensity in the second Muon Storage Ring.

7.4 The Count Rate N(t) versus f_a

The polarization direction of the muons as a function of time is followed by recording the electrons emitted by muon decay in flight. As we have discussed above (see Section 5), these decay electrons are emitted preferentially parallel to the muon spin. Because of the combined effects of the precession of the spin and the electron decay asymmetry, the number of electrons emitted parallel to β_μ varies as $(1 + AP\cos 2\pi f_a t)$, where A is the asymmetry parameter described in Section 5 and P is the polarization of the muon sample. In Section 4 it has been shown that the kinematics of the muon–electron decay provides correlation between the angle $(\beta_e \cdot \beta_\mu)$ and the electron energy in the laboratory frame. If the electrons with energy E greater than a given value E_t are counted, they are emitted from forward decays of muons and the count rate is modulated at the (g – 2) frequency

$$N_e(E > E_t, t) = N_0\, e^{-t/\tau}\,[1 - A(E_t)\cos(2\pi f_a t + \phi)]\,. \tag{7.2}$$

It was pointed out in Section 4 that in order to minimize the statistical error in the angular frequency f_a, the optimum of N_eA^2 must be found. The maximum value of N_eA^2, and therefore the minimum uncertainty in f_a, is obtained at $E_t \approx 0.65E_{max}$, which corresponds to the values E_t = 780 MeV, A = 0.42, and AP = 0.11. The exact value of E_t is not critical because the curve of N_eA^2 has a broad maximum. The full asymmetry data obtained in this experiment are shown in Fig. 19. The initial attempt to fit these data using Eq. (7.2) revealed a dependence of f_a and τ on the particular portion of data chosen, due to the presence of a residual uniform background and to muon losses caused by orbit perturbations. The data were therefore fitted with an eight-parameter fit of the following form:

$$N_e(E > E_t, t) = N_0 e^{-t/\tau}[1 - A(E_t)\cos(2\pi f_a t + \phi)](1 + A_L e^{-t/\tau}) + W\,. \qquad (7.3)$$

The data fitted with this modulated function provide a value of f_a with no statistically significant dependence on the portion of data used in the least-squares fit. The frequency f_a varied by less than ± 0.2 standard deviations as a function of the starting time of the fit.

To calculate a_μ from the data using Eq. (7.3), the value of $(e/mc)\bar{B}$ is required. The quantity $\bar{B}$ is the mean magnetic field experienced by the muon sample. The value of $(e/mc)\bar{B}$ is obtained from the magnetic field measurement in terms of the proton resonance frequency f_p and the known ratio $\lambda = f_s/f_p$ for muon and proton spin precession in the same field [4] [see Eq. (4.9)].

The magnetic field was surveyed in terms of the corresponding proton spin-resonance frequency f_p. The measurement of the magnetic field B versus radius was an integrated value of 288 positions in the azimuth round the ring. At each azimuthal position, the magnetic field was measured for ten different values of radii. During the experiment, the magnetic field was surveyed as a function of both radius and azimuth.

The frequency obtained from fitting the data with the function (7.3) is actually $[\langle f_a \rangle]$, where the time average [ ] reduces to an azimuthal average, and the ensemble average $\langle\ \rangle$ requires the evaluation of the average radius of the stored muons.

As the radial magnetic gradient necessary for vertical focusing implies a field variation of ±0.2% over the full horizontal aperture of the storage ring (8 cm), a major problem is to determine the mean radius of the ensemble of muons that contribute to the data. Since

$$\frac{\Delta B}{B_0} = n\frac{\Delta r}{r_0}, \qquad (7.4)$$

where B_0 is the magnetic field at the central orbit of radius r_0, and since $n = 0.13$, a knowledge of the average radius $\langle r \rangle$ to an accuracy of about 1000 ppm is sufficient to determine $\bar{B} = \langle B \rangle$ to 100 ppm.

Since the proton injection pulse is only 5–10 ns long, and the rotation period of the central momentum of the muons, $T = 2\pi r_0/\beta c$ is 52.5 ns, the muons are initially bunched and the counting rate in the decay electron counters is modulated with a period $T = 52.5$ ns. The low-momentum muons have a shorter rotation period than the high-momentum ones because of the spread in radius, so the bunch begins to overlap itself after about 1.5 μs and the rotational structure of the muon beam disappears after ~ 6 μs. This structure forms the basis of the primary method for finding the mean radius, since the rotation period of a given muon is determined essentially by its equilibrium radius. The analysis of the modulated records yields the mean radius to ±2.7 mm, as shown by tests on Monte Carlo generated data (the mean frequency $f = \beta c/2\pi r$ is determined in the time interval between 1.8 μs and 5.5 μs). The

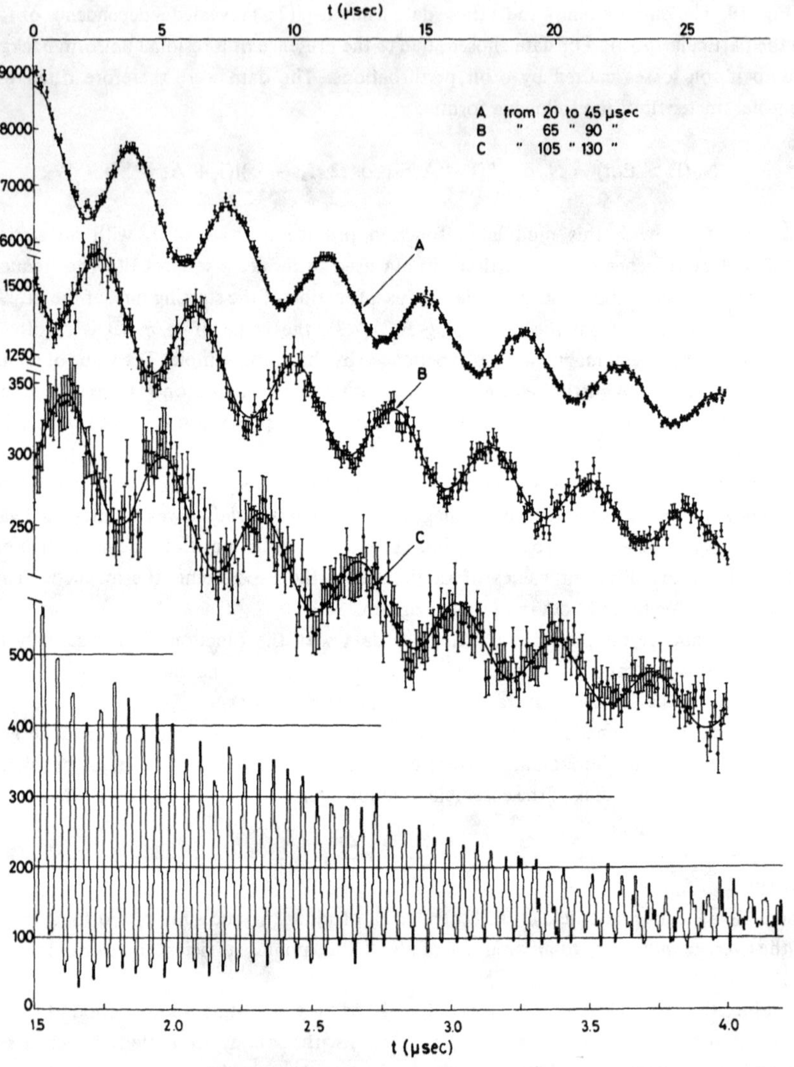

Fig. 19 Muon Storage Ring I: decay electron counts as a function of time after the injected pulse. The lower curve from 2 to 4.75 μs (lower time scale) shows 19 MHz modulation due to the rotation of the bunch of muons around the ring. As it spreads out the modulation dies away. This is used to determine the radial distribution of muon orbits. Curves A, B, and C are defined by the legend (upper time scale); they show various sections of the experimental decay (lifetime 27 μs!) modulated by the (g − 2) precession. The frequency is determined to 215 ppm, $\bar{B}$ to 160 ppm leading to 270 ppm in a_μ.

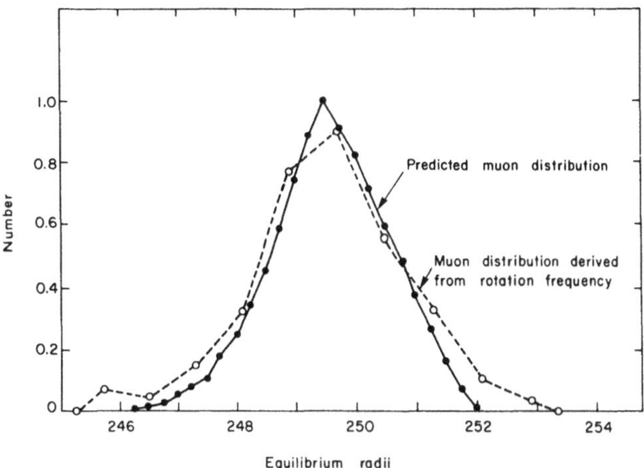

Fig. 20 The distribution of muons in radius (horizontal axis, cm) derived from the analysis of the decay electron events at early time. The muon rotation frequency has been analysed from 1.8 μs to 5.5 μs.

reconstructed muon number versus equilibrium radius may be compared with the radial distribution of muons predicted by the injection calculations. From Fig. 20 we see that the agreement is good. The reconstructed mean radius is:

$$\langle r \rangle \equiv \bar{r} = 2494.3 \pm 2.7\ \text{mm}\,. \tag{7.5}$$

In the time interval in which the mean radius of the muon population was determined, there was an excess of counts; this was due partly to the fact that muons were lost, and partly to the fact that a non-rotating background was produced in the ring as consequence of the injection system (see Fig. 21). Numerous checks [53] indicate that the method of analysis by which the mean radius is determined is valid. First a direct measurement of the rotation frequency at early times was made with reduced intensity, using 1/5 of one-bunch intensity, see Fig. 22. This determination of the mean radius gave

$$\bar{r} = 2492.5 \pm 2.1\ \text{mm}\,. \tag{7.6}$$

and corresponds to the time interval between 0.6–1.6 μs (before the muon bunch overlapped itself) in which the excess count factor was 3.0, compared with an excess count factor of 1.25 at 3 μs. However, the mean radius was in good agreement, in spite of the different excess counting rate. Secondly a loss of particles (~ 30%) occurred between the time interval in which the mean radius was measured (1.8–5.5 μs) and the time interval in which f_a was determined. This raises the possibility of a change of radius between, let us say, 3 μs and 50 μs. To study this,

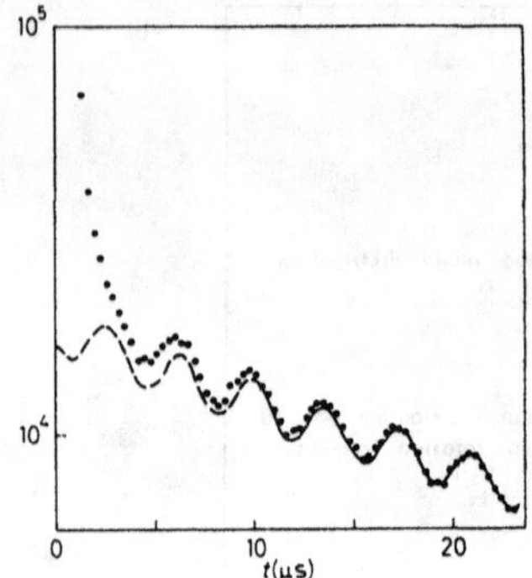

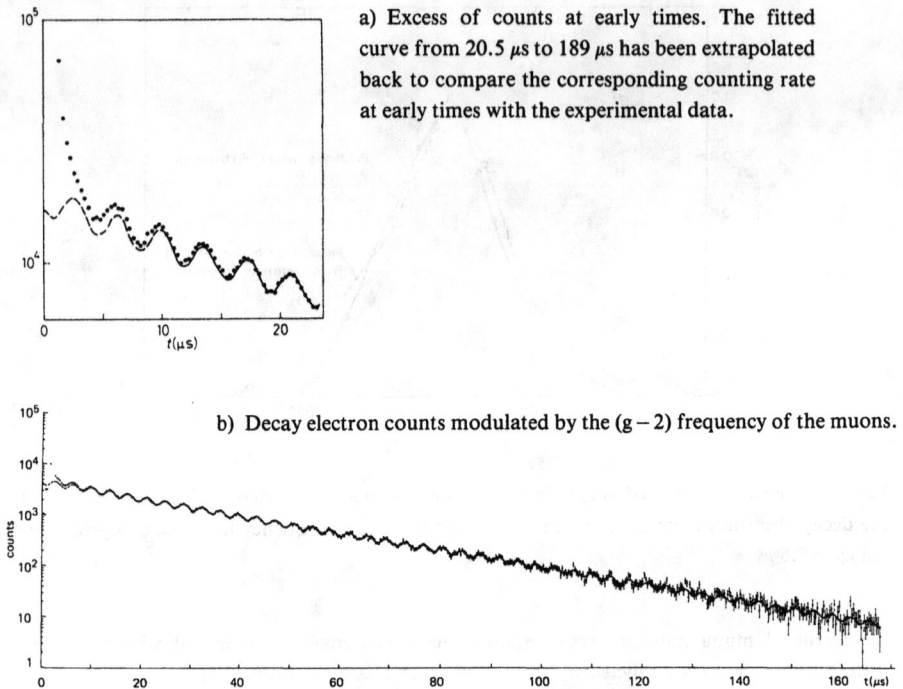

a) Excess of counts at early times. The fitted curve from 20.5 μs to 189 μs has been extrapolated back to compare the corresponding counting rate at early times with the experimental data.

b) Decay electron counts modulated by the (g − 2) frequency of the muons.

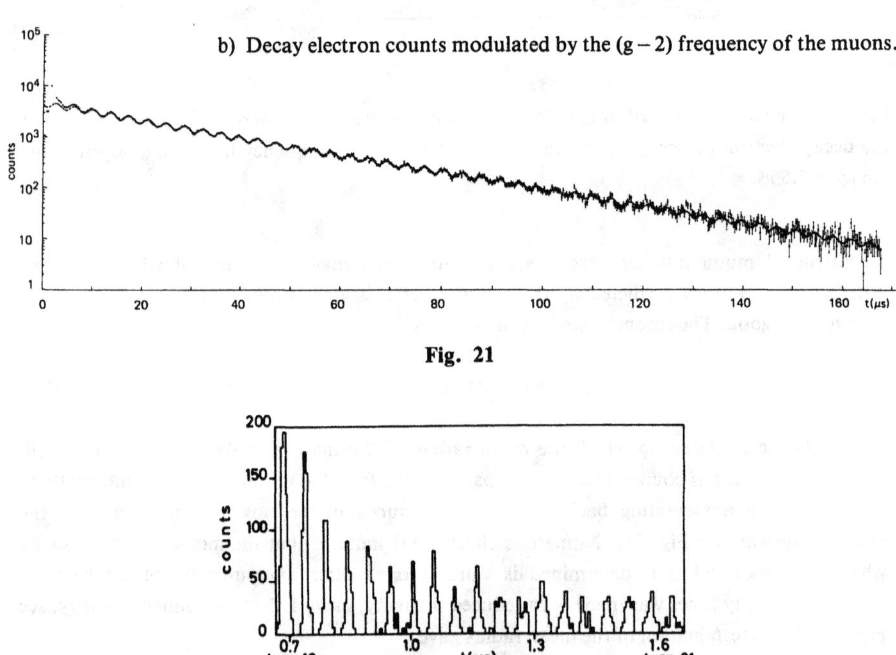

Fig. 21

Fig. 22 Reduced intensity experiment, clearly showing the rotation frequency of the muon bunch.

information was obtained by restricting the horizontal aperture from both the inside and the outside of the ring. These studies indicated that the mean radius did not change with time by more than ±1.1 mm, and a conservative overall error of ±3 mm in radius was assigned, implying an error of 160 ppm in the value of a_μ.

The frequency f_a varied by less than ± 0.2 standard deviations as a function of the starting time of the fit. The statistical error was $\pm 23 \times 10^{-8}$, and the fluctuations of the results of eight different runs about the mean gave $\chi^2 = 7.84$, compared with 6.35 expected. The random error has been increased to $\pm 25 \times 10^{-8}$ by the square root of this ratio, the error in the magnetic field corresponding to ± 3 mm uncertainty in radius was $\pm 19 \times 10^{-8}$ in a. The two errors were combined quadratically to give the overall error in a of $\pm 31 \times 10^{-8}$.

The data-fitting program, applied to the late-time part of the data, yields

$$f_a = 0.270009 \pm 0.000054\,\text{MHz}\,. \tag{7.7}$$

The radius-reconstruction program, applied to the early-time part of the same data, when combined with the magnetic field map of NMR proton frequency as function of radius, gives

$$\bar{f}_p = 72.817 \pm 0.012\,\text{MHz}\,. \tag{7.8}$$

These two frequencies were measured with the same standard crystal, so that even if this standard was in error, it did not affect the ratio that was the final result of the Muon Storage Ring experiment:

$$R = f_a/f_p = (370805 \pm 99) \times 10^{-8}\,. \tag{7.9}$$

This is proportional to the extra-Dirac part of the muon magnetic moment, and the use of the somewhat unfamiliar dimensionless unit MHz/(proton MHz), merely reflects the experimental fact that all precise magnetic field measurements are made in terms of the proton NMR frequency.

We are, of course, most interested in the dimensionless number a_μ, and in order to obtain this we must combine the result of the ratio f_a/f_p with an experimental measurement of the muon magnetic moment. At the time of the second $(g-2)$ experiment, the best such measurement came from the experiment of Hutchinson et al. [51], who stopped μ^+ in water and obtained

$$\lambda = 3.18338 \pm 0.00004\,, \tag{7.10}$$

combined with the calculation of Ruderman [54], who found the diamagnetic correction for muons in water to be

$$(1-\epsilon)^{-1} = 1.000010 \pm 0.000005\,. \tag{7.11}$$

Inserting Eqs. (7.9) and (7.10) in Eq. (4.9) gave

$$a_\mu^{exp} = (116616 \pm 31) \times 10^{-8}\,(270\,\text{ppm})\,. \tag{7.12}$$

Initially, this was nearly two standard deviations higher than the theoretical value, a sign that there was more to discover about the muon. In fact the discrepancy resulted from a defect in the theory. Theorists had originally speculated that the contribution of the photon-photon scattering diagrams to the $(\alpha/\pi)^3$ term in a_μ might be small, or perhaps cancel exactly. The experimental result stimulated Aldins et al. [55] to examine this more carefully, obtaining a coefficient of 18.4! The situation then was

$$a_\mu^{exp} - a_\mu^{th} = 28(31) \times 10^{-8} = 240 \pm 270 \text{ ppm} . \tag{7.13}$$

The difference of ~ 0.9 standard deviation cannot be called a discrepancy; rather, it confirms that the 'μ looks like a heavy electron', but it tests the theory to new levels of precision. For the photon propagator cut-off, this implied $\Lambda_\gamma > 5$ GeV, and for the muon vertex $\Lambda_\mu > 7$ GeV.

One of the by-products of the Muon Storage Ring was the determination of the lifetime of a muon beam decaying in flight [56]. In this experiment the Einstein time dilation in a circular orbit was confirmed to 1%. The muon lifetime was dilated from its value at rest (2.198 ± 0.001) μs to (26.37 ± 0.05) μs, compared with the expected value of 26.69 μs. The statistical error in the fitted lifetime was ~ 0.2%, and the agreement between theory and experiment was at the level of only 1.2%. The expected lifetime corresponds to the assumption that the muon beam was oscillating around the mean equilibrium radius. The shorter measured lifetime was ascribed to a slow loss of muons due to imperfections in the magnetic field. In Section 11 we will describe a more precise confirmation of the Einstein time dilation.

8. 1969–1976: MUON STORAGE RING II

8.1 The Second (g – 2) Experiment at the Proton Synchrotron

By 1969 an electron–electron colliding-beam experiment had demonstrated the point-like nature of the electron ($\Lambda_\gamma > 4$ GeV, $\Delta_e > 6$ GeV) [57], and e^+e^- storage rings were giving useful data on vector-meson production [58]. Experiments on e^+e^- and $\mu^+\mu^-$ pair production, on wide-angle bremsstrahlung, and on muon tridents, and a comparison of e–p and μ–p scattering were all in accord with theory. (For reviews, see Ref. [59].) The pure quantum effects were less satisfactory. The Lamb-shift data [60] were consistently higher than theory, but this was resolved by a recalculation of a small theoretical term by Appelquist and Brodsky [61]. The electron (g – 2) data of Wilkinson and Crane [19] had been rediscussed by Farley [17], by Henry and Silver [20], and by Rich [62], who concluded that $a_e^{exp} - a_e^{th} = -(79 \pm 26)$ ppm. This discrepancy was to be resolved in a new measurement by Wesley and Rich [63]. Thus QED was doing well, but in early 1969, a_μ, a_e, and the Lamb shift all showed uncomfortably large departures from theory. It could have been the beginning of something new.

The major motivations for carrying out a third measurement were therefore as follows:

i) to look for departures from standard QED;
ii) to detect the contribution of strong interactions to a_μ through hadron loops in the vacuum polarization;
iii) to search for new interactions of the muon.

The third experiment [64] had its origins very firmly rooted in the previous one, and the main challenge to a new design lay in the systematic troubles of the last experiment. Its design was an attempt to overcome the major sources of uncertainty arising in that experiment; these sources can be summarized as follows:

i) The radial magnetic gradient required to provide the vertical focusing implied a magnetic field variation of $\pm 0.2\%$ over the aperture in which the muons were stored, and a corresponding radial dependence of f_a. Even if the mean radius was determined precisely after injection, uncertainties in radius could arise from uncontrolled muon losses.

ii) The burst of particles created in the ring at injection upset the counting system and also produced a non-rotating background at early times, which made the interpretation of the data collected in the first few microseconds difficult.

iii) In spite of careful shaping of the magnetic field, there were small muon losses up to at least 100 μs after injection.

The new project overcame the previous systematic troubles by introducing the following improvements:

i) Using a ring magnet with a uniform magnetic field and focusing vertically with an electric quadrupole field. This field defocuses, of course, horizontally, slightly reducing the natural horizontal focusing of the uniform magnetic field.

ii) Injecting a momentum-selected beam of pions, instead of protons, into the ring in order to reduce the background.

iii) Reducing the loss of muons; this can be achieved by a combination of a more precise field shaping and by using an 'electric scraper'.

iv) Increasing the intensity of the stored muons to improve the statistical accuracy in f_a.

v) Increasing the product $B \cdot \gamma$, thus increasing the number of $(g-2)$ cycles per lifetime, and therefore increasing the precision in f_a.

The previous experiment showed that even if the determination of the radial distribution from fast-rotation analysis was very precise, uncertainties about muon losses from the ring would give a non-negligible error in the mean radius. Since we do not know the relative tendency of muons at different radii to be lost, the fundamental problem was to achieve focusing but remove the dependence of f_a on r. *This can be achieved in principle because the forces that hold the muon in its orbit and give focusing for small deviations from equilibrium arise from what appears, in the muon rest frame, as an electric field; but the spin precession is determined by what appears there as a magnetic field. These two fields may therefore be varied independently by applying suitable magnetic and electric fields in the laboratory frame.*

To appreciate the advantages of this method, we rewrite the classical relativistic equations of motion of a charged particle with an anomalous magnetic moment, in the laboratory fields $\vec{B}$ and $\vec{E}$ (using cgs units), already derived above in another form:

$$\frac{d\vec{\beta}}{dt} = \vec{\omega}_c \times \vec{\beta}\,, \qquad \frac{d\vec{\sigma}}{dt} = \vec{\omega}_s \times \vec{\sigma}\,, \tag{8.1}$$

with

$$\vec{\omega}_c = \left(\frac{e}{mc}\right)\left[\frac{\vec{B}}{\gamma} - \left(\frac{\gamma}{\gamma^2 - 1}\right)\vec{\beta} \times \vec{E}\right] \tag{8.2}$$

$$\vec{\omega}_s = \left(\frac{e}{mc}\right)\left[\frac{\vec{B}}{\gamma} - \left(\frac{1}{\gamma + 1}\right)\vec{\beta} \times \vec{E} + a_\mu(\vec{B} - \vec{\beta} \times \vec{E})\right] . \tag{8.3}$$

Here we have assumed that $\vec{\beta} \cdot \vec{E} = \vec{\beta} \cdot \vec{B} = 0$ (the muon charge is $-e$), and we have assumed that the electric dipole moment in the muon rest frame is zero. The precession of the spin relative to the velocity vector is given by

$$\vec{\omega}_a \equiv \vec{\omega}_s - \vec{\omega}_c = \left(\frac{e}{mc}\right)\left[a_\mu \vec{B} + \left(\frac{1}{\gamma^2 - 1} - a_\mu\right)\vec{\beta} \times \vec{E}\right] . \tag{8.4}$$

From this equation it can be seen that the final term giving a dependence of ω_a on the electric field can be made zero if γ is correctly chosen. That is, if one operates at a value of $\gamma = (1 + 1/a_\mu)^{1/2}$ called the magic value $\gamma_{magic} = 29.304$ corresponding to momentum $p = 3.094$ GeV/c. In this case the muons can be stored in a perfectly uniform magnetic field which determines the spin motion, with focusing provided by an electric field which has no influence on the spin. This new principle is the essential basis of the final CERN experiment [65].

The value of the electric field was chosen to give appropriate focusing, but it was not needed for calculating a_μ. Focusing in the simultaneous fields is no problem [24]. If we define $n_B = -(r/B)\, dB/dr$ and

$$n = n_B + \frac{r}{\beta B}\frac{dE_r}{dr} \tag{8.5}$$

then the equations for axial and radial displacements from equilibrium, $z'' + nz = 0$ and $x'' + (1-n)x = 0$, and the focusing condition $0 < n < 1$, are formally unchanged. In the last experiment, we proposed the following, particularly simple solution: $n_B = [E_r]_0 = 0$ and therefore Eq. (8.5) becomes simply

$$n = \frac{r}{\beta B}\frac{dE_r}{dr} . \tag{8.6}$$

The voltage to be applied to electric quadrupoles disposed continuously round the ring with half aperture a (vertical) $\times$ b (horizontal) then becomes

$$V = n\frac{\beta pc}{2e}\frac{a^2 + b^2}{r^2} \sim 22 \text{ kV} \tag{8.7}$$

in our case. In practice the electrodes were not continuous in azimuth and required somewhat higher voltages.

The cancellation of the effect of the electric field on the spin motion is exact only for the central momentum of the muon distribution; for other momenta a correction is required. If the muons uniformly fill the available phase space, the average correction is

$$\Delta\omega_a/\omega_a = 0.2\, n(1 - n)\,(b/r)^2\,. \tag{8.8}$$

This is 1.7 ppm for n = 0.14, b = 6 cm, r = 700 cm.

The system for injecting muons into the ring was designed to give maximum muon polarization, minimum background, and as large an intensity as possible. A high value of the longitudinal polarization can be achieved by starting with a momentum-selected pion beam, and accepting only decay muons whose momenta lie in a narrow band close to that of the pion beam.

It was therefore decided to locate the primary target outside the Muon Storage Ring, and to prepare a momentum-selected pion beam to be guided into the ring by a pulsed inflector. Because of the size of the inflector structure, the pions would make only one turn in the ring, and the useful aperture of the inflector would be very small. The loss of intensity due to these factors could, however, be compensated by using special beam optics, which collected pions over a large solid angle and matched them to the acceptance of the storage ring.

The injector was in the form of a coaxial line in which a 10 μs current pulse of peak value 300 kA produced the required field of about 1.5 T between the inner and outer conductors. The great technical difficulty of this method of injection was outweighed by the increased pion flux and the high longitudinal polarization of the stored muons (95%). This was borne out by the large observed modulation of the decay electron counts. The calculated polarization direction was independent of the muon equilibrium radius; and consequently any possible asymmetric muon losses could cause no significant shift in the measured spin precession frequency f_a. Finally, the background was considerably reduced with respect to the previous experiment (in which the copper target was located in the ring), and therefore the electron detectors could be located all around the ring. Other improvements can be mentioned only briefly in this review.

8.2 Electric Field and Vacuum

The electric quadrupole system has been described by Flegel and Krienen [66]; here we point out only the general features. The quadrupole electrodes were constructed in eight sectors, each sector covering about one tenth of the circumference of the ring. The sectors were arranged in two groups of four with two equal spaces between them. One of these spaces was needed for the pion injection region while the other was required to balance this omission and cancel the first harmonic closed-orbit distortion. The cross-section of the quadrupole and vacuum chamber is shown in Fig. 23. The vacuum was normally better than 10^{-7} Torr and it was obtained by ion getter pumps to avoid any deterioration of the high-voltage behaviour of the electrodes by oil vapour.

The shape of the electrodes approximated to the ideal case of a pair of hyperbolae which would be required to give a perfect quadrupole field, and was calculated by successive approximation to minimize the variations in electric field gradient throughout the storage region. The presence of the vacuum chamber affects the electric field, and so for a given geometry the uniformity of the field gradient depends upon the chosen ratio of potentials applied to the two pairs of electrodes. The ratio which minimized the contribution of other multipoles, in particular the octupole, was found by calculation and later adjusted by measuring the muon losses.

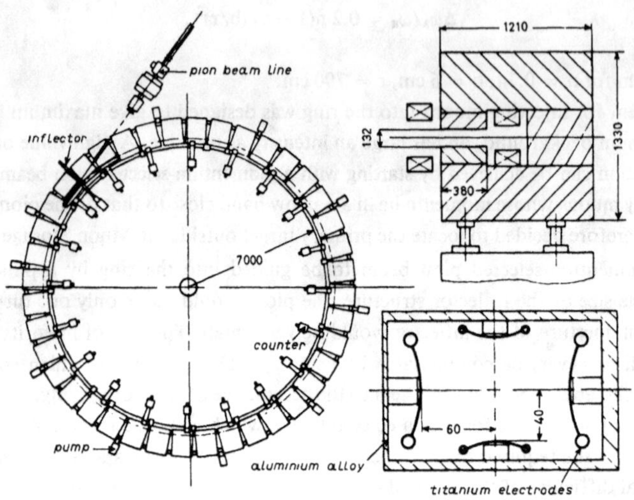

Fig. 23 Muon Storage Ring II, which consists of 40 contiguous magnet blocks. The open side of the C-shaped yoke (upper right) faces the centre of the ring. The cross-section of the vacuum chamber and electric quadrupole is shown at the bottom right. The decay electrons are detected by 20 counters. Dimensions are in mm.

It had been demonstrated in the preparation of the experiment that it was not possible to use steady electric fields for both muon polarities, because trapped low-energy electrons produced too much ionization in the residual gas. These low-energy electrons were trapped, vertically, by the electric potential, and while spiralling up and down the magnetic field lines they slowly drifted along a side electrode in the direction $\vec{E} \times \vec{B}$. At the end of the focusing system the field shape was such that these low-energy electrons passed around the electrode and returned along the reverse side; they were therefore trapped inside the electrode system, with mean energy of 1 keV, and they were able to ionize the residual gas and to cause a voltage breakdown even if the vacuum was better than 10^{-6} Torr. However this process took some milliseconds, so that the danger could be greatly reduced by operating in a pulsed mode. Voltages in the form of flat-topped pulses, about 1 ms long, were applied to the electrodes by an arrangement of spark gaps. The electric field gradient corresponding to $n = 0.185$ was $dE_r/dr = 1.17 \times 10^7\ \mathrm{V \cdot m^{-2}}$ for $B = 1.472$ T, implying about 40 kV applied to the horizontal electrodes.

At the beginning of the experiment the low-voltage pulser used for the vertical electrodes was made with spark gaps which could not be regulated below 5 kV. This was too high to minimize the higher multipole field components, but the resulting muon losses were too low to affect the $(g-2)$ frequency. For the lifetime measurement, however, this mode of operation was inadequate, and the spark gaps were replaced by cascaded thyristors, which allowed the true optimum level to be found and the muon losses to be minimized.

The operation of the full electrode system of the storage ring brought with it some problems which were not present in the test set-up. For the μ^+ polarity, both high and low voltages could be switched on together before the injection of pions and switched off together some 700 μs later, while for the μ^- polarity breakdown ensued unless the low-voltage crowbar was delayed by 16–32 μs after the high-voltage pulse had been terminated.

8.3 Electric Scraping

For the measurement of the muon lifetime it was essential to reduce the late-time muon losses to a minimum. This was done by shifting the muon orbits at early times both vertically and horizontally in order to 'scrape off' those muons which passed near the edge of the aperture and were most likely to be lost. The orbits were shifted by applying voltages asymmetrically to each pair of electrodes, and then gradually bringing them back to normal. While one member of the pair was pulsed normally with 1 μs rise time, the other was fed with a pulse which had a 10 μs rise time. For the side electrodes the asymmetry of the voltages was reversed between the two groups of focusing units on either side of the ring. Thus the muon orbits were shifted radially outwards on one side of the ring and radially inwards on the other; to first order they were just shifted sideways with their shape remaining the same. The bottom electrodes, all the way round the ring, received the longer pulse, and so the median plane of the muon orbits was initially low and returned to its normal position with a 10 μs time constant. The pulses were switched on about 6 μs before the arrival of the pion bunch. The whole question of particle orbits, muon losses, and the effect of scraping is discussed in detail in Section 4 of the final report of the CERN Muon Storage Ring [64]. One minor disadvantage of scraping was that the horizontal perturbation allowed some protons accompanying the π^+ beam to be injected onto stored orbits, and these could hit the vacuum chamber at later times producing background in the counters.

8.4 The Storage Ring Magnet

For a given momentum the maximum accuracy in f_a is obtained by working at the highest magnetic field. The actual field of 1.47 T was chosen to meet this requirement without jeopardizing good field uniformity.

In the preparation of the magnets, the aim was to provide a field as nearly uniform as possible so that the f_a of all the muons would be the same. Because the muons sample the field all the way round the ring this requirement essentially meant that, after averaging in azimuth, the field should be independent of radius. A further aim in the construction of the storage ring was that the field should be stable and reproducible to a few parts per million (ppm). The novel procedure to achieve this spectacular result has been described elsewhere [67]. The success of this procedure is indicated by the vertical field map shown in Fig. 24. This map is a contour line plot for the storage aperture and was obtained by averaging the three-dimensional map in azimuth. The interval between the contours of equal field strength is 2 ppm or 3 μT.

As can be seen from Fig. 23, the Muon Storage Ring consisted of 40 C-shaped bending magnets, each about 1 m long, fitted together to form a regular polygon with open sides facing the centre of the ring. The pole gap was 14 cm high and 38 cm wide. Each magnet was

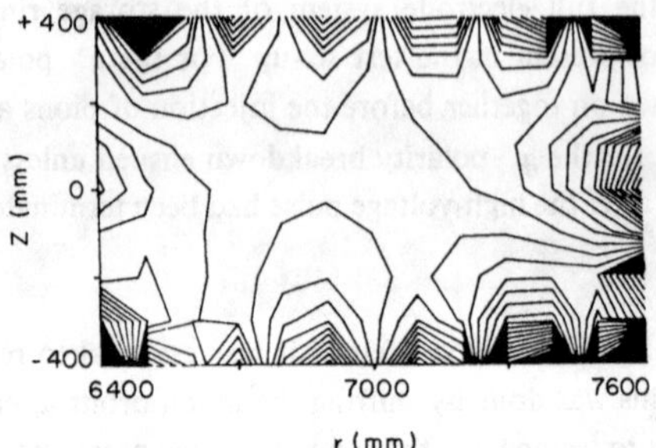

Fig. 24 A contour line plot of the magnetic field strength in the muon storage aperture. This map is obtained by averaging a three-dimensional map in azimuth. The interval between the contours of equal field strength is 2 ppm or 3 μT.

individually supported on a ring-shaped concrete foundation. The pole pieces were cut at an angle of 4.5° at each end so that they fitted together to form a closed polygon. However the yoke was not fitted together and consequently the field at each junction was reduced; this drop in the field, after various adjustments in the geometry of the iron of the magnet, was about 400 ppm. The magnets were energized in series by four large circular coils providing continuous current all the way round the ring although for cooling purposes each coil was divided into five sectors. The field in every sector was monitored with 40 NMR probes, the signal from which was used to set the field at the correct level and to provide its stability via the control of the current through 40 compensating coils wound around the yoke close to each pole. Thus the field at 40 points in the gap just outside the muon storage region was stabilized.

It is worth emphasizing here the extreme insensitivity of the average value of the magnetic field B computed for different assumed radial distributions of muons. Even in extreme cases the average magnetic field was the same within less than 2 ppm, compared with the 160 ppm uncertainty in B in the previous experiment. Thus the uniformity of the magnetic field in the storage ring was such that the (g − 2) frequency was essentially independent of the distribution of muons within the storage region. However, a precise value of the mean radius of the muon population was still needed in order to derive the muon lifetime at rest from the measurement made in flight.

8.5 Stabilization and Reproducibility of the Magnetic Field

To ensure the stability and reproducibility of the magnetic field, each of the 40 magnet blocks was stabilized separately by a control system which had an NMR probe and a pick-up coil as its sensors [68]. The signals from these devices were used to determine the current automatically through additional compensating coils which were wound around the yoke of individual magnets close to each pole tip. The NMR probe contained a flat coil for modulating the field in a small plastic tube, shrunk around a cylindrical RF coil of 2 mm diameter and 4 mm length, and containing a 0.1 molar solution of $NiSO_4$. The applied field modulation, B_{mod}, was a symmetric 30 Hz triangular waveform with a peak-to-peak amplitude of 0.8 mT, and the system operation was such that the proton frequency resonance for $B_{mod} = 0$ equalled the frequency of a crystal oscillator. The signal from the pick-up coil dominated the response to changes in the magnetic field at frequencies above 1 Hz, while the NMR probe counteracted the slower drifts. By using eight different crystal frequencies for the NMR control probes, making local special shimming at the probe position and adjusting the iron shunts, the mean fields in all blocks were as closely equalized as possible. This was done in order to minimize all compensating currents. During the operation of the ring, the currents through the main coils were adjusted automatically so as to keep the average of the 40 compensating currents close to zero.

To bring the magnet into the same operating condition, a special switching-on procedure was adopted; without it the field shape would not have been reproducible, and although the field values at 40 locations of the fixed frequency stabilizer probes were the same, temporary eddy currents or hysteresis in the yoke could so modify the distribution of magnetization in the iron that the overall average field in the storage region could change by as much as 50 ppm. The procedure consisted of three rapid (50 A/s) up and down cycles of the main current to a value some 10% higher than the operating point ($I_0 \sim 4.5$ kA), followed by slow (1 A/s) oscillations about I_0 of a few per cent in amplitude and gradually decreasing. This cycling of the storage ring magnets was controlled automatically and took about half an hour to complete, at which point the stabilization system automatically switched-on.

During data-taking runs the magnetic fields were monitored in 37 of the magnets. This was done with small NMR probes which could be driven into the muon storage ring along a radial line in the median plane without breaking the vacuum. The 37 probes were connected, via multiplexers, to the same eight magnetometers used in the field mapping. The whole procedure was controlled by a computer program and only took 10 minutes. The 400 points at which these plunging probes measurements were made were located within the coordinate system of the full field map by survey and the relationship between the two sets of readings determined. This allowed the drift in the mean field to be followed throughout the periods between full-scale maps.

Between groups of data-taking runs, the whole vacuum chamber was removed and a full map made of the magnetic field with a system of eight NMR probes mounted on a measuring machine driven around the ring under computer control. The probes were connected to automatic magnetometers, and the measurements could be made and recorded whilst the machine was moving at the speed of up to 6 cm/s without significantly distorting the map. About 250000 points were measured throughout the storage volume in steps of 1 cm vertical by 1 cm radial by 2 cm azimuthal. Careful tests had to be carried out on the reproducibility of the field. As a result of the cycling procedure, the field in the magnetic gap followed a predictable pattern; the average value rose by some 5 ppm over the first two days after switching on and then remained constant to within ± 1 ppm. The full maps of the magnetic field truly reflected the field during the data taking, subject to some small systematic effects such as the influence of the vacuum chamber and the electrodes. The effects of different parts of the chamber and electrodes were measured and the overall contribution to the average field value was less than one part per million [68].

The probes used in mapping, monitoring, and stabilizing the magnetic field were calibrated with respect to a special probe of well-defined geometry and susceptibility correction [68]. The crystal clock which was used to measure the proton resonance frequencies was also used to time the muon decays, and as the experimental value of the anomaly depends upon the ratio of these two measurements, the effect of any drift in the clock frequency was largely cancelled out. The radial component of the magnetic field was measured with a Hall-plate pendulum and found to be at the level of 10^{-4} T, sufficient to cause the median plane of the muon orbits to undulate with respect to the geometrically central plane of the storage region over a range of about 2 mm. This shift was confirmed by measurements on the muon distribution.

8.6 Radial Distribution of Muons and Magnetic Field Analysis

Information on the radial distribution was obtained as before by analysis of the bunch structure of the muons in the ring at early times. This fast rotation pattern is shown in Fig. 25. The pion bunch had a width of 10 ns, a time structure which was inherited by the decay muons. Shortly after injection, as the bunch length is much shorter than the rotation period in the ring (147 ns), the decay electrons detected by a particular counter will appear in discrete bunches. The delays of the various counters were adjusted so that the bunched data from different counters could be coherently added, simply by combining single-counter histograms suitably shifted by an integer number of bin widths (10 ns).

Two methods of analysis were available. In the first, which was applicable only to non-overlapping bunches, the centroid in time of each bunch was calculated: by fitting a

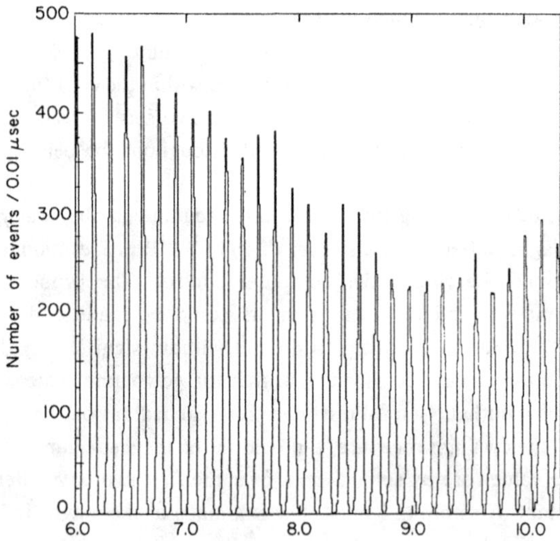

Fig. 25 The fast rotation pattern. This is the count rate at early time which clearly shows the muon bunch rotating around the ring with a period of 147 ns. Horizontal scale in μs.

straight line to these values, the mean rotation frequency f_{rot} was obtained. The muon energy was then deduced from the average γ-parameter $\bar{\gamma}$, which was given by

$$\bar{\gamma} = 2\lambda \bar{f}_p / g f_{rot} , \tag{8.9}$$

where $\lambda = f_s/f_p$ as above; $\bar{f}_p$ is the mean proton magnetic resonance frequency, corrected to vacuum, over the muon orbits; and g is the g-factor of the muon. The mean muon radius is

$$r = \beta c / 2\pi f_{rot} . \tag{8.10}$$

In the second method more detailed information on the radial distribution was obtained by making a least-squares fit to the shape of the bunches as a function of storage time. The program to perform this analysis was also used in the previous CERN (g − 2) experiment where knowledge of the radial distribution was crucial. Although all the muons in the storage ring had essentially the same velocity ~ c, the high-energy ones, because of their longer path length, gradually lagged behind those of lower energy. This resulted in spreading out the bunch as time progressed, so that eventually all bunches overlapped. The rate at which this overlap occurred was sensitive to the radial distribution of the muon orbits. The latter was found by least-squares fitting to the observed time development of the bunches. With one pion bunch injected, the bunches of detected electrons remain resolved until ~ 10 μs, and the analysis was normally carried out in the time range 5 to 20 μs or about 100 revolutions. However, the bunch structure did not disappear completely until about 40 μs after injection. The analysis was also carried out when two bunches, separated by 105 ns, were injected into the ring. In this case the fit was performed entirely on overlapping bunches, and the simple centroid-fitting procedure described above could not be used. A detailed description of the second method used to determine the mean radius is given in the two papers which describe the second and third (g − 2) experiments [53, 64]. The statistical error on the mean radius given by the fit is typically 0.1–0.2 mm. A number of systematic checks of the fitting program were carried out, both by using Monte Carlo data as input and by varying input assumptions and the parameters in the fits to real data. Particular attention was given to systematic errors such as:

i) rate and dead-time effects;
ii) the effect of a flat background between bunches;
iii) the effect of muon losses;
iv) the width of the input pion pulses;
v) the shape of the input pulses.

The fast rotation pattern and the radial muon distribution are shown in Figs. 25 and 26.

The determination of the average magnetic field in terms of the proton magnetic resonance frequency f_p, for each set of data which was analysed separately for f_a, involved three distinct steps:

i) Using the parameters found in the analysis of the radial distribution of the stored muons an ensemble average of the full field map was made. The effect of the electric field for muons not exactly on the nominal radius gave a correction of the order of 2 ppm and was included in this average.

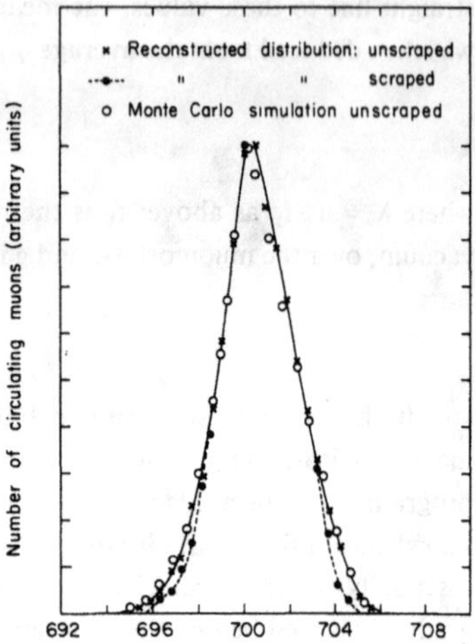

Fig. 26 The reconstructed radial distribution of muons for scraped and unscraped data compared with a Monte Carlo simulation of the latter.

i) Using the parameters found in the analysis of the radial distribution of the stored muons an ensemble average of the full field map was made. The effect of the electric field for muons not exactly on the nominal radius gave a correction of the order of 2 ppm and was included in this average.

ii) Using information from the plunging probe measurements a correction was applied to allow for any changes in the magnetic field between the data collection and the making of the full map.

iii) The measured proton magnetic resonance frequency was corrected to the value in vacuum. These steps are described in Section 5.3 of Ref. [64].

iv) The pitch correction was applied using Eq. (4.36). Assuming that the muons populate the available vertical phase space uniformly the average value of ψ_0^2 is 1/2 $n(a/r)^2$. For $n = 0.135$, $a = 4$ cm, $r = 700$ cm, the pitch correction is then 0.5 ppm.

To achieve the accuracy reached in this last experiment many technical problems had to be solved. The ability to find these solutions constituted part of the beauty of the experiment and we invite the reader to consult in particular the final report on the CERN Muon Storage Ring for all details [64]. Figure 27 shows the combined decay electron counts versus storage time for the whole experiment.

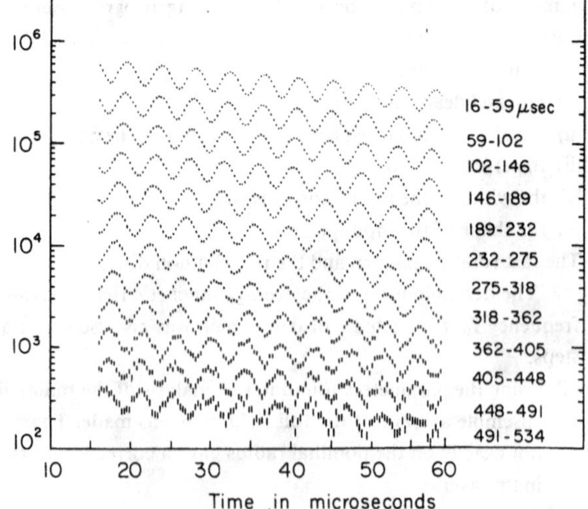

Fig. 27 Muon Storage Ring II: decay electron counts versus time (in microseconds) after injection. Range of time for each line is shown on the right (in microseconds).

8.7 Data Analysis

The raw data for each detected decay electron consisted of a recorded time in bins of 10 ns within the range 0 to 655.35 μs, and labels for counter and pulse height. The time spectrum of the decay electrons was fitted using the function minimization program MINUIT. If we denote the observed spectrum by $H(t_i)$ and the fitted function by $N(t_i)$, the likelihood function L is given, assuming Poissonian errors for $H(t_i)$, by:

$$L = \prod_i \exp[-N(t_i)] \frac{[N(t_i)]^{H(t_i)}}{H(t_i)!} . \tag{8.11}$$

The function actually minimized by the program MINUIT is

$$\ln L = \sum_i [H(t_i) \ln N(t_i) - N(t_i)] , \tag{8.12}$$

where the constant term $\Sigma_i \ln H(T_i)$ is dropped and $N(t_i)$ is taken as

$$N(t_i) = N_0 \{L(t_i) \exp(-t_i/\tau)[1 - A\cos(2\pi f_a t + \phi)] + W\} . \tag{8.13}$$

The function $L(t_i)$ is an empirical correction factor to allow for the effect of muon losses and gain changes at early times. The 'ideal' distribution would have $L(t_i) = 1$ and the background $W = 0$. The loss function $L(t_i)$ is parametrized in the form:

$$L(t_i) = 1 + A_L \exp(-t_i/\tau_L) . \tag{8.14}$$

The total number of parameters used in the fits is then eight: five (N_0, τ, A, f_a, ϕ) from the 'ideal' distribution, plus three (W, A_L, τ_L). As many of the parameters in the fit are correlated with each other, some care is needed in extracting the frequency by such global fits. The only parameter strongly correlated with f_a is ϕ. However, attempting to fit data with early-time loss or gain distortions to a simple six-parameter (N_0, τ, A, f_a, ϕ, W) function leads to an incorrect value of f_a. This effect was noted in the previous $(g-2)$ experiment. For six parameters the fitted value of f_a varies with the starting time of the fit, oscillating about the correct value as a function of time with a frequency of approximately f_a. Including A_L and τ_L in the fit removes these oscillations. The mechanism of this frequency shift is easily understood in terms of 'phase pulling' of the early $(g-2)$ cycles by the excess counts at early times when the simple six-parameter function is used. The final results, quoted in this experiment, were obtained with the starting time chosen for the fits at the zero crossing of the oscillation (21 μs), so that they will be unchanged within statistical error whether the six- or the eight-parameter fit is used.

For μ^+ data with scraping, the background due to protons was at the level of approximately 10^{-3} at early times but had a negligible effect on f_a. For all the μ^- data and those for μ^+ without scraping, W was at the very low level of 2×10^{-5}.

Particular attention was given to possible systematic corrections to the $(g-2)$ frequency, and checks were made with artificial light pulses during the data-taking runs. Among them the more important were:

i) Time slewing effect due to the large counting rates experienced by the counters and the electronics after injection. The effect varied considerably from counter to counter, depending on the position relative to the injected beam. These effects on the frequency shift are very small: 0.9 ppm for a starting time of 15 μs, and $<$ 0.5 ppm for a starting time of 21 μs.

ii) Gain effects. These are included in the empirical function L(t) for the (g − 2) frequency analysis.

iii) Digitron calibration and linearity. Both the linearity and the absolute calibration of all digitrons used in the experiment were found to be good to a small fraction of ppm.

iv) Queuing and dead-time losses. Experimental checks were done with great care and these experiments are discussed in Ref. [64]. The raw frequency queuing-loss correction in the worst case was 1.3 ppm.

8.8 Experimental Results

Nine separate runs were made over a period of two years to measure the (g − 2) precession frequency f_a, the field being determined in terms of the proton resonance frequency f_p. The ratio $R = f_a/f_p$ showed good consistency ($\chi^2 = 7.3$ for eight degrees of freedom) (see Fig. 28). The overall mean value is the principal result of the experiment:

$$R = f_a/f_p = 3.707213(27) \times 10^{-3}\ (7\text{ ppm})\,. \tag{8.15}$$

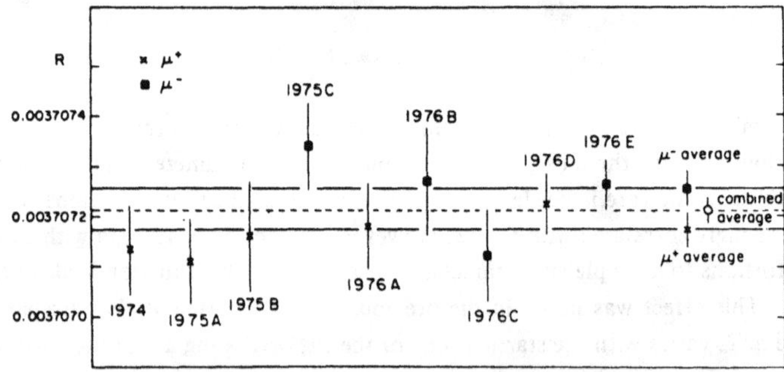

Fig. 28 Individual values of the ratio $R = f_a/\bar{f}_p$ for the nine experimental periods of the third experiment together with the weighted averages.

This error is made up of a 7.0 ppm statistical contribution from f_a and a 1.5 ppm systematic contribution from f_p. This number is the essential result of this experiment. Equation (4.9) allows us to calculate the anomaly if $\lambda = f_s/f_p$ is known.

The magnitude of λ has now been determined directly from measurements of muon precession at rest [13] and indirectly from the hyperfine splitting in muonium [14]. The weighted average value of these measurements [4] is

$$\lambda = 3.18334547(47)\,.$$

This leads to the following results for the anomalous moment, slightly different from those published in Ref. [64] because the value of λ has changed:

$$a_{\mu^+} = 1\,165\,910\,(11) \times 10^{-9}\,(10\,\text{ppm}) \tag{8.16}$$

$$a_{\mu^-} = 1\,165\,936\,(12) \times 10^{-9}\,(10\,\text{ppm}) \tag{8.17}$$

and for μ^+ and μ^- combined

$$a_\mu = 1\,165\,923\,(8.5) \times 10^{-9}\,(7\,\text{ppm})\,. \tag{8.18}$$

Compared with (2.9)

$$a_\mu^{th} = 1\,165\,918\,(2) \times 10^{-9}$$

the result is

$$a_\mu^{exp} - a_\mu^{th} = (5 \pm 8.5) \times 10^{-9}\,. \tag{8.19}$$

Table 1 gives a summary of the measurements of a_μ. It includes an experiment by Henry, Shrank and Swanson [42] using a long solenoid magnet in analogy with the electron (g – 2) experiment of Schupp, Pidd and Crane [41], which we have not described in detail.

Table 1

Experimental results for a_μ

		Ref.	Dates
μ^+	0.001 145 (22)	43	1961
μ^+	0.001 162 (5)	43	1962, 1965
μ^-	0.001 165 (3)	53	1966
$\mu^\pm$	0.001 166 16 (31)	53	1968, 1972
μ^+	0.001 060 (67)	42	1969
μ^+	0.001 165 895 (27)	64	1975
$\mu^\pm$	0.001 165 923 (8.5)	64	1977, 1979

8.9 Comparison Between Theory and Experiment

The main six conclusions that can be drawn from this last measurement of the anomalous magnetic moment of the muon are the following:

1. The QED calculations of the muon anomaly are verified up to the sixth order, the experimental uncertainty being equivalent to 1.2×10^{-6} in A, 3.5×10^{-3} in B, or 4.7% in C [see Eq. (2.1)].

2. The hadronic contribution to the anomaly is confirmed to an accuracy of 20%. The existence of hadronic vacuum polarization has thus been established at the level of five standard deviations.
3. There is no evidence for a special coupling of the muon. The experimental range of possible values of an extra contribution to the moment is

$$-12 \times 10^{-9} < \Delta a_\mu < 22 \times 10^{-9} \tag{8.20}$$

to 95% confidence. The limits implied for unknown boson fields then depend on the nature of the coupling and are given in Fig. 29 [64].

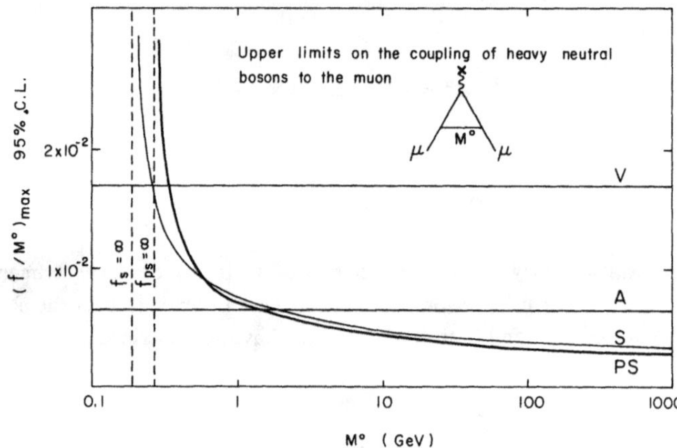

Fig. 29 The upper limits on the coupling constant f of the muon to a heavy neutral boson of mass M^0, for vector (V), axial vector (A), scalar (S) and pseudoscalar (PS).

4. With the advent of renormalizable gauge theories unifying the weak and the electromagnetic interactions, the calculation of the weak interaction contribution to the muon anomaly has become reliable. In general the weak contribution depends upon the parameters of the theory, such as the masses of the Higgs and intermediate vector bosons. To the extent that we do not yet know the correct form of the weak interaction Hamiltonian, the above results for a_μ (and also the result for a_e) can be used to restrict the range of possible models. Only in the simplest of such theories, that of Weinberg [9] and Salam [9], are the parameters sufficiently well determined experimentally to give a firm prediction of the expected value of the weak anomaly. Kinoshita [3] has recently reviewed this subject. If the arguments of Weinberg on the Higgs boson are accepted, and the current limits on $\sin^2\theta_W$ are taken into account, we obtain:

$$1.9 \times 10^{-9} \leq a_\mu\,(\text{weak}) \leq 2.3 \times 10^{-9}\,. \tag{8.21}$$

Clearly the precision of even the latest experiment is inadequate for testing this prediction.

5. The range defined by the inequalities Eq. (8.20) may be used to set limits on single contributions to the muon anomaly from other sources, including various models for breaking QED as discussed in the theoretical section. The following limits apply to 95% confidence:
   (a) The muon may not behave like a point charge, but instead have a finite size, in analogy with the proton. This would show up as a form factor, see Eq. (2.10). The limit imposed on Λ_μ would be $\Lambda_\mu > 36$ GeV.
   (b) The modification in the photon propagator of Eq. (2.11) leads to $\Lambda_\mu > 25$ GeV.
   (c) From the latest experiment it is possible to set a limit to the modification of the muon propagator [69] by a factor $(1 - q^4/\Lambda_{prop}^4)$. The value is $\Lambda_{prop} > 1.5$ GeV.
   (d) A possible new, undiscovered lepton of mass M_L would contribute to the vacuum polarization through a mechanism such as diagrammed in Fig. 1d. The value of the anomaly would depend on the ratio M_L/m_μ. The 95% confidence limit sets the limit $M_L > 210$ MeV/c^2, which is not very interesting. In passing, the recently discovered heavy lepton τ [70], with mass 1.8 GeV/c , gives a contribution, $\Delta a_\mu(\tau) \simeq 0.4 \times 10^{-9}$, well below the present sensitivity.
6. Recently Kadyshevsky [71] has given a new gauge formulation of the electromagnetic interaction theory, containing a 'fundamental length' ℓ as a universal scale constant as important as $h/2\pi$ and c. This new hypothetical constant ℓ, together with $h/2\pi$ and c, is expected to regulate all microscopic phenomena. The quantity $M = h/2\pi\ell c$ plays the role of a fundamental mass. In the new approach the electromagnetic potential becomes a 5-vector associated with the de Sitter group O(4,1). Among the various predictions given is the value of the anomalous moment for a lepton of mass m_ℓ

$$a_{lepton} \simeq m_\ell^2/2M^2$$

and the electric dipole moment (EDM)

$$d_{lepton} \simeq e\ell/2\ .$$

From the present experimental result one then obtains an upper bound for the fundamental length: $\ell < 3.9 \times 10^{-17}$ cm.

8.10 Comparison between the Muon and Electron g-Factor Experiments

It is interesting to consider the following comparison between the recent experiments to measure the anomalous parts of the muon and electron g-factors by the precession and spin resonance methods, respectively [72]. These two experiments are of vastly different scales; the electrons have non-relativistic energies in the region of 1 meV, while the muons are highly relativistic with energy of about 3.1 GeV. The electrons are trapped on cyclotron orbits of radii ~ 1 μm and oscillate vertically over a distance of about 200 μm, while the muons, travelling virtually at the velocity of light, follow near-circular orbits of 7 m radius and execute vertical oscillations over a range of up to 80 mm. There is little doubt about the macroscopic nature of the motion of the many millions of muons which contribute to the measured signal, but in the mono-electron oscillator a single electron spends a sizeable part of its time in the lowest quantum level of the system.

These differences are largely dictated by the need to use the relativistic time dilation of the muon lifetime on the one hand and to minimize the relativistic frequency shift due to the electron trapping potential on the other: in spite of them there are strong similarities between the two experiments; in particular, the use of a type of Penning trap is common to both. The trap consists of an electric quadrupole superimposed upon a uniform magnetic field. Both the configurations used in these two experiments have axial symmetry, and a general form for the electric potential in cylindrical coordinates is

$$V(r,z) = V_0/b^2 \{r^2 - 2r_0^2 [\ln(r) - \ln(r_0)] - r_0^2 - 2z^2\}\,, \tag{8.21}$$

where r_0 is the radius of the circle at which $\partial V/\partial r = 0$. The potential is singular along the symmetry axis except in the case when $r_0 = 0$. At this limit the form of the potential becomes

$$V(r,z) = V_0/b^2\,(r^2 - 2z^2)\,, \tag{8.22}$$

which is just the potential of the electron trap. For the muon experiment, r_0 is 7 m and the particle orbits are restricted by the aperture limits to radii within ± 60 mm of this value and also to within ± 40 mm of the horizontal plane $z = 0$. Thus by writing $r = r_0 + x$ the potential can be approximately reduced to the two-dimensional form

$$V(x,z) = (2V_0/b^2)(x^2 - z^2)\,, \tag{8.23}$$

which completely neglects the effects of curvature. Thus in cross-section the shape of the four electrodes, required to provide the quadrupole field, closely resembles that which satisfies the simple two-dimensional case, but perpendicular to this section the electrodes are curved to follow the circumference of the 7 m radius ring.

In both the electron and muon traps the classical motion of the particles can be described in the same terms as a combination of three frequencies. The first is the relatively fast cyclotron frequency due to the magnetic field, but slightly modified by the trapping potential so that the orbits are not quite circular. The centre of the cyclotron orbits can be considered to slowly drift around the axis of symmetry such that the particle executes an epitrochoidal motion, the frequency of this drift being the so-called magnetron motion. In the case of the muon storage ring the size of the horizontal aperture ensures that the centre of the 7 m radius cyclotron orbit is always within 60 mm of the symmetry axis or centre of the ring. The third frequency is that of the vertical oscillations in the direction of the magnetic field. This latter frequency and the magnetron frequency both go to zero in the absence of a trapping potential.

It should be remembered, however, that strictly speaking this classical approximation is not applicable to the quantum states of the electron in the mono-electron oscillator.

9. ELECTRIC DIPOLE MOMENT

An upper limit for the electric dipole moment (EDM) of the muon has been measured directly in the CERN Muon Storage Ring [73]. For a particle with both magnetic and electric dipole moments the electromagnetic interaction Hamiltonian contains a term $(\vec{\mu}_m \cdot \vec{B} + \vec{d} \cdot \vec{E})$, where $\vec{B}$ and $\vec{E}$ are the magnetic and electric field strengths and $\vec{\mu}_m$ and $\vec{d}$ are the magnetic and electric dipole moment operators. See Eqs. (1.3)–(1.6).

It is well known that the expectation value of the electric dipole moment $\vec{d}$ must be zero for a particle described by a state of well-defined parity. However, Purcell and Ramsey [74] stressed that the existence of an EDM for particles should be treated as a purely experimental question, and they suggested possible physical mechanisms that could lead to a non-vanishing EDM. After the discovery of parity violation in the weak interactions, it was pointed out by Landau [75] that even if P is violated, the existence of an EDM is still forbidden by T invariance, i.e. the existence of a non-vanishing EDM for a particle implies that both P and T are violated. See Field et al. [72] and Jackson [1] for comprehensive reviews of the subject.

The technique used to measure the muon electric dipole moment follows from a suggestion originally made by Garwin and Lederman [76], and the present experiment was carried out at the CERN Muon Storage Ring simultaneously with the measurement of the muon g-factor anomaly $a \equiv 1/2\ (g-2)$. The precession frequency of the muon spin relative to its velocity vector $\vec{\beta}$, which is perpendicular to the magnetic and electric fields $\vec{B}$ and $\vec{E}$, is:

$$\vec{f} = \frac{e}{2\pi mc}\left[a\vec{B} + \left(\frac{1}{\gamma^2 - 1} - a\right)\vec{\beta} \times \vec{E} + \frac{1}{2}\eta\,(\vec{E} + \vec{\beta} \times \vec{B})\right], \tag{9.1}$$

where we have included the effect of an electric dipole moment defined by $|\vec{d}| \equiv (\eta/2)(eh/4\pi mc)$ in analogy with the magnetic moment.

The muon momentum was chosen such that the second term inside the square brackets of Eq. (9.1) vanishes: $\lambda = [1 + (1/a)]^{1/2} = 29.3$. The third term displays the effect of the EDM on the spin motion and, since the laboratory electric field is negligible compared with the magnetic field ($|E| < 10^{-3}$ B), this term reduced to a precession frequency $f_{EDM} = (e/2\pi mc)(1/2 \cdot \eta\vec{\beta} \times \vec{B})$ about an axis radial to the orbit. The origin of this motion is the torque acting on the EDM from the apparent electric field in the muon rest frame. Equation (9.1) is therefore simply the vector sum of f_{EDM} with the normal $(g-2)$ frequency f_a :

$$\vec{f} = \vec{f}_a + \vec{f}_{EDM} = -\frac{e}{mc}\left((a\vec{B} + \frac{1}{2}\eta\vec{\beta} \times \vec{B}\right). \tag{9.2}$$

The effect of the EDM is illustrated in Fig. 30. The plane of the spin precession is tilted such that its normal is at an angle $\delta = f_{EDM}/f_a = \eta\beta/2a$ to the magnetic field. This leads to a vertically oscillating component of the muon polarization with the same frequency as the precession of the horizontal polarization. The observation of this vertical component constitutes the basis of the direct measurement of an EDM. The $(g-2)$ frequency f_a is increased to $f = f_a(1 + \delta^2)^{1/2}$, making an EDM a possible candidate for a discrepancy between the measurement of the anomaly and the theoretical prediction.

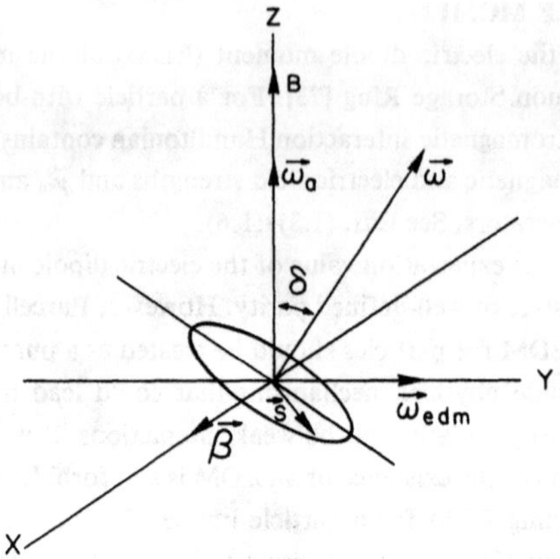

Fig. 30 Precession of the spin relative to the momentum resulting from the combination of an anomalous magnetic moment and an electric dipole moment. The plane of the precession is tilted through the angle $\delta = \beta\eta/2a_\mu$.

The evolution of the muon polarization as a function of storage time is observed through the asymmetry in the angular distribution of the decay electrons with respect to the direction of the muon spin. To be sensitive to the vertical component of the polarization, one has to record whether the decay electron is upward or downward-going. The numbers of decay electrons in these two categories are given by

$$N_{up} = \frac{1}{2} N e^{-t/\tau} [1 - A_\mu \cos(2\pi f_a t + \psi) + A_e \sin(2\pi f_a t + \psi)]$$
$$N_{down} = \frac{1}{2} N e^{-t/\tau} [1 - A_\mu \cos(2\pi f_a t + \psi) - A_e \sin(2\pi f_a t + \psi)] \, . \tag{9.3}$$

The asymmetry A_e is proportional to the magnitude of the EDM and is a function of the energy threshold of the detectors. For a decay-electron energy threshold of 800 MeV, $A_e = (0.164 \pm 0.019)\delta$, for the counter system used in the EDM experiment [73]. Separate measurements on μ^+ and μ^- gave:

$$d_{\mu^+} = (8.6 \pm 4.5) \times 10^{-19} \, e \cdot cm \, , \tag{9.4}$$

$$d_{\mu^-} = (0.8 \pm 4.3) \times 10^{-19} \, e \cdot cm \, . \tag{9.5}$$

Assuming opposite EDMs for the particle and antiparticle, the combined result was

$$d_\mu = (3.7 \pm 3.4) \times 10^{-19} \, e \cdot cm \, . \tag{9.6}$$

For comparison the current upper limits for the electron, proton, and neutron in units of e·cm [77] are electron $\leq 3 \times 10^{-24}$, proton $\leq 2 \times 10^{-20}$, and neutron $\leq 1 \times 10^{-23}$. That these limits are much lower than the limit of the muon largely reflects the fact that, unlike the muons, they are studied in neutral systems. The fundamental length of Kadyshevsky [71] can therefore not be greater than 2×10^{-18} cm (muon evidence) or 10^{-23} cm (electron evidence).

10. MUON LIFETIME IN FLIGHT

Accurate measurements of the muon lifetime in a circular orbit provide a stringent test of Einstein's theory of special relativity. As a bonus it sheds light on the so-called twin paradox, gives an upper limit to the granularity of space–time, and tests the CPT invariance of the weak interaction.

The muon is an unstable particle, and can therefore be regarded as a clock and used to measure the time dilation predicted by special relativity. The existence of cosmic-ray muons at ground level supports the idea of time dilation, for, if the muons lifetime was not lengthened in flight, they would all decay in the upper atmosphere [78]. Experiments verifying the time dilation in a straight path have also been made with high-energy accelerators [79]. Time dilation in a circular path has always seemed more controversial.

Hafele and Keating [80] loaded caesium atomic clocks onto a commercial aircraft on an around-the-world trip and verified the time dilation at low velocity with an accuracy of about 10%.

In the CERN Muon Storage Ring, the muon performs a round trip and so when compared with a muon at rest the experiment mimics closely the twin paradox already discussed in Einstein's first paper [81]. The circulating muons, although they return again and again to the same place, should remain younger than their stay-at-home brothers. It is indeed observed that the moving muons live longer, in agreement to one part in a thousand with the predictions of special relativity. The stationary twin's time scale is given by the muon decay rate at rest determined in a separate experiment.

An accurate measurement of the muon lifetime in a circular orbit at $\gamma = 29.3$ requires high orbit stability in a short time interval (a few hundred microseconds), for any loss of muons will set a limit to the accuracy of the measurement. The reported stability was achieved by using a scraping system that shifted the muon orbits at early times in order to 'scrape off' those muons most likely to be lost.

The experiment consisted of measuring the decay electron counting rate N(t) [see Eq. (8.13)] and the fitting procedure gave the value of $\tau = \gamma\tau_0$. The rotation frequency f_{rot} of the muons obtained from the counting record at early times (see Fig. 25) with the aid of Eq. (8.9) gave

$$\gamma = 29.327(4)\,. \tag{10.1}$$

The best value for the lifetime at rest is 2.19711(8) μs [82], which then gives $\tau_{th} = 64.435(9)$ μs, compared with the experimental result $\tau_{exp} = 64.378(26)$. Thus the transformation of time is validated to an accuracy of $-(0.9 \pm 0.4) \times 10^{-3}$ [79].

In the actual experiment, corrections were made for a residual small loss of stored muons, for variations of the photomultiplier gain accompanying the recovery from the initial flash, and for background counts due to stored protons (in the case of μ^+). In order to measure the muon lifetime with an accuracy of 0.1% it was necessary to study carefully these three effects, which could systematically distort the recorded time spectrum.

Another check on relativity theory can be obtained by comparing (g − 2) measurements carried out at different values of γ. For the electron this has been argued by Newman et al. [83], and discussed by Combley et al. [84] for both e and μ. Inevitably the conclusions are model-dependent, but one can make a plausible case that these results confirm the relativistic transformation laws for magnetic field and mass, as well as for time.

11. VERIFICATION OF THE CPT THEOREM

From CPT it follows that $g_{\mu^+} = g_{\mu^-}$. The measurements in the CERN Muon Storage Ring gave to 95% confidence

$$7 \times 10^{-9} > \frac{g_{\mu^+} - g_{\mu^-}}{g_\mu} > -58 \times 10^{-9}. \tag{11.1}$$

From CPT it follows also that $\tau_{\mu^+} = \tau_{\mu^-}$. The experimental data on the μ^+ and μ^- lifetime in flight give the best test of this equality (as τ_{μ^-} cannot be measured at rest because of muon capture). In this connection it should be noted that the Lorentz γ-factor is the same for μ^+ and μ^- to a much higher precision than the quoted lifetime errors. The limits are

$$3.0 \times 10^{-3} > \frac{\tau_{\mu^+} - \tau_{\mu^-}}{\tau_\mu} > -1.4 \times 10^{-3}. \tag{11.2}$$

Thus the theorem is validated for muons to very high accuracy for the electromagnetic interaction, and rather less accurately for the weak interaction.

12. THE SITUATION TOMORROW: PLANS FOR A NEW MEASUREMENT

The main change since the last (g − 2) experiment has been the firm establishment of the unified electroweak gauge theory [9] which seems to fit all known facts about the electromagnetic and weak interactions. The discovery of the W and Z bosons as free particles [8] has confirmed the major premise of the theory. All present tests of the weak interactions, however, involve only first-order processes, that is the direct exchange of W and/or Z particles. Higher-order processes involving virtual loops of W and Z are an essential part of the theory, but have not so far been detected experimentally. It is important to establish their presence.

The renormalization procedures inherent in the new theory now give definite predictions for the diagrams in Fig. 3b,c,d through which the virtual weak bosons contribute to the muon g-factor. As discussed above the result is a positive contribution from the virtual W, half cancelled by the negative effect of the virtual Z, leaving a resultant of 1.95×10^{-9} that is 1.75 ppm in a_μ. If this could be confirmed it would be a another triumph for the theory.

The above is the result obtained from the standard Weinberg–Salam–Glashow electroweak theory [9]. But a wide variety of alternative gauge theories have been proposed, all compatible with existing data, but giving [10, 85] as it happens different predictions for Δa_μ^{weak}. For most models the (Z + W) effect is in the range (1.5–3.3) $\times$ 10^{-9}. Including the Higgs bosons however can lead to results spreading through (−12 to +17) $\times$ 10^{-9} depending on the Higgs mass. Thus a_μ can provide an excellent laboratory for testing electroweak gauge theories. A new measurement of a_μ at the 0.3 ppm level would provide an important guide to the Higgs masses.

Other current speculations that would alter the value of a_μ include the following:

a) Composite lepton models, leading to finite-size effects or form factors, which would cut off the integrals at high q^2.
b) Lepton substructure, or excited leptons involving new diagrams such as Fig. 31, from Méry et al. in Ref. [86]. A new g − 2 experiment would limit the branching ratio for $Z^0 \rightarrow \mu + \mu^*$ to less than 2% [10].
c) Supersymmetry with many new diagrams and contributions [87].
d) Axions: the standard axion model [10, 88] contributes 10 ppm to a_μ.
e) Theories with multiple W and Z bosons have been proposed and would give their own values of a_μ.

In many of these cases the limits on the exotic effects from a new (g − 2) measurement would be better than those expected from the new generation of electron colliders such as LEP, PEP, and the SLC.

Thus there is now a strong motivation for measuring a_μ with an experimental error much smaller than 1 ppm. But independent of the experiment it will only be possible to determine

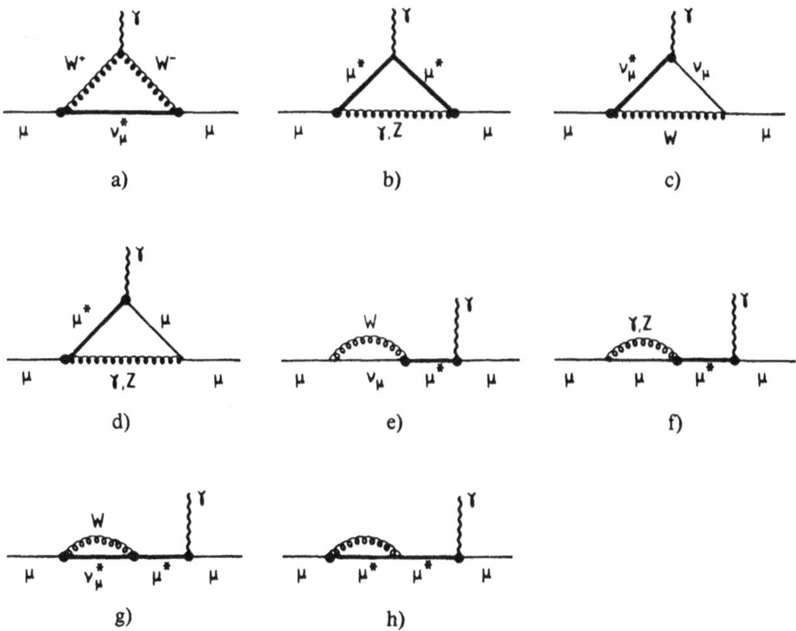

Fig. 31 Excited lepton contributions to a_μ.

Δa_μ^{weak} if the comparatively large hadronic contribution to a_μ can be calculated accurately. The present best result is given in Eq. (2.7), the error being dominated by the experimental uncertainty in the ratio R(s) of pion and muon yields from low-energy electron–positron colliders.

To reduce the error in Δa_μ^{had} to 0.3 ppm, the value of R(s) will need to be known to 0.6%, about 3–4 times better than the present accuracy. New measurements are in preparation at the VEPP collider in Novosibirsk, and Barkov et al. [89] have hopes of reaching the desired accuracy. The measurement of R(s) using high-energy (300 GeV) positron beams annihilating on electrons at rest [90] could very probably also be improved, and is particularly valuable for the lowest values of s just above the pion threshold.

Thus the stage is set for an attempt to measure a_μ to order 0.3 ppm with a view to obtaining the contribution of the weak interaction, and hopefully finding out which of the many alternatives is to be preferred.

12.1 A Fourth (g − 2) Experiment

V.W. Hughes has been considering (g − 2) measurements with a superconducting storage ring (possibly at 5 T) since about 1970. Initially LAMPF was to be the injector, then LAMPF II, but when the intensity of the AGS was upgraded, Brookhaven became the preferred site. In 1984 he organized a discussion meeting there to see whether a new (g − 2) measurement could be initiated, and to work out the general parameters of the experiment. Many of the collaborators in the CERN Muon Storage Ring experiments participated*). Further discussions led to a formal proposal [91]. We describe below the main features of the new experiment as we understand them at present.

The starting point was a muon storage ring working at the 'magic energy', 3.1 GeV, which would as before allow a uniform magnetic field to be used with vertical focusing by means of electric quadrupoles. An improvement in accuracy by a factor of 20 was taken as the target, implying a 400-fold increase in the counting statistics. A large part of this would come from the increased intensity of the Brookhaven AGS, compared to the CERN PS in 1975. (The CERN machine has also been upgraded, but is fully committed as an injector to the SPS and LEP.) The parameters of the storage ring would be re-optimized. In particular one might reduce the radius of the ring and work at a much higher magnetic field B, thus increasing the precession frequency and improving the accuracy. As before the first thought was to inject a prepared beam of pions, and allow them to decay to muons during the first turn of the ring.

The statistical error in finding the frequency f_a from the modulated exponential record of muon decays, Eq. (8.13), is

$$\delta a_\mu / a_\mu = \delta f_a / f_a = \frac{\sqrt{2}}{2\pi f_a A \gamma \tau_0 N_e^{1/2}}, \tag{12.1}$$

*) The main participants were J. Bailey, H.N. Brown, F. Combley, G. Danby, S. Dhawan, F.J.M. Farley, J.H. Field, P. Franzini, M. Giorgi, V.W. Hughes, J.W. Jackson, D. Joyce, F. Krienen, D. Lowenstein, W. Lysenko, W. Marciano, M. May, M. Month, P. Nath, S. Parke, G. Petrucci, E. Picasso and R. Siegel.

where N_e is the total number of decay electrons, A the amplitude of the modulation and τ the lifetime at rest; γ is fixed at the magic value 29.3.

The primary way to reduce the error is to increase the observed number of decays N_e; but increasing f_a by increasing the magnetic field B would clearly be an advantage. As the momentum is to be fixed at the magic value this means

$$\delta a_\mu \propto r/N_e^{1/2} , \tag{12.2}$$

where r is the radius of the orbit, so we must maximize N_e/r^2.

The number N_e is proportional to the number of muons stored N_μ, given by Eq. (7.1). One option for increasing N_μ is to increase the aperture of the storage ring a (vertical) $\times$ b (horizontal). However the pitch correction of Eq. (4.36) is proportional to the square of the vertical angular acceptance, that is to $(a/r)^2$; this should not be increased. Likewise the electric field correction of Eq. (8.8) is proportional to $(b/r)^2$.

If a/r and b/r are held constant to keep these two corrections constant, then Eq. (7.1) implies that N_e is independent of r. However some radial dependent factors are concealed in the parameter F: the number of pions which decay in the first turn of the ring is proportional to r. Also the cross-sectional area of the inflection channel can be larger for larger rings. Therefore one expects

$$N_e \propto N_\mu \propto r^2 \text{ or perhaps } r^3 . \tag{12.3}$$

Substituting in Eq. (12.2) it then appears that a high-field small-radius ring will not after all reduce the statistical error; it might even increase it. So bearing in mind the greater technical difficulties of a high-field ring (field uniformity, inflector design), it was decided to keep the radius at 7 m with B = 1.47 T.

Increasing the decay electron counting rate N_e by a factor 400 turned out not to be trivial. At CERN about 200 electrons, spread over 5 digitrons, were counted per cycle of the PS. With a muon lifetime of 64 μs this implied an initial counting rate of 0.6 per microsecond in each digitron. At rates appreciably higher than this the random overlapping of signals would lead to timing errors which would be increasingly serious as one tried to improve the absolute accuracy of the experiment. One solution is to increase the number of digitrons to, say, 80 (thus reducing the peak counting rate in each), and to repeat the experiment more often per hour. This could be achieved by filling the muon storage ring several times in each cycle of the AGS by ejecting a bunch of protons from the accelerator at intervals of, say, 10 ms during the 'flat top' of the synchrotron. The measurement could thus be repeated several times per AGS cycle. With 80 digitrons and 5 fills per cycle one could handle a factor of 80 in overall counting rate, with no increase in the peak rate per channel. Another factor of 5 could probably come from detailed electronic improvements for detecting and rejecting multiple events which might have been mistimed.

A limiting factor in the CERN experiment was a large perturbing pulse in the decay electron detectors, coinciding with the injection of pions into the ring. This was caused by a strong flash of light in the glass of the photomultipliers, light pipes and scintillators. Light continues to be produced at a lower level for many microseconds after injection, as epithermal neutrons are gradually captured and gamma rays from (n,γ) reactions give rise to Cherenkov light. This phenomenon (also seen in other experiments) is known for convenience as the 'initial flash'. At high injection intensities it can paralyse the detector electronics for excessive periods, leading to counting and timing errors. In the CERN experiment these were controlled to less than 1 ppm in a_μ, negligible compared with the final accuracy of 7 ppm. But in a new measurement at the 0.3 ppm level, the much higher injected intensities could clearly give problems. Unfortunately these effects are hard to study quantitatively without actually building a storage ring!

An innovation which will increase the number of stored muons, while at the same time reducing the initial flash, is to inject muons instead of pions. To do this one sends the prepared beam of pions down a 'muon channel', that is a long (100 m) decay path with focusing elements to collect the muons produced in π-μ decay. At 3 GeV/c the pion decay distance is about 170 m, so about 50% of them decay in the channel, in contrast to only 10% of useful decays if the pions are injected directly into the ring. The muons from almost forward decay, having momentum 1–2% less than the pion momentum, can be separated out at the end of the channel by a bending magnet, giving an almost pure muon beam which is then injected into the muon storage ring.

In this case, however, one must have a full aperture kicker in the ring, to kick the muons onto permanently stored orbits: and this magnetic kick must be switched off within 140 ns, before the bunch of muons makes one turn. There is no difficulty in principle in designing such a kicker, but it must be free of iron or ferrite (which would modify the carefully prepared magnetic field), and the residual field 20 μs later when the muon precession is being recorded must be small and measurable. As the muons continue to pass through the kicker at every turn, its residual field (including that due to any eddy currents induced in the surrounding metal vacuum chamber) must be known to a level corresponding to 0.1 ppm in the average orbit field. This corresponds to about 1:20,000 of the size of the kick, and presents some technical challenge!

In the first Muon Storage Ring the CERN team adopted the technically simple strategy of injecting high-momentum protons onto a target in the ring. The pions produced made a few turns before decaying to muons, some of which fell onto permanently stored orbits. The initial flash effects were severe, and the large number of off-momentum pions decaying at unplanned angles gave stored muons of the wrong polarization, reducing the amplitude of the precession signal. In the second CERN storage ring these difficulties were alleviated by preparing a beam of pions outside the ring and injecting them with a pulsed inflector. Following this a third storage ring in which muons are injected directly is a logical progression. Each step is better for the physics, but presents new technical difficulties.

An overall comparison of pion and muon injection is given in Table 2. With pions, 5000 particles must be injected for every muon stored, so the initial flash will be severe. With muon injection the injected beam is much lower and the flash will be about ten times smaller than in the last CERN experiment. On the other hand, because of the larger decay path, many more muons will be stored. The disadvantage is that the full aperture kicker can perturb the magnetic field in the storage region.

Table 2

Comparison of pion and muon injection

	Pion injection	Muon injection
Particles injected	6×10^7	5×10^4
Muons stored	12,000	30,000
Inflector	Yes	Yes
Full aperture kicker	No	Yes
Initial flash	Severe	Negligible

12.2 The Muon Distribution and the Average Field

Referring to Eq. (4.9) it is clear that an accurate knowledge of the mean magnetic field $\bar{B}$ seen by the muons is necessary. If the field of the storage ring is strictly uniform this is not a problem; but if, as is more likely, the field varies by some parts per million over the storage volume, it is necessary to calculate the average field weighted over the muon distribution, and corrected for the fact that muons on the inside of the ring contribute more decay electron counts than those on the outside. Ideally we should have a detailed knowledge of the distribution of muons in the available aperture, and know how this distribution affects the calculated mean field. Some useful theorems have been developed to simplify the analysis [92].

It is convenient to use a cylindrical-coordinate system, tied to the centre of the storage aperture. Let ϕ be the azimuth angle around the ring, z the height above the median plane and $(x + r_0)$ the radius from the centre of the ring: r_0 is the radius at the centre of the storage aperture, so x measures the radial distance from the centre line of the aperture.

The magnetic field B satisfies Laplace's equation

$$\frac{\delta^2 B}{\delta x^2} + \frac{\delta^2 B}{\delta z^2} + \frac{1}{x + r_0}\frac{\delta B}{\delta x} + \frac{1}{(x + r_0)^2}\frac{\delta^2 B}{\delta \phi^2} = 0 \tag{12.4}$$

For a given (x,z) let us calculate the field B averaged over all azimuths; this is the field seen by a muon circulating round the ring at the given position (x,z) in the aperture:

$$\bar{B}(x,z) = \frac{1}{2\pi}\int_0^{2\pi} B(x,z,\phi)\, d\phi\ ; \tag{12.5}$$

$\bar{B}$ satisfies Eq. (12.4) with the last term omitted, as in carrying out the integral as in Eq. (12.5) the last term cancels out.

Because the curvature of the ring is small it turns out that the third term in Eq. (12.4) is less than 1% of the first two terms, so to a good approximation the Cartesian form of Laplace's equation in two dimensions applies,

$$\frac{\delta^2 \overline{B}}{\delta x^2} + \frac{\delta^2 \overline{B}}{\delta z^2} = 0 . \tag{12.6}$$

It is well known that the field can then be expressed as a series of multipoles, which each satisfy Eq. (12.6). These multipoles are most easily represented by changing to new coordinates (r,θ) in the plane of the aperture and with origin at the centre of the aperture, such that

$$x = r\cos\theta , \quad \text{and} \quad z = r\sin\theta .$$

Then the multipoles of the field are of the form $r^n \cos(n\theta)$ and $r^n \sin(n\theta)$ and in general

$$\overline{B} = \sum_n r^n[c_n \cos(n\theta) + s_n \sin(n\theta)] , \tag{12.7}$$

the c_n and s_n being the multipole coefficients whose values can represent any arbitrary distribution of $\overline{B}$ in the aperture plane. The overall average field is $\overline{B} = c_0$. If the other coefficients are all zero, the field is uniform and no problems arise. How large can they be before they give an appreciable error, and how does one calculate the contribution of each multipole component to $\overline{B}$?

Let the number of muons at point (r,θ) in the aperture be $M(r,\theta)$, the total number being $N = \iint M(r,\theta)\, r\, dr\, d\theta$. Then the average field contributed by the nth cosine multipole will be

$$\begin{aligned}\overline{B}_n &= \frac{1}{N}\iint c_n r^n M(r,\theta)\cos(n\theta)\, r\, dr\, d\theta \\ &= c_n I_n ,\end{aligned} \tag{12.8}$$

where

$$I_n = \frac{1}{N}\iint r^n M(r,\theta)\cos(n\theta)\, r\, dr\, d\theta ; \tag{12.9}$$

I_n may be called the nth moment of the muon distribution; it has no quantities associated with the magnetic field.

Including all multipoles of the field, the overall average field will be expressed as

$$\overline{B} = \sum_n (c_n I_n + s_n J_n) . \tag{12.10}$$

where the skew moments of the muon distribution are given by

$$J_n = \frac{1}{N}\iint r^n M(r,\theta)\sin(n\theta)\, r\, dr\, d\theta . \tag{12.11}$$

As this is a complete specification of the average field it is clear that for this purpose the nth moment I_n only interacts with the nth multipole coefficient c_n; there are no cross terms. This result greatly simplifies the task of analysing the magnetic field.

It was found in the previous storage ring experiments that the muons filled the available phase space in the ring more or less uniformly. While there was no exact experimental check, this can be used as a reasonable basis for planning. With this assumption the moments I_n of the muon distribution have been calculated (a) for a rectangular aperture 10×8 cm^2, (b) for a square aperture 9.14×9.14 cm^2, and (c) for a circular aperture of diameter 11 cm. All three apertures have the same acceptance. (Note that the J_n are theoretically zero because of up-down symmetry.) From the moments, using Eq. (12.10) one can calculate the maximum values of c_n allowable if the mean field is to be affected by less than 0.1 ppm. The corresponding allowable field at the horizontal edge of the aperture is given in Table 3 for each multipole.

Table 3

Allowable field in ppm at the edge of the aperture for 0.1 ppm effect on $\overline{B}$

n	Rectangular	Square	Circular
2	1.12	-	2.8
4	−1.56	0.80	−50
6	−2.00	-	286
8	5.40	0.92	−1014
10	2.06	-	2768
12	25.0	0.62	6446

From the table we see that the rectangular apertures, as used in the CERN experiments, give the muon population significant values of I_8 and I_{10} and therefore require the corresponding field multipole to be small. This is because the corners of the aperture have some muon intensity which is represented by terms in $\cos(4\theta)$, $\cos(8\theta)$, etc. These couple to the 4th, 8th and 12th multipoles which the table shows must be maintained at a very low level. The square aperture is even more sensitive to certain higher multipoles!

With a circular aperture, however, the smoothness of the boundary means that the higher cosine terms in the muon distribution are small, and we can tolerate much larger multipoles in the field. For example the $n = 8$ multipole must be known only to 1014 ppm at the edge of the aperture; this should not be difficult. Therefore it was resolved to put a circular limiting stop in the storage region to define the muon population.

12.3 Pick-Up Electrodes

The above results are based on the assumption that the muons in the ring populate the available phase space uniformly. It will be important at the new level of accuracy to obtain an independent measurement of the actual muon distribution, so that the true average field can be calculated.

The momentum of the muon determines the radius of its closed orbit in the ring, and therefore its orbital rotation frequency. As explained above the muons are bunched at injection, and therefore the rotation frequency is visible in the electron counting data. By analysing the counts one can find the distribution of muons with respect to equilibrium radius. This is an important measurement which has been helpful in finding the average field.

However, the rotation frequency does not determine the muon distribution completely, because a muon will normally oscillate both horizontally and vertically around its theoretical closed orbit. To calculate the average field one needs the distribution of muons including these orbit oscillations. Up to now no direct information on the oscillations has been available, but checks of stored intensity versus aperture size have agreed with predictions based on phase space.

In the new experiment some 15,000 muons will be circulating at 7 MHz implying a muon current of 15 nA, and it is proposed [92] to measure the distribution inside the circular aperture by means of pick-up electrodes. An interesting theorem relates the moments of the muon distribution to the Fourier components of the charge $\sigma(\psi)$ induced on a circular cylinder surrounding the storage aperture, see Fig. 32. The relation between the distribution of charges inside a long cylinder $q(r,\theta)$ and the corresponding distribution of the surface charge $\sigma(\psi)$ can be calculated in static approximation using the method of images. (See Fig. 32 for the definition of the variables.)

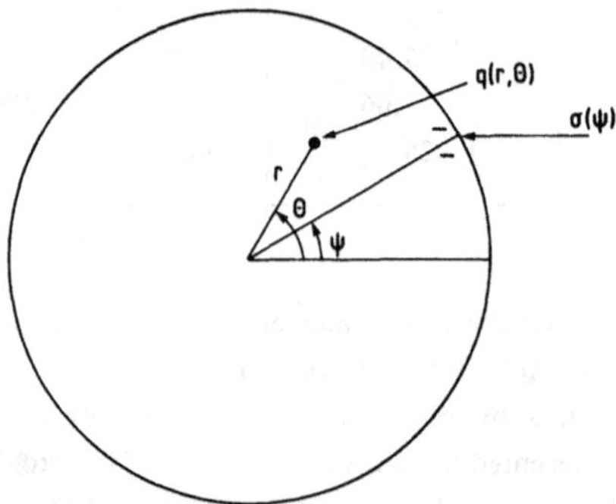

Fig. 32 Definition of the variables needed to calculate the distribution of charge inside a long cylinder.

Consider a long cylindrical pick-up electrode of unit radius. The centre line of the cylinder is taken as the origin for coordinates (r,θ) in the plane transverse to the axis. A distribution of line charges $q(r,\theta)$ per unit volume is placed inside the cylinder. We calculate the corresponding induced charge on the inside surface of the cylinder $\sigma(\psi)$ as a function of azimuth ψ around the cylinder.

In general $\sigma(\psi)$ may be Fourier analysed

$$\sigma(\psi) = \sum_{0}^{\infty} \{u_n \cos(n\psi) + v_n \sin(n\psi)\} . \tag{12.12}$$

It is then found that the amplitude c_n of the nth Fourier component is proportional to the corresponding moment I_n of the charge distribution

$$u_n \propto I_n = \int\int q(r,\theta)\, r^n \cos(n\theta)\, r\, dr\, d\theta\ , \tag{12.13}$$

with a corresponding result for the sine components

$$v_n \propto J_n = \int\int q(r,\theta)\, r^n \sin(n\theta)\, r\, dr\, d\theta\ . \tag{12.14}$$

Thus the Fourier coefficients u_n and v_n of the charge induced on the pick-up cylinder are proportional to the moments I_n, J_n of the muon distribution. This is just the information needed for calculating the average field with the aid of Eq. (12.10).

In translating this theorem into practical reality there are a number of experimental difficulties. The cylinder would be divided into many longitudinal strips, each one connected to a separate amplifier. With only about 3000 muons inside the cylinder at any one time, the induced signal on each strip will be small (of order 1 μV) and this will be below the noise level in the amplifiers. This is not an absolute impediment, however, because the $g-2$ precession signal for a single cycle of the synchrotron is also swamped in noise: by repeating the measurement many times and averaging, significant information is accumulated. In principle the same process can apply to the pick-up electrodes: by digitizing the information with appropriate signal processing, and accumulating over the duration of the experiment, one should end up with accurate position information.

Another method of finding the muon distribution is to measure the decay electron tracks carefully with wire chambers. If the track is extrapolated backwards until it is tangential to the ring this gives a fair estimate of the position of the parent muon. The μ-e decay angle creates some uncertainty, but this turns out to be small.

12.4 Magnet

The general parameters of the new experiment are summarized in Table 4. The radius of the storage ring is slightly larger than in the last CERN experiment, namely 7.11 m instead of 7.00 m. The field is reduced correspondingly to keep the same storage momentum. To have a more uniform field in the storage region the magnet gap is increased from 14 to 18 cm, and the pole width is increased in proportion from 38 to 49 cm. The yoke will be continuous, and the magnet will be energized by superconducting coils. It is expected that the uniform and steady temperature of the coils and the absence of ohmic heating will give a more stable magnet. In addition the iron of the magnet will be lagged and its temperature will be stabilized. By these measures it is hoped that the field shape will be stable to better than 1 ppm.

To increase the field uniformity there will be an air gap of 10 mm between the pole pieces and the yoke. This is a well known technique for decoupling the field in the main gap from irregularities in the iron of the yoke. In addition the field can be adjusted locally by inserting iron sheets into this air gap, instead of shimming or grinding on the pole face itself. Comparatively thick sheets can be used behind the poles to make small changes in the field. Pole face windings will be used to make final corrections to the field.

Table 4

Parameters of the New Experiment

Storage ring radius	7.112 m
Field	1.45 T
Storage momentum	3.094 GeV/c
Muon lifetime	64 μs
Magnet gap	18 cm
Pole width at root	56 cm
Useful aperture inside quadrupole electrodes	10×10 cm^2
Aperture used for muon storage (diameter)	9 cm
No. of AGS protons in each RF bunch (with booster)	4×10^{12}
Length of muon channel	100 m
No. of injected muons per fill	5×10^4
Fraction of muons stored	30%
No. of stored muons per fill	1.5×10^4
Electrons above 1.6 GeV counted per fill (20%)	3000
Fills per AGS cycle	3
Fills per hour	7714
Electron counts per hour	2.3×10^7
Total counts required for 0.3 ppm statistical error	2.85×10^{10}
Running time for 0.3 ppm statistics	1240 hours
Pitch correction	0.40 ppm
Electric field correction	0.69 ppm

As before the field will be surveyed and monitored with proton resonance magnetometers. By scanning the resonance slowly it is expected to achieve an accuracy of 0.1 ppm or better. A remote-controlled trolley which can automatically survey the field inside the vacuum chamber is being developed. This will enable the field to be measured without removing the vacuum chamber. (Removing the chamber normally implies switching off the field, and it may not return to exactly the same value afterwards.) If the trolley can actually move and measure under vacuum that would be an additional bonus because the field could be checked more frequently during the runs.

Another way of calibrating the magnetic field is to inject a narrow (1 ns) pulse of singly charged gold ions [93] with a kinetic energy of 23 MeV. These are available from the tandem Van de Graaff at Brookhaven, which is already linked to the AGS. They would have the same momentum as the stored muons, but would be non-relativistic ($\beta = 0.015$) so the orbit frequency becomes almost independent of radius and is a measure of the magnetic field B averaged over the orbit. The orbit frequency can be measured by kicking the gold ions onto a scintillation counter after many turns. As the orbit frequency of the gold is altered by the radial electric field of the electric quadrupoles, they must be switched off (or operated at very low voltage), but even without any vertical focusing it seems possible to get sufficient turns to make an accurate measurement.

The gold method has the advantage of automatically averaging the field in azimuth. It may also be possible to direct the beam to different parts of the storage aperture to map the field in the x-z plane. At its lowest, the gold method could give an independent check of the NMR measurements. At its best, it might be interleaved with the muon pulses to monitor the field at frequent intervals without losing any running time with the muons.

In summary, the knowledge of the average field seen by the muons will be improved by having a more stable magnet, more frequent surveys and a cross check with the gold ions. The circular defining aperture for the storage volume will reduce the sensitivity to higher multipoles in the field, and the pick-up electrodes and wire chambers will hopefully give precise information about the distribution of muons within the aperture.

12.5 Electronics

With 3000 decay electrons to be detected per fill and timed to an average accuracy of 0.1 ppm, the electronics must be very carefully designed. As the muon lifetime will be 64 μs the average counting rate at the beginning of the fill will be about 50 per microsecond: on the average the pulses are only 20 ns apart.

The forward decays, which have the largest asymmetry parameter, have been selected in the past by requiring a large pulse height in the electron shower calorimeters, implying high electron energy. This will not work properly if the pulses overlap and low-energy electrons are mixed into the high-energy group. The phase of the $(g-2)$ modulation is in fact different for different electron energies, so an admixture of pulses of the wrong height due to overlapping would lead to an error in phase (at early times but not at late times when the counting rate is low). This would create an error in frequency.

Another error arises if two pulses overlap and are processed as a single pulse, with the time taken from the first pulse. This leads to a systematic bias to early times at the beginning of the fill (but not at the end), thus again creating an error in the $(g-2)$ frequency.

The main defence proposed against pulse overlap is spatial separation of the detectors and spatial segmentation within each detector. If, say, 100 independent counters and timing electronics are used, the initial counting rate per channel drops to 0.5 per microsecond. If in addition some information is provided as to where the electron hits the counter, double hits can be detected and rejected. It is better to reject a double event and to lose counts, than to accept it with an error in time. With this strategy it looks as if the necessary timing accuracy can be achieved.

The effective span of time available for measuring the $(g-2)$ frequency superposed on the exponential muon decay is about 3 lifetimes, that is 200 μs. This means that, on average over many events, the timing accuracy must be 20 ps. It is surprising to realise that this can be achieved with a digital timing system that sorts the signals into time bins as much as 10 ns wide. This is discussed in detail in Refs. [53] and [94].

The essential point is that the clock pulses of frequency f_{clock} which determine the channel boundaries must *not* be synchronized with the start pulse which signals the moment of muon injection. This means that for a fixed time interval between 'start' and 'stop' the result of the measurement is statistically distributed between two neighbouring channels. Then from the counts c_n and $c_{(n+1)}$ in the two channels, the exact time interval t can be determined

$$t = f_{clock}^{-1} \{n + c_{(n+1)}/[(c_n + c_{(n+1)})]\} \quad (12.15)$$

subject to the usual statistical uncertainties.

In a test with crystal-controlled pulses with 32,000 counts per channel a timing accuracy of 25 ps was achieved with bins 5 ns wide, a ratio of 1/200 which agrees roughly with what is expected from the statistical factor $[c_n + c_{(n+1)}]^{1/2}$.

In the new experiment one expects 5×10^6 counts per channel at early times, dropping to 2.5×10^5 after 200 μs, so a statistical accuracy of 20 ps will be achieved in each channel. Because many channels contribute to the final frequency determination, this 'binning error' turns out to be entirely negligible.

12.6 Inflector

A new superconducting inflector to guide the incoming particles through the fringing field and into the storage region has been designed [95]. With a carefully structured pattern of parallel conductors carrying d.c. currents, the field of the ring magnet can be cancelled inside the inflector without perturbing the field in the storage region; the return flux is completely contained inside the inflector structure. Model tests have demonstrated the feasibility, and the next step is to test a superconducting model.

12.7 Summary

A measurement of a_μ to about 0.3 ppm has been planned in detail and looks feasible; preparations are now under way in Brookhaven and it is hoped that the experiment will be running in 1992. In the above account we have emphasized some of the new physics ideas that may be incorporated in a new experiment. Inevitably, this is our personal picture; the experiment is still evolving and the final configuration will no doubt be different. To reach this level of accuracy will require great patience and meticulous attention to the many detailed problems that will present themselves en route. In anticipation of this result, considerable theoretical and experimental activity has already been generated.

13. FINAL REMARKS

At its present level of accuracy the muon $(g-2)$ measurement, combined with the theoretical calculation, has made an important contribution to validating QED.

The absence of a discrepancy continues to provide a constraint on the fantasy of theorists, because many new models for fundamental particles turn out at some level to change the theoretical value of a_μ. The experimental value of a_e is now known [96] 2000 times more accurately than a_μ. But new couplings common to the electron and muon usually imply a greater perturbation to a_μ by the factor $(m_\mu/m_e)^2 \sim 42{,}000$. Therefore the muon measurement still, in general, provides a more sensitive limit to exotic speculations.

Another consequence is that, except for possible couplings peculiar to the electron, the muon result ensures that a_e is a 'pure QED quantity' to order 2×10^{-10} in a_e, thus validating a_e as a good measurement of the fine structure constant α. Another good route to α is via the hyperfine splitting in muonium. Here again the experiments on a_μ and a_e ensure that muonium is a 'pure QED system' [97].

We would like to conclude by drawing the reader's attention to the many previous review articles on the muon (g – 2) experiments [98], many of which cover particular topics in further detail.

Acknowledgements

Parts of this chapter have been reproduced, with permission, from the Annual Reviews of Nuclear and Particle Science, Vol. 29, © 1979 by Annual Reviews Inc.

We dedicate this article to all those who participated directly or indirectly in the muon (g – 2) programme. In reviewing the work we inevitably recall again the spirit of constructive interaction which enlivened our debates. It has been our privilege and our pleasure to have a share in this endeavour.

REFERENCES

[1] J.D. Jackson, report CERN 77-17 (1977).

[2] P.A.M. Dirac, Proc. Roy. Soc. **A133,** 60 (1931); Phys. Rev. **74,** 817 (1948).

[3] T. Kinoshita, B. Nižić and Y. Okamoto, Phys. Rev. Lett. **52,** 717 (1984).
T. Kinoshita, IEEE Trans. Instrum. Meas. IM-36, No. 2, 201 (1987).

[4] E.R. Cohen and B.N. Taylor, Rev. Mod. Phys. **59,** 1121 (1987).

[5] V.B. Berestetskii, O.N. Krokhin and A.K. Klebnikov, Zh. Eksp. Teor. Fiz. **30,** 788 (1956) [Transl. Sov. Phys. JETP **3,** 761 (1956)].
W.S. Cowland, Nucl. Phys. **8,** 397 (1958).

[6] C. Bouchiat and L. Michel, J. Phys. Radium **22,** 121 (1961).
L. Durand, Phys. Rev. **128,** 44 (1962).
M. Gourdin and E. de Rafael, Nucl. Phys. **B10,** 667 (1969).
T. Kinoshita and W.B. Lindquist, Phys. Rev. Lett. **47,** 1573 (1981) and Phys. Rev. **D27,** 853 (1983).

[7] T. Kinoshita, B. Nižić and Y. Okamoto, Phys. Rev. **D31,** 2108 (1985).
J.A. Casas, C. López and F.J. Ynduráin, Phys. Rev. **D32,** 736 (1985).
L.M. Barkov, A.G. Chilingarov, S.I. Eidelman, B.I. Khazin, M.Yu. Lelchuk, V.S. Okhapkin, E.V. Pakhtusova, S.I. Redin, N.M. Ryskulov, Yu.M. Shatunov, A.I. Shekhtman, B.A. Schwartz, V.A. Sidorov, A.N. Skrinsky, V.P. Smakhtin and E.P. Solodov, Nucl. Phys. **B256,** 365 (1985).

[8] G. Arnison et al. (UA1), Phys. Lett. **B122,** 103 and **B126,** 392 (1983).
M. Banner et al. (UA2), Phys. Lett. **B122,** 476 (1983).
P. Bagnaia et al. (UA2), Phys. Lett. **B129,** 130 (1983).
C. Rubbia, Rev. Mod. Phys. **57,** 669 (1985) (Nobel lecture 1984).

[9] S. Weinberg, Phys. Rev. Lett. **19,** 1264 (1967).
A. Salam, Proc. 8th Nobel Symposium, Aspenäsgården, 1968, ed. N. Svartholm (Almqvist and Wiksell, Stockholm, 1968), p. 367.
R. Jackiw and S. Weinberg, Phys. Rev. **D5,** 2473 (1972).
G. Altarelli, N. Cabibbo and L. Maiani, Phys. Lett. **B40,** 415 (1972).
I. Bars and M. Yoshimura, Phys. Rev. **D6,** 374 (1972).
K. Fujikawa, B.W. Lee and A.I. Sanda, Phys. Rev. **D6,** 2923 (1972).
W.A. Bardeen, R. Gastmans and B.E. Lautrup, Nucl. Phys. **B46,** 319 (1972).

[10] W. Marciano, Proc. 11th Int. Symp. on Lepton and Photon Interactions at High Energies, Ithaca, 1983, eds. D.G. Cassel and D.L. Kreinick (Cornell Univ. Press, Ithaca, 1983), p. 80.
V.W. Hughes, Phys. Scr. **T22,** 111 (1988).

[11] T.D. Lee and G.C. Wick, Nucl. Phys. **B9,** 209 (1969).

[12] M.E. Peshkin, Proc. 10th Int. Symp. on Lepton and Photon Interactions at High Energies Bonn, 1981 (University, Bonn, 1981), p. 880.
L. Lyons, Progr. Part. Nucl. Phys. **10,** 227 (1983).
E.J. Eichten, K.D. Lane and M.E. Peshkin, Phys. Rev. Lett. **50,** 811 (1983).
M. Suzuki, Phys. Lett. **153B,** 289 (1985).

[13] E. Klempt, R. Schulze, H. Wolf, M. Camani, F.N. Gygax, W. Rüegg, A. Schenk and H. Schilling, Phys. Rev. **D25,** 652 (1982).

[14] F.G. Mariam, W. Beer, P.R. Bolton, P.O. Egan, C.J. Gardner, V.W. Hughes, D.C. Lu, P.A. Sonder, H. Orth, J. Vetter, U. Moser and G. zu Putlitz, Phys. Rev. Lett. **49,** 993 (1982).

[15] J.F. Lathrop, R.A. Lundy, V.L. Telegdi, R. Winston and D.D. Yovanovitch, Nuovo Cimento **17,** 114 (1960).
S. Devons, G. Gidal, L.M. Lederman and G. Shapiro, Phys. Rev. Lett. **5,** 330 (1960).

[16] L.H. Thomas, Phil. Mag **3,** 1 (1927).
A. Sommerfeld, RC Convegno di Fisica Nucleare, Atti Reale Accad. Italia **IX,** 137 (1931).

[17] F.J.M. Farley, Cargèse Lectures in Physics, ed. M. Lévy (Gordon and Breach, New York, 1968), Vol. 2, p. 55.
F.J.M. Farley, Contemp. Phys. **16,** 413 (1975).

[18] H. Mendlowitz and K.M. Case, Phys. Rev. **97,** 33 (1955).
M. Carrassi, Nuovo Cimento **7,** 524 (1958).
V. Bargmann, L. Michel and V.L. Telegdi, Phys. Rev. Lett. **2,** 453 (1959).
M. Fierz and V.L. Telegdi, *in* Quanta, eds. P.G.O. Freund et al. (Univ. Press, Chicago, 1970), p. 196.

[19] D.T. Wilkinson and H.R. Crane, Phys. Rev. **130,** 852 (1963).

[20] G.R. Henry and J.E. Silver, Phys. Rev. **180,** 1262 (1969).

[21] S. Granger and G.W. Ford, Phys. Rev. Lett. **28,** 1479 (1972); Phys. Rev. **D13,** 1879 (1976).

[22] F.J.M. Farley, Phys. Lett. **42B,** 66 (1972).

[23] J.H. Field and G. Fiorentini, Nuovo Cimento **21A,** 297 (1974).

[24] F. Combley and E. Picasso, Phys. Rep. **C14,** 1 (1974).

[25] P.A. Franken and S. Liebes, Jr., Phys. Rev. **104,** 1197 (1956).

[26] R. Karplus and N. Kroll, Phys. Rev. **77,** 536 (1950).

[27] A. Petermann, Helv. Phys. Acta **30,** 407 (1957) and Phys. Rev. **105,** 1931 (1957).

[28] C.M. Sommerfield, Phys. Rev. **107,** 328 (1957).

[29] H. Suura and K. Wichmann, Phys. Rev. **105,** 1930 (1957).

[30] W.H. Louisell, R.W. Pidd and H.R. Crane, Phys. Rev. **91,** 475 (1953).
A.A. Schupp, R.W. Pidd and H.R. Crane, Phys. Rev. **121,** 1 (1962).

[31] J. Mathews, Phys. Rev. **102,** 270 (1956).
A.N. Mitra, Nucl. Phys. **3,** 362 (1957).
S. Hirokawa and H. Komori, Nuovo Cimento **7,** 114 (1958).

[32] J. de Pagter and R.D. Sard, Phys. Rev. **118,** 1353 (1960).

[33] R.P. Feynman, Proc. 12th Solvay Conf., Brussels, 1961 (Interscience, New York and Brussels, 1962), p. 61.

[34] J.S. Schwinger, Ann. Phys. (NY) **2,** 407 (1957).

[35] T.D. Lee and C.N. Yang, Phys. Rev. **104,** 254 (1956).
C.S. Wu, E. Ambler, R.W. Hayward, D.D. Hoppes and R.P. Hudson, Phys. Rev. **105,** 1413 (1957).

[36] R.L. Garwin, L. Lederman and M. Weinrich, Phys. Rev. **105,** 1415 (1957).
J.I. Friedman and V.L. Telegdi, Phys. Rev. **105,** 1681 (1957).

[37] C. Bouchiat and L. Michel, Phys. Rev. **106,** 170 (1957).

[38] L.M. Brown and V.L. Telegdi, Nuovo Cimento **7,** 698 (1958).

[39] G. Feinberg and L. Lederman, Annu. Rev. Nucl. Sci. **13,** 431 (1963).
F.J.M. Farley, Progr. Nucl. Phys. **9,** 259 (1964).

[40] W.K.H. Panofsky, Proc. 8th Int. Conf. on High Energy Physics, Geneva, 1968, ed. B. Ferretti (CERN, Geneva, 1958), p. 3.

[41] A.A. Schupp, R.W. Pidd and H.R. Crane, Phys. Rev. **121,** 1 (1961).

[42] G.R. Henry, G. Schrank and R.A. Swanson, Nuovo Cimento **A63,** 995 (1969).

[43] G. Charpak, F.J.M. Farley, R.L. Garwin, T. Muller, J.C. Sens, V.L. Telegdi and A. Zichichi, Phys. Rev. Lett. **6,** 128 (1961).
G. Charpak, F.J.M. Farley, R.L. Garwin, T. Muller, J.C. Sens and A. Zichichi, Nuovo Cimento **22,** 1043 (1961).
G. Charpak, F.J.M. Farley, R.L. Garwin, T. Muller, J.C. Sens and A. Zichichi, Phys. Lett. **1,** 16 (1962).
G. Charpak, F.J.M. Farley, R.L. Garwin, T. Muller, J.C. Sens and A. Zichichi, Nuovo Cimento **37,** 1241 (1965).

[44] I.Yu. Kobzarev and L.P. Okun, Zh. Eksp. Teor. Fiz. **41,** 1205 (1961) [Transl. Sov. Phys. JETP **14,** 859 (1962)].

[45] G. Danby, J.M. Gaillard, K. Goulianos, L.M. Lederman, N. Mistry, M. Schwartz and J. Steinberger, Phys. Rev. Lett. **9,** 36 (1962).

[46] A. Alberigi-Quaranta, M. DePretis, G. Marini, A.C. Odian, G. Stoppini and L. Tau, Phys. Rev. Lett. **9,** 226 (1962).

[47] G.E. Masek, L.D. Heggie, Y.B. Kim and R.W. Wiliams, Phys. Rev. **122,** 937 (1961).
C.Y. Kim, S. Kaneko, Y.B. Kim, G.E. Masek and R.W. Williams, Phys. Rev. **122,** 1641 (1961).
A. Citron, C. Delorme, D. Fries, L. Goldzahl, J. Heintze, E.G. Michaelis, C. Richard and H. Øverås, Phys. Lett. **1,** 175 (1962).
G.E. Masek, T.E. Ewart, J.P. Toutonghi and R.W. Williams, Phys. Rev. Lett. **10,** 35 (1963).

[48] G. Backenstoss, B.D. Hyams, G. Knop, P.C. Marin and U. Stierlin, Phys. Rev. **129,** 2759 (1963).

[49] V.W. Hughes, P.W. McColm, K. Ziock and R. Prepost, Phys. Rev. Lett. **5,** 63 (1960).

[50] K. Ziock, V.W. Hughes, R. Prepost, J.M. Bailey and W.E. Cleland, Phys. Rev. Lett. **8,**103 (1962).

[51] D.P. Hutchinson, J. Menes, A.M. Patlach and G. Shapiro, Phys. Rev. **131,** 1351 (1963).

[52] F.J.M. Farley, Proposed high precession (g – 2) experiment, CERN Intern. Rep. NP/4733 (1962).

[53] F.J.M. Farley, J. Bailey, R.C.A. Brown, M. Giesch, H. Jöstlein, S. van der Meer, E. Picasso and M. Tannenbaum, Nuovo Cimento **45,** 281 (1966).
J. Bailey, G. von Bochmann, R.C.A. Brown, F.J.M. Farley, H. Jöstlein, E. Picasso and R.W. Williams, Phys. Lett. **B28,** 287 (1968).
J. Bailey, W. Bartl, G. von Bochmann, R.C.A. Brown, F.J.M. Farley, M. Giesch, H. Jöstlein, S. van der Meer, E. Picasso and R.W. Williams, Nuovo Cimento **A9,** 369 (1972).

[54] M.A. Ruderman, Phys. Rev. Lett. **17,** 794 (1966).

[55] J. Aldins, T. Kinoshita, S.J. Brodsky and A.J. Dufner, Phys. Rev. Lett. **23,** 441 (1969).
J. Aldins, S.J. Brodsky, A.J. Dufner and T. Kinoshita, Phys. Rev. **D1,** 2378 (1970).

[56] F.J.M. Farley, J. Bailey and E. Picasso, Nature **217,** 17 (1968).

[57] W.C. Barber, B. Gittelman, G.K. O'Neill and B. Richter, Phys. Rev. Lett. **16,** 1127 (1966).

[58] V.L. Auslander, G.I. Budker, E.V. Pakhtusova, Yu.N. Pestov, V.A. Sidorov, A.N. Skrinskij and A.G. Khabakhaser, Akad. Nauk SSR, Sibirskee Otd., Novosibirsk preprint 243 (1968).
J.E. Augustin, J.C. Bizot, J. Buon, J. Haissinski, D. Lalanne, P. Marin, H. Nguyen Ngoc, J. Perez y Jorba, F. Rumpf, E. Silva and S. Tavernier, Phys. Lett. **B28,** 508 (1969).
J.E. Augustin, A.M. Boyarski, M. Breidenbach, F. Bulos, J.T. Dakin, G.J. Feldman, G.E. Fisher, D. Fryberger, G. Hanson, B. Jean-Marie, R.R. Larsen, V. Lüth,

H.L. Lynch, D. Lyon, C.C. Morehouse, J.M. Paterson, M.L. Perl, B. Richter, R.F. Schwitters, F. Vannucci, G.S. Abrams, D. Briggs, W. Chinowsky, C.E. Friedberg, G. Goldhaber, R.J. Hollebeek, J.A. Kadyk, G.H. Trilling, J.S. Whitaker and J.E. Zipse, Phys. Rev. Lett. **34,** 233 (1975).

[59] S.J. Brodsky, Proc. Brandeis Univ. Summer Inst. in Theoretical Physics, Waltham, 1969, eds. M. Chrétien and E. Lipworth (Gordon and Breach, New York, 1971), vol. 1, p. 91.
S.J. Brodsky and S.D. Drell, Annu. Rev. Nucl. Sci. **20,** 147 (1970).
F.J.M. Farley, Proc. First Int. Conf. Eur. Phys. Soc., Florence, 1969 [Riv. Nuovo Cimento **1** (1969)], p. 59.
E. Picasso, *in* High energy physics and nuclear structure, ed. S. Devons (Plenum, New York, 1970), p. 615.
L.M. Lederman and M.J. Tannenbaum, Adv. Part. Phys. **1,** 1 (1968).

[60] R.T. Robiscoe, Phys. Rev. **168,** 4 (1968) and Cargèse Lect. Phys. **2,** 5 (1968).
R.T. Robiscoe and T.W. Shyn, Phys. Rev. **59,** 559 (1970).

[61] T. Appelquist and S.J. Brodsky, Phys. Rev. Lett. **24,** 562 (1970).

[62] A. Rich, Phys. Rev. Lett. **20,** 967 (1968).

[63] J.C. Wesley and A. Rich, Phys. Rev. **A4,** 1341 (1971).

[64] J. Bailey, K. Borer, F. Combley, H. Drumm, C. Eck, F.J.M. Farley, J.H. Field, W. Flegel, P.M. Hattersley, F. Krienen, F. Lange, G. Petrucci, E. Picasso, H.I. Pizer, O. Rúnolfsson, R.W. Williams and S. Wojcicki, Phys. Lett. **B55,** 420 (1975).
J. Bailey, K. Borer, F. Combley, H. Drumm, C. Eck, F.J.M. Farley, J.H. Field, W. Flegel, P.M. Hattersley, F. Krienen, F. Lange, E. Picasso and W. von Rüden, Phys. Lett. **B68,** 191 (1977).
J. Bailey, K. Borer, F. Combley, H. Drumm, C. Eck, F.J.M. Farley, J.H. Field, W. Flegel, P.M. Hattersley, F. Krienen, F. Lange, G. Lebée, E. McMillan, G. Petrucci, E. Picasso, O. Rúnolfsson, W. von Rüden, R.W. Williams and S. Wojcicki, Nucl. Phys. **B150,** 1 (1979).

[65] J. Bailey, F.J.M. Farley, H. Jöstlein, G. Petrucci, E. Picasso and F. Wickens, Proposal for a measurement of the anomalous magnetic moment of the muon at the level of 10–20 ppm, proposal CERN PH I/COM-69/20 (1969).
J. Bailey and E. Picasso, Progr. Nucl. Phys. **12,** 43 (1970).
F. Combley and E. Picasso, Phys. Rep. **C14,** 1 (1974).
F.J.M. Farley, Contemp. Phys. **16,** 413 (1975).
J.D. Jackson, Classical electrodynamics (Wiley, New York and London, 1975), p. 559.

[66] W. Flegel and F. Krienen, Nucl. Instrum. Methods **143,** 549 (1973).

[67] K. Borer, H. Drumm, C. Eck, G. Petrucci and O. Runólfsson, Proc. 5th Int. Conf. on Magnet Technology, Rome, 1975 (Lab. Naz. CNEN, Rome, 1975), p. 133.
H. Drumm, C. Eck, G. Petrucci and O. Runólfsson, Nucl. Instrum. Methods **158,** 347 (1979).

[68] K. Borer, Nucl. Instrum. Methods **143,** 203 (1977).
K. Borer and F. Lange, Nucl. Instrum. Methods **143,** 219 (1977).

[69] N. Kroll, Nuovo Cimento **45**, 65 (1966).

[70] M.L. Perl, G.S. Abrams, A.M. Boyarski, M. Breidenbach, D.D. Briggs, F. Bulos, W. Chinowsky, J.T. Dakin, G.J. Feldman, C.E. Friedberg, D. Fryberger, G. Goldhaber, G. Hansen, F.B. Heile, B. Jean-Marie, J.A. Kadyk, R.R. Larsen, A.M. Litke, D. Luke, B.A. Lulu, V. Lüth, D. Lyon, C.C. Morehouse, J.M. Paterson, F.M. Pierre, T.P. Pun, P.A. Rapidis, B. Richter, B. Sadoulet, R.F. Schwitters, W. Tanenbaum, G.H. Trilling, F. Vannucci, J.S. Whitaker, F.C. Winkelmann and J.E. Wiss, Phys. Rev. Lett. **35**, 1489 (1975).

[71] V.G. Kadyshevsky, Fermi Lab. Publ. 78/70 THY (1970).

[72] J.H. Field, E. Picasso and F. Combley, Uspekhi Fiz. Nauk **127**, 553 (1979) (in Russian) [Transl. Sov. Phys. Usp. **22**, 199 (1979)].

[73] J. Bailey, K. Borer, F. Combley, H. Drumm, F.J.M. Farley, J.H. Field, W. Flegel, P.M. Hattersley, F. Krienen, F. Lange, E. Picasso and W. von Rüden, J. Phys. **G4**, 345 (1978).

[74] E.M. Purcell and N.F. Ramsey, Phys. Rev. **78**, 807 (1950).

[75] L. Landau, Nucl. Phys. **3**, 127 (1957).

[76] R.L. Garwin and L. Lederman, Nuovo Cimento **11**, 776 (1959).

[77] A. Pais and J.R. Primack, Phys. Rev. **D8**, 3063 (1973).

[78] B. Rossi and D.B. Hall, Phys. Rev. **59**, 223 (1941).

[79] J. Bailey, K. Borer, F. Combley, H. Drumm, F.J.M. Farley, J.H. Field, W. Flegel, P.M. Hattersley, F. Krienen, F. Lange, E. Picasso and W. von Rüden, Nature **268**, 301 (1977).

[80] J.C. Hafele and R.E. Keating, Science **177**, 166 (1972).

[81] A. Einstein, Ann. Phys. **17**, 891 (1905).

[82] M.P. Balandin, V.M. Grebenyuk, V.G. Zinov, A.D. Konin and A.N. Ponomarev, Sov. Phys. JETP **40**, 811 (1974).

[83] D. Newman, G.W. Ford, A. Rich and E. Sweetman, Phys. Rev. Lett. **40**, 1355 (1978).

[84] F. Combley, F.J.M. Farley, J.H. Field and E. Picasso, Phys. Rev. Lett. **42**, 1383 (1979).

[85] S.R. Moore, K. Whisnant and B.L. Young, Phys. Rev. **D31**, 105 (1985).

[86] N. Cabibbo, L. Maiani and Y. Srivastava, Phys. Lett. **139B**, 459 (1984).
F.M. Renard, Phys. Lett. **116B**, 264 (1982) and **139B**, 449 (1984).
P. Méry, S.E. Moubarik, M. Perrottet and F.M. Renard, Constraints on non-standard effects from present and future $g-2$ measurements, preprint Centre of physique théorique Marseille, CPT 89/P2226 (1989), Z. Phys. (in the press).

[87] R. Arnowitt and P. Nath, Supersymmetry corrections to muon $(g-2)$, Proc. Conf. on the Intersections between Particle and Nuclear Physics, Lake Louise, Canada, 1986 (Amer. Inst. Phys., New York, 1986), p. 582.

[88] T. Kinoshita, Phys. Rev. Lett. **47**, 1573 (1981).
S.J. Brodsky, E. Mottola, I.J. Muzinich and M. Soldate, Phys. Rev. Lett. **56**, 1763 (1986).

[89] V.M. Aulchenko et al., Nucl. Instrum. Methods **A252**, 299 (1986).
L.M. Barkov, private communication.

[90] R. Schwitters and K. Strauch, Annu. Rev. Nucl. Part. Sci. **26,** 89 (1978).
S.R. Amendolia et al., Phys. Lett. **138B,** 454 (1984); **146B,** 116 (1984).

[91] V.W. Hughes and T. Kinoshita, Comments Nucl. Part. Phys. **14,** 341 (1985).
C. Heisey, F. Krienen, J.P. Miller, B.L. Roberts, L.R. Sulak, H.N. Brown, E.D. Courant, G.T. Danby, C.R. Gardner, J.W. Jackson, M. May, M. Month, P.A. Thompson, M.S. Lubell, W.P. Lysenko, W. Williams, J.T. Reidy, F. Combley, G. zu Putlitz, S.K. Dhawan, A.A. Disco, F.J.M. Farley, V.W. Hughes, Y. Kuang, J.K. Markey, H. Orth and H. Venkateramania, 'A new precision measurement of the muon $(g-2)$ value at the level of 0.35 ppm', Brookhaven AGS Proposal 821, Sept. 1985, revised Sept. 1986. (See also Design Report for AGS 821, March 1989.)
V.W. Hughes, Phys. Scr. **T22,** 111 (1988) and Proc. 8th Int. Conf. on High Energy Spin Physics, Minneapolis, 1988 (Amer. Inst. Phys., New York, 1989), p. 326.

[92] F.J.M. Farley and W.P. Lysenko, 'The muon distribution and the average field', Brookhaven Laboratory, $g-2$ muon Tech. Note 23 (1985).

[93] M. May, Nucl. Instrum. Methods **A279,** 410 (1989).

[94] H.I. Pizer, Nucl. Instrum. Methods **123,** 461 (1975).
H.I. Pizer, L. Van Köningsveld and H. Verweij, report CERN 76-17 (1976).

[95] F. Krienen, D. Loomba and W. Meng, Nucl. Instrum. Methods **A283,** 5 (1989).

[96] R.S. van Dyck Jr., P.B. Schwinberg and H.G. Dehmelt, Atomic Physics **9,** 53 (1984).

[97] B.E. Lautrup, A. Petermann and E. de Rafael, Phys. Rep. **3C,** 240 (1972).

[98] F.J.M. Farley, Progr. Nucl. Phys. **9,** 259 (1963); Cargèse Lectures in Physics, Vol. 2 M. Lévy, ed.), p. 55 (1965).
J. Bailey and E. Picasso, Progr. Nucl. Phys. **12,** 43 (1970).
F. Combley and E. Picasso, Phys. Rep. **14,** 1 (1974).
F.J.M. Farley, Contemp. Phys. **18,** 413 (1975).
F.H. Combley, Rep. Progr. Phys. **A2,** 1889 (1979).
F.J.M. Farley and E. Picasso, Annu. Rev. Nucl. Part. Sci. **29,** 243 (1979).
F. Combley, F.J.M. Farley and E. Picasso, Phys. Rep. **68,** 2 (1981).
E. Picasso, Il Nuovo Saggiatore **4**(1), 22 (1985).
E. Picasso, 'Elettrodinamica quantistica', *in* Enciclopedia del Novecento, Vol. VIII (Istituto della Enciclopedia Italiana, Rome, 1989).

THEORY OF HYDROGENIC BOUND STATES

Jonathan R. Sapirstein
Department of Physics, University of Notre Dame,
Notre Dame, IN 46556

and

Donald R. Yennie
Laboratory of Nuclear Studies, Cornell University,
Ithaca, NY 14853

CONTENTS

I. INTRODUCTION

The study of hydrogen-like systems has played a major role in the development of modern physics, from the original understanding of the spectra in terms of quantum mechanics, through such refinements as relativistic, spin orbit, and quantum electrodynamic (QED) contributions, to the present concern with precision tests of relativistic two-body theory in such diverse systems as hydrogen, positronium, muonium, and quarkonium. The need for high precision calculations of bound state properties was ushered in with the experimental discovery of the Lamb shift[1] which, together with the concurrent discovery of the electron's anomalous magnetic moment,[2] led to the modern era of quantum electrodynamics.

High precision bound state calculations are different in character from other precision calculations, such as the electron anomaly. The latter requires the highest refinement in perturbative analysis of renormalization, including intricate subtractions of overlapping and nested divergences.[a] Renormalization theory in QED is a perfectly well-defined procedure in terms of an expansion in numbers of photons, but its implementation is extremely complex and lengthy. On the other hand, precision bound state calculations are essentially nonperturbative in the binding potential. Though they involve kernels which can be written in terms of Feynman graphs, the dimensionless parameters, particularly the fine structure constant α, enter the wave functions and the particle propagators (through

[a]See the description of this by Kinoshita in the present volume.

the energy) in a nonperturbative way. Thus one cannot simply count powers of α in a given kernel in order to determine its ultimate contribution to an energy shift. Typically, a given kernel includes some leading order plus an infinite series of smaller terms. Further, kernels can often be grouped together in such a way that their leading order cancels; it is important to take this into account so as to avoid calculating terms of spurious lower order. As a consequence, the formal development of bound state theory is not at all mechanical. There are many routes to a correct answer, but they are not equally felicitous. Fortunately, the tedium involved in exploring various of these paths is partially offset by the fact that the ultimate work involved in algebra and integrals is orders of magnitude simpler than that of the electron anomaly—provided the formal analysis is well chosen. This is because most calculations are at the two-loop level, rather than the four-loops which presently characterize the anomaly calculation. The dual aims of this paper are to provide the reader with an overview of the situation and introduce him or her to the techniques which currently seem most viable.

In spite of the statement in the preceding paragraph that bound state theory is nonperturbative, it is possible to make use of small parameters such as α and m_e/m_N (where m_N is the mass of the nucleus) to develop expressions in increasing orders of smallness. However, the nonperturbative nature of the expansion shows up in nonanalytic dependence on these parameters (such as logarithms). As implied by the preceding paragraph, there is an art in developing a theoretical expression in this manner. At the most elementary level, one must be aware that there are nonanalytic terms and seek them out; the trick is to arrange the expression being calculated into a tractable form so that the analytic part of leading order is easily extracted and terms of higher order isolated. Usually it is best to do this before integrations are carried out, rather than later. We can then make an estimate of the accuracy of our calculation by guessing the size of neglected terms, which have additional factors of the small parameters. Since they can occasionally have large coefficients, such estimates can be very unreliable even though they give the correct order of magnitude; one should be aware of the danger of putting them on the same level as experimental errors.

In principle, the problem of precision analysis of hydrogen-like bound state energy levels is exceedingly complicated. The system consists of a nucleus and an electron, and involves their mutual and self interactions through the electromagnetic field. For some nuclei, the structure of the nucleus also requires analysis of the strong interactions, or the use of other experiments to calibrate such effects. It is desirable, but very difficult, to give a unified treatment of the problem so that different contributions need not be combined in a patchwork way. While we hope to give a glimpse of how such a treatment can be developed, the approach of this paper is more pragmatic. It is fortunate that different physics, such as nuclear motion and radiative corrections, can largely be compartmentalized. We shall not dwell on the reasons here, but it is certainly much simpler than for other systems, such as multi-quark particles. For example, we can largely ignore recoil effects while treating the *radiative corrections*, which refer to corrections due to the emission and absorption of photons by the electron and also to vacuum polarization corrections to the Coulomb interaction. In these, the recoil of the nucleus is ignored except for the dependence of some quantities on the reduced mass m_r rather than m_e. The treatment of renormalization is similar to that for free electrons, except that electron propagators in an external Coulomb field are used (Furry[3]) representation). Similarly, in treating *recoil corrections*, in which truly dynamical effects due to nuclear recoil are incorporated, radiative corrections may be ignored, except to the extent that it may be legitimate to incorporate a particle's anomalous magnetic moment phenomenologically. Until recently, it was quite possible to keep these two categories separate to the order of interest. Thus a worker could be an expert on radiative correction effects without dealing with the complications of recoil, and *vice versa.* Now, the level of accuracy has reached the point where both complications must be treated together as *radiative-recoil* corrections. However, in combination some of the subtleties of each category are absent. [At some level it is necessary to include a fourth category consisting of other small effects such as weak interactions, finite nuclear size, nuclear polarizability, etc.] In this paper, we are concerned with all these effects.

Rather than attempt an organization of the paper which might be more logical from a theoretical viewpoint, we have arranged the topics in a way which is hoped to be more useful to a reader who is interested in the status of various subtopics. We start in Chapter II with a summary of the most important theoretical results in the field and a comparison of them with experimental results. At that point, there is some attempt to give their meaning, but not their derivation. Chapter III discusses those aspects of *radiative corrections* which can be understood without recourse to a complete relativistic two-body formalism. Historically, the one-photon radiative corrections (with all orders of the Coulomb interaction) have been considered in the most detail, and the work required to attain the present level of accuracy has extended over four decades. While the expressions to be calculated are well-known and well-defined, there is as yet no single agreed-upon way to carry through their evaluation. We present three approaches, in partial detail, so that the reader may discern the general nature of such calculations. A brief primer on orders of magnitude of different contributions in this situation is presented; this is one of the most perplexing questions to novices in the field.

In Chapter IV, we turn to the formulation of the relativistic two-body problem. Every physicist knows the trivial analysis for separating the center of mass and relative motion in the nonrelativistic two-body system to find a simplified problem in which the relative motion is described using the reduced mass. When one or both particles are to be treated relativistically, there is no similar simple procedure. In fact, if one starts with a two body system and lets one of the particle masses become infinite, it requires a nontrivial analysis to demonstrate that the result can be expressed in terms of a relativistic equation for the other particle in a central potential.[b] The physics is clear; it is simply the formalism which is awkward. In the following few paragraphs, we trace briefly the history of the development of the understanding of the recoil problem. Our aim is to give a general perspective of the subject rather than to present a complete or critical survey of the

[b]A demonstration of this fact is given by Brodsky[4]) and Gross.[5])

literature. The calculation of recoil corrections is obviously related to the development of the theory of a relativistic two-body system. There is a vast amount of literature concerning this theory. Some of it is purely formal and does not address any particular bound state problem, and some of it is relevant to the treatment of the energy shifts of interest here.

The first main step in treating the recoil problem involving Dirac particles was due to Breit.[6] In that work he presented his famous equation for two electrons interacting through the electromagnetic field. That equation has the form of an instantaneous interaction between the electrons, but it does incorporate retardation effects to the extent that they are important at the v^2/c^2 level. His equation was the only one used for the treatment of recoil effects for over two decades. In fact, there were very few workers in the field, other than Breit and his associates, during that period. For example, the Breit equation was used by Breit and Meyerott[7] to find the correct reduced mass dependence in the hyperfine structure (hfs). It took the stimulation of subsequent experiments to bring on the development of more precise formulations.

The Breit equation does not produce corrections of relative order $\alpha m_e/m_N$ in the fine structure or hyperfine structure. As was pointed out by Salpeter,[8] the reason is that it corresponds essentially to a single electron theory, rather than to hole theory. Thus these contributions, which are present and experimentally relevant, can be found only if the equivalent of hole theory is used. As experimental precision increased, it became necessary to develop a more accurate formalism based on a complete field theoretic treatment of the two-body problem by Schwinger[9] and by Salpeter and Bethe.[10] The result is usually referred to as the Bethe-Salpeter equation. It is now generally recognized that this equation, which is a homogeneous four-dimensional integral equation, is not in a practical form for high precision analysis of energy levels. The first systematic approximation procedure for solving this equation was developed by Salpeter,[8] who showed how to develop a perturbation expansion in which the starting point is a solvable three-dimensional equation. A well-known complication is that it is difficult to obtain the correct result in the limit in which the mass of one of the particles becomes infinite. The correct treatment of that limit requires the use of kernels with crossed photon lines of arbitrarily high order. In most other recent approaches, the calculation is arranged differently, so that the problem of the limit of large mass does not occur.

Although more recent approaches are based on the Bethe-Salpeter philosophy, the practical procedure is rather different. To solve the Bethe-Salpeter bound state integral equation directly to the required accuracy seems to be quite awkward and not at all useful. Present work is generally based on some three-dimensional formulation which incorporates varying amounts of the basic physics. Additional field theoretic effects are then introduced through various perturbation kernels.

Chapter V deals with the problem of recoil, excluding radiative corrections and going beyond simple reduced mass effects. In the case of the Lamb shift, the leading terms are of the same order in α as the term due to radiative corrections on the electron line, but they are smaller in magnitude by an additional factor of m_e/m_N. For the hyperfine structure, introducing the effects of recoil also increases the order in α by one or more powers.

Chapter VI describes radiative-recoil corrections, in which the previously separate lines of investigation are brought together. It was pointed out by Caswell and Lepage[11] that, in addition to the recoil contribution to the hfs due purely to photon exchange, radiative-recoil contributions are important at the present level of accuracy. Their importance arises from the fact that in the one-loop recoil corrections, the characteristic momentum of the loop integration is large compared to the electron mass, causing a significant modification of the photon propagator. Other corrections, of less importance, arise from radiative corrections to the electron leg. Radiative-recoil corrections to the Lamb shift are considerably more subtle in that they are nonperturbative in the Coulomb interaction, while the analysis of the two-body problem is tied in initially to a perturbative discussion of the four-point function. Bhatt and Grotch[12] study this by considering an effective Dirac equation which incorporates the recoil kinematics. They obtain the correct overall reduced mass dependence, as well as the expected reduced mass dependence of the argument of the Bethe logarithm. Then they go on to evaluate the more dynamical consequences of recoil. As the present authors prefer a treatment entirely within the framework of quantum field theory, a description is given of how the Bhatt-Grotch approach is related to such a framework.

One of our goals in this paper is to describe previously calculated contributions in a consistent and systematic way. To our knowledge, this has not always been done before, even though workers in the field have satisfied themselves about the consistency of their results. When extracting leading terms, it is not difficult to make approximations consistently since different kinematic regimes and different analytic structures are easy to identify and isolate. The complication that faces the next round of calculations is that one will often be seeking terms which are small residues of earlier terms. The authors have attempted to outline a consistent formalism in which these calculations can be framed. However, it is clear that a great deal of ingenuity and difficult computation will be required before the terms of one higher order in α or m_e/m_N are all determined.

II. PRESENT STATUS OF THEORY AND COMPARISON WITH EXPERIMENT

It is our purpose in this chapter to present a review of the experimental situation and theoretical status of fine structure and hyperfine structure in one-electron atoms, specifically for positronium, muonium, and hydrogen. We limit ourselves to discussion of ground state hyperfine splitting, and to fine structure of only the $n = 1$ and $n = 2$ levels. We also discuss the Lamb shift in He^+ and other hydrogenic ions. The chapter is intended to be self-contained, so that a reader interested mainly in the status of tests of QED in these systems as of mid-1989 need consult only this chapter. However, we will refer to parts of the following chapters in which various effects are discussed in more theoretical detail, so that a reader interested in either understanding where the formulae come from or in seeing where extensions of theory are required can explore more deeply. Chapter III deals with radiative correction effects that can be treated without considering the recoil of the nucleus. However, it is important to stress that theory is now approaching a level where a unified treatment of the two-body bound state problem is necessary for further progress: simple combination of a non-recoil result with a recoil result can lead to errors in higher orders of α or m_e/m_N. (We use m_N for the mass of the positively charged particle the electron is bound to and refer to that particle as the nucleus in this review so as to be able to describe hydrogen, muonium, positronium, and hydrogenic ions simultaneously.) Such a treatment appears in Chapter IV; it is of course not unique, and future researchers may well find a more convenient one. However, it provides a well-defined framework in which to calculate recoil corrections, and these will be described in Chapter V. Chapter VI provides a discussion of the situation when both radiative and recoil effects must be treated at the same time.

In all the following the 1986 adjustment of physical constants will be used:[13] for convenient reference the relevant constants are:

$$\begin{aligned} c &= 2.99792458 \times 10^{10}\text{cm/sec} \\ \alpha^{-1} &= 137.0359895(61) \\ R_\infty &= 109737.315709(18)\text{cm}^{-1} \\ a_e &= 1.159652193(10) \times 10^{-3} \\ a_\mu &= 1.1659230(84) \times 10^{-3}\ . \end{aligned} \tag{2.1}$$

We have in the above presented a more recent determination of R_∞[14] which disagrees slightly with the 1986 adjustment. The difference is unimportant for most purposes, but is significant for consideration of $n = 2$ to $n = 1$ transitions. It should be noted that the comparison of theory and experiment for several QED systems, most importantly the electron $g - 2$ and muonium hfs, is incorporated in the above adjustment. However, the particular numbers shown in (2.1) are influenced only very slightly by the experiments to be discussed here. Other quantities, such as the ratio of the muon mass to the electron mass, are strongly influenced by the experiments to be discussed. Therefore we should not use such adjusted values in the discussion. We also note that there is a more recent determination of the fine structure constant[15].

Let us start with a few observations about the reduced mass problem and the Dirac equation. There is a trivial rearrangement of the non-relativistic Schrödinger equation to express the kinetic energy in terms of the reduced mass, subsequently leading to a factor of the reduced mass in the Bohr expression for the energy levels. The same procedure cannot be applied directly to the Dirac equation. However, in making refined calculations of energy levels of hydrogenic systems, it seems to us to be desirable to find some procedure which includes as much of the correct mass dependence as is possible from the start in order to avoid treating all effects of the nuclear motion perturbatively. Such a procedure should include some of the obvious physics such as the kinetic energy of the nucleus (possibly in relativistic form) and perhaps the convection current due to its motion. Other effects of the nuclear motion would be regarded as bonafide dynamical recoil effects. Having experimented with various approaches, we find that there is no unique way to do this. Nevertheless, it is possible to incorporate a large amount of the correct mass dependence in the treatment of the initial problem. In particular, if one is considering radiative corrections only in lowest order, the exact reduced mass dependence is known, and we will present formulas valid for arbitrary mass ratio for such terms.

To make this concrete and to describe what has been done so far, let us first consider the energy levels for the Dirac-Coulomb problem without recoil. These are given by[c]

$$E_{nj} = m_e f(n,j) \tag{2.2a}$$

where

$$f(n,j) = \left[\frac{1}{1+\dfrac{(Z\alpha)^2}{(n-\beta)^2}}\right]^{\frac{1}{2}} . \tag{2.2b}$$

Here n is the principal quantum number,

$$\beta = j + \frac{1}{2} - \sqrt{(j+\frac{1}{2})^2 - (Z\alpha)^2} \, ,$$

and j is the total angular momentum. Note that the energy levels are independent of the orbital angular momentum given by l, which is related to j by $l = j \pm \frac{1}{2}$. We may expand f in powers of $(Z\alpha)^2$; the first few terms are

$$\begin{aligned} f(n,j) = 1 &- \frac{(Z\alpha)^2}{2n^2} - \frac{(Z\alpha)^4}{2n^3}\left(\frac{1}{j+1/2} - \frac{3}{4n}\right) \\ &- \frac{(Z\alpha)^6}{8n^3}\left(\frac{1}{(j+1/2)^3} + \frac{3}{n(j+1/2)^2} + \frac{5}{2n^3} - \frac{6}{n^2(j+1/2)}\right) + \dots . \end{aligned} \tag{2.2c}$$

The first term here has the correct mass dependence corresponding to the rest mass of the electron. The second term should be modified by a factor of $m_r/m_e = 1/(1+m_e/m_N)$ to take into account the dominant non-relativistic behavior embodied in the Bohr energy levels. We can see already that it would be awkward to attain this dependence by a perturbative expansion in the kinetic energy of the nucleus. (To exceed the accuracy of the Rydberg, at least three powers of m_e/m_N would be required!) The exact mass dependence of the $(Z\alpha)^4$ term has been worked out, and it introduces a small l-dependence. The result is given below. The exact mass dependence of the $(Z\alpha)^6$ term has not been worked out because true dynamical effects already set in at the level of $\alpha^5 m_r^2/m_N$, making the meaning of the reduced mass dependence of this term ambiguous.

In Chapter IV, we present a discussion of the two-body problem which starts from field theory and succeeds in obtaining much of the correct mass dependence of the first three terms of (2.2c), with the remaining mass dependence through that order coming

[c]In presenting theoretical expressions, we use natural units in which $\hbar = c = 1$ for convenience. For subsequent comparison with experiment, the usual units can easily be restored.

from one photon exchange kernels. An expression for the total energy in which terms of order $\alpha^6 m_e^3/m_N^2$ have been neglected (totally negligible for hydrogen and muonium) and perturbation kernels through order α^4 have been incorporated is

$$E = M + m_r[f(n,j) - 1] - \frac{m_r^2}{2M}[f(n,j) - 1]^2 + \frac{(Z\alpha)^4 m_r^3}{2n^3 m_N^2}\left[\frac{1}{j+\frac{1}{2}} - \frac{1}{l+\frac{1}{2}}\right](1 - \delta_{l0}) \quad (2.3)$$

where $M = m_e + m_N$. Note that the terms depending on $f(n,j)$ retain the degeneracies of the original Dirac-Coulomb result, but the last term contributes a tiny amount, -2 kHz, to the $2S_{1/2} - 2P_{1/2}$ splitting (i.e., to the Lamb shift). While the second term does not contribute to this splitting, we note that it does contribute about 22 MHz to the $n = 2$ to $n = 1$ energy difference in hydrogen. The complete correct mass dependence through order α^4, including the small term which breaks the degeneracy, was obtained originally by Barker and Glover.[16] Several authors also derived the result for positronium energy levels to the same order in α.[17]

A. Lamb Shift

QED was decisively established as the most accurate modern physical theory by two tests, the anomalous magnetic moment of the electron and the Lamb shift (the $2S_{1/2} - 2P_{1/2}$ splitting in hydrogen); these are reviewed in this volume by Kinoshita and Pipkin, respectively. A useful review of the energy levels of one-electron atoms was given about ten years ago by Erickson.[18]

In the following we use a more general terminology for the Lamb shift to include any deviation from the energy levels predicted by the Dirac equation with the binding scaled by the reduced mass of the electron, namely the deviation from $M + m_r[f - 1]$. (Note that this definition includes the effect of finite nuclear size in the Lamb shift: an equally satisfactory definition would include that effect instead in the solution of the Dirac equation. While simply a matter of definition, it is important to note that the physics involved in the finite nuclear size shift is quite different from that of the rest of the Lamb shift). Where appropriate, we refer to the $2S_{1/2} - 2P_{1/2}$ splitting as the 'classic' Lamb shift. At first ignoring nuclear motion, the bulk of the Lamb shift due to one photon radiative corrections to the electron and vacuum polarization corrections to the potential can be expressed by the formula

$$\Delta E_n(\text{one-loop}) \equiv \frac{m_e \alpha (Z\alpha)^4}{\pi n^3} F_n(Z\alpha) \ . \quad (2.4)$$

For large values of Z the function $F_n(Z\alpha)$ must be determined numerically, as will be discussed in Chapter III-C. For smaller values, however, one can expand it in powers of $Z\alpha$.

In the non-recoil limit, this expansion involves constants and the logarithmic factor $\ln(Z\alpha)^{-2}$. Certain consequences of nuclear motion can be incorporated through reduced mass dependence, as shown in Chapter VI-B. For example, most terms acquire an overall factor of $(m_r/m_e)^3$, which arises simply from the scaling of the wave functions. Spin-orbit terms associated with the electron's anomalous magnetic moment have instead a factor $(m_r/m_e)^2$; the difference is a consequence of the proton's convection current. The argument of the logarithm which arises from radiative corrections to the electron is really the ratio of the electron's rest energy and a characteristic binding energy ($\propto m_r(Z\alpha)^2$). Thus the logarithm takes on an argument $(m_e/m_r)(Z\alpha)^{-2} \equiv \sigma(Z\alpha)^{-2}$. (The combined effect of the overall factors and the change in the argument of the logarithm are sometimes Taylor expanded to first order in m_e/m_N: the error induced by this procedure is negligible for hydrogen and muonium, but the expansion cannot of course be used for positronium.) Other nuclear motion corrections which cannot be incorporated with these simple substitutions are recoil corrections and are described later. We then have the expansion

$$\begin{aligned} F_n(Z\alpha) = & A_{40}(n) + A_{41}(n)\ln[\sigma(Z\alpha)^{-2}] + (Z\alpha)A_{50}(n) \\ & + (Z\alpha)^2\{G(n,Z\alpha) + A_{61}(n)\ln[\sigma(Z\alpha)^{-2}] + A_{62}(n)\ln^2[\sigma(Z\alpha)^{-2}]\}\ . \end{aligned} \tag{2.5a}$$

The functions $G(n, Z\alpha)$ represent non-logarithmic terms of order $m_e\alpha(Z\alpha)^6$ plus all higher order corrections. The work involved in calculating these constants and functions has involved many physicists and has extended over almost four decades. The main part of this work, and associated references, are given in Chapter III; but a few references on the more refined contributions are given here. For discussing the classic Lamb shift and $n = 2$ to $n = 1$ transitions, we need the following constants; where convenient, we present them in a general form which applies to other states. Values of the Bethe logarithms $\ln k_0(n)$ are given in (3.13).

$$\begin{aligned} A_{40}(nS) &= \left[\frac{10}{9} - \frac{4}{15} - \frac{4}{3}\ln k_0(nS)\right]\left(\frac{m_r}{m_e}\right)^3 \\ A_{40}(n, l \neq 0) &= \left[\frac{C_{lj}}{2(2l+1)}\right]\left(\frac{m_r}{m_e}\right)^2 + \left[-\frac{4}{3}\ln k_0(n, l\neq 0)\right]\left(\frac{m_r}{m_e}\right)^3 \\ A_{41}(nS) &= \left[\frac{4}{3}\right]\left(\frac{m_r}{m_e}\right)^3 \\ A_{50}(nS) &= 4\pi\left[\frac{139}{128} + \frac{5}{192} - \frac{1}{2}\ln 2\right]\left(\frac{m_r}{m_e}\right)^3 \\ G(n, Z\alpha) &= [G_{SE}(n, Z\alpha) + G_{VP}(n, Z\alpha)]\left(\frac{m_r}{m_e}\right)^3 \\ A_{61}(1S) &= \left[\frac{28}{3}\ln 2 - \frac{21}{20} - \frac{2}{15}\right]\left(\frac{m_r}{m_e}\right)^3 \\ A_{61}(2S) &= \left[\frac{16}{3}\ln 2 + \frac{67}{30} - \frac{2}{15}\right]\left(\frac{m_r}{m_e}\right)^3 \\ A_{61}(2P_{1/2}) &= \left[\frac{103}{180}\right]\left(\frac{m_r}{m_e}\right)^3 \\ A_{61}(2P_{3/2}) &= \left[\frac{29}{90}\right]\left(\frac{m_r}{m_e}\right)^3 \\ A_{62}(nS) &= -[1]\left(\frac{m_r}{m_e}\right)^3\ . \end{aligned} \tag{2.5b}$$

where $C_{lj} \equiv 2(j-l)/(j+1/2)$. The constants A_{41}, A_{50}, and A_{62} vanish for P states. The above expressions combine the self-energy (including the anomalous magnetic moment) and vacuum polarization contributions: the latter comprise the $-\frac{4}{15}$ part of A_{40}, the $\frac{5}{192}$ part of A_{50}, and the $-\frac{2}{15}$ part of $A_{61}(nS)$. The functions G_{VP} and G_{SE} for Z=1 are

$$\begin{aligned}
G_{SE}(1S) &= [-30.3 \pm 1.5] \\
G_{SE}(2S) &= [-32.1 \pm 1.5] \\
G_{SE}(2P_{1/2}) &= [-0.8 \pm 0.3] \\
G_{SE}(2P_{3/2}) &= [-0.5 \pm 0.3] \\
G_{VP}(1S) &= \left[\frac{4}{15}\ln 2 - \frac{1289}{1575} + \left(\frac{19}{45} - \frac{\pi^2}{27}\right) + \mathcal{O}(Z\alpha)\right] \\
G_{VP}(2S) &= \left[-\frac{743}{900} + \left(\frac{19}{45} - \frac{\pi^2}{27}\right) + \mathcal{O}(Z\alpha)\right] \\
G_{VP}(2P_{1/2}) &= \left[-\frac{9}{140} + \mathcal{O}(Z\alpha)\right] \\
G_{VP}(2P_{3/2}) &= \left[-\frac{1}{70} + \mathcal{O}(Z\alpha)\right] .
\end{aligned} \tag{2.5c}$$

The terms in parentheses in the formulas for G_{VP} come from higher order terms in the Wichmann-Kroll expression;[19] the lowest order part, the Uehling potential, gives rise to the remaining terms. A detailed numerical study of vacuum polarization with references to previous calculations has recently been presented by Mohr and Soff.[20] Appendix A contains a discussion of the results for G_{SE}, which come from extrapolating the Z=10, 20, and 30 values calculated by Mohr[21] down to $Z = 1$.

There is one last set of single-loop non-recoil terms to take into account. These are vacuum polarization corrections from heavier mass particles. Generally such corrections are reduced because they are proportional to $1/\text{mass}^2$. For example, the muon contributes $(m_e/m_\mu)^2$ as much as the electron, or about 0.6 kHz to the classic Lamb shift in hydrogen. The hadronic contribution is more complicated, but it can be worked out in principle using the Källen-Lehman representation, whose weight function can be determined from the total cross section for $e^+e^- \to$ hadrons. For many purposes, such as the contribution to $g-2$ or the muonium hyperfine structure, one can get a reasonable estimate by using the ρ^0-resonance contribution. The estimate is smaller than 0.1 kHz in the classic Lamb shift in hydrogen. In the application to hydrogen, it is questionable in principle whether these small contributions should be incorporated. The reason is that the measured proton size determined by electron scattering experiments should be used in the correction described below. Since those experiments would take into account the radiative corrections to the electron and the vacuum polarization from electron pairs, the *apparent* form factor would already incorporate the effect of the muon and hadronic vacuum polarization, so no additional correction is needed in the Lamb shift.

At this point, we have a complete description of the one-loop non-recoil Lamb shift with an important set of reduced mass effects built in. Because we have not assumed a small mass ratio, the expression is correct, though not complete, for positronium. However, in that case the effect of self-energy on the positive particle is just as important, and one must add in the expression[22]

$$\Delta E_n(\text{nucleus}) = \frac{\alpha^5}{\pi n^3}\frac{m_r^3}{m_N^2}\left[\left(\frac{10}{9}+\frac{4}{3}\ln\frac{m_N}{m_r\alpha^2}\right)\delta_{l0}-\frac{4}{3}\ln k_0(n)\right] . \tag{2.6a}$$

We mention this term only for the purposes of having a complete set of formulas to order $m_e\alpha^5$ for positronium. For positronium, we also need the spin-orbit contributions associated with the positron's anomalous magnetic moment. In order to show the symmetry between the electron and positron, we display the complete effective spin-orbit Hamiltonian from the anomalous magnetic moments:

$$\Delta H(\text{spin-orbit}) = \frac{\alpha^2}{4\pi m_e m_N}\left[\frac{m_N}{m_e}\vec{\sigma}_e\cdot\vec{L}+\frac{m_e}{m_N}\vec{\sigma}_N\cdot\vec{L}+(\vec{\sigma}_e+\vec{\sigma}_N)\cdot\vec{L}\right]\frac{1}{r^3} . \tag{2.6b}$$

The terms proportional to the electron spin are already incorporated into (2.5b). The first one is independent of the nuclear motion; and the second arises from the convection current of the nucleus, as described in Chapter VI-B. The combination is responsible for changing the power of m_r/m_e from three to two. For the case of hydrogen, (2.6a) and the nuclear spin dependent part of (2.6b) are inseparable from the structure of the proton, and are in any case extremely small. For muonium the term is larger and unambiguous, but far smaller than present experimental error.

We now turn to the correction of order $(Z\alpha)^5 m_r^3/(m_e m_N)$ derived by Salpeter.[8] This is sometimes referred to as a relativistic recoil correction, but is referred to here as the Salpeter correction. This is an example of a pure recoil correction, which requires the use of a form of the Bethe-Salpeter equation since the previously used Breit equation had missed terms of this order entirely. The analysis of this order is complicated by having parts that involve very different physics. As will be discussed further in Chapter V, the exchange of a transverse photon between the electron and the nucleus with any number of Coulomb interactions occurring between emission and absorption gives rise to an expression closely related to the lowest order self-energy calculation, but reduced by a factor m_e/m_N. This term involves highly nonrelativistic momenta. However, there is also a term of the same order associated with various combinations of two-photon exchange kernels, which involves both non-relativistic and relativistic electron momenta. The following expression is valid for arbitrary masses:

$$\begin{aligned}\Delta E_n(\text{Salpeter}) \equiv \frac{m_r^3}{m_e m_N}\frac{(Z\alpha)^5}{\pi n^3}\Bigg\{&\frac{2}{3}\delta_{l0}\ln\Big(\frac{1}{Z\alpha}\Big)-\frac{8}{3}\ln k_0(n)-\frac{1}{9}\delta_{l0}-\frac{7}{3}a_n\\ &-\frac{2}{m_N^2-m_e^2}\delta_{l0}\left[m_N^2\ln\frac{m_e}{m_r}-m_e^2\ln\frac{m_N}{m_r}\right]\Bigg\} .\end{aligned} \tag{2.7a}$$

The state-dependent constants a_n, which are associated with the relativistic region, are given by the formula[18]

$$a_n \equiv -2\left(\ln\frac{2}{n}+(1+\frac{1}{2}+\ldots+\frac{1}{n})+1-\frac{1}{2n}\right)\delta_{l0}+\frac{1-\delta_{l0}}{l(l+1)(2l+1)} . \tag{2.7b}$$

Nuclear motion effects in the radiative corrections which cannot be incorporated in the reduced mass dependence are known as radiative-recoil contributions. Those of order $\alpha(Z\alpha)^5 m_e^2/m_N$ have recently been calculated by Bhatt and Grotch.[12] We describe the calculation in Chapter VI, but quote the result here

$$\Delta E_n(\text{rad-rec}) = \frac{\alpha(Z\alpha)^5 m_e^2}{n^3 m_N}\left(\frac{35}{4}\ln 2 - \frac{7333}{960} - 0.415 \pm 0.004\right)\delta_{l0}\,. \tag{2.8}$$

The terms given analytically arise from the external field approximation part of the calculation, which is a modification of the formal approach discussed in Chapter III-D. The term given numerically with errors is due to some kernels not included in the external field approximation. The overall contribution to the Lamb shift in hydrogen is small, approximately -2.5 kHz.

We now turn to higher order radiative corrections. The function $F_n(Z\alpha)$ is nonperturbative inasmuch as its evaluation requires consideration of an infinite set of Coulomb exchanges. However, it represents a perturbative quantity in α, because only one radiative photon loop is taken into account. At the present level of accuracy, it is necessary to consider higher order terms in the loop expansion. The term of order α^2 can be written in analogy with the one loop expression as

$$\Delta E_n(\text{two-loop}) = \frac{m_e \alpha^2 (Z\alpha)^4}{\pi^2 n^3} H_n(Z\alpha) \tag{2.9}$$

The complexity of the evaluation of the lowest order term in $F_n(Z\alpha)$ is connected with the fact that the slope of the electron's Dirac form factor, $dF_1(q^2)/dq^2|_{q^2\to 0}$, is infrared divergent in one loop order. However, with two loops this divergence is no longer present,[d] and to lowest order in $Z\alpha$ it suffices to carry out a standard vertex calculation using free propagators and Feynman parameter techniques.[23,24] That calculation is only slightly more complex than the two-loop calculation of the electron $g-2$ factor. It was first correctly carried out numerically by Appelquist and Brodsky[25] and later done analytically.[26]

Combining their result with terms of the same order arising from the two-loop contribution to the electron anomalous magnetic moment and two-loop contributions to vacuum polarization gives[e]

$$\begin{aligned} H_{\text{S-states}}(Z\alpha) &= \left[-\frac{4358}{1296} - \frac{10}{27}\pi^2 + \frac{3}{2}\pi^2\ln 2 - \frac{9}{4}\zeta(3)\right]\left(\frac{m_r}{m_e}\right)^3 \\ H_{\text{non-S-states}}(Z\alpha) &= \left[\frac{197}{72} + \frac{\pi^2}{6} - \pi^2\ln 2 + \frac{3}{2}\zeta(3)\right]\frac{C_{lj}}{2(2l+1)}\left(\frac{m_r}{m_e}\right)^2 . \end{aligned} \tag{2.10}$$

---

[d]The vertex does have an infrared divergence; but when treated consistently with appropriate self energies on external lines taken into account, the infrared divergent term has *two* powers of q^2. Such a term yields a Lamb shift contribution with an additional power of $Z\alpha$ beyond the one under discussion here.

[e]We know of no calculation of the one-loop vertex combined with the one-loop vacuum polarization, which might have been expected to contribute in the same order. The reason it does not is similar to that mentioned in the preceding footnote.

It contributes 101 kHz to the classic Lamb shift. There are clearly two calculations that need to be done to extend this work. The first is the calculation of the slope of the three-loop Dirac form factor, which will enter with another power of α/π. The factor π in the denominator, which is characteristic of perturbative calculations, will probably force this contribution under 1 kHz, so its evaluation is not urgently needed. Nevertheless, the developments in the technology of evaluating multi-loop Feynman diagrams that have been made to carry out the four-loop calculations of $g-2$ described by Kinoshita in this volume make the evaluation of three-loop graphs rather straightforward, so this effect is certainly calculable. A more important effect can be expected from the $Z\alpha$ correction to $H(Z\alpha)$. This is because binding corrections lack the inverse power of π characteristic of perturbation theory, and in fact can have that factor in the numerator, as evidenced by the second term of $F(Z\alpha)$. Thus this term could enter at the few kHz level. It would be desirable for application to high-Z ions to evaluate $H(Z\alpha)$ without approximation, though renormalization of two-loop graphs is considerably more complex than for the one-loop case. However, because the first order correction is order $\alpha^2(Z\alpha)^5$, which is non-analytic in $(Z\alpha)^2$, the techniques of Layzer[27,28] may prove useful in the evaluation of this correction.

We turn now to the effect of nuclear size. The finite distribution of charge in the nucleus affects the binding of S states, and for nonrelativistic systems leads to the well-known shift

$$\Delta E_n(\text{finite-size}) = \frac{2}{3n^3}(Z\alpha)^4 m_r^3 \langle r^2 \rangle \ . \tag{2.11}$$

While in principal this shift should have some dependence on the detailed distribution of charge in the nucleus, as long as the root mean square radius is chosen to be the same, different models of the charge distribution give results that differ negligibly. The above formula assumes a spherical charge distribution. However, it should be noted that quadrupole and higher electric moments possessed by many nuclei can influence that distribution. We note that the effect of finite nuclear size is particularly important for muonic atoms, in which the muonic wave function has a significant overlap with the nucleus. The combination of radiative corrections and finite nuclear size gives a totally negligible contribution (about 10 Hz) of order $\alpha(Z\alpha)^5 m_r^3\langle r^2\rangle$.[29] For highly charged ions, however, where there is sizeable nuclear penetration, this effect is significant and has been calculated with the numerical techniques described in Chapter III-C.[30].

At some level the effect of nuclear polarizability will play a role. This refers to the contribution from intermediate excited states of the nucleus. In Section C below, we discuss the consistent treatment of the finite size and polarizability of the nucleus in some detail for the case of hyperfine splitting. However, the effect should be greatly suppressed for fine structure because of a cancellation between two-photon exchange ladder and crossed contributions, similar to the Caswell-Lepage cancellation for three-photon exchange contributions to the hfs, as described in Chapter V-B(2). We note that in an earlier work on this subject[31] approximations have been made that appear not to be in accord with this mechanism. Until those results are confirmed, we prefer not to incorporate them in the analysis. It is most likely that the effect of nuclear polarizability cannot be detected in hydrogen because of its smallness and the theoretical and experimental uncertainties. Note however, that in high-Z ions the effect is more important, and a significant energy shift has recently been estimated for the ground state of one-electron uranium.[32]

At this point, we have collected all the known formulas that apply to fine structure. Important terms that have not been calculated are, in estimated order of importance, as follows:

i) "Pure recoil" terms of order $m_e^2(Z\alpha)^6/m_N$ Some have been calculated by Erickson and Grotch,[33] but others remain. A complete calculation of these terms is one of the most pressing calculational challenges facing the field today. It is our belief that the techniques described in Chapters IV and V of this review are well suited to these calculations.

ii) Binding corrections to the two-loop Lamb shift. These corrections, which were discussed in connection with the $m\alpha^2(Z\alpha)^4$ terms, may contribute at the few kHz level.

iii) Lowest order three-loop Lamb shift. Probably they give a contribution smaller than one kHz.

It is important to note that the numerical uncertainty in G_{SE} leads to a theoretical error of about 8 kHz in the classic Lamb shift for $Z = 1$. Hence the calculations described above must also be accompanied with a more accurate calculation of that quantity in order to reduce theoretical uncertainties to the 1 kHz level.

We turn now to comparison with experiment. Rather than presenting a set of tables listing the various theoretical contributions, we have instead written a computer program that we now describe, and present only the final theoretical results. A listing of this program can be found in Appendix B. We encourage the interested reader to write his or her own program, which can be modified as the fundamental constants get revised or additional theoretical contributions are found. It is useful to leave the mass ratio and nuclear charge as input parameters so that different atoms can be treated with the same program. The first step is to incorporate the fine structure following (2.3). It is not necessary to use the Taylor expansion given in (2.2c), and if high-Z ions are being considered, the original Dirac formula should be used. A convenient basic unit is

$$\frac{m_e c^2 \alpha^5}{h\pi} = \frac{2cR_\infty \alpha^3}{\pi} = 813.862\,88(11)\ \text{MHz}\ , \tag{2.12}$$

which is 6 times the so-called Lamb constant. One then adds in (2.4), (2.7), (2.8), (2.9), and (2.11). (The last term of (2.7a) is dropped for the same reasons discussed in connection with (2.6)). This does not include hyperfine splitting, but correctly accounts for the centroids of hyperfine multiplets, except for the case of positronium, where annihilation plays an important role to be described below.

1a. Classic Lamb Shift in Hydrogen

The application of the above procedure to the classic Lamb shift immediately encounters the problem that there are two measurements of the charge radius of the proton in the literature that are discrepant by more than the quoted error bars.[34,35] As we are in no position to judge one measurement more reliable than the other, we simply carry out the calculation for each radius, and find

$$\begin{aligned} \mathcal{S} &= 1057.853(9)(10) \text{ MHz for } \langle r^2 \rangle^{1/2} = .805(11) \text{ fm}^{34]} \\ \mathcal{S} &= 1057.871(9)(10) \text{ MHz for } \langle r^2 \rangle^{1/2} = .862(12) \text{ fm}^{35]} \; . \end{aligned} \tag{2.13}$$

where $\mathcal{S} = \Delta E_{S_{1/2}} - \Delta E_{P_{1/2}}$. The first theoretical error refers to the uncertainty in G_{SE} and the proton radius, and the second is an estimate of the size of the uncalculated terms discussed in the previous subsection. The experimental result of Lundeen and Pipkin is 1057.845(9) MHz:[36] references to the extensive experimental literature on this transition can be found in their paper or in the review by Pipkin in this volume. A new measurement of the proton charge distribution is clearly needed before theory can be meaningfully confronted with experiment. Barring that, progress in reducing the theoretical uncertainties discussed above has the potential of providing an atomic physics determination of a fundamental property of the proton.

1b. Classic Lamb Shift in Muonium

Muonium provides a very interesting system in which to study the Lamb shift, both because the muon is pointlike and thus has no finite size uncertainty, and because the mass ratio of the electron to the nucleus is significantly larger, by a factor of 8.9, than in hydrogen. Therefore recoil corrections, which test the basic framework of our understanding of the relativistic two-body problem, are larger and more easily studied. This field is still very new; and the most recent experiments

$$\begin{aligned} \mathcal{S}(\text{muonium}) &= 1042^{+21}_{-23} \text{ MHz}^{37]} \\ \mathcal{S}(\text{muonium}) &= 1070^{+12}_{-15} \text{ MHz}^{38]} \; , \end{aligned} \tag{2.14}$$

while consistent with the theoretical result 1047.49(1)(9) MHz, do not yet provide such a test. We note that the calculation of Bhatt and Grotch,[12] which contributes a basically negligible -2.53kHz to the classic Lamb shift in hydrogen, contributes a much larger -22.3kHz in this atom. This new physics can then be qualitatively tested in muonium if experimental accuracies can be reduced to the order of 10kHz, whereas it is problematical whether the situation in hydrogen will ever allow information about this effect to be obtained.

1c. n=2 Energy Levels in Positronium

The study of this beautiful system of course provides the most rigorous test of recoil, since the mass ratio is now unity. Unfortunately, theory has advanced relatively slowly for two reasons. The first is that one lacks a small expansion parameter to simplify the kernels: reference to Chapter IV will show that a considerable simplification is thus lost. The second is that because of the possibility of annihilation into photons, a large number of extra kernels must be considered. However, the result to order $m_e\alpha^5$ has been known since the mid-1950's, and is given below. There has been recent experimental progress[39] in measuring the $n = 2$ levels. The measurements were of the $2\,^3S_1 \to 2\,^3P_0\ (\nu_0)$, $2\,^3S_1 \to 2\,^3P_1\ (\nu_1)$, and $2\,^3S_1 \to 2\,^3P_2\ (\nu_2)$ transitions, which were were found to be

$$\nu_0 = 18504.1(10.0)(1.7)\ \text{MHz}$$

$$\nu_1 = 13001.3(3.9)(0.9)\ \text{MHz} \tag{2.15}$$

and

$$\nu_2 = 8619.6(2.7)(0.9)\ \text{MHz}$$

where the first errors are statistical and the second systematic.

A discussion of the theory for $n\,^3S_1$ states to order $m\alpha^5$ is given in Subsection 3: for the case n=2 adding in the nonrelativistic energy and fine structure to (2.22) gives

$$E(2\,^3S_1) = m_e\alpha^2\left[-\frac{1}{16} + \frac{65}{3072}\alpha^2 + \frac{\alpha^3}{64\pi}\left(3\ln\alpha^{-2} + \frac{97}{30} + \frac{2}{3}\ln 2 - \frac{16}{3}\ln k_0(2S)\right)\right]. \tag{2.16}$$

The energies of the P states are somewhat simpler to evaluate because they are unaffected by annihilation, and are given by

$$E(2\,^3P_J) = m_e\alpha^2\left[-\frac{1}{16} + \frac{A_J}{3072}\alpha^2 + \frac{\alpha^3}{64\pi}\left(-\frac{7}{18} - \frac{16}{3}\ln k_0(2P) + B_J\right)\right] \tag{2.17}$$

where $A_0 = -95$, $A_1 = -47$, $A_2 = -\frac{43}{5}$, and $B_0 = -1$, $B_1 = -\frac{1}{6}$, and $B_2 = \frac{3}{10}$. These formulas lead to the theoretical splittings $\nu_0 = 18496.1$ MHz, $\nu_1 = 13010.9$ MHz, and $\nu_2 = 8625.2$ MHz, all in reasonable agreement with theory. If the next round of experiments reach the 1 MHz level of accuracy, a considerable challenge to theory to determine the $m\alpha^6$ corrections will be presented.

2. Hydrogenic Ions

Although the classic tests of bound-state QED involve positronium, muonium, and hydrogen, the theory should in principle describe all atomic and molecular systems. Restricting our attention to atoms, we observe that any atom is associated with a number of isoelectronic sequences. The neutral atom is the first member of the primary isoelectronic sequence, which is named after the atom. The once-ionized atom is then the second member of the isoelectronic sequence named after the preceding element, and so on, until finally the atom with only one electron remaining is a member of the hydrogenic isoelectronic sequence. There are then more than 4000 ions that can be studied, and advances in beam-foil spectroscopy and other techniques for forming and studying these ions have opened up a field that, particularly for highly charged ions, involves a fascinating blend of QED and many-body physics. We restrict our attention here to the hydrogen isoelectronic sequence, and refer the reader to recent reviews[44,45] for more information about the field.

It is standard in treating one-electron bound states to leave a factor Z in the Coulomb potential, even if one is working with a singly charged nucleus, as this facilitates distinguishing terms associated with binding, which involve the factor $Z\alpha$, from radiative effects, which involve only the factor α. However, for discussion of the hydrogenic isoelectronic sequence, Z takes on other values; it should also be noted that terms involving nuclear spin and finite size must be treated on a case-by-case basis. We recall that the theoretical interpretation of the hydrogen Lamb shift was made uncertain by conflicting experimental measurements of the finite size of the proton. An advantage of studying hydrogenic ions is that a precise measurement in a system with more completely understood nuclear properties can bypass this problem. In addition, the occurrence of high powers of Z leads to enhanced radiative effects: in particular, the function $G(Z\alpha)$ introduced above, is enhanced relative to the leading order of the Lamb shift by Z^2. Thus, although the experimental error given below for He^+ is very large compared to that of the hydrogenic measurements, the study of that system gives information about G_{SE} competitive with the latter experiments.

The classic Lamb shift in He^+ has recently been measured by Drake, Patel, and van Wijngaarden[40] to be

$$\mathcal{S}(\mathrm{He}^+) = 14042.22(35)\ \mathrm{MHz} \tag{2.18}$$

To apply the fortran program in Appendix B it is important to note that G_{SE}, which is given there for $Z = 1$, must be replaced with its $Z = 2$ values, which are -29.8, -31.4, -.8, and -.5 for the $1S$, $2S$, $2P_{1/2}$, and $2P_{3/2}$ states respectively; these values are determined from the fitting procedure described in Appendix A. There is a very accurate determination of the nuclear radius of 1.673(1)fm for this system coming from an experiment on muonic helium.[41] A note of caution should be added that there has been difficulty in repeating this experiment.[42] The theoretical result is then 14042.2(5)(6) MHz, in excellent agreement with experiment.[f]

There has been considerable work done on measuring the classic Lamb shift in more highly charged ions; references to experiments in lithium, oxygen, flourine, phosphorus, chlorine, and argon can be found in (40). The experimental errors are relatively large, but agree with theory within errors. There has also been work done on two- and three-electron uranium by Gould[43] that agrees within large error bars with theory (the complications of many-electron structure are not a major source of theoretical uncertainty at high Z). This provides a substantial check of the highly nonperturbative behavior of the self-energy function G_{SE} illustrated in Figure 3.

3. $n = 2$ to $n = 1$ transitions

The Lamb shift is generally taken to refer to the $2S_{1/2}$-$2P_{1/2}$ energy splitting in one-electron systems. However, because the bulk of the effect scales as $1/n^3$, the $1S_{1/2}$ energy is shifted from the Dirac value by an amount about 8 times larger than for the $2S_{1/2}$. Despite the larger size of the effect, this shift cannot be measured as directly because

[f] The theoretical errors have the same source as discussed in Subsection 1a, and are larger by a factor of 64 because the dominant theoretical uncertainties scale as Z^6.

there is no nearby $P_{1/2}$ state; and until recently it has been experimentally inaccessible. However, advances in laser technology and the introduction of the method of Doppler-free spectroscopy[46] have now made it possible to directly measure the $2S_{1/2}$-$1S_{1/2}$ transition for hydrogen,[50] muonium,[47,48] and positronium.[49,48] What makes this transition of particular interest is its extremely small linewidth. The accuracy of the measurement of the $n = 2$ Lamb shift is fundamentally limited by the large $2P_{1/2}$ linewidth. Because of the metastability of the 2S state, this problem is absent at the present level of accuracy for $n = 2$ to $n = 1$ transitions, although muonium and positronium have larger linewidths due to the decay rate of the muon for the former and annihilation into three photons for the latter.

Hydrogen:

There are two approaches possible to the interpretation of the hydrogenic measurement recently carried out by Beausoleil et. al.,[50]

$$\Delta\nu = 2\ 466\ 061\ 413.8(1.5)\ \text{MHz}\ . \tag{2.19a}$$

The first is to assume that we understand the theory of the hydrogen spectrum, and to use the measurement to provide a determination of the Rydberg constant. Alternatively, one can use the very high accuracy determinations of the Rydberg constant that rely on exciting the $2S$ state to higher excited states. The measurement then provides a combined test of the $1S$ and $2S$ Lamb shifts. We adopt the latter approach, using the new value of the Rydberg constant given in (2.1). The theoretical prediction of 2 466 061 413.6(1) MHz (using $< r^2 >^{1/2} = .805(11)$ fm) is in excellent agreement with experiment. It is also of interest to consider the Lamb shift for the $1S$ electron. This has been done in reference (50), where the 1S Lamb shift is determined to be

$$\Delta\nu = 8173.3(1.7)\ \text{MHz}\ . \tag{2.19b}$$

This is consistent with the theoretical value 8172.96(7)(8) MHz, though the experimental error is relatively large. Because the natural linewidth of the transition is only 1.3 Hz, however, the prospects for a high accuracy test of QED for this state are clearly very bright.

Muonium:

The $n = 2$ to $n = 1$ splitting between $F = 1$ states in muonium has been measured[47,48] to be 2 455 527 936(120)(140) MHz, where the first error is statistical and the second systematic. Without hyperfine splitting, the theoretical prediction is 2 455 528 935 MHz, with error well under 1 MHz and a further uncertainty of 3.5 MHz arising from the reduced mass. Ignoring the small state dependence in hyperfine splitting, we can account for that effect by subtracting 7/8 of one quarter of the ground state hfs (the $F = 1$ part of the splitting), and find 2 455 527 959 MHz, in very good agreement with experiment.

Positronium:

The experiment measuring the energy difference between the $n = 1$ and $n = 2$ triplet states in positronium, which is described in this volume by Chu and Mills, has determined the splitting[48]

$$\Delta E = 1\ 233\ 607\ 218.9(10.7)\ \text{MHz}\ . \tag{2.20}$$

We review in some detail the theory up to order $m\alpha^5$ in order to illustrate some of the characteristic difficulties of positronium calculations. The bulk of the transition is of course three eighths of a Rydberg, scaled down by a factor of two from the same transition in hydrogen because $m_r = m_e/2$. At this point it is important to use the newer value of the Rydberg given in (2.1), which leads to

$$\frac{3}{8}R_\infty = 1\ 233\ 690\ 735.4(1)\ \text{MHz}\ . \tag{2.21}$$

The fine structure formula (2.3) leads to an energy shift of $-(5/32)\alpha^2 R_\infty$ for the 1S state and $-(21/512)\alpha^2 R_\infty$ for the 2S state. However, to these corrections must be added the equally important effect of hyperfine splitting, which shifts the triplet states we are concerned with here by $\alpha^2 R_\infty/6n^3$, and the effect of annihilation into a single virtual photon, which shifts these states upward by $\alpha^2 R_\infty/2n^3$. Putting these effects together leads to a relativistic correction to the splitting of $-(719/1536)\alpha^2 R_\infty = -82{,}005.6$ MHz. At this point experiment lies 1510.9(10.7) MHz below theory.

The situation in the next order, while quite complex, is correctly described for triplet S-states by the formula

$$E(n^3S_1\ :\ \alpha^5) = \frac{m_e\alpha^5}{8\pi n^3}[-\frac{7}{3}a_n - \frac{109}{15} + \frac{2}{3}\ln 2 - 6\ln\alpha - \frac{16}{3}\ln k_0(n)] \tag{2.22}$$

where the state dependent a_n coefficients are given in (2.7b). It should be noted that this result corrects two errors in a previous publication.[51] This result comes from many sources. Firstly, the self-energy correction on the positron is just as important as for the electron, so both (2.4) and (2.6a) must be added: this procedure also accounts for the vacuum polarization contribution, which is not to be double counted. Secondly, the Salpeter correction is added in; note that the last term in that expression is no longer negligible in this case. Finally, the effect of hyperfine structure must be accounted for. In units of $m\alpha^5/\pi$, ground state hyperfine structure leads to contributions to the triplet energy of -1 from one-loop corrections to one-photon annihilation, -1/8 from the one-loop contribution given in (5.16), 1/12 from the $g-2$ correction to ordinary hyperfine splitting, and -2/9 from the effect of vacuum polarization in the annihilation photon. One finally uses the fact that these contributions scale as $1/n^3$ to obtain the above formula.

The contribution to $n = 1$ to $n = 2$ splitting from this order is now easily seen to be -1501.5 MHz, which leaves theory 9.5 MHz above experiment. It should be clear from the above discussion that the complete evaluation of the $m\alpha^6$ terms will be a very large scale task involving a large number of Feynman diagrams. The nominal order of these terms is 18.6 MHz, so the next round of experiments, which should achieve an order of magnitude reduction in error, will clearly present a major challenge to theory in a system which is most sensitive to our understanding of the relativistic bound state problem.

B. Hyperfine Splitting in Muonium

Muonium hyperfine splitting provides one of the important ingredients for testing QED. This is primarily because, to the present level of accuracy, muonium is essentially a pure QED system. The effect of strong interactions, which enter through the vacuum polarization, is very small and has an uncertainty of less than one part in 10^8. Weak interactions between the electron and muon also give a very small effect. At the same time, the experimental accuracy is excellent and there is no obstacle to working out the theory to the same level; most of that theoretical work has already been accomplished. The principal uncertainties in comparing theory and experiment are currently smaller than 0.3 ppm and are likely to decrease further in the coming years. At the present time, it provides the most strenuous test of relativistic two-body theory in QED.

Hyperfine splitting in muonium is simpler than in hydrogen because it does not suffer the complications of nucleon structure. On the other hand, recoil corrections are relatively more important. We review briefly the results, including radiative correction contributions. The outlook for further work is also described. Here the term hyperfine splitting refers to the energy (or frequency) separation between the spin 0 and spin 1 S-states of the atom. To match the accuracy of the experiment[52]

$$\Delta\nu(\exp) = 4\ 463\ 302.88(16)\ \text{kHz}, \tag{2.23}$$

it is necessary to know all theoretical contributions through 1 ppm (e.g., through relative order α^3 and $\alpha^2 m_e/m_N$).

The first formulation of the theory of the hyperfine splitting (hfs) was due to Fermi.[53] Traditionally the leading contribution to the ground state hfs is referred to as the "Fermi splitting":

$$E_F \equiv \frac{8}{3}\alpha^4 c^2 \frac{m_r^3}{m_e m_N} = \frac{16}{3}\alpha^2 \frac{m_r^3}{m_e^2 m_N} hcR_\infty \ . \tag{2.24}$$

The second form shows the best way to handle the fundamental constants to minimize the uncertainties. The Rydberg is known to ten places, so its uncertainty may be ignored. Planck's constant disappears in the comparison with the frequency measurement. The remaining uncertainties are in the value of α and the ratio of the electron mass to the muon mass. As mentioned below (2.1), it would be inconsistent to use here the value of m_μ/m_e obtained in the adjustment of the fundamental constants since that value is influenced by the muonium hfs measurement itself. However, another part of the same experiment independently determines the magnetic moment of the muon. When this is combined with the value from SIN,[54] and the value of the muon anomaly a_μ, one obtains $m_\mu/m_e = 206.768\,262(62)$. The central value of the mass ratio is the same as that given by the adjustment of constants, but the error is larger by about a factor of two.

To obtain his result, Fermi used the Dirac equation but made an approximate two-component reduction. While ignoring recoil in the dynamical sense, he introduced the reduced mass through the square of the non-relativistic wave function at the (spatial) origin. Breit[55] treated the Dirac wave functions without approximation and derived a

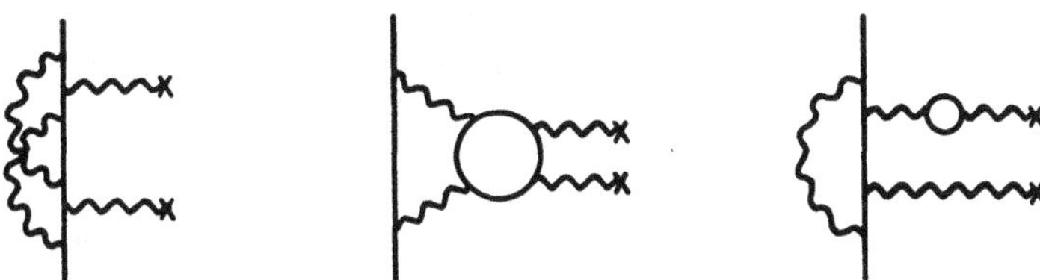

Figure 1: Graphical examples of contributions to D_1. Either of the external interactions is with the nuclear magnetic moment and the other is with its Coulomb potential.

correction factor to (2.24), which we refer to as the Breit relativistic correction; for the ground state, this factor takes the form $(1 + \frac{3}{2}\alpha^2 + ...)$. He could not supply the correct reduced mass factor at that time because of the complications of dealing properly with recoil for the Dirac equation.

We refer to the contributions not involving recoil as the QED contributions. To indicate the separation of binding effects from radiative corrections, we use the convention of including factors of $Z\alpha$ for the former in the following expression:

$$\begin{aligned}\Delta E(\text{hfs; QED}) &= E_F(1+a_\mu)\left\{1 + \frac{3}{2}(Z\alpha)^2 + a_e + \alpha(Z\alpha)\left(\ln 2 - \frac{5}{2}\right)\right. \\ &\left. -\frac{8\alpha(Z\alpha)^2}{3\pi}\ln Z\alpha\left(\ln Z\alpha - \ln 4 + \frac{281}{480}\right) + \frac{\alpha(Z\alpha)^2}{\pi}(15.38 \pm 0.29) + \frac{\alpha^2(Z\alpha)}{\pi}D_1\right\}.\end{aligned} \tag{2.25}$$

Here a_μ represents the anomalous magnetic moment of the muon. Note that we use E_F as defined in (2.24), so that an explicit factor of $(1 + a_\mu)$ appears here; in discussing the hydrogen hfs, however, it is now conventional to incorporate the total proton magnetic moment into the definition of E_F. The $\alpha(Z\alpha)$ radiative correction was calculated originally by Kroll and Pollock[56,57] and Karplus, Klein, and Schwinger.[58] A simpler method is given by Sapirstein, Terray, and Yennie.[59] The $\alpha(Z\alpha)^2$ radiative corrections containing $\ln \alpha$ were first calculated by Layzer[27,28] and Zwanziger.[60,61] Brodsky and Erickson[62] confirmed those terms and estimated the non-logarithmic term. More recently, the non-logarithmic term was evaluated numerically by Sapirstein.[63] The D_1 term represents uncalculated radiative corrections involving two virtual photons, as illustrated in Fig. 1.

The presently known recoil corrections sum to

$$\begin{aligned}\Delta E(\text{hfs; rec}) &= E_F\left\{-\frac{3\alpha}{\pi}\frac{m_e m_\mu}{m_\mu^2 - m_e^2}\ln\frac{m_\mu}{m_e}\right. \\ &\left.+\frac{\gamma^2}{m_e m_\mu}[2\ln\frac{m_r}{2\gamma} - 6\ln 2 + 3\frac{11}{18}]\right\}.\end{aligned} \tag{2.26}$$

where $\gamma \equiv m_r\alpha$. The first term in mass symmetric form is due to Arnowitt.[64] The $(\gamma^2/m_e m_\mu)\ln[1/2\alpha]$ term was calculated by Lepage[65] and confirmed by Bodwin and Yennie.[66] The other terms were first reported in Bodwin, Yennie, and Gregorio;[67] they have been verified numerically by Caswell and Lepage.[68] The muon's anomalous moment does not appear here since its effect should be counted as part of the radiative-recoil correction.

The radiative-recoil contributions, which arise from both lepton lines and from vacuum polarization, are given by

$$\begin{aligned}\Delta E(\text{hfs; rad-rec}) = E_F\left(\frac{\alpha}{\pi}\right)^2\frac{m_e}{m_\mu}&\left[-2\ln^2\frac{m_\mu}{m_e} + \frac{13}{12}\ln\frac{m_\mu}{m_e}\right. \\ &\left.+\frac{21}{2}\zeta(3) + \frac{\pi^2}{6} + \frac{35}{9} + (1.9\pm 0.3)\right].\end{aligned} \tag{2.27}$$

The present importance of these terms was first pointed out by Caswell and Lepage,[11] who evaluated the one proportional to $\ln^2(m_\mu/m_e)$. Terray and Yennie[69] then evaluated the single ln terms; an apparent change from that reference in the coefficient is due to our changed choice of conventions in writing these expressions. Subsequently, the additive non-logarthmic contribution was calculated analytically for the vacuum polarization and numerically for the lepton lines by Sapirstein, Terray, and Yennie.[59] And finally, the lepton line contributions of those terms were evaluated analytically by Eides, *et al.*[70] The agreement between the two approaches is adequate, although somewhat outside the estimated uncertainty of the numerical calculation. We regard the agreement as a confirmation of the quite intricate analytic work; probably this point should be reviewed in the future should the overall accuracy of other contributions improve. The hadronic vacuum polarization contribution is represented by the 1.9±0.3 term.

The complete theoretical result is

$$\Delta\nu(\text{theory}) = 4\,463\,303.11(1.33)(0.40)(1.0)\ \text{kHz}\,. \tag{2.28}$$

The first uncertainty reflects the uncertainty from the measurement of m_μ, the second from that of α, and the third is an order of magnitude estimate of the uncalculated D_1 contribution. Weak neutral current effects, which are estimated to be less than 0.1 ppm [Bég and Feinberg[71] and Bodwin and Yennie[66]], have not been included. The difference between theory and experiment, as well as the experimental uncertainty, is considerably smaller than any of these. The comparison with the experimental result is very satisfactory; in fact, the difference between theory and experiment is much smaller than the various errors. The main job remaining for theorists is the evaluation of the D_1 coefficient. After that, uncalculated QED contributions will be at the level of one part in 10^8. For purposes of comparison with other QED experiments, it is sometimes useful to turn the comparison around to determine a value of the fine structure constant from muonium. This is given in the article by Kinoshita and Yennie in this volume.

C. Hyperfine Splitting in Hydrogen

The hyperfine splitting between the spin-0 and spin-1 levels in the hydrogen ground state is one of the most accurately measured quantities in physics. The most recent experimental determinations of this quantity[72,73] give

$$\nu(\exp) = 1420.405\,751\,766\,7(9)\mathrm{MHz}. \tag{2.29}$$

Various theoretical contributions to the hydrogen ground state hfs have been calculated over the years. The most important of these do not involve proton recoil or the dynamics of the proton structure; these "QED" contributions are given in (2.25), but with the explicit factor of $(1 + a_\mu)$ omitted and a factor of $(1 + \kappa)$ incorporated into the definition of E_F in (2.24). Here κ is the proton's anomalous moment coefficient. The definition used here is more usual for the hydrogen hfs. The expression (2.25) arises from contributions whose characteristic momenta are of order m_e or less. They do not involve the details of the proton structure, but they *are* sensitive to the low energy properties of the proton. This is the reason that the factor $(1 + \kappa)$ representing the proton's total magnetic moment can appear in the Fermi splitting.

We may now evaluate the known QED contributions to the hfs, as given in (2.25), using the latest values for the fundamental constants given in (2.1). The Fermi splitting in frequency units is $\nu_F = 1418.840\,29(14)$ MHz; the dominating uncertainty arises from the value of the fine structure constant. With the corrections given in (2.25) taken into account, the theoretical value of the hyperfine splitting is

$$\nu(\mathrm{QED}) = 1420.451\,99(14)\ \mathrm{MHz}\ . \tag{2.30}$$

The difference between QED theory and experiment is then

$$\frac{\nu(\mathrm{QED}) - \nu(\mathrm{Exp})}{\nu_F} = 32.55(10)\ \mathrm{ppm}. \tag{2.31}$$

In order to take full advantage of the refinements described so far, it would be necessary to calculate the recoil and dynamical corrections involving the proton's structure to about 0.1 ppm relative accuracy. Although such a goal has not yet been achieved, substantial progress has been made in working out these structure-dependent corrections. They are usually expressed as

$$\Delta E(\mathrm{hfs;\ structure}) = E_F\Big[\delta_p(\mathrm{rigid}) + \delta_p(\mathrm{polarizability})\Big]. \tag{2.32}$$

The quantity $\delta_p(\mathrm{rigid})$ is computed by using elastic form factors to approximate the electromagnetic interactions of the proton. The term $\delta_p(\mathrm{polarizability})$, the proton polarizability correction, contains all of the effects of the dynamics of the proton that are not included in $\delta_p(\text{rigid})$.

The most important structure-dependent correction does not involve recoil; it is known as the "nonrelativistic size correction." In computing this correction, one treats the proton as a nonrecoiling particle with a fixed charge-current distribution of finite extent. The nonrelativistic size correction was first analyzed by Zemach;[75] hence, we denote it by $\delta_p(\text{Zemach})$ and write

$$\delta_p(\text{rigid}) = \delta_p(\text{Zemach}) + \delta_p(\text{recoil}), \tag{2.33}$$

where $\delta_p(\text{recoil})$ contains all of the elastic recoil corrections to the hydrogen hfs. The nonrelativistic size correction is given by

$$\delta_p(\text{Zemach}) = -2m_r \alpha R_{pr}, \tag{2.34a}$$

where R_{pr} is a mean radius associated with the proton's charge-current distribution. Accurate calculation of $\delta_p(\text{Zemach})$ requires good knowledge of the elastic form factors from experiment. Because it is a contribution which requires a serious reanalysis, we give here the explicit expression for it in terms of the electric and magnetic form factors

$$\delta_p(\text{Zemach}) = \frac{2\alpha m_r}{\pi^2} \int \frac{d^3p}{p^4} \left[\frac{G_E(-p^2) G_M(-p^2)}{1+\kappa} - 1 \right] . \tag{2.34b}$$

When evaluated with the commonly-used dipole parametrization of the elastic form factors, it gives a contribution of −38.72 ppm, which removes most of the difference between theory and experiment. The uncertainty in the Zemach correction is a rather complex issue. The existing literature contains an estimate of the uncertainty in the total structure-dependent contribution of 0.9 ppm.[76] We have been unable to determine the basis for that estimate.

Now let us turn to the effects of recoil. In hydrogen, the term $\delta_p(\text{recoil})$ replaces the expression (2.26). The one-loop (relative order $\alpha(m_e/m_p)$) contributions to $\delta_p(\text{recoil})$ correspond to the first term of (2.26). These contributions involve very large characteristic momenta and hence they involve the proton's internal structure in an important way. On the other hand, the second term of (2.26), which contains no logarithm of the heavy particle mass, involves momenta of order m_e and less. Thus, in the case of the hydrogen hfs, such contributions would not show a sensitivity to the proton's structure. Hence, in analyzing contributions to the hydrogen hfs of this type, one can make use of any results that were derived originally for the muonium hfs. In addition contributions that arise from the anomalous moment of the heavy particle must also be included. The one-loop recoil corrections were initially estimated by Arnowitt[64] and Newcomb and Salpeter.[77] The first actual calculation was carried out by Iddings and Platzman,[78] who found that the combination of $\delta_p(\text{Zemach})$ and $\delta_p(\text{recoil})$ gave a contribution of −35 ppm. Grotch and Yennie[79] later arrived at a similar total (−34.6 ppm), but were able to carry out the calculation in a way that separated $\delta_p(\text{recoil})$ from the Zemach correction. The small net size for $\delta_p(\text{recoil})$ is deceptive, since individual parts are comparable in size to the Zemach contribution.

These recoil and Zemach corrections are the subject of a paper by Bodwin and Yennie.[80] Their results are

$$\delta_p\Big(\alpha(m_e/m_p)\text{-structure}\Big) \approx (5.22 \pm 0.01)\ \text{ppm}. \tag{2.35}$$

and

$$\delta_p\Big(\alpha^2(m_e/m_p)\Big) \approx 0.46\ \text{ppm}. \tag{2.36}$$

Combining (2.36) and (2.35) to give the total recoil correction to the hfs, excluding the proton polarizability, through relative order $\alpha^2(m_e/m_p)$, we find

$$\delta_p(\text{recoil}) \approx 5.68\ \text{ppm}\ . \tag{2.37}$$

This is a net increase of 1.4 ppm compared with the result of Grotch and Yennie. When the new result (2.37) is combined with the Zemach correction of (-38.72 ± 0.56) ppm, the difference between theory and experiment becomes

$$\frac{\nu(\text{theory}) - \nu(\text{Exp})}{\nu_F} = (-0.47 \pm 0.56 \pm \text{unknown}\ )\text{ppm}. \tag{2.38}$$

The error of 0.56 ppm contains a small contribution from the uncertainty in α, but it arises mainly from the uncertainty in the parameter Λ in the dipole fit to the proton's elastic form factors. The primary effect of this uncertainty in Λ is in the Zemach correction, where it enters through the mean radius of convolution of the proton's electric and magnetic form factors $R_{pr} \propto \Lambda^{-1}$. As seen in Ref. (80), the corresponding uncertainties in δ_p(recoil) are much smaller, both because the Λ-dependence is essentially logarithmic for them, and because there is a partial cancellation between the various terms.

Finally, we turn to the term δ_p(polarizability). Its size was first estimated by Iddings,[81] and it was analyzed further by Drell and Sullivan.[82] Some progress has been made since that initial work. For details we refer the reader to the review of Hughes and Kuti.[83] Here we merely note that deRafael[84] and Gnädig and Kuti[85] have shown how to use data from inelastic electron scattering with polarized beam and target to put a bound on δ_p(polarizability). In their review, Hughes and Kuti give $|\delta_p(\text{polarizability})| < 4$ ppm. The result (2.38) leaves little room for a polarizability correction.

The part of the error labeled "unknown" in (2.38) represents all remaining uncertainties in ν(theory). The most important sources of these uncertainties are the radiative corrections to the structure-dependent contributions, the systematic errors in the parametrization of the proton's form factors, and δ_p(recoil). The radiative corrections to the structure-dependent contributions, which are described in Chapter VI, could potentially contribute at the level of 1 ppm. However, they could probably be calculated to a precision of 0.01 ppm or better. More important are the systematic errors in the parametrization of the proton's form factors, which are evidenced by statistically significant deviations of the scaling assumption and dipole parametrization obtained from the elastic scattering data. It is difficult to estimate how much the use of more precise expressions for the form factors might shift the central value of ν(theory) or how large the remaining statistical uncertainty in that determination might be. As we remarked earlier, a value of 0.9 ppm has appeared in the literature as an estimate of the complete uncertainty in the δ_p(Zemach) + δ_p(recoil). The work of Bodwin and Yennie has shifted the theory by more than that amount (by $\approx +1.5$ ppm). Their subjective impression is that a statistical uncertainty of 1 ppm seems to be a reasonable estimate of what might be achieved through a more precise treatment of existing form factor data, together with an evaluation of radiative corrections that are significant at that level. It would not be surprising if such an analysis were to lead to a shift of as much as 1 ppm in the central value of ν(theory). However, it is unlikely that the analysis would reveal any incompatibility with the above estimate $|\delta_p(\text{polarizability})| < 4$ ppm. In fact, by incorporating the refinements in the computation of ν(theory) that we have already mentioned, one could use the hydrogen hfs to determine δ_p(polarizability) with a precision of roughly 1 ppm.

D. Hyperfine Structure in Positronium

The theoretical expression for the positronium hfs is given by

$$\nu = \alpha^2 \mathrm{R}_\infty \left[\frac{2}{3} + \frac{1}{2} - \frac{\alpha}{\pi}(\ln 2 + \frac{16}{9}) + \frac{5}{12}\alpha^2 \ln \alpha^{-1} + K\alpha^2 + K'(\alpha)\alpha^3 \right] . \qquad (2.39)$$

The α/π term includes radiative corrections as well as a term obtained by taking the equal mass limit of (2.26). The $\alpha^2 \ln \alpha^{-1}$ term comes partially from (2.26), but it has previously noted contributions from annihilation kernels [Barbieri and Remiddi,[86] and Caswell and Lepage[87]]. The α^2 term has been only partially evaluated. Various contributions to K are: two photon annihilation kernels:[88] 1.408; the three photon annihilation kernels:[89] -.098; and the radiative-recoil corrections:[59] -1.192(3). A recent development has been a recalculation of the three-photon annihilation kernels by Adkins, Bui, and Zhu.[90] It was found that although the calculation is infrared finite, divergences in separate parts require a consistent treatment with a regulator. Use of two different regulators, one a photon mass and the other an energy shift led to the same answer, in slight numerical disagreement with the previous calculation: the corrected contribution to K is -.104. The principal set of uncalculated terms involves two-loop corrections to the one-photon annihilation graphs. The K' term is uncalculated, but it could be enhanced by factors of $\ln \alpha$. Recently recoil corrections of the type contained in (2.26) were evaluated by Caswell and Lepage[68] using a novel approach to the bound state problem based on the use of effective Lagrangians. The idea of this technique is to effect a separation of the high-energy part of the calculation, where renormalization is an important issue and Feynman gauge the most convenient gauge, from the low-energy part, which is dominated by the nonrelativistic region and is best calculated in Coulomb gauge. They find a contribution to K of 0.33(7). The most recent experimental result by Ritter *et al.*[91] is

$$\nu = 203\,389.10(74) \text{ MHz}. \qquad (2.40)$$

The present theoretical value [Sapirstein and Kinoshita[92]], including all known contributions along with the corrected three-photon answer and the recoil term calculated by

Caswell and Lepage, is

$$\nu = 203\ 404.5(0.6)(9.3)\ \text{MHz}\ . \tag{2.41}$$

The first error is dominated by the numerical uncertainty of the Caswell and Lepage calculation, and the second corresponds to a possible magnitude of unity for uncalculated contributions to K. The difference between theory and experiment is quite compatible with the expected magnitude of the uncalculated terms.

III. THEORETICAL METHODS IN THE NON-RECOIL LIMIT

In this review, we claim to be giving a unified treatment of the hydrogen-like bound state problem. Logically, this treatment should flow from an analysis of the two-body problem, which would follow the conceptual ideas, if not the precise methods, of Bethe and Salpeter. However, when such a treatment is carried through, it is found to compartmentalize; i.e., it breaks down into pieces which can largely be understood and treated independently of each other. Only at the most refined levels needs the unification play a role. The relatively complex two-body formalism simplifies considerably in the limit in which $m_e/m_N \to 0$, which we refer to as the non-recoil limit. The underlying formalism then becomes invisible, and the only role of the nucleus is to provide an external field—its combined Coulomb and magnetic dipole field. Having made the simplification, one cannot patch back in the features of recoil; it is necessary to return to the parent problem to see how to deal with them. However, some features of recoil can be incorporated by judicious use of the reduced mass dependence. To avoid dealing with all the complications right from the beginning, we elect to take advantage of this compartmentalization. We describe the non-recoil radiative corrections in this chapter.

It turns out that in the detailed evaluation of radiative level shifts, there is not yet a standard preferred method. This is because the calculations involve the Dirac-Coulomb propagator, which is a complex object that can be treated in several different ways, each of which has characteristic advantages and disadvantages. A few of the approaches are described here in the context of the Lamb shift, but the methods are similar for the hyperfine splitting. The first step is the derivation of the computational object, which is the same in all cases. This is described in Section A. Also in that section is a brief discussion of how one identifies orders of magnitude of various expressions; that is, indeed, a very subtle and important question. Although the computational object is gauge invariant, the complications which arise in different approaches are not independent of the choice of gauge. Section B deals with one of the approaches which is based on a rearrangement of the electron propagator internal to the electron self energy so that the renormalizations can easily be extracted. To avoid the problem of spurious terms of lower order than the final result, the Coulomb gauge for the "radiative correction" photon is used. It leads to a powerful method for numerical computation, which is somewhat tied to the expansion in powers of $Z\alpha$. Section C describes an entirely different method in which a partial wave expansion of the electron propagator is used. While it is in principle exact, involving no expansion in $Z\alpha$, it works best for large values of Z. Section D gives a formal treatment of the electron self energy which is not sensitive to the choice of gauge. It is a primarily analytic method which has not been pushed as far computationally as the other two methods. Perhaps it is an approach which could be wedded to numerical analysis to produce useful results. Its greatest purpose may be formal. For example, in Chapter VI-B, it is generalized so that it can be used to discuss the reduced mass effects of recoil and to help identify the radiative-recoil contributions.

A. Introduction

Once the external field approximation has been made, QED is naturally treated in the Furry, or bound interaction representation.[3] In this representation the electron field operator is expanded in terms of Coulomb bound and scattering states, and the electron propagator (Fourier transformed in time) obeys

$$(E\gamma_0 + \frac{Z\alpha}{r}\gamma_0 + i\vec{\gamma}\cdot\vec{\nabla} - m)S_F(\vec{r},\vec{r}\,';E) = \delta(\vec{r}-\vec{r}\,') \tag{3.1a}$$

together with appropriate boundary conditions. We will refer to S_F as the Dirac-Coulomb propagator, and its nonrelativistic counterpart as the Schrödinger-Coulomb propagator. The connection with the two-body treatment is that if S_F is expanded graphically in powers of the Coulomb potential, the n'th term of S_F approximates an n-Coulomb photon ladder graph. The effect of the magnetic moment of the nucleus could be incorporated in this equation by adding a term $e\vec{\gamma}\cdot\vec{A}_{\text{dipole}}$ where $\vec{A}_{\text{dipole}}$ is the vector potential which is produced by the magnetic moment of the nucleus. Since this term can be treated consistently only to first order, it is better to omit it from the equation and treat it by perturbation theory. Higher order effects must be carefully extracted from kernels in the complete two-body formalism.

The function S_F has poles (cuts) in E at all bound (continuum) eigenvalues of the Dirac-Coulomb Hamiltonian. This is seen in its representation in terms of energy eigenfunctions

$$S_F(\vec{r},\vec{r}\,';E) = \sum_n \frac{\psi_n(\vec{r})\bar{\psi}_n(\vec{r}\,')}{E - E_n(1-i\epsilon)}\,, \tag{3.1b}$$

where the sum includes integrations as appropriate. The introduction of the $i\epsilon$ corresponds to the usual hole theory. The eigenfunctions are normalized according to $\int \psi^\dagger\psi d^3x = 1$. One way to derive an expression for the energy shift produced by radiative corrections is to consider the modification of the electron propagator produced by such corrections. These corrections produce self-energy diagrams similar to the usual ones except that the internal electron propagators are in the Furry representation. We use the same symbol Σ for them. With these corrections, the electron propagator becomes

$$S_F + S_F\tilde{\Sigma}S_F + S_F\tilde{\Sigma}S_F\tilde{\Sigma}S_F + \ldots \tag{3.2a}$$

where $\tilde{\Sigma}$ is the electron self energy with the mass renormalization subtraction taken into account. Renormalization works in the same way as in the free representation,[24] and it takes the form

$$\tilde{\Sigma} = BS_F^{-1} + (1 - B)\tilde{\Sigma}_{\text{ren}} \tag{3.2b}$$

where B is the same constant that occurs in the free case and $\tilde{\Sigma}_{\text{ren}}$ is finite. An important difference from the free case is that there $\tilde{\Sigma}_{\text{ren}}$ is proportional to $[S_F^{-1}]^2$, while here it contains terms without these factors. As seen below in Section D, these additional terms are necessary for the existence of the Lamb shift and the anomalous magnetic moment of the electron. Using (3.2b), the series (3.2a) may be resummed to give the renormalized propagator divided by $(1 - B)$

$$\frac{1}{1-B}\left[S_F + S_F\tilde{\Sigma}_{\text{ren}}S_F + S_F\tilde{\Sigma}_{\text{ren}}S_F\tilde{\Sigma}_{\text{ren}}S_F + \ldots\right] \equiv \frac{S_F^{\text{ren}}}{1-B}\,. \tag{3.2c}$$

1. Perturbation Theory of Level Shifts

The energy levels are given by the positions of the poles of (3.2c) as a function of E. The method described here for determining this shift is used again in Chapter IV in the discussion of the two-body problem. To see how it works, consider the first few terms in the series for the renormalized propagator. Taking an expectation value in a given unperturbed state labelled n, we find that the first term becomes (here $|n\rangle \equiv \psi_n$ and $\langle n| \equiv \bar{\psi}_n = \psi_n^\dagger\gamma_0$ is the Dirac adjoint)

$$\langle n|\gamma_0 S_F\gamma_0|n\rangle = (|n\rangle)^\dagger S_F\gamma_0|n\rangle = \frac{1}{E - E_n}\,.$$

The second is

$$\frac{\langle n|\tilde{\Sigma}_{\text{ren}}|n\rangle}{(E - E_n)^2}\,,$$

and the third is

$$\frac{1}{(E - E_n)^2}\sum_m \frac{\langle n|\tilde{\Sigma}_{\text{ren}}|m\rangle\langle m|\tilde{\Sigma}_{\text{ren}}|n\rangle}{E - E_m(1 - i\epsilon)}\,.$$

This is enough for us to see the pattern which is developing. It is important to split off terms where $m = n$ in various sums. The series can then be resummed to give

$$\frac{1}{E - E_n - \sigma_n(E)} \tag{3.3a}$$

where

$$\sigma_n(E) = \langle n|\tilde{\Sigma}_{\text{ren}}(E)|n\rangle + \sum_{m\neq n} \frac{\langle n|\tilde{\Sigma}_{\text{ren}}(E)|m\rangle\langle m|\tilde{\Sigma}_{\text{ren}}(E)|n\rangle}{E - E_m(1 - i\epsilon)} + \ldots\,. \tag{3.3b}$$

It is now an easy matter to find the position of the pole by expanding σ about E_n. This gives for the modified energy $\hat{E}_n$

$$\hat{E}_n = E_n + \sigma_n(E_n) + \sigma_n(E_n)\frac{\partial\sigma_n(E_n)}{\partial E_n} + \ldots\,. \tag{3.3c}$$

In this chapter we will be exclusively concerned with lowest-order effects, the one-loop electron self energy and vacuum polarization graphs of Figs. 2a and Fig. 2b respectively. In

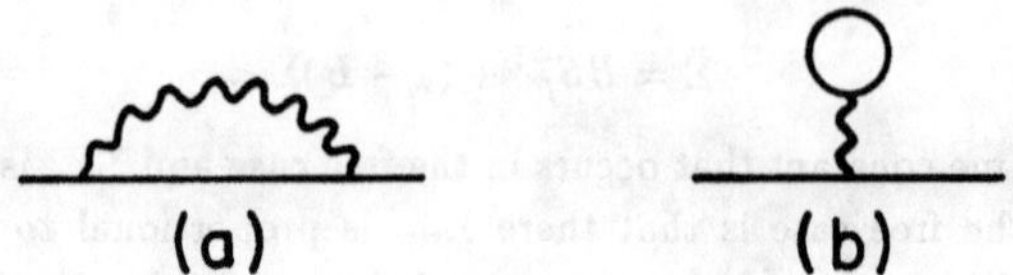

Figure 2: Second order contributions to the Lamb shift. Heavy lines represent propagation in the external potential. (a) Electron self energy (SE). (b) Vacuum polarization (VP).

the free case, Fig. 2b would not be regarded as a self energy diagram; however, since it fits into the algebraic structure just described, we treat it as one here. Using the procedure just outlined, we find that the energy shift from the one loop electron self-energy (expressed in the Feynman gauge) is

$$\Delta E_n(\mathrm{SE}) = -e^2 \int d^3r d^3r' \int \frac{d^4k}{-(2\pi)^4 i} \frac{e^{i\vec{k}\cdot(\vec{r}-\vec{r}')}}{k^2+i\epsilon} \bar{\psi}_n(\vec{r})\gamma_\mu S_F(\vec{r},\vec{r}';E_n-k_0)\gamma^\mu \psi_n(\vec{r}') - \delta m^{(2)} \int d^3r \bar{\psi}_n(\vec{r})\psi_n(\vec{r}) \tag{3.4a}$$

and from vacuum polarization

$$\Delta E_n(\mathrm{VP}) = -e^2 \int d^3r d^3r' \int \frac{d^3k}{(2\pi)^3} \int \frac{dE}{-2\pi i} \bar{\psi}_n(\vec{r})\gamma_0 e^{i\vec{k}\cdot\vec{r}} \psi_n(\vec{r}) \frac{1}{\vec{k}^2} \times e^{-i\vec{k}\cdot\vec{r}'} \left\{ \mathrm{Tr}\left[\gamma^0 S_F(\vec{r}',\vec{r}';E)\right] - Z\Pi_0^{(2)}\delta(\vec{r}')\right\} ; \tag{3.4b}$$

where Π_0 is the charge renormalization constant associated with vacuum polarization. This is conventionally written in terms of an induced charged density $\rho_{VP}(\vec{r})$ via

$$\Delta E_n(\mathrm{VP}) \equiv -\frac{e}{4\pi} \int d^3r d^3r' \bar{\psi}_n(\vec{r})\gamma_0\psi_n(\vec{r}) \frac{1}{|\vec{r}-\vec{r}'|} \rho_{VP}(\vec{r}') \; . \tag{3.4c}$$

Both of these integrals are ultraviolet divergent and must be renormalized. These infinities are most easily dealt with if the principal advantage of the Furry representation, treating the propagation of the electron with an arbitrary number of Coulomb interactions using the Dirac-Coulomb propagator S_F, is undone. For example, in the electron self energy, the first term of the expansion, in which the electron propagates freely, can be treated with standard Feynman parameter techniques, and renormalized in the usual way with the mass correction explicitly subtracted and the wave function renormalization constant identified. Using Ward's identity, the latter is cancelled by the vertex renormalization constant associated with a single Coulomb interaction. At this point no ultraviolet divergences remain. In addition, because the electron is not on the mass shell, there are no infrared divergences present either, so a completely finite expression remains to be evaluated. In Section D, these renormalizations are accomplished for the SE by formal manipulations without expanding in powers of the Coulomb potential. The vacuum polarization contribution includes all numbers of Coulomb interactions within a single closed electron loop. It was originally analyzed in this form by Wichmann and Kroll.[19]

2. A Brief Digression on Orders of Magnitude

To indicate the type of difficulties which can occur with an insufficiently refined approach, suppose one were to expand S_F in (3.4a) in powers of the external potential. The electron propagator becomes

$$S_F = S_F^0 + S_F^0\gamma_0\frac{-Z\alpha}{r}S_F^0 + S_F^0\gamma_0\frac{-Z\alpha}{r}S_F^0\gamma_0\frac{-Z\alpha}{r}S_F^0\ldots\,. \tag{3.5a}$$

It is the first two terms of this expansion which participate in the renormalization discussion of the preceding paragraph. In momentum space, a factor of S_F^0 takes the form

$$S_F^0 = \frac{m_e + E_n\gamma_0 - \vec{p}\cdot\vec{\gamma} - \not{k}}{k^2 - 2E_nk_0 + 2\vec{p}\cdot\vec{k} - \vec{p}^{\,2} - \gamma_n^2 + i\epsilon}\,. \tag{3.5b}$$

Here $\gamma_n^2 \equiv m_e^2 - E_n^2 \approx m_e^2\alpha^2/n^2$, and $\vec{p}$ takes on a numerical value associated with the position of the factor in the product. To avoid the details of discussing renormalization (which, however, must be dealt with in a real calculation), we consider a term with several external potential factors.

The first step in analyzing a term is to estimate its "nominal" order of magnitude. This refers to the order of magnitude which would result if all integrations were to converge in the non-relativistic region characterized by electron momenta of order γ_n. This is the characteristic momentum of the wave function; it is also the reciprocal of the Bohr radius. To study the order of magnitude, we set aside a number of small contributions at first and then return to discuss them later. To analyze our expression, we consider what happens when we carry out the k_0-integration first (of course, it is possible that this may not be the best way actually to evaluate the integral). If we close the k_0-contour below, we obtain contributions either from negative energy electron poles or from the photon pole. The former give a small contribution in which the characteristic momentum is forced to be relativistic; note also the small numerators when $k_0 \approx E_n + \sqrt{m_e^2 + \vec{p}^{\,2}}$ with $|\vec{p}| \sim \gamma_n$. The photon pole occurs at $k_0 = |\vec{k}| \equiv \omega$, so the electron denominators become $-2E_n\omega + 2\vec{p}\cdot\vec{k} - \vec{p}^{\,2} - \gamma_n^2$.

Now, to estimate the order of magnitude, we use a simple dimensional argument. First we extract powers of m_e (note $E_n \approx m_e$) in such a way that the order of magnitude of the integral is a power of $m_e\alpha \equiv \gamma'$. The dominant part of the numerator of (3.5b) is a large component projector times $2m_e$. Each of these factors may be taken with a factor of α from a potential to produce a power of γ', with one excess factor of m_e to be taken outside the integral. To account for the explicit m_e in the denominator, we change variables using $\vec{\xi} = m_e\vec{k}$. Two powers of m_e then appear in the denominator from $\int d^3k/k$ and can be factored out. Now we may let $m_e \to \infty$ inside the integral and obtain a finite result which can only be a power of γ' (of course, it also depends on n). Including the factor of α from the virtual photon, the final order of magnitude is then $\alpha\gamma'^a/m_e$ where the power a

must be 2 in order to give the right dimensions. Note that since the electron numerators are essentially large component projectors, only the $\mu = 0$ term in the sum produces this order. Since the actual order of magnitude of the Lamb shift is two powers of α smaller, this is a very bad estimate; and there must be a delicate cancellation between terms with different powers of the external potential.

If instead we consider spatial values for μ, we obtain a much smaller estimate. This is because the factor $m_e + \gamma_0 E_n$ is approximately $\gamma_n^2/2m_e$ in this situation, increasing the order of magnitude by α^2 for each free electron propagator. Actually, a much larger contribution is obtained by taking a factor of $-\vec{\gamma} \cdot \vec{p}$ next to these spatial γ_μ's, either from the wave function or from the adjacent propagator. In effect, this introduces two additional factors of m_e into the denominator outside the integral, changing the final order to $\alpha\gamma'^4/m_e^3$, which is the actual order of the Lamb shift. Of course, it remains necessary to sum an infinite series.

How about some of the other numerator terms which have been bypassed? An additional factor of $-\vec{\gamma} \cdot \vec{p}$ in the numerator may replace a factor $m_e + \gamma_0 E_n$. Some of the remaining factors of $m_e + \gamma_0 E_n$ then may have the order of magnitude $\gamma_n^2/2m_e$, greatly reducing the magnitude of the result. The dominant term then has two factors of $-\vec{\gamma} \cdot \vec{p}$ in succession, reducing the magnitude by two powers of α. A factor of $\not{k}$ in the numerator is even smaller because the order of magnitude of k is $\alpha^2 m_e$. This kind of discussion can be somewhat tricky because as these expansions are made, the important range of integration may be pushed into the relativistic region. This behavior is most frequently encountered with low orders of expansion in the Coulomb interaction which are also sensitive to renormalization. However, enough has been said to get started on understanding the issues. Rather than continuing to discuss the expansion along these lines, we return to the main discussion and insert remarks about the order of magnitude as appropriate.

3. Elimination of Spurious Orders of Magnitudes

Although the infinite renormalizations can be removed as discussed earlier, in a general gauge there are present terms of a spurious order, specifically $m\alpha(Z\alpha)^2$. As just argued, these can be traced to photon pole terms with $\mu = 0$ in (3.4a). A numerator factor γ_0 when taken between two positive energy states is of order unity, as opposed to numerator terms involving $\vec{\gamma}$, which are of the order of an atomic velocity, i.e. $Z\alpha$. As the true order is actually $m\alpha(Z\alpha)^4$, all such terms must cancel. There have been a variety of ways of dealing with this problem developed in the various Lamb shift calculations carried out over the decades, which we now describe; some of these will be discussed in greater detail in the following sections. If one describes the original calculation of Bethe[93] in the framework of modern calculations, it is seen to correspond to substituting a transverse photon for the Feynman gauge photon in (3.4a) and to retaining only the photon pole contribution in our discussion of orders of magnitude. This calculation accounts correctly for the non-relativistic part of the Lamb shift, but a full relativistic calculation is required to correctly determine the complete answer.

One possibility is simply to ignore the problem. While individual terms certainly have the spurious order, they must cancel in the full calculation. The partial wave expansion approach introduced by Brown, Langer, and Schaefer[94], which was first applied to numerical calculations by Brown and Mayers,[95] and later used by Desiderio and Johnson,[96] has individual terms of this order which come from different parts of the calculation and are cancelled numerically. As those calculations were meant to be applied to high-Z one-electron ions, the cancellation is not a severe one. However, the method is not well adapted to hydrogen, both because of the more severe cancellation and because of other numerical problems. These others are serious enough so that Mohr's[21] calculation, which does not have this cancellation, goes down only to $Z = 10$. The other calculation that ignores the problem is the powerful method used by Layzer[27,28] The basic idea of that approach is to concentrate only on terms nonanalytic in $(Z\alpha)^2$. The troublesome 'deep non-relativistic' region, associated with the photon pole terms, can be shown to contribute only as a power series in $(Z\alpha)^2$. Layzer devised a method that is insensitive to such contributions, but would pick up nonanalytic terms such as $m\alpha(Z\alpha)^5$, or $m\alpha(Z\alpha)^4\ln(Z\alpha)$. These nonanalytic terms arise from low orders in the expansion of S_F in powers of the Coulomb potential. The important contributions of order $m\alpha(Z\alpha)^6\ln(Z\alpha)$ and $m\alpha(Z\alpha)^6\ln^2(Z\alpha)$ were first calculated with this method, and it is probably the best way to attack the evaluation of the unknown $m\alpha(Z\alpha)^7$ terms. It is, of course, not able to pick up the non-logarithmic part of the Lamb shift, or the constants in order $m\alpha(Z\alpha)^6$.

The modern phase of Lamb shift calculations involved the introduction of methods powerful enough to recover the lowest order result without the use of a cutoff, and also to allow the evaluation of the first order binding corrections of order $m\alpha(Z\alpha)^5$. Two Feynman gauge calculations were carried out, the first by Baranger, Bethe, and Feynman,[97] and the second by Karplus, Klein, and Schwinger.[98] The former paper solved the problem of the spurious terms with the use of a lemma, which in essence rearranges the calculation so that they do not occur. The second solves the problem by expanding the electron propagator in a gauge invariant manner instead of the direct expansion in the Coulomb potential described above. This latter technique was extended later by Erickson and Yennie,[99] and will be described in some detail in Section D of this chapter.

Another method of dealing with the terms of spurious order is to abandon the Feynman gauge, and to use either the Coulomb or Fried-Yennie gauge.[100] It is well known that the energy shift is gauge-independent; this is easy to show directly for the one-loop contribution. With the photon propagator expressed as $-G_{\mu\nu}/k^2$, the two gauges are specified by

$$G_{\mu\nu} = g_{\mu\nu} + 2\frac{k_\mu k_\nu}{k^2} \qquad \text{Fried-Yennie} \tag{3.6a}$$

$$G_{\mu\nu} = \left\{ \begin{array}{ll} -\dfrac{k^2}{\vec{k}^2} & \mu = \nu = 0 \quad \text{(Coulomb photon)} \\ 0 & \mu = 0,\ \nu = i \text{ or } \mu = i,\ \nu = 0 \\ -(\delta_{ij} - \dfrac{k_i k_j}{\vec{k}^2}) & \mu = i,\ \nu = j \quad \text{(Transverse photon)} \end{array} \right\} \quad \text{Coulomb}\ . \tag{3.6b}$$

In the Coulomb term of the Coulomb gauge, the photon pole is eliminated so that the multi-potential contributions are associated with negative energy electron poles and are therefore strongly suppressed in the deep NR region as described above. In the Fried-Yennie gauge the combination of terms is chosen so as to provide a suppression factor for the dangerous $\mu = \nu = 0$ term; an explanation of how this removes the difficulty is given in an Appendix to a paper by Sapirstein, Terray, and Yennie.[59] The result in both cases is that the $\mu = 0$ contributions are of the same order of magnitude as for spatial indices. Both gauges are useful, with the principle advantage of the Fried-Yennie gauge being its covariance, and that of the Coulomb gauge being its clean separation of the dominant physics of the atom, the Coulomb interaction, from the smaller magnetic effects associated with the transverse photon interaction. A disadvantage shared by both gauges in comparison to Feynman gauge is a more complicated numerator structure, and this is exacerbated in Coulomb gauge by its also being noncovariant. Despite this latter fact, it is possible to evaluate the Lamb shift complete to order $m\alpha(Z\alpha)^6$ in this gauge, and we now proceed to a description of this calculation.

B. Use of the Schrödinger-Coulomb Green's Function in the Lamb Shift Calculation

As a preliminary to this discussion, we first consider the evaluation of a part of the lowest order Lamb shift that is available only numerically, the so called Bethe logarithm. All calculations except those described in Section C make contact with this quantity. This object arises in a completely NR context in the evaluation of the second order perturbation theory expression[g]

$$\Delta E_n(\Lambda;\ \mathrm{NR}) = \frac{2e^2}{3m^2}\int_0^\Lambda \frac{d^3k}{2\omega(2\pi)^3}\sum_m \frac{\langle n|\vec{p}|m\rangle\cdot\langle m|\vec{p}|n\rangle}{(E_n-\omega-E_m)}\ . \tag{3.7}$$

This expression can be obtained from the Coulomb gauge version of (3.4a) by making several approximations: insert (3.1b), take only the photon pole contribution as described above, use the dipole approximation $e^{i\vec{k}\cdot(\vec{r}-\vec{r}\,')} \to 1$, replace $\vec{\alpha}$ with $\vec{p}/m$, and finally replace all wave functions with their nonrelativistic counterparts. Without negative energy states, this expression diverges linearly with Λ. After removing that linear divergence by subtracting from (3.7) the same expression with denominator $-\omega$ (which corresponds to a portion of the mass renormalization), a logarithmic dependence on Λ remains

$$\begin{aligned}\Delta E_n(\Lambda;\ \mathrm{NR\text{-}sub}) &= \frac{e^2}{6m^2\pi^2}\sum_m |\vec{p}_{nm}|^2(E_m-E_n)\ln\frac{\Lambda}{|E_m-E_n|}\\ &\equiv \frac{e^2}{6m^2\pi^2}\sum_m |\vec{p}_{nm}|^2(E_m-E_n)\ln\frac{\Lambda}{k_0(n)Ry}\\ &= \frac{4m\alpha(Z\alpha)^4}{3\pi n^3}\ln\left(\frac{\Lambda}{k_0(n)Ry}\right)\ .\end{aligned} \tag{3.8}$$

The absolute value of the argument of the logarithm is taken in the first expression to give the real part of the energy shift; the neglected imaginary part corresponds to the decay

[g]Note: In this section, there is a proliferation of different ΔE_n's having different sources in the formalism and using different approximations. We attempt to distinguish these by a descriptive argument.

width of the state. The logarithmic term with the constant $k_0(n)$ is known as the Bethe logarithm. Technically, the second line is non-vanishing only for S states; however, the third line still defines k_0 in other cases, with Λ taking on the value of one Rydberg.

There exists a variety of methods to evaluate the Bethe logarithm. The first calculations were done by summing a number of bound intermediate states and integrating continuum intermediate states on a mesh. We now describe a method that is particularly convenient for later connection with a fully relativistic calculation; in addition, the method allows for the simplest high-accuracy evaluation of this important quantity. As it contributes around 385 MHz to the Lamb shift, an accuracy approaching 1 ppm is required to get under the other theoretical uncertainties discussed in Chapter II. We see below that uncertainties well under this have been achieved.

The nonrelativistic Green's function in a Coulomb potential, which we refer to as the Schrödinger-Coulomb Green's function, satisfies

$$[\nabla^2 + \frac{2mZ\alpha}{r} + 2mE]G_{NR}(\vec{r},\vec{r}\,';E) = \delta(\vec{r}-\vec{r}\,') \tag{3.9a}$$

and has a formal solution in terms of NR Coulomb orbitals

$$G_{NR}(\vec{r},\vec{r}\,';E) = \sum_m \frac{\langle\vec{r}|m\rangle\langle m|\vec{r}\,'\rangle}{2m(E-E_m)}\,. \tag{3.9b}$$

Thus (3.7) can be written in momentum space as

$$\Delta E_n(\Lambda;\ \mathrm{NR}) = \frac{4e^2}{3m}\int_0^\Lambda \frac{d^3k}{2\omega(2\pi)^3}\int\frac{d^3p}{(2\pi)^3}\int\frac{d^3p'}{(2\pi)^3}\vec{p}\cdot\vec{p}\,'\,\phi_n^\dagger(\vec{p})\tilde{G}_{NR}(\vec{p},\vec{p}\,',E_n-\omega)\phi_n(\vec{p}\,'). \tag{3.10}$$

Quite some time ago Schwinger[101] developed an elegant solution for the Schrödinger-Coulomb Green's function in momentum space. This first appeared in print in 1963; at that same time Hostler independently derived the Green's function in coordinate space[102] and in momentum space.[103] The solution breaks naturally into three parts: the first is the free Green's function, the second accounts for the interaction with a single Coulomb potential, and the last sums up in one expression all interactions with two or more Coulomb potentials. Specifically, with $E \equiv E_n - \omega$, $p_0^2 \equiv 2m\omega + (mZ\alpha/n)^2$, and $\nu \equiv mZ\alpha/p_0$,

$$\begin{aligned}\tilde{G}_{NR}(\vec{p},\vec{p}\,',E) = &-\frac{(2\pi)^3\delta(\vec{p}-\vec{p}\,')}{(\vec{p}\,^2+p_0^2)} - 8\pi mZ\alpha\frac{1}{(\vec{p}\,^2+p_0^2)}\frac{1}{|\vec{p}-\vec{p}\,'|^2}\frac{1}{(\vec{p}\,'^2+p_0^2)}\\ &-\frac{32\pi(mZ\alpha)^2p_0}{(\vec{p}\,^2+p_0^2)(\vec{p}\,'^2+p_0^2)}\int d\rho\frac{\rho^{-\nu}}{4\rho|\vec{p}-\vec{p}\,'|^2p_0^2+(1-\rho)^2(\vec{p}\,^2+p_0^2)(\vec{p}\,'^2+p_0^2)}\,.\end{aligned} \tag{3.11}$$

The three parts will be referred to in the following as the zero-potential (0-P), one-potential (1-P), and many-potential (M-P) terms.

Inserting this solution into (3.10), we find that the first two terms can be evaluated analytically. The results for the 1S state are

$$\Delta E_n(\Lambda;\ \text{NR: 0-P}) = -\frac{2\alpha(Z\alpha)^2\Lambda}{3\pi} + \frac{2m\alpha(Z\alpha)^4}{\pi}\ln\left(\frac{2\Lambda}{m(Z\alpha)^2}\right) - \frac{m\alpha(Z\alpha)^4}{\pi}(3+4\ln 2)$$

$$\Delta E_n(\Lambda;\ \text{NR: 1-P}) = \frac{2m\alpha(Z\alpha)^4}{3\pi}\left(\ln\frac{m(Z\alpha)^2}{2\Lambda}+1\right) + \frac{m\alpha(Z\alpha)^4}{\pi}\cdot\frac{10}{3}(\ln 4 - 1) \tag{3.12a}$$

$$\Delta E_n(\Lambda;\ \text{NR: M-P}) \equiv \frac{m\alpha(Z\alpha)^4}{\pi}D_2\ .$$

Removing the linear term in Λ, which corresponds to the subtraction in the definition of (3.8), we can then identify

$$\ln k_0(1S) = \frac{17}{4} - \ln 4 - \frac{3}{4}D_2\ . \tag{3.12b}$$

While the M-P term D_2 has never been evaluated analytically, it is noteworthy that the 0-P and 1-P parts of (3.12b) account for 96% of the complete result.

An elegant and powerful way of evaluating Bethe logarithms with high accuracy that uses the O(4) symmetry present in the Schrödinger-Coulomb Green's function was developed by Lieber[104] and Huff.[105] They were able to reduce the calculation to a set of summations involving various terms of the form $1/l^n$. In this way, with just the use of a hand calculator, the following results were obtained in reference (105):

$$\begin{aligned} \ln k_0(1S) &= \ \ 2.98412\ 85559(3) \\ \ln k_0(2S) &= \ \ 2.81176\ 98932(5) \\ \ln k_0(2P) &= -0.03001\ 67089(3) \end{aligned} \tag{3.13}$$

The accuracy of this approach can easily be extended. However, because the quoted results are already accurate to well under one Hertz, many orders of magnitude under present experimental precision, there is no need for such an exercise.

The reason for emphasizing this particular approach to the calculation of the Bethe logarithm has to do with the special role played by low orders of the potential expansion of both the relativistic and nonrelativistic Coulomb Green's function. In the latter case, as seen above, a linear and a logarithmic dependence on a cutoff exist in the 0-P and 1-P contributions, but the M-P term is finite. We see below that in the relativistic case a similar situation holds, except that the linear divergence is softened to a logarithmic one. The relativistic version of the M-P term leads to an energy shift that to leading order in $Z\alpha$ agrees precisely with the D_2 term in (3.12a).

We now describe a method of calculating the Lamb shift[106] that utilizes the fact that the Dirac-Coulomb propagator can be expanded[102] in terms of a closely related propagator that has the same form as the Schrödinger-Coulomb Green's function. In particular, a breakup into a 0-P and 1-P term exists, and all questions of renormalization can be dealt with while considering this part of the calculation. The M-P term can be shown then to have as its NR limit exactly the M-P term considered above that leads to the contribution D_2. The difference between the two, which is evaluated numerically, is of order $m\alpha(Z\alpha)^6$. The calculation is carried out in Coulomb gauge for the virtual photon so as to avoid contributions of a spurious order. The principle advantage of this method is that the expansion of the Dirac-Coulomb propagator involves a small number of terms provided one neglects terms of order $m\alpha(Z\alpha)^7$ and higher. A disadvantage is that the calculation is numerical; however, the numerical uncertainties can be controlled at the level of experimental uncertainties. This is in contrast to the formal method to be described in Section D, which has so many terms contributing at the order of interest that no complete calculation of the constant terms of order $m\alpha(Z\alpha)^6$ has yet been carried out, although the constants associated with logarithmic terms, which are numerically dominant, were evaluated by Erickson and Yennie.[99]

The Dirac-Coulomb Green's function can be expressed in terms of a Klein-Gordon like propagator $G(\vec{r},\vec{r}\,';E)$ via

$$S(\vec{r},\vec{r}\,';E) \equiv \left[(E+\frac{Z\alpha}{r})\gamma_0 + i\vec{\gamma}\cdot\vec{\nabla} + m\right] G(\vec{r},\vec{r}\,';E) \tag{3.14a}$$

which then must satisfy the equation

$$\left[(E+\frac{Z\alpha}{r})^2 + \vec{\nabla}^2 - m^2 + \frac{iZ\alpha\vec{\alpha}\cdot\hat{r}}{r^2}\right] G(\vec{r},\vec{r}\,';E) = \delta(\vec{r}-\vec{r}\,') \tag{3.14b}$$

If we define $p_0^2 = m^2 - E^2$ and $\nu = \dfrac{EZ\alpha}{p_0}$, this can be rearranged as

$$\left[(\vec{\nabla}^2 + \frac{2\nu p_0}{r} - p_0^2) + \{\frac{(Z\alpha)^2}{r^2} + \frac{iZ\alpha\vec{\alpha}\cdot\hat{r}}{r^2}\}\right] G(\vec{r},\vec{r}\,';E) = \delta(\vec{r}-\vec{r}\,') \tag{3.14c}$$

With the neglect of the inner brace within square brackets, (3.9a) is recovered, with the difference that p_0 and ν are redefined. Therefore, defining $G_0(\vec{r},\vec{r}\,';p_0)$ as the solution of

$$[\vec{\nabla}^2 + \frac{2\nu p_0}{r} - p_0^2]G_0(\vec{r},\vec{r}\,';p_0) = \delta(\vec{r}-\vec{r}\,')\ , \tag{3.14d}$$

we find the integral equation for G

$$G(\vec{r},\vec{r}\,';p_0) = G_0(\vec{r},\vec{r}\,';p_0) - \int d^3x G_0(\vec{r},\vec{x};p_0)\left[\frac{(Z\alpha)^2 + iZ\alpha\vec{\alpha}\cdot\hat{x}}{x^2}\right] G(\vec{x},\vec{r}\,';p_0)\ . \tag{3.14e}$$

This integral equation can be expanded in terms of the expression in brackets to any desired order. The expansion will not be useful, however, if many terms are required to obtain an expression applicable to the order of interest. Denoting the first term as $T0$ and the second as $T1$, we describe how these perturbations contribute. It turns out that the leading order of the Lamb shift, $\alpha(Z\alpha)^4$, requires $T0$ and $T1$, along with the unperturbed piece of G. Next it can be observed that in the very non-relativistic region each additional power of $T0$ increases the order by $(Z\alpha)^2$ and each additional power of $T1$ increases the order by one or two powers of $Z\alpha$, depending on whether the factor of $\vec{\alpha}$ requires the presence of a small component of the wave function. The net effect is that through order $m\alpha(Z\alpha)^6$, the required terms are $T0, T1, T0\ T1, T1\ T1$, and $T1\ T1\ T1$. Perturbations involving $T1$ can be simplified with the use of the following relations:

$$\begin{aligned}\int d^3x G_0(\vec{r},\vec{x};E)\frac{\vec{\alpha}\cdot\hat{x}}{x^2}G_0(\vec{x},\vec{r}';E) &= \frac{1}{2EZ\alpha}\vec{\alpha}\cdot(\vec{\nabla}+\vec{\nabla}')G_0(\vec{r},\vec{r}';E)\\ &\equiv \frac{1}{2EZ\alpha}\vec{\alpha}\cdot(\hat{r}+\hat{r}')G_\Delta(\vec{r},\vec{r}';E)\end{aligned}$$

which can be derived by considering the differential equation satisfied by G_0. Therefore, the perturbation $T1$ can be taken together with G_0, and in addition the remaining terms involving $T1$ can be simplified. The final coordinate space expression that will be used for the electron propagator is

$$G(\vec{r},\vec{r}';E) = \left[G_0(\vec{r},\vec{r}';E) - \frac{i}{2E}(\alpha_r+\alpha_{r'})G_\Delta(\vec{r},\vec{r}';E)\right] + \sum_{i=1}^{4} G_i(\vec{r},\vec{r}';E) \tag{3.15a}$$

with

$$\begin{aligned}G_1(\vec{r},\vec{r}';E) &\equiv -(Z\alpha)^2\int\frac{d^3x}{x^2}G_0(\vec{r},\vec{x};E)G_0(\vec{x},\vec{r}';E)\\ G_2(\vec{r},\vec{r}';E) &\equiv -\frac{Z\alpha}{4E}\int\frac{d^3x}{x^2}\left[G_\Delta(\vec{r},\vec{x};E)G_0(\vec{x},\vec{r}';E)(\alpha_r+\alpha_x)\alpha_x\right.\\ &\qquad\left.+\,G_0(\vec{r},\vec{x};E)G_\Delta(\vec{x},\vec{r}';E)\alpha_x(\alpha_x+\alpha_{r'})\right]\\ G_3(\vec{r},\vec{r}';E) &\equiv \frac{iZ\alpha}{4E^2}\int\frac{d^3x}{x^2}G_\Delta(\vec{r},\vec{x};E)G_\Delta(\vec{x},\vec{r}';E)(\alpha_r+\alpha_x)\alpha_x(\alpha_x+\alpha_{r'})\\ G_4(\vec{r},\vec{r}';E) &\equiv i\frac{(Z\alpha)^2}{2E}\int\frac{d^3x}{x^2}\left[G_\Delta(\vec{r},\vec{x};E)G_0(\vec{x},\vec{r}';E)(\alpha_r+\alpha_x)\right.\\ &\qquad\left.+\,G_0(\vec{r},\vec{x};E)G_\Delta(\vec{x},\vec{r}';E)(\alpha_x+\alpha_{r'})\right]\end{aligned} \tag{3.15b}$$

where we have used the shorthand notation $\alpha_a = \vec{\alpha}\cdot\hat{a}$.

While we have developed the above expansion of the Dirac-Coulomb Green's function in coordinate space, the bulk of the calculation is best carried out in momentum space. For example, for the G_0 terms of (3.15a), one finds for the Coulomb and transverse photon pieces

$$\begin{aligned}\Delta E_n(\mathrm{Coul}) =& 2\pi\alpha\int\frac{d^4k}{-(2\pi)^4 i}\frac{1}{|\vec{k}|^2}\int\frac{d^3p\,d^3p'}{(2\pi)^6}\tilde{G}_0(\vec{p}-\vec{k},\vec{p}'-\vec{k};E-k_0)\\ &\times\left[\bar{\psi}(\vec{p})[2m-k_0\gamma_0+\vec{\gamma}\cdot(2\vec{p}-\vec{k})]\gamma_0\left(1+\frac{\vec{\alpha}\cdot(\vec{p}-\vec{p}')}{2(E-k_0)}\right)\gamma_0\psi(\vec{p}')\right.\\ &\left.+\bar{\psi}(\vec{p})\gamma_0\left(1+\frac{\vec{\alpha}\cdot(\vec{p}-\vec{p}')}{2(E-k_0)}\right)\gamma_0[2m-k_0\gamma_0+\vec{\gamma}\cdot(2\vec{p}'-\vec{k})]\psi(\vec{p}')\right]\end{aligned} \tag{3.16}$$

and

$$\Delta E_n(\text{Trans}) = 2\pi\alpha \int \frac{d^4k}{-(2\pi)^4 i} \frac{\left(\delta_{ij} - \frac{k_i k_j}{|\vec{k}|^2}\right)}{k^2} \int \frac{d^3p d^3p'}{(2\pi)^6} \tilde{G}_0(\vec{p}-\vec{k}, \vec{p}\,'-\vec{k}; E-k_0)$$
$$\times \left[\bar{\psi}(\vec{p})[\not{k}\gamma_i + 2(p-k)_i]\left(1 + \frac{\vec{\alpha}\cdot(\vec{p}-\vec{p}\,')}{2(E-k_0)}\right)\gamma_j\psi(\vec{p}\,')\right.$$
$$\left. +\bar{\psi}(\vec{p})\gamma_i\left(1 + \frac{\vec{\alpha}\cdot(\vec{p}-\vec{p}\,')}{2(E-k_0)}\right)[\gamma_j\not{k} + 2(p'-k)_j]\psi(\vec{p}\,')\right] \tag{3.17}$$

respectively. It should be noted that the forms given here result from some rearrangements using the bound state wave equation. Similar expressions hold for $G_1 - G_4$.

At this point the calculation can be split into three parts, the calculation of the 0-P,1-P, and M-P terms, with all ultraviolet divergences occurring in the first two; these will be regulated with the technique of dimensional regularization.[107] Because one is dealing with free propagators in this part of the calculation, standard Feynman parameter techniques can be used, and it is entirely straightforward to isolate the gauge-invariant electron self-mass term, and further to identify the gauge-variant wave-function renormalization and vertex renormalization infinities, which cancel by Ward's identity and leave a finite quantity of order $m\alpha(Z\alpha)^4$. While the entire calculation is too lengthy to describe here, we discuss a part of it to give the flavor of the method. Specifically, for the 1S state consider the 0-P part of (3.16); the quantity to be evaluated is

$$\Delta E_n(\text{Coul: 0-P}) = 4\pi\alpha \int \frac{d^nk}{-(2\pi)^n i} \int \frac{d^3p}{(2\pi)^3} \frac{\bar{\psi}_{1s}(\vec{p})[2m - k_0\gamma_0 + \vec{\gamma}\cdot(2\vec{p}-\vec{k})]\psi_{1s}(\vec{p})}{[(p-k)^2 - m^2]|\vec{k}|^2} \tag{3.18}$$

where the dimension of space-time is taken to be $n \equiv 4 - \epsilon$, where ϵ is to be taken to zero wherever possible. Note that because of rearrangements using the wave equation, this is not the same form which would be written down directly from the electron self energy.

While this integral is non-covariant, we can still combine the two denominators with a Feynman parameter x to find

$$\Delta E_n(\text{Coul: 0-P}) = -4\pi\alpha \int \frac{d^{n-1}k}{(2\pi)^{n-1}} \int \frac{dk_0}{-2\pi i} \int \frac{d^3p}{(2\pi)^3} \int_0^1 dx$$
$$\times \frac{\bar{\psi}(\vec{p})[2m - k_0\gamma_0 + \vec{\gamma}\cdot(2\vec{p}-\vec{k})]\psi(\vec{p})}{[x(E-k_0)^2 - xm^2 - |\vec{k}-x\vec{p}|^2 - x(1-x)\vec{p}^{\,2}]^2}\,.$$

The k_0 integration can be done, leading to a $\vec{k}$ integration that would diverge logarithmically at large momenta in 4 dimensions. Because we are using dimensional regularization, we find instead

$$\Delta E_n(\text{Coul:. 0-P}) = \frac{\alpha C}{2\pi\epsilon} \int_0^1 \frac{dx}{\sqrt{x}} \int \frac{d^3p}{(2\pi)^3} \bar{\psi}(\vec{p})[2m - E\gamma_0 + \vec{\gamma}\cdot\vec{p}(2-x)]\psi(\vec{p})$$
$$\times [x(1 + \vec{p}^{\,2}/m^2(1-x))]^{-\epsilon/2}$$

where $C \equiv (4\pi/m^2)^{\epsilon/2}\Gamma(1+\epsilon/2)$. Making a Taylor expansion in the small quantity ϵ, we find

$$\begin{aligned}\Delta E_n(\text{Coul: 0-P}) = &\frac{\alpha C}{\pi\epsilon}\int d^3p\bar{\psi}(\vec{p})[2m - E\gamma_0 + \frac{5}{3}\vec{\gamma}\cdot\vec{p}]\psi(\vec{p}) \\ &+\frac{\alpha}{\pi}\int\frac{d^3p}{(2\pi)^3}\bar{\psi}(\vec{p})[2m - E\gamma_0 + \frac{17}{9}\vec{\gamma}\cdot\vec{p}]\psi(\vec{p}) \\ &-\frac{\alpha}{4\pi}\int_0^1\frac{dx}{\sqrt{x}}\int\frac{d^3p}{(2\pi)^3}\bar{\psi}(\vec{p})[2m - E\gamma_0 + \vec{\gamma}\cdot\vec{p}(2-x)]\psi(\vec{p}) \\ &\qquad\times\ln(1+\frac{\vec{p}^2}{m^2}(1-x))\,.\end{aligned} \tag{3.19}$$

The first two integrals of (3.19) are primarily mass renormalization and wave function renormalization terms, and they cancel with the mass renormalization counterterm and terms coming from the 1-P term. The first two terms of the last integral can be rearranged into a term that contributes to the Lamb shift and a term of a spurious order that cancels out against similar terms coming from other parts of the calculation. The $\vec{\gamma}\cdot\vec{p}$ part of the last integral, however, is finite and of the order of interest by the power counting arguments of Section A. We consider it as an example; an integration by parts brings it into the form

$$\Delta E_n(\text{Coul: 0-P; example}) = -\frac{\alpha}{\pi m^2}\int dx\sqrt{x}(1-x/6)\int\frac{d^3p}{(2\pi)^3}\frac{\vec{p}^2\bar{\psi}(\vec{p})\vec{\gamma}\cdot\vec{p}\psi(\vec{p})}{1+\frac{\vec{p}^2}{m^2}(1-x)}\,. \tag{3.20a}$$

At this point if one is interested only in the leading order, the denominator can be replaced with 1, the x integration carried out trivially, and nonrelativistic approximations made on the wave functions to obtain the result

$$\Delta E_n(\text{Coul: 0-P; example: NR}) = -\frac{3m\alpha(Z\alpha)^4}{\pi}\,. \tag{3.20b}$$

Naive power counting would suggest that the correction from considering the actual denominator would be down by a factor $(Z\alpha)^2$. However, because the momentum integration for the nonrelativistic calculation behaves as $\int dp/p^2$, a Taylor expansion of the denominator would give a linearly divergent integral. This is an example of the warning given in Section A of integrals being pushed into the relativistic region. To find the relativistic corrections to (3.20a) one subtracts a term with the approximations that lead to (3.20b), without making approximations on the denominator. The important range of integration is then $p \sim m$. In this range the wave functions simplify considerably, as they involve structures such as $\vec{p}^2 + (mZ\alpha)^2$ which can be replaced with $\vec{p}^2$ for relativistic values of the momentum. It is then possible with such approximations to carry out the remaining integral analytically with the result

$$\Delta E_n(\text{Coul: 0-P; example: } (Z\alpha)^5) = 7m\alpha(Z\alpha)^5 \tag{3.20c}$$

All higher order terms can now be evaluated numerically by forming a second level of subtraction, removing both (3.20b) and (3.20c) from (3.20a). When this procedure is carried out for the entire calculation, a result consistent with the value of $G_{SE}(1S)$ given in (2.5c) is found.[106] It is also possible to evaluate at least some terms of order $m\alpha(Z\alpha)^6$ analytically, but a complete calculation has never been carried out.

C. Numerical Evaluation of the Self Energy for High Z

The treatment of the self-energy just presented, as well as the treatment to be presented in the following section, relies in part on an expansion in the small quantity $Z\alpha$. However, when Z is large an expansion in this quantity will clearly become unreliable unless the coefficients of higher order terms become small, a pattern which is not observed in the orders calculated so far. In fact, as shown in Figure 3, truncating the expansion of $F_{1S}(Z\alpha)$ at fourth, fifth, and sixth order in $Z\alpha$ leads to very different behavior at large Z. For this

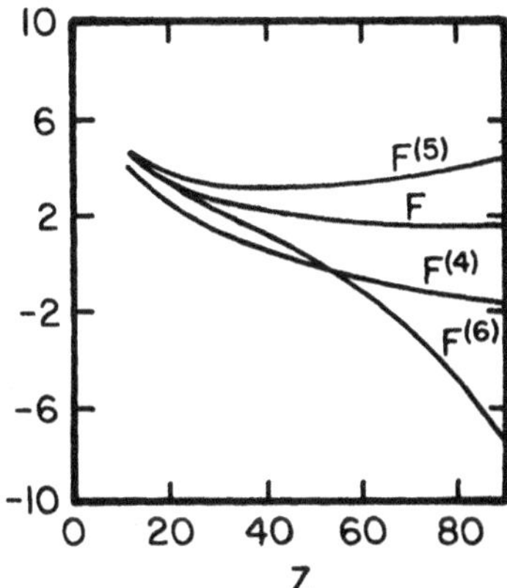

Figure 3: $F_{1S}(Z\alpha)$ and approximations $F^{(n)}(Z\alpha)$ obtained by cutting off the power series expansion at order n

reason it is desirable to evaluate the self-energy with the numerical approach mentioned in Section A when Z is large. The basic idea is to evaluate the self energy-expression numerically; it is nominally a ten-dimensional integral. We proceed to a description of the calculation following references (96), (94), and (95). The first step is to perform the $\int d^3k$ integration in (3.4a), to obtain the seven-dimensional integral

$$\Delta E_n = \frac{e^2}{4\pi}\int \frac{d^3r d^3r'}{|\vec{r}-\vec{r}'|}\int \frac{dk_0}{-2\pi i} e^{ib|\vec{r}-\vec{r}'|}\bar{\psi}_n(\vec{r})\gamma_\mu S_F(\vec{r},\vec{r}';E_n-k_0)\gamma^\mu\psi_n(\vec{r}') - \delta m^{(2)}\int d^3r\bar{\psi}_n(\vec{r})\psi_n(\vec{r}) \tag{3.21}$$

where $b \equiv \sqrt{k_0^2+i\epsilon}$. Note that covariance has been abandoned by preferentially performing the spatial part of the k integration; particular care must be taken in connection with this point. The remaining energy integration is to be carried out numerically. In

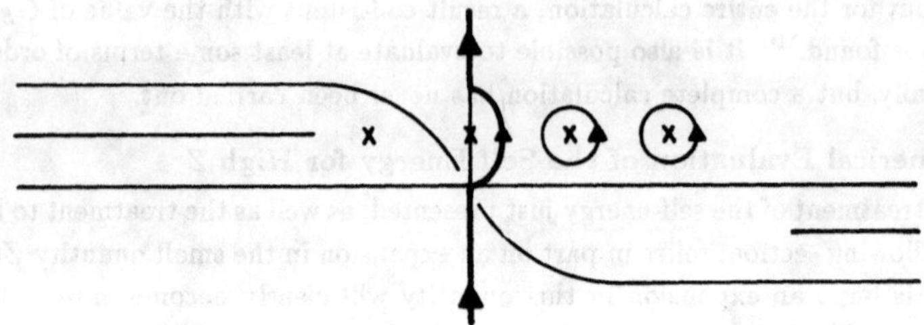

Figure 4: Poles and cuts in the complex k_0 plane and the contour rotation used in the self-energy calculation.

order to avoid dealing with oscillatory functions, the contour is rotated as shown in Figure 4. While the branch cuts associated with the photon propagator can be avoided by the counter-clockwise rotation indicated, poles in the electron propagator will play a role.

For the ground state shift, the pole at the origin forces the contour to describe a half-circle about the origin, which leads to a term R_{1S}, where

$$R_{1S} = \frac{\alpha}{2} \int \frac{d^3r d^3r'}{|\vec{r} - \vec{r}'|} \bar{\psi}_{1S}(\vec{r}) \gamma_\mu \psi_{1S}(\vec{r}) \; \bar{\psi}_{1S}(\vec{r}') \gamma^\mu \psi_{1S}(\vec{r}') \; . \tag{3.22}$$

For excited states, the contour must pass all poles of states of lower energies, leading to full circles about those poles. These contributions give imaginary energy shifts corresponding to the decay widths of the state into these lower energy states. There now remains an integral along the imaginary k_0 axis of the spatial integrals over $\vec{r}$ and $\vec{r}'$. In order to perform the angular part of these integrals the photon propagator is expressed in terms of spherical Bessel functions and spherical harmonics. While this introduces an infinite sum over the partial waves, the angular part of the remaining six-dimensional integral over coordinates can be carried out analytically, which then leaves a two-dimensional integral over the two radial coordinates.

Thus the numerical problem is reduced to evaluating a partial wave sum of three-dimensional integrals. Because we have not explicitly renormalized at this point, an infinite result would come from a direct numerical evaluation. The first step in turning the integral into a finite quantity is to add and subtract the same integral with the Dirac-Coulomb propagator replaced with a free propagator. This integral can be treated with standard Feynman parameter techniques, and in particular the mass renormalization counterterm can be explicitly cancelled at this point. However, because the electron is not on mass shell, an infinite wave function renormalization constant remains to be cancelled. The compensating infinity in this case is found in the subtracted term. After the free propagator is subtracted, the partial wave expansion at a given value of k_0 behaves, after carrying out the r and r' integrations, as $1/l^2$ for large l. A major numerical complication of the approach is that at higher energies the l at which the asymptotic limit is reached gets fairly large: Desiderio and Johnson[96] included seventeen partial waves. Once the coefficients of the leading terms have been identified the series can be summed to infinity, leaving only a single integration over k_0. However, this behaves at large k_0 as $1/k_0$, and it is this divergent behavior that is cancelled by the remaining wave function renormalization term. The cancellation becomes severe at large values of k_0, and coupled with the difficulty of evaluating the partial wave summation at these large values leads to the major source of numerical error. A further problem is that because the calculation is carried out in Feynman gauge, spurious terms of order $\alpha(Z\alpha)^2$ must be cancelled numerically, although at large values of Z this problem is not too severe. It is possible that use of another gauge could eliminate this problem.

Shortly after the work of Desiderio and Johnson appeared, Mohr[21] presented a related calculation that is significantly more accurate. The increased accuracy comes from several sources. Firstly, instead of carrying out the r and r' integrations first, followed by a partial wave summation and then the ω integration, Mohr carried out the l summation for each given value of r, r', and ω. This is facilitated by the use of the representation of the Dirac-Coulomb propagator as a partial wave expansion of products of Whittaker functions, which have well understood asymptotic behavior. Renormalization is also implemented at an earlier stage than in the previous calculation, which led to increased convergence of the partial wave summations and ω integrations. Finally, the contour used is different. Instead of rotating about $\omega = 0$, the rotation is carried out about the electron mass, which adds an extra term in which the contour wraps around the photon cut; this term is the relativistic generalization of (3.10). Even with the increased numerical control made possible by these techniques, the difficulty of the calculations is such that it has to date been applied only to the states $1S_{1/2}$, $2S_{1/2}$, $2P_{1/2}$, and $2P_{3/2}$ for ions with $Z > 10$. The extrapolation of these results to obtain information about the Lamb shift for $Z = 1$ is described in Appendix A.

D. Formal Operator Method

Formal operator techniques for application to QED corrections to bound state energies were pioneered by Karplus, Klein, and Schwinger.[98] Their method used an exponential representation of the propagators. The method was pursued further by Erickson and Yennie,[99] who instead used algebraic expressions. An introduction to the latter approach, which might be useful for a beginner to the field, is given in reference (24). Throughout this section we omit the factor of Z which distinguishes the Coulomb interactions; it is easily restored later.

Formally, we may express the Dirac propagator in an external field as

$$S_F = \langle \vec{r} | \frac{1}{\not{\Pi} - m_e} | \vec{r}' \rangle \tag{3.23a}$$

where the normalization $\langle \vec{r} | \vec{r}' \rangle = \delta(\vec{r} - \vec{r}')$ is used and

$$\Pi_\mu \equiv P_\mu - eA_\mu \; . \tag{3.23b}$$

Here, A_μ represents the time-independent external field, and the four-momentum operator is $P_\mu = (E, \vec{p})$. It is helpful to note that

$$[\Pi_\mu, \Pi_\nu] = -ieF_{\mu\nu} \tag{3.23c}$$

where $F_{\mu\nu} = \partial_\mu A_\nu - \partial_\nu A_\mu$ is the electromagnetic field. We may include the field giving the hyperfine splitting with the proviso that it be used only to first order. Later, in Chapter V-B, this is generalized further and A_μ becomes the quantized electromagnetic field operator, and P_0 has an operator part (time derivative) as well. This gives us the possibility of treating renormalization on the electron line without first committing ourselves to a particular treatment of recoil.

The series in (3.2c) may be replaced by

$$\frac{1}{\not{\Pi} - m_e} + \frac{1}{\not{\Pi} - m_e}\tilde{\Sigma}_{\text{ren}}\frac{1}{\not{\Pi} - m_e} + \frac{1}{\not{\Pi} - m_e}\tilde{\Sigma}_{\text{ren}}\frac{1}{\not{\Pi} - m_e}\tilde{\Sigma}_{\text{ren}}\frac{1}{\not{\Pi} - m_e} + \cdots \tag{3.24}$$

where now $\tilde{\Sigma}_{\text{ren}}$ is to be regarded as an operator. As mentioned above, an important difference from the free case is that $\tilde{\Sigma}_{\text{ren}}$ may contain terms which do not have factors of $(\not{\Pi} - m_e)$. These additional terms arise due to the lack of commutativity of the components of Π_μ.

One can argue that if $\tilde{\Sigma}_{\text{ren}}$ has a term proportional to $\not{\Pi} - m_e$, that term does not contribute in leading order. The important terms then arise from the non-commutivity of the components of Π_μ and occur in the form of field strengths. Such commutator terms are necessary for the Lamb shift and the anomalous magnetic moment of the electron. From now on, we drop terms which contain an external factor of $\not{\Pi} - m_e$. Incidentally, this also means that in leading order we may simply use $\tilde{\Sigma}$ in place of $\tilde{\Sigma}_{\text{ren}}$. The casual reader may wish to skip the rest of this paragraph which justifies these remarks. Consider the effect of a term containing a factor of $\not{\Pi} - m_e$ in one of the factors of (3.24). One (or more) of the denominators are canceled. If an external factor is canceled, the result is to modify the residue of the pole in the discussion of Section A(1). This corresponds to a perturbative change of the wave function, but not directly to an energy correction. If, instead, an internal factor is canceled, the result is a new term in the geometric series and it does produce an energy shift; this new term is essentially second order in $\tilde{\Sigma}_{\text{ren}}$. We would expect such shifts to be very small, since they should contain two factors of the field strength and two explicit factors of α. As far as we know, such contributions have not actually been examined, but their nominal order appears to be smaller than any terms of current interest.

1. Description of Second Order Self Energy

In place of (3.4a), the simplest self energy contribution is given by the operator

$$\tilde{\Sigma}^{(2)} = -e^2 \int \frac{d^4k}{-(2\pi)^4 i}\frac{1}{k^2 + i\epsilon}\gamma_\mu \frac{1}{\not{\Pi} - \not{k} - m_e + i\epsilon}\gamma^\mu - \delta m^{(2)} . \tag{3.25}$$

We want to rearrange this to reveal how the leading terms in the Lamb shift calculation arise. A complete treatment would be prohibitively long so we emphasize the approach without giving a detailed analysis. Before proceeding to this discussion, it is helpful to consider how one identifies contributions of different orders of magnitude, following

Section A. The reason is that if one proceeds too directly, he may find that delicate cancellations are necessary in order to obtain the terms of interest. Even if one does not wish to calculate terms of different order analytically, a numerical calculation would require unprecedented accuracy in order to include the range of terms necessary to compare with present experimental precision.

In order to discuss orders of magnitude in a more useful way for our immediate purposes, let's proceed with a few steps from the formal development. After rationalizing the electron denominator (taking care of operator dependence), we have

$$\tilde{\Sigma}^{(2)} = -e^2 \int \frac{d^4k}{-(2\pi)^4 i} \frac{1}{k^2+i\epsilon} \gamma_\mu \frac{1}{k^2 - 2k\cdot\Pi + \not{\Pi}^2 - m_e^2 + i\epsilon} (\not{\Pi} - \not{k} + m_e)\gamma^\mu - \delta m^{(2)} \ . \tag{3.26}$$

We want to imitate standard procedures for calculating Feynman integrals as far as possible, while taking into account the non-commutativity of quantities in these expressions. Generally, we display only the integrand. From now on, we do not display the $i\epsilon$ term of the denominators explicitly. Since $1/k^2$ is a c-number, the two denominators can be combined with a Feynman parameter integral (introducing an integration $\int_0^1 dx$). The integrand becomes

$$\gamma_\mu \frac{1}{D^2} (\not{\Pi} - \not{k} + m_e)\gamma^\mu \tag{3.27a}$$

where

$$\begin{aligned} D &= k^2 - 2xk\cdot\Pi + x[\not{\Pi}^2 - m_e^2] \\ &= (k - \Pi x)^2 + x(1-x)(\not{\Pi}^2 - m_e^2) + x^2\mathcal{M} - x^2 m_e^2 \end{aligned} \tag{3.27b}$$

where $\mathcal{M} = \not{\Pi}^2 - \Pi^2 = -\frac{e}{2}\sigma_{\mu\nu}F^{\mu\nu}$ is the magnetic moment operator.

Now let us estimate the orders of magnitude obtained by various expansions of the external potentials, using the first form of D. The potential occurs in several ways in the denominator. These are $2exk\cdot A$, $-2exE_nA_0$, $xe^2A\cdot A$, $-x(ep\cdot A + eA\cdot p)$, and $x\mathcal{M}$. Imagine that some combination of these has been expanded up and that the denominators now are independent of the potential. We may now shift k_0 by xE_n so that all the denominators become

$$D_0 = k^2 + 2x\vec{p}\cdot\vec{k} - x[\vec{p}^{\,2} + \gamma_n^2] - x^2[m_e^2 - \gamma_n^2] \ . \tag{3.28}$$

where, in momentum space, $\vec{p}$ takes on a numerical value depending on its position in the product. Also, a numerator factor $-2exk\cdot A$ becomes $-2exk\cdot A - 2ex^2E_nA_0$. Now we may combine all the denominators with a parameter integration. After the the k-integration is carried out, the result will take the form

$$\text{numerator factors} \times \frac{x^m}{\{xm_e^2 - x[\vec{p}(v)^2 + \gamma_n^2] + [\vec{p}^{\,2}(v) + \gamma_n^2]\}^n} \ . \tag{3.29a}$$

where $m \leq n$. Here (v) represents the dependence on parameters, which we need not specify in detail at this point. In the non-relativistic region, the region of small x is important because x has the large coefficient m_e^2 in the denominator. Independently of the size of x, we expect the second term in the denominator to be much smaller than the first so we expand it up and include expanded terms with other numerator factors. Our expression then becomes

$$\text{numerator factors} \times \frac{x^m}{[xm_e^2 + \Delta]^n} \tag{3.29b}$$

where

$$\Delta \doteq \vec{p}^2(v) + \gamma_n^2 \tag{3.29c}$$

and the values of m and n may have changed.

Now we can carry out the x-integration. One way to do this is to write a numerator x as $[(xm_e^2 + \Delta) - \Delta]/m_e^2$. The first term cancels denominators, and the Δ in the second term may be placed with other numerator factors. In the end, no powers of x remain in the numerator. Thus we find terms with the structure

$$\text{numerator factors} \times \frac{1}{m_e^{2m}[xm_e^2 + \Delta]^n} \; . \tag{3.29d}$$

where the value of n may have changed. If $n > 1$, the x-integration gives

$$\text{numerator factors} \times \frac{1}{m_e^{2m+2}} \left(\frac{1}{\Delta^{n-1}} - \frac{1}{[m_e^2 + \Delta]^{n-1}} \right) \; . \tag{3.29e}$$

The first term here is more important than the second in the non-relativistic region. If $n = 1$, a logarithm results

$$\text{numerator factors} \times \frac{1}{m_e^{2m+2}} \ln \left(\frac{m_e^2 + \Delta}{\Delta} \right) \; . \tag{3.29f}$$

Now we are prepared to discuss order of magnitude again. For $\mu = 0$, the numerator of our original expression is of order m_e. The term in the denominator which produces the lowest order when expanded up is $2xE_n eA_0$ since it contributes a factor of γ' and no contribution to m. Consequently, we may factor out $1/m_e$ (including the first factor of (3.29e)) for any power of this term; and with the first term in parenthesis from (3.29e), the integral produces a factor of γ'^2. This agrees with our original estimate; i.e., these are individually spurious terms of too low an order.

Continuing with our discussion of the $2xE_n eA_0$ expansion, what is the result from the second term of (3.29e)? If the contribution continues to arise from the non-relativistic region, the additional inverse powers of m_e^2 require additional positive powers of γ'^2 to keep the dimensions correct. We might call this the "naive" order estimate. However, too many inverse powers of m_e^2 signal that the integral may diverge at the high momentum end unless the Δ in the denominator produces convergence. This is a very complicated subject which cannot be dealt with in a strictly correct way without looking in detail at the integral, but we can give some of the ideas here. Basically, we must see how many powers of γ' can be factored out of the integral. If that power *exceeds* that of the power of m_e in the denominator, the integral *may* converge in the low momentum region and produce inverse powers of γ' in order to agree with the naive order estimate, which is now $\alpha\gamma'^{2n}/m_e^{2n-1}$. In the opposite case, the integral will have momenta characterized by m_e and will produce a power of m_e which balances that of γ' to produce the correct dimensions. It is relatively straighforward to determine the number of powers of γ' which are available. In momentum space, the S-state wave functions contain γ'^5. Each power in the expansion adds a power of γ', giving an overall factor of $\gamma'^{(n+5)}$. Thus, for $n \leq 5$, the naive order estimate for this term should be valid; while for $n > 5$, the order estimate becomes $\alpha\gamma'^{n+5}/m_e^{n+4}$. Either of these estimates is smaller than the present order of importance provided $n > 3$. However, the true upper bound is smaller. If there are more than two Coulomb interactions in succession without a denominator to damp intermediate momentum, divergent integrals are produced. The estimate is valid for $n = 2$ where it is of order of the Lamb shift. For $n = 3$, one would expect a logarithmic factor to occur as well. For $n \geq 4$ intermediate momenta become relativistic, requiring the denominators, while the wave function momenta continue to converge in the non-relativistic region. This local convergence reduces the overall number of γ' factors by 2, so that the likely order of the result is $\alpha\gamma'^{3+n}/m_e^{2+n}$. This is beyond the present order of interest; thus it is not necessary to go beyond three powers of $2xeE_nA_0$ for the second term of (3.29e).

Let's look briefly at some of the other terms which can be expanded up, starting with $x\mathcal{M}$. This may be thought of as introducing only a factor of α, rather than a factor of γ'. We take this factor outside the integral since it does not contribute to the dimensional argument. Thus each of these powers adds a factor of α to the naive order (sometimes more since it may require small components of the wave function to contribute). It also produces a tendency for the naive estimate to fail since the overall integration, or a subintegration, may switch to the relativistic region. The other terms work in a similar manner. The convection term does not actually exist for the Coulomb potential. The term $xe^2A_0^2$ is a seagull term; it adds two powers of α to the overall power count. A term $-2ex^2E_nA_0$ goes like the discussion of the preceding paragraph with the change that the additional power of x generally causes the order in α to increase by two; but the transition to relativistic behaviour is advanced. The $-2xk_0eA_0$ term must be expanded to even orders by symmetry. To second order it produces the structure of (3.29a) with $n = 1, m = 1$, with a factor of α^2 which may be taken outside the integral. The expected order is $\alpha^3\gamma'^2/m_e$, the same as the leading Lamb shift.

We should also verify that expanding up the second term of the denominator in (3.29a) generally increases the order. This would make the m and n of (3.29b) each one larger than originally, which would in turn increase the number of external denominator factors of m_e by 2, adding two powers of α to the naive order estimate. At the point where the integral goes over to the relativistic region, the expansion stops adding powers of α and the term must be retained in the denominator.

It should be clear by now that expanding an expression into terms of different order requires a certain amount of skill and judgment. The next thing is to cope with the terms of spurious lower order. The key to understanding and removing them is to note that we have not yet used the fact that terms with external factors of $\not{\Pi} - m_e$ may be dropped. Thus it is necessary to combine terms with different powers of A_μ to produce that combination, or rather, not to separate them unnecessarily in the first place.

2. Formal Decomposition of the Second Order Self Energy

Returning to (3.27a), we now bring a γ_μ through using

$$\gamma_\mu \frac{1}{D} = \frac{1}{D}\gamma_\mu - \frac{1}{D}[\gamma_\mu, D]\frac{1}{D} = \frac{1}{D}\gamma_\mu - \frac{1}{D}2x[\Pi_\mu, \not{\Pi}]\frac{1}{D} \ . \tag{3.30}$$

The second term here is a field strength term. For the term where the γ_μ went through, we collapse the $\gamma_\mu ... \gamma^\mu$ sum in the usual way. Any resulting factors of $\not{\Pi}$ on the outside can be replaced by m_e as mentioned earlier. This takes care of any contributions to B, as well as some terms which cannot contribute to first order in $\tilde{\Sigma}$.

At this stage, the integrand has been transformed to

$$\frac{1}{D^2}(2m_e + 2\not{k}) - \left[\frac{1}{D}2x[\Pi_\mu, \not{\Pi}]\frac{1}{D^2} + \frac{1}{D^2}2x[\Pi_\mu, \not{\Pi}]\frac{1}{D}\right](2\Pi^\mu - \not{k}\gamma^\mu) \ . \tag{3.31}$$

We start by examining the commutator term, neglecting powers of the potential in the denominators. Its leading part is of the $m = 0, n = 1$ type with no factors of m_e in the numerator, its order of magnitude is then easily seen to be $\alpha^2\gamma'^3/m_e^2$ times a logarithm. All orders of $2xE_neA_0$ produce the same order (but without the logarithm), so one should not expand in powers of that operator. However, the other operators may be expanded up to produce higher order contributions. On the other hand, the first term of (3.31) still has the spurious lower orders; it requires more effort to eliminate them.

Next consider the apparently troublesome terms with $\not{k}$ on the outside. These can be handled with a simple trick. After introducing enough regularization to secure the required convergence (which we do not display explicitly), we note that

$$0 = \int d^4k \frac{\partial}{\partial k^\sigma}\frac{1}{D} = \int d^4k \frac{1}{D}(2k_\sigma - 2x\Pi_\sigma)\frac{1}{D} \ .$$

Hence

$$\frac{1}{D^2}2\not{k} \doteq \frac{1}{D}2x\Pi_\sigma\frac{1}{D}\gamma^\sigma \ . \tag{3.32}$$

Now it is natural to move the γ^σ next to the Π_σ, while creating a field strength term from first commuting it with the denominator. After that, we bring the $\not{\Pi}$ outside and replace it with m_e. The result of all this is

$$\begin{aligned} &\frac{1}{D}2x\not{\Pi}\frac{1}{D} + \frac{1}{D}2x\Pi_\sigma\frac{1}{D}2x[\Pi^\sigma, \not{\Pi}]\frac{1}{D} \\ &\quad = \frac{1}{D^2}2m_e x + \frac{1}{D}\left[2x\not{\Pi}, \frac{1}{D}\right] + \frac{1}{D}2x\Pi_\sigma\frac{1}{D}2x[\Pi^\sigma, \not{\Pi}]\frac{1}{D} \ . \end{aligned} \tag{3.33}$$

The second term contains a field strength and is linear in k. It can be modified by another application of the previous trick. The last term is similar to the one whose order is discussed in the preceding paragraph; however, it has an extra factor of x. This causes a significant change since its leading term is then of the $m = 1$, $n = 1$ type with no factors of m_e in the numerator. This reduces to the form (3.29d) with $m = 1$ and $n = 0$ or 1, giving contributions of nominal order $\alpha^2\gamma'^3/m_e^2$ or $\alpha^2\gamma'^5 \ln\alpha/m_e^4$, respectively (the latter result may be larger, since one wave function may converge in the non-relativistic region making the remaining integrations relativistic and giving the order $\alpha^2\gamma'^4/m_e^3$). Terms obtained by expanding up powers of $2xE_n eA_0$ increase the order in α by 2.

Similarly, the other term of (3.31) with $\not{k}$ becomes

$$\left[\frac{1}{D}2x[\Pi_\mu, \not{\Pi}]\frac{1}{D}x\Pi_\sigma\frac{1}{D} + \frac{1}{D}x\Pi_\sigma\frac{1}{D}2x[\Pi_\mu, \not{\Pi}]\frac{1}{D}\right]\gamma^\sigma\gamma^\mu \,. \tag{3.34}$$

Here we can proceed in a similar manner. Moving the γ^σ or γ^μ into the corresponding factors produces additional field strength terms from commutators with $1/D$. Other field strengths are produced by moving a factor of $\not{\Pi}$ out to act on the wave function. Thus in a few moves this expression breaks down into terms with one, two, or three factors of the field strength. Up to the point where the expression becomes relativistic, the additional field strengths give additional factors of α.

The field strength terms found so far will be considered later. The non-field strength terms are in a sense more troublesome; they are

$$\frac{2m_e(1+x)}{D^2} = \frac{2m_e(1+x)}{[(k-\Pi x)^2 + x(1-x)(\not{\Pi}^2 - m_e^2) + x^2\mathcal{M} - x^2m_e^2]^2} \,. \tag{3.35}$$

Here, if it were legal to ignore field strength terms, we could recover an expression for the $\delta m^{(2)}$ term which is to be canceled by subtraction. First one could ignore the $\mathcal{M}$ in the denominator. Then since they behave as c-numbers, one could shift the origin of the k-integration to make the denominator quadratic. Finally, the $\not{\Pi}^2 - m_e^2$ could be set equal to zero in the expectation value. The result is exactly the expression for the second order contribution to $\delta m^{(2)}$. This suggests a workable procedure. The first step is to expand up powers of $x^2\mathcal{M}$ until terms left over are too small to matter. This rapidly increases the order until the integrals become relativistic. For example, the leading contribution from one power is of the type $m = 1$, $n = 1$ with one power of m_e in the numerator; in addition, because of the structure of $\mathcal{M}$, one small component of the wave function enters. Using the dimensional arguments, we find the order to be $\alpha^2\gamma'^3/m_e^2$. With two powers the type becomes $m = 2$, $n = 2$, with an explicit factor of m_e in the numerator; now only the large components of the wave function are important. The expected order, taking into account that some sub-integrations are relativistic, is $\alpha^3\gamma'^3/m_e^2$. Each additional power increases the order by a factor of α on the average. If in the term with $\mathcal{M}^0$, we formally shift the origin of integration, the result is precisely canceled by the mass renormalization.

It is worth showing the technique for correcting for the operator shift. We want to evaluate (3.35) without the $\mathcal{M}$ minus the same expression with the first term in the denominator replaced by k^2. The direct, but not so useful, way would be to expand up the operator difference in the denominators. Instead, we can write the difference as a single integrand at the expense of introducing a new integration. We have to evaluate

$$\int_0^1 d\lambda \frac{\partial}{\partial\lambda} \frac{2m_e(1+x)}{[(k-\lambda\Pi x)^2 + x(1-x)(\not{\Pi}^2 - m_e^2) - x^2 m_e^2]^2}$$

$$= \int_0^1 d\lambda 2m_e(1+x) \left[\frac{1}{D_\lambda}[2k\cdot x\Pi - 2\lambda x^2\Pi^2]\frac{1}{D_\lambda^2} + \frac{1}{D_\lambda^2}[2k\cdot x\Pi - 2\lambda x^2\Pi^2]\frac{1}{D_\lambda}\right] \tag{3.36a}$$

where

$$D_\lambda = (k-\lambda\Pi x)^2 + x(1-x)(\not{\Pi}^2 - m_e^2) - x^2 m_e^2 \,. \tag{3.36b}$$

The term linear in k can be transformed by a trick similar to the previous ones. Suppressing the integration $\int_0^1 2\lambda d\lambda$, the new integrand is

$$2m_e(1+x)x^2 \left[\frac{1}{D_\lambda}\Pi^2\frac{1}{D_\lambda^2} + \frac{1}{D_\lambda^2}\Pi^2\frac{1}{D_\lambda} - 2\frac{1}{D_\lambda}\Pi_\mu\frac{1}{D_\lambda}\Pi^\mu\frac{1}{D_\lambda}\right] \tag{3.36c}$$

$$= 2m_e(1+x)x^2 \left[\frac{1}{D_\lambda}\Pi_\mu[\Pi^\mu, \frac{1}{D_\lambda}]\frac{1}{D_\lambda} - \frac{1}{D_\lambda}[\Pi_\mu, \frac{1}{D_\lambda}]\Pi^\mu\frac{1}{D_\lambda}\right] \,. \tag{3.36d}$$

The final form reveals the field strength structure. Working out the commutator gives contributions from either the first or second terms of the denominator. We show only the latter here

$$[\Pi_\mu, x(1-x)(\not{\Pi}^2 - m_e^2)] = x(1-x)([\Pi_\mu, \not{\Pi}]\not{\Pi} + \not{\Pi}[\Pi_\mu, \not{\Pi}])$$

Except for the extra factor of $\not{\Pi}$, this is like some terms we've seen before. This extra factor can be commuted to the outside and replaced by m_e; some additional field strength terms are left behind

3. Derivation of Leading Second Order Contributions

We now have a sufficient sampling of terms to be able to discuss some of the methods involved without getting overly bogged down in details. Terms which were given the name L by Erickson and Yennie are discussed first. An example of one of these terms is the first field strength term to occur in (3.31).

$$-\left[\frac{1}{D}2x[\Pi_\mu, \not{\Pi}]\frac{1}{D^2} + \frac{1}{D^2}2x[\Pi_\mu, \not{\Pi}]\frac{1}{D}\right] 2\Pi^\mu \,. \tag{3.37}$$

We want to reduce this to the point where we can identify its leading contribution. This is done by making a number of approximations which must be checked out later. It will become apparent that higher powers of the field strength generally lead to higher orders in α, but higher powers of the potential do not necessarily increase the order in α. It is important to understand this point. In any case, we may start by dropping powers of $\mathcal{M}$ in the denominators. Next, we should like to shift the origin of k so as to make the denominators quadratic. Corrections to this which give higher powers of the field strength are generally smaller, but corrections which are higher derivatives of the one field strength are not necessarily smaller and must be viewed with care. In the present case, it is possible to make such an expansion. After all these approximations, the outside denominators simplify to $k^2 - x^2m_e^2$. Since they are now c-numbers, they may be combined with the inside denominator(s) by introducing a new parametric integral ($\int_0^1 du$). Our integrand then becomes

$$-[\Pi_\mu, \not{\Pi}]\frac{8x}{[k^2 + ux(1-x)(\not{\Pi}^2 - m_e^2) - x^2m_e^2]^3}\Pi^\mu . \tag{3.38}$$

The k-integration is now easily carried out with the result (some factors omitted)

$$-[\Pi_\mu, \not{\Pi}]\frac{8}{-u(1-x)(\not{\Pi}^2 - m_e^2) + xm_e^2}\Pi^\mu . \tag{3.39a}$$

Before evaluating this, it is helpful to discuss the order of magnitude of different terms which it contains. First note that the combination $\not{\Pi}^2 - m_e^2$ appearing in the denominator is zero if it acts on the wave function or of order $(Z\alpha)^2m_e^2$ if it does not (the latter estimate comes from inserting a sum over states, as seen below). If the factor of Π_μ on the right has a constant part, the denominator becomes xm_e^2, but the commutator gives zero because $\not{\Pi}$ can act on the wave function where it is replaced by m_e. For all other terms, the x-integration gives a factor of order $\ln(m_e^2/[(Z\alpha)^2m_e^2]$, which affects the size of the result significantly, but not its order of magnitude. Thus we need consider only the factors in the numerator to decide the order of magnitude. For $\mu = 0$, the factor on the right becomes V_c and the commutator becomes $[V_c, \vec{\gamma}\cdot\vec{p}]$. The γ-matrix structure requires that one small component of the wave function occurs as a factor, and the overall order is seen to be $\alpha(Z\alpha)^6$, which is beyond our immediate interest although it is relevant for comparison with experiment.

For the present discussion, we consider only spatial components of μ and restrict the calculation to the large components of the wave function which may be approximated by Schrödinger wave functions. For consistency, the denominator should also be approximated in the nonrelativistic region. For example, the $\mathcal{M}$ term can be dropped in leading order. If expanded to first order, the naive order of the correction corresponds to having three explicit factors of α and a structure producing $\alpha^3\gamma'^4/m_e^3$, which is at the limit of interest. Similarly, the first order correction from the $e^2A_0^2$ term is estimated to be $\alpha^4\gamma'^3/m_e^2$. Thus we may use

$$\not{\Pi}^2 - m_e^2 \approx -2m_eV_c - \vec{p}^{\,2} - 2m_e(m_e - E_n) = -2m_e(H_{NR} + m_e - E_n) \tag{3.39b}$$

where H_{NR} is the non-relativistic Hamiltonian.

At this stage, our expression has become

$$[\vec{p}, \gamma_0V_c]\frac{8}{u(1-x)2m_e(H_{NR} + m_e - E_n) + xm_e^2}\cdot\vec{p} . \tag{3.39c}$$

Before carrying out the x-integration, it is convenient to insert a factor

$$1 = \frac{[m_e^2 - 2um_e(H_{NR} + m_e - E_n)] + 2um_e(H_{NR} + m_e - E_n)}{m_e^2} \tag{3.40a}$$

between the operators. The second term in the numerator yields a higher order of magnitude than the first,[i] so we set it aside temporarily. The advantage of this rearrangement is that the x-integral of the first term is carried out simply, giving

$$[\vec{p}, \gamma_0 V_c] \frac{8}{m_e^2} \ln \left[\frac{m_e^2}{-2m_e u(H_{NR} + m_e - E_n)} \right] \cdot \vec{p} \, . \tag{3.40b}$$

Now if we insert a sum over a complete set of non-relativistic states next to the logarithm, we are lead immediately to a result of the form (3.8), but with the cutoff parameter replaced by m_e.

Another important term which can arise is the magnetic moment type. For example, suppose we have one power of $\mathcal{M}$ from (3.35). Because this carries two factors of x into the numerator, it turns out that a correction to the subsequent operator shift will increase the order of the result. Therefore we carry out the shift and set aside the correction. At that point, we may replace $\not{\Pi}^2 - m_e^2$ by zero in the denominator, giving us the simplified expression

$$\frac{-4m_e x^2 \mathcal{M}}{(k^2 - x^2 m_e^2)^3} \, . \tag{3.41}$$

The k-integration produces a result of the form $\mathcal{M}/m_e$, which is a contribution to the anomalous magnetic moment. Other contributions arise in a similar manner.

IV. FORMULATION OF AN EQUATION FOR BOUND STATE QED CALCULATIONS

A. General Organization of Bound State Perturbation Theory

There exist many formulations of two-body bound state theory in the literature, mostly based on the Bethe-Salpeter approach. They all give equivalent results if treated to the same level of accuracy, but they differ in considerable detail in overall organization and ease of calculation. The one to be emphasized here seems especially adapted to the situation where there is a large mass ratio, but it can be used in the equal mass case to quite refined levels of contribution. Although much of the material to be presented below is "common knowledge" in the field, it seems important to include it in order to make precise the details of the calculation. We refer to the bound system as hydrogen, but the methods are equally applicable to muonium as well and similar methods can be applied to positronium although some of the details would be different. The actual hydrogen system has additional complications because of nucleon structure, but these are generally quite small and are easily isolated. Throughout, we try to point out which results are also valid for positronium. Our discussion is based on that of Bodwin, Yennie, and Gregorio;[74] since that reference is referred to frequently, we cite it as BYG. In this chapter, we omit the factor of Z associated with exchanged photons; it can easily be restored later.

---

[i]To see this, note that if the operator in the numerator is commuted through $\vec{p}$, it gives zero acting on the wave function. Thus it produces an additional power of the field strength through the commutator. The expected order, provided both wave functions converge separately is $\alpha^3 \gamma'^3 / m_e^2$.

1. The Integral Equation

We wish to choose an unperturbed problem that includes the basic non-relativistic physics of the Schrödinger equation and as much of the relativistic physics as is feasible. There are many ways to accomplish this; all of them make use of the known solutions of the Schrödinger or Dirac equations with a given time-independent external potential. The method developed here is specialized to quantum electrodynamics, but some of the techniques should have a more general utility (for example, to quark-pair bound states). There are many ways to define the unperturbed problem; different choices simply produce different correction kernels. The trick is to make a choice that yields a fairly simple unperturbed problem *and* limits the number and complexity of the perturbation kernels required. Furthermore, one wants the perturbations to yield truly small effects so that one can get reasonable results in finite order. For example, even though one particle mass may be much larger than the other one, it would be poor strategy to start with the infinite mass limit. One should be able to get most of the reduced mass effects at the unperturbed level. This turns out to be a somewhat subtle problem, and in fact there is no precise distinction between reduced mass and genuine dynamical recoil effects. Nevertheless, we will be able to obtain much of the mass dependence exactly, and most of our recoil corrections *look* dynamical.

Following the ideas of Salpeter and Bethe,[10] we arrive at the structure of our unperturbed problem by studying the two-particle four-point function. Schematically, the problem may be described as follows. The four point function G describing the scattering of an electron and a proton satisfies a four-dimensional integral equation; we write this down first in unrenormalized form

$$G_{\rm unr} = S_{\rm unr} + S_{\rm unr} K_{\rm unr} G_{\rm unr} \ , \tag{4.1a}$$

where $S_{\rm unr}$ is the product of two free unrenormalized propagators and $K_{\rm unr}$ is the complete unrenormalized two particle irreducible kernel. The renormalization works exactly as in the usual case. The charge renormalization associated with the photon propagator will not be shown explicitly, but the spurious charge renormalizations will. For example, the propagator product is

$$S_{\rm unr} = \frac{S_{\rm ren}}{Z_e Z_p} \ , \tag{4.1b}$$

where Z_e and Z_p are associated with the electron and proton free propagators, respectively. At the same time, as a consequence of the Ward identity, the kernels satisfy

$$K_{\rm unr} = Z_e Z_p K_{\rm ren} \ . \tag{4.1c}$$

Accordingly, the unrenormalized four point function is related to the renormalized one by

$$G_{\rm unr} = \frac{G_{\rm ren}}{Z_e Z_p} \ , \tag{4.1d}$$

which satisfies the integral equation

$$G_{\rm ren} = S_{\rm ren} + S_{\rm ren} K_{\rm ren} G_{\rm ren} \ . \tag{4.1e}$$

Clearly, the extraction of the overall multiplicative factor cannot affect the positions of the poles. Rather than use complete renormalized propagators in the development of our reference equation, we may incorporate them into perturbative kernels as described below.

Initially, we replace S_{ren} by S, which is the product of two free propagators; in momentum space, it is given by

$$S = \frac{1}{\not{P}_e - m_e} \otimes \frac{1}{\not{P}_p - m_p} \tag{4.2a}$$

where, in the center-of-mass frame,

$$(P_e + P_p)_\mu = E g_{\mu 0} \ .$$

Here E is the total energy of the system. It is convenient to split off a fixed energy piece of each four-momentum and write

$$P_{e\mu} = E' g_{\mu 0} + p_\mu \ , \quad P_{p\mu} = E'' g_{\mu 0} - p_\mu \ . \tag{4.2b}$$

A particularly convenient separation of E into E' and E'' is given below. At this point, the treatment is quite symmetric between the two particles so it would be equally applicable to positronium, provided annihilation diagrams are not to be incorporated into the analysis. If such diagrams are to be incorporated, the sign of P_p should be reversed in (4.2a). This does not affect the discussion of the denominator rearrangement below, but some numerators are changed. Rather than develop a universal notation which applies to all cases, we simply mention the differences for positronium as they occur.

Having split off the radiative corrections contained in S_{ren}, we are first going to study the integral equation

$$G_0 = S + SKG_0 \tag{4.3a}$$

(for notational simplicity, from here on we omit the subscript "ren" on the functions K and G). Writing

$$S_{\text{ren}} = S + \delta S \tag{4.3b}$$

we see that the integral equation for G becomes

$$G = G_0 + \delta S + \delta SKG + G_0 K \delta S + G_0 K \delta SKG \ . \tag{4.3c}$$

The implications of this can be better understood after the development of perturbation theory.

The objective is to find the positions of the poles of G (at first G_0) as a function of the total energy of the four-point function. These give us the energy (mass) eigenvalues of the two-particle system. Usually we are not interested in the wave function, except in some simplified approximation. Thus it is not useful to continue with the analysis of Salpeter and Bethe,[10] which leads to a four-dimensional integral equation for the wave function. In brief, they showed that at the pole, the inhomogeneous term becomes negligible and that G may be replaced by a four-dimensional wave function. We study instead the perturbation theory of the four-point function about some known reference. In fact, it will become clear from our work that the Bethe-Salpeter integral equation separates effects which should be treated together. For example, it is necessary to treat together contributions arising from iterations of the kernels of the Bethe-Salpeter equation and those from higher order kernels. If that is not done, spurious non-recoil effects occur, which cancel only when all contributions are added at the end. An example of this is given in Chapter IV-B(3) below. From either an analytic or a numerical point of view, it is undesirable to calculate spurious terms which are much larger than the ones of interest.

Before proceeding to the discussion of a particular method of calculation of bound state energies, let us present an overview of the general procedure first. As will be seen, there is considerable flexibility in how to proceed through a calculation, and we can compare several different approaches which could be used. A brief description of several of the better known approaches is given in Appendix C. Basically, we must find a reference equation which is easily solved and which incorporates as much of the dominant physics of the original equation as possible. The two known elements of (4.3a) are S and K (K is expressed as a series, of course), and we must make approximations for both of them in order to find a solvable equation. In effect, we have the freedom to choose both the free Hamiltonian and the potential of the reference problem. Thus we write

$$S \equiv N\bar{S} + R\ , \tag{4.4a}$$

where $\bar{S}$ is an effective propagator which must match exactly the behavior of S in the extreme non-relativistic limit. Namely, in that limit it should agree with the two-body Schrödinger propagator. In all practical cases, the relative energy p_0 of the two particles is fixed by some prescription. While the constant N is of course redundant, it usually occurs quite naturally, and it gives us some additional flexibility in putting formulas into a convenient form. (It is also possible for N to be a function of the momentum $|\vec{p}|$; parts of that function can then be attached to adjacent factors of the potential.) Away from the region of non-relativistic momentum, $\bar{S}$ may be chosen in various ways depending on circumstances. Of course, it is desirable to incorporate as much physics into its choice as is feasible.

An understanding of why it is possible to make a decomposition like (4.4a) is provided by studying the pole structure of the two propagators in (4.2a) as a function of p_0. We mention this now, and give the denominators explicitly in (4.9b) below. For non-relativistic spatial components of momentum, the positions of various poles are illustrated in Fig. 5. The dominant feature of the pole structure is that the electron propagator and the proton propagator have poles close to the origin and on opposite sides of the axis, nearly pinching the p_0-contour. Associated with this pinching region, one finds that the leading term in the numerator is the large component projector for each particle. Consequently, the behaviour of the dominant term from this region does agree with that of the Schrödinger equation for the large components. *Any* treatment must incorporate this dominant feature into the definition of $\bar{S}$, and various treatments differ primarily on how this is done and

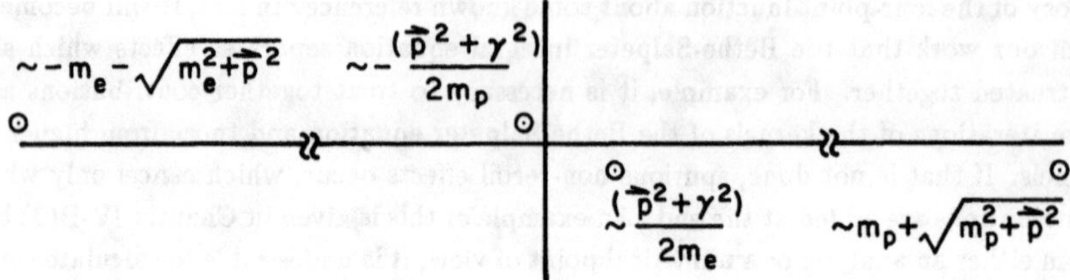

Figure 5: The location of the poles as a function of p_0 in the ladder configuration.

how the other poles are treated. To emphasize that p_0 is fixed at a small value, we write

$$\bar{S} = -2\pi i \delta(p_0)\bar{s}\ , \tag{4.4b}$$

where $\bar{s}$ is now a three-dimensional propagator. It should be emphasized that we do not obtain $\bar{s}$ directly from S by setting $p_0 = 0$; instead, $\bar{s}$ corresponds to an operator residue associated with the dominant pole structure.

In general, we represent (4.4a) diagrammatically as in Fig. 6. The heavy lines indicate the resulting three-dimensional propagator, and R represents the remainder, which is to

$$S = N\bar{S} + R$$

Figure 6: Graphical representation of the decomposition of the propagator product into a three-dimensional piece and a remainder.

be treated as a perturbation. The perturbation scheme for dealing with this separation is presented below.

Having settled on some choice of $\bar{S}$, we next turn our attention to K. Obviously, the most important term in K is the one photon exchange contribution (because of the small value of α). Also, because the numerators in $\bar{S}$ are close to large component projectors, the dominant part of the one photon exchange arises from the time indices of the propagator. In the Coulomb gauge, this dominant term is directly the Coulomb interaction. In a covariant gauge, the discussion of S shows that the dominant values of p_0 are much less than the values of spatial components of momentum. Accordingly, in the leading terms, we may neglect p_0 in the photon propagators, and they effectively become Coulomb interactions. Thus, in all cases, we should approximate K by something close to a Coulomb potential. It turns out that it is sometimes possible to pack a little extra physics into our

choice, so we simply call the approximate form $\mathcal{V}/N$. If we use this approximation and keep only the $N\bar{S}$ piece of S everywhere, we find that G is approximated by $N\bar{G}$ where $\bar{G}$ satisfies an integral equation

$$\bar{G} = \bar{S} + \bar{S}\mathcal{V}\bar{G} \; ; \tag{4.4c}$$

because of (4.4b), this is in reality a three-dimensional integral equation which has been "promoted" to a four-dimensional one. Since the solutions of the three-dimensional equation should be known in principle, this should provide a convenient starting point for the development of a perturbation procedure.

It is our expectation that the bound states of the complete problem are very close to those of the unperturbed problem. If there are other states which are not reached by a perturbation expansion, some other method of identifying them will be necessary. Hence, the next step is to develop an expansion of G in powers of $\bar{G}$, incorporating both effects due to the difference between (4.4c) and (4.3a) and terms arising from (4.3c). This differs from the usual type of expansion since both the free propagator and the interactions differ in all these equations. Intuitively, the result is very clear. First imagine iterating (4.3c) to arbitrarily large order, then in turn iterating the factors of G_0 it contains to arbitrarily high order. In this expansion, replace each factor of S using (4.4a). Between any two factors of $N\bar{S}$ replace K by its approximation $\mathcal{V}/N$ plus a correction $K - \mathcal{V}/N$. Sum up iterations of $\mathcal{V}$ using the integral equation for $\bar{G}$. Taking stock of the situation at this stage, we see that between any two factors of $N\bar{G}$ there now occurs a new kernel $\hat{K}$ defined by

$$\hat{K} = K - \mathcal{V}/N + K[R+\delta S]K + K[R+\delta S]K[R+\delta S]K + \dots \, . \tag{4.5a}$$

For sub-factors not occurring between factors of $\bar{G}$, it is convenient to define another new kernel

$$\tilde{K} = K + K[R+\delta S]K + K[R+\delta S]K[R+\delta S]K + \dots = \hat{K} + \mathcal{V}/N \, . \tag{4.5b}$$

Formally, the entire result may then be written

$$G = R + \delta S + [R+\delta S]\tilde{K}[R+\delta S] + [1 + [R+\delta S]\tilde{K}]N\hat{G}[1 + \tilde{K}[R+\delta S]] \, , \tag{4.5c}$$

where $\hat{G}$ satisfies the integral equation

$$\hat{G} = \bar{G} + \bar{G}N\hat{K}\hat{G} \, . \tag{4.5d}$$

This result can be verified by simple formal manipulations which we need not detail here.

An important consideration in developing these procedures is to avoid defining kernels in such a way that their iterations become increasingly singular. As an example of how this type of difficulty could arise, recall that the dominant part of the hyperfine splitting interaction has the spatial dependence $\delta(\vec{r})$, provided one uses non-relativistic wave functions. If this interaction were to occur directly in K, it could not be iterated. In fact, the kernels as defined here produce no such problems when iterated. In practice, it is sometimes convenient to carry through the preceding analysis in a more refined way by splitting

K into parts which are treated differently in introducing the decomposition (4.4a). We do not introduce that complication at this point. For an example of this procedure, the reader may refer to BYG.

2. The Perturbation Expansion

As a preliminary to describing the perturbation expansion, we consider the three-dimensional Green's function $\bar{G}$. It is usually restricted to some subspace of the full 4×4 spinor space, for example by large component projectors, or by positive energy projectors (defined on a plane-wave basis). We do not wish to deal with these particular details at this point. Rather, we simply note that it is usually expressible in terms of eigenfunctions $|\bar{n}\rangle$ and $\langle\bar{n}|$ through an expression

$$\bar{G} = \sum_n \frac{|\bar{n}\rangle\langle\bar{n}|}{\mathcal{E} - \mathcal{E}_n} \,. \tag{4.6a}$$

The eigenfunctions are restricted to the appropriate subspace and have been "promoted" from three dimensions to four by appending a factor $-2\pi i\delta(p_0)$. $\mathcal{E}$ is a known function of the total energy E; similarly for $\mathcal{E}_n$. $\langle\bar{n}|$ is not necessarily the adjoint of $|\bar{n}\rangle$, but the two functions satisfy an orthonormality relation

$$\langle\bar{n}|L|\bar{n}'\rangle = \delta_{nn'} \,, \tag{4.6b}$$

where L is an appropriate matrix.

In order to describe the calculation of recoil effects precisely, we find it necessary to present the perturbation expansion in some detail. Although our discussion is in the context of the treatment of recoil, many of the details are valid also for fixed potential calculations such as the Lamb shift, as described in Chapter III. The discussion of various kernels is given later in this chapter. Here we describe how one finds a perturbation series for the energy levels. These are given by the poles of G, or equivalently, those of $\hat{G}$.[j] To find them, we consider the series obtained by iterating (4.5d) and substituting the eigenfunction expansion (4.6a). Having done this, we take the expectation value of $\hat{G}$ in the state of interest (labelled $\bar{0}$) to find

$$\begin{aligned}\langle\bar{0}|L\hat{G}L|\bar{0}\rangle &= \frac{1}{\mathcal{E} - \mathcal{E}_0} + \frac{\Sigma}{(\mathcal{E} - \mathcal{E}_0)^2} + \frac{\Sigma^2}{(\mathcal{E} - \mathcal{E}_0)^3} + \dots \\ &= \frac{1}{\mathcal{E} - \mathcal{E}_0 - \Sigma(\mathcal{E})} \,,\end{aligned} \tag{4.7a}$$

where

$$\Sigma(\mathcal{E}) = \langle\bar{0}|N\hat{K}|\bar{0}\rangle + \sum_{n\neq 0} \frac{\langle\bar{0}|N\hat{K}|\bar{n}\rangle\langle\bar{n}|N\hat{K}|\bar{0}\rangle}{\mathcal{E} - \mathcal{E}_n} + \dots \,. \tag{4.7b}$$

[j]The role of the exterior factors in (4.5c) is to build up the four-dimensional Bethe-Salpeter wave functions from the three-dimensional ones. A demonstration of this is given in an Appendix of BYG. These factors do not shift the positions of the poles. [In principle, they might produce entirely new poles which are outside a perturbative approach.]

Since a kernel typically produces a factor of α^4 or smaller, and may have a small mass ratio factor as well, terms which are not exhibited here are generally much too small ($\mathcal{O}(\alpha^8 m_e)$ or smaller) to require further consideration at the present time. By expanding about $\mathcal{E}_0$, we find that the pole in $\mathcal{E}$ is located at

$$\hat{\mathcal{E}}_0 = \mathcal{E}_0 + \Sigma(\mathcal{E}_0) + \Sigma(\mathcal{E}_0)\frac{\partial\Sigma(\mathcal{E}_0)}{\partial\mathcal{E}_0} + \dots . \tag{4.7c}$$

From this, the corrected energy can be found from the known relation between E and $\mathcal{E}$.

3. Two Choices of Bound State Formalisms

We now turn to a particular implementation of the procedure just described. The first step is to analyze the structure of the denominators. Later, we will specialize to two possibilites for the treatment of the numerators of the free propagators, giving us the Schrödinger equation and Dirac equation formulations.

It turns out to be convenient to choose the separation of E into E' and E'' so that

$$E'^2 - m_e^2 = E''^2 - m_p^2 = -\gamma^2 . \tag{4.8a}$$

This is motivated by the fact that in a real scattering process the right hand side would be the incoming $\vec{p}^{\,2}$ in the center of mass. Any other choice would lead to the same final result, of course, but intermediate expressions would be a bit more complicated. Together, these yield

$$\begin{aligned} E' &= \frac{E^2 - m_p^2 + m_e^2}{2E} \\ E'' &= \frac{E^2 - m_e^2 + m_p^2}{2E} . \end{aligned} \tag{4.8b}$$

For QED bound states, it turns out that $\gamma = \mathcal{O}(\alpha m_r)$, so that

$$\begin{aligned} E &= [m_e^2 - \gamma^2]^{1/2} + [m_p^2 - \gamma^2]^{1/2} \\ &\approx m_e + m_p - \frac{\gamma^2}{2m_r} - \frac{\gamma^4}{8}\left[\frac{1}{m_e^3} + \frac{1}{m_p^3}\right] , \end{aligned} \tag{4.8c}$$

where $m_r = m_e m_p/(m_e + m_p)$ is the reduced mass. Final results cannot depend on the momentum routing, of course, but the definition of the unperturbed propagator does.

With the choice (4.8a), (4.2a) takes the form

$$\frac{(m_e + E'\gamma_0 + \not{p})}{D_e(p)} \otimes \frac{(m_p + E''\gamma_0 - \not{p})}{D_p(-p)} , \tag{4.9a}$$

where

$$\begin{aligned} D_e(\pm p) &\equiv p^2 - \gamma^2 \pm 2E'p_0 + i\epsilon \\ D_p(\pm p) &\equiv p^2 - \gamma^2 \pm 2E''p_0 + i\epsilon . \end{aligned} \tag{4.9b}$$

These denominators reveal the near pinching of the p_0-contour of integration referred to earlier and illustrated in Fig. 5: there are poles in (4.9a) at $p_0 \approx (\vec{p}^{\,2} + \gamma^2)/2E' - i\epsilon$ and

$p_0 \approx -(\vec{p}^2 + \gamma^2)/2E'' + i\epsilon$. One way to bring out the large contribution which results is to rearrange the product of denominators

$$\frac{1}{D_e(p)D_p(-p)} = \frac{1}{2E(-p_0 + i\epsilon)} \left\{ \frac{1}{D_e(p)} - \frac{1}{D_p(-p)} \right\} .$$

Note that while the complete expression on the rhs of the equation has no actual pole at $p_0 = 0$, it is convenient to introduce an imaginary part into the denominator to facilitate separating the two terms. Further rearrangement gives

$$\frac{1}{D_e(p)D_p(-p)} = \frac{-2\pi i \delta(p_0)}{-2E(\vec{p}^2 + \gamma^2)} - \frac{1}{2E} \left\{ \frac{1}{(p_0 + i\epsilon)D_e(p)} + \frac{1}{(-p_0 + i\epsilon)D_p(-p)} \right\} . \tag{4.9c}$$

As will become apparent later, the presence of the factor $(2E)^{-1}$ in the δ-function term corresponds to the correct treatment of the reduced mass in the non-relativistic region. The δ-function term of (4.9c) is to be incorporated into the definition of $N\bar{S}$. The remaining terms are part of the definition of R; note that they do not produce a pinch of the contour. We should call attention to a possible problem which could occur if we do not use this result carefully. Term by term, the new form has less convergence at high momenta than the original form. Thus, we should use this separation only when the resulting expressions remain well behaved.

Next consider the numerator structure of (4.9a). We can rewrite the two Dirac factors as

$$E''\gamma_0 + m_p - \not{p} = \tfrac{1}{2}(E'' + m_p)(1 + \gamma_0) + [-\tfrac{1}{2}(E'' - m_p)(1 - \gamma_0) - \not{p}] \tag{4.10a}$$

and

$$E'\gamma_0 + m_e + \not{p} = \tfrac{1}{2}(E' + m_e)(1 + \gamma_0) + [-\tfrac{1}{2}(E' - m_e)(1 - \gamma_0) + \not{p}] . \tag{4.10b}$$

As a matter of principle, it would be desirable to keep these complete numerators in $N\bar{S}$ if the resulting equation were solvable. As noted in the Appendix C, the result is then actually very close to that obtained by rationalizing the Breit propagator. In fact, the latter contains a little more physics. The trouble with either of these choices is that in combination with the Coulomb potential they are too singular. In fact, for large $|\vec{p}|$, they approach a (direction-dependent) constant; when iterated with the Coulomb potential, one finds by simple power counting that the integrals become divergent in a non-renormalizable way. It turns out that we may keep one factor of $\not{p}$, but not two, without encountering this problem.[k] We proceed to possible simplifications which seem practical. The first terms on the right side of (4.10a) and (4.10b) dominate the non-relativistic region, particularly for the heavier particle. We may choose to perturb about one or both of these terms.

At this stage, both the Coulomb gauge and the Feynman gauge lead to the same reference equation in the ladder approximation. Because there is at least one large component

[k]In a recent paper, Malenfant[108] studies the eigenvalue problem for the Breit equation in the positive energy sector for both particles and finds acceptable wave functions. However, the eigenfunctions in the other sectors appear not to be acceptable, and the Breit Hamiltonian is therefore not self-adjoint.

projector on both sides of the γ-matrix in the interaction, only γ_0-terms survive. In the Coulomb gauge, the resulting interaction is directly the Coulomb potential (multiplied by $\gamma_0 \otimes \gamma_0$). For the Feynman gauge, any photon exchanged between the electron and the proton introduces a factor

$$Q_{ph} = e^2 \frac{\gamma_\mu \otimes \gamma^\mu}{q^2} \ . \tag{4.11}$$

This is often separated into a part Q_o with $\mu = 0$ (called an O photon) and a part Q_v with spatial indices (called a V photon). In the case of a ladder photon between two factors of $N\bar{S}$, only the O photon part survives and because of the two δ-functions, it is converted to a Coulomb potential.

Schrödinger Equation Model:

If we take the large component projectors in (4.10a) and (4.10b), we are led immediately to the Schrödinger equation, which is symmetric in the two particles.

$$S \rightarrow N_s \bar{S}_s = N_s \bar{s}_s(\vec{p})(-2\pi i)\delta(p_0) \ , \tag{4.12a}$$

where

$$N_s \equiv \frac{(E'' + m_p)(E' + m_e)}{4Em_r} \approx 1 - \frac{\gamma^2}{4}\left(\frac{1}{m_e} - \frac{1}{m_p}\right)^2 + \ldots \tag{4.12b}$$

and

$$\bar{s}_s(\vec{p}) \equiv \frac{1}{\mathcal{E} - \vec{p}^2/2m_r} \frac{(1+\gamma_0) \otimes (1+\gamma_0)}{4} \ . \tag{4.12c}$$

with

$$\mathcal{E} \equiv -\frac{\gamma^2}{2m_r} \ . \tag{4.12d}$$

The interpretation is obvious. This decomposition takes the form represented by Fig. 6a. Note that the factor N_s is very close to unity, except for very small binding corrections. This small correction should be separated off and treated as a small perturbation (to avoid having the potential term in the reference Hamiltonian dependent on the eigenvalue).

Clearly, this fits into the general scheme described in Chapter IV-A(1,2), with the present $\mathcal{E}$ agreeing with the one there. E may be expressed in terms of $\mathcal{E}$ through (4.8c). Terms not incorporated into $\bar{S}_s$ become part of R. Next, we must make a choice for $\mathcal{V}$. This should be done so as to cancel leading terms from K in (4.5a). Since between factors of $\bar{S}$ the one photon exchange simplifies to the Coulomb interaction, the simplest thing is to choose $\mathcal{V}$ to be the Coulomb potential V_c. Then we are left with an effective perturbation

$$\delta\mathcal{V}_s = N_s V_c - V_c = V_c(N_s - 1) \ , \tag{4.12e}$$

which is to be treated as part of $N\hat{K}$ in the perturbation expansion. With this choice of $\mathcal{V}$, the three-dimensional Schrödinger equation is

$$\left(\frac{\vec{p}^2}{2m_r} + V_c\right)|n\rangle = \mathcal{E}_n|n\rangle \ . \tag{4.13a}$$

The propagator is

$$\bar{S}_s = \frac{1}{4}(1+\gamma_0)\otimes(1+\gamma_0)\sum_n \frac{|\bar{n}\rangle\langle\bar{n}|}{\mathcal{E}-\mathcal{E}_n}\,, \tag{4.13b}$$

where $|\bar{n}\rangle$ represents $-2\pi i\delta(p_0)|n\rangle$.

The Dirac Equation Model:

We take the large component projector term of (4.10a) with the complete (4.10b) to obtain

$$S \to N_d\bar{S}_d = N_d\bar{s}_d(\vec{p})(-2\pi i)\delta(p_0)\,, \tag{4.14a}$$

where

$$N_d \equiv \frac{E''+m_p}{2E} \approx \frac{m_p}{M} \equiv N \tag{4.14b}$$

and

$$\bar{s}_d(\vec{p}) = \frac{\frac{1}{2}(1+\gamma_0^p)}{E'\gamma_0+\not{p}-m_e}\,. \tag{4.14c}$$

Here $M(=m_e+m_p)$ is the total mass. The normalizing factor N_d, which is approximately m_r/m_e, turns out to give the main reduced mass effects. The two-body propagator $\bar{s}_d$ is simply the large component projector for the proton times the Dirac propagator for the electron.

The most obvious choice for $\mathcal{V}$ is $\gamma_0 V_c N$. This is easy to implement and it leads directly to a Dirac equation in which the fine structure constant has been multiplied by m_p/M. It would produce the correct reduced mass dependence of the Bohr energy levels. However, it turns out that we can include a little more physics with a slightly different choice

$$\mathcal{V} \equiv \gamma_0 V_c\left\{1-\frac{\gamma_0 m_e}{M}\right\}\,. \tag{4.14d}$$

For the large components, this choice agrees with the obvious one, but it also includes some effects arising from the exchange of a V photon. Part of the perturbation in $N\hat{K}$ is then

$$N_d\gamma_0 V_c - \mathcal{V} = \gamma_0 V_c' + \delta\mathcal{V}_d\,, \tag{4.14e}$$

where

$$\begin{aligned} V_c' &= V_c\left(\frac{-E'+\gamma_0 m_e}{E}\right) \\ \delta\mathcal{V}_d &= \gamma_0 V_c\left[\frac{m_p-E''}{2E}+\gamma_0 m_e\frac{E-M}{ME}\right] \approx \gamma_0 V_c\frac{\gamma^2}{4m_pM}(1-2\gamma_0)\,. \end{aligned} \tag{4.14f}$$

As will be seen later, the V_c' term is cancelled by one arising from a correction kernel. The $\delta\mathcal{V}_d$ term contributes a term of order $\alpha^4 m_r^3/m_p^2$, and we include it below.

The Dirac equation is

$$(\vec{\gamma}\cdot\vec{p}+\mathcal{V}+m_e)|n\rangle = E_n'\gamma_0|n\rangle\,. \tag{4.15a}$$

This equation can be solved in terms of the known solutions of the usual Dirac-Coulomb problem; this is described below. The corresponding integral equation for bound states is

$$|n\rangle = \bar{s}(E_n')\mathcal{V}|n\rangle \; . \tag{4.15b}$$

The propagator is given by

$$\bar{S}_d = \frac{1}{2}(1+\gamma_0^p)\sum_n \frac{|\bar{n}\rangle\langle\bar{n}|}{E'-E_n'} \tag{4.15c}$$

In our analysis, E' plays the role of the energy to be calculated. I.e., it replaces $\mathcal{E}$ in the discussion of Chapter IV-A(2). A small shift $\Delta E'$ produces a shift in E given by[l]

$$\Delta E = \Delta E' \frac{E}{E''} \; . \tag{4.15d}$$

There are two easily identifiable perturbations on (4.14a). One is the second term of (4.10a) with the original form of (4.9c) (left-hand side). The other is the first term of (4.10a) with the second term of (4.9c). The first of these perturbations is a recoil correction (vanishes as $m_p \to \infty$); the second is not. However, as we show later, the non-recoil part of the perturbation is straightforwardly compensated and so it need not be calculated.

Following the procedure of Grotch and Yennie,[79] we can rearrange (4.15a) into the form of the usual Coulomb-Dirac equation with certain modified parameters:

$$\begin{aligned} \bar{\alpha} &\equiv \alpha\left[1-\frac{m_e^2}{M^2}\right]^{\frac{1}{2}} \\ \bar{m} &\equiv m_e \frac{1-\dfrac{E_n'}{M}}{\left[1-\dfrac{m_e^2}{M^2}\right]^{\frac{1}{2}}} \\ \bar{E}_n' &= \frac{E_n'-\dfrac{m_e^2}{M}}{\left[1-\dfrac{m_e^2}{M^2}\right]^{\frac{1}{2}}} = \bar{m}f(n,j,\bar{\alpha}) \end{aligned} \tag{4.16}$$

where the $f(n,j,\bar{\alpha})$ are just the functions defined in (2.2b). Now the eigenfunctions are

$$\psi_n^{jm} = \left[\frac{1+\gamma_0\dfrac{m_e}{M}}{1+f(n,j,\bar{\alpha})\dfrac{m_e}{M}}\right]^{\frac{1}{2}} \chi_n^{jm}(\bar{m}\bar{\alpha}\vec{r},\bar{\alpha}) \; , \tag{4.17a}$$

[l]The reader may be concerned that this linearized approximation may not be adequate for all situations. It is exact for positronium, and the error made for systems with a large mass ratio is approximately $(\Delta E)^2/(2E'')$, which is several orders of magnitude smaller than any contributions to be considered in the forseeable future.

where χ_n^{jm} are the usual Dirac-Coulomb eigenfunctions and the eigenvalues are given by

$$E'_{nj} = m_e \frac{f(n,j,\bar{\alpha}) + \frac{m_e}{M}}{1 + f(n,j,\bar{\alpha})\frac{m_e}{M}} . \tag{4.17b}$$

Clearly, at this stage the Dirac degeneracy is valid independently of the mass. To simplify some intermediate algebra, it is convenient to rewrite $f(n,j,\bar{\alpha})$ as

$$f = 1 + \left(1 - \frac{m_e^2}{M^2}\right) g . \tag{4.18a}$$

This gives us

$$E'_{nj} \approx m_e \left[1 + \frac{m_p^2}{M^2} g - \frac{m_e m_p^3}{M^4} g^2\right] . \tag{4.18b}$$

To obtain E'', we use (4.8a) and find

$$E''_{nj} \approx m_p + \frac{m_r^2}{m_p} g - \left[\frac{m_r^3}{M m_p} - \frac{m_r^2 m_p}{2M^2} + \frac{m_r^4}{2m_p^3}\right] g^2 . \tag{4.18c}$$

Finally, then, E takes the form

$$E_{nj} \approx m_e + m_p + m_r g - \frac{m_r^2}{2m_p} g^2 . \tag{4.18d}$$

This is an ingredient in our evaluation of the fine structure to order α^4 in Chapter IV-B(4) below.

B. Rearrangements of the Kernels

1. Appending small proton components to the wave function

Before proceeding to a discussion of some particular cases, we describe a useful rearrangement of the kernels when they are treated in lowest order in the Dirac equation model. The following discussion is not always helpful in discussing the sum over states contributions since the rearrangement artificially makes separate terms too singular (i.e., it may lead to divergent expressions which ultimately cancel).

Our original prescription appears somewhat unnatural in the way that the effects of the proton's small components are introduced through perturbation kernels. As emphasized in the discussion of the Dirac equation model, either particle's Dirac properties can be incorporated fully into the reference wave function, but we cannot include those effects for both particles simultaneously. In the discussion leading to (4.15a), we were able to incorporate some effects associated with the proton's convection current into the potential occurring in the basic equation. This is verified below. We now aim to justify that we can incorporate the small proton components in the wave function by introducing factors multiplying the wave functions. For example, for the initial wave function (where the electron has momentum $\vec{p}_i$), we insert the factor

$$\left[1 - \frac{\not{p}_i}{m_p + E''}\right] + \left(\frac{\not{p}_i}{m_p + E''}\right) . \tag{4.19a}$$

This, of course, does not change anything. However, the term in square brackets does incorporate effects associated with the proton's small components. The term in parentheses can be separated off and later be combined with a higher order kernel to compensate certain contributions which arise there. We may think of the proton's small components as having been "borrowed" from the higher order kernel. The similar factor which may be introduced to multiply the final wave function (electron momentum $\vec{p}_f$) is

$$\left[1 - \frac{\not{p}_f}{m_p + E''}\right] + \left(\frac{\not{p}_f}{m_p + E''}\right) . \tag{4.19b}$$

We always have the option of appending these factors, provided we make the proper corrections for doing so.

The simplest example is the one photon kernel $\hat{K}_1$ for which we now sum over four polarizations. The collection of terms from inserting the factors (4.19) may be rearranged (schematically)

$$[\ ]\hat{K}_1[\ ] + [\ ]\hat{K}_1() + ()\hat{K}_1[\ ] + ()\hat{K}_1() \tag{4.20}$$

The first of these terms is studied in the following subsection. Contributions with small components only on one side of the photon interaction yield the dominant hyperfine splitting interaction and the convection current interaction. Certain of the contributions with small components on both sides are too singular, and they may be explicitly cancelled in all the terms at this point. Others, which are less singular, are more conveniently treated as indicated here. We hope that the reader is not overly disconcerted by the seeming indefiniteness in the way various terms may be organized. We have gone through a certain amount of trial and error to discover a felicitous procedure.

The second and third terms of (4.20) will now be seen to compensate certain contributions from $\hat{K}_1 R \hat{K}_1$. Recall that the definition of $N\bar{S}$ included the first term of (4.10a) with the $-2\pi i\delta$-part of the denominator rearrangement. Part of R is associated with the second term of (4.10a) with the complete (4.9c), but here we consider only the δ function part. Using $E'' - m_p = -\gamma^2/(E'' + m_p)$, we see that the piece with a factor of $1 - \gamma_0$ gives a very small contribution since its coefficient is small and also both μ and ν must be spatial. For the fine structure, it is totally negligible (of order $\alpha^6 m_e^3/m_p^2$). It is the same order for the hfs and is incorporated in the final result. For the $-\not{p}$-term, we rearrange the proton factor, taking into account the large component projectors on each side before inserting (4.19), and find

$$\frac{\gamma_\mu[-\not{p}]\gamma_\nu}{2E} \doteq \frac{-g_{\mu 0}\not{p}\gamma_\nu - \gamma_\mu \not{p} g_{\nu 0}}{m_p + E''} N_d . \tag{4.21}$$

Now it is easy to see the partial compensation between these terms and the second and third terms of (4.20). Consider the term with the factor $g_{\nu 0}$. To make the comparison, we note that because of the delta functions the photon propagator with index ν turns into a Coulomb interaction; we incorporate this explicitly along with the γ_0 factor for the electron to give

$$-[\ ]\frac{\gamma_\mu \not{p}}{m_p + E''}\gamma_0^e N_d V_c - ()\frac{\gamma_\mu \not{p}}{m_p + E''}\gamma_0^e N_d V_c , \tag{4.22a}$$

where we have indicated the insertion of the decomposition (4.19b). To put the second term of (4.20) on the same footing, we use the integral equation and relabel the momenta to produce the term

$$[\,]\frac{\gamma_\mu \not{p}}{m_p + E''}\mathcal{V}\,. \tag{4.22b}$$

Other factors in (4.22a) and (4.22b) which are not shown explicitly match each other. From (4.14e), we see that (4.22b) nearly cancels the first term of (4.22a), leaving some small contributions. The one associated with δV_d gives contributions of relative order $\alpha^2 m_e^2/m_p^2$ so it should be negligible for present purposes. The V_c'-term remaining after this cancellation is sometimes cancelled by other contributions; we do not discuss it further here. Note that this result, using the integral equation satisfied by the reference wave function, is valid only for the ground state.

The last term of (4.20) also combines with pieces of higher order kernels to produce cancellations. There are two such contributions like the second term of (4.22a) which are of the opposite sign. The other is the kernel with the structure $\hat{K}_1 R \hat{K}_1 R \hat{K}_1$, which contains a piece with the same sign. Having indicated the general approach, in the following we assume these cancellations.

2. The One-Photon Kernel

First we should discuss the choice of gauge. In the past, it was natural to treat the exchanged photons in the Coulomb gauge and the radiative correction photons in the Feynman gauge. In hydrogen-like systems, two different gauges could be used because the two types of photons are separately gauge invariant. However, in positronium that is not possible because of the annihilation diagrams which link them. In Chapter IV-A, we saw that there is no difference between these two gauges in setting up the bound state problem. The same is true of any covariant gauge, such as the Fried-Yennie gauge. In the study of recoil corrections to the hfs by BYG, it turned out that although the problem was initially set up in the Coulomb gauge, it was greatly simplified by transforming the kernels to the Feynman gauge. In that case, some awkward contributions simply cancelled without the need of being computed. Here we remain neutral about the choice of gauge as long as possible, but illustrate how the two gauges work for the simplest contributions.

The one-photon kernel, including the related $\mathcal{V}$ subtraction is given by

$$N_d \hat{K}(\text{one-photon}) = e^2 \frac{m_p + E''}{2E} \frac{\gamma_\mu \otimes \gamma_\nu G^{\mu\nu}}{q^2} - \mathcal{V}\,, \tag{4.23a}$$

where $G^{\mu\nu}$ defines the gauge of the virtual photon, and $q = p_i - p_f$ is the four-momentum transfer by the photon (in this case, with time component zero). In the Feynman gauge $G^{\mu\nu} = g^{\mu\nu}$, while in the Coulomb gauge, it is

$$G^{\mu\nu} = g^{\mu\nu} + \frac{q^\mu q^\nu - q^\nu q^0 g^{\mu 0} - q^\mu q^0 g^{\nu 0}}{\vec{q}^{\,2}}\,. \tag{4.23b}$$

To evaluate the expectation value of this kernel, we first include the appended proton small components. Then, using the fact that the wave function has large component projectors on both sides of this expression, the proton factor may be written

$$\left[1-\frac{\not p_f}{m_p+E''}\right]\gamma_\nu\left[1-\frac{\not p_i}{m_p+E''}\right]\doteq g_{\nu 0}-\frac{(p_f+p_i)_\nu}{m_p+E''}-\frac{1}{2}\frac{[\not p_f-\not p_i,\gamma_\nu]}{m_p+E''}-\frac{g_{\nu 0}\not p_f\not p_i}{(m_p+E'')^2}\,. \tag{4.24}$$

Of course, ν must be spatial in the second and third terms on the right hand side.

The first term yields the contribution $N_d\gamma_0 V_c$ of (4.14e). The second term represents the convection current of the proton. Its ultimate contribution depends on the choice of gauge. Starting with the Feynman gauge, we find an effective electron interaction

$$\frac{-\not p_f-\not p_i}{2E}V_c=\frac{[m_e-E'\gamma_0-\not p_f]+[m_e-E'\gamma_0-\not p_i]}{2E}V_c-\frac{m_e-E'\gamma_0}{E}V_c\,. \tag{4.25a}$$

The last term here cancels the $\gamma_0 V_c'$ term of (4.14e), as anticipated. The first term can be simplified using the Dirac equation; it gives the effective contribution to $N\hat{K}$

$$-2\frac{V_c\mathcal{V}}{2E}=-2\gamma_0\frac{V_c^2}{2E}(1-\gamma_0\frac{m_e}{M})\,. \tag{4.25b}$$

This term by itself would give a sizable contribution to the Lamb shift, of order $\alpha^4 m_e^2/M$. As is seen below, this leading order piece is precisely cancelled by other terms of the same size.

Next consider the convection term in the Coulomb gauge. To the contributions discussed in the preceding paragraph must be added one from the second piece of (4.23b). This gives an effective electron interaction

$$\frac{(\vec{p}_f{}^2-\vec{p}_i{}^2)(\not p_i-\not p_f)\frac{1}{2}W_c}{2E}=\frac{\frac{1}{2}[\vec{p}^2,W_c]([\not p_i-m_e+E'\gamma_0]-[\not p_f-m_e+E'\gamma_0])}{2E}\,, \tag{4.25c}$$

where in momentum space $\frac{1}{2}W_c$ is $e^2/(\vec{q}^2)^2$ and in coordinate space, W_c is $-\alpha r$.[m] Again with the use of the Dirac equation, this reduces to

$$-\frac{1}{4E}[\mathcal{V},[\vec{p}^2,W_c]]\doteq\gamma_0\frac{V_c^2}{2E}(1-\gamma_0\frac{m_e}{M})\,. \tag{4.25d}$$

This cancels one-half of (4.25b). The difference between the results in the two gauges is compensated by a difference in the two-photon kernel calculation below.

There are some other interesting contributions in (4.24), one of which we shall discuss here. The third term, which is not gauge sensitive, gives the leading contribution to the hyperfine splitting. Now we are able to get the correct reduced mass dependence of that effect. The proton's anomalous moment can also be incorporated at this point.[n] The net

[m]There is no problem in the Fourier transform of such a singular function, provided one takes into account the effect of the factors of $\vec{q}$ in the numerator. The nervous reader may wish to study this by considering $\lim_{\epsilon\to 0}1/(\vec{q}^2+\epsilon^2)^2$.

[n]This arises from the one photon exchange kernel in which the complete proton vertex is approximated by the sum of γ_μ and an anomalous moment interaction whose internal structure is suppressed. The first term is already incorporated in the basic equation, and the second is taken into account here. Corrections to this approximation are described in Chapter II-C.

effect is to multiply the hfs term from (4.24) by $1+\kappa$, where κ gives the anomalous part of the proton's magnetic moment. In momentum space, the hfs part of the one-photon kernel reduces to

$$N_d \hat{K}(\text{hfs}) \doteq \frac{4\pi\alpha(1+\kappa)}{2E} \frac{-i\vec{\sigma}_p \times \vec{q} \cdot \vec{\gamma}_e}{\vec{q}^{\,2}} , \tag{4.26a}$$

where $\vec{q} = \vec{p}_i - \vec{p}_f$. The expectation value of this kernel is to be evaluated using the Grotch-Yennie equation wave functions described in Chapter IV-A(3). Note that $\bar{\alpha}\bar{m}_e \approx \alpha m_r$; the error made in the approximation is of relative order $\alpha^2 m_e/m_p$ and is important only as a correction to the leading term. It need not be considered with terms which are intrinsically smaller.

For S-states, we find that the first order contribution of (4.26a) to the hfs is given by

$$\Delta E(\text{leading-hfs}) = \frac{4\pi\alpha}{2m_p}(1+\kappa)\langle\frac{-i\vec{\sigma}_p \times \vec{q} \cdot \vec{\gamma}_e}{\vec{q}^{\,2}}\rangle , \tag{4.26b}$$

where we use $E'' + m_p \approx 2m_p$, with an error of relative order γ^2/m_p^2, which is not presently of interest for hydrogen, but would be for positronium. Here the expectation value is to be taken with $\psi_{n(l=0)}$ given by (4.17a). Except for the small modifications of the wave function, this is precisely the calculation done by Breit.[6] As an example, for the ground state it gives the result for the hfs[74]

$$\Delta E(\text{leading-hfs}) = E_F(1+\kappa)\left(1 + \frac{3}{2}\alpha^2 + \frac{3}{2}\frac{\gamma^2}{m_e m_p}\right) , \tag{4.26c}$$

where E_F is given by (2.24), and terms of relative order α^4 and $\alpha^2(m_e/m_p)^2$ have been dropped; usually the factor of $(1+\kappa)$ is incorporated in the definition of E_F in the case of hydrogen. Notice the presence of a contribution of relative order $\gamma^2/m_e m_p$. Its coefficient is dependent on our particular choice of reference equation, but of course the complete final result is not.

Incidentally, most earlier treatments of the relativistic two-body problem do not connect with the Dirac equation in a simple way. This means that the $\frac{3}{2}\alpha^2$ term in (3.15b), which arises from the structure of the Dirac wave function (non-integer exponent in the radial dependence), is difficult to obtain. In fact, it appears as a second order (in $\hat{K}$) contribution in other treatments. To our knowledge, the present treatment (from BYG) is the only unified discussion of $\mathcal{O}(\alpha^2)$ recoil *and* non-recoil contributions.

We can indicate here how the radiative corrections can be incorporated into this result. The electron's anomalous moment interaction has a different structure from (4.26a), but its leading contribution can be determined by taking the non-relativistic approximation for the large components of the wave function. This effect can be incorporated by adding the electron anomaly a_e to the parenthesis of (4.26c). Beyond that, the dynamical structure of the magnetic moment operator and other radiative corrections to the electron and photon lines explore the relativistic momentum range for the electron. These add terms of order α^2 to the parenthesis. The work of Sapirstein, *et al.*[59] justifies the reduced mass dependence for these terms.

Because of the two powers of proton mass in the denominator, the last term of (4.24) is very small, provided it is not singular. In fact, there is a spin-independent piece which is well-behaved; its contribution of order α^4 is included with the discussion of those terms below. A spin-dependent piece would contribute a recoil correction to the hfs, but it is too singular so we let those terms cancel within the one-photon kernel as described below (4.20). Contributions of this nature are to be found in higher order kernels where they are treated without separating the $-2\pi i\delta$ parts from other terms. This exhausts the discussions of the one photon kernel.

3. Two-Photon Kernels

In discussing the two photon kernels, we shall concentrate our attention here on contributions which enter in order α^4. However, as we go along, we describe the order of magnitude of other terms and the nature of their contribution. There are two such kernels, which are referred to as ladder and crossed graphs. The ladder contribution is

$$N_d\hat{K}(\text{ladder}) = N_d\hat{K}_1 R\hat{K}_1. \tag{4.27a}$$

Let us recall briefly the meaning of R in the Dirac model. It is defined by taking the first term of (4.10a) with the part of (4.9c) *not* containing $-2\pi i\delta$ plus the second term of (4.10a) with the complete (4.9c). However, in the discussion of (4.21), we have already disposed of the $-2\pi i\delta$-piece of the second term. Hence we are left with the complete (4.10a) with the second part of (4.9c). We append to this the small proton components using the first terms of (4.19). The corrections from the second terms of (4.19) are partially cancelled by pieces of the three photon kernels in a manner similar to the discussion of Chapter IV-B(1); we do not pursue this point here. The resulting proton factor is

$$\left[1 - \frac{\not{p}_f}{m_p + E''}\right]\gamma_\mu \mathrm{R}\left\{\frac{(m_p + E''\gamma_0 - \not{p})}{D_p(-p)}\right\}\gamma_\nu\left[1 - \frac{\not{p}_i}{m_p + E''}\right]. \tag{4.27b}$$

Here the R associated with the middle factor means that the $-2\pi i\delta$-terms are to be dropped as explained above. This corresponds to the substitution

$$\frac{1}{D_p(-p)} \to \frac{1}{2E}\left\{-\frac{1}{(p_0 + i\epsilon)} - \frac{D_e(p)}{(-p_0 + i\epsilon)D_p(-p)}\right\}. \tag{4.27c}$$

Next we rearrange the numerator of this expression in powers of $m_p + E''$ to find (note that $m_p - E'' = \gamma^2/(m_p + E'')$)

$$\begin{aligned}(m_p + E'')g_{\mu 0}g_{\nu 0} &- \{(p_f + p)_\mu + \tfrac{1}{2}[\not{p}_f - \not{p}, \gamma_\mu]\}g_{\nu 0} - \{(p + p_i)_\nu + \tfrac{1}{2}[\not{p} - \not{p}_i, \gamma_\nu]\}g_{\mu 0} \\ &+ p_0\gamma_\mu\gamma_\nu + \frac{\not{p}_f\gamma_\mu(1 + \gamma_0)\gamma_\nu\not{p}_i + 2\not{p}_f\gamma_\mu\not{p}\gamma_\nu + 2\gamma_\mu\not{p}\gamma_\nu\not{p}_i + \gamma^2\gamma_\mu(1 - \gamma_0)\gamma_\nu}{2(m_p + E'')} \\ &+ \frac{\gamma^2[g_{\mu 0}\not{p}_f\gamma_\nu + g_{\nu 0}\gamma_\mu\not{p}_i]}{(m_p + E'')^2} + \frac{\gamma^2 g_{\mu 0}g_{\nu 0}\not{p}_f\not{p}_i}{(m_p + E'')^3}.\end{aligned} \tag{4.27d}$$

Before proceeding to the other terms, we display the similar factor for the crossed graph. Including the small proton components, the complete proton factor is

$$\left[1-\frac{\not p_f}{m_p+E''}\right]\frac{\gamma_\nu(m_p+E''\gamma_0+\not p-\not p_i-\not p_f)\gamma_\mu}{D_p(p-p_i-p_f)}\left[1-\frac{\not p_i}{m_p+E''}\right]. \tag{4.28a}$$

We rearrange the numerator in a similar manner showing only the terms through $1/(m_p+E'')$

$$\begin{aligned}(m_p+E'')g_{\mu0}g_{\nu0}&-\{(2p_f+p_i-p)_\nu+\frac{1}{2}[\not p-\not p_i,\gamma_\nu]\}g_{\mu0}\\&-\{(2p_i+p_f-p)_\mu+\frac{1}{2}[\not p_f-\not p,\gamma_\mu]\}g_{\nu0}-p_0\gamma_\nu\gamma_\mu\\&+\frac{\not p_f\gamma_\nu(1+\gamma_0)\gamma_\mu\not p_i-2\not p_f\gamma_\nu\not Q\gamma_\mu-2\gamma_\nu\not Q\gamma_\mu\not p_i+\gamma^2\gamma_\nu(1-\gamma_0)\gamma_\mu}{2(m_p+E'')},\end{aligned} \tag{4.28b}$$

where $Q=p-p_i-p_f$.

In order to combine the crossed graph contribution conveniently with the one from the ladder graph, we introduce a denominator rearrangement similar to (4.27c)

$$\begin{aligned}\frac{1}{D_p(p-p_i-p_f)}&=\frac{1}{2E(p_0+i\epsilon)}\left\{1-\frac{D_e(p_i+p_f-p)}{D_p(p-p_i-p_f)}\right\}\\&=\frac{1}{2E(p_0+i\epsilon)}\left\{1-\frac{D_e(p_i)+D_e(p_f)-D_e(p)+(p-p_i-p_f)^2+p^2-p_i^2-p_f^2}{D_p(p-p_i-p_f)}\right\}.\end{aligned} \tag{4.28c}$$

We refer to the $D_e(p_i)$ and $D_e(p_f)$ terms in the numerator of the last term in the braces as "artifact terms" because they have the appearance of two-loop contributions which are not actually present in the final calculation. Had we put the proton on mass shell in the definition of the unperturbed problem, they would not have appeared here. However, that would have led to far greater complications in that square roots (such as $(m_p^2+\vec{p}^{\,2})^{1/2}$) would have occurred. To study two-loop contributions properly, it is necessary to consider two-loop graphs from the beginning. If that is done, these artifact terms disappear; from now on, we drop them. A discussion of this type can be carried out to arbitrary order. The final form of this expression is then

$$\frac{1}{D_p(p-p_i-p_f)}\to\frac{1}{2E}\left\{\frac{1}{(p_0+i\epsilon)}-\frac{-D_e(p)+2(p-p_i)\cdot(p-p_f)}{(p_0+i\epsilon)D_p(p-p_i-p_f)}\right\}. \tag{4.28d}$$

The sum of the leading contributions from the ladder and crossed graphs is now

$$\begin{aligned}g_{\mu0}g_{\nu0}N_d\left\{-\frac{D_e(p)}{(-p_0+i\epsilon)D_p(-p)}+\frac{D_e(p)}{(p_0+i\epsilon)D_p(p-p_i-p_f)}\right.\\\left.-\frac{2(p-p_i)\cdot(p-p_f)}{(p_0+i\epsilon)D_p(p-p_i-p_f)}\right\}.\end{aligned} \tag{4.29}$$

Note that the non-recoil pieces have cancelled, as predicted above. Let us consider the terms with the factor of $D_e(p)$ in the numerator; this factor cancels the electron denominator, giving a particularly simple structure. While it would be possible to deal with these terms somewhat more exactly than is done here, we use a simplifying approximation, which is to keep only the term linear in p_0 in the D_p denominators. Corrections to this give extra factors of α. Then we find the combination

$$-\frac{1}{(-p_0+i\epsilon)^2}+\frac{1}{(p_0+i\epsilon)^2}\doteq 2\pi i\frac{\partial\delta(p_0)}{\partial p_0}\,. \tag{4.30}$$

Since the photon propagators are a function of p_0^2 in ether gauge, this contributes only in combination with the term linear in p_0 in the numerator of the electron factor. Approximating $(m_p+E'')/2E''$ by 1,[o] we find the effective contribution to $N\hat{K}$

$$\gamma_0\frac{V_c^2}{2E}N\,. \tag{4.31}$$

This should be compared with (4.25b). For the large components of the wave function, it is exactly half as large and of the opposite sign. For the small components of the wave function, it has two additional powers of α and is accordingly smaller than the terms we are studying in this section. Of course, such terms may be of interest for the Lamb shift.

We set aside the last term of the brace in (4.29) for the moment and consider some of the other terms in (4.27d) and (4.28b). To leading order in $1/M$, the terms which could give a contribution to the hfs compensate exactly as did the first terms of each of these expressions. This type of cancellation is true for all multi-photon kernels. It means that any corrections to the hfs will introduce additional factors of $1/M$ in a manner similar to the ones we have just been describing. It is helpful to combine the convection terms from the ladder and crossed graphs. These yield a part of order $1/M$

$$\begin{aligned}\frac{(p_f+p)_\mu g_{\nu 0}+(p+p_i)_\nu g_{\mu 0}}{2E(p_0+i\epsilon)}&+\frac{-(2p_i+p_f-p)_\mu g_{\nu 0}-(2p_f+p_i-p)_\nu g_{\mu 0}}{2E(p_0+i\epsilon)}\\&=\frac{2(p-p_i)_\mu g_{\nu 0}+2(p-p_f)_\nu g_{\mu 0}}{2E(p_0+i\epsilon)}\,.\end{aligned} \tag{4.32a}$$

Finally, the p_0 terms give the leading effect

$$-\frac{2g_{\mu\nu}}{2E}\,, \tag{4.32b}$$

which is just a seagull term.

Now let us put together the last term of (4.29), (4.32a), and (4.32b), making a simplifying approximation on the last term of (4.29). This gives

$$\frac{2(p_i-p)\cdot(p-p_f)g_{\mu 0}g_{\nu 0}}{2E(p_0+i\epsilon)^2}+\frac{2(p-p_i)_\mu g_{\nu 0}+2(p-p_f)_\nu g_{\mu 0}}{2E(p_0+i\epsilon)}-\frac{2g_{\mu\nu}}{2E}\,. \tag{4.32c}$$

[o]The error is of relative order $\alpha^2 m_r^2/m_p^2$, which is certainly unimportant for hydrogen-like atoms but should be taken into account for positronium.

It is interesting to note that a very similar structure will be obtained for the multi-photon contributions. It is useful to note that this structure is gauge invariant. For example, if we multiply it by $(p_i - p)^\nu$, the result is zero.

Before studying (4.32c), let us first describe some of the remaining terms from (4.27d) and (4.28b). It is clear that all of these lead to factors of order $1/M^2$ or higher. Also, upon examination they are found to produce at least five powers of α. Thus they can be ignored for the fine structure in hydrogen, but might require consideration for positronium fine structure. Terms of order $1/M^2$ are important for the hfs with up to six powers of α. Thus we may ignore the last two terms of (4.27d) and the similar terms of (4.28b) which have not been displayed explicitly. The hfs terms of order $1/M^2$ arise from the corrections to rearranging the denominators in the third and fifth terms of the two numerators and from their sixth terms and from those with the factor $1/(m_p + E'')$. We shall not discuss them further in this section.

Now let us study the contribution of (4.32c), starting in the Feynman gauge. It is useful to note a cancellation between the various terms in either gauge. The temporal part of the dot product in the first term added to the $\mu = \nu = 0$ part of the other terms gives 0. The spatial part of the dot product multiplies the electron factor

$$\frac{\gamma_0(m_e + \gamma_0 E' + \not{p})\gamma_0}{D_e(p)} \ . \tag{4.33a}$$

We note that the numerator term $\not{p}$ leads to contributions of order α^5 but do not study them further in this section. Also for present purposes, we rewrite the electron denominator

$$\frac{1}{D_e(p)} = \frac{1}{2E'(p_0 + i\epsilon)} - \frac{p^2 - \gamma^2}{2E'(p_0 + i\epsilon)D_e(p)} \ . \tag{4.33b}$$

To order α^4, only the first term needs be included at this time. At this point, the product of electron and proton factors becomes (ignoring the difference between E' and m_e which produces a correction of relative order α^2)

$$\frac{2(\vec{p} - \vec{p}_f)\cdot(\vec{p} - \vec{p}_i)}{2E(p_0 + i\epsilon)^3} = \frac{(\vec{p} - \vec{p}_f)\cdot(\vec{p} - \vec{p}_i)}{2E}\frac{\partial^2}{\partial p_0^2}\frac{1}{p_0 + i\epsilon} \ . \tag{4.33c}$$

This is to be multiplied by the photon propagators, which are functions of p_0^2. After an integration by parts, the factor $(p_0 + i\epsilon)^{-1}$ can be replaced using

$$\frac{-1}{p_0 + i\epsilon} = \pi i \delta(p_0) \ . \tag{4.33d}$$

The resulting effective contribution to $N\hat{K}$ is

$$\frac{e^4}{2E}N\int\frac{d^3p}{(2\pi)^3}(\vec{p} - \vec{p}_f)\cdot(\vec{p} - \vec{p}_i)\left\{\frac{1}{[(\vec{p} - \vec{p}_f)^2]^2(\vec{p} - \vec{p}_i)^2} + \frac{1}{(\vec{p} - \vec{p}_f)^2[(\vec{p} - \vec{p}_i)^2]^2}\right\} \ . \tag{4.34}$$

This integral may be transformed by an integration by parts into the same form as (4.31); together they cancel (4.25b), taking into account the approximations described after (4.31).

Alternatively, one may note that in coordinate space it has the same structure as (4.25d) for the large components.

At this point, we have secured the cancellation of terms of order $\alpha^4 m_r^2/M$ in both gauges, but we have to be sure that we have not overlooked any such terms. For example, what happens if we repeat the previous exercise in the Coulomb gauge? In that case, the Coulomb photons are independent of p_0, so the integral of (4.33c) vanishes. The other terms from (4.32c) couple to a spatial γ_μ on the electron side, producing contributions which are necessarily of higher order in α. These are studied in Chapter V. In summary, the leading spin-independent terms of $\mathcal{O}(\alpha^4 m_r^2/M)$ have all cancelled each other.

4. The Mass Dependence of the Fine Structure to Order α^4

Using the results of Chapter IV-A(3), we now work out the fine structure of hydrogen to order α^4, without approximation in the mass ratio. The modified Dirac equation gave the result (4.18d), which we expand to order α^4

$$\begin{aligned} E_{nj} \approx m_e + m_p + m_r \left[\frac{-\alpha^2}{2n^2} - \frac{\alpha^4}{2n^3(j+\frac{1}{2})} + \frac{3\alpha^4}{8n^4} \right] \\ + m_r \frac{\alpha^4}{2n^3(j+\frac{1}{2})} \frac{m_e^2}{M^2} - m_r \frac{\alpha^4}{4n^4} \left[\frac{2m_r^2}{m_p^2} + \frac{m_r}{2M} \right] . \end{aligned} \tag{4.35}$$

There are various perturbation kernels of order α^4 in the Dirac model. Some of these have already been discussed in the one and two photon kernel sections. There it was found that certain contributions of order $\alpha^4 m_r^2/M$ precisely cancelled each other. The first higher order kernel we look at is δV_d; to order α^4, we take the large component contribution and use the non-relativistic approximation for the wave function. Also, this is already a term in $N\hat{K}$, so we multiply the result by M/m_p. Thus we need

$$\left\langle -\frac{\gamma^2 V_c}{4m_p^2} \right\rangle = \left\langle \frac{2m_r \mathcal{E} V_c}{4m_p^2} \right\rangle = \frac{m_r \mathcal{E}^2}{m_p^2} = \frac{m_r^3 \alpha^4}{4n^4 m_p^2} . \tag{4.36}$$

This partially cancels a term in (4.35).

The only term remaining is the last term of (4.24) which has two factors of $\not{p}$. If taken literally with the Dirac wave function for the electron, the resulting expectation value would be too singular. However, it turns out that a correct treatment to order α^4 is equivalent to taking the expectation value of this operator for Schrödinger wave functions. We may also approximate $E'' + m_p \approx 2m_p$ with a completely negligible error. Finally, since this gives a contribution to $\hat{K}$, and the additional factors of N and M/m_p compensate each other to the order of interest, the contribution to the energy shift is

$$\left\langle -\frac{\not{p}}{2m_p} V_c \frac{\not{p}}{2m_p} \right\rangle = \frac{\langle \vec{p}' \cdot \vec{p} V_c \rangle}{4m_p^2} + \frac{\langle i\sigma_p \cdot \vec{p}' \times (\vec{p} - \vec{p}') V_c \rangle}{4m_p^2} , \tag{4.37a}$$

where for convenience we use $\vec{p}'$ to represent a momentum operator to the left of the potential and $\vec{p}$ to represent one to the right. The dot product may be rearranged

$$\vec{p}' \cdot \vec{p} = \frac{1}{2}\vec{p}'^2 + \frac{1}{2}\vec{p}^2 - \frac{1}{2}(\vec{p}' - \vec{p})^2 .$$

The last of these terms is proportional to $\nabla^2 V_c = 4\pi\alpha\delta(\vec{r})$, which is like a Darwin contribution. The last term of (4.37a) is a proton spin-orbit splitting, which we ignore here. The final result from (4.37a) is

$$-\frac{\alpha^4 m_r^3}{2n^3 m_p^2}\left(\frac{1}{j+\frac{1}{2}} - \frac{1}{l+\frac{1}{2}}\right)\delta_{l0} - \frac{\alpha^4 m_r^3}{2n^3(l+\frac{1}{2})m_p^2} + \frac{\alpha^4 m_r^3}{4n^4 m_p^2} \; . \tag{4.37b}$$

When these new results are added to (4.35), we find

$$\begin{aligned} E_{nj} \approx M + m_r &\left[\frac{-\alpha^2}{2n^2} - \frac{\alpha^4}{2n^3(j+\frac{1}{2})} + \frac{\alpha^4}{4n^4}\left(\frac{3}{2} - \frac{m_r}{2M}\right)\right] \\ &+ m_r \frac{\alpha^4 m_e^2}{2n^3 M^2}\left[\frac{1}{j+\frac{1}{2}} - \frac{1}{l+\frac{1}{2}}\right](1-\delta_{l0}) \; . \end{aligned} \tag{4.38}$$

This is a result known from the 1950's.[16] Its derivation here is much more compact. To order α^4, it is incorporated in (2.3); however, that equation also contains some terms of higher order in α.

5. Expanding Higher-Loop Kernels to Order $1/m_N$

Our immediate objective is to expand the kernels to order $1/M$ without restricting the choice of gauge which is ultimately used for the calculation of the energy of concern. Later we can decide which gauge is most convenient. Space does not permit giving the full details here; but we hope that the following provides a useful indication of the method involved.

With the experience gained through the study of the one loop kernels, it is quite straightforward to identify terms of relative order $1/M$ in the higher order kernels; and it is remarkable that when these terms are combined from all kernels with a fixed number of photon exchanges, a simple result emerges. Within a given kernel, suppose a particular proton propagator carries four-momentum Q. This leads to a subfactor in the proton line

$$\ldots \frac{\gamma_\lambda[\frac{1}{2}(E''+m_p)(1+\gamma_0) + \not{Q}]\gamma_\sigma}{D_p(Q)} \ldots \; . \tag{4.39}$$

Here $\gamma_{\lambda(\sigma)}$ are the factors for photon interactions after (before) the given proton line. We have omitted the small component projector from (4.10a) because it produces a term of higher order in $1/M$. Since $D_p(Q)$ is of order M, a term of order M^0 is produced if we take factors of $\frac{1}{2}(E''+m_p)(1+\gamma_0)$ in every numerator together with the 1 in the appended proton factor in the wave function. Since factors $\ldots g_{\lambda 0} g_{\sigma 0} \ldots$ are produced by the large component projectors, all photon polarizations become temporal. To find a term of order $1/M$, it is necessary either to pick up a correction from rearranging the proton denominators or to replace any one of the large component projectors by the $\not{Q}$ term or to use one of the small proton components in the wave function. We frequently rearrange the denominator using

$$\frac{1}{D_p(Q)} = \frac{1}{2E(Q^0+i\delta)}\left[1 - \frac{D_e(-Q)}{D_p(Q)}\right] \; . \tag{4.40}$$

With one factor of $\not{Q}$, we have the structure

$$(1+\gamma_0)\gamma_\lambda \not{Q} \gamma_\sigma (1+\gamma_0) \equiv (1+\gamma_0)[g_{\lambda 0}\not{Q}\gamma_\sigma + g_{\sigma 0}\gamma_\lambda \not{Q} - Q_0 \gamma_\lambda \gamma_\sigma](1+\gamma_0) \; . \tag{4.41a}$$

All other photon polarizations become temporal. It is easy to find the physical meaning of these terms. Let the four-momentum of the proton line preceding (following) the one being studied be labeled Q_r (Q_l). Then associated with the polarization λ there is a structure

$$\gamma_\lambda \not{Q} + \not{Q}_l \gamma_\lambda = (Q_\lambda + Q_{l\lambda}) + \frac{1}{2}[\not{Q}_l - \not{Q}, \gamma_\lambda] \; . \tag{4.41b}$$

The first term obviously corresponds to the convection current of the proton and the second term represents a spin interaction. The four vector $Q_l - Q$ is just the four momentum carried to the proton line by the photon. In the same way, the other vertex yields a convection and spin part. The small proton components in the wave function provide necessary terms for (4.41b). Finally, we have the Q_0 term from (4.41a). If we take the leading term from the denominator rearrangement, the Q_0-dependence cancels. Considering both orders of photon absorption, the product of matrices becomes $2g_{\lambda\sigma}$; this is nothing other than a seagull term.

The analysis of (4.40) is a simple generalization of (4.28c). Normally, Q is expressed in terms of some subset of electron momenta

$$Q = \sum_i s_i p_i \; , \tag{4.42a}$$

where $s_i = \pm 1$ and $\sum_i s_i = -1$. In the non-relativistictic region, the dominant term in $D_e(-Q)$ is the one containing $-2E' \sum s_i p_i^0$. The others contain two powers of momentum and are nominally smaller by one power of α. To single out the pieces which can cancel electron denominators, we write

$$D_e(-Q) = -\sum_i s_i D_e(p_i) + Q^2 + \sum_i s_i p_i^2 \; . \tag{4.42b}$$

Wherever there is a ladder structure adjacent to a wave function, the non-recoil piece of (4.40) has its $-2\pi i \delta(p_0)$ part removed in each factor. Then it is found that the complete non-recoil term arising from all graphs (corresponding to the first term of (4.40) in all factors) vanishes. This is an important result: *there are no net non-recoil contributions from higher kernels.* This also eliminates spurious lower order terms in powers of α, as found by Love.[109] First-order recoil corrections arise from keeping the second term of (4.40) in any one of the factors in each graph.

After a somewhat lengthy analysis, one finds that only in the one-loop kernels does a $D_e(p_j)$ contribution survive. (To show this, one must exercise some care in treating the so-called artifact terms mentioned in connection with the one-loop kernel.) This one-loop kernel is discussed in Chapter IV-B(3). Thus we are left with the terms from the combination $Q^2 + \sum_j s_j p_j^2$. These terms occur only for denominators carrying several electron momenta; for the leading recoil correction, we may replace all $D_p(Q)$ by $2m_p Q_0 +$

$i\epsilon$. While the derivation of the net result from all these contributions is algebraically complicated, the result is very simple. We express the resulting proton factor in terms of electron momenta defined as follows. Let the external momenta be labeled as p_i for the initial wave function and p_f for the final wave function. The internal electron momenta are labeled from p_1 to p_n starting from the one adjacent to the initial wave function. Also let the polarization of the photon of four-momentum $p_i - p_1$ be μ and that of the photon of four-momentum $p_n - p_f$ be ν. It turns out that the polarizations of all other photons are fixed to be 0 by the structure of the proton line in either gauge. Then the result of our analysis is

$$\frac{2(p_i - p_1)\cdot(p_n - p_f)g_{\mu 0}g_{\nu 0}}{2E(p^0 + i\epsilon)^2}\prod_{j=1}^{n-1}(-2\pi i\delta(p_j^0 - p_{j+1}^0))N_d \, , \tag{4.43a}$$

where $p^0 \equiv p_1^0 \equiv p_n^0$. Because of the product of δ-functions, all the internal photon lines become Coulomb interactions. Only the first and last photons retain a loop energy dependence.

Next consider the total contribution from convection contributions. Here the advantage of having incorporated the small proton components in the wave function shows up in unifying the analysis. Because of $1+\gamma_0$ factors, all photons except one have polarization index 0. Again, the rather complicated algebra leads to a simple result. For a photon with polarization index λ, the result depends on where it was emitted from the electron line. It turns out that if this is associated with a photon emitted from an external line the result is non-zero, while if it is emitted from an internal line the result is zero. The result for emission from the initial line is

$$\frac{2(p_n - p_f)_\mu g_{\nu 0}}{2E(p^0 + i\epsilon)}\prod_{j=1}^{n-1}(-2\pi i\delta(p_j^0 - p_{j+1}^0))N_d \, . \tag{4.43b}$$

The similar result for a photon attached to the final electron line is

$$\frac{2(p_1 - p_i)_\nu g_{\mu 0}}{2E(p^0 + i\epsilon)}\prod_{j=1}^{n-1}(-2\pi i\delta(p_j^0 - p_{j+1}^0))N_d \, . \tag{4.43c}$$

Again, all photons except the first and last become Coulomb interactions. Roughly speaking, (4.43b) corresponds to a photon of polarization μ spanning the entire graph from the initial electron line to the final proton line. A similar remark can be made about (4.43c).

The same type of analysis can be made for the seagull contributions. This time two of the photons carry polarization indices μ, ν. It turns out that a non-zero contribution is obtained only if one of these is emitted from the initial electron line and the other from the final electron line. The result in that case is

$$\frac{-2g_{\mu\nu}}{2E}\prod_{j=1}^{n-1}(-2\pi i\delta(p_j^0 - p_{j+1}^0))N_d \, . \tag{4.43d}$$

The sum of terms from (4.43) is contained in (5.1) below.

How about the spin dependent terms in (4.41b)? When all the terms are added, the result is zero. This is a good thing, because the one photon kernel contains all the effects of this order in $1/M$ for proton spin dependence.

V. PURE RECOIL CORRECTIONS

A complete calculation of the presently known recoil corrections to fine structure and hyperfine structure is very lengthy. Therefore, our aim in this chapter is to provide a description of the strategy used, with emphasis on the determination of the terms which have an order of interest. Some details are provided so that the reader can get a sense of the nature of the calculations. Naturally, the authors believe that the approach of this paper is more streamlined and transparent than some of the older approaches. The main reason is that it is less patchwork in nature and some of the approximations are less awkward.

A. Mass-Dependence of Terms of Higher Order in $Z\alpha$ in the Fine Structure

When we consider terms beyond order $(Z\alpha)^4$, there are many sources of mass ratio corrections. Some of these have already been noted in the previous chapter. They are conveniently divided into pure recoil corrections, radiative corrections modified by some reduced mass effects, and radiative-recoil corrections. The first of these refers to kernels involving exchanged photons without any radiative corrections; they are dealt with in this chapter. The exact separation into the second and third categories is not well-defined, but one can give a plausible meaning to it as far as overall factors are concerned. This is further discussed in Chapter VI.

In Chapter IV-B(5), we worked out an expression for multiphoton exchange to order $1/m_N$. We found that all photons between the first and last are Coulomb photons independently of the original choice of gauge. Only the first and last photons need be described explicitly. Let us label one photon by k, λ, μ and the other by k', σ, ν where the first polarization index is for the electron line and the second for the nuclear line. For gauge invariance questions, it is useful to have the flexibility of emitting the photons from the electron line in either order (divide by two later to compensate for double counting). If k is emitted first, the proton factor given by (4.43) is

$$A_{\mu\nu} = \frac{1}{2E}\left[\frac{2k\cdot k' g_{\mu 0} g_{\nu 0}}{(-k_0+i\epsilon)^2} + \frac{2k'_\mu g_{\nu 0}}{-k_0+i\epsilon} - \frac{2k_\nu g_{\mu 0}}{-k_0+i\epsilon} - 2g_{\mu\nu}\right] , \tag{5.1}$$

where the common factors for the intermediate photons are not displayed. If the order of emission is interchanged, a similar result is obtained. It may be expressed in terms of $A_{\mu\nu}$ by

$$B_{\mu\nu} = A_{\mu\nu} + \{-2\pi i\delta(k_0)[-2k'_\mu g_{\nu 0} + 2k_\nu g_{\mu 0}] + 2\pi i\delta'(k_0)2k\cdot k' g_{\mu 0} g_{\nu 0}\}/2E \ . \tag{5.2}$$

This observation is particularly handy since when a gauge insertion is made on the electron line, the part of the proton factor which is the same for either order of emission, namely

$A_{\mu\nu}$, will cancel. Also, a gauge factor on the proton line gives zero directly. We have to associate these expressions with the complete electron and photon factors and arrive at the order of magnitude of various contributions.

1. Relation between Coulomb gauge and Feynman Gauge Multiphoton Contributions

Before discussing the recoil corrections in detail, let us first verify that the difference between the Coulomb gauge and the Feynman gauge in the multiphoton kernels does just account for the difference in the one-photon kernels which appeared in (4.25c). Since the photons between the first and last are Coulomb photons independently of the original choice of gauge, they are unaffected by the gauge difference. Only the first and last photons need be studied explicitly. The gauge terms relating the Coulomb and Feynman gauge necessarily have a factor of k_0, so only the δ'-term of (5.2) can contribute after an integration by parts. Also, it can contribute only for a single gauge term.

Suppose we set up the problem in the Feynman gauge but wish to re-express it in the Coulomb gauge using (4.23b). Using the gauge invariance on the nuclear line for each fixed number of photons, we write $g^{\mu\lambda} \doteq G^{\mu\lambda}_{Coul} + k^\lambda k_0 g^{\mu 0}/\vec{k}^{\,2}$ for the photon labeled k, λ, μ and consider all insertions of the gauge term. The net result arises from the last term of (5.2) where all insertions have k emitted after k'. Taking into account the gauge factor k^λ multiplying the electron line, these have the usual pairwise cancellations between successive insertions, with a piece left over where the kernel contains only two photon exchanges and the electron propagator adjacent to the first photon has been replaced by 1. (Another term where the photon k spans an arbitrarily large number n of Coulomb interactions vanishes in the limit $n \to \infty$.) After examination, one finds that the result agrees with (4.34), without any intervening approximations. Hence the Coulomb gauge and Feynman gauge give the same result to order $1/m_N$, as they must. It is mainly a question of convenience as to which gauge is preferred. After examining that question, we decided that the Coulomb gauge is slightly to be preferred. The reason why is indicated below.

2. Recoil Corrections to the Fine Structure of Order $(Z\alpha)^5 m_e^2/m_N$

Although our present goal is to identify and analyze terms of order $(Z\alpha)^5 m_e^2/m_N$, we also point out the origin of some contributions of higher order in $Z\alpha$. Let us start the consideration of recoil corrections in the Coulomb gauge. Since the spatial part of the dot product in the first term of (5.1) is cancelled by the $\mu = \nu = 0$ parts of the remaining terms, we may replace that expression by

$$A'_{\mu\nu} = \frac{1}{E}\left[\frac{-2\vec{k}\cdot\vec{k}' g_{\mu 0} g_{\nu 0}}{(-k_0+i\epsilon)^2} + \frac{2k'_\mu(1-g_{\mu 0})g_{\nu 0}}{-k_0+i\epsilon} - \frac{2k_\nu(1-g_{\nu 0})g_{\mu 0}}{-k_0+i\epsilon} - 2g_{\mu\nu}(1-g_{\mu 0})(1-g_{\nu 0})\right] . \tag{5.3}$$

At first, we are most interested in identifying multiphoton contributions whose order does not decrease with increasing number of exchanged photons. Consider various possibilities for the types of the two photons. If both are Coulomb, only the first term contributes. It will be seen later that this gives a very small result since when the k_0-contour is closed below only negative energy electron poles can contribute. If both photons are transverse, only the last term contributes. Here the structure of the Dirac matrices gives a smaller result than the leading term, which is found to occur when one photon is a Coulomb interaction and the other is a transverse photon. Therefore we start the discussion with the leading term.

If we take k' to belong to a Coulomb photon and k to belong to a transverse photon, only the second term of (5.3) contributes. We may study this and double the result to take into account the third term. The expression for the energy shift from $n+1$ photons is

$$\begin{aligned}
&-\frac{1}{m_N}\int\frac{d^3p_f}{(2\pi)^3}\bar{\phi}(\vec{p}_f)\int\prod_{j=1}^{n}\frac{d^3p_j}{(2\pi)^3}\int\frac{d^4k}{-(2\pi)^4 i}\frac{4\pi Z\alpha}{(-k_0+i\epsilon)(k^2+i\epsilon)}\frac{-4\pi Z\alpha\gamma_0 N}{\vec{q}_n{}^2}\\
&\times\prod_{j=1}^{n-1}\frac{-4\pi Z\alpha\gamma_0 N}{(\vec{p}_j-\vec{p}_{j+1})^2}\prod_{j=1}^{n}\frac{m_e+E'\gamma_0+\not{p}_j-\not{k}}{k^2-2E'k_0-\vec{p}_j{}^2+2\vec{p}_j\cdot\vec{k}-\gamma^2+i\epsilon}[\vec{q}_n\cdot\vec{\gamma}]_\perp\phi(\vec{p}_i)\,.
\end{aligned}\tag{5.4}$$

where, for notational convenience, $\vec{p}_1\equiv\vec{p}_i$ and $\vec{q}_n\equiv\vec{p}_n-\vec{k}-\vec{p}_f$; intermediate electron lines have four-momenta $(E'-k_0,\vec{p}_j-\vec{k})$. Recall that $\gamma^2\equiv m_e^2-E'^2=\mathcal{O}((Z\alpha)^2m_r^2)$ and $N\equiv m_N/M$. From now on, we ignore the distinction between N and 1 in the recoil corrections to the fine structure because it would be inconsistent to retain it without also improving the treatment of the multiphoton terms to include higher powers of m_e/m_N; the treatment of positronium requires more care. The subscript $\perp$ means that the component of the dot product parallel to $\vec{k}$ is to be removed. The product of electron factors and Coulomb potentials must be properly ordered. The important feature of the Coulomb gauge is that the denominator with $\vec{q}_n{}^2$ has no k_0^2 term; i.e., it has no retardation. We may briefly note how the same term from (5.3) would appear in the Feynman gauge. First, the $\perp$ restriction in the dot product would be dropped. The other change is that the denominator with $\vec{q}_n{}^2$ would have a k_0^2 term, introducing an additional photon pole. The residue of that pole would have less significance because of the factor of $\vec{q}_n$ in the numerator. At the expense of a little complication, one can rearrange the two photon pole contributions with the result that the dominant contribution is the same as (5.4) with the transversality restriction removed. However, a piece of the first term in (5.3), together with the photon propagators which now have retardation, can be shown to restore the transversality condition. In this way, the Feynman gauge reproduces the Coulomb gauge result after some slightly inconvenient approximations.

To begin our analysis of orders of magnitude, we make some simplifying approximations which select out the leading term and later look at the size of the corrections. We notice first that it is convenient to carry out the k_0-integration by closing its contour below. Then the only poles which can contribute are the photon pole at $k_0=|\vec{k}\,|$ and negative energy electron poles. We argue that negative energy electron poles generally give contributions which diminish with increasing numbers of photons. The reason is that then momenta internal to the kernel are dominated by the relativistic region. We can estimate the order of magnitude of the result by factoring out powers of γ from the wave function and explicit factors of $Z\alpha$ from the kernel and claim that the integral produces appropriate powers of m_e to balance the dimensions. If all the momenta in the integral are pushed into the relativistic region, the order produced would be $\gamma^5(Z\alpha)^{n+1}/(m_e^3 m_N)$. Actually, this gives too small an estimate since the $\vec{p}_f$-integration converges at low momentum and produces an inverse power of γ, yielding a final order $\gamma^4(Z\alpha)^{n+1}/(m_e^2 m_N)$ (the $\vec{p}_i$-integration cannot have this behavior because the additional factor of $\vec{\gamma}$ in the integrand brings in small components of the wave function). For $n \geq 2$, this would be negligible at the present time, permitting us to neglect the negative energy electron poles in that circumstance. The $n = 1$ terms of this type should contribute at the level of $(Z\alpha)^6 m_e^2/m_N$, but they have not yet been evaluated. Since the negative energy terms are unimportant, we approximate all electron numerators by $2m_e$ times a large component projector except for one next to the factor $[\vec{q}_n \cdot \vec{\gamma}]_\perp$, where a factor of $-\vec{\gamma} \cdot \vec{p}_i$ on either side is needed. This produces the substitution $[\vec{q}_n \cdot \vec{\gamma}]_\perp \to [\vec{q}_n \cdot 2\vec{p}_i]_\perp/2m_e$.

To produce terms of order $(Z\alpha)^5 m_e^2/m_N$ with $n \geq 2$, the integrals must yield inverse powers of γ. Closing the contour below the axis to avoid positive energy electron poles, we consider the leading contribution from the photon pole at $k_0 = |\vec{k}|$. At this stage, we have the dominant piece

$$\begin{aligned} &\frac{1}{m_N}\int \frac{d^3p_f}{(2\pi)^3}\bar{\phi}(\vec{p}_f)\int \prod_{j=1}^{n}\frac{d^3p_j}{(2\pi)^3}\int \frac{d^3k}{(2\pi)^3}\frac{4\pi Z\alpha}{2\vec{k}^{\,2}}\frac{-4\pi Z\alpha}{\vec{q}_n^{\,2}} \\ &\times \prod_{j=1}^{n-1}\frac{-4\pi Z\alpha}{(\vec{p}_j-\vec{p}_{j+1})^2}\prod_{j=1}^{n}\frac{2m_e}{-2E'k-\vec{p}_j^{\,2}+2\vec{p}_j\cdot\vec{k}-\gamma^2+i\epsilon}\frac{[\vec{q}_n\cdot 2\vec{p}_i]_\perp}{2m_e}\phi(\vec{p}_i)\,, \end{aligned} \tag{5.5}$$

where now we are to use only the large (non-relativistic) components of the wave functions.

We are at first more interested in understanding the integral (5.5) than in developing techniques for calculating it. We note that if n is large enough, the scale of the variables of integration is set by γ. For example, in both the wave functions and in the electron denominator, the natural scale of each $\vec{p}_j$ is γ. On the other hand, for $-2E'k$ to be of order γ^2, we see that the natural scale of k is γ^2/m_e. This implies that $2\vec{p}_j \cdot \vec{k}$ tends to be smaller that the other terms in the electron denominator by relative order γ/m_e. The same is true for each power of the $\vec{k}$-dependence in the $\vec{q}_n^{\,2}$ denominator. Thus, for n not too small, we may estimate the size of the integral by neglecting these small terms in the denominators. To study whether the integrals have adequate convergence properties, we can imagine combining the electron denominators with parametric integrations first. Then the k-integration is easily carried out. At this stage, the combined electron denominators would take the form $1/[(\vec{p})^2(v) + \gamma^2]^{n-1}$, where $(\vec{p})^2(v)$ is a polynomial in the $\vec{p}_j^{\,2}$'s. Taking into account explicit factors from the integrand (including the wave functions), the resulting integral is found to have an overall factor $Z\alpha\gamma^{n+5}/(m_e^2 m_N)$. By rescaling the remaining momentum variables, the integral is seen to produce the factor γ^{-n-1}, giving the overall order $Z\alpha\gamma^4/(m_e^2 m_N)$, which is independent of n. This analysis breaks down for $n = 1$ because the k-integration is logarithmically divergent at the upper limit without the $\vec{k}$-dependence of the denominator $\vec{q}_n{}^2$. For this term, the overall factor is the same, but a logarithm depending on $Z\alpha$ is produced (with upper limit of order $Z\alpha m_e$ and lower limit of order $(Z\alpha)^2 m_e$).

We now consider the effect of the approximations in inverse order. The approximation to neglecting $2\vec{p}_j \cdot \vec{k}$ terms in an electron denominator can be estimated by expanding up those terms; each power is of relative order $Z\alpha$. In addition, a first order term will vanish by symmetry. Thus we get a contribution by expanding up two of these terms. After carrying out the angular integration, we can imagine following the same procedure as in the preceding paragraph. The convergence properties of the k-integration are the same as before, and the same denominator results. Now, however, the resulting integrand has an additional factor like $\vec{p}_l \cdot \vec{p}_j / m_e^2$ from the k-integration. In this case, if the momentum integration converges, it should yield a factor γ^{-n+1}, giving a result which is smaller than the leading one by relative order $(Z\alpha)^2$. The signal for convergence is that the power of γ from the integral is negative. For $n \geq 2$, the integral converges, producing a result which is negligible at the present time. Another term arises from expanding up $2\vec{k} \cdot (\vec{p}_f - \vec{p}_n)$ from the $\vec{q}_n{}^2$ denominator together with a term $2\vec{p}_j \cdot \vec{k}$ from an electron denominator. Going through the same arguments, one again finds that the order is $(Z\alpha)^2$ smaller than the leading term. However, one should note that the k-integration now has a logarithmic divergence for $n = 2$; a more complete treatment would cut this off at $k \approx Z\alpha m_e$, so it is still negligible. On the other hand, if we expand up two powers of $2\vec{k} \cdot (\vec{p}_f - \vec{p}_n)$ or one of $\vec{k}^2$ from the $\vec{q}_n{}^2$ denominator, the estimate becomes unreliable for $n = 2$. Since it produces an inoccuous logarithm for $n = 3$, we may continue to neglect these terms for $n \geq 3$. The lesson is that the $\vec{k}$-dependence in $\vec{q}_n{}^2$ must be dealt with carefully for $n \leq 2$. Even for $n = 2$, its order is expected to be small, namely $(Z\alpha)^6 m_e^2 / m_N$.

Now shift attention to a factor $\not{k}$ in one of the original numerators. Its order of magnitude is smaller than that of a large component projector by a factor of $(Z\alpha)^2$, so it too should be negligible, even though a logarithm could occur for $n = 2$. If it happens to replace a factor of $\not{p}_i$, it is apparently only a factor of $Z\alpha$ smaller. We must then distinguish between $\gamma_0 k$ and $\vec{\gamma} \cdot \vec{k}$. For the former, we must have an additional spatial part of a factor $\not{p}_j$ which adds a factor of $Z\alpha$; for the latter, the odd angular integration requires expanding up a factor of $\vec{k}$ from a denominator, which also adds a factor of $Z\alpha$. In either case, the result is negligible for $n \geq 2$ at the present time.

Continuing to work backwards, we consider other terms dropped in arriving at (5.5) itself. We also neglected factors of $\vec{\gamma} \cdot \vec{p}_j$ except for the ones adjacent to the photon vertex. If we replace one of the latter by one of the former, one or more factors of $m_e - E'$ replace factors of $m_e + E'$. This is smaller by a factor of at least $(Z\alpha)^2$ and does not change the convergence properties of the integral. Hence it should induce a negligible error to drop these terms. Two extra factors of $\vec{\gamma} \cdot \vec{p}_a$ also increase the order by $(Z\alpha)^2$, making them negligible. The conclusion seems to be that we can neglect all these corrections at and beyond the three photon kernel level. This applies only to the contributions arising from the photon pole. However, we have already noted that the contributions from negative energy electron poles are also negligible for $n \geq 2$.

We now turn to the calculation of (5.5), which turns out to be closely related to the Bethe logarithm defined in (3.8). Our strategy is to analyze that expression in momentum space, and to manipulate it into a form which may be compared to (5.5). It is convenient to start with (3.7) and make the subtraction mentioned there to remove part of the mass renormalization (note that $\omega = |\vec{k}|$). The integration over $\vec{k}$ produces (3.8). The integral we consider is

$$\begin{aligned}(3.8) &= \frac{e^2}{m_e^2}\int_0^\Lambda \frac{d^3k}{2\vec{k}^{\,2}(2\pi)^3}\sum_m (E_n - E_m)\frac{[\langle n|\vec{p}|m\rangle\cdot\langle m|\vec{p}|n\rangle]_\perp}{E_n - k - E_m} \\ &= \frac{4\pi\alpha}{m_e^2}\int_0^\Lambda \frac{d^3k}{2\vec{k}^{\,2}(2\pi)^3}\sum_m \frac{[\langle n|[H,\vec{p}]|m\rangle\cdot\langle m|\vec{p}|n\rangle]_\perp}{E_n - k - E_m}\ .\end{aligned} \tag{5.6a}$$

To facilitate the comparison, we have undone the angular integration over $\vec{k}$ and reintroduced the transversality condition in the dot product. The commutator is $-4\pi Z\alpha\vec{q}/\vec{q}^{\,2}$ in momentum space; this $\vec{q}$ corresponds to $\vec{q}_n$ in (5.5), but the two differ in an important way to be elaborated below. At this point we recognize that the sum over m is giving the Schrödinger-Coulomb Green's function, as represented in coordinate space in (3.9b). In momentum space, our expression is

$$\begin{aligned}(3.8) = \frac{1}{m_e}\int_0^\Lambda \frac{d^3k}{(2\pi)^3}\int &\frac{d^3p_i d^3p_n d^3p_f}{(2\pi)^9}\frac{4\pi\alpha}{2\vec{k}^{\,2}}\frac{-4\pi Z\alpha}{\vec{q}^{\,2}} \\ &\times \phi_n^\dagger(\vec{p}_f)2m_e G_{NR}(\vec{p}_n,\vec{p}_i;E_n-k)\frac{[\vec{p}_i\cdot\vec{q}]_\perp}{m_e}\phi_n(\vec{p}_i)\end{aligned} \tag{5.6b}$$

where $\vec{q} \equiv \vec{p}_n - \vec{p}_f$. Now G_{NR} can be expanded in terms of free propagators and Coulomb interactions: the first two terms in this expansion are given in (3.11). Taking the first term, which represents simply free propagation, we have

$$\begin{aligned}(3.8) = \frac{1}{m_e}\int_0^\Lambda \frac{d^3k}{(2\pi)^3}\int &\frac{d^3p_i d^3p_f}{(2\pi)^6}\frac{4\pi\alpha}{2\vec{k}^{\,2}}\frac{-4\pi Z\alpha}{\vec{q}^{\,2}} \\ &\times \phi_n^\dagger(\vec{p}_f)\frac{-2m_e}{2m_e k + \vec{p}_i^{\,2} + \gamma^2}\frac{[\vec{p}_i\cdot\vec{q}]_\perp}{m_e}\phi_n(\vec{p}_i)\ .\end{aligned} \tag{5.6c}$$

Except for an overall factor of m_N/Zm_e, the difference in the definitions of $\vec{q}$ and $\vec{q}_n$, the replacement of E' with m_e, and the presence of the cutoff in this expression, this is precisely the same as the $n = 1$ term in (5.5). It is a simple matter to see that this result generalizes so that the n'th term of (5.5) corresponds to the $(n-1)$'th term in the expansion of the Schrödinger-Coulomb Green's function, so that we conclude

$$(5.5) \sim (Zm_e/m_N)\times(3.8)$$

where $(\sim)$ means that the two expressions have the same structure term by term except for the noted differences. The complete result should be doubled to take into account the contribution from the third term of (5.3).

The integral (5.5) is completely ultraviolet finite, but it is awkward because of the $\vec{k}$-dependence in $\vec{q}_n$. Actually, for all terms except the $n = 1$ one, we can retain accuracy to order $(Z\alpha)^5 m_e^2/m_N$ while dropping this awkward dependence. For $n = 1$, a logarithmic divergence results if the extra k convergence from $1/|\vec{q}_n|^2$ is omitted. To account properly for the $n = 1$ part of (5.5) we first add and subtract $1/|\vec{q}|^2$ from $1/|\vec{q}_n|^2$, introducing an ultraviolet cutoff Λ in intermediate stages with the understanding it is to be taken to infinity at the end of the calculation. The term with $1/|\vec{q}|^2$ can then be expressed in terms of (3.8) as described above. Now consider the remaining difference (for $n = 1$ only). The subtraction causes a suppression of the integrand at low k: all momenta are now of order $m\alpha$. At this point $2m_e k$ is much larger than $\vec{p}_i^{\,2}$ and γ^2, so that the propagator denominator can be replaced with $2m_e k$. After these simplifications, the expression can be Fourier transformed into

$$\Delta E_{\rm sub} = -\frac{(Z\alpha)^2}{4m_e m_N \pi^2}\int\frac{d^3r}{r^3}\int^{\Lambda}\frac{d^3k}{k^3}\phi^{\dagger}(\vec{r})(e^{i\vec{k}\cdot\vec{r}}-1)\vec{\nabla}_r\phi_n(\vec{r})\cdot(\vec{r}-\vec{r}\cdot\hat{k}\hat{k}) \qquad (5.7)$$

The evaluation of this integral gives rise to a logarithmic dependence on Λ that exactly cancels that of the previous term, leaving a state-dependent constant that forms part of a_n introduced in Chapter II.

There are two other sources of $m_e^2(Z\alpha)^5/m_N$ terms that are both associated with one-loop graphs. Using the tools just developed for the Coulomb-transverse photon case, we can easily see how these other cases work. Consider the seagull term in (5.3) which involves only two transverse photons. In place of a factor $1/(-k_0+i\epsilon)$, it has an additional m_e-denominator from having two factors of $\vec{\gamma}$ rather than one. In effect, it is smaller by a factor of k/m_e, which is of order $(Z\alpha)^2$; for the case $n = 2$, it would be expected to have a logarithmic factor as well. Thus it should be negligible for $n \geq 2$. The one loop contribution leads to a logarithmic dependence on α, and also involves state dependent constants. We refer the interested reader to the original literature (Salpeter[8], Fulton and Martin[22]), and the review of Grotch and Yennie[79] for further details.

The double Coulomb term acquires contributions only from the negative energy electron poles. Thus the momenta internal to the kernels scale as m_e, while it is possible for the wavefunction momenta to converge at small momenta. This indicates that the order of magnitude of a term with $n+1$ photons will be $\gamma^3(Z\alpha)^{n+1}/m_e m_N$, which is negligible for $n \geq 3$. The evaluation of the one loop two-Coulomb term is straightforward and is included in (2.7) along with the other contributions described here.

In summary, we conclude that multiphoton terms are important only in the Coulomb-transverse case. For the other cases, and most of the corrections discussed, three photon kernels may contribute at the level of $(Z\alpha)^6 m_e^2/m_N$, which is small, but of possible interest. Two photon kernels also contain terms of that order.

B. Recoil Effects in the Hyperfine Splitting

We have seen in Chapter IV-B(2) how the one photon kernel in our formalism includes the Fermi splitting[53] and the Breit relativistic correction[55] without prejudicing higher order calculations. Now our objective is to study the dynamical recoil corrections to the hyperfine splitting. In the first subsection below, we rederive another historic result in this field, the leading (one loop) recoil correction.[64,77] This is given in enough detail so that the reader should be able to verify the result, if desired. The main features of the two-loop calculation, which introduces an additional power of α into the result, is described briefly in the second subsection.

In contrast to the fine structure which is discussed here in the Coulomb gauge, it turns out to be more convenient to analyze the hyperfine splitting in the Feynman gauge. This might be accomplished by the use of this gauge *ab initio*, or by setting up the kernels in the Coulomb gauge and transforming them into the Feynman gauge. The latter approach was used in BYG. In the following, we assume the Feynman gauge without further comment.

1. The One-Loop Recoil Correction.

Since we aim to make these results accessible to a casual reader, we do not deal with all the complications from the start. On the other hand, the expressions for the leading terms generally contain many higher order contributions as well. We do not wish to calculate these leading terms so crudely that we would find it necessary to deal with extra complications in order to recover higher order terms later on. The actual calculation of higher loop terms is quite lengthy. We refer the reader to BYG for details but try to give the flavor here. Probably the methods described in the present paper could help streamline that earlier analysis somewhat, but we have not redone it. In Chapter IV-B(1), we saw how the small components of the nucleus can be appended to a given kernel by "borrowing" them from a higher order kernel. At the same time, when higher order kernels are considered, we must remember that such terms have already been taken into account. That is, we drop $-2\pi i\delta$-terms from higher kernels unless they include some effect not already incorporated.

We start with the two-photon ladder graph. The nuclear numerator of this graph is given in (4.27d). Since we are interested in hfs terms, we need an even number (other than zero) of spatial γ-matrices in this expression. The terms are arranged there in powers of m_N, and in addition the nuclear denominator is proportional to m_N for non-relativistic nuclear momenta. Since we are interested in a final result of order $1/m_N^2$, we may discard the last line of the numerator. The terms with one power of $1/m_N$ in the numerator involve factors of the wave function momenta or powers of γ. They lead to two powers of α beyond the Fermi energy. While they are of the order of interest, they are smaller than the one-loop terms and are incorporated in the two-loop calculation. Also, from the discussion preceding (4.27d), the other part of the nuclear factor is given by (4.27c). The combination to be considered here appears to give contributions of order $1/m_N$, but when it is combined with the similar contributions from the crossed graph, that order cancels; and the remainder is of order $1/m_N^2$. The numerator terms which contribute to the one loop kernel are

$$-\frac{1}{2}[\not{p} - \not{p}_i, \gamma_\nu] g_{\mu 0} - \frac{1}{2}[\not{p}_f - \not{p}, \gamma_\mu] g_{\nu 0} + p_0 \gamma_\mu \gamma_\nu \; . \tag{5.8a}$$

Technically, the terms with wave function momenta involve two loops, but they can be included here with no additional effort. The corresponding terms from the numerator of the crossed graph given in (4.28b) are

$$-\frac{1}{2}[\not p - \not p_i, \gamma_\nu] g_{\mu 0} - \frac{1}{2}[\not p_f - \not p, \gamma_\mu] g_{\nu 0} - p_0 \gamma_\nu \gamma_\mu \,. \tag{5.8b}$$

For the hfs contribution which we call VO, we single out the first term from each of these. We combine these with the electron numerator, together with the electron and nuclear denominators after they have been rearranged using (4.9c) and (4.28c) and the $-2\pi i\delta$ piece has been dropped. The result is

$$\begin{aligned}\frac{-\frac{1}{2}[\not p - \not p_i, \gamma_\nu] \otimes \frac{1}{2}[\not p - \not p_i, \gamma^\nu]}{2E} \Bigg\{ & -\frac{1}{(-p_0 + i\epsilon) D_p(-p)} + \frac{1}{(p_0 + i\epsilon) D_p(p - p_i - p_f)} \\ & - \frac{(p - p_i - p_f)^2 + p^2 - p_i^2 - p_f^2}{(p_0 + i\epsilon) D_e(p) D_p(p - p_i - p_f)} \Bigg\} \,.\end{aligned} \tag{5.9}$$

Note how the leading non-recoil pieces have cancelled between the ladder and crossed contributions. We have also dropped some "artifact terms" similar to those discussed in Chapter IV-B(3). We can also use a symmetry to cancel the first two terms in the braces. Let $p \to p_i + p_f - p$ in the second term. This interchanges the photon denominators and changes the factor in front of the brace to the one which goes with a contribution we call OV. In the complete expression for VO+OV, the first and second terms then cancel. Carrying out an angular average, we find that the remainder reduces to

$$\frac{\frac{2}{3}\langle \vec\sigma_e \cdot \vec\sigma_p \rangle (\vec p - \vec p_i)^2}{2E} \left\{ \frac{(p - p_i - p_f)^2 + p^2 - p_i^2 - p_f^2}{(p_0 + i\epsilon) D_e(p) D_p(p - p_i - p_f)} \right\} \,. \tag{5.10}$$

We rearrange the numerator of (5.10) using

$$\begin{aligned}(p - p_i - p_f)^2 + p^2 - p_i^2 - p_f^2 &= 2(p - p_f) \cdot (p - p_i) \\ &= 2(p - p_f)^2 + 2(p_f - p_i) \cdot (p - p_f) \,.\end{aligned}$$

It is necessary to proceed with a little caution in treating the wave functions. We see from the first form that there is no difficulty in replacing the Dirac wave functions by the non-relativistic ones. The correction has higher powers of α. After that replacement, we see that the second term of the second form clearly involves wave function momenta, so it does not contribute to the pure one-loop term and we set it aside as part of the two-loop calculation. The first term cancels a photon denominator, decoupling the integration over the final wave function momentum (here there would be a problem had we not previously replaced the Dirac by the non-relativistic wave function). To decouple the other wave function integration, we use the approximation $D_p(p - p_i - p_f) \to D_p(p)$ and also the identity

$$\frac{(\vec p - \vec p_i)^2}{p_0^2 - (\vec p - \vec p_i)^2 + i\epsilon} = -1 + \frac{p_0^2}{p^2 - \gamma^2 + i\epsilon} + p_0^2 \left[\frac{1}{p_0^2 - (\vec p - \vec p_i)^2 + i\epsilon} - \frac{1}{p^2 - \gamma^2 + i\epsilon} \right] \,. \tag{5.11}$$

The first two terms here decouple the wave function, and the square bracket term produces a contribution which is of higher order in α. The introduction of the γ^2 into the denominator is simply so that we can use certain integral tables which have been worked out in BYG. It has the advantage that sometimes the last term has special cancellations which make it even smaller than the order which is straightforwardly estimated. Integration of the wave functions in momentum space produces a factor $\tilde{\phi}_{nr}(0)$ which is the value of the non-relativistic ground state wave function at the spatial origin. This is an adequate approximation for our present purposes. At this point, we are left with the four-dimensional integral

$$\Delta E([\mathrm{VO}]_1 + [\mathrm{OV}]_1) = -(4\pi\alpha)^2 \frac{\frac{2}{3}\langle\vec{\sigma}_e\cdot\vec{\sigma}_p\rangle}{E''+E'} |\tilde{\phi}_{nr}(0)|^2 \int \frac{d^4p}{-(2\pi)^4 i} \frac{1}{D_e(p)D_p(p)} \times \frac{1}{(p_0+i\epsilon)} \left[-1 + \frac{p_0^2}{p^2-\gamma^2+i\epsilon}\right] . \tag{5.12}$$

The evaluation of the integral is simplified by making a further rearrangement of the denominators which separates the integrand into a piece depending on the electron mass and a piece depending on the nuclear mass

$$\frac{1}{D_e(p)D_p(p)} = \frac{1}{2(E''-E')(p_0+i\epsilon)} \left[\frac{1}{D_e(p)} - \frac{1}{D_p(p)}\right] . \tag{5.13}$$

To carry out the integration, one first does the p_0-integration by closing the contour above. The subsequent integration over the magnitude of $\vec{p}$ is then straightforward. The result is

$$\Delta E([\mathrm{VO}]_1 + [\mathrm{OV}]_1) = \frac{E_F}{1+\kappa} \left\{ -\frac{6\alpha}{\pi} \frac{m_e m_N}{m_N^2 - m_e^2} \ln\frac{m_N}{m_e} + \frac{\gamma^2}{m_N m_e} \right\} . \tag{5.14}$$

We may easily include the last term of (5.11) as well; it is known as a "decoupling" correction. It adds a contribution $\gamma^2/(m_N m_e)$ to the brace. Note that the factor involving the nuclear anomalous moment is divided out of the Fermi splitting; it would be incorrect to incorporate it here. Chapter II-C contains a description of how this analysis is modified when the nuclear anomalous moment and structure are taken into account.

The contribution arising from the last term of each of (5.8) is called VV. The electron line introduces a similar factor. The first step in simplification is to use a rearrangement like

$$\frac{\pm p_0}{D_\beta(\pm p)} = \frac{1}{2m_\beta} - \frac{p^2-\gamma^2}{2m_\beta D_\beta(\pm p)} ,$$

where β stands for e or p. Since $\gamma_\mu\gamma_\nu + \gamma_\nu\gamma_\mu = 2g_{\mu\nu}$, the first term does not contribute to the hfs when the ladder and crossed terms are combined (in case it occurs on the electron line, one must also use the symmetry $p \to p_i + p_f - p$). Approximating the nuclear denominator in the crossed term, we are led to the energy shift

$$\Delta E([\mathrm{VV}]_1) = 3(4\pi\alpha)^2 \frac{\frac{2}{3}\langle\vec{\sigma}_e\cdot\vec{\sigma}_p\rangle}{4m_e m_N} \int \frac{d^3p_f d^4p d^3p_i}{-(2\pi)^{10} i} \phi_{nr}^\dagger(p_f)\phi_{nr}(p_i) \times \frac{(p^2-\gamma^2)^2}{[(p_f-p)^2+i\epsilon]D_e(p)[(p-p_i)^2+i\epsilon]} \left(\frac{1}{D_p(p)} - \frac{1}{D_p(-p)}\right) . \tag{5.15}$$

The first step in the evaluation is to rearrange the photon propagators as in (5.11) to separate each photon propagator into a part in which the wave function decouples and a decoupling correction. The piece of (5.15) in which both wave functions decouple is easily evaluated by first doing the p_0 integration. The only subtlety is that the γ^2 dependence in the denominators yields a small contribution which could easily be overlooked. The complete result, including the decoupling corrections, is

$$\Delta E([\mathrm{VV}]_1) = \frac{E_F}{1+\kappa}\left\{\frac{3\alpha}{\pi}\frac{m_e m_N}{m_N^2 - m_e^2}\ln\frac{m_N}{m_e} - \frac{9}{2}\frac{\gamma^2}{m_e m_N}\right\} . \tag{5.16}$$

This completes the calculation of the one-loop terms.

2. The Two-Loop Recoil Correction.

In the discussion of the one-loop recoil correction, we have already seen the appearance of some terms of relative order $\gamma^2/(m_e m_N)$. The natural origin of terms of this order is actually in the two-loop recoil corrections. The complete calculation of them is rather lengthy, so it is not presented here; details are in BYG. Instead, we report some of the more important general features.

In the one-loop recoil corrections, we see the occurrence of terms proportional to $\ln(m_N/m_e)$; such terms do not occur in the two-loop result, although they do appear in the contributions from individual graphs. The presence of these terms in the one-loop case can be understood from (5.12) or (5.15). In the intermediate momentum region where the momentum variables lie between m_e and m_N, either of these integrals tends to be dimensionless after overall powers of mass are factored out where appropriate. This signals the possibility of logarithmic behavior from this region. The same thing happens with the integrals for the two-loop contributions, so we might expect them also to produce these terms. The reason that they do not occur is due to an important compensation between different graphs, which we refer to as the Caswell-Lepage cancellation.[11] This arises because if the photon connections to the electron line are reversed, the hyperfine contribution from the intermediate momentum region changes sign. I.e., there is a pairwise cancellation between different graphs in this region. The result is that if the contributions are combined before integration, the complete integrand is strongly suppressed in this region and the contribution comes primarily from momenta of order m_e and smaller. This is an illustration of the desirability of finding cancellations before calculating integrals rather than later.

From the observations of the preceding paragraph, it is clear that we should group together contributions which represent "gauge invariant" sets. For example, for a given set of attachments of O and V type photons on the electron line, we group together all ways of attaching the three photons to the nuclear line for a total of six graphs. This will automatically give us the pairs necessary for the Lepage-Caswell cancellation. At this point, since the integral arises from the very non-relativistic region for the nucleus, we can make simplifying approximations to the nuclear propagators in which only the denominator term proportional to $2E''$ is kept. (For positronium, something better should be done.)

It turns out that this always produces at least one or two δ-functions of a photon energy. In the case of one δ-function, a different photon energy appears in a denominator, e.g., $1/(p_0 + i\epsilon)$. Some of these terms are discarded if they are already taken into account according to the discussion of Chapter IV-B(1). The remaining ones are easily estimated to give results of relative order $\gamma^2/(m_e m_N)$, although the mass symmetric form is not guaranteed. In case there are two δ-functions of photon energies, a factor of $\ln \alpha$ arises.

Fortuitously, it turns out that the integrals can be carried out analytically to the order of interest by a special trick, which will now be described. One first combines the electron and photon denominators using Feynman paramters and carries out the integrations over all energies and momenta. This produces a rather intractable result in which the integrations over the parameters are not readily carried out analytically, although they could be done numerically using a program such as VEGAS.[112] The next step is to make simplifying approximations using the fact that $\gamma \ll m_e$. At this point, it can be recognized that the same result in terms of Feynman parameters can be obtained from integrals having spatial momentum integrations only. (Sometimes this is delicate and additional tricks are needed.) These integrals can in turn be expressed in terms of coordinate space integrations having only one space variable, and they become trivial. In this way, integral tables are constructed to do all the integrals needed. A few are a bit more complicated since they involve three and four loops.

Another source of terms of relative order $\gamma^2/(m_e m_N)$, aside from those which are already observed in the one-loop calculation, is the second order kernels in (4.7b). It turns out that in the Feynman gauge these are second order in the hfs; i.e., there are no contributions in which only one factor is a hyperfine interaction. Although the algebra is slightly laborious, these terms are easily worked out using the Dalgarno-Lewis[113] method.

VI. RADIATIVE-RECOIL CORRECTIONS

In Chapter IV-A, we describe how radiative corrections can be introduced perturbatively as contributions to $\hat{K}$. For example, in (4.5a) the two-particle irreducible kernels K include radiative corrections to exchanged photon lines and to the electron line with virtual photons spanning one or more exchanged photons, and δS includes radiative corrections to the electron propagator between photon exchanges. Any of these may occur in combination with R's in adjacent loops. We recall that the term radiative-recoil corrections refers to effects that involve recoil and radiative corrections simultaneously. Sometimes it is natural to think of these as primarily one type or the other with modifications produced by the other type. For example, in the hfs calculation, it seems more natural to start with the leading (one-loop) recoil calculation and append the effects of radiation. However, the discussion of the Lamb shift in Chapter III indicates that in the non-recoil limit the separation of the calculation into different numbers of photon exchanges introduces spurious orders which cancel between them. It is desirable to keep all these terms together in a single expression. This is a situation in which we tend to think of recoil as a modification of the basic radiative corrections. A more subtle approach which tries to deal with both issues simultaneously seems necessary, and a possible approach of this type is described in Section B below.

A. Hyperfine Structure

1. Muonium

In Chapter V-B, we saw that the leading recoil correction arose from one-loop kernels and that the characteristic integration momentum range was between m_e and m_μ, giving rise to a logarithm in the mass ratio. Recognizing this, Caswell and Lepage[11] pointed out that radiative corrections to this would be rather important because they would contain terms proportional to the square of the logarithm of the mass ratio. It is easy to understand this intuitively. As they pointed out, the most important corrections of this type arise from the insertion of a vacuum polarization correction into an exchanged-photon line. In the region where the photon four-momentum p is very large compared to the electron mass, the vacuum polarization introduces a factor

$$\frac{\alpha}{3\pi}\ln\frac{-p^2}{m_e^2}\,. \tag{6.1}$$

It is straightforward to estimate the size of the effect. In the most important terms, the p-integration is approximately logarithmic over the range m_e to m_μ, so (6.1) yields a factor

$$\frac{2\alpha}{3\pi}\ln\frac{m_\mu}{m_e} \tag{6.2}$$

in conjunction with such terms. In (6.2) we have taken into account a factor of 2 for the two ways of inserting the vacuum polarization and a factor of $\frac{1}{2}$ that arises from the integration. An estimate of the size of the largest radiative-recoil correction may then be obtained by multiplying the logarithmic terms of Chapter V-B by the factor (6.2) to produce terms of relative order $(\alpha/\pi)^2(m_e/m_\mu)\ln^2(m_\mu/m_e)$.

Because the correction discussed is relatively important for muonium, it is important to work out the singly logarithmic and constant terms as well. Our description follows that of Sapirstein, Terray and Yennie,[59] which we refer to here as STY. To take into account vacuum polarization, either of the photon propagators is replaced as follows

$$\frac{1}{p^2+i\epsilon}\rightarrow\frac{\alpha}{\pi}\int_{s_{th}}^{\infty}\frac{\rho(s)ds}{p^2-s+i\epsilon}\,. \tag{6.3}$$

The function $\rho(s)$ depends on the ingredient of the vacuum polarization, the electron closed loops being the most important. It turns out to be convenient to work out the contributions for the electron and muon closed loops together, there being some convenient combinations of terms which arise. The combined contribution to the muonium hfs is included in (2.27). The hadronic contributions to the vacuum polarization are also easily incorporated, but they require numerical evaluation. They contribute about 0.05 ppm and introduce an uncertainty smaller than one part in 10^8. The non-hadronic contribution for positronium can be calculated analytically; the hadronic contribution is presently quite negligible.

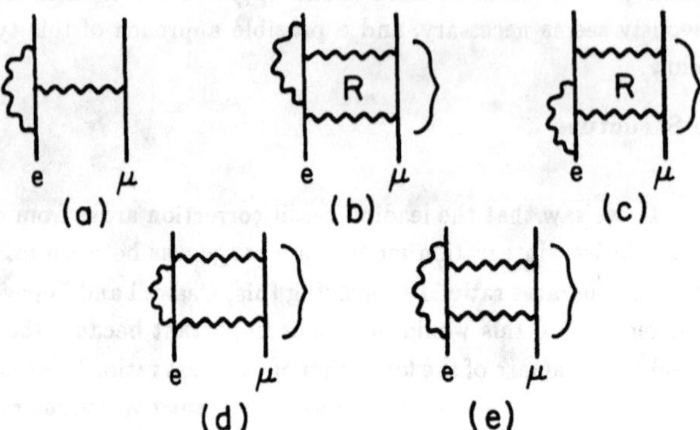

Figure 7: Graphical representation of electron line radiative corrections to the hyperfine structure. The large brace represents contributions in which the photon connections to the muon line are permuted in all ways to generate other contributions.

Next we describe radiative corrections to the electron and muon lines. Following the discussion of Chapter IV-A, kernels giving radiative corrections to order $\alpha^2(m_e/m_\mu)E_F$ on the electron line are shown in Fig. 7. For our present purposes, it turns out that it is helpful to use the Fried-Yennie gauge (3.6a) in which contributions with the virtual photon spanning more than two exchanged photons are of higher order in α (see Ref. (59)). This provides an example of a situation in which the calculation simplifies with a particular choice of gauge, while of course the final result is independent of gauge.[p] Following the discussion of Chapter IV-A, it is understood that the self-energy and vertex subgraphs represent the *renormalized* functions $\tilde{\Sigma}_{\text{ren}}$ and $\Lambda^{\mu}_{\text{ren}}$, respectively.

There are some properties of the individual contributions from Fig. 7 which should be pointed out. For example, the first one, if approximated by a static anomalous moment term and taken together with Dirac-Coulomb wave functions, would give an infinite contribution because of the δ-function singularity at the origin. This simply means that the internal structure of the vertex cannot be ignored. However, the leading term of the result is given by using the static anomalous moment with Schrödinger-Coulomb wave functions, and the corrections have additional powers of $Z\alpha$. Thus we cannot find the analog of the Breit relativistic correction for the anomalous moment term; this explains why we cannot simply factor out that correction. When an electron self-energy correction occurs outside a recoil correction loop (not shown), its contribution is smaller than those to be discussed here because $\tilde{\Sigma}_{\text{ren}}$ contains two factors of $\not{p} - m_e$. The actual calculation

[p]The situation differs from that in the Lamb shift, where the use of the Fried-Yennie gauge eliminates terms of spurious lower order, but not the need to take into account an arbitrary number of Coulomb interactions.

was handled slightly differently in STY. First, the self-energy and vertex functions were not renormalized from the beginning. To compensate for the renormalization constant which is introduced in this way, we can rearrange the present expression by adding one term where an electron self-energy is outside the recoil loop. Second, the R's are initially ignored, and the non-recoil part is removed at a later stage, as described below.

Because higher order terms in α can be neglected, it is possible to treat these contributions in a simple way by setting the spatial part of their external momentum equal to zero, with the effect of atomic structure being accounted for by simply introducing a factor of the square of the nonrelativistic wave function at the origin. Because of the binding energy, the illustrated amplitudes are slightly off-mass-shell. This serves to regulate the infrared divergence, which ultimately cancels between the five contributions. The calculation then reduces to a free-particle Feynman diagram analysis. At this stage, the gauge choice is immaterial in principle because after the infrared cancellations, the result is independent of gauge. However, an additional advantage of the Fried-Yennie gauge is that it turns out to be straightforward to make manifest certain compensations between individual amplitudes which eliminate terms involving $\ln^2(m_\mu/m_e)$, and make it very straightforward to extract the singly logarithmic terms analytically. The compensation facilitates an improvement in the numerical accuracy in the work of STY, and helps avoid complications in analytic work. It should be observed, however, that a future researcher desiring to evaluate higher order corrections in α will find that the situation is more complicated since amplitudes where the virtual photon spans an arbitrary number of exchanged photons will be important. This is a relatively common occurence in QED, and is one of the reasons we believe that a unified treatment such as the one presented in this review will be absolutely necessary for the next round of higher order calculations in this field.

Following the procedure just described, one finds after considerable algebraic manipulations that if the loop momentum of the exchanged photons is denoted as p, the energy shift can be written as

$$\Delta E_F = -i\alpha^2 E_F \frac{m_e m_\mu}{\pi^3} \int \vec{p}^{\,2} dp \int dp_0 \frac{1}{(p^2)^2} [(2\vec{p}^{\,2} - 3p_0^2)T_1 + 3p_0 T_2] \times \left[\frac{1}{p^2 - \gamma^2 - 2m_\mu p_0 + i\epsilon} + \frac{1}{p^2 - \gamma^2 + 2m_\mu p_0 + i\epsilon} \right], \tag{6.4}$$

where T_1 and T_2 are integrals over two Feynman parameters that come from carrying out the radiative loop integral on the electron line, and involve extensive cancellations and regroupings of the various graphs. Interestingly, in this form the non-recoil radiative corrections are retained to relative order $\alpha(Z\alpha)$. We can recover them by replacing the last bracket with $-2\pi i\delta(p_0)/2m_\mu$, after which the momentum integration and Feynman parameter integrations can be easily carried out to give the known result.[56,57,58] This provides a useful check on the algebra and suggests that we can remove the non-recoil result by subtracting the δ-function term from the square bracket. However, that would not be convenient for numerical integrations, so STY devised a simple trick to accomplish the same result. The first step is to carry out the p_0 integration without further approximation with the aid of a Wick rotation $p_0 \to ip_4$. While the resulting integration will include the non-recoil result, it is now straightforward to subtract this by introducing a counter term in which we set $p_4 = 0$ in all factors of the integrand except the last bracket and also drop p_4^2 terms in the denominators of that bracket. The subtracted term just reproduces the non-recoil calculation mentioned above. After the subtraction, the resulting four dimensional integral can be evaluated with standard Monte-Carlo integration techniques,[112] or if one can make an expansion in m_e/m_μ, be carried out analytically. The numerical approach was pursued by STY, and was improved analytically by Eides, *et al.*,[70] although the details of their calculation are different from the one indicated here. A similar calculation can be carried out when the radiative correction is on the other fermion line, though of course in that case there is no non-recoil result to be subtracted out. The results for muonium and positronium are included in (2.27) and the discussion following (2.39), respectively.

2. Remarks about Radiative Corrections Involving the Proton Structure in Hydrogen hfs

In the case of radiative corrections that do not involve the proton's structure, the most important contributions are contained in (2.25). As was the case for the muonium radiative corrections, we expect the largest ones involving the structure to arise from vacuum polarization insertions into contributions whose characteristic momentum is large compared to the electron mass. From Chapter II-C, we see that the most important of these is the Zemach contribution. From (2.34b), one can see that this contribution comes from a broad range of momenta characterized by Λ^2. Since the logarithm in (6.1) varies slowly in that range, we obtain a crude estimate of the size of the correction by replacing $-p^2$ by Λ^2 and doubling the result to allow for the two ways of inserting the vacuum polarization. This yields a correction factor

$$\frac{4\alpha}{3\pi}\ln\frac{\Lambda}{m_e} \approx 0.023\,, \tag{6.5}$$

which would lead to a correction of about -1 ppm. The total result for all the radiative corrections to the Zemach contribution would probably be somewhat smaller, since the corrections on the electron line tend to contribute with the opposite sign.

In the case of radiative corrections to δ_p(recoil), the largest individual correction is then about 1 ppm (in magnitude). However, the individual terms are likely to cancel as in Chapter V, giving a contribution at the level of 0.03 ppm. The radiative corrections that do not contain the leading logarithm of (6.1), including corrections on the electron line, are potentially more problematic, since they would likely affect the various recoil corrections in different ways. However, their size is intrinsically smaller than that of the leading logarithmic correction. Hence, the corresponding radiative-recoil corrections would probably contribute less than 0.1 ppm.

B. Recoil Effects in the Lamb Shift

For the Lamb shift, it is desirable to keep all the radiative correction terms together in a single expression. This is a situation in which we tend to think of recoil as a modification of the basic radiative corrections. However, the goal of keeping these terms together seems to be incompatible with a field-theoretical treatment of recoil which appears to require discussion of individual kernels with definite numbers of exchanged photons. The problem is that the kernels with a small number of exchanged photons inside the radiative correction loop (including none) must be treated exactly in order to renormalize correctly, while the multiphoton exchange terms are very non-relativistic and should be treated differently.

After a lot of trial and error, we were not able to find a straightforward way of treating recoil before coping with renormalization. If we use the original denominator rearrangement of (4.9c) and (4.28d), it turns out to be very difficult to get the correct reduced mass dependence of the argument of the Bethe logarithm. One can expand the logarithm in powers of m_e/m_N, but there is no simple way even to get the first order term correctly, and the expansion would fail completely for positronium. The trouble is that those rearrangements are fortuitously simple for the particular situation discussed in Chapter IV and do not adequately take into account the actual positions of the poles in the present situation when a radiative correction photon loop is present. Since these difficulties are associated with the very non-relativistic region for the "spectator" particle, a possible way out is to first pick out the leading non-relativistic terms in the propagator for that particle. This is, in essence, the approach followed by Bhatt and Grotch,[12] which extends the work of Grotch and Yennie,[79] who used an effective Hamiltonian which treated the proton non-relativistically but did not incorporate radiative corrections. Bhatt and Grotch thus remove most of the dynamical details for the proton and incorporate some of them in modifications of the electron propagator. This is referred to as the external field approximation (EFA).

We start with a brief description of the Bhatt-Grotch approach and then try to relate it to the present framework. Their first step is to work out the leading terms in the EFA. Effects not included in the EFA are later incorporated through various perturbation kernels. The EFA evolved from the equation studied by Grotch and Yennie, which takes the form of the Dirac equation with the redefinition of variables

$$\Pi_0 = E_n - V_c - \frac{\vec{p}^2}{2m_N}, \quad \vec{\Pi} = \vec{p} - e\vec{A} \tag{6.6a}$$

where

$$e\vec{A} = -\frac{1}{2m_N} V_c[\vec{p} + \hat{r}\hat{r}\cdot\vec{p}] \,. \tag{6.6b}$$

This equation reproduces the fine structure result (2.3) correctly to first order in $1/m_N$. Next, these expressions are substituted into the formal treatment of Chapter III-D. This gives a very simple derivation of the reduced mass dependence of the leading term of the Lamb shift, which we describe further below. This is incorporated in the first three lines of (2.5b) and in the first logarithm of (2.5a). They also find the reduced mass dependence of the "relativistic" contribution to the Lamb shift, given in the fourth line of (2.5b). In addition, the EFA gives some new terms not anticipated by simple intuitive arguments. Some of these arise from the fact that in addition to the effects of the scaling of the wave function, the actual integrals encountered are modified because of the proton mass dependence. Others arise because the wave function from the Grotch-Yennie treatment is slightly different from the Dirac-Coulomb wave function. These further corrections,

$$\frac{\pi\alpha(Z\alpha)^5 m_e^2}{\pi n^3 m_N}\left(\frac{35}{4}\ln 2 - 8 + \frac{1}{5} + \frac{31}{192}\right)\delta_{l0}\ , \tag{6.7}$$

are included in (2.8). The final step is to identify kernels which are corrections to the EFA. These are essentially the contributions associated with Fig. 7b–e, where now R means to remove contributions already included in the EFA. Techniques similar to those of Sapirstein, *et al.*[59] are used to work out the algebra and evaluate the resulting integrals using VEGAS.[112] The result is

$$\frac{\pi\alpha(Z\alpha)^5 m_e^2}{\pi n^3 m_N}(-0.415 \pm 0.004)\delta_{l0}\ , \tag{6.8}$$

which is also included in (2.8).

Since we have repeatedly stressed the importance of obtaining all results from a single basic formalism in order to avoid gaps or overlaps in results, we wish to indicate how the Bhatt-Grotch procedure may be derivable from the formulation discussed in Chapter IV. We should stress that we have not gone through the complete analysis, but we have proceeded far enough to see that it is quite plausible. We start by extending the formal treatment of the Lamb shift discussed in Chapter III-D. In this extension, we study the radiative corrections on the electron line first, and include the effects of the interaction with the proton later. This is accomplished by generalizing A_μ to be the second quantized electromagnetic field with the understanding that it has no pairings within the electron line. Such pairings are already taken into account explicitly through the electron self energies. Also, the p_0 operator (time derivative) is included so that we can subsequently study recoil corrections by assigning it numerical values for loop integrations. After formal reduction of the self-energy operator, one may introduce pairings with the proton line to include exchanged photons. The advantage of this approach is that the renormalization discussion may be carried out first independently of the discussion of recoil.

We can now take over much of the work of Chapter III-D without change. We start with (3.24) in this new context; this is to be combined with a similar factor for the proton line (we ignore the complications of strong interactions here and use only the first term in that expansion). To see what this means, consider the first term of the series and make all photon pairings between the electron and proton lines. As mentioned above, p_0 is assigned a value for each electron line and becomes a set of loop integration variables. The result corresponds to the analysis of Chapter IV for the four-point function without radiative corrections, where a bound state corresponds to a first order pole as a function of $(E - E_0)$.

To obtain a contribution to first order in the radiative corrections, we must consider the term linear in $\tilde{\Sigma}_{\rm ren}$ and seek a second order pole in $(E - E_0)$, with the coefficient of that pole representing the energy shift. The two poles can be produced by the factors of $1/(\not{\Pi} - m_e)$, keeping only pairings between the electron and proton lines in which there are limited overlaps between $\tilde{\Sigma}_{\rm ren}$ and the other factors so that the two poles can develop. As in the discussion of Chapter III-D, a factor of $\not{\Pi} - m_e$ arising from $\tilde{\Sigma}_{\rm ren}$ cancels one of these poles and no energy shift results. Overlaps which eliminate factors of $\bar{G}$ may produce single poles or no poles. As usual, the modification of the residue of a single pole factors out of the whole series and represents a wave function correction. Thus we can carry through the formal analysis along the lines of Chapter III-D and drop such terms as they arise. In this way, $\tilde{\Sigma}_{\rm ren}$ evolves into a radiative correction kernel $K_{\rm rad}$ which includes exchanged photon and proton propagators. A careful description is necessary to make $K_{\rm rad}$ well defined. We incorporate in it everything except adjacent factors of $\bar{G}$ so that it may be used directly in the perturbative analysis of Chapter IV-A(2) just like any other kernel. This means that it may include ladder structure outside the radiative correction photon if that structure has had its non-relativistic pieces removed (this is a slight modification of the meaning of R for the present circumstances). In this way, related contributions involving remainders of ladders together with crossed photon lines are kept together so that, as we have seen, cancellations often occur. Whenever such structures occur, they correspond to some recoil corrections.

Higher orders in perturbation theory involving $K_{\rm rad}$ together with other kernels generally produce results much smaller than the leading contribution, which is $\mathcal{O}(\alpha(Z\alpha)^4 m_r)$. To see this, note that the order of magnitude of other kernels is typically smaller than $\mathcal{O}((Z\alpha)^4 m_r^2/M)$, while the energy denominator occurring in the sum is of $\mathcal{O}((Z\alpha)^2 m_r)$, resulting in a correction factor smaller than $\mathcal{O}((Z\alpha)^2 m_r/M)$. This is unlikely to be of any importance in hydrogen or muonium and is still unimportant in positronium. Some care must be taken in this argument. In the discussion of Chapter IV-B(2,3) where the cancellation of kernels of $\mathcal{O}((Z\alpha)^4 m_r^2/M)$ was demonstrated in the expectation value, it was necessary to use the equation satisfied by the bound state. In the present case, if the operator $[\vec{\gamma}\cdot\vec{p} + \mathcal{V} + m_e - E_0'\gamma_0]$ acts on a state $|\bar{n}\rangle$, it becomes $(E_n' - E_0')\gamma_0$ rather than 0. Thus the energy denominator is cancelled and closure (with an appropriate adjustment for the term with $n = 0$) can be used to carry out the sum. This gives an effective wave function correction which may be combined with other contributions to produce the correction described below.

We still appear to be facing the dilemma that to treat the Lamb shift properly, it is necessary to keep all orders of the interaction between the electron and the proton, while to study the recoil, we need to make an expansion in the number of photons exchanged. However, we also have the experience of Chapter IV that the dominant contributions come from considering ladder graphs only, and in those we need keep only the effects of putting the proton on mass shell (approximately). In the absence of radiative corrections, these contributions can be summed up to give one of the solvable models discussed there, including reduced mass dependence; the remaining terms give real dynamical recoil corrections. Here we follow a similar strategy. Namely, we first expand up the operators A_μ from the electron denominators and pair them with similar operators from the proton line in various configurations. Next we make suitable approximations to reproduce the leading effects of taking into account the near pinching of the p_0-integrations by the poles in the electron and proton denominators. This amounts to making the non-relativistic approximation for the proton and treating the corrections perturbatively as dynamical recoil corrections. Finally, we resum the leading terms of the resulting series to recover the wave function and operator dependence of Chapter III.

Basically, our aim is to change the proton factor from a Dirac structure to a Schrödinger one with certain well-defined corrections which can be identified and treated as perturbations. Except for effects which can be absorbed into exact reduced mass dependence, we are willing here to make an expansion in powers of $1/m_N$. These corrections are typically of relative order $\alpha^{1,2}m_e/m_N$ and we refer to them as dynamical radiative-recoil corrections; some of these are identified as we proceed. The distinction is not totally unique, but it is practical for hydrogen and muonium. For positronium, the corrections have higher powers of α at least. For convenience, we replace m_N in the propagator numerator by E'' with an error of relative order γ^2/m_N^2, which is totally unimportant for hydrogen and small for positronium. Next we look at the large-large components of any of the resulting free proton propagators, since these give the dominant structure. We concentrate on the proton poles which are very close to the origin as a function of the various p_0's. In this region, we may make the approximation for a free proton propagator

$$\frac{2E''+Q_0}{2E''Q_0+Q_0^2-\vec{Q}^{\,2}-\gamma^2+i\epsilon}\rightarrow\frac{2E''}{2E''Q_0-\vec{Q}^{\,2}-\gamma^2+i\epsilon}\,, \tag{6.9}$$

where $Q=\sum_i s_i p_i$, as in the discussion of (4.42a). The error made is of second order in $1/m_N$ in the non-relativistic region. It may be treated as a perturbation which would produce additional powers of α as well. Similarly, quadratic powers of p_0's in photon denominators produce higher powers of $1/m_N$ and α so we set them aside to be treated perturbatively. As seen below, some such approximations are also made on the electron line after the reduced mass effects are taken into account.

In the case of a ladder structure where $Q=-p$ for some electron momentum, we define $K(p)\equiv 2E''p_0+\vec{p}^{\,2}+\gamma^2$ and make the rearangement

$$\frac{2E''}{-K(p)+i\epsilon}=-2\pi i\delta\left(p_0+\frac{\vec{p}^{\,2}+\gamma^2}{2E''}\right)-\frac{2E''}{K(p)+i\epsilon}\,. \tag{6.10}$$

The δ-function is an approximation to putting the proton on mass shell. For proton propagators involving several p_i's, the right side of (6.9) becomes

$$\frac{2E''}{2E''Q_0-\vec{Q}^{\,2}-\gamma^2+i\epsilon}=\frac{2E''}{\sum_i s_iK(p_i)-[\vec{Q}^{\,2}+\sum_i s_i\vec{p}_i^{\,2}]+i\epsilon}\,. \tag{6.11}$$

The term in brackets can be expanded up and treated as a recoil correction, much as in the discussion of Chapter IV-B(5). In the present treatment, there are no artifact terms to deal with.

Now consider contributions where all photons have the time polarization. Then we may apply (6.9). The leading contribution arises from ladder graphs in which the δ-function part of (6.10) is used everywhere so that p_0 is replaced by $-(\vec{p}^{\,2}+\gamma^2)/2E''$. This corresponds to the definition of Π_0 in the EFA as given in (6.6a). The apparent difference due to the γ^2 term is a consequence of the way we define the routing of the constant part of the total energy; it changes nothing. Correction terms arise from the second term of (6.10) or from various crossings of photon lines. When these terms are properly grouped together, they are found to sum to zero (see the remarks following (4.42)). The corrections from expanding up the brackets to first order may be treated as in Chapter IV-B(5) and give a result like (4.43a), with only the spatial part of the dot product. In combination with radiative corrections on the electron line, we estimate that with two Coulomb interactions in the kernel this adds a power of $Z\alpha$ and one of m_e/m_N to the lowest order Lamb shift order of magnitude. Each additional Coulomb interaction increases the order by one power of $Z\alpha$. We define these to be dynamical recoil corrections.

If all photons except one have time polarization, the spatial index photon produces a factor (4.41b). Since this is smaller than the dominant term by one power of $1/m_N$, we obtain the leading result by taking the δ-function parts of (6.10) with all the proton propagators in the ladder configuration. The corrections from the remainder to (6.10) and from crossed photon diagrams produce structures similar to (4.43b,c); their expected order is $Z\alpha m_e/m_N$ smaller than that of the leading contribution to the Lamb shift. If two photons have spatial indices, they must be adjacent on the proton line and they produce a seagull.

To summarize the proton line results, in the dominant terms time index photons have become Coulomb interactions and spatial index photons give an effective vector potential acting on the electron. It turns out that in the Coulomb gauge the spin independent part of the effective vector potential takes the form

$$e\vec{A} = -\frac{1}{2m_N}\{\vec{p}V_c + V_c\vec{p}\} - \frac{1}{4m_N}[\vec{p},[\vec{p}^{\,2}, W_c]] \,, \tag{6.12}$$

where W_c is the quantity defined in (4.25c). This is equivalent to (6.6b). We have also noted some contributions which are to be regarded as dynamical recoil corrections.

Now we return to the electron line and try to resum contributions to the greatest extent possible. Suppose that p is the momentum in an electron line outside $K_{\rm rad}$. That line contributes a factor

$$\begin{aligned} \frac{m_e + \gamma_0(E'+p_0) - \vec{\gamma}\cdot\vec{p}}{2E'p_0 + p_0^2 - \vec{p}^{\,2} - \gamma^2 + i\epsilon} &\to \frac{m_e + \gamma_0(E'+p_0) - \vec{\gamma}\cdot\vec{p}}{2E'p_0 - \vec{p}^{\,2} - \gamma^2 + i\epsilon} \\ &\to \frac{m_e + \gamma_0 E' - \vec{\gamma}\cdot\vec{p}}{-(E/E'')[\vec{p}^{\,2} + \gamma^2]} + \frac{\gamma_0}{2E} \,. \end{aligned} \tag{6.13}$$

The first step involves the approximation of neglecting p_0^2 in the denominator. The corrections to this approximation add powers of $1/m_N$ and α, and are to be regarded as dynamical recoil corrections. The second step substitutes from the δ-function fixing p_0. The first term of the overall result is equivalent to that obtained from (4.9c) with the numerator structure giving the Dirac model. The second term is already incorporated in the discussion of the compensation of terms of $\mathcal{O}((Z\alpha)^4 m_r^2/M)$ in the kernels; it corresponds to the discussion preceding (4.31). The expansion in the Coulomb interaction can be resummed to produce the Dirac model propagator.

Since terms of second order in the kernels can be ignored, we consider the expectation value of the radiative correction kernel, including the wave function corrections mentioned earlier. This should incorporate everything in the Bhatt-Grotch analysis, albeit with a different organization of details. There is still considerable flexibility in choosing the point in the formal analysis at which the proton propagator analysis is to be invoked. It appears obvious that this point should be after the initial reductions have been carried out so that we have only terms with explicit field strength factors. As an actual analysis has not yet been carried through, the best way to optimize it is unknown. In any case, our purpose here is only to give the flavor of the analysis and an indication of the flavor of the calculation of Bhatt and Grotch.

As a simple example, let's start with (3.37) from Chapter III-D; other terms can be studied in a similar manner. For convenient reference, we rewrite it here;

$$-\left[\frac{1}{D}2x[\Pi_\mu, \not{\Pi}]\frac{1}{D^2} + \frac{1}{D^2}2x[\Pi_\mu, \not{\Pi}]\frac{1}{D}\right] 2\Pi^\mu \ , \tag{6.14}$$

where D is defined in (3.27b), and the operator Π_μ now has the components $(E' - eA_0 + p_0, \vec{p} - e\vec{A})$. We may now follow exactly the same steps which led to (3.39a). The terms set aside all produce higher powers of α in the present context, but not necessarily higher powers of $1/m_N$. Each correction can be examined in turn to find its recoil corrections. As justified earlier, after dropping the terms $\mathcal{M}$ in the denominators and making the formal shift of k, the terms $\not{\Pi}^2 - m_e^2$ in the external denominators may be replaced by zero. At this point, the electron denominators may be combined and the k-integration carried out. The resulting expression is (3.39a), with the change in the meaning of Π_μ. Now we are ready to study the effects of recoil.

The denominator in (3.39a) is

$$u(1-x)[\not{\Pi}^2 - m_e^2] - xm_e^2 \ , \tag{6.15a}$$

where

$$\begin{aligned}\not{\Pi}^2 - m_e^2 = &-2E'eA_0 + e^2A_0^2 - \vec{p}^{\,2} - \gamma^2 + \mathcal{M} \\ &+ \{2E'p_0 + p_0^2 - p_0eA_0 - eA_0p_0 + \vec{p}\cdot e\vec{A} + e\vec{A}\cdot\vec{p} - e^2\vec{A}^2\} \ .\end{aligned} \tag{6.15b}$$

Here we group the principal terms which depend on the proton recoil inside the brace; in addition, $\mathcal{M}$ has such terms through the vector potential. If we use the δ-function part of (6.10), it is clear that the most important term in the brace is the first one. The p_0^2 term is formally of order $1/m_N^2$, but the rearrangement (6.10) is not appropriate for such a term. We simply expand it up and set it aside from present considerations. The p_0eA_0 and $e^2\vec{A}^2$ terms are probably beyond the order of interest so we ignore them. Now if we use the δ-function from (6.10) and replace $\vec{A}$ according to (6.12), we have the EFA, which gives

$$
\begin{aligned}
&u(1-x)\left[\not{\Pi}^2 - m_e^2\right] - x m_e^2 \\
&\quad \to u(1-x)\Big[-\frac{E}{E''}(\vec{p}^2+\gamma^2) - 2E'eA_0 + \vec{p}\cdot e\vec{A} + e\vec{A}\cdot\vec{p} + e^2A_0^2 + \mathcal{M}\Big] - x m_e^2 \\
&\quad \approx u(1-x)\left[\frac{M}{m_N}[\not{\Pi}'^2 - m_e^2]\right] - x m_e^2 \,,
\end{aligned}
\tag{6.15c}
$$

where $\Pi_0' = E' - \frac{m_N}{M} eA_0$, $\vec{\Pi}' = \vec{p} - \frac{m_N}{M} e\vec{A}$. A small error has been made in the $e^2A_0^2$ term; this could be incorporated in another small correction. Also, since our discussion of the proton factor is only to first order in $\vec{A}$, the $e^2\vec{A}^2$ term is to be dropped here.

Following through the steps leading to (3.39c), we find that the argument of the logarithm in (3.40b) becomes

$$
\frac{m_e}{\dfrac{\vec{p}^2+\gamma^2}{2m_r} + V_c} \,. \tag{6.16}
$$

There is no special complication when one proton denominator carries the same momentum as two electron denominators, as can happen with the factor Π^μ of (3.39a), for example (however, not with the factor $[\Pi_\mu, \not{\Pi}]$, which of necessity represents an exchange photon vertex). From our treatment of the proton factor, eA_0 has become a Coulomb interaction and $\vec{A}$ takes the form (6.12).

We are also interested in the overall factor in this contribution to the Lamb shift. It is easy to trace this through in our analysis. Each of the propagators which produces a power of $1/(E' - E_0')$ has an overall factor of $E''/E \approx N$. One of these factors out of the series, and the other is compensated by the factor from (4.15d). The result is that the energy shift is given directly by the expectation value of the operator in (3.40b) in the state under consideration. It should be emphasized that the quantity m_e appearing in that expression is in fact the electron mass and not the reduced mass. Now it is easy to see how the reduced mass occurs in the final result. At this point, the only scale in the wave function is γ; the electron mass makes an explicit appearance only in the operator. The operator has an explicit power of $\alpha(Z\alpha)$ from the virtual photon and the factor of V_c. Thus the evaluation of the expectation value must produce an overall power of γ^3 to provide the correct dimensions; the result is an overall factor $\alpha(Z\alpha)^4 m_r^3/m_e^2$. Also, the denominator of the argument of the logarithm must be proportional to $\alpha^2 m_r$, giving the expected reduced mass dependence.

The reduced mass dependence of the anomalous magnetic moment works out slightly differently and requires some attention to detail. If one carries out the preceding analysis using only Coulomb interactions, the operator $\mathcal{M}/m_e$ couples the large and small components of the wave function. Using (4.17a), it can be seen that to a reasonable approximation the small component of the wave function is $\vec{\sigma}\cdot\vec{p}/(2m_e)$ acting on the large component. The resulting effective operator between large components has terms of the form $\nabla^2 V_c/m_e^2$ and $\vec{\sigma}_e \cdot \vec{L}e^2/(m_e^2 r^3)$. The first applies to S-states and the second is a spin-orbit coupling which applies to other orbital angular momentum states. Thus the dimensional argument leads to the same conclusion. However, the transverse interaction also contributes directly between the large electron components and it has only a spin-orbit structure. It adds a contribution such that its effect is to multiply the previous spin-orbit term by $1 + \frac{m_e}{m_N} = \frac{m_e}{m_r}$.

So far, we have sketched the Bhatt-Grotch discussion for the reduced mass dependence of the leading term. Various recoil terms have been dropped in passing from (6.15a) to (6.16). We can argue that they are all too small to be of present interest; of course, our nominal order estimates require care because orders can be reduced due to the properties of the integrals. If the correction due to dropping p_0^2 is considered, we expect it to become very relativistic and to have an additional two factors of $1/m_N$. The dimensional arguments then suggest the order $\alpha(Z\alpha)^6 m_r^5/m_e^2 m_N^2$. However, this is probably too small since the natural scale of some subintegrations is likely to be m_e rather than γ; nevertheless, our estimate is still small, namely $\alpha(Z\alpha)^5 m_e^3/m_N^2$. All the terms from $p \cdot eA$ have an explicit factor of $Z\alpha$ in the numerator and a factor of $1/m_N$; their contribution is expected to be of order $\alpha(Z\alpha)^6 m_r^4/m_e^2 m_N$, which is beyond the present level of interest. The $e^2 \vec{A}^2$ contribution is expected to be even smaller.

Having dispensed with these recoil terms in the operator, Bhatt and Grotch then note that the wave function corrections alluded to earlier do contribute at the level of interest within the EFA. These take the form of $-[\vec{p}^2, W_c]/4m_N$ acting on the wave function. Although the nominal order of these contributions appears to be $\alpha(Z\alpha)^6 m_r^4/m_e^2 m_N$, its actual order is less by one factor of $Z\alpha$. The reason is that both wave function integrations can converge in the non-relativistic region producing inverse powers of γ while the internal integration is relativistic for the electron.

Bhatt and Grotch identify other terms of order $\alpha(Z\alpha)^5 m_e^2/m_N$ in the EFA. These arise from contributions of higher order in $Z\alpha$ in the original formal reduction. Certain kernels have two Coulomb exchanges and (electron) relativistic internal momenta, but the wave function integrations converge separately. Their contribution without recoil corrections is of order $\alpha(Z\alpha)^4 m_r^3/m_e^2$, with the reduced mass dependence coming from the wave functions. In the kernels, the main effect of the recoil is in the value of p_0. The calculation is quite tedious, but has been carried through analytically by them.

They also identify terms of the same order in kernels which are not taken into account within the EFA, as mentioned above. They would appear in a different way here, and we see no way to make a direct connection. For example, the term from expanding up the bracket of (6.11) is not included in the EFA. Since a correction of this sort would make the electron momenta within the kernel relativisitic, at most two powers of the Coulomb interaction would be important and it might not be very difficult to carry through the calculation this way. In any case, the result of Bhatt and Grotch is that these non-EFA contributions are very small.

Acknowledgements

The preparation of this review extended over a three year period. Most of it was done at the authors' home institutions, where it was supported in part at Notre Dame by NSF Grant PHY-86-08101, and at Cornell by the National Science Foundation. Some of it was also carried out at other places. A particularly fruitful period was spent at the program on *Relativistic, Quantum Electrodynamic, and Weak Interaction Effects in Atoms*, which took place at the Institute for Theoretical Physics at Santa Barbara during the first half of 1988. We gratefully acknowledge the hospitality provided by the Institute and its support through NSF Grant PHY82-17853, with supplemental funds provided by the National Aeronautics and Space Administration. One of us (DRY) also benefited from the hospitality of the Theory Groups at the Fermi National Accelerator Laboratory (late spring, 1988) and at the Stanford Linear Accelerator Center (winter, 1989). Several colleagues also deserve special thanks. We drew heavily from two papers that one of us (DRY) had coauthored with G. T. Bodwin. We benefited greatly from the special insights of G. P. Lepage in this field. In many ways the review was improved by discussions with G. S. Adkins, G. Drake, W. R. Johnson, and P. Mohr. We thank T. Kinoshita for inviting us to prepare this review and for his gentle encouragement to complete it.

APPENDICES

A. Determination of the Function $G_{SE}(n, Z\alpha)$.

The values for the non-recoil part of the Lamb shift presented in Chapter II involved logarithmic terms with analytically known coefficients, Bethe logarithms known to extremely high accuracy numerically, and finally the quantities G_{SE} and G_{VP}. One of the largest theoretical uncertainties in the Lamb shift comes from the large error in G_{SE}, and the purpose of this appendix is to give the details of its determination for low values of the nuclear charge Z.

Any of the theoretical approaches to the non-recoil Lamb shift described in Chapter III should be able to account for this contribution. However, the methods which rely on expanding in the quantity $Z\alpha$ become quite complex in higher orders, and to date only the approach described in Chapter III-B has been applied to a direct calculation of this term for the ground state. A dominant part of the term that includes large constants associated with the logarithmic terms was calculated by Erickson and Yennie[99] using the approach of Chapter III-D. However, it is the remaining approach of Chapter III-C that gives the most accurate determination of these terms. In principle, that numerical approach could be used for any Z, and would be superior to the expansion techniques because all orders of $Z\alpha$ are accounted for. Unfortunately, as Z becomes of order 1, the numerical difficulty of this kind of calculation increases rapidly, and the functions $G_{SE}(n, Z\alpha)$ have been calculated only for Z=10, 20, 30, ..., 110.[21] The actual calculation is of the self energy contribution to the function $F(Z\alpha)$. In Table A-1, we reproduce the results for the first three values of Z for the $n = 1$ and $n = 2$ states. It is clear from that table that the numerical error for S-states is greatest for Z=10. This error, together with the systematic error of extrapolation, is responsible for the relative inaccuracy of the results presented in Chapter II.

Table A-1. Values of the self-energy part of $F_n(Z\alpha)$

State	$1S_{1/2}$	$2S_{1/2}$	$2P_{1/2}$	$2P_{3/2}$
Z=10	4.654(2)	4.893(2)	-.1145(4)	.1303(4)
Z=20	3.246(1)	3.5063(4)	-.0922(4)	.1436(4)
Z=30	2.5519(5)	2.8391(3)	-.0641(4)	.1604(4)

For values of Z larger than 30, $F_n(Z\alpha)$ is a relatively smoothly varying function, as can be seen from Fig. 3, and ordinary interpolating techniques can be used to evaluate the function for any Z. This has been done in a very useful compilation by Johnson and Soff.[114] For lower values, however, the function changes fairly rapidly, and the following procedure is applied. Because a great deal of that variation comes from the known terms of low order in $Z\alpha$, one subtracts from the above table the terms involving the self-energy parts of $A_{40}, A_{41}, A_{50}, A_{61}$, and A_{62} given in (2.5b). This result is then divided by $(Z\alpha)^2$ to determine the values of G_{SE} shown in Table A-2.

Table A-2. Values of $G_{SE}(Z\alpha)$

State	$1S_{1/2}$	$2S_{1/2}$	$2P_{1/2}$	$2P_{3/2}$
Z=10	-26.63(38)	-27.58(38)	-.715(75)	-.383(75)
Z=20	-23.62(5)	-24.15(2)	-.585(19)	-.290(19)
Z=30	-21.114(10)	-21.468(6)	-.433(8)	-.206(8)

One next fits the entries in Table A-2 to the function

$$G_{SE}(Z\alpha) = A_{60} + (Z\alpha)[A_{71}\ln(Z\alpha)^{-2} + A_{70}]$$

The three constants in the fit are uncertain because of the errors in the table. The error estimate quoted in Chapter II comes from assuming extremes in the value at Z=10, which dominates the error. Once the constants are determined, of course, predictions for low-lying members of the isoelectronic sequence can be immediately determined. It is clear that it would be very desirable to recalculate the Z=10 value with higher precision, or, even better, to find some numerical improvement that would allow direct evaluations of G_{SE} without any of the uncertainties associated with extrapolation.

B. Fortran Listing of Fine Structure Program

The following program was used to generate the theoretical numbers quoted in the text (the quoted error estimates were, however, derived separately). This program is used for hydrogen: it can be applied to muonium or hydrogenic ions by adjusting the nuclear radius, mass, and charge appropriately. Four states are treated together, the $1S_{1/2}$, $2S_{1/2}$, $2P_{1/2}$, and $2P_{3/2}$ states in that order. The array *bl* holds the Bethe logarithms (3.12b), the arrays *a*40-*a*62 represent the constants given in (2.5b), with *a*60 dependent on the function G_{SE}

given in (2.5c) and discussed in the previous appendix. The array *en* represents the principal quantum number, the array *at* the state dependent constants (2.7b), *sal* the Salpeter contribution (2.7a) *ab* the two-loop contribution (2.10), *fns* the effect of finite nuclear size, *bg* the Bhatt-Grotch correction (2.8), and *fs* the fine structure given in (2.3).

```
program qed(input=tty,output=tty)
implicit double precision(a-h,o-z)
dimension bl(4),a40(4),a41(4),a50(4),a62(4),a61(4),a60(4),
. en(4),at(4),ans(4),sal(4),ab(4),fns(4),gse(4),fs(4)
data gse/-30.3, -32.1, -0.8, -0.5/
data en/1.,2.,2.,2./
data bl/2.9841285559,2.8117698932,-.0300167089,-.0300167089/
data pi/3.14159265358979/, zeta/1.202056903159594/, two/2./
data eme/.51099906/, hc/197.32705/, alin/137.0359895/
data c/2.99792458d10/, ryd/109737.315709/
z=1.
rnuc=.805                                 Proton charge, radius, and mass
emp=938.27231
del=eme/emp
emr=1./(1.+del)                           mr/me
al=1./alin
zal=z*al
zlog=dlog(1./zal**2/emr)                  Build in reduced mass into logarithm
zlog0=dlog(1./zal**2)
fac= 2.*c*ryd*al**3/pi*1.d-6              Constant for Lamb shift
facp=2.*c*ryd*al**2*1.d-6                 Constant for fine structure
f1=-zal**2/2.-zal**4/8.-zal**6/16.
f2=-zal**2/8.-5.*zal**4/128.-21.*zal**6/1024.
f3=f2                                                   (2.2c)
f4=-zal**2/8.-zal**4/128.-zal**6/1024.
fs(1)=facp*(emr*f1-emr**3*del/2.*f1**2)/al**4
fs(2)=facp*(emr*f2-emr**3*del/2.*f2**2)/al**4
fs(3)=facp*(emr*f3-emr**3*del/2.*f3**2)/al**4+          (2.3)
. facp*z**4*emr**3*del**2/48.
fs(4)=facp*(emr*f4-emr**3*del/2.*f4**2)/al**4-
. facp*z**4*emr**3*del**2/96.
a40(1)=(10./9.-4./15.-4./3.*bl(1))*emr**3
a40(2)=(10./9.-4./15.-4./3.*bl(2))*emr**3               (2.5b)
a40(3)=-1./6.*emr**2-4./3.*bl(3)*emr**3
a40(4)=1./12.*emr**2-4./3.*bl(4)*emr**3
a41(1)=4./3.*emr**3
a41(2)=4./3.*emr**3
```

```
      a41(3)=0.                                                  "
      a41(4)=0.
      a50(1)=4.*pi*(139./128.+5./192.-.5*dlog(two))*emr**3
      a50(2)=a50(1)
      a50(3)=0.                                                  "
      a50(4)=0.
      a60(1)=(gse(1,iflag)-1289./1575.+4./15.*dlog(two)+19./45.-
     . pi**2/27.)*emr**3
      a60(2)=(gse(2,iflag)-743./900.+19./45.-pi**2/27.)*emr**3
      a60(3)=(gse(3,iflag)-9./140.)*emr**3                       " + (2.5c)
      a60(4)=(gse(4,iflag)-1./70.)*emr**3
      a61(1)=(28./3.*dlog(two)-21./20.-2./15.)*emr**3
      a61(2)=(16./3.*dlog(two)+67./30.-2./15.)*emr**3
      a61(3)=103./180.*emr**3                                    "
      a61(4)=29./90.*emr**3
      a62(1)=-emr**3
      a62(2)=a62(1)
      a62(3)=0.                                                  "
      a62(4)=0.
      at(1)=-2.*(dlog(two)+1.5)
      at(2)=-2.*2.25
      at(3)=1./6.                                                (2.7b)
      at(4)=1./6.
      sal(1)=4.*z/3.*del*(.25*zlog0-1./12.-2.*bl(1)-7./4.*at(1))*emr**3
      sal(2)=4.*z/3.*del*(.25*zlog0-1./12.-2.*bl(2)-7./4.*at(2))*emr**3
                                               Eqn. 2.7a (last term dropped)
      sal(3)=4.*z/3.*del*(-2.*bl(3)-7./4.*at(3))*emr**3
      sal(4)=4.*z/3.*del*(-2.*bl(4)-7./4.*at(4))*emr**3
      ab(1)=(-4358./1296.-10./27.*pi**2+3./2.*pi**2*dlog(two)-2.25*zeta)*emr**3
      ab(2)=ab(1)
      ab(3)=(-197./432.-pi**2/36.+pi**2/6.*dlog(two)-.25*zeta)*emr**2
      ab(4)=-.5*ab(3)                                            (2.10)
      bg(1)=zal*del*(35./4.*dlog(two)-7333./960.-.415)
      bg(2)=bg(1)                                                (2.8)
      bg(3)=0.
      bg(4)=0.
      fns(1)=2.*pi/3./al*(eme*rnuc/hc)**2*emr**3
      fns(2)=fns(1)
      fns(3)=0.                                                  (2.11)
      fns(4)=0.
      do 10 i=1,4
      ans(i)=a40(i)+a41(i)*zlog+zal*a50(i)+zal**2*(a62(i)*zlog**2+
```

```
     a61(i)*zlog+a60(i))+sal(i)+al/pi*ab(i)+bg(i)+fns(i)
     ans(i)=fac/en(i)**3*z**4*ans(i)+fs(i)
10   continue
     print *, ans(1)-fs(1), ans(2)-fs(2),ans(3)-fs(3)
     print *, ans(2)-ans(3), ans(2)-ans(1)
     stop
     end
```

C. Comparison of Some Bound State Formalisms

As mentioned before, modern precision calculations of bound state energies are based on three-dimensional formulations of the basic wave equation, with corrections inserted as perturbation kernels of various sorts. We indicate here some of the highlights of the development of these formulations without attempting to be complete. As emphasized by Lepage,[65] an arbitrariness in the choice of a three-dimensional formalism arises from the fact that there is no unique choice for the free propagator which occurs in the zeroth order integral equation. Different choices of propagator simply lead to different perturbation kernels. Thus there is freedom to choose the initial approximation in any convenient way for the particular problem under consideration. Many of these approaches contain an awkward feature that relativistic energies involving square roots occur in the denominator of the propagator. This makes it difficult to find closed form analytic solutions as the starting point of a perturbation treatment and leads to intractable integrals in low orders of perturbation theory. A well-known approach was developed by Blankenbecler and Sugar,[115] but it was oriented more toward the multichannel scattering problem. Using certain assumptions [on-shell unitarity and the linearity of the inverse propagator in $\vec{p}^2$], Todorov[116] arrives at an equation which has many features in common with ours. Austen and de Swart[117] give a field theoretic derivation of an equation — which is also similar to ours — in which the potential takes into account an important part of the physics of planar and crossed two-photon exchanges. In all of these three-dimensional formalisms, the loop energy p_0 is fixed by some procedure; and the difference between the full kernel and the fixed one is treated with the other perturbations. There is freedom of choice both on the value at which p_0 is fixed and the terms which are kept in the residue of the product of the two particle propagators.

Our particular procedure retains the mass symmetry in the choice of the point at which p_0 is fixed, but introduces mass asymmetry in the residue by retaining the full Dirac numerator structure for the lighter particle but not for the heavy one. Thus we have some of the dynamics necessary for obtaining mass symmetric results. Our method works best when there is a large mass ratio, but it has utility for positronium as well.

For the purposes of this Appendix, it is convenient to rewrite (4.2a) in a slightly different form

$$S = \frac{1}{E' + p_0 - H_e} \, \frac{1}{E'' - p_0 - H_p} \gamma_0 \otimes \gamma_0 \tag{C.1}$$

where $H_e = \vec{\alpha}_e \cdot \vec{p} + \beta_e m_e$ and $H_p = -\vec{\alpha}_p \cdot \vec{p} + \beta_p m_p$. Recall that each m is accompanied by $-i\epsilon$. The various model equations can be understood in terms of various ways of approximating (C.1)

The Breit Equation

The first main step in treating the recoil problem involving Dirac particles was due to Breit.[6] In that work he presented his famous equation for two electrons interacting through the electromagnetic field. That equation has the form of an instantaneous interaction between the electrons, but it does incorporate retardation effects to the extent that they are important at the v^2/c^2 level. He was guided by the classical result but arrived at the equation by a somewhat laborious analysis. His equation was the only one used for the treatment of recoil effects for over two decades. In fact, there were very few workers in the field, other than Breit and his associates, during that period. It took the stimulation of subsequent experiments to bring on the development of more precise formulations.

Breit assumed that the free part of the two-particle Hamiltonian is simply the sum of the electron and proton free Hamiltonians. In effect, this corresponds to replacing the first factor of (C.1) by (we also omit the factor of $-2\pi i\delta(p_0)\gamma_0 \otimes \gamma_0$ in this and subsequent propagators)

$$\frac{1}{E - H_e - H_p} = \frac{(E' + H_e)(E'' + H_p) - \vec{p}^2 - \gamma^2}{-2E(\vec{p}^2 + \gamma^2)} . \tag{C.2}$$

It is interesting to note that the numerator is very close to the one obtained by keeping the complete numerator with the first term of (4.9c). The difficulties of that literal definition were emphasized there. The complete definition of Breit's equation also requires a choice of potential. This was taken to be the one photon exchange interaction in the Coulomb gauge.

How is the Breit equation related to the two-particle propagator? We can obtain it from (C.1) by first redefining the positions of the poles. If we replace the $-i\epsilon$ accompanying the masses by $+i\epsilon$ in each denominator, we can close the p_0 contour (ignoring photon poles) and obtain (C.2). As pointed out by Salpeter, this makes it clear that the Breit equation corresponds to a one-electron one-proton picture, rather than to hole theory. Incidentally, if one continues to assume that a multiparticle Hamiltonian is simply the sum of individual Hamiltonians, worse difficulties arise as the number of particles increase. This phenomenon has been labeled "continuum dissolution" by Sucher.[118] Of course, it is not a true physical difficulty. Rather, it is a consequence of trying to treat a field theoretical problem in too simple a manner.

The Breit equation does give contributions of order α^4 correctly, but it is not well adapted toward calculation of smaller terms. To do so, one would have to undo the approximations made in arriving at (C.2) from (C.1). These "undoing corrections" would have to combine with other terms in order to avoid the difficulties inherent in the equation in higher order. We have not tried to pursue this systematically, but it looks awkward.

The Salpeter Equation

In the case where (C.1) lies between Coulomb interactions, one may carry out the p_0

integration explicitly. To do this, insert the identity in the form $[\Lambda_+^e(p)+\Lambda_-^e(p)][\Lambda_+^p(-p)+\Lambda_-^p(-p)]$ next to the denominators, where $\Lambda_\pm(\pm p) = [E(p) \pm H_d(\pm p)]/2E(p)$ with $E(p) = [m^2 + \vec{p}^{\,2}]^{1/2}$. Taking into account that the Coulomb potential is independent of p_0, one finds that the result of the contour integration is

$$\frac{1}{E - H_e - H_p}[\Lambda_+^e(p)\Lambda_+^p(-p) - \Lambda_-^e(p)\Lambda_-^p(-p)] = \frac{1}{E - H_e - H_p}[1 - \Lambda_-^e(p) - \Lambda_-^p(-p)] \quad (C.3)$$

In the second form, one may easily see the connection with the Breit equation, which corresponds to the first term in the square brackets. The Salpeter equation is a viable approach to bound state problems, but the result is quite complicated and not too useful for the refined calculations that are now being done. The square root expressions are not very tractable in such refined terms. In simple terms, one may approximate them by expanding in powers of $\vec{p}^{\,2}$, treating it as a small quantity until one reaches divergent integrals. At that point, one retains an exact remainder. The leading terms of such a procedure are usually quite straightforward, but later terms are quite complicated, and there are unexpected compensations between them. One calculation which reveals these complications was carried through by Bodwin and Yennie;[66] however, it was not possible to push the method as far as later approaches (BYG[74] and Caswell and Lepage[68]). With further approximations, it can be reduced to a form that leads to the Schrödinger equation with reduced mass. One deficiency of the Salpeter equation is that it does not reduce to the Dirac equation as one mass goes to infinity. Rather, it contains a positive-energy (plane wave) projection operator for the lighter particle. One can show that crossed Coulomb kernels restore the physics of the Dirac equation to given order in α, but the procedure is quite cumbersome.

The Grotch-Yennie Equation

In this equation, which is similar to the Breit equation, the proton is treated non-relativistically. The form of the free propagator is

$$\frac{1}{E - m_p - H_d^e - \dfrac{\vec{p}^{\,2}}{2m_p}} \frac{1 + \gamma_0^p}{2} \quad (C.4)$$

At some point, this is bound to become a poor approximation; it is necessary to treat terms where the proton becomes relativistic with care. One advantage of this equation is that when the interactions are taken into account (including the spatial part of the Breit interaction), a very simple equation is obtained which can be solved in terms of the usual Dirac-Coulomb problem. This is the basis of our treatment in Chapter IV-A(3). This equation produces quite a good account of the hydrogen bound state energies to relative order m_e/m_p.

The Gross Equation

This equation differs from the Salpeter one by putting only the proton on mass shell.

I.e., the projection operator in (8) is replaced by the positive-energy proton projector. Its original application was in the treatment of the deuteron problem.

The Caswell-Lepage Equation

Caswell and Lepage have developed a new renormalization group strategy for analyzing bound state problems in QED and other field theories.[68] It employs a nonrelativistic effective Lagrangian to separate effects due to relativistic loop momenta from those coming from nonrelativistic momenta, thereby avoiding much of the complexity inherent in traditional analyses. Applications to muonium and positronium hyperfine splitting are mentioned in Chapter II. A description of the approach is contained in an article by Lepage and Kinoshita in this volume.

REFERENCES

1. W. E. Lamb, and R. C. Retherford, Phys. Rev. **72**, 241 (1947).
2. J. E. Nafe, E. B. Nelson, and I. I. Rabi, Phys. Rev. **71**, 914 (1947); D. E. Nagle, R. S. Julian, and J. R. Zacharias, Phys. Rev. **72**, 971 (1947).
3. W. H. Furry, Phys. Rev. **81**, 115 (1951).
4. S. J. Brodsky, *Atomic Physics and Astrophysics*, Vol. I, edited by M. Chrétien and E. Lipworth (Gordon and Breach Science, New York), p. 91 (1969)
5. F. Gross, Phys. Rev. C**26**, 2203 (1982).
6. G. Breit, Phys. Rev. **29**, 553 (1929).
7. G. Breit, and R. E. Meyerott, Phys. Rev. **72**, 1023 (1947).
8. E. E. Salpeter, Phys. Rev. **87**, 328 (1952).
9. J. Schwinger, Proc. Nat. Acad. Sci. USA **37**, 452, 455 (1951).
10. E. E. Salpeter, and H. A. Bethe, Phys. Rev. **84**, 1232 (1951).
11. W. E. Caswell and G. P. Lepage, Phys. Rev. Lett. **41**, 1092 (1978).
12. G. Bhatt and H. Grotch, Phys. Rev. A**31**, 2794 (1985); Phys. Rev. Lett. **58**, 471 (1987); Ann. Phys. (NY) **178**, 1 (1987).
13. E. Richard Cohen and Barry N. Taylor, Rev. Mod. Phys. **59**, 1121 (1987).
14. F. Biraben, J. C. Garreau, L. Julien, and M. Allegrini, Phys. Rev. Lett. **62**, 621 (1989).
15. B.N. Taylor and T.J. Witt, Metrologia **26**, 47(1989).
16. W. A. Barker and F. N. Glover, Phys. Rev. **99**, 317 (1955).
17. J. Pirenne, Arch. Sci. Phys. Nat. **29**, 121, 207, 265 (1947); V. Berestetski and L. Landau, J. Exp. Theor. Phys. USSR **19**, 673, 1130 (1949); R. A. Ferrell, Phys. Rev. **84**, 858 (1951).

18. G. W. Erickson, J. Phys. Chem. Ref. Data, **6**, 833 (1977).

19. E.H. Wichmann and N.M. Kroll, Phys. Rev. **101**, 8431 (1956). The vacuum polarization numbers in (2.5c) have been calculated using the formulas of this paper by P. Mohr (private communication).

20. Gerhard Soff and Peter J. Mohr, Phys. Rev. **A38**, 5066 (1988).

21. P.J. Mohr, Phys. Rev. **A23**, 2338 (1982).

22. Fulton, T., and P. C. Martin, Phys. Rev. **95**, 811 (1954).

23. R. Mills and N. Kroll, Phys. Rev. **98**, 1489 (1955).

24. J. A. Fox and D. R. Yennie, Ann. Phys. (N.Y.) **81**, 438 (1973).

25. T. W. Appelquist and S. J. Brodsky, Phys. Rev. **A2**, 2293 (1970).

26. B.E. Lautrup, A. Peterman, and E. deRafael, Phys. Lett. **B31**, 577 (1970); R. Barbieri, J.A. Mignaco, and E. Remiddi, Nuovo Cim. Lett. **3**, 588 (1970).

27. A. J. Layzer, Bull. Am. Phys. Soc. **6**, 514 (1961).

28. A. J. Layzer, Nuovo Cimento **33**, 1538 (1964).

29. G. P. Lepage, D. R. Yennie, and G. W. Erickson, Phys. Rev. Lett. **47**, 1640 (1981).

30. K.T. Cheng and W.R. Johnson, Phys. Rev. **A14**, 1943 (1976); G. Soff, P. Schlüter, B. Müller, and W. Greiner, Phys. Rev. Lett. **48**, 1465 (1982).

31. V. A. Petrun'kin, and S. F. Semenko, Sov. J. Nucl. Phys. **3**, 355 (1966).

32. G. Plunien, B. Müller, W. Greiner and G. Soff, GSI preprint GSI-88-45.

33. G. W. Erickson and H. Grotch, Phys. Rev. Lett. **60**, 2611 (1988); erratum: Phys. Rev. Lett. **63**, 1326 (1989).

34. D.J. Drickey and L.N. Hand, Phys. Rev. Lett. **9**, 521 (1962); L.N. Hand, D.J. Miller, and R. Wilson, Rev. Mod. Phys. **35**, 335 (1963).

35. G.G. Simon, Ch. Schmidt, F. Borkowski, and V.H. Walther, Nucl. Phys. **A333**, 381 (1980).

36. S. R. Lundeen and F. M. Pipkin, Phys. Rev. Lett. **46**, 232 (1981); S. R. Lundeen and F. M. Pipkin, Metrologia **22**, 9 (1986).

37. K. A. Woodle *et al.*, Phys. Rev. **A41**, 93 (1990).

38. C.J. Oram et. al., Phys. Rev. Lett. **52**, 910 (1984).

39. S. Hatamian, R.S. Conti, and A. Rich, Phys. Rev. Lett. **58**, 1833 (1987).

40. G.W.F. Drake, J. Patel, and A. van Wijngaarden, Phys. Rev. Lett. **60**, 1002 (1988).

41. E. Borie and G.A. Rinker, Phys. Rev. **A18**, 324 (1978).

42. M. Eckhause, P. Guss, D. Joyce, J.R. Kane, R.T. Sidgel, W. Vulcan, R.E. Wlesh, and R. Whyley, Phys. Rev. **A33**, 1743 (1986).

43. C.T. Munger and H. Gould, Phys. Rev. Lett. **57**, 2927 (1986) and private communication.

44. I. Martinson, Rep. Prog. Phys. **52**, 157 (1989).

45. A. van Wijngaarden, J. Patel, and G.W.F. Drake, in "Atomic Physics 11", edited by S. Haroche (World Scientific, Singapore, 1989).

46. L.S. Vasilenko, V.P. Chebotaev, and A.V. Shishaev, JETP Lett. **12**, 113 (1970); and T.W. Hänsch, S.A. Lee, R. Wallenstein, and C. Wieman, Phys. Rev. Lett. **34**, 307(1975).

47. Steven Chu, A.P. Mills, Jr., A.G. Yodh, K. Nagamine, Y. Miyake, and T. Kuga, Phys. Rev. Lett. **60**, 101 (1988).

48. K. Danzmann, M.S. Fee, and Steven Chu, Phys. Rev. **A39**, 6072 (1989).

49. Steven Chu, Allen P. Mills, Jr., and John L. Hall, Phys. Rev. Lett. **52**, 1689 (1984).

50. R. G. Beausoleil, D. H. MacIntyre, C. J. Foot, E. A. Hildum, B. Couillaud, and T. W. Hänsch, Phys. Rev. **A35**, 4878 (1987).

51. T. Fulton, Phys. Rev. **A26**, 1794 (1982). This error has also been noted by Suraj N. Gupta, Wayne W. Repko, and Casimir J. Suchyta, III, Phys. Rev. **D40**, 4100 (1989).

52. F. G. Mariam, W. Beer, P. R. Bolton, P. O. Egan, C. J. Gardner, V. W. Hughes, D. C. Lu, P. A. Souder, H. Orth, J. Vetter, U. Moser, and G. zu Putlitz, Phys. Rev. Lett. **49**, 993 (1982).

53. E. Fermi, Z. Phys. **60**, 320 (1930).

54. E. Klempt et. al., Phys. Rev. **D25**, 652 (1982).

55. G. Breit, Phys. Rev. **35**, 1447 (1930).

56. N. Kroll and F. Pollock, Phys. Rev. **84**, 594 (1951).

57. N. Kroll and F. Pollock, Phys. Rev. **86**, 876 (1952).

58. R. Karplus, A. Klein, and J. Schwinger, Phys. Rev. **84**, 597 (1951).

59. J. R. Sapirstein, E. A. Terray, and D. R. Yennie, Phys. Rev. Lett. **51**, 982 (1983); Phys. Rev. **D29**, 2290 (1984).

60. D. Zwanziger, Bull. Am. Phys. Soc. **6**, 514 (1961).

61. D. Zwanziger, Nuovo Cimento **34**, 77 (1964).

62. S. J. Brodsky and G. W. Erickson, Phys. Rev. **148**, 26 (1966).

63. J. R. Sapirstein, Phys. Rev. Lett. **51**, 985 (1983).

64. R. Arnowitt, Phys. Rev. **92**, 1002 (1953).

65. G. P. Lepage, Phys. Rev. **A16**, 863 (1977).

66. G. T. Bodwin and D. R. Yennie, Phys. Rep. **43C**, 267 (1978).

67. G. T. Bodwin, D. R. Yennie, and M. Gregorio, Phys. Rev. Lett. **48**, 1799 (1982).

68. W. E. Caswell and G. P. Lepage, Phys. Lett. **167B** 437 (1986).

69. Terray, E. A., and D. R. Yennie, Phys. Rev. Lett. **48**, 1803 (1982).

70. M.I. Eides, S.G. Karshenboim, V.A. Shelyuto, Phys. Lett. **202B**, 572 (1988); S.G. Karshenboim, V.A. Shelyuto, and M.I. Eides, Zh. Eksp. Teor. Fiz. **92**, 1188 (1987) [Eng. transl.: Sov. Phys. JETP **65**, 664 (1987).

71. M. A. B. Bég, and G. Feinberg, Phys. Rev. Lett. **33**, 606 (1974).

72. H. Hellwig, R. F. C. Vessot, M. W. Levine, P. W. Zitzewitz, D. W. Allan, and D. J. Glaze, IEEE Trans. Instrum. **IM-19**, 200 (1970).

73. L. Essen, R. W. Donaldson, M. J. Bangham, and E. G. Hope, Nature **229**, 110 (1971).

74. G. T. Bodwin, D. R. Yennie, and M. Gregorio, Rev. Mod. Phys. **57**, 723 (1985).

75. A. C. Zemach, Phys. Rev. **104**, 1771 (1956).

76. S. J. Brodsky and S. D. Drell, Ann. Rev. Nuclear Science **20**, 147 (1970).

77. W. A. Newcomb and E. E. Salpeter, Phys. Rev. **97**, 1146 (1955).

78. C. K. Iddings and P. M. Platzman, Phys. Rev. **113**, 192 (1959).

79. H. Grotch and D. R. Yennie, Z. Phys. **202**, 425 (1967); Rev. Mod. Phys. **41**, 350 (1969).

80. G. T. Bodwin, and D. R. Yennie, Phys. Rev. D**37**, 498 (1988).

81. C. K. Iddings, Phys. Rev. **138**, B446 (1965).

82. S. D. Drell and J. D. Sullivan, Phys. Rev. **154**, 1477 (1967).

83. V. W. Hughes and J. Kuti, Ann. Rev. Nucl. Part. Sci. **33**, 611 (1983).

84. E. de Rafael, Phys. Lett. **37B**, 201 (1971).

85. P. Gnädig and J. Kuti, Phys. Lett. **42B**, 241 (1972).

86. R. Barbieri and E. Remiddi, Phys. Lett. **65B**, 258 (1976); R. Barbieri and E. Remiddi, Nucl. Phys. **B141**, 413 (1978).

87. W. E. Caswell and G. P. Lepage, Phys. Rev. A**20**, 36 (1979).

88. V. K. Cung, A. Devoto, T. Fulton, and W. W. Repko, Phys. Lett. **78B** 116 (1978).

89. V. K. Cung, A. Devoto, T. Fulton, and W. W. Repko, Phys. Lett. **68B**, 474 (1977); V. K. Cung, A. Devoto, T. Fulton, and W. W. Repko, Nuovo Cimento **43A**, 643 (1978).

90. G. S. Adkins, M. H. T. Bui, and D. Zhu, Phys. Rev. A**37**, 4071 (1988).

91. M. Ritter, P. O. Egan, V. W. Hughes, and K. A. Woodle, Phys. Rev. A**30**, 1331 (1984).

92. T. Kinoshita and J. Sapirstein, in Atomic Physics **9**, Eds. Robert S. Van Dyck, Jr. and E. Norval Fortson, World Scientific (Singapore, 1984).

93. H. A. Bethe, Phys. Rev. 72, 339 (1947).

94. G.E. Brown, J.S. Langer, and G.W. Schaefer, Proc. Roy. Soc. (London) **A251**, 92 (1959).

95. G.E. Brown and D.F. Mayers, Proc. Roy. Soc., **A251**, 105 (1959).

96. A. M. Desiderio and W. R. Johnson, Phys. Rev. A**3**, 1267 (1971).

97. M. Baranger, H. A. Bethe, and R. P. Feynman, Phys. Rev. **92**, 482 (1953).

98. R. Karplus, A. Klein, and J. Schwinger, Phys. Rev. **86**, 288 (1952).

99. G. W. Erickson, and D. R. Yennie, Ann. Phys. (N.Y.) **35**, 271 (1965); G. W. Erickson, and D. R. Yennie, Ann. Phys. (N.Y.) **35**, 447 (1965).

100. H. M. Fried and D. R. Yennie, Phys. Rev. **112**, 1391 (1958).

101. J. Schwinger, J. Math. Phys. **5**, 1606 (1964).

102. L. Hostler, J. Math. Phys. **5**, 591 (1964).

103. L. Hostler, J. Math. Phys. **5**, 1235 (1964).

104. M. Lieber, Physical Review **174**, 2037 (1968).

105. R. W. Huff, Physical Review **186**, 1367 (1969).

106. J. Sapirstein, Phys. Rev. Lett. **47**, 1773 (1981).

107. G. 't Hooft and M. Veltman, Nucl. Phys. **B44**, 189 (1972).

108. Jerome Malenfant, Phys. Rev. D**38**, 3295 (1988).

109. S. Love, Ann. Phys. **113**, 153 (1978).

110. L. H. Chan, *et al.*, Phys. Rev. **141**, 1298 (1966).

111. P. N. Kirk, *et al.*, Phys. Rev. **D8**, 66 (1973).

112. G. Peter Lepage, J. Comput. Phys. **27**, 192 (1978); G. P. Lepage, Cornell University preprint CLNS-80/477, March 1980.

113. A. Dalgarno, and J. T. Lewis, Proc. Roy. Soc. (London) **A233**, 70 (1955).

114. W. R. Johnson and Gerhard Soff, Atomic Data and Nuclear Data Tables **33**, 405 (1985).

115. R. Blankenbecler and R. Sugar, Phys. Rev. **142**, 1051 (1966).

116. I. T. Todorov , Phys. Rev. D**3**, 2351 (1971).

117. G. J. M. Austen and J. J. de Swart, Phys. Rev. Lett. **50**, 2039 (1983).

118. J. Sucher, Phys. Rev. A**30**, 703 (1980).

ATOMIC HYDROGEN HYPERFINE STRUCTURE EXPERIMENTS

Norman F. Ramsey
Lyman Physics Laboratory
Harvard University
Cambridge, MA 02138, USA

CONTENTS

1. Early History

The first experimental observation of structure in the spectral lines of atomic hydrogen beyond that accounted for by the Balmer series goes back to Michelson and Morley[1], who observed additional spectral lines that were later described as fine structure and attributed to relativistic and electron spin effects in accordance with the Dirac theory. The Lamb-Retherford experiment[2] showed slight departures of the experimental fine structure from the predictions of the Dirac theory and led to the the development of quantum electrodynamics (QED).

Although an additional structure, called hyperfine structure, was observed in the optical spectra of many heavier atoms; it was not seen in the early atomic hydrogen spectra. The hyperfine structure was attributed by Pauli[3] to the magnetic interaction of a small nuclear magnetic moment with the magnetic field from the electrons in the atom. The first observations of this interaction in hydrogen atoms was by the atomic beam deflection experiments described below. Fermi[4] showed from the Dirac theory that the magnetic interaction energy $(h\Delta\nu)_F$ should be given by

$$h(\Delta\nu)_F = (8\pi/3)[(2I+1)/I]\mu_B\mu_I\mu_N|\Psi_{no}(0)|^2 \qquad (1)$$

$$= (4/3)[(2I+1)/I](m_r/m)^3(m/M)\mu_I\alpha^2 hcR_{y\infty}Z^3/n^3,$$

where μ_I is the nuclear magnetic moment in nuclear magnetons, μ_N is the nuclear magneton given by μ_N = eh/4πMc with M being the mass of the proton, μ_B is the Bohr magneton given by μ_B = eh/4πmc, $\Psi_{no}(0)$ is the atomic hydrogen wave function at the nucleus, m_r is the reduced mass mM/(m + M) of the electron, $R_{y\infty}$ is the Rydberg constant $2\pi^2 me^4/h^3c$, and the subscript F indicates values calculated by the Fermi's Eq. (1) including the Breit[5] reduced mass correction $(m_r/m)^3$.

2. Atomic Beam Deflection Experiments

In 1931 Breit and Rabi[6,7] developed the important Breit-Rabi formula, which shows how the energy of an atom and its effective magnetic moment vary with the strength of the external magnetic field. These changes occur because the atomic configuration at low magnetic fields corresponds to the electron angular momentum, **J**, being primarily coupled to the nuclear angular momentum, **I**, to give the resultant angular momentum, **F** = **I** + **J** whereas at high external field, **J**, is primarily coupled to the external field; angular momenta are measured in units of $h/2\pi$. The variation of a hydrogen atom's energy and effective magnetic moment with external magnetic field are shown in Fig. 1. One can see from this figure that the effective component of

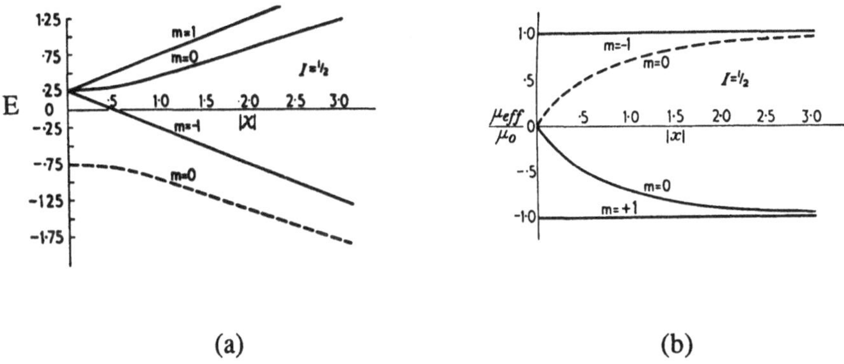

(a) (b)

Fig. 1. Variation with magnetic field of (a) the energy and (b) the effective component of the atomic magnetic moment along the field[6]. The variable $x = (-\mu_J/J + \mu_I/I)H_0/h\Delta\nu$. The dashed lines are for the the F = J - 1/2 state when the nuclear moment is assumed positive. If the nuclear moment were negative all curves would be inverted, i. e. reflected in the horizontal coordinate axis.

the atomic magnetic moment depends on the hyperfine separation energy, $h\Delta\nu$, so by measuring the deflection of the atom in an inhomogeneous magnetic field the hyperfine energy can be determined.

Rabi, Kellogg and Zacharias[8] first used this method for atomic hydrogen. They deflected the atoms with a two wire inhomogeneous magnetic field and detected the deflected beam by the visible trace on a molybdenum oxide coated plate. Since they were using the experiment to determine the magnetic moment of the proton, they did not directly report their value of $h\Delta\nu$ but instead expressed their result in terms of their measured proton magnetic moment, 3.25 ± 0.32 nuclear magnetons. From this result and Eq. (1) it is easily seen that their measured value of $\Delta\nu = (1645 \pm 150)$ MHz.

In a closely related experiment, the same authors[8] measured the hyperfine structure of the deuteron (then called the deuton) and found it corresponded to a deuteron magnetic moment of 0.77 ± 0.2 nuclear magnetons or $\Delta\nu = (293 \pm 76)$ MHz.

3. Atomic Beam Refocussing Experiments

The above deflection experiments could not be very accurate because the velocity distribution of the molecules produced broad deflection patterns. However, the same experimenters[9] soon devised a refocussing technique which largely overcame this difficulty and produced much more accurate results.

The collimator slit was placed in the center of the apparatus with magnetic fields of opposite inhomogeneity before and after the collimator so the deflection produced by the first field could be exactly compensated by the second field to provide refocussing of the beam as in the resonance method discussed below and illustrated in Fig. 2. Since the refocussing condition is velocity independent the refocussed peaks are much sharper and more intense. The authors arranged that the first inhomogeneous deflecting field should be longer but weaker than the second with the strengths being such that μ_{eff} in the first field was approximately (1/2) μ_B but in the second refocussing field it was approximately μ_B. From the relative strengths of the two fields to provide maximum refocussing and from the relative lengths of the two regions one could determine the value μ_{eff} in the weak field region and consequently from Fig. 1 could determine the value $|x|$ in that region which in turn determines the value $\Delta\nu$ if the external field H_0 is measured. In this manner Kellogg, Rabi and Zacharias[9] found the magnetic moment of the proton to be 2.85 ± 0.15, corresponding to $\Delta\nu_H = 1442 \pm 75$ and for the deuteron 0.85 ± 0.15 and $\Delta\nu_H = 323 \pm 11$.

From the figure caption of Fig. 1 it is apparent that the sign of the nuclear moment and therefore of $\Delta\nu$ can be determined if one can decide whether the $F = 0$ state has a positive or negative atomic magnetic moment at high fields. The direction of the deflection gives the sign of the atomic magnetic moment and the identity of whether that is the $F = 0$ or $F = 1$ state can be made by passing the atom through a region in which the direction of a weak magnetic field is successively reversed. In such a region non-adiabatic transitions are induced to other $F = 1$ states but no such transitions are possible for an atom in the $F = 0$ state. In this manner Kellogg, Rabi and Zacharias[9] determined that the magnetic moments of the proton and deuteron were both positive and that both $\Delta\nu$'s are therefore also positive.

4. Atomic Beam Magnetic Resonance Experiments

The first successful magnetic resonance experiments[10] were for the purpose of measuring nuclear magnetic moments in external magnetic fields and were with $^1\Sigma$ molecules possessing no electronic angular momentum. However, it soon became apparent that the method provided a general means for studying radiofrequency spectroscopy[11] which could be applied to atoms[12] as well.

The basic atomic beam magnetic resonance method[7] is illustrated in Fig. 2. An atomic beam emerging from a source S is deflected by one inhomogeneous magnetic field and refocussed by a similar field with its inhomogeneity in the opposite direction. In passing between the two fields the atoms are subjected to a weak oscillatory magnetic field at frequency ν. When ν equals the Bohr frequency $\nu_o = (W_f - W_i)/h$, transitions can take place with a consequent refocussing failure and a reduction in beam intensity. By measuring the beam intensity as a function of frequency one can thereby determine the spacing of the atomic energy levels.

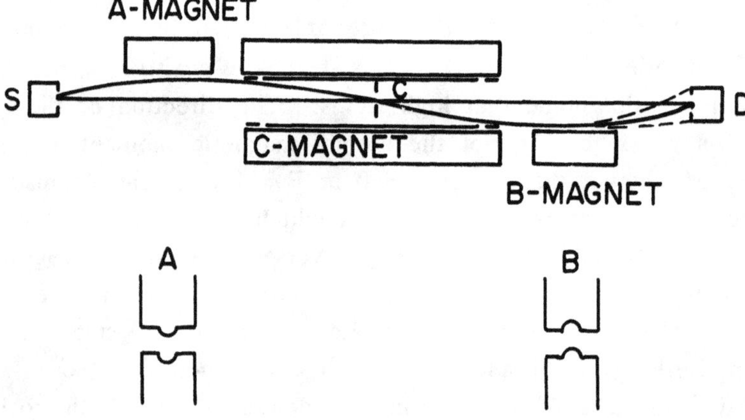

Fig. 2. Schematic diagram of magnetic resonance apparatus. S is the source and D the detector. The cross sections of the magnet pole tips are shown.

Due to the interruption of World War II, no attempts were made to apply the resonance method to the atomic hydrogen hyperfine structure until 1947. In that year Nafe, Nelson and Rabi[13] made the first accurate measurement of the atomic hydrogen hyperfine structure and found for hydrogen, deuterium and tritium the values

$$\begin{aligned} \Delta\nu_H &= 1{,}420.410 \pm 0.006 \text{ MHz} \\ \Delta\nu_D &= 327.384 \pm 0.003 \text{ MHz} \\ \Delta\nu_T &= 1{,}516.702 \pm 0.010 \text{ MHz} \\ \Delta\nu_H / \Delta\nu_D &= 4.33867 \pm 0.00004. \end{aligned} \qquad (2)$$

Shortly thereafter, with a similar but independent atomic beam apparatus, Nagle, Julian and Zacharias[14] found similar values.

At the time of these experiments it was assumed by theorists that the value of the electron magnetic moment was exactly one Bohr magneton and the value of the proton magnetic moment was measured[15] in terms of the electron magnetic moment. When these assumptions were used in the theoretical Eq. (1), the predicted result was

$$\Delta\nu_H = 1{,}416.97 \pm 0.54 \text{ MHz} \qquad (3)$$

which differs from the experimental result by far more than the estimated errors. It was just this disagreement which first suggested to Breit[5] and Schwinger[16] the possibility of an anomalous value for the magnetic moment of the electron since a change in the value of the electron magnetic moment would change the value of Eq. (1) both directly through the electron moment and indirectly through the proton moment since the two appear as products in Eq. (1) whereas they entered as ratios in experiments by which the proton's magnetic moment at that time had been measured. Schwinger's development of a relativistic QED started with his efforts to understand the anomalous hyperfine structure of atomic hydrogen. The agreement between the experimental value of $\Delta\nu$ and the theoretical value from the new QED was sufficiently close that it stimulated active programs to increase the accuracy of both and to measure related quantities such as the electron magnetic moment and the magnetic moment of the atom.

Prodell and Kusch[17] improved the precision of the atomic hydrogen hyperfine structure measurements by designing the oscillatory field region to reduce Doppler effects and by operating at a much lower magnetic field to increase stability. By measuring three of the transitions allows on Fig.1, they obtained

$$\begin{aligned} \Delta\nu_H &= 1{,}420.40573 \pm 0.00005 \text{ MHz} \\ \Delta\nu_D &= 327.384302 \pm 0.000030 \text{ MHz} \\ \Delta\nu_H / \Delta\nu_D &= 4.33864947 \pm 0.00000043 \end{aligned} \tag{4}$$

Although the largest quantum electrodynamic corrections to the atomic hydrogen hyperfine structure are those due to the anomalous magnetic moment of the electron there are also important quantum electrodynamic and further reduced mass corrections to the Fermi relation in Eq. (1) as well. When all of these theoretical corrections and a correctly calibrated value for the proton magnetic moment were included, the agreement between theory and experiment for $\Delta\nu_H$ was so good that the main limitation to the comparison was the uncertainty in the fine structure constant, α, so the experiment and theory were used in the evaluation of α. In subsequent years the accuracy of the measurements of α have been markedly increased, as discussed in the next section. The theoretical accuracy has also been improved by QED calculations to higher order and by further refinements to recoil and other corrections as reviewed by Bodwin and Yennie[18] and as discussed below in Section 10.

The theoretical value for $(\Delta\nu_H / \Delta\nu_D)_F$ from Eq. (1) is 4.3393876 ± 0.0000008 which differs markedly from the experimental value for $(\Delta\nu_H / \Delta\nu_D)_{exp}$ from Eq. (4). The discrepancy can be expressed as

$$\Delta = 1 - [(\Delta\nu_H / \Delta\nu_D)_{exp} / (\Delta\nu_H / \Delta\nu_D)_F] = (1.703 \pm 0.005) \times 10^{-4} \tag{5}$$

Low and Salpeter[19] could account for some, but not all of this discrepancy, by introducing an additional reduced mass correction which allowed for the nuclear recoil that occurs when virtual photons are exchanged. One of the most fruitful proposals for the source of the remaining discrepancy was due to A. Bohr[20] who pointed out that the electron charge near the nucleus follows the proton in its motion in the deuteron. Hence, in a calculation of the magnetic interactions of the neutron with the electrons, the neutron can not be considered to be at the exact center of the atom. The combination of these two corrections gave a theoretical value for Δ of $(2.28 \pm 0.20) \times 10^{-4}$. The difference between this theoretical value and the experimental value was attributed to the structure of the nucleons and further recoil corrections[6,18] as discussed in Section 10.

In a comparison of the hyperfine structure separations of hydrogen and tritium, one would expect with merely the Breit[5] reduced mass correction in Eq. (1) that

$$\begin{aligned}(\Delta\nu_T)_F &= (\Delta\nu_H)_F\,(\mu_t/\mu_p)\,(m_T/m_H)^3 \\ &= 1516.709 \pm 0.015 \text{ MHz}\end{aligned} \tag{6}$$

This result is in excellent agreement with the value of Nelson and Nafe[13] in Eq. (2), which is 1516.702 ± 0.010 MHz. At first sight it is surprising that this excellent agreement should obtain in the case of tritium, whereas for deuterium there was no such agreement until allowances were made for the structure of the nucleus and for nuclear recoil in the exchange of virtual photons. However, Fermi[22], Teller[22] and others[23,24] showed that the tritium anomaly should be much smaller than for deuterium.

5. Anomalous Magnetic Moment of the Electron

The first evidence for an anomalous magnetic moment of the electron came from the above measurements of the atomic hyperfine structure of atomic hydrogen. This led to an early atomic beam magnetic resonance experiment by Kusch and Foley[25] to measure the spin magnetic moment μ_e of the electron in terms of the orbital magnetic moment μ_L of a P state atom. They assumed that either magnetic moment might depart from μ_B by $\mu_e / \mu_B = 1 + \delta_S$ and $\mu_L / \mu_B = 1 + \delta_L$ and then measured Ga atoms in the $^2P_{3/2}$ and $^2P_{1/2}$ states, Na in the $^2S_{1/2}$ state, and In in the $^2P_{1/2}$. None of these measurements depended on μ_L alone since both orbital and spin magnetic moments were involved in each of the P states, but from the combination of measurements with the spin and orbital moments appropriately added in the $^2P_{3/2}$ case and subtracted in the $^2P_{1/2}$ the effect of μ_L could be determined so the magnetic field could be calibrated in terms of μ_L. For the reasonable assumption $\delta_L = 0$, they found

$$\mu_e / \mu_B = 1 + \delta_S = (1.00119 \pm 0.00005) \qquad (7)$$

in agreement with later more accurate values as discussed elsewhere in this volume.

6. Atomic Hydrogen Maser Experiments

By far the most accurate measurements of the atomic hydrogen hyperfine structure are those with the atomic hydrogen maser, which was invented by Ramsey and Kleppner[26]. The basic principles of this maser are illustrated in the schematic diagram shown in Fig. 3. An intense electrical discharge in the source converts commercially available molecular hydrogen (H_2) into atomic hydrogen (H). The atoms emerge from the source into a region that is evacuated to a pressure below 10^{-6} torr. A beam of diverging atoms enters the state-selecting magnet which has north poles alternating in a circle with three south poles. By symmetry the magnetic field is zero on the axis of such a magnet and increases in magnitude away from the axis. From Fig. 1(a) it is apparent that the low energy F = 0 state (dashed line) decreases in energy with increasing field or distance off the axis. Since a mechanical system always falls to lower potential energy state these atoms will move away from the axis and be defocussed as in Fig. 3. Inversely, from Fig. 1(a), two of the three high energy F = 1

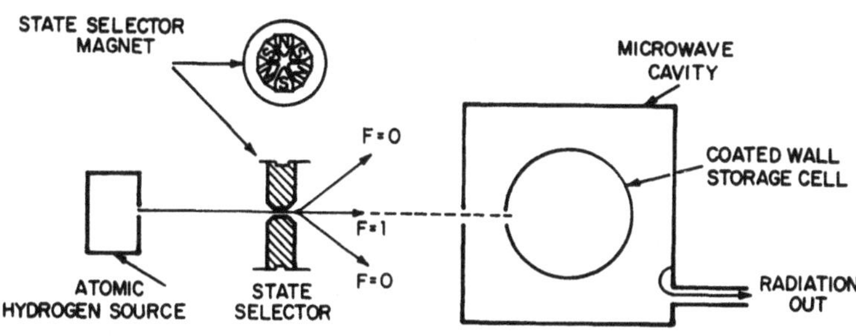

Fig. 3. Schematic diagram of atomic hydrogen maser.

states (full lines) are focussed onto the small aperture of the storage cell, a 15 cm diameter bulb coated with teflon on the inside. If these atoms are exposed to microwave radiation at the hyperfine frequency, more atoms will be stimulated to go from the higher energy state to the lower so the released energy will make the microwave radiation stronger -- that is the device will be an amplifier, a maser. If the storage cell is placed inside a tuned cavity, an oscillation at the resonance frequency will be increased in magnitude until an equilibrium value is reached, at which level the oscillation will continue indefinitely. The energy to maintain the oscillation comes from the continuing supply of hydrogen atoms in the high energy hyperfine state.

The atomic hydrogen maser has unprecedentedly high stability due a combination of favorable features. The atoms typically reside in the storage bulb for 10 seconds, which is much longer than in an atomic beam apparatus so the resonance line is much narrower. The atoms are stored at low pressure so they are relatively free and unperturbed while radiating. The first order Doppler shift is removed, since the average velocity is extremely low for atoms stored for 10 seconds in a bottle with the same aperture for entrance and exit. Masers have very low noise levels, especially when the amplifying elements are isolated single atoms. Over periods of several hours, the hydrogen maser's stability is better than 1×10^{-15}.

The major disadvantage of the hydrogen maser is that the atoms collide with the walls at intervals, changing slightly the hyperfine frequency and giving rise to a wall shift of two parts in 10^{11}. The wall shifts can be experimentally determined by measurements with bulbs of different diameters or with a deformable bulb whose surface to volume ratio can be altered.

The atomic hydrogen hyperfine structure $\Delta\nu$ was first measured with a hydrogen maser by Goldenberg, Kleppner and Ramsey[26] and has subsequently been measured by many observers, chiefly in connection with its use as a highly stable clock or frequency standard. All of these values have been in complete agreement to limits far beyond the significance of any theoretical calculation. In obtaining the best values, the primary emphasis has to be on measuring the wall shift for the storage bottle actually used and in comparing it to a cesium beam standard of high accuracy. The second is defined in terms of the cesium frequency and the potential errors of the cesium measurement are comparable to those of hydrogen. The values that are usually considered of greatest accuracy are those of Hellwig[27], Vessot[27], Essen[28] and their associates[27, 28]. Their combined value for the hyperfine frequency of the ground state of atomic hydrogen is

$$\Delta\nu_H = 1{,}420{,}405{,}751.7667 \pm 0.0009 \text{ Hz} \qquad (8)$$

The theoretical interpretations of $\Delta\nu_H$ are discussed in Sections 1, 4 and 10. The accuracies of the measurement and of the theory are sufficient that for a number of years the principal uncertainty in comparing theory with experiment was the value of the fine structure constant α. During that time the hyperfine structure measurements provided one of the best means for determing α. When the Josephson effect finally became available for determing the value of α, the agreement between the values of α determined from hyperfine structure measuremnts and from the Josephson effect was excellent. As discussed in greater detail in Section 10, the presnt best theoretical value for $\Delta\nu_H$ agrees with the above experimental value to within the uncertainty of the theoretical calculation which is approximately one part per million.

The $\Delta\nu$ for the deuterium atom has been measured by Ramsey, Crampton, Kleppner, and others[29,30]. Since $\Delta\nu_D$ is much smaller than $\Delta\nu_H$, a full size deuterium maser is much larger than a hydrogen maser; so the experiment was done by spin exchange collisions in a hydrogen maser. They operated a hydrogen maser on its usual (F,m) (1,0) → (1,-1) transition at 1420 MHz but with both hydrogen and deuterium atoms going into the storage bulb. Spin exchange collisions between D and H relax the oscillation of the hydrogen magnetic moment by an amount depending on the deuterium electron polarization as well as on the spin exchange collision rate. When a radiation field is applied at the deuterium resonance frequency, the electron polarization of the deuterium decreases and there is a corresponding diminution in the hydrogen maser output power level, which serves to detect the deuterium resonance. Later a full size deuterium maser was constructed by Wineland and Ramsey[31], who obtained a result consistent with the first measurements but much more accurate. Their result for the ground state of atomic deuterium is

$$\Delta\nu_D = 327{,}384{,}352.5219 \pm 0.0017 \qquad (9)$$

The above number differs slightly from their published value, since, at the time of experiment, the ratio of $\Delta\nu_H$ to $\Delta\nu_D$ was measured more accurately than $\Delta\nu_H$ could be calibrated in terms of cesium; the above value has therefore been adjusted to be consistent with Eq. (8). A discussion of the theories of the ratio $\Delta\nu_H / \Delta\nu_D$ is given in Section 4.

The $\Delta\nu$ for the tritium atom has been measured by Mathur, Crampton, Kleppner and Ramsey[32]. Since the hydrogen and tritium hyperfine frequencies are approximately the same, a regular hydrogen maser was used with tritium gas substituted for hydrogen. The result including a readjustment to normalize to the latest value for $\Delta\nu_H$, is

$$\Delta\nu_T = 1{,}516{,}701{,}470.7731 \pm 0.0072 \text{ Hz} \qquad (10)$$

The theories of the ratio $\Delta\nu_H / \Delta\nu_T$ are also discussed in Section 4.

7. Dependence of Hyperfine Structure on External Magnetic Field

The dependence of $\Delta\nu_H$ on magnetic field strength has been studied by Brenner and Ramsey[33]. Theoretically it was expected that the energies should vary with the magnetic field as in the Breit-Rabi formula shown in Fig. 1, but it was not known how accurately this curve would be followed. Brenner and Ramsey[33] found that the Breit-Rabi equation was followed to an accuracy of 7 parts in 10^9.

Larson, Valberg, and Ramsey[26] and Hughes and Robinson[34] have measured the ratio of the magnetic moment of the hydrogen atom to that of deuterium. They looked at (F,m) (1,1) → (1,0) and (1,0) → (1,-1) transitions in various magnetic fields. It is apparent from Fig. 1 that these measurements will give the ratio of atomic magnetic moments. More accurate values have been found by Walther, Phillips and Kleppner[35]. Similar measurements have been made with a tritium maser by Larson and Ramsey[36]. The best results from these measurements of the ratios of the atomic magnetic moments μ_J in the atomic ground states are

$$\mu_J(H) / \mu_J(D) = 1.000\ 000\ 007\ 22 \pm 0.000\ 000\ 000\ 03$$
$$\mu_J(H) / \mu_J(T) = 1.000\ 000\ 010\ 7 \pm 0.000\ 000\ 002\ 0 \quad (11)$$

When these experiments were first started, the QED theories predicted results that were even closer to 1 than the above measurements. However, Hegstrom[37] and Grotch[37] found additional QED terms that accounted for the difference. Their theoretical value for $\mu_J(H) / \mu_J(D) = 1.000\ 000\ 007\ 22$, in excellent agreement with experiment.

Myint, Kleppner, Robinson, Ramsey, Winkler and Walther[38,39] have made a hydrogen maser that operates in a stable high magnetic field and have used it to study the hyperfine spectrum at 3,500 Gauss. From Fig. 1, it is apparent that for the (F,m) transition (1,0) → (1,-1) the frequency is primarily dependent on the electron magnetic moment whereas the (1,+1) → (1,0) transition depends to a considerable extent on the magnetic moment of the proton. From the ratio of these two frequencies a value is obtained for the ratio of the magnetic moment, $(\mu_J)_H$, of the hydrogen atom to the apparent magnetic moment, $(\mu_p)_H$, of the proton shielded inside the hydrogen atom. For both of these quantities it is necessary to make small known corrections[39] to obtain the ratio μ_e/μ_p of the proton. The value of this ratio, as found from these hyperfine structure measurements[39], is

$$\mu_e/\mu_p = -658.210\ 706 \pm 0.000\ 006 \qquad (12)$$

From the above result and from the known value of the electron moment, the magnetic moment of the free proton in Bohr magnetons is

$$\mu_p = 0.001\ 521\ 032\ 181 \pm 0.000\ 000\ 000\ 015\ \mu_B \qquad (13)$$

The value of μ_p has also been obtained from NMR measurements with the H_2 molecule, but the magnetic shielding corrections are much more uncertain for molecules than for atoms. Nevertheless, the two values agree when the molecular shielding factor is obtained from Ramsey's[40] shielding theory; thus providing the first experimental test of the molecular shielding theory.

The same high field maser has been used by Philips, Cook and Kleppner[41] to measure the ratio the proton NMR resonance frequency in water to the electron resonance frequency at the same magnetic field in an atomic hydrogen maser. Although there is little interest in this number by itself, it is of great value in other magnetic moment measurements for which the magnetic fields are often calibrated in terms of an NMR signal with a water sample. For example it is used in obtaining the value of the neutron magnetic moment in Bohr magnetons[42].

8. Dependence of Hyperfine Structure on External Electric Field

Since the hydrogen atom in its ground state has no electric dipole moment, there should be no first order effect of an external electric field. However, due to the electric polarizability of the atom a small effect proportional to E^2 might be expected. This was first looked for by Fortson[43] with an ordinary hydrogen maser to which an ordinary electric field could be applied. Gibbons and Ramsey[44] modified the maser to use higher electric fields and to permit the magnetic field and hence the axis of quantization to be either parallel or perpendicular to the electric field. Stuart, Larson and Ramsey [45] used a cylindrical geometry to obtain a still higher field. The theory for the electric field dependence has been developed by Schwarz[46], Sandars[47] and Gibbons[44]. The hyperfine separation is proportional to the probability density of the electron at the nucleus. The change in this density is approximately proportional to the fraction of excited electronic states mixed in with the ground state by the electric field. This quantity in turn is proportional to the electronic polarization energy[46] divided by the excitation energy. The experimental and theoretical results for the shifts $\delta\nu$ with E in volts per meter are

$$\begin{array}{lll} \mathbf{E} \parallel \mathbf{B} & \text{Expt.} & \delta\nu = (-8.37 \pm 0.84) \times 10^{-14} E^2 \\ & \text{Theory} & \delta\nu = -8.47 \times 10^{-14} E^2 \\ \mathbf{E} \perp \mathbf{B} & \text{Expt.} & \delta\nu = (-8.45 \pm 0.20) \times 10^{-14} E^2 \\ & \text{Theory} & \delta\nu = -8.24 \times 10^{-14} E^2 \end{array} \quad (14)$$

Measurements of the frequency shifts at different strengths of electric field confirmed that the shifts were proportional to $\mathbf{E}^2$.

9. Hyperfine Structure Experiments in Higher Electronic States

The most precise atomic hydrogen hyperfine structure experiments have been those in the ground electronic state of the atom as discussed above so these are the quantities of greatest interest to discussions of QED. However, hyperfine structure measurements of less accuracy have been made in excited atomic states, especially the metastable $2^2S_{1/2}$ state.

The first observations were those of of Lamb[2] with the apparatus used to discover the Lamb shift in the fine structure of atomic hydrogen. Much more accurate measurements of the hyperfine structure of the $2^2S_{1/2}$ were made by Heberle, Reich and Kusch[48] with an atomic beam apparatus especially built for the purpose. Their results can most easily be interpreted in terms of the ratio, R, of the hyperfine frequency in the $2^2S_{1/2}$ state to that in the $1^2S_{1/2}$ state, in which case the first order dependence on the value of α and the largest theoretical corrections both cancel and R should be approximately n^{-3} or 1/8. they found that there experimental result could be expressed as

$$8R - 1 = 0.000\,034\,6 \pm 0.000\,000\,3 \qquad (15)$$

Most of this number can be accounted for in terms of the relativistic Breit[21] correction but there remains a discrepancy about four times the experimental error. Mittelman[49] and Zwanziger[50] have accounted for the discrepancy in terms of QED corrections of order α^3. The theoretical value for Eq. 14 including these corrections is 0.000 034 5 ± 0.000 000 3. Similar excellent agreement between theory and experiment is obtained with atomic deuterium in the $2^2S_{1/2}$ state[44, 47].

Subsequent to the above measurements of Heberle, Reich and Kusch[48], many measurements have been made of the atomic fine structure in excited states but the corresponding hyperfine structure has usually not been accurately measured. Lundeen, Jessop and Pipkin[51], however, have measured the hyperfine separation in the $2^2P_{1/2}$ state of atomic hydrogen and have found the the value 59.22 ± 0.14 MHz in good agreement with the theoretical prediction 59.1501 ± 0.0001 MHz.

10. Further Theoretical Discussion

In most cases the theories of the above atomic hydrogen hyperfine structure experiments have been discussed briefly along with the experiments. However, in the case of the basic values for $\Delta\nu_H$ there have been extensive QED calculations and further discussion is given here.

The primary constituent to the the QED corrections to the atomic hyperfine structure is the electron's anomalous moment coefficient, a_e, defined by $\mu_e = \mu_B (1 + a_e)$. This coefficient is discussed extensively elsewhere in this volume. The coefficient a_e has been very accurately measured as well as accurately calculated by QED, with the two values being in close agreement. As a result, in discussing the remainder of the atomic hydrogen hyperfine QED corrections, the experimental value for a_e is ordinarily used.

Many papers have been written on the remaining QED corrections to the atomic hydrogen hyperfine structure. These papers have recently been reviewed, with extensive references, by Bodwin and Yennie[18] and the reader is referred to that article. The QED value for the hydrogen hyperfine separation, $h\Delta\nu_{QED}$, can be written as follows, including the Fermi factor $h\Delta\nu_F$ from Eq. (1), the electron's anomalous moment coefficient a_e and other QED terms:

$$\begin{aligned} h\Delta\nu_{QED} = h\Delta\nu_F\{&1+(3/2)(Z\alpha)^2+ae+\alpha(Z\alpha)(\ln 2-[5/2]) \\ &-[8\alpha(Z\alpha)2/3\pi]\ \ln(Z\alpha)[\ln(Z\alpha)-\ln 4+(281/480)] \\ &+[\alpha(Z\alpha)^2/\pi]\ (15.38\pm 0.29) + [\alpha^2(Z\alpha)/\pi]D_1\} \\ = {}& 1420.451\ 95 \pm 0.000\ 14\ \text{MHz}. \end{aligned} \tag{16}$$

The difference between QED theoretical calculations and experimental measurements is usually expressed as

$$\{(\Delta\nu_{QED} - \Delta\nu_{expt}) / \Delta\nu_F\} = 32.56 \pm 0.10\ \text{ppm} \tag{17}$$

The remaining discrepancy is attributed to proton structure corrections, $\delta\Delta\nu$(structure), which are usually written as

$$\delta\Delta\nu(\text{structure}) / \Delta\nu_F = \delta_p(\text{Zemach}) + \delta_p(\text{recoil}) + \delta_p(\text{polarizability}) \quad (18)$$

Bodwin and Yennie[18] discuss the theoretical evaluation of these terms and conclude that the finite size or Zemach correction δ_p(Zemach) = -38.72 ±0.56 ppm, δ_p(recoil) = 5.68 ppm and δ_p(polarizability) < 4 ppm. With these corrections, with $\Delta\nu_{theory}$ representing the theoretical value for $\Delta\nu_H$, and with all known contributions included,

$$\{(\Delta\nu_{theory} - \Delta\nu_{expt}) / \Delta\nu_F\} = (-0.48 \pm 0.56 \pm \text{unknown}) \text{ ppm} \quad (19)$$

RFERENCES

1. Michelson and Morley, Phil. Mag. (5) 24, 46 (1887).
2. Lamb, W. E. and Retherford, R. C., Phys. Rev. 72, 741 (1947) and Reports on Progress in Physics 14, 19 (1951).
3. Pauli, W., Naturwiss. 12, 741 (1924).
4, Fermi, E., Zeits. f. Physik, 60, 320 (1930).
5. Breit, G., Phys. Rev. 72, 984 (1947).
6. Breit, G. and Rabi, I. I., Phys. Rev. 38, 2082 (1931).
7. Ramsey, N. F., *Molecular Beams* , Oxford University Press (1956 and 1985) and *History of Atomic Clocks*, J. Res. NBS 88, 301 (1983)
8. Rabi, I. I., Kellogg, J. M. B. Kellogg and Zacharias, J. R., Phys. Rev. 46, 157 and 163 (1934).

9. Kellogg, J. M. B., Rabi, I. I. and Zacharias, J. R., Phys. Rev. 50, 472 (1936).
10 Rabi, I. I., Zacharias, J. R., Millman and Kusch, P., Phys. Rev. 53, 318 (1938) and 55, 526 (1939).
11. Kellogg, J. M. B., Rabi, I. I., Ramsey, N. F. and Zacharias, J. R., Phys. Rev. 55, 318 (1939); 56, 728 (1939); 57, 677 (1940) and 58, 226 (1940).
12. Kusch, P., Millman, S. and Rabi, I. I., Phys. Rev. 57, 765 (1940)
13. Nafe, J. E. , Nelson, E. B. and I. I. Rabi, Phys. Rev. 71, 914 (1947); 73, 718 (1948); 75, 1194 (1949) and 76, 1858 (1949).
14. Nagle, D. E., Julian, R. S. and Zacharias, J. R., Phys. Rev.72, 971 (1947) and 76, 847 (1949).
15. Millman, S. and Kusch, P., Phys. Rev. 60, 91 (1941).
16. Schwinger, J., Phys. Rev. 73, 416 (1948) and 76, 790 (1949).
17. Prodell, A. G. and Kusch, P., Phys. Rev. 79, 1009 (1950) and 88, 184 (1952).
18. Bodwin, G. T. and Yennie, D. R., Phys. Rev. D37, 498 (1988).
19. Low, F. and Salpeter, E. E.,Phys. Rev. 83, 478 (1951) and 77, 361 (1950).
20. Bohr, A., Phs. Rev. 72, 1109 (1948); 77, 94 (1950) and 90, 717 (1953).
21. Breit, G., Phys. Rev. 35, 1447 (1930).
22. Fermi, E. and Teller, E., Notes on Pocono Conference of Physics, sponsored by National Academy of Sciences, April 1, 1948.
23. R. Avery and R. G. Sachs, Phys. Rev.74, 433 and1320 (1948).
24. Sessler, A. M. and Foley, H. M., Phys. Rev. 98, 6 (1955).
25. Kusch, P. and Foley, H. M., Phys. Rev. 72, 1256 (1947) and 74, 250 (1948).
26. Goldenberg, H. M., Kleppner, D. and Ramsey, N. F., Phys. Rev. Lett. 5, 361 (1960); Phs. Rev. 126, 603 (1962); 138A, 1972 (1965).

27. Hellwig, H., Vessot, R. F. C., Levine,M. W., Zitzewitz, P. W., Allan, D. W. and Glaze, D. J., IEEE Trans. Instrum. Meas. IM-19, 200 (1970).
28. Essen, L., Donaldson, M. J., Bangham, M. J., and Hope, E. G., Nature 229, 110 (1971).
29. Crampton, S. B., Robinson, H. G.,Kleppner, D. and Ramsey, N.F., Phys. Rev. 141, 55 (1966).
30. Larson, D. J., Valberg, P. A. and Ramsey, N. F., Phys. Rev. Lett. 23, 1369 (1969) and A3, 554 (1971)..
31. Wineland, D. J. and Ramsey, N. F., Phys. Rev. A5, 821 (1972).
32. Mathur, B. S., Crampton, S. B., Kleppner, D. and Ramsey, N. F., Phys. Rev. 158, 14 (1967).
33. Ramsey, N. F., Physics of One- and Two- Electron Systems, page 218 (North Holland, 1969).
34. Hughes, W. M. and Robinson, H. G., Phys. Rev. Lett. 23, 1369 (1969).
35. Walther, F. G., Phillips, W. D. and Kleppner, D., Phys. Rev. 28, 1159 (1972).
36. Larson, D. J. and Ramsey, N. F., Phy. Rev. A9, 1543 (1974).
37. Hegstrom, R. A. and Grotch, H., Phys. Rev. 184, 17 (1969); A2, 1605 (1970); and A4, 1543 (1974).
38. Myint, T., Kleppner, D., Ramsey, N. F. and Robinson, H. G., Phys. Rev. Lett., 17, 405 (1966).
39. Winkler, P. J., Kleppner, D., Myint, T., and Walther, J. G., Phys. Rev.A5, 83 (1972).
40. Ramsey, N. F., Phys. Rev. 78, 699 (1950).
41 Phillips, W. D., Cook, W. E. and Kleppner, D., Phys. Rev. Lett., 35, 1619 (1975).
42. Greene, G. L., Ramsey, N. F., Mampe, M., Pendlebury, Smith, K., Dress, W. B., Miller, P. D. and Perrin, P., Phys. Rev. D20, 2139 (1979).
43. Fortson, E. N., Phys. Rev. Lett. 13, 22 (1964).
44. Gibbons, P. C. and N. F. Ramsey, Phys. Rev. A5, 73 (1972).

45. Stuart, J. G., Larson, D. J. and Ramsey, N. F., Phys. Rev. A22, 2092 (1980).
46. Schwartz, C., Ann. Phys. (N. Y.) 6, 156 (1959).
47. Sandars, P. G. H., Proc. Roy. Soc. 92, 857 (1967).
48. Heberle, J. W., Reich, H. A. and Kusch, P., Phys. Rev. 101, 612 (1956) and Phys. Rev. 104, 1585 (1956).
49. Mittelman, M. H., Phys. Rev. 107, 1170 (1957).
50. Zwanziger, D. E., Phys. Rev. 121, 1128 (1961).
51. Lundeen, S. R., Jessop, P. E. and Pipkin, F. M., Phys. Rev. Letters 34, 377 (1975).

Lamb Shift Measurements

Francis M. Pipkin
Lyman Laboratory of Physics
Harvard University
Cambridge, Massachusetts 02138

Contents

1. INTRODUCTION

One of the early triumphs of quantum mechanics was the successful prediction of the energy levels for the hydrogen atom.[1-3] The initial step was the demonstration by Schroedinger that the non-relativistic wave equation yielded energy levels for an electron in a Coulomb field that agreed with those predicted by the Bohr theory. The addition using perturbation theory of the relativistic expression for the kinetic energy, the spin and anomalous magnetic moment of the electron, and the spin-orbit interaction predicted the observed doublet fine structure of the Balmer radiation. A major achievement was Dirac's invention of a relativistic equation which correctly predicted the spin and anomalous magnetic moment of the electron. The Dirac theory indicated that for the n=2 state of hydrogen the $^2S_{1/2}$ and $^2P_{1/2}$ levels would be degenerate. It was realized early that since the $^2S_{1/2}$ electron had a high probability and the $^2P_{1/2}$ electron had a low probability of being at the nucleus, the degeneracy of these two levels could be used to study the behavior of the force between the electron and proton at small distances.[4-7] In the decade of the 1930's several spectroscopic measurements were carried out to confirm the fine structure predicted by the Dirac theory.[8-13] These measurements were made difficult by the Doppler width of the spectral lines and the fact one had to deal with the fine structure for both of the states connected by the optical transition. The measurements suggested, however, that the fine structure predicted by the Dirac theory was not correct. In 1938 Pasternack[14] summarized the status of the measurements and concluded that the data indicated that the binding energy for the $2^2S_{1/2}$ electron was roughly 0.03 cm^{-1} less than that for the $2^2P_{1/2}$ electron. This was not confirmed by measurements carried out by Drinkwater, et al. in 1940.[15]

After World War II, Lamb and Retherford undertook a series of experiments to resolve this question.[16-18] Lamb realized that if the two levels were not degenerate, in the absence of an electric field, the $2^2S_{1/2}$ level would be metastable and one could form an easily detectable beam of atoms in this state. A radiofrequency field could then be used to drive transitions to the $2^2P_{1/2}$ state. Since the lifetime of the $2^2P_{1/2}$ state is 1.6 ns, it would decay immediately and the resonance could be detected by the decrease in the number of atoms in the $2^2S_{1/2}$ state. They reported the success of this experiment and the preliminary measurement of the $2^2S_{1/2}$-$2^2P_{1/2}$ interval, which is now called the Lamb shift, at the Shelter Island Conference in 1947.[1] This result stimulated theoretical physicists to overcome the infinities encountered in quantum field theory and predict the shift in energy levels. The first calculation was carried out by Bethe.[19] Subsequently Feynman, Schwinger, and Tomanaga independently developed techniques for using quantum electrodynamics to calculate corrections to all orders in the fine structure constant $\alpha = e^2/\hbar c$. This theory uses the technique of renormalization to obtain finite measureable predictions by replacing the uncalcuable quantities such as the charge and mass of the electron by their measured values. The divergent integrals encountered in perturbation theory are made convergent by expressing observable quantities in terms of physical masses and charges.[20-21]

In this article we will review the measurements of the Lamb shift and summarize the current status of the agreement between theory and experiment. The presentation will be in part based on the techniques used in the measurements and in part historical. The sections deal in turn with a summary of the theory, the calculation of the signals, slow beam experiments, bottle experiments, fast beam experiments, laser spectroscopy measurements, measurements on two electron systems, and a comparison of theory and experiment. A table summarizes the results of the reported measurements. Earlier general reviews of the theory[21-26] and measurements[27-28] of the Lamb shift are given in the references.

2. SUMMARY OF THEORY

The Lamb shift is due to the change in the energy levels from that predicted by the Dirac equation resulting from quantum electrodynamics. We shall call this the QED shift and denote the QED shift of a particular level by $\Delta E_Q\,(n^2L_J)$. Fig. 1 shows the energy levels for the n=2 state of a hydrogenic atom and the notation used to denote the intervals. In this notation

$$S(n=2) = \Delta E_Q(2^2S_{1/2}) - \Delta E_Q(2^2P_{1/2}). \tag{2.1}$$

There are two dominate contributions to the Lamb shift -- the electron self energy and the vacuum polarization. Feynman graphs for these two contributions are shown in Fig. 2. The self energy term, which is due to emission and reabsorption of virtual photons, can be pictured as the interaction of the electron with the zero point fluctuations of the electromagnetic field.[29] This tends to smear the electron charge over a mean square radius of

$$\langle \delta r^2 \rangle = \frac{2\alpha^3 a_o^2}{\pi} \ln(Z\alpha)^{-1} = (5.838 \times 10^{-10}\ \mathrm{cm})^2 \ \text{for}\ Z=1. \tag{2.2}$$

Table 1 summarizes the symbols and the values of the fundamental constants that will be used in this paper.(30,31)

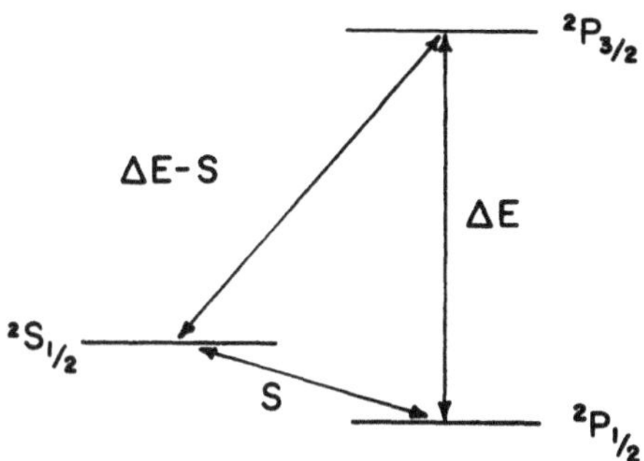

Figure 1. A level diagram for the n=2 state of a hydrogenic atom showing the notation that will be used to describe the levels.

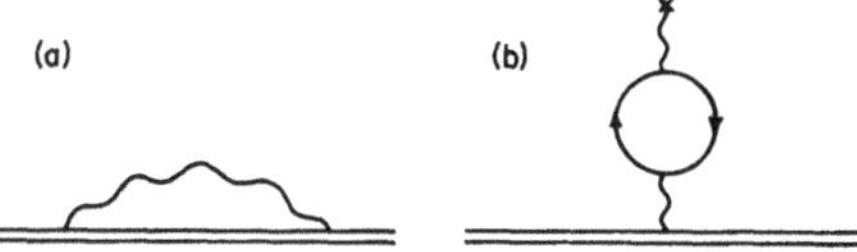

Figure 2. Feynman diagrams for the lowest order condtributions to the Lamb shift.

The vacuum polarization correction can be pictured as arising from a polarization of the vacuum due to creation of virtual charged particle pairs. The physically measured charge is smaller than the bare charge since it is screened by the polarized vacuum.

The theoretical formulas for the Lamb shift are discussed in the article in this volume by Sapirstein and Yennie. In this section we will simply summarize the

Table 1. Symbols and values for fundamental constants used in this paper. With the exception of the Rydberg, these constants were taken from the 1986 adjustment.[30] The value for the Rydberg comes from the recent measurement by Biraben, et al.[31]

Symbol	Quantity	Value	Units
m_e	mass of electron	9.109 389 7(54)	10^{-31} kg
		5.485 799 03(13)	10^{-4} u
m_p	mass of proton	1.672 623 1(10)	10^{-27} kg
e	charge on electron	1.602 177 33(49)	10^{-19} C
$\hbar$	Planck constant/2π	1.054 572 66(63)	10^{-34} Js
c	velocity of light	2.997 924 58	10^{8} ms^{-1}
R_∞	Rydberg constant	1.097 373 157 09(18)	10^{7} m^{-1}
α^{-1}	inverse fine structure constant	1.370 359 895(61)	10^{2}
$\lambda\!\!\!^{-}_C$	electron Compton wavelength/2π	3.861 593 23(35)	10^{-13} m
a_o	Bohr radius	0.529 177 249(24)	10^{-10} m

important features of the theory. The dependence on the reduced mass is not simple and requires a sophisticated analysis.

For a nucleus with infinite mass, the energy levels given by the Dirac theory for an electron in a Coulomb field are

$$E_{nj} = m_e c^2 f(n,j) \tag{2.3}$$

where

$$f(n,j) = \left[1 + \frac{(Z\alpha)^2}{(n-\beta)^2}\right]^{-1/2} \tag{2.4}$$

$$\beta = j + \frac{1}{2} - \sqrt{\left(j + \frac{1}{2}\right)^2 - (Z\alpha)^2} \tag{2.5}$$

Here n is the principal quantum number and j the total angular momentum. For a nucleus with finite mass m_N, the energy levels are given by

$$E/c^2 = M + m_r\,(f\,(n,j) - 1) - \frac{m_r^2}{2M}\,(f\,(n,j) - 1)^2$$

$$+ \frac{(Z\alpha)^4\, m_r^3}{2n^3\, m_N^2}\left[\frac{1}{j + \frac{1}{2}} - \frac{1}{l + \frac{1}{2}}\right](1 - \delta_{l0}), \qquad (2.6)$$

where l is the angular momentum,

$$M = m_e + m_N, \qquad (2.7)$$

and the reduced mass m_r is given by

$$m_r = \frac{m_e m_N}{m_e + m_N}. \qquad (2.8)$$

The accepted method for calculating the QED shift is to evaluate successively higher terms in powers of α, $Z\alpha$ and m_e/m_N. The QED shift can be expressed in the form

$$\Delta E_Q(n^2L_J) = \left[\frac{\alpha}{\pi}\frac{(Z\alpha)^4}{n^3} m_e c^2\right] F(Z\alpha, n^2L_J) = \left[\frac{2chR_\infty \alpha^3}{\pi} Z^4\right] F(Z\alpha, n^2\, L_J) \qquad (2.9)$$

where for a particular angular momentum state

$$F(Z\alpha) = F_{SE}(Z\alpha) + F_{VP}(Z\alpha) + F_{FS}(Z\alpha) + F_{HO}(Z\alpha) + F_{RR}(Z\alpha) + F_{RRM}(Z\alpha).$$

Here the successive terms are respectively due to self energy, vacuum polarization, finite size of the nucleus, higher order contributions, relativstic recoil, and relativistic reduced mass. The self energy term for an S state is

$$F_{SE}(Z\alpha) = A_4 + (Z\alpha)A_5 + (Z\alpha)^2 A_6 + \frac{4}{3}(Z\alpha)^2\, G_{SE}(Z\alpha) \qquad (2.10)$$

where

$$A_4 = A_{40} + \frac{4}{3}\ln\,(Z\alpha)^{-2}, \qquad (2.11)$$

$$A_5 = A_{50}, \tag{2.12}$$

$$A_6 = A_{61} \ln (Z\alpha)^{-2} + A_{62} \ln^2 (Z\alpha)^{-2}. \tag{2.13}$$

The coefficients A_{40}, A_{50}, A_{61}, and A_{62} have been calculated analytically and the major uncertainty is incorporated in $G_{SE}(Z\alpha)$. The analogous expression for the vacuum polarization is

$$F_{VP}(Z\alpha) = B_4 + (Z\alpha)B_5 + (Z\alpha)^2 B_6 + \frac{4}{3} (Z\alpha)^2 G_{VP}(Z\alpha), \tag{2.14}$$

where

$$B_4 = -\frac{4}{15}, \tag{2.15}$$

$$B_5 = B_{50}, \tag{2.16}$$

$$B_6 = B_{60} + B_{61} \ln (Z\alpha)^{-2}. \tag{2.17}$$

The principal dependence on n comes from the Bethe logarithm which is included in A_{40}. Explicit expressions for the coefficients are given in the article by Sapirstein and Yennie. Sapirstein and Yennie advocate a formulation for the theoretical expressions in which the reduced mass dependence is included in the individual terms and the argument of the logarithmic terms is $((m_e/m_r)(Z\alpha))^{-2}$ rather than $(Z\alpha)^{-2}$. They also include the factor (4/3) in the G_{SE} and G_{VP}. We follow here the convention used in the earlier literature. Alternate summaries of the formulae are given in refs. 38 and 208.

With the assumption that the nuclear charge is distributed uniformly in a sphere, to a good approximation for Z < 30 the nuclear size correction for the S states is

$$F_{FS}(Z\alpha) = \left(\frac{2}{3}\right)\left(\frac{\pi Z^2}{\alpha^3}\right)(1 + 1.70(Z\alpha)^2)\left(\frac{Z\alpha R_N}{\lambda_C}\right)^{2\gamma}, \tag{2.18}$$

where $\gamma = \sqrt{1-(Z\alpha)^2}$, $ƛ_C = \hbar/m_e c$, and R_N is the rms charge radius of the nucleus. As Z increases the finite size correction becomes increasingly dependent on n and the finite size contribution for the $P_{1/2}$ state must be taken into account when computing the Lamb shift.

There are similar expressions for the shift in the states with angular momentum. The shifts are largest for the S states and decrease as the angular momentum increases. For hydrogen the dominate contributions for n=2 are

$$\Delta E_{SE} = \frac{\alpha}{\pi}\frac{(Z\alpha)^4}{n^3} m_e c^2 \left(\frac{4}{3}\ln (Z\alpha)^{-2} - 2.6379\right) \rightarrow 1066 \text{ MHz}, \tag{2.19}$$

$$\Delta E_{VP} = \frac{\alpha}{\pi}\left(\frac{(Z\alpha)^4}{n^3}\right) m_e c^2 \left(-\frac{4}{15}\right) \rightarrow -27 \text{ MHz}. \tag{2.20}$$

For a discussion of the calculation of the Lamb shift and tables summarizing the formulae for calculating the Lamb shift, the reader should see the article by Sapirstein and Yennie in this volume. Erickson[32] has provided a discussion of the expressions for the fine structure and the dependence on the reduced mass. He has also given tables of the calculated values for the Lamb shift and the other hydrogenic fine structure intervals. There is an error in Erickson's calculation of $G_{SE}(Z\alpha)$ which should be corrected when using these tables.[33] The best values for $G_{SE}(Z\alpha)$ are those given by Mohr.[34-37]

Johnson and Soff[38] have provided a table of calculated values of the QED shifts for the n=1, $S_{1/2}$ and n=2, $S_{1/2}$, $P_{1/2}$, $P_{3/2}$ levels with Z=1 to 110. They give in their tables the relative size of the various contributions to the Lamb shift. As Z increases the $G_{SE}(Z\alpha)$ term becomes increasingly important. Figure 3 shows in graphical form the total QED shift for the $2S_{1/2}$ state and the major contributions to the shift as a function of Z.

A major limitation on experiments to determine the Lamb shift is due to the lifetime of the states involved. If one observes a transition between two states such as the $2^2S_{1/2}$ and $2^2P_{1/2}$ states, the natural linewidth is determined by the sum of the decay probabilities of the states through the expression

$$\Delta\nu = \frac{1}{2\pi}(\gamma_1 + \gamma_2). \tag{2.21}$$

The $2^2P_{1/2}$ and $2^2P_{3/2}$ states decay through electric dipole radiation to the ground state with the decay rate[39]

$$\gamma(2^2P_{1/2,3/2}) = 6.265 \times 10^8\, Z^4\, s^{-1}. \tag{2.22}$$

The $2^2S_{1/2}$ state decays primarily through two photon radiation for low Z and magnetic dipole radiation for high Z. The decay rate is given by the approximate expression[40]

$$\gamma(2^2S_{1/2}) = (8.2293\, Z^6\, (2E1) + (2.50 \times 10^{-6})Z^{10}\, (M1))\, s^{-1}. \tag{2.23}$$

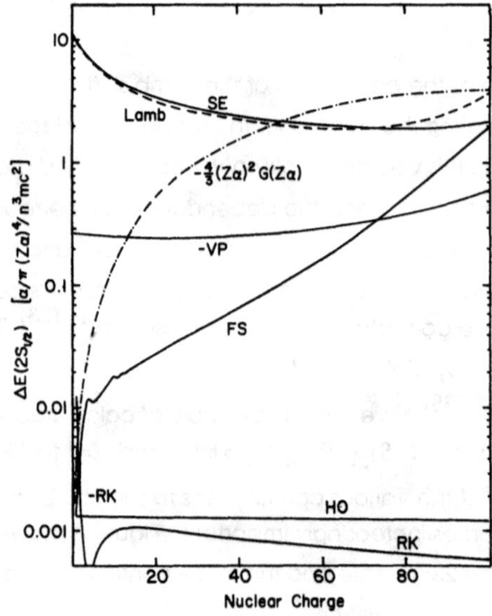

Figure 3. Contributions to the QED shift for the $2^2S_{1/2}$ state in hydrogenic atoms as a function of the nuclear charge. The curve labeled RK is the sum of the relativistic recoil and relativsitc reduced mass terms. Most of the contributions as well as the total Lamb shift are repulsive. Attractive contributions are indicated by a minus sign.

Since the lifetime of the $2^2P_{1/2}$ state has roughly the same dependence on Z as the Lamb shift, the ratio of the linewidth to the $^2S_{1/2}$ - $^2P_{1/2}$ Lamb shift interval, which is 1/10 for hydrogen, increases slowly as Z increases. Figure 4 shows graphically the dependence of the n=2 Lamb shift and the decay probabilities of the S and P states on Z.

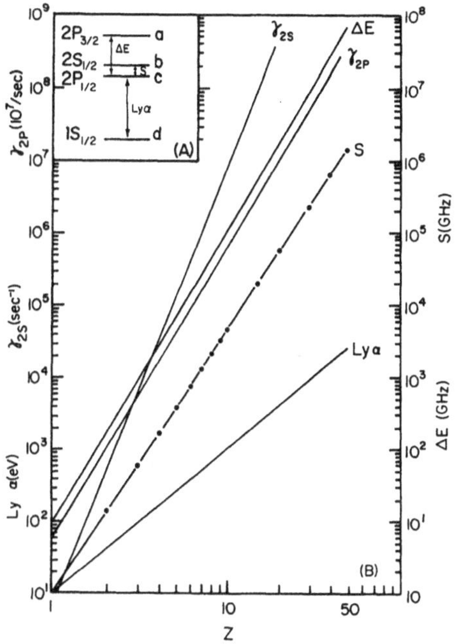

Figure 4. A graph showing the dependence on Z of the spontaneous decay rates for the $2S_{1/2}$ and $2P_{1/2}$ states, the Lamb shift S, the fine structure ΔE , and the energy of the Lyman-α transition.

A second potential source for precision measurements of the Lamb shift is two electron systems such as helium.(25) Figure 5 shows the energy level diagram for the low-lying levels of a helium like atom. The intervals of greatest interest are those between the metastable $^3S_1$ level and the $^3P_{0,1,2}$ levels. For low Z ions the methods of laser spectroscopy can be used to measure this interval with a precision of 1 in 10^{10}. For helium the Lamb shift is roughly 19 parts in 10^6 so a measurement could give the Lamb shift with a precision of 5 in 10^6. At present the major limitation in using this interval to test QED is the calculation of the energy levels without the inclusion of QED effects.

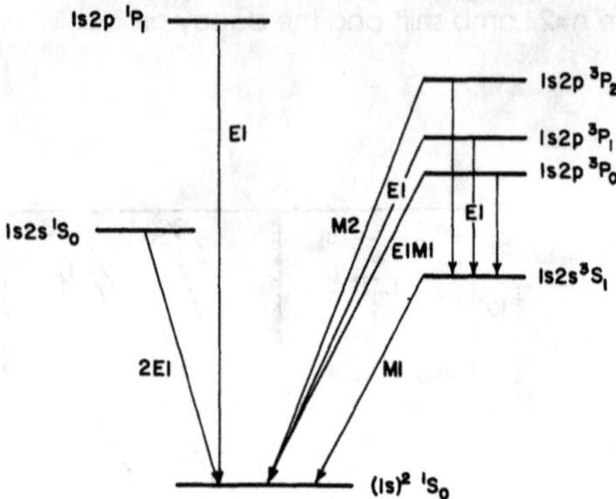

Figure 5. The energy level diagram for the lowest S and P states of a helium like atom. Also shown is the character of the electromagnetic transitions through which the states decay.

3. CALCULATION OF SIGNALS

Three basic techniques have been used to measure the Lamb shift -- quenching by an external electric field, radiofrequency or laser induced transitions between two states, level crossing-anticrossing experiments. Since the natural linewidth for $n^2S_{1/2}$-$n^2P_{1/2}$ transitions is roughly 10% of the Lamb shift, a precise measurement of the Lamb shift requires measuring the center of the resonance line to a small fraction of the line width. This entails a detailed knowledge of the line shape and the factors which can shift the line or make it asymmetric. In this section, we will summarize the calculation of the signals and the factors common to all the experiments.

A static electric field will mix the S and P levels producing two new eigenstates whose lifetimes are intermediate between those of the initial S and P states. [39] By forming a beam in the $^2S_{1/2}$ state and observing the lifetime as a function of electric field, one can measure the Lambshift. If we represent the hydrogen atom by a system with two levels, the Hamiltonian in an external electric field is

$$H = H_o + H_D - \vec{\mu}_E \bullet \vec{E}_{DC}, \tag{3.1}$$

where H_o is a hermetian operator yielding the energy levels

$$H_o|1> = \hbar\omega_1|1>, \tag{3.2a}$$

$$H_o|2> = \hbar\omega_2|2>; \tag{3.2b}$$

H_D is a phenomenological non-hermition operator accounting for the decay

$$H_D|1> = -i\frac{1}{2}\hbar\gamma_1|1>, \tag{3.3a}$$

$$H_D|2> = -i\frac{1}{2}\hbar\gamma_2|2>; \tag{3.3b}$$

$\vec{\mu}_E$ is the electric dipole moment operator and $\vec{E}_{DC}$ the applied electric field. The two levels are described by the wave function

$$|\psi> = c_1(t)|1> + c_2(t)|2>; \tag{3.4}$$

and the evolution of the system is given by the Schrodinger equation

$$i\hbar\frac{\partial|\psi>}{\partial t} = H|\psi>. \tag{3.5}$$

The combination of Eqs. 3.1, 3.4, and 3.5 gives

$$i\dot{c}_1 = \omega_1 c_1 - i\frac{1}{2}\gamma_1 c_1 + V_{DC}c_2, \tag{3.6a}$$

$$i\dot{c}_2 = \omega_2 c_2 - i\frac{1}{2}\gamma_2 c_2 + V_{DC}c_1, \tag{3.6b}$$

where

$$V_{DC} = -\frac{1}{\hbar}<1|\vec{\mu}_E \bullet \vec{E}_{DC}|2>. \tag{3.7}$$

It is straight forward to solve these equations and obtain the time dependence for an atom initially in the $|1\rangle$ state which suddenly finds itself in an electric field.[39,41] For the situation where $\gamma_2 >> \gamma_1$ to a good approximation the atom decays exponentially with the decay constant

$$\gamma = \gamma_1 + \gamma_2 \frac{|V_{DC}|^2}{(\omega_1 - \omega_2)^2 + \left(\frac{\gamma_2}{2}\right)^2} \tag{3.8}$$

For a radiofrequency or laser field the interaction Hamiltonian is

$$H_I = - \vec{\mu}_E \bullet \vec{E}_{RF} \cos(\omega t + \delta). \tag{3.9}$$

The Schroedinger equation for two coupled states is

$$i\dot{c}_1 = \omega_1 c_1 - i\frac{1}{2}\gamma_1 c_1 + 2\, V_{RF}\, c_2 \cos(\omega t + \delta), \tag{3.10a}$$

$$i\dot{c}_1 = \omega_1 c_2 - i\frac{1}{2}\gamma_2 c_2 + 2\, V_{RF}\, c_1 \cos(\omega t + \delta), \tag{3.10b}$$

where

$$V_{RF} = -\frac{1}{2\hbar} \langle 1 | \vec{\mu}_E \bullet \vec{E}_{RF} | 2 \rangle. \tag{3.11}$$

These equations do not have simple analytic solutions. By neglecting the counterrotating component of the rf field, which does not contribute to the resonance, one obtains the following equations:

$$i\dot{c}_1 = \omega_1 c_1 - i\frac{1}{2}\gamma_1 c_1 + V_{RF}\, c_2\, e^{-i(\omega t + \delta)}, \tag{3.12a}$$

$$i\dot{c}_2 = \omega_2 c_2 - i\frac{1}{2}\gamma_2 c_2 + V_{RF}\, c_1\, e^{i(\omega t + \delta)}. \tag{3.12b}$$

The analytic solution for these equations[42,43] with the initial conditions at $t = 0$

$$c_1(t) = c_1(0), \tag{3.13a}$$

$$c_2(t) = c_2(0), \tag{3.13b}$$

are

$$c_1(t) = \exp(-\frac{1}{4}(\gamma_1 + \gamma_2)t - i(\frac{1}{2})(\omega + \omega_1 + \omega_2)t)$$

$$\times ((\cos\frac{1}{2}at - i\cos\theta\sin\frac{1}{2}at)\,c_1(0)$$

$$- e^{-i\delta}(i\sin\theta\sin\frac{1}{2}at)\,c_2(0)), \tag{3.14a}$$

$$c_2(t) = \exp(-\frac{1}{4}(\gamma_1 + \gamma_2)t + i(\frac{1}{2})(\omega - \omega_1 - \omega_2)t)$$

$$\times (-e^{i\delta}(i\sin\theta\sin\frac{1}{2}at)c_1(0)$$

$$+ (\cos\frac{1}{2}at - i\cos\theta\sin\frac{1}{2}at)\,c_2(0)), \tag{3.14b}$$

where

$$a = (4V_{RF}^2 + (\Omega + iQ)^2)^{1/2}; \tag{3.15}$$

$$\omega_0 = \omega_1 - \omega_2; \tag{3.16}$$

$$\Omega = \omega - \omega_0; \tag{3.17}$$

$$Q = \frac{1}{2}(\gamma_1 - \gamma_2); \tag{3.18}$$

$$\sin\theta = 2V_{RF}/a; \tag{3.19}$$

$$\cos\theta = (\Omega + iQ)/a. \tag{3.20}$$

These solutions are the generalization to decaying states of the Rabi formula for rf induced transitions.[44] In general the line shape and linewidth for a beam experiment is determined by the transit time through the rf field, the rf power, the decay constants for the states, and the average over the velocity distribution. The linewidth can be less than the optical natural line width $(\gamma_1 + \gamma_2)$; for some fast beam experiments $(\gamma_2 - \gamma_1)$, where $\gamma_2 > \gamma_1$, plays the role of the natural linewidth.

For the situation where the time spent in the rf field is long compared to the lifetime of the short lived state and the perturbation is weak, it can be shown that one has an effective decay probability for the long lived state of the form[45]

$$\gamma = \gamma_1 + \gamma_2 \frac{|V_{RF}|^2}{(\omega-\omega_o)^2 + \left(\frac{\gamma_2}{2}\right)^2} . \qquad (3.21)$$

This gives a quenching analogous to that for a DC field and shows that the line profile is Lorentzian with a width determined by the lifetime of the short lived state. An approximate expression which takes into account the saturation due to the rf field is

$$\gamma = \gamma_1 + \frac{(\gamma_1 + \gamma_2)\ |V_{RF}|^2}{\left((\omega-\omega_o)^2 + \left(\frac{\gamma_1 + \gamma_2}{2}\right)^2 + \frac{(\gamma_1 + \gamma_2)^2}{\gamma_1\gamma_2}|V_{RF}|^2\right)} . \qquad (3.22)$$

Since it is technically difficult to keep the microwave electric field constant when tuning over the wide band required to measure the line profile, the early experiments were carried out with a fixed frequency and a magnetic field was used to tune through resonance. This introduces complications since the wavefunctions for the initial and final states depend on the field. The electric field seen by the atom due to motion through the magnetic field must also be taken into account and an average made over the velocity distribution of the atoms in the beam. Figure 6 shows an energy level diagram for the n=2 manifold of atomic hydrogen as a function of an external magnetic field. The vertical lines and circles indicate fields used in measurements of the fine structure. This figure also shows the notation introduced by Lamb to designate the different sublevels.

Rather than use a microwave field, one can use a DC electric field and tune the levels through the fields where the S and P levels cross. The DC electric field mixes the levels and the long lived state is quenched. This is in essence a resonance at zero frequency and has a Lorentzian line profile.

An alternate method applicable in some situations is level anticrossing.[46] The levels involved are prevented from crossing by an internal perturbation which couples the two magnetic substates. The substates repel each other and their wavefunctions interchange their identities as the magnetic field is tuned through the region of closest approach. At the field strength of closest approach the wavefunction of each state is a mixture of the wavefunctions of the uncoupled states.

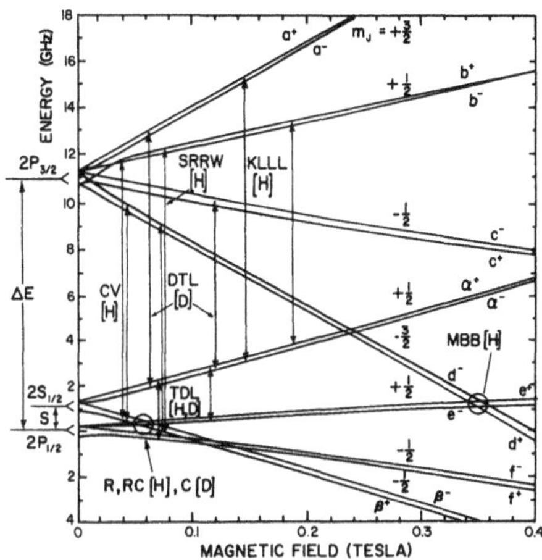

Figure 6. The energy levels of a hydrogen atom in a magnetic field.

In bottle experiments in which the environment is a gas and one does not have information about the time of production and detection, it is advantageous to use the density matrix to describe the signals. This situation has been analyzed in detail by Lamb and Sandars.(47) They showed that, for two states 1 and 2 with production rates r_1, r_2 which undergo spontaneous transitions to lower states with decay rates γ_1, γ_2, and detection fractions f_1 and f_2, the signal is

$$S = -(f_1 - f_2)\left(\frac{r_1}{\gamma_1} - \frac{r_2}{\gamma_2}\right) \times \frac{(\gamma_1 + \gamma_2)\,|V_{RF}|^2}{\left((\omega-\omega_0)^2 + \frac{1}{4}(\gamma_1 + \gamma_2)^2 + \frac{(\gamma_1+\gamma_2)^2}{\gamma_1\gamma_2}\,|V_{RF}|^2\right)} \tag{3.23}$$

where V_{RF} is as defined earlier. This gives a Lorentzian profile whose width is determined by the optical natural line width.

4. SLOW BEAM MEASUREMENTS

Figure 7 shows a schematic diagram of the apparatus used by Lamb and Retherford for their measurements of the Lamb shift.[16,41,48,49] The key element in the design is the metastability of the $2^2S_{1/2}$ state. A tungsten oven was used to dissociate molecular hydrogen and form a beam of hydrogen atoms. A transverse

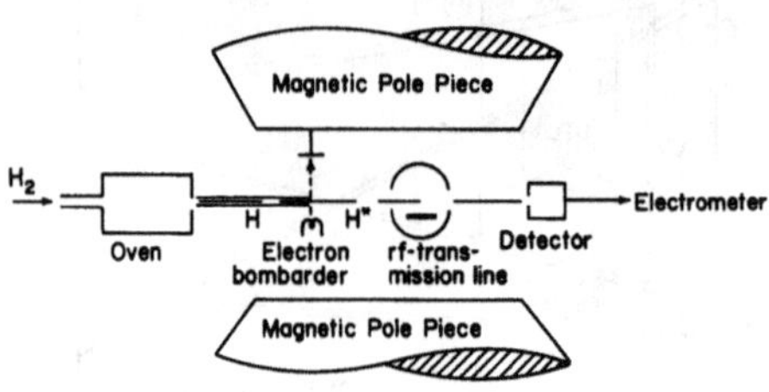

Figure 7. A diagram of the apparatus used by Lamb and Retherford for their measurements of the Lamb shift.

electron beam excited the atoms into the $2^2S_{1/2}$ state. The beam then passed through a radiofrequency field located in the center of a magnet and was detected through the electrons emitted when the atoms struck a tungsten plate. The energy levels were tuned through resonance by varying the magnetic field with the rf frequency constant. The observed resonances were actually close lying doublets due to the hyperfine structure resulting from the interaction of the proton magnetic moment with the magnetic field due to the electron.

After the initial series of measurements by Lamb and Rutherford, Triebwasser, Dayhoff and Lamb[50,51] improved the apparatus and made a further series of measurements. Figure 8 shows the resonance profiles for the αf transitions in hydrogen and deuterium at respectively 707.77 G and 704.57 G with an rf frequency of 2395 MHz. The two components in the scan for H are due to the hyperfine structure. In deuterium, where the hyperfine structure is smaller, the line width is only slightly greater than the 100 MHz natural linewidth due to the 1.6 ns lifetime of the $^2P_{1/2}$ state. Lamb carried out an extensive analysis of the signals and the sources of systematic corrections. These publications are very informative and

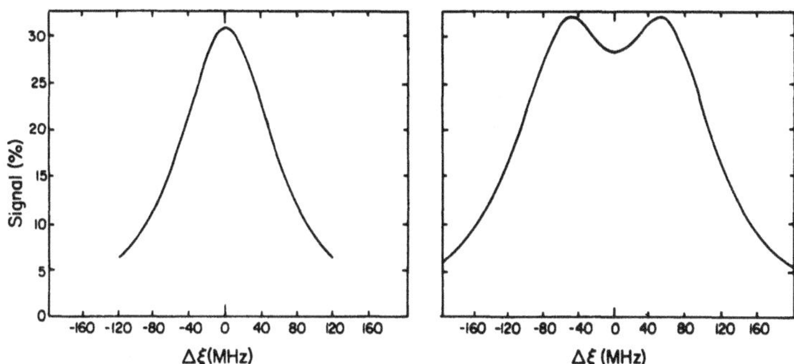

Figure 8. Line profiles for αf transitions in hydrogen and deuterium obtained by Triebwasser, et al.

recommended reading for individuals interested in precision measurements. The outcome of this series of experiments was precise measurements of the $2^2S_{1/2}$ - $2^2P_{1/2}$ Lamb shift interval for hydrogen and deuterium and the $2^2P_{3/2}$ - $2^2S_{1/2}$ fine structure interval in deuterium. Brodsky and Parsons[52] carefully analyzed the theory for the Zeeman effect used to determine the zero field fine structure intervals from the measured transitions. Taylor, Langenberg, and Parker[53] have given a thoughtful critique of the Lamb measurements in the light of subsequent work and reevaluated their errors so they are equivalent to one standard deviation. The precision of the final results, taken as one standard deviation, was 60 kHz. This corresponds to 6/10000 of the 100 MHz natural linewidth. The results are summarized in Table 7 along with the results from subsequent experiments.

The second series of slow beam measurements were those carried out by Robiscoe.[54] Robiscoe used a static electric field to induce transitions from the $2S_{1/2}$ to the $2P_{1/2}$ state at the field near 600G where the levels cross. Figure 9 shows a schematic diagram of the apparatus used by Robiscoe. In order to eliminate the asymmetric line obtained when both hyperfine components were present, Robiscoe used a second level crossing to prepare a beam of atoms in the β^- hyperfine state. The electron gun was located in a field of 575 G at the β-e crossing point so that all atoms formed in the β state were immediately quenched. A non-adiabatic transition near zero field or a rf field where used, respectively, to prepare a beam in the β^- or β^+

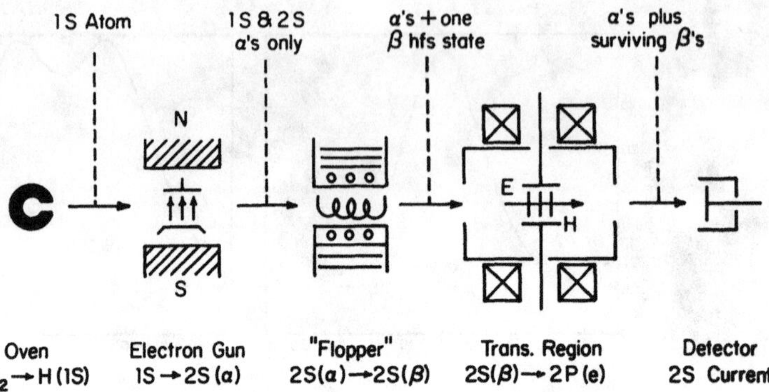

Figure 9. A schematic drawing of the apparatus used by Robiscoe for the level crossing measurements of the Lamb shift in hydrogen

states. He then observed either the $\beta^- - e^-$ crossing at 605G or the $\beta^+ - e^+$ crossing at 538G. Figure 10 shows the line profile obtained by Robiscoe and Cosens(55) in an improved version of the level crossing experiment. The observed linewidth of 61.1G is only slightly greater than the 54G Lorentz width expected from the 1.6 ns lifetime of the $2P_{1/2}$ state. Robiscoe's initial result disagreed with that obtained by Lamb and coworkers. Robiscoe(56) subsequently realized that he had failed to make the proper correction for the distortion of the beam velocity distribuiton by preferential quenching of high velocity atoms due to motional electric fields. When the proper correction was made, it brought the results of the crossing experiments into agreement with those of Lamb. Robiscoe's Cargese lectures(57) provide a thorough treatment of the history and testify to the care required to measure the center of a line to high precision. In a subsequent communication, Robiscoe and Shyn(58) reported calculations and measurements of the effective velocity distribution for metastable hydrogen atoms in a typical slow beam experiment and showed it was characterized by $U^n \exp(-U^2)$ with $n \sim 4$ rather than $n=2$ as normally assumed. They concluded that this changed the motional field-asymmetry correction in the Cosens-Robiscoe experiments and raised the Lamb shift by 0.040 MHz. This correction also called into doubt the final value reported by Triebwasser, Dayhoff and Lamb. Taylor, Parker, and Langenberg(53) discuss these measurements in a note added in proof. The final result had a precision comparable to that obtained by Lamb et al.

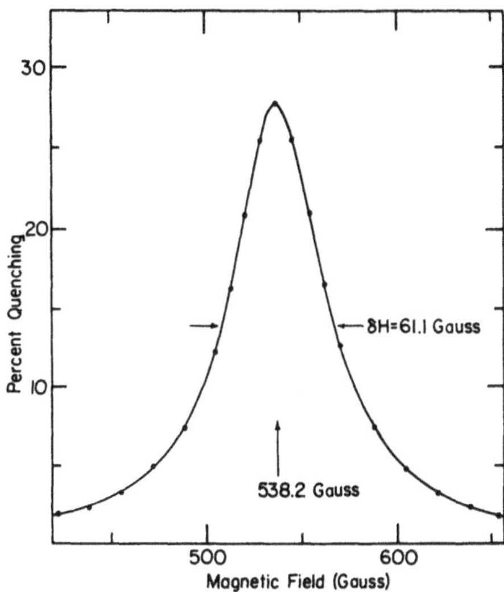

Figure 10. Typical quenching line profile obtained by Robiscoe and Cosens near 538G with an 0.6 V/cm static electric field. The solid line is the velocity-averaged line shape derived from the Bethe Lamb theory of the lifetime of the 2S state in an external field.

Cosen[59] subsequently carried out a similar level crossing measurement of the Lamb shift in deuterium. The result for this measurement differs significantly from that of Triebwasser, et al.

Three groups have used a slow beam-rf field method to measure the $2^2P_{3/2}$ - $2^2S_{1/2}$ fine structure interval. Dayhoff, Triebwasser and Lamb[51] measured the interval in deuterium using the αa and αc transitions. Cosens and Vorburger[60,61] measured the transition in hydrogen. They used a modification of the apparatus employed for the level crossing experiments to select either the β^+ or β^- hyperfine state and observed rf transitions near 10 GHz to the b and d states. They obtained a result with a precision of 4 ppm. Shyn, Rebane, Robiscoe, and Williams [63,63] used the β^+d and β^+b transitions near 800 Gauss to measure the $2^2P_{3/2}$ - $2^2P_{1/2}$ transition in hydrogen. A low field crossing near 578 gauss followed by an rf transition near 180 MHz was employed to form a beam of atoms in the β^+ state. A microwave wave field at 8-11 GHz was used to drive the transitions of interest. They obtained a precision of 6 ppm.

5. BOTTLE EXPERIMENTS

A major limitation of the slow beam experiments is that they depend on the metastability of the $2^2S_{1/2}$ state and cannot easily be used to measure the fine structure of states with $n \geq 3$. The magnetic field used to tune through the resonance makes it difficult to work with a slow ion beam since the ion beam will be deflected by the transverse magnetic field normally used. The bottle technique was introduced to overcome both these difficulties.[27,64]

Figure 11 shows a schematic diagram for a bottle experiment. The electron excitation, resonance transition, and detection all take place in the same region of

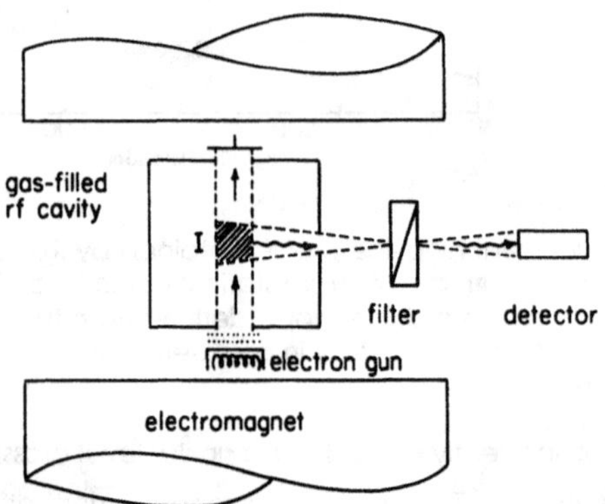

Figure 11. A schematic diagram of a bottle experiment in which production, resonance transition, and detection all take place in the interaction volume I.

space where there is a dilute gas of the atoms to be investigated. An electron beam parallel to the magnetic field excites the atoms in the gas to the n manifold of interest and the light emitted by the excited atoms or ions is observed. The intensity, polarization, and spectral distribution of the emitted light depend on the excitation and decay paths and in general change when the atom is tuned through resonance. The large variation in the decay rates for the different angular momentum levels in a given n manifold results in markedly different populations for the states.

Major sources of difficulty in the bottle experiments are the perturbations due to collisions and the electric fields due to ions and electrons in the plasma. Atoms in high n excited states are very sensitive to electric fields. The electric fields mix the states and distort the energy levels. Some experimenters have pulsed the electron beam and observed the resonances in a subsequent time interval to reduce the electric field perturbations. A second source of difficulty is cascades from higher n states. There are often overlapping resonance in higher n states which saturate at a low rf power and distore the resonance in the manifold of interest.

The first bottle experiment was carried out by Skinner and Lamb[65,66] to measure the Lamb shift in the n=2 state of He^+. An electron beam ionized the helium atoms and excited them to the $2^2S_{1/2}$ metastable state. A resonant rf field drove the atoms to the $2^2P_{1/2}$ state whose lifetime is 10^{-10} s and which decayed to the $1^2S_{1/2}$ ground state with emission of radiation at 304Å. The resonance was observed by photodetection of the 304Å radiation. The large background was a major problem which limited the precision of the measurement. Subsequently, Novick, Lipworth, and Yerkin[67] and later Lipworth and Novick[68] refined the technique and measured the Lamb shift in He^+ n=2 with a precision of 120 ppm. Independently Narasimham and Strombotne[69] used this technique to measure the Lamb shift in He^+. They obtained a result with a precision of 85 ppm but which differed by 276 ppm from the result of Novick, et al.

An example of the difficulties encountered in bottle experiments is provided by the measurement by Mader, et al.[70] of the Lamb shift in He^+ n=3. Figures 12 and 13 show, respectively, the resonance transitions and the power shift for the αe transition of interest. There are overlapping transitions whose amplitudes have a different dependence on the microwave power. By including the $7\beta c$ signal in the data analysis they could understand the power shift and were able to measure the Lamb shift interval with an uncertainty of $\pm$0.5 MHz (100 ppm).

Kaufman et al.[71,72] used the bottle technique to measure the $2^2P_{3/2}$ - $2^2S_{1/2}$ interval in H by observing transitions αa, αb and αc. They obtained excellent signals and reported a final value with an uncertainty of 26 kHz (26 ppm). This result , however, disagrees significantly with the other measurements of this interval. This illustrates the basic difficulty with bottle experiments. It is not obvious when one has eliminated residual sources of systematic corrections due to the environment.

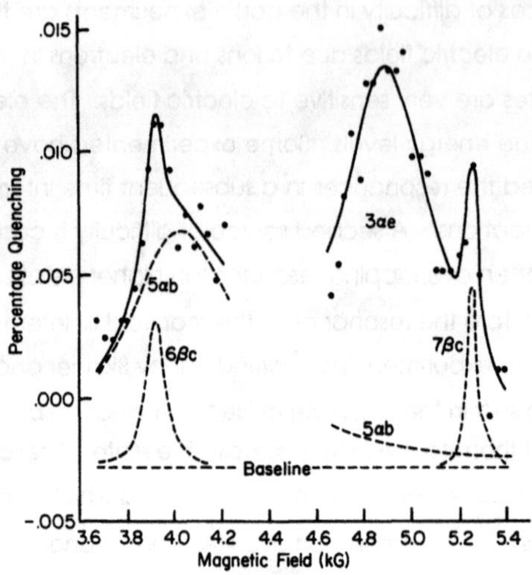

Figure 12. A very low power scan taken at 9 GHz with 0.0005W. Each point represents 1h of signal averaging. Solid curve is a theoretical fit. The main resonance is ae with a height about 1/1000 of its saturated height.

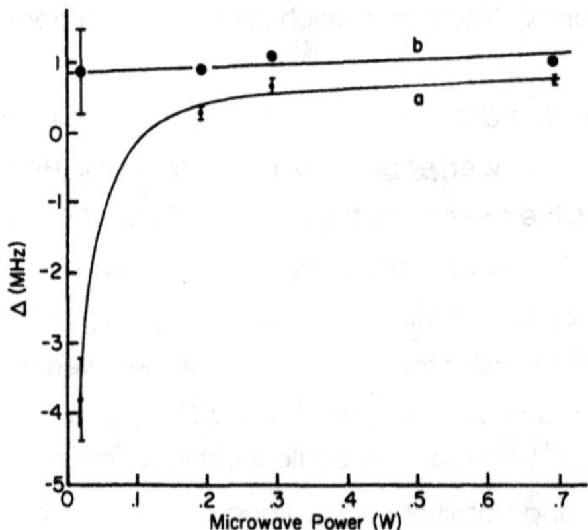

Figure 13. (a) Observed power shift for the αe transition.
(b) Power shift for αe transition after correction for the large shift at low power due to the 7βc transition.

Leventhal and Harvey[73,74] employed the bottle technique to measure the Lamb shift in the n=2 state of Li^{++}. They used pulsed excitation and observed the βc transition. They measured the interval with a precision of 330 ppm.

Kleinpoppen[75] carried out a measurement of the Lamb shift in H, n=3 through an experiment which combined the bottle and beam techniques. He formed a beam of H atoms and then used Lyman β radiation to excite the atoms to the 3p state. Roughly 10% of the atoms in the $3^2P_{1/2}$ state decayed to the $2^2S_{1/2}$ metastable state which was detected in a metastable detector. When the atoms were tuned through resonance they were driven to the 3S state which does not decay to the 2S state. The experiment was difficult and the precision obtained was 1%.

Another application of bottle experiments has been the detection of S-D and S-F intervals through multistep level crossing signals.[46] These resonances are narrower than the S-P resonances because of the longer lifetimes of the terminal states. They are also very sensitive to Stark shifts and require a careful extrapolation to zero electric field.

Table 2 lists other bottle experiments not discussed in the text. A more complete review is given by Beyer.[27] Only the 100 ppm measurement of Mader, et al., for He^+ n=3 and the measurements for He^+ n=2 have a precision comparable with that obtained for H, n=2.

6. FAST BEAM EXPERIMENTS

Fast beam experiments utilize an energetic atomic or ion beam to overcome the limitations of the slow beam and bottle experiments. The speed of the beam makes it possible to work with short lived states. Accelerators can be used to prepare intense well collimated, nonenergetic beams of atoms and ions in the metastable $2^2S_{1/2}$ state. Accelerator based techniques can be used to make a beam of hydrogen and helium like ions with high Z. Several different techniques have been used to measure the Lamb shift. In this section, we shall summarize the measurements and briefly describe the major methods.

6.1 Stark Quenching

The first fast beam measurements of the Lamb shift were carried on Li^{++} by Sellin[100] and later refined by Fan, Garcia-Munoz, and Sellin.[101,102] Figure 14 shows

Table 2. A summary of the bottle experiments not discussed in the text

System	Experimenters	Reference
H, n=3	Lamb, Sanders	47
	Glass-Maujean, Descoubes	78
	Richardson, Hughes	79, 80
	Glass-Maujean	98
D, n=3	Lamb, Sandars	47
H, n=4	Brown, Pipkin	81
D, n=4	Wilcox, Lamb	77
He^+, n=3	Leventhal, et al.	82
	Bauman, Eibofner	83,85,93
He^+, n=4	Lea, et al.	86
	Hatfield, Hughes	87
	Beyer, Kleinpoppen	88
	Jacobs, et al.	89
	Eibofner	90,91
	Eck, Huft	92
He^+, n=5	Beyer, Kleinpoppen	95,96,97
	Bauman, Eibofner	93
	Eibofner	91,94
He^+, n=6	Eibofner	91,94,99

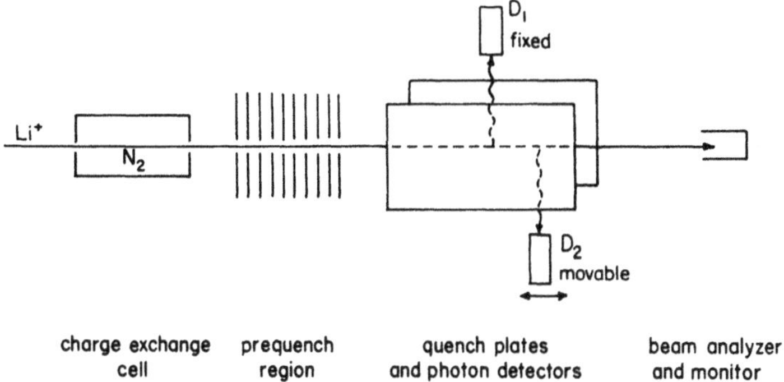

Figure 14. A schematic diagram of the apparatus used to determine the Lamb shift in Li^{++} by measuring the lifetime of atoms initially in the $2^2S_{1/2}$ state as a function of the applied electric field.

a schematic diagram of the apparatus. A beam of Li^{++} in the metastable $2^2S_{1/2}$ state was prepared by passing a 8 MeV beam of $^6Li^+$ through a N_2-filled gas cell. The radiation emitted as a function of distance along the beam was used to measure the lifetime as a function of the external electric field. The Lamb shift was determined from the dependence of the lifetime on electric field by the use of a refinement of Eq. 3.8 which took into account higher order terms, the $2^2P_{3/2}$ state and the hyperfine structure.[102-103] A prequench region with alternating 12 kV and earth potential was employed to modulate the population in the metastable state and improve separation of the $2^2S_{1/2}$ signal from the background. The net uncertainty was ±0.5% which corresponds to 1/25 of the natural linewidth.

A refinement of this technique was used first by Leventhal and Murnick and subsequently by Kugel, Leventhal and Murnick[105-107] to measure the Lamb shift in $^{12}C^{5+}$. They first formed a beam of C^{6+} and then used charge capture in a cell to form C^{5+} $2^2S_{1/2}$ ions. They used the motion in a transverse magnetic field to create the electric quench field. A serious source of uncertainty was the variation in the detection efficiency due to the change in the beam trajectory with the magnetic field. The final uncertainty was ±1% which corresponds to 1/16 of the natural linewidth.

A similar technique was used by Leventhal, Murnick and Kugal[108] and by Lawrence, Fan and Bashkin[109] to measure the Lamb shift in O^{7+}, by Zacek, et al.[110] to measure the Lamb shift in S^{15+} and by Gould and Marrus[111,112] to measure the Lamb shift in Ar^{17+}. Figure 15 shows a schematic diagram of the apparatus used by Gould and Marrus. A fully stripped Ar^{18+} beam was passed

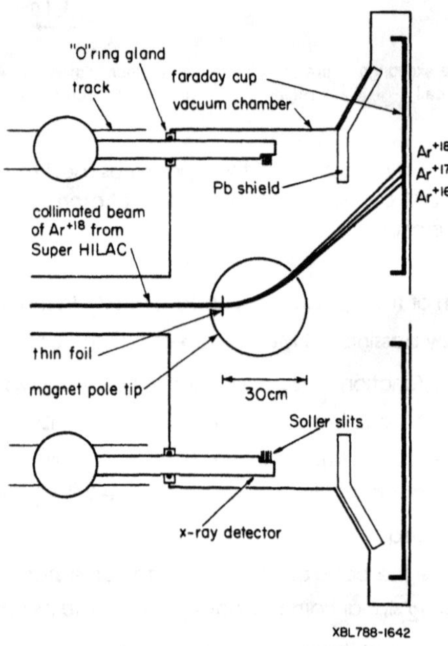

Figure 15. Schematic diagram of the apparatus used to measure the lifetime of the quenched $2^2S_{1/2}$ state of hydrogen like argon. Special precaution was taken to understand the changes in count rate due to beam dynamics.

through a thin carbon foil to form hydrogenlike Ar^{17+} in the $2^2S_{1/2}$ state. The lifetime was measured as a function of a transverse magnetic field. A symmetrical arrangement of x-ray detectors was employed to measure the intensity of emitted x-rays as a function of position.

Additional discussions of the radiative decay of a hydrogen like atom in an electric field are given in refs. 241 and 242.

6.2 Anisotropy of Quench Radiation

Fite, Kauppila, and Ott [113,114] observed that the Lyman-α radiation emitted by $2^2S_{1/2}$ hydrogen atoms in a electric field is polarized. In an analysis of the polarization, Drake and Grimley[104] showed that this quench radiation is spatially anisotropic, and that the anisotropy, which is roughly proportional to the Lamb shift, could be used to measure the Lamb shift. The anisotropy is defined as

$$R_O = \frac{I(0) - I(\frac{\pi}{2})}{I(0) + I(\frac{\pi}{2})}, \qquad (6.2.1)$$

where I(0) and $I(\frac{\pi}{2})$ are, respectively, the intensity of the radiation emitted parallel and perpendicular to the electric field. It can be shown that for a hydrogen-like atom with no hyperfine structure

$$R_O \simeq \frac{3}{2}\frac{S}{\Delta E}, \qquad (6.2.2)$$

and that to first order R_O is independent of the electric field strength. Drake and his coworkers[115-117] have refined the theory so as to take into account the hyperfine structure, P states with n>2, relativistic corrections to the matrix elements, and the interference between electric dipole and magnetic quadrupole decay channels.

van Wijngaarden, Drake, and Farago[118,119] used the anisotropy method to measure the Lamb shift in the n=2 state of hydrogen. They obtained a result with a precision of 850 ppm in agreement with other measurements. van Wijngaarden _et al._[120] also measured the Lamb shift in deuterium with a precision of 150 ppm. The value agreed with, but was less precise than, earlier measurements.

Van Wijngaarden _et al._[121,122,123] have carried out a series of experiments using the anisotropy method to measure the Lamb shift in the n=2 state of He^+. Figure 16 shows a schematic diagram of the apparatus used to measure the anisotropy. A He^+-ion beam containing about 0.5% metastable ions was formed by passing a beam of 121.5 keV ground-state He^+ ions from a magnetic mass analyzer through a gas cell. The emerging beam passed through a magnetic lens, a prequencher, and a collimator and entered the observation cell, where a transverse static electric field was produced by supplying opposite polarities to two pairs of cylindrical conducting

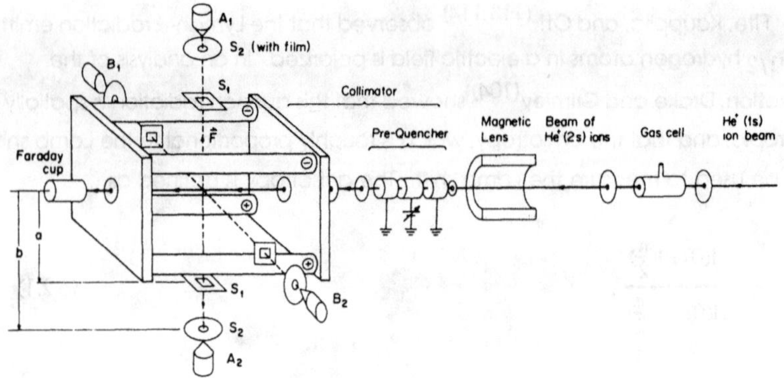

Figure 16. Schematic diagram of the apparatus used to measure the anisotropy for He^{+}.

rods. Two diode photodetectors with an absolute photon-detection efficiency of about 20% were used to measure the number of photons emitted along and at right angle to the electric field. The polarity of the rods was switched so that the roles of the two photodetectors were periodically interchanged. In their most recent measurement, they reported a value for the Lamb shift in He^{+} with a precision of 25 ppm. This corresponds in a resonance experiment to 2/10000 of the natural linewidth.

Curnette, Clocke, and Dubois[124] used the anisotropy method to measure the Lamb shift in O^{7+}. A 32 MeV O^{5+} beam was converted by passage through a thin (<10 $\mu g/cm^2$) carbon foil to O^{7+} in the metastable $2^2S_{1/2}$ state. The Stark mixing field was the motional electric field due to passage through a magnetic field. The Lamb shift was measured to 0.7%.

6.3 Quantum Beat Experiments

The passage of a fast beam through a thin foil or a localized electric field can be used to excite the atoms to a coherent superposition of fine or hyperfine structure states. If these states subsequently decay to the same final state, the spatial distribuiton of the emitted light can be used to measure the frequency difference between the states. An external electric field can be used to mix states of different parity so the beats can be observed. Similar experiments can be carried out using a laser or electron beam to excite the atoms and observing the emitted radiation as a function of time. These techniques provide a method for observing the natural frequencies of the atomic system.

These methods have been explored extensively and experiments carried out on both hydrogen and He^+. They have not, however, been used to make precise measurements of the Lamb shift. Details of the quantum beat method and further references can be found in the review articles by Andrä,[125,126] Beyer,[27] and Haroche.[127]

6.4 Atomic Interferometer

Sokolov and his coworkers[128-134] used a variation of the quantum beat technique to measure with high precision the Lamb shift in H, n=2. Figure 17 shows the

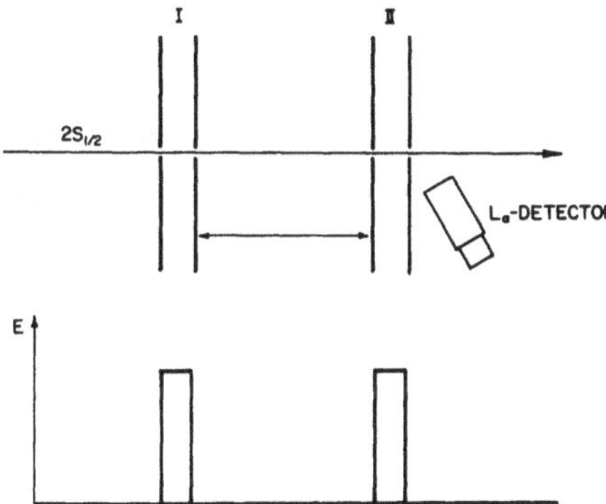

Figure 17. The arrangement of the electric fields used by Sokolov for the measurement of the Lamb shift in hydrogen.

two electric field region interferometer used by Sokolov. A 20 keV proton beam forms through charge exchange in H_2 a well collimated, highly monochromatic beam of hydrogen atoms in the $2^2S_{1/2}$ state. A weak magnetic field removes residual protons and rf fields at 1147 and 1088 MHz quench atoms in the $2^2S_{1/2}$ (F=1) hyperfine state. The atoms in the $2^2S_{1/2}$ (F=0) state then pass with nonadiabatic entry and exit through the two short regions with a longitudinal electric field separated by the distance L. A photodetector is used to detect the radiation from atoms in the $2^2P_{1/2}$ state when they exit the second electric field region. The signal of interest is the difference in the number of atoms in the $^2P_{1/2}$ state when the direction of the electric field in the second region is reversed. The signal is measured as a function of the separation of the two regions and a fit made to determine the Lamb shift in terms of the beam velocity. A separate experiment is employed to measure the beam velocity by observing the spatial decay of the Lyman-α radiation after passage of the metastable atoms through one electric field region. One in essence measures the Lamb shift in terms of the lifetime of the $2^2P_{1/2}$ state. Yakovlev[128] has calculated a precise value for this lifetime and found it to be

$$\tau(2^2P_{1/2}) = 1.59619946(48)\text{ns}. \qquad (6.4.1)$$

Using this method Sokolov, et al. found a value of the Lamb shift with an uncertainty of 1.8 parts in 10^6. This corresponds to 2 in 10^5 of the natural linewidth for a resonance experiment. These measurements can be regarded as a determination of the Lamb shift if one assumes the calculation of the lifetime is correct or a determination of the lifetime of the $2^2P_{1/2}$ state if one assumes the theoretical value for the Lamb shift is correct or uses an independent experimental value for the Lamb shift.

6.5 Fast Beam Radiofrequency Experiments

6.5.1 Single Field

With a fast beam it is difficult to use a magnetic field to tune the atoms to resonance bcause of the large perturbations due to the motional electric field. It is advantageous to use zero magnetic field and to sweep the frequency to observe the resonances. This requires a detailed understanding of the rf system and instrumentation for monitoring the rf field as the line is scanned. Advances in microwave technique have made this feasible.

Another problem encountered is overlapping resonances due to hyperfine structure. In general the hyperfine structure produces lines with several components of unequal amplitude which are not suitable for precision measurements of the line centers. To overcome this problem additional rf fields are used to quench the hyperfine states which are not of interest and obtain a line with a single hyperfine component.

The signals for a fast beam experiment are best calculated by solving the Schrodinger equation with the configuration of fields used in the experiment. These calculations can be carried out numerically so as to take into account the other states in the n manifold of interest and the spatial distribution of the microwave field. The signals are not well represented by the solution used in the slow beam work and the widths are not in general limited by the optical natural line width.

The first fast beam rf experiment was employed by Fabjan and Pipkin[42,135] to measure the Lamb shift in the n=3 state of hydrogen. They used a 20 keV proton beam and obtained a beam of atoms in the $3^2S_{1/2}$ state through charge capture in a carbon foil or a N_2 charge exchange cell. The excited atoms passed through the rf spectroscopy region and then through an rf hyperfine-state selection field before passing in front of a photomultiplier which observed selectively Balmer-α radiation. They measured the $3^2S_{1/2}$ (F=0) → $3^2P_{1/2}$ (F=1) transition and obtained a value for the Lamb shift with a precision of 200 ppm (1/500 of the natural linewidth). Subsequently, Fabjan, Pipkin, and Silverman[136] used this apparatus to measure all the fine structure intervals in the hydrogen n=3,4 manifolds and all but the F-G intervals in the n=5 manifold.

The most precise single field measurement was made by Andrews, Newton, and Unsworth[137-138] to determine the Lamb shift in the n=2 state of hydrogen. They used a calculable 'slab line' geometry for the microwave interaction region which consisted of a cylindrical center conductor midway between two flat plates. A precision, wide-band coaxial to slab line converter was used to match the interaction region to 50Ω coaxial lines. The rf power was monitored with a power meter which was calibrated using reference standards directly traceable to U.K. prime standards. They obtained lines with a width of 120 MHz and were able to determine the center to 20 kHz. This provided a measurement of the Lamb shift with a precision of 19 ppm.

Churassy, Gaillard, and Silver[139] used the fast beam rf method to measure the Lamb shifts in the n=4-7 manifolds of He^+. They used beams of 300, 400 and 500 keV and employed carbon foils to excite atoms in the $^4He^+$ ground state. These measurements were of relatively low precision (2.5%) but demonstrated the potential of the technique.

The detection of the two photon $3^2S_{1/2}$ - $3^2D_{5/2}$ transition in hydrogen was first reported by Kramer, Lundeen, Clark, and Pipkin.[140] Subsequently, van Baak, Clark, and Pipkin[141] developed a theoretical basis for understanding this transition and Clark, van Baak, Lundeen, and Pipkin[142,143] made a measurement of the $3^2S_{1/2}$ - $3^2D^{5/2}$ interval in hydrogen with a precision of 19 ppm (1/128 of the natural linewidth).

Brandenburger, Lundeen, and Pipkin[144] observed one-, two-, three- and four-photon resonances in the n=6,7, and 9 manifolds of He^+. The spectrum was quite complicated because of the differences in the saturation character and the shift with rf power of the different lines. This exploratory experiment suggested the potential for precision measurements of the multiphoton transitions in higher n manifolds.

Dietrich, Lebrow, de Zafra and Metcalf [145] used a fast beam microwave resonance experiment to measure the Lamb shift in $^6Li^{2+}$. Li^{2+} ions of 3-7 MeV were obtained from the Stoney Brook tandem Van de Graff accelerator and metastable ions were formed via ion-atom collisions in a gas or fail target. Fixed frequency millimeter wave radiation at 55 GHz and Zeeman tuning in the range 2-8 kG was used to observed the resonance curve. The magnetic field was produced by an iron clad solenoid with the axis of the solenoid parallel to the ion beam. The metastables were detected by observation of the Lyman-a photons emitted when an electric field was used to mix the $^2S_{1/2}$ and $^2P_{1/2}$ states. The uncertainty in the measured Lamb shift was 1100 ppm.

Dewey and Dunford[146] measured the Lamb shift in He^+ n=2 using a method which combined features of the fast and slow beam experiments. A 170-eV beam of $^4He^+$ ions, some of which were in the $2^2S_{1/2}$ state, was extracted from an electron bombardment ion source and directed along the magnetic field lines of a large solenoid through two microwave resonators into a metastable detector. The metastable detector was a large-solid angle photodetector sensitive to the 304Å radiation emitted when a large electric field quenched the metastable He^+ atoms. The first resonator drove atoms in the β state to the 2p state where they decayed. The second resonator, which was located in the central homogeneous part of the electric field, drove the remaining atoms from the α state to the f state. The resonance was observed by keeping the microwave frequency fixed and scanning the magnetic field. The interplay between the mode structure of the resonator and the helical trajectories of the ions provided a serious source of potential uncertainty. To correct for this effect they put the resonator on a translation stage and in each run obtained resonance curves at several transverse positions of the resonator. The Lamb shift was obtained from a fit in which the centroids were characterized by a sinusoidal function of position. They measured the Lamb shift with a precision of 85 ppm which is 7.5/10000 of the natural linewidth.

6.5.2. Separated Oscillatory Fields

The main gain achieved with fast beam radiofrequency spectroscopy comes with the use of two separated oscillatory fields. With two separated oscillatory fields, one can obtain a line whose width depends on the time spent between the fields rather than the natural width and whose width does not increase with the rf power. Figure 18 shows schematically the origin of the interference signal.

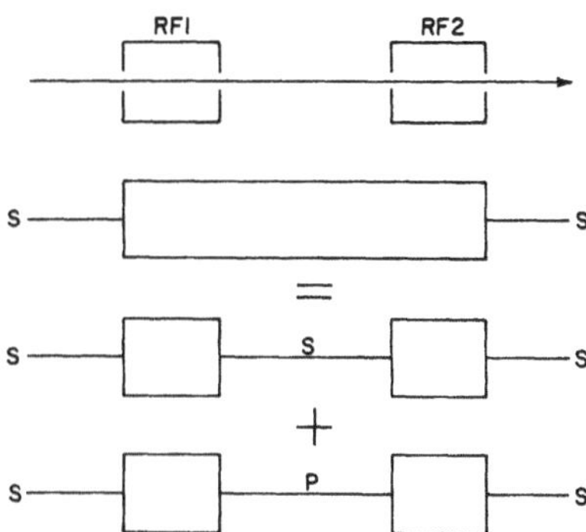

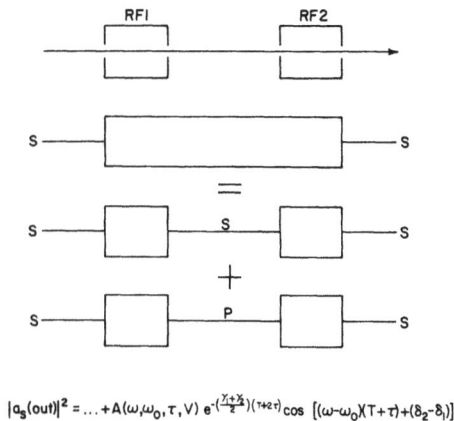

$$|a_s(out)|^2 = \ldots + A(\omega,\omega_0,\tau,V)\, e^{-\left(\frac{\gamma_1+\gamma_2}{2}\right)(T+2\tau)} \cos\, [(\omega-\omega_0)(T+\tau)+(\delta_2-\delta_1)]$$

Figure 18. A schematic diagram showing the origin of the narrowed interference signal. γ_1 and γ_2 are decay constants for the S and P states, respectively, ω_0 is the fine structure interval $(\omega_1-\omega_2)$, δ_1 and δ_2 are the phases of the radiofrequency fields in the two oscillatory field regions, τ is the length in time of each rf field region, T is the separation in time of the two regions.

There are two quantum mechanical amplitudes for an atom initially in an S state to pass through the two rf regions and emerge in the S state -- one in which it is in an S state in the intermediate region and one in which it is in a P state in the intermediate region. The probability is the square of the absolute value of the sum of the two amplitudes; the interference term comes from the product of the S and P path amplitudes. The sign of the interference signal is sensitive to the relative phase of the rf fields in the two regions. By subtracting the quench signals obtained with the relative phase 180° from that with the relative phase 0°, the interference signal can be isolated. Figure 19 shows the calculated SOF signals for one particular situation. The interference signal has the approximate analytic form

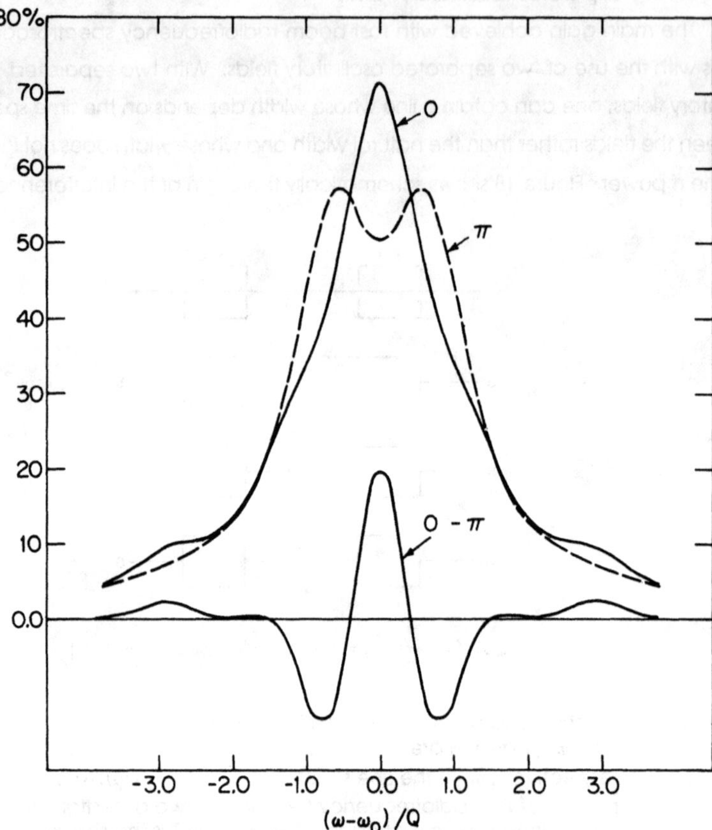

Figure 19. The calculated SOF signals for one particular situation.

$$I(\omega,\omega_O,T,\tau,V_{RF}) = A(\omega,\omega_O,\tau,V_{RF})e^{-(1/2)(\gamma_2-\gamma_1)(T+2\tau)} \cos((\omega-\omega_O)T+\tau) \quad (6.5.2.1)$$

Here τ is the atom-field interaction time in each rf region, V_{RF} the transition matrix element, T the time spent in the immediate field free region, ω the applied microwave frequency, ω_O the resonance frequency, and γ_i the inverse lifetime of state i. The factor A gives the single field quench envelope. For a detailed treatment for states which do not decay, see ref. 44; for a treatment for states which can decay see, refs. 42 and 43.

The first fast beam separated oscillatory field experiment was carried out by Fabjan and Pipkin and used to observe the Lamb shift transition in H, n=3.(42,147) Subsequently, Lundeen and Pipkin(43,148-150) refined the method and in two separate series of experiments used it to measure the Lamb shift in the n=2 state of hydrogen. Figure 20 shows a schematic diagram of the apparatus. A beam of

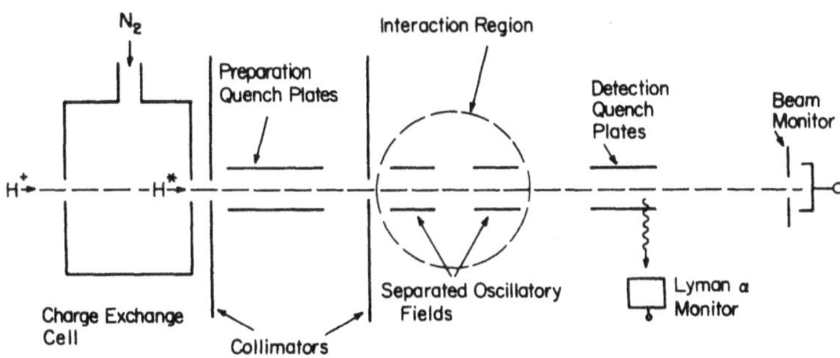

Figure 20. Schematic diagram of the fast beam apparatus used for the measurements of the Lamb shift in H, n=2.

atoms in the $2^2S_{1/2}$ state was formed from a 50 or 100 keV proton beam through charge capture in a N_2 cell. The $2^2S_{1/2}$ (F=0) hyperfine state was selected by the combination of a 1110 MHz rf prequench field ($2^2S_{1/2}$ (F=1) $\rightarrow$ $^2P_{1/2}$ (F=0)) and a 910 MHz rf quench field ($^2S_{1/2}$ (F=0) $\rightarrow$ $^2P_{1/2}$ (F=1)) in the detector region. Figure 21 shows the energy levels for the n=2 manifold of hydrogen when one includes the hyperfine structure. Figure 22 gives a series of plots of the average quench and interference signals showing the hyperfine state selection due to continuous rf fields in the state-selection region and in the detection region. Figure 23 shows examples of the average quench and interference signals obtained with a 55.1 keV beam and increasing separation for the rf regions. The narrowing and decrease of the signal as the separation increases is apparent. The useful reduction of the linewidth from the natural linewidth was a factor of 2.5 to 3. Lundeen and Pipkin were able to measure the Lamb shift to 9 kHz. This gives a precision of 9 ppm and it is 9 parts in 10^5 of the 100 MHz natural linewidth. The limiting factors in this experiment were the determination of the strength of the rf field over the line and the reproducibility of the rf connections.

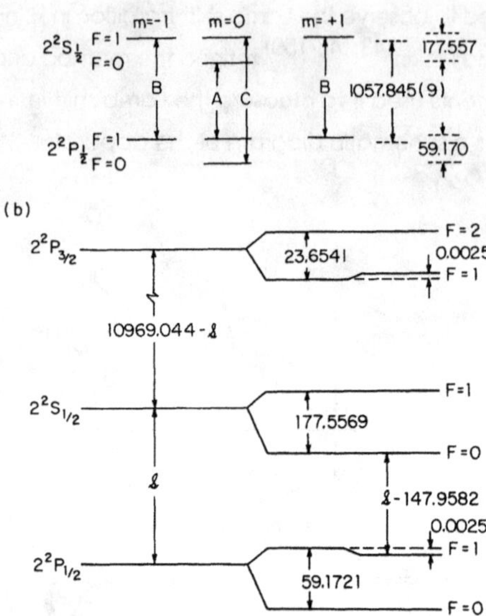

Figure 21. Energy levels for the n=2 manifold when hyperfine structure is included.

Lundeen, Jessop, and Pipkin[151] used the method of separated oscillatory fields to measure the hyperfine structure in the $2^2P_{1/2}$ state of hydrogen. The additional resolution provided by the interference signal made it possible to distinguish the normally overlapping hyperfine transitions which are separated by less than the natural linewidth.

van Baak, Clark, Lundeen, and Pipkin[152,153] used the method of separated oscillatory fields to measure the $3^2S_{1/2}$ - $3^2D_{5/2}$ interval in hydrogen. To obtain the 180° phase shift between these two regions, they used two rf sources. The phase of one source was alternated between 0° and 180° so as to isolate the separated-oscillatory field interference signal; the frequency of the other source was varied to sweep over the resonance. The precision of the measured value was 14 ppm (1/200 of the natural linewidth). The measurement was limited by the background count rate in the detector.

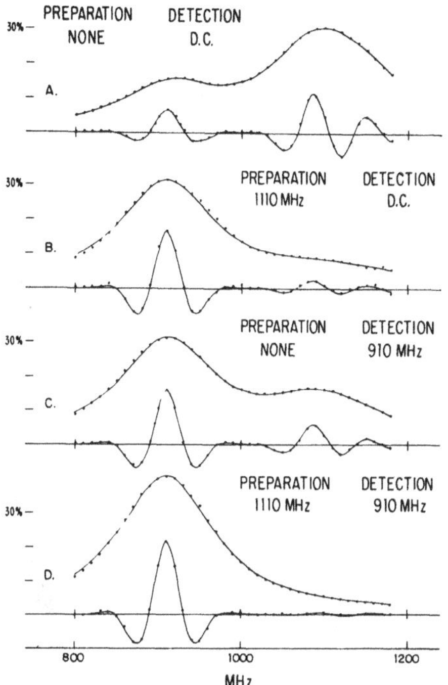

Figure 22. Plots of the average quench and interference signals showing the hyperfine state selection due to continuous rf fields in the state-selection region and in the detection region.

Safinya, Chan, Lundeen, and Pipkin[154] used the method of separated oscillatory fields to measure the $2^2P_{3/2} - 2^2S_{1/2}$ interval in hydrogen. This experiment used X-band waveguide for the interaction regions. Figure 24 shows the observed line profile for a 49 keV beam. The confined extent of the microwave electric field in each region gives an interference curve with more oscillations than for the rf plates used for the Lamb shift measurements. They measured the ΔE-S interval with a precision of 4 ppm which is 1/2400 of the natural linewidth.

Bollinger, Lundeen, and Pipkin[155] used the separated oscillatory field technique to measure the $4^2S_{1/2} - 4^2P_{1/2}$ Lamb shift transition in He^+. A 125 keV beam of He^+ was excited to the $4^2S_{1/2}$ state by passage through a cell filled with N_2 and detected downstream by using a large solid angle gas proportional counter with

MgF_2 windows to observe the 121.5 nm uv radiation emitted in the decay of the $4^2S_{1/2}$ state to the $2^2P_{1/2,3/2}$ states. A 21 GHz prequench field was used to modulate the

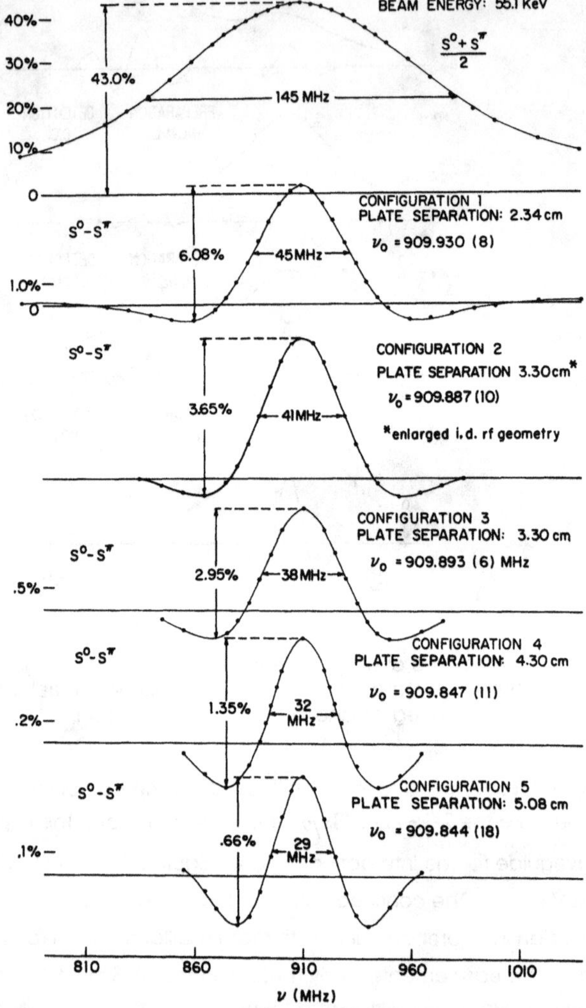

Figure 23. Examples of the average quench and interference signals obtained for hydrogen with a 55.1 keV beam and increasing separation of the rf regions.

population of the $4^2S_{1/2}$ state by driving the $4^2S_{1/2} \rightarrow 4^2P_{3/2}$ transition. The measurement was limited by the dependence of the center of the resonance on the beam intensity and overlapping transitions in higher n states detected through cascade transitions to the n=4 manifold. The final precision was 475 ppm (1 in 240 of the natural linewidth).

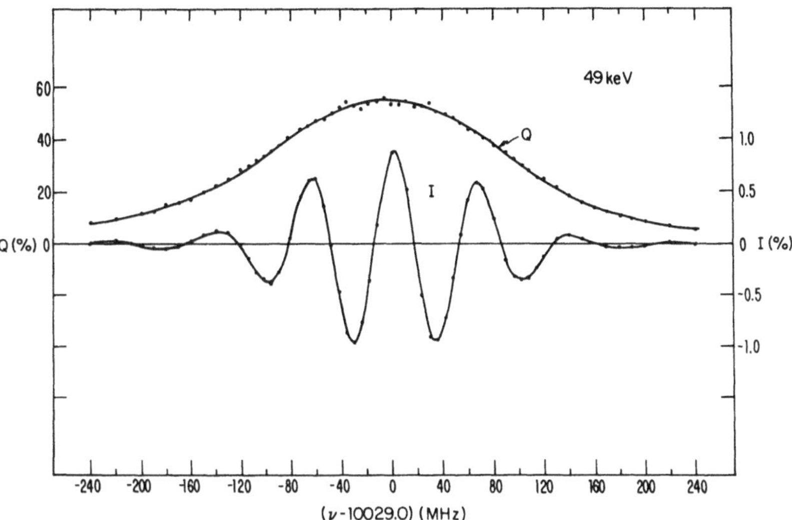

Figure 24. Line profile obtained for separated oscillatory field measurement of the $2^2S_{1/2}$-$2^2P_{3/2}$ transition in hydrogen near 10 GHz.

Majumder and Pipkin[156] used the separated oscillatory field method to measure the three photon $4^2F_{5/2}$ - $4^2S_{1/2}$ transition in He^+. This transition is especially attractive since, due to the relatively long lifetimes of the $4^2S_{1/2}$ (14.4 ns) and $4^2F_{5/2}$ (4.6 ns) states, it has a fast beam natural width of 8 MHz at the 9000 MHz microwave frequency required to drive the transition. Figure 25 shows a schematic diagram of the apparatus used to observe the transition. The arrangement was essentially the same as that employed in the earlier measurements of the He^+ n=4 Lamb shift. Microwave cavities tuned with a dielectric rod were used to produce the large microwave electric field (300 V/cm) required to drive the transition.

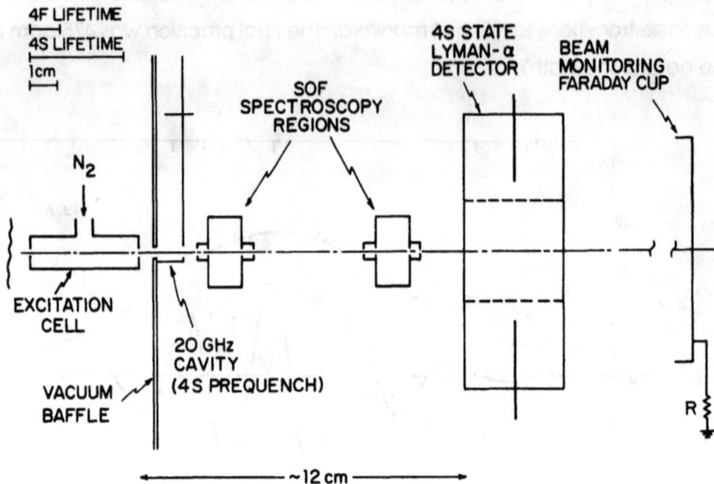

Figure 25. Schematic diagram of apparatus used for measurement of the $4^2S_{1/2}$ - $4^2F_{5/2}$ three photon transition in He^+.

Measurements were made for three separations of the oscillatory fields and at each separation data were taken as a function of microwave power. A numerical integration of the Schroedinger equation which included all 32 levels in the n=4 manifold, included both rotating components of the microwave electric field, and took for the fields in the cavities a numerical solution of Maxwell's equations, was used to simulate the signals and extrapolate the measured line centers to zero microwave power. Figure 26 shows the calculated average quench and interference signals over a wide range in frequency. Due to the power shift, the center of the quench curve is not the same as the center of the interference curve. Figure 27 shows the central lobe of the interference curve for three values of microwave power at the narrow separation. Figure 28 shows the measured center frequencies as a function of microwave power for the three separations at which data were taken. The solid curves are the simulations. The simulations were fit to the data, taking as a free parameter the intercept at zero power, to determine the fine structure interval. The precision of the measurement of the $4^2F_{5/2}$ - $4^2S_{1/2}$ fine structure interval was 5 ppm (1/64 of the fast beam natural linewidth). The measured frequency was used to determine a value for the He^+ n=4 Lamb shift with a precision of 70 ppm. This

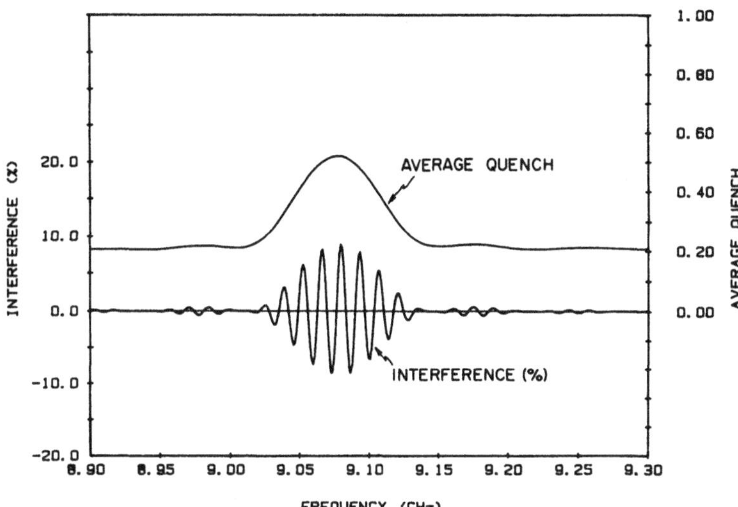

Figure 26. The calculated separated oscillatory field line profiles for the quench and interference signals for the $4^2S_{1/2}$-$4^2F_{5/2}$ three photon transition in He^+.

measurement was limited by the signal to noise ratio due to background counts in the detector.

6.6 Fast Beam Laser Resonance

A laser can be used in conjunction with a fast beam to extend the analog of the microwave fine structure measurements to high Z ions. The laser can be brought into resonance through angle tuning by using the Doppler shift or by varying the frequency of the laser.

The first laser measurement was carried out by Kugel, Leventhal, Murnick, Patel and Wood[(157)]. Figure 29 shows a schematic diagram of the geometry of the laser beam interaction. $^{14}F^{7+}$ beams at 50 or 64 GeV from the Rutgers tandem accelerator were either first post-stripped to obtain beams of bare fluorine nuclei and then passed through a thin carbon foil to produce ions in the hydrogenic $2^2S_{1/2}$ state or passed directly through a thin carbon foil. Two thin-window gas flow proportional counters above and below the beam were used to detect two photon spontaneous emission from the hydrogenic $2^2S_{1/2}$ state. Optical radiation to drive the

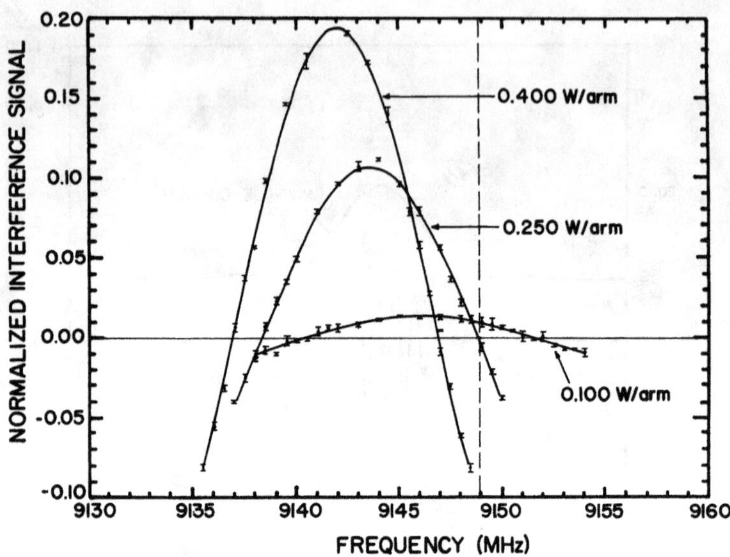

Figure 27. The measured center lobe of the interference signal for three values of microwave power at the narrow separation of the fields. The dashed line gives the center extrapolated to zero microwave power.

$2^2S_{1/2}$ - $2^2P_{3/2}$transition was provided by a pulsed HBr chemical laser. This laser operated at numerous frequencies and appropriate lines were Doppler tuned into resonance. The laser output was directed into the interaction chamber through a KCl window by a rotatable mirror system consisting of mirrors M_2 and M_3 and lens L so as to intersect the particle beam on the axis of the rotating mirror system between the two proportional counters. A time coincidence between the laser pulses and the detected x-rays was used to reduce the background. The counting rate was quite low with 1-2 counts per 10 laser pulses and 1.5 laser pulses per second. Figure 30 shows a typical resonance curve for a 64 MeV beam. The variation of the laser power form point to point and the unresolved hyperfine structure complicate the analysis. The final precision was 0.05% in (ΔE-S) and 1% in S.

Wood, et al.[158] used a pulsed CO^2 laser to measure directly the Lamb shift transition in $C1^{16+}$. A 190 MeV Cl^{14+} beam from the Brookhaven National Laboratory tandem accelerator was post-stripped to bare chlorine nuclei, Cl^{17+}, by passage through 60 μg/cm^2 carbon foils. The beam (2-3 particle nA) was then partially promoted into the metastable $2^2S_{1/2}$ state of Cl^{16+} by electron pickup in a thin

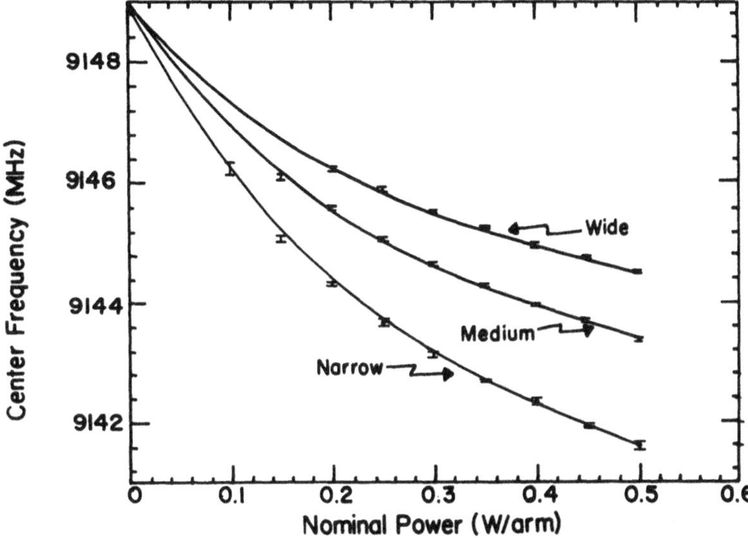

Figure 28. The measured center frequencies as a function of microwave power for the three separations of the rf field at which data were taken.

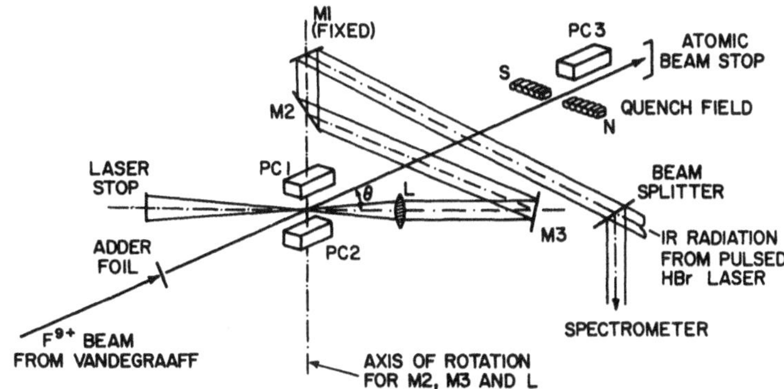

Figure 29. Schematic drawing of particle beam-laser beam interaction. A rotating-mirror system was used for Doppler-shift tuning.

moveable carbon foil. The CO^2 laser beam intersected the particle beam at an angle of roughly 130°. The resonance was scanned using eight different laser

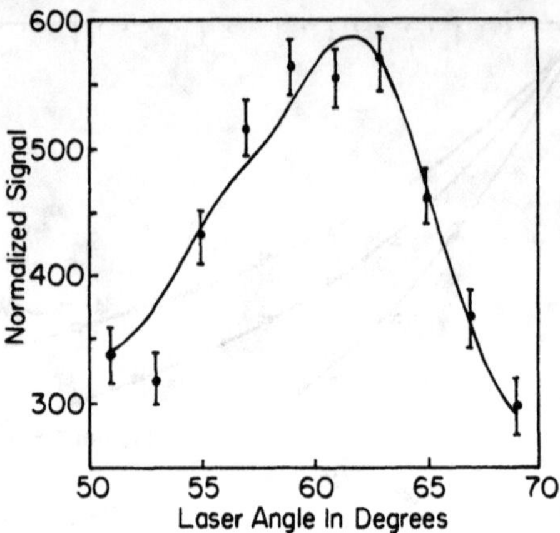

Figure 30. Resonance curve for ΔE-S transition in n=2 hydrogen-like fluorine.

frequencies from within the 4-THz tunability range of the CO^2 laser. They determined the Lamb shift with a precision of 0.7%.

Pellegrin, El Masri and Palffy[159] used a tunable pulsed laser (near 556.0 nm) to measure the $2^2P_{3/2}$ - $2^2S_{1/2}$ interval in hydrogenic phosphorous. 87.1 MeV P^{5+} ions from the Louvain-1a-Neuve Isochronus Cyclotron were stripped in a carbon foil to produce P^{15+}; a second foil yielded hydrogenic phosphorus P^{14+} (~20 nA) of which part was in the metastable $2^2S_{1/2}$ state. The laser beam intersected the ion beam at 90° and a fast, multiwire, large solid angle gas counter monitored the 2.3 keV X-ray from the decay of the $2^2P_{1/2}$ state to the ground state. The 7 ns 50 Hz laser pulses were synchronized with the12.02 MHz cyclotron pulses to maximize the overlap. The wavelength of the laser radiation was measured with an Ebert-Fastie monochrometer located remotely. Figure 31 shows a typical line scan and a fit to a single Lorentzian with the assumption the hyperfine splitting is negligible. They measured (ΔE-S) with a precision of 370 ppm; this yielded S with a precision of 1%.

Georgiadis, et al.[160] used laser spectroscopy to measure the $2^2S_{1/2}$ - $2^2P_{3/2}$ transition in hydrogenic sulfur. A beam of 130 MeV $^{32}S^{9+}$ ions from the Strasburg MP trandem accelerator was further stripped by a 100 μg/cm^2 carbon foil, and the magnetically separated 15+ component was made to intersect the beam of a tunable pulsed dye laser (428.6 nm). Transitions were induced in $^{32}S^{15+}$ from the metastable $2^2S_{1/2}$ state produced in the stripping to the $2^2P_{3/2}$ state. The resonance

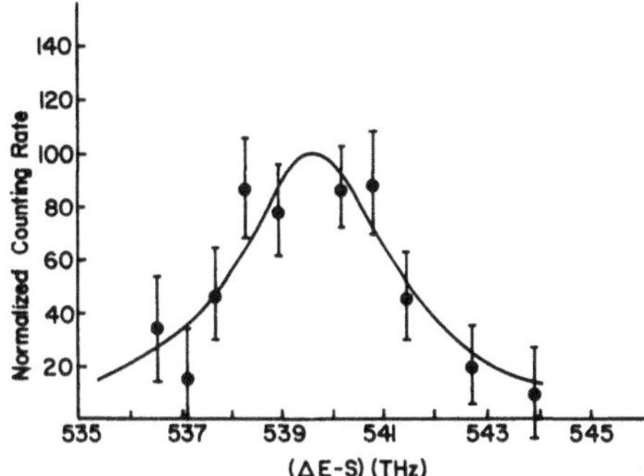

Figure 31. A typical resonance curve for the laser measurement of the (ΔE-S) transition in hydrogenic phosphorus.

was observed by monitoring the count rate in two gas flow proporational counters which detected the 2.6 keV decay of the $2^2P_{3/2}$ state to the ground state.

A geometry with counter-propagating beams with an intersection angle close to 180° was chosen so the ions would spend a longer time in the laser field. 150 scans were made in three beam periods and fit to a Lorentzian curve to obtain the $2^2S_{1/2}$ - $2^2P_{3/2}$ interval. The (ΔE-S) interval was measured with a precision of 90 ppm; S was determined with a precision of 0.25%.

von Brentano, Müller and Gassen[161,162] independently used the laser resonance method to measure the $2^2P_{3/2}$ - $2^2S_{1/2}$ fine structure interval in hydrogenic phosphorus. 100 MeV P^{8+} ions from the Strassburg MP-Tandem accelerator were passed through a carbon foil to produce a beam with hydrogenic phosphorus (P^{14+}) in the $2^2S_{1/2}$ state. The counterpropagating laser beam from a flash lamp pumped dye laser intersected the ion beam at a small angle. The wavelength of the laser was controlled by tilting an interference filter in the laser cavity. Two proportional counters monitored the 2.3 keV x-ray from the decay of the $2^2P_{3/2}$ state to the ground state. Gated time windows were used to minimize background counts. Figure 32 shows a typical resonance curve. They were able to measure (ΔE-S) with a precision of 72 ppm and to determine the Lamb shifts with a precision of 0.2%.

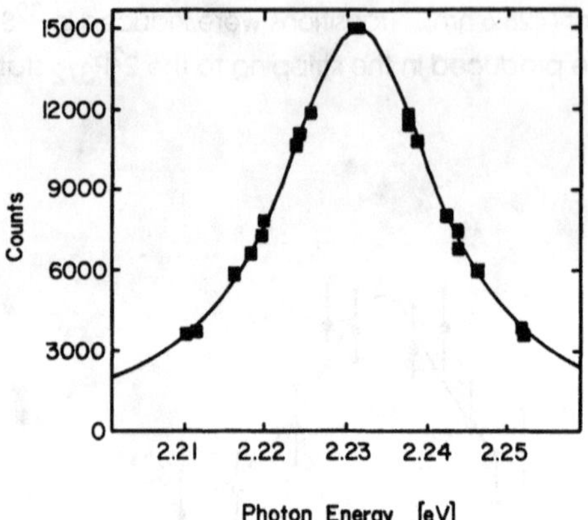

Figure 32. Typical resonance curve for (ΔE-S) transition in hydrogenic phosphorus obtained by von Brentano, et al.

7. LASER SPECTROSCOPY

The advent of high resolution laser spectroscopy has made it feasible to measure the Lamb shift in the ground state of hydrogen and deuterium.[162-168] A laser with a wavelength of 243 nm can be used to drive a two photon transition from the $1^2S_{1/2}$ ground state to the $2^2S_{1/2}$ metastable state. The transition can be observed by detecting the metastable atoms or the Lyman-α radiation emitted when an electric field is used to mix the $2^2S_{1/2}$ and $2^2P_{1/2}$ states. A configuration in which the atoms absorb a photon from each of two counterpropagating beams gives a Doppler free signal. Due to the 1/7 s lifetime for the $2^2S_{1/2}$ state, the natural linewidth for this transition is 1.3 Hz. This gives a line Q of 2×10^{16}; the potential for precision measurements is enormous. To extract the Lamb shift requires, however, an independent precise measurement of the Rydberg or a precise measurement of another interval in the same atom so the ratio can be used to reduce the sensitivity to the Rydberg.

A great deal of effort has been devoted to measuring this transition. It is technically difficult because of the laser frequency required and the need to measure the wavelength absolutely. The first experiments used pulsed lasers to drive the transition.[169,170] The pulsed lasers were subsequently replaced by pulse amplification[171,172] of a continuous wave dye laser. A major difficulty with the use of pulsed amplifiers was the frequency chirp associated with the transient nature of the gain. This effect was reduced by spectrally narrowing the radiation in a confocal filter cavity.[172,173] The Southhampton group has undertaken a program to improve the performance of the pulsed amplifier experiments. Pulsed lasers may have advantageous features for absolute frequency measurements.[164,174]

A major advance was the development of continuous wave sources at 243 nm. Two methods have been used. The Stanford group summed the frequency of a 351 nm argon ion laser and the 790 nm radiation from a dye laser in a KDP crystal.[175,176] The Oxford gorup used a beta-barium borate doubling crystal mounted intra-cavity in a krypton-pumped Coumarin 102 ring dye laser.[177] The Oxford source is advantageous in that the primary 486 nm radiation is near Balmer-β and thus makes feasible a simultaneous measurement to high precision of the Balmer-β transition.

To date all the experiments have used a cell containing $^{130}Te_2$ as a reference standard for frequency measurements. Selected tellurium lines near 486 nm have been calibrated at the National Physical Laboratory by comparison with the iodine stabilized helium-neon standard at 633 nm.[178] This frequency measurement is a major source of uncertainty.

Figure 33 shows the set-up used by the Oxford group.[179] Figure 34 shows the resonance signal obtained by the Oxford group for the F = 3/2 transition in deuterium. Comparable signals were obtained by the Stanford group.[177] Table 3 summarizes the $1s \rightarrow 2s$ centroid frequencies obtained by the two groups. These measurements can be used in conjunction with the measured value for the Rydberg to determine the Lamb shift in the ground state of H and D.

At this time the most precise value for the Rydberg comes from the measurements of Biraben et al.[31] They measured, in a metastable beam, the wavelengths of the 2S-8D, 2S-10D, and 2S-12D transitions for atomic hydrogen and deuterium. Their I_2-stabilized He-Ne reference laser at 633 nm was compared to the standard lasers of the Bureau International des Poids et Measures via an intermediate standard He-Ne laser of the Institut National de Métrologic. The final value for the Rydberg was

$$R_\infty = 109\,737.315\,709(18)\ \mathrm{cm}^{-1}$$

The precision of the measurement was limited by the uncertainty in the reference standard (1.6×10^{-10}) rather than the uncertainties inherent in the experiment.

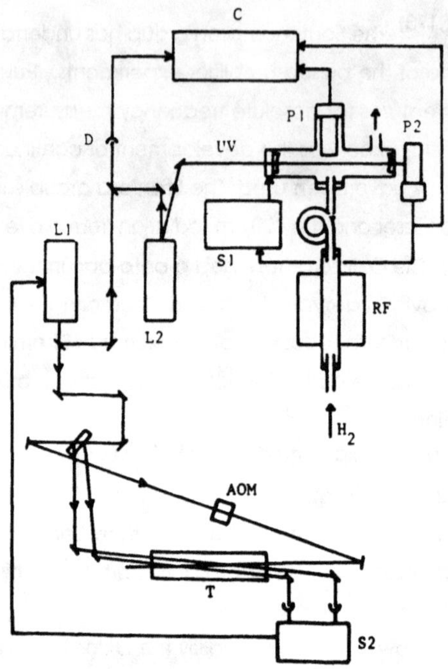

Figure 33. Diagram of the apparatus used in the measurements by the Oxford group. L1, L2 tunable dye lasers; UV, ultra violet radiation (243 nm); RF, radio-frequency dissociation of flowing molecular hydrogen; PM1, signal photomultiplier (Lyman-α detector); PM2, photomultiplier for cavity 1 locking and signal normalization; S1, cavity length servo-control; C, comuter; AOM, acoustic-opto modulator; T, heated quartz cell containing tellurium; S2, laser frequency servo-control; D, fast photodiode.

Using this value for the Rydberg, the Oxford and Stanford experiments give values for the Lamb shift in the $1^2S_{1/2}$ state of hydrogen with a precision, respectively, of 103 ppm and 232 ppm. Work is in progress at Oxford,[163] Southhampton,[164] and Munich[165] to reduce further the uncertainty. The major obstacle is the limitation in the techniques and standards for absolute frequency measurements. A direct comparison with other transitions in H made with the experimental uncertainties encountered by Biraben, et al. would yield a determination of the Lamb shift with a precision of 100 kHz (12 ppm). This is comparable with the precision of the measurements of the Lamb shift in the $2^2S_{1/2}$ state.

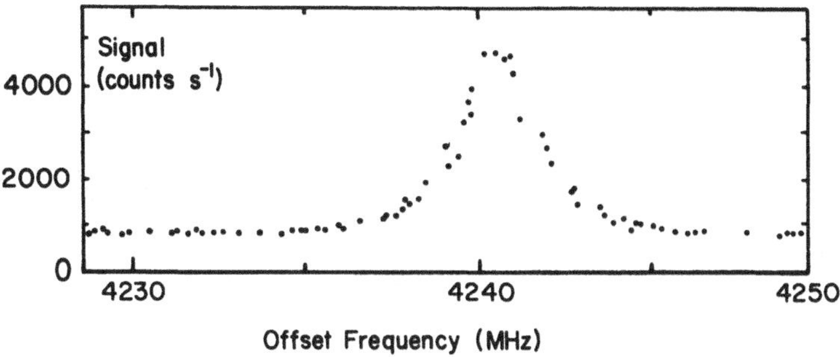

Figure 34. The F= 3/2 component for the 1s-2s two photon transition in deuterium. The signal is the normalized Lyman-α fluorescence observed as a function of the frequency difference between lasers L1 and L2 (Figure 33).

Table 3. The measurements of the centroid for the 1S → 2S interval in H and D

Isotope	Group	$2S_{1/2}$ - $1S_{1/2}$ (MHz)	Ref.
H	Stanford	2 466 061 413.19(175)	179
	Oxford	2 466 061 414.12(75)	177
	Southhampton	2 466 061 397.(25)	174
D	Oxford	2 466 732 408.45(69)	177

8. DIRECT MEASUREMENT OF LYMAN-α RADIATION

The direct measurement of the wavelength of Lyman-α radiation can be used to measure the Lambshift in the ground state of high Z ions. For hydrogenic Fe^{25+} a 1 ppm measurement of the Lyman-α radiation will yield the $^1S_{1/2}$ Lamb shift to 0.17%. The radiation is in the x-ray region and can be measured using a grating or a crystal spectrometer. Silver[180] has reviewed the work in this area.

In recent years several groups have tried to make precise measurements of the 1S Lamb shifts in hydrogenic argon,[181,182] chlorine,[183,185] iron,[186,187]

germanium,[188] and krypton.[189] Fast beam sources, recoil-ion sources, and tokamak sources have been used. It is expected that the newly developed ion sources such as the Electron Cyclotron Resonance (ECR) source, the Electron Beam Ion Source (EBIS)[190] and the Electron Beam Ion Trap (EBIT)[191] will provide good opportunities for spectroscopic measurements on high Z ions.

One technique which has been suggested is to use a spectrometer to observe nearby in different orders Lyman-α and Balmer-β.[180] This provides self calibration and independence of the Doppler shift which make absolute measurements difficult with fast beam sources.

At this time none of these measurements have been sufficienlty precise to provide an "interesting" test of QED.

9. EXOTIC ATOMS

Positronium[192] and muonium,[193] which are, respectively, the bound states of an electron and a positron and of an electron and a muon, are atoms formed from particles which, in present theories, are believed to be elementary. The energy levels for these exotic atoms should be completely calculable from Quantum Electrodynamics and the Unified Electro-weak theory. The calculations for positronium are technically difficult; due to the equal mass of the two constituents, one cannot use an expansion in m_e/m_p as is used for hydrogen.

Figure 35 shows the energy level diagram for positronium. There is no pure QED analog to the $2^2S_{1/2}$ - $2^2P_{1/2}$ separation in hydrogen; the $^3S_1$ - $^3P_{0,1,2}$ and the $^1S_0$ - $^1P_1$ intervals have contributions from the Lamb shift and other QED effects. Three groups have measured precisely intervals in positronium. The fine structure in the triplet n=2 state was first measured by Mills, Berko, and Cantor[194] at Brandeis and subsequently by Hatamian, Conte, and Rich[195] at Michigan. Two photon laser absorption has been used at Bell Laboratories by Chu, Mills, and Hall[196,197] to measure the 1^3S_1 - 2^3S_1 interval. The results are summarized in Table 4. The theoretical values for these intervals are not sufficiently precise to make a definitive comparison between theory and experiment.

Figure 36 shows the energy level diagram for muonium. The largest contribution to the $2S_{1/2}$ - $2P_{1/2}$ Lamb shift interval in muonium is due to vacuum polarization. It is sufficiently large that the $^1S_{1/2}$ state is more tightly bound than the $^2P_{1/2}$ state. Two groups have measured the Lamb shift in muonium. Oram et al. at TRIUMP[198] and Badertscher, et al. [199] at LAMPF. The experimental values were

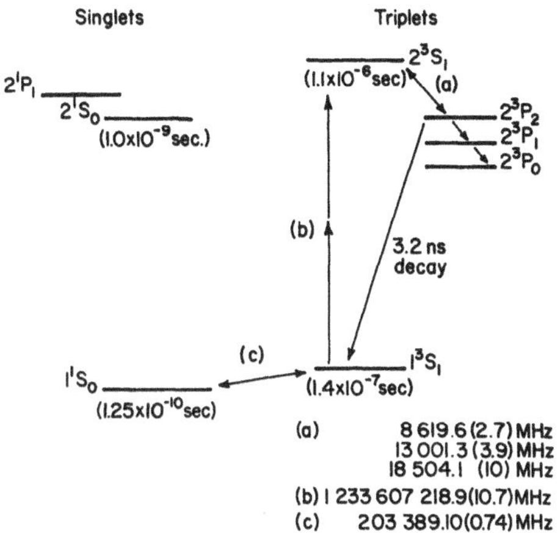

Figure 35. Energy level diagram for lower lying levels of positronium.

Table 4. Measured energy level splittings in positronium.

Interval	Measured Value (MHz)	Reference
$2^3S_1 - 2^3P_0$	18 504.1±10.0±1.7	Hatamian[195]
$2^3S_1 - 2^3P_1$	13 001.3±3.9±0.9	Hatamian[195]
$2^3S_1 - 2^3P_1$	8 628.4±2.8	Mills[194]
	8 619.6±2.7±0.9	Hatamian[195]
$1^3S_1 - 2^3S_1$	1 233 607 218.9±10.7	Chu[196] Correction, Danzmann[197]

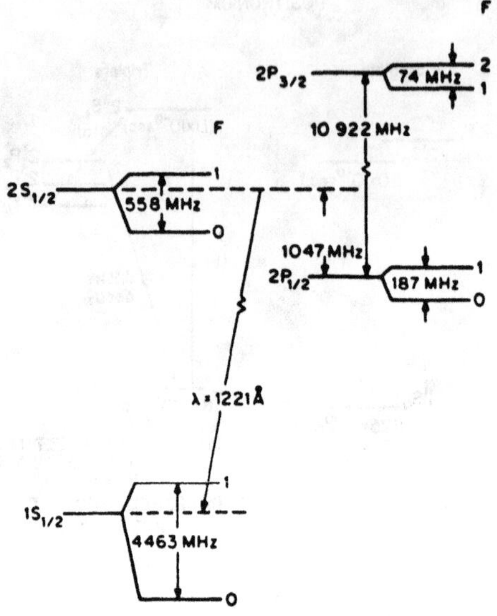

Figure 36. Energy level diagram for low lying levels of muonium

TRIUMPF: $S\,(\text{muonium}) = 1070^{+12}_{-15}$ MHz,

LAMPF: $S\,(\text{muonium}) = 1054 \pm 22$ MHz.

These results agree satisfactorily with the theoretical value

$$S(e^-\mu^+) = 1047.578(300)\ \text{MHz}.$$

The agreement between theory and experiment is satisfactory but the measurements are not sufficiently precise to provide an interesting test of theory.

A CERN-PISA collaobration[200-202] used a laser to measure the Lamb shift in μ^-He^+ in the n=2 state. The results agreed satisfactorily with theory if one used the value of the radius of the alpha particle obtained in electron scattering experiments.[203] The results were also inverted to obtain a precise value for the radius of the alpha particle.[204] Two groups[205,206] have tried to repeat this measurement and have had difficulty in producing long lived μ^-He^+. It is found that at the pressure used by the CERN-PISA Group, the lifetime is much shorter than that reported by the CERN-PISA Group.

Two photon laser absorption has been used to detect the $1^2S_{1/2}$ - $2^2S_{1/2}$ transition in muonium but the results are not sufficienlty precise to confront definitely the theory.[207]

10. TWO ELECTRON ATOMS

A second potential source for precision measurements of the Lamb shift is two electron helium like atoms.[208-211] Figure 5 shows the low lying S and P states of helium-like atoms. The intervals of interest are those between the metastable $^1S_0$ and $^3S_1$ levels and, respectively, the $^1P_1$ and $^3P_{0,1,2}$ levels. Figure 37 shows a plot[211] of the QED contribution to the 2^2P_0 - 2^2S_1, 2^3P_2 - 2^3S_1, and 2^1P_1 - 1^1S_0 intervals in two-electron atoms as a function of Z. In helium the Lamb shift for the $^3S_1$ state is roughly 5320 MHz; this is to be compared with 14022 MHz for the n=2 state of He^+. The 10^{-7} second lifetime for the $^3P_{0,1,2}$ states gives a natural linewidth of 1.6 MHz for the $^3P_{0,1,2}$ - $^3S_1$ transitions so one should be able to measure the transitions with an uncertainty of 1.6 kHz. This would give, in principle, a 0.3 ppm measurement of the

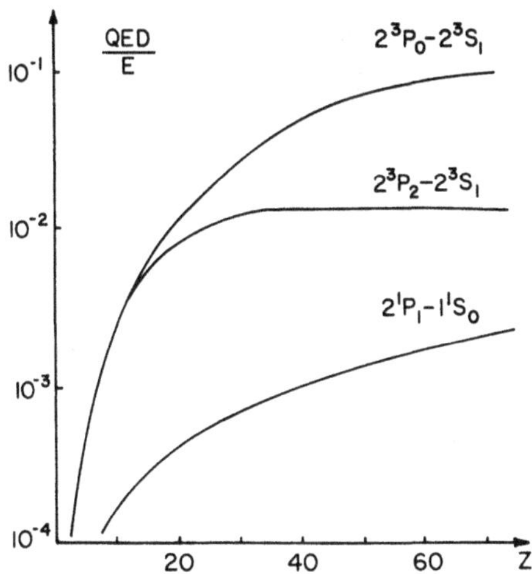

Figure 37. A plot of the QED contributions to the 2^3P_0 - 2^2S_1, 2^3P_2 - 2^2S_1, and 2^1P_1 - 1^1S_0 intervals in two-electron atoms as a function of Z.

Lamb shift. At present the major limitation in the determination of the Lamb shift for the $^3S_1$ state is the calculation of the dominant part of the ^3P - ^3S interval due to non QED quantum mechanics. The total interval is roughly 2.769×10^8 MHz so a 0.3 ppm determination of the Lamb shift requires a calculation of the non-QED component with a precision of 6 in 10^{12}. Recent developments suggest that this precision will be attainable in the future.[212-215]

Table 4 summarizes the measurements of the ^3P - $^3S_1$ transitions that have been reported in the literature. The measurements with lower Z have used laser spectroscopy to attain high precision. The measurements for the high Z ions were made using beam foil spectroscopy.

Table 5. The measured values for the 1s 2s $^3S_1$-1s 2p $^3P_2$ interval in helium-like atoms. Units are cm^{-1}. E_{QED} is the Lamb shift for the hydrogenic atom in the n=2 state. (After ref. 162).

Ion	Theory	Experiment	E_{QED}	$\frac{\Delta E_{exp}}{E_{QED}}$	Ref.
He	9 230.792 3	9 230.792 2(5)	0.46	1×10^{-3}	216
		9 230.792 08(8)		2×10^{-4}	217
Li	18 228.19(1)	18 228.197 99(12)	2.02	5×10^{-3}	218,248
		18 228.199 6(8)		4×10^{-4}	219,239,240
		18 228.199 35(19)			238
N	52 720.3(3)	52 720.2(7)	43.4	2×10^{-2}	220
O	61 589.3(5)	61 589.7(5)	69.8	7×10^{-3}	221
F	70 700.3(8)	70 705.8(3)	106.0	3×10^{-3}	222
Ne	80 122.4(1.2)	80 123.33(83)	154.0	5×10^{-3}	221
Mg	100 253.7(2.6)	100 263.(6)	292.0	2×10^{-3}	223
Al	111 153.0(3.6)	111 157.(6)	386.0	2×10^{-2}	223
Si	122 744.3(4.8)	122 746.(3)	499.0	6×10^{-3}	224
P	135 153.4(6.3)	135 153.(1.8)	634.0	3×10^{-2}	225
S	148 499.1(8.2)	148 493.(5)	793.0	6×10^{-3}	224
Cl	162 926.(10)	162 923.(6)	978.0	6×10^{-3}	224
Ca	214 179.(18)	214 225.(45)	1710.0	3×10^{-2}	226
Ti	256 693.(26)	256 746.(46)	2370.0	2×10^{-2}	227
Fe	368 756.(57)	368 976.(125)	4188.0	3×10^{-2}	228
Cu	483 761.(42)	483 910.(200)	6073.0	3×10^{-2}	229

A novel method has been used by Munger and Gould[230] to determine the Lamb shift for helium like U^{90+}. Figure 38 shows the energy levels

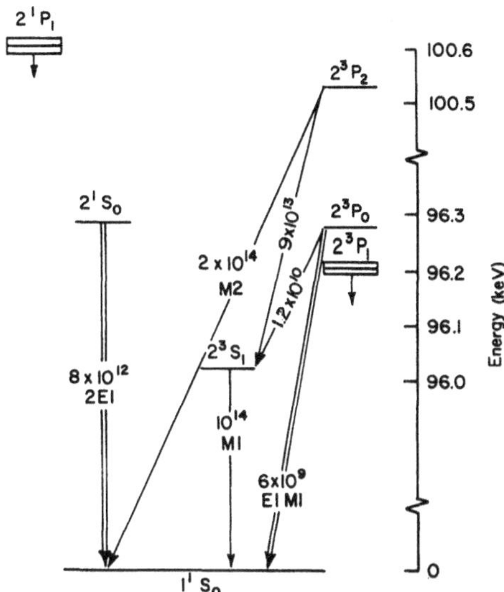

Figure 38. A diagram of the energy levels for helium like uranium showing the decay channels.

for helium like uranium together with the decay paths. A 218 MeV/u beam of hydrogen like uranium was converted to the 1s2p $^3P_0$ state of helium like uranium in a 0.9 mg/cm^2 Pd foil. Downstream from the palladium foil they observed the 96.01 keV X-ray from the decay of the 1s2s $^3S_1$ state to the $(1s^2)^1S_0$ ground state. They assumed the 54.4±3.4 ps observed lifetime was due to the decay of the 1s2p $^3P_0$ state and through use of the calculated E1M1 0.564(5) x10^{10} s^{-1} decay rate for the 2^3P_0 - 1^1S_0 transition determined the E1 decay rate for the 2^3P_0 state. The dipole length formula was then used to calculate the $^3P_0$ - $^3S_1$ interval. By subtracting the calculated Coulomb contribution they determined the Lamb shift to be 70.4(79) eV.

11. COMPARISON OF THEORY AND EXPERIMENT

At present the best tests of the theory can be made using the Lamb shift in hydrogen, helium, and high Z nuclei. We shall discuss each of these comparisons briefly.

11.1 Hydrogen

The most precise experimental values for the Lamb shift have been obtained for hydrogen. The theoretical value, however, is limited by the knowledge of the rms radius of the proton. The radius is determined from electron scattering measurements. The early measurements made by workers centered around Stanford give[231,232]

$$<r_p^2>^{1/2} = 0.805(11)\text{fm}.$$

The more recent determination carried out by workers centered around Bonn gives[233]

$$<r_p^2>^{1/2} = 0.862(12)\text{ fm}.$$

If the additional recoil terms calculated by Erickson and Grotch[235] and by Doncheski, Grotch and Owen[249] are not included, the theoretical values obtained using these radii are

$$S(H,n{=}2) = 1057.853(11)\text{ MHz} \qquad \text{if } <r_p^2> = 0.805(11)\text{ fm};$$
$$S(H,n{=}2) = 1057.871(11)\text{ MHz} \qquad \text{if } <r_p^2> = 0.862(12)\text{ fm}.$$

The two most precise experimental values are:

$$S(H,n{=}2) = 1057.845(9)\text{ MHz} \qquad (9\text{ ppm})\ (\text{Fast beam});$$
$$S(H,n{=}2) = 1057.8514(19)\text{ MHz} \qquad (2\text{ ppm})\ (\text{Atomic interferometer}).$$

The additional recoil terms decrease the theoretical value for the Lamb shift by roughly 3 kHz. We have not included these additional recoil terms here or in Table 7 because it is not clear the calculations are complete. For further discussion of the recoil corrections see the article in this volume by Sapirstein and Yennie.

Experiment and theory are in satisfactory agreement if the old value for the proton radius is used; the agreement is less satisfactory if the new value for the proton radius is used. Because of the discrepant values for the proton radius, one can not make a definitive comparison between theory and experiment.

11.2 Helium

The most precise value for the Lamb shift in helium is that obtained by the measurent of the anisotropy of the quench radiation

$$S(He^4, n=2) = 14042.220 \pm 0.35 \qquad (25\ \text{ppm})$$

Two determinations have been made of the radius of the He^4 nucleus. Sick, et al.[203] found from electron scattering measurement

$$<r_\alpha^2>^{1/2} = (1.676 \pm 0.008)\ \text{fm}.$$

Taking as input the measurement of the Lamb shift in μ^- - He^+ by the CERN-PISA group, Borie and Rinker obtained for the radius[204]

$$<r_\alpha^2>^{1/2} = (1.673 \pm 0.001)\ \text{fm}.$$

The theoretical values obtained using these two radii are respectively

$$S(He^4, n=2) = 14042.27 \pm 0.50\ \text{MHz},$$

and

$$S(He^4, n=2) = 14042.24 \pm 0.50\ \text{MHz}.$$

The agreement between theory and experiment is excellent. In this case it is limited by the uncertainty in the calculated value for $G_{SE}(Z\alpha) = -22.9 \pm 1.0$. This theoretical value does not includes the correction to the relativistic recoil terms recently calculated by Erickson and Grotch,[235] and Doncheski, Grotch, and Owen.[249]

11.3 High Z Atoms

Because of the increase with $(Z\alpha)^2$ of the coefficient for the $G_{SE}(Z\alpha)$ term, high Z hydrogenic atoms provide a sensitive way to test the calculation of $G_{SE}(Z\alpha)$. Figure 39 shows a plot of the difference between experiment and theory as a function of Z. The scaling of the vertical axis by Z^6 emphasizes the $G_{SE}(Z\alpha)$ term. Table 6 shows a comparison between experiment and theory in which the experimental value for $G_{SE}(Z\alpha)$ has been extracted from the measurements. The overall agreement between theory and experiment is quite satisfactory. The only indication of any

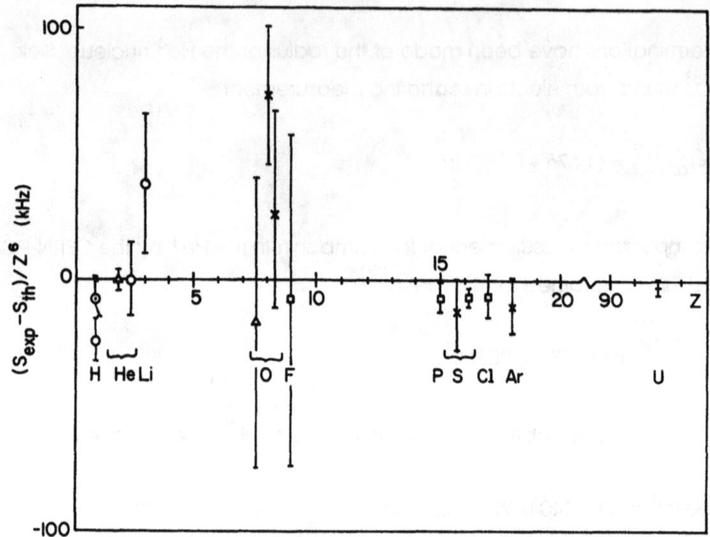

Figure 39. Z dependence of the difference between the experiment and theoretical values of the Lamb shift. The two values for hydrogen are for the two measured values of the rms radius of the proton.

discrepancy is that for higher Z the experimental value is consistently roughly one standard deviation smaller than the theoretical value.

12. PROSPECTS

To make further progress requires an independent determination of the radius of the proton, improved measurements of the Lamb shift in high Z atoms, and more precise calculations of $G_{SE}(Z\alpha)$. The general agreement between theory and experiment is quite encouraging and suggests that there will be no deviations which can not be accounted for within the QED, electro-weak , QCD framework.

13. ACKNOWLEDGEMENTS

The preparation of this review was supported by National Science Foundation Grant PHY-8704527 and Department of Energy Contract DE-AC02-76ER03064.

Table 6. Comparison of theory and experiment for the total Lamb shift, and the derived electron self-energy term G_{SE} $(Z\alpha)$ in Eq. (2.10). R_N is the nuclear radius used. (Taken from ref. 236)

Ion	R_N (fm)	$S_{exp.}$	$S_{theo.}$	$(G_{SE})_{exp.}$ [a]	$(G_{SE})_{theo.}$ [c]	$(G_{VP})_{theo.}$ [b]
^{1}H	0.862(20)	1057.845(9)[d]	1057.866(11) MHz	-26.19±1.25±0.93	-23.4(1.2)	-0.517(20)
^{4}He	1.673(1)	14042.22(35)[e]	14042.22(50)	-22.91±0.76±0.05	-22.9(1.0)	-0.508(17)
^{6}Li	2.56(5)	62765.(21)[f]	62737.(6)	-17.12±4.0±0.77	-22.49(88)	-0.500(16)
^{16}O	2.711(14)	2192.(15)[g]	2196.13(21) GHz	-22.90±7.9±0.03	-20.72(45)	-0.473(12)
		2215.6(7.5)[h]		-10.44±4.0±0.03		
		2203.(11)[i]		-17.09±5.8±0.03		
^{19}F	2.900(15)	3339.(35)[j]	3343.0(1.8)	-21.46±9.0±0.03	-20.42(39)	-0.469(7)
^{31}P	3.197(5)	20.17(7)[k]	20.25(1) THz	-19.82±0.85±0.004	-18.81(14)	-0.449(2)
^{32}S	3.247(4)	25.266(63)[l]	25.373(13)	-19.45±0.52±0.004	-18.57(11)	-0.446(2)
^{35}Cl	3.335(18)	31.19(22)[m]	31.34(2)	-19.23±1.3±0.01	-18.34(8)	-0.444(1)
^{40}A	3.428(8)	37.89(38)[n]	38.24(2)	-19.55±1.6±0.0.005	-18.11(6)	-0.442(1)
^{238}U	5.751(50)	70.4(8.3)[o]	75.3(4) eV	-8.32±0.46±0.02	-8.050(4)	-0.600(8)

[a]The first uncertainty listed is due to the experimental uncertainty in S, and the second to the nuclear radius uncertainty. Nuclear size corrections to the self-energy and vacuum polarization terms have been subtracted for the high-Z ions.

[b]Includes Uehling and Wichmann-Kroll vacuum polarization contributions. The total $G(Z\alpha)$ for a point nucleus is $G_{SE} + G_{VP}$.

[c]reference 35, 36 [f]reference 74 [i]reference 108 [l]reference 160
[d]reference 43 [g]reference124 [j]reference 157 [m]reference158
[e]refernece 123 [h]reference 109 [k]reference 161 [n]reference 111
[o]reference230

Table 7. A summary of Lamb-shift and fine-structure measurements which determine the position of the S states with respect to the states with $l \neq 0$ which have a smaller QED shift. The uncertainties for the experimental values are one standard deviation. The uncertainties for the theoretical values are estimates of uncertainties due to the uncertainties in R_N, G_{SE}, G_{VP}, and the uncalculated terms.

Atom	Interval	Calculated Value	Measured Value	Meas.-Calc. / σ(exp)	Method	Ref.
H	$S(^1S_{1/2})$	8172.94(9) MHz	8600. ± 800 MHz	0.5	two photon absorption	169
		[8173.08(9)]	8220. ±100	0.5	two photon absorption	170
			8175. ±30	0.1	two photon absorption	172
			8184.8 ±5.4	2.2	two photon absorption	173
			8182. ±25	0.4	two photon absorption	174
			8173.9 ±1.9	0.5	two photon absorption	179
			8172.94 ±0.0	0.0	two photon absorption	163,177
	$2S_{1/2} - 2P_{1/2}$	1057.853(11)	1057.772 ±0.063	-1.3	slow beam rf	50,53
		[1057.871(11)]	1057.90 ±0.100	0.5	slow beam, crossing	58
			1057.893 ±0.020	2.0	fast beam, sep rf	149
			1058.05 ±0.260	0.8	fast beam, simulated rf	243
			1057.862 ±0.020	0.5	fast beam, rf	138
			1058.3 ±0.500	0.9	fast beam, interfero	166
			1057.3 ±0.900	-0.6	fast beam, anisotropy	119
			1057.85 ±0.440	0.0	fast beam, anisotropy	244
			1057.845 ±0.009	-0.9	fast beam, sep rf	43
			1057.8514 ±0.0019	0.8	fast beam, interfero	133,134

$2P_{3/2} - 2S_{1/2}$	9911.190(11) MHz	9911.173	±0.042 MHz	-0.5	slow beam, rf	61
		9911.250	±0.063	0.9	slow beam, rf	63
		9911.377	±0.026	7.1	bottle, rf	72
		9911.117	±0.041	-1.9	fast beam, rf	154
$3S_{1/2} - 3P_{1/2}$	314.879(3)	314.819	±0.048	-1.2	fast beam, rf	42
		315.11	±0.89	0.3	fast beam, rf	136
		314.9	±0.9	0.0	bottle, crossing	78
		313.6	±2.9	-0.4	slow beam, rf	75
$3P_{3/2} - 3S_{1/2}$	2935.210(3)	2933.5	±1.2	-1.4	fast beam, rf	136
$3D_{3/2} - 3S_{1/2}$	2929.880(3)	2930.3	±0.8	0.5	bottle, crossing	98
$3D_{5/2} - 3S_{1/2}$	4013.219(3)	4013.8	±0.8	0.7	bottle,crossing	
		4013.204	±0.078	-0.2	fast beam, rf	143
		4013.106	±0.057	-2.0	fast beam, rf	153
$4S_{1/2} - 4P_{1/2}$	133.078(1)	132.53	±0.68	-0.8	bottle, rf pulsed	81
		133.18	±0.59	0.7	fast beam, rf	136
$4P_{3/2} - 4S_{1/2}$	1238.051(1)	1237.79	±0.27	-1.0	bottle, rf pulsed	81
		1235.9	±1.3	-1.7	fast beam, rf	136
$4D_{3/2} - 4S_{1/2}$	1235.765(1)	1235.0	±2.1	-0.4	bottle, crossing	245
$5S_{1/2} - 5P_{1/2}$	68.195(1)	64.6	±5.0	-0.7	fast beam, rf	136
$5P_{3/2} - 5S_{1/2}$	633.823(1)	622.4	±10.1	-1.1	fast beam, rf	136

D	$S(^1S_{1/2})$	8183.92(9) MHz	7900	± 1100 MHz	-0.3	two photon absorbtion	237
			8300	±300	0.4	two photon absorbtion	169
			8260	± 110	0.7	two photon absorbtion	170
			8189	±30	0.2	two photon absorbtion	172
			8184.10	±0.80	0.1	two photon absorbtion	163,177
	$2S_{1/2} - 2P_{1/2}$	1059.233(15)	1059.28	± 0.06	0.8	slow beam, ac	59,53
			1058.996	± 0.064	-3.7	slow beam, rf	50,53
			1058.7	± 1.1	-0.5	fast beam, anisotropy	119
	$2P_{3/2} - 2S_{1/2}$	9912.805(15)	9912.607	± 0.056	-3.6	slow beam, rf	51,53
			9912.803	± 0.094	0.0	slow beam, rf	51,53
	$3S_{1/2} - 3P_{1/2}$	315.288(4)	315.3	± 0.4	0.0	bottle, rf	77
	$4S_{1/2} - 4P_{1/2}$	133.250(2)	133.0	± 5.0	-0.1	bottle, rf	77
$^4He^+$	$2S_{1/2} - 2P_{1/2}$	14042.24(50)	14046.2	± 1.2	3.3	bottle, rf	69
			14040.2	± 1.8	-1.1	bottle, rf	68
			14040.9	± 2.9	-0.5	fast beam, anisotropy	121,123
			14042.22	± 0.35	-0.1	fast beam, anisotropy	123
			14042.0	± 1.20	-0.2	fast beam, rf	146
	$3S_{1/2} - 3P_{1/2}$	4183.63(15)	4183.17	± 0.54	-0.9	bottle, rf	70
			4184.2	±2.4	0.2	bottle, ac	84
	$3P_{3/2} - 3S_{1/2}$	47844.21(15)	47844.05	±0.48	-0.3	bottle, rf	70

$4S_{1/2}-4P_{1/2}$	1768.77(5) MHz	1769.0	±2.0 MHz	0.1	bottle, rf	90
		1768.0	±5.	-0.2	bottle, rf	89
		1766.0	±7.5	-0.4	bottle, rf	87
		1751.0	±13.	-1.4	bottle, rf	88
		1755.0	±44.	-0.3	fast beam, rf	139
		1769.16	±0.84	0.5	fast beam, sep rf	155
$4P_{3/2}-4S_{1/2}$	20180.38(6)	20180.6	±0.8	0.3	bottle, rf	91
		20179.7	±1.2	-0.6	bottle, rf	89
$4D_{3/2}-4S_{1/2}$	20143.71(6)	20145.3	±3.9	0.4	bottle, crossing	96
$4D_{5/2}-4S_{1/2}$	27459.46(6)	27455.4	±3.7	-1.1	bottle, crossing	96
$4F_{5/2}-4S_{1/2}$	27446.49(6)	27435.7	±6.9	-1.6	bottle, crossing	95
		27446.490	±0.124	-0.1	fast beam, sep rf	156
$4F_{7/2}-4S_{1/2}$	31104.28(6)	31098.8	±5.2	-1.1	bottle, crossing	95
$5S_{1/2}-5P_{1/2}$	906.54(3)	902	±15	-0.3	fast beam, rf	139
		909.2	±7	0.4	bottle, rf	93
		905.2	±1.0	-1.3	bottle, rf	94
$5P_{3/2}-5S_{1/2}$	10331.37(3)	10332.9	±1.4	1.1	bottle, rf	91
$5D_{3/2}-5S_{1/2}$	10312.41(3)	10309.8	±7.3	-0.4	bottle, crossing	97
$5D_{5/2}-5S_{1/2}$	14058.07(3)	14055.4	±6.3	-0.4	bottle, crossing	97
$5F_{5/2}-5S_{1/2}$	14051.38(3)	14058.2	±6.8	1.0	bottle, crossing	97

	$5F_{7/2}-5S_{1/2}$	15924.17(3) MHz	15924.5	±5.4 MHz	0.1	bottle, crossing	97
	$5G_{7/2}-5S_{1/2}$	15920.67(3)	15924.4	±5.9	0.6	bottle, crossing	246
	$5G_{9/2}-5S_{1/2}$	17044.34(3)	17040.5	±8.0	-0.5	fast beam, rf	246
	$6S_{1/2}-6P_{1/2}$	524.92(2)	530.	±10	0.5	fast beam, rf	139
			524.	±13	-0.1	fast beam, rf	139
			526.7	±2.7	0.7	bottle, rf	94
	$6P_{3/2}-6S_{1/2}$	5978.47(2)	5979.1	±1.2	0.5	bottle, rf	91
	$6D_{5/2}-6S_{1/2}$	8135.06(2)	8133.4	±1.0	-1.7	bottle, rf	99
	$7S_{1/2}-7P_{1/2}$	330.68(1)	333.0	±8	0.3	fast beam, rf	139
$^{6}Li^{2+}$	$2S_{1/2}-2P_{1/2}$	62737.(6) GHz	62765.	±21GHz	1.3	fast beam, rf	74
			62709.	±70	-0.4	fast beam, rf	145
			63031.	±327	0.9	fast beam, lifetime	102
$^{12}C^{5+}$	$2S_{1/2}-2P_{1/2}$	781.98(23)	780.1	±8.0 GHz	-0.2	fast beam, lifetime	107
$^{16}O^{7+}$	$2S_{1/2}-2P_{1/2}$	2196.19(21)	2215.6	±7.5	3.6	fast beam, lifetime	109
			2192.	±15	-0.3	fast beam, anisotropy	124
			2203.	±11	0.6	fast beam, lifetime	108
$^{19}F^{8+}$	$2P_{3/2}-2S_{1/2}$	68850.0(24)	68853.9	±35	0.1	fast beam, laser	157
$^{31}P^{14+}$	$2P_{3/2}-2S_{1/2}$	539.444(11) THz	539.60	±0.20	0.8	fast beam, laser	159
			539.550	±0.038	2.8	fast beam, laser	161

$^{32}S^{15+}$	$2S_{1/2}-2P_{1/2}$	25.374(17) THz	25.14	±0.24 THz	-1.0	fast beam, lifetime	110
	$2P_{3/2}-2S_{1/2}$	699.922(14)	700.070	±0.062	2.4	fast beam, laser	160
$^{35}Cl^{16+}$	$2S_{1/2}-2P_{1/2}$	31.35(2)	31.19	±0.22	-0.7	fast beam, laser	158
$^{40}A^{17+}$	$2S_{1/2}-2P_{1/2}$	38.25(2)	37.89	±0.38	-0.9	fast beam, lifetime	112

REFERENCES

1. For an overview of the history, see. Pais, A., Inward Bound, (Oxford U.P., New York, 1986).
2. For a detailed account of the history, see Mehra, J. and Rechenberg, H., The Historical Development of Quantum Theory, (Springer-Verlag, New York, 1982-1987), Vols. 1-5.
3. For an interesting treatment of the history, see Schweber, S. S., Am. J. Phys. 57, 299 (1989).
4. Kent, N. A., Taylor, L. B., and Pearson, H., Phys. Rev. 30, 266 (1927).
5. Houston, W. V., Astrophys. J. 64, 81 (1926).
6. Spedding, F. H., Shane, C. D., and Grace, N. S., Phys. Rev. 44, 58 (1933).
7. Kemble, E. C., and Present, R. D., Phys. Rev. 44, 1031 (1933).
8. Houston, W. V. and Hsieh, Y. M. , Phys. Rev. 45, 263 (1934).
9. Williams, R. C. and Gibbs, R. C., Phys. Rev. 45, 475 (1934).
10. Kopfermann, H., Naturwissenshaffen 22, 218 (1934).
11. Spedding, F. H., Shane, C. D., and Grace, N. S., Phys. Rev. 47, 38 (1935).
12. Houston, W. V., Phys. Rev. 51, 446 (1937).
13. Williams, R. C., Phys. Rev. 54, 558 (1938).
14. Pasternack, S., Phys. Rev. 54, 1113 (1938).
15. Drinkwater, J. W., Richardson, O., and Williams, W. E., Proc. Roy. Soc. London 174, 164 (1940).
16. Lamb, W. E., Sr. and Retherford, R. C., Phys. Rev. 72, 241 (1947).
17. Lamb, W. E., Jr., Rep. Prog. Phys, 14, 19 (1951).
18. Lamb, W. E., Jr., in The Birth of Particle Physics, edited by L. M. Brown, and L. Hoddeson, (Cambridge U.P., Cambridge, 1983), p. 311.
19. Bethe, H. A., Phys. Rev. 72, 339 (1947).
20. For a recent text dealing with quantum field theory and renormalization see, Itzykson, C. and Zuber, J. B., Quantum Field Theory (McGraw Hill, New York, 1980).
21. For a review of the theoretical calculation of the Lamb shift see the article in this volume by Sapirstein, J. and Yennie, D.
22. Lautrup, B. E., Peterman, A., and deRafael, E., Phys. Rep. 36, 193 (1972).
23. Ermolaev, A. M., in Physics of Atoms and molecules, Part A, edited by W. Hanle and H. Kleinpoppen (Plenum Press, New York, London, 1978) p. 149.

24. Lepage, G. P. and Yennie, D. R., in Precision Measurements and Fundamental Constants, edited by B. N. Taylor and W. D. Phillips (Natl. Bur. Stand. (U.S.), Spec. Publ. 617, 1984), p. 185.
25. Drake, G. W. F., in Atomic Physics of Highly Ionized Atoms, edited by R. Marrus (Plenum Press, New York, London, 1983), p. 113.
26. Kinoshita, T., and Sapirstein, J., in Atomic Physics 9, edited by R. S. van Dyck, Jr., and E. N. Fortson (World Scientific, Singapore, 1984), p. 38.
27. Beyer, H.-J., in Physics of Atoms and Molecules, Part A., edited by W. Hanle and H. Kleinpoppen, (Plenum Press, New York, London, 1978), p. 529.
28. Kugel, H. W. and Murnick, D. E., Rep. Prog. Phys. 40, 297 (1977).
29. For a simple explanation of the origin of the Lamb shift in terms of the zero point fluctuations of the radiation field, see Welton, T.A., Phys. Rev. 74, 1157 (1948).
30. Cohen, E. R. and Taylor, B. N., Rev. Mod. Phys. 59, 1121 (1987).
31. Biraben, F., Garreau, J. C., Julien, L. , and Allegrini,M., Phys. Rev. Lett. 62, 621 (1989).
32. Erickson, G. W., J. Phys. Chem. Ref. Data, 6, 831 (1977).
33. Erickson, G. W., communication to B. N. Taylor.
34. Mohr, P. J., Ann. Phys. (NY) 88, 26 (1974); 88, 54 (1974).
35. Mohr, P. J., in Beam Foil Spectroscopy, edited by I. Sellin and D. J. Pegg (Plenum, New York, 1976), Vol. 1, p. 89.
36. Mohr, P. J., Phys. Rev. A 26, 2338 (1982).
37. Mohr, P. J., At. Data and Nucl. Data Tables 29, 453 (1983).
38. Johnson, W. R. and Soff, G., At. Data and Nucl. Data Tables 33, 405 (1985).
39. Bethe, H. A., and Salpeter, E. E., Quantum Mechanics of One- and Two-Electron Atoms, (Springer-Verlag, Berlin, 1957).
40. Klarsfeld, S., Phys. Lett. 30A, 382 (1969).
41. Lamb, Jr., W. E. and Retherford, R. C., Phys. Rev. 79, 549 (1950).
42. Fabjan, C. W. and Pipkin, F. M. , Phys. Rev. A6, 556 (1972).
43. For an alternate formulation in terms of matrices, see Lundeen, S. R. and Pipkin, F. M. , Metrologia 22, 9 (1986).
44. Ramsey, N. F., Molecular Beams (Oxford U. P., London, 1956), p. 124.
45. Lamb, Jr., W. E., Phys. Rev. 85, 259 (1952).

46. Beyer, H.-J., and Kleinpoppen, in Physics of Atoms and Molecules, Part A, edited by H. Hanle and H. Kleinpoppen, (Plenum Press, New York, 1978), p. 607.
47. Lamb, Jr., W. E. and Sanders, Jr., T. E., Phys. Rev. 119, 1901 (1960).
48. Lamb, Jr., W. E. and Retherford, R. C., Phys. Rev. 81, 222 (1951).
49. Lamb, Jr., W. E. and Retherford, R. C., Phys. Rev. 86, 1014 (1952).
50. Triebwasser, S., Dayhoff , E. S., and Lamb, Jr., W. E., Phys. Rev. 89, 98 (1953).
51. Dayhoff, E. S., Triebwasser, S., and Lamb, Jr., W. E., Phys. Rev. 89, 106 (1953).
52. Brodsky, S. J. and Parsons, R. G., Phys. Rev. 163, 134 (1967); Phys. Rev. 176, 423 (1968)
53. Taylor, B. N. Parker, W. H., and Langenberg, D. N., Rev. Mod. Phys. 41, 375 (1969).
54. Robiscoe, R. T., Phys. Rev. 138, A22 (1965).
55. Robiscoe, R. T. and Cosens, B. L., Phys. Rev. Lett. 17, 69 (1966).
56. Robiscoe, R. T., Phys. Rev. 168, 4 (1968).
57. Robiscoe, R. T., in Cargése Lectures in Physics, Vol. 2, edited by M. Lévy (Gordon and Breach, New York, 1968), p. 3.
58. Robiscoe, R. T. and Shyn, T. W., Phys. Rev. Lett. 24, 559 (1970).
59. Cosens, B. L., Phys. Rev. 173, 49 (1968).
60. Cosens, B. L., and Vorburger, T. V., Phys. Rev. Lett. 23, 1273 (1969).
61. Cosens, B. L. and Vorburger, T. V., Phys. Rev. A2, 16 (1970).
62. Shyn, T. W., Williams, W. L., Robiscoe, R. T., and Rebane, T., Phys. Rev. Lett. 22, 1273 (1969).
63. Shyn, T. W., Rebane, T., Robiscoe, R. T., and Williams, W. L., Phys. Rev. A3, 116 (1971).
64. Wing, W. H. and MacAdam, K. B., in Physics of Atoms and Molecules, Part A, edited by W. Hanle and M. Kleinpoppen (Plenum Press, New York, 1978) p. 441.
65. Skinner, M. and Lamb, Jr., W. E., Phys. Rev. 75, 1325A (1949).
66. Lamb, Sr., W. E., and Skinner, M., Phys. Rev. 78, 539 (1950).
67. Novick, R., Lipworth, E., and Yergin, P. F., Phys. Rev. 100 1153 (1955).
68. Lipworth, E. and Novick, R., Phys. Rev. 108, 1434 (1957).
69. Narasimham, M. A. and Strombotne, R. I., Phys. Rev. A4, 14 (1971).

70. Mader, D. L., Leventhal, M. and Lamb, Jr., W. E. Phys. Rev. A3, 1832 (1971).

71. Kaufman,S. L., Lamb, Jr., W. E., Lea, K. R., and Leventhal, M., Phys. Rev. Lett. 22, 507 (1969).

72. Kaufman,S. L., Lamb., Jr., W. E., Lea, K. R. and Leventhal, M., Phys. Rev. A4, 2128 (1971).

73. Leventhal, M., and Harvey, P. E., Phys. Rev. Lett. 32, 808 (1974).

74. Leventhal, M. , Phys. Rev. A 11, 427 (1975).

75. Kleinpoppen, H., Z. Phys. 164, 174 (1961).

76. Lamb, Jr., W. E. and Sanders, T. M. Phys. Rev. 103, 313 (1956).

77. Wilcox, L. R., and Lamb, Jr., W. E. , Phys. Rev. 119, 1915 (1960).

78. Glass-Maujean, M., and Descoubes, J. P., C. R., Acad., Sci. 273B, 721 (1971).

79. Richardson, C. B. and Hughes, R. H., Bull. Am. Phys. Soc. II 15, 45 (1970).

80. Richardson, C. B., Mitchell, D. A. and Hughes, R. H., Bull. Am. Phys. Soc.15, 1509 (1970).

81. Brown, R. A. and Pipkin, F. M. , Ann. Phys. (NY) 80, 479 (1973).

82. Leventhal, M., Lea, K. R., and Lamb, Jr., W. E., Phys. Rev. Lett. 15, 1013 (1965).

83. Baumann, M. and Eibofner, A., Phys. Lett. 34A, 421 (1971).

84. Eibofner, A., Z. Phys. 249, 58 (1971).

85. Baumann, M., and Eibofner, A., Phys. Lett. 33A, 409 (1970).

86. Lea, K. R., Leventhal, M., and Lamb, Jr., W. E., Phys. Rev. Lett. 16, 163 (1966).

87. Hatfield, L. L. and Hughes, R. H., Phys. Rev. 156, 102 (1967).

88. Beyer, H.-J. and Kleinpoppen H., Z. Phys. 206, 177 (1967).

89. Jacobs, R. R., Lea, K. R., and Lamb, Jr., W. E. , Phys. Rev. A3, 884 (1971).

90. Eibofner, A., Phys. Lett. 58A, 219 (1976).

91. Eibofner, A., Z. Phys. A277, 225 (1976).

92. Eck, T. G., and Huff, R. J., Phys. Rev. Lett. 22, 319 (1969).

93. Baumann, M., and Eibofner, A., Phys. Lett. 43A, 105 (1973).

94. Eibofner, A., Phys. Lett. 47A, 399 (1974).

95. Beyer, H.-J., and Kleinpoppen, H. J. ,J. Phys. B4, L129 (1971).

96. Beyer, H.-J., and Kleinpoppen, H. J., J. Phys. B5, L12 (1972).

97. Beyer, H.-J., and Kleinpoppen, H., J. Phys. B8, 2449 (1975).

98. Glass-Maujean, M., Opt. Comm. 8, 260 (1973).
99. Eibofner, A., Phys. Lett. 61A, 159 (1977).
100. Sellin, I. A., Phys. Rev. 136, A1245 (1964).
101. Fan, C. Y., Garcia-Munoz, M., and Sellin, I. A., Phys. Rev. Lett. 15, 15 (1965).
102. Fan, C. Y., Garcia-Munoz, M., and Sellin, I. A., Phys. Rev. 161, 6 (1967).
103. Holt, H. K. and Sellin, I. A., Phys. Rev. A6, 508 (1972).
104. Drake, G. W. F. and Grimley, R. B., Phys. Rev. A8, 157 (1973).
105. Leventhal, M. and Murnick, D. E., Phys. Rev. Lett. 25, 1237 (1970).
106. Murnick, D. E., and Leventhal, M., and Kugel, H. W., Phys. Rev. Lett. 27, 1625 (1971).
107. Kugel, H. W., Leventhal, M., and Murnick, D. E., Phys. Rev. A6, 1306 (1972).
108. Leventhal, M., Murnick, D. E., and Kugel, H. W., Phys. Rev. Lett. 28, 1609 (1972).
109. Lawrence, G. P., Fan, C. Y. and Bashkin, S., Phys. Rev. Lett. 28, 1612 (1972).
110. Zacek, V., Bohn, H., Brum, H., Faestermann, T., v. Feilitzsch, F., Giorginis, G., Kienle, P., and Schunbeck, S., Z. Phys. A318, 7 (1984).
111. Gould, M. , and Marrus, R., Phys. Rev. Lett. 41, 1457 (1978).
112. Gould, M., and Marrus, R., Phys. Rev. A28, 2001 (1983).
113. Fite, W. L., Kauppila, W. E., and Ott., W. R., Phys. Rev. Lett. 20, 409 (1968).
114. Ott, W. R., Kauppila, W. E., and Fite, W. L., Phys. Rev. A1, 1089 (1970).
115. Drake, G. W. F., and Grimley, R. B., Phys. Rev. A11, 1614 (1975).
116. Drake, G. W. F., and Lin. C.-P., Phys. Rev. A14, 1296 (1976).
117. Drake, G. W. F., J. Phys. B, Atom. Molec. Phys. 10, 775 (1977).
118. van Wijngaarden, A., Drake, G. W. F., and Farago, P. S., Phys. Rev. Lett. 33, 4 (1974).
119. Darke, G. W. F., Farago, P. S., and van Wijngaarden, A., Phys. Rev. A11, 1621 (1975).
120. van Wijngaarden, A. and Drake, G. W. F., Phys. Rev. A17, 1366 (1978).
121. Drake, G. W. F., Goldman, S. P., and van Winjgaarden, A., Phys. Rev. A20, 1299 (1979).
122. Patel, J., van Winjgaarden, A., and Drake, G. W. F., Phys. Rev. A36, 5130 (1987).

123. Drake, G. W. F., Patel, J. and van Wijngaarden A., Phys. Rev. Lett. 60, 1002 (1988).

124. Curnutte, B., Cocke, C. L., and Dubois, R. D., Nucl. Instr. Meth. 202, 119 (1982).

125. Andrä, H. J., Phys. Scr. 9, 257 (1974).

126. Andrä, H. J., in Physics of Atoms and Molecules, Part B, edited by W. Hanle and H. Kleinpoppen (Plenum Press, New York, 1978), p. 829.

127. Häroche, S., in High-Resolution Laser Spectroscopy, edited by K. Shimoda (Springer-Verlag, Berlin, 1976), p. 252.

128. Sokolov, Yu. L., Zh. Eksp. Theo. Fiz. 63, 461 (1972); Sov. Phys. JETP 36, 243 (1973).

129. Sokolov, Yu. L, in Atomic Physics 6, edited by R. Damburg (Plenum Press, New York 1979), p. 207.

130. Sokolov, Yu. L. and Yakovlev, V. P., Zh. Exp. Theor. Fiz. 83, 15 (1982); Sov. Phys. JETP 56, 7 (1982).

131. Polchikov, V. G., Sokolov, Yu. L., and Yakovlev, V. P., Lett. Jour. Tech. Phys. 38, 347 (1983).

132. Sokolov, Yu. L., in Precision Measurements and Fundamental Constants I, edited by B. N. Taylor and W. D. Phillips (Natl. Bur. Stand. (U.S.), Spec. Publ. 617, 1984), p. 135.

133. Palchikov, V. G., Sokolov, Yu. L., and Yakovlev, V. P., Metrologia 21, 99 (1985).

134. Sokolov, Yu. L., in The Hydrogen Atom, edited by G. F. Bassani, M. I Inguscio, and T. W. Hänsch (Springer-Verlag, Berlin, 1989), p. 16.

135. Fabjan, C. W., and Pipkin, F. M. Phys. Rev. Lett. 25, 421 (1970).

136. Fabjan, C. W., Pipkin, F. M., and Silverman, M., Phys. Rev. Lett. 26, 347 (1971).

137. Andrews, D. A., and Newton, G. Phys. Rev. Lett. 37, 1254 (1976).

138. Newton, G. Andrews, D. A., and Unsworth, P. J., Philos. Trans. R. Soc. London 290, 373 (1979).

139. Churassy, S. Gaillard, M. L., and Silver, J. D., Phys. Rev. Lett. 33, 185 (1974).

140. Kramer, P. B., Lundeen, S. R., Clark, B. O., and Pipkin, F. M. , Phys. Rev. Lett. 32, 635 (1974).

141. van Baak, D. A., Clark, B. O., and Pipkin, F. M.., Phys. Rev. A19, 787 (1979).

142. Clark, B. O., van Baak, D. A., Lundeen, S. R., and Pipkin, F. M., Phys. Lett. A64, 172 (1977).

143. Clark, B. O., van Baak, D. A., Lundeen, S. R., and Pipkin, F. M., Phys. Rev. A19, 802 (1979).

144. Brandenberger, J. R., Lundeen, S. R., and Pipkin, F. M., Phys. Rev. A14, 341 (1976).

145. Dietrich, D. D. , Lebow, P., de Zafra, D., and Metcalf, M., Bull. Am. Phys. Soc. 21, 625 (1976).

146. Dewey, M. S., and Dunford, R. W., Phys. Rev. Lett. 60, 2014 (1988).

147. Fabjan, C. W., and Pipkin, F. M. , Phys. Lett. 36A, 69 (1971).

148. Lundeen, S. R., Yung, Y. L., and Pipkin, F. M., Nucl. Instr. Meth. 110, 355 (1973).

149. Lundeen, S. R. and Pipkin, F. M., Phys. Rev. Lett. 34, 1368 (1975).

150. Lundeen, S. R., and Pipkin, F. M. , Phys. Rev. Lett. 46, 232 (1981).

151. Lundeen, S. R, Jessop, P. E., and Pipkin, F. M., Phys. Rev. Lett. 34, 377 (1975).

152. Clark, B. O., van Baak, D. A., and Pipkin, F. M., Phys. Lett. A62, 78 (1977).

153. van Baak, D. A., Clark, B. O., Lundeen, S. R., and Pipkin, F. M., Phys. Rev. A22, 591 (1980).

154. Safinya, K. A., Chan, K. K., Lundeen, S. R., and Pipkin, F. M., Phys. Rev. Lett. 45, 1934 (1980).

155. Bollinger, J. J., Lundeen, S. R., and Pipkin, F. M., Phys. Rev. A30, 2170 (1984).

156. Majumder, P. K., and Pipkin, F. M., Phys. Rev. Lett. 63, 372 (1989).

157. Kugel, H. W. Leventhal, M. , Murnick, D. E., Patel, C. K. N., and Wood, II, O. R., Phys. Rev. Lett. 35, 647 (1975).

158. Wood, II, O. R., Patel, C. K. N., Murnick, D. E., Nelson, E. T., Leventhal, M., Kugel, H. W., and Niv, Y., Phys. Rev. Lett. 48, 398 (1982).

159. Pellegrin, P., El Masri, Y., and Palffy, L., Phys. Rev. A31, 5 (1985).

160. Georgiadis, A. P. Müller, D., Sträter, H.-D., Gassen, J., von Brentano, P., Sens, J. C., and Pope, A., Phys. Lett. A115, 108 (1986).

161. Muller, D., Gassen, J. , Kremer, L., Pross, H. J., Schwer, F. Sträter, H. D. von Brentano, P., Pope, A., and Sens, J. C., Europhys. Lett. 5, 503 (1988).

162. von Brentano, P., Müller, D., and Gassen, J., in Spectroscopy and Collisions of Few Electron Ions, (World Scientific, Singapore, 1988).

163. Stacey, D. N. in The Hydrogen Atom, edited by G. F. Bassani, M. I. Inguiscio, and T. W. Hänsch (Springer-Verlag, Berlin, Heidelberg, 1989), p. 68.

164. Ferguson, A. I., Tolchard, J. M., and Persaud, M. A., in The Hydrogen Atom, edited by G. F. Bassani, M. Inguscio, and T. W. Hänsch (Sprigner-Verlag, Berlin, Heidelbergf, 1989), p. 81.

165. Hänsch, T. W., in The Hydrogen Atom, edited by G. F. Bassani, M. I. Inguscio, and T. W. Hänsch (Springer-Verlag, Berlin, Heidelberg, 1989), p. 93.

166. Bloembergen, N. and Levenson, M. D., in High Resolution Laser Spectroscopy, edited by K. Shimoda (Springer-Verlag, Berlin, Heidelberg, 1976), p. 314.

167. Balkanov, E. V., and Chebotayev, V. P., Opp. Comm. 12, 312 (1974); Opt. Spectrosc. 38, 215 (1975).

168. Cagnac, B., Grynberg, G., and Biraben, F., Jour. de. Phys. 34, 845 (1973).

169. Hänsch, T. W., Lee, S. A., Wallenstein, R., and Weiman, C., Phys. Rev. Lett. 34, 307 (1975).

170. Lee, S. A., Wallenstein, R. and Hänsch, T. W., Phys. Rev. Lett. 35, 1262 (1975).

171. Weiman, C. E., and Hänsch, T. W., Phys. Rev. Lett. 36, 1170 (1976).

172. Weiman, C. E., and Hänsch, T. W., Phys. Rev. A22, 192 (1980).

173. Hildum, E. A., Boesl, U., McIntyre, D. H., Beausoleil, R. G., and Hänsch, T. W., Phys. Rev. Lett. 56, 576 (1986).

174. Barr, J. R. M., Girkin, J. M., Tolchard, J. M., Ferguson, A. I., Phys. Rev. Lett. 56, 580 (1986).

175. Foot, C. J., Couillaud, B., Beausoleil, R. G., and Hänsch, T. W., Phys. Rev. Lett. 54, 1913 (1985).

176. Beausoleil, R. G., McIntyre, D. H., Foot, C. J., Couillaud, B., Hildum, E. A., and Hänsch, T. W., Phys. Rev. A35, 4878 (1987).

177. Boshier, M. G., Baird, P. E. G., Foot, C. J., Hinds, E. A., Plimmer, M. D., Stacey, D. N., Swan, J. B., Tate, D. A., Warrington, D. M., Wookgate, G. K., Nature 330, 463 (1987).

178. Barr, J. R. M., Girkin, J. M., Ferguson, A. I., Barwood, G. P., Gill, P. Rowley, W. R. C., and Thompson, Opt. Comm. 54, 217 (1985).

179. McIntyre, D. H., Beausoleil, R. G., Foot, C. J., Hildum, E. A., Couillaud, B., and Hänsch, T. W., Phys. Rev. A39, 4591 (1989).

180. Silver, J. A., in The Hydrogen Atom, edited by G. F. Bassani, M. Inguscio, and T. W. Hänsch (Springer-Verlag, Berlin, 1989), p.221.

181. Beyer, H. F., Deslattes, R. D., Folkmann, F., and LaVilla, R. E., J. Phys. B18, 207 (1985).

182. Marmar, E. S., Rice, J. E.,Källne, E., Källne, J. and LaVilla, R. E., Phys. Rev. A33, 774 (1986).

183. Richard, P., Stöckli, M., Deslattes, R. D., Cowan, P., LaVilla, R. E., Johnson, B., Jones, K., Meron, M., Mann, R., and Schartner, K., Phys. Rev. A29, 2939 (1984).

184. Deslattes, R. D., Schuch, R. and Justiniano, E., Phys. Rev. A32, 1911 (1985).

185. Källne, E., Källne, J., Richard, P., and Stöckli, M., J. Phys. B17, L115 (1984).

186. Briand, J. P., Tavernier, M., Indelicato, P., Marrus, R., and Gould, H., Phys.Rev. Lett. 50, 832 (1983).

187. Silver, J. D., McClelland, A. F., Laming, J. M., Rosner, S. D., Chandler, G. C., Dietrich, D. D., and Egan, P. O., Phys. Rev. A36, 1515 (1987).

188. Laming, J. M., Chandler, C. T., Silver, J. D., Dietrich, D. D., Finch, E. C., Mokler, P. H., and Rosner, S. D. , Nucl. Instr. and Meths. B31, 21 (1988).

189. Tavernier, M., Briand, J. P. Indelicato, P., Liesen, D., and Richard, P., J. Phys. B18, L327 (1985).

190. Dunets, E. D., and Ousyannikov, V. P., JETP 53, 466 (1981).

191. Levine, M. A., Marrs, R. E., Henderson, J. R., Knapp, D. A., and Schneider, M. B., Phys. Scr. T22, 157 (1988).

192. For reviews of positronium see: Rich, A., Rev. Mod. Phys. 53, 127 (1981); Berko, S. and Pendleton, H. N., Ann. Rev. Nucl. Part. Sci. 30, 543 (1980).

193. For a recent overview of muonium see: Hughes, V. W., in The Hydrogen Atom, edited by G. F. Bassani, M. Inguscio, and T. W.Hänsch, , (Springer-Verlag, Berlin, 1989), p. 171.

194. Mills, A. P., Berko, S., and Canter, K. F., Phys. Rev. Lett. 34, 1541 (1975).

195. Hatamian, S., Conti, R. S., and Rich, A., Phys. Rev. Lett. 58, 1833 (1987).

196. Chu, S., Mills, A. P., and Hall, J. L., Phys. Rev. Lett. 52, 1689 (1984).

197. Danzmann, K., Fee, M. S., Chu, S., Phys. Rev. A39, 6072 (1989).

198. Oram, C. J., Bailey, J. M., Schmor, P. W., Fry, C. A., Kiefl, R. F., Warren, J. B., Marshall, G. M., and Olin, A., Phys. Rev. Lett., 52, 910 (1984).

199. Badertscher, Dhawan, S., Egan, P. O., Hughes, V. W., Lu, D. C., Ritter, M. W., Woodle, K. A., Gladisch, M. Orth, H., zu Putlitz, G., Eckhause, M., Kane, J., Mariam, F. E., and Reidy, J., Phys. Rev. Lett. 52, 914 (1984).

200. Carboni, G., Gorini, G., Torelli, G., Palffy, L., Palmonari, F., and Zavottini, E., Nucl. Phys. A278, 381 (1977).

201. Carboni, G., Gorini, G., Iacopini, E., Palffy, L., Palmonari, F., Torelli, G., and Zavattini, Phys. Lett., B73, 229 (1978).

202. Carboni, G., Gastaldi, V., Neri, G., Pitzurra, O., Polacco, E., Torelli, G., Bertin, A., Gorini, G., Placci, A., Zavattini, E., Vitale, A., Duclos, J., and Picard, J., Nuovo Cimento 34A, 493 (1976).

203. Sick, I., McCarthy, J. S., and Whitney, R. R., Phys. Lett. 64B, 33 (1976); Sick, S., Phys. Lett. B116, 213 (1982).

204. Borie, F. and Rinker, G. A. Phys. Rev. A18 , 324 (1978).

205. von Arb, H. P., Dittus, F., Heeb, H., Hofer, H., Kottman, F., Niggli, S., Schaerin, R., Taqqu, D., Unternäher, J., and Egelhof, P., Phys. Lett. 136B, 232 (1984).

206. Eckhause, M., Guss, P., Joyce, D., Kane, J. R., Siegel, R. T., Vulcan, W., Welsh, R. E. Whyley, R., Dietlicher, R., and Zehnder, A., Phys. Rev. A33, 1743 (1986).

207. Chu, S., Mills, Jr., A. P., Yodh A. G., Nagamine, K., Miyaki, Y., and Kuga, T., Phys. Rev. Lett. 60, 101 (1988).

208. Drake, G. W. F., in Advan. in At. and Mol. Phys. 18, edited by D. Bates and B. Bederson (Academic Press, New York, 1982), p. 399.

209. Goldman, S. P., and Drake, G. W. F., J. Phys. B: At. Mol. Phys. 17, L197 (1984).

210. Mohr, P. J., Nucl. Instr. Meth. B31, 1 (1988).

211. Desesquelles, J., Nucl. Instr. Meth. B31, 30 (1988).

212. Drake, G. W. F., Phys. Rev. Lett. 59, 1549 (1987).

213. Drake, G. W. F., Nucl. Instr. Meth. B31, 7 (1988).

214. Drake, G. W. F. and Makowski, A. J., J. Opt. Soc. Am. B5, 2207 (1988).

215. Baker, J., Hill., R. N., and Morgan III, J. D., to be published.

216. Sansonetti, C. J., and Martin, W. C., Phys. Rev. A29, 159 (1984).

217. Zhao, Ping, Lawall, J. R., Kam, A. W., Lindsay, M. D., Pipkin, F. M., and Lichten, W., to be published.

218. Holt, R. A., Rosner, S. D., Gaily, T. D., and Adam, A. G., Phys. Rev. A22, 1563 (1980).

219. Kowalski, J., Neumann, R. Noehte, S., Schwartzwald, R., Suhr., H. and zu Putlitz, G. in Laser Spectroscopy VI, edited by H. P. Weber and W. Lüthy, Springer Series in Optical Sciences, Vol. 40, (Springer-Verlag, Berlin, 1983).

220. Stamp, M. F., Ph.D. Thesis, University of Oxford, 1983.

221. Peacock, N. J., Stamp, M. F. and Silver, J. D., Phys. Scr. T8, 10 (1984).

222. Stamp, M. F., Armour, I. A., Peacock, N. J. and Silver, J. D., J. Phys. B14, 3551 (1981).

223. Klein, H. A., Moscatelli, F., Myers, E. G., Pinnington, E. H., Silver, J. D. and Träbert, E. J., Phys. B18, 1483 (1985).

224. DeSerio, R., Berry, H. G., Brooks, R. L., Hardis, J., Livingston, A. E. and Hinterlong, S. J., Phys. Rev. A24, 1872 (1981).

225. Livingston, A. E. and Hinterlong, S. J., Nucl. Instr. Meth. 202, 103 (1982).

226. Hinterlong, S. J. and Livingston, A. E., Phys. Rev. A33, 4378 (1986).

227. Galvez, E. J., Livingston, A. E., Mazure, A. J., Berry, H. G., Engström, L., Hardis, J. E., Sommerville, L. P. and Zei, D., Phys. Rev. A33, 3667 (1986).

228. Buchet, J. P., Buchet-Poulizac, M. C., Denis, A., Desesquelles, J., Druetta, M., Grandin, J. P. and Husson, X. Phys. Rev. A23, 3354 (1981).

229. Buchet, J. P., Buchet-Poulizac, M. C., Denis, A., Desesquelles, J. Druetta, M., Grandin, J. P., Husson, X., Lecler, D. and Beyer, H. F., Nucl. Instr. Meth. B9, 645 (1985).

230. Munger, C. T. and Gould, H., Phys. Rev. Lett. 57, 2927 (1986).

231. For a summary and references, see ref. 43.

232. Hand, L. M., Miller, D. G., and Wilson, R., Rev. Mod. Phys. 35, 335 (1963).

233. Simon, G. G., Schmitt, Ch., Borkowski, F., and Walther, V. W., Nucl. Phys. A333, 381 (1980).

234. Blatt, G. C. and Grotch, H., Ann. Phys. (N.Y.) 178, 1 (1987).

235. Erickson, G. W., and Grotch, H., Phys. Rev. Lett. 60, 2611 (1988); errata 63, 1326 (1989).

236. van Wijngaarden, A., Patel, J., and Drake, G. W. F., in Atomic Physics II, edited by S. Haroche, J. C. Gay and G. Grynberg (World Scientific, Singapore, 1989), p. 355.

237. G. Herzberg, Proc. R. Soc. London, Ser A 234, 516 (1956).

238. Riis, E., Berry, H. G., Poulsen, O., Lee, S. A., and Tang, S. Y., Phys. Rev. A33, 3023 (1986).

239. Bayer, R., Kowalski, J., Neumann, R., Nochte, S., Suhr, H., Winkler, K., and zu Putlitz, G., Z. Phys. A 292, 329 (1979).

240. Englert, M., Kowalski, J. Mayer, F., Neumann, R., Nochte, S., Schwarzwald, R., Suhr, H. Winkler, K., and zu Putlitz, G., Appl. Phys. B 28, 81 (1982).

241. Kelsey, E. J. and Macek, J., Phys. Rev. A16, 1322 (1977).

242. Hillery, M., and Mohr, P. J., Phys. Rev. A21, 24 (1980).

243. Newton, G., Unsworth, P. J., and Andrews, D. A., J. Phys. B8, 2928 (1975).

244. Bentz, E., Liedtke, T., and Salzborn, Verhandl DPG (VI) 11, 89 (1976).

245. Glass-Maujean, M., Thesis, Université de Paris (1974).

246. Beyer, H.-J., Kleinpoppen, H., and Woolsey, J. M., Phys. Rev. Lett. 28, 263 (1972).

247. Lamb, Jr., W. E., Schlicher, R. R., and Schully, M. O., Phys. Rev. A 36, 2763 (1987).

248. Holt, R. A., Rosner, S. D., and Gaily, T. D., Phys. Rev. A 29, 1544 (1984).

249. Doncheski, M. A., Grotch, H., Owen, D. A., Penn State Preprint PSU/TH/57 (1989).

PRECISION MEASUREMENTS IN POSITRONIUM

Allen P. Mills, Jr.
AT&T Bell Laboratories,
Murray Hill, NJ 07974

and

Steven Chu
Physics Department, Stanford University
Stanford, CA 94305-4060

CONTENTS

1. INTRODUCTION

Positronium ($e^{+}-e^{-}$) is the lightest of the exotic hydrogen-like atoms made of a lepton and an anti-lepton. Since the leptons are considered to be structureless particles, careful measurements of the lifetimes and energy levels of positronium atoms give us direct information about the electromagnetic interactions of a lepton-anti-lepton pair. The positronium atom provides an excellent testing ground for studying the relativistic bound-state two-body system and the quantum electrodynamic corrections to that system. The corrections to the energy levels of positronium are of particular interest because they contain virtual annihilation terms not found in hydrogen or muonium. Positronium can also provide a test for non-perturbative calculations of two-body systems such as charmonium where the strength of the quark-quark interaction makes a perturbative solution impossible. For sufficiently accurate measurements, the influence of the weak interactions may be observable as well.

A study of simple atoms is a good way to search for possible new forces or particles signaled by discrepancies with theory, to check the validity of certain conservation laws, to further our knowledge of the fundamental constants, and to foster the technical advances that are the vanguard at the frontiers of solid state and atomic physics. The present article is a summary of recent progress in the study of the positronium energy levels using microwave and laser spectroscopy, the measurement of the lifetimes of the singlet and triplet ground states of the atom, and the search for exotic decays or decays that would break selection rules based on symmetries such as parity and charge conjugation. Other properties of positronium such as the cross sections for excitation and ionization by photons and electrons, the electron affinity, the properties of the positronium negative ion, and interactions of positronium with other atoms, molecules, surfaces, and in condensed matter will be discussed in this review only insofar as they pertain to the experiments that test the more fundamental properties. The reader is referred to a representative set of other reviews concerning positronium and positrons.[1–20]

To date, the annihilation rates of positronium (Ps) in its triplet and singlet ground states[21,22], the n=1 hyperfine interval,[23,24] the n=2 fine structure intervals,[25,26] and the triplet 1S-2S interval[27,28] of Ps have been measured with a precision sufficient to require QED correction terms to be included in the theory. A term diagram showing the energy levels and decay rates is presented in Fig 1. The study of the negative ion[29,30], which is the analogue of the H^{-} ion, should eventually reach comparable precision. Still awaiting techniques for their production and observation are the larger composites first envisioned by Wheeler,[31] such as the positronium molecule (Ps_2), as well as the various phases of the $e^{+}e^{-}$ plasma,[32] and the Bose-condensed positronium fluid.[33]

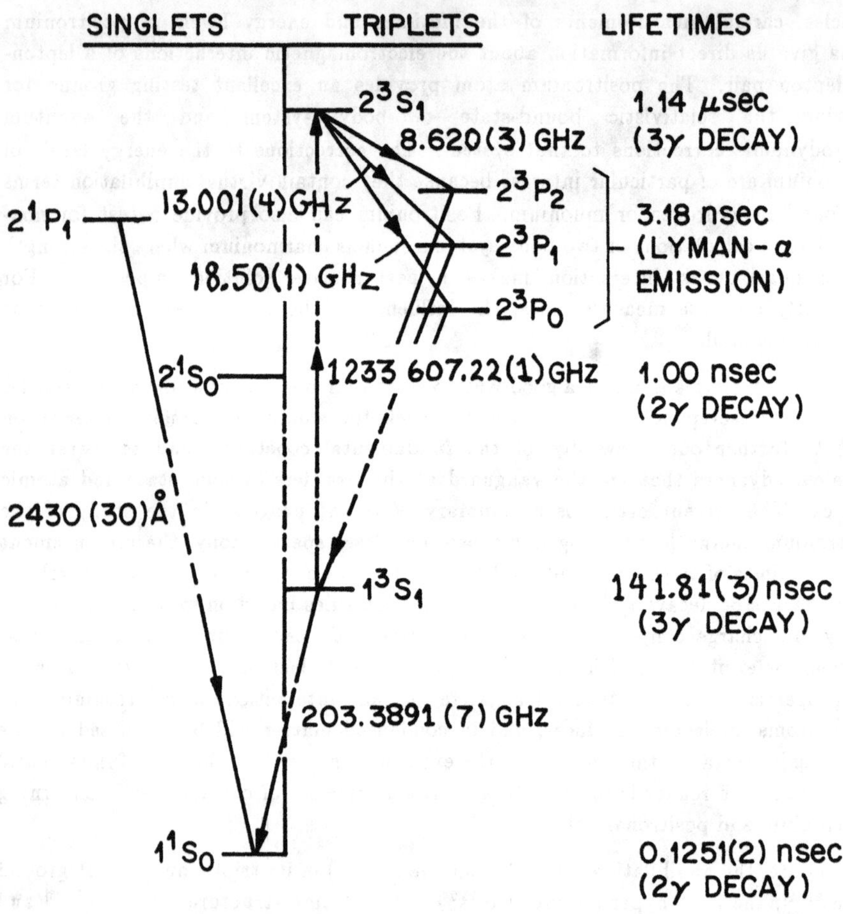

Fig. 1 Energy levels of the n=1 and n=2 states of positronium. The quantities with error estimates in parentheses are measured values.

2. METHODS OF POSITRONIUM FORMATION

The positronium atom (Ps) in its ground state and first excited states and the negative ion have been produced in the laboratory by techniques that include Ps formation in a gas[34] or finely divided solid in which β^+ particles are stopping,[35–38] slow positron beam charge exchange in a thin foil or a gas target,[39–41] and the interaction of slow positrons with the surface of a solid in vacuum.[42,43]

2.1 Positronium Formation in Gases and Powders

The pioneering experiments of M. Deutsch in 1951 demonstrated that positronium can be formed by stopping the energetic positrons from a radioactive source in a low density gas.[34] Since that time, the measurements of the ground state properties of Ps have relied principally on Deutsch's discovery for the production of Ps atoms. Fast positrons that are scattering in a gas will slow down to a few eV kinetic energy before capturing an electron. If the kinetic energy T of the positron is larger than E_{exc}, where E_{exc} is the lowest excitation energy of the stopping gas, inelastic collisions with the gas are the dominant interaction. When T is less than $E_i - E_{Ps}$, where E_i is the ionization energy of the gas atom or molecule and E_{Ps} is the binding energy of positronium, it is energetically impossible to form Ps. Also, for $T > E_i$, the kinetic energy of any Ps formed will exceed E_{Ps} and the Ps has a high probability for dissociation in subsequent collisions. Thus, there is a narrow energy range $E_{exc} \geq T \geq E_i - E_{Ps}$, or $E_i \geq T \geq E_i - E_{Ps}$ called the Ore gap[44,15] where Ps formation is likely. A more realistic treatment includes the effect of the energy dependence of the cross sections for positronium formation and break-up, and for positron inelastic scattering.[45] In all but the simplest noble gases, such as Ar and Ne, the physics is even more complicated because the interactions of the positron with the electrons and ions of the fast positron's ionization trail or "spur" cannot be neglected, as pointed out by Mogensen.[46] Thus, the actual yield in most cases depends on the gas density and impurities in a complicated manner. In any case, positronium is typically formed with probability of order 50% at gas pressures of one atmosphere.

After the positronium is formed it thermalizes at a rate that depends on the gas pressure and whether there are low lying vibrational and rotational excited states of the gas molecules. Collisions of the Ps increase the decay rate because the positron can annihilate with a foreign electron, or the triplet and singlet substates can be perturbed if there are unpaired electrons on the gas molecules. Collisions also perturb the Ps hyperfine interval proportionally to the gas density. Except in the case of helium, the hyperfine pressure shift is dominated by the van der Waals interaction which tends to pull the positron and electron apart. Collision effects are eliminated from the final measurements of the Ps lifetime and hyperfine interval by extrapolation to zero gas density. The excited states are easily quenched by collisions and must be studied in vacuum or in a very low density environment.

Positronium has also been produced in solids and liquids,[47–51] but the atoms are highly perturbed by their environment so that they are more useful as a material probe than for atomic measurements. Finely divided solids are also used to produce positronium. Paulin and Ambrosino showed that nearly free Ps is formed in various oxide powders when irradiated by energetic positrons from a radioactive source.[35] The effect was explained by Brandt and Paulin[52] as being due to Ps forming within a powder grain and rapidly diffusing to the grain surface where it escapes into the vacuum space between grains. A more careful analysis suggests that Mogensen's spur model[46] again is applicable, and that some of the Ps actually forms outside the grain surfaces with electrons liberated by the positron when it was slowing down. Once outside the grains, the Ps can make many thousands of collisions with the grains during its lifetime and can come into thermal equilibrium with the powder.[53] Any precision measurement of Ps properties must be extrapolated to zero powder density to compensate for the effects of collisions.[54] The discovery that nearly free Ps can be formed in powders has been exploited in measurements of the annihilation decay rates and is the basis for one method for producing Ps in vacuum.[55,56]

2.2 Positronium Formation in Vacuum

Frequently it is desirable to study Ps in vacuum, free of the perturbations introduced by gases or powders. Positronium formed near the surface of a layer of powder can escape into the vacuum[55,56] where it subsequently has no collisions. Although a thin powder layer makes Ps with kinetic energies of the order of 1 eV, Ps from deep inside a powder layer has been found to be thermalized[53] and is a possible source of 4 K Ps.[57] A similar powder layer has been used to make thermal muonium (the $\mu^+ - e^-$ bound state) at room temperature.[58,59] It was shown by Canter, Mills, and Berko that beams of slow e^+ can be used to form Ps with $\sim$1 eV kinetic energy efficiently at the surface of a solid target in vacuum without the background radiation associated with radioactive sources.[42] Another important step in the development of vacuum Ps sources was the realization that thermal Ps can also be emitted from a surface.[60–64] Extensive work on the interactions of positrons with surfaces has led to the development of intense sources of thermal Ps.[19,20]

Presently, a vacuum Ps source begins with a source of energetic positrons, either a radioactive source or a pair production source at an electron accelerator beam dump. A moderator is used to reduce the positron energy spread from MeV to eV widths. Once the brightness of the positron source has been increased by the moderator, the positrons are transported by magnetic fields or electrostatic lenses to a secondary target where thermal Ps can be made with high efficiency. Slow positron beams (see Fig 2) are now available with intensities as high as $10^8 sec^{-1}$,[65,66] or in pulses of 10^5 positrons in bunches a few nsec long,[66,67] or with a spot size as small as 10 μm,[68] or with an energy width as small as 27 meV.[69,70] We have yet to combine all of these qualities in a single beam, the purpose of which would be to allow the study of systems

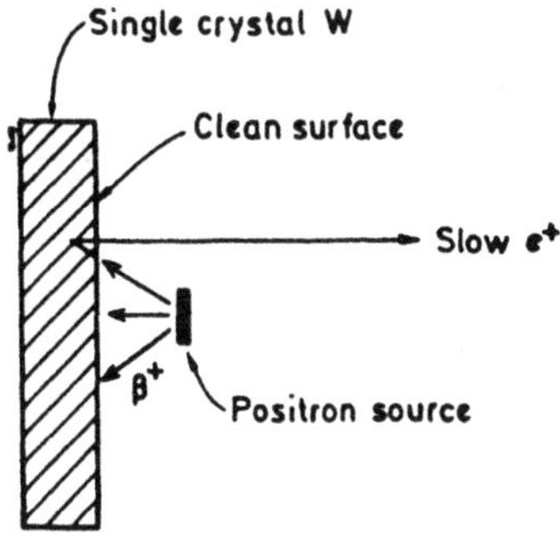

Fig. 2 Production of slow positrons. Energetic positrons from a radioactive source or from pair production are stopped in a clean single crystal moderator, typically made of W or Cu. Some of the positrons can diffuse to the surface where they are ejected into the vacuum because of their negative affinity for the metal.

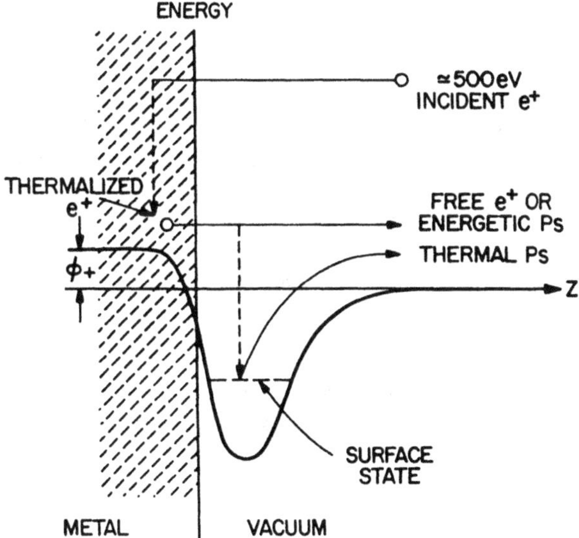

Fig. 3 Schematic representation of some of the processes that occur when slow positrons are near a surface.

containing many interacting positrons.[71]

Progress in developing low energy sources of positrons in vacuum has been possible because of our increasing understanding of positron interactions with surfaces.[19,20] For example, a positron implanted into a metal quickly thermalizes with the electrons and can diffuse back to the surface over a distance of about 10^3Å. At the surface the positrons see a dipole potential that is caused by the electron gas spilling out beyond the ion cores. This potential that holds electrons in pushes positrons out, giving positrons a negative affinity for some metals. Positrons with a thermal energy spread and a total energy of a few eV are thus emitted from certain metal surfaces exposed to energetic positrons from a radioactive source. The fraction of the implanted positrons that are reemitted from the surface is approximately the ratio of the thermalized positron diffusion length to the implantation depth of the energetic positrons in the solid. Yields of up to 0.2% have been obtained using a Co^{58} positron source and a well-annealed clean single crystal tungsten slow-positron moderator in ultra-high vacuum.[71,72] Even larger yields have been obtained using a solid Ne moderator. In a rare gas solid, the mean free path of a positron becomes hundreds of angstroms when T becomes less than $E_{exc}=10-20$eV. Low energy positrons may thus be emitted, not because of a negative affinity, but because they have a long diffusion length and can escape while T is still several eV.[73]

Although the number of slow positrons is much less than the total flux of fast positrons from the source, there are many advantages in having a beam of slow particles. 1) At the distant target, the background from nuclear γ-rays is very much reduced. 2) The positrons may be injected into a sample with controlled energy to achieve different conditions. For example, the moderator material is chosen to maximize the positron diffusion and emission probabilities. Since the target is separate from the moderator, it may be chosen to produce the most Ps. 3) Since the target is in vacuum, the Ps so produced suffers no collisions.

Positron beams are governed by the same principles used for electron optics, but extraordinary care is needed to avoid the loss of flux associated with the use of aberration-reducing beam stops. A trivial positron beam uses the uniform magnetic field of a solenoid to transport the positrons, and transverse electric fields to bend their paths by ExB drift. An electrostatically transported beam will introduce no Zeeman shifts in a precision experiment, and also facilitates the manipulation of the beam phase space. Remoderation of the energetic positrons of a focussed beam can be used to increase the beam brightness by many orders of magnitude[74] for use as a positron microscope.[68] A positron beam focussed to a 1 micron spot will also find use in the study of Ps at high density, and possibly for making a bright beam of cold Ps atoms.[75]

Since the production of cold Ps is important for precision experiments, we will now consider in detail one production means, the thermal desorption of surface positrons.

It has been known for some time that positronium atoms are formed with high probability when a metal surface is bombarded with slow positrons and that the probability of the Ps formation increases when the surface is heated[42]. Studies carried out under ultra-high vacuum conditions showed that the Ps is formed from positrons implanted into a sample at relatively high energies which then thermalize and diffuse to the sample surface[61]. The Ps observed at low temperatures has a maximum kinetic energy equal to minus the Ps work function given by

$$-\phi_{Ps} = \tfrac{1}{4}R_\infty - \phi_- - \phi_+, \tag{1}$$

where ϕ_- and ϕ_+ are the electron and positron work functions of the metal, and $\frac{1}{4}R_\infty$ is the positronium binding energy. For an Al(111) sample this Ps has a velocity distribution consistent with the metal being left in a one-hole excited state due to the sudden removal of an electron from the solid.[76]

At elevated temperatures, however, a different Ps formation occurs in which the fraction f of the incident positrons forming Ps has a temperature dependence characteristic of a thermally activated process.[60–63] The velocity distribution of the thermally activated Ps is consistent with the Ps being thermally desorbed from the sample.[64] This phenomenon is explained by a model in which positrons bound in their "image" potential well at the surface are thermally desorbed as Ps when sufficient energy is supplied by thermal fluctuations. Thermal activation measurements have been analyzed using thermodynamic arguments to obtain the positron surface state binding energies, E_b, and estimates of the positronium formation rate constants.

The scenario for the trapping of positrons in surface states is shown schematically in Fig 3, and is outlined as follows. Slow positrons penetrate into a metal and quickly lose their energy by plasmon and phonon scattering. While some of the positrons annihilate with electrons in the bulk crystal, most of them diffuse to the surface where they encounter the surface dipole layer that may give the positron a negative work function. As the positron leaves the surface it sees an effective potential well due to a modified image potential at large distances from the metal surface and electron correlation at short distances. Either non-thermal free positrons or positronium will emerge, or else the positron may fall into the surface well as a result of inelastic collisions. Once in the well the positron can either annihilate, principally into two γ rays, or else escape from the surface as free Ps if sufficient energy is available from thermal fluctuations.

The fraction f of Ps produced from an incident positron is typically 0.5 at room temperature and increases to nearly unity at elevated temperatures. The temperature dependence of f fits an activation curve that can be described by a Ps formation rate of the form $z = z_0\exp(-E_a/kT)$, where the activation energy E_a is dictated by energy balance

$$E_a = E_b + \phi_- - \tfrac{1}{4}R_\infty, \tag{2}$$

where $E_b > 0$ is the binding energy of the positron at the surface. If we let f_0 denote the fraction of positrons that directly form Ps in the low temperature limit, and f_s be the fraction that becomes trapped in the surface well, then the fraction f_t of positrons that are thermally desorbed from the surface as free positronium will be

$$f_t = \frac{z f_s}{\gamma + z} = \frac{z_0 e^{-E_a/kT} f_s}{\gamma + z_0 e^{-E_a/kT}}, \tag{3}$$

where γ is the 2γ annihilation rate of the positron surface state which has a unique lifetime characteristic of a low electron density.[77,78] The fraction f_t can be related to the experimentally determined high-temperature limit $f_\infty = f_0 + f_t$ at $T \rightarrow \infty$. If we assume that $\gamma << z_0$, and f_s is only a weak function of temperature, the Ps formation fraction f is

$$f = f_0 + f_t = \frac{f_0 + (z/\gamma) f_\infty}{1 + (z/\gamma)}. \tag{4}$$

If we may consider that the surface positrons form a classical 2D gas, thermodynamic arguments[79] show that the factor z_0 is given by

$$z_0 = \frac{4kT}{h}(1 - \bar{r}), \tag{5}$$

where $r(v_{Ps})$ is the reflection coefficient for a positronium atom of velocity v impinging on the surface, and $\bar{r}$ is an average over a thermal distribution of velocities. Essentially, $(1 - \bar{r})$ is the analogue of the familiar emissivity of a body emitting thermal radiation.

The above model has been tested by time of flight measurements of the energy distribution of Ps thermally desorbed from Al(111) surfaces.[80] The thermally desorbed Ps energy spectrum from Al(111) is an exponentially decreasing function of energy (see Fig 4) and is consistent with Ps having a velocity-independent reflection coefficient, contrary to earlier measurements on the Cu(111)+S surface[64]. The rate coefficient deduced from the Al(111) thermal activation[80] measurements (Fig 5) turns out to be consistent with a surface state model due to Platzman and Tzoar[81] in which the entity bound at the surface is a positronium atom rather than a bare positron. If this model is correct, the data implies that the reflection coefficient is in fact zero at thermal energies, and that the Al(111) surface is a "black body" for Ps emission.

The important finding in the present context is that the Ps energy spectrum extends to zero perpendicular energy without diminution, and thus the low energy tail of the Ps beam-Maxwellian distribution is of sufficient intensity to be of use in precision experiments requiring low velocities. It is known that the activation energy for thermal desorption of Ps may be tuned by the addition of a partial monolayer of a material that will change the surface potential.[63,82,83] It is thus possible to cause Ps to

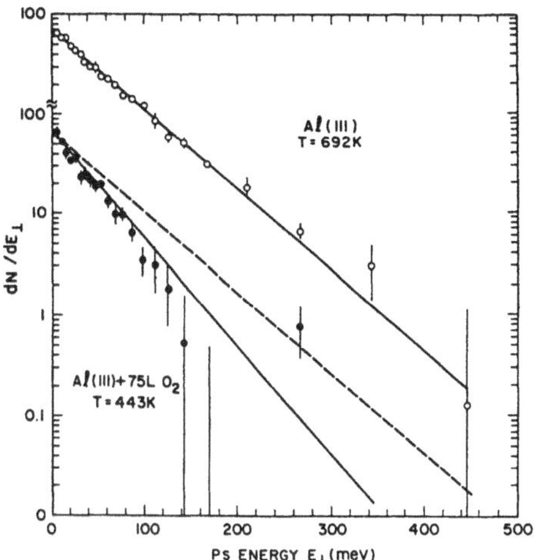

Fig. 4 Spectra of the perpendicular energy component of the positronium thermally desorbed from Al(111) surfaces at two different temperatures. The fitted exponential curves imply positronium temperatures of 464(51)K and 636(64)K for the two cases. [From Ref 80]

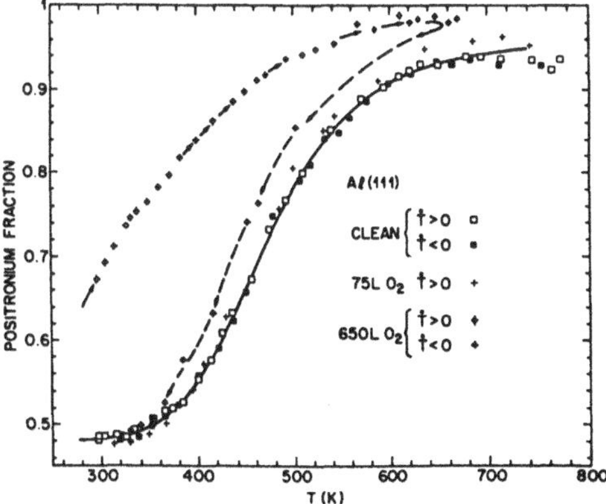

Fig. 5 Positronium fraction *vs.* temperature for clean and oxygen-exposed Al(111) surfaces. The solid line is a fitted activation curve for the clean Al(111) data. [From Ref 80]

be desorbed at temperatures below 300K. Recent experiments[84] show that the Ps energy spectrum for Al(111) with a partial layer of oxygen retains its beam-Maxwellian shape and has a significant intensity at temperatures as low as 230 K. Work is in progress to find out how low the activation energy may be tuned without causing a significant loss of the low energy component. Possible loss mechanisms would include an energy dependence of the reflection coefficient which might tend to make $r \rightarrow 1$ as $v \rightarrow 0$, or an unevenness of the oxygen surface coverage causing a reduction of the thermal component of the Ps.

3. MEASUREMENTS OF THE PHYSICAL PROPERTIES

3.1 Positronium Mass

Measurements of the total energy of the annihilation γ-rays imply that the total energy of Ps is very nearly the same as that of two electron masses. Unfortunately, a scheme for a precision bent-crystal determination of $m_e c^2$ using Ps formed in He vapor at 5 K has not come to pass.[85,86] The inertial mass of Ps has not been measured directly with very great precision either, but we may infer from the agreement of the Ps time-of-flight energy spectrum in Fig 4 with a beam-Maxwellian that M_{Ps} is within 10% of being equal to two electron masses.[80] The present agreement between theory and experiment of the $1^3S_1 - 2^3S_1$ interval implies that the Ps Rydberg constant and the hydrogen Rydberg constant are in agreement, and thus that the positron and electron inertial masses are equal to within a roughly 40 ppm (1 σ) level of uncertainty.[28] A brief discussion of the possible measurement of the passive gravitational mass of Ps will be given in Section 4.4.

3.2 Annihilation Rates

A unique feature of the particle-antiparticle pair atoms is that they annihilate. Positrons and electrons annihilate into one or more photons, the total energy of which equals the sum of the rest energies $2m_e c^2$, less the total binding energy, plus the kinetic energy of the pair, and less any energy taken up by recoiling particles. In vacuum, conservation of C-parity requires that the singlet state of Ps decay into an even number of photons, and the triplet state into an odd number. Tests of C-parity non-conservation by searches for the violation of this rule[87,88] are hopelessly less sensitive than required to observe effects predicted by the standard model of the weak interaction.[89] For further discussion, see section 4.1.

The decay rates, λ_s and λ_t, of singlet and triplet ground state Ps have been measured to a precision sufficient to test the first order corrections to the calculated rates by Gidley, Rich, Sweetman, and West,[21] and by Westbrook, Gidley, Conti, and Rich.[22] Positronium is formed in a gas in a uniform magnetic field using positrons from a ^{68}Ge or ^{22}Na radioactive source. [Fig. 6] The start signal is generated by the fast positrons passing through a thin scintillator that is coupled to a photomultiplier

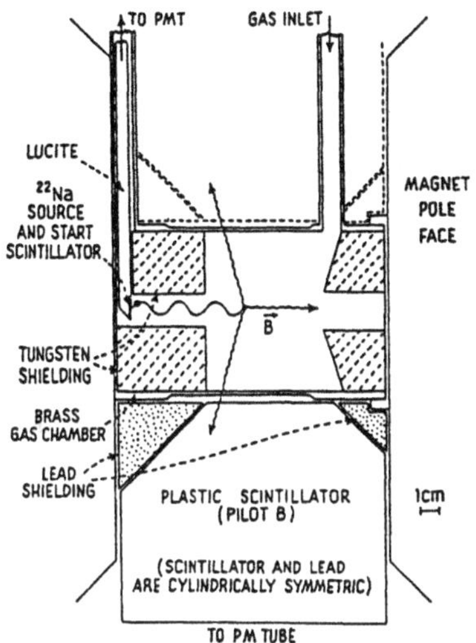

Fig. 6 Positronium formation chamber and detector arrangement. [From Ref 97]

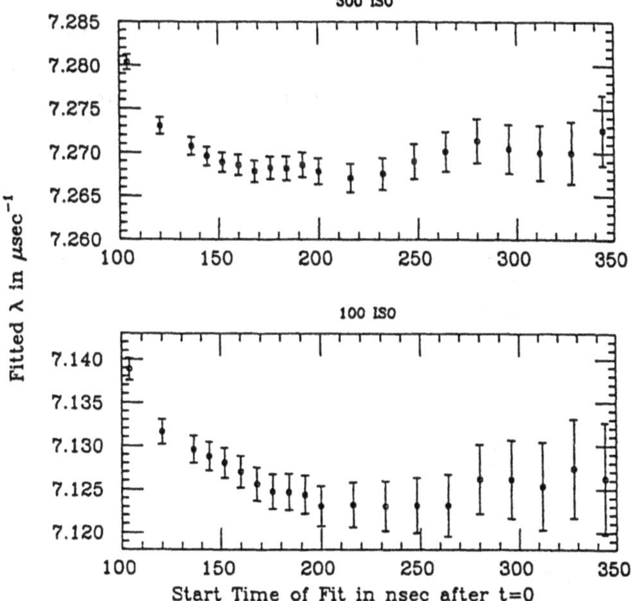

Fig. 7 Fitted decay rates for Ps formation in 100 and 300 torr of isobutane. [From Ref 97]

tube. The annihilation gamma rays are detected by an annular plastic scintillator. The time interval between the birth of the positron and its annihilation is measured by time-to-digital converters that are calibrated using precision quartz oscillators. Histograms of numbers of counts vs time delay, containing more than 10^8 events, are acquired using several types of gas at various densities. The decay curves are fitted to exponential forms that include the effect of the non-zero stop rate.

The magnetic field confines the positrons to the axis of the apparatus so that the volume in which Ps is formed is not too dependent on gas pressure. The triplet $|m|=1$ states are unaffected by the field, but the m=0 state decays at a more rapid rate λ_t' due to mixing with the singlet m=0 state. >From the theory of positronium in a magnetic field[90] Gidley *et al.*[21] compute λ_s from their measurements of λ_t'. Writing $\underline{\lambda}=\frac{1}{2}(\lambda_s-\lambda_t)$, and $\bar{\lambda}=\frac{1}{2}(\lambda_s+\lambda_t)$, we have[91,92]

$$\lambda_t'=\bar{\lambda}-\sqrt{\tfrac{1}{2}\sqrt{(1-\underline{\lambda}^2+x^2)^2+4\underline{\lambda}^2}-\tfrac{1}{2}(1-\underline{\lambda}^2+x^2)} \tag{6}$$

In this expression $x=2g'\mu_0 B_z/\Delta\nu$, $g'=g_e(1-(\frac{5}{24})\alpha^2)$ is the bound state g-factor of an electron in positronium,[93–96] and B_z is the magnetic field along the z axis.

Lack of thermalization of the positronium could possibly cause systematic effects, but by waiting for about 150 nsec the decay rate approaches an asymptotic value[97] that appears to be representative of thermalized Ps. [Fig 7] The most serious systematic effect in the measurement is caused by the increase in the decay rate due to the collisions of the Ps with the gas. An extrapolation of λ_t to zero gas pressure is shown in Fig 8. The singlet Ps decay rate is measured to be 7.994(11) ns^{-1} in agreement with the Harris and Brown[98] calculation with an $\alpha^2\ln\alpha$ contribution calculated by Caswell and Lepage:[99]

$$\lambda_s=\tfrac{1}{2}\alpha^5\frac{mc^2}{\hbar}\left[1-\alpha(\frac{5}{\pi}-\frac{\pi}{4})-\frac{2}{3}\alpha^2\ln\alpha+\cdots\right]=7.986654(1)ns^{-1}. \tag{7}$$

The theoretical error indicated includes only the contributions from the 0.045 ppm uncertainty in the fine structure constant,[100] $\alpha^{-1}=137.0359895(61)$. The triplet decay rate is 7.0516(13) μs^{-1} in significant disagreement with theory:[99,101,102]

$$\lambda_t=\alpha^6\frac{mc^2}{\hbar}\frac{2(\pi^2-9)}{9\pi}\left[1-(3.273\pm0.001)\alpha+\frac{1}{3}\alpha^2\ln\alpha+\cdots\right]=7.03831(7)\mu s^{-1}. \tag{8}$$

In this case the uncertainty is from the numerical integrations. The experiment is in agreement with less precise determinations using positronium formation in SiO_2 powder[56] and in vacuum.[103] The reason for the discrepancy with theory is not known at this time. One possibility is that the coefficient of the order α^2 term in the theoretical expansion is an order of magnitude larger than one might guess, namely 35±4. Non-QED mechanisms that would make the Ps decay faster cannot be ruled out, but decay into light, neutral, hitherto undetected particles such as axions does

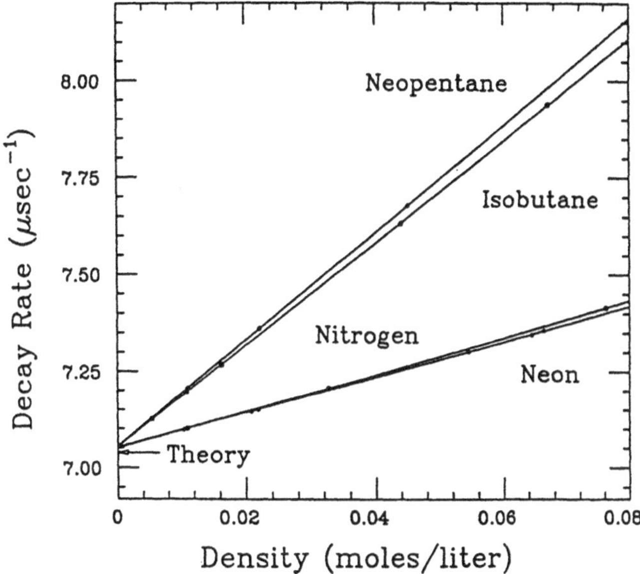

Fig. 8 Plot showing extrapolations of the triplet decay rate in four gases. [From Ref 97]

not appear likely.[104]

Singlet and triplet positronium can also decay into four and five photons respectively. The rates have not been measured, but the branching ratios of these higher order processes are calculated to be about $0.01\alpha^2$, or about one part per million,[105,106] too small to have any significant effect on the measured decay rates. Also see section 4.2 on exotic decays.

3.3 Positronium Energy Levels

The gross structure of the Ps energy levels is similar to H, except that the reduced mass of Ps is $\frac{1}{2}m_e$ instead of approximately m_e. The bound state energies are thus about half of the corresponding energies in H. (See Fig 1.) The fine structure of the Ps bound states is qualitatively different from that of H because the magnetic moment of the positron is about 10^3 times greater than that of the proton. Thus the ground state hyperfine interval is 203GHz for Ps, whereas it is only 1.4GHz in H, the latter being the source of the famous 21cm line of radio astronomy. Similarly, the Lamb shift in H represents a purely QED energy splitting between a 2S and a 2P state that are perfectly degenerate when described by the Dirac equation for an electron in a Coulomb potential; in Ps the analogous levels are split by much more than the Lamb-shift terms.

3.3.1 Hyperfine Interval

Shortly after his discovery of positronium, M. Deutsch and collaborators made a measurement of the energy difference between the singlet and triplet ground states[107] with an accuracy sufficient to confirm the existence of radiative corrections. Whereas the analogous hyperfine splitting in H is one of the most accurately known quantities in all of physics, primarily because of the long lifetime of the hyperfine transition and the availability of atomic H, the Ps hyperfine interval is difficult to measure because of the short $^1S_0$ lifetime and the problems of producing the Ps atoms in quantity. Nevertheless, the Ps hyperfine interval has now been measured to a remarkable accuracy in a series of experiments done principally by Hughes and collaborators at Yale.[23,24]

The Zeeman levels of positronium are shown in Fig 9. Since the hyperfine interval is at an inconvenient frequency, all measurements of this quantity have depended on the measurement of the triplet m=0 to m=1 splitting in a magnetic field, *ie.* the transition f_0 in Fig 9. Positronium is formed in a microwave cavity in a uniform magnetic induction B. When the resonant condition of B and f_0 occurs, there is an increase in the 2γ annihilation yield because the m=1 states, ordinarily decaying only via 3γ's, are mixed with the triplet and singlet m=0 states by the rf magnetic field. Since large rf fields are required, the experiments use a tuned microwave cavity, and one moves through the resonance condition by varying the magnetic field, as shown in Fig 10. Using the Breit-Rabi formula,

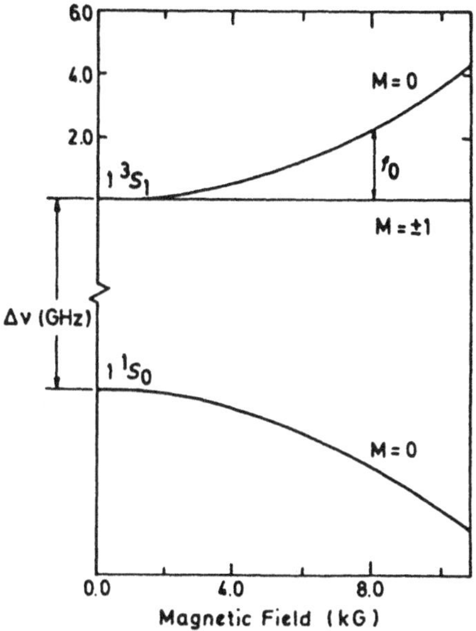

Fig. 9 Zeeman energy levels for positronium in the ground state. [From Ref 23]

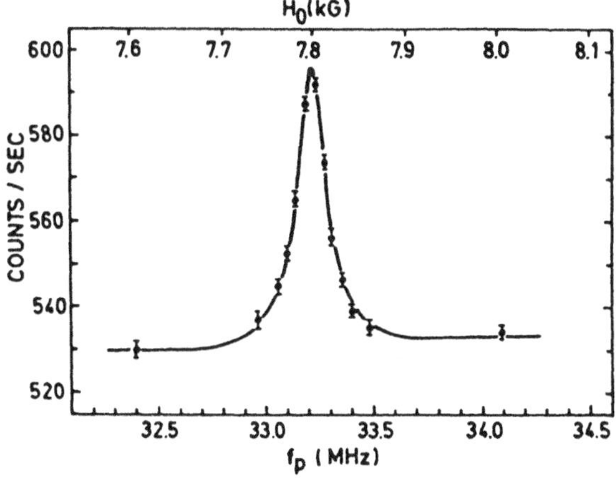

Fig. 10 Typical Zeeman resonance used to determine the positronium hyperfine interval. [From Ref 23]

$$f_0 = \tfrac{1}{2}\Delta\nu\left[(1+x^2)^{\frac{1}{2}}-1\right], \tag{9}$$

with x defined as in Eq. 6, one may derive a value for the hyperfine interval from the position of the peak of the resonance:

$$\Delta\nu = (g'/g_e)^2(\mu_e/\mu_p')^2(f_p^2/f_0) - f_0, \tag{10}$$

where f_p is the proton resonance frequency at the line center value of the magnetic field, and μ_e/μ_p' is the ratio of the electron magnetic moment to that of the proton in a spherical sample of water. If a spherical water sample is not being used, then corrections for the geometry of the NMR sample and its diamagnetic environment must be made. Since the magnetic field enters as the square in Eq 10, its homogeneity must be established with great precision. The field map from Fig 9 of Ref 23 shows that B in this measurement was constant to better than ± 1 ppm.

It is to be noted that Eq 10 represents a simplification that is not necessarily justified at the few ppm precision level attained by the $\Delta\nu$ measurements. Decay terms may lead to energy shifts that must be accounted for accurately.[108] A fully interacting model of positronium with its decay channels, many sublevels, and virtual radiative interactions would be prohibitively complicated. In analyzing Ps hyperfine interval measurements, experimenters have assumed that only the four ground state sublevels are important, and that decay and radiative corrections to the energy levels may be introduced by small complex additions to the unperturbed energy levels. Thus Ps in a magnetic field is described by a four component state vector, and its interactions by a 4x4 non-Hermitian Hamiltonian.[90] An effective Hamiltonian is a good approximation if the decay energy, in this case of order mc^2, is large compared to the energy splittings.[109,110] While this condition is amply satisfied by the Ps ground state, care must be exercised in the actual calculations because a complex Hamiltonian does not necessarily have orthogonal left and right eigenvectors. A calculation of the Ps hyperfine resonance in the effective Hamiltonian approximation shows that the resonance peak position differs from the Breit-Rabi prediction Eq 10 by only 0.52 ppm, but that the line shape is slightly asymmetric.[111–113] If a Lorentzian line shape is used to fit the measurements, the value deduced for $\Delta\nu$ will be about 2 ppm low. The most precise experiment includes an analysis of the data using the full line shape.[23]

A substantial shift in the resonance occurs due to Ps collisions with the molecules of the buffer gas used to form the Ps. Since the theory of the Ps hyperfine pressure shift is not sufficiently advanced,[114] the systematic shift is removed by extrapolating to zero gas density. Fig. 11 shows the latest result of the Yale group,[23] with an extrapolated 3.6 ppm value for $\Delta\nu$,

$$\Delta\nu(1^3S_1 - 1^1S_0) = 203.38910(74)\text{GHz}, \tag{11}$$

in agreement with the earlier 8 ppm Brandeis[24] measurement, corrected for the line shape asymmetry.[112]

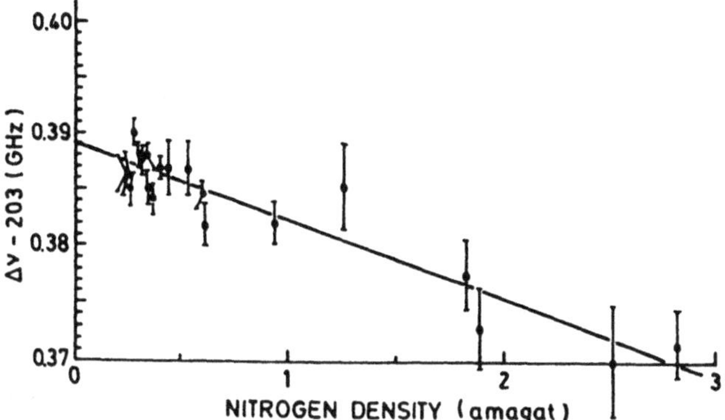

Fig. 11 Extrapolation of the positronium hyperfine interval measurements to zero buffer gas density. [From Ref 23]

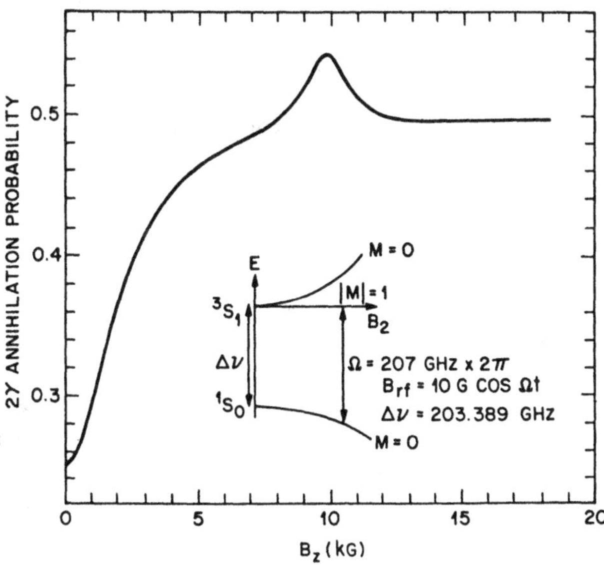

Fig. 12 Calculated 2γ annihilation probability for positronium in a variable magnetic field in the z-direction and excited by an rf magnetic field in the x-direction with constant frequency.

Unfortunately, the theory is very difficult, even though Ps is seemingly so simple. In spite of advances in the understanding of the Bethe-Salpeter equation describing the equal mass two body system,[115] the $\alpha^4 R_\infty$ corrections to $\Delta\nu(1^3S_1-1^1S_0)$ are incomplete, leaving us with essentially the original Karplus and Klein[116] calculation good to order $\alpha^3 R_\infty$:

$$\Delta\nu=\alpha^2 cR_\infty\left[\frac{7}{6}-\frac{\alpha}{\pi}(\frac{16}{9}+\ln 2)-\frac{5}{12}\alpha^2\ln\alpha+\cdots\right]=203.40029(2)GHz. \tag{12}$$

The error in the theory includes only the 0.09 ppm uncertainty in α^2. The next higher order terms are of order 10 MHz.

One might ask if there is any possibility for substantial improvements in the measurement. The real difficulty is the ≈0.5% width of the resonance in Fig 10. While counting statistics was the primary limiting factor in all the measurements, no experiment has yet made use of the slow positron beam. The pulsed nature and high flux of electron accelerator based beams[66,67] would give us a low background and sufficient Ps atoms to permit the use of line-narrowing techniques.[13] The magnetic field inhomogeneities that limit the Zeeman resonance experiments to a 1ppm uncertainty[23] could be overcome by exciting the hyperfine interval directly. [See Fig 12.] To avoid tuning the microwaves, the experiment would have to be done in a small magnetic field to permit Zeeman tuning; however, the precision and uniformity of that field could be an order of magnitude poorer than in the usual Zeeman resonance experiment.

3.3.2 Positronium n=2 Fine Structure

The production of excited states of positronium and the measurement of its fine structure was a long standing goal[117] finally reached in 1975 thanks to the introduction of the slow positron beam.[118–120] The discovery that positronium can be formed in vacuum by slow positrons impinging on a metal surface[42] made it seem likely that a small fraction of the atoms would be formed in excited states. Whereas any Ps formed in excited states in the presence of a dense buffer gas would be rapidly quenched, the excited states of Ps in vacuum would have a relatively long lifetime in which to perform spectroscopic measurements. Fig 13 shows the slow positron beam apparatus for observing Ps Lyman-α radiation.[43] Slow positrons from a MgO moderator[121] were guided by a bent solenoid to a target that is viewed by a uv-sensitive photomultiplier behind one of three interference filters centered on the expected Lyman-α wavelength of 2430Å and at 30Å on either side. Fig 14 shows that there is a significant increase in the uv detector count rate detected in coincidence with an annihilation γ-ray when the filter is chosen to be at the Lyman-α wavelength. The signal goes away when the positron energy is increased from 25eV to 400eV. When the positrons are implanted at higher energies, they diffuse back to the surface with a more nearly thermal energy distribution.[60] Evidently, formation of the 2S state

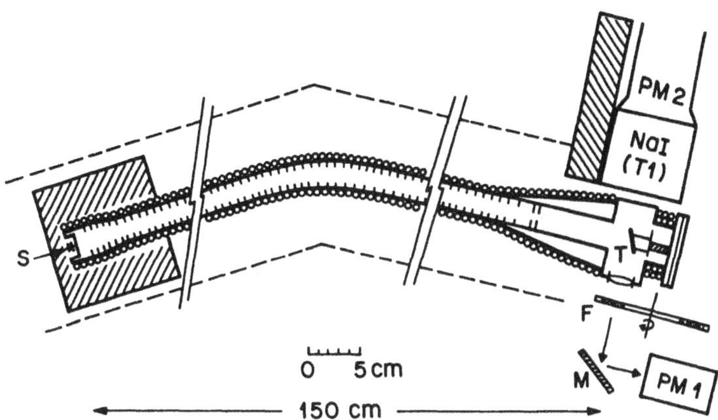

Fig. 13 Slow positron beam apparatus for observing positronium Lyman-α radiation. S, ^{58}Co source; T, target; F, ultraviolet filter wheel; M, mirror; PM1, Lyman-α photon detector; PM2-NaI(Tl), annihilation γ-ray detector. The slow positrons were produced by a MgO powder moderator and guided to the target in vacuum by a long bent solenoid. [From Ref 43]

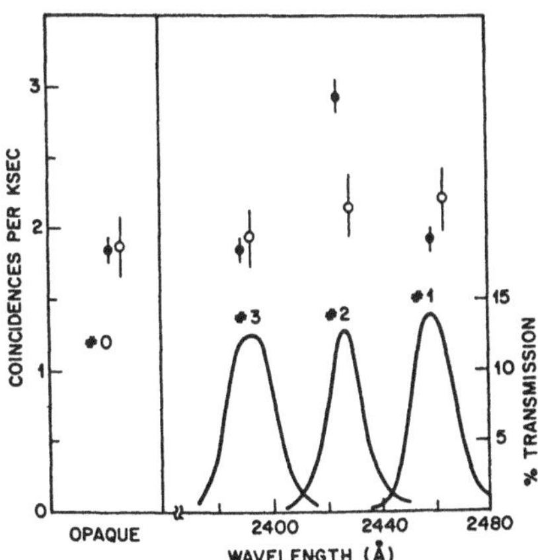

Fig. 14 Photon detector-annihilation γ-ray coincidence count rate versus uv filter. There is a Lyman-α signal at 2430Å with 25eV positrons (solid circles), but no signal with 400eV positrons (open circles). [From Ref 43]

requires positrons to reach the surface with several eV of kinetic energy.

The yield of 2S positronium in this experiment was about 10^{-4} 2S atoms per incident positron, or about 10 sec^{-1}, sufficient to allow measurement of a fine structure interval in the n=2 state.[25] The $2^3S_1-2^3P_2$ interval was chosen for the measurement because it was predicted to lie in the convenient X band and involves the relatively long-lived triplet S state (see Fig 1). The measurement was based on the observation of an enhanced Lyman-α emission when $2^3S_1-2^3P_2$ transitions were induced by an rf electric field at the resonance frequency. Fig 15 shows the target of the slow positron beam apparatus used in the measurement. The positrons impinge on the back of a small microwave cavity. Lyman-α photons are detected in coincidence with 1S annihilation gammas, and the count rate is measured as a function of the microwave frequency. The resonance and its derivative shown in Fig 16 lead to a value for the fine structure interval $\Delta\nu(2^3S_1-2^3P_2) = 8.628(5)\text{GHz}$.

The energies of the triplet n=2 levels of positronium have been calculated to order $\alpha^3 R_\infty$ by Ferrell[122] and Fulton and Martin:[123]

$$E(2^3S_1)=\frac{1}{8}R_\infty\left\{-1+\frac{65}{192}\alpha^2-(\frac{3}{2\pi})\alpha^3\ln\alpha-0.8995\alpha^3+\cdots\right\} \tag{13}$$

$$E(2^3P_2)=\frac{1}{8}R_\infty\left\{-1-\frac{43}{960}\alpha^2+0.0057\alpha^3+\cdots\right\} \tag{14}$$

$$E(2^3P_1)=\frac{1}{8}R_\infty\left\{-1-\frac{47}{192}\alpha^2-0.0315\alpha^3+\cdots\right\} \tag{15}$$

$$E(2^3P_0)=\frac{1}{8}R_\infty\left\{-1-\frac{95}{192}\alpha^2-0.0978\alpha^3+\cdots\right\}. \tag{16}$$

The $2S_1-2P_J$ intervals ν_J are

$$\nu_2=8625.2MHz \tag{17}$$

$$\nu_1=13010.9MHz \tag{18}$$

$$\nu_0=18496.1MHz. \tag{19}$$

The uncalculated higher order terms are expected to contribute several MHz.

The positronium "Lamb-shift" experiment has been extended by Hatamian, Conti and Rich[26] to include all three triplet 2S-2P intervals. As in the earlier experiment, the 2^3P_J states were detected by the emission of Lyman-α photons. An electrostatically guided beam of positrons was focused onto a polycrystalline molybdenum target in a section of waveguide carrying the rf electric field. A fraction 3×10^{-4} of the positrons were emitted from the target in the n=2 states of Ps. The improvements in the measurement included the use of a more stable microwave oscillator, a traveling wave to excite the transitions, and a zero magnetic field which eliminated motional stark shifts in the transitions. A simultaneous analysis of all three transitions shown in Fig 17 yielded

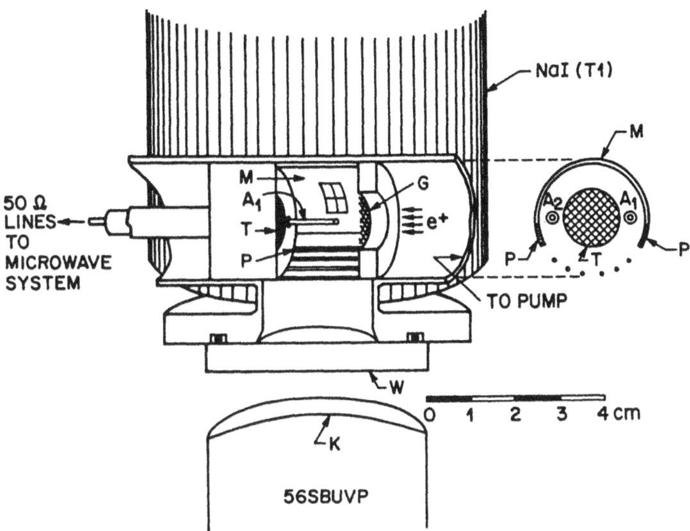

Fig. 15 Positron target chamber and microwave cavity for measuring fine structure interval in the n=2 state of positronium. G, grid; T, copper target; M, mirror; W, window; K, CsTe photocathode of the uv photon detector; A, antennae; NaI(Tl), annihilation γ-ray detector. [From Ref 25]

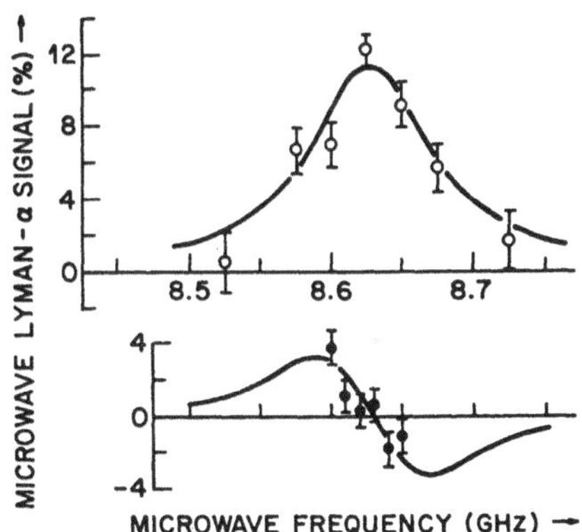

Fig. 16 Lyman-α signal and its logarithmic first difference as a function of microwave frequency. The inset is the schematic term diagram for the n=1 and n=2 Ps states indicating the relevant transitions and the lifetimes for each level. [From Ref 25]

$\nu_2 = 8619.6 \pm 2.7 \pm 0.9$ MHz,
$\nu_1 = 13001.3 \pm 3.9 \pm 0.9$ MHz, and
$\nu_0 = 18504.1 \pm 10.0 \pm 1.7$ MHz.
The first error is statistical, and the second is an estimate of the systematic uncertainties. The experiment is in agreement with Eqs 13-19, and the ν_2 measurement agrees with Ref 25 and is about twice as precise.

The natural line width of the triplet 2S-2P fine structure intervals in Ps is 50MHz due to the 3.2 nsec 2^3P radiative lifetime. If the hyperfine interval measurements are any guide, we may expect that eventually one could improve the fine structure measurements by at least a factor of 100 given a sufficient count rate. While there have been many improvements in positron beam techniques in the last ten years, and small increases in the efficiency of n=2 state production,[124] these have not yet resulted in an improved measurement.[125]

3.3.3 1S-2S Interval

The largest QED level shifts combined with the narrowest line widths are to be found in the triplet S states, with the shifts scaling roughly as $1/n^3$. A measurement of the $1^3S_1-2^3S_1$ interval is especially attractive. The 142 nsec and 8×142 nsec lifetimes of the n=1 and 2 triplet states introduce a 1.26 MHz line width to the transition, and since the 2S state is optically metastable, there is no further broadening of the natural line width. On the other hand, in the hyperfine and fine structure measurements, the comparison 1^1S_0 or 2^3P level has a large width.

Measurement of the 1S-2S interval permits one to use the two-photon Doppler-free technique first introduced by Vasilenko, Chebotaev, and Shishaev.[126] The light mass of the atom and concomitant high velocity would normally introduce a Doppler broadening that would be orders of magnitude more than the natural line width. By irradiating the atom with two counter-propagating laser beams of equal frequency, any first order Doppler shift of one photon due to the motion of the atom will be canceled by an equal but opposite shift in the other beam. The two-photon technique is especially suited to the spectroscopy of Ps, given the limited number of atoms available in the experiments. In other Doppler-free techniques, such as saturation spectroscopy,[127-129] the fraction of atoms excited is typically given by the ratio of the natural line width to the Doppler width, roughly 10^{-4} for room temperature Ps. On the other hand, all the atoms in the velocity distribution are in two-photon resonance (apart from second-order Doppler shifts due to time dilation) if the laser frequency is tuned to one-half the energy interval. The two-photon line shape caused by a Boltzmann velocity distribution has been discussed by Minogin.[130]

The optical excitation of Ps to the 2S state was demonstrated by Chu and Mills[27] and followed by a precise measurement of the $1^3S_1-2^3S_1$ interval by Chu, Mills, and Hall.[28] Fig 18 shows the transitions used in the optical measurements, namely the 1S-

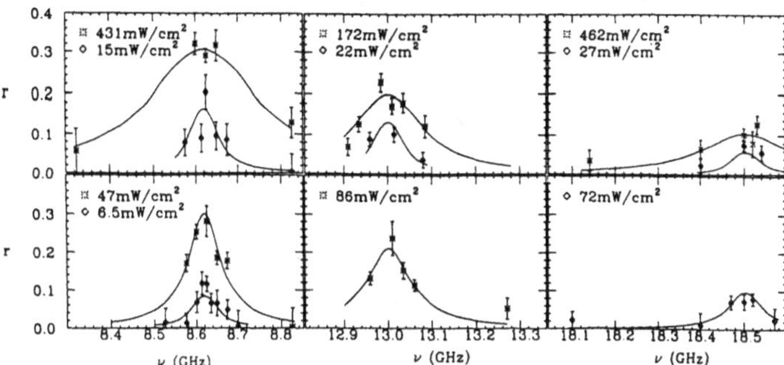

Fig. 17 Microwave-induced $2^3S_1 - 2^3P_J$ signal *vs.* rf frequency for several rf intensities. A single function fitted to all the data is shown by the solid line. [From Ref 26]

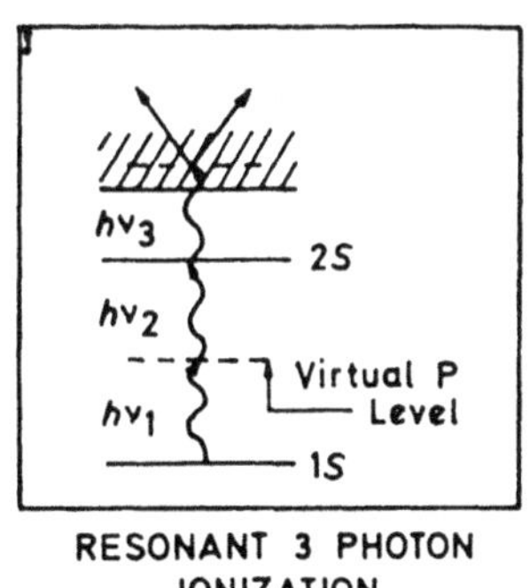

Fig. 18 1S-2S resonant three photon ionization of positronium.

2S two-photon resonant, three-photon ionization of Ps. A high energy pulsed laser was used to excite the interval, since the 1S-2S transition is highly forbidden. The Ps atoms must be formed in vacuum to avoid collisions, and this dictated the use of a positron beam. The excitation of the atoms was observed by ionization from the 2S state by another photon of the same frequency and subsequent detection of the positive ion fragments (positrons). Because of the low duty cycle of the pulsed laser, a pulsed source of Ps was developed that created a high instantaneous flux of atoms that was compatible with the 10 nsec pulse width of the laser.[131–133]

Slow positrons for the pulsed source of Ps were obtained from a ^{58}Co radioactive source and single crystal metal negative work function positron moderator. The slow positrons were injected into a "bottle" consisting of a pinched field magnetic mirror on the entrance and an electrostatic grid at the other end. Positrons that entered the bottle were prevented from leaving by exciting their transverse degree of freedom with a rf cavity driven at the cyclotron resonance frequency. After loading the bottle for a time significantly longer than its 100 μsec storage time, the low energy positrons were collected into a short pulse using a suddenly switched on harmonic potential as shown in Fig 19. With a 300 mCi ^{58}Co source, a 1 kHz repetition rate source of $\approx$100 positrons in 10 nsec pulses over an area of 0.7 cm^2 was achieved. The positrons struck a clean Al(111) target surface in ultrahigh vacuum. The target was kept at 300°C to desorb the surface state positrons as free thermal Ps in vacuum.[64,80] Yields were as high as 20 Ps atoms per pulse.

The Ps was excited from the 1^3S_1 state to the 2^3S_1 state by two counter-propagating 486nm laser pulses. As shown in Fig 20, the light was narrowed in frequency by a Fabry-Perot interferometer in the vacuum system that surrounded the Al target. Ps atoms from the target were ionized by the light and collected by an electron multiplier detector. Fig 21 shows a single 5 min scan of the Ps resonance along with a simultaneously recorded Te_2 reference line and the frequency marker signal. While the line center relative to the Te_2 reference line can be determined to 5ppb in one scan, accounting for systematic effects limited the final accuracy of the experiment to about 12 ppb.

The major systematic effects are as follows. The ac Stark shift introduced by the intense laser beam was accounted for by extrapolating the Ps resonance line center measurement to zero laser power as shown in Fig 22. The second-order Doppler shift (time dilation) due to the Ps motion relative to the laser frame of reference was accounted for by measuring the two-photon excitation frequency as a function of the Ps velocity. The velocity v of the atoms excited was chosen by delaying the laser pulse relative to the time of creation of the Ps, with the laser beams intersecting the Ps at a fixed distance from the target. The measurements were extrapolated linearly to zero v^2/c^2 as shown in Fig 23. The frequency offset between the cw dye laser oscillator and the high-powered pulsed amplifier chain was measured directly using a scanning

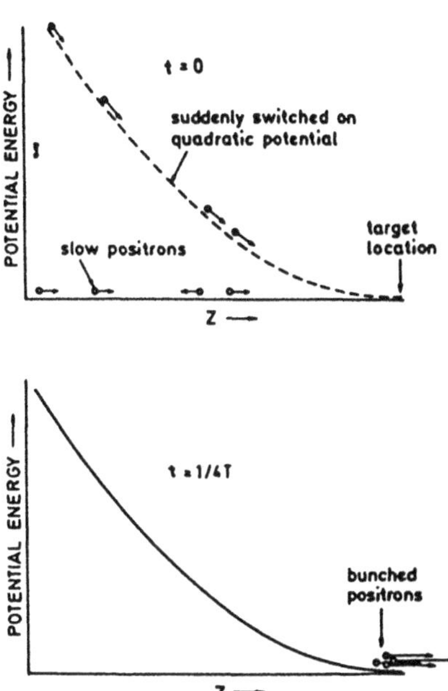

Fig. 19 Quadratic potential bunching of a slow positron beam.

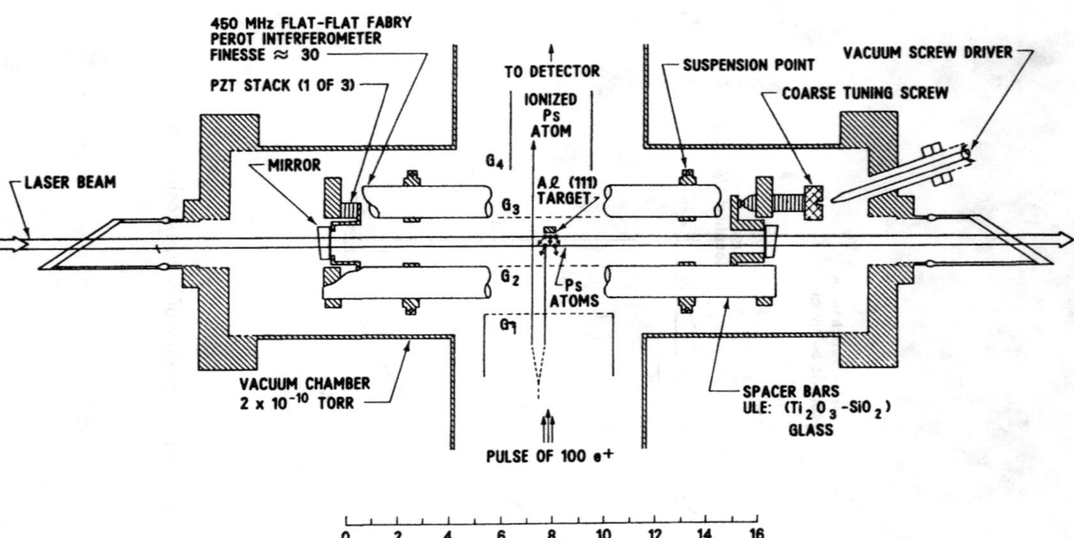

Fig. 20 Thermal positronium-laser beam interaction region. Positronium is formed by a bunch of positrons that is stopped by a clean Al surface in ultrahigh vacuum. Positronium atoms thermally desorbed from the surface are ionized by the laser and the e^+ fragments are collected by a single particle detector. The laser pulse is narrowed in frequency by the Fabry-Perot interferometer. [From Ref 28]

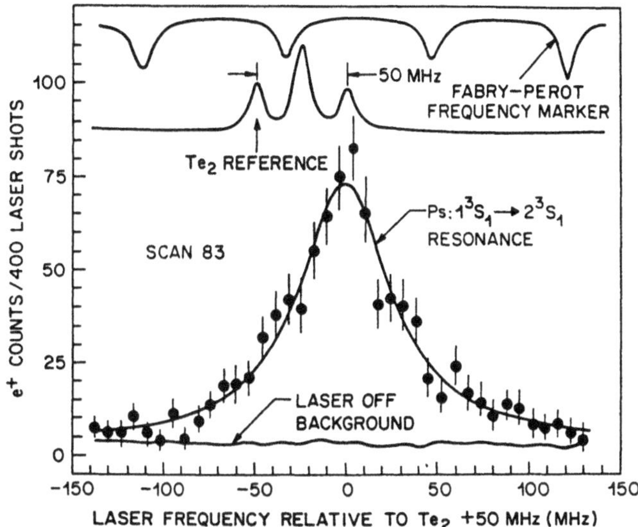

Fig. 21 Resonant three-photon ionization of positronium due to $1^3S_1 + 3h\nu \rightarrow 2^3S_1 + h\nu \rightarrow e^+ + e^-$. The Te_2 reference line has been split into three lines by acousto-optically modulating the cw dye laser at 50MHz. For this scan, the line center was 25.9±2.7 MHz above the Te_2 line. [From Ref 28]

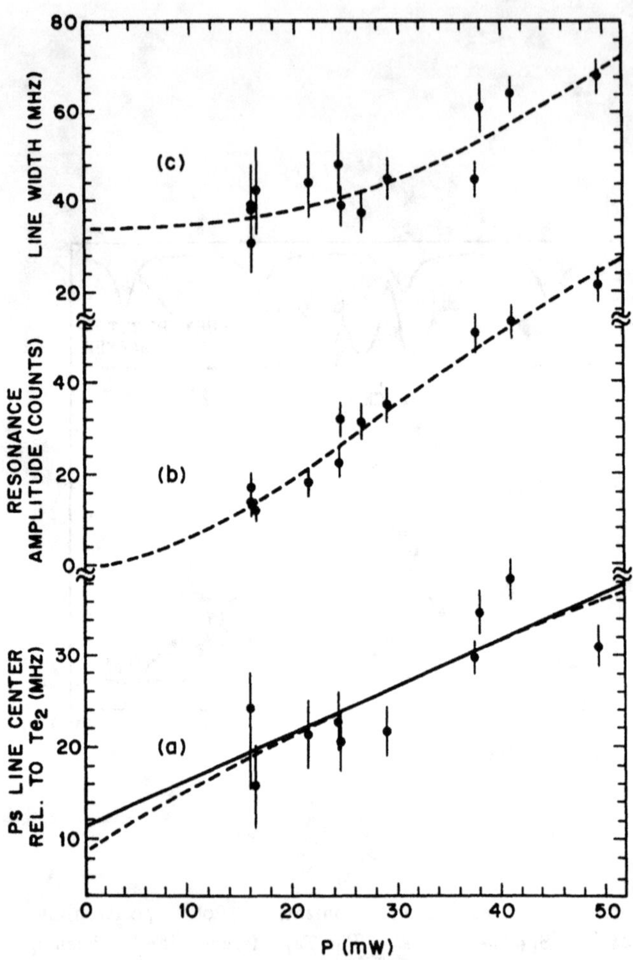

Fig. 22 Frequency shift, linewidth, and resonance amplitude plotted vs laser power transmitted through the Fabry-Perot cavity. The simultaneous fit of three theoretical curves to the data is given by the dashed lines. The simple quadratic Stark shift is shown by the solid line. [From Fig 28]

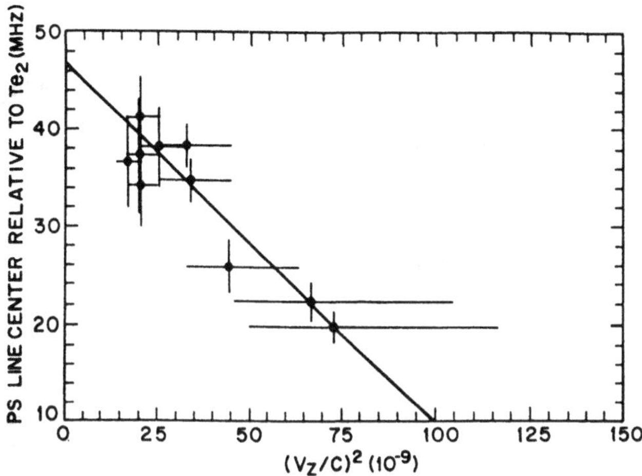

Fig. 23 Ps line center relative to the Te_2 reference line plotted vs v_z^2/c^2 showing the second-order Doppler shift. [From Ref 28]

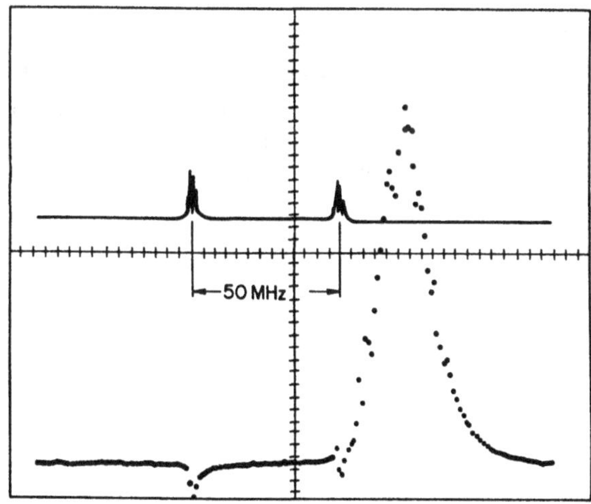

Fig. 24 Frequency offest of the pulsed laser (dotted curve) relative to the acousto-optically modulated cw laser beam (solid curve). The right-hand peak of the solid curve is due to the cw laser, and the left-hand peak is from an rf side band of the carrier frequency introduced by an acousto-optic modulator.

Fabry-Perot interferometer. The pulsed laser was believed to be about 20MHz lower than the cw frequency (Fig 24).

The measurements were reduced to an absolute value for the $1^3S_1-2^3S_1$ interval by measuring the frequency of the Te_2 reference line relative to the deuterium $2S_{1/2}-4P_{3/2}$ Balmer-β line. The absolute value of the deuterium line was obtained from the calculation by Erickson[134] updated using a newer value of the Rydberg constant.[135] The final 1984 result was

$$\Delta\nu(1^3S_1-2^3S_1)=1\ 233\ 607\ 185\pm 15\text{MHz}, \tag{20}$$

differing from the calculation of Fulton[136] by $13\pm 10.5\pm 10.6$ MHz, where the first error is from the Ps measurement relative to the Te_2 calibration line, and the second is from the determination of the absolute position of the Te_2 line.

Since the publication of the above result in 1984, two important corrections have been made. First, the molecular tellurium reference line has been recalibrated in a separate measurement by McIntyre and Hänsch. [137] Second, the sign of the frequency offset effect was discovered to be incorrect while developing an improved technique to measure the frequency of a pulsed laser beam for the next generation positronium experiment.[138] Using the better reference calibration, and inverting the frequency offset, the experimental value is now

$$\Delta\nu(1^3S_1-2^3S_1)=1\ 233\ 607\ 218.9\pm 10.7\text{ MHz}. \tag{21}$$

The theoretical prediction is from Ferrell[122] and from the work of Fulton[136] as recalculated by Sapirstein and Yennie,[139] whose corrections are reported in a companion article in this volume. The recalculation has also been done by Gupta, Repko, and Suchyta,[140] and by Fell.[141] The energy of the triplet levels are given by

$$\begin{aligned}\nu(n\,^3S_1)=&-\tfrac{1}{2}R_\infty c n^{-2} \\ &+2R_\infty c\alpha^2 n^{-3}\{\frac{1}{12}+\frac{11}{64n}\} \\ &+\frac{1}{4\pi}R_\infty c\alpha^3 n^{-3}\{-\frac{7}{3}a_n-\frac{109}{15}+\frac{2}{3}\ln 2+6\ln\alpha^{-1}-\frac{16}{3}\ln k_o(n)\} \\ &+\frac{1}{6}R_\infty c\alpha^4 n^{-3}\ln\alpha+\cdots,\end{aligned} \tag{22}$$

where a_n are the Salpeter terms[134], $a_{1S}=-2\ln 2-3$, $a_{2S}=-9/2$, ..., and $\ln k_o(n)$ are the Bethe logarithms,[142] $\ln k_o(1)=$ 2.9841285558..., $\ln k_o(2)=$ 2.8117698931.... Fell[141] has computed the $\alpha^4R_\infty\ln\alpha$ terms, bringing the theory to 1 233 607 221.8 MHz, 2.9 MHz higher than experiment. It is to be noted that like all Ps calculations so far, the theory does not include α^4R_∞ terms that might contribute about 10MHz, so that we may conclude at this time that the theory and experiment are in agreement to within about 20 MHz.

The precision of the 1S-2S measurement can certainly be improved greatly. With some advances in laser metrology and positronium sources, one can expect an eventual measurement that splits the natural line width by a factor of 10^3. As one approaches such precision, a new set of systematic effects will plague the measurement. As an example, consider the uncertainties caused by an imperfect laser source. Present narrow-band pulsed lasers constructed by injecting a cw laser beam into a series of amplifiers do not have a frequency spectrum that is the Fourier transform of the pulse envelope. If such a laser is used to excite a one-photon transition, the line shape of the resonance is the convolution of the atomic line shape with the power spectrum of the laser radiation. Since the spectrum $I(\omega)=E(\omega)E^*(\omega)$ can be measured with a Fabry-Perot interferometer, a distorted line shape caused by an imperfect laser can be readily accounted for. However, if the transition is non-linear as is the case for the two-photon 1S-2S transition, changes in the "instantaneous frequency" defined as $d\phi/dt$, where ϕ is the accumulated phase of the excitation light, distort the shape and position of the resonance in a way that cannot necessarily be corrected from knowledge of $I(\omega)$ alone. Danzmann, Fee, and Chu have recently developed a heterodyne technique to measure the complex electric field $E(t)$ directly.[143] Given $E(t)$, the Schrödinger equation can be numerically integrated to find the line shape and resonance position.

In order to test the analysis and heterodyne technique, the complex pulse of the cw dye laser driven, excimer-pumped amplifier chain used in the 1S-2S muonium experiment[59] was measured. A similar laser system was used in the measurement of the 1S-2S interval in both Ps and hydrogen.[144] If one assumes that the frequency chirp of the lasers used in the hydrogen and muonium experiments were similar, both the previously unexplained asymmetric line shape and the discrepancy between the pulsed measurement and the subsequent and more precise cw measurement can be resolved. The non-linear chirp corrections were also applied to the Ps experiment, but since the light was not doubled to 244 nm as in the case of hydrogen and muonium, the corrections were only on the order of 1 MHz. We have not included a chirp correction in the final value for $\Delta\nu(1^3S_1-2^3S_1)$, since the exact correction depends on the specific laser used, but we have added in quadrature to the other errors an error estimate of ± 1 MHz to account for chirp effects. A complete analysis of two-photon transitions using pulsed lasers with frequency chirps and a heterodyne technique for measuring these chirps has been discussed by Fee, Danzmann, and Chu.[146]

Ideally, the optical spectroscopy of Ps should be measured using a well-characterized cw laser. In the first experiments, the limited number of atoms demanded that the probability of exciting the two photon transition be close to unity, so that a high-power pulsed laser was needed. Because of the dramatic improvement in Ps sources afforded by electron accelerators,[66,67] one can now work with lower excitation probabilities. Furthermore, a cw dye laser can now be stabilized and locked to a high finesse Fabry-Perot cavity with the aid of high frequency feedback. Finesses over 100,000 have been reported in so-called "super-cavities", and the circulating powers can exceed one kilowatt.[143,147] Thus a cw Ps 1S-2S measurement is now possible in which the dominant line width will be the transit time broadening due to the atoms traversing the laser beam. An intense room temperature source of Ps should yield a line width of ≈10 MHz and good enough signal-to-noise to split the line by a factor of 1000. If combined with a significant advance in the theory, this measurement would provide one of the best tests of relativistic bound-state two-body theory and a very precise value for the positronium Rydberg constant.

3.3.4 Rydberg Levels

Recently Howell and coworkers have reported the optical excitation of the $1^3S_1-2^3P_J$ transitions in positronium followed by a second excitation from the n=2 level to higher n states.[148] Positronium was produced by focusing the Livermore positron beam[66,149] onto a clean, 1000 K Cu target in ultrahigh vacuum. The authors were able to saturate the n=2 level and to create a substantial population in the Rydberg levels. The Ps was observed by means of its annihilation γ's, resulting in a low signal-to-noise compared to experiments using single particle detection.[27] Nevertheless, the signal (see Fig 25) was sufficient to obtain evidence for excitation into the n=13-19 states, and permitted an accurate measurement of the resonance wavelength for the n=2 to n=14 transition: (744.040±0.035) nm; and for the n=2 to n=15 transition: (741.993±0.040) nm. The wavelengths are in reasonable agreement with theory, and the large measured widths are consistent with the first order Doppler spread in the Ps combined with the line profile of the laser.

3.4 Positronium Negative Ion

A bound system of two electrons and one positron was first demonstrated theoretically by Wheeler,[31] and eventually produced in the laboratory by a beam-foil method.[29] Fig 26 shows the annihilation line of the positronium negative ion, Ps^-, Doppler-shifted by various amounts corresponding to different acceleration potentials. The non-relativistic part of the binding energy of Ps^- has been found to arbitrary accuracy by a long series of variational calculations[150] that followed Wheeler's work. The most interesting properties of Ps^- are its annihilation decay rate, which has been determined[30] (see Fig 27) to be (2.09±0.09) $nsec^{-1}$ in agreement with theory,[151] and the photodetachment cross section,[152,153] which is predicted to exhibit an infinite series of Feshbach resonances and a single $^1P^o$ shape resonance.[154,155] The lifetime offers an important independent way to examine 2γ decay, since the annihilation of Ps^- is dominated by singlet overlap annihilation. Accuracies of parts in 10^4 should be possible eventually.[156] Since the lowest $^1P^o$ resonance is quite narrow (≈10 meV), its position can in principle be measured with a precision capable of testing radiative corrections in the three-body system. Measurement of the rate for single photon annihilation[157] would be a difficult but sensitive test of the accuracy of the three-body wave function. Finally, precision measurements of the Doppler shift of the annihilation γ's could in principle be used to extend accurate voltage references to the 100 kV range.

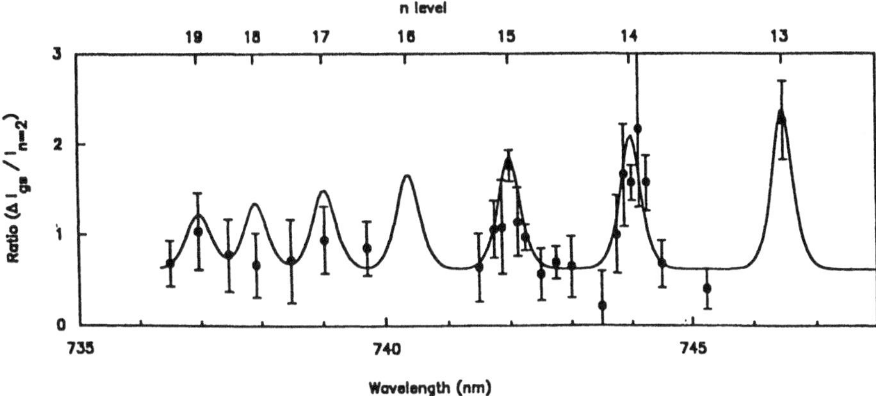

Fig. 25 Excitation from the 2P states to the n=13 and higher Rydberg states of Ps. [From Ref 148]

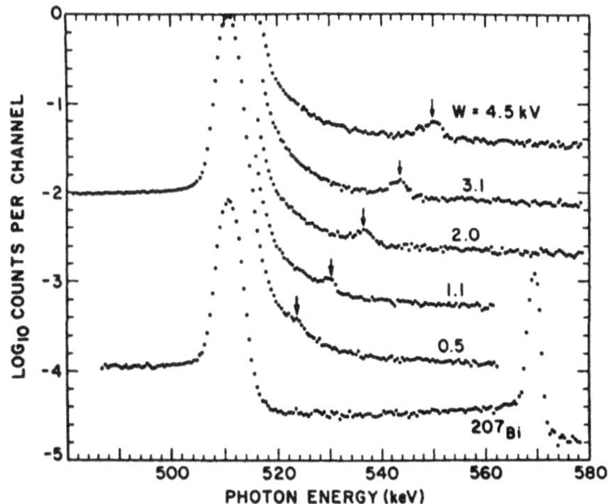

Fig. 26 Annihilation γ energy spectra obtained by accelerating Ps^- ions through different potential drops. [From Ref 29]

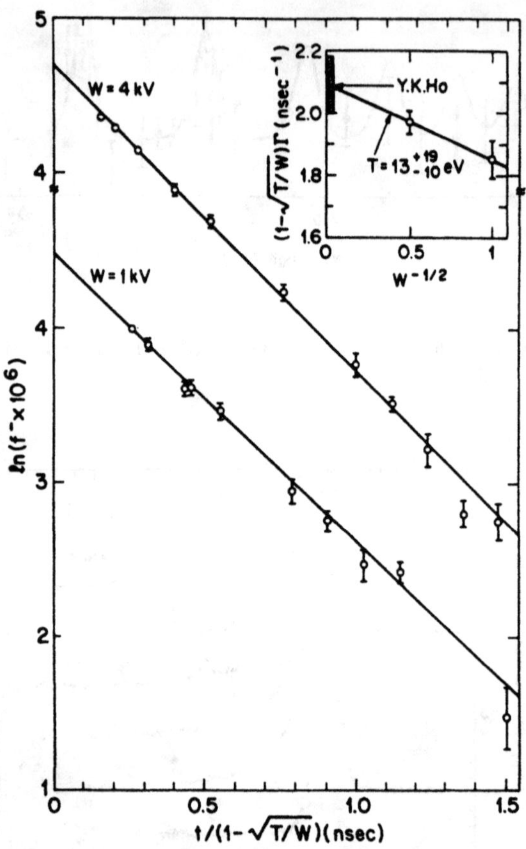

Fig. 27 Natural logarithm of the relative amount of Ps^- surviving for a time $t/(1-\sqrt{T/W})$, where T is the Ps^- kinetic energy at the time of formation in a carbon film, and W is the acceleration potential. The inset shows the extrapolation of the decay rate to infinite acceleration potential. [From Ref 30]

4. EXOTICA

4.1 Tests of Discrete Symmetries

Positronium is the only easily accessible leptonic system that has eigenstates of charge conjugation (C), parity (P), and charge conjugation and parity (CP). Thus, the atom offers a unique opportunity to search for violations of the discrete symmetries.

There are several CPT tests that can be done with positronium. An accurate measurement of the 1S-2S energy difference will yield a precise value for the positronium Rydberg constant, R_{Ps}. The equality of R_{Ps} and R_∞ at the present level of experimental and theoretical uncertainty implies that the inertial masses of the electron and positron are equal to roughly 40 ppb, as expected from CPT conservation. The anticipated experimental uncertainty at the 10^{-11} level, if matched by an equally precise theory and R_∞ measurement, would set an upper limit to the particle-antiparticle inertial mass difference that could exceed the accuracy expected eventually for mass difference measurements that can be done in electromagnetic traps. By comparison, the mass difference of the $\overline{K}^0$ and K^0 is less than one part in 10^{18} of the kaon mass.[158]

A general analysis of one-photon transitions of the form Ps→ Ps' + γ, or Ps + γ→ Ps' as tests of C and/or P and/or T has been given by Conti, Hatamian, and Rich.[159] The possible experiments may be classified into (1) searches for symmetry-forbidden transitions in the absence of external electric and magnetic fields, and (2) measurements of asymmetries due to the interference of a symmetry-violating amplitude with a symmetry-conserving amplitude. For the most part, other experiments such as heavy atom parity tests, limits to the static electric dipole moments of atoms, and searches for forbidden annihilation paths of Ps set better upper limits than the one-photon transition tests. One-photon transitions that test CPT-violating interactions may set better limits than various alternatives. Arbic, *et al.*[161] have examined the angular correlation of the quantity $\vec{S}\cdot(\vec{k}_1\times\vec{k}_2)$, where $\vec{S}$ is the Ps spin, and $\vec{k}_i$ are the momenta of the the three annihilation γ's in the decay of o-Ps with the requirement that $k_1>k_2>k_3$. No CPT violation was observed at the 2% level.

Within the context of the Weinberg-Salam-Glashow standard model, Bernreuther and Nachtmann[89] have examined possible effects of weak interactions in positronium. Since Ps is presumed to be an eigenstate of CP, many of the possible weak-electromagnetic interference effects are forbidden and all the effects are feeble. The parity mixing of the 2^3S_1 state with the nearby 2^3P_1 state has a predicted amplitude of $\delta_{2,2}\approx-1.7\times10^{-14}$, and the corresponding energy shifts will scale as the square of this ratio. Perhaps the most promising experiments would be to look for an interference effect in the two-photon transition between the 1^3S_1 and the 2^3P_1 states in a weak magnetic field, B. The perturbed P-state is

$$|2^3P_1(m)\rangle = |2^3P_1(m)\rangle + i\delta_{2,2}|2^3S_1(m)\rangle + m\frac{eB}{2^{1/2} m_e(E_{2^3P_1} - E_{2^1P_1})}|2^1P_1(m)\rangle, \quad (23)$$

where $m=\pm 1,0$. From a practical point of view, the two-photon transition makes it possible to make use of the entire distribution of Ps velocities using a well established technique, and would have the narrow line width needed to filter out the effects of the nearby allowed transitions. In the two-photon transition

$$\gamma(\vec{k},\vec{\epsilon}_1) + \gamma(-\vec{k},\vec{\epsilon}_2) + 1^3S_1(m) \rightarrow 2^3P_1(m), \quad (24)$$

small admixtures of the 2^3S_1 state due to the weak interaction (at the 10^{-14} level) and the 2^1P_1 state due to the magnetic field will interfere and give a P-violating asymmetry between the absorption rate of two right-circularly polarized photons and two left-circularly photons. Bernreuther and Nachtmann[89] estimate that the asymmetry will be on the order of $4\times10^{-4}/B$ [Gauss]. Unfortunately, the signal is proportional to B^2, and the effect at small values of B would tend to be a large asymmetry with very little count rate. The signal-to-noise ratio would be independent of B.

CP-violating effects due to various models of milli-, micro-, and super-weak forces in leptonic systems have been reviewed by Cheng.[160] Experimental tests of interactions that violate C but conserve P have been in the form of searches for the C-violating decays $1^1S_0 \rightarrow 3\gamma$,[87] and $1^3S_1 \rightarrow 4\gamma$.[88] The 1σ upper limits to the branching ratios are

$R(1^1S_0 \rightarrow 3\gamma)/R(1^1S_0 \rightarrow 2\gamma) < 2.8\times10^{-6}$, and
$R(1^3S_1 \rightarrow 4\gamma)/R(1^3S_1 \rightarrow 3\gamma) < 8\times10^{-6}$.

As mentioned above, these limits are much smaller than the 10^{-27} effects predicted by the standard model of the weak interaction.[89]

4.2 Exotic Decays

There have been a number of searches for exotic particle decay of positronium, mostly spurred on by the apparent discrepancy between theory and experiment of the lifetime of o-Ps, or by the mystery surrounding the discovery of correlated back-to-back emission of e^+e^- pairs in heavy ion collisions.[161] Samuel has argued that the decay of Ps into γ plus a neutral particle X with a branching ratio $R(1^3S_1 \rightarrow X)/R(1^3S_1 \rightarrow 3\gamma) \approx 2\times10^{-3}$ can explain the o-Ps lifetime discrepancy without conflicting with the electron g-2 experiments if the particle has a mass below 5.7 keV.[162] In response, Orito, *et al.*[104] measured the upper limit to the branching ratio and found that for any pseuscalar particle of mass less than 30 keV the ratio is less than 6.4×10^{-5} at a 90% confidence level. Atoyan, *et al.*[164] have set an upper limit to the branching ratio $R(1^3S_1 \rightarrow \text{nothing})/R(1^3S_1 \rightarrow 3\gamma) < 5.8\times10^{-4}$, ruling out the possibility that the o-Ps lifetime is shortened significantly by the Ps vanishing into undetectable particles.

The discovery of the still baffling resonances in heavy ion collisions has prompted speculation about other mystery particles. If the new particle has a scalar coupling $g\phi\bar{\psi}\psi$, the measured electron g-factor limits the coupling constant to $g^2/4\pi \lesssim 10^{-8}$.[165] Given this upper limit, the 2S-2P splittings can be off from the QED value by 100 kHz, and the 1S-2S interval by 1 MHz, if the mystery particle has a point-like structure. If the mystery particle has structure, its form factor would tend to suppress its contribution to the electron g-2 value relative to its contribution to the Ps energy level shifts or effects in Bhabha scattering. Schäfer, *et al.*[166] have pointed out that the agreement with theory in the ground state Ps hyperfine interval sets good limits on possible vector, axial vector, and pseudo scalar couplings, but does not restrict a possible scalar coupling significantly. The possible extended structure of the mystery particle and its influence on a variety of experiments has been reviewed by Graf, *et al.*[167]

4.3 Cryogenic Positronium

Since the development of laser cooling techniques,[168,169] researchers have considered laser cooling Ps.[170,171] Applications of cold Ps include the reduction of the transit time broadening in a cw laser measurement of the 1S-2S interval, quantum reflection of atoms with huge deBroglie wave lengths, and more speculative suggestions such as the study of the Bose condensation of a dilute gas,[32,33,172] the formation of a medium for gamma-ray amplification via stimulated annihilation,[173–177] and free-fall measurements on a particle-antiparticle system.

The laser cooling of Ps presents certain interesting challenges. The natural choice for the cooling transition is the 1S-2P Lyman-α transition at 243 nm which has a natural lifetime of 3.2 nsec and a natural line width of 50 MHz. Because the atom has such a low mass, only 50 scattered photons are required to bring Ps to rest from room temperature. We calculate that an atom with an initial velocity corresponding to a temperature of 100 K can be cooled in the 142 nsec lifetime of the triplet state. Note that since the atom spends some time in the 2P state, the effective lifetime is increased.[178] The velocity change that results from a single photon recoil gives a Doppler shift of 6 GHz, so that the atom recoils out of resonance with any monochromatic laser. Also, the initial Doppler width for 100 K atoms is $\approx$280 GHz. These boundary conditions eliminate the possibility of conventional swept-frequency laser cooling,[179] or cooling in an inhomogeneous magnetic field,[180] which in any case would shorten the o-Ps lifetime due to the mixing of p-Ps states.

One solution to the problem would be to use "white light" cooling based on a modification of a scheme proposed by Hoffnagle.[181] If a quasi-continuous spectrum of light is tuned to be on the red-wavelength side of the absorption line, and the atoms are illuminated in the usual "optical molasses" configuration,[182] the atoms might be expected to cool down to temperatures corresponding roughly to the recoil velocity of a single photon, $k_B T = \hbar^2 k^2/4m_e = 0.15K$. Here, $k = \omega/c$ is the wave vector of the photon, and the expression has been evaluated for the case of 1S-2P cooling. On the other hand, the limiting temperature could be strongly modified by the hyperfine structure of the Ps and the gradients in the polarization of the light field.[183,184] Also, the strong recoil effects should be included in a consistent manner by expressing the atomic wave function in terms of both the internal degrees of freedom and the momentum states of the center of mass.[185]

A source of ultra-cold Ps would open up the possibility of measuring the acceleration due to gravity of a particle-antiparitcle system complimentary to the proposed anti-hydrogen free-fall measurements.[186] The short annihilation lifetime presents a problem, but the atom can be excited into a circular Rydberg state[187,188] where it would have the necessary long radiative lifetime and insensitivity to electric fields. Such an experiment would be similar to the Witteborn-Fairbank[189] falling electron experiment, but the problems associated with electromagnetic shielding might not be so severe.

5. ACKNOWLEDGMENTS

The work of S. Chu was supported in part by a grant from the National Science Foundation, and a Precision Measurements Grant from the National Institute of Standards and Technology. We would like to thank M. S. Fee for a critical reading of the manuscript, and R. Fell and J. Sapirstein for useful discussions. We would also like to thank T. Kinoshita whose gentle arm-twisting made this article possible.

REFERENCES

[1] S. deBenedetti and H. C. Corben, *Ann. Rev. Nucl. Sci.* **4**, 191 (1954).

[2] M. Deutsch and S. Berko, in *Alpha-, Beta- and Gamma-Ray Spectroscopy*, Vol. 2, edited by Kai Siegbahn (North-Holland, Amsterdam, 1968) p. 1583.

[3] V. W. Hughes, in *Atomic Physics* Vol.3 (Plenum, New York, 1972)

[4] "Discovery of Positronium", in *Adventures in Experimental Physics*, Vol. 4, edited by B. Maglic (World Science Education, Princeton, NJ, 1975) pp. 64-127.

[5] M. A. Stroscio, *Phys. Lett. C* **22**, 215 (1975).

[6] T. C. Griffith and G. R. Heyland, *Nature* **269**, 109 (1977)

[7] S. Berko, K. F. Canter, and A. P. Mills, Jr. in *Progress in Atomic Spectroscopy*, Part B, edited by W. Hanle and H. Kleinpoppen (Plenum, New York, 1979) p. 1427.

[8] G. T. Bodwin and D. R. Yennie, *Phys. Lett. C* **43**, 267 (1978).

[9] S. Berko and H. N. Pendleton, *Annu. Rev. Nucl. Part. Sci.* **30**, 543 (1980).

[10] A. Rich, *Rev. Mod. Phys.* **53**, 127 (1981).

[11] D. W. Gidley, A. Rich, and P. W. Zitzewitz, in *Positron Annihilation*, edited by P. G. Coleman, S. C. Scharma, and L. M. Diana (North-Holland, Amsterdam, 1982) p. 11.

[12] *Positron Scattering in Gases*, edited by J. W. Humberston and M. R. C. McDowell (Plenum, New York, 1983).

[13] V. W. Hughes in *Precision Measurements and Fundamental Constants II*, B. N. Taylor and W. D. Phillips, eds. (National Bureau of Standards (U.S.), Spec. Publ. 617, Washington DC, 1984) p 237.

[14] *Positron (Electron)-Gas Scattering*, edited by W. E. Kaupilla, T. S. Stein, and J. M. Wadehra (World Scientific, Singapore, 1985).

[15] M. Charlton, *Rep. Prog. Phys.* **48**, 737 (1985).

[16] *Positron Studies of Solids, Surfaces, and Atoms*, edited by A. P. Mills, Jr., K. F. Canter, and W. S. Crane (World Scientific, Singapore, 1986).

[17] A. P. Mills, Jr., in *The Spectrum of Atomic Hydrogen: Advances*, edited by G. W. Series (World Scientific, Singapore, 1988) p. 447.

[18] *Annihilation in Gases and Galaxies*, edited by R. J. Drachman (NASA Conference Publication 3058, Greenbelt, MD, 1990).

[19] A. P. Mills, Jr. in *Positron Solid State Physics*, edited by W. Brandt and A. Dupasquier (North-Holland, Amsterdam, 1983) p. 432.

[20] P. J. Schultz and K. G. Lynn, *Rev. Mod. Phys.* **60**, 701 (1988).

[21] D. W. Gidley, A. Rich, E. Sweetman, and D. West, *Phys. Rev. Lett.* **49**, 525 (1982).

[22] C. I. Westbrook, D. W. Gidley, R. S. Conti, and A. Rich, *Phys. Rev. Lett.* **58**, 1328 (1987).

[23] M. W. Ritter, P. O. Egan, V. W. Hughes, and K. A. Woodle, *Phys. Rev. A* **30**, 1331 (1984).

[24] A. P. Mills, Jr. and G. H. Bearman, *Phys. Rev. Lett.* **34**, 246 (1975).

[25] A. P. Mills, Jr., S. Berko and K. F. Canter, *Phys. Rev. Lett.* **34**, 1541 (1975).

[26] S. Hatamian, R. S. Conti, and A. Rich, *Phys. Rev. Lett.* **58**, 1833 (1987).

[27] S. Chu and A. P. Mills, Jr., *Phys. Rev. Lett.* **48**, 1333 (1982).

[28] S. Chu, A. P. Mills, Jr., and J. L. Hall, *Phys. Rev. Lett.* **52**, 1689 (1984).

[29] A. P. Mills, Jr., *Phys. Rev. Lett.* **46**, 717 (1981).

[30] A. P. Mills, Jr., *Phys. Rev. Lett.* **50**, 671 (1983).

[31] J. A. Wheeler, *Annals N. Y. Acad. Sci.* **48**, 219 (1946).

[32] P. M. Platzman, in *Positron Studies of Solids, Surfaces, and Atoms*, edited by A. P. Mills, Jr., W. S. Crane, and K. F. Canter (World Scientific, Singapore, 1986) p. 84.

[33] A. P. Mills, Jr., P. M. Platzman, S. Berko, K. F. Canter, K. G. Lynn, and L. O. Roellig, *Bull. Amer. Phys. Soc.* **34**, 588 (1989).

[34] M. Deutsch, *Phys. Rev.* **82**, 455 (1951).

[35] R. Paulin and G. Ambrosino, *J. de Physique* **29**, 263 (1968).

[36] H. Morinaga, *Phys. Lett.* **68A**, 105 (1978).

[37] H. Morinaga and Y. Matsuoka, *Phys. Lett.* **71A**, 103 (1979).

[38] U. Zimmerman, F. Stucki, and F. Heinrich, *Phys. Lett.* **74A**, 346 (1979).

[39] A. P. Mills, Jr. and W. S. Crane, *Phys. Rev. A* **31**, 593 (1985).

[40] M. H. Weber, *et al.*, *Phys. Rev. Lett.* **61**, 2542 (1988).

[41] B. L. Brown, in *Positron Annihilation*, edited by P. C. Jain, R. M. Singru, and K. P. Gopinathan (World Scientific, Singapore, 1985), p. 328.

[42] K. F. Canter, A. P. Mills, Jr. and S. Berko, *Phys. Rev. Lett.* **33**, 7 (1974).

[43] K. F. Canter, A. P. Mills, Jr. and S. Berko, *Phys. Rev. Lett.* **34**, 177 (1975).

[44] A. Ore, Universitet Bergen Arbok, Naturvitenskapelig rekke No. 9 (1949).

[45] D. M. Schrader and R. E. Svetic, *Can. J. Phys.* **60**, 517 (1982).

[46] O. E. Mogensen, *J. Chem. Phys.* **60**, 998 (1974).

[47] R. L. Garwin, *Phys. Rev.* **91**, 1571 (1953).

[48] M. Dresden, *Phys. Rev.* **93**, 1413 (1954).

[49] L. A. Page and M. Heinberg, *Phys. Rev.* **102**, 1545 (1956).

[50] W. Brandt, G. Cussot, and R. Paulin, *Phys. Rev. Lett.* **23**, 522 (1969).

[51] A. L. Greenberger, A. P. Mills, Jr., A. Thompson, and S. Berko, *Phys. Lett. A* **32**, 72 (1970).

[52] W. Brandt and R. Paulin, *Phys. Rev. Lett.* **21**, 193 (1968).

[53] T. Chang, M. Xu, and X. Zeng, *Phys. Lett. A* **126**, 189 (1987).

[54] D. W. Gidley, K. A. Marko, and A. Rich, *Phys. Rev. Lett.* **36**, 395 (1976).

[55] S. M. Curry and A. L. Schawlow, *Phys. Lett. A* **37**, 5 (1971).

[56] D. W. Gidley and P. W. Zitzewitz, *Phys. Lett.* **69A**, 97 (1978).

[57] A. P. Mills, Jr., E. D. Shaw, R. J. Chichester, and D. M. Zuckerman, *Phys. Rev. B* **40**, 2045 (1989).

[58] G. A. Beer, *et al.*, *Phys. Rev. Lett.* **57**, 671 (1986).

[59] S. Chu, A. P. Mills, Jr., A. G. Yodh, K. Nagamine, Y. Miyake, and T. Kuga, *Phys. Rev. Lett.* **60**, 101 (1988).

[60] A. P. Mills, Jr., P. M. Platzman, and B. L. Brown, *Phys. Rev. Lett.* **41**, 1076 (1978).

[61] A. P. Mills, Jr., *Phys. Rev. Lett.* **41**, 1828 (1978).

[62] K. G. Lynn, *Phys. Rev. Lett.* **43**, 391, 803 (1978).

[63] A. P. Mills, Jr., *Solid State Commun.* **31**, 623 (1979).

[64] A. P. Mills, Jr. and L. Pfeiffer, *Phys. Rev. Lett.* **43**, 1961 (1979).

[65] K. G. Lynn, M. Weber, L. O. Roellig, A. P. Mills, Jr., and A. R. Moodenbaugh, in *Atomic Physics with Positrons*, edited by J. W. Humberston and E. A. G. Armour (Plenum, New York 1987) p. 161.

[66] R. H. Howell, R. A. Alverez, and M. Stanek, *Appl. Phys. Lett.* **40**, 751 (1982).

[67] A. P. Mills, Jr., E. D. Shaw, R. J. Chichester, and D. M. Zuckerman, *Rev. Sci. Instrum.* **60**, 825 (1989).

[68] G. R. Brandes, K. F. Canter, and A. P. Mills, Jr., *Phys. Rev. Lett.* **61**, 492 (1988).

[69] B. L. Brown, W. S. Crane, and A. P. Mills, Jr., *Appl. Phys. Lett.* **48**, 739 (1986).

[70] E. M. Gullikson and A. P. Mills, Jr., *Phys. Rev. B* **36**, 8777 (1987).

[71] A. P. Mills, Jr., in *Positron Scattering in Gases*, edited by J. W. Humberston and M. R. C. McDowell (Plenum, New York, 1984) p. 121.

[72] A. Vehanen, K. G. Lynn, P. J. Schultz, and M. Eldrup, *Appl. Phys. A* **32**, 163 (1983).

[73] A. P. Mills, Jr. and E. M. Gullikson, *Appl. Phys. Lett.* **49**, 1121 (1986).

[74] A. P. Mills, Jr., *Appl. Phys.* **23**, 189 (1980).

[75] A. P. Mills, Jr., *Hyperfine Interactions* **44**, 107 (1988).

[76] A. P. Mills, Jr., L. Pfeiffer, and P. M. Platzman, *Phys. Rev. Lett.* **51**, 1085 (1983).

[77] K. G. Lynn, W. E. Frieze, and P. J. Schultz, *Phys. Rev. Lett.* **52**, 1137 (1984).

[78] W. S. Crane and A. P. Mills, Jr., *Rev. Sci. Instrum.* **56**, 1723 (1985).

[79] S. Chu, A. P. Mills, Jr., and C. A. Murray, *Phys. Rev. B* **23**, 2060 (1981).

[80] A. P. Mills, Jr. and L. Pfeiffer, *Phys. Rev. B* **32**, 53 (1985).

[81] P. M. Platzman and N. Tzoar, *Phys. Rev. B* **33**, 5900 (1986).

[82] K. G. Lynn, *Phys. Rev. Lett.* **44**, 1330 (1980).

[83] D. W. Gidley, A. R. Köyman, and T. W. Capehart, *Phys. Rev. B* **37**, 2465 (1988).

[84] A. P. Mills, Jr., E. D. Shaw, M. Leventhal, R. J. Chichester, D. M. Zuckerman, and R. Lee, to be published.

[85] W. C. Sauder and R. D. Deslattes, *J. Res. Nat. Bur. Stand.* **71A**, 347 (1967).

[86] E. G. Kessler, Jr., R. D. Deslattes, A. Henins, and W. C. Sauder, *Phys. Rev. Lett.* **40**, 171 (1978).

[87] A. P. Mills, Jr. and S. Berko, *Phys. Rev. Lett.* **18**, 420 (1967).

[88] K. Marko and A. Rich, *Phys. Rev. Lett.* **33**, 980 (1974).

[89] W. Bernreuther and O. Nachtmann, *Z. Phys. C.* **11**, 235 (1981).

[90] O. Halpern, *Phys. Rev.* **94**, 904 (1954).

[91] A. Rich, PhD dissertation, Univ. Michigan, 1965 (available from University Microfilms, Ann Arbor, MI) p. 80.

[92] A. P. Mills, Jr., *Phys. Rev. A* **41**, 502 (1990).

[93] H. Grotch and R. A. Hegstrom, *Phys. Rev. A* **4**, 59 (1971).

[94] E. R. Carlson, V. W. Hughes, M. L. Lewis, and I. Lindgren, *Phys. Rev. Lett.* **29**, 1059 (1972).

[95] H. Grotch and R. Kashuba, *Phys. Rev. A* **7**, 78 (1973).

[96] M. L. Lewis and V. W. Hughes, *Phys. Rev. A* **8**, 625 (1973).

[97] C. I. Westbrook, D. W. Gidley, R. S. Conti, and A. Rich, *Phys. Rev. A* **40**, 5489 (1989).

[98] I. Harris and L. Brown, *Phys. Rev.* **105**, 1656 (1957).

[99] W. G. Caswell and G. P. Lapage, *Phys. Rev. A* **20**, 36 (1979).

[100] E. R. Cohen and B. N. Taylor, *Rev. Mod. Phys.* **59**, 1121 (1987).

[101] A. Ore and J. L. Powell, *Phys. Rev.* **75**, 1696 (1949).

[102] G. S. Adkins, *Ann. Phys. (N.Y.)* **146**, 78 (1983).

[103] D. W. Gidley, A. Rich, P. W. Zitzewitz, and D. A. L. Paul, *Phys. Rev. Lett.* **40**, 737 (1978).

[104] S. Orito, K. Yoshimura, T. Haga, M. Minowa, and M. Tsuchiaki, *Phys. Rev. Lett.* **63**, 597 (1989).

[105] A. Billoire, R. Lacaze, A. Morel, and H. Navelet, *Phys. Lett.* **78B**, 140 (1978).

[106] G. S. Adkins and F. R. Brown, *Phys. Rev. A* **28**, 1164 (1983).

[107] M. Deutsch and S. C. Brown, *Phys. Rev.* **85**, 1047 (1952).

[108] A. Rich, *Phys. Rev. A* **23**, 2747 (1981).

[109] P. R. Fontana and D. J. Lynch, *Phys. Rev. A* **2**, 347 (1970).

[110] F. H. M. Faisal and J. V. Moloney, *J. Phys. B: At. Mol. Phys.* **14**, 3603 (1981).

[111] A. P. Mills, Jr., *J. Chem. Phys.* **62**, 2646 (1975);

[112] A. P. Mills, Jr., *Phys. Rev. A* **27**, 262 (1983).

[113] F. H. M. Faisal and P. S. Ray, *Phys. Rev. A* **30**, 2316 (1984).

[114] G. H. Bearman and A. P. Mills, Jr., *J. Chem. Phys.* **65**, 1841 (1976)

[115] G. P. Lepage, in *Atomic Physics 7,* D. Kleppner and F. M. Pipkin, eds. (Plenum, N.Y., 1981), p. 297.

[116] R. Karplus and A. Klein, *Phys. Rev.* **87**, 848 (1952).

[117] H. W. Kendall, Ph. D. thesis, MIT (1954).

[118] W. Cherry, PhD dissertation, Princeton Univ. (1958) (Available from University Microfilms, Ann Arbor, MI).

[119] D. E. Groce, D. G. Costello, J. W. McGowan, and D. F. Herring, *Bull. Amer. Phys. Soc.* **13**, 1397 (1968).

[120] D. G. Costello, D. E. Groce, D. F. Herring, and J. W. McGowan, *Phys. Rev. B* **5**, 1433 (1972).

[121] K. F. Canter, P. G. Coleman, T. C. Griffith, and G. R. Heyland, *J. Phys. B: Atom. Molec. Phys.* **5**, L167 (1972).

[122] R. A. Ferrel, *Phys. Rev.* **84**, 858 (1951).

[123] T. Fulton and P. C. Martin, *Phys. Rev.* **95**, 811 (1954).

[124] D. C. Schoepf, S. Berko, K. F. Canter, and A. H. Weiss, in *Positron Annihilation,* P. G. Coleman, S. C. Sharma, and L. M. Diana, eds. (North-Holland, Amsterdam, 1982) p. 165.

[125] S. Berko, K. F. Canter, B. O. Clark, and D. C. Schoepf, in *Positron Annihilation* R. R. Hasiguti and K. Fugiwara, eds. (Japan Inst. of Metals, Sendai, 1979) p. 531.

[126] L. S. Vasilenko, V. P. Chebotaev, and A. V. Shishaev, *JETP Lett.* **12**, 113 (1970).

[127] C. Borde, *Compt. Rend.* **271**, 371 (1970).

[128] P. W. Smith and T. W. Hänsch, *Phys. Rev. Lett.* **26**, 740 (1971).

[129] T. W. Hänsch, in *Atomic Physics 8,* I. Lindgren, A. Rosen, and S. Svanberg, eds. (Plenum, N.Y., 1983) p. 55.

[130] V. G. Minogin, *Kavantovaya Elektronika* **3**, 2061 (1976).

[131] A. P. Mills, Jr., *Appl. Phys.* **22**, 273 (1980).

[132] A. P. Mills, Jr. and S. Chu, in *Atomic Physics 8,* I. Lindgren, A. Rosen, and S. Svanberg, eds. (Plenum, N.Y., 1983) p. 83.

[133] S. Chu, A. P. Mills, Jr., and J. L. Hall, in *Laser Spectroscopy VI,* H. P. Weber and W. Luthy, eds. (Springer-Verlag, Heidelberg, 1983) p. 28.

[134] G. W. Erickson, *J. Phys. Chem. Ref. Data* **6**, 831 (1977).

[135] S. R. Amin, C. D. Caldwell, and W. Lichten, *Phys. Rev. Lett.* **47**, 1234 (1981).

[136] T. Fulton, *Phys. Rev. A* **26**, 1794 (1982).

[137] D. H. McIntyre and T. W. Hänsch, *Phys. Rev. A* **36**, 4115 (1987).

[138] K. Danzmann, M. S. Fee, and S. Chu, *Phys. Rev. A* **39**, 6072 (1989).

[139] J. R. Sapirstein and D. R. Yennie, *Theory of Hydrogenic Bound States*, in this volume.

[140] S. N. Gupta, W. W. Repko, and C. J. Suchyta, *Phys. Rev. D* **40**, 4100 (1989).

[141] R. Fell, private communication.

[142] S. Klarsfeld and A. Maquet, *Phys. Lett.* **43B**, 201 (1973).

[143] K. Danzmann, M. S. Fee, and S. Chu, in *Laser Spectroscopy IX*, edited by M. S. Feld, J. E. Thomas and A. Morradian, (Academic, New York, 1989) p. 328.

[144] E. A. Hildum, *et al.*, *Phys. Rev. Lett.* **56**, 576 (1986).

[145] R. G. Beausoleil, *et al.*, *Phys. Rev. A* **35**, 4878 (1987).

[146] A complete description of this work has been submitted to Phys. Rev. A by M. S. Fee, K. Danzmann, and S. Chu.

[147] J. Bergquist, private communication.

[148] R. H. Howell, K. P. Ziock, F. Magnotta, C. D. Dermer, and R. A. Failor, in *Annihilation in Gases and Galaxies*, edited by R. J. Drachman (NASA Conference Publication 3058, Greenbelt, MD, 1990) p. 201.

[149] R. H. Howell, M. J. Fluss, I. J. Rosenberg, and P. Meyer, *Nucl. Inst. Meth. in Phys. Res. B* **10/11**, 373 (1985).

[150] For a review see Y. K. Ho, in *Annihilation in Gases and Galaxies*, edited by R. J. Drachman (NASA Conference Publication 3058, Greenbelt, MD, 1990) p. 243.

[151] A. K. Bhatia and R. J. Drachman, *Phys. Rev. A* **28**, 2523 (1983).

[152] A. K. Bhatia and R. J. Drachman, *Phys. Rev. A* **32**, 3745 (1985).

[153] S. J. Ward, M. R. C. McDowell, and J. W. Humberston, *Europhysics Lett.* **1**, 167 (1986).

[154] J. Botero and C. H. Greene, *Phys. Rev. Lett.* **56**, 1366 (1986).

[155] J. Botero, *Phys. Rev. A* **35**, 36 (1987).

[156] A. P. Mills, Jr., P. G. Friedman, and D. M. Zuckerman, in *Annihilation in Gases and Galaxies,* edited by R. J. Drachman (NASA Conference Publication 3058, Greenbelt, MD, 1990) p. 213.

[157] M.-C. Chu and V. Pönisch, *Phys. Rev. C* **33**, 2222 (1986).

[158] Particle Data Group, *Phys. Lett. B* **204**, 46 (1988).

[159] R. S. Conti, S. Hatamian, and A. Rich, *Phys. Rev. A* **33**, 3495 (1986).

[160] H.-Y. Cheng, *Phys. Rev. D* **28**, 150 (1983).

[161] B. K. Arbic, S. Hatamian, M. Skalsey, J. Van House, and W. Zheng, *Phys. Rev. A* **37**, 3189 (1988).

[162] T. Cowan, *et al., Phys. Rev. Lett.* **56**, 1463 (1986).

[163] M. A. Samuel, *Mod. Phys. Lett.* **3**, 1117 (1988).

[164] G. S. Atoyan, S. N. Gninenko, V. I. Razin, and Yu. V. Ryabov, *Phys. Lett. B* **220**, 317 (1989).

[165] S. J. Brodsky, E. Mottola, I. Muzinich, and M. Soldate, *Phys. Rev. Lett.* **56**, 1763 (1986).

[166] A. Schäfer, J. Reinhardt, W. Greiner, and B. Müller, *Mod. Phys. Lett. A* **1**, 1 (1986).

[167] S. Graf, S. Schramm, J. Reinhardt, B. Müller, and W. Greiner, *J. Phys. G: Nucl. Part. Phys.* **15**, 1467 (1989).

[168] See for example, the Special Issue on "The Mechanical Effects of Light", edited by P. Meystre and S. Stenholm, *J. Opt. Soc. B* **2** (1985).

[169] See also the Special Issue on "Laser Cooling and Trapping", edited by S. Chu and C. Weiman, *J. Opt. Soc. B* **6** (1989).

[170] E. P. Liang and C. D. Dermer, *Optics Commun.* **65**, 419 (1988).

[171] K. Danzmann, *et al.*, in *Laser Spectroscopy IX.*

[172] A. Loeb and S. Eliexer, *Laser and Particle Beams* **4**, 3 (1986).

[173] C. M. Varma, *Nature* **267**, 686 (1977).

[174] M. Bertolotti and C. Sibilia, *Appl. Phys.* **19**, 127 (1979).

[175] R. Ramaty, J. M. McKinley, and F. C. Jones, *Ap. J.* **256**, 238 (1982).

[176] A. Loeb and S. Eliezer, *Laser and Particle Beams* **4**, 577 (1986).

[177] G. Kurizki and A. Friedman, *Phys. Rev. A* **38**, 512 (1988).

[178] M. H. Mittleman, *Phys. Rev. A* **33**, 2840 (1986).

[179] W. Ertmet, R. Blatt, J. L. Hall, and M. Zhu, *Phys. Rev. Lett.* **54**, 996 (1985).

[180] J. V. Prodan, A. Mignall, W. D. Phillips, I. So, H. Metcalf, and J. Dalibard, *Phys. Rev. Lett.* **54**, 992 (1985).

[181] J. Hoffnagle, *Opt. Lett.* **13**, 102 (1988).

[182] S. Chu, L. Hollberg, J. E. Bjorkholm, A. Cable, and A. Ashkin, *Phys. Rev. Lett.* **55**, 48 (1985).

[183] J. Dalibard and C. Cohen-Tannoudji, *J. Opt. Soc. B* **6**, 2023 (1989).

[184] P. J. Ungar, D. S. Weiss, E. Riis, and S. Chu, *J. Opt. Soc. B* **6**, 2072 (1989).

[185] A. Aspect, E. Arimondo, R. Kaiser, N. Vansteenkiste, and C. Cohen-Tannoudji, *J. Opt. Soc. B* **6**, 2122 (1989).

[186] G. Gabrielse, *et al.*, *Physica Scripta* **22**, 36 (1988).

[187] R. G. Hulet and D. Kleppner, *Phys. Rev. Lett.* **51**, 1430 (1983).

[188] J. Hare, M. Gross, and P. Goy, *Phys. Rev. Lett.* **61**, 1938 (1988).

[189] W. M. Fairbank, F. C. Witteborn, J. M. J. Madey and J. M. Lockhart, in *Experimental Gravitation, Proc. Int. Sch. Phys. "Enrico Fermi"* Course LVI, edited by B. Bertotti (Academic, New York, 1974) p. 310.

MUONIUM

Vernon W. Hughes
Department of Physics, Yale University
New Haven, CT 06511

and

Gisbert zu Putlitz
Physikalisches Institut der Universität Heidelberg
Federal Republic of Germany

CONTENTS

I. INTRODUCTION

Muonium (μ^+e^-, M) is the bound state of a positive muon and an electron.[1,2,3] From the viewpoint of quantum electrodynamics muonium is of great interest because it provides the simplest bound system in which the interaction of the muon and the electron can be studied. In the standard model of modern particle physics the electron and the muon are treated as structureless point leptons with electromagnetic and weak interactions but no strong interactions. Indeed the studies of muonium have provided strong confirmation of modern quantum electrodynamics and of the behaviour of the muon as a heavy electron. The fact that the muon appears to have no structure or hadronic interactions makes muonium a more ideal atom for testing quantum electrodynamics than hydrogen, where the proton has structure and undergoes hadronic interactions.

The energy intervals which have been measured with the highest precision and have provided most information to date are the hyperfine structure interval $\Delta\nu$ in the ground $1^2S_{1/2}$ state and the Zeeman energy intervals in the ground state. The hfs interval $\Delta\nu$ is a sensitive measure of the magnetic interaction of the muon and electron, and the Zeeman energy intervals determine the muon spin and the muon magnetic moment. The effect of weak interactions on $\Delta\nu$ is below the present experimental and theoretical accuracies. In addition the fine structure and Lamb shift of the n=2 state have been measured with much less accuracy, and the 1S-2S transition has been observed.

Muonium is also a useful atom for the study of certain aspects of the weak interactions of the muon, in particular, the possible conversion of muonium to antimuonium (μ^-e^+).[4,5] Muonium can be considered an isotope of hydrogen in which the proton is replaced by the positive muon, and its interactions in gases and in condensed matter are of importance to atomic and molecular collisions and to condensed matter physics.[6,7,8] These topics will not be treated in this article, except as they relate to our principal topic of the energy levels of muonium.

The possibility of studying muonium experimentally depended initially on parity nonconservation in the weak interactions involving the muon. Parity nonconservation in pion decay

$$\pi^+ \to \mu^+ + \nu_\mu \tag{1.1}$$

provides muons (μ^+) polarized in a direction opposite to their momenta i.e. with negative helicity, and parity nonconservation in muon decay

$$\mu^+ \to e^+ + \nu_e + \bar{\nu}_\mu \tag{1.2}$$

provides through the asymmetry in the positron angular distribution the means of determining the muon spin direction at the instant of decay.[1,9]

The discovery of muonium[10] in 1960 followed by several years the discovery of parity nonconservation in the weak interactions of the pion and muon[11,12] and was made through the observation of the Larmor precession of muonium at its characteristic Zeeman frequency. Following the initial measurement of $\Delta\nu$ by a microwave magnetic resonance method[13], there have occurred over the past 25 years a series of increasingly more precise measurements of the hfs levels of muonium in weak and strong static magnetic fields. These microwave spectroscopy experiments all utilized parity non-conservation as the tool for producing polarized muonium and for detecting an induced transition.

By the mid 1980's the Lamb shift[14, 15] and then later the fine structure[16] in the n = 2 state of muonium were measured in experiments similar to recent hydrogen Lamb shift experiments, which did not utilize the parity non-conservation phenomena. Finally, quite recently the 1S-2S transition in muonium was observed as a two photon transition in a laser spectroscopy experiment.[17]

II. THEORY OF MUONIUM ENERGY LEVELS

The basic theory of the energy levels of muonium starts from the Bethe-Salpeter equation for the bound state of two Dirac particles, the electron and the heavier muon.[18] The energy levels of muonium are very similar to those of hydrogen, and it is useful to discuss the energy levels with the several approximate equations conventionally applied to hydrogen.[19]

II.1 Schroedinger Equation

The time-independent nonrelativistic Schroedinger equation for muonium reads

$$\left[-\frac{\hbar^2}{2m_e}\nabla_1^2 - \frac{\hbar^2}{2m_\mu}\nabla_2^2 - \frac{e^2}{r_{12}}\right]\psi_0(\vec{r}_1,\vec{r}_2) = E_o\psi_0(\vec{r}_1,\vec{r}_2) \tag{2.1}$$

in which 1 (or e) refers to the electron and 2 (or μ) to the muon and the symbols have their usual meanings. The discrete energy eigenvalues are

$$E_o = -2\pi^2 m_r e^4/h^2n^2 = -R_M/n^2 \tag{2.2}$$

in which n is the principal quantum number, and $m_r[=m_e m_\mu/(m_e + m_\mu)]$ is the reduced mass of the electron in muonium. From the values for the Rydberg constant in cm^{-1} $R_\infty(=2\pi^2 m_e e^4/h^3c)$ and the ratio m_μ/m_e, (see Eq. 2.14) we obtain the value of the Rydberg constant for muonium:

$$R_M = 109{,}209.144(03)\ cm^{-1} \tag{2.3}$$

The binding energy of the ground $1^2S_{1/2}$ state of muonium is 13.539 eV.

II.2 Breit Equation

The Breit equation for muonium gives energy levels accurate to order $\alpha^2 R_M$:

$$\{\beta_1 m_e c^2 + c\vec{\alpha}_1\cdot\vec{p}_1 + \beta_2 m_\mu c^2 + c\vec{\alpha}_2\cdot\vec{p}_2 - \frac{e^2}{r_{12}}$$

$$+ \frac{e^2}{2r_{12}} [\ \vec{\alpha}_1\cdot\vec{\alpha}_2 + \frac{(\vec{\alpha}_1\cdot\vec{r}_{12})(\vec{\alpha}_2\cdot\vec{r}_{12})}{r_{12}^2}]\}\ \Psi = E\Psi \tag{2.4}$$

The quantities $\vec{\alpha}_1$, β_1, $\vec{\alpha}_2$, β_2 are the usual Dirac matrices. The wavefunction Ψ is a 16-component wavefunction.

The Pauli approximation to the Breit equation can be derived by a treatment similar to the Pauli approximation to the Dirac equation for a single electron or by a generalized Foldy-Wouthysen transformation. In the center of mass coordinate system where $\vec{p}_1 + \vec{p}_2 = 0$, the equation reads

$$W\psi = (H_0 + H_1 + H_2 + H_3 + H_4 + H_5)\psi \tag{2.5}$$

where

$$\begin{aligned}
H_0 &= -\frac{e^2}{r} + \frac{1}{2}\left(\frac{1}{m_e} + \frac{1}{m_\mu}\right) p^2 \\
H_1 &= -\frac{1}{8c^2}\left(\frac{1}{m_e^3} + \frac{1}{m_\mu^3}\right) p^4 \\
H_2 &= \frac{-e^2}{2m_e m_\mu c^2}\ \frac{1}{r}\ (p^2 + p_r^2) \\
H_3 &= -\frac{\vec{r}\times\vec{p}}{r^3}\cdot\left[\frac{\mu_1 e_2}{m_e c}\vec{s}_1 + \frac{\mu_2 e_1}{m_\mu c}\vec{s}_2 + \frac{2\mu_1 e_2}{m_\mu c}\vec{s}_1 + \frac{2\mu_2 e_1}{m_e c}\vec{s}_2\right] \\
H_4 &= -\frac{ie^2\hbar}{(2c^2)}\left(\frac{1}{m_e^2} + \frac{1}{m_\mu^2}\right)\vec{p}\cdot\nabla\frac{1}{r} \\
H_5 &= 4\mu_1\mu_2\left[-\frac{8\pi}{3}(\vec{s}_1\cdot\vec{s}_2)\,\delta^3(\vec{r}) + \frac{1}{r^3}(\vec{s}_1\cdot\vec{s}_2 - 3s_{1r}s_{2r})\right]
\end{aligned} \tag{2.5a}$$

and in which

$$\vec{p} = \vec{p}_1 = -\vec{p}_2, \qquad \vec{r} = \vec{r}_{12} = \vec{r}_1 - \vec{r}_2$$

$$\mu_1 = -\mu_B^e = -\frac{e\hbar}{2m_e c}, \qquad \mu_2 = \mu_B^\mu = \frac{e\hbar}{2m_\mu c}$$

$$s_{1r} = \left|\vec{s}_1\cdot\vec{r}\right|/r, \qquad p_r^{\,2} = \vec{r}\cdot(\vec{r}\cdot\vec{p})\vec{p}/r^2$$

The wavefunction ψ consists of the largest 4 components of Ψ.

Evaluation of the energy levels accurate to order $\alpha^2 R_M$ can be made by perturbation theory. The zeroth-order approximation is the nonrelativistic Schroedinger equation

$$H_0\psi_0 = E_0\psi_0 \tag{2.6}$$

The energy eigenvalues are given in Eq.(2.1). The energy eigenstates are the nonrelativistic Schroedinger wavefunctions $\psi_0(\vec{r})$ in the center of mass system, and are chosen as the degenerate eigenfunctions characterized by the quantum numbers n,L,J,M_J for principal quantum number, orbital angular momentum, total angular momentum, and z component of total angular momentum, respectively. The energy term E_1^{fs} corresponding to fine structure in hydrogen is obtained as the diagonal matrix element:

$$\langle\psi_0\big| H_{fs}\big|\psi_0\rangle = E_1^{fs} \tag{2.7}$$

where

$$H_{fs} = -\frac{1}{8c^2}\frac{1}{m_e^3}p^4 + H_2 - \frac{\vec{r}\times\vec{p}}{r^3}\cdot\left(\frac{\mu_1 e}{m_e c}\right)\vec{s}_1 - \frac{ie^2\hbar}{(2c^2)}\frac{1}{m_e^2}\vec{p}\cdot\nabla\frac{1}{r} \tag{2.8}$$

The Hamiltonian term H_{fs} neglects terms in $H_1 + H_2 + H_3 + H_4 + H_5$ that are of relative order $(m_e/m_\mu)H_{fs}$ or smaller. Hence we obtain

$$E_1^{fs} = -\frac{\alpha^2 R_M}{n^3}\left(\frac{1}{j+\frac{1}{2}} - \frac{3}{4n}\right) \tag{2.9}$$

The leading term for the hfs energy contribution E_1^{hfs} is of order $(m_e/m_\mu)\alpha^2 R_M$ and is contained as a portion of H_3 and H_5:

$$\mathcal{H}_{hfs} = -\left(\frac{4\mu_1\mu_2}{r^3}\right)(\vec{r} \times \vec{p})\cdot\vec{s}_2 + \mathcal{H}_5 \tag{2.10}$$

We compute the contribution of $\mathcal{H}_{hfs}$ by perturbation theory using degenerate eigenfunctions characterized by the quantum numbers n,L,J, I,F,M_F, where $I = \frac{1}{2}$ is the muon spin quantum number, $F(\vec{F} = \vec{I} + \vec{J})$ the total angular momentum quantum number, and M_F the quantum number for the z component of total angular momentum:

$$E_1^{hfs} = \langle n,L,J,F,M_F | \mathcal{H}_{hfs} | n,L,J,F,M_F \rangle$$

$$E_1^{hfs} = \alpha^2 R_M \left(\frac{m_e}{m_\mu}\right)\frac{2}{n^3}\left[\frac{F(F+1) - I(I+1) - J(J+1)}{J(J+1)(2L+1)}\right] \tag{2.11}$$

II.3 Hyperfine Structure Interval in Ground State

Under the assumption that both the muon and the electron are Dirac particles with the conventional coupling to the electromagnetic field, the hfs interval $\Delta\nu$ of the ground n = 1 state of muonium can be calculated with modern quantum electrodynamic theory from the Bethe-Salpeter relativistic bound-state two-body equation. The result is expressed as a series expansion in the small parameters α and m_e/m_μ.

We write the formula for the muonium hfs energy interval in the form given by Sapirstein and Yennie.[18] The leading contribution to the hfs interval is given by the Fermi term:

$$E_F = \frac{8}{3}\alpha^4 c^2 \frac{m_r^3}{m_e m_\mu} = \frac{16}{3}\alpha^2 \frac{m_r^3}{m_e^2 m_\mu} hcR_\infty \tag{2.12}$$

in which $m_r = \frac{m_e m_\mu}{m_e + m_\mu}$. The second form is useful for evaluating E_F from the known values of the fundamental constants in order to minimize the error associated with their uncertainties.

The full expression for ΔE (hfs) including all calculated terms follows:

$$\Delta E(\text{hfs}) = h\Delta\nu = \Delta E\,(\text{hfs; QED}) + \Delta E\,(\text{hfs; rec}) + \Delta E\,(\text{hfs; rad-rec}) \tag{2.13}$$

The term ΔE (hfs; QED) includes all terms not involving recoil. The separation of binding effects from radiative corrections is indicated by including the factors $Z\alpha$ to refer to the binding effects.

$$\Delta E(\text{hfs; QED}) = E_F(1 + a_\mu)\left\{1 + \frac{3}{2}(Z\alpha)^2 + a_e + \alpha(Z\alpha)\left(\ln 2 - \frac{5}{2}\right) - \frac{8\alpha(Z\alpha)^2}{3\pi}\ln Z\alpha\left(\ln Z\alpha - \ln 4 + \frac{281}{480}\right) + \frac{\alpha(Z\alpha)^2}{\pi}(15.38 \pm 0.29) + \frac{\alpha^2(Z\alpha)}{\pi}D_1\right\} \tag{2.13a}$$

The factor $(1+a_\mu)$ provides for the muon anomalous magnetic moment. The D_1 term represents uncalculated radiative corrections involving two virtual photons.

The recoil corrections are given by:

$$\Delta E(\text{hfs; rec}) = E_F\left\{-\frac{3\alpha}{\pi}\frac{m_e m_\mu}{m_\mu^2 - m_e^2}\ln\frac{m_\mu}{m_e} + \frac{\gamma^2}{m_e m_\mu}\left[2\ln\frac{m_r}{2\gamma} - 6\ln 2 + 3\frac{11}{18}\right]\right\} \tag{2.13b}$$

where $\gamma = m_r\alpha$.

The last term gives radiative-recoil contributions which arise from both lepton lines and from vacuum polarization.

$$\Delta E(\text{hfs; rad-rec}) = E_F\left(\frac{\alpha}{\pi}\right)^2\frac{m_e}{m_\mu}\left[-2\ln^2\frac{m_\mu}{m_e} + \frac{13}{12}\ln\frac{m_\mu}{m_e} + \frac{21}{2}\zeta(3) + \frac{\pi^2}{6} + \frac{35}{9} + (1.9 \pm 0.3)\right] \tag{2.13c}$$

The hadronic vacuum polarization contribution is represented by the (1.9 ± 0.3) term.

For evaluation of $\Delta\nu$ from (2.13) values of the relevant fundamental constants are needed. With some exceptions to be indicated we use the values of the constants given in "The 1986 Adjustment of the Fundamental Constants."[20]

$$
\begin{aligned}
c &= 2.997\ 924\ 58 \times 10^{10}\ \text{cm/sec (exact)} \\
\alpha^{-1} &= 137.035\ 989\ 5\ (61)\ (0.045\ \text{ppm}) \\
R_\infty &= 109\ 737.315\ 73\ (2)\ \text{cm}^{-1}\ (0.000\ 2\ \text{ppm}) \\
a_e &= 1.159\ 652\ 193\ (10) \times 10^{-3}\ (0.009\ \text{ppm}) \\
a_\mu &= 1.165\ 923\ 0\ (84) \times 10^{-3}\ (7.2\ \text{ppm}) \\
\mu_p/\mu_B^e &= 1.521\ 032\ 202\ (15) \times 10^{-3}\ (0.01\ \text{ppm}) \\
\mu_\mu/\mu_p &= 3.183\ 345\ 5\ (9)\ (0.3\ \text{ppm}) \\
m_\mu/m_e &= 206.768\ 260\ (60)\ (0.3\ \text{ppm})
\end{aligned}
\tag{2.14}
$$

The value for R_∞ differs slightly from that of Cohen and Taylor and takes into account more recent measurements.[21,22,23] The value for μ_μ/μ_p is obtained from two experiments; the measurement of the Zeeman effect in the ground state of muonium[24] and a measurement of the μSR frequency of μ^+ in liquid bromine.[25] Cohen and Taylor give a somewhat more precise value for μ_μ/μ_p obtained by equating the experimental and theoretical values for the hfs interval $\Delta\nu$ of muonium. However for this article we wish to compare the theoretical and experimental values for $\Delta\nu$ and hence need an independent determination of μ_μ/μ_p. The mass ratio m_μ/m_e is determined from μ_μ/μ_p by the relation.

$$
\frac{m_\mu}{m_e} = \left(\frac{g_\mu}{2}\right)\left(\frac{\mu_p}{\mu_\mu}\right)\left(\frac{\mu_B^e}{\mu_p}\right) \tag{2.15}
$$

A convenient form for evaluating the Fermi factor E_F or $E_F(1+a_\mu)$ is

$$
\frac{E_F(1+a_\mu)}{h} = \frac{16}{3}\,\alpha^2 c R_\infty \left(\frac{\mu_\mu}{\mu_p}\right)\left(\frac{\mu_p}{\mu_B^e}\right)\left(1 + \frac{m_e}{m_\mu}\right)^{-3} \tag{2.16}
$$

Using Eqn. (2.13) and values of the fundamental constants from Eq. (2.14), we obtain

$$
\Delta\nu(\text{theory}) = 4\ 463\ 303.11(1.33)(0.40)(1.0)\ \text{kHz}\ (0.4\ \text{ppm}) \tag{2.17}
$$

The first uncertainty arises from the uncertainty in μ_μ/μ_p, the second from that in α, and the third is an order of magnitude estimate of the uncalculated D_1 contribution. If an accurate evaluation of the D_1 coefficient is made, then uncalculated contributions will be at the level of 1 part in 10^8.

A weak interaction contribution to $\Delta\nu$ arises from the exchange of the weakly interacting Z boson between electron and muon.[26,27] This is an axial vector-axial vector coupling due to the Hamiltonian term of Eq. (2.18)

$$H_{M\to\overline{M}} = \frac{G}{\sqrt{2}} \bar{\psi}_\mu \gamma_\lambda (1 + \gamma_5)\psi_e \bar{\psi}_\mu \gamma^\lambda (1 + \gamma_5)\psi_e + \text{H.C.} \tag{2.18}$$

in which G_F is Fermi coupling constant. Its contribution is

$$\Delta\nu(\text{weak}) = 0.07 \text{ kHz} \tag{2.19}$$

This interaction has been quantitatively studied with high energy e^+e^- storage rings through the forward-backward asymmetry in the reaction $e^+e^-\to\mu^+\mu^-$.[28]

Various speculations beyond the standard model involving new interactions or new particles have been made which would alter the theoretical value for $\Delta\nu$ given above. The theoretical formula for $\Delta\nu$ includes many higher order radiative processes and hence involves lepton and photon propagators, vertex functions, and bound state relativistic recoil terms. Hence its verification tests many aspects of QED and, most importantly, the muon-electron interaction. Possible causes of a breakdown have been discussed extensively in terms of a modification of lepton or photon propagators or vertex functions[29]; in more modern terms this question must be treated within the standard model with the electroweak interaction and its possible breakdowns.[30]

II.4 Zeeman Effect in $1^2S_{1/2}$ Ground State

The part of the Hamiltonian for muonium that is relevant for the energy splittings of the ground n = 1 state is[2]

$$H = a\vec{I}\cdot\vec{J} + \mu_B^e g_J \vec{J}\cdot\vec{H} - \mu_B^\mu g'_\mu \vec{I}\cdot\vec{H} \tag{2.20}$$

in which a is the hfs coupling constant ($\equiv h\Delta\nu$), I the muon spin operator, J the electron total angular momentum operator, g_J the electron gyromagnetic ratio in muonium, g'_μ the muon gyromagnetic ratio in muonium, and H the external static magnetic field. The quantities g_J and g'_μ are related to the free electron and muon g values g_e and g_μ by the equations[31]

$$g_J = g_e\left(1 - \frac{1}{3}\alpha^2 + \frac{\alpha^2}{2}\frac{m_e}{m_\mu} + \frac{\alpha^3}{4\pi}\right) \qquad g'_\mu = g_\mu\left(1 - \frac{1}{3}\alpha^2 + \frac{\alpha^2}{2}\frac{m_e}{m_\mu}\right) \tag{2.21}$$

The differences of g_J and g'_μ from the free values g_e and g_μ are due to relativistic binding contributions.

The energy eigenvalues associated with the Hamiltonian of Eq.(2.20) are[2]

$$W_{F=\frac{1}{2}\pm\frac{1}{2},M_F} = -\frac{1}{4}\Delta W - \mu_B^\mu g'_\mu M_F H \pm \frac{1}{2}\Delta W\,(1 + 2M_F x + x^2)^{\frac{1}{2}} \tag{2.22}$$

in which $a = \Delta W = h\Delta\nu$; $x = (g_J\mu_B^e + g'_\mu\mu_B^\mu)H/\Delta W$. From Eqs. (2.17) and (2.21), we find $x = H/1585$. The energy level diagram corresponding to Eq.(2.22) is shown in Fig. 2.1 with the states labeled by numbers 1-4. The strong field quantum numbers M_J (magnetic quantum number for component of electronic angular momentum) and M_μ (magnetic quantum number for z component of muon spin) are indicated.

The energy eigenfunctions can be written

$$\psi = \phi(r)\chi_{F,M_F}(H) \tag{2.23}$$

in which $\phi(r)$ is the spatial part of the wavefunction and $\chi_{F,M_F}(H)$ is the spin eigenfunction. The spin eigenfunctions can be expressed in terms of the strong-field Pauli spin eigenfunctions:

$$\chi_{1,1}(H) = \alpha_e\alpha_\mu, \qquad \chi_{1,0}(H) = c\alpha_e\beta_\mu + s\beta_e\alpha_\mu$$
$$\chi_{1,-1}(H) = \beta_e\beta_\mu, \qquad \chi_{0,0}(H) = c\beta_e\alpha_\mu - s\alpha_e\beta_\mu \tag{2.23a}$$

where α_μ, β_μ are the normalized spin eigenfunctions of μ^+, α_μ corresponds to spin orientation along the positive z direction ($M_\mu = +\frac{1}{2}$) and

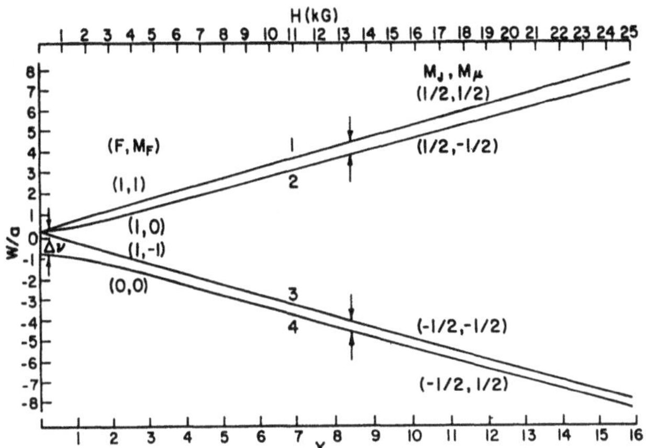

Figure 2.1. Breit-Rabi energy level diagram for muonium in its $1^2S_{1/2}$ ground state in a magnetic field.

β_μ to the opposite spin direction ($M_\mu = -\frac{1}{2}$); α_e, β_e are similarly defined spin eigenfunctions of the electron. The quantities s(H) and c(H) are given by

$$s = \frac{1}{\sqrt{2}}\left[1 - \left(\frac{x}{1+x^2}\right)^{1/2}\right]^{1/2} = \sin\left(\frac{1}{2}\,\text{arc cot}\ x\right)$$

$$c = \frac{1}{\sqrt{2}}\left[1 + \left(\frac{x}{1+x^2}\right)^{1/2}\right]^{1/2} = \cos\left(\frac{1}{2}\,\text{arc cot}\ x\right) \tag{2.23b}$$

where $s^2 + c^2 = 1$.

Corrections to the Breit-Rabi equation (2.22) arise because of excited states of higher n. The hfs operator $a\,\vec{I}\cdot\vec{J}$ in Eq.(2.20) is not correct in the sense that it does not have matrix elements to higher n states; however, this correction is of relative order $(a/Ry)^2 \sim 10^{-12}$ and can be neglected. There will also be off-diagonal matrix elements of the Zeeman operator to higher n states and associated corrections to the Breit-Rabi equation of order $(\mu_B^e H/Ry)^2$, but for practically achievable magnetic fields of 50 kG or less, the modifications are negligible.

II.5 Fine Structure and Lamb Shift in n = 2 Excited State

The energy level separations in the n = 2 state involve the fine structure, the Lamb shift and hyperfine structure terms. For the pure quantum electrodynamics problem involved in muonium the levels are in principle exactly calculable (apart from relatively small virtual effects arising from hadronic interactions) and do not as for hydrogen involve the uncertainty of proton structure. The energy levels of muonium in the n = 2 state are very similar to those of hydrogen.

Figure 2.2 shows the energy-level diagram for muonium in the n = 1 and n = 2 states.[32] In the absence of hyperfine interaction and Zeeman splitting, the unperturbed states can be specified by quantum numbers n,L,J,m_J with the Lamb-shift and fine-structure energy intervals designated $\mathbf{S}_M = E(2^2S_{1/2}) - E(2^2P_{1/2})$ and $\Delta E = E(2^2P_{3/2}) - E(2^2P_{1/2})$. The theoretical expression for $\mathbf{S}_M$ (in MHz) is[18,32]

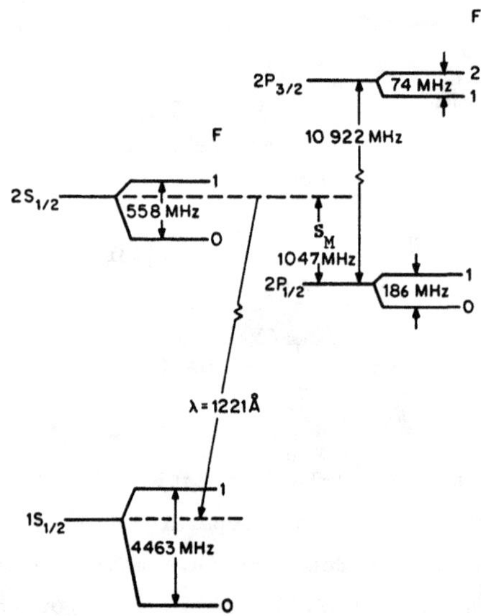

Figure 2.2. Energy level diagram for muonium in its n = 1 and n = 2 states, including fine structure, Lamb shift, and hfs terms.

$$\mathcal{S}_M = \frac{\alpha^3 R_\infty c}{3\pi}\ \{[\ln(\alpha^{-2})-2.208\ 45-\ln(\frac{m_r}{m_e})][\frac{m_r}{m_e}]^3 + \frac{1}{8}[\frac{m_r}{m_e}]^2 + 2.2962\pi\alpha[\frac{m_r}{m_e}]^3$$
$$+\alpha^2[-\frac{3}{4}\ln^2(\alpha^{-2})+3.9184\ \ln(\alpha^{-2})+G(\alpha)]\ [\frac{m_r}{m_e}]^3$$
$$+0.323\frac{\alpha}{\pi}\ [\frac{m_r}{m_e}]^3 + [\frac{1}{4}\ln(\alpha^{-2})+2.399\ 77+\frac{3\pi}{4}\alpha[3-\ln(2/\alpha)]]\frac{m_e}{m_\mu}$$
$$-4.684(9)\alpha\frac{m_e}{m_\mu}\} - \frac{\alpha^2 R_\infty c}{24}\ [\frac{m_e}{m_\mu}]^2$$
$$= 1047.490(300)\ \text{MHz}. \qquad (2.24)$$

In evaluating Eq. (2.24) we use $G(\alpha) = -24 \pm 1.2$ for the binding correction. The contributions of the various terms to the Lamb shift in muonium are summarized in Table 2.1 The reduced mass and relativistic recoil terms, which involve the factor m_e/m_μ, are of course larger for muonium than for hydrogen. Indeed, the present theoretical uncertainty in $\mathcal{S}_M$ arises from uncalculated terms of order $\alpha^3 R_\infty\ (m_e/m_\mu)^2$ and $\alpha^4 R_\infty\ (m_e/m_\mu)$.

The theoretical value of the fine-structure (FS) interval (in MHz) is given by[33]

$$\Delta E = \frac{\alpha^2 R_\infty c}{16}\ [(1+\frac{5}{8}\alpha^2)[\frac{m_r}{m_e}] - [\frac{m_e}{m_\mu}]^2\ [\frac{m_r}{m_e}]^3\ +2a_e\ [\frac{m_r}{m_e}]^2\ - \frac{\alpha^3}{\pi}[\ln(\alpha^{-2})+\delta_{FS}]]$$
$$= 10\ 921.832(2)\ \text{MHz}, \qquad (2.25)$$

where $a_e = (g_e - 2)/2$ is the electron-magnetic-moment anomaly and $\delta_{FS} = \frac{11}{24} \pm \frac{3}{2}$ is a radiative correction. The 2-kHz uncertainty in ΔE is due mainly to the uncertainty in δ_{FS}.

The hyperfine interaction has the form

$$\mathcal{H}_{hfs}(n,L,J) = a(n,L,J)\vec{I}\cdot\vec{J}, \qquad (2.26)$$

and the hyperfine-structure interval for a fine-structure term n,L,J is the difference of the expectation values

Table 2.1

Contributions to the theoretical value of the Lamb shift in muonium. The uncertainty in the theoretical value is due to uncalcaulated terms of higher order in m_e/m_μ, i.e., terms (m_e/m_μ)(reduced mass term) and α(reduced mass term).

Correction	Order (mc^2)	Value (MHz)
Self-energy	$\alpha(Z\alpha)^1[\ln(Z\alpha)^{-2},1,Z\alpha,\ldots]$	1085.812
Vacuum polarization	$\alpha(Z\alpha)^4(1,Z\alpha,\ldots)$	-26.897
Fourth order	$\alpha^2(Z\alpha)^4$	0.102
Reduced mass	$\alpha(Z\alpha)^4(m_e/m_\mu)[\ln(Z\alpha)^{-2},1]$	-14.493
Relativistic recoil	$(Z\alpha)^5(m_e/m_\mu)[\ln(Z\alpha)^{-2},1,Z\alpha]$	3.159
Higher-order recoil	$(Z\alpha)^4(m_e/m_\mu)^2$	-0.171
Radiative recoil	$\alpha(Z\alpha)^5 m_e/m_\mu$	-0.022
		1047.490(300)

$$\langle H_{hfs}(n,L,J)\rangle = \frac{g_\mu m_e}{n^3 m_\mu}\,\alpha^2 hcR_\infty\,\frac{F(F+1)-I(I-1)-J(J+1)}{(2L+1)J(J+1)} \tag{2.27}$$

with $F=J+I$ and $F=J-I$, where $I=\frac{1}{2}$ for the muon spin. For the 1S ground state the hfs interval $\Delta\nu$ has been measured and calculated to high precision, and $\Delta\nu_{expt}$ = 4 463 302.88(16) kHz, in good agreement with theory. The leading term for $\Delta\nu$ in the 2S state scales with $1/n^3$; hence $\Delta\nu(2^2S_{1/2})\approx\frac{1}{8}\Delta\nu(1^2S_{1/2})$. Similar considerations of the leading terms for the P states give $\Delta\nu(2^2P_{1/2})\approx\frac{1}{3}\Delta\nu(2^2S_{1/2})$ and $\Delta\nu(2^2P_{3/2})\approx\frac{2}{15}\Delta\nu(2^2S_{1/2})$. These expressions give the energy intervals between the F eigenstates (Fig.2.2) to a precision sufficient for our present purpose. In our experiment the magnetic field was less than 1G, and hence the Zeeman effect was negligible.

II.6 Muonium Decay

Muonium is an unstable atom because of the decay of the muon due to the weak interaction. Its dominant mode of decay is due to that of the free positive muon: [34,35]

$$\mu^+ \rightarrow e^+ + \nu_e + \bar{\nu}_\mu \tag{2.28}$$

The free muon decay rate γ or the mean life τ_μ ($\gamma \equiv 1/\tau_\mu$), given approximately by[36]

$$\gamma = 1/\tau_\mu = m_\mu{}^5 G_F{}^2/192\pi^3 \tag{2.29}$$

in which G_F is the Fermi coupling constant. The current experimental value for τ_μ is[37]

$$\tau_\mu = 2.197\ 03(4) \times 10^{-6}\ \text{sec (18 ppm)} \tag{2.30}$$

The momentum and angle spectrum for the positrons is given with high accuracy upon neglect of terms for radiative corrections and of terms involving m_e/m_μ by the equation

$$N(y,\theta) \propto 4y^2\{3(1-y)+2\rho(\tfrac{4}{3}\ y-1) - P\xi[1 - y + 2\delta(\tfrac{4}{3}\ y - 1)]\cos\theta\} \tag{2.31}$$

for positive muons with degree of polarization P, in which y is the positron momentum in units of $\frac{1}{2}\ m_\mu c$; θ is the angle between the muon spin direction and the positron momentum; ρ is the spectrum shape parameter (Michel parameter) ($=\frac{3}{4}$); ξ is the asymmetry parameter ($=-1$); δ is the energy dependence of the asymmetry parameter ($=\frac{3}{4}$). The positron spectrum extends up to a maximum total energy of 52.8 MeV. The positron asymmetry spectrum or angular distribution, $N(\theta)$, is obtained by integrating $N(y,\theta)$ from y=0 to 1:

$$N(\theta) \propto 1 - \frac{1}{3}\ A\cos\theta, \tag{2.32}$$

in which $A = P\xi$. Figure 2.3 shows the theoretically predicted momentum and asymmetry spectrum for the decay positrons from positive muon decay based on the standard V-A theory.[36] Figure 2.4 shows a comparison of the experimental and theoretical momentum spectra.[34]

Other then the ordinary muon decay of Eq.(2.28) radiative muon decay is the only muon decay so far observed[37]

$$\mu^+ \to e^+ + \nu_e + \bar{\nu}_\mu + \gamma \qquad \text{(branching ratio} \simeq 10^{-3}) \quad (2.33)$$

Small upper limits have been set on various possible decay modes:[37,38]

$$\mu^+ \to e^+ + \gamma,\ \mu^+ \to e^+ + e^+ + e^-,\ \mu^+ \to e^+ + \gamma + \gamma \quad (2.34)$$

Another possible mode of disappearance of muonium is the reaction

$$\mu^+ + e^- \to \nu_e + \bar{\nu}_\mu \quad (2.35)$$

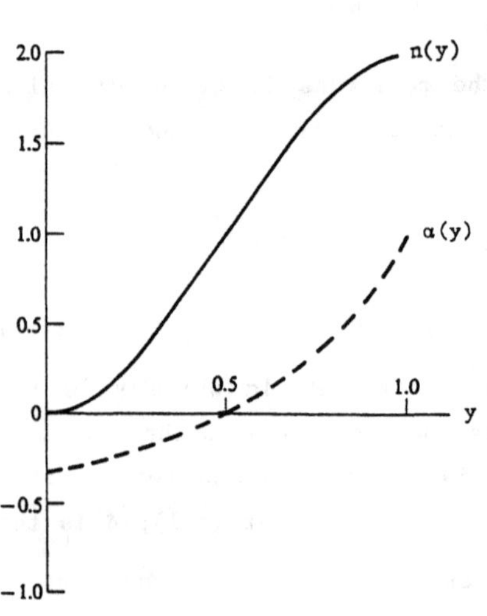

Figure 2.3. Normalized positron spectrum $n(y)=2y^2(3-2y)$ and asymmetry $\alpha(y) = \frac{2y-1}{3-2y}$ for fully polarized μ^+ decay, as predicted by the V-A theory, in the approximation $m_e = 0$. $N(y,\theta) \propto n(y)[1+\alpha(y)\cos\theta]$.

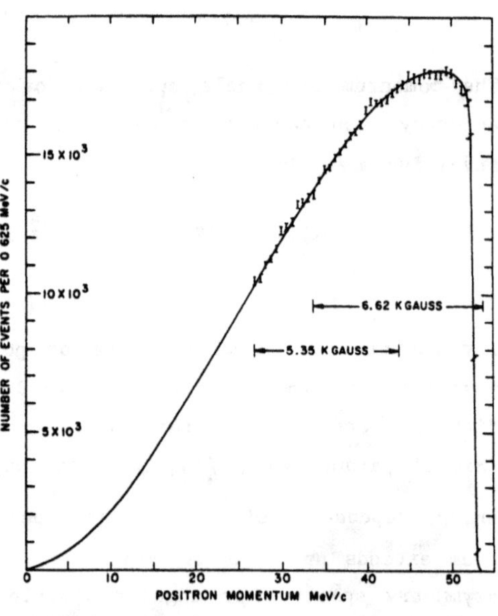

Figure 2.4. Momentum spectrum of positrons from positive muon decay.[34] The experimental points are compared to the theoretical curve for $\rho = \frac{3}{4}$, modified for radiative corrections and experimental resolution.

which is the decay reaction of Eq. (2.28) in which the muonium electron partakes. The probability of the reaction Eq. (2.35) relative to the dominant mode of muon decay Eq. (2.28) is about 10^{-10}.[39]

III. DISCOVERY OF MUONIUM AND MUONIUM FORMATION

III.I Theoretical Considerations on Muonium Formation

In the discovery of muonium and for measurements of the energy levels of the n = 1 state muonium atom it has been important to form muonium in a gas, and indeed to utilize as low a gas pressure as possible to minimize the perturbing effects of neighboring atoms on the energy levels of muonium. To avoid chemical reactions and other depolarizing reactions of muonium, which is an isotope of hydrogen and a highly reactive paramagnetic atom, it has been necessary for formation and ground state energy level measurements to use pure inert gases; argon and krypton have been used prinicpally thus far. For the explicit study of the reaction of muonium with other atoms and molecules, these atoms or molecules were introduced in small fractional amounts with the inert gas.[40]

Muonium can be formed in a collision of a positive muon with an atom by an electron capture reaction, e.g., with argon,

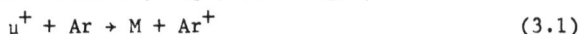

$$\mu^+ + Ar \rightarrow M + Ar^+ \qquad (3.1)$$

The muonium M can be in any one of its bound states, including the ground n = 1 state.

An energy level diagram that is useful for a discussion of muonium formation in helium and argon is shown in Fig. 3.1 With reference to helium the zero energy level for an electron is taken to be that of an electron bound in the ground state of helium. The ionization energy of helium $E_I(He)$ is 24.58 eV and that of muonium $E_I(M)$ is 13.54 eV. Hence, in order that a muon can capture an electron from a helium atom to form muonium in its ground state, the kinetic energy of the muon-helium system in its c.m. coordinate system must be greater than the threshold value

$$E_t(CM) = E_I(He) - E_I(M) \qquad (3.2)$$

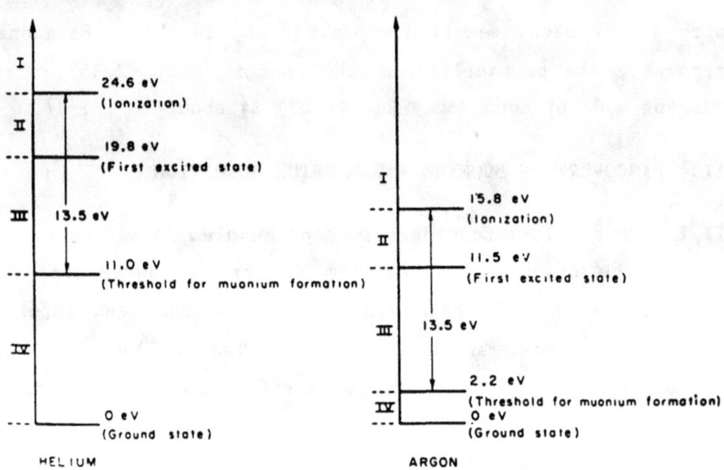

Figure 3.1. Energy level diagram relevant to muonium formation in helium and argon.

which equals 11.04 eV. In the laboratory coordinate system with the helium atom initially at rest, the kinetic energy of the muon required for muonium formation is given by

$$E_t(\mathrm{lab}) = E_t(\mathrm{CM})m_\mu/m_r \tag{3.3}$$

in which $m_r = m_\mu m_{He}/(m_\mu + m_{He})$ where m_{He} is the mass of a helium atom. Hence the threshold kinetic energy of the muon is $E_t(\mathrm{lab}) = 11.3$ eV. Formation of muonium in its excited n=2 state by a charge capture reaction with helium will require a kinetic energy for the muon-helium system of 21.2 eV. Threshold energies for formation of muonium for several atoms and molecules are shown in Table 3.1. Note that for some cases, e.g. xenon, E_t is negative, and hence muonium formation is always possible energetically.

The theory of the charge capture reaction as an example of a rearrangement collision has received much attention since the early days of the quantum theory of atomic collisions, but accurate values for inelastic charge capture cross sections have not been calculated due to the mathematical complexity of the problem.[41,42] However, the general features of the charge capture cross section involved in muonium formation can be predicted theoretically. Since the mass of the muon is

TABLE 3.1
Threshold Energies E_t(CM) for Muonium Formation

Atom or molecule	E_t(CM)(eV)
He	+11.04
Ne	+8.02
Ar	+2.22
Kr	+0.46
Xe	-1.41
SF_6	+2.1
N_2	+2.0
CH_4	-0.6
N_2O	-0.7
O_2	-1.5

207 times the mass of the electron, the muon can be treated as a heavy particle similar to the proton in atomic reactions. In the so-called adiabatic region above the threshold energy, the capture cross section for a reaction such as Eq. (3.1) rises rapidly with muon kinetic energy E_μ and has approximately an exponential dependence on the inverse of the muon velocity. The cross section has a relatively flat maximum for muon velocities v_μ that satisfy the condition

$$a|\Delta E|/\hbar v_\mu \simeq 1 \qquad (3.4)$$

where $|\Delta E|$ is the magnitude of the change in internal energy involved in the reaction ($|\Delta E|$ = 2.2 eV for the argon and muonium in its ground n=1 state) and a is a distance parameter that characterizes the range of the muon-atom interaction potential ($a \simeq 8 \times 10^{-8}$ cm). In the high-energy region, as E_μ increases the cross section decreases monotonically toward zero with a high inverse power of v_μ.

Theoretical estimates of the cross section for muonium formation in its ground state have been made for the case of electron capture from helium:[10]

$$\mu^+ + He \rightarrow M(1S) + He^+ \qquad (3.5)$$

The results shown in Fig. 3.2 agree with the general features of the capture cross section just described.

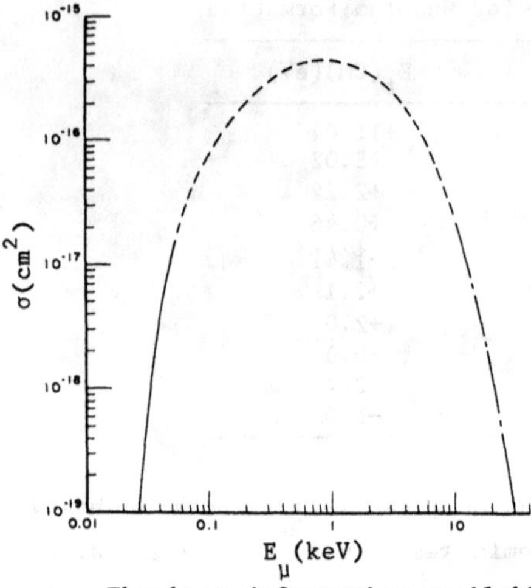

Figure 3.2. Calculated capture cross section for the reaction $\mu^+ + He \rightarrow M + He^+$, in which muonium M is formed in its ground state by electron capture from helium, as a function of muon kinetic energy E_μ. The three portions of the theoretical curve are based on the following theoretical approximations:——, at low energy E_μ, the adiabatic approximation or the perturbed stationary state method; ---, at intermediate energy E_μ, an impact parameter method; ——.——.——, at high energy E_μ, a Born approximation.[10]

The best information available on charge capture reactions relevant to muonium formation in gases is obtained from experimental data on charge capture reactions of protons to form hydrogen atoms.[42] Argon has been an important gas for muonium studies and Fig. 3.3 shows the measured capture cross section for the reaction

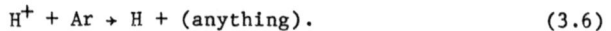

$$H^+ + Ar \rightarrow H + (\text{anything}). \qquad (3.6)$$

The measurements yield the total cross section for producing neutral H, including its excited states, and do not determine the final charge state of the argon ion. Direct capture into the ground state probably accounts for about 0.9 of the total cross section. From our theoretical understanding of the charge capture reaction we know that the cross section curve for the corresponding muon charge capture reaction will be similar in form to Fig. 3.3, and the cross section values will be approximately equal when the velocity of the muon equals the velocity of the proton.

Information relevant to the question of what fraction of the muons stopping in a gas will form muonium is provided by measurements of the equilibrium charge states of protons passing through a gas. Figure 3.4 shows experimental data on the equilibrium fractions of the beam in the charge states H^+, H, and H^- (F_{+1}, F_0, and F_{-1}, respectively) as a function of proton kinetic energy in argon starting from the lowest

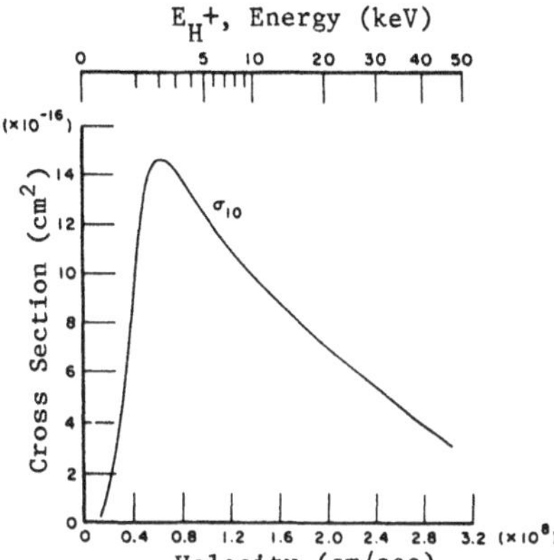

Figure 3.3. Measured cross section for the reaction H^{+} Ar → H + (anything), as a function of the velocity or kinetic energy of the incident proton.

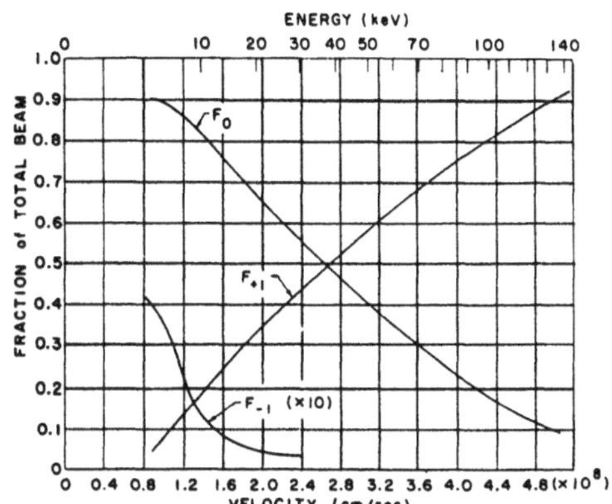

Figure 3.4. Measured equilibrium fractions of the hydrogen charge states H^{+}, H, and H^{-} (F_{+1}, F_0, and and F_{-1}) in argon as a function of hydrogen velocity or kinetic energy.

energy of about 5 keV at which data are available. The ratio of the fraction of the beam in the charge state +1 to that in 0, F_1/F_0, at a given energy is approximately equal to the ratio of the cross sections for ionization and charge capture, σ_{01}/σ_{10}. The measurements determine only the charge state of the proton and for the neutral form H do not establish directly that H is in its ground state; however, the neutral form is expected to be dominantly in the ground state. At a kinetic energy of 5 keV a fraction of the beam greater than 0.9 is in the neutral state H. The muon and muonium cross sections will be about the same as the H^{+} and H cross sections at the same velocities.

Although these data on equilibrium charge states of protons in argon and the expected similar data for muons in argon suggest that a large fraction of the muons stopping in argon will form muonium in the ground state, a quantitative calculation of this fraction requires a simultaneous numerical integration of the differential equations for the charge fractions of the muon beam and the energy loss (dE/dx) from high energies to thermal energy.[43] Many atomic processes will be important and indeed molecular formation of μ^{+}Ar could also be significant.[44] At energies below about 5 keV these cross section data and the dE/dx curve for protons, as well as the applicability of proton results to muons, are not known well enough to allow accurate predictions of muonium formation. Current estimates are that the fraction of positive muons stopping in argon that form muonium will be between 0.7 and 0.9.

Since the incident muons are polarized because of their origin from pion decay, muonium atoms are formed with a net polarization, and indeed it is essential for many experiments with muonium that the muonium atoms be polarized. We shall specialize our discussion to the formation of the n=1 ground state, including its four hfs substates. The electron capture reaction by which muonium is formed is dominated by the Coulomb interaction, and magnetic interactions can be neglected in the formation process.

In general, a superposition of the four hfs energy eigenstates is formed and the appropriate description for an ensemble of muonium atoms involves the density matrix ρ.[45,46] We assume that the incident muons have a polarization P in the z direction and that the atomic electrons are unpolarized. The density matrices for the muons and for the electrons in the basis states M_μ and M_J respectively are

$$\rho_\mu = \frac{1}{2}\begin{pmatrix} 1+P & 0 \\ 0 & 1-P \end{pmatrix}; \quad \rho_e = \frac{1}{2}\begin{pmatrix} 1 & 0 \\ 0 & 1 \end{pmatrix} \tag{3.7}$$

Hence calculation of the combined muon and electron, or muonium, density matrix ρ for magnetic field $H = 0$ in the basis states (F,M_F) yields

$$\rho(0) = \frac{1}{4}\begin{pmatrix} 1+P & 0 & 0 & 0 \\ 0 & 1 & 0 & +P \\ 0 & 0 & 1-P & 0 \\ 0 & +P & 0 & 1 \end{pmatrix} \tag{3.8}$$

The symbol $\rho(0)$ is the density matrix ρ at time t=0. The matrix element is $\rho_{nm}(0) = \langle a_n^*(0)a_m(0)\rangle$, where the ensemble average is indicated and $a_n(0)$ and $a_m(0)$ are the amplitudes of the zero magnetic field eigenstates in the wavefunction for an atom in the ensemble. The diagonal elements $\rho_{ii}(0)$ give the probabilities of formation of the hfs states 1-4:

$$p_1 = (1+P)/4, \; p_2 = \frac{1}{4}, \; p_3 = (1-P)/4, \; p_4 = \frac{1}{4} \tag{3.9}$$

>he time-dependent density matrix elements are given by

$$\rho_{nm}(t) = \rho_{nm}(0)\ \exp[-i(\omega_n - \omega_m)t] \tag{3.10}$$

in which $\omega_n = W_n/\hbar$, where W_n is given in Eq.(2.22) with H=0.

It is useful to have the density matrix also for the case in which there is a magnetic field H along the z direction. Corresponding to Eq. (3.8), we have

$$\rho(0) = \frac{1}{4}\begin{pmatrix} 1+P & 0 & 0 & 0 \\ 0 & 1+P(s^2 - c^2) & 0 & 2csP \\ 0 & 0 & 1-P & 0 \\ 0 & 2csP & 0 & 1+P(c^2 - s^2) \end{pmatrix} \tag{3.11}$$

where c and s are defined by Eq.(2.23b). The basis states are the energy eigenstates in the magnetic field H as given in Eq.(2.23a). The formation probabilities of hfs states 1-4 are

$$\begin{aligned} p_1 &= 1 + P, & p_2 &= 1 + P(s^2 - c^2), \\ p_3 &= 1 - P, & p_4 &= 1 + P(c^2 - s^2). \end{aligned} \tag{3.12}$$

III.2 Discovery of Muonium and Other Muonium Formation Experiments

The discovery of muonium was made by studies of the polarization of muons stopped in gases.[10] The basic idea was to observe the characteristic Larmor precession frequency of muonium in an external magnetic field perpendicular to the z direction of the incident muon polarization. Classically, a body with magnetic moment $\vec{\mu}$ and angular momentum $\vec{I}$ will precess in a magnetic field perpendicular to $\vec{H}$ with the frequency

$$f = \mu H/2\pi I \tag{3.13}$$

For muonium, the relevant initial state is state 1 with $(F,M_F) = (1,1)$, which has $\mu = \mu_B^\mu - \mu_B^e \simeq -\mu_B^e$, $I = \hbar$, and hence

$$f = \mu H/Fh = 1.40\ H(G)\ \text{MHz}. \tag{3.14}$$

The muon magnetic moment is collinear with the total atomic magnetic moment μ in this state. Since the muon moment direction can be observed through the angular asymmetry of the decay positron distribution as given by Eq.(2.31), the direction of the total atomic magnetic moment is also thus determined.

The quantum mechanical treatment requires the evaluation of the expectation value of the muon polarization P and the use of the density matrix. It is sufficient to consider the z component of the muon polarization:

$$\begin{aligned} P_z(t) &= P_{z_0}(t)e^{-\gamma t} = \langle\psi|2I_z|\psi\rangle \\ &= \sum_{i,j=1}^{4} a_i^* a_j \langle\chi_i|2I_z|\chi_j\rangle \exp(i\omega_{ij}t) \end{aligned} \tag{3.15}$$

in which

$$\psi = \phi(r) \sum_{i=1}^{4} a_i(t)\chi_i \exp(-iW_i t/\hbar) \tag{3.16}$$

and χ_i are the zero magnetic field eigenstates given in Eq.(2.23a) with $H = 0$. The state amplitudes $a_i(t)$ satisfy the Schroedinger equation with the Hamiltonian of Eq.(2.20) with $H = H_x$:

$$\mathcal{H} = a\vec{I}\cdot\vec{J} + \mu_\beta^e\, g_J J_x H - \mu_B^\mu\, g'_\mu I_x H \tag{3.17}$$

Since the basis states are eigenstates of the Hamiltonian term $a\vec{I}\cdot\vec{J}$, the state amplitudes satisfy the equations

$$\dot{a}_i = -\frac{i}{\hbar} \sum_j a_j V_{ij} \exp\left[-\frac{i}{\hbar}(W_j - W_i)t\right] - \frac{\gamma}{2} a_i \tag{3.18}$$

where

$$V = \mu_B g_J^e J_x H - \mu_B^\mu g'_\mu I_x H \tag{3.18a}$$

The matrix elements V_{ij} are given in Table 3.2. Evaluation of the ensemble average $\langle P_z(t)\rangle$ using Eqs. (3.15-3.18) and Table 3.2 and assuming that the initial muon polarization is equal to 1 gives

$$\langle P_z(t)\rangle = \frac{1}{2}\sin[\partial\omega_{12}/\partial H]Ht \tag{3.19}$$

In the limit of very weak magnetic field, Eq.(3.19) gives the classical Larmor precession of Eq.(3.14).

The experimental arrangement used in the discovery of muonium[10] is shown in Fig. 3.5. Simple Helmholtz coils were used to provide a homogeneous magnetic field of about 4 G. The argon gas in the target was initially of high-purity grade, and the argon was recirculated over hot titanium for further purification, since it was expected and subsequently verified that muonium could be depolarized in collisions, e.g. with paramagnetic O_2 molecules. The time distribution of the decay positrons with respect to the stopping time of the muons (defined by the $12\bar{3}$ counts) was measured with fixed counters (45 counts). Fourier analysis of the data indicated a strong frequency component at the characteristic muonium Larmor precession frequency as indicated in Fig. 3.6. Analysis showed that between 50 and 100% of the muons stopped in argon form muonium in agreement with the theoretical expectation.

Because the value of the muon spin is involved of course in determining the muonium precession frequency of Eqs.(3.13) and (3.14), this experiment can be interpreted as a measurement of the muon spin[47] and establishes that I = 1/2.

Abundant muonium formation in krypton was subsequently observed in microwave magnetic resonance measurements of muonium energy levels.[48] Later studies of muonium formation in noble gases and in noble gas mixtures have been made by observation of the characteristic muonium Larmor precession [Eq.(3.14)].[43] The results are given in Table 3.3, in which $F_\mu(F_M)$ is the fraction of muons in the gas that remain free (form muonium). These results are understood qualitatively in terms of the electron capture and loss cross sections. The small admixture of xenon serves as an electron donor to the positive muon even if the muon is at rest. Also recent experiments with positive

Table 3.2

Matrix Elements of $(1/2\hbar)(g_J\mu_B^e \vec{J} - g'_\mu\mu_B^\mu \vec{I})\cdot\vec{H}$ Between Muonium States of Eq.(2.23a)[a]

Final state	1	2	3	4
1	$(B_j + B_I)H_z$	$(sB_J + cB_I)(H_x - iH_y)$	0	$(cB_J - sB_1)(H_x - iH_y)$
2	$(sB_j + cB_1)(H_x + iH_y)$	$(c^2 - s^2)(B_j - B_I)H_z$	$(cB_J + sB_I)(H_x - iH_y)$	$-2sc(B_J - B_I)H_z$
3	0	$(cB_J + sB_I)(H_x + iH_y)$	$-(B_J + B_I)H_z$	$(-sB_J + cB_I)(H_x + iH_y)$
4	$(cB_J - sB_I)(H_x + iH_y)$	$-2sc(B_J - B_I)H_z$	$(-sB_J + cB_I)(H_x - iH_y)$	$(s^2 - c^2)(B_J - B_I)H_z$

[a] s and c are defined by Eq.(2.23a) and x is defined for Eq.(2.22). The quantities B_J and B_I are defined by $B_J = -\frac{1}{4}(g_J\mu_B^e/\hbar)$ and $B_I = -\frac{1}{4}(g'_\mu\mu_B^\mu/\hbar)$.

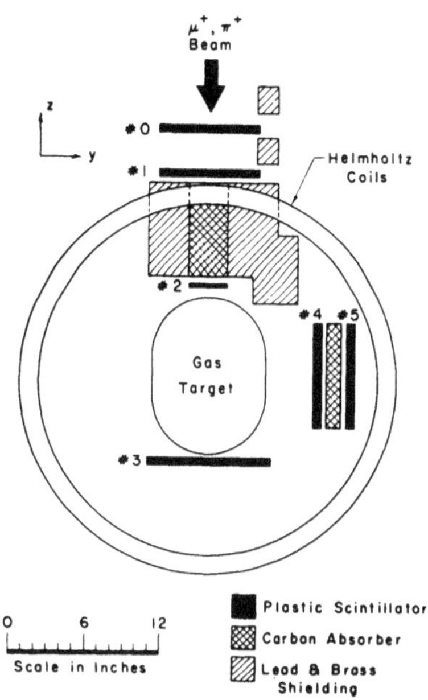

Figure 3.5. Schematic diagram of the experimental arrangement used for the study of the depolarization of muons in gases and of the Larmor precession of muonium.[10]

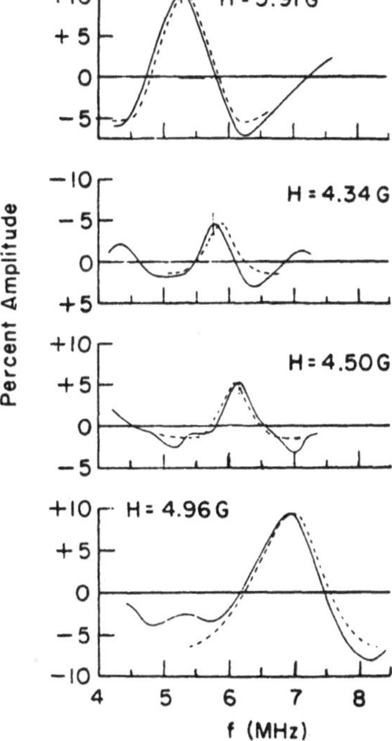

Figure 3.6. Frequency analysis, illustrating the precession of muonium when polarized muons are stopped in argon gas, for several magnetic field values H. The percentage amplitude of a frequency component is plotted versus frequency (——, experimental curve; ---theoretical curve assuming Larmor precession of muonium).[10]

TABLE 3.3

Experimental Results on Muonium Formation in Noble Gases[43]

Target Gas	F_{μ} (%)	F_{M} (%)
He	99(5)	1(5)
He + 0.015% Xe	83(15)	———
He + 0.09% Xe	25(9)	75(9)
Ne	100(2)	0(2)
Ne + 0.15% Xe	19(3)	81(3)
Ar	35(5)	65(5)
Xe	10(5)	100 [a]

[a]No error estimate given.

muons in liquid helium observed the free muon Larmor precession but no muonium formation.[49]

IV. HYPERFINE STRUCTURE AND ZEEMAN EFFECT IN GROUND STATE MUONIUM

Precise measurements have been made of the hfs energy levels of muonium in its ground state by the microwave magnetic resonance method. For muonium the possibility of performing a magnetic resonance measurement of the hfs energy intervals in the ground state depends on the formation of polarized muonium and on the angular asymmetry of the decay positrons as indicator of the muon spin direction. Transitions by absorption or stimulated emission can be induced by applied rf or microwave power at resonant frequencies of muonium and can be detected through the resulting change in the decay positron angular distribution.

The quantities in the Hamiltonian of Eq.(2.20) for the hfs and Zeeman effect of muonium in its ground state that are regarded as the unknowns to be measured are those involving the muon, and include the muon spin $\vec{I}$, the muon magnetic moment in muonium $g'_{\mu}\mu_B^{\mu}\,\vec{I}$, and the hfs constant a (or the hfs interval $\Delta\nu$). The quantities associated with the electron only, namely, its spin $\vec{J}$ and its magnetic moment in muonium (the same as in hydrogen) $g_J\mu_B^e\,\vec{J}$, are known from other measurements. The unknowns can be determined in various ways by measuring various transitions between the hfs levels in different magnetic fields as will be discussed below.

IV.1 Transition Frequencies

The transition frequencies can be obtained from the energy eigenvalues Eq.(2.22). The four transition frequencies ν_{12}, ν_{14}, ν_{24}, and ν_{34} are of particular interest; these frequencies are given below and are plotted in Figure 4.1.

$$
\begin{aligned}
\nu_{12} &= \frac{W_{1,1} - W_{1,0}}{h} = \frac{-\mu_B^\mu g'_\mu H}{h} + \frac{\Delta\nu}{2}\left[(1+x) - (1+x^2)^{1/2}\right] \\
\nu_{14} &= \frac{W_{1,1} - W_{0,0}}{h} = \frac{-\mu_B^\mu g'_\mu H}{h} + \frac{\Delta\nu}{2}\left[(1+x) + (1+x^2)^{1/2}\right] \\
\nu_{24} &= \frac{W_{1,0} - W_{0,0}}{h} = \Delta\nu\left[(1+x^2)^{1/2}\right] \\
\nu_{34} &= \frac{W_{1,-1} - W_{0,0}}{h} = \frac{+\mu_B^\mu g'_\mu H}{h} + \frac{\Delta\nu}{2}\left[(1-x) + (1+x^2)^{1/2}\right]
\end{aligned}
\qquad (4.1)
$$

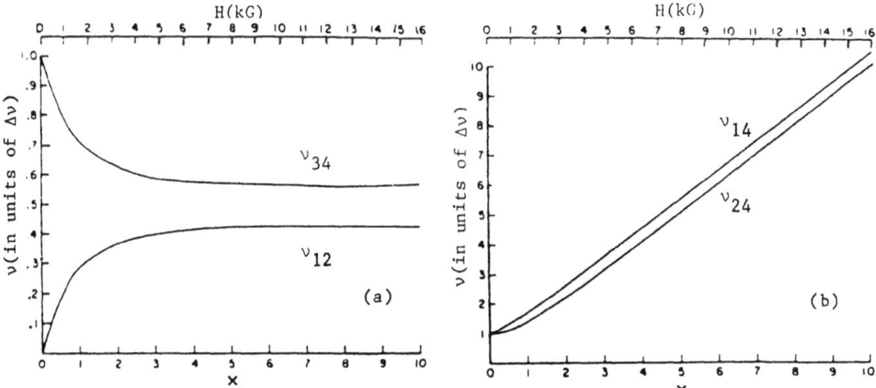

Figure 4.1. Plots of resonance frequencies (a) ν_{12} and ν_{34} and (b) ν_{14} and ν_{24}, in units of $\Delta\nu$ versus x or H, as given in Eq.(4.1).[50]

In the limit of strong magnetic field (x>>1) the transition frequencies from Eq. (2.22) are given approximately by the expressions

$$\nu_{12} = \frac{\Delta\nu}{2} - \frac{\mu_B^\mu g_\mu' H}{h} , \qquad \nu_{14} = \frac{\Delta\nu}{2} + \frac{\mu_B^e g_J H}{h}$$
$$\nu_{24} = \frac{\mu_B^e g_J H}{h} + \frac{\mu_B^\mu g_\mu' H}{h} , \qquad \nu_{34} = \frac{\Delta\nu}{2} + \frac{\mu_B^\mu g_\mu' H}{h} \tag{4.2}$$

In the limit of weak magnetic field (x<<1) the transition frequencies from Eq.(2.22) are given approximately by the expressions

$$\begin{aligned} \nu_{12} &= \frac{g_J \mu_B^e H}{2h} - \frac{g_\mu' \mu_B^\mu H}{2h} \\ \nu_{14} &= \Delta\nu + \frac{g_J \mu_B^e H}{2h} - \frac{g_\mu' \mu_B^\mu H}{2h} \\ \nu_{24} &= \Delta\nu \\ \nu_{34} &= \Delta\nu - \frac{g_J \mu_B^e H}{2h} + \frac{g_\mu' \mu_B^\mu H}{2h} \end{aligned} \tag{4.3}$$

Measurements at strong magnetic field have been used to determine both $\Delta\nu$ and μ_μ, and measurements at weak magnetic field determine $\Delta\nu$.

IV.2 Theory of Resonance Line Shape

Transitions are induced between the hfs magnetic substates of muonium by application of a microwave magnetic field. The associated time-dependent Hamiltonian term is given by

$$\mathcal{H}' = (g_J \mu_B^e \vec{J} - g_\mu' \mu_B^\mu \vec{I}) \cdot \vec{H}_1 \cos \omega t = \mathcal{H}_0' \cos \omega t \tag{4.4}$$

in which $\vec{H}_1$ is the vector amplitude of the applied microwave field and ω its angular frequency. The time-dependent muonium wavefunction is given by Eq.(3.16) in which $a_i(t)$ are the state amplitudes and the other quantities are defined in Eqs.(2.22)-(2.23b). The theory of the observed resonance line shape is based on the time-dependent equations for the state amplitudes including the term for muon decay. We obtain a solution for these amplitudes, calculate the muon polarization and decay positron angular distribution, and finally integrate over the observation time and detector solid angle to obtain the line shape as a function of the applied microwave frequency ω or the static magnetic field H.

The theory of resonance line shapes is a central problem in precision atomic spectroscopy and has been discussed extensively.[51,52]
For muonium, several idealized cases that can be treated analytically have been important, including two that employ line-narrowing techniques. These cases are

(1) Only two muonium states are involved, the static magnetic field H is arbitrary, and a single constant-amplitude oscillatory field $\vec{H}_1 \cos\omega t$ is applied.

(2) Three muonium states are involved at zero magnetic field, and a single constant-amplitude oscillatory field is applied.

(3) Three muonium states are involved at zero magnetic field, and two pulses of microwave power separated in time are applied (separated oscillating fields method).

(4) Only two muonium states are involved, the static magnetic field H is arbitrary, a single constant-amplitude oscillatory field $\vec{H}_1 \cos\omega t$ is applied, and decay positrons are observed only following a certain time interval after muonium formation ("old" muonium method). The "old" muonium method has been applied to transitions at zero magnetic field for which the three state problem can be reduced to a two level case.

Figure 4.2 shows the timing diagram for these resonance methods.

We treat briefly only case 1. We require the use of the density matrix for muonium since we are dealing with an ensemble of muonium atoms. The incident muons have a polarization P in the z direction and the atomic electrons, which are captured by the muons to form muonium, are unpolarized. The initial muonium density matrix at time $t = 0$ of muonium formation for an external static magnetic field H in the z direction is given in Eq.(3.11).

For case 1, the time-dependent equations for the state amplitudes are given by

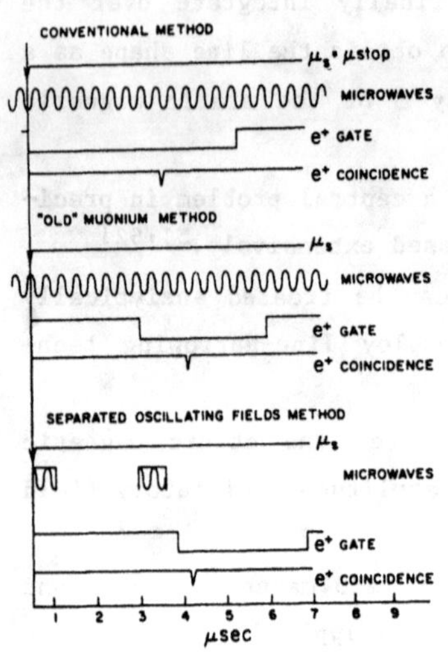

Figure 4.2. Timing diagram for various methods of observing muonium resonance transitions.[53]

$$\dot{a}_1 = -ia_2 b \exp\left[+i(\omega_{12} - \omega)t\right] - \frac{1}{2} a_1 \gamma \qquad (4.5a)$$

$$\dot{a}_2 = -ia_1 b^* \exp\left[-i(\omega_{12} - \omega)t\right] - \frac{1}{2} a_2 \gamma \qquad (4.5b)$$

The numbers 1 and 2 refer to any two muonium states, and

$$\omega_{12} = (W_1 - W_2)/\hbar, \qquad b = \langle \chi_1 | H_0' | \chi_2 \rangle/(2\hbar) \qquad (4.5c)$$

$$\gamma = \text{muon decay rate } \left[\text{Eq.2.30}\right]$$

Other quantities are defined in Eqs.(2.22)-(2.23b), (3.16), and (4.4). It is assumed that the diagonal matrix elements of H_0' in states 1 and 2 are zero. The time-dependent factor in H' of Eq.(4.4) is written $\cos\omega t = (e^{+i\omega t} + e^{-i\omega t})/2$, and the nonresonant term in the solutions for the state amplitudes-either $e^{+i\omega t}$ or $e^{-i\omega t}$ -is omitted. Solution of Eq.(4.5) with the initial conditions

$$a_1(0) = 1, \qquad a_2(0) = 0 \qquad (4.6)$$

gives

$$|a_1(t)|^2 = [\cos^2(\tfrac{1}{2}\Gamma t) + [(\omega_{12} - \omega)/\Gamma]^2 \sin^2(\tfrac{1}{2}\Gamma t)]e^{-\gamma t}$$
$$|a_2(t)|^2 = (4|b|^2/\Gamma^2)\sin^2(\tfrac{1}{2}\Gamma t)e^{-\gamma t} \tag{4.7}$$

in which $\Gamma = [(\omega - \omega_{12})^2 + 4|b|^2]^{1/2}$.

For any time t, the quantity $|a_2(t)|^2$ attains its maximum value at resonance ($\omega = \omega_{12}$) when $|b|t = \pi/2$. The matrix element b determines the selection rules and probabilities for transitions between states. The selection rules are the usual ones for a magnetic dipole transition, and are given in the general case in Table 3.2.

The quantity observed as the resonance signal is the number of decay positrons emitted in a particular direction (e.g., z direction), which is proportional to the z component of the muon polarization [Eq.(3.15)]:

$$\begin{aligned} P_z(t) &= P_{z0}(t)e^{-\gamma t} = \langle\psi|2I_{\mu z}|\psi\rangle \\ &= |a_1|^2\langle\chi_1|2I_{\mu z}|\chi_1\rangle + |a_2|^2\langle\chi_2|2I_{\mu z}|\chi_2\rangle \\ &\quad + a_1^* a_2\langle\chi_1|2I_{\mu z}|\chi_2\rangle\exp(i\omega_{12}t) \\ &\quad + a_2^* a_1\langle\chi_2|2I_{\mu z}|\chi_1\rangle\exp(-i\omega_{12}t). \end{aligned} \tag{4.8}$$

The resonance signal S is defined as the difference between the number of positrons observed with the microwave magnetic field on and with the microwave magnetic field off, normalized to 1 for a single muon. The observation time is taken from t = 0, when the muonium atom is formed, to t = ∞. We use Eq. (4.7) for the state amplitudes and Eq. (2.32) for the positron angular distribution. Furthermore, we neglect γ compared to ω_{12}, which leads to the elimination of the interference terms involving a_1 and a_2 in Eq. (4.8)

$$S = (a\cos\theta\, d\Omega/12\pi)\left[-\langle\chi_1|2I_{\mu z}|\chi_1\rangle + \langle\chi_2|2I_{\mu z}|\chi_2\rangle\right] 2|b|^2/(\Gamma^2+\gamma^2) \tag{4.9}$$

in which $d\Omega$ is the solid angle subtended by the positron detector at the position of the muonium atom and θ the polar angle to the positron detector. This line shape is Lorentzian with regard to variation of the applied frequency ω and has its maximum value when $\omega = \omega_{12}$. The full width between half-intensity points (FWHM) is given by

$$\Delta\omega_{1/2} = 2\pi\Delta f_{1/2} = 2(4|b|^2 + \gamma^2)^{1/2} \tag{4.10}$$

In the limit of zero microwave power ($|b|^2 = 0$), we obtain the natural linewidth

$$\Delta f_{1/2} = \gamma/\pi = 0.145 \text{ MHz} \tag{4.11}$$

A nonzero value of $|b|^2$ contributes power broadening of the line. At resonance,

$$S_{max} \propto |b|^2/(4|b|^2 + \gamma^2) \tag{4.12}$$

The three-level problem of case 2, which involves the (F,M_F) levels (1, 1), (1, -1), and (0, 0) at $H = 0$ and with $\vec{H}_1 = H_{1x}\hat{i} + H_{1y}\hat{j}$ [Eq.(4.4)], can be reduced to a two-level problem[46,54]. The signal is proportional to

$$S \propto |b|^2(\gamma^2 + 2|b|^2)/[(\gamma^2 + 2|b|^2)^2 + \gamma^2(\omega - \omega_{12})^2] \tag{4.13}$$

The line shape is a Lorentzian in ω with the FWHM value

$$\Delta f_{1/2} = (1/\pi\gamma)(\gamma^2 + 2|b|^2) \tag{4.14}$$

As in case 1, the natural linewidth is γ/π. However, the broadening of the line by microwave power is proportional to the power or to $|b|^2$ at high power, whereas in case 1 this broadening is proportional to the square root of the microwave power.

The separated oscillating fields method, case 3, involves the use of two pulses of microwave power separated in time (Fig. 4.2). It is advantageous because it can yield a resonance line narrower than those obtained by the conventional single oscillatory field method discussed in cases 1 and 2, and indeed narrower than the natural linewidth γ/π given in Eq.(4.11).[52, 53, 54, 55, 56, 57] The qualitative idea is simply to obtain the resonance curve using muonium atoms which live longer than the mean life of the muon. Also microwave power broadening of the resonance line is relatively unimportant. If the two pulses of microwave power are each of duration τ and separated by a time interval T, then for the two-level case the linewidth of the central interference peak is

$$\Delta f_{1/2} = \frac{1}{3}\,\pi[1/(T + \tau)] \tag{4.15}$$

Another line-narrowing method is the "old" muonium method, case 4, which involves the application of relatively low microwave power and the observation of decay positrons only subsequent to a time interval T (Fig. 4.2). In the limit where $T \gg \tau_\mu$ and $|b| \ll \gamma$, the linewidth for this method is

$$\Delta f_{1/2} = 2\,|b|/\pi \tag{4.16}$$

Although the statistical counting error and various systematic errors are most important in determining the precision achieved in a resonance experiment, the resonance linewidth by itself is an important criterion for the expected precision of an experiment. In principle, the idealized line shape is exactly calculable, and there is much experience in precision atomic spectroscopy to indicate that a resonance line center can be chosen to at least $1/10^3$ of its linewidth.[51,52]

IV.3 Measurements at Strong Magnetic Field

The first observation of a microwave resonance transition for muonium[13] and the first precise measurement of an energy interval[58,59] were done with the Columbia Nevis synchrocyclotron by a Yale group. The transition ν_{12} [Eq.(4.1)] was observed at a static magnetic field of 5000 to 6000 G, where the strong field quantum numbers for the transition $(M_J, M_\mu) = (\frac{1}{2}, \frac{1}{2}) \rightarrow (\frac{1}{2}, -\frac{1}{2})$ are applicable.

The experimental arrangement is shown in Fig. 4.3. Positive muons were stopped in the high-pressure aluminum vessel containing highly purified argon gas at pressures from 10 to 60 atm, which was continuously purified by recirculation over titanium at a temperature of about 750°C. The external static magnetic field was provided by a split solenoid and had a homogeneity over the target volume of about 2 parts in 10^4. The magnetic field was measured with a proton NMR signal. The high-Q microwave cavity operated in the TM_{110} mode with a microwave field perpendicular to the static magnetic field. Frequency stability to better than 1 part in 10^6 and power stability to 2% were achieved. The stopping of a positive muon in the target is indicated by the scintillation counter coincidence-anticoincidence count $0123\bar{4}$. Decay positrons detected as $\bar{2}45$ were counted over a time interval of about 4 μsec following the stopping of a muon. Data were taken as a function of the external static magnetic field with the microwave power on and off.

Typical resonance curves obtained at different gas pressures are shown in Fig. 4.4. The signal is the ratio of the number of decay positron counts with the microwave power on to the number with the microwave power off minus 1, and the error bars are one standard deviation statistical counting errors. The solid curves are least square fits to the experimental points of the theoretical line shape of Eq.(4.9). The linewidths are about four times the natural linewidth and are due principally to microwave power broadening.

The resonance values of microwave frequency and magnetic field were used to determine $\Delta\nu$, which had not been previously measured except by an imprecise static magnetic field method[60] and hence was the least well-known quantity in the Hamiltonian of Eq.(2.20).

The dependence of $\Delta\nu$ on gas pressure or density was observed in this experiment. This density shift in $\Delta\nu$ is due to the effect of collisions of muonium with argon atoms, which distort the muonium

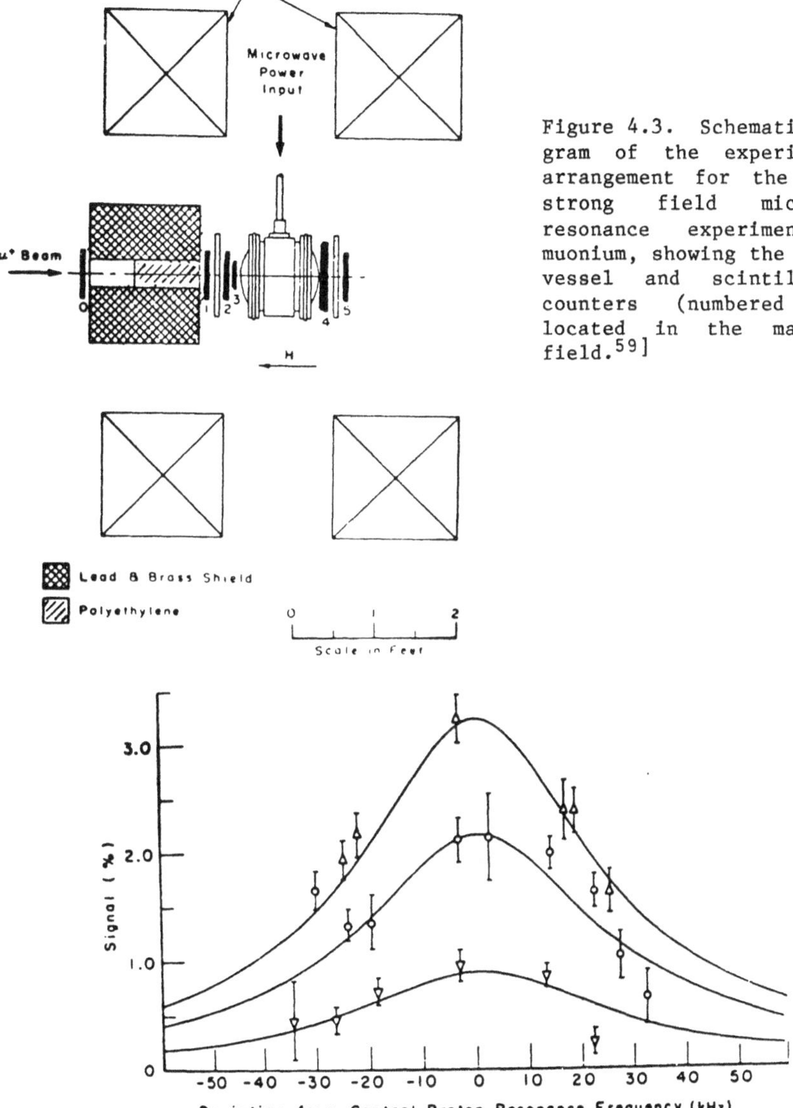

Figure 4.3. Schematic diagram of the experimental arrangement for the first strong field microwave resonance experiment on muonium, showing the target vessel and scintillation counters (numbered 0-5) located in the magnetic field.[59]

Figure 4.4. Muonium forward signal (counts $\overline{2}45$) as a function of magnetic field for several values of argon gas density. Each curve is taken with an input microwave power to the cavity of 24W. The three sets of data have been shifted to a common line center for ease of comparison. The solid curves are the calculated best-fit curves to the data with the line center and the peak signal intensity as variables. Pressure: Δ, 33.9; O, 23.9; ∇, 9.5 atm.[59]

wavefunction and hence alter $\Delta\nu$. Density shifts have been observed and studied particularly in precision optical pumping measurements of atomic hfs intervals.[61, 62] At the gas pressures used, the effective collision rate is some 10^5 - 10^6 times the muon decay rate γ, and hence the resonance measurement determines a value for $\Delta\nu$ averaged over many collisions. Since a high precision is sought in the determination of $\Delta\nu$ and since quite high gas pressure is needed to stop an adequate number of muons in the gas target, the density dependence of $\Delta\nu$ is important and must be studied carefully. It can be expressed as

$$\Delta\nu(D) = \Delta\nu\left[1 + aD + bD^2\right] \qquad (4.17)$$

in which D is the gas density. The quantity a is the coefficient of linear density dependence and arises from two-body collisions of one muonium atom with one argon atom. The quantity b is the coefficient of quadratic density dependence and arises from three body collisions of one muonium atom with two argon atoms. Values for a and b are determined by fitting the expression for $\Delta\nu(D)$ of Eq.(4.17) to the observed points $\Delta\nu(D)$. [Figure 4.15 shows data on $\Delta\nu(D)$ obtained in weak-field experiments.]

In the initial reported precision result for $\Delta\nu$ from the strong-field measurement, the linear fit to Eq.(4.17) with b = 0 was chosen. Subsequently, more precise measrements of $\Delta\nu$ indicated that the quadratic fit should have been chosen, although it was somewhat less favored by its χ^2 value. The final value for $\Delta\nu$ from this experiment was accurate to 0.12 MHz or 27 ppm.[59] Table 4.1 gives the final results of the various measurements of $\Delta\nu$.

Other improved measurements of muonium transitions at strong magnetic field were made by the Chicago group, which introduced several new approaches.[63,64,65] One was the observation that at a certain magnetic field H_m of about 11.35 kG the transition frequencies ν_{12} and ν_{34} have maximum and minimum values so that $\partial\nu_{12}/\partial H|_{H_m} = \partial\nu_{34}/\partial H|_{H_m} = 0$. Hence at H_m magnetic field inhomogeneities have only a second order effect on the transition frequency. The experimental arrangement is shown in Fig. 4.5, where proportional wire chambers are used in the

TABLE 4.1

History of $\Delta\nu$ Measurements

Time	Method	Group	$\Delta\nu$
1961	Static H	Yale-Nevis,[10,60]	5500^{+2900}_{-1500} MHz
1962	Strong H	Yale-Nevis,[13,50]	4 461.3(2.0)MHz(450ppm)
1964	Strong H	Yale-Nevis,[58,59]	4 463.24(0.12)MHz(27ppm)
1966	Weak H	Yale-Nevis,[1]	4 463.18(0.12)MHz(27 ppm)
1969	Weak H	Yale-Nevis,[48]	4 463.26(0.04)MHz(9.0ppm)
1969	Strong H	Chicago,[63,65]	4 463.317(0.021)MHz(4.7ppm)
1970	Strong H	Chicago,[64]	4 463.302 2(89)MHz(2.0 ppm)
1971	Weak H	Yale-Nevis,[46]	4 463.308(11)MHz(2.5ppm)
1971	Weak H, SOF	Chicago,[54]	4 463.301 2(23)MHz(0.5ppm)
1973	Weak H, SOF	Chicago-SREL,[72,73]	4 463 304.0(1.8)kHz
1975	Weak H, "old" μ^+e^-, SOF	Yale-Heidelberg-LAMPF,[69]	4 463 302.2(1.4)kHz(0.3ppm)
1977	Strong H	Yale-Heidelberg-LAMPF,[68]	4 463 302.35(0.52)kHz (0.12ppm)
1982	Strong H	Yale-Heidelberg-LAMPF,[24]	4 463 302.88(0.16)kHz (0.036ppm)

pressure vessel to reduce background from muon stops in the vessel walls. The muon stopping distribution was measured in a separate experiment with a small scintillation counter probe. A resonance curve for the transition ν_{12} of signal versus microwave frequency is shown in Fig. 4.6a. The signal at the resonance frequency versus time is shown in Fig. 4.6b, fit to the form expected from Eq.(4.7). The value of $\Delta\nu$ obtained was accurate to 18 kHz or 4ppm.[65]

Measurement of two transitions at strong magnetic field allows the determination of both $\Delta\nu$ and the muon magnetic moment μ_μ, or μ_μ/μ_p, as is clear from the discussion in Section IV.1. The Chicago group have measured both transitions ν_{12} and ν_{34} at $H = H_m$ in a double resonance experiment, which involved the simultaneous

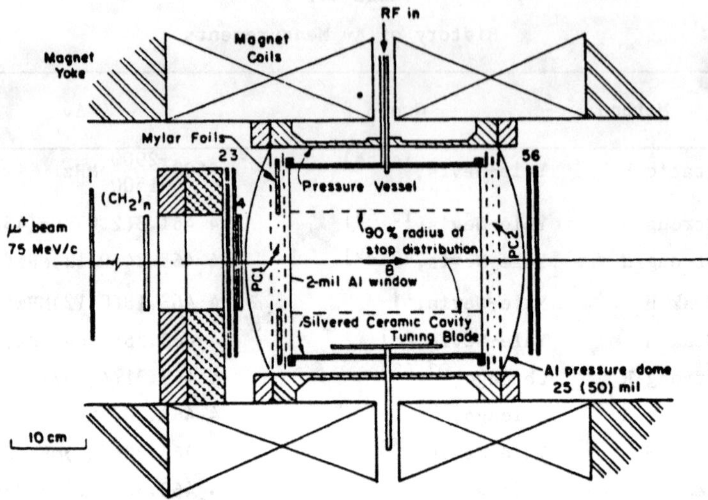

Figure 4.5. Schematic cross section through the experimental setup for a strong field microwave resonance experiment at a field where the transition frequency is a maximum. The setup uses proportional wire chambers inside the pressure vessel and indicates the result of an independent measurement of the muon stopping distribution.[65]

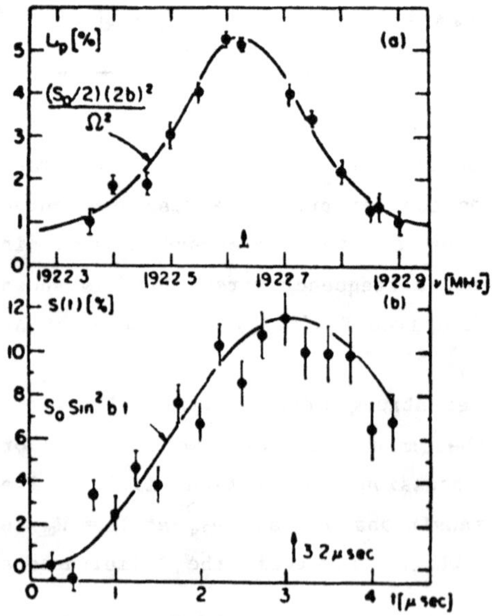

Figure 4.6. (a) Resonance curve for the transition ν_{12} obtained with the experimental setup of Fig. 4.5. (b) Signal S_o at the resonance frequency ν_{12} versus the time following the muon stop.[65]

observation of both transitions.[64] A value for $\Delta\nu$ was obtained accurate to 8.9 kHz or 2 ppm. Because of the density shift of g_J, which was not measured and could only be estimated theoretically,[66] the accuracy in the determination of μ_μ/μ_p was largely limited by this theoretical correction of 8 ppm, whose precision is difficult to estimate. The overall precision in μ_μ/μ_p from this experiment is μ_μ/μ_p in a precession experiment on μ^+ in liquids[67]; the two values of μ_μ/μ_p are in satisfactory agreement.

Subsequently the results of new measurements at strong magnetic field by the Yale-Heidelberg group at LAMPF have become available.[68] The transitions $(M_J, M_\mu) = (+\frac{1}{2},+\frac{1}{2}) \leftrightarrow (+\frac{1}{2},-\frac{1}{2})$ or ν_{12}, and $(-\frac{1}{2},-\frac{1}{2})\leftrightarrow(-\frac{1}{2},+\frac{1}{2})$ or ν_{34} were measured at a static magnetic field of 13.6 kG using a constant-amplitude oscillatory field. With the LAMPF proton linac providing a proton beam of 100 μA (average) intensity at a duty factor of 6% and with the 50 MeV/c μ^+ beam from the muon channel, a stopping rate of 2.5×10^4 sec^{-1} (average) was obtained in a 0.2 gm/cm^2 krypton (1.7 atm) gas target. A high-precision eighth-order solenoid electromagnet provided a magnetic field stable to ~1 ppm and homogeneous to ~5 ppm over the region of the microwave cavity (19 cm diameter, 19 cm length). Gas proportional wire chambers as well as plastic scintillators were used to detect the muons stopping in the gas and the decay positrons.[69] Data were taken at krypton gas pressures of 1.7 and 5.3 atm. The hfs interval $\Delta\nu$ is given by $\Delta\nu = \nu_{12} + \nu_{34}$, and the ratio μ_μ/μ_p is obtained from the difference frequency $\nu_{12} - \nu_{34}$ together with the NMR measurement of the static magnetic field. Linear extrapolation of the data to zero gas density gives a value for $\Delta\nu$ accurate to 0.12 ppm and of μ_μ/μ_p accurate to 1.4 ppm.

$$\Delta\nu = 4\,463\,302.35\,(52)\ \text{kHz}(0.12\ \text{ppm})$$
$$\mu_\mu/\mu_p = 3.183\,340\,3(44)(1.4\ \text{ppm})$$

The latest and most precise measurement of $\Delta\nu$ and of μ_μ/μ_p was made at LAMPF in a strong field experiment[24] with the experimental setup just described.[68] The increased precision was due principally to the use of the high-intensity, low-momentum "surface" muon beam.

A diagram of the experimental apparatus is shown in Fig. 4.7. The LAMPF stopped-muon channel was tuned to accept and transmit μ^+ originating from the decay of π^+ stopped near the surface of the pion production target in the primary proton beam. With an 800-MeV proton beam of 300 μA average, the μ^+ beam had an instantaneous flux of $3\text{x}10^7$ s^{-1} ($2\text{x}10^6$ s^{-1} average) after collimation. The muon momenta were between 25 and 28 MeV/c, and the longitudinal polarization was close to 1. The ratio of μ^+ to e^+ in the beam was about 6/1. The μ^+ flux was monitored with a thin (0.12 mm) plastic scintillator S1 by integrating the anode current over the 600-μs proton beam pulse. The target vessel was filled with Kr gas at pressures of $\frac{1}{2}$ or 1 atm, and about half of the incident μ^+ stopped in the gas. Scintillation counters S2 and S3 detected e^+ from μ^+ decays, and the plastic and aluminum moderator downstream from the cavity helped reduce the background from the e^+ contamination in the beam. A central element of the experiment was the high-precision solenoid electromagnet which provided the magnetic field of 13.6 kG, which was homogeneous over the volume of the microwave cavity to about 3 ppm rms and had a long term stability of better than 0.3 ppm. The microwave cavity was resonant in the TM_{110} mode at 1.918 GHz (ν_{12} at 13.6 kG) and in the TM_{210} mode at 2.545 GHz (ν_{34}). It was excited with an input power of ~20 W switched on and off at a repetition period of 160 ms, alternating between modes for successive microwave-on periods.

Typical resonance curves are shown in Fig. 4.8 and were observed by varying the magnetic field H in small steps with fixed microwave frequency and power. The signal at each value of H is defined by $S = \left\{(S2 \cdot S3/S1)_{\text{rf on}} / (S2 \cdot S3/S1)_{\text{rf off}}\right\} - 1$. The theoretical line shape, which includes as free parameters essentially a resonance line center, a linewidth, and a height, is fitted to the experimental points and determines the resonance magnetic field value. The theoretical line shape incorporates the measured distribution of the magnetic field

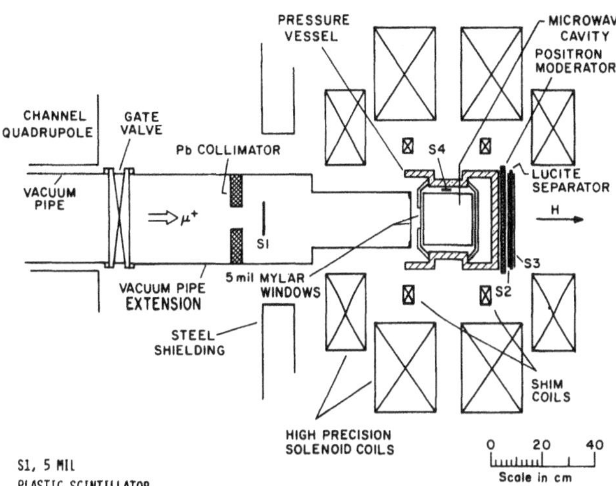

Figure 4.7. Experiment at LAMPF in which the latest precision measurement of the hyperfine structure interval $\Delta\nu$ in muonium was made. S1, S2, S3, and S4 are plastic scintillation counters of thicknesses 0.005, 0.25, 0.25, and 0.25 in., respectively. Counter S4 was used to indicate maximum μ^+ stopping rate in center of microwave cavity.

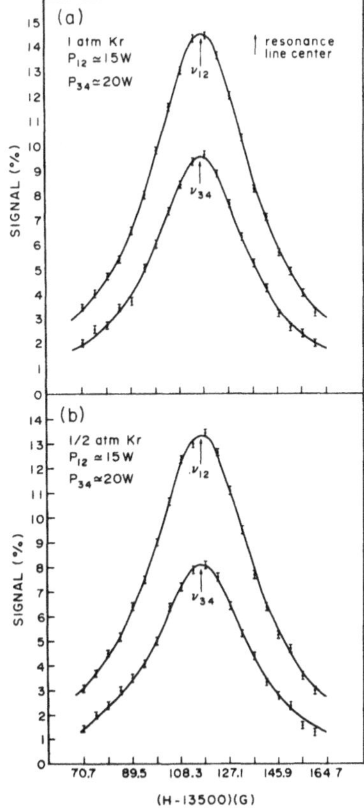

Figure 4.8. Typical resonance lines with theoretical line shapes (solid lines) fitted to the data points obtained with the apparatus shown in Fig. 4.7. Data points were taken alternately on opposite sides of the line center. Data-taking time for each pair of resonance lines was less than 2 h.

H, the measured μ^+ stopping distribution, the microwave power distribution over the cavity, and the solid angle for detection of an e^+ from μ^+ decay. A total of 184 resonance lines obtained in about 600 h of data taking (102 at 0.5 atm and 54 at 1 atm in the present experiment, and 18 at 1.7 atm and 10 at 5.2 atm from a previous experiment) were analyzed. After adjustment of all transition frequencies to correspond to a magnetic field H of 13 616.0 G and correction for a small measured quadratic density shift, the transition frequencies were extrapolated to zero gas density with use of the linear density dependence $\nu(D) = \nu(0)(1 + aD)$. Using the resonance frequencies of Eq.(4.1), we then obtain

$$\Delta\nu = 4\,463\,302.88(16)\ \text{kHz}\ (0.036\ \text{ppm})$$
$$\mu_\mu/\mu_p = 3.183\,346\,1(11)\ (0.36\ \text{ppm}) \tag{4.18}$$

in which the one-standard-deviation total error including systematic and statistical errors is given. Table 4.2 lists the sources of error.

The value of $\Delta\nu$ given in Eq. (4.18) agrees with the earlier measurement[68] of $\Delta\nu$, but the error is less by a factor of 3.3. Our value of μ_μ/μ_p agrees with the most recent muon spin rotation measurement[25] done in liquid bromine which gave $\mu_\mu/\mu_p = 3.183\,344\,1(17)$ (0.53 ppm).

The hfs and g_J fractional density shifts in Kr can be evaluated from the experimental data. They are:

$$(1/\Delta\nu)(\partial\Delta\nu/\partial D) = -\ 10.57(4)\times10^{-9}\ \text{Torr}^{-1}\ (0°\text{C,Kr});$$
$$(1/g_J)(\partial g_J/\partial D) = -\ 1.83(32)\times10^{-9}\ \text{Torr}^{-1}\ (0°\text{C,Kr}). \tag{4.19}$$

Use of our value of μ_μ/μ_p in Eq. (4.18) together with that of Ref. 25, and of experimental values[20] of g_{μ^+} and of μ_p/μ_B^e determines the most precise value for m_μ/m_e, as follows:

$$m_\mu/m_e = (g_\mu/2)(\mu_p/\mu_\mu)(\mu_B^e/\mu_p) = 206.768\,259(62)\ (0.30\ \text{ppm}). \tag{4.20}$$

TABLE 4.2

Corrections and errors for $\Delta\nu$ and μ_μ/μ_p

	Source	$\delta\Delta\nu$(kHz)	$\delta(\mu_\mu/\mu_p)$ (ppm)
1.	Statistical error (e^+ counts)	0.000±0.073	0.000±0.196
2.	Random error associated with μ^+ beam monitor	0.000±0.031	0.000±0.084
3.	Muon stopping distribution and detector solid angle distribution	0.000±0.008	0.000±0.119
4.	Magnetic field measurement (±0.31 ppm)	0.000±0.000	0.000±0.093
5.	Microwave power averaging	+0.021±0.035	-0.102±0.104
6.	Gas density	0.000±0.065	0.000±0.001
7.	Temperature dependence of a^*	-0.073±0.073	-0.013±0.013
8.	Quadratic density shift	0.000±0.041	0.000±0.005
9.	Field-dependent line-shape systematics†	+0.037±0.083	+0.356±0.217
10.	Bloch-Siegert term and nonresonant states††	+0.004±0.000	-0.005±0.000
11.	Small approximations in line fitting	0.000±0.000	0.000±0.047
	Total correction and one-standard-deviation error	-0.011±0.160	+0.236±0.359

*Data of Ref. 68 and of Ref. 24 were taken at two temperatures differing by 2.5°C. Corrections to the data of Ref. 68 were made for the dependence of a on temperature based on experimental (Refs. 61 and 62) and theoretical (Ref. 70) information on hydrogen density shifts.

†Based on measurements of broadened resonance lines at high microwave power.

††Calculation with Ref. 71.

IV.4 Measurements at Weak Magnetic Field

Microwave magnetic resonance measurements at weak magnetic field determine the hfs interval $\Delta\nu$ [Eq.(4.3)]. The earliest measurements[48] done by the Yale group utilized the hfs transitions $\Delta F = \pm 1$, $\Delta M_F = \pm 1$, first at $H \simeq 3$ G and then at $H \simeq 0$. For $H \simeq 3$ G only two levels are involved in the transition (case 1 of Section IV.2) and for $H \simeq 0$ the three states $(F, M_F) = (1, +1), (1, -1)$, and $(0, 0)$ are involved (case 2 of Section IV.2). The experimental setup shown in Fig. 4.9 differs from the strong field setup of Fig. 4.3 principally in that a longitudinal magnetic field of a few milligauss is required at the microwave cavity for the $H \simeq 0$ case and it is achieved with the use

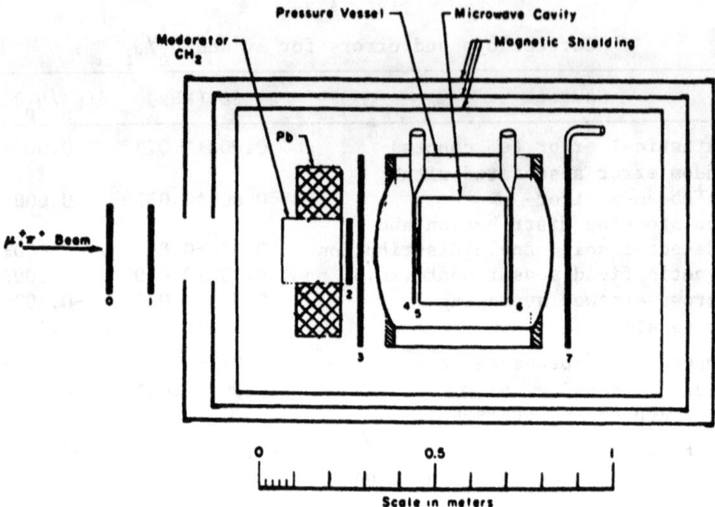

Figure 4.9. Diagram of experimental apparatus used for very weak field $\Delta F = \pm 1$ transitions, showing the magnetic shields and the target with its internal scintillation counters.[46]

of three moly-permalloy shields including endcaps. The microwave cavity operated in the TM_{220} mode, and a constant amplitude oscillatory field was applied. Observed resonance curves are shown in Fig. 4.10b-f for very weak field transitions ($H \sim 10$ mG), with argon and krypton at different pressures. In addition, the transition ν_{14} at $H = 2.85$ G is shown in Fig. 4.10a and is seen to be a weak signal as expected. We note that the linewidths of the very weak field transitions are about 600 kHz, due predominantly to microwave power broadening. The precision achieved in the determination of $\Delta\nu$ was 2.5 ppm.[46] (Table 1.)

An important advance was made by the Chicago group by introducing the method of separated oscillating fields for the observation of the very weak field transition.[54] Figure 4.11 shows the central interference signal obtained by taking the difference between the positron counts when the phase of the second pulse with respect to the first pulse is $+\pi/2$ and $-\pi/2$. The narrow resonance line of 47 kHz allowed a very precise determination of $\Delta\nu$ to 1.8 kHz or 0.4 ppm.[72,73]

In addition, a precise determination of $\Delta\nu$ at weak field has been

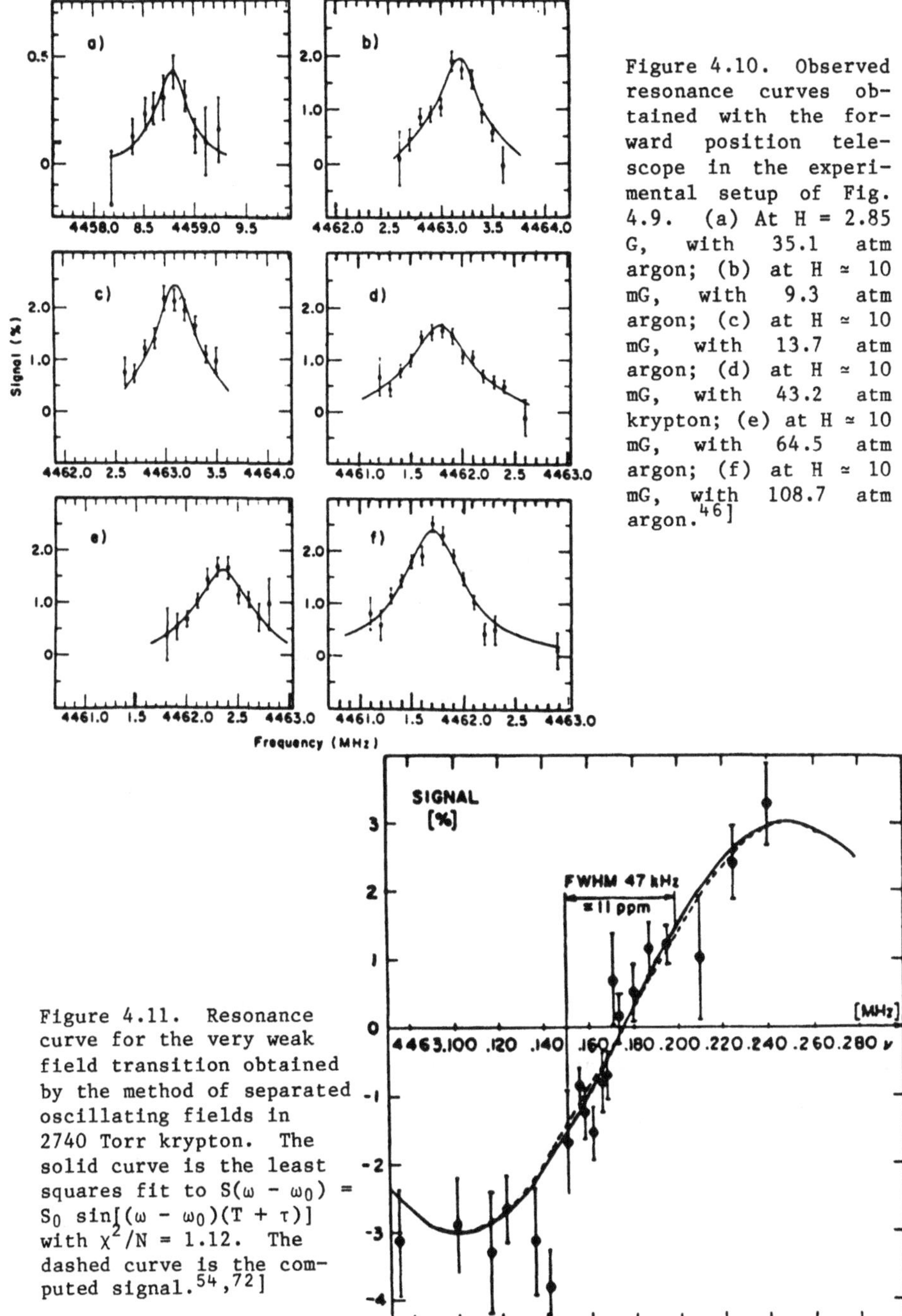

Figure 4.10. Observed resonance curves obtained with the forward position telescope in the experimental setup of Fig. 4.9. (a) At H = 2.85 G, with 35.1 atm argon; (b) at H ≈ 10 mG, with 9.3 atm argon; (c) at H ≈ 10 mG, with 13.7 atm argon; (d) at H ≈ 10 mG, with 43.2 atm krypton; (e) at H ≈ 10 mG, with 64.5 atm argon; (f) at H ≈ 10 mG, with 108.7 atm argon.[46]

Figure 4.11. Resonance curve for the very weak field transition obtained by the method of separated oscillating fields in 2740 Torr krypton. The solid curve is the least squares fit to $S(\omega - \omega_0) = S_0 \sin[(\omega - \omega_0)(T + \tau)]$ with $\chi^2/N = 1.12$. The dashed curve is the computed signal.[54,72]

done by the Yale-Heidelberg group at LAMPF.[69] Figure 4.12 shows the experimental arrangement. With the external primary proton beam of 10 μA average on a production target of 6.0 cm carbon, the muon stopping rate in a 1.7-atm krypton target was 2000/sec average. Both the "old" muonium technique and the separated oscillating fields technique were used, and resonance lines observed by the two techniques are shown in Figs. 4.13 and 4.14, respectively. The overall statistical powers of the two techniques were approximately the same in this experiment. The determination of $\Delta\nu$ requires extrapolation of the measured values of $\Delta\nu(D)$ to zero gas density using Eq.(4.17). Figure 4.15 shows all of the Yale/Heidelberg group's weak-field data for argon and krypton and the fitted curves. The final result from these data is

$$\Delta\nu = 4{,}463{,}302.2(1.4) \text{ kHz} \quad (0.3 \text{ ppm}) \tag{4.21}$$

in which the dominant error is the statistical counting error.[69]

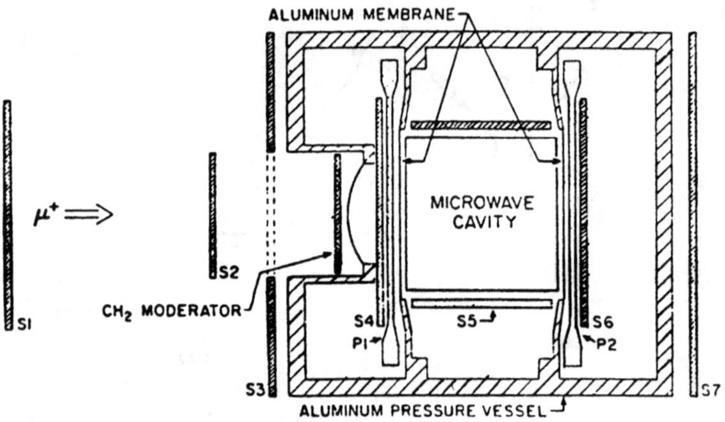

Figure 4.12. Diagram of experimental arrangement of the target vessel for the weak field measurement at LAMPF. P1 and P2 are proportional wire chambers. S1-S7 are plastic scintillation counters. A muon stopping in the active region of the target is defined as $\mu_S = S1\cdot S2\cdot S4\cdot \overline{S6}\cdot P1\cdot \overline{P2}$. A forward decay positron defined as $e_F = \overline{S1}\cdot S6\cdot S7\cdot \overline{P1}\cdot P2$ and a backward decay positron as $e_B = \overline{S1}\cdot S3\cdot S4\cdot P1\cdot \overline{P2}$. The cylindrical counter S5 surrounds the microwave cavity and is used to detect positrons that emerge normal to the beam axis.[69]

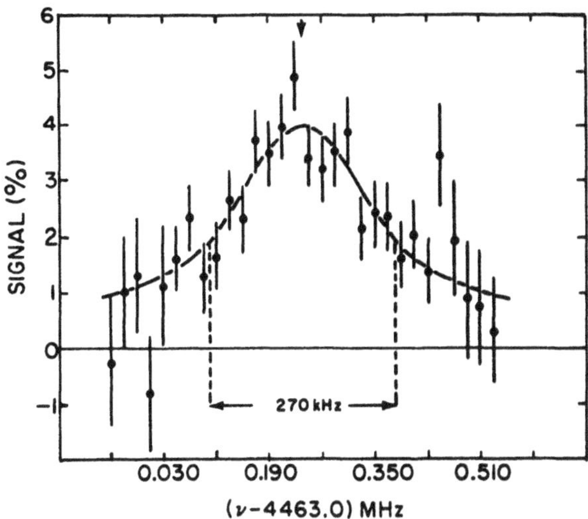

Figure 4.13. "Old" muonium resonance curve for a very weak field transition in 1.65 atm krypton. The theoretical line shape is approximately a Lorentzian.[69]

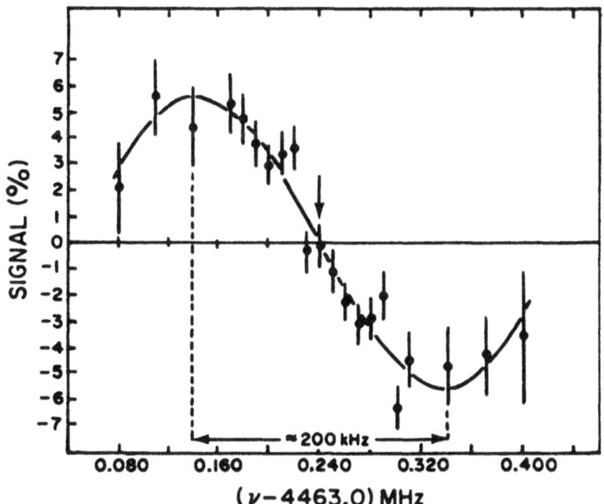

Figure 4.14. Separated oscillating fields resonance curve for a very weak field transition in 1.64 atm krypton. The data are fitted to the expression $\sin[(\omega - \omega_0)(T + \tau)]$ with $T + \tau = 2.55\ \mu\text{sec}$.[69]

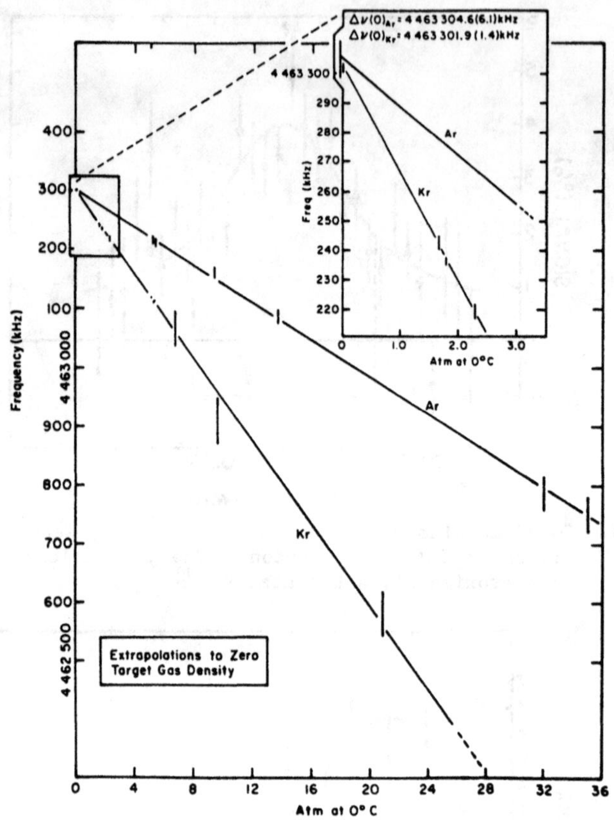

Figure 4.15. $\Delta\nu(D)$ versus density for argon and krypton for all the weak-field measurements done by the Yale/Heidelberg group. Data are fit by Eq. (4.17). The inset figure shows the low-density data and the extrapolated values of $\Delta\nu(0)$. The values of a and b for argon and krypton are $a_{Ar} = -4.85(17) \times 10^{-9}$/Torr, $b_{Ar} = 6.7(2.0) \times 10^{-15}$/Torr^2, $a_{Kr} = -10.66(10) \times 10^{-9}$/Torr, $b_{Kr} = 9.7(2.0) \times 10^{-15}$/Torr^2.

IV.5 Summary and Prospectives

The measurements of $\Delta\nu$ are listed in Table 4.1 from both the strong field and weak field experiments and Fig. 4.16 presents a bar chart of the history of $\Delta\nu$ measurements. Note that the measured values have agreed well over their history and have improved in precision by a factor of about 10^7 over this period. Since the precision of the latest value given in Eq.(4.18) is so much higher than those of the earlier measurements we take the latest value as the measured value.

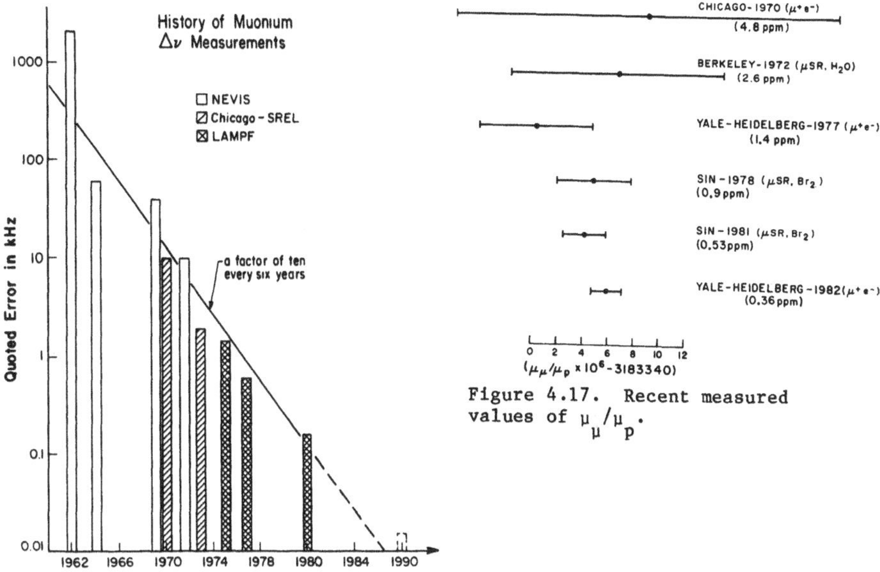

Figure 4.17. Recent measured values of μ_μ/μ_p.

Figure 4.16. History of measurements of $\Delta\nu$.

The experimental value for $\Delta\nu$ agrees well with the theoretical value of Eq.(2.17).

$$\begin{aligned} \Delta\nu_{th} - \Delta\nu_{exp} &= 0.23\ (1.7)\ \text{kHz} \\ \frac{\Delta\nu_{th} - \Delta\nu_{exp}}{\Delta\nu_{exp}} &= (0.05 \pm 0.4)\ \text{ppm} \end{aligned} \tag{4.22}$$

Since $\Delta\nu_{exp}$ of Eq.(4.18) is known with much higher precision than $\Delta\nu_{th}$ due in substantial part to uncertainty in our knowledge of the fundamental constant μ_μ/μ_p, it is useful to regard the theoretical formula of Eq.(2.13) as correct and use it together with $\Delta\nu_{exp}$ of Eq.(4.18) to determine μ_μ/μ_p. This gives

$$\mu_\mu/\mu_p = 3.183\ 345\ 47(47)\ (0.15\ \text{ppm}) \tag{4.23}$$

It is also useful to determine a value for the fine structure constant α from $\Delta\nu_{exp}$ and $\Delta\nu_{th}$. This gives

$$\alpha^{-1} = 137.035\ 993(22)\ (0.16\ \text{ppm}) \qquad (4.24)$$

in good agreement with the current value for α of Eq.(2.14) which comes principally from the electron g-2 value and the quantized Hall effect.

A new experiment is being undertaken at LAMPF[74] to measure $\Delta\nu$ and μ_μ/μ_p to precisions of about 10 ppb and 50 ppb, respectively. The measurement will be done at strong magnetic field with a new precision superconducting solenoid. Line narrowing will be employed using the old muonium technique with a chopped and separated surface muon beam derived from a 1 mA (time average) primary proton beam.

V. THE LAMB SHIFT IN MUONIUM

V.1 Introduction

The Lamb shift in hydrogen was not only a discovery which led to the development of modern quantum electrodynamic field theory, but it has been also a sensitive and crucial testing ground for the elaboration and confirmation of the theory.[33,75,76] A precise measurement of the Lamb shift in the n = 2 state of M would provide an ideal test of quantum electrodynamics free of the effect of proton structure which now complicates the interpretation of the Lamb shift in hydrogen.

At present the theoretical and experimental values for the Lamb shift in hydrogen $\mathbf{S}_H$ agree within about 1 standard deviation or 2 parts in 10^5.[18] The contribution of proton structure to $\mathbf{S}_H$ is 145 kHz or 1.4 parts in 10^4, based on the latest measurement of the proton form factor, or $\langle r_p^2\rangle$, through e-p elastic scattering. The estimate of the uncertainty in knowledge of this contribution is 4 parts in 10^6, but it should be noted that the previous measurement of $\langle r_p^2\rangle$ disagreed with the latest value and implied a proton structure contribution of 127 kHz rather than 145 kHz. These values differ by 1.7 parts in 10^5.

For muonium the theoretical expression for S_M is very similar to S_H except for the absence of a nuclear size term (Table 2.1). Of course the terms involving reduced mass and relativistic recoil which involve the factor m_e/m_μ are larger for muonium than for hydrogen. Indeed the present theoretical uncertainty in S_M arises from the uncalculated terms of order $\alpha^5(m_e/m_\mu)^2$.

As for hydrogen the 2S atomic state of muonium is metastable with respect to spontaneous emission, and decays to the 1S state by 2 photon emission with a mean life of 1/7 s. However, of course, the muon mean life τ_μ of about 2.2 μs associated with its decay due to the weak interaction determines the lifetime of M(2S). The 2P state decays spontaneously by electric dipole radiation to the ground 1S state with a mean lifetime of 1.6 ns and with emission of a Lyman α photon of 1221Å wavelength.

In principle an experiment to measure the Lamb shift in muonium can be done in a similar way to the measurement of the Lamb shift in hydrogen[76]. A beam of M(2S) is formed, a microwave transition is induced from the 2S state to the 2P state, and the 2P→1S photons are detected as the signal of the transition. As for hydrogen, because of the Stark quenching of M(2S) in an atomic collision (see Sec. V.3) it is necessary to have the M(2S) atoms in vacuum.

V.2 Muonium Production in Vacuum by the Beam Foil Method

In an experiment[78] using the stopped muon channel (SMC) at the Los Alamos Meson Physics Facility (LAMPF) it was shown that muonium in its ground 1S state is formed by passing a μ^+ beam through a thin foil in vacuum. A slightly degraded "surface" muon beam of 28 MeV/c was used for these studies.[79] It was found that about 1/3000 of the incoming μ+ beam produced M(1S) downstream of the foil and that this production efficiency is substantially independent of the foil target material. The observed production efficiency agrees with our general understanding of the physical phenomena involved and with relevant Monte Carlo calculations. The result of a Monte Carlo calculation of the degradation of an incident μ^+ beam with mean energy of 3.7 MeV in a Be foil of 148 mg/cm^2 thickness is shown in Figure 5.1. The Monte Carlo calculation is based on extensive data on p-Be and H-Be atomic scattering cross sections, scaled through the Born approximation to μ-Be and M-Be cross sections. We note the broad spread in energy of the emerging μ^+ beam. Since only muons with kinetic energies up to about 20 keV have an appreciable probability of capturing an electron as they emerge from the surface (Figure 5.2), the order of magnitude of the M production efficiency is to be expected.

The formation of M(2S) atoms should also occur through the electron capture process as the μ^+ emerge from the foil. From proton data[80] we anticipate that the ratio M(2S)/M(1S) will be about 0.1. Higher excited states will also be formed but with lower probability, and most of these excited state atoms will decay rapidly.

In order to obtain a useful intensity of M(2S) atoms in a sufficiently background-free environment a substantial effort was devoted to the development of a very low-momentum, high purity positive muon beam.[81,82] To obtain a relatively pure μ^+ beam a static $\vec{E}\times\vec{B}$ separator was developed and installed in the beam line to transmit μ^+ and to deflect e^+ out of the beam. This device had an e^+ suppression factor of 10^4, and our resulting μ^+ beam at 10 MeV/c had $e^+/\mu^+ < 0.1$.

Since M formation occurs for μ^+ emerging from the target foil only in the kinetic energy range up to about 20 keV, it is clearly desirable to have a μ^+ beam with as low a momentum as possible. This was achieved by tuning the muon channel to transmit low momenta μ^+ which originate from π^+ stopped below the surface of the primary proton target (subsurface muons). Using the range-momentum relationship for muons in carbon, we find that the muon intensity transmitted by the muon channel, for which the fractional momentum acceptance $\Delta p/p$ is independent of p, is proportional to

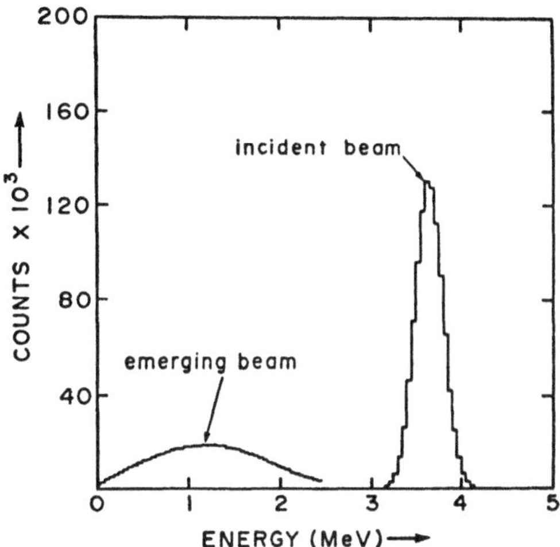

Figure 5.1. Calculated energy distribution for muons emerging from a Be degrader of 148 mg/cm^2 for an incident muon beam of momentum 28 MeV/c.

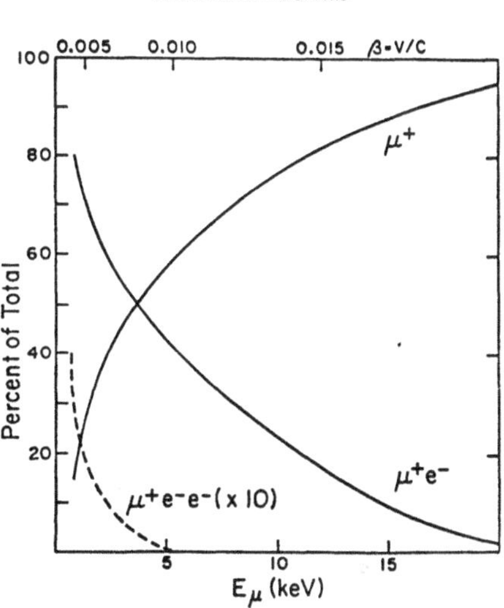

Figure 5.2. Expected charge-state distribution for muons emerging from foil targets.

$$p^{3.5} e^{-m_\mu \ell/p\tau} \qquad (5.1)$$

in which ℓ = channel length, m_μ and τ are muon mass and mean life. The exponential factor accounts for muon decay in the channel. Figure 5.3 shows the measured muon rates and the solid curve is Eqn.(5.1) normalized to the muon rate at p = 20 MeV/c. Although the muon rate r decreases rapidly with decrease in p, Monte Carlo calculations indicate that the probability of M formation by a μ^+ increases as p decreases. Overall the total M rate downstream of the foil target is approximately independent of p. Figure 5.4 shows the calculated output μ+ energy spectrum for an incident momentum of 10 MeV/c and a Au foil target of thickness 8.7 mg/cm^2.

The muonium ground state yield and the ratio of M(1S)/μ^+ was measured with the apparatus shown in Figure 5.5. A 20 μm thick muon scintillation counter (NE 102A) served as a degrader, and muonium was formed in a thin Aℓ foil (0.2 mg/cm^2) at ground potential immediately downstream of the scintillator. Comparison between the number of e^+ from μ^+ decay detected at the NaI(Li) detector with the sweeping magnetic field B on and off gives the number of neutrals passing the field (i.e. μ^+e^-) compared to the number of μ^+ getting downstream. Assuming an isotropic angular distribution of M emerging from the foil, the M yield per incoming μ^+ is 4% at its maximum (Figure 5.6) as expected from Monte Carlo calculations. New measurements with even thinner targets (0.5 mg/cm^2) indicate that the measured M yield at 7 MeV/c is still higher (~10%), as expected.

V.3 Observation of Muonium in the 2S State by Static Electric Field Quenching

It is well known that the metastable 2S state of M will be quenched by application of a static electric field $\vec{E}$, i.e. it will decay to the ground 1S state with emission of a uv photon of 1221 Å wavelength. The

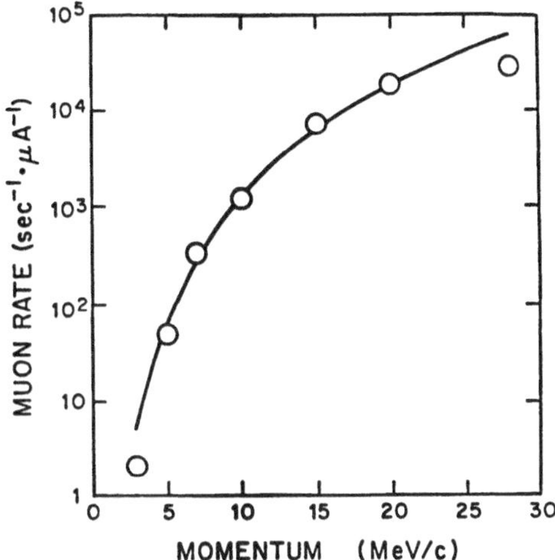

Figure 5.3. Measured muon rates vs. beam momentum. The circles show the measurements (statistical errors are a diameter or smaller). The solid curve represents the theoretically expected rates, normalized to the point at 20 MeV/c.

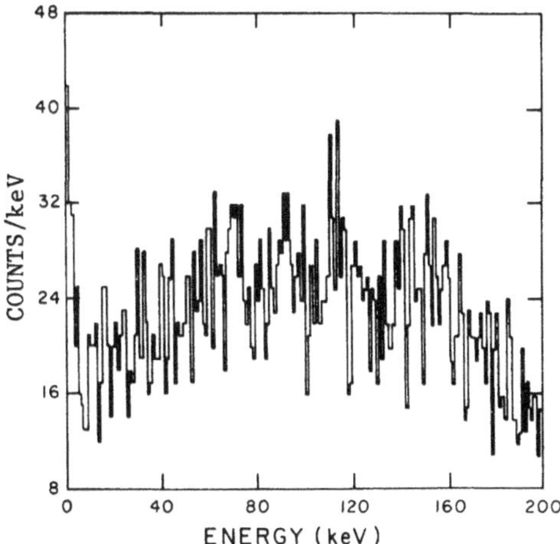

Figure 5.4. Calculated energy distribution for muons emerging from a Au degrader of 8.7 mg/cm^2 (4.5 μm) with an incident muon beam of 10 MeV/c.

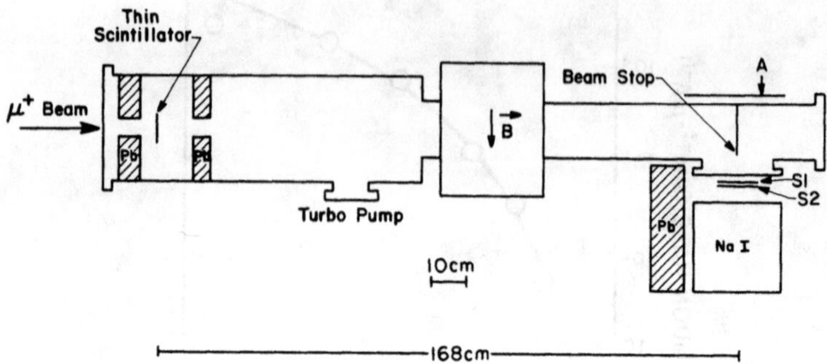

Figure 5.5. Apparatus used in the detection of muonium from a target foil. The thin Al foil is attached to the thin scintillator.

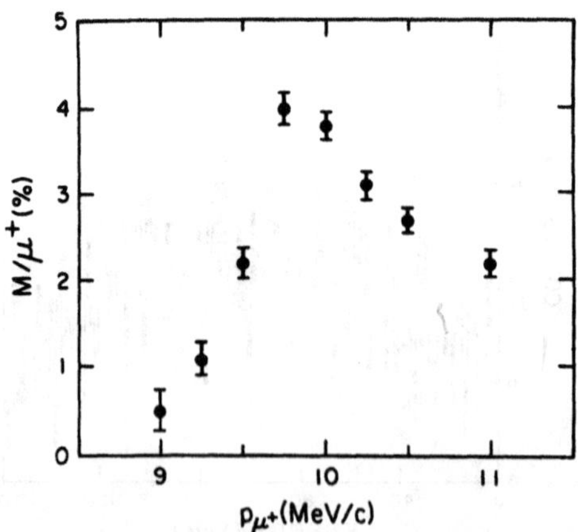

Figure 5.6. Measured fraction of incident μ^+ which form M downstream of the Al foil.

process involves the admixture of the 2P state due to E and an electric dipole transition from 2P to 1S. The quenching rate γ_{Stark} for small electric fields is given by:[83]

$$\gamma_{Stark} = \frac{\gamma_p |V|^2}{\hbar^2(\omega^2 + \frac{1}{4}\gamma_p^2)} \tag{5.2}$$

in which $\omega = (E_0(2P) - E_0(2S))/\hbar$, γ_p = 2P decay rate = $6.3\times10^8\ s^{-1}$, and $V = (2P|e\vec{E}\cdot\vec{r}|2S)$.

Observation of M(2S) atoms can be done by detecting the static electric field quenching of M(2S).[81] The apparatus used at LAMPF both for the static electric field quenching and for the microwave-induced 2S→2P transitions discussed in the next section is shown in Fig. 5.7. It consists basically of:

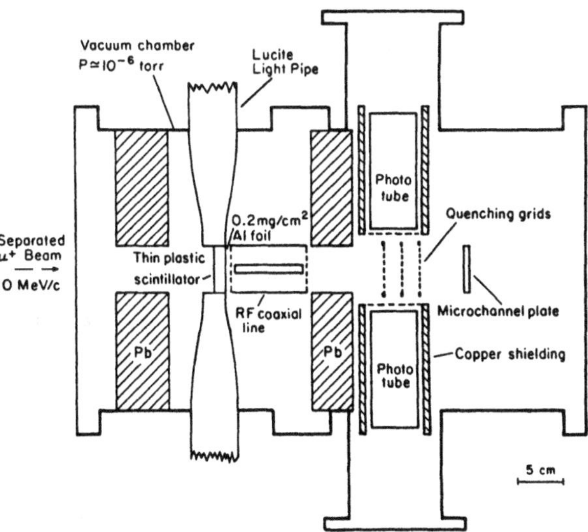

Figure 5.7. Experimental apparatus used in the LAMPF muonium Lamb shift experiment.

1. A thin scintillation counter and aluminum foil to detect the incident μ^+ and to form M. The scintillation counter (NE 102A) was 20 μm thick and served also as a degrader. The Al foil (0.2 mg/cm^2) immediately downstream was at ground potential and served as the target foil for M formation.
2. Phototubes to detect Lyman-α uv photons. The four photomultiplier tubes were the Hamamatsu type R2050 tube with a MgF_2 window of 5 cm diameter and a quantum efficiency of 10% at 1221Å. The total solid angle subtended by the 4 tubes was 30% of 4π.
3. A microchannel plate of 7.5 cm plate diameter just downstream of the phototubes to detect M(1S) atoms and μ^+.
4. Three quenching grids located at the axial position of the phototubes produced an axial static electric field of 600 V/cm. The center grid was at - 1200V and the two outer grids were grounded.

The search for M(2S) involved modulating the static electric field between on (600 V/cm) and off and looking for a difference in the event rate due to Lyman-α photons emitted during the field-on phase. A typical incident μ^+ rate was 70 x 10^3 s^{-1} average with a 9% duty factor. The $\vec{E}$ field was large enough to quench all M(2S) atoms during their transit time in the field. An event E required a triple coincidence between an incoming muon, a delayed Lyman-α photon in one of the four uv phototubes, and a further delayed M(1S) atom detected in the microchannel plate (see Figure 5.8). The distance between the scintillator and the center plane of the uv phototubes was 20 cm and that between the scintillator and the microchannel plate 27.5 cm; the time gates were set to accept M(2S) atoms with kinetic energies between 4 and 26 keV. In two days of data-taking 174 events were observed with the quenching field on and 72 events with the field off. Assigning the difference to events due to Lyman α photons from quenched M(2S) gives a Lyman-α event rate induced by the static electric field of (4.7 ± 0.7)/h. Within the uncertainties of detection efficiencies this value agrees with the

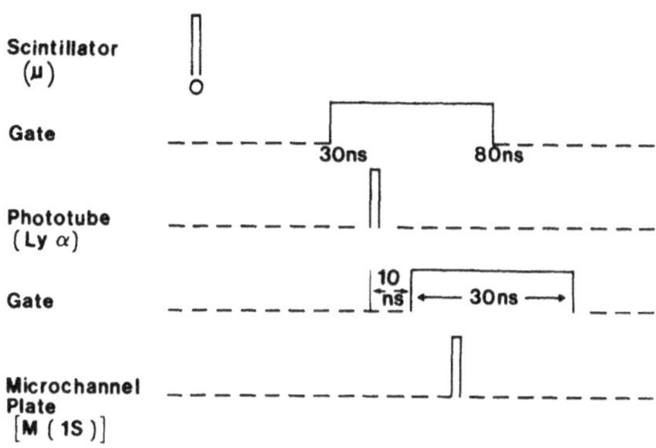

Figure 5.8. Timing diagram of logic used to define signal events by a triple delayed coincidence.

expected rate of M(2S). The normalized difference signal S between quenching field on and field off, $S=[(E/\mu)_{on}-(E/\mu)_{off}]/[(E/\mu)_{on}+(E/\mu)_{off}]$, was S = (41.5 ± 5.8)%. For delayed coincidences with delay times outside the expected flight times of M(2S) atoms, the signal is consistent with zero, namely S = (-1.7 ± 6.4)%. The signal versus the M(2S) kinetic energy and time-of-flight is shown in Figure 5.9. Formation and static electric field quenching of M(2S) has also been reported from an experiment at TRIUMF.[84]

V.4 Observation and Measurement of the Lamb Shift Transition

The principle of the method is to induce a 2S→2P transition with an applied microwave electric field and to observe the accompanying decrease in the 2S beam intensity. Transition to the 2P state is followed rapidly by decay to the 1S state with emission of a Lyman α photon. The center of the resonance line determines the Lamb shift.[85]

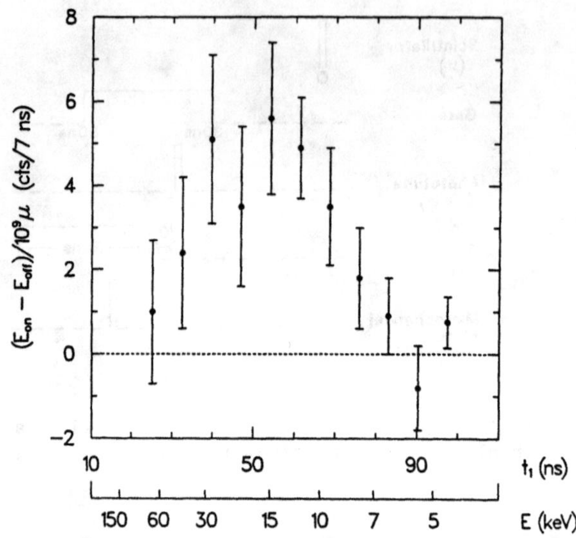

Figure 5.9. Measured M(2S) time-of-flight distribution. The data from the static quenching measurements are plotted vs. the time-of-flight and vs. the M(2S) kinetic energy E.

The transition probability or resonance line shape for the induced transition is given by:

$$P\ (2S \rightarrow 2P) = (2\pi e^2 \gamma_p S_0 / c\hbar^2)\ \frac{|(2P|\hat{e}\cdot\vec{r}|2S)|^2}{(\omega - \omega_0)^2 + (\gamma_p/2)^2} \tag{5.3}$$

in which S_0 is the incident flux density of the radiation having angular frequency ω and electric polarization $\hat{e}$, $\omega_0=(E_0(2S)-E_0(2P))/\hbar$ and $(|\vec{r}|)$ is the matrix element of the coordinate vector $\vec{r}$ for the transition. The quantity $\gamma_p = 1/\tau_p$ is the optical decay rate of the 2P state. At resonance $\omega = \omega_0$ and

$$P = (8\pi e^2 S_0 / c\hbar^2 \gamma_p)(|\hat{e}\cdot\vec{r}|)^2. \tag{5.3a}$$

The matrix elements are given by:

$$|(2^2P_{1/2},m_J=-1/2|\ x\ |2^2S_{1/2},m_J=+1/2)|^2 = 3a_0{}^2$$
$$|(2^2P_{1/2},m_J=+1/2|\ z\ |2^2S_{1/2},m_J=+1/2)|^2 = 3a_0{}^2 \qquad (5.3b)$$

The fraction of 2S atoms quenched by the rf field is

$$\phi = 1 - \exp(-\frac{P\ell}{v}) \qquad (5.4)$$

in which ℓ = length of rf region, v = speed of atoms, and P is the transition probability from Eq.(5.3).[81]

In the experimental setup at LAMPF shown in Figure 5.7 the rf interaction region consisted of a coaxial line. It was 8.3 cm long and had tungsten meshes with a transmission of 80% each at the entrance and exit planes to allow the muon beam to pass through. The diameter of the inner conductor was 1 cm and the inner diameter of the outer conductor was 5.1 cm, which resulted in a variation of a factor of 25 in the power density between the inner and the outer radii.

Data were taken with an input power of 25 W to the coaxial line. The microwave power was modulated between on and off with a frequency of 1 Hz. The static electric field in the quenching grids was held constant at 600 V/cm. With microwave power on, M(2S) atoms should be quenched in the rf region, while with power off quenching will take place only in the region of the static field quenching grids viewed by the phototubes. Event counts were triple coincidences as defined in section V.3 and the signal was the reduction in event counts due to the microwave power. The signal S defined as the normalized difference in the number of events between rf power on and power off was measured as a function of frequency between 800 and 1400 MHz, which covered the range where the two transitions 2S(F=1)→2P(F=1) → and 2S(F=1)→2P(F=0) should occur.

The signal and a fitted line shape are shown in Figure 5.10. The maximum signal event rate was about 4 events/hr, and the data shown were obtained in about 200 h of running. The maximum signal was about 25%. No signals were observed with delayed coincidences outside the expected flight times of muonium atoms.

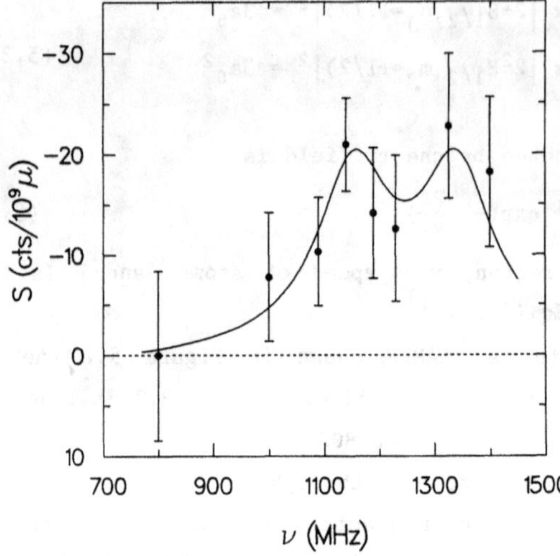

Figure 5.10. Microwave-resonance data and the best fit line shape.

If the reference z-axis for an atom is chosen to be along the direction of $\vec{E}_{rf}$ in the coaxial line, then only transitions $\Delta m_J=0$, $\Delta M=0$ are allowed (M is the magnetic quantum number associated with F). Hence the relative intensity at resonance for the two transitions 2S(F=1) → 2P(F=1) and 2S(F=1) → 2P(F=0) is 2 to 1, if we assume an equal distribution in all sublevels of the initial $2^2S_{1/2}$ state. Then the theoretical curve is of the form:

$$S = \frac{A}{(\omega - \omega_{01})^2 + (\gamma/2)^2} + \frac{3A/7}{(\omega - \omega_{02})^2 + (\gamma/2)^2} \qquad (5.5)$$

Theoretical values for the hfs intervals in the $2^2S_{1/2}$ and $2^2P_{1/2}$ states in which ω_{01} = resonance frequency for the transition ($2^2S_{1/2}$, F=1) → ($2^2P_{1/2}$, F=1) and ω_{02} for ($2^2S_{1/2}$, F=1) → ($2^2P_{1/2}$, F=0); γ is the line width factor associated with γ_p and rf power broadening. In one analysis of the data S is fit to the data points taking the Lamb shift $\mathbf{S}_M$, A and γ as fitting parameters. In another more complete analysis measured values of S_0 and of the M(2S) atom transit times are used, and the fitting parameters are $\mathbf{S}_M$ and the heights of the two resonance lines. The result is $\mathbf{S}_M = 1042^{+21}_{-23}$ MHz.[32] A similar experiment has been done at TRIUMF[86] with the reported result $\mathbf{S}_M = 1070^{+12}_{-15}$ MHz. Systematic errors are small compared to the statistical errors in these measurements of the Lamb shift.

These two experimental values for S_M are in satisfactory agreement and also agree with the theoretical value given in Table 2.1.

Very recently the fine structure transition $2^2S_{1/2} - 2^2P_{3/2}$ in muonium has been observed for the first time in an experiment at LAMPF similar to the Lamb shift measurement.[16] Its measured value is 9895^{+35}_{-30} MHz, which is in reasonable agreement with the value of 9874.3(3) MHz predicted from the theoretical values for the fine structure and Lamb shift intervals.

A more accurate measurement of the Lamb shift to a precision of about 0.1% may be possible with improvements in the type of experiment that has been done. These could include: (1) use of a lower pressure gas multiwire proportional detector as the muon counter to allow use of a lower momentum muon beam of 7 to 8 MeV/c where the muonium rate is the same but the muon rate is lower and pileup is less troublesome. (2) use of M(2S) formed at about 30° to the incident μ^+ beam where the background from the μ^+ is greatly reduced but the M(2S) rate is only reduced by a factor of about 2. (3) use of a high frequency resonant microwave cavity for the $2^2S_{1/2}-2^2P_{3/2}$ fine structure transition where the Lyman-α photons following the transition can be directly observed, and the first order Doppler effect can be avoided, (4) use of a higher efficiency and larger acceptance Lyman-α detector, perhaps achieved with uv reflecting mirrors and a gas counter. A still higher precision measurement of the Lamb shift probably requires the development of a more intense source of low energy muons by some method of phase space compression such as stochastic cooling or muon moderation.

In addition, the analysis of more precise experiments will require a more careful treatment of systematic effects including the Bloch-Siegert effect, rf Stark effect, Zeeman effect, first-and second-order Doppler effects, variation of microwave power with frequency, spatial variation of microwave power, and velocity distribution of M(2S).

The precise measurement of the optical 1S-2S transition discussed in section VI is another way of determining the Lamb shift.

VI. THE 1S-2S TRANSITION IN MUONIUM

The 1S-2S transition in muonium[87] has become of considerable interest only recently because of its observation in a pioneering experiment at KEK.[17] We comment briefly on the theory of this transition, on the initial experiment[17] and on an experiment in progress to make a precise measurement of the transition frequency.[88]

The basic treatment of the theoretical value of the 1S-2S energy interval is given by J.R. Sapirstein and D.R. Yennie in this volume.[18] Table 6.1 gives the contributions to this interval.

The uncertainties in the theoretical result are due to the sources listed in Table 6.2. The uncertainties due to the Rydberg constant (0.67 MHz) and the muon/electron mass ratio (3.43 MHz) enter through the Dirac energy and its reduced mass correction, respectively. Several groups are working on new measurements of the Rydberg constant, and its uncertainty can be expected to decrease during the next few years. A new measurement of the muon magnetic moment is also underway at LAMPF[74] which aims to improve the precision with which m_μ/m_e is known by a

TABLE 6.1

Contributions to the muonium 1S-2S interval

Contribution	Frequency (MHz)
Dirac energy	2,467,411,582.59(67)
Reduced mass correction	-11,875,786.91(343)
Relativistic two-body correction	195.73
Self-energy	-6,967.40(6)
Anomalous magnetic moment	-356.06
Vacuum polarization	188.28
QED recoil and reduced mass corrections	79.96(60)
Higher-order radiative corrections	-0.88(4)
Total interval M(1S-2S)	2,455,528,935.31(355)

TABLE 6.2

Contributions to the uncertainty in the muonium 1S-2S interval

Contribution	Frequency (MHz)
Muon/electron mass ratio	3.43
Uncalculated QED recoil/reduced mass corrections	0.6
Rydberg constant	0.67
Numerical estimate of QED binding corrections	0.06
Uncalculated QED radiative corrections	0.04
Total uncertainty	3.55

factor of five. Uncalculated recoil and reduced mass corrections (about 1 MHz) are the dominant source of the QED uncertainty in the 1S-2S interval. There has recently been considerable theoretical interest in the calculation of terms of this type and it is believed that this uncertainty can be reduced considerably. It is clear that the increased sensitivity of muonium to reduced mass and recoil corrections makes it an ideal system for testing the calculation of these terms. The uncertainty in the calculation of the QED binding corrections (0.06 MHz) is a numerical uncertainty that results from an extrapolation of accurate results calculated for high Z back to Z = 1. This uncertainty will not limit the interpretation of measurements of M(1S-2S) in the near future. Similarly, the uncalculated radiative corrections will not be important for some time.

A measurement of M(1S-2S) can be interpreted in a number of ways. Firstly, by taking values for the Rydberg constant and the muon mass determined by other experiments, one can subtract off the reduced Dirac energy to leave the difference between the 1S and 2S Lamb shifts, which is about 7 GHz. A measurement of M(1S-2S) with 10 MHz uncertainty would thus test the calculation of the Lamb shift in muonium at a level of about 1 part in 10^3, a factor of about 10 better than the current best tests, which are obtained from rf measurements of the 2S Lamb shift. Alternatively, the measurement of M(1S-2S) gives the muon/electron mass ratio.

The pioneering observation[17] of the 1S-2S transition in a two-photon laser spectroscopy experiment was done at KEK, the National High Energy Laboratory in Japan. A schematic diagram of the experiment is shown in Fig. 6.1. The pulsed proton beam from the 500 MeV booster ring of the 12 GeV proton synchrotron with 50ns pulses spaced by 50ms was used to produce a pulsed muon beam and hence a pulsed thermal ground state muonium source from a SiO_2 powder surface. The 1S(F=1)-2S(F=1) transition was induced by two photons from counter-propagating laser beams of 244nm obtained from a cw dye-laser using pulsed amplification and frequency doubling. (Fig. 6.2) The two photon transition is free of first order Doppler effect and inherently capable of very high precision, as has been achieved for the hydrogen 1S-2S transition.[23] The transition is detected by ionization of M(2S) which occurs in the strong laser field and then detection of the slow μ^+ after transport to a microchannel plate detector. The laser frequency was calibrated against a nearby and known Te_2 spectral line. The signal is shown in Fig. 6.3, but it is weak due to the low muon beam intensity available at KEK. The observed 1S(F=1) to 2S(F=1) transition frequency is within 300 MHz or 1 part in 10^7 of the QED prediction of Table 6.1.

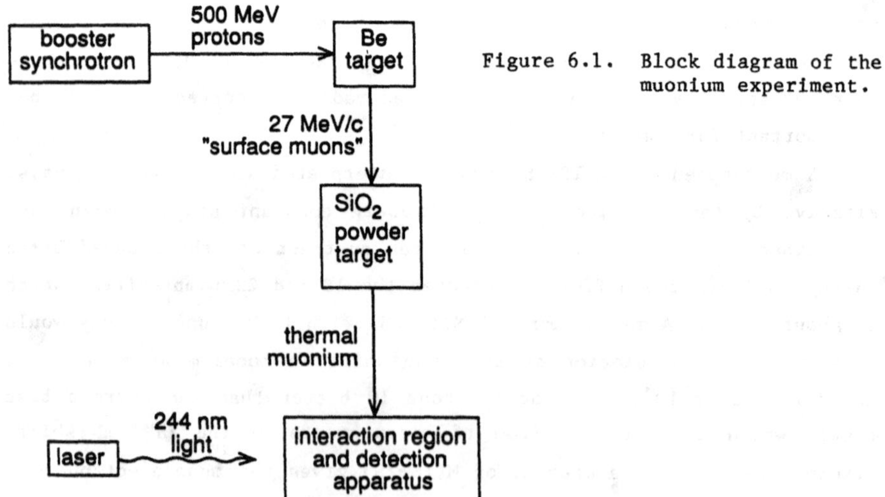

Figure 6.1. Block diagram of the muonium experiment.

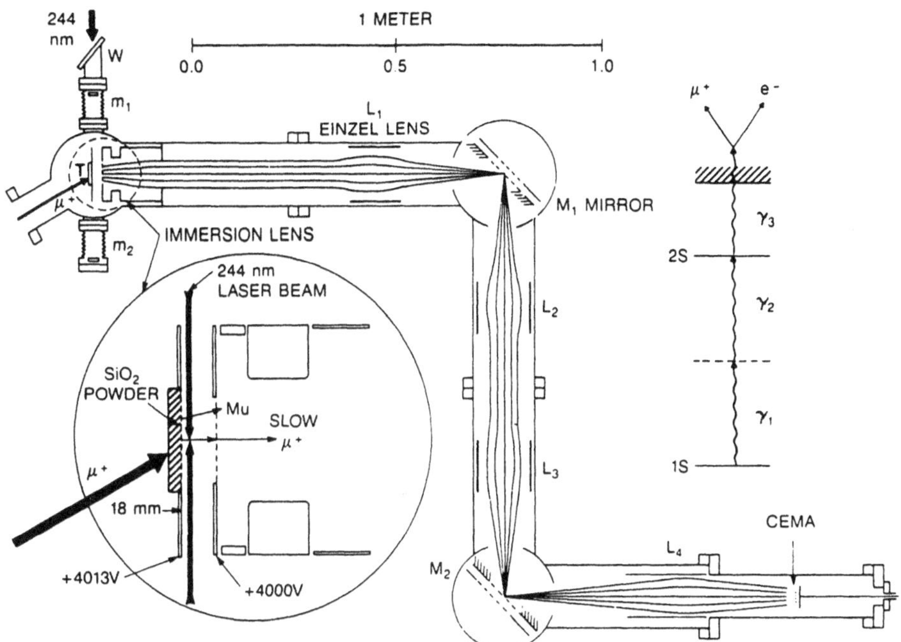

Figure 6.2. Target chamber and electrostatic slow-muon collection optics. Inset: Two-photon-resonant, three-photon ionization scheme.

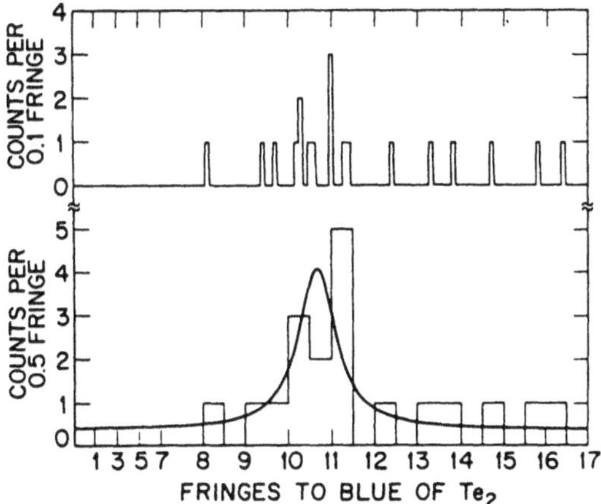

Figure 6.3. Frequency spectrum for all the runs (16h). A Lorentzian is fitted to the individual events shown at the top of the figure.

A new experiment based on the same experimental approach is in progress at Rutherford Laboratory by a Heidelberg-Oxford-Rutherford-Yale group.[88] The proton synchrotron produces a pulsed proton beam of 1 GeV at a repetition rate of 60 Hz and a pulse width of 50 ns. The muon beam intensity at Rutherford is a factor of 100 greater than at KEK. It is expected that a precision of about 10^{-9} in the measurement of the 1S-2S transition frequency can be achieved and that such a result can be interpreted as a measurement of the Lamb shift in the n = 1 state to a precision of about 1 part in 10^3.

VII. MUONIC HELIUM ATOM

The muonic helium atom $^4He\mu^-e^-$ is the simple basic atom in which one of the two electrons in a normal helium atom is replaced by a negative muon. In its structure the muonic helium atom is similar to hydrogen, with the relatively small muonic helium ion "pseudonucleus" $(^4He\mu^-)^+$ corresponding to the proton. The radius of the electron orbit is large compared to the radius of the μ^- in the pseudonucleus, and thus the electron sees a pseudonucleus with a unit positive charge and a magnetic moment equal to that of a negative muon (Fig. 7.1). Hence the muonic helium atom is similar to muonium, and indeed its hfs interval $\Delta\nu$ in the ground state will be approximately equal to that of muonium, but inverted because of the different signs of the magnetic moments of μ^+ and μ^-. The Breit-Rabi energy level diagram is shown in Fig. 7.2.

The $^4He\mu^-e^-$ atom has been formed by stopping polarized negative muons in helium gas at 14 atm with a xenon admixture of 2%.[89] When a negative muon stops in helium gas, it is captured by a helium atom in an Auger process, and then as a result of further Auger and radiative processes will form $(^4He\mu^-)^+$ in its ground 1S state. In a collision with helium at thermal energy, $(^4He\mu^-)^+$ cannot capture an electron to form $^4He\mu^-e^-$ because of the 11.0 eV excess in binding energy of an electron in helium compared to $^4He\mu^-e^-$. Xenon with an ionization energy of 12.1 eV can serve as an electron donor to form $^4He\mu^-e^-$ in the reaction

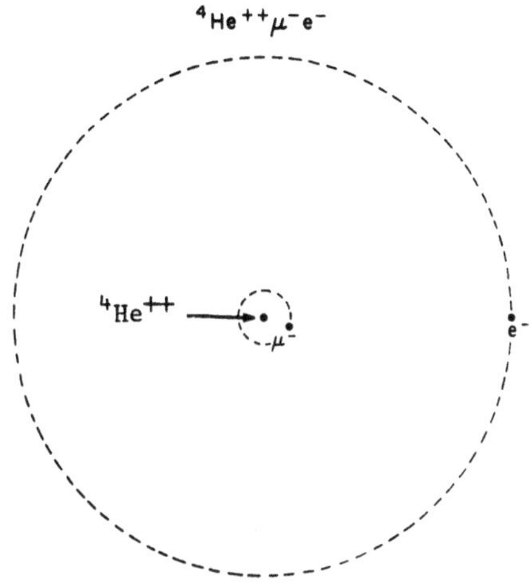

Figure 7.1. The muonic helium atom. Bohr picture.

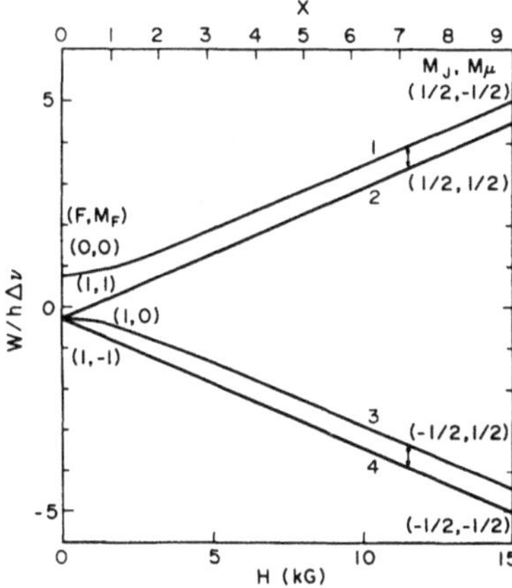

Figure 7.2. The muonic helium atom. Breit-Rabi energy level diagram. $\Delta\nu \simeq 4465$ MHz.

$$({}^4\text{He}\mu^-)^+ + \text{Xe} \rightarrow {}^4\text{He}\mu^- e^- + \text{Xe}^+. \tag{7.1}$$

The residual polarization of μ^- in ${}^4\text{He}\mu^-(1S)$ has been measured to be $P_\mu = 0.06 \pm 0.01$ for μ^- stopped in pure helium gas at a pressure of 14 atm by observation of the amplitude of Larmor precession at the free-muon precession frequency of 13.5 kHz/G. This anomalously low value for P_μ is believed to be associated with collisional Stark mixing of different orbital (L) levels of ${}^4\text{He}\mu^-$ during its cascade from high n states to the 1S state.

Despite the low residual polarization it was possible to observe the Larmor precession characteristic of ${}^4\text{He}\mu^- e^-$ at the frequency of 1.4 MHz/G when μ^- were stopped in the He-Xe mixture in an experiment similar to that in which muonium was discovered[10] (Section III.2). Figure 7.3a shows the amplitude of the Fourier component $A(\gamma)$, where $\gamma = f/H$, obtained from data at four magnetic fields of 3.10, 3.42, 3.73, and 4.64 G, respectively. Figure 7.3b shows corresponding data for muonium formation.

Precision measurements of the hfs interval $\Delta\nu$ and Zeeman effect in ${}^4\text{He}\mu^- e^-$ similar to those for muonium have been made both at weak and strong magnetic fields. The hfs transition was first observed at SIN at weak magnetic field.[90] The experimental arrangement was very similar to that for muonium weak field transitions and is shown in Fig. 7.4. A resonance curve is shown in Fig. 7.5. Transitions at strong magnetic field were studied[91] at LAMPF with the experimental arrangement shown in Fig. 4.7, and resonance curves are shown in Fig. 7.6. The results of the measurements of $\Delta\nu$ at weak and strong magnetic fields are the following:

$$\begin{aligned} \Delta\nu &= 4\ 464.95(6)\ \text{MHz}\ (13\ \text{ppm})\quad (\text{weak field}) \\ \Delta\nu &= 4\ 465.004(29)\ \text{MHz}\ (6.5\ \text{ppm})\quad (\text{strong field}). \end{aligned} \tag{7.2}$$

In addition, the strong field experiment determined

$$\mu_{\mu^-}/\mu_p = 3.183\ 28(15)\ (47\ \text{ppm}) \tag{7.3}$$

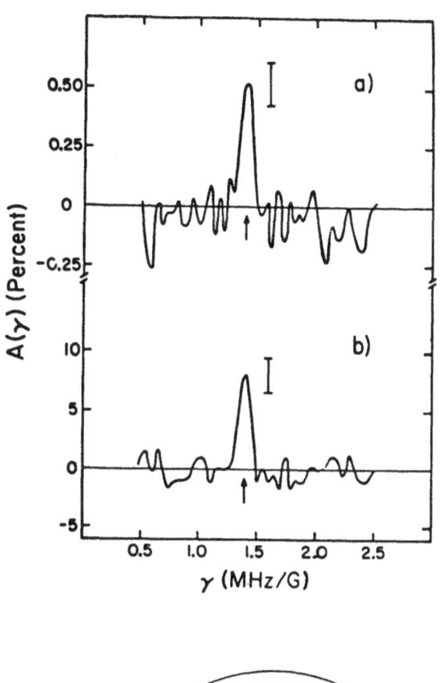

Figure 7.3. Observed Larmor precession amplitudes $A(\gamma)$ versus gyromagnetic ratio $\gamma=f/H$. (a) μ^- stopped in He + 2% Xe, and forming $^4He\mu^-e^-$ ($3\times10^8\mu_s$); (b) μ^+ stopped in He+2% Xe, and forming μ^+e^-($1.8 \times 10^6\mu_s$). The arrows indicate the expected gyromagnetic ratio $\gamma=1.4$ MHz/G.[89]]

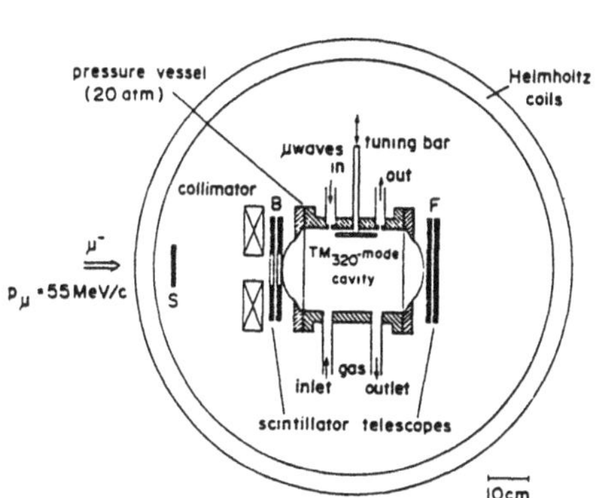

Figure 7.4. Schematic view of the apparatus used to measure weak field transitions in $^4He\mu^-e^-$. The Helmholtz coils are used for muon-spin rotation. A cylindrical high-permeability metal shield (diameter 50 cm, length 100 cm) was installed (not shown in the figure) during the microwave magnetic-resonance experiment to reduce the stray magnetic fields.

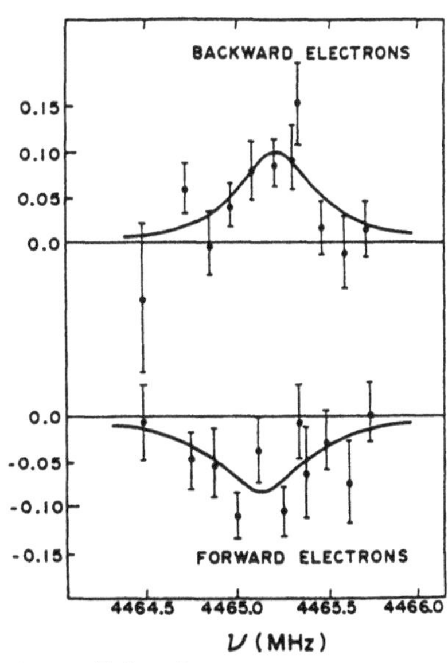

Figure 7.5. Resonance curves for the $\Delta F = \pm1$, $\Delta M_F=\pm1$ hfs transitions in $^4He\mu^-e^-$, simultaneously observed in the backward (upper graph) and forward (lower graph) electron telescopes as a function of the microwave resonance frequency.

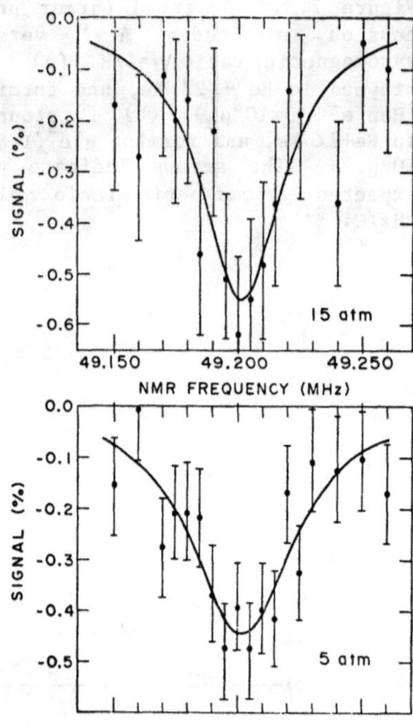

Figure 7.6. Typical resonance curves for the ν_{12} strong field transition in $^4He\mu^-e^-$ obtained with the forward telescope at 15 atm and 5 atm. The data for these curves were obtained in 24h and 100h respectively.

The values of $\Delta\nu$ from the weak and strong field measurements are in good agreement. Also the value for μ_{μ^-}/μ_p agrees within its limited accuracy with the much more accurately determined value for μ_{μ^+}/μ_p given in Eq. (4.18).

The theoretical value of $\Delta\nu$ for the muonic He atom can be written:[92]

$$\Delta\nu = \Delta\nu_F\ (1 + \delta^{rel} + \delta^{rad} + \delta^{rec}) \qquad (7.4)$$

in which $\Delta\nu_F$ is the Fermi value, δ^{rel} the relativistic contribution, δ^{rad} the radiative contribution, and δ^{rec} the radiative recoil contribution. Apart from the simplest reduced mass correction, the principal difference between $\Delta\nu$ for muonium and $\Delta\nu$ for $^4He\mu^-e^-$ is a term of order m_e/m_μ due to the finite size or structure of the pseudonucleus $^4He\mu^-$ and the penetration of the electron inside this structure. Calculation of $\Delta\nu$ requires accurate Schroedinger wave functions in the region of the μ-e overlap. The results of different calculations[92-98] have been compiled in Table 7.1.

Table 7.1
Comparison of Theoretical Results with Experiment

Method	$\Delta\nu$ (MHz)
Variational calculation[94]	4 465.0(0.3)
Perturbation theory coordinate representation[96]	4 464.3(1.8)
Born-Oppenheimer theory, global operator technique[97]*	4 450
Perturbation theory momentum representation[98]*	4 462.9
Experiment[91]	4 465.004(29)

*Only leading term $\Delta\nu_F$ is evaluated.

Another interesting atomic system is $^3He\mu^-e^-$ where three different particles each with spin 1/2 form the atom. Coupling of the muon to the ^3He nucleus results in a large muonic hfs splitting of 332 THz between the two states with angular momentum quantum numbers G = 0 and G = 1. Addition of the electron to the $(^3He\mu^-)^+$ pseudonucleus results in a state F = 1/2 for G = 0 and two states F = 1/2 and F = 3/2 for G = 1 (see Fig. 7.7). The latter F states are separated by the hfs interval $\Delta\nu \simeq$ 4166 MHz, which is similar to the $^4He\mu^-e^-$ hfs but about 10% smaller because of the coupling of the ^3He nuclear magnetic moment to the muon magnetic moment.

Hfs transitions have been observed at weak magnetic field (H<20mG) between the G = 1, F = 1/2 and F = 3/2 states (see Fig. 7.8). The result of the measurement is[99]

$$\Delta\nu = 4166.41(5) \text{ MHz}$$

in agreement with the measurements in $^4He\mu^-e^-$ and with theory.[92,96]

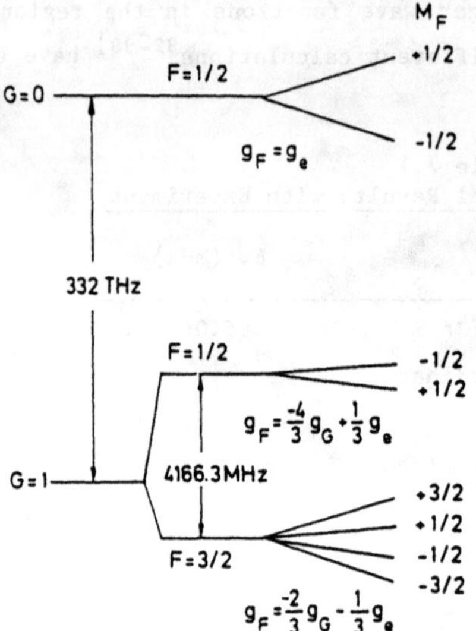

Figure 7.7. Ground state hfs of $^3\mathrm{He}\mu^-e^-$.

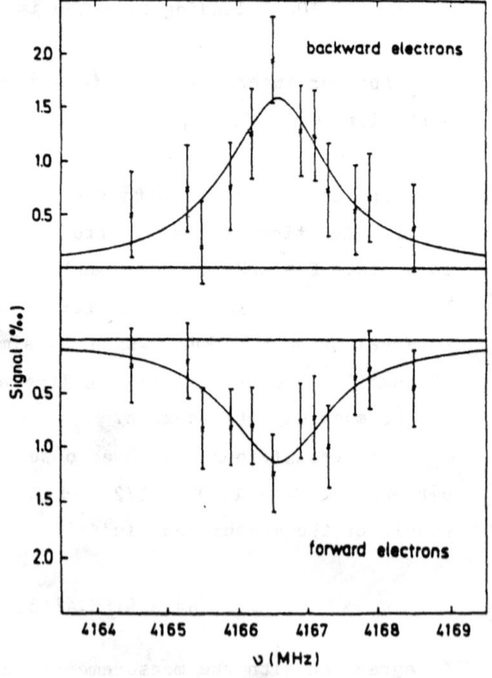

Figure 7.8. Hfs resonance signal obtained from μ-decay electrons in and opposite to the direction of the polarization of the muon.

If a suitable high power laser were available, it would of course be interesting to measure the high frequency transition between the G=0 and G=1 states. This would provide precise information about the nuclear form factor of ^{3}He.

Research is supported in part by the U.S. Department of Energy and the Federal Minister of Scientific Research and Technology, FRG. One of us (VWH) is grateful for support through an award by the Alexander von Humboldt Foundation, and one of us (G.z.P) for support by the NATO Office of Scientific Affairs.

REFERENCES

1. Hughes, V.W., Ann. Rev. Nucl. Sci. 16, 445 (1966).
2. Hughes, V.W. and Kinoshita, T., Muon Physics I, ed. by V.W. Hughes and C.S. Wu, (Academic Press, New York, 1977), p. 11.
3. Hughes, V.W., Fundamental Symmetries, ed. by P. Bloch, P. Pavlopoulos, and R. Klapisch (Plenum Publishing Corp, 1987) p. 287.
4. Hughes, V.W., et al., Nuclear Weak Process and Nuclear Structure, ed. by M. Morita, H. Ejiri, H. Ohtsubo, and T. Sato, (World Scientific, Singapore, 1989) p. 157.
5. Olin, A., et al., Nuclear Weak Process and Nuclear Structure, ed. by M. Morita, H. Ejiri, H. Ohtsubo, and T. Sato, (World Scientific, Singapore, 1989) p. 164.
6. Brewer, J.H., et al., Muon Physics III, ed. by V.W. Hughes and C. S. Wu, (Academic Press, New York, 1975) p. 3.
7. Walker, D., Muon and Muonium Chemistry, (Cambridge University Press, Cambridge, 1983).
8. Schenck, A., Muon Spin Rotation Spectroscopy, (Adam Hilger Ltd., England, 1985).
9. Feinberg, G. and Lederman, L.M. Ann. Rev. Nucl. Sci. 13, 431 (1963).
10. Hughes, V.W., et al., Phys. Rev. Lett. 5, 63 (1960); Phys. Rev. A1, 595 (1970).
11. Garwin, R.L., Lederman, L.M., and Weinrich, M., Phys. Rev. 105, 1415 (1957).
12. Friedman, J.I. and Telegdi, V.L., Phys. Rev. 105, 1681 (1957).
13. Ziock, K., et al., Phys. Rev. Lett. 8, 103 (1962).
14. Oram, C.J., Atomic Physics 9, ed. by R.S. Van Dyck, Jr. and E. N. Fortson, (World Scientific, Singapore, 1985) p. 75.

15. Badertscher, A., et al., Atomic Physics 9, ed. by R.S Van Dyck, Jr. and E.N. Fortson, (World Scientific, Singapore, 1985) p. 83.

16. Kettell, S.H., "Measurement of the $2^2S_{1/2}$-$2^2P_{3/2}$ Fine Structure Interval in Muonium", Ph.D. Thesis, Yale University, 1990. (Unpublished)

17. Chu, S., et al., Phys. Rev. Lett. 60, 101 (1988).

18. Sapirstein, J.R. and Yennie, D.R., This volume.

19. Bethe, H.A. and Salpeter, E.E., Quantum Mechanics of One-and Two-Electron Atoms, (Plenum, New York, 1977).

20. Cohen, E.R. and Taylor, B.N., Rev. Mod. Phys. 59, 1121 (1987).

21. Zhao, P., et al., Phys. Rev. Lett. 58, 1293 (1987).

22. Biraben, F., et al., Phys. Rev. Lett. 62, 621 (1989).

23. Boshier, M.G., et al., Phys. Rev. A40, 6169 (1989).

24. Mariam, F.G., et al., Phys. Rev. Lett. 49, 993 (1982).

25. Klempt, E., et al., Phys. Rev. D25, 652 (1982).

26. Bég, M.A.B. and Feinberg, G., Phys. Rev. Lett. 33, 606 (1974); 35, 130(E) (1975).

27. Bodwin, G.T. and Yennie, D.R., Phys. Rep. 43C, 267 (1978).

28. Wu, S.L., Nucl. Phys. B (Proc. Suppl.) 3, 138 (1988).

29. Brodsky, S.J. and Drell, S.D., Ann. Rev. Nucl. Sci. 20, 147 (1970).

30. Amaldi, U., et al., Phys. Rev. D36, 1385 (1987).

31. Grotch, H. and Hegstrom, R.A., Phys. Rev. A4, 59 (1971).

32. Woodle, K.A., et al., Phys. Rev. 41, 93 (1990).

33. Lepage, G.P. and Yennie, D.R., Precision Measurement and Fundamental Constants II, Natl. Bur. Stand, (U.S.) Spec. Publ. No. 617, ed. by B.N. Taylor and W.D. Phillips (U.S. GPO, Washington, D.C., 1984) p. 185.

34. Sachs, A.M. and Sirlin, A., Muon Physics II, ed. by V.W. Hughes and C.S. Wu, (Academic Press, New York, 1975) p. 49.

35. Lee, T.D. and Wu, C.S., Ann. Rev. Nucl. Sci. 15, 381 (1965).

36. Commins, E.D., Weak Interactions, (McGraw-Hill, Inc., New York, 1973); Morita, M., Beta Decay and Muon Capture, (W.A. Benjamin, Inc., Reading Massachusetts, 1973).

37. Review of Particle Properties, Phys. Lett. B204, 1 (1988).

38. Mélése, P.L., Comments Nucl. Part. Phys., 19, 117 (1989); Bryman, D., et al., Rare Decay Symposium (World Scientific, Singapore, 1988).

39. Pontecorvo, B., Zh. Eksp. Teor. Fiz. 33, 549 (1957) [English transl.: Sov. Phys. - JETP 6 429 (1958)].

40. Mobley, R.M. et al., J. Chem Phys. 44, 4354 (1966); 47, 3074 (1967). For a review of more recent works in Muonium Chemistry see: Roduner, E., Lecture Notes in Chemistry 49, ed. by G. Berthier et al., (Springer-Verlag, N.Y., 1988).

41. Mott, N.F., and Massey, H.S.W., The Theory of Atomic Collisions, 3rd. ed. (Oxford, Clarendon, London, 1965).

42. Massey, H.S.W. and Burhop, E.H.S., Electronic and Ionic Impact Phenomena, (Clarendon Press, Oxford, 1969), 2nd ed.

43. Stambaugh, R.D., et al., Phys. Rev. Lett. 33, 568 (1974).

44. Hughes, V.W., Phys. Rev. 108, 1106 (1957).

45. Breit, G. and Hughes, V.W., Phys. Rev. 106, 1293 (1957).

46. Thompson, P.A., et al., Phys. Rev. A8, 86 (1973).

47. Kibble, T.W.B., Phys. Rev. 155, 1554 (1967).

48. Thompson, P.A., et al., Phys. Rev. Lett. 22, 163 (1969).

49. Crane, T.W., et al., Phys. Rev. Lett. 33, 572 (1974).

50. Bailey, J.M., et al., Phys. Rev. A3, 871 (1971).

51. Kusch, P. and Hughes, V.W., Handbuch der Physik, Vol. 37/1, ed. by S. Flugge/Marburg, (Springer-Verlag, Berlin, 1959), p. 1.

52. Ramsey, N.F., Molecular Beams, (Oxford, Clarendon, London, 1956).

53. Hughes, V.W., High-Energy Physics and Nuclear Structure 1975, ed. by D.E. Nagle et al., AIP Conf. Proc. No. 26, p. 515.

54. Favart, D., et al., Phys. Rev. Lett. 27, 1336 (1971); Kells, W., et al., Nuovo Cimento 35A, 289 (1976), gives an extension to a strong field transition.

55. Hughes, V.W., Quantum Electronics, ed. by C.H. Townes, (Columbia Univ. Press, New York, 1960) p. 582; Thompson, P.A., et al., Phys. Rev. Lett. 22, 163 (1969).

56. zu Putlitz, G., Comments Atomic Molecular Phys. I, 74 (1969).

57. Ramsey, N.F., Phys. Rev. 76, 996 (1949).

58. Cleland, W.E., et al., Phys. Rev. Lett. 13, 202 (1964).

59. Cleland, W.E., et al., Phys. Rev. A5, 2338 (1972).

60. Prepost, R. et al., Phys. Rev. Lett. 6, 19 (1961).

61. Clark, G.A., J. Chem. Phys. 36, 2211 (1962); Pipkin, F.M. and Lambert, R.H., Phys. Rev. 127, 787 (1962).

62. Ensberg, E.S. and Morgan, C.L., Phys. Lett. 28A, 106 (1968); Ensberg, E.S. and zu Putlitz, G., Phys. Rev. Lett, 22, 1349 (1969); Morgan, C.L. and Ensberg, E.S., Phys. Rev. A7, 1494 (1973).

63. Ehrlich, R.D., et al., Phys. Rev. Lett. 23, 513 (1969).

64. De Voe, R., et al., Phys. Rev. Lett. 25, 1779 (1970).

65. Ehrlich, R.D., et al., Phys. Rev. A5, 2357 (1972).

66. Jarecki, J. and Herman, R.M., Phys. Rev. Lett. 28, 199 (1972).

67. Crowe, K.M., et al., Phys. Rev. D5, 2145 (1972).

68. Casperson, D.E., et al., Phys. Rev. Lett. 38, 956; 1504 (1977).

69. Casperson, D.E., et al., Phys. Lett. 59B, 397 (1975).

70. Rao, B.K., et al., Phys. Rev. A2, 1411 (1970).

71. Shirley, J.H., Phys. Rev. 138B, 979 (1965); Salwen, H. Phys. Rev. 99, 1274 (1955).

72. Favart, D., et al., Phys. Rev. A8, 1195 (1973); Phys. Rev. Lett. 27, 1340 (1971).

73. Kobrak, G.E., et al., Phys. Lett. 43B, 526 (1973).

74. LAMPF Research Proposal 1054, "Ultrahigh Precision Measurements on Muonium Ground State: Hyperfine Structure and Muon Magnetic Moment," Hughes, V.W., zu Putlitz, G., Souder, P.A., et al. (1986).

75. Lautrup, B.E., et al., Phys. Rep. Phys. Lett. C (Netherlands) 3C, 193 (1972); Quantum Electrodynamics, G. Kallen, (Springer-Verlag, New York, 1972); Quantum Electrodynamics, V.B. Berestetskii, E.M. Lifshitz, and Pitaevskii (Pergamon Press, New York, 1982).

76. Pipkin, F.M., This volume.

77. Owen, D.A. Phys. Lett. 44B, 199 (1973).

78. Bolton, P.R., et al., Phys. Rev. Lett. 47, 1441 (1981); Gladisch, M., Atomic Physics 8, ed. by I. Lindgren, A. Rosen and S.

Svanberg, (Plenum Press, New York, 1984), p. 197; Bolton, P.R., Ph.D. Thesis, Yale Univ., 1982 (unpublished).

79. Reist, H.-W., et al., Nucl. Instrum. and Meth. 153, 61 (1978).
80. Gabrielse, G., Phys. Rev. A23, 775 (1981).
81. Badertscher, A., et al., Phys. Rev. Lett. 52, 914 (1984).
82. Badertscher, A., et al., Nucl. Instr. and Meth. A238, 200 (1985).
83. Lamb, W.E., Jr. and Retherford, R.C., Phys. Rev. 79, 549 (1950).
84. Oram, C.J., et al., J. Phys. B At. Mol. Phys. 14, L789 (1981).
85. Lamb, W.E., Jr. Phys. Rev. 85, 259 (1952).
86. Oram, C.J., et al., Phys. Rev. Lett. 52, 910 (1984).
87. Chu, S., The Hydrogen Atom, ed. by G.F. Bassani, M. Inguscio, and T.W. Hänsch, (Springer-Verlag, Heidelberg, 1989), p. 144.
88. Rutherford Appleton Laboratory Experiment, "High Resolution Two Photon Spectroscopy of the Muonium 1S-2S Transitions," Jungmann, K., zu Putlitz G., et al. (1988).
89. Souder, P.A., et al., Phys. Rev. Lett. 34, 1417 (1975); Phys. Rev. A22, 33 (1980).
90. Orth, H., et al., Phys. Rev. Lett. 45, 1483 (1980).
91. Gardner, C.J., et al., Phys. Rev. Lett. 48, 1168 (1982).
92. Huang, K.N., et al., High-Energy Physics and Nuclear Structure, ed. by G. Tibell, (North-Holland Publ., Amsterdam, 1974), p. 312; Hughes, V.W. and Penman, S., Bull. Am. Phys. Soc. 4, 80 (1959).
93. Otten, E.W., Z. Physik 225, 393 (1969).
94. Huang, K.-N. and Hughes, V.W., Phys. Rev. A20, 706 (1979); Phys. Rev. A21, 1071(E) (1980); Phys. Rev. A26, 2330 (1982).
95. Borie, E., Z. Phys. A291, 107 (1979).
96. Lakdawala, S.D. and Mohr, P.J. Phys. Rev. A22, 1572 (1980); Phys. Rev. A24, 2224 (1981); Phys. Rev. A29, 1047 (1984).
97. Drachman, R.J. Phys. Rev. A22, 1755 (1980); J. Phys. B14, 2733 (1981).
98. Amusia M. Yu, et. al., J. Phys. B16, L71 (1983).
99. Arnold, K.P. et al., Abstracts, Eighth International Conference on Atomic Physics, A58 (1982); and to be published.

THE FINE STRUCTURE IN THE 2^3P STATE OF HELIUM

Francis M. J. Pichanick
Department of Physics and Astronomy,
University of Massachusetts,
Amherst MA 01003,

and

Vernon W. Hughes
Gibbs Laboratory,
Physics Department,
Yale University,
New Haven CT 06520.

CONTENTS

1. INTRODUCTION

The fine structure of the 2^3P state of helium has played an important role in the development of spectroscopy, quantum mechanics, and quantum electrodynamics. For a brief period it yielded the most accurate value of the fine-structure constant α. At the present time much more accurate values for α are available from other measurements, but the helium fine structure is still very important as a precision test of the quantum electrodynamics of the two-electron atom. We believe that both the theoretical and experimental precisions can be improved to the extent that it can once again be competitive both as a test of fundamental quantum electrodynamics and as a source of the fine-structure constant α.

2. EXPERIMENT

2.1 The first optical and microwave experiments

The earliest measurements of the fine-structure were done by optical spectroscopy, the most recent being done by Brochard *et al*[1)]. The first direct measurements of transitions within the 2^3P state using a microwave technique were done by Lamb and his associates at Stanford[2)]. In these experiments the 2^3P state was produced with a net magnetic polarization by electron bombardment of ground-state helium in a cell, (see Fig.1 for the energy levels, although this diagram refers specifically to the atomic beam experiments described below). Microwave magnetic-resonance transitions between fine-structure levels were observed through the resultant changes in intensity and polarization of the fluorescent radiation when the 2^3P state decayed back to the 2^3S. The Stanford result for the 2^3P_1-2^3P_2 interval was 2291.42 (36) MHz, (157ppm).

2.2 Level crossing measurements

Optical level-crossing experiments have been important in the measurement of the fine structure. Indeed the technique[3)] was first seen

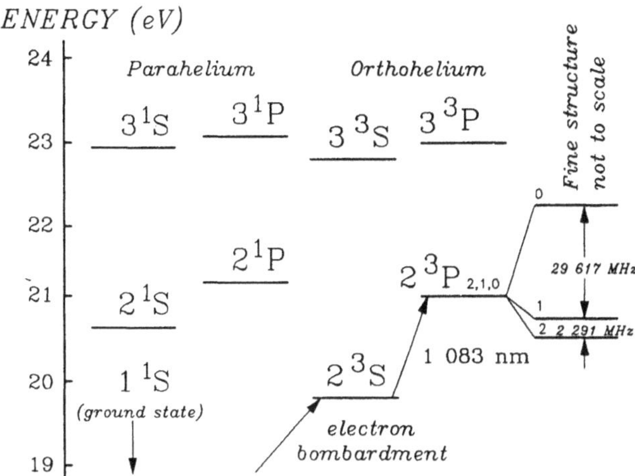

Fig.1 : Helium energy levels of interest

in the 2^3P state of helium when it was excited from the metastable 2^3S state. In this type of experiment observations are made when two Zeeman sublevels, from different fine or hyperfine structure sublevels of an optically excited state, become degenerate, or "cross" at a particular value of magnetic field. If the m values of the crossing sublevels have the appropriate differences, they can be excited in coherence. In such cases their fluorescent radiation will show interference effects when they decay back to the lower state. In the Zeeman diagram (see Fig. 4) two such crossings can be seen just below 600 Gauss. They are between the two pairs of (J,m_J) sublevels (1,0)-(2,2) and (1,-1)-(2,1), both of which have been observed experimentally. The condition for coherent excitation of the crossing sublevels is that they must be allowed to "share" the same photon, meaning that they must be excitable from the same sublevel of the lower state. The selection rule $\Delta m = \pm 1$ for electric dipole radiation then imposes the condition that the m value difference of the two crossing sublevels must be 2, 1, or 0. When the difference is 2 the interference causes a change in both the angular distribution and polarization of the fluorescence, and the effect is quite easy to observe as a change in intensity when the magnetic field is modulated at the crossing point. In fact the effect was first seen by accident during an optical pumping experiment. When the m value difference is 1 the interference causes a change in the polarization only, and the apparatus has to be a little more elaborate. A difference of 0 does not occur because sublevels of the same m cannot cross. The fine structure can be deduced from measurements of the values of the magnetic fields at which the crossings occur.

Level-crossing measurements do not, of course, require the use of a radiofrequency (rf) magnetic field. This gives them an advantage over rf techniques for states with very short lifetimes, when inordinately large rf fields would be required. They do not have any resolution advantage over rf techniques, since the linewidth, as the field is swept through the crossing, is determined by the lifetime, as it is for rf techniques, although there may be some rf broadening. For a time the level-crossing experiments were the source of the most accurate numbers for the fine-structure intervals in helium. This technique suffers, however, from the disadvantage that measurements of the fine-structure separations and the Zeeman g-factors are not independent. Other problems are that they are done in a gas, with collisions affecting the lineshape, and a discharge is needed to excite the metastable state giving further possibilities of systematic error. Their most important contributions[4)] to measurements on the 2^3P state, in both ^4He and ^3He have been to yield g-factors, using fine and hyperfine structure measurements obtained by the atomic beam technique described below.

The level crossing result for the 2^3P_1-2^3P_2 interval was 2291.200 (22) MHz (9.6ppm).

2.3 Atomic Beam Measurements

The most accurate measurements have been made at Yale, using optical excitation in conjunction with microwave magnetic resonance in an atomic beam apparatus[5),6),7),8)]. The optical-microwave atomic beam method was first used at Columbia[9)] to measure hyperfine structure in the 3^2P state of ^{23}Na. The atomic-beam method has the advantage that the atoms are in the free state, uncomplicated by collisional effects, and transitions can be observed in any value of external magnetic field.

The theory of the experiment was described in Ref. 5 as well as the measurement of the J =1 to 2 interval. In Ref. 6 we reported the theory and measurement of the Zeeman g-factors, and quoted a more precise value for the J = 1 to 2 interval because of the improved knowledge of the g-factors. The measurement of the J = 0 to 1 interval was reported in Ref. 7, and of the J = 0 to 2 interval in Ref. 8. This chapter is devoted to the fine structure measurements on the isotope ^{4}He which has no hyperfine structure. We mention briefly that a series of measurements was made also on the 2^3P state of the isotope ^{3}He[10]. In ^{3}He one has the rare instance of fine and hyperfine intervals being comparable and considerably more theory is needed to evaluate the experimental data[11]. The hyperfine constants were evaluated with the aid of the fine structure measurements on ^{4}He, and the effect of 2^1P admixture had to be included explicitly in the data analysis, which was not necessary in the case of the measurements on ^{4}He[12]

2.3.1 General description of the experiments. A beam of helium atoms was excited from the ground state to the 2^3S_1 (Fig.1) metastable state by electron bombardment. Experiments using metastable helium beams also originated at Columbia[13], where a Wood's tube discharge was used for the excitation to the metastable state. We give some discussion of electron gun design in Section 5.1 below. A conventional atomic-beam magnetic-resonance apparatus (Fig.2), with two deflecting magnets A, B was used to separate the $2^3S(m_S)$ sublevels into different trajectories. The apparatus had A- and B-field gradients in opposite directions which, together with a collimator, defined the m_S = +1, 0, and -1 trajectories as shown in Fig.3. Between the A and B magnets was a uniform magnetic field, the C-field, wherein the atoms were excited simultaneously by resonant optical and microwave radiation. The optical radiation, from a helium discharge lamp, excited the atoms to the 2^3P state. The process of excitation and decay had the effect of transferring some of the atoms from one m_S sublevel to another, thereby altering the trajectories of these atoms in the B-region.

The dotted trajectory in Fig.3 is for atoms which have transferred from m_S = 0 to +1. We were able to single out this trajectory using an off-axis stop before the A-magnet, and a detector location as shown in Fig.3. The resultant signal due to excitation and decay was called the "light flop".

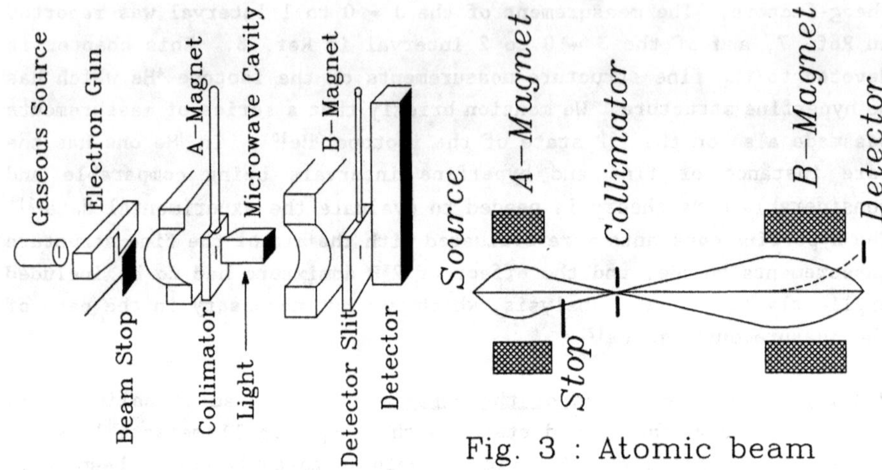

Fig. 2 : Schematic diagram of the atomic beam apparatus.

Fig. 3 : Atomic beam trajectories

In the C-field region the atoms also passed through a microwave magnetic field which was used to induce transitions during the time the 2^3P state was being excited. A magnetic-resonance transition between (J,m_J) sublevels within the 2^3P state will alter the light flop, since this will tend to equalize the populations of the two relevant sublevels, which will have different branching ratios to the ^3S sublevels. This change in the light flop, which is discussed in more detail in Section 5.2 of this chapter, constituted the resonance signal for observing fine-structure transitions in the ^3P state. It was measured as a fraction of the light flop using a double modulation of both the optical and microwave sources.

It was necessary, as explained below, for the measurements to be made in the presence of a uniform magnetic field. The Zeeman Hamiltonian for the 2^3P states can be written in the form

$$H = H_{fs} + g_s\mu_B \mathbf{S}.\mathbf{B} + g_L\mu_B \mathbf{L}.\mathbf{B} \quad ....(1).$$

H_{fs} contains the spin-orbit and spin-spin operators which lead to the fine structure, and is given in detail in equations 3 below in the theory section. It is diagonal in $\mathbf{J}(=\mathbf{L}+\mathbf{S})$, and for purposes of experimental analysis we were only concerned with its eigenvalues E_J, with J = 0,1 and 2, in a representation in which **J** is diagonal. The fine structure intervals which were measured were the differences between the E_J. **S** and **L** are respectively the total spin and orbital angular momentum operators, and **B** is the magnetic field operator. The symbols g_s and g_L are respectively the spin and orbital electronic g-factors, and μ_B is the Bohr magneton. The g-factors had relativistic corrections of order $\alpha^2\mu_B B$ which were both measured and calculated. In addition there were corrections to g_L of order m/M[14], where m and M are the electron and nuclear masses respectively, due to the effects of nuclear motion.

The Zeeman Hamiltonian **H** is not diagonal in J because the field-dependent terms couple sublevels having the same m_J and different J, where m_J is the quantum number associated with J_z, the component of **J** in the direction of the magnetic field. For this reason the $m_J=0$ sublevels, which would be field-independent without this coupling, have a second-order field dependence. While it was therefore necessary to measure the magnetic field, the accuracy required was much less than the quoted precisions of the fine-structure intervals. In fact the field-dependence was convenient for two reasons. A magnetic resonance lineshape could be swept through by variation of the field. The only other alternative would have been to sweep the frequencies of the sources of microwave magnetic field. With the high-power oscillators and tuned cavities that we used, this would have been difficult to accomplish without changing the intensity of the microwave field, thereby distorting the lineshapes of the resonance transitions. The other reason for the presence of a magnetic field was to make observable the J = 0 to 2 transition as described below.

The variation of the levels (J,m_J) as a function of magnetic field, (Zeeman diagrams), are illustrated in Figs. 4 and 5. Fig 4 shows only the J = 1 and 2 levels at relatively low fields, while Fig. 5 shows the J = 0, 1, and 2 levels extended to higher fields. The microwave transitions measured were always between fine-structure sublevels both having $m_J = 0$, which made the transition frequencies independent of magnetic field to first order. The optical decay rate of the 2^3P state is $10^7 s^{-1}$, and for a good signal the microwave transition rate had to be comparable.

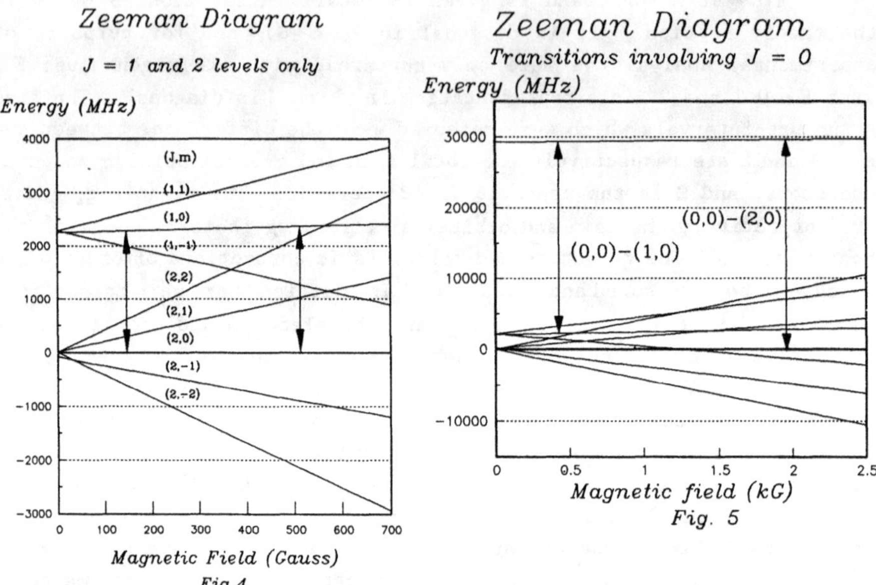

Fig. 4

Fig. 5

This required a microwave field amplitude of about 2 Gauss. We measured all three possible combinations of fine-structure intervals. Hereafter we shall refer to these intervals as ν_{01}, ν_{12} and ν_{02}, where the subscript refers to the appropriate values of J. The interval ν_{12} was measured by two series of runs at the values of magnetic field shown by the arrows in Fig. 4. The sources of error for the two series were somewhat different. At low field the main source of error was the presence of an unresolved transition which distorted the lineshape. At the higher magnetic field this was resolved, but the transition became more field-dependent, and the uncertainty in field measurement restricted the precision. In this way the measurements complemented each other.

An example of a resonance curve is shown in Fig. 6a for the transition $2^3P(J,m_J)$ (1,0) to (2,0). The microwave source was maintained at constant power and frequency, and so the curve was plotted by varying the magnetic field making use of the second order Zeeman effect, as has been mentioned above. The same curve is shown in Fig. 6b as a function of transition frequency units.

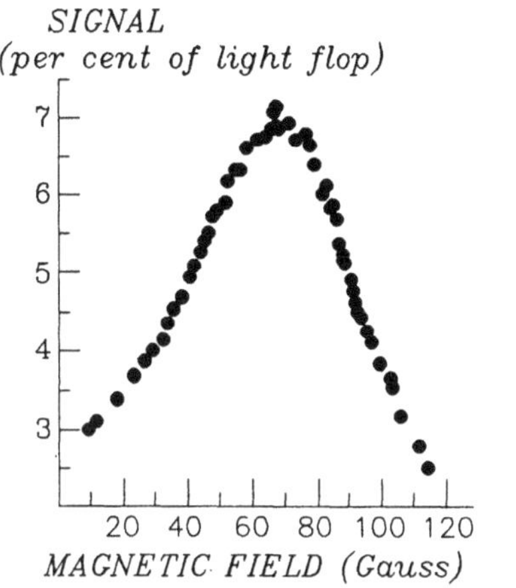

Fig. 6a : J = 1 to 2 resonance plotted as a function of magnetic field

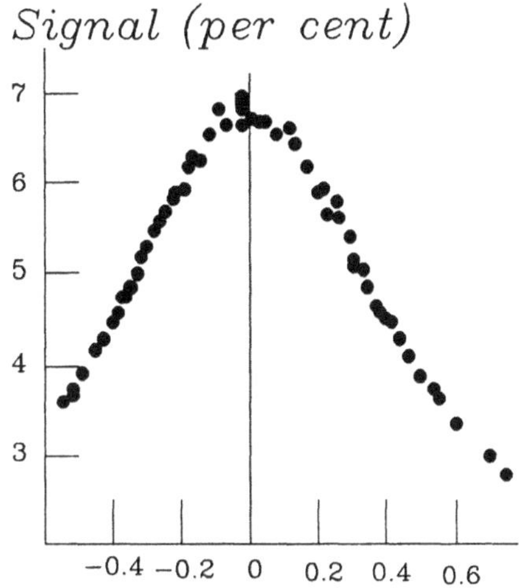

Fig. 6b : J = 1 to 2 resonance plotted against frequency. The units are linewidth fractions.

The latter curve had the theoretical profile given by the Lorentzian:

$$S(\alpha,\beta)=\frac{2(\mu_B H_1)^2|V|^2}{(\omega_{\alpha\beta}-\omega)^2+4(\mu_B H_1)^2|V|^2+\gamma^2}\;.....(2)$$

In this expression $S(\alpha,\beta)$ is the magnitude of the signal between sublevels α and β, $2H_1$ is the amplitude of the microwave magnetic field, $|V|$ is the magnitude of the time-independent part of the microwave transition probability, $\omega_{\alpha\beta}$ is the sublevel energy separation in angular frequency units, ω the microwave magnetic field frequency, and γ is the decay rate (reciprocal of the lifetime). Since the resonance curve illustrated in Fig. 6a was obtained by sweeping the magnetic field, it is in essence a plot of $S(\alpha,\beta)$. In practice the lineshape had to modified because the value of H_1 was not uniform over the microwave cavity. This affected the maximum value of the resonance signal but the overall shape was indistinguishable from that given by Eq. 2. Exhaustive tests were made for distortions and in particular asymmetries in the lineshape, but nothing significant was found, and the fits were within statistical limits. The precision of the experiments would only have been affected by systematic asymmetries.

The "natural" linewidth (width at half-maximum), in the limit of vanishing H_1, is given from Eq. 2 by 2γ, (in effect each sublevel contributes one γ), where the decay rate γ has a value of about 10^7 per second. The natural linewidth in frequency units is

$$\Delta\nu = \frac{2\gamma}{2\pi} Hz \approx 3.2 MHz.$$

The optimum amount of power broadening is obtained when the ratio of the height to width of the lineshape is maximum, because this maximizes the accuracy with which the center of the line can be found. This situation exists when the linewidth is broadened to $2\sqrt{2}\gamma/\pi$, and the amplitude of the required oscillatory field is $2H_1 = \sqrt{2}\gamma/2\pi\mu_B|V| \approx 3$ Gauss for the transitions studied.

The fine-structure interval was obtained by extrapolating to zero field using measured values of the appropriate g-factors[6)]. For the other two intervals[7),8)] the signal-to-noise ratios were not as good as that indicated in Figs. 6a and 6b. The accuracy in the measurement of the interval ν_{12} was limited mostly by uncertainties in field-dependent effects, and by the overlapping of a nearby microwave transition between the $m_J = \pm 1$ sublevels of the same interval. The uncertainties in the measurements of the ν_{01} and the ν_{02} intervals were almost entirely due to counting statistics. Examples of resonance curves for the transitions J = 0 to 1 and J = 0 to 2 are shown respectively in Figs. 7 and 8.

In the latter case, as mentioned above, transitions were allowed between J = 0 and 2 because the experiment was done in a uniform magnetic field, and the Zeeman Hamiltonian admixes the effective J value of all three fine-structure levels.

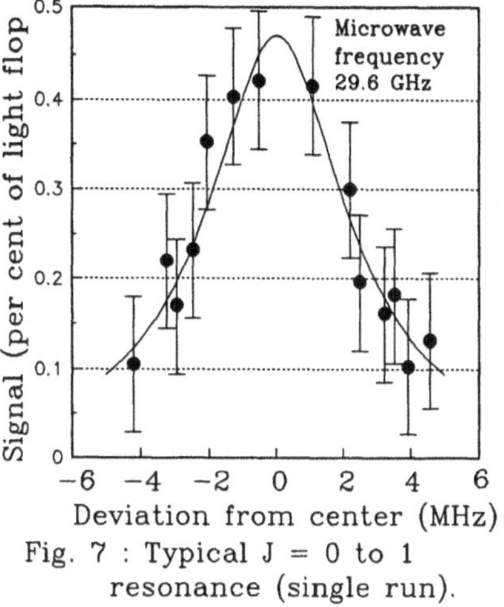

Fig. 7 : Typical J = 0 to 1 resonance (single run).

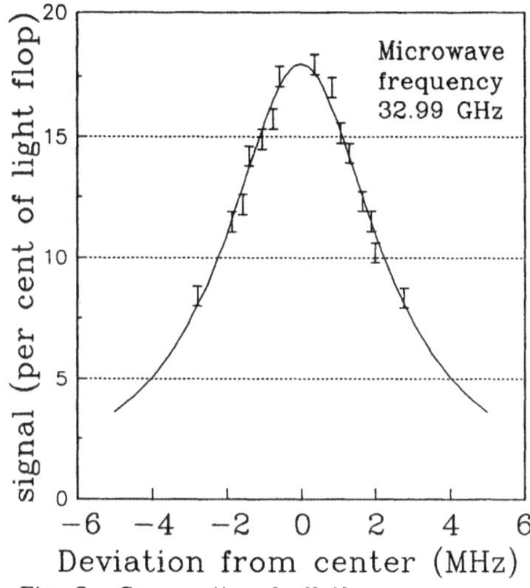

Fig. 8 : Composite of all the resonance data for the J = 0 to 2 transition.

2.3.2 <u>Optical excitation</u>. The metastable beam was excited to the 2^3P state by means of a helium discharge lamp. In the first experiment[5)] a modified commercial lamp was used, and this was connected, in series with a ballast, to a DC power supply. In the subsequent experiments a home-made rf discharge system was used. The lamp, sketched as an inset in Fig. 9, was made out of 3/8 in. Pyrex tubing slightly enlarged at the ends. Contaminants in the tube seriously affected lamp quality, and so it was thoroughly cleaned, baked under vacuum, and then given a final purge with a discharge before being filled and sealed. The typical operating pressure for the helium in the lamp was about 3 Torr. A typical lamp spectrum is illustrated in Fig. 9.

The lamp was energized by placing it in the tank circuit of a 40W 100MHz oscillator. The light interacted with the atomic beam by entering the microwave cavity through small slots in the cavity bottom. A shutter was placed between the lamp and cavity so that the lamp could be modulated in order to measure the light flop.

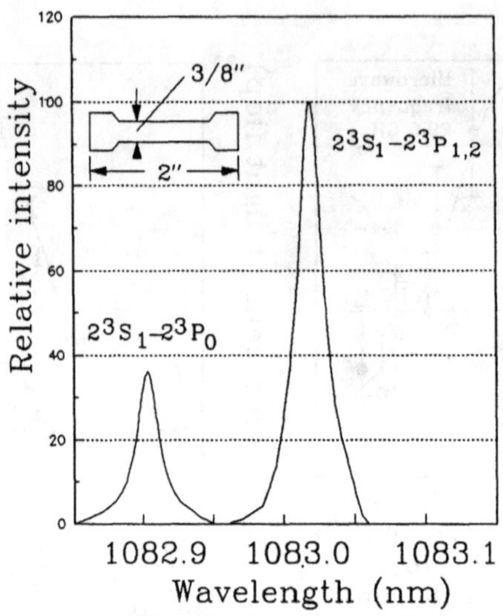

Fig. 9 : Lamp spectrum and dimensions.

2.3.3 <u>Microwave systems</u>. The measurements of the J = 1 to 2 transition were done over a frequency range of 2295 to 2420 MHz. In this case the "cavity" design was that of a U-shaped, shorted co-axial transmission line illustrated in Fig. 10. This low-Q arrangement had the virtue of staying in tune, but it required about a kilowatt of power to provide the necessary microwave magnetic field amplitude. Getting so much power into the vacuum system necessitated very careful design and fabrication of the transmission line, since any reflection along the line would lead to a runaway burn-up. The line entered the vacuum system by means of a carefully matched cylindrical Teflon plug, which fitted tightly around the inner conductor and pressed an O-ring around the outer conductor. The inner and outer conductors were made of brass, and the ratio of their radii gave the line an impedance of 50Ω. The length of the line was designed so that there would be a maximum of the microwave magnetic field at the region of the light-atomic-beam interaction. The atomic beam entered and exited through narrow slots in the elbows of the U, and travelled along the horizontal portion of the U just under the center conductor. The light entered through a slot in the bottom of the outer conductor.

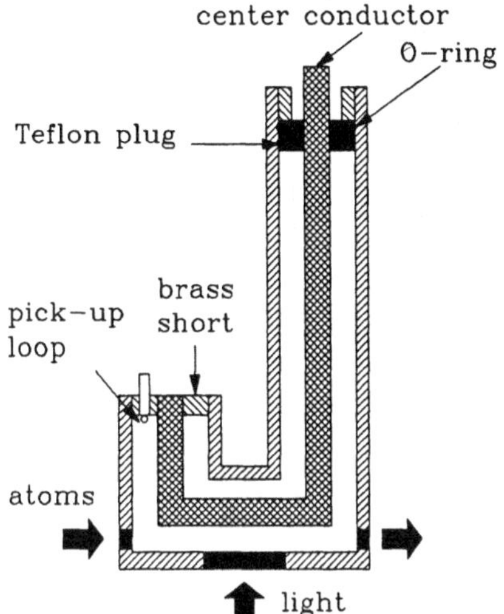

Fig. 10 : Microwave structure for J = 1 to 2.

The high-power microwaves were provided by a 1kw, four-cavity tunable klystron amplifier, which had a power gain of about a million. The input to the klystron was a harmonic of the output of a 0-50MHZ frequency synthesizer, which had a stability of 3 parts in 10^9. The harmonic was generated by a step-recovery diode and amplified by a travelling-wave tube. Square-wave modulation of the input to the klystron was accomplished by means of a diode switch activated by an audio oscillator, which in turn was triggered by the detector channel switching circuit described below.

The intensity of the microwave magnetic field inside the transmission line was monitored by a small pick-up loop, the rectified output of which activated a power-levelling circuit which controlled the input level to the klystron.

The measurement of the J=0 to 1 fine structure was done at a frequency of 29.6 GHz using a cavity made of silver RG 96/U waveguide, (1cm x 0.5cm), which is illustrated in Fig. 11. This was bent into a U shape to provide entrance and exit for the atomic beam as shown. It was shorted at one end and coupled by an iris to the power source at the other. The length of the cavity (10.15cm) was such that it sustained the $TE_{10n=14}$ mode. There was a tuning screw at the shorted end of the cavity. The slits for the atomic beam and the light were 0.07cm wide. The cavity was water cooled to provide stability in the cooling.

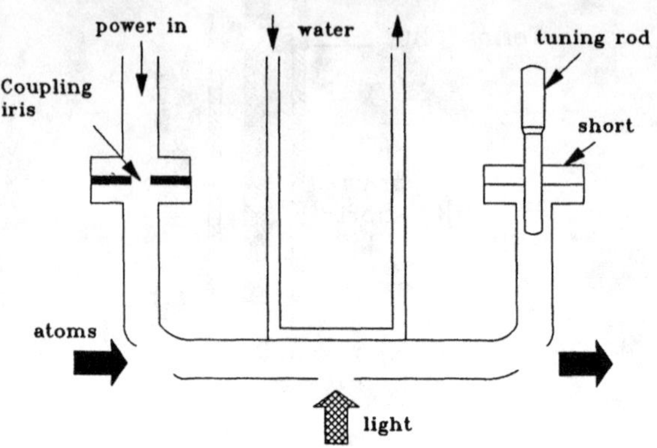

Fig. 11 : The microwave structure
for J = 0 to 1

The microwave power was provided by 10 watt two-cavity klystron specially developed for us by Varian Associates. The local frequency standard was the aforementioned frequency synthesizer. A 9.9 GHz low-power klystron was phase-locked to this standard, and its third harmonic was mixed with part of the output of the 29.6 GHz klystron. The frequency difference was amplified and fed to an FM discriminator, which generated a dc voltage that was fed back to the power supply of the 29.6 GHz klystron to provide frequency stabilization. The output of the 29.6 GHz klystron was square-wave modulated by a ferrite switch which was activated by the channel switch of the detector circuit.

The microwave cavity that was used for the J=0 to 2 measurement was a rectangular copper structure operating in the TE_{410} mode at 32.99 GHz with a Q of 6500, (Fig. 12). The upper part of the diagram is a plan view of the cavity showing the magnetic field lines. On each side of the cavity were co-axial lines feeding the microwaves used to measure the 2^3S Zeeman frequency (5.7 GHz) for magnetic field calibration. The cavity was clamped between water-cooled blocks to prevent thermal drifts of the resonant frequency. The geometry of the arrangement was very similar to that shown below in Section 5, (Fig. 15). The inside of the cavity was silvered and polished to improve the optical reflectivity thereby making the optimal use of the resonant radiation entering the cavity. The entire arrangement could be translated along the beam line, and the 2^3S Zeeman resonances were used to map the magnetic field.

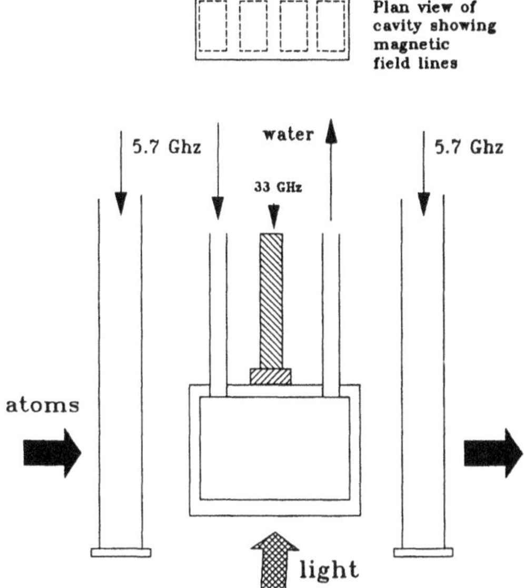

Fig. 12 : microwave structure for J = 0 to 2.

The microwave power was provided by a Varian extended-interaction oscillator which typically supplied 50W during the course of its life. Its frequency was locked to a frequency synthesizer in a manner similar to that described above for the two-cavity 29.6 GHz klystron.

At the time that these experiments were done, obtaining high power oscillators in the 30 GHz frequency region was difficult. Future experiments will be done with travelling-wave tube amplifiers, readily and inexpensively available because this frequency range is now used for satellite transmission of commercial television.

2.3.4 <u>Detector systems</u>. The metastable atoms with the appropriate trajectory impinged on the tungsten cathode of a continuous dynode electron multiplier. With an overall efficiency that we estimated to be about 10 per cent, the metastable atoms ejected Auger electrons from the cathode[15)] which were amplified into a detectable pulse by the multiplier. After pre-amplification and noise discrimination the pulses were processed by a logic network directing them into the proper channel.

The modulation logic grew in sophistication as the experiments progressed. Counts were recorded in separate channels with all combinations of light on/off and microwaves on/off. In the latest experiment[8], the measurement of the J = 0 to 2 interval, a computer was used to generate square-wave modulation signals for both the light and the microwaves, sending the appropriate signals to the microwave output and the light shutter. The logic of the modulations was designed to be as effective as possible to suppress drifts in lamp intensity and microwave power. The 2^3P microwave signal was recorded, as noted before, as a fraction of the light flop. Less time was needed to record the light flop since it was relatively much larger than the microwave signal. Counts were not recorded during the modulation switching because of transient effects. The readout of each data point was the average of 30 sets of counts, and the computer generated a standard deviation which was compared with that to be expected on the basis of counting statistics. In this way we could ensure that the suppression of drifts and fluctuations was satisfactory.

2.3.5. Magnetic field measurement. The value of the magnetic field was measured by inducing a Zeeman resonance in the 2^3S_1 beam. For the J = 1 to 2 and 0 to 1 intervals this was accomplished by inserting small radiofrequency loops in the slits where the light entered the microwave cavities. The field measurements were made between the data acquisition of individual points on the 2^3P resonance curve. At the end of each point the field measurement was repeated before starting another, and the data were abandoned if unacceptable drift had occurred in the field.

The measurement of the J = 0 to 2 interval was made at a much higher magnetic field (2 kGauss) than the others for the reason mentioned above. In this case the field was continuously monitored by means of a nuclear magnetic resonance (NMR) probe, the signal from which was also used to send a stabilization feedback signal to the power supply for the field coils. The NMR measurements were calibrated against Zeeman transitions in the 2^3S_1 atomic beam. The Zeeman transitions were induced at a frequency of 5.7 GHz, the power for which was introduced to the beam on either side of the 2^3P cavity as shown in Fig. 12. The entire microwave structure was translatable along the atomic beam line so that the magnetic field could be mapped.

The g-factor for the 2^3S_1 state was known to much more than sufficient accuracy for the magnetic field measurements[16].

2.3.6 Atomic beam. The intensity of the measured metastable beam varied during the course of the different experiments. For example after the completion of the J = 1 to 2 experiment a new apparatus was built, and this was twice as long as the old machine in order to accommodate a large external C-magnet with 15-inch pole faces, but the additional length reduced the beam intensity. The data given below are typical.

The measured beam at the center of the detector region consisted of about 45 per cent metastable atoms, the remainder being presumably UV photons from the optical decay of short-lived states that were excited by the electron gun. The fraction of the beam that consisted of 2^1S_0 metastables was measured by optically exciting the beam with the lamp. This had the effect of quenching only the singlets since they decayed predominantly to the ground state, while the triplets decayed back to the metastable 2^3S_1 state. The Stern-Gerlach effect was used to measure the fraction of the beam consisting of 2^3S_1 metastables. We found the triplet/singlet ratio to be 3:1 as would be expected from their statistical weights.

The detected beam at the center was about 2 x 10^5 pulses per second (pps). This was consistent with the following numbers which are order-of-magnitude estimates:

flow of atoms past the gun filaments : 10^{14} per sec;
metastable excitation efficiency : 10^{-5};
solid angle subtended by detector : 10^{-3} sr; and
efficiency of detector/discriminator : 0.1.

The light flop at our off-axis detector position was about 200 pps in a background of about 1000 pps. The background varied considerably during the course of the experiments, and it was believed to be due to the tail of the central beam and scattered UV photons.

3. THEORY

The first quantum-mechanical calculation of the helium fine-structure was made by Heisenberg[17] in 1926. Next Breit[18] derived his famous equation using a relativistic retarded interaction and a sixteen component wave-function. He later[19] used the Pauli approximation to reduce the wave-function to four components, and the Hamiltonian to seven sets of terms, of which only the following two contribute to the fine-structure:

$$H_3 = \frac{\mu_0}{mc}\left\{\left[(\mathbf{E}_1\times\mathbf{p}_1)+2\frac{e}{r_{12}^3}\,\mathbf{r}_{12}\times\mathbf{p}_2\right].\mathbf{s}_1+\left[(\mathbf{E}_2\times\mathbf{p}_2)+2\frac{e}{r_{12}^3}\,\mathbf{r}_{21}\times\mathbf{p}_1\right].\mathbf{s}_2\right\}\ldots.(3a)$$

$$H_5 = 4\mu_0^2\left\{-8\frac{\pi}{3}\,\mathbf{s}_1.\mathbf{s}_2\delta(\mathbf{r}_{12})+\frac{1}{r_{12}^3}\left[\mathbf{s}_1.\mathbf{s}_2-\frac{3(\mathbf{s}_1.\mathbf{r}_{12})(\mathbf{s}_2.\mathbf{r}_{12})}{r_{12}^2}\right]\right\}\ldots(3b)$$

Equation (3a) contains the spin-orbit and spin-other orbit interactions while Equation (3b) contains the spin-spin interactions. The improvements in the theory over the decades have been in three areas which we describe below in the following order:

i) more sophisticated wave-functions;
ii) quantum electrodynamic contributions; and
iii) higher order corrections.

3.1 Wave-functions

In Breit's original work[18,19] a two-term wave-function was used with a $\cos\theta_{12}$ term to provide correlation between the electrons. Bethe[20] used a wave-function that was not fully antisymmetric. Araki, Ohta and Mano[21] used a four-component variational wave-function which included pd admixture into the 1s2p configuration. Traub and Foley[22] used an 18-term Hylleraas wave-function.

In the 1950's and 1960's, when powerful computers had become more readily available, ambitious programs were launched, by Pekeris with his co-workers in Israel[23] and Schwartz in the United States[24], to develop Hylleraas wave-functions with hundreds of parameters. Once this was accomplished, the accuracy of the non-relativistic wave-functions was no longer a limitation on the precision of the theoretical fine-structure.

3.2 Quantum Electrodynamic Contributions

These have been the subject of monumental endeavors by Daley[25], and by Douglas and Kroll[26]. They based their investigation on the co-variant Bethe-Salpeter equation[27] including an external potential to take into account the Coulomb field of the nucleus. They included all contributions, to order α^4Ry, which arise from Feynman[28] diagrams involving the exchange of one, two, and three photons, as well as radiative corrections to the electron magnetic moment.

The fully covariant Bethe-Salpeter equation is currently regarded as the most rigorous starting point for the treatment of the two-electron atom. It was derived by Salpeter and Bethe from the Feynman form of quantum electrodynamics, and also from field theory by Gell-Mann and Low[29]. It has been shown by Salpeter[30], and by Sucher[31] that the Bethe-Salpeter equation can be cast into a form similar to the Breit equation. Douglas and Kroll acknowledge that much of their procedure and notation follow that of Sucher. Parts of the calculation by Douglas and Kroll were in disagreement with an earlier work by Kim[32], who used an extension of the Breit technique.

In the Douglas and Kroll procedure with the Bethe-Salpeter equation, the instantaneous Coulomb interaction is separated from the remainder, and a perturbation procedure is developed for calculating the contributions of other interactions in terms of an "equal times" wave-function. In addition, the equation involving the "equal times" wave-function is transformed to a Pauli-Schroedinger type of equation. They developed a procedure for evaluating Dirac operators as Pauli-type matrix elements in terms of the transformed wave-function. Energy shifts for the fine structures were calculated in this way to order α^4 Ry.

The effect of single photon exchange gives rise to a term of order α^2 Ry which they call the "retardation" fine structure. There are numerous correction terms to this of order α^4 Ry. The effect of exchange of two transverse photons gives rise to corrections of order α^4 Ry. The radiative corrections are treated phenomenologically in the sense that they are taken into account by ascribing to the electron a modified charge form factor and a modified magnetic moment. Initially the expectation values were presented in terms of wave-functions in momentum space. For purposes of numerical calculation these were transformed into co-ordinate space.

They point out that, in addition to the fact that only spin-dependent operators need be evaluated, there are two factors which produce considerable simplification of the calculation. One is that contact terms of the form $\langle \phi_0 | \delta^3(r_1 - r_2) | \phi_0 \rangle$ are zero for the spatially antisymmetric triplet states. The other is that operators which contribute terms of order α^4 Ry can be evaluated in a non-relativistic approximation, also because of the spatial antisymmetry of the triplet states.

3.3 Higher Order Contributions

There are a number of terms not included in Eqs. 3a and 3b. The motion of the nucleus is taken into account by inclusion of the usual mass-polarization term

$$H_{MP} = \frac{m}{M}(\mathbf{p}_1 \cdot \mathbf{p}_2) \ldots (4)$$

where m, M, respectively, are the electron and nuclear masses. This operator, when mixed with the spin-dependent operators that contribute to the fine structure, give terms of order $\alpha^2(m/M)$ Ry.

There are second order contributions to the fine structure energies due to the admixtures of intermediate states. These corrections were originally calculated by Hambro[33)] using the method of Dalgarno and Lewis[34)]. In this method an inhomogeneous Schroedinger equation is solved variationally, giving corrections to the wave-function for each operator in Eqs. 3a and 3b, enabling the second-order energies to be evaluated. Hambro computed the second-order energies using intermediate states with symmetries ^{3}P, ^{1}P, and ^{3}D. This did not prove accurate enough, and the second-order energies converged poorly as the number of wave-function parameters increased.

Lewis and Serafino[35)] re-computed the second order energies, including ^{1}D and ^{3}F symmetries for the intermediate states as well as those used by Hambro. In the calculations of the energies they used Hylleraas wave-functions with up to 455 terms for ^{3}P symmetry, and up to

286 terms for the others. Each energy was extrapolated to the limit of an infinite number of terms in the wave-function. The ^3P, ^1P, and ^3D intermediate states contributed to all the fine-structure intervals, while the ^3F and ^1D contributed only to the interval ν_{12}.

4. RESULTS

The fundamental constants used in Table I were[36]:

$$R_\infty = 109737.31534(13) cm^{-1}; \quad \alpha^{-1} = 137.0359895(61); \quad c = 2.99792458 cm/s.$$

The most accurate theoretical number is for ν_{01}, while the most accurate experimental value is for ν_{02}, so that the best comparison with theory is to subtract the experimental value of ν_{12} from that of ν_{02}, obtaining

ν_{01} = 29 616.914 (43) MHz (1.5ppm), theory, and
ν_{01} = 29 616.844 (21) MHz (0.7ppm), experiment.

The agreement is satisfactory. These numbers can be used to obtain a value for the fine-structure constant α. In accordance with common practice we quote our number in terms of the inverse of the fine-structure constant

$$\alpha^{-1} = 137.03612\ (11)\ (0.8ppm),$$

to be compared with the current least-squares adjusted global value (Ref. 36)

$$\alpha^{-1} = 137.0359895\ (61)\ (0.045ppm).$$

Table I : Theoretical and Experimental Results.

Interval	ν_{01}	ν_{12}	ν_{02}
Theory			
α^2 Ry	29 564.577 ± 0.006 (0.20ppm)	2 317.203 ± 0.0018 (0.79ppm)	31 881.780 ± 0.006 (0.19ppm)
α^3 Ry	54.708	-22.548	32.160
$(m/M)\alpha^3$ Ry	-10.707 ± 0.00044 (0.015ppm)	1.952 ± 0.00088 (0.38ppm)	-8.755 ± 0.0010 (0.031ppm)
Second order	11.657 ± 0.042 (1.42ppm)	-6.866 ± 0.081 (35.3ppm)	4.791 ± 0.091 (2.85ppm)
α^4 Ry	-3.331 ± 0.0039 (0.13ppm)	1.542 ± 0.0068 (3.0ppm)	-1.789 ± 0.0078 (0.24ppm)
ν_{theory}	29 616.904 ± 0.043 (1.44ppm)	2 291.283 ± 0.081 (35ppm)	31 908.187 ± 0.091 (2.9ppm)
Experiment			
$\nu_{experiment}$	29 616.864 ± 0.036 (1.2ppm)	2 291.196 ± 0.005 (2.2ppm)	31 908.040 ± 0.020 (0.63ppm)

5. OUTLOOK FOR THE FUTURE

It would be highly desirable for the helium fine structure to provide a test of quantum electrodynamics to a precision ofabout 20 parts per billion. We would also like it to provide an independent value for α competitive with the others. On the experimental side we have mentioned that the accuracies of the measured values of ν_{01} and ν_{02} were limited by counting statistics. We propose to improve this by introducing two major improvements :

i) a more intense 2^3S metastable beam; and
ii) selective excitation of the 2^3P using an infrared laser.

5.1 Improvement in Beam Intensity

An important improvement will be the use of a more efficient electron gun to excite the metastable state from the 1^1S ground state. The gun is of the general type used by Lamb and Retherford[37)] for the production of atomic hydrogen in the metastable 2^2S state, and also used by us in the measurement of ν_{12} [5)]. The gun, illustrated in Fig. 13, is basically a diode with a linear tungsten filament as cathode and a U-shaped anode with a linear slit near, and parallel to the cathode. A ribbon of electrons is emitted by the filament, which is approximately co-planar with the atomic beam, passing through the U-shaped gap in the anode. An important feature of the design is a magnetic field in the direction of the electron motion, which prevents divergence of the electron beam before it encounters the atoms.

An important advance in this gun design, originally made by Lichten[38)] and further developed by Eyler and Pipkin[39)], was the reduction in the slit width by a factor of about 10, and an increase of the magnetic field to about 0.3T. These improvements have led to more than an order of magnitude increase in electron current density at the atomic beam. A gun of this design has been constructed and partially tested. The spread in electron energy due to space-charge should be only about a volt, and by setting the electron energy at the peak in the 2^3S cross-section just above threshold, we should be able to discriminate against excitation of the 2^1S metastable state which provides an unwanted background.

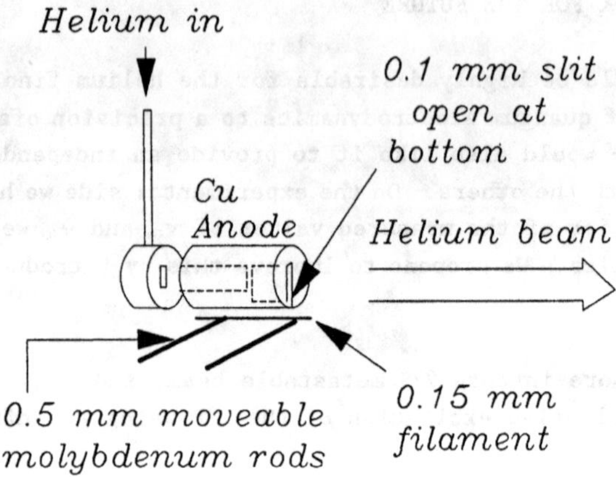

Fig. 13 : Sketch of the new electron gun.

5.2 Selective Excitation of the 2^3P State

In the previous experiments the 2^3P state was excited by means of a helium discharge lamp. The relevant emission lines are in the region of 1083 nm. We have mentioned earlier that our signal depends on the change, due to a microwave 2^3P fine-structure transition, of the light flop, the number of atoms transferring from $2^3S(m_S = 0)$ to $(m_S = +1)$ sublevels. The discharge lamp excited all of the 2^3P sublevels, thereby considerably increasing the background of atoms that are not affected by the microwave transition.

We illustrate this in Fig. 14a for a sample of 1000 atoms originating in the $(m_S = 0)$ sublevel, and how they populate the various 2^3P sublevels when excited by the discharge lamp. The diagram has taken roughly into account the spectral intensity distribution of a typical discharge lamp that we used. With the lamp only, the decay rates are such that 311 of these return to the $(m_S = +1)$ sublevel, (light flop). When a microwave transition is induced between two 2^3P sublevels, the optimal condition is for their populations to be approximately equalized, as shown in the middle diagram of Fig. 14a for the transition m_J=0, J=0 to 1, (01 transition) and in the lower diagram for m_J=0, J=0 to 2 (02 transition). We see that for the 01 transition the number of atoms in (m_S=+1) is changed to 315.5 (the signal is 4.5/311 or 1.5% of the light flop), and for the 02 transition to 351.5 (13% signal).

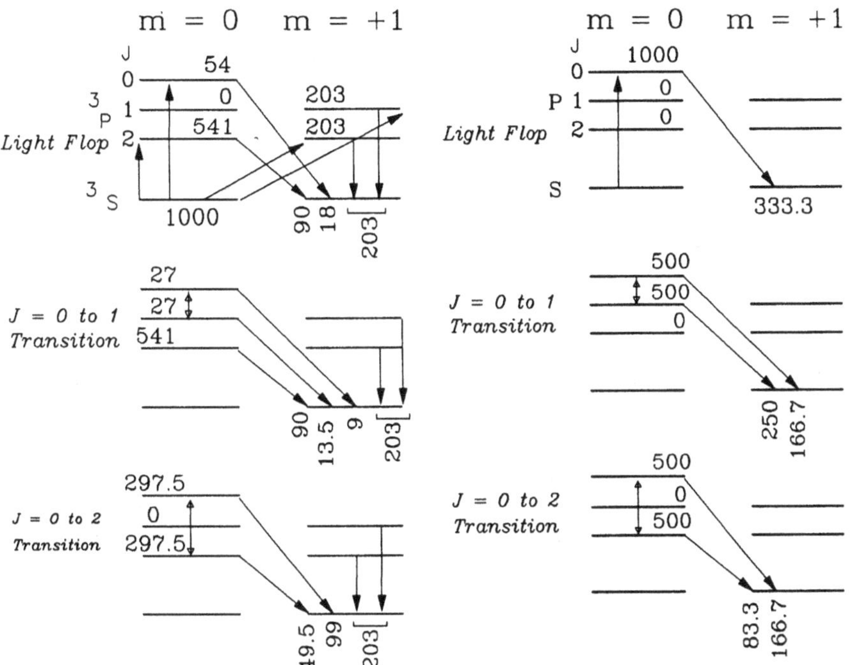

Fig. 14a : Sublevel populations using lamp excitation.

Fig. 14b : Sublevel populations using laser excitation

In Fig. 14b we have assumed that all 1000 atoms have been selectively excited to the $2^3P_0(m_J=0)$ sublevel, of which 333.3 constitute the light flop. This number changes to 416.7 for the 01 transition (25% signal) and to 250 for the 02 transition (also a 25% signal). The most dramatic improvement is the factor of 17 increase in the signal-to-light-flop for the 01 transition. Furthermore there is a 50% increase in the light flop itself, and therefore in the signal intensity, which will add to the improvements in beam intensity. The 01 transition should prove to be the most important, since this fine-structure separation has thus far been the least subject to inaccuracies in the higher order theoretical corrections. While statistically the 02 transition can be measured to the same accuracy, it requires a relatively large magnetic field to give a reasonable transition probability between the J = 0 and

2 states. The additional field-dependence of the transition frequency could limit the accuracy of the 02 measurement, and so below we emphasize the 01 measurement.

Infra-red lasers have been used for some time now to excite the 2^3S-2^3P transition. Solid-state lasers have been developed using a variety of crystals doped with Nd^{3+} as the lasing medium. They are stable, easily tunable, operate cw, and are relatively inexpensive. The fluorescent transition of interest is $^4F_{3/2}$ to $^4I_{11/2}$ in the $4f^3$ configuration of Nd^{3+}, and it occurs at a wavelength around 1050 nm. This line is split and broadened by the lattice crystal. Until a few years ago the standard host crystal for the wavelength region was yttrium-aluminum-garnet (YAG), but its fluorescent spectrum is too narrow for the laser to be tunable to our desired transition, 2^3S_1 to 2^3P_0, (hereafter called D_0), which has a wavelength of 1082.9 nm. The first crystal to be used with success for D_0 was lanthanum neodymium hexa-aluminate (LNA)[40], which has other advantages over YAG, including six times more Nd concentration, and a much wider tuning range because of its broader fluorescence spectrum. The laser has been used to optically pump gaseous helium with several hundreds milliwatts of power. Two other crystals are potentially useful. One is yttrium-aluminum-perovskite (YAP)[41] which has been used successfully to pump the D_0 line, even though the line falls in between the crystal's two main fluorescent peaks. The other possible crystal is the readily available lithium niobate.

Our requirements for laser excitation of the atoms in a beam are somewhat different from those for optical pumping in a gas, where the Doppler width is about 1 GHz. In an atomic beam experiment the Doppler width would be about 10 MHz, determined by the angular divergence of the atomic beam and the geometry with which the light beam crosses it. Ideally, for maximum efficiency of excitation, the Doppler width should be matched or exceeded by the spectral width of a single laser mode. Usually the laser spectral width is much narrower, and it would have to be chirped in some way, such as using a piezoelectric device to oscillate one of the mirrors. On the other hand care must be taken that this does not interfere with the requirements of tuning stability, to minimize light shifts as explained below. The geometry of the proposed arrangement is illustrated in Fig. 15.

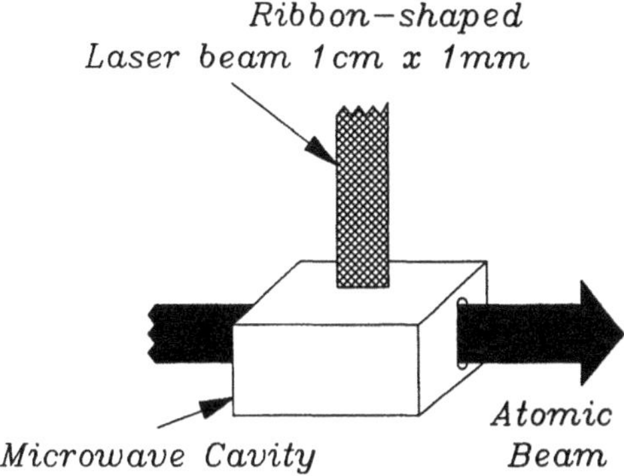

Fig. 15 : Geometry of the laser and atomic beams intersecting in the microwave cavity.

5.3 Projected errors

The projected errors for the proposed experiment are summarized below in Table II. The overall error corresponds to the determination of the resonance line center to 1/8000 of the width. Our experience when measuring ν_{12} has convinced us that we understand the line-shapes adequately to accomplish this.

Table II : Projected errors for re-measurement of ν_{01}.

Source	Magnitude (kHz)
Statistical Error (1000 hrs. data-taking)	0.4
Light shift	0.4
Bloch-Siegert effect	0.045
Doppler Effect	<0.2
Magnetic Field	<0.2
	Overall error 0.6 kHz (20ppb)

A more detailed discussion of systematic errors follows.

Bloch-Siegert Effect

This is due to the counter-propagating component of the oscillating microwave field, and is given by[42)]

$$\delta\nu \approx \frac{\gamma^2}{8\pi^2 f} \quad \text{......(5)}$$

where γ is the decay rate of the 2^3P state ($10^7 s^{-1}$) and f is the microwave frequency. The shift would be approximately 45 Hz for the optimal strength of the microwave magnetic field. This is small and calculable.

Light Shifts

There will be a shift in the energy of the 2^3P_0 sublevel due to the radiation field which excites it from the 2^3S metastable state. The shift is given by

$$\delta E = Re \frac{\langle g | p.\epsilon | e \rangle \langle e | p.\epsilon | g \rangle}{(E_g + \hbar\omega) - (E_e - i\hbar/2) + \hbar k.v}$$

where the numerator is the product of electric dipole matrix elements with the labels g,e representing the lower (2^3S) and upper (2^3P) states respectively, and E_g, E_e are the corresponding energies. Other parameters are ω, the angular frequency of the optical radiation, γ, the decay rate of the upper state, **v**, the atomic velocity, and **k**, the wave vector of the laser radiation. The term **k.v** must be averaged over the Doppler width of the atomic beam, which in our case would be determined by geometric factors in the apparatus.

Using analytical techniques similar to those developed by Happer and Mathur[43)] we have obtained the following expression for the light shift of our microwave resonance in frequency units

$$\delta\nu = \frac{8}{3} ln2 \frac{\lambda r_0 I f}{2\pi h} \cdot \frac{1}{\Delta_D^2} \cdot \frac{\Omega}{2\pi},$$

where Ω is the detuning of the laser and r_0 ($= e^2/mc^2$) is the classical electron radius. The other parameters, followed in parenthesis by the numerical values appropriate for our experiment, are the Doppler width, Δ_D (10 MHz), the oscillator strength f (0.06), the wavelength λ (1μ), and the laser intensity I (1 mW/cm^2). This gives a value for $\delta\nu$ of 7 kHz per MHz of laser detuning, assuming negligible frequency width in the laser. The effect of finite laser width is to lessen the shift. We have estimated that for a laser width comparable to the Doppler, which is desirable for excitation efficiency, the shift is reduced by about 25 per cent. Thus our overall estimate is a light shift of about 5 kHz per Mhz of laser detuning.

We plan to measure the light shift as a function of both detuning and light intensity. Accurate measurements of light shifts in an excited state for a simple atomic system are of interest in their own right.

Doppler Shifts

The atoms see the microwave field as two counter-propagating waves whose frequencies differ by about 200 khz. Thus our resonance line is actually two Lorentzian lines whose centers are separated by 200 kHz. A Doppler shift in the overall line center occurs if the two counter-propagating waves are of unequal amplitude. For a microwave cavity with a Q of about 1000, which is typical in our case, the Doppler shift will be 200 Hz.

6. OTHER FINE-STRUCTURE INTERVALS OF HELIUM AND TWO-ELECTRON IONS

Many fine-structure intervals have been measured for other states of helium and for two-electron ions. Because of the limited experimental and theoretical accuracies, however, these intervals are not useful for a determination of the fine-structure constant α. In addition they are not of comparable value to the 2^3P_J fine-structure intervals in helium as a test of the quantum electrodynamic theory of the two-electron atom. Hence we do not give details of these other fine-structure intervals in this chapter.

We give a few references to other fine-structure intervals for the simpler and more precisely known cases. Extensive microwave measurements of fine-structure intervals in higher excited states have been made, together with some field-crossing measurements[44]. In some cases the experimental accuracies are as good as 10 ppm, but for various reasons the theoretical precision is limited to the parts per thousand level[45].

The 2^3P_J fine-structure intervals in Li^+ have been measured to accuracies of 100 ppm using laser spectroscopy[46], but theory and experiment are in poor agreement. Measurements have been made also on the 2^3P_J levels on higher Z two-electron ions using beam foil spectroscopy[47], and in the case of F^{7+} the experimental precision is about 10ppm[48]. The agreement between theory and experiment, however, is no better than the level of parts per thousand for the fine-structure intervals[49]

7. ACKNOWLEDGEMENTS

We are grateful to Professor Edward A. Hinds for providing a critical reading of the manuscript for this chapter which yielded several very helpful suggestions.

The research of one of us, (VWH), is currently supported by the Department of Energy. In the past the helium 2^3P experiments were supported by the Air Force Office of Scientific Research and the National Science Foundation.

References

1. J. Brochard, R. Chabbal, H. Chantret, and P. Jacquinot, J. Phys. Radium 13, 433 (1952).

2. W. E. Lamb, Jr., and T. H. Maiman, Phys. Rev. 105, 573 (1957); I. Wieder and W. E. Lamb, Jr., *ibid.*, 107, 125 (1957).

3. F. D. Colegrove, P. A. Franken, R. R. Lewis, and R. H. Sands, Phys. Rev. Letters, 3, 420 (1959); J. Lifsitz, Ph.D. thesis, *University of Michigan*, 1965 (unpublished).

4. C. Lhuillier, J. P. Faroux, and N. Billy, J. Phys. (Paris), 37, 335 (1976); E. A. Hinds, J. D. Prestage, and F. M. J. Pichanick, Phys. Rev. A33, 68 (1986).

5. F. M. J. Pichanick, R. D. Swift, C. E. Johnson, and V, W. Hughes, Phys. Rev., 169, 55 (1968).

6. S. A. Lewis, F. M. J. Pichanick, and V. W. Hughes, Phys. Rev. A2, 86 (1970).

7. A. Kponou, V. W. Hughes, C. Johnson, S. A. Lewis, and F. M. J. Pichanick, Phys. Rev. A24, 264 (1981).

8. Frieze, E. A. Hinds, V. W. Hughes, and F. M. J. Pichanick, Phys. Rev. A24, 279 (1981).

9. M. L. Perl, I. I. Rabi, and B. Senitsky, Phys. Rev. 98, 611 (1955); see also P. Kusch and V. W. Hughes, in *Encyclopaedia of Physics*, edited by S. Flügge (Springer-Verlag, Berlin, 1959), Vol. 37/1. The application of the technique to the 2^3P state of helium, as part of the atomic beam helium program at Yale, was first suggested by B. B. Aubrey, L. Y. Chow, V. W. Hughes, and R. Swift, Bull. Am. Phys. Soc. 6, 248 (1961).

10. J. D. Prestage, E. A. Hinds, and F. M. J. Pichanick, Phys. Rev. Lett., 50, 828 (1983); J. D. Prestage, C. E. Johnson, E. A. Hinds, and F. M. J Pichanick, Phys. Rev. A 32, 2712 (1985).

11. E. A. Hinds, J. D. Prestage, and F. M. J. Pichanick, Phys. Rev. A 32, 2615 (1985).

12. W.E. Lamb, Jr., Phys. Rev. 105, 559 (1957).

13. V. W. Hughes, G. Tucker, E Roderick, and G. Weinreich, Phys. Rev. 91, 828 (1953).

14. M. Phillips, Phys. Rev. 76, 1803 (1949).

15. G. J. Schulz and R. E. Fox, Phys. Rev. 106, 1179 (1957); H. D. Hagstrum, *ibid.* 96, 336 (1954).

16. C. W. Drake, V. W. Hughes, A. Lurio, and J. A. White, Phys. Rev. 112, 1627 (1958); B. E. Zundell and V. W. Hughes, Phys. Lett. 59A, 381 (1976).

17. W. Heisenberg, Z.Phys. 39, 499 (1926).

18. G. Breit. Phys. Rev. 34, 553 (1929); Phys. Rev. **36**, 383 (1930).

19. G. Breit, Phys. Rev. 39, 616 (1932).

20. H. A. Bethe, *Hanbuch der Physik*, Vol. 24/1, Springer Verlag, Berlin, 1933.

21. G. Araki, M. Ohta, and K. Mano , Phys. Rev. 116, 651 (1959).

22. J. Traub and H. M. Foley, PHys. Rev. 116, 914 (1959).

23. C. Pekeris, B. Schiff, and H Lifson, Phys. Rev., 126, 1057 (1952); B. Schiff, C. L. Pekeris, and H. Lifson, Phys. Rev., 137, A1672 (1965).

24. C. Schwartz, Phys. Rev., 134, A1181, (1964).

25. J. Daley, Ph.D. Thesis, *University of California at Berkeley*, 1972, (unpublished).

26. M. Douglas and N. M. Kroll, Annals of Physics, 82, 89 (1974).

27. E. E. Salpeter and H. A. Bethe, Phys. Rev. 84, 1232 (1951).

28. R. P. Feynman, Phys. Rev. 76, 749 (1949); *ibid.*, 76, 769(1949).

29. M. Gell-Mann and F. Low, Phys. Rev. 84, 350 (1951).

30. E. E. Salpeter, Phys. Rev. 87, 328 (1952).

31. J. Sucher, Phys. Rev. 109, 1010 (1958); Ph.D. thesis, *Columbia University*, 1958 (unpublished).

32. K. Y. Kim, Phys. Rev. 140, A1498 (1965).

33. L. Hambro, Phys. Rev. 5, A2027 (1972); Phys. Rev. 6, 865 (1972); Phys. Rev. 7, 479 (1973).

34. A. Dalgarno and J. T. Lewis, Proc. Roy. Soc. (London), A 233, 70 (1956).

35. M. L. Lewis and P. H. Serafino. Phys. Rev. A18, 867 (1978).

36. E. R. Cohen and B. N. Taylor, *The 1986 Adjustment of the Fundamental Physical Constants*, (CODATA), Pergammon Press, (Paris), 1986; E. R. Cohen and B. R. Taylor, Rev. Mod. Phys. 59, 1121 (1987).

37. W. E. Lamb and R. C. Retherford, Phys. Rev. 79, 549 (1950); A. V. Haeff, Proc. IRE 27, 586 (1939).

38. W. Lichten, Phys. Rev. 120, 848 (1960).

39. E. E. Eyler and F. M. Pipkin, Phys. Rev. Letters, 50, 828 (1981); E. E. Eyler, Ph. D. Thesis, *Harvard University*, 1982 (unpublished).

40. L. D. Schearer *et al.*, IEEE J of Quant. Electr. QE-22, 713 (1986).

41. L. Schearer and M. Leduc, IEEE J. Of Quant. Electr. QE-22, 756 (1986).

42. F. Bloch and A. Siegert, Phys. Rev. 57, 552 (1940); Jon H. Shirley, *ibid.* 138, B979 (1965).

43. E. W. Happer and B. Mathur, Phys. Rev. 163, 12 (1967).

44. B. MacAdam and W. H. Wing, Phys. Rev. A13, 2163 (1976); J. W. Farley, K. B. MacAdam, and W. H. Wing, Phys. Rev. A20, 1754 (1979).

45. E. S. Chang, Phys. Rev. A35, 2777 (1987); W. C. Martin, Phys. Rev. A29, 1883 (1984), and references cited therein.

46. R. Bayer, J. Kowalski, R. Neumann, S. Noehte, H. Suhr, K. Winkler, and G. zu Putlitz, Z. Physik A292, 329 (1979).

47. J. D. Silver, Physica Scripta 37, 720 (1988), and referneces cited therein.

48. J. D. Silver, *private communication*.

49. O. Corveix, P. Indelicato, and J. P. Desclaux, J. Phys. B20, 639 (1987); P. Indelicato, O. Corveix, and J. P. Desclaux. J. Phys. B20, 651 (1987).

APPENDIX
HISTORICAL REVIEW AND BIBLIOGRAPHY OF QUANTUM ELECTRODYNAMICS*

Kan-ichi YOKOYAMA and Reijiro KUBO

Research Institute for Theoretical Physics

Hiroshima University, Takehara, Hiroshima 725, Japan

Contents

---

* Translated by R.Kubo from *"Quantum Electrodynamics", New Series of Selected Papers on Physics No.65*,(ed.K.Yokoyama) published by the Physical Society of Japan, under the permission of the authors and the Physical Society of Japan.

INTRODUCTION

In this Appendix we make a historical review of the development of quantum electrodynamics(QED) on the basis of text books and articles which are collected as Bibliography in the latter part of this review. We do not make a thorough historical analysis of QED, but we restrict ourselves to discussing topics indicated in the table of contents, since this review was originally written about a decade ago. In fact, QED was almost established as a typical renormalizable quantum field theory by the 1950's. We lay our emphasis upon earlier history of QED rather than its recent development. Various important problems remained unsolved are also discussed.

The authors would like to thank Professor T.Kinoshita for valuable comments. In particular, they wish to thank Professor N.Nakanishi for critical reading the manuscript and for useful suggestions. They are also greatly indebted to Miss K.Kanda for typewriting the manuscript.

1. CONSTRUCTION OF QED

1.1 Lorentz Invariant Formulation

Quantum electrodynamics (QED) as a typical quantum field theory began with the famous work by Dirac[1]. Important works in the pioneering era of QED are found in [1-7]. Quantum field theory in a time-dependent canonical formalism was established by Heisenberg and Pauli in [4,5]. Tomonaga[8] and Schwinger[12] independently reformulated Heisenberg and Pauli's formalism in a Lorentz covariant way. Tomonaga and Schwinger's theory was called the super-many-time theory. It was applied to QED for the first time in [9,10]. Refs.[13-15] are a series of elaborate works by Schwinger along the line of [12]. A Lorentz covariant framework of QED was founded by Tomonaga and Schwinger's works.

On the other hand, Feynman proposed the method of path integration in [16], on the basis of which he formulated Feynman rules with Feynman diagrams[17,18]. Mathmatical foundation of Feynman's theory was established in [19-21]. Tomonaga, Schwinger and Feynman's theories were dealt with in a unified way by Dyson in [68, 69]. The notion of asymptotic fields is inevitable in view of the Heisenberg representation and it was shown in [22] that the notion of asymptotic fields actually leads to the Yang-Feldman equation.

Bound state problems are investigated in [23,24]. In particular, it is shown in [23] that Green's functions in the Heisenberg representation are given by solutions to Gell-Mann and Low's equation. Refs.[25-37] have played important and fundamental roles in the development of the modern quantum field theory. Usefulness of Green's functions is discussed in [29,30]. Construction of quantum field theory by means of the variational principle was attempted in a series of Schwinger's works[31-37], in which invariance of the theory under Poincaré group and the relation between spin and statistics are extensively investigated.

Various considerations together with interesting memoirs and episodes in the development of QED are found in the Nobel lectures [38-40].

1.2 CPT Invariance

Invariance properties of physical quantities under discrete trnansformations such as charge conjugation (C), space inversion (P) and time reversal (T), together with invariance under Lorentz transformation have been often discussed in various stages of the development of QED. Considerations about these invariance properties have eventually led us to the unified notion of the CPT theorem. Various invariance properties are studied in [41-62]. In particular, Refs.[57-60] are extensive review articles on invariance principles of physical laws.

2. PROBLEMS ON ULTRAVIOLET DIVERGENCES

2.1 General Renormalization Theory

The problem of ultraviolet divergence is regarded as an essential dificulty in quantum field theory. One can not get around this difficulty in a Lorentz-covariant framework of quantum field theory. The renormalization theory gives us a prescription to find a finite result by a formal subtraction of the ultraviolet divergence.

On the basis of perturbation theory, Dyson [68,69] formulated renormalization theory in a self-contained form in QED. Earlier works on renormalization theory in Japan on the basis of the Tomonaga theory are given in [63-67]. Wick's theorem appeared in [70]. An extension of Dyson's theory to the meson theory was attempted by Matthews in [71-78]. The method of T^*-product is introduced in [71] for a Lagrangian which contains a derivative term. The Ward identity $Z_1 = Z_2$, was derived in [80-82]. As a result of the Ward identity genuine divergent quantities appearing in a S-matrix in QED are reduced to two quantities, that is, the selfmass and charge of the electron.

The problem of overlapping divergences, owing to which it is impossible to renormalize self-energies of the electron and photon by simply subtracting infinity from calculated amplitudes was investigated in detail by Salam [83,84]. Bypassing the difficulty of overlapping divergences, Ward [85,86] proposed an elegant way to find renormalized physical quantities by extending the result obtained in [81]. This idea was induced by Ward's identity but Ward's identity itself is not necessarily essential to guarantee renormalizability of a theory. Thus the prescription for renormalization in perturbation theory was almost completed in QED, and it was shown that elimination of divergences appearing in S-matrix is possible in each order of perturbation expansion. Some further problems left to be studied are as follows: (1) Is it also possible to eliminate divergences present in local quantities such as Heisenberg operators? (2) How is the convergence property of perturbation series after renormalization? These problems were studied by Dyson in [88-91], in which it was shown in a skillful manner that if perturbation series converges after renormalization then the answer to (1) is affirmative. However, the convergence of a perturbation series is very questionable. As a consequence Dyson expressed a pessimistic opinion to (2) in [92]. The convergence problem of perturbation series still remains unsolved.

Conditions of renormalizability for various interaction forms of quantized fields were investigated in [93,94]. It was shown in [95-98] that renormalization is also possible in a way different from Dyson's. Ref.[98] is a review article of renormalization theory. Refs.[93-103] built a road to the BPHZ theory. In Valatin's theory [99-102] a limiting procedure for the space-time point was introduced so as to deal with only renormalized quantities.

An asymptotic behavior for a propagator at high energies was investigated by Gell-Mann and Low in [104], in which they also investigated a relation between a bare charge and renormalized one in order to gain an insight into applicability of renormalization theory. Similar problems are studied in [105-117]. Gell-Mann and Low's paper played an important role in the discovery of renormalization groups. Works done in the Landau school are given in [109,112]. For a recent development of renormalization groups, see a review article given in *quantum field theory III, New Series of Selected Papers on Physics No.189*,ed.K.Nishijima and N.Nakanishi(Physical Society of Japan, 1975).

The Ward-Takahashi identity(WT) plays a very important role in quantum field theory in view of invariance under gauge transformation. WT is a generalization of the Ward identity. Logically, however, WT should have been discovered prior to the Ward identity: One would obtains the Ward identity as a byproduct of WT.

Green [118] derived WT for the first time. Takahashi derived WT in a simplified form in [120]. Various problems related to WT are dealt with in [121-130]. A general formulation useful to study WT was demonstrated in [122-125]. Characteristic properties of WT were discussed in [131,132],thereby the authors showing that specific field equations and commutation relations are unnecessary to obtain WT. Ref.[133] is an extensive review article on renormalization theory.

2.2 Feynman Integral

The theory of Feynman integrals began originally with Feynman's paper[18]. Feynman-parametric integrals were investigated to a great extent in [131-141]. A graph-theoretical consideration was made in [141]. Analyticity property in the Feynman integrals is investigated in detail in [142,143]. The Landau singularities and problems on an anomalous threshold were dealt with in [144-151]. Analyticity of an amplitude was most generally discussed in terms of duality diagrams in [145,151,152]. Ref.[153] is a review article on integral representations of the Feynman-parametric integrals. Subsequent studies on this problem are seen in [154-159].Refs.[160-162] are extensive review articles on Feynman integrals. Some further technical problems are discussed in [163,164].

A proof of the power-counting theorem which was proposed by Dyson[69] is given in [138]. A more rigorous proof is given in [165]. Refs.[166,167] give a proof in the case in which four-momenta are Euclidean. Its extension to the case of Minkowski metric is given in [168]. The power-counting theorem is studied in a more general case in [169,170]. An analytic regularization of Feynman integrals was attempted in [161,171-174], in which finite parts of Feynman integrals were extracted by applying the method of analytic continuation. The method of dimensional regularization was developed in [175-179].

2.3 BPHZ Theory

An exact verification of the proposition that a renormalized Feynman integral actually converges was originally given by Bogoliubov and Parasiuk [180]. Finding some defects in their formulation, Hepp improved it and completed the proof [181]. Zimmermann[182] made similar consideration on the basis of Dyson's formalism. Their theories are well known as the BPHZ theory.

A method of subtraction by means of usual differential calculus was developed in [185], and in [186] renormalization was performed in terms of an integral representation with parameters. More rigorous proofs of convergence are given in [187,188]. Refs.[189-196] discuss how to extract renormalized quantities automatically. Attempt at finding a prescription to eliminate ultraviollet divergences as well as infrared divergences at the same time is done in [197-201]. Ref.[202] is useful to see the present status of the BPHZ theory.

3. MODIFICATION OF THE FORMALISM (INTRODUCTION OF INDEFINITE METRIC AND GAUGE PROBLEMS)

3.1 Gupta-Bleuler Formalism and Related Problems

When one attempts to quantize the classical electromagnetic field in a relativistically covariant way in the Lorentz gauge, it has been well known that the Lorentz condition is incompatible with commutation relations between electromagnetic fields. In this circumstance Fermi proposed a subsidiary condition, which imposes the Lorentz condition upon the physical states, as an equation defining the physical states. This is called Fermi's subsidiary condition (FSC). Tomonaga [9], Schwinger [12] and some other authors [152-156] investigated the method to accomodate FSC in the interaction representation. However, the contradiction between FSC and commutation relations was by all means unavoidable. Gupta [209] was the first to succeed in the Lorentz-covariant quantization of the electromagnetic field in the Feynman gauge by introducing indefinite metric into a state-vector space and proposing Gupta's subsidiary condition(GSC), which extracts the positive-frequency part of FSC. Bleuler completed Gupta's method of quantization in [210]. Refs.[203-208] discussed various problems on subsidiary conditions prior to Gupta.

The Gupta-Bleuler formalism was extensively applied to quantum field theory in [213-218]. The reason why the Gupta-Bleuler formalism is most successful in the covariant quantization of the electromagnetic field was clarified in [219-225] on the basis of an axiomatic consideration. It is shown in [219] that introduction of indefinite metric is inevitable as far as the electromagnetic potential A_μ satisfies $\Box A_\mu = 0$ in a manifestly covariant QED. The Maxwell equations do not hold as they stands as operator equations in QED irrespective of whether there exists the Lorentz condition or not. This fact was substantiated in [220-225] in connection with the necessity of indefinite metric and GSC. Mathews,Seetharaman and Simon[226] discussed the necessity of the indefinite metric on the basis of the structure of a little group of the Poincare group.

GSC is investigated in detail in [227-232]. As a consequence it has become clear that whatever formalism one adopts for QED should satisfy the following facts:(1) The Maxwell equation does not hold as operator equations. However,(2) The Maxwell equations are reproduced as equations of expectation values with respect to physical states by introducing a subsidiary condition. (3) Introduction of indefinite metric is indispensable. Haller and Landovitz[228-230] discussed the relation between GSC and the adiabatic asymptotic conditions as $t \to \infty$, and they showed that the formalism in the Feynman gauge and that in the Coulomb gauge are equivalent to each other if one applies Bleuler's subsidiary condition to them without adiabatic switching, though their claim was misleading[231,232].

3.2 Källén Formalism and Related Problems

Källén [233] proposed a renormalized formalism of QED in the Feynman gauge in which only renormalized field operators appear as dynamical variables. This formalism can be regarded as an extension of the Gupta-Bleuler Formalism, although some new devices such as the Källén-Lehmann spectral representation are incorporated in itself. In [234] Källén claimed the impossibility of making all the renormalization constants finite. However, some authors [236-238] pointed out that Källén's proof was incorrect and that there are some other possibilities to get finite renormalization constants. Rollnik, Stech and Nunnemann [239] showed in connection with [233] that asymptotic fields for the Heisenberg operators A_μ do not satisfy canonical commutation relations as they stand [240]. Refs.[241-243] deal with similar problems as in [233].

3.3 Gauge Transformation

We present papers which deal with problems on gauge invariance of the S-matrix and related problems. The Landau-Khalatnikov gauge and the Fried-Yennie gauge are introduced in [244] and in [245], respectively. The fact that $Z_1 = Z_2$ is gauge-dependent was shown by Johnson and Zumino in [246], and that it becomes finite at 2nd order of perturbation theory in Landau gauge is shown in [247]. Ref.[248] gives a good review of gauge transformations.

A study to find out a gauge that makes $Z_1 = Z_2$ finite at all orders of perturbation theory was attempted by Bialynicki-Birula in [250]. Renormalized transition probabilities are calculated in [255], by rigorously verifying gauge invariance of the S-matrix. It is shown in [256] that renormalization procedure gives rise to a shift in gauge generally. Some examples of noncovariant gauges are shown in [257-259]. Q-number gauge transformations are dealt with in [260-264].

3.4 Covariant-Gauge Formalism and Related Problems

It is not straightforward to derive the photon propagator from the operator formalism in covariant gauges other than the Feynman gauge. The photon propagator depends on a gauge parameter, which had been introduced by hand into the propagator because gauge invariance guarantees the gauge-parameter independence of the S-matrix.

Nakanishi [265,266] formulated quantum theory of electromagnetic fields in covariant gauges by introducing a dipole ghost. Independently, Lautrup [267] constructed a canonical formalism in covariant gauges. Lautrup did not introduce a dipole ghost and as a consequence Lorentz covariance was not manifest in his formalism. On the other hand, Nakanishi's formalism is not canonical but manifestly Lorentz covariant. However, field equations and commutation relations are essencially equivalent in both formulations. Synthesizing these two, we obtain the Nakanishi-Lautrup formalism (NL). For details, see Ref.(A) and [350]. The Gupta-Bleuler formalism is comprised in NL as a particular case of the Feynman gauge.

It should be remarked that it is inevitable to introduce a dipole ghost in any covariant-gauge formalism except for that of the Feynman gauge. When a dipole ghost exists, we can not use positive frequency parts of field operators to define the vacuum properly. This difficulty is resolved in [268]. An asymptotic condition in NL is studied in [270]. An auxiliary field is introduced as a Lagrange multiplier in NL. Utiyama[271] was the first to apply this method to classical field theory by extending canonical formalism. Quantum field theory using a Lagrange multiplie is formulated also in [272]. NL is investigated from some different point of view in [273-275].

If one takes two gauges which are different from each other, then one obtains two distinct theories of QED corresponding to the two adopted gauges in NL, and it is impossible to connect the two theories of QED by an appropriate gauge transformation without destroying Lorentz covariance. That is, two theories with different gauges are inequivalent to each other.On the other hand, a gauge parameter is renormalized and shifted by renormalization procedure in NL. As a consequence renormalization gives rise to a gauge transformation, which, however, can not be carried out in a manifestly covariant way except for the Landau gauge in NL. This was pointed out by Nakanishi in [350]. If one restricts oneself to the case of the Landau gauge, then there arises no trouble associated with renormalization. However, this restriction leads us to a rigid formulation of QED, in which gauge transformation does not play an important role.

The problem underlying between gauge transformation and renormalization is resolved to some extent in [276], in which a state space is extended by introducing another dipole ghost, called gaugeon, in addition to the original one in NL. It should be noted that all the gauges appearing in [276] are equivalent to one another. A gauge structure is studied in this system in [277]. This covariant formalism is applied to the case involving a neutral-vector field in [278-286].

As is well known, the propagator for the massive neutral vector field is singular at the massless limit of the vector field in an ordinary sense. Up to that time there existed no neutral vector field theory having a nonsingular massless limit,and hence one can not apply Johnson's theorem naively to such vector field theories [304,305]. Also associated with the infrared divergence problem we did not have a sufficient theoretical background which guaranteed introduction of a fictitious photon mass in QED. It was also known that a neutral-vector field theory is unrenormalizable unless a neutral vector field interacts only with a conserved current.

The above shortcommings associated with the massive neutral vector field theory are improved in [278] by extending QED in the Landau gauge. The Landau-gauge QED is reproduced by taking a massless limit of a neutral vector field in [278]. Ghose and Das [279] studied a similar problem as in [278]. Although they did not define the massless limit clearly, after applying an appropriate transformation [286], essential ingredients in [279] become almost the same as those obtained in [278]. Extension of the neutral vector field theory formulated in [278] is attempted in [280-282]. In this way we have arrived at a synthesized formalism of abelian gauge field theory including QED, which is manifestly renormalizable in a covariant gauge.

A four-dimensional momentum representation is investigated in covariant-gauge QED in [283], in which the vacuum is defined by making use of the massless limit of neutral vector fields. It is pointed out in [284] that the scale dimension of a neutral vector field becomes different from the canonical one in neutral-vector field theories. It is shown in [285,286] that there is no serious problem associated with the scale dimension in the formalisms in [278-281]. For more detailed exposition of matelials on neutral-vector field theory, see [287-317]. Application of neutral vector field theory to gravity theory is attempted in [318-321]. In particular, Kimura [320,321] extended the ingredients in [280-281] to a case incorporating gravitons.

3.5 Indefinite Metric

Refs.[322-357] deal with state-vector spaces with indefinite metric and related problems in QED. Dirac was the first to introduce indefinite metric into quantum field theory[322-323], intending to eliminate ultraviolet divergences. Gupta[326] formulated Pauli-Villars' regularization method as quantum field theory with indefinite metric. Originally the operator η was introduced to realize indefinite metric in a Hilbert space. However, the use of η gives rise to confusion in QED. It is pointed out by Sunakawa[330] that the commutation relations between η and A_μ break manifestly covariant nature of physical quantities with respect to Lorentz transformation. It is shown in [331,332] that a state-vector space with indefinite metric can be relevantly constructed without introducing of the indefinite-metric operator η.

Froissart's model, a typical model in which a massive dipole ghost is incorporated, was demonstrated in [339]. It should be remarked that there is an essencial difference between massive and massless cases because the massless limit of the ghost does not exist in Froissart's model. In the $m \neq 0$ case, from the coupled equations, $(\Box - m^2)A = B$ and $(\Box - m^2)B = 0$, one can express A in terms of B, but this is not the case for the $m = 0$ case. This circumstance is discussed in detail in [350,273]. For a review of the problem on indefinite metric, see [338,342,349,350]. There are some misprints in [349], which are corrected in [350].

4. PROBLEMS ON INFRARED DIVERGENCES

The investigation of infrared divergences in QED has a long history, since infrared divergences appeared at an early stage of the development of QED. Actually, infrared divergences cancel out completely, and they do not exist in reality like a mirage [384]. However, when we attempt to verify that infrared divergences really do not exist and infrared divergences can,in fact, be eliminated completely,there arise various complicated problems. There are several methods to verify the cancellation of the infrared divergence, that is, the method in terms of coherent states, perturbation theory, extraction of infrared phase factors, *etc.*

4.1 Coherent State

Bloch and Nordsieck (BN) showed that the infrared divergence which arises from the contribution of the soft-photon emission can be eliminated by constructing coherent states of photons. Their paper [358], which appeared in 1937, is well known and Ref.[359,360] were written on the basis of BN's theory. Since then, the BN theory was improved by many authors[361-367]. In particular, the BN theory was rewritten in a covariant form by Thirring and Touschek[362].

Glauber gave a general theory of the coherent state in [363,367]. It is irrelevant to taking the usual Fock space to deal with a set of soft photons, and it is necessary to rearrange physical states so as not to give rise to infrared divergences. Chung[369] succeeded in constructing a physical state appropriate to deal with the infrared divergence on the basis of the method developed in [396]. Various devices have been developed in a series of papers [370-376] in constructing final states.

Phenomena associated with emission of soft photons are brought about by a long-range effect of Coulomb interaction. This interaction Hamiltonian does not vanish sufficiently fast as $|t| \to \infty$, and we can not apply the usual asymptotic condition as it stands. In this circumstance it is necessary to redefine the interaction Hamiltonian. Dollard[368] showed for the first time this peculiar property of the Coulomb force concretely. This important concept was further developed by Kulish and Faddeev [377], and by Zwanziger [378-380]. The results obtained in [369] were generalized in [377] by using a more tractable prescription. Papanicolau developed an elegant method in [381] in which he applied the renormalization group to the infrared problem. Ref.[382] is an exposition of the infrared problem on the basis of his investigation. For details of the theory of the coherent states, see Ref.(O).

4.2 Cancellation in Perturbation Theory

It can be shown that infrared divergences appearing in each term in perturbation expansion cancel out if one sums up all the contribution from respective

Feynman diagrams. This fact is explicitly shown in a simple example in [384]. That Heitler's method of radiation damping can not eliminate infrared divergences is pointed out in [383]. Nakanishi proposed a generalized theory of cancellation of infrared divergences by making use of the K-diagrams in [390]. Kinoshita took into account the fact that the infrared divergence has its origin in a characteristic property of massless particles, and he generalized the theory of the infrared divergence to a wider theory of mass singularities in [392].

4.3 Extraction of Phase Factors

All the contributions of infrared divergences can be extracted together as an infrared phase factors from all orders of perturbation expansion. Schwinger proposed this idea in [14]. Yennie, Frautschi and Suura extensively studied this method in [396]. Their work was improved and further developed by Grammer and Yennie in [401]. A simple method of extracting infrared singularities was proposed in [395].

4.4 Other Topics

Infrared divergences in experiment are discussed in [402-407]. Infrared divergences in gravitation theory are explored by Weinberg in [408]. The dimensional regularization is applied to infrared problems in [410,411].

5. PROBLEMS ON THE ANOMALY

5.1 Goto-Imamura-Schwinger Term

The spectral representation for the vacuum expectation value of four- dimensional commutation relations of current densities is obtained on the basis of general axiomatic requirements, and the representation is independent of specific theories. The vacuum expectation value of the equal-time commutation relations between the temporal-component of the current and its spatial-component does not vanish identically, whereas if one evaluate the same thing on the basis of the canonical commutation relations one finds that it is identically zero. This contradiction was pointed out for the first time by Goto and Imamura[412],and then by Schwinger[413]. We call the terms added to the results obtained from canonical commutation relations the GIS terms. The GIS terms generally appear in commutation relations of current algebras.

The consequence of general considerations[416,417] is that the equal-time commutation relations between temporal component of the current and its spatial component does not vanish provided that the conservation law holds. Quantities which are given by product forms of some operators at the same space-time point are in principle not well defined in the sense of distributions. Hence, it is dangerous to apply canonical commutation relations naively to quantities such as current components.

In order to resolve the GIS problem it is necessary to define a current density rigorously, as is seen in [418-420]. Although there are still many unsolved problems associated with the GIS terms, it is in principle necessary from an axiomatic view point to take into account the GIS terms to remove contradiction in quantum field theory[416-427].

Apart from pure theoretical problems, however, the GIS terms seem to bring about only troublesome complication and play no physically relevant roles in quantum field theory. It is shown in [415] that actually the GIS terms can be eliminated by means of an appropriate regularization. Some people do not take the problem of the GIS terms seriously, since this is not the problem of physics but it is the problem of pure mathematics. Orzalesi made an extensive survey of the GIS terms in [432].

Supposing that this problem arised from the same origin as that of the indefiniteness of the photon self-mass, Källén treated a GIS term as the photon self-mass [433]. However, the incorrectness of his treatment was pointed out in [256,434]. It can not be easily concluded whether a GIS term is a c-number or q-number. It is shown in [428,431] that the GIS terms in spinor QED can be treated as c-numbers without any contradiction, and in [435] that they are in fact c-numbers.

5.2 The PCAC Anomaly

The PCAC anomaly arises when one deals with axial vector currents in quantum field theory, and it is deeply connected with the $\pi^0 \rightarrow 2\gamma$ decay problem and other problems on strong interaction. If one constructs an axial-vector current in terms of the Dirac fields and computes the divergence of the axial current by applying the Dirac equation naively, then one only obtains a quantity proportional to the mass of the Dirac particle. If this were the case, then π^0 would never decay into 2γ. This fact is well known as the PCAC puzzle.

A careful computation shows in fact that an additional anomalous term appears in the divergence of axial-vector currents. The existence of this anomalous term was pointed out by Schwinger in [15]. The PCAC anomaly in a sigma model is studied in [436], in which a typical method is developed to deal with this kind of problems. The axial-vector vertex is extensively investigated within the framework of perturbation theory by Adler in [438] and it is shown that the result is exact to all orders.

The PCAC puzzle originated in an ill-defined products of distributions. Jackiw and Johnson gave a prescription for a proper treatment of distributions in [440]. Various problems associated with anomaly are discussed in [441,442]. Review articles [443,444] are very instructive as an introduction to anomaly theories.

6. OTHER TOPICS

6.1 Low-Energy Theorems

The behavior of transition probabilities of reactions of elementary particles and that of scattering amplitudes in the neighborhood of threshold energies are investigated in [445-449], in which low-energy theorems are derived. Low-energy theorems play important roles not only in QED but also in treating strong interactions on the basis of renormalization theory, and they yield various simple formulas involving renormalized coupling constants, magnetic moments, *etc.*. As a consequence, we have a direct connection between theory and experiment.

A tactful method of deriving a low-energy theorem was proposed by Kroll and Ruderman[445] in a problem of production of π mesons by photons. Low-energy theorems in an electron-photon scattering and in a bremsstrahlung are studied in [446,447] and [449], respectively. A systematic study on the derivation of low-energy theorems in various reactions of elementary particles and their usefulness was made by Klein in [448].

6.2 Experiments

We summarize some important low-energy experiments in QED together with theories corresponding to the experiments . The theory of the Lamb shift is found in [450-459]; in particular, Bethe[450] showed that the Lamb shift is given rise to as electromagnetic reaction and it is understandable on the basis of renormalization theory. Experiments of the Lamb shift are reported in [468-470].

A review on the fine structure of the hydrogen atom was given by Grotch and Yennie in [460], in which the level difference $2P_{3/2} - 2P_{1/2}$ was studied in detail. Experiments on the fine structure of the hydrogen atom are reported in [468-470], and experiments on the fine structure of the positronium together with the hyperfine structure of the muonium are reported in [457-477].Theories on these problems are in [471-474]. Theories on hyperfine structure of the ground state of the hydrogen atom are found in [461-465]. Comparison between theory and experiment in QED is extensively reviewed in [505,506].

Measurements of the anomalous magnetic moments of an electron and muon have played an especially important role in the establishment of perturbation theory and renormalization theory in QED. Experiments on the anomalous magnetic moment are reported in [493-496], and theories in [478-492]. In particular, the anomalous magnetic moment of the electron is calculated to sixth order of perturbation expansion by Cvitanovic and Kinoshita in [491] by making use of elaborate technique of Feynman integrals.

Problems on radiation corrections to experimental formulas for various processes in QED are dealt with in [497-503]. Recent experimental values of the fine structure constant α are given in [504].

6.3 Magnetic Monopoles

Problems on a magnetic monopole are of interest in connection with the recently developed non-abelian gauge theory. A magnetic monopole was introduced in field theory by Dirac in [507]. Here the notion of the Dirac string was introduced and it was shown that the electric charge can be quantized owing to the existence of the magnetic monopole.

Zwanziger[508] discussed that the analyticity of a S-matrix becomes complicated if there exists a monopole. The theory of magnetic monopole in a system of the electromagnetic field combined with gravity has been developed by Weinberg in [509]. The relationship between spin and a magnetic monopole was studied in [510]. Quantum field theory of a monopole was developed by Schwinger in [511].

BIBLIOGRAPHY

TEXTBOOKS

(A) N. Nakanishi: *Quantum Field Theory,* New Series of Texts in Physics 19 (Baifukan Publ., Tokyo, 1975).

(B) K. Nishijima: *Relativistic Quantum Mechanics,* New Series of Texts in Physics 13 (Baifukan Publ., Tokyo 1973) ; *Quantum Field Theory* (Kiinokuniya Publ., Tokyo, 1987).

(C) Y. Takahashi: *Quantum Field Theory for Solid-state Physicist I, II,* New Series of Texts in Physics 16, 17 (Baifukan Publ., Tokyo, 1974).

(D) G. Takeda and H. Miyazawa: *Elementary Particle Physics,* Selected Texts in Physics 9 (Shokabo Publ., Tokyo, 1965).

(E) H. Yukawa: *Elementary Particle Physics,* Foundations of Modern Physics 11 (Iwanami Book Co., Tokyo, 1974).

(F) A.I. Akhiezer and V.B. Berestetskii: *Elements of Quantum Electrodynamics* (Oldbourne Press, London, 1962).

(G) A.I. Akhiezer and V.B. Berestetskii: *Quantum Electrodynamics*, Interscience Monographs and Texts in Physics and Astronomy, Vol. 11 (Interscience Publ., New York, 1965).

(H) V.B. Berestetskii, E.M. Lifshitz and L.P. Pitaevskii: *Relativistic Quantum Theory*, Vol.4 of Course of Theoretical Physics (Pergamon Press, Oxford, 1971).

(I) J.D. Bjorken and S.D. Drelll: *Relativistic Quantum Mechanics*, International Series in Pure and Applied Physics (McGraw-Hill Book Co., New York, 1964).

(J) J.D. Bjorken and S.D. Drelll: *Relativistic Quantum Fields*, International Series in Pure and Applied Physics (McGraw-Hill Book Co., New York, 1965).

(K) N.N. Bogoliubov and D.V. Shirkov: *Introduction to the Theory of Quantized Fields*, Interscience Monographs and Texts in Pure and Applied Physics and Astronomy, Vol.3 (Interscience Publ., New York, 1959).

(L) P.A.M. Dirac: *The Principles of Quantum Mechanics, 3rd ed.*, The International Series of Monographs on Physics (Clarendon Press, Oxford, 1947).

(M) R.P. Feynman: *Quantum Electrodynamics* (Benjamin Inc., New York, 1961).

(N) W. Heitler: *The Quantum Theory of Radiation*, The International Series of Monographs on Physics (Oxford Univ. Press, Oxford, 1949).

(O) J.M. Jauch and F. Rohrlich: *The Theory of Photons and Electrons, 2nd expanded ed.*, Texts and Monographs in Physics (Springer-Verlag, New York, 1976).

(P) G. Källén: *Quantum Electrodynamics* (Springer-Verlag, Berlin, 1972).

(Q) K. Nishijima: *Fields and Particles*, Lecture Notes and Supplements in Physics (Benjamin Inc., New York, 1969).

(R) J.J. Sakurai: *Invariance Principles and Elementary Particles* (Princeton Univ. Press, 1964).

(S) J.J. Sakurai: *Advanced Quantum Mechanics* (Addison-Wesley Publ. Co., London, 1967).

(T) S.S. Schweber: *An Introduction to Relativistic Quantum Field Theory* (Harper and Row Publ., New York, 1962).

(U) J. Schwinger: *Selected Papers on Quantum Electrodynamics* (Dover Publ., New York, 1958).

(V) J. Schwinger: *Particles, Sources, and Fields, Vol.I, II, III* (Addison-Wesley Publ. Co., London, 1973).

(W) Y. Takahashi: *An Introduction to Field Quantization* (Pergamon Press, Oxford, 1969).

(X) W.E. Thirring: *Principle of Quantum Electrodynamics*, Pure and Applied Physics, Vol.3 (Academic Press, New York, 1958).

(Y) H. Umezawa: *Quantum Field Theory*, Series in Physics (North-Holland Publ. Co., Amsterdam, 1956).

(Z) G. Wentzel: *Quantum Theory of Fields* (Interscience Publ., New York, 1949).

1. CONSTRUCTION OF QED

1.1 Lorentz Invariant Formulation

[1] P.A.M. Dirac: The Quantum Theory of the Emission and Absorption of Radiation, *Proc. Roy. Soc.* **A114** (1927) 243.

[2] P. Jordan and W. Pauli: Zur Quantenelektrodynamik Ladungsfreier Felder, *Z. Phys.* **47** (1928) 151.

[3] P. Jordan und E. Wigner: Über das Paulische Äquivalenzverbot, *Z. Phys.* **47** (1928) 631.

[4] W. Heisenberg und W. Pauli: Zur Quantendynamik der Wellenfelder, *Z. Phys.* **56** (1929) 1.

[5] W. Heisenberg und W. Pauli: Zur Quantentheorie der Wellenfelder, II, *Z. Phys.* **59** (1930) 168.

[6] E. Fermi: Quantum Theory of Radiation, *Rev. Mod. Phys.* **4** (1932) 87.

[7] W. Pauli: Relativistic Field Theories of Elementary Particles, *Rev. Mod. Phys.* **13** (1941) 203.

[8] S. Tomonaga: On a Relativistically Invariant Formulation of the Quantum Theory of Wave Fields, *Prog. Theor. Phys.* **1** (1946) 27.

[9] Z. Koba, T. Tati and S. Tomonaga: On a Relativistically Invariant Formulation of the Quantum Theory of Wave Fields, II, *Prog. Theor. Phys.* **2** (1947) 101.

[10] Z. Koba, T. Tati and S. Tomonaga: On a Relativistically Invariant Formulation of the Quantum Theory of Wave Fields, III, *Prog. Theor. Phys.* **2** (1947) 198.

[11] Z. Koba: Note on a Lorentz-invariant Integration in the Quantum Field Theory, *Prog. Theor. Phys.* **5** (1950) 696.

[12] J. Schwinger: Quantum Electrodynamics, I — A covariant formulation, *Phys. Rev.* **74** (1948) 1439.

[13] J. Schwinger: Quantum Electrodynamics, II — Vacuum polarization and self-energy, *Phys. Rev.* **75** (1949) 651.

[14] J. Schwinger: Quantum Electrodynamics, III — The electromagnetic properties of the electron — Radiative corrections to scattering, *Phys. Rev.* **76** (1949) 790.

[15] J. Schwinger: On Gauge Invariance and Vacuum Polarization, *Phys. Rev.* **82** (1951) 664.

[16] R.P. Feynman: Space-time Approach to Non-relativistic Quantum Mechanics, *Rev. Mod. Phys.* **20** (1948) 367.

[17] R.P. Feynman: The Theory of Positrons, *Phys. Rev.* **76** (1949) 749.

[18] R.P. Feynman: Space-time Approach to Quantum Electrodynamics, *Phys. Rev.* **76** (1949) 769.

[19] R.P. Feynman: Relativistic Cut-off for Quantum Electrodynamics, *Phys. Rev.* **74** (1948) 1430.

[20] R.P. Feynman: Mathematical Formulation of the Quantum Theory of Electromagnetic Interaction, *Phys. Rev.* **80** (1950) 440.

[21] R.P. Feynman: An Operator Calculus Having Applications in Quantum Electrodynamics, *Phys. Rev.* **84** (1951) 108.

[22] C.N. Yang and D. Feldman: The S-Matrix in Heisenberg Representation, *Phys. Rev.* **79** (1950) 972.

[23] M. Gell-Mann and F. Low: Bound States in Quantum Field Theory, *Phys. Rev.* **84** (1951) 350.

[24] E.E. Salpeter and H.A. Bethe: Relativistic Equation for Bound State Problem, *Phys. Rev.* **84** (1951) 1232.

[25] Y. Nambu: On Lagrangian and Hamiltonian Formalism, *Prog. Theor. Phys.* **7** (1952) 131.

[26] R.E. Peierls: The Commutation Laws of Relativistic Field Theory, *Proc. Roy. Soc.* **A214** (1952) 143.

[27] K. Nishijima: On Lagrangian Formalism, *Prog. Theor. Phys.* **8** (1952) 401.

[28] T. Imamura, S. Sunakawa and R. Utiyama: On the Construction of S-matrix in Lagrangian Formalism, *Prog. Theor. Phys.* **11** (1954) 291.

[29] J. Schwinger: On the Green's Functions of Quantized Fields, I, *Proc. Nat. Acad. Sci.* **37** (1951) 452.

[30] J. Schwinger: On the Green's Functions of Quantized Fields, II, *Proc. Nat. Acad. Sci.* **37** (1951) 455.

[31] J. Schwinger: The Theory of Quantized Fields, I, *Phys. Rev.* **82** (1951) 914.

[32] J. Schwinger: The Theory of Quantized Fields, II, *Phys. Rev.* **91** (1953) 713.

[33] J. Schwinger: The Theory of Quantized Fields, III, *Phys. Rev.* **91** (1953) 728.

[34] J. Schwinger: The Theory of Quantized Fields, IV, *Phys. Rev.* **92** (1953) 1283.

[35] J. Schwinger: The Theory of Quantized Fields, V, *Phys. Rev.* **93** (1953) 615.

[36] J. Schwinger: The Theory of Quantized Fields, VI, *Phys. Rev.* **94** (1954) 1362.

[37] J. Schwinger: A Note on the Quantum Dynamical Principle, *Phil. Mag.* **44** (1953) 1171.

[38] S. Tomonaga: Development of Quantum Electrodynamics, *Phys. Today* **19, No.9** (1966) 25.

[39] J. Schwinger: Relativistic Quantum Field Theory, *Phys. Today* **19, No.6** (1966) 27.

[40] R.P. Feynman: The Development of the Space-time View of Quantum Electrodynamics, *Phys. Today* **19, No.8** (1966) 31.

1.2 CPT Invariance

[41] W. Heisenberg: Bemerkungen zur Diracshen Theorie des Positrons, *Z. Phys.* **90** (1934) 209. [Erratum:, *ibid.* **92** (1934) 692]

[42] W. Furry: A Symmetry Theorem in the Positron Theory, *Phys. Rev.* **51** (1937) 125.

[43] W. Pauli: Théorie quantique relativiste des particules obeissant *à* la statistique de Einstein-Bose, *Ann. Inst. H. Poincáre* **6** (1936) 137.

[44] W. Pauli: The Connection between Spin and Statistics, *Phys. Rev.* **58** (1940) 716.

[45] W. Pauli and F.J. Belinfante: On the Statistical Behaviour of Known and Unknown Elementary Particles, *Physica* **7** (1940) 177.

[46] L. Eisenbud and E.P. Wigner: Invariant Forms of Interaction between Nuclear Particles, *Proc. Nat. Acad. USA* **27** (1941) 281.

[47] T.D. Newton and E.P. Wigner: Localized States for Elementary Systems, *Rev. Mod. Phys.* **21** (1949) 400.

[48] S. Watanabe: Reversibility of Quantum Electrodynamics, *Phys. Rev.* **84** (1951) 1008.

[49] F. Coester: Principle of Detailed Balance, *Phys. Rev.* **84** (1951) 1259.

[50] F. Coester: The Symmetry of the S-Matrix, *Phys. Rev.* **89** (1953) 619.

[51] A. Pais and R. Jost: Selection Imposed by Charge Conjugation and Charge Symmetry, *Phys. Rev.* **87** (1952) 871.

[52] G. Lüders: Zur Bewegungsumkehr in Quantisierten Feldtheorien, *Z. Phys.* **133** (1952) 325.

[53] G. Lüders, R. Oehnse and W.E. Thirring: π-mesonen und Quantisierte Feldtheorien, *Z. Naturforsch.* **7a** (1952) 213.

[54] L. Michel: Selection Rules Imposed by Charge Conjugation, *Nuovo Cimento* **10** (1953) 319.

[55] G. Lüders: On the Equivalence of Invariance Under Time Reversal and Under Particle-antiparticle Conjugation for Relativistic Field Theories, *K. Danske Vidensk. Selsk. Mat.-fis. Medd.* **28, No.5** (1954).

[56] H. Umezawa, S. Kamefuchi and S. Tanaka: On the Time Reversal in the Quantized Field Theory, *Prog. Theor. Phys.* **12** (1954) 383.

[57] W. Pauli: Exclusion Principle, Lorentz Groups and Reflection of Space-time and Charge, *Niels Bohr and the Development of Physics* (Pergamon Press, London, 1955) p.30.

[58] S. Watanabe: Symmetry of Physical Laws, Pt. I — Symmetry in space-time and balance theorems, *Rev. Mod. Phys.* **27** (1955) 26.

[59] S. Watanabe: Symmetry of Physical Laws, Pt. II — Q-number theory of space-time inversions and charge conjugation, *Rev. Mod. Phys.* **27** (1955) 40.

[60] W. Pauli: Lectures on Continuous Groups and Reflections in Quantum Mechanics, *UCRL-8213* (Univ. of California, 1958).

[61] J. Schwinger: Spin, Statistics, and the TCP Theorem, *Proc. Nat. Acad. Sci.* **44** (1958) 223. [Addendum: *ibid.* 617]

[62] R.F. Streater and A.S. Wightman: *PCT, Spin and Statistic and All That* (Benjamin Inc., New York, 1964).

2. PROBLEMS ON ULTRAVIOLET DIVERGENCES

2.1 General Renormalization Theory

[63] Z. Koba and S. Tomonaga: On Radiation in Collision Processes, I, *Prog. Theor. Phys.* **3** (1948) 290.

[64] Z. Koba and G. Takeda: Radiation Reaction in Collision Processes, II, *Prog. Theor. Phys.* **3** (1948) 407.

[65] Z. Koba and G. Takeda: Radiation Reaction in Collision Process, III, *Prog. Theor. Phys.* **4** (1949) 60; 130.

[66] T. Tati and S. Tomonaga: A Self-consistent Subtraction Method in the Quantum Field Theory, I, *Prog. Theor. Phys.* **3** (1948) 391.

[67] H. Fukuda, Y. Miyamoto and S. Tomonaga: A Self-consistent Subtraction Method in the Quantum Field Theory, II, *Prog. Theor. Phys.* **4** (1949) 47; 121.

[68] F.J. Dyson: The Radiation Theories of Tomonaga, Schwinger and Feynman, *Phys. Rev.* **75** (1949) 486.

[69] F.J. Dyson: The S Matrix in Quantum Electrodynamics, *Phys. Rev.* **75** (1949) 1736.

[70] G.C. Wick: The Evaluation of the Collision Matrix, *Phys. Rev.* **80** (1950) 268.

[71] P.T. Matthews: The Application of Dyson's Methods to Meson Interactions, *Phys. Rev.* **76** (1949) 684.

[72] P.T. Matthews: The S-matrix for Meson-nucleon Interactions, *Phys. Rev.* **76** (1949) 1254.

[73] P.T. Matthews: The S-matrix for Meson-nucleon Interaction, *Phil. Mag.* **41** (1950) 185.

[74] P.T. Matthews: Spinless Mesons in the Electromagnetic Field, *Phys. Rev.* **80** (1950) 292.

[75] F. Rorhlich: Quantum Electrodynamics of Charged Particles without Spin, *Phys. Rev.* **80** (1950) 666.

[76] P.T. Matthews: Renormalization of the Meson-photon-nucleon interaction, *Phil. Mag.* **42** (1951) 221.

[77] P.T. Matthews: Renormalization of Neutral Mesons in Three-field Problems, *Phys. Rev.* **81** (1951) 936.

[78] P.T. Matthews and A. Salam: The Renormalization of Meson Theory, *Rev. Mod. Phys.* **23** (1951) 311.

[79] R. Jost und J.M. Luttinger: Vacuumpolarisation und e^4-Ladungsrenormalization für Electronen, *Helv. Phys. Acta* **23** (1950) 201.

[80] J.C. Ward: The Scattering of Light by Light, *Phys. Rev.* **77** (1950) 293.

[81] J.C. Ward: An Identity in Quantum Electrodynamics, *Phys. Rev.* **78** (1950) 182.

[82] A. Salam: Differential Identities in Three-field Renormalization, *Phys. Rev.* **79** (1950) 910.

[83] A. Salam: Overlapping Divergences and the S-matrix, *Phys. Rev.* **82** (1951) 217.

[84] A. Salam: Divergence Integrals in Renormalizable Field Theories, *Phys. Rev.* **84** (1951) 426.

[85] J.C. Ward: On the Renormalization of Quantum Electrodynamics, *Proc. Phys. Soc.* **A64** (1951) 54.

[86] J.C. Ward: Renormalization Theory of the Interaction of Nucleon, Mesons and Photons, *Phys. Rev.* **84** (1951) 897.

[87] R. Utiyama, S. Sunakawa and T. Imamura: On the Theory of the Green-functions in Quantum Electrodynamics, *Prog. Theor. Phys.* **8** (1952) 77.

[88] F.J. Dyson: Heisenberg Operators in Quantum Electrodynamics, I, *Phys. Rev.* **82** (1951) 428.

[89] F.J. Dyson: Heisenberg Operators in Quantum Electrodynamics, II, *Phys. Rev.* **83** (1951) 608.

[90] F.J. Dyson: The Renormalization Method in Quantum Electrodynamics, *Proc. Roy. Soc.* **A207** (1951) 395.

[91] F.J. Dyson: The Schrödinger Equation in Quantum Electrodynamics, *Phys. Rev.* **83** (1951) 1207.

[92] F.J. Dyson: Divergence of Perturbation Theory in Quantum Electrodynamics, *Phys. Rev.* **85** (1952) 631.

[93] S. Sakata, H. Umezawa and S. Kamefuchi: Applicability of the Renormalization Theory and the Structure of Elementary Particles, *Phys. Rev.* **84** (1951) 154.

[94] S. Sakata, H. Umezawa and S. Kamefuchi: On the Structure of the Interaction of the Elementary Particles, I – The renormalizability of the interaction, *Prog. Theor. Phys.* **7** (1952) 377.

[95] S.N. Gupta: On the Elimination of Divergencies from Quantum Electrodynamics, *Proc. Phys. Soc.* **A64** (1951) 426.

[96] G. Takeda: On the Renormalization Theory of the Interaction of Electrons and Photons, *Prog. Theor. Phys.* **7** (1952) 359.

[97] S. Kamefuchi and H. Umezawa: On the Renormalization in Quantum Electrodynamics, *Prog. Theor. Phys.* **7** (1952) 399.

[98] P.T. Matthews and A. Salam: Renormalization, *Phys. Rev.* **94** (1954) 185.

[99] J.G. Valatin: Singularities of Electron Kernel Functions in an External Electromagnetic Field, *Proc. Roy. Soc.* **A222** (1954) 93.

[100] J.G. Valatin: On the Dirac-Heisenberg Theory of Vacuum Polarization, *Proc. Roy. Soc.* **A222** (1954) 228.

[101] J.G. Valatin: On the Definition of Finite Operator Quantities in Quantum Electrodynamics, *Proc. Roy. Soc.* **A226** (1954) 254.

[102] J.G. Valatin: On the Propagation Functions of Quantum Electrodynamics, *Proc. Roy. Soc.* **A226** (1954) 535.

[103] N.N. Bogoliubow und D.W. Schirkow: Probleme der Quantenfeld-theorie, *Forts. Phys.* **4** (1956) 438.

[104] M. Gell-Mann and F.E. Low: Quantum Electrodynamics at Small Distances, *Phys. Rev.* **95** (1954) 1300.

[105] L.D. Landau. A.A. Abrikosov and I.I. Halatnikov: On the Removal of Infinities in Quantum Electrodynamics, *Dok. Akad. Nauk. SSSR* **95** (1954) 497.

[106] L.D. Landau, A.A. Abrikosov and I.M. Halatnikov: An Asymptotic Expression for the Green Function, *Dok. Akad. Nauk SSSR* **95** (1954) 733.

[107] L.D. Landau, A.A. Abrikosov and I.M. Halatnikov: An Asymptotic Expression for the Greeen Function, *Dok. Akad. Nauk SSSR* **95** (1954) 1177.

[108] L.D. Landau, A.A. Abrikosov and I.M. Halatnikov: The Mass of the Electron in Quantum Electrodynamics, *Dok. Akad. Nauk SSSR* **96** (1955) 261.

[109] L.D. Landau: On the Quantum Theory of Fields,*Niels Bohr and the Development of Physics* (Pergamon Press, London, 1955) p.52.

[110] I. Pomerancuk: Vanishing of the Renormalized Charge in Electrodynamics and in Meson Theory, *Nuovo Cimento* **3** (1956) 1186.

[111] I.Y. Pomeranchuk, V.V. Sudakov, and K.A. Ter-Martirosyan: Vanishing of Renormalized Charges in Field Theories with Point Interaction, *Phys. Rev.* **103** (1956) 784.

[112] L.D. Landau, A. Abrikosov and I. Halatnikov: On the Quantum Theory of Fields, *Nuovo Cimento Suppl.* **3** (1956) 80.

[113] N.N. Bogoliubov and D.V.Shirkov: The Multiplicative Renormalization Group in the Quantum Theory of Fields, *Soviet Phys. JETP* **3** (1956) 57. [*Zh. Eksper. Teor. Fiz.* **30** (1956) 77]

[114] N.N. Bogoliubov and D.V. Sirkov: Charge Renormalization Group in Quantum Field Theory, *Nuovo Cimento* **3** (1956) 845.

[115] K.E. Eriksson: Partial Summation of the Perturbation Expansion in High-energy Quantum Electrodynamics, *Nuovo Cimento* **27** (1963) 178.

[116] E.R. Caianiello, M. Marinaro and F. Guerra: Form-invariant Renormalization, *Nuovo Cimento* **60A** (1969) 713.

[117] J.H. Lowenstein: Differential Vertex Operations in Lagrangian Field Theory, *Commm. Math. Phys.* **24** (1971) 1.

[118] H.S. Green: A Pre-renormalized Quantum Electrodynamics, *Proc. Phys. Soc.* **A66** (1953) 873.

[119] E.S. Fradkin: Concerning Some General Relations of Quantum Electrodynamics, *Soviet Phys. JETP* **2** (1956) 361. [*Zh. Eksper. Teor. Fiz.* **29** (1955) 258.

[120] Y. Takahashi: On the Generalized Ward Identity, *Nuovo Cimento* **6** (1957) 371.

[121] E. Kazes: Generalized Current Conservation and Low Energy Limit of Photon Interactions, *Nuovo Cimento* **13** (1959) 1226.

[122] K. Nishijima: Asymptotic Condition and Perturbation Theory, *Phys. Rev.* **119** (1960) 485.

[123] K. Nishijima: Time-ordered Green's Functions and Electromagnetic Interactions, *Phys. Rev.* **122** (1961) 298.

[124] K. Nishijima: Renormalization of Time-ordered Green's Functions, *Phys. Rev.* **124** (1961) 255.

[125] M. Muraskin and K. Nishijima: Time-ordered Green's Functions and Perturbation Theory, *Phys. Rev.* **122** (1961) 331.

[126] K. Just and K. Rossberg: Ward Relation for Gravity, *Nuovo Cimento* **40** (1965) A1077.

[127] R.J. Rivers: Gauge Approximation in Mesodynamics, *J. Math.* **7** (1966) 385.

[128] S. Coleman and R. Jackiw: Why Dilatation Generators Do Not Generate Dilatations, *Ann. Phys.* **67** (1971) 552.

[129] K. Nishijima and R. Sasaki: Dispersion Approach to Anomalies in the Axial-vector Ward-Takahashi Identities, *Prog. Theor. Phys.* **53** (1975) 261.

[130] B. deWit and D.Z. Freedman: Combined Supersymmetric and Gauge-invariant Field Theories, *Phys. Rev.* **D12** (1975) 2286.

[131] K. Nishijima and R. Sasaki: Ward-Takahashi Identities in Quantum Electrodynamics, *Prog. Theor. Phys.* **53** (1975) 829.

[132] Y. Takahashi and T. Goto: General Ward-like Relation in Canonical Field Theory,, *Prog. Theor. Phys.* **57** (1977) 1732.

[133] E.R. Caianiello: *Renormalization and Invariance in Quantum Field Theory* (Plenum Press, New York 1974).

2.2 Feynman Integrals

[134] R. Chisholm: Calculation of S-matrix Elements, *Proc. Cam. Phil. Soc.* **48** (1952) 300 [Addendum: *ibid.* 518].

[135] C.A. Hurst: The Enumeration of Graphs in the Feynman-Dyson Technique, *Proc. Roy. Soc.* **A214** (1952) 44.

[136] W. Thirring: On the Divergence of Perturbation Theory for Quantized Fields, *Helv. Phys. Acta* **26** (1953) 33.

[137] A. Petermann: Renormalisation dans les séries Divergentes, *Helv. Phys. Acta* **26** (1953) 291.

[138] N. Nakanishi: General Integral Formula of Perturbation Term in the Quantized Field Theory, *Prog. Theor. Phys.* **17** (1957) 401. [Soryushiron Kenkyu **12** (1956) 217; **13** (1956) 89].

[139] Y. Nambu: Parametric Representations of General Green's Functions, *Nuovo Cimento* **6** (1957) 1064.

[140] K. Symanzik: Dispersion Relations and Vertex Properties in Perturbation Theory, *Prog. Theor. Phys.* **20** (1958) 690.

[141] Y. Shimamoto: Graph Theory and Parametric Representations of Feynman Amplitudes, *Nuovo Cimento* **25** (1962) 1292.

[142] R.C. Hwa and V.L. Teplitz: *Homology and Feynman Integrals* (Benjamin, Inc., New York, 1966).

[143] R.J. Eden, P.V. Landshoff, D.I. Olive and J.C. Polkinghorne: *The Analytic S-matrix* (Cambridge Univ. Press, London, 1966).

[144] N. Nakanishi: Ordinary and Anomalous Thresholds in Perturbation Theory, *Prog. Theor. Phys.* **22** (1959) 128.

[145] L.D. Landau: On Analytic Properties of Vertex Parts in Quantum Field Theory, *Nucl. Phys.* **13** (1959) 181.

[146] J.C. Polkinghorne and G.R. Screaton: The Analytic Properties of Perturbation Theory I, *Nuovo Cimento* **15** (1960) 289.

[147] R.E. Cutkosky: Singularities and Discontinuities of Feynman Amplitudes, *J. Math. Phys.* **1** (1960) 429.

[148] R. Karplus, C.M. Sommerfield and E.H. Wichmann: Spectral Representations in Perturbation Theory, I — Vertex funcion, *Phys. Rev.* **111** (1958) 1187.

[149] Y. Nambu: Dispersion Relations for Form Factors, *Nuovo Cimento* **9** (1958) 610.

[150] R. Oehme: Vertex Function in Quantized Field Theories, *Phys. Rev.* **111** (1958) 1430.

[151] R. Karplus, C.M. Sommerfield and E.H. Wichmann: Spectral Representations in Perturbation Theory, II — Two-particle scattering, *Phys. Rev.* **114** (1959) 376.

[152] J.C. Taylor: Analytic Properties of Perturbation Expansions, *Phys. Rev.* **117** (1960) 261.

[153] N. Nakanishi: Parametric Integral Formulas and Analytic Properties in Perturbation Theory, *Prog. Theor. Phys. Suppl.* **18** (1961) 1 [Erratum: *ibid.* **26** (1961) 806].

[154] T.T. Wu: Perturbation Theory of Pion-pion Interaction I — Renormalization, *Phys. Rev.* **125** (1962) 1436.

[155] A.A. Logunov, I.T. Todorov and N.A. Chernikov: Generalization of Symanzik's Theorem on the Majorization of Feynman Graphs, *Soviet Phys. JETP* **15** (1962) 891 [*Zh. Eksper. Teor. Fiz.* **42** (1962) 1285].

[156] Y. Chow and D.J. Kleitman: Some Theorems on the U-functions of Parametrized Feynman Amplitudes, *Prog. Theor. Phys.* **32** (1964) 950.

[157] Y. Chow: On Some Topological Properties of Feynman Graphs and Their Application to Formulas Related to the Feynman Amplitudes, *J. Math. Phys.* **5** (1964) 1255.

[158] A. Jaffe: Divergence of Perturbation Theory for Bosons, *Comm. Math. Phys.* **1** (1965) 127.

[159] C.S. Lam and J.P. Lebrun: Feynman-parameter representations for Momentum- and Configuration-space Diagrams, *Nuovo Cimento* **59A** (1969) 397.

[160] I.T. Todorov: *Analytic Properties of Feynman Diagrams in Quantum Field Theory* (Pergamon Press, Oxford, 1971).

[161] E.R. Speer: *Generalized Feynman Amplitudes* (Princeton Univ. Press, New Jersey, 1969).

[162] N. Nakanishi: *Graph Theory and Feynman Integrals* (Gordon and Breach, New York, 1971).

[163] P. Cvitanovic and T. Kinoshita: Feynman-Dyson Rules in Parametric Space, *Phys. Rev.* **D10** (1974) 3978.

[164] P. Cvitanovic and T. Kinoshita: New Approach to the Separation of Ultraviolet and Infrared Divergences of Feynman-parametric Integrals, *Phys. Rev.* **D10** (1974) 3991.

[165] N. Nakanishi: Fundamental Properties of Perturbation-theoretical Integral Representations, II, *J. Math: Phys.* **4** (1963) 1385.

[166] S. Weinberg: High-energy Behavior in Quantum Field Theory, *Phys. Rev.* **118** (1960) 838.

[167] Y. Hahn and W. Zimmermann: An Elementary Proof of Dyson's Power Counting Theorem, *Commm. Math. Phys.* **10** (1968) 330.

[168] W. Zimmermann: The Power Counting Theorem for Minkowski Metric , *Comm. Math. Phys.* **11** (1968) 1.

[169] M.C. Bergère and Y.-M.P. Lam: Asymptotic Expansion of Feynman Amplitudes Part I — The convergent case, *Comm. Math. Phys.* **39** (1974) 1.

[170] J.H. Lowenstein and W. Zimmermann: The Power Counting Theorem for Feynman Integrals with Massless Propagators, *Commm. Math. Phys.* **44** (1975) 73.

[171] P. Breitenlohner and H. Mitter: Analytic Regularization and Gauge Invariance, *Nucl. Phys.* **B7** (1968) 443.

[172] E.R. Speer: Analytic Renormalization, *J. Math. Phys.* **9** (1968) 1404.

[173] F. Guerra: On Analytic Regularization in Quantum Field Theory, *Nuovo Cimento* **1A** (1971) 523.

[174] E.R. Speer: Seminars on Renormalisation Theory, 1 — Lectures on analytic renormalisation, *Tech. Rep. No.73-067* (Univ. Maryland 1972).

[175] E.R. Speer: Renormalization and Ward Identities using Complex Space-time Dimension, *J. Math. Phys.* **15** (1974) 1.

[176] G. 'tHooft and M. Veltman: Regularization and Renormalization of Gauge Fields, *Nucl. Phys.* **B44** (1972) 189.

[177] C.G. Bollini and J.J. Giambiagi: Dimensional Regularization. The number of dimensions as a regularizing parameter, *Nuvo Cimento* **12B** (1972) 20.

[178] G. 'tHooft: An Algorithm for the Poles at Dimension Four in the Dimensional Regularization Procedure, *Nucl. Phys.* **B62** (1973) 444.

[179] J.F. Ashmore: On Renormalization and Complex Space-time Dimensions, *Commm. Math. Phys.* **29** (1973) 177.

2.3 BPHZ Theory

[180] N.N. Bogoliubov and O.S. Parasuik: Über die Multiplikation der Kausalfunktion in der Quantentheorie der Felder, *Acta Math.* **97** (1957) 227.

[181] K. Hepp: Proof of the Bogoliubov-Parasiuk Theorem on Renormalization, *Comm. Math. Phys.* **2** (1966) 301.

[182] W. Zimmermann: Convergence of Bogoliubov's Method of Renormalization in Momentum Space, *Comm. Math. Phys.* **15** (1969) 208.

[183] O.I. Zav'yalov and B.M. Stepanov: Asymptotics of Diverging Feynman Diagrams, *Soviet J. Nucl. Phys.* **1** (1965) 658 [*Yad. Fiz.* **1** (1965) 922].

[184] R.A. Brandt: Bogoliubov-Parasiuk-Hepp Renormalization Theorem and Space-like Regularization, *J. Math. Phys.* **8** (1967) 1112.

[185] P.K. Kuo and D.R. Yennie: Renormalization Theory, *Ann. Phys.* **51** (1969) 496.

[186] T. Appelquist: Parametric Integral Representations of Renormalized Feynman Amplitudes, *Ann. Phys.* **54** (1969) 27.

[187] S.A. Anikin, O.I. Zav'yalov and M.K. Polivanov: A Simple Proof of the Bogoliubov-Parasyuk Theorem, *Theor. Math. Phys.* **17** (1974) 1082 [*Teor. Mat. Fiz.* **17** (1973) 189].

[188] M.C. Bergère and J.B. Zuber: Renormalization of Feynman Amplitudes and Parametric Integral Representation, *Comm. Math. Phys.* **35** (1974) 113.

[189] R.A. Brandt: Derivation of Renormalized Relativistic Perturbation Theory from Finite Local Field Equations, *Ann. Phys.* **44** (1967) 221.

[190] R.A. Brandt: Gauge Invariance in Quantum Electrodynamics, *Ann. Phys.* **52** (1969) 122.

[191] W. Zimmermann: Local Operator Products and Renormalization in Quantum Field Theory. *Lectures on Elementary Particles and Quantum Field Theory, 1* (M.I.T. Press, Cambridge, 1970) p.395.

[192] J.H. Lowenstein: Normal-Product Quantization of Currents in Lagrangian Field Theory, *Phys. Rev.* **4D** (1971) 2281.

[193] J. Lowenstein: Seminars on Renormalisation Theory, 2. Normal product methods in renormalized perturbation theory, *Tech. Rep. No.73-068* (Univ. Maryland 1972).

[194] M. Gomes and J.H. Lowenstein: Linear Relations Among Normal-product Fields, *Phys. Rev.* **D7** (1973) 550.

[195] W. Zimmermann: Composite Operators in the Perturbation Theory of Renormalizable Interactions, *Ann. Phys.* **77** (1973) 536.

[196] M. Gomes, J.H. Lowenstein and W. Zimmermann: Generalization of the Momentum-space Subtraction Procedure for Renormalized Perturbation Theory, *Comm. Math. Phys.* **39** (1974) 81.

[197] J.H. Lowenstein and W. Zimmermann: On the Formulation of Theories with Zero-mass Propagators, *Nucl. Phys.* **B86** (1975) 77.

[198] J.H. Lowenstein and W. Zimmermann: Infrared Convergence of Feynman Integrals for the Massless A^4-model, *Comm. Math. Phys.* **46** (1976) 105.

[199] J.H. Lowenstein and E. Speer: Distributional Limits of Renormalized Feynman Integrals with Zero-mass Denominators, *Comm. Math. Phys.* **47** (1976) 43.

[200] J.H. Lowenstein: Convergence Theorems for Renormalized Feynman Integrals with Zero-mass Propagators, *Comm. Math. Phys.* **47** (1976) 53.

[201] M.C. Bergère and Y.-M.P. Lam: Zero-mass Limit in Perturbative Quantum Field Theory, *Comm. Math. Phys.* **48** (1976) 267.

[202] G. Velo and A.S. Wightman: Renormalization Theory, *Proceedings of the NATO Advanced Study Institute* (D. Reidel Publ., Dordrecht, Holland, 1976).

3. MODIFICATION OF THE FORMALISM (INTRODUCTION OF INDEFINITE METRIC AND GAUGE PROBLEMS)

3.1 Gupta-Bleuler Formalism and Related Problems

[203] F.J. Belinfante: On the Vanishing of divE-$4\pi p$ in Quantum Electrodynamics, *Physica* **12** (1946) 17.

[204] F.J. Belinfante: On the Part Played by Scalar and Longitudinal Photons in Ordinary Electromagnetic Fields, *Phys. Rev.* **76** (1949) 226.

[205] S.T. Ma: Relativistic Formulation of the Quantum Theory of Radiation, *Phys. Rev.* **75** (1949) 535.

[206] S.T. Ma: Quantum Theory of the Longitudinal Electromagnetic Field, *Phys. Rev.* **80** (1950) 729.

[207] K. Husimi and R. Utiyama: Canonical Theory of Quantum Electrodynamics, *Prog. Theor. Phys.* **5** (1950) 718.

[208] R. Utiyama, T. Imamura, S. Sunakawa and T. Dodo: Note on the Longitudinal and Scalar Photons, *Prog. Theor. Phys.* **6** (1951) 587.

[209] S.N. Gupta: Theory of Longitudinal Photons in Quantum Electrodynamics, *Proc. Phys. Soc.* **A63** (1950) 681.

[210] K. Bleuler: Eine neue Methode zur Behandlung der Longitudinalen und Skalaren Photonen, *Helv. Phys. Acta* **23** (1950) 567.

[211] K. Bleuler and W. Heitler: The Reversal of Time and the Quantization of the Longitudinal Field in Quantum Electrodynamics, *Prog. Theor. Phys.* **5** (1950) 600.

[212] F.J. Belinfante: Direct Proof of the Covariance of Gupta's Indefinite Metric in Quantum Electrodynamics, *Phys. Rev.* **96** (1954) 780.

[213] S.N. Gupta: On Stueckelberg's Treatment of the Vector Meson Field, *Proc. Phys. Soc.* **A64** (1951) 695.

[214] S.N. Gupta: Quantization of Einstein's Gravitational Field — Linear approximation, *Proc. Phys. Soc.* **A65** (1952) 161.

[215] S.N. Gupta: Quantization of Einstein's Gravitational Field — General treatment, *Proc. Phys. Soc.* **A65** (1952) 608.

[216] K. Just: Quantization Problem of Gravity, *Nuovo Cimento* **34** (1964) 567.

[217] K. Just and K. Rossberg: Ward Relations for Gravity, *Nuovo Cimento* **40A** (1965) 1077.

[218] K. Just and K. Rossberg: The Gauge Dependence of the Graviton Propagator, *Nuovo Cimento* **40** (1965) A1088.

[219] A.S. Wightmann and L. Garding: Fields as Operator-Valued Distributions in Relativistic Quantum Theory, *Ark. f. Fis.* **28** (1964) 129.

[220] F. Strocchi: Gauge Problem in Quantum Field Theory, *Phys. Rev.* **162** (1967) 1429.

[221] F. Strocchi: Gauge Problem in Quantum Field Theory II — Difficulties of combining Einstein equations and Wightman theory, *Phys. Rev.* **166** (1968) 1302.

[222] F. Strocchi: Gauge Problem in Quantum Field Theory III — Quantization of Maxwell equation and weak commutativity, *Phys. Rev.* **D2** (1970) 2334.

[223] R. Ferrari, L.E. Picasso and F. Strocchi: Some Remarks on Local Operators in Quantum Electrodynamics, *Comm. Math. Phys.* **35** (1974) 25.

[224] J. Bertrand: Poincaré Covariance and Quantization of Zero-mass Fields I — The electromagnetic field, *Nuovo Cimento* **A1** (1971) 1.

[225] F. Strocchi and A.S. Wightman: Proof of the Charge Superselection Rule in Local Relativistic Quantum Field Theory, *J. Math. Phys.* **15** (1974) 2198.

[226] P.M. Mathews, M. Seetharaman and M.T. Simon: Indecomposability of Poincaré-group Representations over Massless Fields and the Quantization Problem for Electromagnetic Potentials, *Phys. Rev.* **D9** (1974) 1706.

[227] K. Just: The Lorentz Condition in Quantum Theory, *Nuovo Cimento* **38** (1965) 400.

[228] K. Haller and L.F. Landovitz: Subsidiary Condition in Quantum Electrodynamics, *Phys. Rev.* **171** (1968) 1749.

[229] K. Haller and L.F. Landovitz: Equivalence of the Coulomb Gauge and the Reformulated Lorentz Gauge, *Phys. Rev.* **182** (1969) 1922.

[230] K. Haller and L.F. Landovitz: Renormalization Constants, Wave Functions, and Energy Shifts in the Coulomb and Lorentz Gauges, *Phys. Rev.* **D2** (1970) 1498.

[231] S.N. Gupta: Comment on Quantum Electrodynamics, *Phys. Rev.* **180** (1969) 1601.

[232] S.P. Tomczak and K. Haller: The Generalized-Lorentz-gauge Description of Quantum Electrodynamics, *Nuovo Cimento* **8B** (1972) 1.

3.2 Källén Formalism and Related Problems

[233] G. Källén: On the Definition of the Renormalization Constants in Quantum Electrodynamics, *Helv. Phys Acta* **25** (1952) 417.

[234] G. Källén: On the Magnitude of the Renormalization Constants in Quantum Electrodynamics, *K. Danske Vidensk. Selsk. Mat.-fis. Medd.* **27** (1953) No.12.

[235] G. Källén: Charge Renormalization and the Identity of Ward, *Helv. Phys. Acata* **26** (1953) 755.

[236] K.A. Johnson: Consistency of Quantum Electrodynamics, *Phys. Rev.* **112** (1958) 1367.

[237] S.G. Gasiorowicz, D.R. Yennie and H. Suura: Magnitude of Renormalization Constants, *Phys. Rev. Letters* **2** (1959) 513.

[238] B. Zumino: The Renormalization Constants in Quantum Electrodynamics, *Nuovo Cimento* **17** (1960) 547.

[239] H. Rollnik, B. Stech und E. Nunnemann: Quantenelektrodynamik und Asymptotenbedingung, *Z. Phys.* **159** (1960) 482.

[240] R.S. Willey: Asymptotic Condition and Covariant Gauges in Quantum Electrodynamics, *Ann. Phys.* **45** (1967) 167.

[241] L. Evans, G. Feldman and P.T. Matthews: Gauge Invariance and Renormalization Constants, *Ann. Phys.* **13** (1961) 268.

[242] V.D. Skarzhinski: An Axiomatic Formulation of Quantum Electrodynamics, *Soviet Phys. JETP* **25** (1967) 601 [*Zh. Eksper. Teor. Fiz. USSR* **52** (1967) 910] .

[243] Y.A. Gol'fand: Quantum Electrodynamics without Bare Constants and Divergences, *Soviet J. Nucl. Phys.* **7** (1968) 133 [*Yad. Fiz.* **7** (1968) 183].

3.3 Gauge Transformation

[244] L.D. Landau and I.M. Khalatnikov: The Gauge Transformation of the Green's Function for Charged Particles, *Soviet Phys. JETP* **2** (1956) 69 [*Zh. Eksper. Teor. Fiz.* **29** (1955) 89].

[245] H.M. Fried and D.R. Yennie: New Techniques in the Lamb Shift Calculation, *Phys. Rev.* **112** (1958) 1391.

[246] K. Johnson and B. Zumino: Gauge Dependence of the Wave-function Renormalization Constant in Quantum Electrodynamics, *Phys. Rev. Letters* **3** (1959) 351.

[247] B. Zumino: Gauge Properties of Propagators in Quantum Electrodynamics, *J. Math. Phys.* **1** (1960) 1.

[248] B. Zumino: The Gauge Transformation of Propagators in Quantum Electrodynamics, *Lectures on Field Theory and the Many-body Problem* (Academic Press, London, 1961) p.27.

[249] S. Okubo: The Gauge Properties of Green's Functions in Quantum Electrodynamics, *Nuovo Cimento* **15** (1960) 949.

[250] I. Bialynicki-Birula: Finite Value of the Wave Function Renormalization Constant in Quantum Electrodynamics, *Nuovo Cimento* **17** (1960) 122.

[251] I. Bialynicki-Birula: On the Gauge Properties of Green's Functions, *Nuovo Cimento* **17** (1960) 951.

[252] H. Rollnik: Operatoreichtransformationen in der Quanten-electrodynamik, *Z. Phys.* **161** (1961) 370.

[253] M. Zulauf: On Two Gauge Classes and the Invariance of the S Matrix in Quantum Electrodynamics, *Helv. Phys. Acta* **39** (1966) 439.

[254] I. Bialynicki-Birula: Gauge Transformation in the S-matrix Theory, *Phys. Rev.* **166** (1968) 1505.

[255] I. Bialynicki-Birula: Renormalization, Diagrams, and Gauge Invariance , *Phys. Rev.* **D2** (1970) 2877.

[256] M. Hayakawa and K. Yokoyama: Gauge, Renormalization and the Goto-Imamura-Schwinger Term in Quantum Electrodynamics, *Prog. Theor. Phys.* **44** (1970) 533.

[257] J.G. Valatin: On Quantum Electrodynamics, *K. Danske Vidensk. Selsk. Mat.-fis. Medd.* **26 No.13** (1951) .

[258] L.E. Evans and T. Fulton: Asymptotic Conditions in Quantum Electrodynamics, *Nucl. Phys.* **21** (1960) 492.

[259] Y-P. Yao: Quantization of Electrodynamics in the Axial Gauge, *J. Math. Phys.* **5** (1964) 1319.

[260] A. Peres: Commutation Relations for Electromagnetic Potentials, *Nuovo Cimento* **34** (1964) 346.

[261] J. Lukierski: Electromagnetic Potentials in Landau Gauge, *Bull. Acad. Polon.* **14** (1966) 569.

[262] J. Lukierski: The Formulation of Quantum Electrodynamics with Strong Lorentz Condition, I — Heisenberg picture, *Acta Phys. Polon.* **31** (1967) 63.

[263] J. Lukierski: The Formulation of Quantum Electrodynamics with Strong Lorentz Condition, II — Interaction picture, *Acta Phys. Polon.* **31** (1967) 905.

[264] Y. Nambu: Quantum Electrodynamics in Nonlinear Gauge, *Prog. Theor. Phys. Supp. Extra Number* (1968) 190.

3.4 Covariant-Gauge Formalism and Related Problems

[265] N. Nakanishi: Covariant Quantization of the Electromagnetic Field in the Landau Gauge, *Prog. Theor. Phys.* **35** (1966) 1111.

[266] N. Nakanishi: Quantum Electrodynamics in the General Covariant Gauge, *Prog. Theor. Phys.* **38** (1967) 881.

[267] B. Lautrup: Canonical Quantum Electrodynamics in Covariant Gauges, *K. Danske Vidensk. Selsk. Mat.-fis. Medd.* **35 No.11** (1967) 1.

[268] N. Nakanishi: A Way Out of a Formal Difficulty Encountered in the Landau-gauge Quantum Electrodynamics, *Prog. Theor. Phys.* **51** (1974) 952.

[269] N. Nakanishi: Ward-Takahashi Identities in Quantum Field Theory with Spontaneously Broken Symmetry, *Prog. Theor. Phys.* **51** (1974) 1183.

[270] N. Nakanishi: The Lehmann-Symanzik-Zimmermann Formalism for Manifestly Covariant Quantum Electrodynamics, *Prog. Theor. Phys.* **52** (1974) 1929.

[271] R. Utiyama: Theory of Invariant Variation and the Generalized Canonical Dynamics, *Prog. Theor. Phys. Suppl.* **9** (1959) 19.

[272] J. Schwinger: Non-Abelian Gauge Fields, Lorentz Gauge Formulation, *Phys. Rev.* **130** (1963) 402.

[273] J. Lukierski: The Operator Formalism of Quantum Electrodynamics in Landau Gauge, *Nuovo Cimento Suppl.* **5** (1967) 739.

[274] T. Goto and T. Obara: The Canonical Quantization of the Free Electromagnetic Field in the Landau Gauge, *Prog. Theor. Phys.* **38** (1967) 871.

[275] K. Yokoyama: Photon Propagator in Quantum Electrodynamics, *Prog. Theor. Phys.* **40** (1968) 160.

[276] K. Yokoyama: Canonical Quantum Electrodynamics with Invariant One-parameter Gauge Families, *Prog. Theor. Phys.* **51** (1974) 1956.

[277] K. Yokoyama and R. Kubo: Gauge Structure of Canonical Quantum Electrodynamics, *Prog. Theor. Phys.* **52** (1974) 290.

[278] N. Nakanishi: Massive Vector Field and Electromagnetic Field in the Landau Gauge, *Phys. Rev.* **D5** (1972) 1324.

[279] P. Ghose and A. Das: A Variant of the Stuckelberg Formalism for Massive Gauge Fields and Applications, *Nucl. Phys.* **B41** (1972) 299.

[280] K. Yokoyama: Neutral-vector Field Theory with Invariant Gauge Families, *Prog. Theor. Phys.* **52** (1974) 1669.

[281] R. Kubo and K. Yokoyama: Gauge Structure of Neutral-vector Field Theory, *Prog. Theor. Phys.* **53** (1975) 871.

[282] R. Kubo and K. Yokoyama: Amount of Gauge Transformations in Neutral-vector Field Theory, *Prog. Theor. Phys.* **53** (1975) 911.

[283] K. Yokoyama and S. Yamagami: Momentum-space Operators in Covariant-gauge Quantum Electrodynamics — Massless limit of neutral-vector field theory, *Prog. Theor. Phys.* **55** (1976) 910.

[284] Y. Takahashi: Note on the Scale Transformation, *Phys. Rev.* **D3** (1971) 622.

[285] A. Das and P. Ghose: Scale Transformations and Massive Neutral Vector Fields, *Phys. Rev.* **D8** (1973) 3706.

[286] R. Kubo, Y. Takahashi and K. Yokoyama: Scale Transformation and Massless Limit in Neutral-vector Field Theory, *Phys. Rev.* **D11** (1975) 2335.

[287] A. Proca: Sur la Théorie Ondulatoire des Electrons Positifs et Négatifs, *J. Phys. Radium* **7** (1936) 347.

[288] R.J. Duffin: On the Characteristic Matrices of Covariant Systems, *Phys. Rev.* **54** (1938) 1114.

[289] N. Kemmer: The Particle Aspect of Meson Theory, *Proc. Roy. Soc.* **A173** (1939) 91.

[290] E.C.G. Stueckelberg: Die Wechselwirkungskräfte in der Elektrodynamik und in der Feldtheorie der Kernkräfte, Teil II und III, *Helv. Phys. Acta* **11** (1938) 299.

[291] Y. Miyamoto: On the Interaction of the Meson and Nucleon Field in the Super-many-time Theory, *Prog. Theor. Phys.* **3** (1948) 124.

[292] P.T. Matthews: The S-matrix for Meson-nucleon Interactions, *Phys. Rev.* **76** (1949) 1254.

[293] H. Umezawa: On the Structure of the Interactions of the Elementary Particles, II, *Prog. Theor. Phys.* **7** (1952) 551.

[294] S. Kamefuchi and H. Umezawa: On the Structure of the Interaction of the Elementary Particles, III, *Prog. Theor. Phys.* **8** (1952) 579.

[295] R.J. Glauber: On the Gauge Invariance of the Neutral Vector Meson Theory, *Prog. Theor. Phys.* **9** (1953) 295.

[296] O. Hara and H. Okonogi: On Gauge Invariance in Electrodynamics and the Self-energy of the Photon, *Prog. Theor. Phys.* **10** (1953) 191.

[297] R.J. Glauber: Note on the Neutral Vector Meson Theory, *Prog. Theor. Phys.* **10** (1953) 690.

[298] E.C.G. Stueckelberg: Théorie de la Radiation de Photons de Masse Arbitrairement Petite, *Helv. Phys. Acta* **30** (1957) 209.

[299] Y. Fujii: On the Analogy Between Strong Interaction and Electromagnetic Interaction, *Prog. Theor. Phys.* **21** (1959) 232.

[300] A. Salam: An Equivalence Theorem for Partially Gauge-invariant Vector Meson Interactions, *Nucl. Phys.* **18** (1960) 681.

[301] S. Kamefuchi: On Salam's Equivalence Theorem in Vector Meson Theory, *Nucl. Phys.* **18** (1960) 691.

[302] A. Salam and J.C. Ward: On a Gauge Theory of Elementary Interactions, *Nuovo Cimento* **19** (1961) 165.

[303] V.I. Ogievetskii and I.V. Polubarinov: A Gauge Invariant Formulation of Neutral Vector Field Theory, *Soviet Phys. JETP* **14** (1962) 179 [*Zh. Eksper. Teor. Fiz.* **41** (1961) 247].

[304] K. Johnson: Relation between the Bare and Physical Masses of Vector Mesons, *Nucl. Phys.* **25** (1961) 435.

[305] J. Schwinger: Gauge Invariance and Mass, *Phys. Rev.* **125** (1962) 397.

[306] J. Schwinger: Gauge Invariance and Mass, II, *Phys. Rev.* **128** (1962) 2425.

[307] D.G. Boulware and W. Gilbert: Connection between Gauge Invariance and Mass, *Phys. Rev.* **126** (1962) 1563.

[308] T.D. Lee and C.N. Yang: Theory of Charged Vector Mesons Interacting with the Electromagnetic Field, *Phys. Rev.* **128** (1962) 885.

[309] G. Feldman and P.T. Matthews: Massive Electrodynamics, *Phys. Rev.* **130** (1963) 1633.

[310] J.A. Young and S.A. Bludman: Electromagnetic Properties of a Charged Vector Meson, *Phys. Rev.* **131** (1963) 2326.

[311] S. Bonometto: On Gauge Invariance for a Neutral Massive Vector Field, *Nuovo Cimento* **28** (1963) 309.

[312] S. Kamefuchi and H. Umezawa: The Mass of Gauge Particles and Self-consistent Method of Quantum Field Theory, *Nuovo Cimento* **32** (1964) 448.

[313] Y. Fujii and S. Kamefuchi: A Generalization of the Stueckelberg Formalism of Vector Meson Fields, *Nuovo Cimento* **33** (1964) 1639.

[314] Y. Fujii: Properties of a Massive Neutral Gauge Particle, *Phys. Rev.* **138** (1965) B423.

[315] M.P. Fry: Gauge Invariance and Mass in Scalar Electrodynamics, *Phys. Rev.* **178** (1969) 2389.

[316] J.H. Lowenstein and B. Schroer: Gauge Invariance and Ward Identities in a Massive-vector Meson Model, *Phys. Rev.* **D6** (1972) 1553.

[317] R.A. Brandt and Ng Wing-Chiu: Gauge Invariance and Mass, *Phys. Rev.* **10** (1974) 4198.

[318] R. Kubo and K. Yokoyama: Massless Tensor Fields and Tripole Ghosts, *Prog. Theor. Phys.* **40** (1968) 421.

[319] K. Yokoyama: Massless Tensor Fields and Tripole Ghosts II, *Prog. Theor. Phys.* **41** (1969) 1384.

[320] T. Kimura: Canonical Quantization of Massless Tensor Field in General Covariant Gauge, *Prog. Theor. Phys.* **55** (1976) 1259.

[321] T. Kimura: Note on Massive Tensor Field in General Covariant Gauges, *Prog. Theor. Phys.* **55** (1976) 1328.

3.5 Indefinite Metric

[322] P.A.M. Dirac: The Physical Interpretation of Quantum Mechanics, *Proc. Roy. Soc.* **A180** (1942) 1.

[323] W. Pauli: On Dirac's New Method of Field Quantization, *Rev. Mod. Phys.* **15** (1943) 175.

[324] W. Pauli: On the Connection Between Spin and Statistics, *Prog. Theor. Phys.* **5** (1950) 526.

[325] W. Pauli and F. Villars: On the Invariant Regularization in Relativistic Quantum Theory, *Rev. Mod. Phys.* **21** (1949) 434.

[326] S.N. Gupta: Quantum Electrodynamics with Auxiliary Fields, *Proc. Phys. Soc.* **A66** (1953) 129.

[327] W. Heisenberg: Lee Model and Quantisation of Nonlinear Field Equations, *Nucl. Phys.* **4** (1957) 532.

[328] R. Ascoli and E. Minardi: On Quantum Theories with Indefinite Metric, *Nucl. Phys.* **9** (1958) 242.

[329] R. Ascoli and E. Minardi: On the Unitarity of the S-matrix in quantum Field Theories with Indefinite Metric, *Nuovo Cimento* **8** (1958) 951.

[330] S. Sunakawa: Quantum Electrodynamics with the Indefinite Metric — Non-lorentz-invariance of the Gupta formalism, *Prog. Theor. Phys.* **19** (1958) 221.

[331] S.N. Gupta: Quantum Mechanics with an Indefinite Metric, *Canad. J. Phys.* **35** (1957) 961.

[332] S.N. Gupta: Lorentz Covariance of Quantum Electrodynamics with the Indefinite Metric, *Prog. Theor. Phys.* **21** (1959) 581.

[333] G. Konishi and T. Ogimoto: Quantum Theory in Pseudo-Hilbert Space, *Prog. Theor. Phys.* **20** (1958) 868.

[334] G. Konishi and T. Ogimoto: Quantum Theory in Pseudo-Hilbert Space II, *Prog. Theor. Phys.* **21** (1959) 727. 3.3 The Gauge Transformation

[335] K.L. Nagy: On a Possibility for the Elimination of the Non-physical Consequences of the Indefinite Metric, *Nuovo Cimento* **10** (1958) 1071.

[336] A. Uhlmann: Schema Einer Quantenmechanik mit Indefiniter Metrik, *Nucl. Phys.* **9** (1958) 588.

[337] A. Uhlmann: Untersuchungen über Quantentheorien mit Indefiniter Metrik, *Nucl. Phys.* **12** (1959) 103.

[338] L.K. Pandit: Linear Vector Spaces with Indefinite Metric, *Nuovo Cimento Suppl.* **11** (1959) 157.

[339] M. Froissart: Covariant Formulation of a Field with Indefinite Metric, *Nuovo Cimento Suppl.* **14** (1959) 197.

[340] K.L. Nagy: Dipole Ghost Contribution to Propagators, *Nuovo Cimento* **15** (1960) 993.

[341] K.L. Nagy: Tripole Ghosts in Field Theory, *Nuovo Cimento* **17** (1960) 384.

[342] K.L. Nagy: Indefinite Metric in Quantum Field Theory, *Nuovo Cimento Suppl.* **17** (1960) 92.

[343] S. Schlieder: Indefinite Metrik im Zustandsraum und Wahrscheinlichkeitsinterpretation, *Z. Naturforsch.* **15a** (1960) 448; 460; 555.

[344] K.L. Nagy: A Model with Multipole-type Ghosts, *Acta Phys. Hung.* **14** (1962) 11.

[345] J.J. Bowman and J.D. Harris: Green's Distributions and the Cauchy Problem for Iterated Klein-Gordon Operator, *J. Math. Phys.* **3** (1962) 396.

[346] J. Lukierski: Theory of Free Relativistic Multipole Field I — Classical part, *Acta Phys. Polon.* **32** (1967) 551.

[347] J. Lukierski: Theory of Free Relativistic Multipole Field II — Quantum Part, *Acta Phys. Polon.* **32** (1967) 771.

[348] K. Yokoyama and R. Kubo: Vector Spaces with Indefinite Metric and Pole-type Ghosts, *Prog. Theor. Phys.* **41** (1969) 542.

[349] K.L. Nagy: *State Vector Spaces with Indefinite Metric in Quantum Field Theory* (P. Noordhoff, Groningen, Netherland 1966).

[350] N. Nakanishi: Indefinite-metric Quantum Field Theory, *Prog. Theor. Phys. Suppl.* **No.51** (1972) 1.

3.6 Other Topics

[351] J.D. Bjorken: A Dynamical Origin for the Electromagnetic Field, *Ann. Phys.* **24** (1963) 174.

[352] I. Bialynicki-Birula: Quantum Electrodynamics without Electromagnetic Field, *Phys. Rev.* **130** (1963) 465.

[353] H.P. Dürr, W. Heisenberg, H. Yamamoto and K. Yamazaki: Quantum Electrodynamics in the Nonlinear Spinor Theory and the Value of Sommerfeld's Fine-structure Constant, *Nuovo Cimento* **38** (1965) 1220.

[354] T-T. Chou and M. Dresden: S-matrix Theory of Electromagnetic Interactions, *Rev. Mod. Phys.* **39** (1967) 143.

[355] I. Brevik and B. Lautrup: Quantum Electrodynamics in Material Media, *K. Danske Vidensk. Selsk. Mat.-fis. Medd.* **38** (1970) No.1.

[356] J. Gomatan: Coherent States and Indefinite Metric — Applications to the Free Electromagnetic and Gravitational Fields, *Phys. Rev.* **D3** (1971) 1292.

[357] M. Namiki and K. Yokoyama: Gauge Invariant Quantum Electrodynamics with the Mass-changing Minimal Current, *Prog. Theor. Phys.* **48** (1972) 2093.

5. PROBLEMS ON INFRARED DIVERGENCES

4.1 Coherent State

[358] F. Bloch and A. Nordsieck: Note on the Radiation Field of the Electron, *Phys. Rev.* **52** (1937) 54.

[359] W. Pauli and M. Fierz: Zur Theorie der Emission Langwelliger Lichtquanten, *Nuovo Cimento* **15** (1938) 167.

[360] W. Braunbek und E. Weinmann: Die Rutherford-Streuung mit Berucksichtigung der Ausstrahlung, *Z. Phys.* **110** (1938) 360.

[361] R. Jost: Compton Scattering and the Emission of Low Frequency Photons, *Phys. Rev.* **72** (1947) 815.

[362] W. Thirring and B. Touschek: A Covariant Formulation of the Bloch-Nordsieck Method, *Phil. Mag.* **42** (1951) 244.

[363] R.J. Glauber: Some Notes on Multiple-boson Processes, *Phys. Rev.* **84** (1951) 395.

[364] J.M. Jauch and F. Rohrlich: The Infrared Divergence, *Helv. Phys. Acta* **27** (1954) 613.

[365] C.R. Hagen: Radiation Gauge Electrodynamics I — The two-point function, *Phys. Rev.* **130** (1963) 813.

[366] C.R. Hagen: Radiation Gauge Electrodynamics II — The asymptotic condition, *Nuovo Cimento* **28** (1963) 970.

[367] R.J. Glauber: Coherent and Incoherent States of the Radiation Field, *Phys. Rev.* **131** (1963) 2766.

[368] J.D. Dollard: Asymptotic Convergence and the Coulomb Interaction, *J. Math. Phys.* **5** (1964) 729.

[369] V. Chung: Infrared Divergence in Quantum Electrodynamics, *Phys. Rev.* **140** (1965) B1110.

[370] M. Greco and G. Rossi: A Note on the Infrared Divergence, *Nuovo Cimento* **50A** (1967) 168.

[371] J.K. Storrow: Photons in S-matric Theory, *Nuovo Cimento* **54A** (1968) 15.

[372] J.K. Storrow: Photons in S-matric Theory II — Systems with many charged particles, *Nuovo Cimento* **57A** (1968) 763.

[373] T.W.B. Kibble: Coherent Soft-Photon States and Infrared Divergences II — Mass shell singularities, *Phys. Rev.* **173** (1968) 1527.

[374] T.W.B. Kibble: Coherent Soft-Photon States and Infrared Divergences III — Reduction formula, *Phys. Rev.* **174** (1968) 1882.

[375] T.W.B. Kibble: The Scattering Soft-photon States and Infrared Divergences IV — The scattering operator, *Phys. Rev.* **175** (1968) 1624.

[376] T.W.B. Kibble: Coherent Soft-photon States and Infrared Divergences I — Classical currents, *J. Math. Phys.* **9** (1968) 315.

[377] P.P. Kulish and L.D. Faddeev: Asymptotic Condition and Infrared Divergence in Quantum Electrodynamics, *Theor. Math. Phys.* **4** (1970) 745 [*Teor. Mat. Fiz.* **4** (1970) 153].

[378] D. Zwanziger: Reduction Formulas for Charged Particles and Coherent States in Quantum Electrodynamics, *Phys. Rev.* **D7** (1973) 1083.

[379] D. Zwanziger: Scattering Theory for Quantum Electrodynamics I — Infrared renormalization and asymptotic fields, *Phys. Rev.* **D11** (1975) 3481.

[380] D. Zwanziger: Scattering Theory for Quantum Electrodynamics II — Reduction and cross-section formulas, *Phys. Rev.* **D11** (1975) 3504.

[381] N. Papanicolaou: On the Infrared Singularities of Green's Function in Quantum Electrodynamics, *Ann. Phys.* **89** (1975) 423.

[382] N. Papanikolaou: Infrared Problems in Quantum Electrodynamics, *Phys. Reports* **24C** (1976) 229.

4.2 Cancellation of IR in Perturbation Theory

[383] H.A. Bethe and J.R. Oppenheimer: Reaction of Radiation on Electron Scattering and Heitler's Theory of Radiation Damping, *Phys. Rev.* **70** (1946) 451.

[384] T. Kinoshita: Note on the Infrared Catastrophe, *Prog. Theor. Phys.* **5** (1950) 1045.

[385] M.R. Schafroth: Höhere Strahlungstheoretische Näherungen zur Klein-Nishina-Formel, *Helv. Phys. Acta* **23** (1950) 542.

[386] L.M. Brown and R.P. Feynman: Radiative Corrections to Compton Scattering, *Phys. Rev.* **85** (1952) 231.

[387] K. Baumann: Die Infrarotkatastrophen der Quantenelektrodynamik, *Acta Phys. Austr.* **7** (1953) 248.

[388] R.G. Newton: Radiative Corrections to Electron Scattering, *Phys. Rev.* **97** (1955) 1162 [Erratum: *ibid.* **98** (1955) 1514].

[389] M. Chrétien: Radiative Corrections to Electron Scattering, *Phys. Rev.* **98** (1955) 1515.

[390] N. Nakanishi: General Theory of Infrared Divergence, *Prog. Theor. Phys.* **19** (1958) 159 [Soryushiron Kenkyu **15** (1957) 344].

[391] E.L.Lomon: Radiative Corrections for Nearly Elastic Scattering, *Phys. Rev.* **113** (1959) 726.

[392] T. Kinoshita: Mass Singularities of Feynman Amplitudes, *J. Math. Phys.* **3** (1962) 650.

4.3 Extraction of Phase Factors

[393] H. Suura: Radiative Correction to High-energy Electron Scattering, *Phys. Rev.* **99** (1955) 1020.

[394] D.R. Yennie and H. Suura: Higher Order Radiative Corrections to Electron Scattering, *Phys. Rev.* **105** (1957) 1378.

[395] T. Murota: On Radiative Corrections due to Soft Photons, *Prog. Theor. Phys.* **24** (1960) 1109.

[396] D.R. Yennie, S.C. Frautschi and H. Suura: The Infrared Divergence Phenomena and High-energy Processes, *Ann. Phys.* **13** (1961) 379.

[397] K.E. Erikson: On Radiative Corrections due to Soft Photons, *Nuovo Cimento* **19** (1961) 1010.

[398] K.T. Mahanthappa: Multiple Production of Photons in Quantum Electrodynamics, *Phys. Rev.* **126** (1962) 329.

[399 T.D. Lee and M. Nauenberg: Degenerate Systems and Mass Singularities, *Phys. Rev.* **133** (1964) B1549.

[400] G.Q. Hassoun and D.R. Yennie: Infrared Divergence of the Angular Momentum of Bremsstrahlung and the Physical Structure of the Electron, *Phys. Rev.* **134** (1964) B436.

[401] G. Grammer, Jr. and D.R. Yennie: Improved Treatment for the Infrared-divergence Problem in Quantum Electrodynamics, *Phys. Rev.* **D8** (1973) 4332.

4.4 Other Topics

[402] L.I. Schiff: Radiative Correction to the Angular Distribution of Nuclear Recoils from Electron Scattering, *Phys. Rev.* **87** (1952) 750.

[403] F. Rohrlich: Infrared Divergence in Bound State Problems, *Phys. Rev.* **98** (1955) 181.

[404] R.V. Polovin: Radiative Corrections to the Scattering of Electrons by Electrons and Positrons, *Soviet Phys. JETP* **4** (1957) 385 [*Zh. Eksper. Teor. Fiz.* **31** (1956) 449].

[405] F.E. Low: Bremsstrahlung of Very Low-energy Quanta in Elementary Particle Collisions, *Phys. Rev.* **110** (1958) 974.

[406] J.D. Bjorken, S.D. Drell and S.C. Frautschi: Wide-angle Pair Production and Quantum Electrodynamics at Smalll Distances, *Phys. Rev.* **112** (1958) 1409.

[407] P.I. Fomin: Radiation Corrections to Bremsstrahlung, *Soviet Phys. JETP* **35** (1959) 491 [*Zh. Eksper. Teor. Fiz.* **35** (1958) 707].

[408] S. Weinberg: Infrared Photons and Gravitons, *Phys. Rev.* **140** (1965) B516.

[409] J. Frölich: On the Infrared Problem in a Model of Scalar Electrons and Massless Scalar Bosons, *Ann. Inst. H. Poincaré* **19** (1973) 1.

[410] R. Gastmans and R. Meuldermans: Dimensional Regularization of the Infrared Problem, *Nucl. Phys.* **B63** (1973) 277.

[411] W.J. Marciano and A. Sirlin: Dimensional Regularization of Infrared Divergences, *Nucl. Phys.* **B88** (1975) 86.

6. PROBLEMS ON THE ANOMALY

5.1 Goto-Imamura-Schwinger Term

[412] T. Goto and T. Imamura: Note on the Non-Perturbation-approach to Quantum Field Theory, *Prog. Theor. Phys.* **14** (1955) 396.

[413] J. Schwinger: Field Theory Commutators, *Phys. Rev. Letters* **3** (1959) 296.

[414] S. Okubo: Inconsistency of Canonical Commutation Relations Among Current Densities, *Nuovo Cimento* **44A** (1966) 1015.

[415] J.W. Moffat: Regularized Vacuum Expectation Values in Quantum Field Theory, *Nucl. Phys.* **16** (1960) 304.

[416] K. Johnson: Current-charge Density Commutation Relations, *Nucl. Phys.* **25** (1961) 431.

[417] J. Schwinger: Commutation Relation and Conservation Laws, *Phys. Rev.* **130** (1963) 406.

[418] H.J. Borchers: Field Operators as C^∞ Functions in Spacelike Direction, *Nuovo Cimento* **33** (1964) 1600.

[419] D. Kastler, D.W. Robinson and A. Swieca: Conserved Currents and Associated Symmetries — Goldstone's theorem, *Comm. Math. Phys.* **2** (1966) 108.

[420] B. Schroer and P. Stichel: Current Commutation Relations in the Framework of General Quantum Field Theory, *Comm. Math. Phys.* **3** (1966) 258.

[421] L.S. Brown: Gauge Invariance, Lorentz Covariance, and Current Correlation Functions, *Phys. Rev.* **150** (1966) 1338.

[422] D.G. Boulware: Gauge Invariance and Current Definition in Quantum Electrodynamics, *Phys. Rev.* **151** (1966) 1024.

[423] D. Boulware and S. Deser: Necessity of Field Dependence for Interacting Currents, *Phys. Letters* **22** (1966) 99.

[424] D.G. Boulware and S. Deser: Necessary Dependence of Currents on Fields They Generate, *Phys. Rev.* **151** (1966) 1278.

[425] W.S. Hellman and P. Roman: Schwinger Terms from Local Currents, *Nuovo Cimento* **52** (1967) 1341.

[426] R.A. Brandt and C.A. Orzalesi: Equal-time Commutator and Zero-energy Theorem in the Lee Model, *Phys. Rev.* **162** (1967) 1747.

[427] R.A. Brandt: Approach to Equal-time Commutators in Quantum Field Theory, *Phys. Rev.* **166** (1968) 1795.

[428] T. Nagylaki: Current Commutators in Quantum Electrodynamics, *Phys. Rev.* **158** (1967) 1534.

[429] S.G. Brown and S.A. Bludman: Application of the Dirac-Schwinger Covariance Condition in Quantum Electrodynamics, *Phys. Rev.* **161** (1967) 1505.

[430] R. Jackiw and G. Preparata: Probes for the Constituents of the Electromagnetic Current and Anomalous Commutators, *Phys. Rev. Letters* **22** (1969) 975.

[431] M.S. Chanowitz: Schwinger Terms in Fermion Electrodynamics, *Phys. Rev.* **D2** (1970) 3016.

[432] C.A. Orzalesi: Lecture on Field-theoretic Aspects of Currrent Algebra, *Tech. Rep. No.833* (Univ. Maryland, 1968).

[433] G. Källén: Gradient Terms in Commutators of Currents and Fields, *Acta Phys. Austr. Suppl.* **5** (1968) 268.

[434] P. Ghose and A. Das: On the Goto-Imamura-Pradhan-Schwinger Term in Quantum Electrodynamics, *Prog. Theor. Phys.* **46** (1971) 1623.

[435] K. Nishijima and R. Sasaki: Nature of the Schwinger Term in Spinor Electrodynamics, *Prog. Theor. Phys.* **53** (1975) 1809.

5.2 PCAC Anomaly

[436] J.S. Bell and R. Jackiw: A PCAC Puzzle — $\pi^o \to \gamma\gamma$ in the σ-model, *Nuovo Cimento* **60A** (1969) 47.

[437] C.R. Hagen: Derivation of Adler's Divergence Condition from the Field Equations, *Phys. Rev.* **177** (1969) 2622.

[438] S.L. Adler: Axial-vector Vertex in Spinor Electrodynamics, *Phys. Rev.* **177** (1969) 2426.

[439] I.S. Gerstein and R. Jackiw: Anomalies in Ward Identities for Three-point Functions, *Phys. Rev.* **181** (1969) 1955.

[440] C.R. Jackiw and K. Johnson: Anomalies of the Axial-Vector Current, *Phys. Rev.* **182** (1969) 1459.

[441] S.L. Adler and W.A. Bardeen: Absence of Higher-order Corrections in the Anomalous Axial-vector Divergence Equation, *Phys. Rev.* **182** (1969) 1517.

[442] S.L. Adler and D.G. Boulware: Anomalous Commutators and the Triangle Diagram, *Phys. Rev.* **184** (1969) 1740.

[443] S.L. Adler: Perturbation Theory Anomaly, *Lectures on Elementary Particles and Quantum Field Theory, Vol. 1* (M.I.T. Press, Cambridge, 1970) p.1.

[444] R. Jackiw: *Field Theoretic Investigations in Current Algebra. Current Algebra and Its Applications*, Princeton Series in Physics (Princeton Univ. Press, New Jersey, 1972).

7. QED EXPERIMENTS

6.1 Low-Energy Theorems

[445] N.M. Kroll and M.A. Ruderman: A Theorem on Photomeson Production Near Threshold and the Suppression of Pairs in Pseudo-scalar Meson Theory, *Phys. Rev.* **93** (1954) 233.

[446] M. Gell-Mann and K.L. Goldberger: Scattering of Low-energy Photons by Particles of Spin 1/2, *Phys. Rev.* **96** (1954) 1433.

[447] F.E.Low: Scattering of Light of Very Low Frequency by Systems of Spin 1/2, *Phys. Rev.* **96** (1954) 1428.

[448] A. Klein: Low-energy Theorems for Renormalizable Field Theories, *Phys. Rev.* **99** (1955) 998.

449. F.E. Low: Bremsstrahlung of Very Low-energy quanta in Elementary Particle Collisions, *Phys. Rev.* **110** (1958) 974.

6.2 Experiment

[450] H.A. Bethe: The Electromagnetic Shift of Energy Levels, *Phys. Rev.* **72** (1947) 339.

[451] Y. Nambu: The Level Shift and the Anomalous Magnetic Moment of Electron, *Prog. Theor. Phys.* **4** (1949) 82.

[452] R. Karplus, A. Klein and J. Schwinger: Electrodynamic Displacement of Atomic Energy Levels II — Lamb shift, *Phys. Rev.* **86** (1952) 288.

[453] R.L. Mills and N.M. Kroll: Fourth-order Radiative Corrections to Atomic Energy Levels II, *Phys. Rev.* **98** (1955) 1489.

[454] A.J. Layzer: New Theoretical Value for the Lamb Shift, *Phys. Rev. Letters* **4** (1960) 580.

[455] H.M. Fried and D.R. Yennie: Higher Order Terms in the Lamb Shift Calculation, *Phys. Rev. Letters* **4** (1960) 583.

[456] Ya.A. Smorodinski: Limits of Quantum Electrodynamics and Accuracy of Global Constants, *Soviet Phys. Uspekhi* **11** (1968) 130 [*Uspekhi Fiz. Nauk* **94** (1968) 359] .

[457] T. Appelquist and S.J. Brodsky: Order α^2 Electrodynamic Corrections to the Lamb Shift, *Phys. Rev. Letters* **24** (1970) 562.

[458] A. Peterman: Analytic 4th Order Crossed Ladder Contribution to the Lamb Shift, *Phys. Lett.* **35B** (1971) 325.

[459] J.A. Fox and D.R. Yennie: Some Formal Aspects of the Lamb Shift Problem, *Ann. Phys.* **81** (1973) 438.

[460] H. Grotch and D.R. Yennie: Effective Potential Model for Calculating Nuclear Corrections to the Energy Levels of Hydrogen, *Rev. Mod. Phys.* **41** (1969) 350.

[461] R. Karplus, A. Klein and J. Schwinger: Electrodynamic Displacement of Atomic Energy Levels, *Phys. Rev.* **84** (1951) 597.

[462] R.Karplus and A. Klein: Electrodynamic Displacement of Atomic Energy Levels I — Hyperfine structure, *Phys. Rev.* **85** (1952) 972.

[463] D.E. Zwanziger: α^2 Corrections to Hyperfine Structure in Hydrogenic Atoms, *Phys. Rev.* **121** (1961) 1128.

[464] S.D. Drell and J.D. Sullivan: Polarizability Contribution to the Hydrogen Hyperfine Structure, *Phys. Rev.* **154** (1967) 1477.

[465] D.de Rafael: The Hydrogen Hyperfine Structure and Inelastic Electron Proton Scattering Experiments, *Phys. Letters* **37B** (1971) 201.

[466] W.E. Lamb and R.C. Retherford: Fine Structure of the Hydrogen Atom by a Microwave Method, *Phys. Rev.* **72** (1947) 241.

[467] R.T. Robiscoe: Reconciliation of Experimental Lamb Shifts, *Phys. Rev.* **168** (1968) 4.

[468] S.L.Kaufman, W.E. Lamb, Jr. K.R. Lea and M. Leventhal: Measurement of the $2^2S_{1/2} - 2^2P_{3/2}$ Interval in Atomic Hydrogen, *Phys. Rev. Letters* **22** (1969) 507.

[469] T.W. Shyn, W.L. Williams, R.T. Robiscoe and T. Rebane: Experimental Value of $\Delta E_H - S_H$ in Hydrogen, *Phys. Rev. Lettters* **22** (1969) 1273.

[470] B.L. Cosens and T.V. Vorburger: Remeasurement of $\Delta E - S$ in Atomic Hydrogen, *Phys. Rev. Letters* **23** (1969) 1273.

[471] T. Fulton: Corrections to the Positronium Hyperfine Structure of Order $\alpha^2 \ell n(1/\alpha)$, *Phys. Rev. Letters* **24** (1970) 1035 [Erratum: *ibid.* **25** (1970) 782].

[472] D.A. Owen: Fourth-order Vacuum Polarization Correction to the Positronium Hyperfine Structure, *Phys. Rev. Letters* **30** (1973) 887.

[473] R. Barbieri and P. Christillin: On the Theoretical Value of Positronium Ground State Splitting, *Phys. Letters* **43B** (1973) 411.

[474] M.A. Stroscio: Positronium, *Phys. Reports* **22** (1975) 215.

[475] R. De Voe et al.: Measurement of the Muonium hfs Splitting and of the Muon Moment by "Double Resonance" and a New Value of α, *Phys. Rev. Letters* **25** (1970) 1779.

[476] D. Favart et al.: Precision Measurement of the Hyperfine Interval of Muonium by a Novel Technique Ramsey Resonance in Zero Field, *Phys. Rev. Letters* **27** (1971) 1336.

[477] E.R. Carlson et al.: High-precision Determination of the Fine-structure Interval in the Ground State of Positronium and the Fine-structure Density Shift in Nitrogen, *Phys. Rev. Letters* **29** (1972) 1059.

[478] J. Schwinger: On Quantum-electrodynamics and the Magnetic Moment of the Electron, *Phys. Rev.* **73** (1948) 416.

[479] R. Karplus and N.M. Kroll: Fourth-order Corrections in Quantum Electrodynamics and the Magnetic Moment of the Electron, *Phys. Rev.* **77** (1950) 536.

[480] C.M. Sommerfield: Magnetic Dipole Moment of the Electron, *Phys. Rev.* **107** (1957) 328.

[481] A. Petermann: Fourth Order Magnetic Moment of the Electron, *Helv. Phys. Acta* **30** (1957) 407.

[482] T. Kinoshita and A. Sirlin: Radiative Corrections to Fermi Interactions, *Phys. Rev.* **113** (1959) 1652.

[483] B.E. Lautrup and E.de Rafael: Calculation of the Sixth-order Contribution from the Fourth-order Vacuum Polarization to the Difference of the Anomalous Magnetic Moments of Muon and Electron, *Phys. Rev.* **174** (1968) 1835.

[484] J. Aldins, S.J. Brodsky, A.J. Dufner and T. Kinoshita: Photon-photon Scattering Contribution to the Sixth-order Magnetic Moments of the Muon and

Electron, *Phys. Rev.* **D1** (1970) 2378.

[485] S.J. Brodsky and T. Kinoshita: Vacuum-polarization Contributions to the Sixth-order Anomalous Magnetic Moment of the Muon and Electron, *Phys. Rev.* **D3** (1971) 356.

[486] T. Kinoshita and P. Cvitanovic: Sixth-order Radiative Corrections to the Electron Magnetic Moment, *Phys. Rev. Letters* **29** (1972) 1534.

[487] B.E. Lautrup: On the Order of Magnitude of 8th Order Corrections to the Anomalous Magnetic Moment of the Muon, *Phys. Letters* **38B** (1972) 408.

[488] M.J. Levine and J. Wright: Anomalous Magnetic Moment of the Electron, *Phys. Rev.* **D8** (1973) 3171.

[489] M.J. Levine and R. Roskies: New Technique for Vertex Graphs, *Phys. Rev. Letters* **30** (1973) 772.

[490] R. Carroll and Y.P. Yao: α^3 Contributions to the Anomalous Magnetic Moment of an Electron in the Mass-operator Formalism, *Phys. Letters* **48B** (1974) 125.

[491] R. Cvitanovic and T. Kinoshita: Sixth-order Magnetic Moment of the Electron, *Phys. Rev.* **D10** (1974) 4007.

[492] R. Barbieri and E. Remidi: Electron and Muon $(g-2)/2$ from Vacuum Polarization Insertions, *Nucl. Phys.* **B90** (1975) 233.

[493] P. Kusch and H.M. Foley: Precision Measurement of the Ratio of the Atomic 'g Values' in the $^2P_{32}$ and $^2P_{12}$ States of Gallium, *Phys. Rev.* **72** (1947) 1256.

[494] J. Bailey et al.: Precision Measurement of the Anomalous Magnetic Moment of the Muon, *Phys. Letters* **28B** (1968) 287.

[495] J.F. Hague et al.: Precision Measurement of the Magnetic Moment of the Muon, *Phys. Rev. Letters* **25** (1970) 628.

[496] J.C. Wesley and A. Rich: High-field Electron g-2 Measurement, *Phys. Rev.* **A4** (1971) 1341.

[497] J. Schwinger: On Radiative Corrections to Electron Scattering, *Phys. Rev.* **75** (1949) 898.

[498] J. Schwinger: The Quantum Correction in the Radiation by Energetic Accelerated Electrons, *Proc. Nat. Acad. Sci.* **40** (1954) 132.

[499] S.M. Berman: Radiative Corrections to μ and Neutron Decay, *Phys. Rev.* **112** (1958) 267.

[500] L. Matsson: On Radiative Corrections to Muon Decay, *Nucl. Phys.* **B13** (1969) 647.

[501] A.C. Hearn, R.K. Kuo and D.R. Yennie: Radiative Corrections to an Electron-positron Scattering Experiment, *Phys. Rev.* **187** (1969) 1950.

[502] M. Roos and A. Sirlin: Remarks on the Radiative Corrections of Order α^2 to $\mu-$decay and the Determination of G_μ, *Nucl. Phys.* **B29** (1971) 296.

[503] D.A. Ross: Radiative Corrections to Muon Decay, *Nuovo Cimento* **10A** (1972) 475.

[504] B.N. Taylor, W.H. Parker and D.N. Langenberg: Determination of e/h, Using Macroscopic Quantum Phase Coherence in Superconductors Implication for QED and the Fundamental Physical Constants, *Rev. Mod. Phys.* **41** (1969) 375.

[505] B.E. Lautrup, A. Peterman and E.de Rafael: Recent Developments in the Comparison Between Theory and Experiments in Quantum Electrodynamics, *Phys. Reports* **3** (1972) 193.

[506] T. Kinoshita: Present Status of QED, *J. Phys. Soc. Japan* **29** (1974) 471.

6.3 Magnetic Monopole

[507] P.A.M. Dirac: The Theory of Magnetic Poles, *Phys. Rev.* **74** (1948) 817.

[508] D. Zwanziger: Dirac Magnetic Poles Forbidden in S-matrix Theory, *Phys. Rev.* **137** (1965) B657.

[509] S. Weinberg: Photons and Gravitons in Perturbation Theory. Derivation of Maxwell's and Einstein's Equations, *Phys. Rev.* **138** (1965) B988.

[510] A.S. Goldhaber: Role of Spin in the Monopole Problem, *Phys. Rev.* **140** (1965) B1407.

[511] J. Schwinger: Magnetic Charge and Quantum Field Theory, *Phys. Rev.* **144** (1966) 1087.

8. Articles added

1.1

[512] S. Hayakawa, Y. Miyamoto and S. Tomonaga: On the Elimination of the Auxiliary Condition in the Quantum Electrodynamics I, *J. Phys. Soc. Japan* **2** (1947) 172.

[513] S. Hayakawa, Y. Miyamoto and S. Tomonaga: On the Elimination of the Auxiliary Condition in the Quantum Electrodynamics III, *J. Phys. Soc. Japan* **2** (1947) 199.

[514] S. Kanesawa and S. Tomonaga: On a Relativistically Invariant Formulation of the Quantum Theory of Wave Fields V, *Prog. Theor. Phys.* **3** (1948) 1 [Addendum: *ibid.* 101].

[515] Z. Koba, Y. Oisi and M. Sasaki: Auxiliary Condition and Gauge Transformation in the "Super-many-time Theory" I, *Prog. Theor. Phys.* **3** (1948) 141.

[516] Z. Koba, Y. Oisi and M. Sasaki: Auxiliary Condition and Gauge Transformation in the "Super-many-time Theory" II, *Prog. Theor. Phys.* **3** (1948) 229.

3.1

[517] D. Maison and D. Zwanziger: On the Subsidiary Condition in Quantum Electrodynamics, *Nucl. Phys.* **B91** (1975) 425.

3.4

[518] D. Zwanziger: Physical States in Quantum Electrodynamics, *Phys. Rev.* **D14** (1976) 2570.

3.6

[519] H.P. Dürr and E. Rudolph: Good and Bad Ghosts in Quantum Electrodynamics, *Nuovo Cimento* **62A** (1969) 411.

[520] H.P. Dürr and E. Rudolph: Indefinite Metric in Massless Quantum Field Theories of Arbitrary Spin, *Nuovo Cimento* **65A** (1970) 423.

6.3

[521] A.S. Goldhaber: Connection of Spin and Statistics for Charge Monopole Composites, *Phys. Rev. Letters* **36** (1976) 1122.

www.ingramcontent.com/pod-product-compliance
Lightning Source LLC
Chambersburg PA
CBHW052113230326
41598CB00079B/3664